Springer Proceedings in Physics

Volume 290

Indexed by Scopus

The series Springer Proceedings in Physics, founded in 1984, is devoted to timely reports of state-of-the-art developments in physics and related sciences. Typically based on material presented at conferences, workshops and similar scientific meetings, volumes published in this series will constitute a comprehensive up to date source of reference on a field or subfield of relevance in contemporary physics. Proposals must include the following:

- Name, place and date of the scientific meeting
- A link to the committees (local organization, international advisors etc.)
- Scientific description of the meeting
- List of invited/plenary speakers
- An estimate of the planned proceedings book parameters (number of pages/articles, requested number of bulk copies, submission deadline).

Please contact:

For Americas and Europe: Dr. Zachary Evenson; zachary.evenson@springer.com
For Asia, Australia and New Zealand: Dr. Loyola DSilva; loyola.dsilva@springer.com

Abdul Basit · Erma Yulihastin ·
Sri Yudawati Cahyarini · Heru Santoso ·
Widodo S. Pranowo · Lilik Slamet S. ·
Halda Aditya Belgaman
Editors

Proceedings of the International Conference on Radioscience, Equatorial Atmospheric Science and Environment and Humanosphere Science

INCREASE 2022, 22–23 Nov, Indonesia

Springer

Editors
Abdul Basit
National Research and Innovation Agency
Bandung, Indonesia

Sri Yudawati Cahyarini
National Research and Innovation Agency
Bandung, Indonesia

Widodo S. Pranowo
The Indonesian Naval Postgraduate
Military Service School (STTAL)
Jakarta, Indonesia

Halda Aditya Belgaman
National Research and Innovation Agency
Bandung, Indonesia

Erma Yulihastin
National Research and Innovation Agency
Bandung, Indonesia

Heru Santoso
National Research and Innovation Agency
Bandung, Indonesia

Lilik Slamet S.
National Research and Innovation Agency
Bandung, Indonesia

ISSN 0930-8989 ISSN 1867-4941 (electronic)
Springer Proceedings in Physics
ISBN 978-981-19-9767-9 ISBN 978-981-19-9768-6 (eBook)
https://doi.org/10.1007/978-981-19-9768-6

This Springer imprint is published by the registered company Springer Nature Singapore Pte Ltd.
The registered company address is: 152 Beach Road, #21-01/04 Gateway East, Singapore 189721, Singapore

Preface

This book highlights research on climate and atmospheric dynamics over the maritime continent and its impact on sectors. It is part of the 2nd International Conference on Radioscience, Equatorial Atmospheric Science and Environment (INCREASE) organized by the Research Center for Climate and Atmosphere (PRIMA) of National Research and Innovation Agency (BRIN). The symposium aims to provide a scientific platform for university students, school teachers and scientists to discuss ideas and current issues as well as to design the solutions.

The symposium was held online via Zoom and ran over two days from November 22 to 23, 2022. The first day of the symposium was opened by presentations with Keynote speakers, Dr. Albertus Sulaiman, Prof. Edvin Aldrian, Prof. Greg McFarquhar, Prof. Manabu Yamanaka and followed by ten invited speakers as climate weather experts from Indonesia, Japan, Philippine and USA. On the second day, the presentations were delivered by selected participants (86 people) who were divided into five groups. The time allocation for each presenter was about 10 minutes followed by the question and answer sessions after all presenters finished delivering their talks. Along with the symposium, we have also facilitated the presenters to submit their manuscripts to be able to publish in a peer-reviewed journal. The review process includes (1) the first screen of manuscripts by the chief editor and the decision for full peer review, (2) the review of manuscripts by one or two reviewers who are experts in the field, (3) final decision of the editors based on reviewers' comments, (4) decision for accepted or rejected. We have received 135 manuscripts, and only 83 manuscripts were finally selected for the further scientific publication process.

High appreciation would be dedicated to the participants and speakers who have attended and supported the 2nd INCREASE. We also would like to thank the scientific committee and reviewers for their contributed time and thorough comments to improve the manuscripts. Lastly, many thanks should be addressed to the organizing committee for their hard work so that the symposium can be accomplished as planned. See you all in the next INCREASE.

Bandung, Indonesia

Dr.rer.nat. Abdul Basit
Editor in Chief of 2nd INCREASE
Proceeding 2022

Contents

Contributors

Fildzah 'Adany Research Center for Chemistry, National Research and Innovation Agency (BRIN), South Tangerang, Banten, Indonesia

Prayitno Abadi Research Center for Climate and Atmosphere, National Research and Innovation Agency, Bandung, Indonesia

Muhammad Rais Abdillah Atmospheric Sciences Research Group, Faculty of Earth Sciences and Technology, Institut Teknologi Bandung, Bandung, Indonesia

Emmanuel Adetya Research Center for Climate and Atmosphere, National Research and Innovation Agency, Bandung, Indonesia ;
Research Center of Climate and Atmosphere, National Research and Innovation Agency, Central Jakarta, Indonesia

Zahwa Vieny Adha Department of Physics, Universitas Andalas, Padang, Indonesia

Khanifah Afifi Master Program in Earth Sciences, Faculty of Earth Sciences and Technology, Institut Teknologi Bandung, Bandung, Indonesia

Yuliana Eva Agasi Department Agriculture Engineering, Faculty of Agricultural Industrial Technology, Padjadjaran University, Sumedang, Indonesia

Suaidi Ahadi Indonesian Meteorological, Climatological and Geophysical Agency (BMKG), Kemayoran, DKI Jakarta, Indonesia

Wan Alamsyah Universitas Samudra, Langsa-Aceh, Indonesia

Mustofa Amirullah Research Center for Smart Mechatronics, National Research and Innovation Agency (BRIN), Bandung, West Java, Indonesia

Dava Amrina Meteorology Climatology and Geophysics Agency, Jakarta Pusat, Indonesia

Dita Fatria Andarini Research Center for Climate and Atmosphere, National Research and Innovation Agency (BRIN), Bandung, West Java, Indonesia;
National Research and Innovation Agency, Jakarta, Indonesia

B. Moh. Andi Aris Research Center for Space, Bandung, West Java, Indonesia

Yoga Andrian Research Center for Space, Bandung, West Java, Indonesia

Ega A. Anggari Research Center for Satellite Technology, Bogor, West Java, Indonesia

Titi Anggono Research Center for Geological Disasters, National Agency of Research and Innovation (BRIN), Jawa Barat, Indonesia

Yayan Apriyana Research Centre for Climate and Atmosphere, National Research and Innovation Agency, Bandung, Indonesia

M. Arif Setyo Aji Bandung Institute of Technology, Bandung, Indonesia

Lely Q. Avia Research Center for Atmospheric Science and Technology, Indonesia National Agency of Research and Innovation, Bandung, Indonesia

Praditya Avianto National Research and Innovation Agency of Indonesia (BRIN), Jakarta, Indonesia

Awaluddin Research Center for Climate and Atmosphere, National Research and Innovation Agency (BRIN), Jawa Barat, Indonesia

Asif Awaludin BRIN—Research Organization for Earth Sciences and Maritime, Bandung, Indonesia;
Research Center for Climate and Atmosphere, National Research and Innovation Agency (BRIN), Bandung, West Java, Indonesia

Sholehhudin A. Ayubi National Research and Innovation Agency, Jakarta, Indonesia

Prita Ayuningtyas Pontianak Observatory for Space and Atmosphere, Pontianak, West Kalimantan, Indonesia

Nabila Afifah Azuga Department of Earth Sciences, Faculty of Earth Sciences and Technology, Institut Teknologi Bandung, Bandung, Indonesia

Harry Bangkit Research Center for Smart Mechatronics, Bandung, Indonesia

Abdul Basit National Research and Innovation Agency of Indonesia (BRIN), Jakarta, Indonesia

Halda A. Belgaman National Research and Innovation Agency, Jakarta, Indonesia

Rikha Bramawanto Research Center for Climate and Atmosphere, National Research and Innovation Agency (BRIN), Bandung, Indonesia

Eldian Yosua Budiarto Faculty of Earth Sciences and Technology-Institut Teknologi Bandung, Bandung, Indonesia

Asep Sandra Budiman National Research and Innovation Agency of Indonesia (BRIN), Jakarta, Indonesia

Rinaldi Bursan Faculty of Economics, University of Lampung, Bandar Lampung, Indonesia

Sri Yudawati Cahyarini Research Center of Climate and Atmosphere, Research Group of Paleoclimate and Paleoenvironment, National Research and Innovation Agency, Jakarta, Indonesia

Waluyo Eko Cahyono Research Center for Climate and Atmosphere, National Research and Innovation Agency, Bandung, Indonesia

Nani Cholianawati Research Center for Climate and Atmosphere, National Research and Innovation Agency (BRIN), Bandung, Indonesia;
Research Center of Climate and Atmosphere, National Research and Innovation Agency, Central Jakarta, Indonesia

A. Dariah National Research and Innovation Agency, Jakarta, Indonesia

Arief Darmawan Research Center for Climate and Atmosphere, National Research and Innovation Agency (BRIN), Jawa Barat, Indonesia

Syukri Darmawan BRIN—Research Organization for Electronics and Informatics, Bandung, Indonesia

Saktyanu Kristyantoadi Dermoredjo Research Centre for Economics of Industry, Services, and Trade, BRIN, Jakarta, Indonesia

Dodi Devianto Department of Mathematics and Data Science, Andalas University, Padang, West Sumatera, Indonesia;
Department of Mathematics, UNAND, West Sumatera, Indonesia

Cinantya Nirmala Dewi Research Center for Geological Disasters, National Agency of Research and Innovation (BRIN), Jawa Barat, Indonesia

E. R. Dewi National Research and Innovation Agency, Jakarta, Indonesia

Saraswati Dewi National Research and Innovation Agency, Jakarta, Indonesia

Karunika Diwyacitta Research Center for Remote Sensing, National Research and Innovation Agency, Jakarta, Indonesia

Mohamad Rahman Djuwansah Research Centre for Environment and Clean Technology, National Agency for Research and Innovation, Bandung, Indonesia

Sri Ekawati Research Center for Climate and Atmosphere, National Research and Innovation Agency, Bandung, Indonesia

Erlansyah Research Center for Space, Bandung, West Java, Indonesia

Woro Estiningtyas Research Centre for Climate and Atmosphere, National Research and Innovation Agency, Bandung, Indonesia;
Research Center for Climate and Atmosphere, BRIN, Jakarta, Indonesia

Faiz Rohman Fajary Atmospheric Sciences Research Group, Faculty of Earth Sciences and Technology, Institut Teknologi Bandung, Bandung, Indonesia;
Graduate Program of Earth Sciences, Faculty of Earth Sciences and Technology, Institut Teknologi Bandung, Bandung, Indonesia

Y. R. Fanggidae National Research and Innovation Agency, Jakarta, Indonesia

Yudi Riadi Fanggidae The Research Center for Climate and Atmosphere, National Research and Innovation Agency, Bandung, Indonesia

Ibnu Fathrio Research Center for Climate and Atmosphere, National Research and Innovation Agency (BRIN), Bandung, West Java, Indonesia;
National Research and Innovation Agency, Jakarta, Indonesia;
Indonesian National Research and Innovation Agency, Jakarta, Indonesia

Dita Fatria National Research and Innovation Agency, Bandung, Indonesia

Febty Febriani Research Center for Geological Disasters, National Agency of Research and Innovation (BRIN), Jawa Barat, Indonesia;
Research Center for Geological Disaster, BRIN, Bandung, Indonesia

Naffisa Adyan Fekranie Department of Oceanography, Faculty of Earth Sciences and Technology, Institut Teknologi Bandung, Bandung, Indonesia

Abriani Fensionita Directorate of Food Crops Protection, Directorate General of Food Crops, Indonesia Ministry of Agriculture, Jakarta, Indonesia

Hadi Ferdiansyah Regional Research and Development Planning Agency of Sumedang Regency, Sumedang, Indonesia

Hana Listi Fitriana Research Center for Remote Sensing, National Research and Innovation Agency, Jakarta, Indonesia

Nurul Fitriya Research Center for Oceanography, National Research and Innovation Agency, Jakarta Utara, Indonesia

Gammamerdianti National Research and Innovation Agency, Jakarta, Indonesia;
Research Center for Atmospheric Science and Technology, Indonesia National Agency of Research and Innovation, Bandung, Indonesia

R. D. Goenawan National Research and Innovation Agency, Jakarta, Indonesia

Hidayat Gunawan Research Center for Remote Sensing, National Research and Innovation Agency, Jakarta, Indonesia

Niruti Gupta Malaviya National Institute of Technology, Jaipur, India

Dessy Gusnita National Research and Innovation Agency, Bandung, Indonesia

Muhammad N. Habibi Meteorology Climatology and Geophysics Agency, Jakarta Pusat, Indonesia

Tri Wahyu Hadi Graduate Program of Earth Sciences, Faculty of Earth Sciences and Technology, Institut Teknologi Bandung, Bandung, Indonesia

Trisni Hadiningrum Department of Atmospheric and Planetary Science, Sumatera Institute of Technology, Lampung, Indonesia

Saipul Hamdi Research Center for Climate and Atmosphere, National Research and Innovation Agency-BRIN, Bandung, Indonesia

Cicik Oktasari Handayani Research Center for Horticultural and Estate Crops, National Research and Innovation Agency, Cibinong Science Center, Cibinong, Bogor, Indonesia

Muhammad Hanif Geoinformatics, Department of Computer Science, College of Computing, Khon Kaen University, Khon Kaen, Thailand

Teguh Harjana Research Center for Climate and Atmosphere, National Research and Innovation Agency (BRIN), Bandung, West Java, Indonesia;
Indonesian National Research and Innovation Agency, Jakarta, Indonesia

Wendi Harjupa Research Center for Climate and Atmosphere, National Research and Innovation Agency (BRIN), Bandung, West Java, Indonesia

Wahyudi Hasbi Research Center for Satellite Technology, Bogor, West Java, Indonesia

Mohammad Hasib Research Center for Geological Disasters, National Agency of Research and Innovation (BRIN), Jawa Barat, Indonesia

Bernadetta Rina Hastiestari Research Centre for Genetic Engineering, National Research and Innovation Agency, Indonesia (BRIN), Jakarta, Indonesia

Rahaden Bagas Hatmaja National Research and Innovation Agency of Indonesia (BRIN), Jakarta, Indonesia;
Research Centre for Climate and Atmosphere, National Research and Innovation Agency, Bandung, Indonesia

Marfasran Hendrizan Research Center for Climate and Atmosphere-National Research and Innovation Agency, Bandung, Indonesia;
Research Center of Climate and Atmosphere, Research Group of Paleoclimate and Paleoenvironment, National Research and Innovation Agency, Jakarta, Indonesia

Eddy Hermawan Research Center for Climate and Atmosphere, National Research and Innovation Agency (BRIN), Bandung, West Java, Indonesia

Sendi Purnama Hidayat Department Agriculture Engineering, Faculty of Agricultural Industrial Technology, Padjadjaran University, Sumedang, Indonesia

Yukiharu Hisaki Department of Physics and Earth Sciences, Graduate School of Engineering and Science, University of the Ryukyus, Nishihara, Japan

Lambok M. Hutasoit Graduate Program of Earth Sciences, Faculty of Earth Sciences and Technology, Institut Teknologi Bandung, Bandung, Indonesia

Candra N. Ihsan Research Center for Atmospheric Science and Technology, Indonesia National Agency of Research and Innovation, Bandung, Indonesia

Andy Indradjad Research Center for Remote Sensing, National Research and Innovation Agency, Jakarta, Indonesia

Asri Indrawati Research Center for Climate and Atmosphere, National Research and Innovation Agency-BRIN, Bandung, Indonesia;
Laboratory Management, Research Facilities and Science, and Technology Park, National Research and Innovation Agency, Central Jakarta, Indonesia;
Directorate of Laboratory Management, Research Facilities, and Science and Technology Park, National Research and Innovation Agency, Jakarta, Indonesia

Mochamad Furqon Azis Ismail National Research and Innovation Agency of Indonesia (BRIN), Jakarta, Indonesia;
Humboldt Fellow, GEOMAR Helmholtz Centre for Ocean Research Kiel, Kiel, Germany

Luthfiyah Jannatunnisa Master Program in Earth Sciences, Faculty of Earth Science and Technology, Institut Teknologi Bandung, Bandung, Indonesia

Ina Juaeni Research Center for Climate and Atmosphere, National Research and Innovation Agency (BRIN), Bandung, West Java, Indonesia

Neneng S. Juariah National Research and Innovation Agency, Jakarta, Indonesia

Kiki Kartikasari Carbon and Environmental Research Indonesia, Bogor, Indonesia

Amanda Pasa Kencana Bandung Institute of Technology, Bandung, Indonesia

Dwi Rustam Kendarto Department Agriculture Engineering, Faculty of Agricultural Industrial Technology, Padjadjaran University, Sumedang, Indonesia

Timothy Kenoly Marine Science Program, Faculty of Marine Sciences and Fisheries, Udayana University, Bali, Indonesia

Ninong Komala Research Center for Climate and Atmosphere, National Research and Innovation Agency, Bandung, Indonesia;
Research Center of Climate and Atmosphere, National Research and Innovation Agency, Central Jakarta, Indonesia

Prawira Yudha Kombara Research Center for Climate and Atmosphere, National Research and Innovation Agency, Bandung, Indonesia;
Research Center of Climate and Atmosphere, National Research and Innovation Agency, Central Jakarta, Indonesia

Osamu Kozan Center for Southeast Asian Studies (CSEAS), Kyoto University, Kyoto, Japan;
Research Institute for Humanity and Nature (RIHN), Kyoto, Japan

Prawira Yuda Kumbara National Research and Innovation Agency, Bandung, Indonesia

Andriarti Kusumawardani Directorate of Food Crops Protection, Directorate General of Food Crops, Indonesia Ministry of Agriculture, Jakarta, Indonesia

Farid Lasmono Center for Data and Information, National Research and Innovation Agency (BRIN), Jakarta, Indonesia;
Indonesian National Research and Innovation Agency, Jakarta, Indonesia

Deni Okta Lestari Department of Atmospheric and Planetary Science, Sumatera Institute of Technology, Lampung, Indonesia

Retno Puji Lestari The Center for Standardization of Environmental Quality Instruments, The Ministry of Environment and Forestry, Jakarta, Indonesia

Sopia Lestari Research Center for Climate and Atmosphere, National Research and Innovation Agency, Bandung, West Java, Indonesia;
National Research and Innovation Agency, Jakarta, Indonesia

U. A. Linarka Meteorology, Climatology, and Geophysical Agency, Jakarta, Indonesia;
National Research and Innovation Agency, Jakarta, Indonesia

D. N. Lintangsasi Undergraduate School of Faculty of Geography, Gadjah Mada University, Yogyakarta, Indonesia

Indah Listiana Faculty of Agriculture, University of Lampung, Bandar Lampung, Indonesia

Niken Ida Lovita Department Agriculture Engineering, Faculty of Agricultural Industrial Technology, Padjadjaran University, Sumedang, Indonesia

Ellida Novita Lydia Universitas Samudra, Langsa-Aceh, Indonesia

Olivia Maftukhaturrizqoh Research Center for Remote Sensing, National Research and Innovation Agency, Jakarta, Indonesia

Sindy Maharani I Gusti Ngurah Rai Meteorological Office, Indonesian Agency for Meteorology Climatology and Geophysics, Bali, Indonesia

Gandhi Mahendra Sentani Meteorological Station, Indonesian Agency for Meteorology Climatology and Geophysics, Sentani, Jayapura Region, Indonesia

Dyah Rahayu Martiningrum Research Center for Climate and Atmosphere, National Research and Innovation Agency, Bandung, Indonesia

Martono Research Center for Climate and Atmosphere, National Research and Innovation Agency, West Java, Indonesia

Edy Maryadi Research Center for Data Science and Information, National Research and Innovation Agency, Jakarta Pusat, Indonesia;
BRIN—Research Organization for Electronics and Informatics, Bandung, Indonesia;
Research Center for Data and Information Sciences, National Research and Innovation Agency, West Java, Indonesia

Marzuki Marzuki Department of Physics, Universitas Andalas, Padang, Indonesia

Yutaka Matsumi Nagoya University, Nagoya, Japan

Nata Miharja Research Center for Space, Bandung, West Java, Indonesia

Misnawati The Research Center for Climate and Atmosphere, National Research and Innovation Agency, Bandung, Indonesia;
National Research and Innovation Agency, Jakarta, Indonesia

Yash Kumar Mittal Malaviya National Institute of Technology, Jaipur, India

Robi Muharsyah Agency for Meteorology, Climatology and Geophysics of Republic Indonesia, Jakarta, Indonesia

Subekti Mujiasih Research Center for Climate and Atmosphere, National Research and Innovation Agency (BRIN), Bandung, West Java, Indonesia;
GeoHydrodynamics and Environment Research (GHER), Department of Astrophysics, Geophysics and Oceanography (AGO), University of Liège, Liège, Belgium;
Research Center for Climate and Atmosphere, National Research and Innovation Agency (BRIN), Jakarta, Indonesia

Hanung Agus Mulyadi Research Center for Oceanography, National Research and Innovation Agency, Jakarta Utara, Indonesia;
Department of Marine Science and Technology, Faculty of Fisheries and Marine Science, IPB University, Bogor, Indonesia

Muhammad Arif Munandar Indonesian Agency for Meteorological, Climatological, and Geophysics, Jakarta, Indonesia

La Ode M. Musafar Kilowasid Pontianak Observatory for Space and Atmosphere, Pontianak, West Kalimantan, Indonesia

Weny Anggi Mustika Regional III Office, Indonesian Agency for Meteorology Climatology and Geophysics, Bali, Indonesia

Eka Mutia Universitas Samudra, Langsa-Aceh, Indonesia

Abdul Mutolib Graduate Program, University of Siliwangi, Tasikmalaya, Indonesia

Muzirwan Research Center for Climate and Atmosphere, Bandung, Indonesia

Aisya Nafiisyanti Research Center for Climate and Atmosphere, National Research and Innovation Agency (BRIN), Bandung, West Java, Indonesia;
Indonesian National Research and Innovation Agency, Jakarta, Indonesia

Gandhi Napitupulu Department of Earth Sciences, Faculty of Earth Sciences and Technology, Institut Teknologi Bandung, Bandung, Indonesia

Ida Narulita Research Centre for Climate and Atmospheric Science, National Agency for Research and Innovation, Bandung, Indonesia;
Faculty of Engineering, University of Indonesia, Depok, Indonesia

Fadli Nauval Research Center for Climate and Atmosphere, National Research and Innovation Agency (BRIN), Bandung, West Java, Indonesia

Sushmi Nimje Malaviya National Institute of Technology, Jaipur, India

Takeshi Nishimura Department of Geophysics, Tohoku University, Sendai, Japan

Noersomadi National Research and Innovation Agency, Jakarta, Indonesia;
Research Center for Climate and Atmosphere, National Research and Innovation Agency (BRIN), Bandung, Indonesia

Acep Catur Nugraha Research Center for Climate and Atmosphere, National Research and Innovation Agency (BRIN), Bandung, West Java, Indonesia;
Indonesian National Research and Innovation Agency, Jakarta, Indonesia

Reva Edra Nugraha Department of Chemical Engineering, Faculty of Engineering, Universitas Pembangunan Nasional "Veteran" Jawa Timur, Surabaya, East Java, Indonesia

Dwiyoga Nugroho Ministry of Marine Affairs and Fisheries Republic of Indonesia, Jakarta Pusat, Indonesia

Ginaldi Ari Nugroho BRIN—Research Organization for Earth Sciences and Maritime, Bandung, Indonesia

Fitri Nuraeni Research Center for Space, Bandung, West Java, Indonesia

Candra Nuraini Faculty of Agriculture, University of Siliwangi, Tasikmalaya, Indonesia

Susanna Nurdjaman Oceanography Research Group, Faculty of Earth Sciences and Technology, Institut Teknologi Bandung, Bandung, Indonesia

Erti Nurfindarti Regional Development Planning Agency of Serang City, Serang, Indonesia

Ayu Utami Nurhidayati Research Center of Climate and Atmosphere, Research Group of Paleoclimate and Paleoenvironment, National Research and Innovation Agency, Jakarta, Indonesia

Amalia Nurlatifah Research Center for Climate and Atmosphere, National Research and Innovation Agency, Jakarta Pusat, Indonesia;
Research Center of Climate and Atmosphere, National Research and Innovation Agency, Central Jakarta, Indonesia;
Research Center of Climate and Atmosphere, National Research and Innovation Agency, West Java, Indonesia;

Research Center for Climate and Atmosphere, National Research and Innovation Agency, Jakarta, Indonesia

Siti Nurzakiah Research Center for Food Crops, National Research and Innovation Agency, Cibinong, Bogor, Indonesia

Mariko Ogawa Center for Southeast Asian Studies (CSEAS), Kyoto University, Kyoto, Japan

Shin-Ya Ogino Japan Agency for Marine-Earth Science and Technology, Yokosuka, Japan

Koko Ondara Research Center for Oceanography, National Research and Innovation Agency, Jakarta, Indonesia

Dewangga Palguna Komodo Meteorological Station, Indonesian Agency for Meteorology Climatology and Geophysics, West Manggarai, Indonesia

Tri Astuti Pandansari Research Center for Remote Sensing, National Research and Innovation Agency, Jakarta, Indonesia

Hansan Park Korea-Indonesia Marine Technology Cooperation Research Center, Bandung Institute of Technology, Bandung, Indonesia

Namira N. Perdani Faculty of Fishery and Marine Science, Universitas Padjadjaran, West Java, Indonesia

Namira Nasywa Perdani Marine Sciences Department, Faculty of Fisheries and Marine Sciences, Padjajaran University, Bandung, Indonesia

Perdinan IPB University, Bogor, Indonesia

Teguh A. Pianto Research Center for Geospatial, National Research and Innovation Agency, Geostech Building, PUSPIPTEK, South Tangerang, Banten, Indonesia

Mahardika Jalu Pradana Agency for Meteorology, Climatology, and Geophysics of the Republic of Indonesia (BMKG), Jakarta, Indonesia

Alfan Sukmana Praja Research Center for Climate and Atmosphere, National Research and Innovation Agency (BRIN), Bandung, West Java, Indonesia

Aris Pramudia Research Center for Climate and Atmospheric, National Research and Innovation Agency, Kota, Bandung, West Java, Indonesia;
Innovation Center for Tropical Science, Bogor, Indonesia

Setyanto C. Pranoto Research Center for Space, Bandung, West Java, Indonesia

Widodo Setiyo Pranowo National Research and Innovation Agency of Indonesia (BRIN), Jakarta, Indonesia

Aditya Dwi Prasetio Research Center for Geological Disasters, National Agency of Research and Innovation (BRIN), Jawa Barat, Indonesia

Alvin Pratama Department of Atmospheric and Planetary Science, Sumatera Institute of Technology, Lampung, Indonesia

Khafid Rizki Pratama Department of Earth Sciences, Faculty of Earth Sciences and Technology, Institut Teknologi Bandung, Bandung, Indonesia

Ridho Pratama Research Center for Climate and Atmosphere, National Research and Innovation Agency (BRIN), Bandung, West Java, Indonesia

Danil Rahmad Priatna Universitas Samudra, Langsa-Aceh, Indonesia

Dendi Rona Purnama Public Weather Services, Indonesian Agency for Meteorology Climatology and Geophysics, Kemayoran, Central Jakarta, Indonesia

Narizka Nanda Purwadani Master Program in Earth Sciences, Faculty of Earth Sciences and Technology, Institut Teknologi Bandung, Bandung, Indonesia

Adi Purwandana National Research and Innovation Agency of Indonesia (BRIN), Jakarta, Indonesia

Anis Purwaningsih Research Center for Climate and Atmosphere, National Research and Innovation Agency (BRIN), Bandung, West Java, Indonesia

Asis Purwoko Directorate of Food Crops Protection, Directorate General of Food Crops, Indonesia Ministry of Agriculture, Jakarta, Indonesia

Wasana Putklang Geoinformatics, Department of Computer Science, College of Computing, Khon Kaen University, Khon Kaen, Thailand

Angga Yolanda Putra Pontianak Observatory for Space and Atmosphere, Pontianak, West Kalimantan, Indonesia;
National Research and Innovation Agency, Bandung, Indonesia

I. Dewa Gede Arya Putra Center for Research and Development, Meteorological, Climatological, and Geophysical Agency (BMKG), Jakarta Pusat, Indonesia

I. Nyoman Agus Astina Putra Kendari Marine Meteorological Station, Indonesian Agency for Meteorology Climatology and Geophysics, Kendari, Indonesia

Arif Rachman Directorate of Laboratory Management, Research Facilities, and Science and Technology Park, National Research and Innovation Agency, Jakarta, Indonesia

Arief Rachman Research Center for Oceanography, National Research and Innovation Agency, Jakarta Utara, Indonesia;
Directorate of Laboratory Management, Research Facilities, and Science and Technology Park, National Research and Innovation Agency, Jakarta, Indonesia

Atep Radiana Research Center for Climate and Atmosphere, National Research and Innovation Agency-BRIN, Bandung, Indonesia

Ivonne M. Radjawane Oceanography Research Group, Faculty of Earth Sciences and Technology, Institut Teknologi Bandung, Bandung, Indonesia;
Korea-Indonesia Marine Technology Cooperation Research Center, Bandung Institute of Technology, Bandung, Indonesia

Yunita Fauziah Rahim Directorate of Food Crops Protection, Directorate General of Food Crops, Indonesia Ministry of Agriculture, Jakarta, Indonesia

A. Rahmadya Research Center for Limnology and Water Resources, National Research and Innovation Agency (BRIN), Jakarta Pusat, Indonesia

Ali Rahmat Research Center for Limnology and Water Resources, National Research and Innovation Agency, Bogor, Indonesia;
Research Center for Limnology and Water Resources, National Research and Innovation Agency, Jakarta Pusat, Indonesia

Fahmi Rahmatia Research Center for Climate and Atmosphere, National Research and Innovation Agency (BRIN), Bandung, West Java, Indonesia;
Agam Space and Atmospheric Observation Station, National Research and Innovation Agency, West Sumatera, Indonesia

M. Rizqi Ramadhan Oceanography Department, Faculty of Earth Sciences and Technology, Bandung Institute of Technology, Bandung, Indonesia

Ravidho Ramadhan Department of Physics, Universitas Andalas, Padang, Indonesia

F. Ramadhani National Research and Innovation Agency, Jakarta, Indonesia

Andri Ramdhani Agency for Meteorology, Climatology, and Geophysics of the Republic of Indonesia (BMKG), Jakarta, Indonesia

Dwi Ratnasari Mataram University, Lombok, Indonesia

Herlina Ika Ratnawati National Research and Innovation Agency of Indonesia (BRIN), Jakarta, Indonesia

Dian Nur Ratri Meteorological, Climatological, and Geophysical Agency (BMKG), Jakarta, Indonesia;
Droevendaalsesteeg, Wageningen University and Research, Wageningen, The Netherlands

Findy Renggono National Research and Innovation Agency, Jakarta, Indonesia

Edi Riawan Bandung Institute of Technology, Bandung, Indonesia

Ainur Ridho Cerdas Antisipasi Risiko Bencana Indonesia (CARI), Bandung, Indonesia

I. Ridwansyah Research Center for Limnology and Water Resources, National Research and Innovation Agency (BRIN), Jakarta Pusat, Indonesia

Risyanto Research Center for Climate and Atmosphere, National Research and Innovation Agency (BRIN), Bandung, West Java, Indonesia;
Laboratory Management, Research Facilities and Science, and Technology Park, National Research and Innovation Agency, Central Jakarta, Indonesia

Syahril Rizal Bina Darma University, Palembang, Indonesia

Moch S. Romadhon Center for Climate and Atmosphere, National Research and Innovation Agency, Jakarta Pusat, Indonesia

Nugrahana Fitria Ruhyana Regional Research and Development Planning Agency of Sumedang Regency, Sumedang, Indonesia;
Universitas Sebelas April, Sumedang, Indonesia

Sirapop Rukhajee Geoinformatics, Department of Computer Science, College of Computing, Khon Kaen University, Khon Kaen, Thailand

Thanadon Brahmasakha Na Sakolnagara Geoinformatics, Department of Computer Science, College of Computing, Khon Kaen University, Khon Kaen, Thailand

Toni Samiaji Research Center for Climate and Atmosphere, National Research and Innovation Agency, Bandung, West Java, Indonesia

Wilin Julian Sari Research Center for Quantum Physics, National Research and Innovation Agency, Jakarta, Indonesia

Didi Satiadi Research Center for Climate and Atmosphere, National Research and Innovation Agency (BRIN), Bandung, West Java, Indonesia;
Indonesian National Research and Innovation Agency, Jakarta, Indonesia

I Kadek Mas Satriyabawa I Gusti Ngurah Rai Meteorological Office, Indonesian Agency for Meteorology Climatology and Geophysics, Bali, Indonesia

Haries Satyawardhana National Research and Innovation Agency, Jakarta, Indonesia;
Research Center for Atmospheric Science and Technology, Indonesia National Agency of Research and Innovation, Bandung, Indonesia

Elfira Saufina Research Center for Climate and Atmosphere, National Research and Innovation Agency (BRIN), Bandung, West Java, Indonesia

Kiky Corneliasari Sembiring Research Center for Chemistry, National Research and Innovation Agency (BRIN), South Tangerang, Banten, Indonesia

Harry Septanto Research Center for Smart Mechatronics, National Research and Innovation Agency, KST Samaun Samadikun, Bandung, Indonesia

F. Setiawan Research Center for Limnology and Water Resources, National Research and Innovation Agency (BRIN), Jakarta Pusat, Indonesia

Rachmad Setijadi Department of Geological Engineering, Universitas Jenderal Soedirman, Purbalingga, Indonesia

Wiwiek Setyawati Research Center for Climate and Atmosphere, National Research and Innovation Agency-BRIN, Bandung, Indonesia;
Research Center for Climate and Atmosphere, National Research and Innovation Agency, Jakarta, Indonesia

Takashi Shibata Nagoya University, Nagoya, Japan

Tiin Sinatra BRIN—Research Organization for Earth Sciences and Maritime, Bandung, Indonesia;
Research Center for Climate and Atmosphere, National Research and Innovation Agency (BRIN), Bandung, West Java, Indonesia

Sinta B. Sipayung Research Center of Climate and Atmosphere, National Research and Innovation Agency, West Java, Indonesia

Sinta Berliana Sipayung Research Center for Climate and Atmosphere, National Research and Innovation Agency, West Java, Indonesia;
Research Center for Climate and Atmosphere, National Research and Innovation Agency, Jakarta Pusat, Indonesia

Yogina Lestari Ayu Situmorang Department Agriculture Engineering, Faculty of Agricultural Industrial Technology, Padjadjaran University, Sumedang, Indonesia

Lilik Slamet Research Center for Climate and Atmosphere, National Research and Innovation Agency, Jakarta Pusat, Indonesia

Iis Sofiati Research Center for Climate and Atmosphere, National Research and Innovation Agency, Bandung, Indonesia;
Research Center for Atmospheric Science and Technology, Indonesia National Agency of Research and Innovation, Bandung, Indonesia

Tubagus Solihuddin Research Center of Climate and Atmosphere, Research Group of Paleoclimate and Paleoenvironment, National Research and Innovation Agency, Jakarta, Indonesia

A. Sopaheluwakan Meteorology, Climatology, and Geophysical Agency, Jakarta, Indonesia

Ardhasena Sopaheluwakan Indonesia Agency for Meteorology, Climatology, and Geophysics, Jakarta, Indonesia

Jakkrapong Srisuwan Geoinformatics, Department of Computer Science, College of Computing, Khon Kaen University, Khon Kaen, Thailand

Suaydhi Research Centre for Climate and Atmosphere, National Research and Innovation Agency (BRIN), Bandung, Indonesia

Luki Subehi Research Center for Limnology and Water Resources, National Research and Innovation Agency, Bogor, Indonesia

Suciantini The Research Center for Climate and Atmosphere, National Research and Innovation Agency, Bandung, Indonesia;
National Research and Innovation Agency, Jakarta, Indonesia

Bambang Suhandi Research Center for Climate and Atmosphere, National Research and Innovation Agency, Bandung, Indonesia

Akas Pinaringan Sujalu Universitas, Samarinda, Indonesia

Sukamto Hokkaido University, Sapporo, Japan

Sukarjo Research Center for Horticultural and Estate Crop, National Research and Innovation Agency, Cibinong Science Center, Cibinong, Indonesia

Albertus Sulaiman Research Center for Climate and Atmosphere, National Research and Innovation Agency (BRIN), Jawa Barat, Indonesia;
Research Center for Climate and Atmosphere, BRIN, Bandung, Indonesia

Reni Sulistyowati Research Center for Geospatial, National Research and Innovation Agency, Geostech Building, PUSPIPTEK, South Tangerang, Banten, Indonesia;
Research Center for Geospatial, National Research and Innovation Agency (BRIN), Gedung GEOSTECH, Serpong, Tangerang Selatan, Indonesia

Kadek Sumaja I Gusti Ngurah Rai Meteorological Office, Indonesian Agency for Meteorology Climatology and Geophysics, Bali, Indonesia

Sumaryati Research Center for Climate and Atmosphere, National Research and Innovation Agency-BRIN, Bandung, Indonesia;
Research Center of Climate and Atmosphere, National Research and Innovation Agency, Central Jakarta, Indonesia

Lilik S. Supriatin Research Center of Climate and Atmosphere, National Research and Innovation Agency, West Java, Indonesia

Lilik Slamet Supriatin Research Center for Climate and Atmosphere, National Research and Innovation Agency, West Java, Indonesia

Djoko Suprijanto Faculty of Mathematics and Natural Sciences, Combinatorial Mathematics Research Group, Institut Teknologi Bandung, Bandung, Indonesia

Totok Suprijo Oceanography Research Group, Faculty of Earth Sciences and Technology, Institut Teknologi Bandung, Bandung, Indonesia

Jen Supriyanto Environmental Agency, Balikpapan, Indonesia

Dewi Surinati National Research and Innovation Agency of Indonesia (BRIN), Jakarta, Indonesia

E. Surmaini National Research and Innovation Agency, Jakarta, Indonesia

Elza Surmaini Research Centre for Climate and Atmosphere, National Research and Innovation Agency, Bandung, Indonesia

Arief Suryantoro Research Center for Climate and Atmosphere, National Research and Innovation Agency (BRIN), Bandung, West Java, Indonesia

E. Susanti National Research and Innovation Agency, Jakarta, Indonesia

Erni Susanti The Research Center for Climate and Atmosphere, National Research and Innovation Agency, Bandung, Indonesia

Indah Susanti Directorate of Repository, Multi Media, and Scientific Publishing, National Research and Innovation Agency, West Java, Indonesia;
Directorate of Repository, Multimedia, and Scientific Publishing, National Research and Innovation Agency, Jakarta Pusat, Indonesia

Dewi Syahidah Research Centre for Veterinary, National Research and Innovation Agency, Indonesia (BRIN), Jakarta, Indonesia

M. R. Syahputra Bandung Institute of Technology, Bandung, Indonesia

Muhammad Ridho Syahputra Atmospheric Sciences Research Group, Faculty of Earth Science and Technology, Institut Teknologi Bandung, Bandung, Indonesia

Syahrul BRIN—Research Organization for Electronics and Informatics, Bandung, Indonesia

Fadli Syamsudin Faculty of Fishery and Marine Science, Universitas Padjadjaran, West Java, Indonesia

Syuhada Research Center for Geological Disasters, National Agency of Research and Innovation (BRIN), Jawa Barat, Indonesia

Muhammad Takdir Mulyadi Directorate of Food Crops Protection, Directorate General of Food Crops, Indonesia Ministry of Agriculture, Jakarta, Indonesia

Dyah Aries Tanti Research Center for Climate and Atmosphere, National Research and Innovation Agency-BRIN, Bandung, Indonesia;
Research Center for Climate and Atmosphere, National Research and Innovation Agency, Jakarta, Indonesia

Ankiq Taofiqorahman Marine Sciences Department, Faculty of Fisheries and Marine Sciences, Padjajaran University, Bandung, Indonesia

Opik Taopik Directorate of Laboratory Management, Research Facilities, and Science and Technology Park, National Research and Innovation Agency, Jakarta, Indonesia

Laras Toersilowati Research Center of Climate and Atmosphere, National Research and Innovation Agency, West Java, Indonesia

Eko Triarso Ministry of Marine Affairs and Fisheries, Jakarta, Indonesia

Nurjanna Joko Trilaksono Atmospheric Sciences Research Group, Faculty of Earth Sciences and Technology, Institut Teknologi Bandung, Bandung, Indonesia

Trinugroho Research Center for Geological Disaster, BRIN, Bandung, Indonesia

Trismidianto Research Center for Climate and Atmosphere, National Research and Innovation Agency (BRIN), Bandung, West Java, Indonesia

Rainer Arief Troa Ministry of Marine Affairs and Fisheries, Jakarta, Indonesia

Laras Tursilowati Research Center for Climate and Atmosphere, National Research and Innovation Agency, Jakarta Pusat, Indonesia

Satriya Utama Research Center for Satellite Technology, Bogor, West Java, Indonesia

Dwi Amanda Utami Research Center of Climate and Atmosphere, Research Group of Paleoclimate and Paleoenvironment, National Research and Innovation Agency, Jakarta, Indonesia

Mutya Vonnisa Department of Physics, Universitas Andalas, Padang, Indonesia

Nur Wakhid Research Center for Ecology and Ethnobiology, National Research and Innovation Agency, Cibinong, Bogor, Indonesia

Trinah Wati Graduate Program of Earth Sciences, Faculty of Earth Sciences and Technology, Institut Teknologi Bandung, Bandung, Indonesia;
Indonesia Agency for Meteorology, Climatology, and Geophysics, Jakarta, Indonesia

Visca Wellyanita Research Center for Space, Bandung, West Java, Indonesia

Sigit Kurniawan Jati Wicaksana Research Centre for Oceanography, National Research and Innovation Agency, Jakarta, Indonesia

Raden Ajeng Diana Widyastuti Faculty of Agriculture, University of Lampung, Bandar Lampung, Indonesia

Tirtha Wijaya Agency for Meteorology, Climatology, and Geophysics of the Republic of Indonesia (BMKG), Badung, Bali, Indonesia

Yusuf Jati Wijaya Department of Oceanography, Faculty of Fisheries and Marine Science, Diponegoro University, Semarang, Indonesia

Ulung Jantama Wisha Department of Physics and Earth Sciences, Graduate School of Engineering and Science, University of the Ryukyus, Nishihara, Japan;
Research Center for Oceanography, National Research and Innovation Agency (BRIN), Jakarta, Indonesia

Adi Witono Research Center for Climate and Atmosphere, National Research and Innovation Agency, Jakarta Pusat, Indonesia;
Research Center of Climate and Atmosphere, National Research and Innovation Agency, West Java, Indonesia;
Research Center for Climate and Atmosphere, National Research and Innovation Agency (BRIN), Bandung, West Java, Indonesia

Eka P. Wulandari Research Center for Atmospheric Science and Technology, Indonesia National Agency of Research and Innovation, Bandung, Indonesia

Eka Putri Wulandari National Research and Innovation Agency, Jakarta, Indonesia

Manabu D. Yamanaka Research Institute for Humanity and Nature, Kyoto, Japan; Professor Emeritus, Kobe University, Kobe, Japan;
Japan Agency for Marine-Earth Science and Technology, Yokosuka, Japan

Helvi Yanfika Faculty of Agriculture, University of Lampung, Bandar Lampung, Indonesia

Clara Y. Yatini Research Center for Space, Bandung, West Java, Indonesia

Mutia Yollanda Department of Mathematics and Data Science, Andalas University, Padang, West Sumatera, Indonesia;
Department of Mathematics, UNAND, West Sumatera, Indonesia

Erma Yulihastin Research Center for Climate and Atmosphere, National Research and Innovation Agency (BRIN), Bandung, West Java, Indonesia;
National Research and Innovation Agency of Indonesia (BRIN), Jakarta, Indonesia;
Research Center for Atmospheric Science and Technology, Indonesia National Agency of Research and Innovation, Bandung, Indonesia

Helmi Yusnaini Department of Physics, Universitas Andalas, Padang, Indonesia

Teti Zubaidah Mataram University, Lombok, Indonesia

Rina Zuraida Research Center for Climate and Atmosphere-National Research and Innovation Agency, Bandung, Indonesia

Hidayatuz Zu'amah Research Center for Horticultural and Estate Crops, National Research and Innovation Agency, Cibinong Science Center, Cibinong, Bogor, Indonesia

Indonesian Coastlines Controlling the Whole Earth's Atmosphere

Manabu D. Yamanaka and Shin-Ya Ogino

Abstract A coastline is the triple boundary among land, ocean, and atmosphere constructing the Earth system. We have revealed that the diurnal cycle (sea-land breeze circulation) along the world's longest coastline surrounding major islands of the Indonesian maritime continent (IMC) is the most robust mode of cloud-rainfall generation. This process produces latent heat compensating the global radiational-dynamical energy imbalance, and any modifications of the IMC land surface may change the local diurnal cycle and finally the global climate. The diurnal cycle circulation consists of equi-amplitude, bidirectional (sea- and land-ward propagating), internal gravity waves, and our recent analysis (Yamanaka and Ogino in Tropical coastlines robustizing the stratospheric quasi-biennial oscillation. (in preparation) (2022)) shows that they propagate also upward and maintain the stratospheric most robust mode: quasi-biennial oscillation (QBO). As so far shown by many studies, QBO affects the whole middle and upper atmospheric processes, and therefore the IMC coastlines control the whole Earth's atmosphere.

1 Introduction

Robust periodic phenomena in the atmosphere should be forced by fixed (e.g., astronomical) sources. However, any such sources have been known for the quasi-biennial oscillation (QBO; discovered by [4, 28] reviews by e.g., [1, 2, 8, 9] maintained robustly with around 2.5 years periodicity in the equatorial lower stratosphere. Almost zonally-uniform zonal-wind amplitude and downward phase progression are around $\pm$ 20 m/s and (20–35 km)/(27 months/2) ≈ -1.1 km/month, respectively,

M. D. Yamanaka (✉)
Research Institute for Humanity and Nature, Kyoto, Japan
e-mail: mdy@chikyu.ac.jp; mdy@kobe-u.ac.jp

Kobe University, Kobe, Japan

M. D. Yamanaka · S.-Y. Ogino
Japan Agency for Marine-Earth Science and Technology, Yokosuka, Japan

A. Basit et al. (eds.), *Proceedings of the International Conference on Radioscience, Equatorial Atmospheric Science and Environment and Humanosphere Science*, Springer Proceedings in Physics 290, https://doi.org/10.1007/978-981-19-9768-6_1

and the easterly phase is somewhat stronger and slower than the westerly phase. QBO has been explained by the wave-mean flow interaction (e.g., [19, 25, 26], but requested bidirectional (east- and west-ward propagating) waves generated in the troposphere have not been identified.

Another robust phenomenon confirmed recently in the equatorial lower troposphere is the coastal diurnal cycle (CDC) of sea-land breeze circulations (SLBC) (e.g., reviews by [34, 37]). The diurnal cycle is more dominant than seasonal (annual and semiannual) cycles and on land also than the intraseasonal/interannual variations, and the latter phenomena appear as amplitude modulations of the former. Those climatological modulations (as well as anthropogenic land modifications) cause the sea-land temperature contrast that forces the CDC-SLBC. SLBC is a sum of up- and downward propagating internal gravity waves in latitudes lower than about 30° ([29], Section 6.5 of [17]). Dominance of CDC is the reason why the Indonesian maritime continent (IMC) with the longest coastlines has the most active convection on Earth [27, 34].

These two robust phenomena appear dominantly in the latitudes lower than about 15°, which is narrower than the internal range of diurnal waves, and both of them are still not accurately simulated in an operational numerical model. These suggest that they are related to each other. The equatorial SLBC is somewhat larger horizontally and vertically than the other (higher) latitude regions, which corresponds to horizontal and vertical wavelengths of 10^3 km and 10 km, respectively, of internal gravity waves with 1-day periodicity of 10^2 times of Väisälä-Brunt period (as expected from the dispersion relation).

In this paper, we analyze a global objective reanalysis dataset of horizontal wind and pressure velocity and obtain components of 1-day periodicity and zonal scale shorter than 30° in longitude. They are concentrated along coastlines of IMC and African and South American continents at the surface, and their local-time behaviors correspond clearly to SLBC. Upward fluxes of their zonal momenta are distributed in zonal-wind vertical shears of the same direction, as expected from the QBO theory.

2 Data and Evidence of Coastal Diurnal Cycle (CDC)

Major features of CDC-SLBC analyzed from campaign observations over major islands of IMC have been shown in the INCREASE symposium last year [35]. SLBC corresponds to internal gravity waves, with sea-/land-ward phase velocity $c = \omega/k$, where the frequency $\omega = 2\pi/1$ day, and the horizontal wavenumber $k \sim 2\pi/(10^2-10^3$ km). Usually, $c = (1-3) \times 10^2$ km/10 h $= 3-7$ m/s, which is faster on sea than on land and is much slower than cumulonimbus gust (~20 m/s). The amplitude is proportional to the sea-land temperature contrast $k\Delta T$, where ΔT is the temperature difference determined mainly by the land surface temperature with daytime heating and nighttime cooling. Inland sea propagations between major islands and re-excitation at small islands on the way have been observed.

In small islands, propagation directions are unclear, and the periodicities are often semi-diurnal (with a secondary peak in the seaside of a major island).

In order to study CDC-SLBC over the whole equatorial region, we [36] used a 1.25° latitude/longitude grid dataset of the Japanese 55-year Reanalysis (JRA-55; [14]) for 20-years during 00 UTC 1 January 2000–18 UTC 31 December 2019. In the vertical direction and time, data are given for the surface and 37 isobaric surfaces (1000, 975, 950, 925, 900, 875, 850, 825, 800, 775, 750, 700, 650, 600, 550, 500, 450, 400, 350, 300, 250, 225, 200, 175, 150, 125, 100, 70, 50, 30, 20, 10, 7, 5, 3, 2, and 1 hPa) every six hours (00, 06, 12, and 18 UTC). We have used the same dataset of 1981–2010 for analyzing the coastal concentration along coastlines mainly due to CDC [23]. The JRA-55 dataset has been used in QBO studies (e.g., [1]). The 20-year mean of surface horizontal wind anomaly at each UTC from the diurnal mean and its divergence are shown in Fig. 1 for IMC and African and South American continents. We confirm CDC-SLBC is distributed along all the coastlines, and is most active over IMC with the longest coastlines [34, 37].

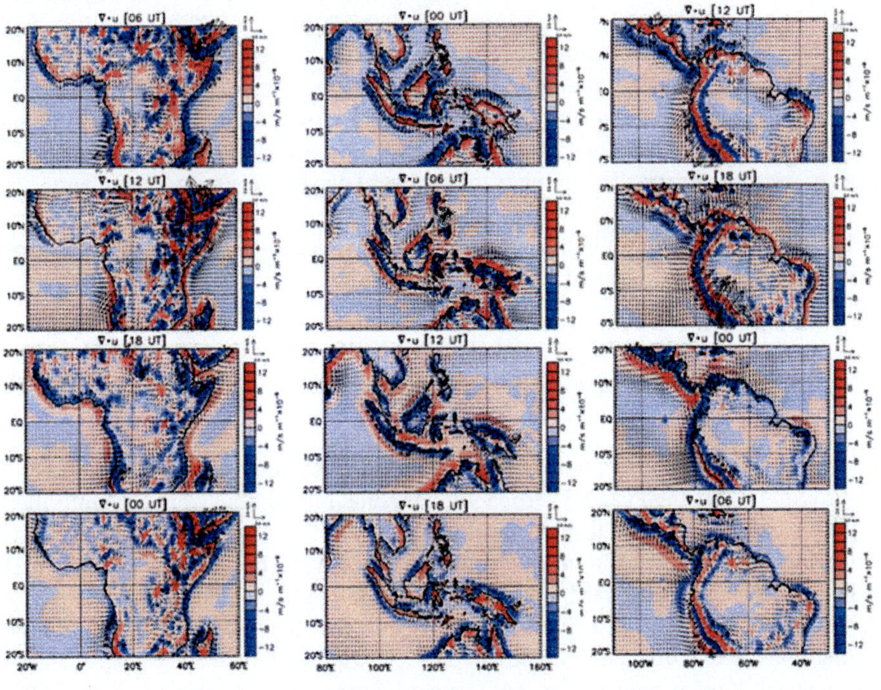

Fig. 1 Diurnal cycles analyzed from JRA-55 (2000–2019) for equatorial regions of Africa (left), IMC (middle), and South America (right) [36]. The 20-year mean of surface horizontal wind anomaly (vectors) at each UTC from the daily mean and its divergence (contours) are shown. Results for Africa and South America are displaced by +6 and −12 h, respectively, from those for IMC (LT = UTC + (7−9) h), showing the morning (top), afternoon (second row), evening (third row), and midnight (bottom)

3 Result on Gravity Waves and Zonal Wind

3.1 Coastal Diurnal Cycle (CDC)-Induced Gravity Waves

The diurnal cycle consists of local CDC-SLBC and global tide (see discussions later). An example of zonal wind anomaly from daily mean is shown in Fig. 2. A westward super-sonic (40,000 km/day = 463 m/s) propagation of the diurnal tide is found, but it is too fast to interact with the mean flow at least up to the middle atmosphere. The other disturbances are also diurnally reversed, but much shorter, vertical, and almost stationary (in such a global plot) in the troposphere below 100 hPa (about 17 km) level, which is reasonable as SLBCs. Between 100 and 30 hPa levels, where the zonal mean wind was westerly for the time of Fig. 3 (January 2000), contours seem rather horizontal, suggesting typical stratospheric gravity waves (with diurnal periodicity in this case), which is consistent to their interactions with the mean zonal flow and generation of QBO shown later in this paper.

In order to study internal gravity waves generated as the CDC-SLBC seen clearly in Fig. 1, we make a monthly mean and a high-pass filter of 30° in longitude, and obtain 20-year mean zonal-vertical plots along the equator of zonal-vertical flow and vertical flux of zonal momentum (divided by atmospheric density) as shown in Fig. 3, in which p-velocity in the p-coordinate are converted into vertical velocity in the $\ln p$ coordinate. Westward (eastward) waves are emitted dominantly from the west (east) coast to the troposphere. This implies that seaward waves are stronger than landward waves, and the bidirectional waves are balanced in the both coasts of each continent, rather than the day and night at each coast. Because of 20-year mean, stratospheric values of Fig. 3 are very small.

3.2 Mean Zonal Wind and QBO

Figure 4 shows the interannual-vertical variations of the zonal-mean zonal wind and longitudinally 30°-high passed vertical flux of zonal momentum (roughly multiplied by gravity acceleration). In the lower troposphere the zonal flow is very weak easterly, and an annual cycle of northern winter-peaked nearly 850 hPa eastward and summer-to-winter ascending westward momentum fluxes is found. In the upper troposphere an annual cycle of northern summer easterly and winter westerly is found, but the momentum flux is unclear. The stratospheric part is enlarged in Fig. 5. QBO is dominant as known well so far, and upward fluxes of zonal momenta of CDC-SLBC-waves distributed upward-decreasingly in zonal-wind vertical shears of the same directions are known newly in this paper. We have also confirmed that zonal momentum fluxes with lower cutoff and no filters show much worse correlations with QBO than 30°-high passed results shown in Figs. 4 and 5. Therefore, the CDC-SLBC waves make the largest contribution (at least among the whole 1-day period waves) to interaction with QBO. It has been reported that QBO was disrupted around 2016

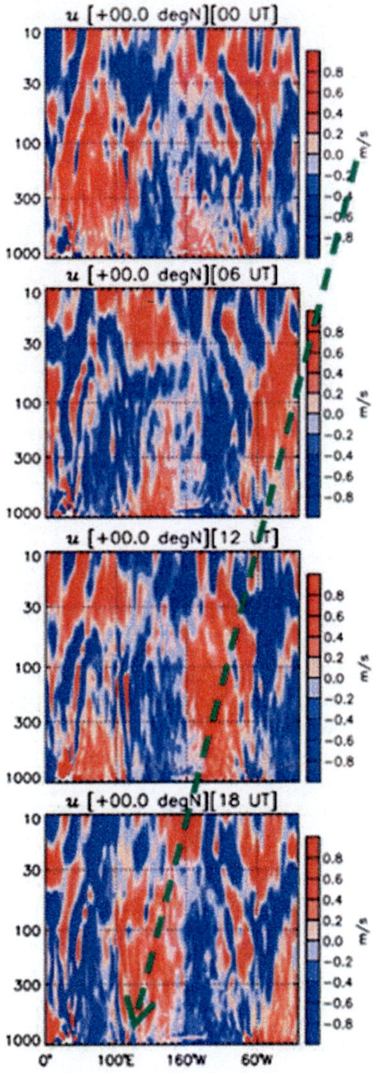

Fig. 2 Example of monthly-mean zonal-vertical plot along the equator at each UTC of zonal wind anomaly from daily mean of JRA-55 data in January 2000 [36]. In addition to local diurnal cycles, a fast westward propagating diurnal tide with a peak at 06 LT (indicated by a green arrow)

(e.g., [24]) (and also 2020 but is not seen well in the data until 2019). During this period, the CDC-SLBC wave momentum flux is also somewhat unusual.

3.3 *Discussions and Comparison with Theoretical Studies*

We have shown observationally that the both robust phenomena CDC-SLBC and QBO are related with each other. Theoretically, the CDC is forced by sea and land

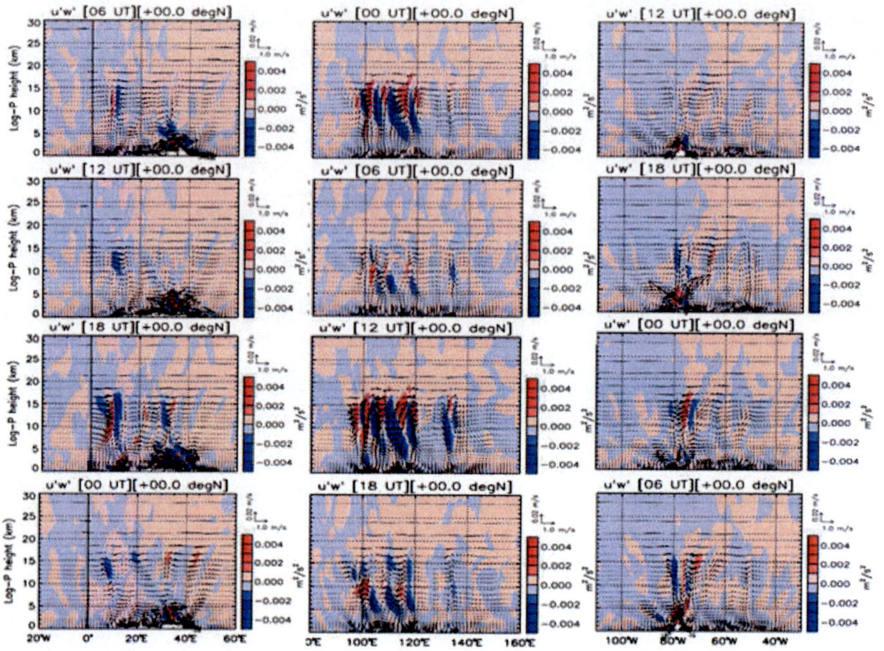

Fig. 3 Same as Fig. 1 but for zonal-vertical plots along the equator of longitudinally 30°-high passed monthly-mean zonal-vertical wind anomaly (vectors) from the daily mean and their product (contours) [36]

surface temperature difference as shown in Fig. 6. Coastal sea-surface temperature is changed by ~± 1 °C/year or weaker for interannual or intraseasonal variations, whereas the land surface temperature is changed by ~± 3 °C/half day or larger every day. Therefore, the bottom boundary condition of CDC-SLBC for temperature θ is:

$$\theta = \theta_0 - \theta_1(\sin kx + 1)\sin \omega t$$
$$= \theta_0 - \frac{\theta_1}{2}[\cos(kx - \omega t) - \cos(kx + \omega t) + 2\sin \omega t] \text{ at } z = 0, \qquad (1)$$

where x, z, and t are the distance from coast, altitude, and time, $k \sim 2\pi/(10^2-10^3)$ km, and $\omega \sim 2\pi/1$ day are horizontal wavenumber (a component perpendicular to the coastline) and frequency, θ_0 and $\theta_0 \pm 2\theta_1$ are sea-side and land-side (highest/lowest) temperature, respectively. The first two terms of the bracket of (1) excite bidirectional waves, and the last term corresponds to the diurnal tide with the circum-equatorial wavelength (= 4×10^4 km). The bidirectional waves make a standing wave in the troposphere, which is SLBC, and are separated by the mean zonal flow in the stratosphere: one wave with the same propagation direction as the vertical shear of the mean zonal flow is amplified and absorbed, and the other wave is propagated into the upper layer with the opposite shear of the mean zonal flow. The diurnal tide

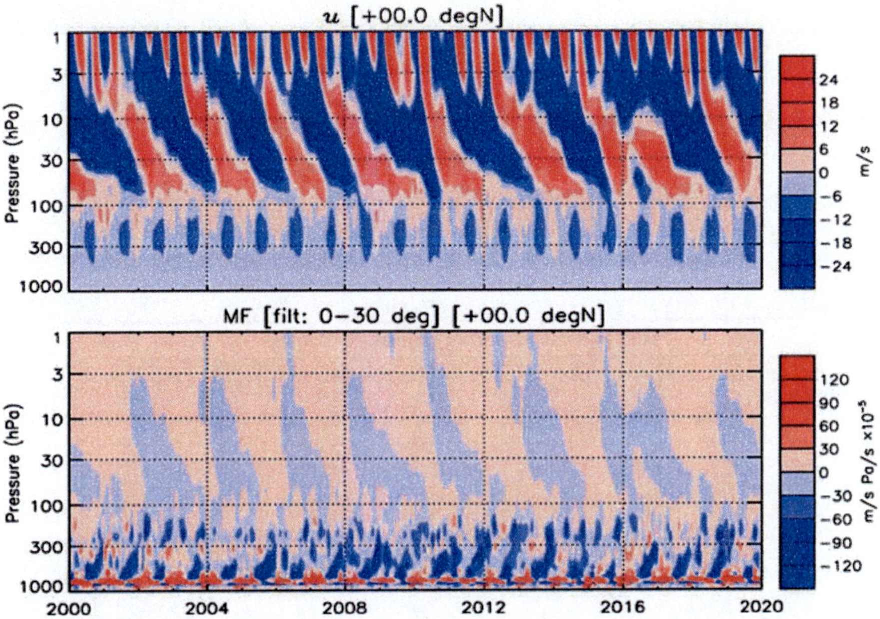

Fig. 4 Temporal-vertical sections along the equator of the zonal-daily-mean zonal flow (top) and product of longitudinally 30°-high passed monthly-mean zonal wind and –p-velocity anomalies from the daily means (bottom) analyzed from the 20-year JRA-55 data [36]

has a supersonic propagation velocity, which is far from and has no interaction with the mean flow (except the top of the atmosphere just below the turbopause; see, e.g., [22]). Note that shorter waves generated with individual clouds cannot propagate upward (i.e., trapped and ducting near the ground), as discussed by Scorer [30]. It should be noted that Kawatani et al. [12] studied contributions of various waves in a high-resolution general circulation model, but the diurnal component was filtered out.

The situations mentioned above are quite similar to Plumb's (1977) model and Plumb and McEwan's (1978) experiment which simulated QBO with standing waves forced at the bottom. The wave with the phase velocity (either ω/k or $-\omega/k$) with the same direction of vertical shear of the mean zonal flow is separated just below the level of the phase velocity approaching (but somewhat slower than) the mean zonal velocity, because vertical wavelength and group velocity become much smaller than the other (opposite direction) wave and small enough to cause local instability ("breaking") and/or radiative dissipation ("damping"). Thus, the mean zonal shear flow at somewhat lower level (somewhat slower than the wave) absorbs the wave momentum, and the mean zonal shear flow descends:

$$\frac{\partial U}{\partial t} = -\frac{1}{\rho}\frac{\partial}{\partial z}(\text{wave momentum flux})$$

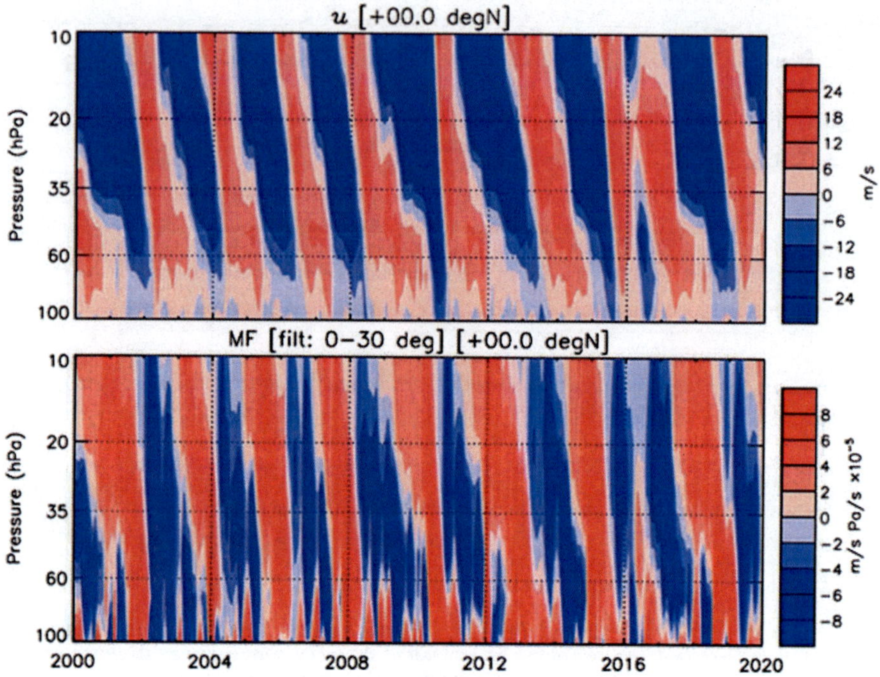

Fig. 5 Same as Fig. 4 but enlarged for the stratosphere [36]

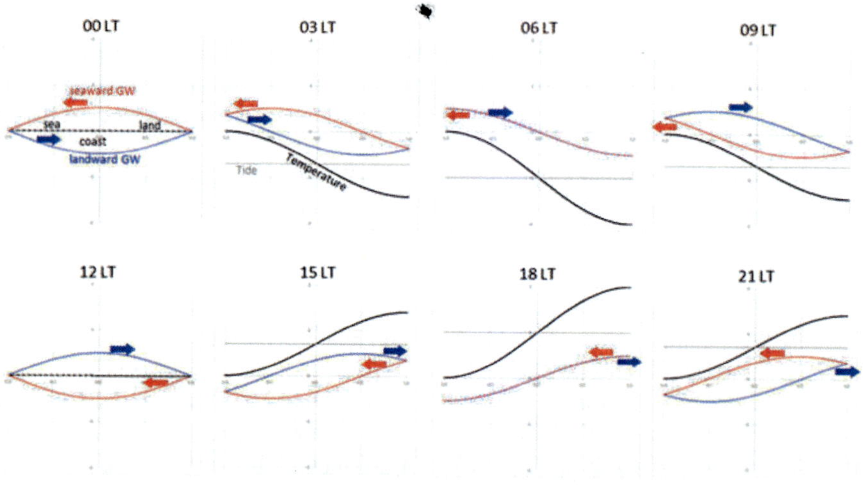

Fig. 6 Schematic plots of the surface temperature (black) for each local time (LT) near a coast (left: sea; right: land), consisting of seaward (red) and landward (blue) gravity waves and global tide (gray). The sea-surface temperature is almost constant (actually varying slightly with interannual variations). The land-surface temperature varies largely by daytime solar heating and nighttime cooling (mainly by sprinkler-like effect of rainfall; see, e.g., [34])

$$= -\frac{1}{\rho}\frac{\partial}{\partial U}(\text{wave momentum flux})\frac{\partial U}{\partial z} \equiv -W\frac{\partial U}{\partial z}, \tag{2}$$

where U is the mean zonal velocity, ρ is the mean density, and the wave momentum flux is given by $\rho \times$ product of zonal and vertical velocity anomalies and is $\propto 1/(\text{zonal phase velocity}-U)$ (see e.g., Section 4.6 of Andrews et al., 1987). The advection velocity $W = \frac{1}{\rho}\frac{\partial}{\partial U}$ (wave momentum flux) = (wave momentum flux)/(zonal phase velocity$-U$) is always negative, which gives descending velocity of the QBO pattern. The zonal phase speed $|\omega/k|/\cos\alpha$ (dependent on the coastline direction α oblique to the meridian) restricts the zonal mean flow amplitude of QBO, and the wave momentum flux controls $|W|$ and the shortness of QBO period (as so far known). Interannual and intraseasonal variations may modulate CDC-SLBC amplitudes, wave momentum flux and then QBO. Waves associated with CDC-SLBC also have the meridional and vertical components, which may be related to the Brewer–Dobson circulation (e.g., [6, 11]).

The other (opposite-direction) wave is propagated much higher up to the opposite-direction shear region of the mean zonal flow, otherwise up to a self-breaking level due to amplification compensating atmospheric density decrease, and is absorbed in a similar way. The diurnal tide should be self-breaking just below the turbopause [22], and this should determine the turbopause (homopause for the atmospheric constituents) level [16, 33].

Some results using Plumb's (1977) model have been presented in the INCREASE symposium last year [35]. The general dispersion relation of CDC-SLBC waves in viscous atmosphere on the surface for each of bidirectional modes is a cubic equation for vertical wavenumber squared, of which a steady case with large eddy viscosity/diffusivity corresponds to a local circulation convection such as a heat island [13, 31]. When the atmospheric density stratification (scale height) is stronger than any damping, atmospheric motions (circulation layers or waves) appear also above the circulation at the bottom. In case of weaker viscosity/diffusivity/damping with a mean flow apart from the surface, three roots forced at the top (amplifying upward) are omitted, and two roots are degenerated as amplitude damping of the remaining one root under so-called WKB-approximation, which is the solution used by Plumb (1977). The period and structure of calculated "QBO" may be dependent on the number and location of coastlines, that is, the distribution of lands. For example, if coastlines are less (more), the period of "QBO" becomes longer (shorter). IMC with the longest coastline is most responsible to produce QBO. Because CDC-SLBC is quite robust as seen in Fig. 1, QBO generated by bidirectional waves separated from SLBC must be robust, which is different from foregoing discussions on generation of QBO with broadband waves (e.g., [3]).

4 Conclusions

In this paper, we show observational evidence that the equatorial tropospheric CDC-SLBC plays a principal role to maintain the stratospheric QBO robustly and explain it from theories of SLBC and the wave-mean-flow interaction. The coastlines are the triple boundary among land, sea, and air constructing the Earth system, and our present results suggest even the QBO is not occurred incidentally but is controlled under the Earth system. As in the distributions of surface rainfall and tropospheric convective cloud activity, the strongest CDC along the world's longest coastline surrounding major islands of IMC is most responsible in the generation of bidirectional (sea- and land-ward) gravity waves interacting with QBO. The longer equatorial (Kelvin and mixed Rossby-gravity) waves induced by a larger-scale ocean-continent or open-ocean heating contrast [5, 20] may modulate (cluster) the bidirectional gravity waves in the stratosphere, just as land distributions and some intraseasonal variations do in the troposphere [7]. This Matsuno-Gill heating is equivalent to meridional gradient of zonal forcing, and zonally uniform case makes Hadley circulation and subtropical trade wind zones [15]. The other non-annual variations such as semiannual oscillations (SAOs) and mesospheric QBO may be considered by CDC-SLBC bidirectional waves similarly.

In the extratropics the orographic (stationary) internal gravity waves robustize the middle-stratospheric weak wind layers [32], which permits both east- and westward traveling gravity waves propagating upward and producing the reversal of the middle-atmospheric zonal circulation and winter-poleward meridional circulation (balanced with annually-reversed both hemispheric difference of radiational budget) at the mesopause [10, 18, 21]. Both the extratropical orography and the equatorial coastlines have been changed in the long history of Earth, and the paleoclimate may be discussed as applications of those gravity-wave processes. The robust 1-day periodicity of equatorial CDC is another effect of Earth's rotation like the extratropical Coriolis force.

As Himalaya-Tibet in the mid-latitudes, IMC in the tropics is the geologically most important product of plate tectonics in Mesozoic-Cenozoic (recent 100 million years). They govern the global atmosphere, through the process mentioned in this paper for the equatorial lower-to-upper atmosphere. Therefore, more observational and theoretical understanding of CDC-SLBC may contribute to further development of atmospheric sciences in Indonesia.

Acknowledgements This study has been supported partly by JSPS KAKENHI Grant Number JP19H04248.

References

1. Anstey, J.A., Osprey, S.M., Alexander, J., Baldwin, M.P., Butchart, N., Gray, L., Kawatani, Y., Newman, P.A., Richter, J.H.: Impacts, processes and projections of the quasi-biennial oscillation. Nat. Rev. Earth Environ. **3**, 588–603 (2022). https://doi.org/10.1038/s43017-022-00323-7
2. Baldwin, M.P., Gray, L.J., Dunkerton, T.J., Hamilton, K., Haynes, P.H., Randel, W.J., Holton, J.R., Alexander, M.J., Hirota, I., Horinouchi, T., Jones, D.B.A., Kinnersley, J.S., Marquardt, C., Sato, K., Takahashi, M.: The quasi-biennial oscillation. Rev. Geophys. **39**, 179–229 (2001). https://doi.org/10.1029/1999RG000073
3. Dunkerton, T.J.: The role of gravity waves in the quasi-biennial oscillation. J. Geophys. Res. **102**, 26053–26076 (1997). https://doi.org/10.1029/96JD02999
4. Ebdon, R.A.: Notes on the wind flow at 50 mb in tropical and sub-tropical regions in January 1957 and January 1958. Quart. J. Roy. Meteor. Soc. **86**, 540–542 (1960). https://doi.org/10.1002/qj.49708637011
5. Gill, A.E.: Some simple solutions for heat-induced tropical circulation. Quart. J. Roy. Meteor. Soc. **106**, 447–462 (1980). https://doi.org/10.1002/qj.49710644905
6. Hasebe, F., Aoki, S., Morimoto, S., Inai, Y., Nakazawa, T., Sugawara, S., Ikeda, C., Honda, H., Yamazaki, H., Komala, N., Putri, F.A., Budiyono, A., Soedjarwo, M., Ishidoya, S., Toyoda, S., Shibata, T., Hayashi, M., Eguchi, N., Nishi, N., Fujiwara, M., Ogino, S.-Y., Shiotani, M., Sugidachi, T.: Coordinated upper-troposphere-to-stratosphere balloon experiment in Biak. Bull. Amer. Meteor. Soc. **99**, 1213–1230 (2018). https://doi.org/10.1175/BAMS-D-16-0289.1
7. Hayashi, Y.-Y., Sumi, A.: The 30–40 day oscillations simulated in an "aqua planet" model. J. Meteor. Soc. Jpn. **64**, 451–467 (1986). https://doi.org/10.2151/jmsj1965.64.4_451
8. Haynes, P., Hitchcock, P., Hitchman, M., Yoden, S., Hendon, H., Kiladis, G., Kodera, K., Simpson, I.: The influence of the stratosphere on the tropical troposphere. J. Meteor. Soc. Jpn. **99**, 803–845 (2021). https://doi.org/10.2151/jmsj.2021-040
9. Hitchman, M.H., Yoden, S., Haynes, P.H., Kumar, V., Tegtmeier, S.: An observational history of the direct influence of the stratospheric quasi-biennial oscillation on the tropical and subtropical upper troposphere and lower stratosphere. J. Meteor. Soc. Jpn. **99**, 239–267 (2021). https://doi.org/10.2151/jmsj.2021-012
10. Holton, J.R.: The role of gravity wave induced drag and diffusion in the momentum budget of the mesosphere. J. Atmos. Sci. **39**, 791–799 (1982). https://doi.org/10.1175/1520-0469(1982)039%3c0791:TROGWI%3e2.0.CO;2
11. Kawatani, Y., Hamilton, K.: Weakened stratospheric quasibiennial oscillation driven by increased tropical mean upwelling. Nature **497**, 478–481 (2013). https://doi.org/10.1038/nature12140
12. Kawatani, Y., Sato, K., Dunkerton, T.J., Watanabe, S., Miyahara, S., Takahashi, M.: The roles of equatorial trapped waves and internal inertia-gravity waves in driving the quasi-biennial oscillation, Part I: zonal mean wave forcing. J. Atmos. Sci. **67**, 963–980 (2010). https://doi.org/10.1175/2009JAS3222.1
13. Kimura, R.: Dynamics of steady convections over heat and cool islands. J. Meteor. Soc. Jpn. **53**, 440–457 (1975). https://doi.org/10.2151/jmsj1965.53.6_440
14. Kobayashi, S., Ota, Y., Harada, Y., Ebita, A., Moriya, M., Onoda, H., Onogi, K., Kamahori, H., Kobayashi, C., Endo, H., Miyaoka, K., Takahashi, K.: The JRA-55 reanalysis: general specifications and basic characteristics. J. Meteor. Soc. Jpn. **93**, 5–48 (2015). https://doi.org/10.2151/jmsj.2015-001
15. Kosaka, Y., Matsuda, Y.: Roles of Rossby and gravity waves on circulation associated with tropical and subtropical heating. J. Meteor. Soc. Jpn. **83**, 481–498 (2005). https://doi.org/10.2151/jmsj.83.481
16. Leovy, C.B.: Control of the homopause level. Icarus **50**, 311–321 (1982). https://doi.org/10.1016/0019-1035(82)90128-2
17. Lin, Y.-L.: Mesoscale dynamics. Cambridge University Press, pp. 630, ISBN-978-0-521-80875-0 (2007)

18. Lindzen, R.S.: Turbulence and stress owing to gravity wave and tidal breakdown. J. Geophys. Res. **86**, 9707–9714 (1981). https://doi.org/10.1029/JC086iC10p09707
19. Lindzen, R.S., Holton, J.R.: A theory of the quasi-biennial oscillation. J. Atmos. Sci. **25**, 1095–1107 (1968). https://doi.org/10.1175/1520-0469(1968)025%3c1095:ATOTQB%3e2.0.CO;2
20. Matsuno, T.: Quasi-geostrophic motions in the equatorial area. J. Meteor. Soc. Jpn. **44**, 25–43 (1966). https://doi.org/10.2151/jmsj1965.44.1_25
21. Matsuno, T.: A quasi one-dimensional model of the middle atmosphere circulation interacting with internal gravity waves. J. Meteor. Soc. Jpn. **60**, 215–226 (1982). https://doi.org/10.2151/jmsj1965.60.1_215
22. Miyahara, S.: Zonal mean winds induced by vertically propagating atmospheric tidal waves in the lower thermosphere. J. Meteor. Soc. Jpn. **56**, 86–97 (1978). https://doi.org/10.2151/jmsj1965.56.2_86
23. Ogino, S.Y., Yamanaka, M.D., Mori, S., Matsumoto, J.: Tropical coastal dehydrator in global atmospheric water circulation. Geophys. Res. Lett. **44**(1), 636–11643 (2017). https://doi.org/10.1002/2017GL075760
24. Osprey, S.M., Butchart, N., Knight, J.R., Scaife, A.A., Hamilton, K., Anstey, J.A., Schenzinger, V., Zhang, C.: An unexpected disruption of the atmospheric quasi-biennial oscillation. Science **353**, 1424–1427 (2016). https://doi.org/10.1126/science.aah4156
25. Plumb, R.A.: The interaction of two internal waves with the mean flow: implications for the theory of the quasi-biennial oscillation. J. Atmos. Sci. **34**, 1847–1858 (1977). https://doi.org/10.1175/1520-0469(1977)034<1847:TIOTIW>2.0.CO;2
26. Plumb, R.A., McEwan, A.D.: The instability of a forced standing wave in a viscous stratified fluid: a laboratory analogue of the quasi-biennial oscillation. J. Atmos. Sci. **35**, 1827–1839 (1978). https://doi.org/10.1175/1520-0469(1978)035%3c1827:TIOAFS%3e2.0.CO;2
27. Qian, J.H.: Why precipitation is mostly concentrated over islands in the maritime continent. J. Atmos. Sci. **65**, 1428–1441 (2008). https://doi.org/10.1175/2007JAS2422.1
28. Reed, R.J., Campbell, W.J., Rasmussen, L.A., Rogers, D.G.: Evidence of a downward-propagating, annual wind reversal in the equatorial stratosphere. J. Geophys. Res. **66**, 813–818 (1961). https://doi.org/10.1029/JZ066i003p00813
29. Rotunno, R.: On the linear theory of the land and sea breeze. J. Atmos. Sci. **40**, 1999–2009 (1983). https://doi.org/10.1175/1520-0469(1983)040%3c1999:OTLTOT%3e2.0.CO;2
30. Scorer, R.: Theory of waves in the lee of mountains. Quart. J. Roy. Meteor. Soc. **75**, 41–56 (1949). https://doi.org/10.1002/qj.49707532308
31. Stommel, H., Veronis, G.: Steady convective motion in a horizontal layer of fluid heated uniformly from above and cooled non-uniformly from below. Tellus **9**, 401–407 (1957). https://doi.org/10.3402/tellusa.v9i3.9100
32. Tanaka, H., Yamanaka, M.D.: Atmospheric circulation in the lower stratosphere induced by the mesoscale mountain wave breakdown. J. Meteor. Soc. Jpn. **63**, 1047–1054 (1985). https://doi.org/10.2151/jmsj1965.63.6_1047
33. Yamanaka, M.D.: Homopause control by gravity wave breaking in the planetary atmospheres. Adv. Space Res. **15**(4), 47–50 (1995). https://doi.org/10.1016/0273-1177(94)00063-7
34. Yamanaka, M.D.: Physical climatology of Indonesian maritime continent: an outline to comprehend observational studies. Atmos. Res. **178–179**, 231–259 (2016). https://doi.org/10.1016/j.atmosres.2016.03.017
35. Yamanaka, M.D.: EAR construction motivation revisited: Indonesian coastline representing earth. An invited talk at EAR 20-Year Anniversary—INCREASE 2021, 20–21 Sept 2021. (to be printed in this volume) (2021)
36. Yamanaka, M.D., Ogino, S.-Y.: Tropical coastlines robustizing the stratospheric quasi-biennial oscillation. (in preparation) (2023)
37. Yamanaka, M.D., Ogino, S.-Y., Wu, P.-M., Hamada, J.-I., Mori, S., Matsumoto, J., Syamsudin, F.: Maritime continent coastlines controlling Earth's climate. Prog. Earth Planet Sci. **5**(21), 1–28 (2018). https://doi.org/10.1186/s40645-018-0174-9

EAR Construction Motivation Revisited: Indonesian Coastline Representing Earth

Manabu D. Yamanaka

Abstract EAR was constructed to target the vertical coupling at a single station. The equatorial region with maximum insolation and moisture in the lower atmosphere was less observed and less understood because of its ageostrophic, sub-synoptic cloud-dominant features. For the middle and upper atmosphere, vertically-temporally high-resolution radar sounding was much anticipated to clarify waves generating stratospheric quasi-biennial oscillation (QBO), for example. During these two decades after the construction of EAR, the significant role of the Indonesian maritime continent (IMC) for the global lower-atmospheric climate has been recognized. Many "mini-EARs" (wind profilers and weather radars), as well as satellites and cloud-resolving model, have revealed that the diurnal cycle (sea-land breeze circulation) along the world's longest coastline surrounding major islands of IMC is the most robust mode of cloud-rainfall generation. This process produces latent heat (amount of roughly 10–20% of the greenhouse effect) compensating the global radiational-dynamical energy imbalance, and causes floods in stronger land-sea temperature gradients during remarkable monsoon (cold surge) and/or La Niña periods. Any anthropogenic modification of the land surface-hydrologic conditions such as urbanization and cultivation also may change the diurnal cycle, local weather, and also the global climate. The diurnal-cycle circulation consists of equi-amplitude bidirectional (sea- and land-ward) internal gravity waves likely to cause QBO. Indeed, the coastline of IMC is the triple boundary maintained among land, sea, and atmosphere of the earth, and the EAR and many mini-EARs listen to sounds of the whole earth system.

A paper presented at INCREASE 2021

M. D. Yamanaka (✉)
Research Institute for Humanity and Nature, Kyoto, Japan
e-mail: mdy@chikyu.ac.jp; mdy@kobe-u.ac.jp

Professor Emeritus, Kobe University, Kobe, Japan

A. Basit et al. (eds.), *Proceedings of the International Conference on Radioscience, Equatorial Atmospheric Science and Environment and Humanosphere Science*, Springer Proceedings in Physics 290, https://doi.org/10.1007/978-981-19-9768-6_2

1 Introduction

It was 32 years ago when Prof. Susumu Kato and Prof. Shoichiro Fukao of Kyoto University, Pak Alex Sudibyo of LAPAN, Pak Chandra Manan Mangan and Bu Tien Sribimawati of BPPT, and others including myself climbed Bukit Kototabang for the first time. It was 4 years after Kyoto University and LAPAN started site surveys for "Equatorial Radar" in Biak, followed in Pontianak, and was among more than ten locations in Sumatera Barat (Fig. 1). Considering ground conditions for construction of a large VHF radar, infrastructure, and support for operation, and accessibility from Jakarta and overseas, Kototabang was selected [25]. After very quick permission of Prof. Bacharuddin Jusuf Habibie, followed soon by BMKG's GAW station, we needed 12 years until constructing the EAR in 2001.

On the first proposal of "Equatorial Radar" construction [25], establishment of "International Center for Equatorial Research (ICEAR)" was also planned for (i) effective management of international observational activities and data archives, (ii) development and integration of ground-based profiling techniques including various types of radars, and (iii) construction of a prototype of new-generation atmospheric science. Prof. Fukao considered (iii) based on (ii), that is, "radar atmosphere physics" [5–7].

Atmospheric radars providing temporally-vertically high-resolution observations almost automatically obtained innovative results for the extratropical middle and upper atmosphere in 1980s and were much anticipated to do so in the tropics with more special features and less observations (Fig. 2). This dream came true by EAR for these two decades, as many other papers presented in this symposium. In this paper, the background before the EAR construction and the progress of tropical lower-atmospheric observations in parallel with the EAR observations are overviewed.

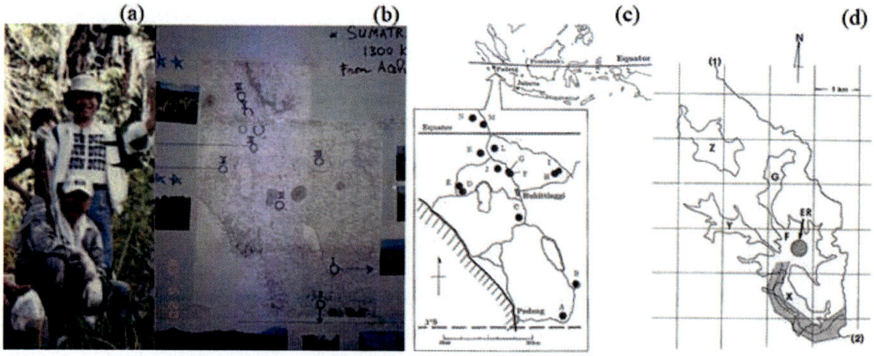

Fig. 1 **a** Professor Susumu Kato (1928–2020) and Professor Shoichiro Fukao (1943–2014) on a site survey at Bukit Kototabang in June 1989. Candidate site maps **b** during and **c** after site surveys. **d** Map of Bukit Kototabang [25]

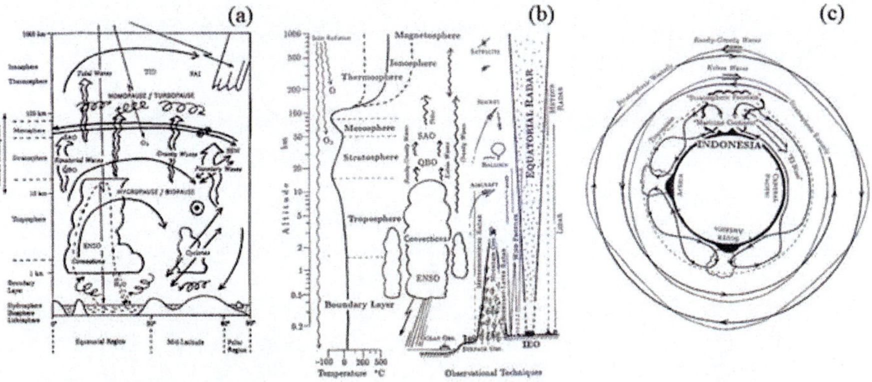

Fig. 2 Schematic **a** meridional-vertical cross-section, and equatorial **b** vertical and **c** horizontal structures (Overview by Fukao, in [25])

2 Result and Discussion

2.1 Middle/Upper-Atmospheric Motivation of EAR Before *Construction*

The first use of the name Sumatera occurred in 1017 as *Samudra*, which meant "sea-land" [31], that is, the maritime continent! "Great voyage" re-discovered trade wind, monsoon, and diurnal cycles [49], reprinted report of de Houtman's first arrival in Enggano and Banten in 1596, [9] which were discovered originally by ancient Greek and Arab sailors. Before WW2, the IMC was one of the densest meteorological observation regions over the world (Fig. 3). *Nederlands-Indië* (*Hindia Belanda*) installed more than 2000 rain gauge stations in 19C [22, 26, 30] which were used by the first climatological classification by Köppen [27].

Telegram network was first established by Lord Kelvin during 1866–70, which communicated the 1883 eruption of Krakatau (misspelled as "*Krakatoa*") over the world within a couple of days [65]. The volcanic ashes went around the equatorial stratosphere westward for two weeks, suggesting easterly of about 30 m/s [13, 14]. At Batavia (Jakarta), van Bemmelen [61, 62] did many balloon observations to confirm high tropopause, both easterly and westerly in the stratosphere, tropospheric monsoons, and diurnal cycles of sea-land breeze circulations.

Operational meteorological observations were continued under Japanese occupation (1942–45) [69] and were partly supported by Japanese war reparation (1958–70) after independence of Indonesia. They were organized as a national institute (Lembaga MG) in 1955, an agency (Badan MG) in 1980, and the present BMKG in 2008. Middle/upper-atmospheric (large balloon and rocket) and space (ionospheric and satellite) observations were started by LAPAN established in 1963, which

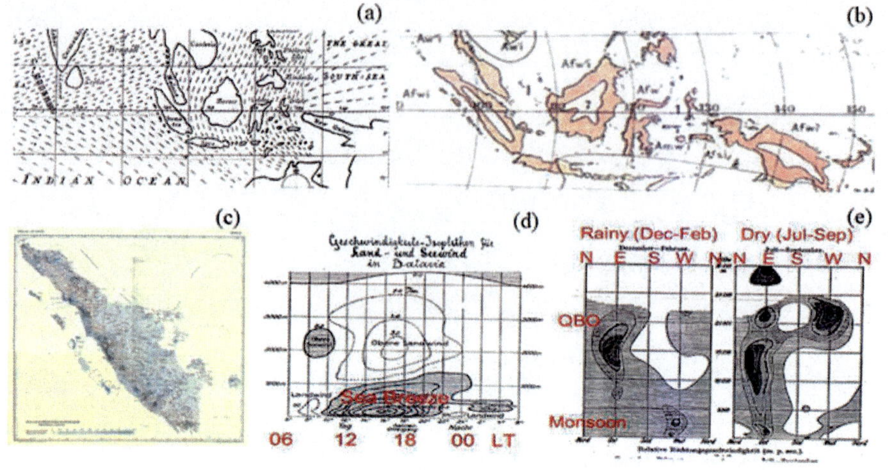

Fig. 3 Parts of first global maps of **a** surface wind distribution [9] and **b** climatological classification [27]. **c** Annual rain amount map for Sumatera [30], and vertical distributions of **d** local time-mean wind [61, 62] and **e** wind direction histograms for rainy and dry seasons [61, 62] at Batavia

becomes an organization under BRIN in 2021, together with other agencies such as BPPT and LIPI.

The middle atmosphere (stratosphere and mesosphere) is interdisciplinary between meteorology and upper atmospheric/space physics, and its dynamics have been understood as the wave-mean flow interaction during 1970–80s. The extratropical mesopausal zonal-wind reversal is explained by east/westward gravity waves propagating from the troposphere [29, 32], which results from stationary (orographic) gravity waves breaking completely in the middle-stratospheric weak wind [58] (Fig. 4). The equatorial-stratospheric quasi-biennial oscillation (QBO) has been explained by equi-amplitude zonally-bidirectional equatorial waves (e.g., [21, 44], but what waves and how they are generated are still controversial [3, 20]. Probably, gravity waves contribute greatly, and are observed by sufficiently high-resolution observations. Balloons have been operated by LAPAN at Watukosek, Jawa Timur (Fig. 4c), but their time resolution (every three hours, i.e., eight times per day at maximum) and vertical coverage (up to the stratosphere and often below the tropopause) are limited (e.g., [59, 60]). Therefore, the atmospheric radars are keenly anticipated to clarify these unsolved wave issues at least, as succeeded already for the extratropical middle-atmosphere.

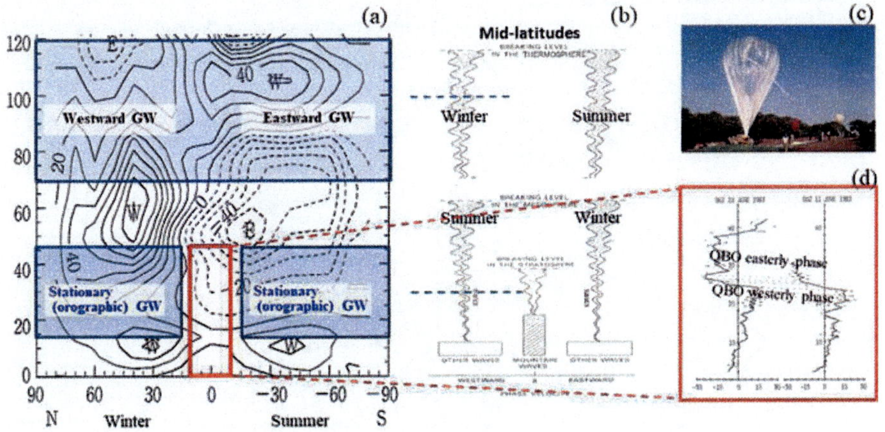

Fig. 4 **a** Meridional-vertical distribution of zonal wind in January (CIRA, 1986) [47], and **b** breaking levels of internal gravity waves [58]. **c** Balloon observations and **d** zonal wind vertical profiles at Watukosek, Jawa Timur [72]

2.2 Climatological Importance of IMC Recognized in These Two Decades

During these two decades, the significant role of the Indonesian maritime continent (IMC) for the global lower-atmospheric climate has been recognized. Interannual (ENSO, IOD), annual (monsoons, rainy/dry seasons), and intraseasonal (MJO) variations have been known (e.g., [4, 10–12, 17–19, 28, 37, 38, 42, 54, 63, 64, 71, 74], but the most dominant mode is the diurnal cycle [71, 74]. GMS cloud data clearly show that intraseasonal variations dominant over the ocean are replaced by the diurnal cycle near the coastline (left two panels of Fig. 5).

Diurnal cycle as the tropospheric coastal most dominant mode

The diurnal cycle along the world's longest coastline surrounding many islands of IMC is the most robust feature of cloud-rainfall generation [1, 2, 17, 18, 23, 33–37, 50–52, 55–57, 66–68]. It is associated with the updraft (surface convergence) zone of a sea-land breeze circulation. Figure 6 shows a long-time average of morning/evening surface wind obtained from so-called "objective analysis" (a numerical model calculation with integrating dynamical equations based on available observational data) calculated by Japan Meteorological Agency (JMA) almost two decades ago. At that time both observational and modeling capabilities were rather limited, but the diurnal cycle of sea-land breeze was clearly seen, because of the diurnal cycle itself is a very basic mode of atmosphere. Recent analysis is done much more accurately, and the diurnal cycle appears as observed without averaging (see, e.g., [46]).

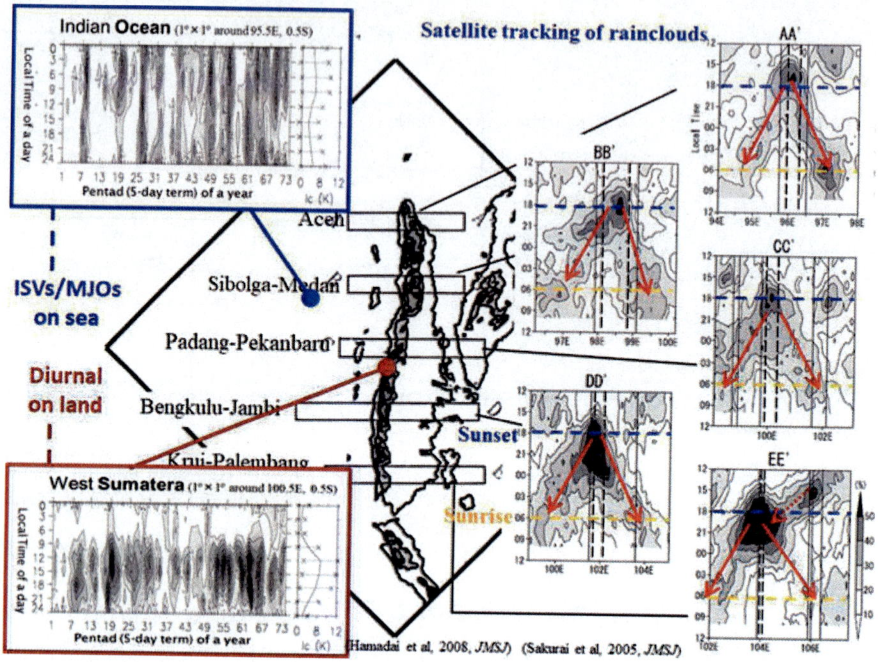

Fig. 5 Intraseasonal and diurnal dominance over open ocean and near coast, respectively, (left two local time-pentad diagrams; [11], and diurnal march along five cross-sections (right-hand side five Hovmöller (local time-longitude) diagrams, [52]) of cloud activity over Sumatera

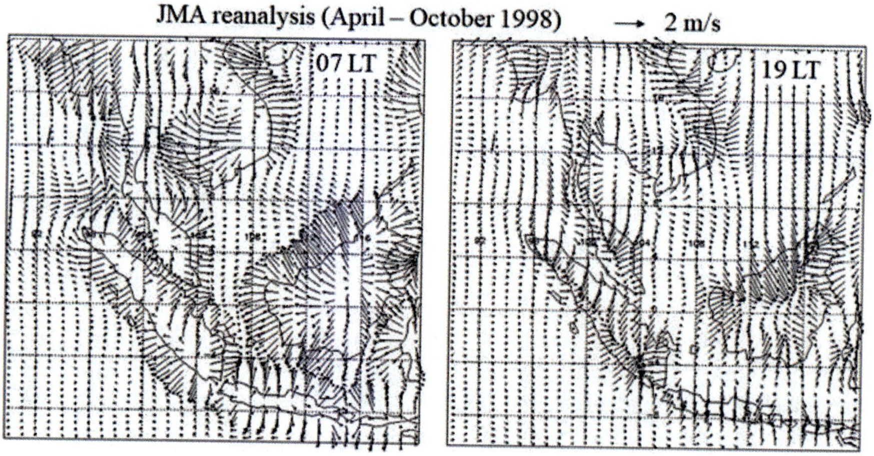

Fig. 6 Examples of long-term (one year of 2001) average of objective analysis data of Japan Meteorological Agency (unpublished MS Thesis work in 2003)

Radar and satellite observations as well as numerical models (Figs. 5, 6 and 7) have shown that the characteristic horizontal scale of the diurnal cycle zone around a coastline is of the order of some hundreds of kilometers, which is somewhat broader than in the extratropics. In Sumatera (Fig. 5) active rainclouds are generated near the central (Barisan) mountains around sunset, migrated both coastward, and arrived offshore before sunrise. In the northern (inland sea) coast of Jawa (Fig. 7), morning seaside to evening mountainside and opposite routes have been observed. A smaller island in the open-ocean side of a large island has two (evening and morning) rainfall peaks [12, 57, 68] and those in the inland (Karimata) sea re-excite the diurnal-cycle migration between two large (Kalimantan and Sumatera) islands [2].

Observational evidence of the diurnal cycle mentioned above is obtained mainly by weather radars as horizontal raindrop distributions. Profilers including EAR and WPRs may provide vertical wind profiles including vertical components and clear areas without raindrops. Complementary uses of both types of radars, as well as collaborations with high-resolution numerical models, are anticipated.

Energy/Water cycles controlled by coastal diurnal cycle

The diurnal cycle appears also in the river water [55], and may cause floods in stronger land-sea temperature gradients during remarkable monsoon (cold surge) and/or La Niña periods. Such applications are fruitful for disaster prevention.

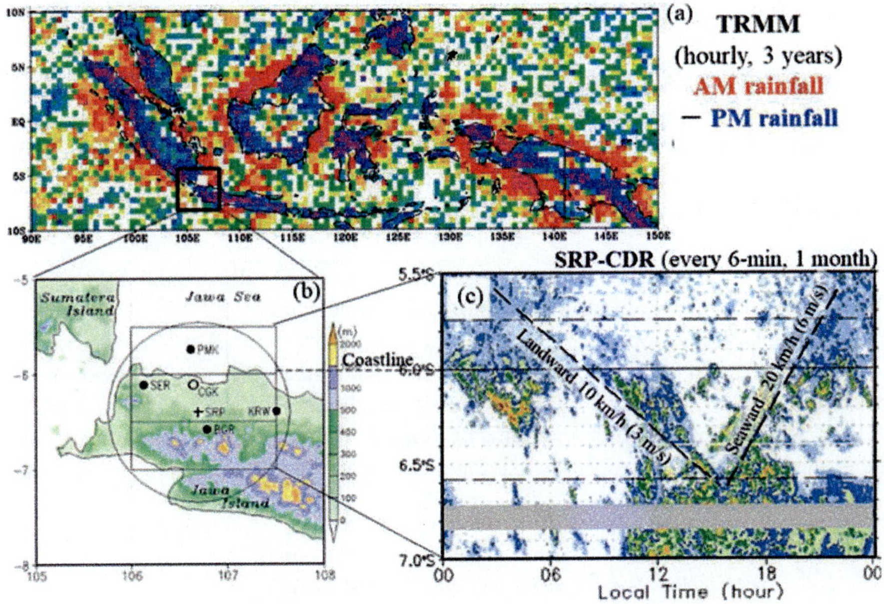

Fig. 7 **a** Diurnal cycle zones along IMC coastlines are shown as AM–PM differences (worm and cold colors indicate positive and negative) of TRMM rainfall [35]. **b** Surface/radiosonde network (●) during 16 Jan–15 Feb 2010 around a C-band radar (CDR) (+) providing **c** Hovmöller (latitude-local time) diagram of diurnal-cycle rainfall over west Jawa [33]

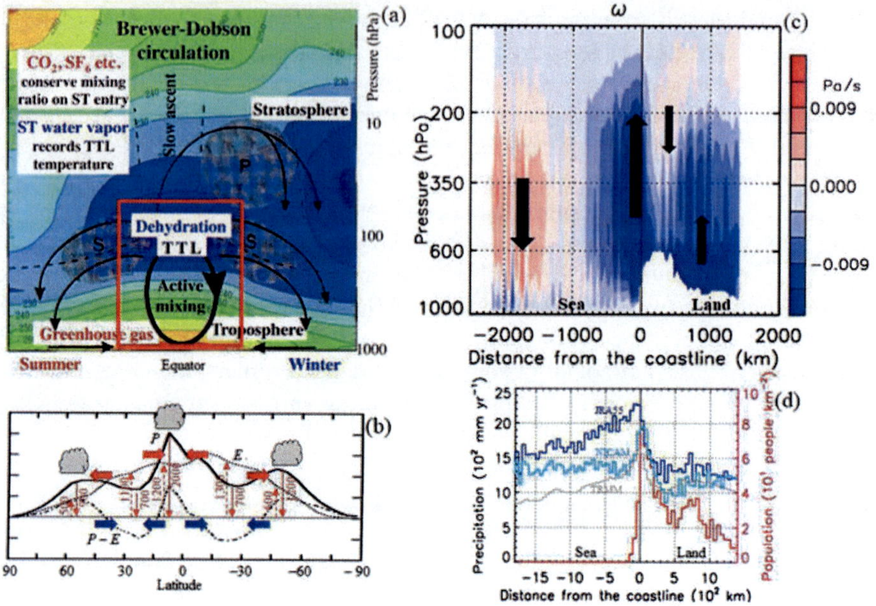

Fig. 8 **a** Brewer-Dobson circulation [16] and **b** surface water budget [15] in the meridional plot. **c** Vertical velocity and **d** precipitation (and population) referenced to coastline (modified from [40, 41])

The annual rainfall amount in the IMC coastal regions is about 2500 mm/year, which is about 2.5 times of the global mean amount 1000 mm/year corresponding to latent heat release 80 W/m^2 (about 1/3 of the greenhouse effect). Thus, the diurnal cycle (producing about 2/3 of the total rainfall amount, or roughly 10–20% of the greenhouse effect) compensates the global radiational-dynamical energy imbalance [74].

Concerning the water budget, rainwater of 1×10^{14} m^3 falling annually over the IMC coastal region is about 1/3 of the tropical total and about 1/6 of the global total [74]. Thus the global atmospheric water transport (so-called Brewer-Dobson circulation or "cold trap" mechanism) concentrated in the tropics in the zonal-mean image (Fig. 8a, b) is more localized near the coastal regions (Fig. 8c, d) in particular of IMC [40, 41]. EAR in the mountains of Sumatera is included in the coastal region mentioned here, and may observe just the global water and atmosphere circulations.

2.3 *Momentum Budget and Vertical Coupling*

The diurnal-cycle sea-land breeze circulation consists of horizontally (sea- and land-ward) and vertically (upward) propagating internal gravity waves (Fig. 9) (e.g., [48]), which is a different category from the global tides [24]. In the lower troposphere, both up/downward propagating modes make a circulation cell satisfying zero vertical velocity at the ground, and large viscosity makes it almost steady and reversed between day and night. Above the cell, waves are almost inviscid and propagating sea- or land-ward, and the top boundary (radiative) condition selects only upward mode which is amplified with upward-decreasing density to satisfy energy (action) conservation. Note that shorter waves corresponding to individual clouds cannot propagate upward (i.e., trapped and ducting near the ground), which is just the same as Scorer's (1949) [53] cutoff for stationary (mountain) waves.

The internal gravity waves generated above the diurnal-cycle sea-land breeze circulation cell are equi-amplitude horizontally-bidirectional waves, which are likely to cause the quasi-biennial oscillation (QBO) [43–45] as the most robust mode in the stratosphere. It is reasonable that the meridional range of QBO is narrower than the internal (vertical propagation) region of inertio-gravity waves because the solar diurnal cycle is larger than the annual cycle within 10°-latitudes similar to the meridional range of the IMC. The horizontal propagation direction are from mountains in the evening with destination to the maritime destination in the morning, or from seas before sunrise to the mountainous destination in the evening. The horizontal phase velocity is 100 km/half a day = 10 km/h = 3 m/s on land, and faster off the coast, and its zonal component becomes much master if the coastline is inclined from the meridional direction.

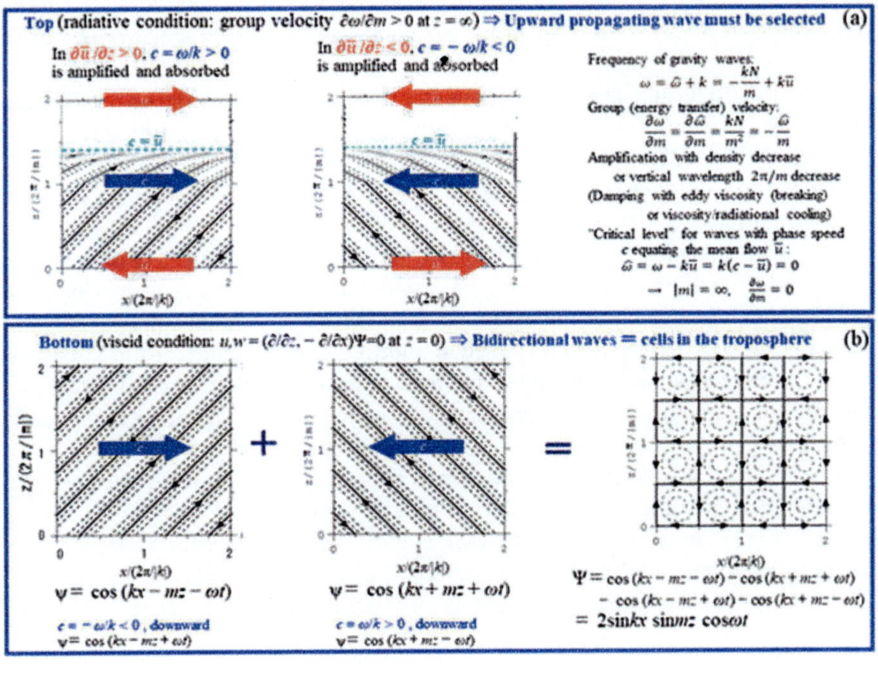

Fig. 9 Gravity wave solutions for **a** stratosphere with the radiative top boundary condition and **b** troposphere with the rigid bottom boundary condition

The wave amplitude is dependent on the sea-land temperature contrast, which is modified by climatological sea-surface variations (ENSO, IOD, and MJO) and anthropogenic land-surface modifications. The period of QBO decreases with an increase of the wave amplitudes, and becomes biperiodic for too strong amplitudes (Fig. 9). The QBO period becomes longer for a weaker diurnal cycle, and decayed if too weak. Longitudinal locations of Africa (−90°) and South America (+180°) and rapid rotation of the earth homogenize the local solar heating zonally.

The sea-land breeze circulations have been observed successfully by single-stationed lower-atmospheric UHF radars (WPRs) [1, 8, 17, 18, 56]. Connecting them with upper-tropospheric/stratospheric internal gravity waves observed by larger VHF radars such as EAR is strongly expected (Fig. 10).

2.4 Recent Whole IMC Tropospheric Observations

Many "mini-EARs" (wind profilers and weather radars at BMKG stations) have been installed and operated successfully over IMC (Figs. 11, 12 and 13). Scientific projects become multi-national, multi-stationed, multi-instrumented and/or multi-disciplinary, and the BMKG operational radar network covers almost all the Indonesian territory. Collaborations and contributions of G20 (G7 + EU + China, India, Brazil, Russia, Korea, Australia, Mexico, Indonesia, Saudi Arabia, Turkey, Argentina, and South Africa) covering our planet more densely are strongly anticipated. In Indonesia National Research and Innovation Agency (BRIN) has been established in 2020, and integrates various scientific activities covered by BPPT, LIPI, and LAPAN, which is noted internationally. As an example of multi-disciplinary approach, any modification of the land surface-hydrologic conditions such as urbanization and plantationization may change the diurnal cycle, local weather, and also the global climate.

Radars have been developed for long years in the extratropical countries with cooler and drier climate and cleaner environment. This is one reason why radar operations are not so easy in the tropics such as IMC. For example, the HARIMAU C-band radar was damaged by a connector shorting due to humid dust in 2014, and finally destroyed by circuit damages due to thunderbolt current in 2015. Establishment of Indonesian radar makers and development of Indonesian radars are strongly anticipated.

New applications of radars in particular for special targets in the tropics are also expected. For example mixed layer top detection initiated by an *L*-band radar at Serpong [17, 18] may be applied also by a *C*-band weather radar for watching smog caused by wildfire [46]. The future (Equatorial MU Radar) project as the next step of EAR [70] is also expected in the common directions mentioned here.

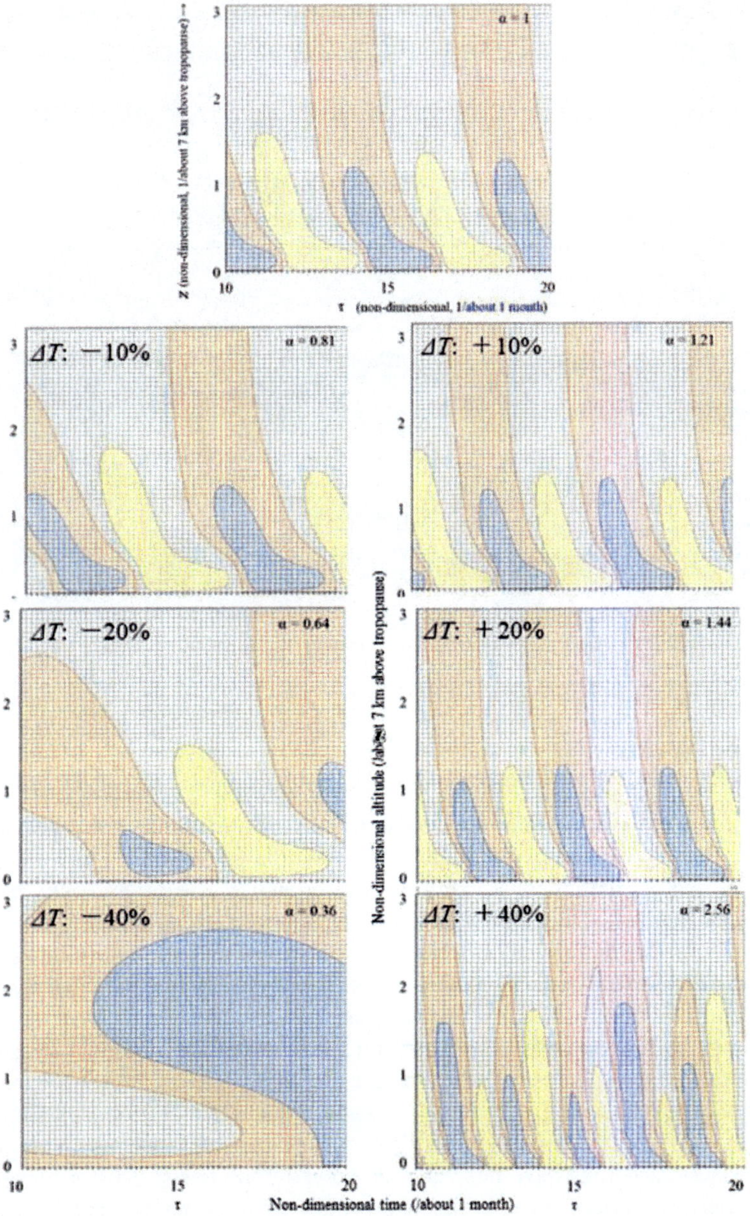

Fig. 10 Plumb's [44] QBO analogue applied for internal gravity waves generated by diurnal-cycle sea-land breeze circulations

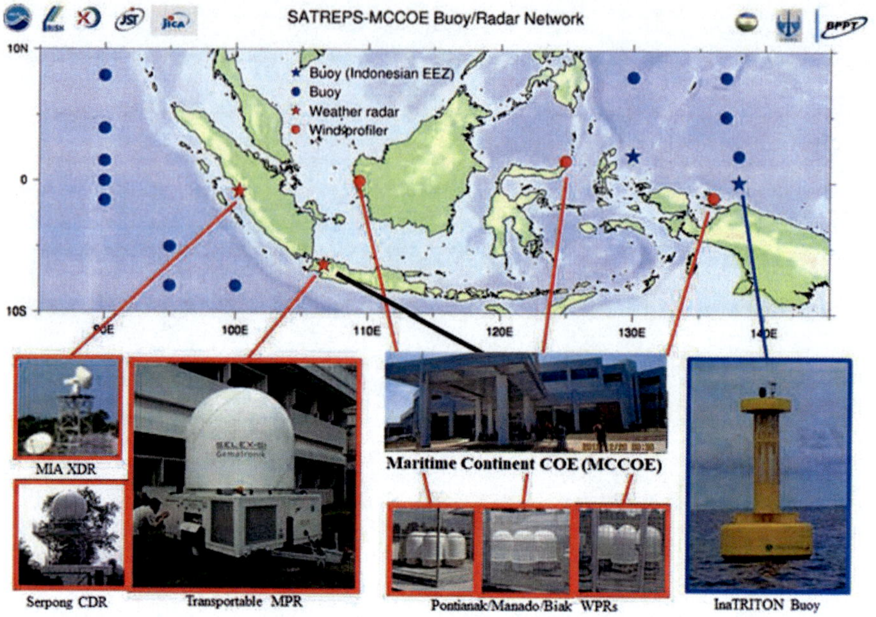

Fig. 11 The JEPP/HARIMAU-SATREPS/MCCOE radar-profiler-buoy networks (see [73, 74] for details)

3 Conclusion

Successful twenty years of EAR proved correctness of the initial motivation of EAR construction. The geographical blank of the middle atmosphere, in particular the equatorial region, before construction is covered now vertically by EAR, and horizontally by satellites. Operational meteorological observations by BMKG have been promoted greatly (more than 40 weather radars, more than 20 radiosonde stations, etc.) and their data are opened on the internet. Those observational progresses lead to improvement of the global objective analyses, and they may complement the vertical observation by EAR. However, the dynamical mystery on equi-amplitude zonally-bidirectional wave sources in particular for the QBO generation is still controversial. Similarly, relationships or interactions among interannual (tropospheric ENSO, IOD, global warming, stratospheric QBO, etc.), seasonal (monsoons), and intraseasonal (MJO) variations have not yet been solved completely. This is serious also for disaster prevention, because superimposition of these phenomena may induce extreme phenomena (flood or drought).

The dominance of diurnal cycles along coastlines of IMC may be a key also for the unsolved issues mentioned above [71, 74]. A sea or land breeze circulation cell (of which the updraft portion is associated with active convective clouds) is equivalent to the sum of a pair of sea- and land-ward propagating internal gravity waves satisfying zero velocity at the ground. Those waves resemble the simple two-wave model of

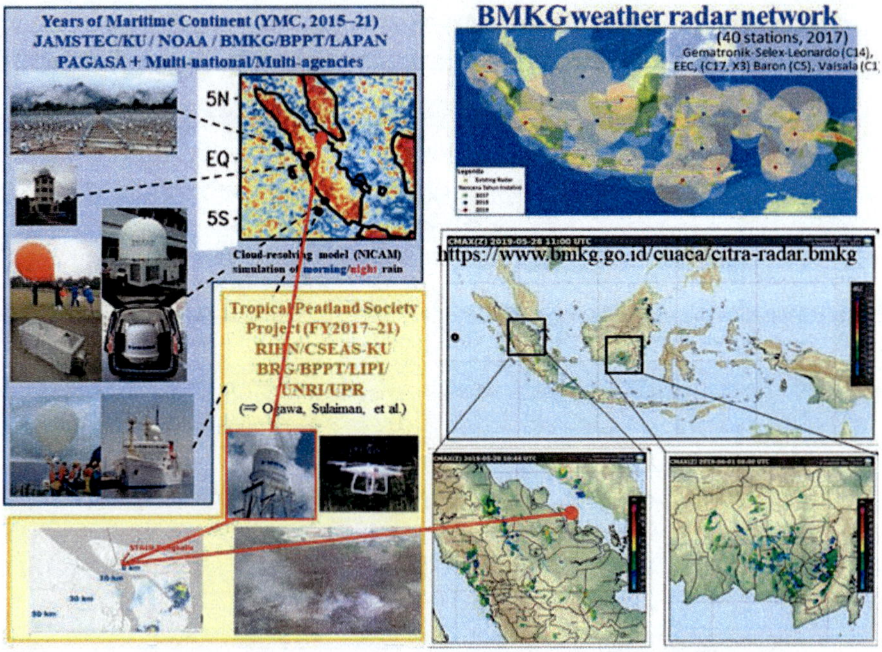

Fig. 12 YMC [75], RIHN Tropical Peatland Society Project [39], and BMKG C-band weather radar network

QBO by Plumb (1977) and propagate also vertically satisfying the energy conservation through the upward decreasing density and the radiative condition at the top. The cell intensity and wave amplitude are proportional to the land-sea surface temperature gradient, modified with the interannual, annual, and intraseasonal variations, as observed.

Indeed, the coastline of IMC is the triple boundary maintained among land, sea, and atmosphere of the earth, and the EAR and many mini-EARs listen to sounds of the whole earth system. The coastal region has a large gradient of physical quantities such as temperature, ground viscosity, and mountain orography vs flat sea surface. The biosphere and its diversity are maintained to fit/relax this gradient, and the anthroposphere (including megacities, peatland plantations, etc.) is made inside this biosphere. Therefore, more utilization/application of radars in particular coastal regions (some of which are not covered by the BMKG radars installed mainly near the major airports) may be important for the future of Indonesia.

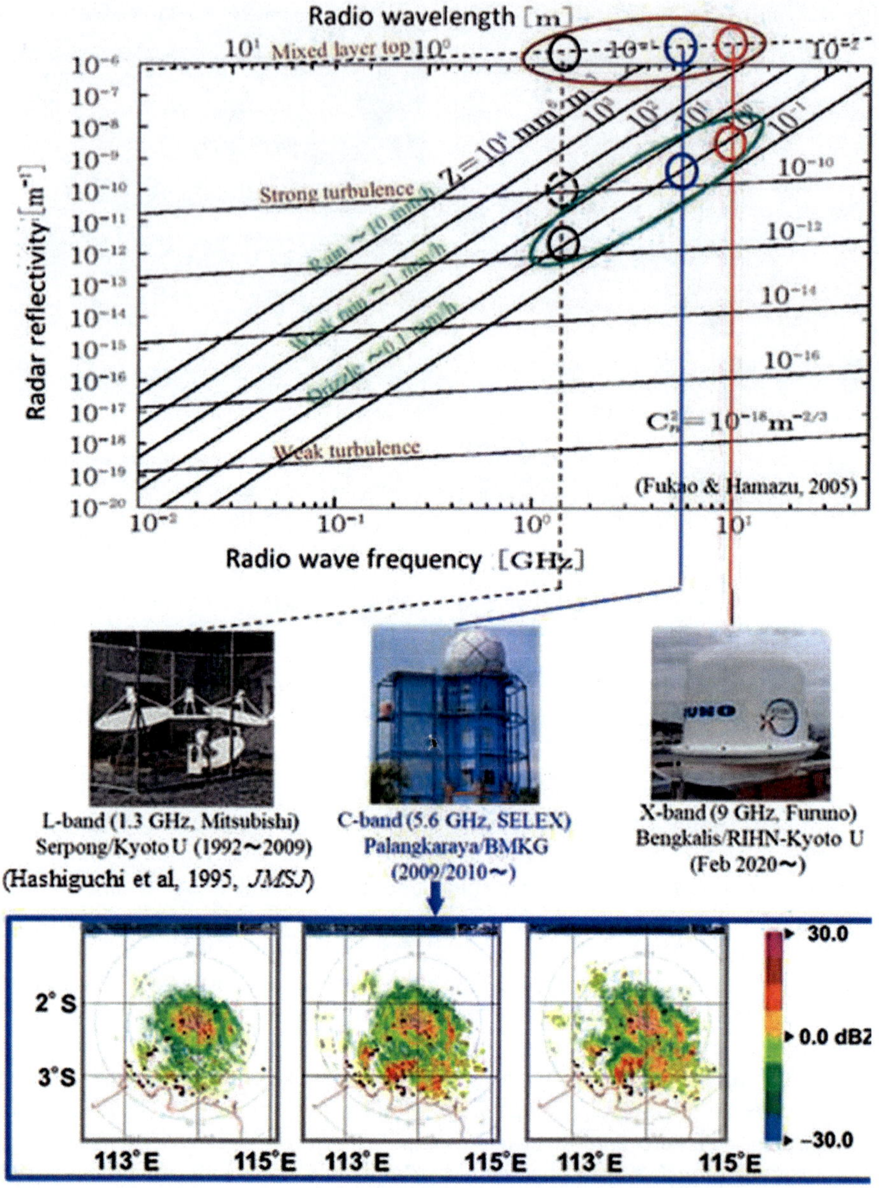

Fig. 13 Radar reflectivity-frequency diagram for raindrops (labeled by Z) and turbulence layers (by Cn^2) (top) [7], radars used by our collaborative studies in Indonesia (middle), and the mixed layer top detected by Palangkaraya BMKG C-band weather radar (bottom) [46]

Acknowledgements This study has been supported partly by JSPS KAKENHI Grant Number JP19H04248.

References

1. Araki, R., Yamanaka, M.D., Murata, F., Hashiguchi, H., Oku, Y., Sribimawati, T., Kudsy, M., Renggono, F.: Seasonal and interannual variations of diurnal cycles of local circulation and cloud activity observed at Serpong, West Jawa, Indonesia. J. Meteor. Soc. Jpn. **84A**, 171–194 (2006). https://doi.org/10.2151/jmsj.84A.171
2. As-syakur, A.R., Imaoka, K., Ogawara, K., Yamanaka, M.D., Tanaka, T., Kashino, Y., Nuarsa, I.W., Osawa, T.: Analysis of spatial and seasonal differences in the diurnal rainfall cycle over Sumatera revealed by 17-year TRMM 3B42 dataset. SOLA, **15**, 216–221 (2019). https://doi.org/10.2151/sola.2019-039
3. Baldwin, M.P., Gray, L.J., Dunkerton, T.J., Hamilton, K., Haynes, P.H., Randel, W.J., Holton, J.R., Alexander, M.J., Hirota, I., Horinouchi, T., Jones, D.B.A., Kinnersley, J.S., Marquardt, C., Sato, K., Takahashi, M.: The quasi-biennial oscillation. Rev. Geophys. **39**, 179–229 (2001). https://doi.org/10.1029/1999RG000073
4. Fujiwara, M., Tomikawa, Y., Komala, N., Saraspriya, S., Manik, T., Suripto, A., Kita, K., Kondo, Y., Kawakami, S., Ogawa, T., Kelana, E., Suhardi, B., Sribimawati, T., Kudsy, M., Harijono, S.W.B., Yamanaka, M.D.: Ozonesonde soundings in the Indonesian maritime continent in September-October 1998 and in August-September 1999. Atmos. Environ. **37**, 353–362 (2003). https://doi.org/10.1016/S1352-2310(02)00891-9
5. Fukao, S.: Coupling processes in the equatorial atmosphere (CPEA): a project overview. J. Meteor. Soc. Jpn. **84A**, 1–18 (2006). https://doi.org/10.2151/jmsj.84A.1
6. Fukao, S.: Recent advances in atmospheric radar study. J. Meteor. Soc. Jpn. **85B**, 215–239 (2007). https://doi.org/10.2151/jmsj.85B.215
7. Fukao, S., Hamazu, K.: Radar for Meteorological and Atmospheric Observations, pp. 537. Springer (2014)
8. Hadi, T.W., Horinouchi, T., Tsuda, T., Hashiguchi, H., Fukao, S.: Sea-breeze circulation over Jakarta, Indonesia: a climatology based on boundary layer radar observations. Mon. Wea. Rev. **130**, 2153–2166 (2002). https://doi.org/10.1175/1520-0493(2002)130<2153:SBCOJI>2.0.CO;2
9. Halley, E.: An historical account of the trade winds, and monsoons, observable in the seas between and near the Tropicks, with an attempt to assign the physical cause of the said winds. Phil. Trans. **16**, 153–168 (1686). https://doi.org/10.1098/rstl.1686.0026
10. Hamada, J.-I., Yamanaka, M.D., Matsumoto, J., Fukao, S., Winarso, P.A., Sribimawati, T.: Spatial and temporal variations of the rainy season over Indonesia and their link to ENSO. J. Meteor. Soc. Jpn. **80**, 285–310 (2002). https://doi.org/10.2151/jmsj.80.285
11. Hamada J.-I., Yamanaka, M.D., Mori, S., Tauhid, Y.I., Sribimawati, T.: Differences of rainfall characteristics between coastal and interior areas of central western Sumatera, Indonesia. J. Meteor. Soc. Jpn. **86**, 593–611 (2008). https://doi.org/10.2151/jmsj.86.593
12. Hamada J.-I., Mori, S., Yamanaka, M.D., Haryoko, U., Lestari, S., Sulistyowati, R., Syamsudin, F.: Interannual rainfall variability over northwestern Jawa and its relation to the Indian Ocean dipole and El Niño southern-oscillation events. SOLA, **8**, 69–72 (2012). https://doi.org/10.2151/sola.2012-018
13. Hamilton, K.: Observations of tropical stratospheric winds before World War II. Bull. Amer. Meteor. Soc. **79**, 1367–1371 (1998). https://doi.org/10.1175/1520-0477(1998)079<1367:OOTSWB>2.0.CO;2
14. Hamilton, K.: Sereno Bishop, Rollo Russell, Bishop's ring and the discovery of the "Krakatoa easterlies". Atmos.-Ocean, **50**, 169–175 (2012). https://doi.org/10.1080/07055900.2011.639736

15. Hartmann, D.L.: Global Physical Climatology, pp. 411. Academic (1994)
16. Hasebe, F., Aoki, S., Morimoto, S., Inai, Y., Nakazawa, T., Sugawara, S., Ikeda, C., Honda, H., Yamazaki, H., Komala, N., Putri, F.A., Budiyono, A., Soedjarwo, M., Ishidoya, S., Toyoda, S., Shibata, T., Hayashi, M., Eguchi, N., Nishi, N., Fujiwara, M., Ogino, S.-Y., Shiotani, M., Sugidachi, T.: Coordinated upper-troposphere-to-stratosphere balloon experiment in Biak. Bull. Amer. Meteor. Soc. **99**, 1213–1230 (2018). https://doi.org/10.1175/BAMS-D-16-0289.1
17. Hashiguchi, H., Fukao, S., Tsuda, T., Yamanaka, M.D., Tobing, D.L., Sribimawati, T., Harijono, S.W.B., Wiryosumarto, H.: Observations of the planetary boundary layer over equatorial Indonesia with an L-band clear-air Doppler radar: initial results. Radio Sci. **30**, 1043–1054 (1995a). https://doi.org/10.1029/95RS00653
18. Hashiguchi, H., Tsuda, T., Fukao, S., Yamanaka, M.D., Harijono, S.W.B., Wiryosumarto, H.: Boundary layer radar observations of the passage of the convection center over Serpong, Indonesia (6°S, 107°E) during the TOGA-COARE intensive observation period. J. Meteor. Soc. Jpn. **73**, 535–548 (1995b). https://doi.org/10.2151/jmsj1965.73.2B_535
19. Hashiguchi, N.O., Yamanaka, M.D., Ogino, S.-Y., Shiotani, M., Sribimawati, T.: Seasonal and interannual variations of temperature in tropical tropopause layer (TTL) over Indonesia based on operational rawinsonde data during 1992–1999. J. Geophys. Res. **111**, D15110 (2006). https://doi.org/10.1029/2005JD006501
20. Hitchman, M.H., Yoden, S., Haynes, P.H., Kumar, V., Tegtmeier, S.: An observational history of the direct influence of the stratospheric quasi-biennial oscillation on the tropical and subtropical upper troposphere and lower stratosphere. J. Meteor. Soc. Jpn. **99**, 239–267 (2021). https://doi.org/10.2151/jmsj.2021-012
21. Holton, J.R., Lindzen, R.S.: An updated theory for the quasi-biennial cycle of the tropical stratosphere. J. Atmos. Sci. **29**, 1076–1080 (1972). https://doi.org/10.1175/1520-0469(1972)029<1076:AUTFTQ>2.0.CO;2
22. Kajita, R., Yamanaka, M.D., Kozan, O.: Reconstruction of rainfall records at 24 observation stations in Sumatera, Colonial Indonesia, from 1879 to 1900. J. Hydrometeor. **23**, 1627–1643 (2022). https://doi.org/10.1175/JHM-D-20-0245.1
23. Kamimera, H., Mori, S., Yamanaka, M.D., Syamsudin, F.: Modulation of diurnal rainfall cycle by the Madden-Julian oscillation based on one-year continuous observation with a meteorological radar in west Sumatera. SOLA, **8**, 111–114 (2012). https://doi.org/10.2151/sola.2012-028
24. Kato, S.: Diurnal atmospheric oscillation, 1. Eigenvalues and Hough functions. J. Geophys. Res. **71**, 3201–3209 (1966). https://doi.org/10.1029/JZ071i013p03201
25. Kato, S. (ed.): Report on "Feasibility Study on the Equatorial Radar in Indonesia". STP Sub-Committee, Committee on Geoelectromagnetism, Science Council of Japan (1989). https://www.researchgate.net/publication/359861088
26. Können, G.P., Jones, P.D., Kaltofen, M.H., Allan, R.J.: Pre-1866 extensions of the southern oscillation index using early Indonesian and Tahitian meteorological readings. J. Clim. **11**, 2325–2339 (1998). https://doi.org/10.1175/1520-0442(1998)011<2325:PEOTSO>2.0.CO;2
27. Köppen, W.: Klassifikation der Klimate nach Temperatur, Niederschlag und Jahresablauf. Petermanns Geogr. Mitt., **64**, 193–203, 243—248 (1918). http://koeppen-geiger.vu-wien.ac.at/pdf/Koppen_1918.pdf
28. Lestari, S., Hamada J.-I., Syamsudin, F., Sunaryo, Matsumoto, J., Yamanaka, M.D.: ENSO influences on rainfall extremes around Sulawesi and Maluku Islands in the eastern Indonesian maritime continent. SOLA, **12**, 37–41 (2016). https://doi.org/10.2151/sola.2016-008
29. Lindzen, R.S.: Turbulence and stress owing to gravity wave and tidal breakdown. J. Geophys. Res. **86**, 9707–9714 (1981). https://doi.org/10.1029/JC086iC10p09707
30. LMG (Lembaga Meteorological dan Geofisika): Rainfall atlas of Indonesia, Vol. II: Sumatra, Kalimantan, Sulawesi, Nusatenggara, Maluku, Irian Jaya (1911–1940). In: Meteorological Note No. 9, Department of Communications, Meteorological and Geophysical Institute Republic of Indonesia, 47pp. (1973)
31. Loeb, E.M.: Sumatra; its history and people, Universitat Wien (reprinted at Oxford University Press in 1972), 350pp. (1935)

32. Matsuno, T.: A quasi one-dimensional model of the middle atmosphere circulation interacting with internal gravity waves. J. Meteor. Soc. Jpn. **60**, 215–226 (1982). https://doi.org/10.2151/jmsj1965.60.1_215
33. Mori, S., Hamada J.-I., Hattori, M., Wu, P.-M., Katsumata, M., Endo, N., Ichiyanagi, K., Hashiguchi, H., Arbain, A.A., Sulistyowati, R., Lestari, S., Syamsudin, F., Manik, T., Yamanaka, M.D.: Meridional march of diurnal rainfall over Jakarta, Indonesia, observed with a C-band Doppler radar: an overview of the HARIMAU2010 campaign. Prog. Earth Planet Sci. **5**(47), 1–23 (2018). https://doi.org/10.1186/s40645-018-0202-9
34. Mori, S., Hamada J.-I., Sakurai, N., Fudeyasu, H., Kawashima, M., Hashiguchi, H., Syamsudin, F., Arbain, A.A., Sulistyowati, R., Matsumoto, J., Yamanaka, M.D.: Convective systems developed along the coastline of Sumatera Island, Indonesia observed with an X-band Doppler radar during the HARIMAU2006 campaign. J. Meteor. Soc. Jpn. **89A**, 61–81 (2011). https://doi.org/10.2151/jmsj.2011-A04
35. Mori, S., Hamada J.-I., Tauhid, Y.I., Yamanaka, M.D., Okamoto, N., Murata, F., Sakurai, N., Sribimawati, T.: Diurnal rainfall peak migrations around Sumatera Island, Indonesian maritime continent observed by TRMM satellite and intensive rawinsonde soundings. Mon. Wea. Rev. **132**, 2021–2039 (2004). https://doi.org/10.1175/1520-0493(2004)132<2021:DLRPMO>2.0.CO;2
36. Mori, S., Hamada J.-I., Yamanaka, M.D., Kodama, Y.-M., Kawashima, M., Shimomai, T., Shibagaki, Y., Hashiguchi, H., Sribimawati, T.: Vertical wind characteristics in precipitating cloud systems over west Sumatera, Indonesia, observed with Equatorial Atmosphere Radar: case study on 23-24 April 2004 during the first CPEA Campaign Period. J. Meteor. Soc. Jpn. **84A**, 113–131 (2006). https://doi.org/10.1029/2005JD006501
37. Murata, F., Yamanaka, M.D., Fujiwara, M., Ogino, S.-Y., Hashiguchi, H., Fukao, S., Kudsy, M., Sribimawati, T., Harijono, S.W.B., Kelana, E.: Relationship between wind and precipitation observed with a UHF radar, GPS rawinsonde and surface meteorological instruments at Kototabang, West Sumatera during September—October 1998. J. Meteor. Soc. Jpn. **80**, 347–360 (2002). https://doi.org/10.2151/jmsj.80.347
38. Murata, F., Yamanaka, M.D., Hashiguchi, H., Mori, S., Kudsy, M., Sribimawati, T., Suhardi, B., Emrizal: Dry intrusions following eastward-propagating synoptic-scale cloud systems over Sumatera Island. J. Meteor. Soc. Jpn. **84**, 277–294 (2006). https://doi.org/10.2151/jmsj.84.277
39. Ogawa, M., Yamanaka, M.D., Awaluddin, Darmawan, A., Sulistyowati, R., Sulaiman, A., Kozan, O.: X-band weather radar observation over Eastern Sumatra coastal area: statistical analysis of diurnal cycle of rainfall. (in this issue) (2023)
40. Ogino, S.-Y., Yamanaka, M.D., Mori, S., Matsumoto, J.: How much is the precipitation amount over the tropical coastal region? J. Clim. **29**, 1231–1236 (2016). https://doi.org/10.1175/JCLI-D-15-0484.1
41. Ogino, S.Y., Yamanaka, M.D., Mori, S., Matsumoto, J.: Tropical coastal dehydrator in global atmospheric water circulation. Geophys. Res. Lett. **44**, 11636–11643 (2017). https://doi.org/10.1002/2017GL075760
42. Okamoto, N., Yamanaka, M.D., Ogino, S.-Y., Hashiguchi, H., Nishi, N., Sribimawati, T., Numaguti, A.: Seasonal variation of tropospheric wind over Indonesia: comparison between collected operational rawinsonde data and NCEP reanalysis for 1992–99. J. Meteor. Soc. Jpn. **81**, 829–850 (2003). https://doi.org/10.2151/jmsj.81.829
43. Otobe, N., Sakai, S., Yoden, S., Shiotani, M.: Visualization and WKB analysis of the internal gravity wave in the QBO experiment. Nagare: Jpn. Soc. Fluid Mech. **17** (1998). http://www2.nagare.or.jp/mm/98/otobe/index.htm
44. Plumb, R.A.: The interaction of two internal waves with the mean flow: implications for the theory of the quasi-biennial oscillation. J. Atmos. Sci. **34**, 1847–1858 (1977). https://doi.org/10.1175/1520-0469(1977)034<1847:TIOTIW>2.0.CO;2
45. Plumb, R.A., McEwan, A.D.: The instability of a forced standing wave in a viscous stratified fluid: a laboratory analogue of the quasi-biennial oscillation. J. Atmos. Sci. **35**, 1827–1839 (1978). https://doi.org/10.1175/1520-0469(1978)035<1827:TIOAFS>2.0.CO;2

46. Rahman, M.A., Nugroho, D.S., Yamanaka, M.D., Kawasaki, M., Kozan, O., Ohashi, M., Hashiguchi, H., Mori, S.: Weather radar detection of planetary boundary layer and smoke layer top of peatland fire in Central Kalimantan, Indonesia. Sci. Rep. **11**, 367 (2021). https://doi.org/10.1038/s41598-020-79486-6
47. Rees, D., Barnett, J.J., Labitzke, K. (eds.): COSPAR international reference atmosphere (CIRA) 1986, Part II: middle atmosphere models. Adv. Space Res. **10**(12), 1–525 (1990). https://www.sciencedirect.com/journal/advances-in-space-research/vol/10/issue/12 (Some data are freely available at https://ccmc.gsfc.nasa.gov/modelweb/atmos/cospar1.html)
48. Rotunno, R.: On the linear theory of the land and sea breeze. J. Atmos. Sci. **40**, 1999–2009 (1983). https://doi.org/10.1175/1520-0469(1983)040<1999:OTLTOT>2.0.CO;2
49. Rouffaer, C.P., en IJzerman (uitg. en toeg.): De Eerste Schipvaart der Nederlanders naar Oost-Indië onder Cornelis de Houtman, 1595–1597. Martinus Nijhoff, Bladz. 248 (1915). https://archive.org/details/deeersteschipvaa01rouf
50. Sakurai, N., Kawashima, M., Fujiyoshi, Y., Hashiguchi, H., Shimomai, T., Mori, S., Hamada J.-I., Murata, F., Yamanaka, M.D., Tauhid, Y.I., Sribimawati, T., Suhardi, B.: Internal structure of precipitating cloud system migrating with diurnal cycle over Sumatera Island during CPEA-I campaign observations. J. Meteor. Soc. Jpn. **86**, 157–170 (2009). https://doi.org/10.2151/jmsj.87.157
51. Sakurai, N., Mori, S., Kawashima, M., Fujiyoshi, Y., Hamada J.-I., Shimizu, S., Fudeyasu, H., Tabata, Y., Harjupa, W., Hashiguchi, H., Yamanaka, M.D., Matsumoto, J., Emrizal, Syamsudin, F.: Migration process and 3D wind field of precipitation systems associated with a diurnal cycle in west Sumatera: Dual Doppler radar analysis during the HARIMAU 2006 campaign. J. Meteor. Soc. Jpn. **89**, 341–361 (2011). https://doi.org/10.2151/jmsj.2011-404
52. Sakurai, N., Murata, F., Yamanaka, M.D., Hashiguchi, H., Mori, S., Hamada J.-I., Tauhid, Y.I., Sribimawati, T., Suhardi, B.: Diurnal cycle of migration of convective cloud systems over Sumatera Island. J. Meteor. Soc. Jpn. **83**, 835–850 (2005). https://doi.org/10.2151/jmsj.83.835
53. Scorer, R.: Theory of waves in the lee of mountains. Quart. J. Roy. Meteor. Soc. **75**, 41–56 (1947). https://doi.org/10.1002/qj.49707532308
54. Shibagaki, Y., Shimomai, T., Kozu, T., Mori, S., Fujiyoshi, Y., Hashiguchi, H., Yamamoto, M.K., Fukao, S., Yamanaka, M.D.: Multiscale aspects of convective systems associated with an intraseasonal oscillation over the Indonesian maritime continent. Mon. Weather Rev. **134**, 1682–1696 (2006). https://doi.org/10.1175/MWR3152.1
55. Sulistyowati, R., Hapsari, R.I., Syamsudin, F., Mori, S., Oishi, S.T., Yamanaka, M.D.: Rainfall-driven diurnal variations of water level in the Ciliwung River, West Jawa, Indonesia. SOLA, **10**, 141–144 (2014). https://doi.org/10.2151/sola.2014-029
56. Tabata, Y., Hashiguchi, H., Yamamoto, M.K., Yamamoto, M., Yamanaka, M.D, Mori, S., Syamsudin, F., Manik, T.: Lower tropospheric horizontal wind over Indonesia: a comparison of wind-profiler network observations with global reanalyses. J. Atmos. Solar Terr. Phys. **73**, 986–995 (2011a). https://doi.org/10.1016/j.jastp.2010.09.016
57. Tabata, Y., Hashiguchi, H., Yamamoto, M.K., Yamamoto, M., Yamanaka, M.D, Mori, S., Syamsudin, F., Manik, T.: Observational study on diurnal precipitation cycle in equatorial Indonesia using 1.3-GHz wind profiling radar network and TRMM precipitation radar. J. Atmos. Solar Terr. Phys. **73**, 1031–1042 (2011b). https://doi.org/10.1016/j.jastp.2010.10.003
58. Tanaka, H., Yamanaka, M.D.: Atmospheric circulation in the lower stratosphere induced by the mesoscale mountain wave breakdown. J. Meteor. Soc. Jpn. **63**, 1047–1054 (1985). https://doi.org/10.2151/jmsj1965.63.6_1047
59. Tsuda, T., Murayama, Y., Wiryosumarto, H., Harijono, S.W.B., Kato, S.: Radiosonde observations of equatorial atmosphere dynamics over Indonesia: 1. Equatorial waves and diurnal tides. J. Geophys. Res. **99**, 10491–10505 (1994a). https://doi.org/10.1029/94JD00355
60. Tsuda, T., Murayama, Y., Wiryosumarto, H., Harijono, S.W.B., Kato, S.: Radiosonde observations of equatorial atmosphere dynamics over Indonesia: 2. Characteristics of gravity waves. J. Geophys. Res. **99**, 10507–10516 (1994b). https://doi.org/10.1029/94JD00354
61. van Bemmelen, W.: Die Erforschungdes tropischen Luftoszeans in Niederländische-Ost-Indien. Luftfahrt und Wissenschaft **5**, 1–50 (1913)

62. van Bemmelen, W.: Land- und Seebrise in Batavia. Beitr. Phys. Frei. Atmos. **10**, 169–177 (1922)
63. Widiyatmi, I., Yamanaka, M.D., Hashiguchi, H., Fukao, S., Tsuda, T., Ogino, S.-Y., Harijono, S.W.B., Wiryosumarto, H.: Quasi-four-day mode observed by the boundary layer radar at Serpong (6°S, 107°E), Indonesia. J. Meteor. Soc. Jpn. **77**, 1177–1184 (1999). https://doi.org/10.2151/jmsj1965.77.6_1177
64. Widiyatmi, I., Hashiguchi, H., Fukao, S., Yamanaka, M.D., Ogino, S.-Y., Gage, K.S., Harijono, S.W.B., Diharto, S., Djojodihardjo, H.: Examination of 3–6 day disturbances over equatorial Indonesia based on boundary layer radar observations during 1996–1999 at Serpong, Biak and Bukittinggi. J. Meteor. Soc. Jpn. **79**, 317–331 (2001). https://doi.org/10.2151/jmsj.79.317
65. Winchester, S.: Krakatoa: The Day the World Exploded: August 27, 1883. Harper Collins Pub., New York, 416pp. (2003)
66. Wu, P.-M., Hamada J.-I., Mori, S., Tauhid, Y.I., Yamanaka, M.D., Kimura, F.: Diurnal variation of precipitable water over a mountainous area in Sumatera Island. J. Appl. Meteorol. **42**, 1107–1115 (2003). https://doi.org/10.1175/1520-0450(2003)042<1107:DVOPWO>2.0.CO;2
67. Wu, P.-M., Hara, M., Hamada J.-I., Yamanaka, M.D., Kimura, F.: Why a large amount of rain falls over the sea in the vicinity of western Sumatera Island during nighttime. J. Appl. Meteor. Climatol. **48**, 1345–1361 (2009). https://doi.org/10.1175/2009JAMC2052.1
68. Wu, P.-M., Mori, S., Hamada J.-I., Yamanaka, M.D., Matsumoto, Kimura, F.: Diurnal variation of rainfall and precipitable water over Siberut Island off the western coast of Sumatera Island. SOLA, **4**, 125–128 (2008). https://doi.org/10.2151/sola.2008-032
69. Yamamoto, H.: Weather Observations at Empire of Japan, II: The Army Weather Part. Agriculture and Forestry Statistics Publishing Inc., 531pp. (in Japanese) (2015)
70. Yamamoto, M.: Future project. (presented at INCREASE 2021) (2021)
71. Yamanaka, M.D.: Physical climatology of Indonesian maritime continent: an outline to comprehend observational studies. Atmos. Res. 178–179, 231–259 (2016). https://doi.org/10.1016/j.atmosres.2016.03.017
72. Yamanaka, M.D., Akiyama, H., Koma, Y., Okabe, Y., Nishimura, J., Soegijo, J.: Vertical microstructure of equatorial stratospheric wind on the zonal wind reversal of QBO. Abstracts of Papers Presented at the Meteorological Society of Japan, 44, 8. (in Japanese) (1983)
73. Yamanaka, M.D., Mori, S., Wu, P.-M., Hamada J.-I., Sakurai, N., Hashiguchi, H., Yamamoto, M.K., Shibagaki, Y., Kawashima, M., Fujiyoshi, Y., Shimomai, T., Manik, T., Erlansyah, Setiawan, W., Tejasukmana, B., Syamsudin, F., Djajadihardja, Y.S., Anggadiredja, J.T.: HARIMAU radar-profiler network over Indonesian maritime continent: a GEOSS early achievement for hydrological cycle and disaster prevention. J. Disaster Res. **3**, 78–88 (2008). https://doi.org/10.20965/jdr.2008.p0078
74. Yamanaka, M.D., Ogino, S.-Y., Wu, P.-M., Hamada J.-I., Mori, S., Matsumoto, J., Syamsudin, F.: Maritime continent coastlines controlling Earth's climate. Prog. Earth Planet Sci. **5**(21), 1–28 (2018). https://doi.org/10.1186/s40645-018-0174-9
75. Yoneyama, K., Zhang, C.: Years of the maritime continent. Geophys. Res. Lett. **47**, e2020GL087182 (2020). https://doi.org/10.1029/2020GL087182

Study on Diurnal Variation of Rainfall Observed by X-band Polarimetric Radar in Peatlands Over Bengkalis Island, Eastern Sumatra, Indonesia

Mariko Ogawa, Manabu D. Yamanaka, Awaluddin, Arief Darmawan, Albertus Sulaiman, Reni Sulistyowati, I. Dewa Gede Arya Putra, and Osamu Kozan

Abstract Understanding the precipitation systems over tropical peatlands is essential for flood and forest fire risk management. The diurnal variation of rainfall over the Indonesian maritime continent has been analyzed previously by satellite, and more recently, by analyzing data from rain gauge network across Sumatra. In the complicated topography of Bengkalis Island located between Sumatra Island and the Malay Peninsula, the ground observation network is unevenly distributed. Consequently, analyzing the regional characteristics of diurnal variations of rainfall over the region is difficult. This paper presents preliminary results from observations for five rain gauges over Bengkalis Island and high spatiotemporal resolution compact X-band polarimetric radar whose data were collected from February 2020 to June 2021. The average hourly rainfall estimates by radar was compared with rain gauge data. The average hourly rainfall mentioned above during the observation period were re-averaged for each time zone. Rainfall occurred over both sea and land areas along the coast from 20:00 to 06:00 local time (LT), and heavy rain (≥50 mm/h) occurred across Bengkalis Island from west to east after midnight from 00:00 to

M. Ogawa (✉) · O. Kozan
Center for Southeast Asian Studies (CSEAS), Kyoto University, 46 Shimoadachi-Cho, Yoshida Sakyo-Ku, Kyoto 606-8501, Japan
e-mail: ogwmari@cseas.kyoto-u.ac.jp

M. D. Yamanaka · O. Kozan
Research Institute for Humanity and Nature (RIHN), 457-4 Motoyama, Kamigamo, Kita-Ku, Kyoto 603-8047, Japan

Awaluddin · A. Darmawan · A. Sulaiman
Research Center for Climate and Atmosphere, National Research and Innovation Agency (BRIN), Jalan Dr. Djunjunan No.133, Jawa Barat 40173, Indonesia

R. Sulistyowati
Research Center for Geospatial, National Research and Innovation Agency (BRIN), Gedung GEOSTECH, Jalan Kawasan Puspitek, Serpong, Tangerang Selatan 15314, Indonesia

I. D. G. A. Putra
Center for Research and Development, Meteorological, Climatological, and Geophysical Agency (BMKG), Jalan Angkasa I, No.2 Kemayoran, Jakarta Pusat 10610, Indonesia

A. Basit et al. (eds.), *Proceedings of the International Conference on Radioscience, Equatorial Atmospheric Science and Environment and Humanosphere Science*, Springer Proceedings in Physics 290, https://doi.org/10.1007/978-981-19-9768-6_3

03:00 LT. On the other hand, rainfall was distributed over land areas in the daytime from 10:00 to 16:00 LT. These findings showed that sea-land breeze circulation is the dominant wind system in this area.

1 Introduction

Tropical peatlands are distributed in lowlands along the coast of Southeast Asia. It is essential for understanding the water balance and fire hazards in the peatlands to obtain accurate real-time rainfall data for each region. Such information can only be collected by radar network in Indonesia (e.g. [16]) because the remoteness area makes rain gages maintenance difficult. Compact X-band polarimetric radars have been installed in Indonesia for disaster prevention against heavy rains and volcanic debris flows [2, 14]. These compact X-band polarimetric radar systems cover a smaller area than Global Satellite Mapping of Precipitation (GSMaP) [6] and the Indonesian Meteorological, Climatological, and Geophysical Agency (BMKG) C-band radar (single polarization) network. The compact X-band radar systems have temporal and spatial resolutions of approximately 1 min and 100 m, respectively. Consequently, more accurate rainfall with a high spatiotemporal resolution can be expected through observations of the compact X-band polarimetric radar.

Tropical coastal rainfall has a regular diurnal cycle which is typical in Indonesia [15, 17]. Satellite observations of diurnal rainfall cycles in Indonesia have been performed as part of the tropical rainfall measuring mission (TRMM) [9]. The findings of the TRMM have shown that more rain falls in the evening than in the morning over the relatively large islands and that there is more rainfall in the morning than in the evening over coastal sea regions surrounding the islands. Furthermore, rainfall moves from the central mountains to the coastal areas on either side of Sumatra Island.

On the other hand, the east coast of Sumatra, where peatlands are widespread, has a complex topography and the Malay Peninsula and numerous small islands lie to the north. Mori et al. [9] showed almost no difference in the annual mean rainfall between the mornings and evenings in this region.

Kozan [5] investigated the diurnal cycle of rainfall and the maximum hourly rainfall after midnight using one rain gauge installed in Bukit Batu village on the east coast of Sumatra. More recently, Marzuki et al. [8] studied the regional characteristics of the diurnal cycle in rainfall over Sumatra. They used 229 BMKG rain gauges deployed throughout Sumatra to perform cluster analyses of rainfall amount, intensity, duration, and frequency. However, since the density of the BMKG rain gauge network in eastern Sumatra is lower than that in the western part, the regional characteristics of the diurnal cycle of rainfall in the east region contain uncertainties.

The purpose of this study is to examine the regional characteristics of the diurnal cycle of rainfall and the movement of rain by using a compact X-band polarimetric radar system from February 2020 to June 2021 over Bengkalis Island, Riau Province. This is the first report on weather radar observations from the east coast of Sumatra.

Firstly, this study verifies the accuracy of radar-based rainfall estimates by comparing the average hourly rainfall calculated from the radar data with accumulated rainfall measured by five rain gauges. Secondly, to clarify the regional characteristics of the diurnal rainfall cycle, the average hourly rainfall mentioned above was re-averaged to each local time zone from 00:00 to 23:00.

2 Instruments and Methodology

2.1 Furuno Radar

A compact type X-band polarimetric radar system (WR-2100, Furuno Electric Co., Ltd), hereafter referred to as the Furuno radar, was installed at Bengkalis State Islamic College (STAIN Bengkalis) on Bengkalis Island, Riau Province (1° 29' N, 102° 07' E) in February 2020 (Fig. 1). Figure 2 shows the study area. Table 1 shows the specifications of the Furuno radar system.

Fig. 1 Furuno radar was installed at STAIN Bengkalis in February 2020

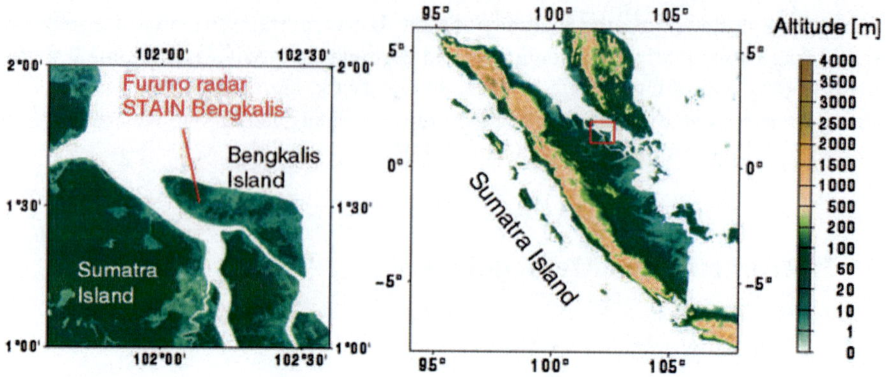

Fig. 2 Study area. The line indicates the location of Furuno radar installed at STAIN Bengkalis on Bengkalis Island, Riau Province

2.2 Radar Preprocessing and Rainfall Estimation Algorithms

Radar data were converted from polar to cartesian coordinates. The rainfall data within a 100 m grid cell is determined by calculating the weighted average using all data values within a search radius of 1 km. Estimates of radar rainfall data in this study used radar parameters, Z_H and K_{DP}, and the method of Park et al. [11], where Z_H is the horizontal reflectivity factor and K_{DP} is the specific differential phase. Before calculating K_{DP}, the observed phase information was smoothed using an iterative filter method [3] to obtain Φ_{DP}, which is the cross-polarization difference phase.

In addition, attenuation correction using the method of Bringi and Chandrasekar [1] was performed before estimating rainfall using the radar data.

2.3 Average Hourly Rainfall

Scanning radar data obtained at 10 min intervals were used to calculate the average hourly rainfall only when six radar scans could be performed per hour. These values may differ from accumulated hourly rainfall measured using a rain gauge because tropical rainfall intensity can change at scales of <1 km (this study used data within a search radius of 1 km) and <10 min (e.g., [4]). This study used the plan position indicator (PPI) radar mode with a 4° elevation angle. Furthermore, to characterize the regional characteristics of the diurnal rainfall cycle, average hourly rainfall data >10 mm during the observation period were re-averaged for each local time zone from 00:00 to 23:00 when three or more samples for each radar grid were obtained. In this study, we limited the precipitation threshold to events with relatively heavy rainfall of 10 mm/hr or more. This is to make it easier to understand the regional characteristics of rainfall areas. For example, rainfall of 5 mm/hr or less occurred

Table 1 Specifications of the WR-2100 Furuno radar system

Parameter	Description
Operating frequency	9.4 GHz
Maximum distance display	50 km
Maximum range fixed observation level	30 km
Transmitter	Solid-state 100 W per channel (H, V)
Size	Radome size: Ø1085 mm, 1025 mm height
Weight	68 kg
Polarity	Dual polarimetric horizontal (H) and vertical (V)
Pulses	PRF 600–2500 Hz, Width 0.1–50 μs
Antenna	Ø 750 mm, 2.7° beam width
Antenna gain	33.0 dBi
Wavelength	3.3 cm
Scan mode	PPI, CAPPI, RHI
Data output	Reflectivity factor: Z_H (dBZ), Differential reflectivity: Z_{DR} (dB), Doppler velocity: V (m/s), Doppler velocity spectrum width: W (m/s), Specific differential phase: K_{DP} (°/km), Copolar correlation coefficient: ρ_{HV}, Rainfall intensity: R (mm/h), Cross polarization difference phase: Φ_{DP} (°)

Source Furuno (WR-2100) operator's manual

everywhere within the radar range, and the regional characteristics of rain areas were not very clear (No image).

3 Observation and Data

This study analyzed 69-day rainfall data collected from February 15, 2020, to June 3, 2021; no data were available for 269 days when radar operations were stopped due to electrical and/or PC system problems. In November 2020, lead-acid batteries and

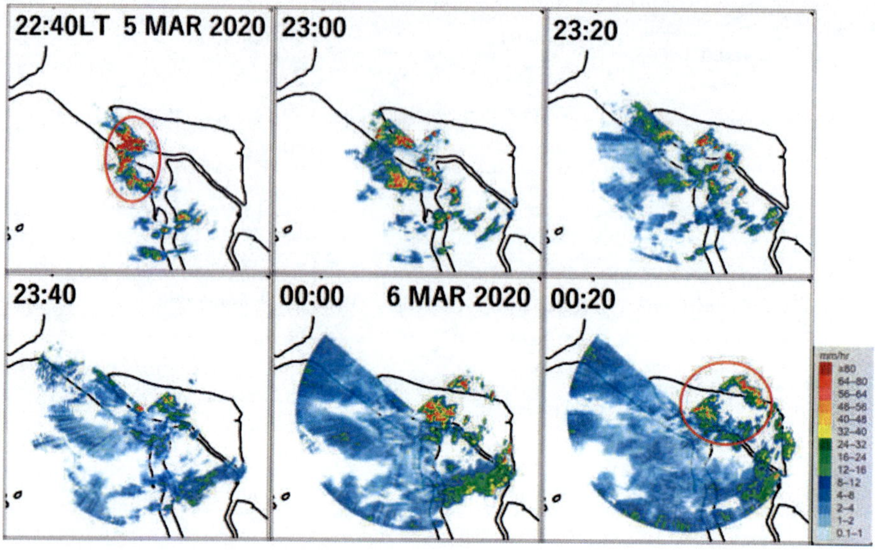

Fig. 3 Instantaneous values of rainfall were detected by the Furuno radar from 22:40 LT on March 5, 2020 to 00:20 LT on March 6, 2020. Rainfall exceeding 50 mm/h (circled in red) estimated using the method of Marshall and Palmer [7] moved from west to east over Bengkalis Island. Since the roof of the building exists on the north side of the Furuno radar (Fig. 1), the radar is unobservable on the direction

a generator were installed with an automatic transfer switch to ensure a continuous power supply.

Figure 3 shows a rainfall event detected by the Furuno radar system from 22:40 LT on March 5, 2020 to 00:20 LT on March 6, 2020. Rainfall exceeding 50 mm/h estimated using the method of Marshall and Palmer [7] in systems moving from west to east over Bengkalis Island was recorded. Figure 4 shows examples of hourly rainfall data obtained from GSMaP from 20:00 LT on March 5, 2020 to 04:59 LT on March 6, 2020, including the rainfall event detected by the Furuno radar. As shown in Fig. 4, rainfall moved across Sumatra from west to east. These movements in rainfall systems over the large spatial scale of Sumatra and the small scale over Bengkalis correspond to the measurements of Mori et al. [9] and Kozan [5], respectively.

4 Results and Discussion

4.1 *Evaluation of Rainfall Estimation by the Furuno Radar*

Average hourly rainfall estimated using the Furuno radar was compared to accumulated rainfall data measured at five rain gauge stations (Table 2 and Fig. 5). Station No. 5 is the farthest from the Furuno radar with about 45 km. Therefore, the maximum

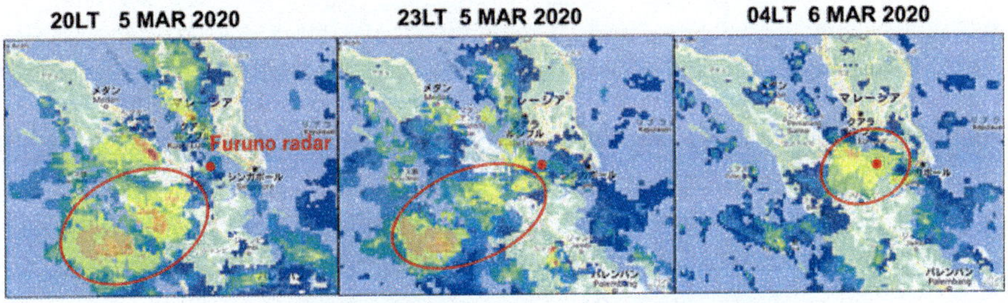

Fig. 4 Examples of hourly rainfall data were obtained from GSMaP from 20:00 LT on March 5, 2020 to 04:00 LT on March 6, 2020. The dots indicate the location of the Furuno radar. Rains ≥10 mm/h moved across Sumatra from west to east (yellow and orange areas circled in red). *Source* https://sharaku.eorc.jaxa.jp/GSMaP/index_j.htm

radar range used in this study is 45 km. Point data were extracted from PPI data with the exact latitude and longitude as the rain gauges on the ground.

Rainfall measurements at two stations (No. 6 and 7) were considerably underestimated. Since the rainfall moves faster and the heavy rain of about 30 mm/hr or more is localized (see Fig. 3), it is suggested that the radar data analyzed every 10 min misses the heavy rain. For example, No. 6 is located near the coast where heavy rains often occur (as confirmed later in Fig. 7 at 22:00 local time), so Furuno

Table 2 Locations of rain gauge stations

No	Name of rain gauge stations	Name of district	Location (Longitude, Latitude)
1	BRG_147205_01	MEDANG KAMPAI	101.645537, 1.613610
2	BRG_140301_01	BENGKALIS	102.099167, 1.519444
3	BRG_140302_02	BANTAN	102.433056, 1.516389
4	BRG_140312_01	SIAK KECIL	102.062222, 1.240833
5	BRG_140312_02	SIAK KECIL	102.039343, 1.114076
6	Perapat Tunggal	BENGKALIS	102.029137, 1.547426
7	Selatbaru	BENGKALIS	102.231427, 1.564642

Notes No. 1 and 2 are out of the radar coverage area. Rainfall data are collected every 10 min. Data are sent to the Peatland Restoration Agency in Indonesia (BRG) (No. 1–5) server and the SESAME (No. 6–7) server. SESAME: Sensory data transmission service assisted by Midori Engineering Laboratory Co., Ltd.

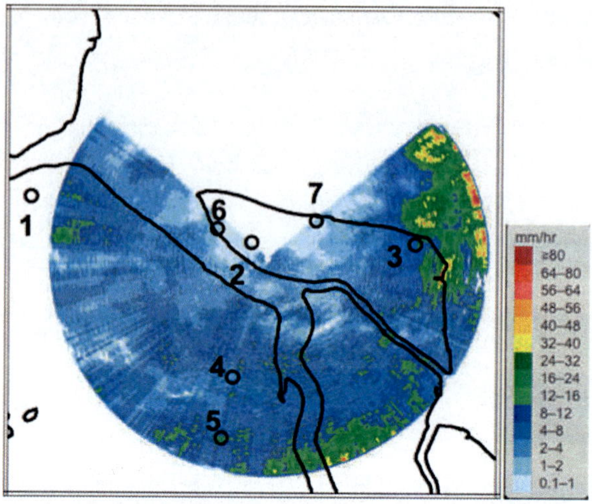

Fig. 5 Locations of rain gauges. The numbers on the map correspond to rain gauges, No. 1–7 depicted in Table 2. Exactly No. 1 and 2 are out of the radar coverage area (It is seen that No. 2 is located almost on the border of the area covered by the radar). Radar echo shows an example of instantaneous values of rainfall observed by the Furuno radar at 01:50 LT on March 5, 2020. The rainfall data were estimated using the method of Marshall and Palmer [7]

radar is significantly underestimated. Except for these two stations (No. 6 and 7), the coefficient of determination varied from 0.4 to 0.9 (Fig. 6).

The slopes of the regression lines were ranged from 0.6 to 2 at the other three stations (No. 3–5). The coefficient of determination did not correspond to the distance from the radar, for example, at station No. 3, which is located approximately 30 km from the radar site, the radar only detects precipitation particles at an altitude of 2 km and above. This evaluation of radar rainfall data did not consider the relationship between the precipitation system and the time resolution of the data, in terms of the different altitude of the extracted radar data. Syarifuddin et al. [13] showed that the correlation coefficient was >0.9 between the amount of rain measured by rain gauge and by radar estimated rainfall using the method of Park et al. [11]. They interpolated Furuno radar data using the scanning mode with a constant altitude plan position indicator (CAPPI) composed of multiple PPI data with different elevation angles. They then extracted the radar grid at a constant altitude of 500 m, and confirmed the existence of a precipitation system and melting layer through radiosonde data and the radar parameters, Z_{DR} and ρ_{HV}, where Z_{DR} is the differential reflectivity and ρ_{HV} is the copolar correlation coefficient.

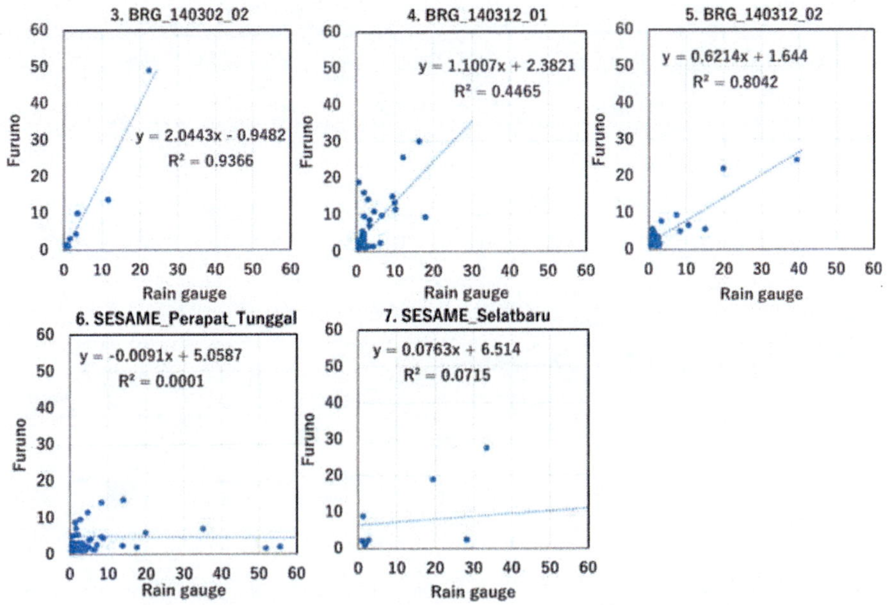

Fig. 6 Scatterplots of data from radar and five rain gauges during the observation period from February 15, 2020 to June 3, 2021. The numbers in the titles of scatterplots correspond to the station numbers shown in Table 2. The X-axis shows 1 h rainfall accumulation (mm) measured by rain gauges. Y-axis shows average hourly rainfall (mm) calculated by the Furuno radar using the method of Park et al. [11]. The coefficient of determination varied from 0.4 to 0.9, excluding two stations (No. 6 and 7) where heavy rainfalls of 30 mm/hr or more were frequently observed

4.2 *Regional Characteristics of the Diurnal Rainfall Cycle*

Figure 7 shows the average hourly rainfall during the observation period in each time zone from 00:00 to 23:00 LT. Rainfall occurred over both sea and land areas along the coast from nighttime to sunrise (i.e., 20:00–06:00 LT). Heavy rains approximately ≥50 mm/h occurred across Bengkalis Island from west to east after midnight (i.e., 00:00–03:00 LT). The timing of heavy rainfalls was similar to that observed by Kozan [5]. On the other hand, rainfall was distributed over land before noon to before sunset (i.e., 10:00–16:00 LT), with heavy rains ≥50 mm/h moving from north to south across Bengkalis Island during that time. These findings show that land-sea breeze circulation is dominant over Bengkalis Island and the surrounding area. Here, it should be noted that the rainfall estimated by the Furuno radar can overestimate compared to the rain gauge. According to Fig. 6, when the rainfall estimated by the Furuno radar was about 50 mm/hr at station No. 3, the rain gauge was about 23 mm/hr.

Although previous studies using rain gauges have reported that the rainfall in eastern Sumatra has a diurnal cycle [5, 8], this study confirmed the regional characteristics of this diurnal rainfall cycle in eastern Sumatra using radar data.

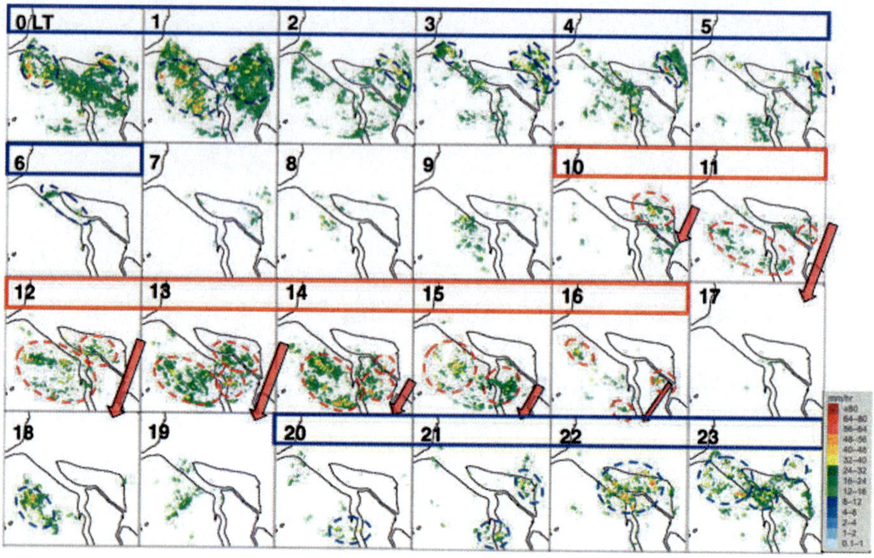

Fig. 7 Average hourly rainfall derived from Furuno radar during the observation period from February 15, 2020 to June 3, 2021 in each time zone from 00:00 to 23:00 LT (modified from Ogawa et al. [10]). Rainfall data were estimated using the method of Park et al. [11]. Rainfall occurred over both sea and land areas along the coast from nighttime to sunrise (the area surrounded by the blue dotted line from 20:00 to 6:00 LT). Heavy rains approximately ≥50 mm/h occurred across Bengkalis Island from west to east after midnight (the area was surrounded by the blue dotted line from 00:00 to 03:00 LT). On the other hand, rainfall was distributed over land before noon to before sunset, with heavy rains ≥50 mm/h (the area surrounded by the red dotted line from 10:00 to 16:00 LT) moving from north to south across Bengkalis Island during that time (down arrow)

5 Conclusions

This paper aims to clarify the regional characteristics of the diurnal cycle of rainfall and the movement of rain over the east coast of Sumatra, Indonesia. Furuno radar, a compact X-band polarimetric radar with a high spatiotemporal resolution, was newly installed and operated from February 2020, to June 2021 on Bengkalis Island, Riau Province.

Firstly, the average hourly rainfall observed by Furuno radar every 10 min was compared to accumulated rainfall data obtained from five rain gauge stations. The coefficient of determination varied from 0.4 to 0.9, excluding two stations where heavy rainfalls of 30 mm/hr or more were frequently observed. These results include uncertainties related to the time resolution of the data and the altitude of the extracted radar data, which are associated with the precipitation system. On the other hand, rainfall measurements at two stations were considerably underestimated. This is because the rainfall moves faster and the heavy rain of about 30 mm/hr or more is localized, so that the radar data analyzed every 10 min might miss the heavy rain.

Secondly, this study qualitatively clarified the regional characteristics of the diurnal cycle of rainfall and the rainfall movement over the island. Using average hourly rainfall data obtained from the Furuno radar during the observation period in each time zone (i.e., 00:00–23:00 LT), the land-sea rainfall and the movement of rainfall were explained in terms of their diurnal characteristics, i.e., whether they occurred at night time (20:00–06:00 LT) or daytime (10:00–16:00 LT) and in which time zone. The findings indicated that land-sea breeze circulation is dominant in the Bengkalis Island area.

Radar can also be applied to monitor the upper bounds of smoke layers when there is no precipitation [12], such as the smoke associated with peatland fires which have caused extensive environmental issues both locally and globally [5].

Acknowledgements This study was supported in part by a grant from JSPS KAKENHI (No. 19KK0106) and RIHN research project entitled "Toward the Regeneration of Tropical Peatland Societies: Building an International Research Network on Paludiculture and Sustainable Peatland Management." Rain gauge data from stations at Perapat Tunggal and Selatbaru were provided by Assoc. Prof. Koichi Yamamoto, Yamaguchi University.

References

1. Bringi, V.N., Chandrasekar, V.: Polarimetric Doppler Weather Radar: Principles and Applications. Cambridge University Press, United Kingdom (2001)
2. Hapsari, R.I., Oishi, S., Syarifuddin, M., Asmara, R.A., Legono, D.: X-MP Radar for developing a lahar rainfall threshold for the Merapi Volcano using a Bayesian approach. J. Disaster Res. **14**(5), 811–828 (2019). https://doi.org/10.20965/jdr.2019.p0811
3. Hubbert, J., Bringi, V.N.: An iterative filtering technique for the analysis of copolar differetial phase and dual-frequency radar measurements. J. Atmos. Oceanic Technol. **12**(3), 643–648 (1995). https://doi.org/10.1175/1520-0426(1995)012%3c0643:AIFTFT%3e2.0.CO;2
4. Kiguchi, M., Oki, T.: Point precipitation observation extremes in the world and Japan. J. Jpn. Soc. Hydrol. Water Resour. **23**(3), 231–247 (2010). https://doi.org/10.3178/jjshwr.23.231
5. Kozan, O.: Precipitation and groundwater variations in peat swamps. Stud. Sustain. Humanosphere **4**, 271–287 (2012). (in Japanese)
6. Kubota, T., Aonashi, K., Ushio, T., Shige, S., Takayabu, Y.N., Kachi, M., Arai, Y., Tashima, T., Masaki, T., Kawamoto, N., Mega, T., Yamamoto, M. K., Hamada, A., Yamaji, M., Liu, G., Oki, R.: Global Satellite Mapping of Precipitation (GSMaP) products in the GPM era. In: Satellite Precipitation Measurement, Springer (2020). https://doi.org/10.1007/978-3-030-24568-9_20
7. Marshall, J.S., Palmer, W.M.: The distribution of raindrops with size. J. Meteor. **5**(4), 154–166 (1948). https://doi.org/10.1175/1520-0469(1948)005%3c0165:TDORWS%3e2.0.CO;2
8. Marzuki, M., Suryanti, K., Yusnaini, H., Tangang, F., Muharsyah, R., Vonnisa, M., Devianto, D.: Diurnal variation of precipitation from the perspectives of precipitation amount, intensity and duration over Sumatra from rain gauge observations. Int. J. Climatol. **41**(8), 4386–4397 (2021). https://doi.org/10.1002/joc.7078
9. Mori, S., Jun-Ichi, H., Tauhid, Y.I., Yamanaka, M.D., Okamoto, N., Murata, F., Sakurai, N., Hashiguchi, H., Sribimawati, T.: Diurnal land-sea rainfall peak migration over Sumatera Island, Indonesian maritime continent observed by TRMM satellite and intensive rawinsonde soundings. Mon. Wea. Rev. **132**(8), 2021–2039 (2004). https://doi.org/10.1175/1520-0493(2004)132<2021:DLRPMO>2.0.CO;2

10. Ogawa, M., Yamanaka, M. D., Awaluddin, A., Darmawan, A., Sulistyowati, R., Sulaiman, A., Kozan, O.: X-band weather radar observations in the east coast of Sumatra: statistical analysis of diurnal cycle of rainfall. In: Proceedings of the 15th Symposium on MU Radar and Equatorial Atmosphere Radar (2021)
11. Park, S.G., Bringi, V.N., Chandrasekar, V., Maki, M., Iwanami, K.: Correction of radar reflectivity and differential reflectivity for rain attenuation at X band. Part I: theoretical and empirical basis. J. Atmos. Oceanic Technol. **22**(11), 1621–1631 (2005). https://doi.org/10.1175/JTECH1803.1
12. Rahman, M.A., Nugroho, D.S., Yamanaka, M.D., Kawasaki, M., Kozan, O., Ohashi, M., Hashiguchi, H., Mori, S.: Weather radar detection of planetary boundary layer and smoke layer top of peatland fire in Central Kalimantan Indonesia. Sci. Rep. **11**(367), 1–9 (2021). https://doi.org/10.1038/s41598-020-79486-6
13. Syarifuddin, M., Ogawa, M., Nakamichi, H., Oishi, S., Iguchi, M.: Performance of a small-compact X-MP radar to monitor extreme rainfall event on 7 July 2018. In: Proceedings of the 22nd IAHR-APD Congress 2020, pp. 1–8 (2020)
14. Syarifuddin, M., Oishi, S., Hapsari, R.I., Shiokawa, J., Mawandha, H.G., Iguchi, M.: Estimating the volcanic ash fall rate from the mount sinabung eruption on february 19, 2018 using weather radar. J. Disaster Res. **14**(1), 135–150 (2019). https://doi.org/10.20965/jdr.2019.p0135
15. Yamanaka, M.D.: Physical climatology of Indonesian maritime continent: an outline to comprehend observational studies. Atmos. Res. **178–179**, 231–259 (2016). https://doi.org/10.1016/j.atmosres.2016.03.017
16. Yamanaka, M.D., Hashiguchi, H., Mori, S., Wu, P.M., Syamsudin, F., Manik, T., Hamada, J.I., Yamamoto, M.K., Kawashima, M., Fujiyoshi, Y., Sakurai, N., Ohi, M., Shirooka, R., Katsumata, M., Shibagaki, Y., Shimomai, T., Erlansyah, Setiawan, W., Tejasukmana, B., Djajadihardja, Y.S., Anggadiredja, J.T.: HARIMAU radar-profiler network over the Indonesian maritime continent: a GEOSS early achievement for hydrological cycle and disaster prevention. J. Disaster Res. **3**(1), 78–88 (2008). https://doi.org/10.20965/jdr.2008.p0078
17. Yamanaka, M.D., Ogino, S.-Y., Wu, P.-M., Hamada, J.-I., Mori, S., Matsumoto, J., Syamsudin, F.: Maritime continent coastlines controlling Earth's climate. Prog. Earth Planet Sci. **5**(21), 1–28 (2018)

Detrended Fluctuation Analysis (DFA) of Gunungsitoli Geomagnetic Station to Assess the Possibility of the Earthquake Precursor

Febty Febriani, Cinantya Nirmala Dewi, Suaidi Ahadi, Titi Anggono, Syuhada, Mohammad Hasib, and Aditya Dwi Prasetio

Abstract Nias Island is one of the active seismic regions in Indonesia. We analyzed three components of geomagnetic data recorded at the Gunungsitoli (GSI) station to identify possible geomagnetic anomalies before earthquakes on Nias Island in 2018. We selected earthquakes with a magnitude > 5 and an epicenter ≤ 100 km from the GSI. We calculated the local seismicity index (K_{LS}) and the daily local earthquake energy (Es). We applied detrended fluctuation analysis (DFA) to investigate the possibility of the earthquake precursor related to the analyzed earthquake. We also used the disturbance storm time (Dst) index to analyze geomagnetic storm activity. The result presents a decrease in the DFA value, which exceeds the DFA's value threshold during the analyzed period. The decrease occurred about two months and one month prior to the Mw5.4 (06/09/2018) and Mw5.2 (20/08/2018) earthquakes. The Dst index also shows the quiet period where the DFA value decreases. It confirms that the anomalies were not related to geomagnetic storms. The Mw5.4 earthquake had a larger magnitude and the shallowest depth. It can be the main contributor to the appearance of geomagnetic anomalies. Therefore, we suggest that the decrease in the DFA value is possibly related to the Mw5.4 earthquake.

Keywords Geomagnetic anomaly · Detrended fluctuation analysis · Earthquake forecast

---

F. Febriani (✉) · C. N. Dewi · T. Anggono · Syuhada · M. Hasib · A. D. Prasetio
Research Center for Geological Disasters, National Agency of Research and Innovation (BRIN), Jl. Sangkuriang, Kompleks BRIN, Jawa Barat 40135, Indonesia
e-mail: febt001@brin.go.id

C. N. Dewi
e-mail: cina001@brin.go.id

S. Ahadi
Indonesian Meteorological, Climatological and Geophysical Agency (BMKG), Jl. Angkasa I No. 2, Kemayoran, DKI Jakarta 10720, Indonesia

A. Basit et al. (eds.), *Proceedings of the International Conference on Radioscience, Equatorial Atmospheric Science and Environment and Humanosphere Science*, Springer Proceedings in Physics 290, https://doi.org/10.1007/978-981-19-9768-6_4

1 Introduction

Nias is a small Island on the western coast of Sumatra, Indonesia, and is geologically located in the Sunda megathrust. The collision between the Indo-Australian Plate affects the geological structure of Nias Island and causes the Island as one of the most active seismic regions in Indonesia. As a result, some earthquakes occurred in the surrounding Nias Island in the past. For example, the M8.6 Nias earthquake occurred on March 28, 2005, and caused many casualties, loss, and damaged buildings [1].

Many studies have observed that geomagnetic precursors may appear before moderate-large earthquakes worldwide e.g., [2–8]. In Indonesia, the research related to the earthquake precursor by investigating anomalies in the geomagnetic signals has evolved. Furthermore, the geomagnetic anomalies before the earthquakes occurred on Sumatra Island have been reported e.g., [3, 9–13]. Many methods have been applied to investigate geomagnetic anomalies prior to an earthquake. One of them is detrended fluctuation analysis (DFA) e.g., [7, 14–17].

Therefore, it is essential to assess the precursor associated with earthquakes based on the geomagnetic data as a tool for earthquake hazards and risk mitigation on Sumatra Island. In this study, we apply DFA as the analysis method. We also calculated the certain earthquake epicenter distance and local earthquake energy released by earthquakes in observing geomagnetic anomalies to investigate the anomalies related to the earthquakes.

2 Data and Method

Figure 1 presents the location of the Gunungsitoli (GSI) station and all Mw5 earthquake epicenters during 2018 based on the United States Geological Survey (USGS) catalog. The magnetometer system is operated by the Indonesian Meteorological, Climatological, and Geophysical Agency (BMKG) with an averaged sampling frequency of 1 Hz. It was developed by Magnetic Data Acquisition System (MAGDAS) project belonging to Space Environment Research Center (SERC), Kyushu University. It recorded H, D, Z, and total (F) components with the observation range $\pm$ 70,000 nT.

In this study, we chose an earthquake based on criteria $M > 5$ and the distance between the epicenter and the station within 100 km. The analyzed data period in this study is from January 25, 2018, to December 31, 2018. There is no data from January 1, 2018, to January 24, 2018. We use X, Y, and Z components in this study. Therefore, we convert the H, D, and Z to X, Y, and Z components. We also calculated the local seismicity index (K_{LS}) and the daily sum of the local earthquake energy (E_s) by applying Eqs. 1 and 2 [18, 19].

$$K_{LS} = \frac{10^{0.75M}}{R + 100} \quad (1)$$

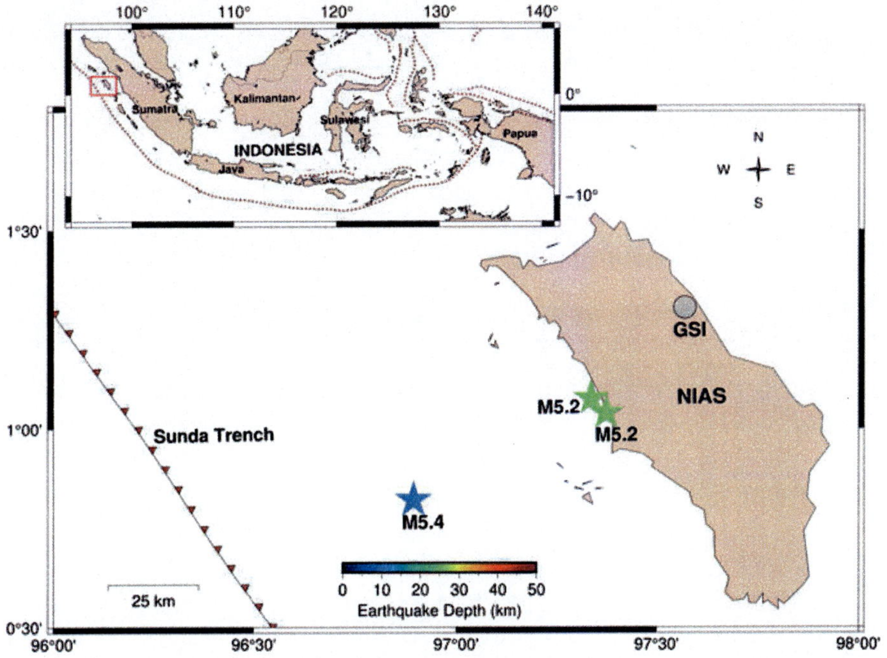

Fig. 1 The red square in Indonesia's map inset indicates the location of the study area. The gray circle is the Gunung Sitoli (GSI) station. The stars are the epicenter of the earthquakes, which is indicated in Table 1. The color scale describes the depth of the earthquakes

$$E_s' = \frac{10^{4.8+1.5M}}{r^2} \tag{2}$$

R is the distance between the earthquake epicenter and the station, r is the hypocenter distance, and M is the earthquake's magnitude.

The local seismicity index (K_{LS}) represents the correlation between the magnitude and distance of an earthquake [18]. It can be applied to investigate the primary source of the detected precursor in an earthquake series [20]. The E_s parameter considers the hypocenter distance and the magnitude of the selected earthquake events [19]. The detailed parameters of all earthquakes analyzed in this study are listed in Table 1.

In this study, we use the night data (17:00–21:00 UTC) since they have less contamination from artificial and natural noises [7, 21, 22].

Then, we apply detrended fractal analysis (DFA) for further analysis. This analysis is carried out to define the data scaling behavior. Even though we do not know the detail of the data origin and shape, we can investigate the possible trends of the data by applying DFA [23]. We assume $y(k)$ is a time series of interevent $x(i)$ and x_{ave} indicates the mean interevent time. Then, we can calculate $y(k)$as follows:

Table 1 A list of earthquakes that occurred during the analyzed data period based on the USGS global earthquake catalog

No	Lat (*N*)	Long (*E*)	Depth (km)	Mag (Mw)	Date	Time (UTC)	Epicenter distance *R* (km)	Hypocenter distance *r* (km)	Local seismicity index K_{LS}	Daily sum of the local earthquake energy E_s
1	1.041	97.376	25.69	5.2	20/08/2018	10:51:05	36.403	44.555	58.23	2.0E + 09
2	0.821	96.895	10	5.4	06/09/2018	00:13:50	92.424	92.963	58.31	9.2E + 08
3	1.077	97.338	25.91	5.2	29/12/2018	08:15:27	36.085	44.424	58.37	2.2E + 09

$$y(k) = \sum_{i=}^{k} [x(i) - x_{\mathrm{ave}}] \tag{3}$$

with $k = 1 \ldots N$.

Then, the $y(k)$ is divided into boxes with equal length, n, and a least-square polynomial $y_n(k)$ of degree p is applied in each box. It will determine the trend of the order p. In the next step, we find $F(n)$ through Eq. (4):

$$F(n) = \left(\sum_{k=1}^{N} [y(k) - y_n(k)]^2 \right)^{1/2} \tag{4}$$

If we calculated $F(n)$ for all box sizes, the relationship between $F(n)$ and n could be determined as follows:

$$F(n) \propto n^{\alpha} \tag{5}$$

According to the above formula, we define the α value. In the last step, we calculated the mean $\pm$ 2σ of all α values during the analyzed data period as the threshold with σ is the standard deviation. We determined the anomaly if there is data that exceeds the threshold. This assumption refers to the normal distribution of data that can be assumed as the peculiarity of the data trend if data surpasses the threshold.

3 Results and Discussion

Figure 2 shows the X, Y, and Z components of the geomagnetic data recorded at the GSI station. The data was recorded on November 25, 2018. The figure shows that the daytime data of the X, Y, and Z components are higher than that of nighttime data. It presents typical geomagnetic data.

Figure 3 indicates the correlation between $F \sim n$ for the X, Y, and Z components of November 25, 2018, geomagnetic data. We calculated the correlation for all available data by applying Eq. (5) to obtain the α value of each component during the analyzed data period. We present them in Fig. 4.

The first, second, and third panels in Fig. 4 indicate the α value during the analyzed data period for the X, Y, and Z components. The last panel in Fig. 4 is the Dst index during the analyzed data period. It describes the geomagnetic activity, especially in the low latitude region [24]. We used the Dst index to confirm that the geomagnetic anomalies which appeared prior to the earthquake does not origin from external disturbance or solar-terrestrial interaction (solar wind, interplanetary magnetic field). The Dst ≤ -100 nT, -100 nT < Dst ≤ -50 nT, and -50 nT < Dst $\leq$ -30 nT indicates intense, moderate, and small storms, respectively [25].

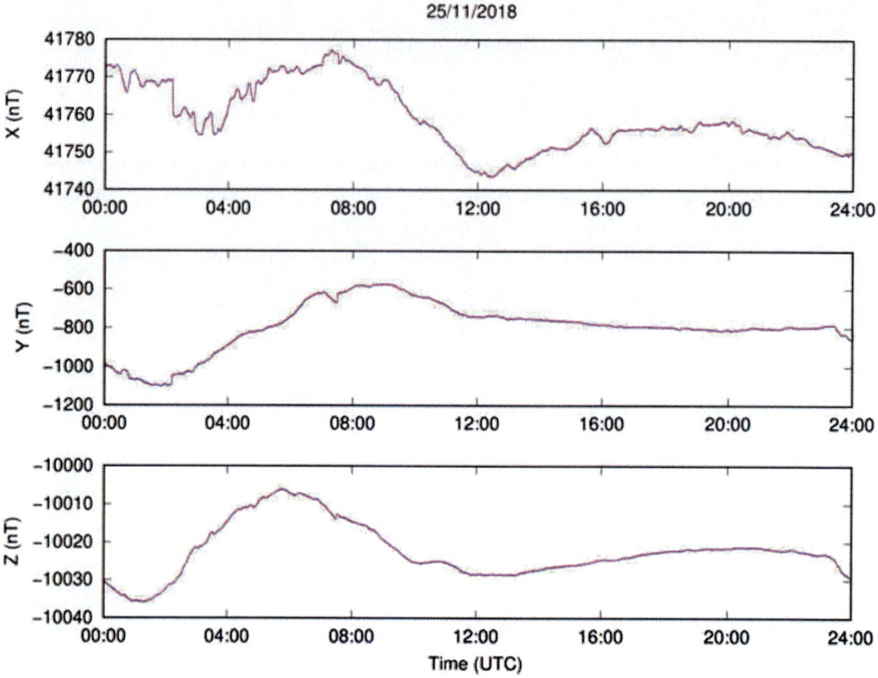

Fig. 2 The *X*, *Y*, and *Z* components of the November 25, 2018, geomagnetic data were recorded at the GSI geomagnetic station. The *X*-axis indicates the recorded data time in UTC. The Y axes are magnetic fields for each component in nano Tesla (nT)

Figure 4 presents global geomagnetic storms from January 25, 2018, to December 31, 2018. They are indicated by the green circle in the last panel of Fig. 4. The highest storm occurred before the Mw5.4 earthquake, as indicated by number 9 in the last panel of Fig. 4. However, there is no α value of the three components that surpassed the threshold when the ninth storm occurred. The storm prior to an earthquake is an exciting character since a recent publication suggested that the probability of a geomagnetic storm prior to a shallow earthquake is high [26].

In addition, Fig. 4 also shows that the decrease of αvalue exceeded the threshold when the second, third, fourth, fifth, tenth, and eleventh geomagnetic storms appeared. The anomalies related to a geomagnetic storm are very clear in the α value of the *Y* component.

The decrease of the αvalue surpassing the threshold also appeared about 30 and 50 days prior to the Mw5.2 (20/08/2018) and Mw5.4 (06/09/2018) earthquakes. The anomaly period ends approximately 10 and 25 days before the Mw5.2 (20/08/2018) and Mw5.4 (06/09/2018) earthquakes. The anomalies emerge when the quiet period of the Dst index. The previous studies revealed that the α value could be increased or decreased before an earthquake [26, 27]. Figure 4 also presents there is no anomaly prior to the Mw5.2 (29/12/2018).

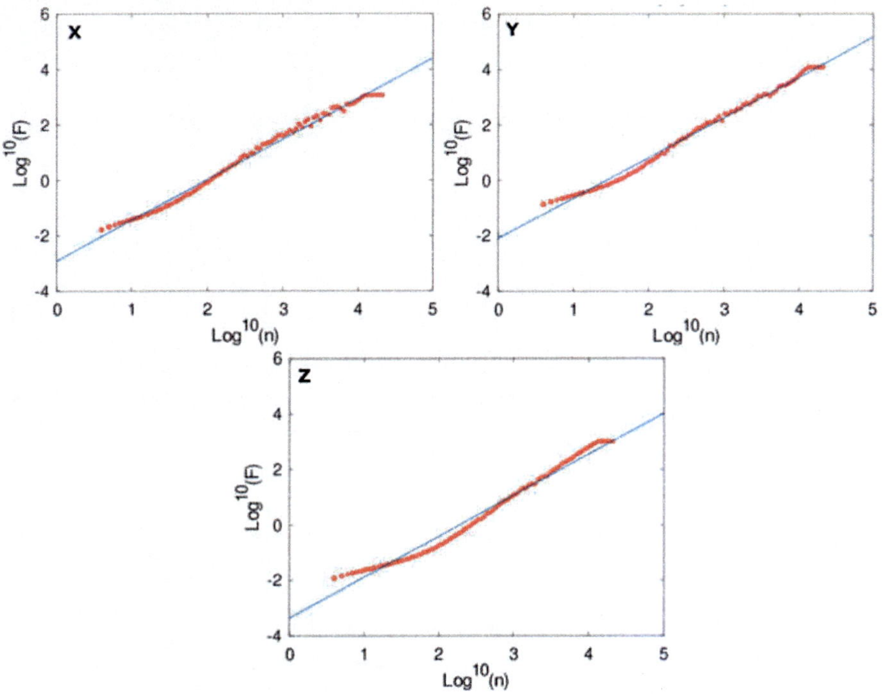

Fig. 3 An example of the correlation between $F \sim n$ for the X, Y, and Z components of the November 25, 2018, geomagnetic data

During the analyzed data period, the Mw5.4 earthquake (06/09/2018) is the earthquake that has the smallest depth. The K_{LS} value of the Mw5.4 (06/09/2018) earthquake is larger than that of the Mw5.2 (20/08/2018) earthquake. Previous research has reported that an earthquake with a high K_{LS} value in a series of earthquakes is considered the primary source of every detected precursor [21]. Therefore, we assume the anomalies are possibly related to the Mw5.4 earthquake. It also indicated that the critical factors related to the geomagnetic precursor before an earthquake is possibly the depth and magnitude of the earthquake.

This study supports the recent research that revealed electromagnetic/geomagnetic anomaly could be one candidate of the short-term precursor before an earthquake [28]. Therefore, since Indonesia is an active seismic country in the world, studying the geomagnetic precursor related to an earthquake must be carried out intensively in the future.

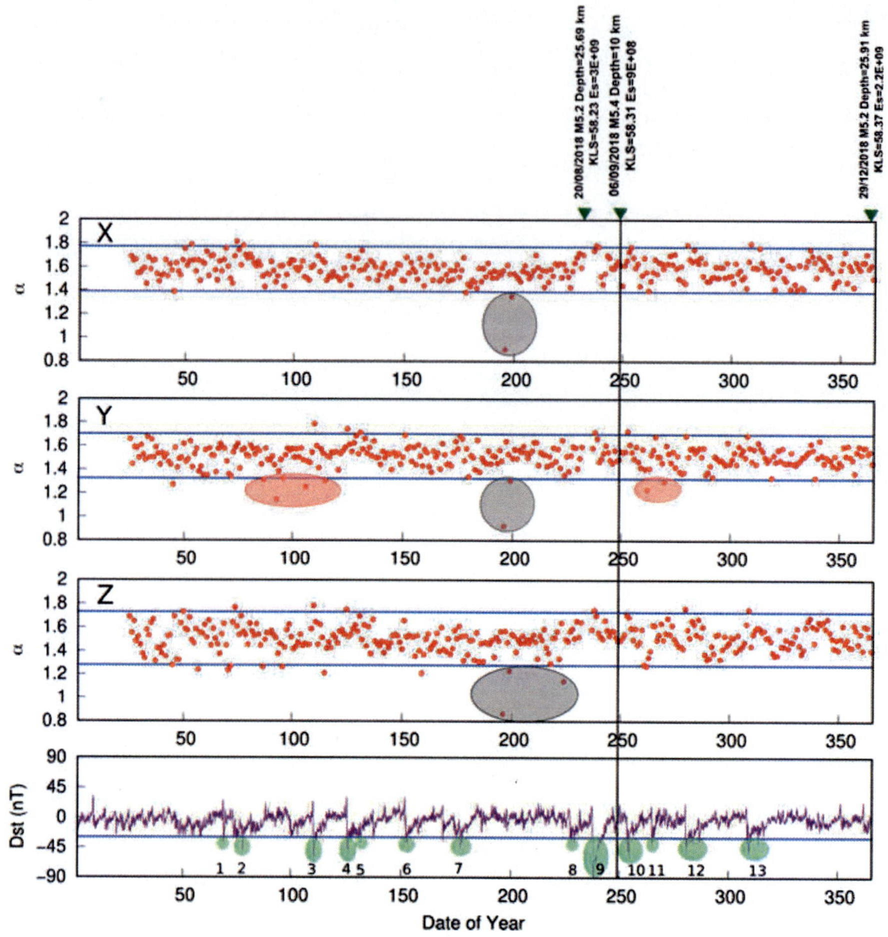

Fig. 4 The result of the DFA analysis of the geomagnetic data and Dst index during the analyzed data period. The reverse triangles indicate that all M5 earthquakes occurred during 2018. The numbers in the last panel indicate the geomagnetic storm that occurred in 2018. The blue lines indicate the threshold mean $\pm 2\sigma$, with σ as the standard deviation

4 Conclusion

We analyzed the geomagnetic data recorded in the Guningsitoli station to assess the possibility of the geomagnetic anomalies associated with the Mw $\geq$ 5 earthquake in 2018. We considered the distance between the station and the epicenter less than 100 km. We applied detrended fluctuation analysis (DFA) in this study. The results present that the α value decreased, surpassing from 30–10 days and 50–25 days prior to the Mw5.2 (20/08/2018) and Mw5.4 (06/09/2018) earthquakes, respectively. Since the Mw5.4 (06/09/2018) earthquake has the smallest depth and bigger K_{LS} than that of Mw5.2 (20/08/2018), the anomalies are possibly related to the Mw5.4 earthquake. This study also revealed that the possibility of the earthquake's depth and magnitude are important factors for the geomagnetic precursor to appear before an earthquake.

Funding We thank you for funding support from L'oreal-UNESCO for Women in Science Program and the Disaster Research Program (Grant No. SP DIPA-124.01.1.690501/2022) of the National Research and Innovation Agency (BRIN) for this research. We also thank the U.S. Geological Survey (USGS) for the earthquake database and World Data Center for Geomagnetism, Kyoto University, for the Dst index (http://wdc.kugi.kyoto-u.ac.jp/dstdir/). The Generic Mapping Tools 7 is applied in this paper to make some figures (http://www.soest.hawaii.edu/gmt/; Wessel and Smith 1998).

References

1. Setiyono, U., Gunawan, I., Proyobudi, Yatimantoro, T., Imananta, R.T., Ramdhan, M., Hidayanti, Anggraini, S., Rahayu, R.H., Hawati, P., Yogaswara, D.S., Julius, A.M., Apriani, Harvan, M., Simangunsong, G., Kriswinarso, T.: Significant and destructive earthquakes catalog 1821–2018. Indonesian Meteorological, Climatological, and Geophysical Agency (BMKG), Jakarta (2019) (in Indonesia)
2. Febriani, F., Han, P., Yoshino, C., Hattori, K., Nudiyanto, B., Effendi, N., Maulana, I., Suhardjono, Gaffar, E.: Ultra low frequency (ULF) electromagnetic anomalies associated with large earthquakes in Java Island, Indonesia by using wavelet transform and detrended fluctuation analysis. Nat. Hazards Earth Syst. Sci. **14**, 789–798(2014)
3. Ahad, S., Puspito, N.T., Ibrahim, G., Saroso, S., Yumoto, K., Yoshikawa, A.: Muzli: anomalous ULF emissions and their possible association with the strong earthquakes in Sumatra, Indonesia, during 2007–2012. J. Math. Fundam. Sci. **47**, 84–103 (2015)
4. Han, P., Hattori, K., Huang, Q., Hirooka, S., Yoshino, C.: Spatiotemporal characteristic of the geomagnetic diurnal variations prior to the 2011 off the Pacific coast of Tohuku earthquake (Mw9.0). J. Asian Earth Sci. **129**, 13–21 (2016)
5. Han, P., Zhuang, J., Hattori, K., Chen, C.-H., Febriani, F., Chen, H., Yoshino, C., Yoshida, S.: Assesing the potential earthquake precursory information in ULF magnetic data recorded in Kanto, Japan, during 2000–2010: Distance and Magnitude Response. Entropy **22**(8), 859 (2020)
6. Swati, Singh, B., Pundhir, D., Sinha, A.K., Rao, K.M., Guha, A., Hobara, Y.: Ultra-low frequency (ULF) magnetic field emissions associated with some major earthquake occurred in Indian Subcontinent. J. Atmos. Solar Terr. Phys. **211**, 105469 (2020)
7. Febriani, F., Ahadi, S., Anggono, T., Dewi, C.N., Prasetio, A.D.: Applying wavelet analysis to assess the ultra low frequency (ULF) geomagnetic anomalies prior to M6.1 Banten earthquake (2018). In: Subehi, L., Sugeha, H.Y. (eds), International Conference on the Ocean and Earth Sciences, vol. 789, 012064. IOP Publishing Ltd (2021)
8. Yusof, K.A., Abdullah, M., Hamid, N.S.A., Ahadi, S., Yoshikawa, A.: Correlations between earthquake properties and characteristic of possible ULF geomagnetic precursor over multiple earthquake. Universe **7**(1), 20 (2021)
9. Ahyar, A.S., Sunardi, B.: Correlation of Z/H magnetic polarization for earthquake precursor identification around Pelabuhan Ratu. Spektra: Jurnal Fisika dan Aplikasinya **2**, 179–186 (2017) (in Indonesia)
10. Takla, E.M., Yumoto, K., Liu, J.Y., Kakinami, Y., Uozumi, T., Abe. S., Ikeda, A.: Anomalous geomagnetic variations possibly linked with the Taiwan earthquake (Mw= 6.4) on 19 December 2009. Int. J. Geophys. **2011**, 848467 (2011)

11. Ramadhani, A.F., Syafriani, Triyono, R., Ubaya, T.: Analysis of magnetic field anomalies before earthquakes based on ULF (Ultra Low Frequency) method using magnetic sensor data in Sumatra. IOP Conf. Ser. J. Phys. Conf. Ser. **1317**, 012044 (2019)
12. Sokacana, I., Rosid, M.S., Ahadi, S.: Optimum frequency identification of anomalous geomagnetic signals related to earthquake precursor in Sumatra Island. IOP Conf. Ser. Earth Environ. Sci. **406**, 012021 (2019)
13. Marzuki, M., Hamidi, M., Ahadi, S., Putra, A., Afdal, A., Harmadi, H., Karnawati, D., Suprihatin, H.S., Syirojudin, M., Marsyam, I.: ULF geomagnetic anomaly associated with the Sumatra-Pagai Island earthquake swarm during 2020. Contribution Geophys. Geodesy **52**(2), 185–207 (2022)
14. Ida, Y., Hayakawa, M.: Fractal Analysis for the ULF data during 1993 Guam earthquake to study prefracture critically. Nonlinear Process. Geophys. **13**, 409–412 (2006)
15. Telesca, L., Hattori, K.: Non-uniform scaling behavior in ultra-low frequency (ULF) earthquake-related geomagnetic signals. Phys. A **384**, 522–528 (2007)
16. Telesca, L., Lapenma, V., Macchiato, M., Hattori, K.: Earth Planet. Sci. Lett. **268**, 219–224 (2008)
17. Hayakawa, M., Ida, Y.: Fractal (mono- and multi-) analysis for the ULF data during the 1993 Guam earthquake for study of prefracture criticality. Curr. Dev. Theory Appl. Wavelets **2**, 159–174 (2008)
18. Molchanov, O.A., Hayakawa, M.: Seismo-Electromagnetics and Related Phenomena: History and Latest Results, vol. 189. Terrapub, Tokyo (2008)
19. Hattori, K., Serita, A., Yoshino, C., Hayakawa, M., Isezaki, N.: Singular spectral analysis and principal component analysis for signal discrimination of ULF geomagnetic data associated with 2000 Izu Island Earthquake Swarm. Phys. Chem. Earth **31**, 281–291 (2006). https://doi.org/10.1016/j.pce.2006.02.034
20. Yusof, K.A., Hamid, N.S.A., Abdullah, M., Ahadi, S., Yoshikawa, A.: Assessment of signal processing methods for geomagnetic precursor of the 2012 M6.9 Visayas, Philippines earthquake. Acta Geophysica **67**, 1297–1306 (2019)
21. Potirakis, S.M., Hayakawa, M., Schekotov, A.: (2017) Fractal analysis of the ground-recorded ULF magnetic fields prior to the 11 March 2011 Tohoku earthquake (Mw = 9): discriminating possible earthquake precursors from space-sourced disturbances. Nat. Hazards **85**, 59–86 (2017)
22. Peng, C.K., Buldyrev, S.V., Havlin, S., Simons, M., Stanley, H.E., Goldberger, A.L.: Mosaic organization of DNA nucleotides. Phys. Rev. E **49**, 1685–1689 (1994)
23. Nose, M., Iyemori, T., Sugiura, M., Kamei, T.: Geomagnetic Dst index world data center for geomagnetism Kyoto (2015).https://doi.org/10.17593/14515-74000
24. Gonzalez, W.D., Joselyn, J.A., Kamide, Y., Kroehl, H.W., Rostoker, G., Tsurutani, B.T., Vasyliunas, V.M.: What is a geomagnetic storm? J. Geophys. Res. **99**, 5771 (1994)
25. Chen, H., Wang, R., Miao, M., Liu, X., Ma, Y., Hattori, K., Han, P.: A statistical study of the correlation between geomagnetic storms and M≥7.0 global earthquake during 1957–2020. Entropy **22**(11), 1270 (2020)
26. Ida, Y., Yang, D., Li, Q., Sunvand, H., Hayakawa, M.: Fractal analysis of ULF electromagnetic emissions in possible association with earthquakes in China. Nonlinear Process. Geophys. **19**, 577–583 (2012)
27. Potirakis, M., Hayakawa, M., Schekotov, A.: Fractal analysis of the ground-recorded ULF magneticfields prior to the 11 March 2011 Tohoku earthquake (Mw = 9): discriminating possible earthquake precursors from space-sourced disturbances. Nat. Hazards **85**, 59–86 (2017)
28. Wang, J.H.: Piezoelectricity as a mechanism on generation of electromagnetic precursors before earthquakes. Geophys. J. Int. **224**(1), 682–700 (2021)

Low-Latitude Fluctuation of Ionospheric Magnetic Field Measured by LAPAN-A3 Satellite

Fitri Nuraeni, La Ode M. Musafar Kilowasid, Clara Y. Yatini, Visca Wellyanita, Satriya Utama, Yoga Andrian, Teti Zubaidah, Wahyudi Hasbi, B. Moh. Andi Aris, Setyanto C. Pranoto, Harry Bangkit, Muzirwan, Ega A. Anggari, Erlansyah, Nata Miharja, Dwi Ratnasari, Prita Ayuningtyas, and Angga Yolanda Putra

Abstract The fluctuation of the magnetic field measured by satellite LAPAN-A3 is analyzed for low-latitude data. The low-latitude magnetic field characteristic has been long well understood from the ground-based observation that the magnetic variation shows a peak around local time and it is strongly related to the amount of ionospheric ionization, which affects current flow in the ionosphere. The LAPAN-A3 satellite orbits at an ionospheric altitude where the low-latitude geomagnetic field lines lie almost horizontally in the ionosphere so that variations in the ionospheric magnetic field at low latitudes will show a pattern similar to local time variations at ground-based stations at low latitudes. To examine the magnetic field fluctuation measured by LAPAN-A3 satellite, we are analyzing the local time ionospheric magnetic variation under quiet conditions.

F. Nuraeni · C. Y. Yatini · V. Wellyanita · Y. Andrian · B. Moh. Andi Aris · S. C. Pranoto · Erlansyah · N. Miharja
Research Center for Space, Bandung, West Java, Indonesia

L. O. M. Musafar Kilowasid (✉) · P. Ayuningtyas · A. Y. Putra
Pontianak Observatory for Space and Atmosphere, Pontianak, West Kalimantan, Indonesia
e-mail: laod002@brin.go.id

S. Utama · W. Hasbi · E. A. Anggari
Research Center for Satellite Technology, Bogor, West Java, Indonesia

T. Zubaidah · D. Ratnasari
Mataram University, Lombok, Indonesia

H. Bangkit
Research Center for Smart Mechatronics, Bandung, Indonesia

Muzirwan
Research Center for Climate and Atmosphere, Bandung, Indonesia

A. Basit et al. (eds.), *Proceedings of the International Conference on Radioscience, Equatorial Atmospheric Science and Environment and Humanosphere Science*, Springer Proceedings in Physics 290, https://doi.org/10.1007/978-981-19-9768-6_5

1 Introduction

The study of geomagnetic fields and related phenomena in space can be carried out using the measurements of ground-based magnetometers and satellites. Measurement of the geomagnetic field on the earth's surface has been carried out for a long time, and several magnetometer networks have been established to study and understand dynamic processes in space [1]. From the measurements on the ground, various phenomena related to the magnetic field in space have been identified and characterized that can describe the dynamics and processes that occur in the magnetosphere and/or ionosphere [2]. In addition, several satellites have also been launched to carry out in-situ measurements to confirm the gains through measurements of the ground-based magnetometer.

One of the earliest observations regarding the characteristics of the geomagnetic field was that geomagnetic variation is the variation in solar quiet that depends on solar activity [3]. Thus, an increase in solar radiation will cause an increase in the amount of ionization of the upper atmosphere and change the flow of ionospheric currents whose strength depends on the conductivity known as the ionospheric dynamo effect. Ionospheric currents at geomagnetically quiet conditions are commonly referred to as Sq currents. At low latitudes, this current system results in a magnetic field variation pattern having a maximum value at noon local time.

LAPAN-A3 satellite is the first experimental satellite launched by Indonesia to measure magnetic fields in space. The satellite has a polar orbit at an altitude of 500–550 km. However, the LAPAN-A3 satellite magnetic field measurement experiment does not carry out magnetic field measurements continuously for 24 h because the satellite is not specifically intended to perform magnetic field measurements. The longest continuous measurement time span was 6 h. This paper aims to validate the magnetic field measured by LAPAN-A3 satellite. This is carried out by analyzing the pattern of magnetic field variations at low latitudes where in the morning local time the magnetic field variation increases and reaches a maximum around midday local time and then decreases around the afternoon.

2 Data and Methods

In this paper, we used the magnetic data from magnetometer measurements carried by the LAPAN-A3 satellite where the satellite orbits at the height of the ionosphere. The magnetometer mounted on the satellite records three components of the magnetic field with a resolution of 128 Hz in each direction of the sensor that is mounted orthogonal to each other. Because the attitude of the satellite is always changing, the three components of the magnetic field recorded by the satellite are representations of the magnetic field vector components in an arbitrary orthogonal coordinate system.

The satellite only carries one magnetometer, therefore the magnetic field correction related to the satellite's attitude is carried out using the IGRF model, and in this

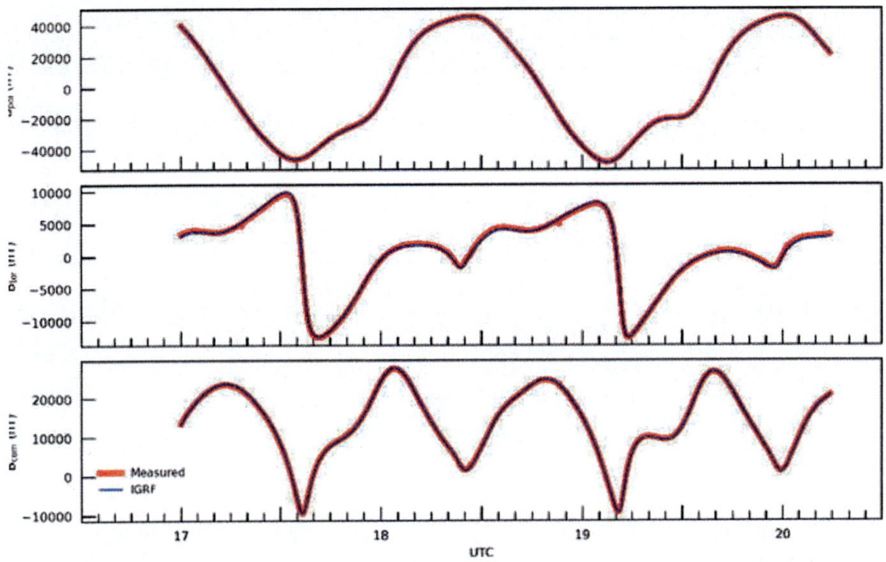

Fig. 1 Fit of the measured magnetic field by LAPAN-A3 satellite with IGRF model

case, the IGRF-13th model has been used. The component of the magnetic field vector is calculated at each time and position of the satellite [4]. Furthermore, the correction results are transformed into the NEC coordinate system using the GEOPACK software package [5–7]. Examples of magnetic field correction results from LAPAN-A3 satellite measurements and the IGRF model are given in Fig. 1. The panels in Fig. 1 represent the poloidal, toroidal, and compressional components of the magnetic field in the satellite's local coordinate system, respectively.

In this paper, the magnetometer data of LAPAN-A3 satellite during 2021 were analyzed. The positions of the satellite are expressed in a magnetic coordinate system, and the data are selected when the satellite is around $\theta = 10°N$ and $\theta = 10°S$ and latitudinal width of $\Delta\theta = \pm 2°$. So, we analyzed the data in the latitudinal range between 8°N to 12°N and 8°S to 12°S. During 2021, in that range of latitude, the altitude of LAPAN-A3 is between 500 and 520 km, as shown in Figs. 2 and 3 for the northern and southern hemisphere, respectively. The figures can be seen that the altitude distribution of LAPAN-A3 is almost similar.

The magnetic field variation is then calculated as the difference between the measured magnetic field component and the corresponding magnetic field component which is calculated using the 13th IGRF model for 2015. In addition, the position of the satellite in the magnetic coordinate system is expressed in local magnetic time where the calculation is done by utilizing the GEOPACK module written in Python.

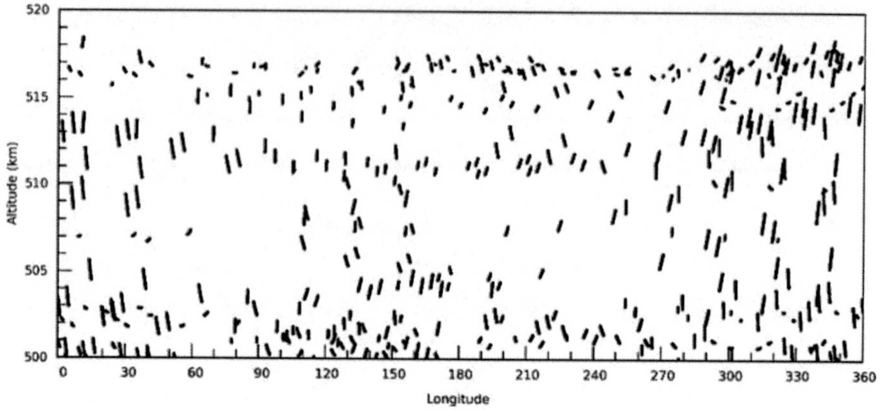

Fig. 2 Distribution of LAPAN-A3 altitude in the northern hemisphere of 8° to 12° latitude

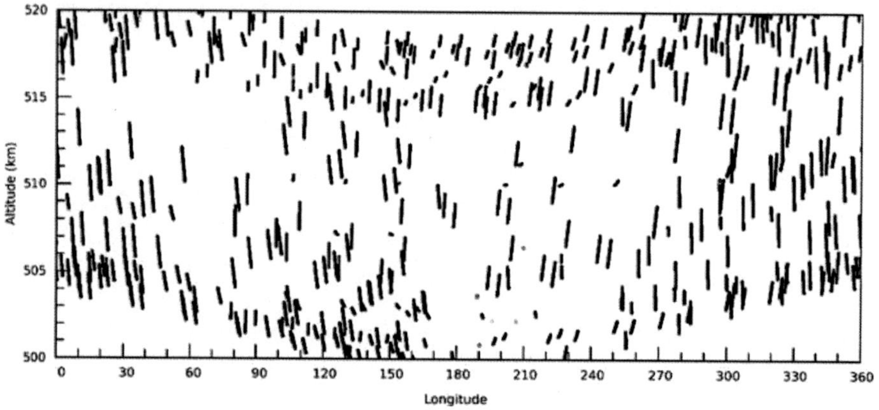

Fig. 3 Distribution of LAPAN-A3 altitude in the southern hemisphere of 8° to 12° latitude

3 Result and Discussions

The H (north–south) component of the low-latitude magnetic field increases at midday local time. This increase in the H component occurs due to ionospheric currents flowing at the height of the ionosphere. An example of such an increase in the H component is shown in Fig. 4. The data in the figure are selected on an international geomagnetically quiet day, $K_p \leq 1$. In addition to these properties, the *D* component (east–west) has anti-symmetric properties to the equatorial plane, while the *Z* component has symmetry properties to the equatorial plane.

Because the LAPAN-A3 satellite orbits at the height of the ionosphere, the magnetic field measured by the magnetometer on the satellite experiences the effect of ionospheric currents. Magnetometer data are presented in poloidal, toroidal, and

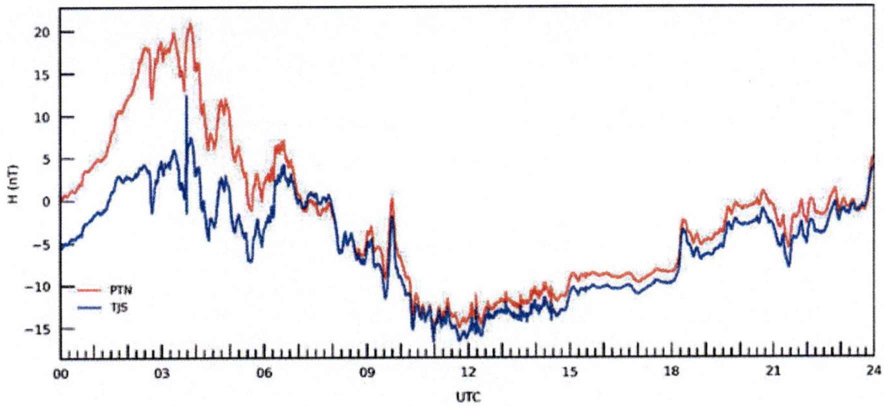

Fig. 4 H component of magnetic field recorded by ground-based magnetometer at Tanjungsari (blue) and Pontianak (red) stations

compressional representations. At low latitudes, the magnetic field lines are almost parallel to the ionosphere plane so that the poloidal component of the magnetic field almost has a direction parallel to the H component of the ground magnetic field. Therefore, the properties of the low-latitude ionospheric magnetic field can be considered similar to the character of H component of the ground magnetic field. The toroidal and compressional components can also consider having a similar character with the *D* and *Z* components of the geomagnetic field.

Figures 5, 6 and 7 show the behavior of the poloidal, toroidal, and compressional components as a result of magnetic field measurements by the LAPAN-A3 satellite at low latitudes during 2021. The top panel of each image represents the magnetic field in the northern hemisphere, while the bottom panel represents the magnetic field in the northern hemisphere for the southern hemisphere. Due to the limited measurement, the LAPAN-A3 satellite only recorded data in a fairly short period namely at 07–09 MLT and 20–22 MLT. So the complete daily behavior of each component of the geomagnetic field is difficult to characterize. However, the poloidal components of the magnetic field in the northern and southern hemispheres have opposite trends, as shown in Fig. 5 for data on 07–09 MLT. That is, in the northern hemisphere, there is an increase toward local noon (MLT), while in the southern hemisphere, there is a decrease. These may reflect the magnetic field anti-symmetry of low-latitude poloidal components. The toroidal component also showed the same thing. Meanwhile, in every component around 20–22 MLT, there was no clear trend showed.

Figure 7 shows compressional components for the same measurement range as Figs. 5 and 6. At 07–09 MLT, there was an increase of this component in both hemispheres. Data at 20–22 MLT also showed the same thing. This behavior shows evidence that the compressional component is symmetrical about the equatorial plane. From the results mentioned above, we can state that although it is difficult to observe clearly, measurements of LAPAN-A3 satellite are still consistent with the behavior of the geomagnetic field.

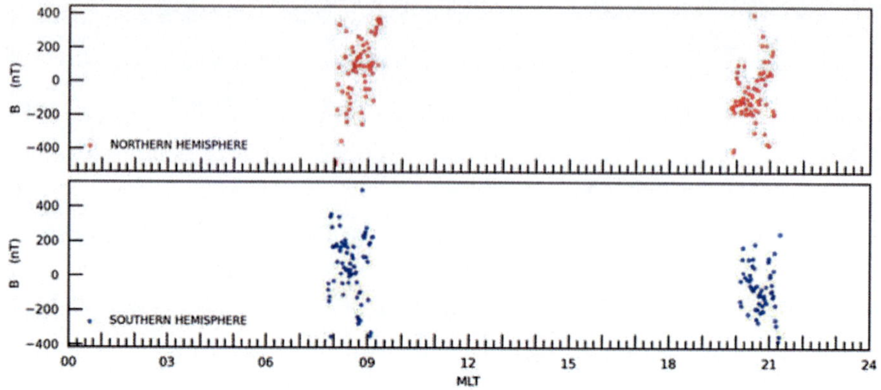

Fig. 5 Average poloidal component of the magnetic field was measured by the LAPAN-A3 satellite at geomagnetically quiet conditions during 2021

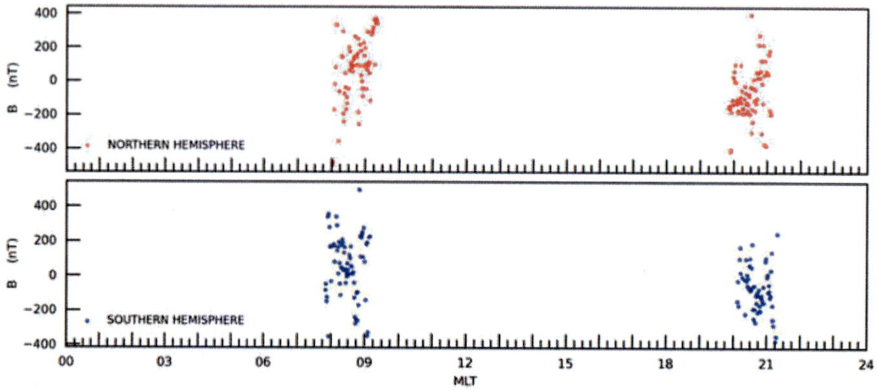

Fig. 6 Average toroidal component of the magnetic field was measured by the LAPAN-A3 satellite at geomagnetically quiet conditions during 2021

The asymmetry of the poloidal and toroidal of ionospheric magnetic field component has an important relationship with the direction of the ionospheric current flow. In the low latitude of northern hemisphere, the ionospheric currents flow eastward during the daytime, and based on asymmetry of magnetic field, the current flows eastward in the southern hemisphere. As a result, asymmetry of poloidal and toroidal components is strongly related to the two vortices ionospheric current systems at low-to mid-latitude in the northern and southern hemispheres. And, the symmetry of compressional components is related to the radial component of the magnetic field and can have an effect on ionospheric plasma transport. However, this statement requires further confirmation.

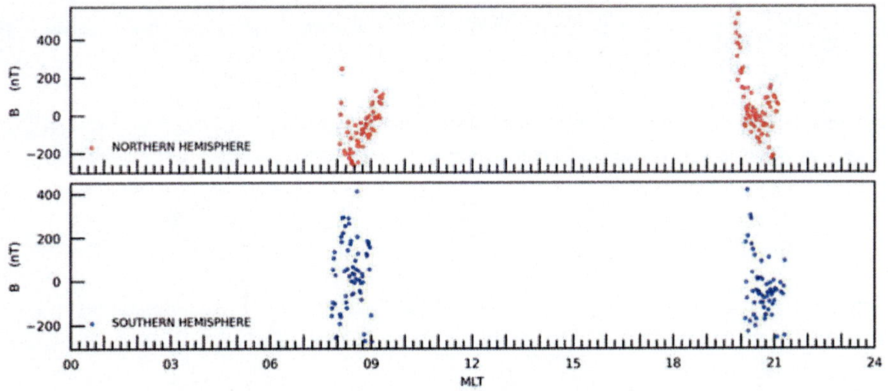

Fig. 7 Average compressional component of the magnetic field measured by the LAPAN-A3 satellite at geomagnetic quiet conditions during 2021

4 Conclusion

The LAPAN-A3 satellite orbits at an ionospheric altitude where the low-latitude geomagnetic field lines lie almost horizontally in the ionosphere so that variations in the ionospheric magnetic field at low latitudes will show a pattern similar to local time variations at ground-based stations at low latitudes. Based on the result, the poloidal and toroidal components of the magnetic field in the northern and southern hemispheres have opposite trends that may reflect the magnetic field anti-symmetry of low-latitude poloidal and toroidal components, while the compressional component is symmetrical about the equatorial plane. Therefore, measurements of LAPAN-A3 satellite are still consistent with the behavior of the low-latitude ionospheric geomagnetic field.

Acknowledgements This research is funded by Space research center BRIN of 2022 fiscal year.

References

1. Kamide, Y., Balan, N.: The importance of ground magnetic data in specifying the state of magnetosphere–ionosphere coupling: a personal view. Geosci. Lett. **3**(1), 10 (2016)
2. Motoba, T., Kikuchi, T., Okuzawa, T., Yumoto, K.: Dynamical response of the magnetosphere-ionosphere system to a solar wind dynamic pressure oscillation. J. Geophys. Res. Space Phys. **108**, A5 (2003)
3. Yamazaki, Y.: Solar and lunar daily geomagnetic variations and their equivalent current systems observed by Swarm. Earth, Planets Space **74**, 99 (2022)
4. Musafar, L.M., Yatini, Y., Nuraeni, F., Wellyanita, V., Winarko, A., Juangsih, M., Ratnasari, E.A., Hasbi, W., Utama, S.: Pre-processing data magnetometer satelit LAPAN-A3. Jurnal Sains Dirgantara **19**(1), 1–12 (2021)

5. Hapgood, M.A.: Space physics coordinate transformations: a user guide. Planet. Space Sci. **40**(5), 711–717 (1992)
6. Tsyganenko, N.A.: A new data-based model of the near magnetosphere magnetic field: 1. Mathematical structure. 2. Parameterization and fitting to observations (2001)
7. Tsyganenko, N.A., Sitnov, M.I.: Modeling the dynamics of the inner magnetosphere during strong geomagnetic storms. J. Geophys. Res. **110**(A3) (2005)

Study of the Low Latitude Ionosphere Irregularities Using Multi-instrument Observations

Dyah Rahayu Martiningrum, Sri Ekawati, Prayitno Abadi, and Bambang Suhandi

Abstract Equatorial plasma bubbles (EPB) or in other term known as equatorial spread *F*(ESF) are extended zones of depleted *F*-region plasma density that grow from irregularities caused by the generalized Rayleigh–Taylor instability mechanism in the post-sunset equatorial sector. Study of equatorial plasma bubbles is interesting because plasma density depletions generate rapid changes in both the amplitude and phase of radio signals, which propagate through the bubble, producing ionospheric scintillations that degrade communications and navigation signals. Scintillations come from many sources. Therefore, this study was conducted to understand characteristics of the EPB/ESF and to confirm the source of interference in the case study scintillation events at any given time. In this research, scintillations have analyzed using scintillation index (S4) from GPS Ionospheric Scintillation and TEC Monitor (GISTM) at Kototabang (0.20°S, 100.32°E; 10.36°S dip lat) and Pontianak (0.05°N, 109.34°E; 8.9°S dip lat) for the case March 1 and March 8, 2011. The field aligned irregularities (FAI) obtained from Equatorial Atmosphere Radar (EAR) data. Analysis of ionosonde data from CADI Pontianak and FMCW Kototabang also showed that spread *F* occurred on March 1 and March 8, 2011. Scintillation index showed strong value at the appropriate time. Further, analysis of EAR and other data will help understanding of equatorial plasma bubbles (EPBs) and scintillation.

1 Introduction

Equatorial plasma bubbles (EPBs) or equatorial spread *F* (ESF) are plasma density irregularities in nighttime low latitude ionosphere with a large range of scale sizes and amplitudes at almost all altitudes along the latitude and longitude sectors. The ESF predominantly occurs at nighttime. The complex dynamical phenomena and

D. R. Martiningrum (✉) · S. Ekawati · P. Abadi · B. Suhandi
Research Center for Climate and Atmosphere, National Research and Innovation Agency, Bandung, Indonesia
e-mail: dyah003@brin.go.id

A. Basit et al. (eds.), *Proceedings of the International Conference on Radioscience, Equatorial Atmospheric Science and Environment and Humanosphere Science*, Springer Proceedings in Physics 290, https://doi.org/10.1007/978-981-19-9768-6_6

plasma instability play an important role in development of the formation of EPB/ESF included the background ionospheric conditions [1].

The horizontal geomagnetic field lines at the magnetic equator perpendicular to gravity, and also prevailing natural wind and background electric field, compose unique phenomena to develop the plasma irregularities. These plasma irregularities are generally magnetic field aligned. They have zonal widths of typically a few tens of km and extend along the magnetic field lines for hundreds to thousands of km depending on the peak altitude of the irregularity (bubble) development [2], while their vertical heights range from a few tens of km to several hundred km [3]. Vertical plasma density gradient at the bottom side of ionosphere F-region also controls the growth rate of EPB/ESF. Some previous studies have reported occurrence of EPB/ESF under different geophysical conditions which potentially change the ionospheric parameters [4, 5].

When radio signals propagate through these disturbed regions, they cause scintillation [6]. This results in a fade in received signal power, meaning a loss of signal. The region of equatorial scintillations extends 30° on either side of the earth's magnetic equator, and the strongest effects are found at approximately 10° N and S [7]. Another ionospheric irregularities can also find in the nighttime. Under certain condition, we can find the irregularity of electron density which is aligned along the earth's magnetic field. It is known as field aligned irregularity (FAI) that related to EPB/ESF. Another research pointed out that frequent and stronger scintillations associated with EPB during solar maximum period and a drastic reduction in frequency and intensity of scintillation during solar minimum period [8].

This work has been studied the characteristics of the equatorial spread F in associated with the scintillation and field aligned irregularities (FAI) occurrences over Kototabang by using GPS Ionospheric Scintillation and TEC Monitor (GISTM), Equatorial Atmosphere Radar (EAR), and ionosonde data. This study also has been done to understand variability of plasma irregularities of ionosphere, and therefore, it gives general pattern of the relationship between scintillation and plasma irregularities of ionospheric layer.

2 Data and Method

Equatorial Atmosphere Radar (EAR) has operated at Kototabang, West Sumatra, and Indonesia (0.20°S, 100.32°E; 10.36°S dip lat). EAR is the large monostatic radar which operate at 47.0 MHz. This instrument developed to study dynamic of lower and upper atmosphere [9] (Fig. 1).

Under certain condition, amplitude fading and phase scintillation effects can cause loss of carrier lock and intermittent GPS receiver operation [10]. GPS scintillations generally occur shortly after sunset and may persist until just after local midnight. GISTM used to obtain scintillation index (S_4) which indicated occurrences of scintillation. Ionosonde is another instrument was used in this study to confirm the equatorial spread F (ESF) occurrences over Kototabang and Pontianak (Fig. 2).

Frequency	47.0 MHz
Output Power	Peak: 100 kW, Ave.: 5 kW
Antenna System	Quasi-circular active phased array (110 m-diameter, 560 three-element Yagis)
Beam Width	3.4°
Beam Direction	Anywhere (within 30°zenith angles)
Observation Range	1.5 km - 20 km (Atmospheric turbulence) over 80 km (Ionospheric irregularity)

Fig. 1 Equatorial atmosphere radar specification and location

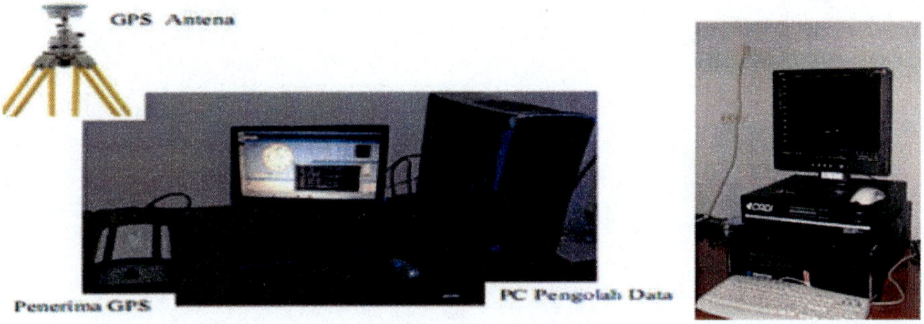

Fig. 2 GPS ionospheric scintillation and TEC monitor (left) and ionosonde (right)

The Equatorial Atmosphere Radar (EAR) unable to observe the ionosphere region in directions perpendicular to the geomagnetic field for studying F-region FAI associated with equatorial spread F (ESF). Intense FAI echoes have been observed in the nighttime F-region over Kototabang during 2011–2013 and classified in post-sunset and post-midnight occurrences. Furthermore, we have taken two cases of ionosphere irregularity and analyzed possibility relationship between ESF and scintillation between Kototabang and Pontianak.

3 Result and Discussion

3.1 Seasonal Variability of Field Aligned Irregularities (FAI)

Characteristics of the EPB/ESF have already been observed as daily percentage of occurrence of field aligned irregularities (FAI). We have used a threshold SNR value of −5 dB. Percentage obtained from ratio between number of signal of particular range and time to total number of signal. Figure 3 is result of radar data analysis during 2011–2013 shows that the higher percentage of *F*-region post-sunset FAI occurred around March (~30%) and September–October (~40%), while *F*-region post-midnight FAI occurs around June–July (~35–50%). The result clearly confirms dependency of ionosphere irregularities to solar activity.

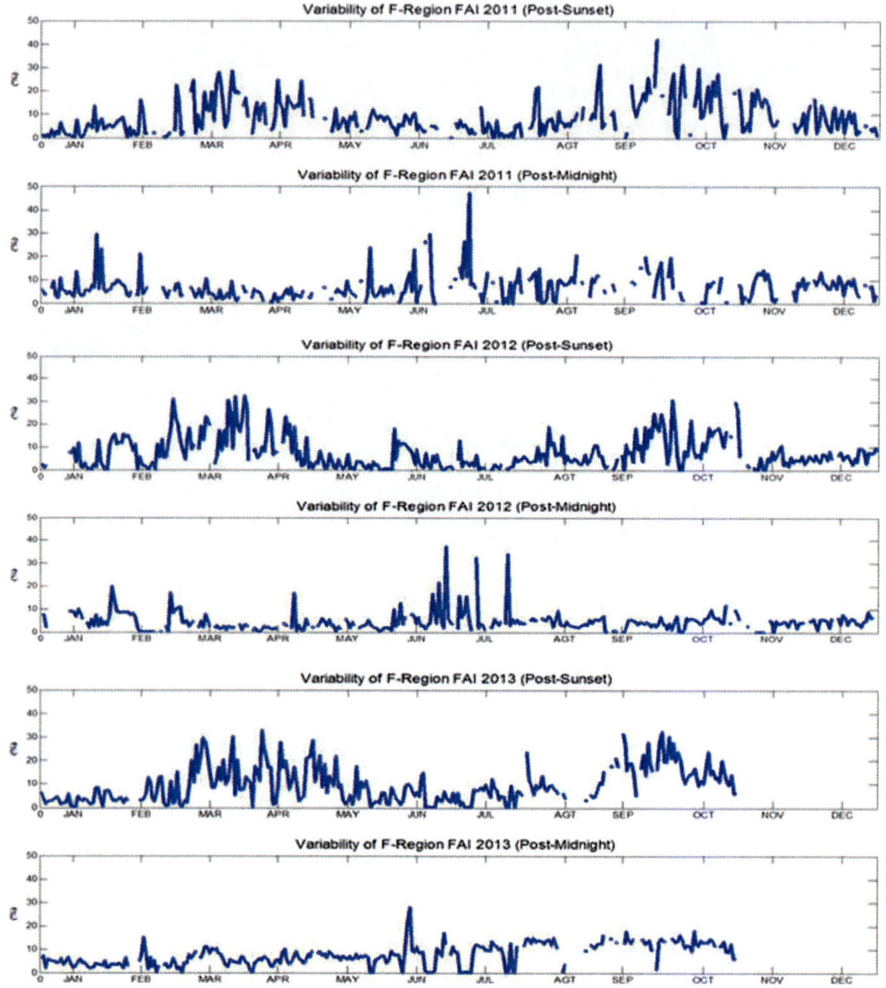

Fig. 3 Daily percentage of occurrence of *F*-region echoes

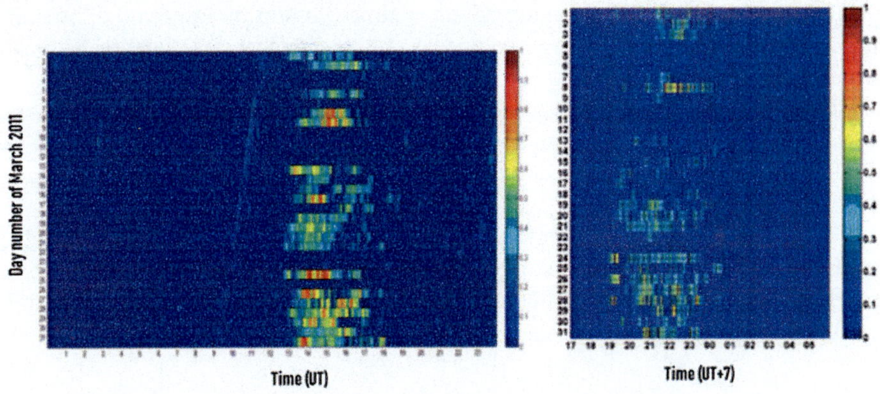

Fig. 4 Scintillation index (S4) of Kototabang (left) and Pontianak (right)

3.2 *Ionospheric Scintillation Occurrences During March 2011*

The corrected scintillation index (S_4) obtained from amplitude scintillation data of GISTM which has already been reduced multipath effects and ambient noise described as

$$S_4 = \sqrt{S_{4\,\text{tot}}^2 - S_{4\,\text{corr}}^2}$$

where the S_4 is a metric for indicating the amount of variation in the amplitude of a signal; $S_{4\,\text{tot}}$ is total value of S_4 and $S_{4\,\text{corr}}$ is correction value of S_4.

Figure 4 shows scintillation occurrences during March 2011 over Kototabang and Pontianak. Those results have been compared with FAI occurrences from EAR (Kototabang), shows the strong correlation of both data.

3.3 *Relationship Between Field Aligned Irregularities (FAI), Scintillation, and Equatorial Spread* F *(ESF)*

During March 2011, we founded strong relationship between FAI and scintillation events over Kototabang. We also observed scintillation event which possible associated with ionospheric irregularities over Pontianak. Pontianak placed about 1058.17 km eastern of Kototabang.

For further analysis, we took two cases, March 1, 2011, and March 8, 2011. The echoes scatter were analyzed to obtain zonal distance by using distance approximation of two location at the certain height over the earth's surface. Each of bubbles of SNR plot named as A, B, C, D, and E (Fig. 5). Drift velocity determined from zonal distance of each bubble and gradient of time. Possibility of period of bubbles propagate from Kototabang to Pontianak was obtained from simple relation between distance and drift velocity. Table 1 shows the result of all bubbles.

The echoes scatter of Bubble A shows the fastest drift velocity and Bubble B is the slowest. By using the similar relation between distance, velocity, and time, will be obtained period/time which needed all of bubbles for propagating from Kototabang to Pontianak (Table 2).

The virtual height over Kototabang and Pontianak from ionosonde data also increased at that time. It means that electron density of F-region of ionosphere layer over Kototabang and Pontianak fluctuated or changed mainly for March 1 and March 8, 2011 (red line in Fig. 6).

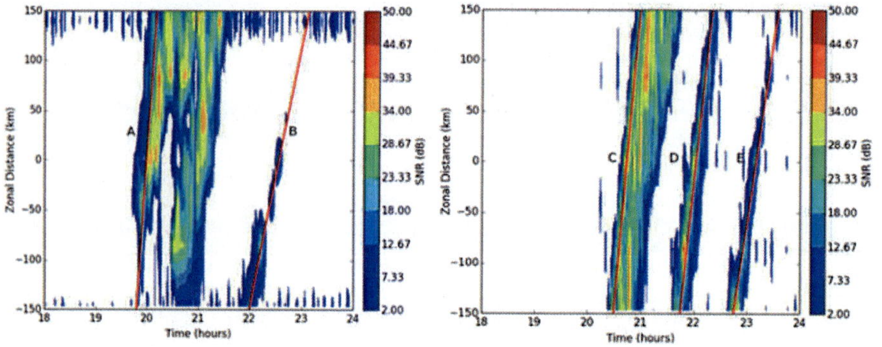

Fig. 5 Zonal distance from SNR plot of echoes scatter EAR for March 1, 2011, (left) and March 8, 2011 (right)

Table 1 Date and time of FAI from EAR

Date	Time Occurrence	Date	Time Occurrence
1 March 2011	19:30 – 23:00	19 March 2011	20:00 – 23:00
2 March 2011	21:00 – 01:00	20 March 2011	20:00 – 22:00
7 March 2011	21:00 – 23:00	21 March 2011	19:30 – 22:00
8 March 2011	20:00 – 24:00	24 March 2011	19:30 – 22:00
13 March 2011	19:30 – 23:00	25 March 2011	19:30 – 24:00
14 March 2011	20:30 – 02:00	26 March 2011	20:00 – 24:00
15 March 2011	20:00 – 01:00	27 March 2011	20:00 – 23:00
16 March 2011	20:00 – 24:00	28 March 2011	19:30 – 24:00
17 March 2011	22:30 – 23:30	30 March 2011	20:00 – 24:00
18 March 2011	20:00 – 24:00	31 March 2011	20:00 – 23:00

Table 2 Characteristic of each bubble

No.	Date	Bubble's name	Drift velocity (m/s)	Period/time to propagate (h:m)
1	March 1, 2011	A	208.3	1:24
2	March 1, 2011	B	72.46	4:36
3	March 8, 2011	C	166.67	1:48
4	March 8, 2011	D	138.89	2:06
5	March 8, 2011	E	98.04	3:00

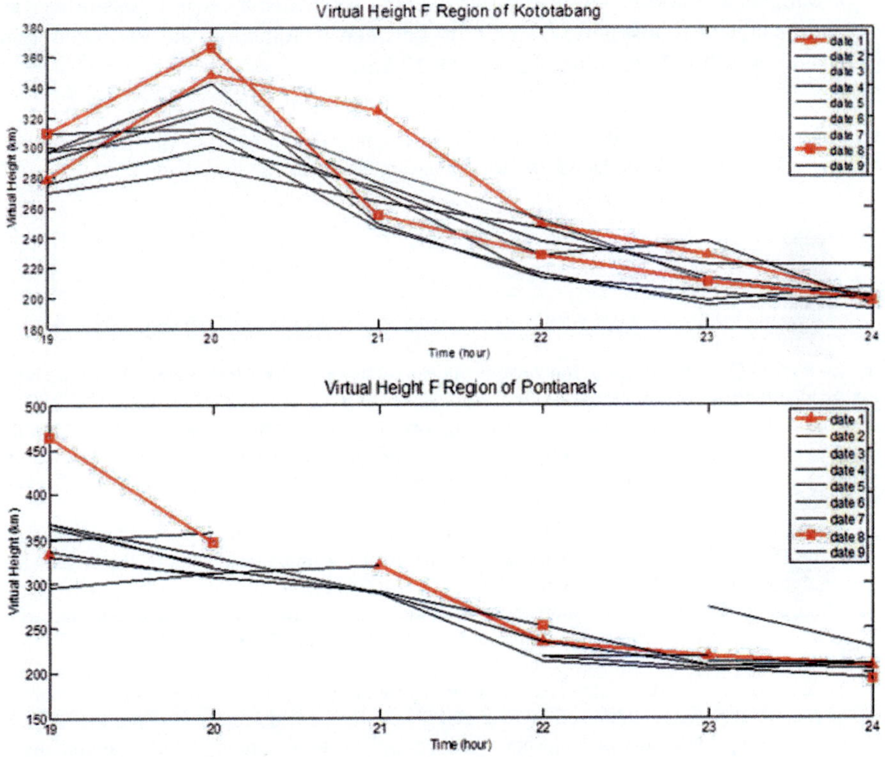

Fig. 6 Virtual height of *F*-region for Kototabang and Pontianak

4 Conclusion

We studied characteristic of ionospheric irregularity by using three different instruments that is Equatorial Atmosphere Radar Kototabang, GPS Ionospheric Scintillation and TEC Monitor (GISTM) at Kototabang and Pontianak, and also ionosonde at Kototabang and Pontianak. The results obtained from present study suggest that.

- Radar data analysis during 2011–2013 shows that the higher percentage of *F*-region post-sunset *F*AI occurred around March (~30%) and September–October (~40%), while *F*-region post-midnight FAI occurred around June–July (~35–50%). The result clearly confirms dependency of ionosphere irregularities to solar activity.
- The relationship between the scintillation and ionospheric irregularities (ESF) found from analysis GISTM and ionosonde data over Kototabang and Pontianak. Variability of virtual height also shows increasing of *F*-region's virtual height when equatorial spread *F* and scintillation occurred.
- The further analysis of echoes scatter of EAR for certain case shows characteristics of bubbles at certain height (350 km), their zonal distance and possibility the relationship of ionospheric irregularity between Kototabang and Pontianak that have longitudinally spacing about 1058.17 km.

Acknowledgements The authors thank Research Institute for Sustainable Humanosphere and Yamamoto Laboratory, Kyoto University for supporting EAR data.

References

1. Kelley, M.C.: The Earth's Ionosphere: Plasma Physics and Electrodynamics, 2nd edn. Academic Press, Amsterdam (2009)
2. Sobral, J.H.A., Abdu, M.A., Takahashi, H., Taylor, M.J., de Paula, E.R., Zamlutti, C.J., de Aquino, M.G., Borba., G.L.: Ionospheric plasma bubble climatology over Brazil based on 22 Years (1977–1998) of 630 nm airglow observations. J. Atmos. Terr. Phys. **64**, 1517–1524 (2002)
3. Labelle, J., Jahn, J.M., Pfaff, R.F., Swartz, W.E., Sobral, J.H.A., Abdu, M.A., Muralikrishna, P., de Paula., R.: The Brazil/guara equatorial spread F campaign: results of the large-scale measurements. Geophys. Res. Lett. **24**(13), 1691–1694 (1997)
4. Abdu, M.A.: Day to day and short-term variabilities in the equatorial plasma bubble/spread F irregularity seeding and development. Prog. Earth Planet. Sci. (2019). https://doi.org/10.1186/s40645-019-0258-1
5. Yamamoto, M., Otsuka, Y., Jin, H., Miyoshi, Y.: Relationship between day-to-day variability of equatorial plasma bubble activity from GPS scintillation and atmospheric properties from ground-to-topside model of atmosphere and ionosphere for aeronomy (GAIA) assimilation. Prog. Earth Planet. Sci. (2018). https://doi.org/10.1186/s40645-018-0184-7
6. Huang, N., Cheng, K., Huang, W.T.: Seasonal and solar cycle variation of spread F at the equator anomaly crest zone. J. Geomag. Geoelectr. **39**, 639–657 (1987)
7. Wanninger, L.: Effects of equatorial ionosphere on GPS. GPS World **4**(7), 48–52, 54 (1993). ISSN 1048-5104
8. Basu, S., Basu, S., Stubbe, P., Kopka, H., Waaramaa, J.: Daytime scintillations induced by high-power HF Waves at Tromso, Norway. J. Geophys. Res. Space Phys. **92**(A10), 11149–11157 (1978)
9. Fukao, S., Hashiguchi, H., Yamamoto, M., Tsuda, T., Nakamura, T., Yamamoto, M.K., Sato, T., Hagio, M., Yabugaki, Y.: Equatorial atmosphere radar (EAR): system description and first results. Radio Sci. **38**(3), 1053 (2003). https://doi.org/10.1029/2002RS002767
10. Klobuchar, J.A.: Ionospheric effects on GPS. GPS World **2**, 48–51 (1991)

Distribution of Peat Soil Carbon Under Different Land Uses in Tidal Swampland

Nur Wakhid and Siti Nurzakiah

Abstract Indonesia has the largest area of tropical peatlands in Southeast Asia and is an essential component of the global carbon (*C*) pool through its large *C* deposit inside the peat. Peat soil in Indonesia was usually associated with swampland both in freshwater and tidal swampland. However, the assessment of peat *C* deposit on swampland remains uncertain. Therefore, this study aimed to assess the *C* deposit of peat soil established on tidal swampland. The study was conducted in 2012 at Pulang Pisau, Central Kalimantan, under three different land uses, intercropping of rubber and pineapple, rubber plantation mixed with shrubland, and shrubland. The depth of peat soil was in the range of 4 to 8 m, dominated by sapric and hemic. The mean of *C* deposit was 3963 ± 561, 6666 ± 1065, and 4325 ± 605 t ha^{-1}, respectively, in intercropping of rubber and pineapple, rubber plantation mixed with shrubland, and shrubland. *C* deposit in a rubber plantation mixed with shrubland was high because the peat depth was deeper than other land uses. We expect this finding could be used as a baseline information for sustainable peatland management cultivated for agricultural plantation.

1 Introduction

Tropical peatlands are an important component of the global carbon (*C*) deposit. They have a huge *C* deposit both above–ground through plant biomass and below-ground underlying in the soil. It has been estimated that the tropical peatlands store up to 104.7 Petagrams of *C* below the soil, even though only about 17% of global peatland areas [1, 2]. More than 50% of the tropical peatland area is located in Southeast

N. Wakhid (✉)
Research Center for Ecology and Ethnobiology, National Research and Innovation Agency, Cibinong, Bogor 16911, Indonesia
e-mail: nurw007@brin.go.id

S. Nurzakiah
Research Center for Food Crops, National Research and Innovation Agency, Cibinong, Bogor 16911, Indonesia

A. Basit et al. (eds.), *Proceedings of the International Conference on Radioscience, Equatorial Atmospheric Science and Environment and Humanosphere Science*, Springer Proceedings in Physics 290, https://doi.org/10.1007/978-981-19-9768-6_7

Asia, with the vast majority in Indonesia and Malaysia [2]. The total peatland area in Indonesia was estimated to be 13.4 million ha distributed in Sumatra, Kalimantan, Sulawesi, and Papua islands [3].

In Indonesia, peat soil is usually associated with swampland under water-logged conditions, both in fresh water and or tidal swampland. The logged conditions bring through the slow decomposition of organic matter under an anaerobic situation. Today's *C* deposit on tropical peat soil is the result of organic matter accumulation over thousands of years from the forest that grows on the top of soil under these logged conditions. However, the huge *C* deposit underlying peat soil is now vulnerable due to land conversion. During the last decades, many areas of tropical peatlands have been converted mainly for plantations [4].

The land conversion is usually accompanied by drainage construction to shift the water-logged conditions to low groundwater levels. As a result, the peat soil characteristic would be changed [5]. Also, the peat soil *C* deposit is potentially decreased drastically. Therefore, it is crucial to quantify the *C* deposit of peat soil to determine the effects of land conversion. The objective of this study was to quantify the distribution of peat soil *C* deposits under different land uses in tidal swampland.

2 Methods

2.1 Study Site

The study was conducted on three different land uses (1) an intercropping of rubber and pineapple, (2) rubber mixed with shrubland, and (3) shrubland, on the same area established on peat soil in Pulang Pisau, Central Kalimantan, Indonesia (see Figs. 1 and 2). The distance between each land use is about 100 m. The study site originally was a peat swamp forest that was degraded due to the failure of the mega rice project in the late 1990s.

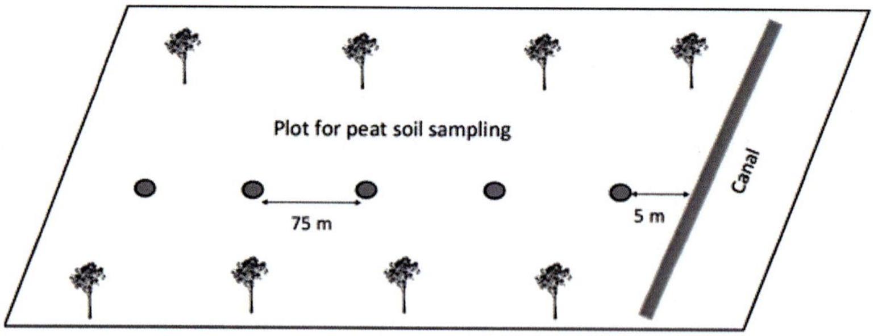

Fig. 1 Illustration of peat soil sampling plots

Fig. 2 Study plots for peat soil sampling, **a** intercropping of rubber and pineapple, **b** rubber mixed with shrubland, and **c** shrubland

2.2 Carbon Deposit Measurement

C deposit was estimated from soil sampling that collected in 2012 within five plots on each land use, using a peat auger (Eijkelkamp). The plots were on the same perpendicular transect at 75 m intervals between each plot to cover the different depths of the peat (see Fig. 1). In each plot, the samples were collected deeply until the peat auger reached the substratum. The samples were stratified based on the maturity of the peat. Thus, the depth interval of the samples was different between each plot. Also, the groundwater level in each plot was measured simultaneously during peat soil sampling. Peat maturity was estimated directly in the field using three indicators: decomposition stage, color, and squeezing the peat soil in the palm [6]. The peat maturity was divided into sapric (mature), hemic (medium), and fibric (immature). After stratification, each soil sample at different depth interval was separated, tagged, and stored in plastic bags for the next analysis in laboratory.

The *C* deposit was quantified from *C* content, peat bulk density, and peat depth. The peat depth was determined directly in the field during peat soil sampling. Whereas *C* content and BD were estimated at laboratory for each soil sample at different depth intervals. The *C* content was determined using a conversion factor of 0.58 from the percent of organic matter to organic *C* [6]. The percent of organic matter was measured using the loss on ignition (LOI) method at 550 °C in a muffle furnace. BD was estimated using gravimetric methods by weighing the oven dry weight of peat [6]. The *C* deposit (CS, t ha^{-1}) was calculated following the equation [6]:

$$CS = CD \times PV \quad (1)$$

$$CD = BD \times (CC/100) \quad (2)$$

$$PV = PD \times 10000 \quad (3)$$

where CD is *C* density (t m^{-3}), PV is peat volume in each layer (m^3 ha^{-1}), BD is bulk density (g cm^{-3}), CC is *C* content (% in mass), and PD is peat depth or peat thickness (m).

2.3 *Data Analysis*

The difference between each plot was compared using Tukey's HSD following analysis of variance (ANOVA). Furthermore, the linear regression and Pearson correlation methods were used to analyze the relationship between peat depth and *C* deposit. All data analysis was conducted using Microsoft Excel and Real Statistics Resource Pack software (Charles Zaiontz, www.real-statistics.com).

3 Results and Discussion

The peat soil in all study plots was dominated by sapric and hemic. The peat depth varied from 5.23–5.74, 6.16–8.50, and 4.95–5.46 m, respectively, in intercropping of rubber and pineapple, rubber mixed with shrubland, and shrubland (see Fig. 3a). The deepest peat soil was recorded in rubber mixed with shrubland up to 8.50 m. Mean of peat depth was 5.50, 7.12, and 5.16 m in intercropping of rubber and pineapple, rubber mixed with shrubland, and shrubland, respectively. The peat depth in rubber mixed with shrubland was significantly different with intercropping of rubber and pineapple, and shrubland ($p < 0.05$).

The lowest average of bulk density (BD) was found in rubber and pineapple with a range of BD from 0.12 to 0.18 g cm^{-3} (see Fig. 3b). The BD in rubber and pineapple was lower than that of BD of peat soil in a nearby rubber plantation that measured from soil samples collected by soil ring sample [7]. In this study, BD was measured from soil samples collected using a peat soil auger. Soil sampling method is likely to affect the result of BD measurement. The mean of BD was 0.16, 0.19, and 0.20 g cm^{-3}, respectively, in intercropping of rubber and pineapple, rubber mixed with shrubland, and shrubland. BD of peat soil in this study was in the range of

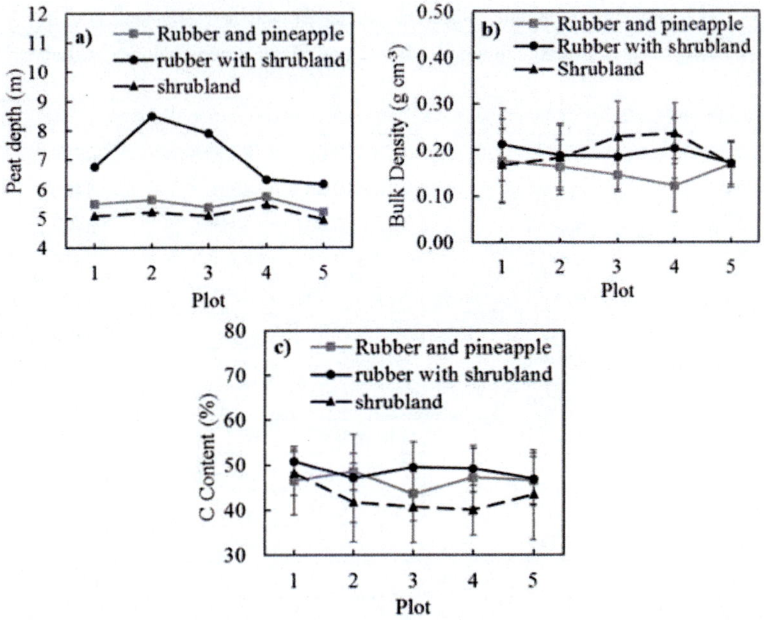

Fig. 3 Variations of **a** Peat depth, **b** Bulk density (BD), and **c** Carbon (*C*) content, at different plots in **a** intercropping of rubber and pineapple (gray square with solid line), **b** rubber plantation mixed with shrubland (black circle with solid line), and **c** shrubland (black triangle with dashed line). Vertical bars in (**b**) and (**c**) denote 1 standard deviation

BD in Sumatra and Kalimantan of 0.02–0.03 g cm^{-3} [6]. BD in intercropping of rubber and pineapple was significantly different from BD in shrubland ($p < 0.05$), but not for rubber mixed with shrubland. In intercropping of rubber and pineapple and rubber mixed with shrubland, the highest BD was measured near the canal, but not for shrubland. BD in a degraded tropical peatland was reported to increase near the canal [8]. In shrubland, as an abandoned peatland, the drainage canal is likely not intense as in intercropping of rubber and pineapple and rubber mixed with shrubland. Thus, the effect of drainage on BD was less.

Mean of *C* content in all plots was similar in the range of 46, 49, and 43%, in intercropping of rubber and pineapple, rubber mixed with shrubland, and shrubland, respectively. The mean *C* content was in the range of peat *C* content across Southeast Asian peatlands (41.6–62%) [2, 9]. The *C* content varied from 40 to 51% in all plots (see Fig. 3c). The *C* content in rubber mixed with shrubland showed a significant difference with shrubland ($p < 0.05$), but in intercropping of rubber and pineapple was not significantly different. Shrubland has the highest BD with the lowest *C* content (see Table 1). The highest value of BD is usually associated with low *C* content [2]. Also, high BD with low *C* content may indicate a more advanced degree of peat decomposition at the site [10].

Table 1 *C* deposit distribution under different land uses (mean ± 1 SD)

Plot	Peat depth (m)	Sample number	BD (g cm^{-3})	Carbon content (%)	*C* deposit (t/ha)
Rubber and pineapple					
1	5.50	$n = 11$	0.18 ± 0.09	47 ± 8	4494
2	5.64	$n = 13$	0.16 ± 0.06	49 ± 4	4465
3	5.38	$n = 11$	0.15 ± 0.03	44 ± 6	3424
4	5.74	$n = 12$	0.12 ± 0.06	47 ± 7	3319
5	5.23	$n = 11$	0.17 ± 0.05	47 ± 5	4113
					3963 ± 561
Rubber with shrubland					
1	6.75	$n = 14$	0.21 ± 0.08	51 ± 3	7288
2	8.5	$n = 12$	0.19 ± 0.07	47 ± 10	7540
3	7.89	$n = 11$	0.18 ± 0.08	50 ± 6	7225
4	6.32	$n = 12$	0.20 ± 0.07	49 ± 5	6333
5	6.16	$n = 12$	0.17 ± 0.05	47 ± 6	4945
					6666 ± 1065
Shrubland					
1	5.09	$n = 10$	0.17 ± 0.06	48 ± 5	4092
2	5.21	$n = 11$	0.18 ± 0.06	42 ± 9	3978
3	5.09	$n = 5$	0.23 ± 009	41 ± 8	4744
4	5.46	$n = 4$	0.24 ± 0.09	40 ± 6	5149
5	4.95	$n = 10$	0.17 ± 0.06	43 ± 10	3664
					4325 ± 605

Mean of *C* deposit was estimated to be 3963 ± 561, 6666 ± 1065, and 4325 ± 605 t ha^{-1} (mean ± 1 standard deviation (SD)), respectively, in intercropping of rubber and pineapple, rubber mixed with shrubland, and shrubland (see Table 1). The largest *C* deposit was estimated in rubber mixed with shrubland. The *C* deposit in rubber mixed with shrubland was significantly different with intercropping of rubber and pineapple and shrubland. Deep peat soil would result in a higher peat *C* deposit (see Fig. 3a and Table 1). The relationship between peat depth and *C* deposit was significant if using all data ($n = 15$, $p < 0.05$) from three plots. However, if only using data from each plot ($n = 5$), the relationship between peat depth and *C* deposit was not significant ($p > 0.05$).

The peat depth is probably the most factor that affected the amount of peat *C* deposit estimations on tropical peatland. However, the peat depth is likely also the most contributor to uncertainty on peat *C* deposit estimations since the peat depth measurements are often estimated from a few samples that may not represent the heterogeneity of the peat ecosystem [11]. Peat *C* deposit estimations may also be overestimated, as the assessment has not yet considered the mineral layer that may exist under the peat layers. Peat C deposit in this study (3963 ± 561, 6666 ± 1065, 4325 ± 605 t ha^{-1}) was higher than the peat C deposit of rubber plantations on freshwater swampland in South Kalimantan (2025.21 ± 132.77 t ha^{-1}, peat depth: 3.1–4.7 m) [12]. The lower C deposit in that study was probably due to the shallow peat depth compared to our study.

4 Conclusion

Peat carbon (C) deposit was estimated in a tidal swampland under three different land uses, intercropping of rubber and pineapple, rubber mixed with shrubland, and shrubland. Rubber mixed with shrubland has the deepest peat soil with the highest peat C deposit up to 6666 ± 1065 t ha^{-1}. The peat C deposit estimations on swampland are facing some uncertainties as the heterogeneity of peatland was high. Also, the peat depth measurement as the most factor for peat C estimations required a lot of effort and was expensive. Therefore, further study is needed to obtain a new method or model for inexpensive and fast peat C deposit assessment.

References

1. Dargie, G.C., Lewis, S.L., Lawson, I.T., Mitchard, E.T.A., Page, S.E., Bocko, Y.E., Ifo, S.A.: Age, extent and carbon storage of the central Congo Basin peatland complex. Nature **542**, 86–90 (2017)
2. Page, S.E., Rieley, J.O., Banks, C.J.: Global and regional importance of the tropical peatland carbon pool. Glob. Change Biol. **17**, 798–818 (2011)
3. Anda, M., Ritung, S., Suryani, E., Sukarman, S., Hikmat, M., Yatno, E., Mulyani, A., Subandiono, R.E., Suratman, S., Husnain, H.: Revisiting tropical peatlands in Indonesia: semi-detailed mapping, extent and depth distribution assessment. Geoderma **402**(115235), 1–14 (2021)
4. Gaveau, D.L.A., Sheil, D., Husnayaen, Salim, M.A., Arjasakusuma, S., Ancrenaz, M., Pacheco, P., Meijaard, E.: Rapid conversions and avoided deforestation: examining four decades of industrial plantation expansion in Borneo. Sci. Rep. **6**, 32017 (2016)
5. Furukawa, Y., Inubushi, K., Ali, M., Itang, A.M., Tsuruta, H.: Effect of changing groundwater levels caused by land-use changes on greenhouse gas fluxes from tropical peatlands. Nutr. Cycl. Agroecosyst. **71**, 81–91 (2005)
6. Agus, F., Hairiah, K., Mulyani, A.: Measuring Carbon Stock in Peat Soils: Practical Guidelines. World Agroforestry Centre (ICRAF) Southeast Asia Regional Program, Indonesian Centre for Agricultural Land Resources Research and Development, Bogor, Indonesia, 60 pp. (2011)
7. Wakhid, N., Hirano, T., Okimoto, Y., Nurzakiah, S., Nursyamsi, D.: Soil carbon dioxide emissions from a rubber plantation on tropical peat. Sci. Total Environ. **581–582**, 857–865 (2017)
8. Sinclair, A.L., Graham, L.L.B., Putra, E.I., Sahaijo, B.H., Applegate, G., Grover, S.P., Cochrane, M.A.: Effects of distance from canal and degradation history on peat bulk density in a degraded tropical peatland. Sci. Total Environ. **699**, 134199 (2020)
9. Page, S.E., Wüst, R.A.J., Weiss, D., Rieley, J.O., Shotyk, W., Limin, S.H.: A record of Late Pleistocene and Holocene carbon accumulation and climate change from an equatorial peat bog (Kalimantan, Indonesia): implications for past, present and future carbon dynamics. J. Q. Sci. **19**, 625–635 (2004)

10. Anshari, G.Z., Afifudin, M., Nuriman, M., Gusmayanti, E., Arianie, L., Susana, R., Nusantara, R.W., Sugardjito, J., Rafiastanto, A.: Drainage and land use impacts on changes in selected peat properties and peat degradation in West Kalimantan Province, Indonesia. Biogeosciences **7**, 3403–3419 (2010)
11. Warren, M.W., Kauffman, J.B., Murdiyarso, D., Anshari, G., Hergoualc'h, K., Kurnianto, S., Purbopuspito, J., Gusmayanti, E., Afifudin, M., Rahajoe, J., Alhamd, L., Limin, S., Iswandi, A.: A cost-efficient method to assess carbon stocks in tropical peat soil. Biogeosciences **9**, 4477–4485 (2012)
12. Nurzakiah, S., Noor, M., Nursyamsi, D.: Carbon stock stratification of peat soils in South Kalimantan, Indonesia. J. Wetlands Environ. Manag. **2**(2), 55–59 (2014)

Study of Air Quality and Pollutant Distribution Patterns in Balikpapan Using the WRF-Chem Model

Dessy Gusnita, Prawira Yuda Kumbara, Waluyo Eko Cahyono, Angga Yolanda Putra, Fahmi Rahmatia, and Jen Supriyanto

Abstract Indonesia faces serious problems related to air quality, in line with the plan to move the capital to East Kalimantan, so it is necessary to study of air quality in the area around the new capital. This paper aims to study air quality using in-situ data from the Balikpapan Environmental Service (DLH Balikpapan) and pollutant distribution using the Weather Research Forecasting/Chemistry (WRF-Chem) model. This model combines topographic, atmospheric, and emission data that can produce a simulation of the distribution and concentration of pollutants in an area at a time. Simulations were carried out in January (14–16), April (21–23), July (16–18), and October (5–7) 2020, with spatial resolution: 5 km (domain 1) and 1 km (domain 2) and hourly time. In-situ DLH Balikpapan data was obtained from AQMS tools installed at 2 points, namely Balikpapan Baru (BB-1.2422744, 116.8596525) and Plaza Balikpapan (PB-1.2771924, 116.8382858) covering pollutant parameters PM2.5, NO_2, and SO_2 using hourly data. The results of the 2020 AQMS monitoring at BB station, indicated that PM2.5 concentration reached a peak concentration at 22.00 at 21.2 $\mu g/m^3$ in April. At the PB station, the PM2.5 peak concentration was lower at 16.0 $\mu g/m^3$ in January. The peak of SO_2 in the PB occurred in July at 07.30 Central Indonesia time with a concentration of 229.3 $\mu g/m^3$ and in the BB, the peak concentration of SO_2 was at 48.9 $\mu g/m^3$, which occurred in October at 14:30 Central Indonesia time. Meanwhile, the peak of NO_2 occurred at 07:00 Central Indonesia time in PB on July 2020, with concentrations reaching 85.18 $\mu g/m^3$. The results of the WRF-Chem model show that the distribution of pollutants with the highest concentration is found in the northern region, which is around Tenggarong, both for PM2.5, SO_2, NO_2.

D. Gusnita (✉) · P. Y. Kumbara · W. E. Cahyono · A. Y. Putra · F. Rahmatia
National Research and Innovation Agency, Bandung, Indonesia
e-mail: dessy.gusnita@brin.go.id

J. Supriyanto
Environmental Agency, Balikpapan, Indonesia

A. Basit et al. (eds.), *Proceedings of the International Conference on Radioscience, Equatorial Atmospheric Science and Environment and Humanosphere Science*, Springer Proceedings in Physics 290, https://doi.org/10.1007/978-981-19-9768-6_8

1 Introduction

Nowadays, the city of Balikpapan is growing quite rapidly, both in the construction of infrastructure, buildings, companies, the number of residents, the number of immigrants, the number of motorized vehicles circulating on the highway every day, especially private cars [1]. As the population grows, the demand increases, giving rise to an increase in urban activity. This increase causes the movement of people and goods which is called transportation [2]. According to the Directorate General of Traffic of the East Kalimantan Police, the number of vehicles in Balikpapan has increased by an average of 6.8%, so that in 2017 the number of vehicles in Balikpapan was 572,976 vehicles [3]. The high consumption of fuel oil in the transportation sector results in a higher potential for air pollution because the air will be polluted by exhaust gases resulting from combustion [4].

The distribution of atmospheric pollutants depends on the direction and speed of the wind, wind speed, air temperature, and humidity are part of the meteorological parameters that can affect the concentration of pollutant gases in the air [5]. One of the models commonly used to model ambient air conditions is Weather Research and Forecasting with Chemical (WRF-Chem). WRF-Chem is a regional scale weather model that models meteorological conditions and their interactions with chemical compounds in the atmosphere spatially and temporally. WRF-Chem simulation using regional emission inventories can show better results than global scale models [6]. To get the distribution of pollutant emissions in an area at a time, a dynamic model can be used. Weather Research Forecasting Chemistry (WRF-Chem) is a combined model of atmospheric modelling with atmospheric chemistry. The distribution of pollutant emissions is influenced by various factors, one of which is meteorological factors such as wind, temperature, atmospheric stability, and so on. Several studies in Indonesia using the WRF-Chem model to analyse the dispersion pattern of particulates and Sulphur dioxide using the WRF-Chem model around the industrial areas of Tangerang and Jakarta [7] and research on the analysis of the effect of seasonal variations on NO_2 dispersion in the city of Tangerang [8].

Many studies use the WRF-Chem model to simulate and predict the distribution of pollutant emissions in an area over a certain period of time [9–11] WRF-Chem is a mesoscale dynamic model that combines atmospheric models with chemical models that include calculations of atmospheric chemical processes, photolysis, deposition, dispersion and mixing of emissions, aerosol interactions, and radiation [9, 10, 12, 13]. This model combines topographic, atmospheric, and emission data to produce a simulation of the distribution and concentration of emissions in an area at a time.

This paper aims to analyse in-situ pollutant data in the city of Balikpapan which consists of PM2.5, SO_2, and NO_2 based on seasons, then to determine the pattern of pollutant distribution using the WRF-Chem model. It is hoped that the results of this study will provide input for the local government in supporting sustainable development.

2 Data and Method

This study uses in-situ data from the Balikpapan Environmental Service (DLH Balikpapan) from measurements of Air Quality Monitoring System (AQMS) consisting of: PM2.5, SO_2, and NO_2 in the city of Balikpapan in 2020. AQMS is installed at two measurement points, namely Balikpapan Baru (BB-1.242744, 116.8596525) and Plaza Balikpapan. (PB-1.2771924, 116.8382588). In-situ data were analysed based on seasonal data, namely the rainy season, transition season, and dry season. Furthermore, to complete the in-situ data, a pollutant distribution model was run using the WRF-Chem model to see the direction of the distribution of pollutants from the city of Balikpapan.

Simulation Setup

This study did a simulation by using dynamical modelling to learn chemical emission concentration and dispersion. We use the WRF-Chem model to run the simulation. WRF-Chem is a mesoscale dynamic model that combines atmospheric models with chemical models that include calculations of atmospheric chemical processes, photolysis, deposition, emission dispersion and mixing, aerosol interactions, and radiation [9, 10, 12]. This model combines topographic, atmospheric, and emission data so that it can produce a simulation of the distribution and concentration of emissions in an area at a time. This study used two nested domains for the simulation and take study area in East Borneo. The spatial resolution for both domains is 5 and 1 km with temporal resolution is hourly. The domain design can be seen at Fig. 1.

For meteorological input, this study used Final Reanalysis (FNL) data that can be downloaded at: http://rda.ucar.edu/datasets/ds083.2. The spatial resolution FNL data is 0.25 degree by 0.25 degree temporal resolution is six hourly. Then, for the emission data input, this study used two kinds of data, both of EDGAR-HTAP for anthropogenic emission and MEGAN for biogenic emission. Furthermore, this study used some parameterization schemes for simulation setup that can be seen in Table 1.

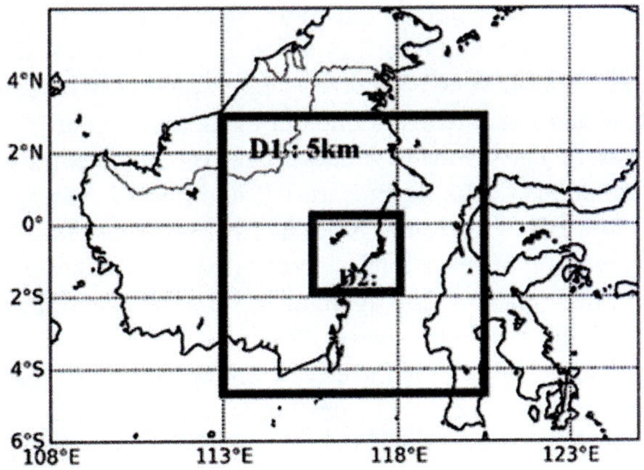

Fig. 1 The domain design for simulation study

Table 1 The parameterization schemes for simulation setup

Parameterization	Domain 1	Domain 2
Cumulus	Grell 3	–
Microphysics	Purdue Lin	Purdue Lin
Boundary layer	YSU	YSU
Radiation (shortwave and longwave)	RRTMG	RRTMG
Surface	NOAH MP	NOAH MP
Chemical	RADM2 and MADE/SORGAM aerosols	RADM2 and MADE/SORGAM aerosols

The simulation was conducted in January (14–16), April (21–23), July (16–18), and October (5–7) 2020.

3 Results and Discussion

3.1 Concentration of Air Pollutants in the City of Balikpapan

Balikpapan City DLH AQMS equipment is installed in 2 separate places, namely in the Balikpapan Baru area (BB-1.242744, 116.8596525) and in the Plaza Balikpapan area (PB-1.2771924, 1168382858). AQMS BB station is installed in residential areas, while AQMS PB is located on the Balikpapan city road, adjacent to shopping centres. The air pollutant parameters measured by AQMS DLH Balikpapan are PM2.5, SO_2, and NO_2.

PM2.5 Concentration Pattern in the observation area of Balikpapan Baru (BB) and Plaza Balikpapan (PB)

Based on seasonal data, namely: rainy season (JF), transitional season (MAM), dry season (JJA), and transitional season (SON), analysis of three pollutant parameters, namely PM2.5, SO_2, and NO_2 based on hourly data, was carried out. This pollutant data analysis intends to find out the pattern of daily pollutant concentrations in each season. The results of monitoring PM2.5 concentrations from AQMS equipment at two points (Balikpapan Baru/BB and Plaza Balikpapan /PB) are shown in Fig. 2.

At BB observation station the maximum PM 2.5 concentration occurred in January, concentrations peaked at 07.00 to 08.00 central Indonesia time in the morning, then peaked again at 16.00–17.00 central Indonesia time in the afternoon. The high concentration of PM2.5 in January is thought to have come from transportation activities in the city of Balikpapan in the morning and evening. In addition, the peak of PM2.5 concentration occurs again at 21.00–22.00 central Indonesia time at night.

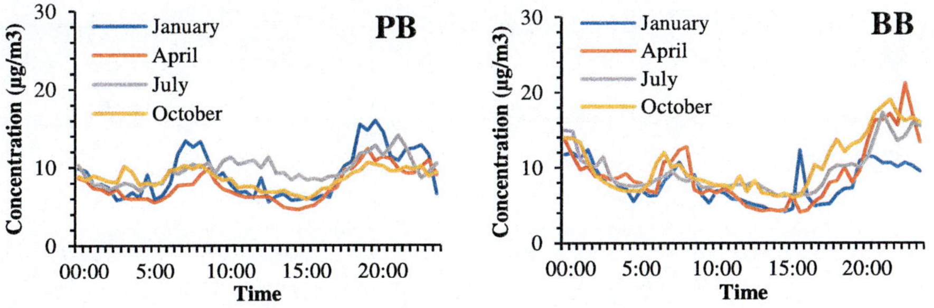

Fig. 2 PM2.5 concentration at the Balikpapan Baru (BB) and Balikpapan Plaza (PB) observation stations. *Source* Data from the Balikpapan Environmental Service

At the PB observation station, the maximum PM2.5 concentration occurred in April, peaking at 07.00–08.00 central Indonesia time in the morning, peaked again at 16.00–17.00 central Indonesia time in the afternoon. This is presumably, due to the transportation activities of the people of Balikpapan in the morning and evening. In addition, the peak of PM2.5 concentration occurred again at 19.00 central Indonesia time at night.

The Pattern of NO_2 Concentration in the observation area of Balikpapan Baru (BB) and Plaza Balikpapan (PB)

The concentration of NO_2 in PB is much higher than at the BB observation station, this is presumably, because PB's location is in a fairly busy shopping centre, so the mobility of people who emit NO_2 pollutants is much higher than BB's located in residential areas (Fig. 3).

The pattern of SO_2 concentration in the observation area of Balikpapan Baru (BB) and Plaza Balikpapan (PB)

In Fig. 4, the concentration of SO_2 reaches 229.3 $\mu g/m^3$ in the dry season, it is suspected that this is due to the lack of cleaning of pollutants during the dry season.

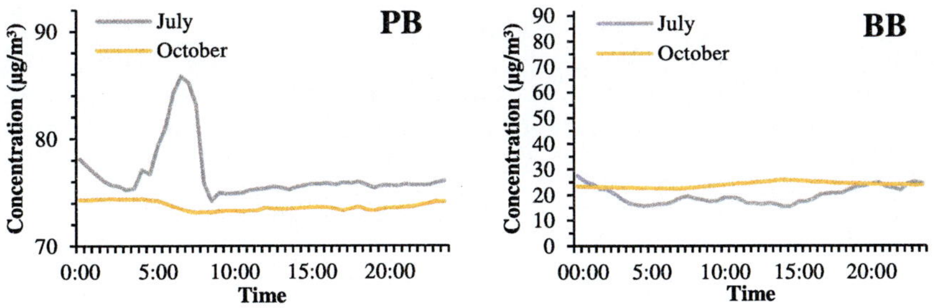

Fig. 3 NO_2 concentration at the Balikpapan Baru (BB) and Balikpapan Plaza (PB) Observation stations. *Source* data from the Balikpapan Environmental Service

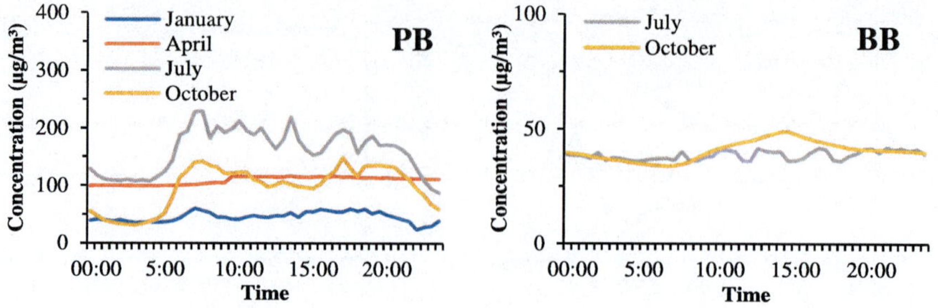

Fig. 4 SO_2 concentration at the Balikpapan Baru (BB) and Balikpapan Plaza (PB) Observation stations. *Source* data from the Balikpapan Environmental Service

As a result, there is an increase in pollutant concentrations. Meanwhile, in the rainy season (January) there was a significant decrease in SO_2 concentration due to the *washing out* process of rainwater. However, in general, both BB and PB stations peak at 07.00 central Indonesia time in the morning.

3.2 *Pollutant Distribution in Dry Season, Transition Season, and Wet Season*

The WRF-Chem running model was carried out on 14 Jan, 21 April, 16 July, and 05 October 2020 at 07.00, 13.00, and 19.00 Central Indonesia time for three pollutant parameters PM10, SO_2, and NO_2. The results of running the WRF-Chem model to analyse the distribution of pollutants in the city of Balikpapan are presented in Fig. 5.

In April, the highest PM2.5 concentration occurred in the morning when, atmospheric conditions tended to be stable, and during the day the concentration reaches 18 $\mu g/m^3$. In July, the concentration of PM2.5 was still quite high, especially in the Tenggarong area towards the north and its surroundings. The highest concentrations of SO_2 and NO_2 in the morning are in the northern area of Balikpapan, especially the Tenggarong area, where there is a coal mining centre which emits high levels of NO_2, SO_2, and PM10 pollutants from their activities. During the day, NO_2 and SO_2 did not show high concentrations, and were more vertically dispersed due to unstable atmospheric conditions. The height of the stable layer during the day is higher than at night due to turbulence in the mixed layer, which causes vertical transport in the atmosphere [8].

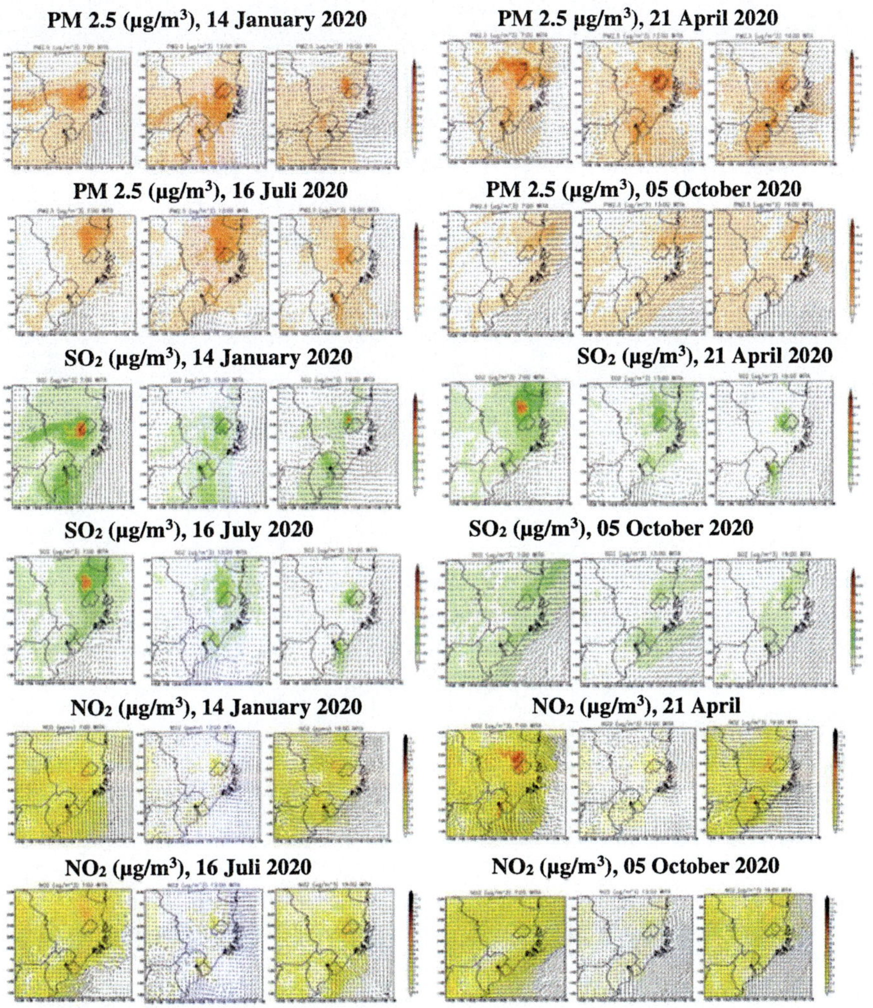

Fig. 5 Pollutant distribution patterns in Balikpapan City and surrounding areas result of running WRF-Chem model

3.3 Wind Characteristics of Balikpapan City in Balikpapan Baru and Plaza Balikpapan

Balikpapan Baru (BB)

Wind direction conditions on 14–16 January 2020 (rainy season) show that the dominant wind comes from the north. The wind speed is relatively low because to the north of Balikpapan there are hills and mountains. 21–23 Apr 2020 (transition season). As in January, April 21–23, the dominant wind comes from the north. In the dry season,

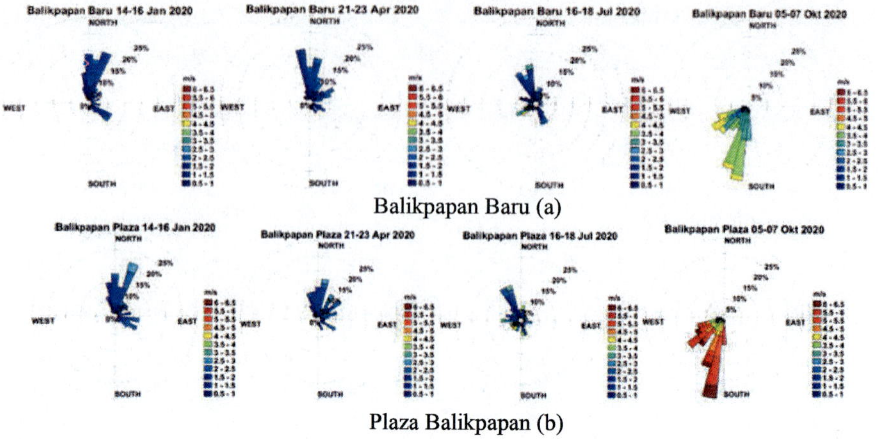

Fig. 6 Balikpapan City wind characteristics in BB (**a**) and PB (**b**) the results of running the WRF-Chem model

July 16–18, the direction of the wind in July is more varied. The most dominant winds are from the North-northwest and southeast. On 05–07 October 2020, the dominant wind direction comes from the south-southwest. The wind speed is greater because the wind directly comes from the sea area (Fig. 6).

Balikpapan Plaza (PB)

Wind conditions on 14–16 Jan 2020, dominant wind comes from the north, indicating the influence of the Asian monsoon. The wind speed is relatively low because to the north there are hills and mountains. 21–23 April 2020. As in January, April 21–23, the dominant wind comes from the north. 16–18 July 2020, the direction of the wind in July is more varied. And the most dominant comes from the North-northwest and southeast. On 05–07 Oct 2020. The dominant wind direction is from the south-southwest. The wind speed is significant because the wind direction comes from the sea. The rate is higher than the BB point because the BP point is closer to the sea.

4 Summary and Conclusion

Air quality simulations in the dry season, transition season, and wet season in the city of Balikpapan show that the distribution of PM10, NO_2, and SO_2 pollutants is strongly influenced by meteorological parameters, including wind direction and speed, atmospheric stability, and rainfall. In general, the dispersion of pollutants in the city of Balikpapan follows the pattern of local wind direction movements that affect the wet season, transition season, and dry season. Based on the distribution simulation from WRF-Chem, it shows that the Tenggarong and surrounding areas have the potential to have high pollutant concentrations. This information on air quality and pollutant distribution is expected to be an input for local governments to mitigate air quality in their area.

The conclusion of the in-situ AQMS data analysis at the BB and PB Balikpapan stations is that, the pollutant concentrations of SO_2 and NO_2 at the Balikpapan PB station have higher concentrations than those at the Balikpapan Baru station, while the highest PM2.5 concentration is at the Balikpapan Baru observation point.

References

1. Winarni, Sistem Transportasi Massal, Kebutuhan Mutlak. Kaltimpost, hal. 25 [Online] tersedia di http://kaltim.prokal.co/read/news/298347-sistem-transportasi-massal-kebutuhan-mutlak.html diakses pada tanggal 26 September 2021 (2017)
2. Al Hadar, A.: Analisis Kinerja Jalan dalam Upaya Mengatasi Kemacetan Lalu Lintas pada Ruas Simpang Bersinyal di Kota Palu. SMARTek. **9**(4), 327–366 (2011)
3. BPS Kota Balikpapan.: Kota Balikpapan dalam Angka 2018. Balikpapan. Badan Pusat Statistik Kota Balikpapan. [Online] bisa diakses di: Desain Rute Jaringan Moda Bus Kota sebagai Antisipasi Kemacetan di Kota Balikpapan59 https://balikpapankota.bps.go.id/publication/2018/08/16/8df63ffd73378cb97b4fc3eb/kota-balikpapan-dalam-angka-2018.html. diakses pada 10 Sept 2018 (2018)
4. Ismiati., et al.: Pencemaran Udara Akibat Emisi Gas Buang Kendaraan Bermotor. Jurnal Manajemen Transportasi & Logistik (JMTransLog) **1**(3) (2014)
5. Neiburger, M.: Understanding our Atmospheric environment, diterjemahkan Ardina Purbo. Memahami Lingkungan Atmosfer Kita, Edisi II, Bandung: ITB Bandung (1995)
6. Powers, J., Klemp, J., Skamarock, W., Davis, C., Dudhia, J., Gill, D., Coen, J., Gochis, D., Ahmadov, R., Peckham, S., Grell G., Michalakes, J., Trahan, S., Benjamin, S., Alexander, C., Diego, G., Wang, W., Schwartz, C., Romine, G., Liu, Z., Snyder, C., Chen, F., Barlage, M., Yu, W., dan Duda, M.: The weather research american meteorological society and forecasting (WRF) model: overview, system efforts, and future directions. Bull. Am. Meteor. Soc. https://doi.org/10.1175/BAMS-D-15-00308.1, in press (2017)
7. Turyanti, A., et al.: Analisis Pola Dispersi Partikulat dan sulfurdioksida menggunakan model WRF Chem di sekitar wilayah industri Tangerang dan Jakarta. Manusia dan lingkungan **23**(2), 169–178 (2016)
8. Faisal, I., Sofyan, A.: Analisis pengaruh variasi musiman terhadap dispersi NO_2 di kota Tangerang dengan menggunakan model WRF-CHEM. Jurnal Teknik Lingkungan 25(1), Hal 1–14 (2019)
9. Fast, J.D., Gustafson, W.I., Easter, R.C., Zaveri, R.A., Barnard, J.C., Chapman, 46. E.G., et al.: Evolution of ozone, particulates, and aerosol direct radiative forcing in the vicinity of Houston using a fully coupled meteorology-chemistry-aerosol model. J. Geophys. Res.: Atmos. **111**(D21) (2006).
10. Grell, G.A., Peckham, S.E., Schmitz, R., McKeen, S.A., Frost, G., Skamarock, W.C., et al.: Fully coupled "online" chemistry within the WRF Model. Atmos. Environ. **39**(37), 6957–6975 (2005)
11. Zhang, Q., Tong, P., Liu, M., Lin, H., et al.: A WRF-chem model-based future vehicle emission control policy simulation and assessment for the Beijing-Tianjin-Hebei region, China. J. Environ. Manag. (2020)

12. Baklanov, A., et al.: Online coupled regional meteorology chemistry models in Europe: current status and prospects. J. Atmos. Chem. Phys., **14**, 317–398 (2014). www.atmos-chem-phys.net/14/317/2014/, https://doi.org/10.5194/acp-14-317
13. Gusnita, D.: Effect of meteorology parameters on air pollutant standard index in the urban area (case study in Jakarta). In: Proceedings of the International Conference on Radio science, Equatorial Atmospheric Science and Environment and Humanosphere Science, vol. 275, pp. 705–716 (2021)

Automatic Processing for Aerosol, Snow/Ice, Cloud, and Volcanic Ash Imagery (ASCI) Products Based on NOAA-JPSS Satellites Data

Olivia Maftukhaturrizqoh, Andy Indradjad, Tri Astuti Pandansari, Hidayat Gunawan, and Karunika Diwyacitta

Abstract Routine air quality monitoring is important to anticipate potential climate changes and other consequences including human health implications. Remote sensing observations can measure air quality estimation indirectly and complement temporally and spatially limited ground-based observation. Aerosol, Snow/Ice, Cloud, and Volcanic Ash (ASCI) is one of the satellite imagery products from the National Oceanic and Atmospheric Administration (NOAA) Joint Polar Satellite System (JPSS) series satellite data. Research Center for Remote Sensing BRIN acquired NOAA-JPSS satellite data via direct receive from the BRIN ground station. The satellite passes above Indonesia four times per day, thus can provide daily observations. An automation system to generate ASCI data products is developed to provide and ensure the continuity and availability of the data. Inputs of the system are from Sensor Data Records (SDR).h5 file format of Visible Infrared Imaging Radiometer Suite (VIIRS) instrument carried by NOAA-JPSS Satellite. The inputs are processed using an open-source module, Community Satellite Processing Package (CSPP) ASCI version 1.2. The output will be in NetCDF (.nc) file format including Aerosol Optical Depth (AOD), Aerosol detection product (ADP), Cloud Mask, Cloud Phase, Cloud Cover Layer, Volcanic Ash, Surface Albedo, Ice concentration (if any), and also generate Quicklook imageries in PNG file format. The automation system consists of several functions to crawl files and look for the newest input data available, process, and save it to the storage. The automation system has been implemented and can automatically produce ASCI imageries. Measurement of processing time was conducted and discovered that it corresponds to the input folder size. The average processing time was 19.82 and 15.31 minutes for daytime and nighttime data, respectively. The near-real-time ASCI products generated from this automation system are expected to be utilized by users to assist their research, forecast, and policy decisions related to air quality, climate, or other corresponding research.

O. Maftukhaturrizqoh (✉) · A. Indradjad · T. A. Pandansari · H. Gunawan · K. Diwyacitta
Research Center for Remote Sensing, National Research and Innovation Agency, Jakarta, Indonesia
e-mail: olivia.maftukhaturrizqoh@brin.go.id

A. Basit et al. (eds.), *Proceedings of the International Conference on Radioscience, Equatorial Atmospheric Science and Environment and Humanosphere Science*, Springer Proceedings in Physics 290, https://doi.org/10.1007/978-981-19-9768-6_9

1 Introduction

Aerosols are suspended particles in the air that can be either liquid or solid and semi-solid, the main component of industrial smog and pollutant in natural disasters such as volcanic eruptions, dust, biomass burning, and forest fires [1, 2]. Aerosols leads to deteriorated air quality [2], and exposure especially to those smaller than 2.5 μm can cause health consequences including heart disease, strokes [1], respiratory problem [3], acute cardiovascular morbidity and mortality, lung cancer, and others that can shorten life expectancy [4] and also have harmful impacts for the environment [2]. Therefore, monitoring of aerosol distributions and studying how they are changing is important, especially for aerosols with large spatial and temporal variability, such as smoke, sand storms, and dust [5, 6]. Studies of aerosol characteristics allow for the anticipation of prospective climate changes, long-term ecological effects [7], and other consequences. Those valuable data can assist forecasters, meteorologists, and other stakeholders in decision-making related to air quality, long-term climate change assessment, and others.

Currently, in Indonesia, there are already in-situ stations for air quality monitoring using Air Quality Monitoring System (AQMS) [8]. However, the time–space dynamics air assessment are constrained by the monitoring's limited network [9]. A large spatial view and resolution of aerosol and air quality estimation [9, 10] can be achieved by spaceborne or remote sensing observation that complements temporally and spatially limited ground-based or in-situ sampling [11] and airborne observations [12]. Recently, remote sensing has become frequently used to monitor the worldwide spatial and temporal distributions of aerosols and offer long-term and continuous coverage [7]. Electromagnetic radiation from the earth's surface is captured and recorded in satellite imageries [9] while aerosol directly scatter and absorb solar and thermal infrared radiation and indirectly change the Earth-leaving radiation [1]. Related to it, Aerosol optical depth (AOD) is a measure of the radiation extinction caused by aerosols due to absorption and scattering [13], the retrieval of AOD is from visible and near IR bands solar radiation reflection [1].

Retrieval of AOD and other aerosol products measured by satellite began in the 1970s that was made initially with sensor in Landsat satellite series, then followed by a variety sensors [1]. Aerosol data more recently obtained by the Visible Infrared Imaging Radiometer Suite (VIIRS) sensor carried by the Suomi National Polar-orbiting Partnership (SNPP) satellite and the retrievals continued to the next generation Joint Polar Satellite System (JPSS) from National Oceanic and Atmospheric Administration (NOAA) [1].

NOAA-JPSS satellite provides continuous access to environmental data worldwide by carrying four instruments, one of those is the VIIRS instrument [14]. Aerosol, Snow/Ice, Cloud, and Volcanic Ash (ASCI) is one of the satellite imagery products from the VIIRS sensor of NOAA-JPSS satellite data. The VIIRS Atmospheric composition aerosol products are the aerosol optical depth (or aerosol optical thickness, AOD, AOT) and the aerosol detection product (ADP). The AOD is a quantitative measure of aerosol loading that requires spectral reflectance measurements from a variety of VIS and NIR bands and many ancillary data sets. The ADP is information on the aerosol type in a qualitative form that identifies volcanic ash, dust, and smoke [14].

In terms of that, Research Center for Remote Sensing BRIN has been acquired satellite data of VIIRS sensor data from SNPP and NOAA-JPSS via direct receive from BRIN ground station since early 2018 [15]. The NOAA-JPSS satellite passes above Indonesia four times per day [16], two times in the daytime and 2 times other are passed in the nighttime, thus it can provide daily observations with national coverage. To provide and ensure the continuity and availability of the data, this paper proposed a near-real-time automation system to generate ASCI imagery data products.

The automation system generates the ASCI imagery automatically right after the new input data from the ground station acquisition is detected. The system can reduce the workload of the human workforce and also can avoid human errors. From the automation system, the ASCI imagery data product will available in near-real-time. The automation system also can be used to process NOAA-JPSS older data from the beginning of input data available in BRIN storage (2018), so that multitemporal or historical data can be provided for long-term assessment, analysis or research. The near-real-time ASCI products generated from this automation system are expected to be utilized by users to assist their research, forecast, and policy decisions related to air quality, climate, or other corresponding researches.

2 Data and Method

The automation system design consists of four major functions. Those functions are the files crawling function, comparing function, processing function, and storing function as shown in Fig. 1a. To be able to run the automation system requires three basic parts, those are input, process, and output. The inputs are from Sensor Data Records (SDR) of the Visible Infrared Imaging Radiometer Suite (VIIRS) instrument carried by NOAA-JPSS Satellite.

NOAA-JPSS Satellite data is acquired via direct receive by BRIN ground station in Parepare, South Sulawesi. All the raw data from the direct receive are transferred further and stored in the BRIN cloud. The ASCI automation system gets the SDR input data to generate ASCI products from the BRIN cloud through BRIN internal network and also all data acquisition and transfer work under the network as shown in Fig. 1b.

The Inputs data of the automation system are all VIIRS M-bands (SVM*) and I-bands (SVI*), the I-Band (GITCO*), and M-band geolocation files (GMTCO*) and Day/Night Band (SVDNB*, GDNBO*) files as shown in Fig. 2. The inputs are then processed in the next part, the process part. In the process part, the CSPP ASCI module from the Space Science and Engineering Center (SSEC) at University of Wisconsin-Madison [17] is utilized. It is an open-source module as an outcome of collaboration between Space Science and Engineering Center (SSEC), Cooperative Institute for Meteorological Satellite Studies

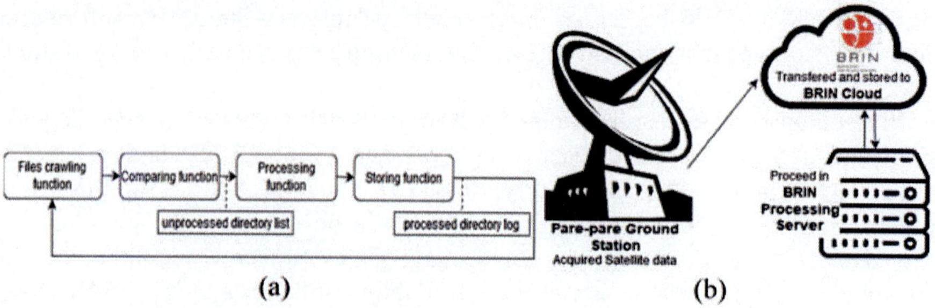

Fig. 1 **a** Four major functions of the automation system design. **b** The process of Satellite data acquisition, transfer, and processing in the Remote Sensing Research Center, National Research and Innovation Agency (BRIN)

(CIMSS), and NOAA Community Satellite Processing Package (CSPP) [18]. The inputs will be processed by the CSPP ASCI version 1.2 to generate several types of output in NetCDF (.nc) extension format including JRR-ADP*.nc, JRR-AOD*.nc, JRR-CloudBase*.nc, JRR-CloudMask*.nc, JRR-CloudPhase*.nc, JRR-CloudCoverLayers.nc, JRR-VolcanicAsh*.nc, SURFALB*.nc, etc. as mentioned in the Fig. 2 output list. The processing module will also generate Quicklook imageries in PNG extension format, those are JRR-AOD*AOD550.png, JRR-CloudMask*.png, JRR-VolcanicAsh*.png, JRR-IceConcentration*.png (if any), and SURFALB*ALBEDO_EDR.png. JRR-AOD and JRR-ADP imagery could only be generated by using SDR VIIRS daytime input data, otherwise, the other product can be generated in the day and nighttime data.

The four major functions as shown in Fig. 1 have a purpose as follows: The first one is a function to perform files crawling so that the system scan and search for the

Fig. 2 Three basic parts of the automation system

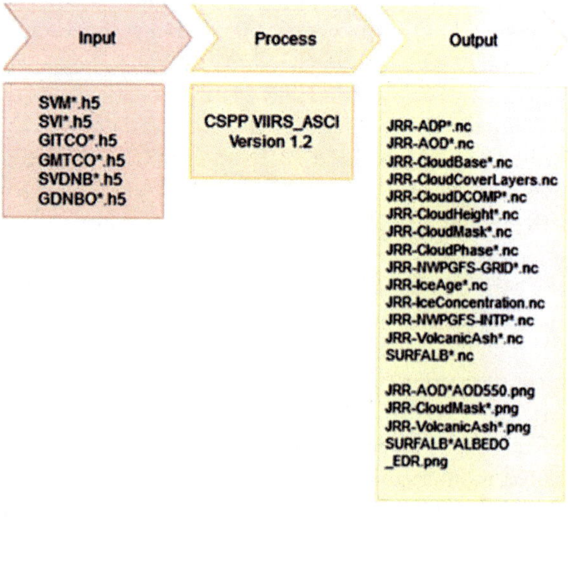

unprocessed input files. The second function compares the file crawling results to the processed directory log to check whether an input directory has been processed. When there is a directory not processed yet, the system activates the next function to process the input directory. After processing the directory, the last storing function store the output in the corresponding directory based on the date and time of the data acquisition.

The module was implemented by using Python programming language. Python programming language is one of the most popular ones that have various libraries to support automation functions. Some automation implementations that use the python programming language as their core program are automatic modeling [19, 20] automatic analysis [21, 22] automatic computation [23], and automatic classification [24].

The flowchart of the automatic system is shown in Fig. 3. below. First, the system will open and read a log file and store it as an array. The log file has contained a list of file directories which already been processed. The files crawling function then scan all files in directories. To determine whether an input directory had already been processed or still has to be processed further, the comparing function performs a comparison between the scanned filename to the log list. When the file name is not in the log yet, the processing function will process the input directories file further using the CSPP ASCI module. The module consists of two main processing, the first will produce ASCI products in .nc format. Afterward, the system run the next processing which will generate a Quicklook of ASCI product in PNG format.

After finishing processing, the storing function check whether the storage directory path based on year/month/date and product name of the input data acquisition

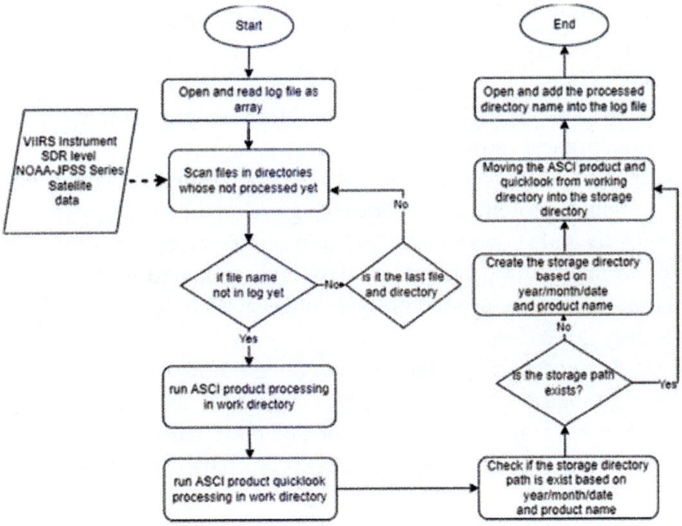

Fig. 3 Flowchart of the automatic system

exists. If the storage path has the not existed yet, the system will create the appropriate storage directory according to acquisition time of the input data and store all the output files in. The directory that has finally been processed will then be written in the processed directory log. The system also computes the time required to process each input directory in generating ASCI products to find out the performance of this automation system. The log file also stores this processing time information of each data. The processes in the automation system will be run each time Sensor Data Records (SDR) from the NOAA-JPSS satellite VIIRS sensor is available from new acquisition activity. The system reduces the workload of the human workforce and also can avoid human errors.

3 Results

As mentioned in the previous section, the ASCI output product that will be generated from the system are VIIRS Aerosol Optical Depth, VIIRS CloudMask, VIIRS Volcanic Ash, and VIIRS Surface Albedo in both.nc and Quicklook.png format. The ASCI output products are at VIIRS M-Band Spatial Resolution which is 750 m. Figure 4 shows four types of Quicklook output in PNG format generate from the automation system using the CSPP ASCI version 1.2 module, the satellite data acquired on May 15th 2022 at 05:31 (UTC).

The Quicklook already contains a legend or color indication for each product. Figure 4a is the Quicklook PNG image of Aerosol Optical Depth (AOD) at 550 nm shown in colors ranging from dark violet to red that indicates the highest value. Figure 4b is the Quicklook PNG image of Cloud Mask shown in four kinds of color indicator, green color indicates clear, light blue indicates probable clear, red indicates probable cloudy, and white indicates cloudy. Figure 4c is the Quicklook of Volcanic Ash confidence product shown ranging from high to not ash. Figure 4d is the Quicklook of Surface Albedo shown in colors ranging from dark violet to red that indicates the highest ratio albedo.

The automation system has been included in operational activities and integrated into the remote sensing satellite data processing system in the BRIN Research Center of Remote Sensing and shows satisfactory and stable performances. The system has been able to automatically perform file crawling, compared to the log, processing, and storing functions. The automation system has already been run for three months from May to July 2022 (Fig. 5).

Table 1 gives a summary of the processing time data on May and June 2022 product data. The measurement of the processing time of each product generation was conducted to know the performance of this automation system. From Table 1 can be discovered that the time required to process data is corresponds to the input folder size. The larger the size of the input data, the longer time is required to process the data. All data in Table 1 are plotted in Fig. 6 graphics which show the correlation of input data size to processing time.

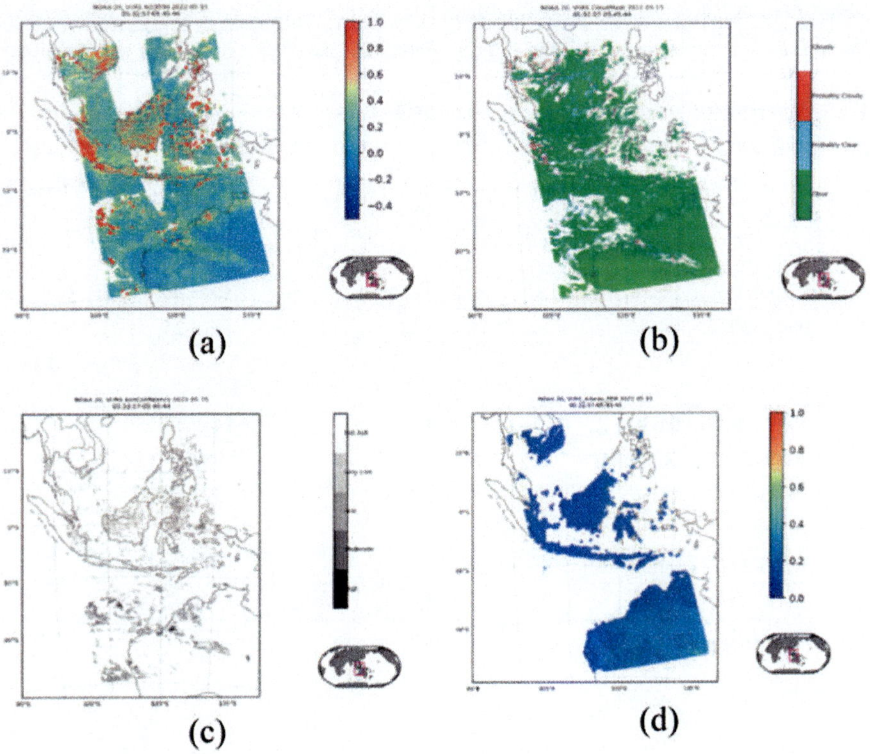

Fig. 4 As a sample of ASCI product Quicklook in.png format acquired on May 15th 2022 at 05:31 (UTC). **a** VIIRS Aerosol Optical Depth (JRR-AOD*550.png). **b** VIIRS CloudMask (JRR-CloudMask*.png). **c** VIIRS Volcanic Ash (JRR-VolcanicAsh*.png). **d** VIIRS Surface Albedo (JRR-SURFALB*.png)

Fig. 5 Scatter plot of correlation between input file size and processing time required

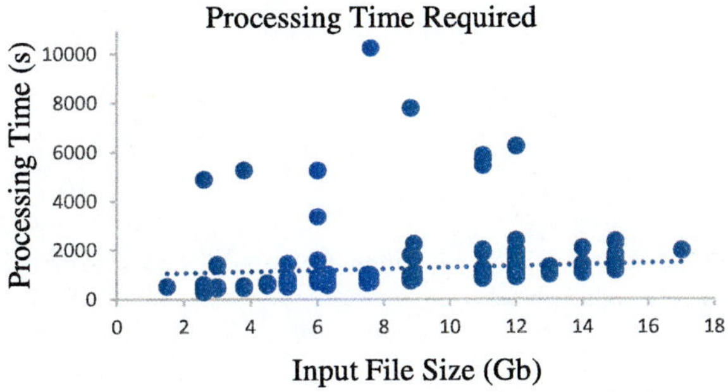

Table 1 Processing time data on May and June 2022 product data

No.	Date	Acquisition time (UTC)	Input file size (Gb)	Processing time required (seconds)
1	01/05/2022	04:54	15	1379.62
2	01/05/2022	06:37	8.9	818.05
3	01/05/2022	16:03	5.1	569.24
4	01/05/2022	17:37	13	1057.26
5	02/05/2022	04:37	12	1142.86
6	02/05/2022	06:18	12	1043.20
7	03/05/2022	04:18	12.0	1113.25
: :	: :	: :	: :	: :
219	29/06/2022	04:49	14	1412.15
220	29/06/2022	06:30	8.9	998.56
221	29/06/2022	15:56	5.1	859.60
222	29/06/2022	17:30	13	1253.72
223	30/06/2022	04:30	12	1617.93
224	30/06/2022	06:11	12	1276.91
225	30/06/2022	17:12	13	1324.17

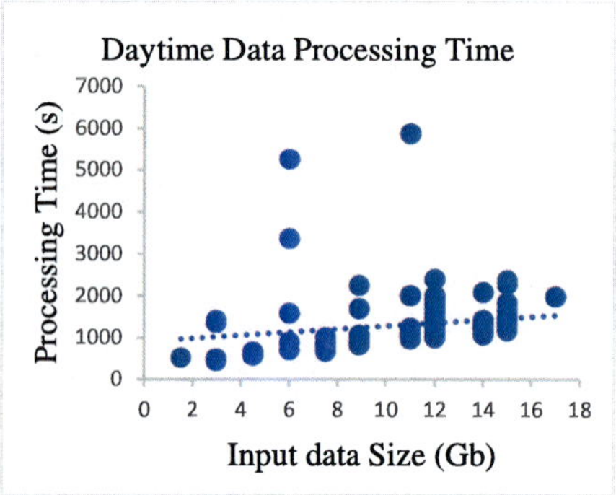

(a)

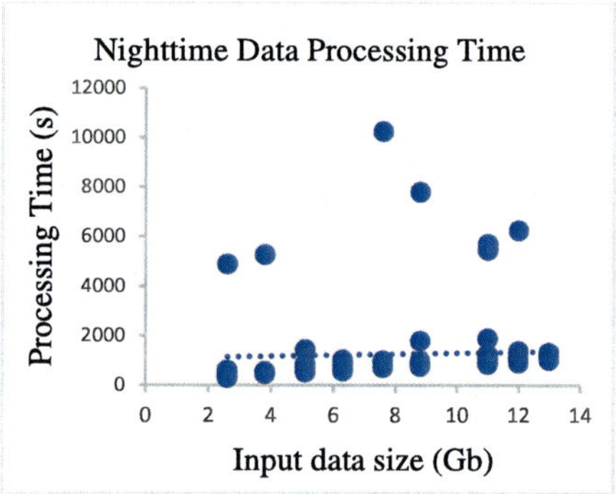

(b)

Fig. 6 Scatter plot of correlation between input file size and processing time required. **a** Daytime data. **b** Nighttime data

4 Discussion

Daytime and Nighttime data then were plotted separately in Fig. 7a and b graphics because as mentioned before the product of Aerosol Optical Depth which was JRR-AOD and JRR-ADP imagery could only be generated by using SDR VIIRS daytime input data. Therefore, daytime and nighttime data processing generate different amounts of output data and may take different average processing times. The average processing time for daytime data was 1333.79 s or 22.23 min. While the average processing time for nighttime data was 1284.66 or 21.41 min. However, in addition to the input requirement, the processing module also required ancillary data. The ancillary data that match the date and time of the input will automatically be downloaded as needed at module runtime. This ancillary data takes time to download before the input data can be processed further by the module. Therefore, the Internet connection network while downloading the ancillary also affects the processing time. Accordingly, further analysis of the time-processing data is in the next section.

In Fig. 7a and b there are several data points which are can be considered outliers. Those could be because of the slow Internet connection during ancillary data downloading at runtime. Figure 7a and d are shown the scattered plot result without the outliers. The removed outliers are interpreted using statistics as the data point which has a processing time longer than the upper fence of processing time data. The upper fence threshold of daytime processing time data was 2016.75 s, thus the data longer than that were removed and plotted into Fig. 7a. Likewise, nighttime processing data were plotted in Fig. 8b graphics after removing outliers that were longer than the upper fence of the data, 1609.46 s.

The trendline of Fig. 7a and b confirm the results that input file size corresponds to the processing time. The processing time has a linear correlation to the input data size. The larger the size of input data, the longer time is required to process the data. While from the filtered processing time data above the average processing time were 1189.14 and 918.43 s or 19.82 and 15.31 min for daytime and nighttime data, respectively. Based on the calculation, this automation system has fulfilled the requirement to produce near-real-time ASCI products because the products can be generated no longer than 30 min after acquisition time. Most of the NOAA-JPSS imagery including ASCI products also available on the NOAA Comprehensive Large Array-Data Stewardship System (CLASS) website, but can be publicly accessed only after 6 h from the acquisition time [25], therefore this automation implementation,

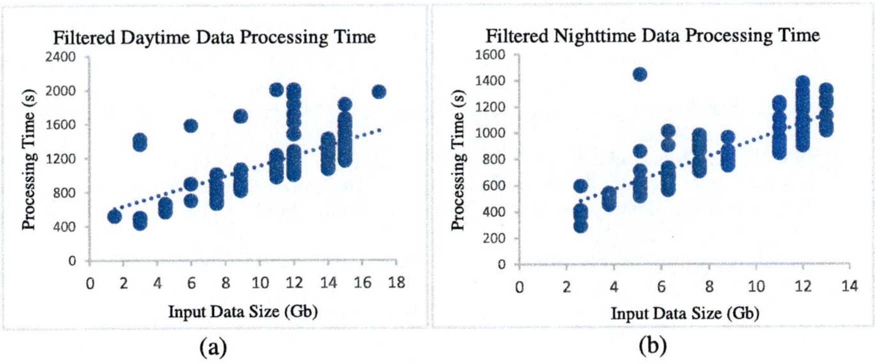

Fig. 7 Scatter plot of correlation between input file size and processing time required after removing outliers. **a** Daytime data. **b** Nighttime data

can provide ASCI products faster with national daily coverage. It is also confirmed that daytime data processing has a longer average time than nighttime data processing because they generate different amounts of output data.

5 Conclusion

A near-real-time automation system for ASCI imagery products has been successfully designed and implemented in the remote sensing satellite data processing system in the BRIN Research Center of Remote Sensing and shows satisfactory and stable performances. The system has been executed to run automatically to generate ASCI imagery products from the NOAA-JPSS satellite data. The processes in the automation system will be run each time Sensor Data Records (SDR) from the NOAA-JPSS satellite VIIRS sensor are available from new acquisition activity. The system can reduce the workload of the human workforce and also can avoid human errors.

This automation system can provide ASCI imagery for daily observations with national coverage and ensure the continuity and availability of the data in near-real-time data. Measurement of processing time was conducted and discovered that it corresponds to the input folder size. The average processing time was 19.82 and 15.31 min for daytime and nighttime data, respectively. The near-real-time ASCI products generated from this automation system are expected to be utilized by users to assist their research, forecast, and policy decisions related to air quality, climate, or other corresponding researches.

References

1. Nesdis, N.: NOAA NESDIS Center for Satellite Applications and EPS Aerosol Optical Depth (AOD) Algorithm Theoretical Basis Document (2016)
2. GOES-R Aerosols/Air Quality Applications fact sheet
3. Talib, N., et al.: Relationship between particulate matter (PM2.5) concentration and aerosol optical depth (AOD) throughout Peninsular Malaysia. IOP Conf. Ser. Earth Environ. Sci. **1019** (2022)
4. Zhang, H., Kondragunta, S.: Daily and hourly surface PM2.5 estimation from satellite AOD. Earth Sp. Sci. **8** (2021)
5. Morgan, M.R.: Climate change 2007. Weather **59** (2007)
6. Star, N.N.: Center for satellite applications and research algorithm theoretical basis document. In: JPSS Enterprise Processing System Aerosol Detection Product Pubu Ciren and Shobha Kondragunta, pp. 1–101 (2022)
7. Filonchyk, M. et al.: Author correction: combined use of satellite and surface observations to study aerosol optical depth in different regions of China (Scientific Reports, (2019), 9, 1, (6174), https://doi.org/10.1038/s41598-019-42466-6). *Sci. Rep.* **9**, 1–15 (2019)
8. Zheng, H.: Design and application for environmental air quality automatic monitoring system. In: ICEMI 2009—Proceeding 9th International Conference Electron Measured Instruments, pp. 2218–2221 (2009). https://doi.org/10.1109/ICEMI.2009.5274611
9. Kumara, N., Chub, A., A.F.: An empirical relationship between PM2.5 and aerosol optical depth in Delhi Metropolitan. C. Model. Eng. Sci. **1**, 119–131 (2007)

10. Chu, D.A., et al.: Global monitoring of air pollution over land from the Earth observing system-terra moderate resolution imaging spectroradiometer (MODIS). J. Geophys. Res. Atmos. **108** (2003)
11. Kaufman, Y.J., Tanré, D.: Satellite remote sensing|aerosol measurements. Encycl. Atmos. Sci. 1941–1956 (2003). https://doi.org/10.1016/b0-12-227090-8/00347-x
12. JGR Atmospheres—2013—Suomi-NPP VIIRS aerosol algorithms and data products.pdf.
13. List, A., et al.: Visible Infrared Imaging Radiometer Suite (VIIRS) Enterprise Aerosol Optical Depth and Aerosol Particle Size Products User's Guide Version 1, February 2020 (2020)
14. Zhou, L., Divakarla, M., Liu, X., Layns, A., Goldberg, M.: An overview of the science performances and calibration/validation of joint polar satellite system operational products. Remote Sens. **11** (2019)
15. Gustiandi, B., Monica, D., Indradjad, A.: Sistem Pengolahan Data Satelit Seri Noaa Jpss Untuk Produksi Informasi Titik Panas Secara Otomatis (Automatic NOAA JPSS Satellite Series Data Processing System To Produce Active Fires Information) **17**, 43–55 (2020)
16. Soleh, M., Suprijanto, A., Mahatmanto, B.P.A.: Preliminary design of remote sensing ground station system for The JPSS-1 (joint polar satellite system) data acquisition and processing. In: Proceedings of The 2nd International Conference of Indonesian Society for Remote Sensing, pp. 62–77. Indonesian Society for Remote Sensing (2016)
17. Space Science and Engineering Center
18. Webmaster, C.: Community Satellite Processing Package. Retrieved from Cooperative Institute for Meteorogical Satellite Studies (2021)
19. K., V.S., K., R.: A python automation script that generates UVM_RAL (Register Abstraction Layer) register model. Mater. Today Proc. **37** (2020)
20. Straatsma, M.W., Kleinhans, M.G.: Flood hazard reduction from automatically applied landscaping measures in RiverScape, a Python package coupled to a two-dimensional flow model. Environ. Model. Softw. **101**, 102–116 (2018)
21. Loganathan, K., Sarath Kumar, R., Nagaraj, V., John, T.J.: CNN & LSTM using python for automatic image captioning. Mater. Today Proc. (2020). https://doi.org/10.1016/j.matpr.2020.10.624
22. Stuckner, J., Frei, K., McCue, I., Demkowicz, M.J., Murayama, M.: AQUAMI: an open source Python package and GUI for the automatic quantitative analysis of morphologically complex multiphase materials. Comput. Mater. Sci. **139**, 320–329 (2017)
23. Estévez Schwarz, D., Lamour, R.: InitDAE: computation of consistent values, index determination and diagnosis of singularities of DAEs using automatic differentiation in Python. J. Comput. Appl. Math. **387**, 112486 (2021)
24. Lamy, J.-B.: Owlready: ontology-oriented programming in Python with automatic classification and high level constructs for biomedical ontologies. Artif. Intell. Med. **80**, 11–28 (2017)
25. List, A., et al.: Suomi National Polar-Orbiting Partnership (SNPP) Visible Infrared Imaging Radiometer Suite (VIIRS) Aerosol Products User's Guide, pp. 1–34 (2013)

The Seasonal Composition of Inorganic Aerosol in an Urban Region of Bandung, Indonesia

Wiwiek Setyawati, Dyah Aries Tanti, Saipul Hamdi, Asri Indrawati, Atep Radiana, Sumaryati, Risyanto, and Retno Puji Lestari

Abstract Bandung is a basin-shaped region surrounded by high mountains with a dense population vulnerable to air quality degradation. The haze can potentially occur in a supportive meteorological condition. The study investigates the influence of local meteorology on the inorganic aerosol composition in an urban area of Bandung in 2020–2021. The ambient air sample is collected weekly using a filter pack and analyzed by ion Chromatography (IC). Local meteorological data is obtained from the Campbell weather station installed on the site. The data analysis is classified into seasons: November to March (NDJFM) for wet and May to September (MJJAS) for dry. The sulfate, nitrate, and ammonium (SNA) contributed more than 75% of the inorganic aerosol composition, and their concentrations were higher in the dry than in the wet season. Non-parametric Spearman's rho correlation (*r*) shows that the relative humidity promoted NH_4^+ stronger in the dry than in the wet season. Therefore, NH_4^+ had a more significant possibility to increase under higher relative humidity in the dry season, thus increasing the possibility of haze formation. The sources of Na^+, Cl^-, and NH_4^+ aerosols are long-distance transport, but K^+ is local. Heterogeneous oxidation under high relative humidity was more critical for NH_4^+ generation than photochemical oxidation was, but for Cl^- it was vice versa.

1 Introduction

The tiny particles suspended in the air, called aerosol, are formed in the atmosphere by direct release from sources (primary aerosol) or indirectly through chemical processes involving precursor gases (secondary aerosol). Aerosol is essential in

W. Setyawati (✉) · D. A. Tanti · S. Hamdi · A. Indrawati · A. Radiana · Sumaryati · Risyanto
Research Center for Climate and Atmosphere, National Research and Innovation Agency-BRIN, Jl. Cisitu, Sangkuriang, Bandung 40175, Indonesia
e-mail: wiwi007@brin.go.id

R. P. Lestari
The Center for Standardization of Environmental Quality Instruments, The Ministry of Environment and Forestry, Puspiptek Serpong, 15310 Jakarta, Indonesia

A. Basit et al. (eds.), *Proceedings of the International Conference on Radioscience, Equatorial Atmospheric Science and Environment and Humanosphere Science*, Springer Proceedings in Physics 290, https://doi.org/10.1007/978-981-19-9768-6_10

climate, health, and ecosystem [1, 2]. Their direct impacts on climate are by scattering and absorbing solar radiation and indirect by altering clouds' properties. The health impact of aerosols is their ability to penetrate deep into the human respiratory system, resulting in the dysfunction of human organs. The aerosols' removal from the atmosphere to the earth's surface can alter the air, water bodies, and soil properties.

Secondary inorganic aerosols (SIA) of sulfate, nitrate, and ammonium (SNA) are the dominant components in particulate matter with sizes less than 2.5 microns (PM2.5) [2, 3]. They are responsible for haze formation and a decrease in visibility during heavily polluted events in urban regions [1, 2, 4]. Furthermore, the increased inorganic fraction in the aerosol particles led to frequent haze days. Meteorological factors have important roles in inorganic aerosol formation. Lower relative humidity favors aerosol hygroscopic formation due to more ammonium nitrate aerosol and the enhancement of the light-scattering ability of atmospheric aerosol. A high concentration of NOx can promote the conversion of SO_2 to form sulfate aerosol via aqueous-phase oxidation during intensive pollution periods [3].

Indonesia lies in the equatorial region, where sunlight is available all year round. Bandung is one of the megapolitan cities in Indonesia, located in a basin area surrounded by high mountains. The unique climatic and topographic conditions and the magnitude of the emission sources can cause dangerous haze, such as those in China [3, 4] and India [2]. Therefore, it is necessary to study the meteorological factors dominant in influencing the inorganic aerosol formation in the atmosphere. This study aims to investigate the role of local meteorology on the inorganic aerosol composition and their possible source in an urban area of Bandung.

2 Methodology

2.1 Sampling Site

The sampling site in Bandung city representing urban areas is located on the 5th floor of the Climate and Atmospheric Research Center (PRIMA), BRIN building at Jl. Dr Djundjunan 133, Bandung, West Java, Indonesia (6.8951710 S, 107.588370 E) as shown in Fig. 1. Bandung has a cool climate, high humidity, large wind speed variabilities, low sunshine duration, and high precipitation [5].

2.2 Data Used

We used weekly dry deposition monitoring data from 2020–2021 carried out by the Research Center for Climate and Atmosphere, BRIN (previously the Center for Atmospheric Science and Technology, LAPAN). There was no sampling from March 9th–June 29th, 2020, June 14th–August 7th, 2021, Aug 14th–7th, 2021, and

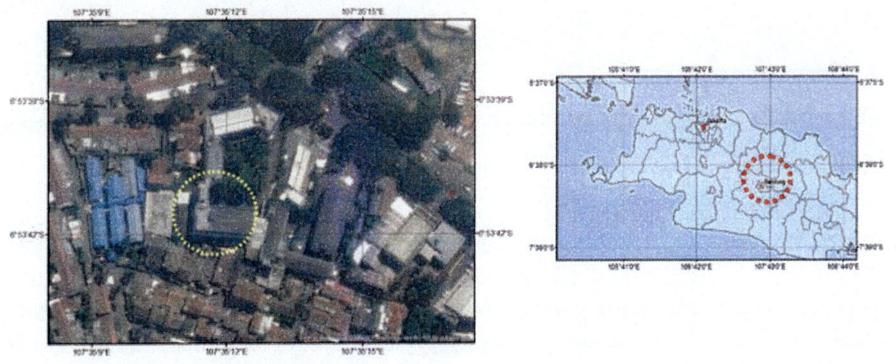

Fig. 1 The Sampling site is marked by a yellow broken circle (Left). The broken red line is the Bandung city administrative border (right)

September 27th–October 4th, 2021, because of the lockdown during the COVID-19 pandemic. The dry deposition monitoring is a part of the East Asia Network (EANET) for acid deposition monitoring collaboration. The ambient air is collected using a filter pack. The sampling duration is one week, and the airflow rate is 1.5 L/min. The filter pack sampling method refers to the manual guide published by Asia Center for Air Pollution Research (ACAP) (EANET, 2010). We focus on the aerosol data: Sulfate (SO_4^{2-}), nitrate (NO_3^-), chloride (Cl^-), ammonium (NH_4^+), calcium (Ca^{2+}), sodium (Na^+), potassium (K^+), and magnesium (Mg^{2+}) ions analyzed by Ion Chromatography (DIONEX ICS-1600/09071855 for anions and DIONEX ICS-1600/09071855 for cation analysis).

The 10-min synoptic data is obtained from the Campbell weather station adjacent to the filter pack. The data is then averaged according to the weekly filter pack sampling duration. The meteorological data used for further analysis are temperature (T) in degrees Celsius, relative humidity (RH) in %, wind speed (WS) in m/s, wind direction (WD) in degrees, precipitation amount (PR) in mm, and solar radiation (SR) in W/m^2.

2.3 Data Analysis

We grouped data into wet and dry seasons to study seasonal variabilities of inorganic aerosol concentrations and their related meteorological condition. In Java Island, the dominant northwest monsoon from November to March (NDJFM) leads to the wet season, and the southeast monsoon from May to September (MJJAS) to the dry season [6]. Spearman's rho correlation coefficient with a 95% significance level ($\alpha = 0.05$) is used to study the relationship between inorganic aerosol concentrations and corresponding selected meteorological factors. The formulae for calculating Spearman's rho correlation coefficient are as follows:

$$r = 1 - \frac{6\sum di^2}{n(n^2 - 1)} \tag{1}$$

where r is Spearman's rho correlation coefficient, d is the difference between ranks, and n is the number of observations.

We draw wind rose diagrams to get information about the wind direction and speed in Bandung during wet (NDJFM) and dry (MJJAS) seasons using a freeware named WRPLOT developed by Lakes Software (https://www.weblakes.com/software/freeware/wrplot-view/). The software uses wind speed and direction as inputs. The output is a wind rose figure that illustrates the frequency of wind occurrence in the specified wind direction sectors and wind speed classes for a given location and time.

We investigate the long-distance source of particles to Bandung during wet and dry seasons by utilizing The HYSPLIT trajectory model developed by Air Resources Laboratory, NOAA (https://www.ready.noaa.gov/hypub-bin/trajtype.pl?runtype=archive). The model vertical trajectories velocity was run for three heights: 100, 300, and 500 m above ground level by selecting GFS meteorological data having 0.25° spatial resolution.

3 Result and Discussion

3.1 Seasonal Variabilities

We report the descriptive statistics of seasonal inorganic aerosol in ambient air by averaging the weekly concentrations for corresponding months in 2020–2021 (see Table 1 left). The average concentration of inorganic aerosols in the dry (MJJAS) was higher than in the wet (NDJFM) season, except for Mg^{2+}. The $SO_4{}^{2-}$, $NO_3{}^{-}$, and $NH_4{}^{+}$ (SNA) had the three highest concentrations with average ± standard deviation in wet were 5.8 ± 2.3 μg/m^3 (51.0%), 1.8 ± 0.98 μg/m^3 (15.5%), and 1.5 ± 0.8 μg/m^3 (13.2%), respectively and in dry were 6.9 ± 2.7 μg/m^3 (50.6%), 2.1 ± 0.7 μg/m^3 (15.3%), and 1.8 ± 0.8 μg/m^3 (13.3%), respectively. SNA's compositions made up more than 75% of total inorganic aerosols. The SNA also dominated the haze pollution in Beijing, China, with $NH_4{}^{+}$ as the highest, followed by $NO_3{}^{-}$ and $SO_4{}^{2-}$ [7]. The secondary aerosols of SNA were produced by the chemical process of their precursors in the atmosphere [8] as follows:

$$2NH_{3(g)} + H_2SO_{4(g)} \rightleftharpoons 2NH_4^+ + SO_4^{2-} \tag{2}$$

$$NH_{3(g)} + HNO_{3(g)} \rightleftharpoons NH_4^+ + NO_3^- \tag{3}$$

The restricted mobilization imposed in Indonesia in 2020–2021 due to the COVID-19 pandemic led to fewer vehicles on the road and cutting down on most

Table 1 The descriptive statistics of inorganic aerosol concentrations in an urban region of Bandung

Inorganic aerosols	Wet	Dry	Meteorology	Wet	Dry
SO_4^{2-} (µg/m^3)	$5.8^{a} \pm 2.3^{b}$ $1.1^{c} - 11.6^{d}$ 40^{e}	$6.9^{a} \pm 2.7^{b}$ $0.4^{c} - 13.1^{d}$ 28^{e}	*T* (°C)	$23.5^{a} \pm 0.5^{b}$ $22.4^{c} - 24.9^{d}$ 40^{e}	$23.8^{a} \pm 0.4^{b}$ $22.9^{c} - 24.5^{d}$ 27^{e}
NO_3^- (µg/m^3)	$1.8^{a} \pm 0.98^{b}$ $0.3^{c} - 5.6^{d}$ 40^{e}	$2.1^{a} \pm 0.7^{b}$ $0.1^{c} - 3.3^{d}$ 28^{e}	RH (%)	$82.9^{a} \pm 3.8^{b}$ $73.7^{c} - 89.4^{d}$ 40^{e}	$76.0^{a} \pm 6.2^{b}$ $66.4^{c} - 90.8^{d}$ 27^{d}
Cl^- (µg/m^3)	$0.3^{a} \pm 0.1^{b}$ $0.1^{c} - 0.7^{d}$ 40^{e}	$0.4^{a} \pm 0.2^{b}$ $0.02^{c} - 0.7^{d}$ 28^{e}	WS (m/s)	$2.2^{a} \pm 0.6^{b}$ $1.1^{a} - 3.5^{b}$ 40^{e}	$1.9^{a} \pm 0.3^{b}$ $1.4^{c} - 2.6^{d}$ 27^{e}
NH_4^+ (µg/m^3)	$1.5^{a} \pm 0.8^{b}$ $0.4^{c} - 3.5^{d}$ 40^{e}	$1.8^{a} \pm 0.8^{b}$ $0.2^{c} - 4.1^{d}$ 28^{e}	PR (mm)	$0.04^{a} \pm 0.03^{b}$ $0.0^{c} - 0.12^{d}$ 40^{e}	$0.03^{a} \pm 0.04^{b}$ $0.0^{c} - 0.2^{d}$ 27^{e}
21.7bNa^+ (µg/m^3)	$0.4^{a} \pm 0.3^{b}$ $0.1^{c} - 1.4^{d}$ 40^{e}	$0.6^{a} \pm 0.2^{b}$ $0.01^{c} - 1.1^{d}$ 28^{e}	SR (W/m^2)	$146.3^{a} \pm 25.7^{b}$ $93.4^{c} - 220.7^{d}$ 40^{e}	$135.6^{a} \pm 21.7^{b}$ $84.5^{c} - 175.7^{d}$ 27^{e}
K^+ (µg/m^3)	$0.3^{a} \pm 0.1^{b}$ $0.1^{c} - 0.6^{d}$ 40^{e}	$0.5^{a} \pm 0.2^{b}$ $0.04^{c} - 1.2^{d}$ 28^{e}			
Mg^{2+} (µg/m^3)	$0.1^{a} \pm 0.1^{b}$ $0.02^{c} - 0.4^{d}$ 40^{e}	$0.1^{a} \pm 0.04^{b}$ $0.02^{c} - 0.2^{d}$ 28^{e}			
Ca^{2+} (µg/m^3)	$1.1^{a} \pm 0.4^{b}$ $0.1^{c} - 2.4^{d}$ 40^{e}	$1.2^{a} \pm 0.4^{b}$ $0.2^{c} - 1.97^{d}$ 28^{e}			

Note superscript a = average, b = standard deviation, c = minimum, d = maximum, e = number of data

industrial activities. The SNA is expected to be lower than the years before the pandemic because their precursor emissions are also lower. Further study is needed to justify this statement.

The seasonal meteorological data was calculated by averaging the data of the corresponding months in 2020–2021, and the results are in Table 1 right. We reported that the wet season in urban Bandung was slightly more humid, windier, wetter, and sunnier than the dry season. However, the dry season was slightly hotter than the wet season. Indonesia experienced a weak to moderate La Nina in 2020–2021, resulting in a wetter dry season.

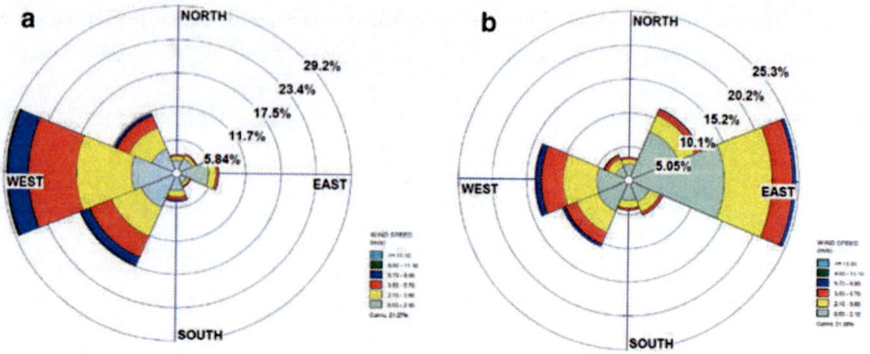

Fig. 2 Wind rose for wind speed and wind direction in **a** wet and **b** dry seasons

3.2 Meteorological Factors Versus Inorganic Aerosol Concentration

Fig. 2 shows the wind rose plot of wind direction and speed in wet and dry seasons. During the wet season, the dominant winds, about 28.4% of the time, came from western Bandung, where a toll road connecting Bandung–Jakarta and very dense industrial and residential areas are located (see Fig. 2a). The wind blew from the west dominantly at speeds 0.5–2.10 m/s (8.18% of the time), 2.10–3.60 m/s (8.22% of the time), and 3.60–5.70 m/s (8.1% of the time). On the contrary, in the dry season, the dominant winds, about 25% of the time, came from eastern Bandung, where also there are heavy transportation, industrial, and residential areas (see Fig. 2b). The wind blew from the east dominantly at speed 0.5–2.10 m/s (14.1% of the time), 2.10–3.60 m/s (7.1% of the time), and 3.60–5.70 m/s (3.0% of the time).

Table 2 presents the non-parametric Spearman's rho correlation coefficient (r) values between seasonal inorganic aerosol concentrations and the corresponding average of temperature (T), relative humidity (RH), wind speed (WS), precipitation (PR), and solar radiation (SR). T significantly inhibited Cl^- during the wet ($r = -0.33$, $p < 0.05$) but enhanced K^+ during the dry ($r = -0.4$, $p < 0.01$) seasons. RH significantly promoted SO_4^{2-} ($r = 0.37, p < 0.05$), NO_3^- ($r = 0.31, p < 0.05$), K^+ ($r = 0{,}38, p < 0.05$), and Ca^{2+} ($r = 0.41, p < 0.05$) in the wet season, but inhibit Cl^- ($r = -0.44, p < 0.05$), Na^+ ($r = -0.42, p < 0.01$), Mg^{2+} ($r = -0.28, p < 0.05$) in dry seasons. RH promoted NH_4^+ stronger in dry ($r = 0.52$, $p < 0.01$) than in wet ($r = 0.44, p < 0.01$) and led to a higher concentration of NH_4^+ in dry than in wet seasons. Low T and high RH are more favorable to forming ammonium nitrate (NH_4NO_3) [9, 10].

PR enhanced NH_4^+ ($r = 0.43$, $p < 0.01$) but reduced Na^+ ($r = -0.38, p < 0.05$) in the dry season. Significant negative correlation between NH_4^+ and SR ($r = -0.31$, $p < 0.05$), but a positive between NH_4^+ and RH ($r = 0.52$, $p < 0.01$) in the dry season showed that heterogeneous oxidation under high relative humidity was more critical for NH_4^+ generation than photochemical oxidation was [10]. On the

Table 2 Spearman rho's correlation coefficient values (r) between inorganic aerosol concentrations and selected meteorological variables

Ion concentration	Meteorological data				
	T (°C)	RH (%)	WS (m/s)	PR (mm)	SR (W/m^2)
SO_4^{2-} (µg/m^3)	$0.09^{(1)}$ $0.11^{(2)}$	$0.37*^{(1)}$ $0.24^{(2)}$	$-0.25^{(1)}$ $-0.22^{(2)}$	$0.24^{(1)}$ $0.23^{(2)}$	$0.09^{(1)}$ $-0.19^{(2)}$
NO_3^- (µg/m^3)	$-0.08^{(1)}$ $0.09^{(2)}$	$0.31*^{(1)}$ $0.1^{(2)}$	$0.14^{(1)}$ $-0.18^{(2)}$	$-0.06^{(1)}$ $0.09^{(2)}$	$0.17^{(1)}$ $0.00^{(2)}$
Cl^- (µg/m^3)	$-0.33*^{(1)}$ $-0.15^{(2)}$	$0.02^{(1)}$ $-0.44**^{(2)}$	$0.42**^{(1)}$ $0.48**^{(2)}$	$-0.3^{(1)}$ $-0.25^{(2)}$	$-0.25^{(1)}$ $0.40**^{(2)}$
NH_4^+ (µg/m^3)	$0.01^{(1)}$ $0.05^{(2)}$	$0.44**^{(1)}$ $0.52**^{(2)}$	$-0.25^{(1)}$ $0.43*^{(2)}$	$0.19^{(1)}$ $0.43**^{(2)}$	$0.10^{(1)}$ $-0.31*^{(2)}$
Na^+ (µg/m^3)	$0.07^{(1)}$ $0.1^{(2)}$	$-0.12^{(1)}$ $-0.42**^{(2)}$	$0.48**^{(1)}$ $0.34*^{(2)}$	$-0.24^{(1)}$ $-0.38*^{(2)}$	$0.21^{(1)}$ $0.26^{(2)}$
K^+ (µg/m^3)	$-0.14^{(1)}$ $0.40**^{(2)}$	$0.38*^{(1)}$ $-0.13^{(2)}$	$-0.5**^{(1)}$ $0.04^{(2)}$	$0.25^{(1)}$ $-0.11^{(2)}$	$-0.26^{(1)}$ $0.24^{(2)}$
Mg^{2+} (µg/m^3)	$0.1^{(1)}$ $0.2^{(2)}$	$0.06^{(1)}$ $-0.28*^{(2)}$	$0.26^{(1)}$ $0.23^{(2)}$	$-0.13^{(1)}$ $-0.27^{(2)}$	$0.3^{(1)}$ $0.16^{(2)}$
Ca^{2+} (µg/m^3)	$-0.07^{(1)}$ $0.06^{(2)}$	$0.41*^{(1)}$ $0.08^{(2)}$	$-0.31^{(1)}$ $-0.09^{(2)}$	$0.01^{(1)}$ $-0.01^{(2)}$	$-0.10^{(1)}$ $-0.15^{(2)}$

Note * correlation is significant at the 0.05 level (2-tailed)
**correlation is significant at the 0.01 level (2-tailed)
Superscript (1) = wet, (2) = dry

contrary, the significant positive correlation between Cl^- and SR ($r = 0.40, p < 0.01$) but negative between Cl^- and RH ($r = -0.44$, $p < 0.01$) in the dry season showed that photochemical oxidation was more critical to Cl^- formation than heterogeneous oxidation under high relative humidity [4]. The reduced ventilation associated with the reduction in SR amount and lower temperature increased the SNA concentrations at the surface [11].

WS significantly supported Cl^- in wet ($r = 0.42, p < 0.01$) and dry ($r = 0.48, p < 0.01$) seasons. WS also enhanced Na^+ in the wet ($r = 0.48, p < 0.01$) and dry season ($r = 0.34, p < 0.05$). The source of Na^+ and Cl^- are mainly sea-salt aerosol. The strong wind can easily transport sea-salt aerosol hundreds of kilometers away from the sea to Bandung. The wet northwest monsoon strongly influenced the wet season and the southeast monsoon during the dry season [6]. We run backward particle trajectories using HYSPLIT by selecting January 1st, 2020, and July 1st, 2020, to represent the wet and dry seasons, respectively. As shown in Fig. 3, the result indicated that sea salt origin in the wet season was the northwestern part of the Java Sea (Fig. 3a) and, in the dry season, the southeastern part of the Indian Ocean (Fig. 3b).

WS inhibited the wet season's K^+ ($r = -0.5$, $p < 0.05$) and supported the dry season's NH_4^+ ($r = 0.43$, $p < 0.05$). The negative correlation between WS and K^+ could indicate the local source. On the contrary, the positive correlation between WS and NH_4^+ could indicate long-distance transport.

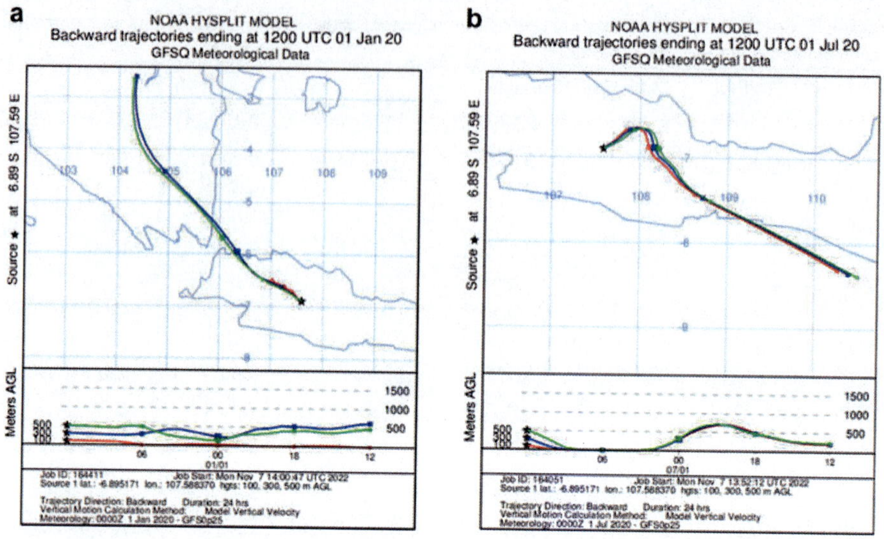

Fig. 3 The output of the backward trajectory model running by using HYSPLIT to simulate the long-distance source of Na^+ and Cl^- to Bandung in **a** wet and **b** dry seasons. A black star indicates the Bandung site

4 Conclusion

Sulfate, nitrate, and ammonium (SNA) dominated the inorganic aerosol composition collected during the wet and dry seasons in 2020–2021. The SNA contributed more than 75% of secondary inorganic aerosol, and their concentrations were higher in the dry than in the wet season. Relative humidity had an essential role in forming SNA in the dry season and promoted NH_4^+ stronger in the dry than in the wet season. Therefore, NH_4^+ had a higher possibility to increase under higher relative humidity in the dry season, thus increasing the possibility of haze formation. Heterogeneous oxidation under high relative humidity was more critical for NH_4^+ generation than photochemical oxidation was, but for Cl^- it was vice versa. The sources of Na^+, Cl^-, and NH_4^+ aerosols are long-distance transport, but K^+ is local. Sea salt originated from the northwestern part of the Java Sea in the wet season and, in the dry season, the southeastern part of the Indian Ocean.

Acknowledgements We thank the chemistry and observation laboratories—BRIN, for providing this study's acid deposition and surface meteorological data.

Author Contributions Wiwiek Setyawati performed the data processing and analysis and wrote the manuscript; Dyah Aries Tanti and Asri Indrawati conducted sampling in the field and testing in the laboratory; Saipul Hamdi, Atep Radiana and Risyanto contributed to meteorological data observation and processing, Sumaryati and Retno Puji Lestari assisted in data quality assurance.

References

1. Allen, S.A., Ana Godson, R.E.E., Micheal Ayodeji, S., Deborah, S.E., Ejike, O.M.: Secondary inorganic aerosols: impacts on the global climate system and human health. Biodiv. Int. J. **3**(6), 245–259 (2019). https://doi.org/10.15406/bij.2019.03.00152
2. Chen, Y., Wang, Y., Nenes, A., Wild, O., Song, S., Hu, D., Liu, D., He, J., Ruiz, L.H., Apte, J.S., Gunthe, S.S., Liu, P.: Ammonium chloride associated aerosol liquid water enhances haze in Delhi, India. Environ. Sci. Technol. **56**, 7163−7173 (2022). https://doi.org/10.1021/acs.est.2c00650
3. Wang, X., Wang, W., Yang, L., Gao, X., Nie, W., Yu, Y., Xu, P., Zhou, Y., Wang, Z.: The secondary formation of inorganic aerosols in the droplet mode through heterogeneous aqueous reactions under haze conditions. Atmos. Environ. **6**, 68–76 (2012). https://doi.org/10.1016/j.atmosenv.2012.09.029
4. Wang, Y., Wang, Y., Wang, L., Petäjä, T., Zha, Q., Gong, C., Li, S., Pan, Y., Hu, B., Xin, J., Kulmala, M.: Increased inorganic aerosol fraction contributes to air pollution and haze in China. Atmos. Chem. Phys. **19**, 5881–5888 (2019). https://doi.org/10.5194/acp-19-5881-2019
5. Setyawati, W., Aries Tanti, D., Indrawati, A. (2022). Air quality in the bandung basin of Indonesia as measured by Passive Sampler. In: Yulihastin, E., Abadi, P., Sitompul, P., Harjupa, W. (eds.) Proceedings of the International Conference on Radioscience, Equatorial Atmospheric Science and Environment and Humanosphere Science, 2021. Springer Proceedings in Physics, vol. 275. Springer, Singapore. https://doi.org/10.1007/978-981-19-0308-3_36
6. Aldrian, E., Dwi Susanto, R.: Identification of three dominant rainfall regions within Indonesia and their relationship to sea surface temperature. Int. J. Climatol. **23**, 1435–1452 (2003). https://doi.org/10.1002/joc.950
7. Wang, Y., Zhuang, G., Tang, A., Yuan, H., Sun, Y., Chen, S., Zheng, A.: The ion chemistry and the source of PM2.5 aerosol in Beijing. Atmos. Environ. **39**, 3771–3784 (2005). https://doi.org/10.1016/j.atmosenv.2005.03.013
8. Pitts, F., Pitts, J.N.: Atmospheric chemistry: fundamentals and experimental techniques. A Willey—Interscience publication, John Willey & Sons, Inc. USA (1986)
9. Fu, X., Wang, S., Chang, X., Cai, S., Xing, J., Hao, J.: Modeling analysis of secondary inorganic aerosols over China: pollution characteristics, and meteorological and dust impacts. Sci. Rep. **6**, 35992 (2016). https://doi.org/10.1038/srep35992
10. Squizzato, S., Masiol, M., Brunelli, A., Pistollato, M., Tarabotti, E., Rampazzo, G., Pavoni, B.: Factors determining the formation of secondary inorganic aerosol: a case study in the Po Valley (Italy). Atmos. Chem. Phys. **13**, 1927–1939 (2013). https://doi.org/10.5194/acp-13-1927-2013
11. Jung, J., Souri, A.H., Wong, D.C., Lee, S., Jeon, W., Kim, J., Choi, Y.: The impact of the direct effect of aerosols on meteorology and air quality using aerosol optical depth assimilation during the KORUS-AQ campaign. Geophys Res Atmos. **124**(14), 8303–8319 (2019). https://doi.org/10.1029/2019JD030641

The Development of DSS SRIKANDI Fifth Version

Emmanuel Adetya, Nani Cholianawati, Prawira Yudha Kombara, Sumaryati, and Ninong Komala

Abstract The Research Center for Climate and Atmosphere (PRIMA) of the National Research and Innovation Agency (BRIN) is developing an information system based on a website application for the composition and air quality of Indonesia's atmosphere. This application is called the Indonesia Atmospheric Composition Information System (DSS SRIKANDI). The version development of the DSS SRIKANDI that is being carried out is the fifth version. In this fifth version, the appearance of the web will be changed entirely with the addition of several new features. Some of these features are the prediction menu for atmospheric chemical parameters (O_3, CO, SO_2, $PM_{2.5}$, PM_{10}, and NO_2) derived from the WRF-Chem model has been run independently by PRIMA. Furthermore, in the observation menu present the aerosol optical depth (AOD) information. The AOD was obtained from the GEMS instrument. The GEMS instrument is the South Korean satellite, which provides aerosol optical depth measurement.

1 Introduction

In today's digital era, a piece of information spreads very rapidly. There are many platforms can be used to disseminate information. One of these platforms is a website-based application. Currently, many website-based applications that have been developed by various organizations to present information systems [1–3]. As one of the research organizations, the Research Center for Climate and Atmospheric (PRIMA) has developed an information system based on a website application called the Decision Supporting System of the Indonesian Atmospheric Composition Information

E. Adetya · N. Cholianawati · P. Y. Kombara (✉) · Sumaryati · N. Komala
Research Center for Climate and Atmosphere, National Research and Innovation Agency, Bandung, Indonesia
e-mail: praw005@brin.go.id

E. Adetya
e-mail: emma002@brin.go.id

A. Basit et al. (eds.), *Proceedings of the International Conference on Radioscience, Equatorial Atmospheric Science and Environment and Humanosphere Science*, Springer Proceedings in Physics 290, https://doi.org/10.1007/978-981-19-9768-6_11

System (DSS SRIKANDI). This DSS SRIKANDI displays information about the composition of the atmosphere in Indonesia.

The atmosphere is composed of several chemical elements. The most dominant component is nitrogen and oxygen, then the rest are several other components including pollutant gases in it [4]. Pollutant gases are usually dominant in the atmosphere near the surface, especially in large cities. Compounds in gaseous form that include pollutants include Carbon Monoxide (CO), Ozone (O_3), Sulfur Dioxide (SO_2), and Nitrogen Dioxide (NO_2). In addition to these gaseous compounds, there are also micro-sized solid particles such as $PM_{2.5}$ and PM_{10}.

Pollutant gases in the atmosphere will pollute the air usually inhaled by living things every day. If you continue to inhale it, it will accumulate in the body and can cause diseases, especially respiratory system diseases [5, 6]. Therefore, PRIMA developed the DSS SRIKANDI to monitor and present information on the concentration of these pollutant gases in the Indonesian atmosphere.

Currently, the development of the DSS SRIKANDI has reached the fifth version. The development of the DSS SRIKANDI involves the Ministry of Environment and Forestry to share information and data on air quality observations for making decisions related to forest fires. The development of the fifth version was carried out to update some of the features that existed in the previous version. In this paper, we describe the updates that are in the fifth version. In this fifth version, an overhaul is implemented in the website's appearance. In addition, updating and adding several features such as predictive information on pollutant concentrations sourced from the results of the operational running of the WRF-Chem model run by PRIMA and visualization of aerosol optical depth (AOD) parameters from South Korea's GEMS satellite imagery. In previous version, prediction information was obtained from operationally downloaded CAMS-Chem global model data. However, the spatial resolution of the CAMS-Chem prediction data is still roughly about 70 km, while the prediction data of the WRF-Chem model independently run by PRIMA has a higher spatial resolution of about 15 km. The update of several features is expected to improve the quality of this DSS SRIKANDI.

2 The DSS SRIKANDI Fifth Version Architecture's

The DSS SRIKANDI is built through several parts, as shown in Fig. 1. The first part is the database that organizes the files and data will be displayed on the website. The database application used in this fifth version is MySQL which is installed on the XAMPP application. The next part is the service module or website application. The website application used is Lokomedia CMS. Lokomedia CMS is an open-source CMS created and developed by an Indonesian named Lukmanul Hakim and his team. Initially, this CMS was intended to practice creating websites using a CMS. However, because it is in great demand, Lokomedia finally becomes a CMS officially that can be used to create websites with various needs such as e-commerce, agencies, personal websites, and so on. The Lokomedia CMS can be downloaded on the page: https://members.phpmu.com/contribution/detail/download-cms-lokomedia-161-versi-codeigniter. Lokomedia CMS is in great demand by website developers because it is PHP-based, easy to install on several OS, easy to use, and there is an online forum to discuss if you have problems during website development.

The next section is the Application Programming Interface (API). The fifth version of SRIKANDI DSS also uses an API to transfer files and data from the database to the interface for display. Then the last part is the frontend or the face of the fifth version of the DSS SRIKANDI website. The website's appearance has changed entirely compared to the previous version. A comparison of the appearance of versions four and five are shown in Fig. 2.

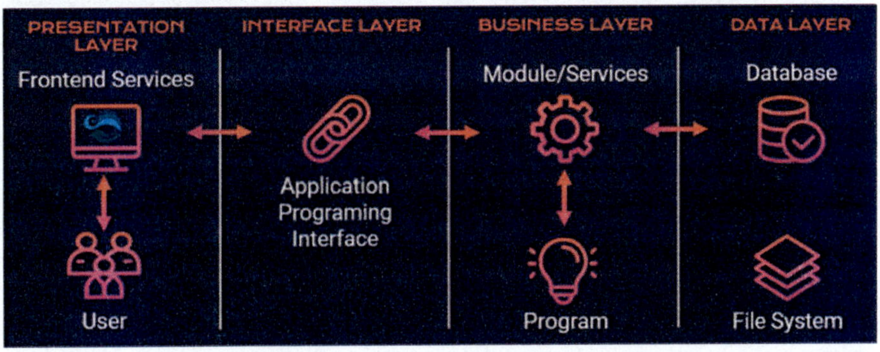

Fig. 1 The architecture of DSS SRIKANDI fifth version

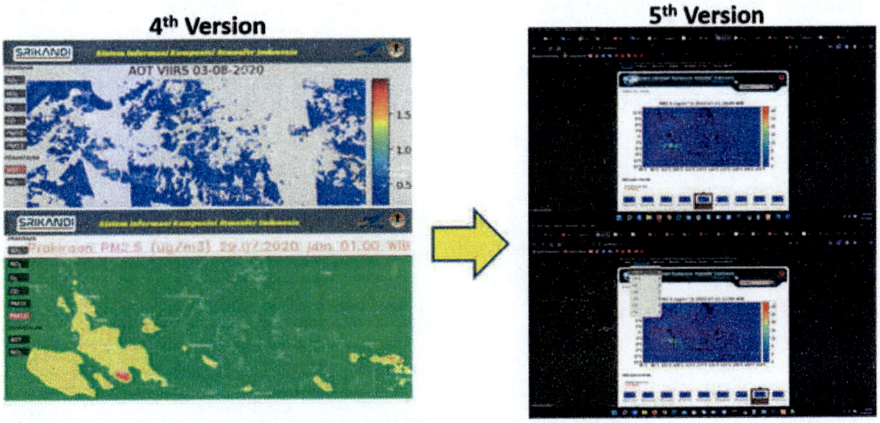

Fig. 2 The comparison of the DSS SRIKANDI version fourth and fifth

3 The Interface of DSS SRIKANDI Fifth Version

The fifth version of the DSS SRIKANDI interface change is shown at a glance in Fig. 2. For more details about the new interface, shown in Fig. 3.

The DSS SRIKANDI version five is still in the prototype stage, so it cannot be accessed by the public. For a display, as shown in Fig. 3, DSS SRIKANDI version 5 has several main menus consist of Home, Forecast, Observation, Karhutla, Knowledge Sharing, and Tentang Srikandi. The main display or when the Home menu is accessed, will display the forecast for the $PM_{2.5}$ parameter, as shown in Fig. 3a.

The Forecast menu has several sub-menus, which are the parameters of the WRF-Chem model output as shown in Fig. 3b. The WRF-Chem model is a mesoscale numerical model that combines the weather and chemical components of the atmosphere [6–9]. The parameters of WRF-Chem output are $PM_{2.5}$, PM_{10}, CO, O_3, NO, NO_2, and SO_2. The forecasts of these parameters displayed are the results of predictions for the next 24 h for all parts of Indonesia with a spatial resolution of about 15 km and time intervals per hour. In previous version, the forecast data was obtained from the downloaded CAMS-Chem model. The purpose of replacing the data source for forecasting with PRIMA's self-running WRF-Chem model is because the CAMS-Chem model has a lower spatial resolution. In addition, other goals are to reduce dependence on external models and increase independence.

Then, the observation menu also has a sub-menu in the form of GEMS (Fig. 3c). The fifth version of the SRIKANDI DSS displays observation data from South Korea's GEMS satellite imagery for AOD parameters. The satellite image displayed has a spatial resolution of 7 km $\times$ 8 km and an hourly time interval from 07.45 to 14.45 in Indonesian time. Actually, the GEMS satellite imagery has other parameters besides AOD, such as SO_2, O_3, cloud cover, and so on. However, this GEMS satellite is still relatively new and is still being developed. So, for the beginning, the AOD parameters are presented first. The reason of GEMS satellite imagery was chosen to be presented because the GEMS satellite is the first geostationary satellite to measure atmospheric chemistry. Therefore, it can produce images in the tropical area, especially in Indonesia with a higher time resolution. Furthermore, the menu about DSS Srikandi (Fig. 3d) has several sub-menus, such as DSS Srikandi, Tim Srikandi, and Contact Us.

4 Further Development

The fifth version of the DSS SRIKANDI website can still be developed further. For further plans, in the observation menu section, especially in the GEMS satellite section, other parameters will also be displayed, such as O_3, SO_2, and cloud cover. In addition to the GEMS satellite, observation data from other instruments, such as the AQMS station owned by KLHK, and the surface observations from the PANDORA instrument will also be displayed. In addition, the Karhutla menu will also display

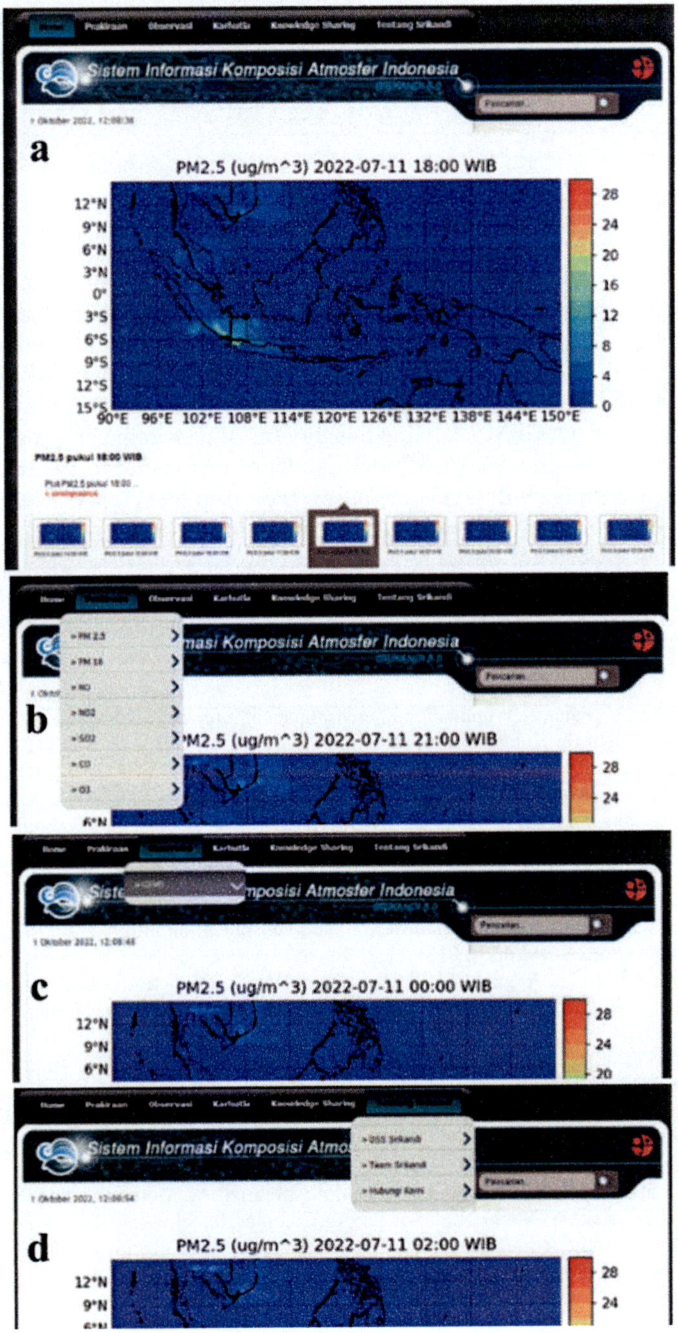

Fig. 3 The interface of DSS SRIKANDI fifth version

the predicted distribution of $PM_{2.5}$ and CO gas obtained from the WRF-Chem model when a forest fire occurs. In the initial development, it will display predictions for the Sumatra area first.

5 Summary

The DSS SRIKANDI website-based application has been developed up to the fifth version. In this fifth version, a comprehensive display update was carried out using the Lokomedia CMS. Several new data are presented in the fifth version of the SRIKANDI DSS, such as prediction of pollutant parameters such as $PM_{2.5}$, PM_{10}, CO, O_3, NO, NO_2, and SO_2 obtained from the WRF-Chem model. In addition, observation data from GEMS satellite imagery for AOD parameter is also displayed on the observation menu. This DSS SRIKANDI can still be developed further. The next development plan will add some observation data from other instruments and predict the distribution of $PM_{2.5}$ and CO when forest fires occur.

References

1. Elekné Fodor, V., Pájer, J.: Application of environmental information systems in environmental impact assessment (in Hungary). Acta Silv. Lignaria Hungarica. **13**, 55–67 (2017). https://doi.org/10.1515/aslh-2017-0004
2. Karayannis, V.G., Kagawa, T., Motosugi, G., Takeda, Y., Ichimura, M., Sasaki, A., Heterostructures, M., Saku, T., Horikoshi, Y., Tokura, Y.: Open Access Proceedings Journal of Physics_Conference Series_Enhanced Reader.pdf (2018)
3. Tudela, J., Martínez, M., Valdivia, R., Romo, J., Portillo, M., Rangel, R.: Enhanced Reader.pdf (2010)
4. Lutgens, F.K., Tarbuck, E.J., Herman, R.L.: The atmosphere: an introduction to meteorology. Pearson Education, New York (2018)
5. Fitri, D.W., Afifah, N., Anggarani, S.M.D., Chamidah, N.: Prediction concentration of PM2.5 in Surabaya using ordinary Kriging method. AIP Conf. Proc. **2329** (2021). https://doi.org/10.1063/5.0042284
6. Ismiyati, Marlita, D., Saidah, D.: Pencemaran Udara Akibat Emisi Gas Buang Kendaraan Bermotor. J. Manaj. Transp. Logistik. **1**, 241–248 (2014)
7. Baklanov, A., Schlünzen, K., Suppan, P., Baldasano, J., Brunner, D., Aksoyoglu, S., Carmichael, G., Douros, J., Flemming, J., Forkel, R., Galmarini, S., Gauss, M., Grell, G., Hirtl, M., Joffre, S., Jorba, O., Kaas, E., Kaasik, M., Kallos, G., Kong, X., Korsholm, U., Kurganskiy, A., Kushta, J., Lohmann, U., Mahura, A., Manders-Groot, A., Maurizi, A., Moussiopoulos, N., Rao, S.T., Savage, N., Seigneur, C., Sokhi, R.S., Solazzo, E., Solomos, S., Sørensen, B., Tsegas, G., Vignati, E., Vogel, B., Zhang, Y.: Online coupled regional meteorology chemistry models in Europe: current status and prospects. Atmos. Chem. Phys. **14**, 317–398 (2014). https://doi.org/10.5194/acp-14-317-2014

8. Fast, J.D., Gustafson, W.I., Easter, R.C., Zaveri, R.A., Barnard, J.C., Chapman, E.G., Grell, G.A., Peckham, S.E.: Evolution of ozone, particulates, and aerosol direct radiative forcing in the vicinity of Houston using a fully coupled meteorology-chemistry-aerosol model. J. Geophys. Res. Atmos. **111**, 1–29 (2006). https://doi.org/10.1029/2005JD006721
9. Grell, G.A., Peckham, S.E., Schmitz, R., McKeen, S.A., Frost, G., Skamarock, W.C., Eder, B.: Fully coupled "online" chemistry within the WRF model. Atmos. Environ. **39**, 6957–6975 (2005). https://doi.org/10.1016/j.atmosenv.2005.04.027

Characteristics of PM2.5 Concentration at Bandung and Palembang from December 2019 to November 2021 Measured by Low-Cost Sensor

Saipul Hamdi, Sumaryati, Asri Indrawati, Atep Radiana, Syahril Rizal, Ridho Pratama, Fahmi Rahmatia, Yutaka Matsumi, and Takashi Shibata

Abstract This paper discusses the daily characteristics of PM2.5 at Bandung and Palembang measured by using low-cost sensor. Two low-cost sensors developed by Nagoya University has been being operated in Bandung and Palembang continuously and simultaneously. The data recorded every minute and has been grouped considering the season of DJF, MAM, JJA, and SON. To get the potential dispersion of PM2.5 concentration in Bandung and Palembang, the surface meteorological data have been used. It is found the different primary peak at both of locations. In Bandung, the primary peak appeared in the morning whereas the primary peak in Palembang appeared in the evening. In general, it is also found the annual concentration of PM2.5 at Bandung higher than in Palembang and both are overvalued of upper limit of Indonesia Government rule.

1 Introduction

We use the term of particulate to describe the dispersion of solid and liquid particles at the atmosphere in normal condition. This particulate has dimension of more than a molecule size ($\pm$ 0.0002 μm) and less than 500 μm. The particle in this range of size has residence time in suspension from several seconds to several months. The source of atmospheric particulate is natural source like volcanic eruption, forest fire, sea spray, and anthropogenic sources like industrial activities and transportation.

S. Hamdi (✉) · Sumaryati · A. Indrawati · A. Radiana · R. Pratama · F. Rahmatia
National Research and Innovation Agency, Jl DR Djundjunan No 133, Bandung, Indonesia
e-mail: saipulh@yahoo.com

S. Rizal
Bina Darma University, Jl Jendral Ahmad Yani No 3, Palembang, Indonesia

Y. Matsumi · T. Shibata
Nagoya University, Nagoya, Japan

A. Basit et al. (eds.), *Proceedings of the International Conference on Radioscience, Equatorial Atmospheric Science and Environment and Humanosphere Science*, Springer Proceedings in Physics 290, https://doi.org/10.1007/978-981-19-9768-6_12

About 50% of total emission of dust in the atmosphere is PM10. This emission contributes significantly for atmospheric warming. This emission not only disperses the atmosphere but also absorbs the solar radiation [6]. The dispersion of PM10 in the atmosphere is determined by physical condition of the source and physical and chemical processes of the source in the atmosphere. The contaminants follow the wind and accumulate at the destination. When the particulate enters the human respirator for longer time, it will damage the functions of human respiratory system and possibility make severe damage. The raising of PM10 concentration can correspond to season or weather. Muhaimin [13] reported a higher total concentration of particulate during the dry season than the rainy season. It is suspected by removal processes of contaminant or by washing out by rainwater [14].

Chaloulakou et al. [3] say that PM10 negatively correlates to wind speed with a value of −0.43 that value shows an inverse correlation between the average wind speed and the average of PM10 concentration, which means the fastest wind speed will cause a decrease in the concentration of PM10. This value also agrees with the results of Maraziotis et al. [12], who found a correlation between particulate concentration and wind speed that correlation value can be determined to enough correlation, which means the level concentration of PM10 is not only dependent on wind speed but also the other factors, i.e., residence time, temperature, and intensity of rain occurred on the observed days. Jallad et al. [8] say that a positive correlation between PM10 and surface temperature shows dependence of surface temperature on the processes of formation of new particle from gas to be particle. Generally, the concentration of particulates is higher on a hot day cause of the high activity of photochemical reactions on the day with high intensity of solar radiation [3].

Forest and land fires events which are causing an increase in the air pollution index, dominantly produces CO or carbon dioxide [7], CO_2, hydrocarbon, and particulate. As amount as 85% of greenhouse gas emission in Indonesia was produced in peatland fires [9]. Most of the area in Indonesia is forest, and potential of forest fire events in this area is inversely proportional to the area of forest. Sumatera Selatan is one of the ten provinces with the most hotspot for forest fire events in 2013 [2]. In the forest fire events in 2015 at South Sumatera, it was found that the longest distance of dispersion of PM10 is about 100 km for the highland natural primary forest, 60 km for secondary lowland natural forest, 170 km for community forests, and 50 km for secondary natural peatland [17]. Besides that, it is known that the height of plume has strong and inversely correlation to wind speed, which means that the low speed of wind will cause higher plume. Increasing wind speed will cause the atmospheric condition became more neutrally [1].

The monitoring of PM2.5 at Bandung basin using low-cost sensor shows the result that decreasing PBL at nighttime will cause local pollutants and transboundary to have mixed and tend to go down to the ground surface. This phenomenon can be correlated to higher concentration of PM2.5 at nighttime [16]. From sample test of black carbon content (BC) in Bandung and Lembang shows concentration of BC in Bandung (urban area) is higher than 60–70% compared to Lembang (suburban area), and the peak of BC concentration occurred in the rainy season [10]. This research aims to study about different PM2.5 characteristics at Bandung and Palembang.

Table 1 Classification of wind speed, Beaufort scale [15]

No	Windspeed (m/s)	Category	Color
1	0 – 0.3	Calm	
2	0.3 – 1.5	Light air	
3	1.5 – 3.3	Light breeze	
4	3.3 – 5.5	Gentle breeze	
5	5.5 – 8	Moderate breeze	
6	≥ 8	Fresh breeze	

2 Data and Method

Two PM2.5 palm-sized optical sensors [18] have been being operated at Bandung (6°53′35.52″ S 107°35′12.16″ E) and Palembang (2°59′59.71″ S 104°47′38.75″ E) continuously and simultaneously. In Bandung City, the sensor is located at near road site which has heavy traffic during office hours. In Palembang City, the sensor is located inside the residence area of Perumahan Srimas Buana Rindang, about 1 km southern side of Musi River, or around 2300 m from Bina Darma University. Both sensors are doing the recording of data every ten minutes. We calculate the daily average of PM2.5, then separate it into four groups correlated to the season in Indonesia: DJF (rainy season), MAM, JJA (dry season), and SON. The period of data is taken from December 2019 to November 2021.

To support the primary data, the wind speed and wind direction for Palembang were taken from the URL https://www.ogimet.com, whereas Bandung was taken from in-situ measurement by AWS. The wind speed classification used in this study has shown in Table 1 refers to Beaufort scale. Wind speed and wind direction data were analyzed by using freeware WRPLOT to find the frequency and dominant speed.

3 Results and Discussions

The surface PM2.5 daily concentration at Bandung and Palembang in the period of 2019–2021 is shown in Fig. 1. The horizontal axis is the Day Number (DN) or Julian Day and started from December 1, 2019, (DN = 335) to November 30, 2021 (DN = 334). In general, it is shown that the PM2.5 concentration at Bandung is higher than in Palembang in two years monitoring but have similar pattern. Statistically, the average concentration of PM2.5 is shown in Table 2. The annual PM2.5 concentration in Bandung is higher than the PM2.5 concentration in Palembang. With the upper limit of PM2.5 as high as 15 $\mu g/m^3$ then the PM2.5 concentration for both locations have over the limit by Indonesian Government rule of PP no 22/2021 about maintenance of protection and management of environment. Meanwhile, it was found that six days of event in 2020 had a daily concentration of more than 55 $\mu g/m^3$ in August, one day in July, also one day in April. It was not found that the daily concentration at Palembang over the upper limit of concentration.

Marsidi et al. [11] reported the average concentration of PM2.5 was valued as 10.77 μg/m^3 at Plaju (Palembang) in fourth week of 2018 that result seems lower compared to the results of this study. The differences can appear correlated to corresponding factors such as relative humidity, surface temperature, and other artificial factors. The increasing temperature of environment will cause decreasing in particulate exposure and increasing of relative humidity will decreasing the particulate exposure. The effect of temperature and relative humidity correlating to changes in particulate concentration has been reported by Jallad et al. [8] and Csavina et al. [4].

Figure 2 shows the PM2.5 daily average concentration at Bandung for four seasons in two years of monitoring. It can be seen the appearance of sharp primary peak in the morning about 8 AM, with a variation in height. The secondary peak appears at the evening/nighttime, and this shape is very slope; also, the value is not as high as the primary peak. At both of MAM and JJA seasons, the concentration of primary peak

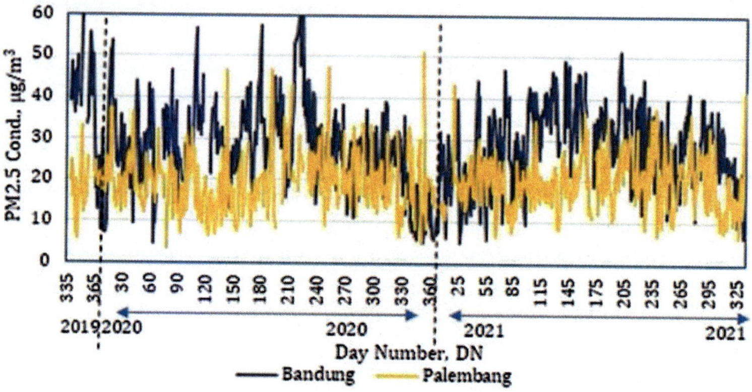

Fig. 1 Time series of daily concentration of PM2.5 at Bandung and Palembang from Dec 1, 2019, to November 30, 2021. The horizontal axis is day number (December 1 = day number 335, and November 30 = day number 334)

Table 2 Statistical of PM2.5 daily concentration measured by using low-cost sensor

		Bandung	Palembang
2020	Average	27.7	19.7
	Maximum	66.4	50.9
	Minimum	4.4	3.5
	Count	351	331
2021	Average	28.8	19.9
	Maximum	51.3	43.3
	Minimum	4.9	6.3
	Count	257	255

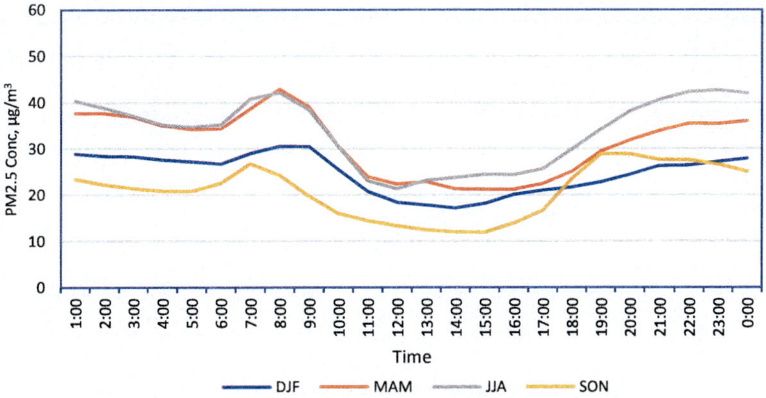

Fig. 2 Characteristics of PM2.5 concentration in Bandung at four seasons (DJF, MAM, JJA, and SON), calculated from daily concentration of December 2019 to November 2021

is relatively higher than the concentration in rest of the others season. MAM means a transition season from rainy season to dry season. The primary peak of PM2.5 at JJA season in Bandung is like MAM season, but PM2.5 concentration at JJA season is relatively higher than MAM season especially after mid-day. The temperature profile in March and July has similar characteristics, and maximum temperature differences at noon as amount as 1 °C. This difference can explain the difference of primary daily average (on 8–9 AM) not corresponding to temperature increment and not corresponding directly to increasing of the planetary boundary layer.

Figure 3 shows the PM2.5 daily average concentration at Palembang for four seasons in two years of monitoring. The profiles were found by calculating PM2.5 concentration from December 2019 to November 2021. Different from concentration of PM2.5 in Bandung, the primary peak in Palembang is seen and sharp after sunset, or around 7 PM. In each seasons, the primary peak of PM2.5 has higher value compared to the secondary peak, and the primary peak appears in the morning around 7 AM. Referring to result about study for surface temperature profile in Palembang [5] says that the highest temperature at Palembang generally does not happen when sun reach culminates. However, it is delayed for around 2 h, then it is suspected that has a good correlation between PM2.5 concentration and temperature of the environment.

The characteristics of wind speed and wind direction in Bandung are shown in Fig. 4, and the wind speed in Fig. 4 can be referred to as color in Table 1. The data on wind have been grouped into four seasons (DJF, MAM, JJA, and SON). For Fig. 4, wind data were measured from December 2019 to November 2021 using automatic weather station operated in the exact location. Generally, wind direction in Bandung for DJF and MAM flows from the west-northwest, so the surround area for this direction must have a dominant role in dispersing PM2.5 from and heading to Bandung. The dominant wind speed in DJF season is 1.5–3.3 m/s which amount to

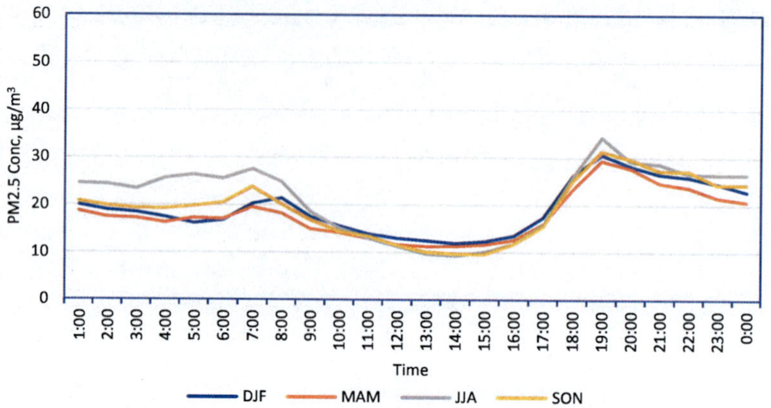

Fig. 3 Characteristics of PM2.5 concentration at Palembang for four seasons (DJF, MAM, JJA, and SON), calculated from daily concentration of December 2019 to November 2021

38.6%, and the dominant wind direction is flowing from the west-northwest, amount 43.4%. The dominant wind direction in MAM is looks like DJF, and wind speed is 0.3–3.3 m/s. The opposite condition has occurred in JJA and SON seasons. In these seasons, the wind is distributed randomly from western and eastern and wind class 1.5–3.3 m/s dominantly. It was not found that the direction of wind heading blew from the North and South direction in JJA and SON. Tangkuban Parahu Mountain at northern side of Bandung does not contribute to PM2.5 concentration in Bandung. As well as southern site of Bandung, there are some industrial area and rice fields area do not contribute to PM2.5 concentration in Bandung.

Figure 5 shows the distribution of wind speed and direction in Palembang for four seasons. Generally, it can be seen that the wind blows from northwest and southeast. These areas which have potentially sent particulate from southeast area are Kayu Agung, Pampangan, and Tulung Selapan. These areas are known as frequent forest fire especially in JJA and SON seasons. Consequently, if these areas get forest fired events in JJA and SON seasons, then these areas have high potential risk to disperse the pollutants reach Palembang City and contribute to increasing the PM2.5 concentration in Palembang.

Meanwhile, in DJF season, which is marked as the rainy season, the wind blows from northwest Palembang. The areas located in northwest of Palembang are Sungai Lilin and Banyung Lencir, also neighboring country Singapore and Malaysia. However, Zhang [19] said wind shear seems more critical than wind speed for modulating ground surface PM2.5 concentration.

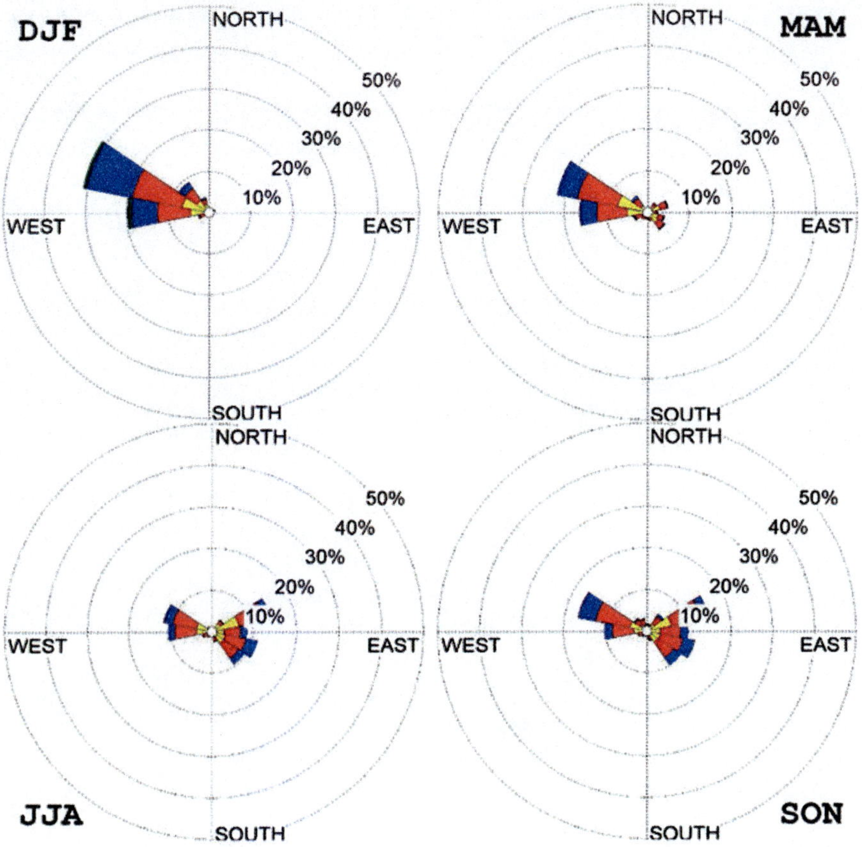

Fig. 4 Pattern of wind speed and direction in Bandung for four seasons calculated using data from December 2019 to November 2021. *Source of data* LAPAN

The wind speed in Palembang is commonly in range of 1.5–3.3 m/s or category of a light breeze, and a small concentration of wind speed is categorized as a gentle breeze. In light breeze category, wind fell on the face, leaves rustled, wind vane moved by wind, and small wavelets on sea. In gentle breeze category, some leaves and small twigs move constantly, light flags extended, and large wavelets on the sea.

4 Conclusions

By the monitoring of PM2.5 concentration at Bandung and Palembang, it has been found different diurnal pattern of PM2.5 concentration for these two locations. It can be seen the primary peak in each locations are contrarily. The primary peak of PM2.5 concentration at Bandung occurred in the morning after sun rises. In contrast, the

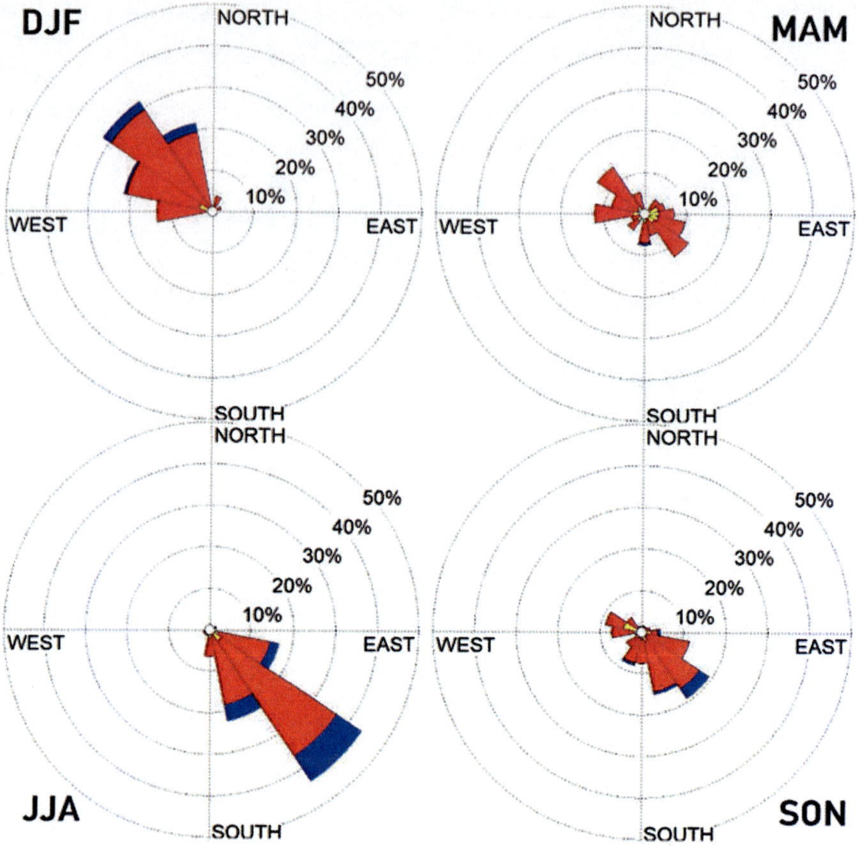

Fig. 5 Pattern of wind speed and wind direction at Palembang for four seasons calculated using data from December 2019 to November 2021. *Source of data* https://www.ogimet.com

primary PM2.5 concentration at Palembang occurred in the evening after sun sets. Meanwhile, the long-time series of monitoring shows that the daily average of PM2.5 concentration (24 h) at Palembang is lower compared to the daily average of PM2.5 concentration at Bandung. The yearly average concentration of PM2.5 in Bandung and Palembang has over the threshold value 15 $\mu g/m^3$ (determined by Indonesian Government). The characteristics of seasonal concentration of PM2.5 does not show significant differences for these two locations. Primary peak of seasonal variation of PM2.5 concentration in Bandung is higher concentration than in Palembang.

Acknowledgements Center for Atmospheric Research and Technology of The Indonesia National Research and Innovation Agency (BRIN) have fully supported this research. The appreciation is delivered to Nagoya University which has provided two PM2.5 palm-size optical sensors.

References

1. Alex, D.V.: Air dispersion modeling. John & Sons Inc., New Jersey (2014)
2. Badan Nasional Penanggulangan Bencana: Rencana Kontinjensi Nasional Menghadapi Ancaman Bencana Asap Akibat Kebakaran Hutan dan Lahan. Badan Nasional Penanggulangan Bencana, Jakarta (2013)
3. Chaloulakou, A., Kassomenos, P., Spyrellis, N., Demokritou, P., Koutrakis, P.: Measurements of PM10 and PM2,5 particle concentrations in Athens, Greece. J. Atmos. Environ. **37**, 649–660 (2003)
4. Csavina, J., Field, J., Félix, O., Corral-Avitiac, A.Y., Sáeza, A.E., Bettertond, E.A.: Effect of wind speed and relative humidity on atmospheric dust concentrations in semi-arid climates. Sci. Total Environ. **15**(487), 82–90 (2013)
5. Hamdi, S.: Karakter suhu udara kota Palembang berdasarkan pengukuran di Kampus Universitas Bina Darma. Sains Atmosfer Teknologi dan Aplikasinya, CV Andira Bandung, halaman, pp 13–25. ISBN 978-979-1458-68-9 (2013)
6. Haywood, J., Boucher, O.: Estimates of the direct and indirect radiative forcing due to tropospheric aerosols: a review. Rev. Geophys. **38**(4), 513–543 (2000)
7. Heriyanto, E., Syaufina, L., Sobri, M.: Forecasting simulation of smoke dispersion from forest and land fires in Indonesia. Procedia Environ. Sci. **24**, 111–119 (2015)
8. Jallad, F.A., Katheeri, E.A., Omar, M.A.: Concentrations of particulate matter and their relationships with meteorological variables. Sustain. Environ. Res. **23**(3), 191–198 (2013)
9. Kementerian Lingkungan Hidup: Pengkajian Baku Mutu Kualitas Udara Ambien Lampiran PP No. 41 Tahun 1999. Jakarta (2011)
10. Lestiani, D.D., Santoso, M., Hidayat, A.: Karakteristik black carbon partikulat udara halus PM2.5 di Bandung dan Lembang 2004–2005. In: Prosiding Seminar Nasional Sains dan Teknologi Nullar PTNBR_BATAN, Bandung (2007)
11. Marsidi, Kamaluddin, M.T., Kurdi, F.N., Novrikasari: Correlation of climate to particulate matter in Palembang. Pollut. Res. **37**(2), 279–284 (2018)
12. Maraziotis, E., Sarotis, L., Marazioti, C., Marazioti, P.: Statistical analysis of inhalable (PM10) and fine particles (PM2,5) concentrations in urban region of Patras, Greece. Glob. NEST J. **10**(2), 123131 (2008)
13. Muhaimin: Permodelan dispersi polutan udara dari aktivitas PLTU Cirebon pada musim kemarau dan hujan serta penggunaan 2 cerobong asap [tesis]. Universitas Gajah Mada, Yogyakarta (2014)
14. Puspitasari, A.D.: Pola spasial pencemaran udara dari sumber pencemar PLTU dan PLTGU Muara Karang [skripsi]. Universitas Indonesia, Depok (2011)
15. Stewart, R.H.: Introduction to Physical Oceanography. Department of Oceanography Texas A & M University, Texas (2008)
16. Sya'bani, A., Chandra, I., Majid, L.I., Vaicdan, F., Barus, R.A.A., Abdurrachman. A., Salam, R.A.: Pemantauan konsentrasi PM2.5 dan CO_2 berbasis low-cost sensor secara real-time di cekungan Udara Bandung Raya. Jurnal Teknologi Lingkungan **21**(1), 009–015 (2020)
17. Tampubolon, A.P.C., Boedisantoso, R.: Analisis persebaran polutan karbon monoksida dan partikulat dari kebakaran hutan di Sumatera Selatan. Jurnal Teknik ITS **5**(2), C160–C165 (2016)
18. Nakayama, T., Matsumi, Y., Kawahito, K., Watabe, Y.: Development and evaluation of a palm sized optical PM2.5 sensor. Aerosol Sci. Technol. **52**(1), 2–12. https://doi.org/10.1080/02786826.2017.1375078 (2018)
19. Zhang, Y., Guo, J., Yang, Y., Wang, Y., Yim, S.H.L.: Vertical wind shear modulates particulate matter pollutions: a perspective from radar wind profiler observations in Beijing, China. Rem. Sens. **12**(3), 546. https://doi.org/10.3390/rs12030546 (2020)

Air Pollution Impact During Forest Fire 2019 Over Sumatra, Indonesia

Prawira Yudha Kombara, Waluyo Eko Cahyono, Wiwiek Setyawati, Hana Listi Fitriana, Emmanuel Adetya, and Alvin Pratama

Abstract In 2019, there was a strong positive Indian Ocean Dipole (IOD) phenomenon and forest and land fire in several areas of Sumatra. The positive IOD phenomenon makes the Sumatra region drier, and the rainfall decreases from normal conditions so that the forest and land fires that occur are getting worse. The environmental impacts, especially the pollution impact of the forest and land fires and the 2019 positive IOD in Sumatra, have been studied in this study by analyzing the spatial pattern of the aerosol optical depth (AOD), particulate matter $(PM)_{2.5}$, and carbon monoxide (CO) parameters. Spatial analysis was carried out from August to November 2019. Then, the general results obtained that the maximum concentrations for the parameters AOD, $PM_{2.5}$, and CO occurred in September 2019. The three parameters spread to several provinces in Sumatra following the wind direction. Several areas affected by the 2019 forest and land fires are the provinces of Riau, Jambi, and South Sumatra. In October, the South Sumatra region became the area with the highest concentrations of $PM_{2.5}$ and CO compared to other regions. This indicates that the forest and land fires incident was centered in South Sumatra Province in October 2019.

P. Y. Kombara (✉) · W. E. Cahyono · W. Setyawati · E. Adetya
Research Center for Climate and Atmosphere, National Research and Innovation Agency, Bandung, Indonesia
e-mail: praw005@brin.go.id

H. L. Fitriana
Research Center for Remote Sensing, National Research and Innovation Agency, Jakarta, Indonesia

A. Pratama
Department of Atmospheric and Planetary Science, Sumatera Institute of Technology, Lampung, Indonesia

A. Basit et al. (eds.), *Proceedings of the International Conference on Radioscience, Equatorial Atmospheric Science and Environment and Humanosphere Science*, Springer Proceedings in Physics 290, https://doi.org/10.1007/978-981-19-9768-6_13

1 Introduction

In 2019, there were forest and land fires in several areas of Sumatra [1]. The impact of the 2019 forest fires caused a smoke hazard and increased the concentrations of aerosols in the form of $PM_{2.5}$ and toxic gases such as carbon monoxide (CO). However, unlike in 2015, when the El Nino phenomenon also occurred, in the 2019 forest fires, the Indian Ocean Dipole (IOD) phenomenon occurred. IOD is a phenomenon that shows the variability of sea surface temperature (SST) between the western and eastern parts of the Indian Ocean. The name Dipole refers to the two sides of the Indian Ocean. When this phenomenon occurs, it will affect the climate and weather around the Indian Ocean. When the west side, of the Indian Ocean experiences an increase in SST, on the east side, there will be a decrease in the value of SST and is called a positive IOD and vice versa. While the positive IOD is active, the west side of the Indian Ocean will receive an abundant supply of water vapor from normal conditions due to the presence of a low-pressure center and make the wind blow and carry water vapor, resulting in excessive precipitation and vice versa on the east side will experience a shortage of water vapor supply and less precipitation. The opposite will happen when the negative IOD is active. The latest IOD phenomenon occurred in 2019 when it was autumn in the Northern Hemisphere. At that time, there was a strong positive IOD [2–4].

The strong positive IOD phenomenon that occurred in 2019 has resulted in the western part of Indonesia, especially some areas in Sumatra, experienced drought [5, 6]. This drought is exacerbated by the occurrence of forest and land fires. However, it should be emphasized that the forest fires incident was first caused by land acquisition using the burning method and caused forest fires occurred. Therefore, the IOD phenomenon only exacerbates the impact of forest fires.

Several previous studies have examined the impacts of the 2019 IOD phenomenon on the weather in the Indonesian Maritime Continent (BMI), especially in Sumatra. Those previous studies explain the impact of drought and reduced rainfall [5, 6]. However, environmental conditions such as the variability of aerosol concentration and the distribution of smoke haze during forest fires, which coincide with the IOD phenomenon, have not been thoroughly studied. Therefore, in this paper, we try to examine the environmental impact of the atmosphere during the forest fires and the 2019 IOD phenomenon. The environmental impacts of the atmosphere that will be investigated are an increase in the concentration of aerosols, $PM_{2.5}$, and toxic gases such as CO.

2 Data and Method

In this study, we used several data such as

1. Data The Modern-Era Retrospective Analysis for Research and Applications, Version 2 (MERRA-2) can be downloaded from: https://disc.gsfc.nasa.gov/datasets/M2IMNXGAS_5.12.4/summary?keywords=aod. This data is used to visualize aerosol optical depth (AOD) and carbon monoxide (CO) parameters with a monthly and spatial–temporal resolution of 0.5° × 0.625°. MERRA-2 is a data reanalysis developed by NASA through the GMAO project [7]. In this study, we used MERRA-2 data from August–November 2019.
2. Sea surface temperature data using data from COBE-SST which can be downloaded on the page: https://psl.noaa.gov/data/gridded/data.cobe.html. COBE-SST data is the input data for sea surface temperature parameters used by JMA to run the climate data assimilation system (JCDAS) model. The COBE-SST data has a spatial resolution of one degree by one degree and a monthly temporal resolution.
3. ECMWF Atmospheric Composition Reanalysis 4 (EAC4) data to visualize Particulate Matter 2.5 μm (PM2.5) parameters which can be downloaded on the page: https://ads.atmosphere.copernicus.eu/cdsapp#!/dataset/cams-global-reanalysis-eac4-monthly?tab=form. This data is the fourth version of global reanalysis data released by ECMWF. The data we use has a spatial resolution of 0.75° and a monthly temporal resolution.
4. Rainfall data from CPC Global Unified Gauge-Based Analysis of Daily Precipitation can be downloaded at https://psl.noaa.gov/data/gridded/data.cpc.globalprecip.html. The data has a monthly temporal and spatial resolution of 2.5° by 2.5°.

In this study, we used two main methods, consist of the calculation of the IOD index and exploratory data analysis (EDA). The determination of the IOD index in 2019 was carried out by calculating the difference in SST anomalies in the western (50E-70E, 10S-10N) and eastern (90E-110E, 10S-0S) of Indian Ocean [8]. Furthermore, analyze the spatial pattern of SST anomalies, rainfall anomalies, AOD, $PM_{2.5}$, and CO using the EDA method when the 2019 strong positive IOD forest fires occurred.

3 Result

3.1 The Strong Positive IOD and Forest Fire 2019

By using the method of Saji et al. [8] can be obtained an index that shows the condition of IOD in 2019. Figure 1 shows the IOD index with a positive value above 1.5 for the months of October–November. The strengthening of the index value started in September and reached its peak in October with a value of 1.8. This indicates that there was a strong positive IOD phenomenon in 2019. The spatial pattern of the SST anomaly as shown in Fig. 2 confirms the results shown by the IOD index.

Figure 2 shows the spatial pattern of SST anomalies from August to November 2019. In August, positive SST anomalies were still around the central and eastern

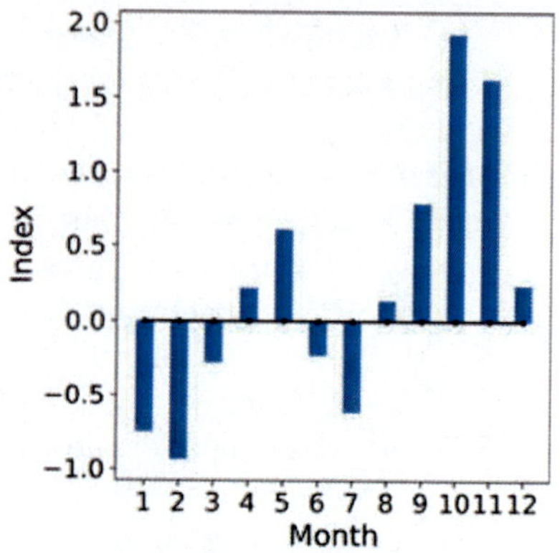

Fig. 1 IOD 2019 index based on Saji et al. [8] method

areas of the Indian Ocean. In September, the positive anomaly area shifted to the west of the Indian Ocean. Then, from October to November, the area of the positive anomaly shifted in the western part of the Indian Ocean, resulting in a positive IOD phenomenon. The impact felt on the eastern part of the Indian Ocean is becoming drier due to the cooling of SST, resulting in decreased convective activity and reduced precipitation as shown in Fig. 3.

Figure 3 shows precipitation anomalies in some areas of Sumatra, Indonesia. It can be seen there are some regions have negative values in September and October 2019. A negative value means some areas of Sumatra experienced a precipitation deficit as previously the study result was done by Lestari et al. [6]. This deficit triggers a drought in some regions of Sumatra. Along with the occurrence of positive IOD 2019, in some Sumatra regions, there were forest fires.

Based on Fig. 4, we can see at several red points, there is smoke (orange circle) due to forest fire. The smoke moves follow the wind direction to other regions that do not experience forest fire. Figure 4 is one example that shows a forest fire event on September 22, 2019. The red points in the figure represent the hot spots when forest fires occurred. The negative anomaly of precipitation makes the forest fire last longer. In addition, the dry air due to positive IOD increases the intensity of forest fires. All of this can cause the degradation of air quality.

3.2 *Environment Atmosphere Impact*

When forest fire occurs, the smoke will appear and emit an amount of aerosol into the atmosphere. Besides aerosol, this smoke contains particulate, and toxic gases that can harm human health, especially the respiratory system [9]. Figure 5 shows aerosol is emitted from smoke when forest fire 2019 occurs in Sumatra. Here, the aerosol is represented by aerosol optical depth (AOD) parameter. In August, the AOD in most parts of Sumatra had a magnitude of around 0.5 and moved toward the north. However, in September, the AOD magnitude increased drastically to reach the value

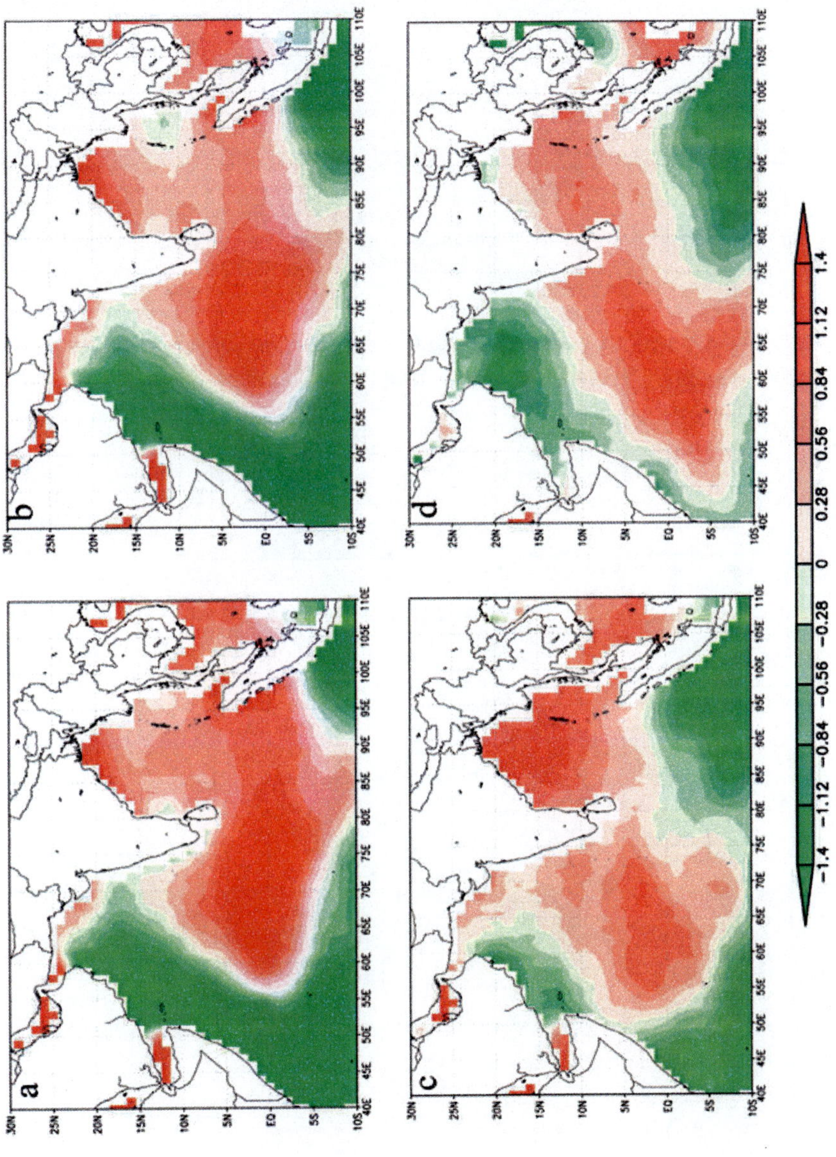

Fig. 2 SST anomaly during strong IOD positive 2019, **a** August, **b** September, **c** October, and **d** November

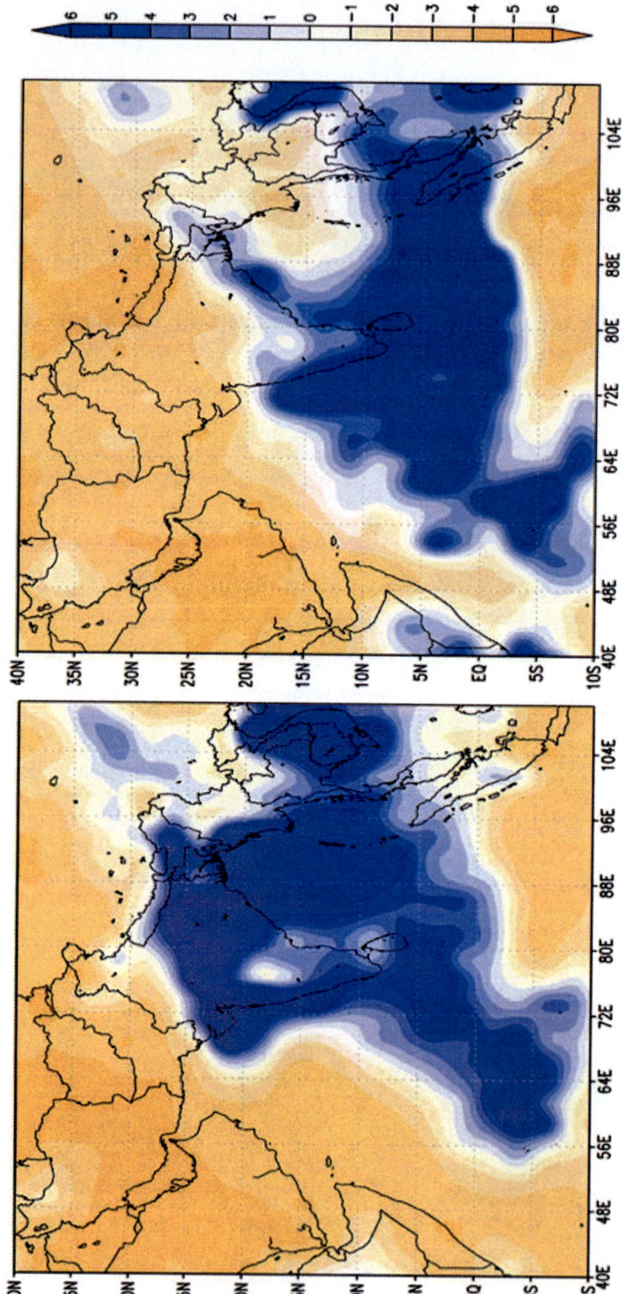

Fig. 3 Precipitation anomaly on September (left) and October (right) 2019

Fig. 4 Forest fire on September 22, 2019, on Sumatra based on Suomi/VIIRS satellite capture (https://worldview.earthdata.nasa.gov/)

of one. The Sumatran area that has a high AOD value is Riau Province. We can say that there were severe forest fires during September 2019 as shown in Fig. 4. This high magnitude of AOD in September resembles the forest fire event in a few years ago when strong El Nino in 2015. At that time, the AOD magnitude reached greater than one based on Cahyono et al. [10].

As shown in Fig. 5b, the high AOD value moved to the north and northwest toward the Malacca Peninsula and West Sumatra Province due to was being carried away by the wind. In September, the background wind conditions still showed the Australian monsoon pattern which means the prevailing wind was southeasterly [11]. Therefore, neighboring countries such as Malaysia and then West Sumatra Province experienced a decrease in air quality and visibility. A high AOD value can be interpreted as smoke due to forest fires. In the next month, the AOD value decreased as in August. This decrease in the AOD value could have been caused by the rain in October. If we see the rainfall anomaly in Fig. 3 on the right, most of the Sumatra region has a positive rainfall anomaly in October. Then in November, the condition of AOD showed a decline in value than in October. We can say that in November, the forest fires started to subside.

Previously, Fig. 5 shows the AOD pattern during the 2019 forest fires in Sumatra. This section shows the distribution pattern of one type of aerosol, namely $PM_{2.5}$ during the 2019 forest fires in Sumatra, as shown in Fig. 6. In August (Fig. 6a), $PM_{2.5}$ concentrations began to increase, and the maximum magnitude of $PM_{2.5}$ occurs in the province of Jambi. $PM_{2.5}$ spread to Riau Province and a little to Malaysia. In

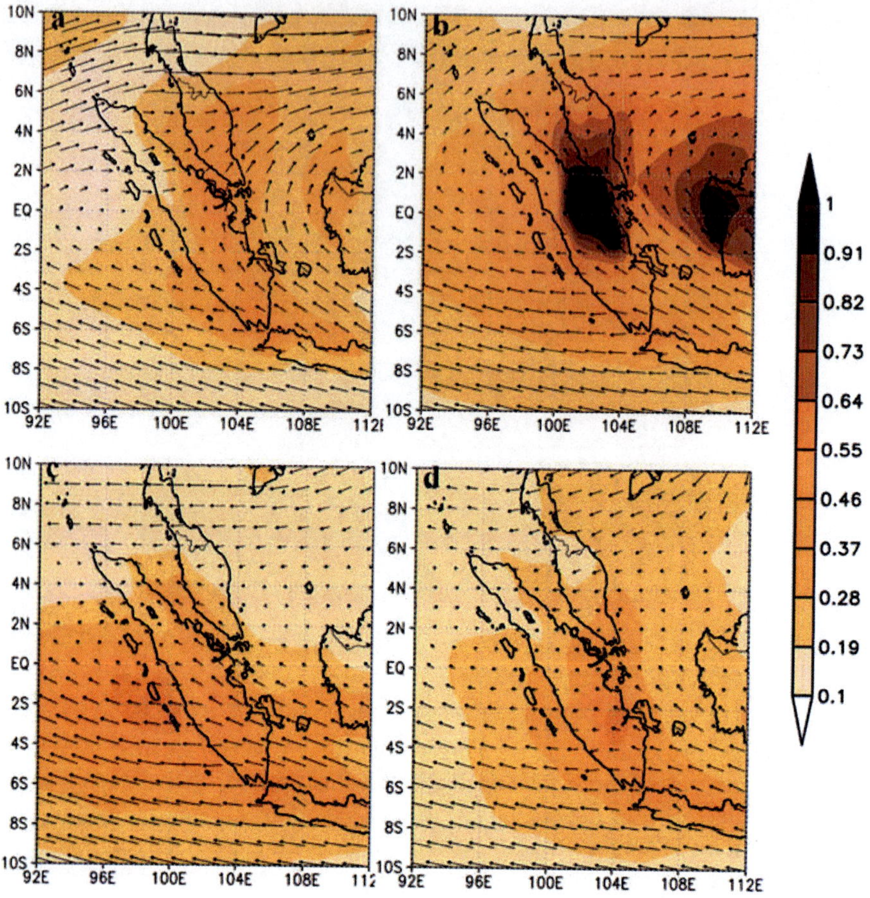

Fig. 5 Aerosol optical depth (AOD) is overlayed by vector wind at 850 hPa during a forest fire and strong positive IOD 2019, **a** August, **b** September, **c** October, and **d** November

the following month of September, it can be said that the forest fires reached its peak. In September (Fig. 6b), there are two centers of magnitude $PM_{2.5}$. The two centers occurred in Jambi and Riau Provinces. The monthly mean of $PM_{2.5}$ in both provinces was exceeding by 300 ug/m^3. From these two provinces, $PM_{2.5}$ spread to surrounding provinces such as Lampung, Bengkulu, North Sumatra, West Sumatra, and South Sumatra. The areas near to the forest fire location will be higher than usual [12]. Then in October (Fig. 6c), the center of the $PM_{2.5}$ magnitude moved to South Sumatra Province, while in Jambi and Riau Provinces it had subsided. From South Sumatra Province, $PM_{2.5}$ spread to surrounding areas such as Lampung, Jambi, and Bengkulu Provinces. In October, it can be said that more forest fires occurred in South Sumatra Province. Then in November (Fig. 6d), the concentration of $PM_{2.5}$ was seen to be quite reduced but still concentrated in South Sumatra Province. As in

Fig. 5d, based on Fig. 6d can also be interpreted that in November, the forest fires had started to subside.

Figure 7 shows the distribution of CO gas during the 2019 forest fires in Sumatra. In August (Fig. 7a), CO concentrations increased and were concentrated in Jambi and Riau Provinces. Then in September (Fig. 7b), CO concentrations increased drastically with the center of magnitude in Jambi Province. This increase shows that forest fires occur significantly. CO gas spreads to the provinces around Jambi because it is carried away by the wind. Other provinces affected include the provinces of Riau, South Sumatra, Bengkulu, to the provinces of North Sumatra. CO gas from these forest fires also spread to Malaysia. In the following month, October (Fig. 7c), CO gas concentrations began to decline, and the center of magnitude was in Jambi and South Sumatra Provinces. For October, there is a difference in the pattern between the parameters AOD, $PM_{2.5}$, and CO. This month, $PM_{2.5}$ and CO parameters still

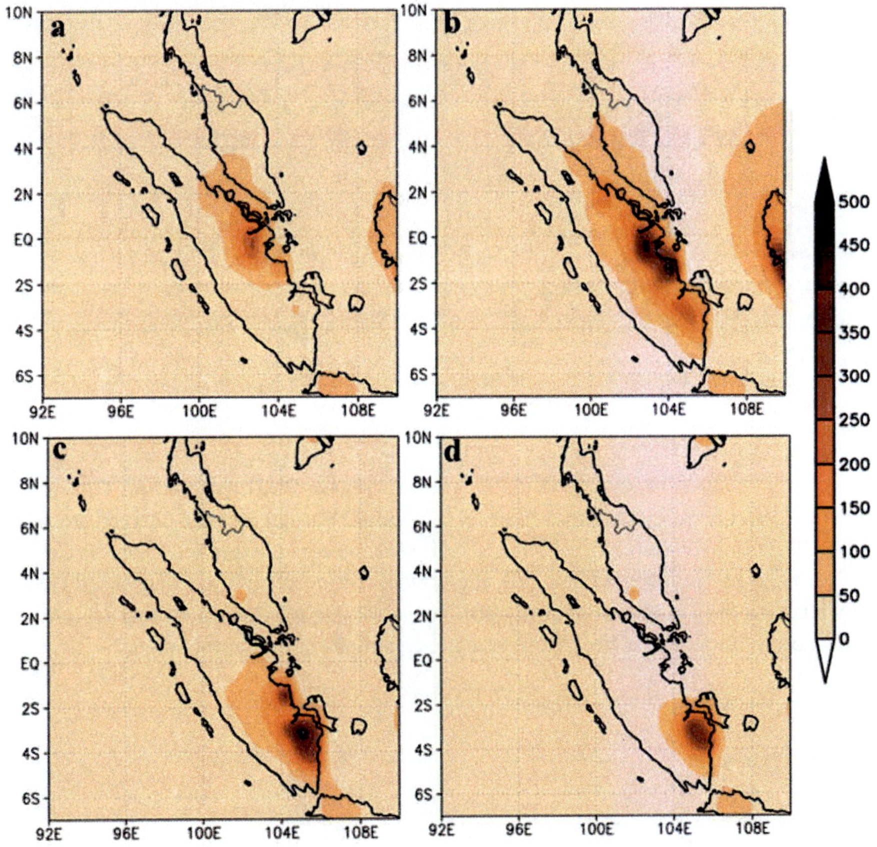

Fig. 6 $PM_{2.5}$ dispersion during forest fire 2019, **a** August, **b** September, **c** October, and **d** November

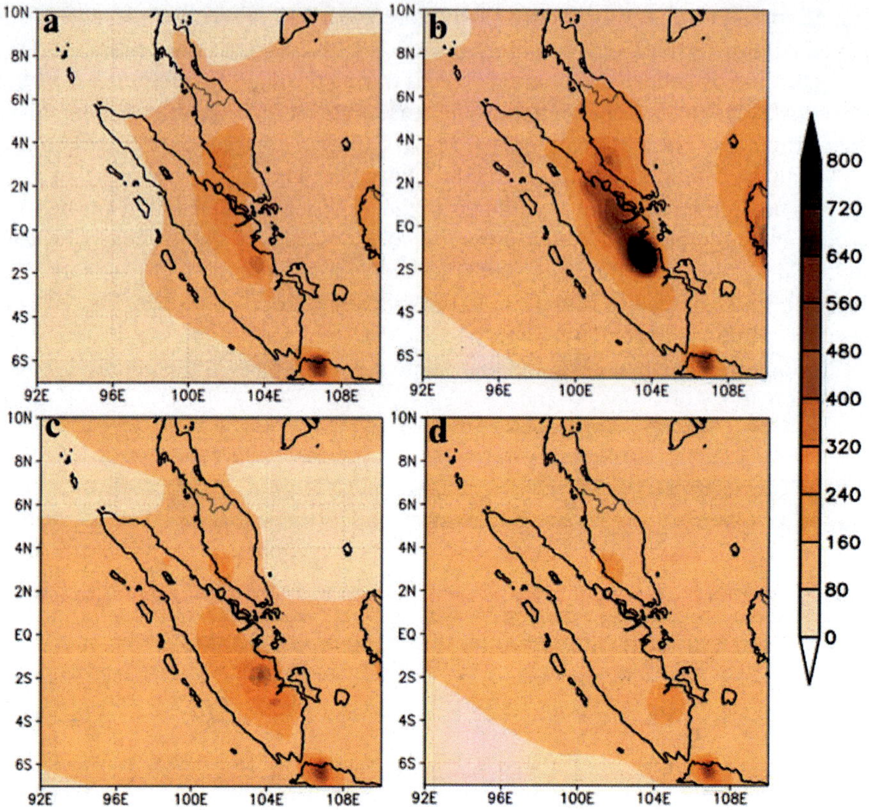

Fig. 7 Carbon monoxide (CO) dispersion during forest fire 2019, **a** August, **b** September, **c** October, and **d** November

showed high concentrations, but the AOD parameter showed the opposite. This difference may be caused by different models. The AOD parameters are derived from the MERRA-2 model, whereas the $PM_{2.5}$ and CO parameters are derived from the CAMS model. Then in November (Fig. 7d), the concentration of CO gas decreased drastically and was only concentrated in South Sumatra. As in Figs. 5d, 6d and 7d can also be interpreted that forest fires were starting to subside in November.

4 Conclusion

In 2019, strong positive IOD and forest and land fires occurred, and there was an increase in the concentration of AOD, $PM_{2.5}$, and CO gas in several provinces of Sumatra. The increase in concentration began to occur in August and reached its peak from September to October. The concentration centers of AOD, $PM_{2.5}$, and CO occurred in the provinces of Riau, Jambi, and South Sumatra. At the peak of the forest fire event, aerosols in the form of $PM_{2.5}$ and CO gas spread to Bengkulu, Lampung, West Sumatra, North Sumatra, and even Malaysia. In September, most forest and land fires occurred in the provinces of Riau and Jambi, while in October most fires occurred in South Sumatra. As soon as November entered, the concentrations of AOD, $PM_{2.5}$, and CO began to decline so it can be interpreted that in this month the forest and land fires began to subside.

Acknowledgements We want to thank Rumah Program Kebencanaan DIPA OR KM BRIN No. SP DIPA-124.01.1.690501/2022 for funding research support. Besides, we acknowledge the use of imagery from the NASA Worldview application (https://worldview.earthdata.nasa.gov), part of the NASA Earth Observing System Data and Information System (EOSDIS).

References

1. Indonesia, C.: Membandingkan Karhutla di Indonesia Pada 2015 dan 2019. https://www.cnnindonesia.com/teknologi/20190918104533-199-431485/membandingkan-karhutla-di-indonesia-pada-2015-dan-2019%0A
2. Lu, B., Ren, H.L.: What caused the extreme Indian Ocean Dipole event in 2019? Geophys. Res. Lett. **47**, 1–8 (2020). https://doi.org/10.1029/2020GL087768
3. Ratna, S.B., Cherchi, A., Osborn, T.J., Joshi, M., Uppara, U.: The extreme positive Indian Ocean Dipole of 2019 and associated Indian summer monsoon rainfall response (2021)
4. Chang, D.Y., Yoon, J., Lelieveld, J., Park, S.K., Yum, S.S., Kim, J., Jeong, S.: Direct radiative forcing of biomass burning aerosols from the extensive Australian wildfires in 2019–2020. Environ. Res. Lett. **16** (2021). https://doi.org/10.1088/1748-9326/abecfe
5. Mardiansyah, W., Setiabudidaya, D., Khakim, M.Y.N., Yustian, I., Dahlan, Z., Iskandar, I.: On the Influence of Enso and IOD on rainfall variability over the Musi Basin, South Sumatra. Sci. Technol. Indones. **3**, 157 (2018). https://doi.org/10.26554/sti.2018.3.4.157-163
6. Lestari, D.O., Sutriyono, E., Sabaruddin, S., Iskandar, I.: Respective influences of Indian Ocean Dipole and El Niño-Southern Oscillation on Indonesian precipitation. J. Math. Fundam. Sci. **50**, 257–272 (2018). https://doi.org/10.5614/j.math.fund.sci.2018.50.3.3
7. Services, G.E.S.D. and I., DISC), 151 Center (GES: Global Modeling and Assimilation Office (GMAO): MERRA-2 tavgM_2d_chm_Nx: 2d, 149 monthly mean, time-averaged, single-level, assimilation, carbon monoxide and ozone diagnos150 tics V5.12.4. https://disc.gsfc.nasa.gov/
8. Saji, N.H., Vinayachandran, P.N.: A dipole mode in the tropical Indian Ocean. Lett. Nat. **401**, 360–364 (1999)
9. Fitri, D.W., Afifah, N., Anggarani, S.M.D., Chamidah, N.: Prediction concentration of PM2.5 in Surabaya using ordinary Kriging method. AIP Conf. Proc. **2329** (2021). https://doi.org/10.1063/5.0042284
10. Eko Cahyono, W., Setyawati, W., Hamdi, S., Cholianawati, N., Yudha Kombara, P., Julian Sari, W.: Observations of aerosol optical properties during tropical forest fires in Indonesia. Mater. Today Proc. **63**, S445–S450 (2022). https://doi.org/10.1016/j.matpr.2022.04.113
11. Xian, P., Reid, J.S., Atwood, S.A., Johnson, R.S., Hyer, E.J., Westphal, D.L., Sessions, W.: Smoke aerosol transport patterns over the Maritime Continent. Atmos. Res. **122**, 469–485 (2013). https://doi.org/10.1016/j.atmosres.2012.05.006
12. Sharma, A., Valdes, A.C.F., Lee, Y.: Impact of wildfires on meteorology and air quality (PM2.5 and O_3) over Western United States during September 2017. Atmosphere (Basel) **13** (2022). https://doi.org/10.3390/atmos13020262

Influence Impregnation Method in the Structure of Bimetallic Ni-Zn/ZrO_2 Catalyst

Fildzah 'Adany, Kiky Corneliasari Sembiring, Mustofa Amirullah, and Reva Edra Nugraha

Abstract Ni-based catalysts are an alternative to precious metal catalysts. However, monometallic Ni catalysts still have weaknesses, such as being easily deactivated and being able to agglomerate if the Ni concentration is too high. This can be overcome by adding a second metal (bimetallic catalyst) in which the structure and morphology of the bimetallic catalyst are strongly influenced by the method used. The wet-impregnation method is used to obtain a supported bimetallic catalyst where the order of metal carrying will affect the structure and morphology and is specific for each material. On the Ni-Zn/ZrO_2 catalyst, which has been synthesized by co-impregnation (Cat-1) and sequentially (Cat-2) method, XRD results show that both catalysts have specific peaks of ZrO_2, NiO, and ZnO. However, the Cat-2 catalyst has lower crystallinity and smaller NiO crystallite size (14.79 nm) than Cat-1 (19.45 nm). Micrograph results show that both catalysts have irregular particle shapes. Cat-1 has a smaller particle size and a more homogeneous metal dispersion than Cat-2. The elemental analysis shows that metal concentration at the surface is close to the theoretical (10%). Further research with further characterization and application to a specific chemical reaction is needed to determine which of the two catalysts perform better. This is because each reaction has specific requirements, and each catalyst will produce different results in each reaction even though they have the same properties.

F. 'Adany (✉) · K. C. Sembiring
Research Center for Chemistry, National Research and Innovation Agency (BRIN), South Tangerang, Banten, Indonesia
e-mail: Fildzah.adany@brin.go.id

M. Amirullah
Research Center for Smart Mechatronics, National Research and Innovation Agency (BRIN), Bandung, West Java, Indonesia

R. E. Nugraha
Department of Chemical Engineering, Faculty of Engineering, Universitas Pembangunan Nasional "Veteran" Jawa Timur, Surabaya, East Java, Indonesia

A. Basit et al. (eds.), *Proceedings of the International Conference on Radioscience, Equatorial Atmospheric Science and Environment and Humanosphere Science*, Springer Proceedings in Physics 290, https://doi.org/10.1007/978-981-19-9768-6_14

1 Introduction

Catalysts are materials that have an essential role in various chemical reactions so that reactions can run more quickly and selectively. Because of this, catalysts have an essential role in various chemical industries, and more than 80% of the chemical industry uses catalysts [1]. In addition to having a vital role in the industry, catalysts also play an essential role in efforts to mitigate global warming and climate change. Catalysts play a role in producing clean fuels such as hydrogen or biogas [2].

There are two types of catalysts based on the reaction system's phase: homogeneous and heterogeneous catalysts. Heterogeneous catalysts have advantages compared to homogeneous catalysts, which are easy to separate from products and reactants, have high stability, and are more environmentally friendly [3]. Heterogeneous catalysts include zeolite, ion exchange resin, carbon-based catalysts, mineral–clay, and metal oxide [4].

Metal oxides are important as catalysts because they are essential in most chemical reactions, such as petroleum refining, energy conversion, hydrogenation, and biomass conversion. Due to their potential application in most industrial chemical processes, metal oxides are of great interest today to researchers. Metal oxide catalysts used in various chemical industries include TiO_2, ZnO, ZrO_2, porous and mesoporous metal oxides, multi-component mixed oxides, polyoxometallates (POMs), phosphates, and perovskites [1].

Metal oxide catalysts are usually modified to have high activity and selectivity in a reaction. Modification of the catalyst usually aims to improve its characteristics, such as hydrophobicity, specific surface area, basicity, and acidity [5]. The catalyst can be modified by depositing a metal on the surface of the support. This modification can increase selectivity and activity by increasing the catalyst's active site [6]. One of the catalyst preparation methods is wet-impregnation, which is often used because it is easy and does not require high costs [7].

Nickel-based catalysts, such as Ni/Al_2O_3, $Ni/MgAl_2O_4$, Ni/ZrO_2, and Ni/SBA15, are well-established catalysts that can be used as an alternative to precious metal-based catalysts (Pd, Pt, and Ru) due to their high catalytic activity, low cost, and widespread availability [8–10]. On the other hand, the Ni-based catalyst is easily deactivated during the catalysis process, necessitating a relatively large amount of H_2 in the hydrogenation reaction. Furthermore, the high Ni concentration in the support causes agglomeration that will inhibit the reaction process [11].

However, research shows that the ability of this monometallic Ni catalyst can be improved by adding other metals (bimetallic catalyst). Because of their much higher activity than monometallic catalysts, current Ni-based bimetallic catalysts are receiving special attention. Tanakabe et al. discovered that a homogeneous alloy of Co and Ni, as well as a low Ni substitution of Co, can increase the activity and stability of the $NiCo/TiO_2$ catalyst, which is due to the catalyst's high resistance to unwanted metal oxidation and coke [12]. Zhang et al. also found that the bimetallic NiCo/MgO catalyst outperforms other monometallic or NiMe metals (Me = Fe, Cu, Mn) in CO_2 reforming gas in activity and stability [13].

The activity of Ni-based bimetallic catalysts has increased significantly in several reactions. However, to obtain a stable catalyst with high activity, the composition, pretreatment process, preparation method, and support characteristics must all be considered [14–16]. Ferari et al. discovered that the sequence of Co and Mo metal impregnation on carbon supports affects metal dispersion, the interaction between the oxide and carrier phases, and the catalytic activity of the hydrodeoxygenation reaction [17]. The order of metal bearing in the impregnation method also affects the structure of the catalyst, which affects its catalytic activity and is specific to each catalyst, as explained by Salerno et al. and Bacariza et al. [18, 19].

As a result, in this study, the Ni-Zn/ZrO_2 catalyst will be prepared using a wet-impregnation method with an ethanolic solvent. The addition of the two metals in the support will be carried out simultaneously (co-impregnation) and sequentially impregnation. Furthermore, the impact of impregnation methods on the structure and morphology of the catalyst was investigated using X-ray diffraction (XRD) and scanning electron microscopy–energy dispersive X-ray (SEM–EDX).

2 Materials and Methods

2.1 *Materials*

The materials used in this study include commercial zirconia (ZrO_2) from Kanto Chemical, $Ni(NO_3)_2{\cdot}6H_2O$, $Zn(NO_3)_2{\cdot}4H_2O$, and ethanol absolute was purchased from Merck. All materials used in this work were analytical grade and were used directly without prior purification.

2.2 *Catalyst Preparations*

Co-impregnation. The solid catalyst Ni-Zn/ZrO_2 was synthesized with the following procedure with some modifications [20]. One gram of ZrO_2 is added with the metal precursor (Ni 10% wt and Zn 10% wt), which has been dissolved in ethanolic solution (ethanol: aquadest = 1:1) slowly. Next, the mixture was stirred for a few hours at room temperature and heated at 65 °C to form a slurry. The slurry was dried at 100 °C overnight and calcined at 400 °C for 4 h. The solid formed is labeled Cat-1.

Bimetallic catalyst impregnated in sequential steps. The bimetallic catalyst impregnated on ZrO_2 was carried out in sequential steps but under the same conditions as the co-impregnation method. The first Ni solution (10% wt), dissolved in the ethanolic solution, is added to 1 g ZrO_2 under stirring. The mixture was stirred for a few hours at room temperature and heated at 65 °C until a slurry was formed. Then, the slurry was dried at 100 °C overnight. The solid formed was then calcined at 400 °C for 4 h. Subsequently, a Zn solution (10% wt) was impregnated, dried, and calcined under the same conditions. So, in this case, the catalyst is double calcination. Impregnation in sequential steps was used to synthesize the Ni–Zn bimetallic catalyst. The solid made in two steps was designed as Cat-2.

2.3 Catalyst Characterizations

X-ray Diffraction (XRD). The crystal structure of the synthesized catalyst was characterized by XRD using PANalytical EMPYREAN. The pattern was run with copper radiation (Cu K = 1.54056 Å) with a second monochromator at 40 kV and 30 mA. The step size is 0.0260, and an angle of 2θ is between 20 and 100°. The crystallite size was calculated using the Scherrer equation shown in (1).

$$D = \frac{K\lambda}{\beta \cos\theta} \tag{1}$$

D is the average value of the crystal diameter, K is a constant (0.9), λ is the wavelength, θ is the diffraction angle, and β is the full width at half maximum diffraction peak (FWHM) [21].

Scanning Electron Microscopy–Energy Dispersive X-ray (SEM-EDX). SEM-EDX was used to determine the morphology and particle distribution of the catalyst. The equipment used was an SEM Hitachi SU350 scanning microscope, which was additionally equipped with a chemical composition analysis system based on energy dispersion scattering—the EDX.

3 Result and Discussion

The crystal structure of the synthesized catalyst was determined using XRD. The XRD results are shown in Fig. 1, which shows that both catalysts have a diffractogram with the same pattern. Cat-1 and Cat-2 consist of ZrO_2, NiO, and ZnO phases. ZrO_2 monoclinic is indicated by the appearance of a typical peak at 2θ = 17.45; 24.07; 24.46; 28.20; 31.49; 34.18° (PDF 00-024-1165). In addition, the peak of the NiO phase with the shape of a cube is also visible at 2θ = 37.24; 43.27; 74.99° (●) and ZnO with a hexagonal shape at 2θ = 36.33; 67.96° (PDF 00-005-0664) (■) [22]. The diffractogram in Fig. 1 shows that the typical peaks of both NiO and ZnO have low intensity, which indicates that NiO and ZnO are dispersed on the surface of ZrO_2. From the diffractogram in Fig. 1, the crystallinity of Cat-2 synthesized by the sequential impregnation method has lower crystallinity when compared to Cat-1 synthesized by the co-impregnation method. The same thing was also explained in the Romero-Sáez study (2018) that the catalyst prepared by the sequential impregnation

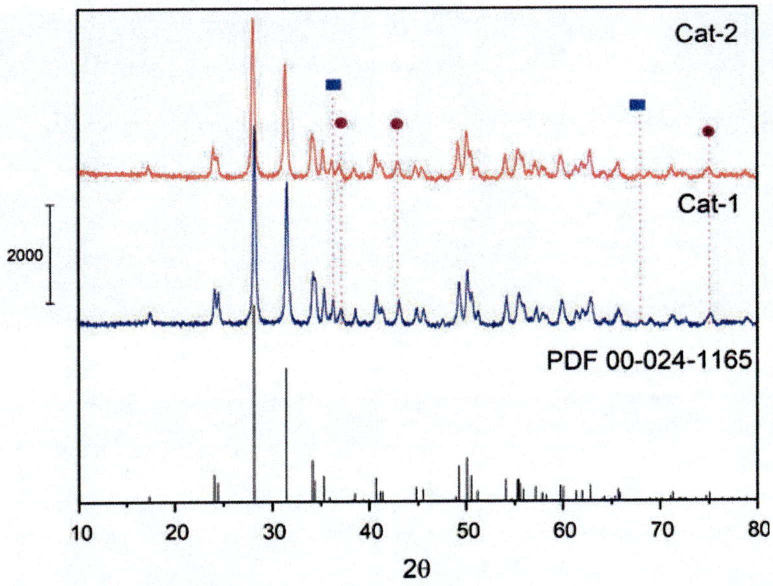

Fig. 1 XRD diffractogram of Cat-1 and Cat-2

method had lower crystallinity than the catalyst prepared by the co-impregnation method [23, 24].

The crystallite size of NiO and ZnO was measured using the Scherrer equation shown in Table 1. Crystallite size measurements were carried out at the dominant peaks of NiO at $2\theta = 43.27°$ and ZnO at $2\theta = 36.25°$ [25, 26]. The calculation results show that the NiO crystallite on the catalyst prepared by sequential impregnation (Cat-2) has smaller crystallite sizes. Meanwhile, the ZnO crystal size tends to be the same in Cat-1 and Cat-2, around 30.20 nm. This indicates that the smaller the crystallite size, the better the dispersion on the support surface.

Figure 2 shows the surface morphology obtained from SEM of Cat-1 (Fig. 2a, b) and Cat-2 (Fig. 2c, d). From Fig. 2, the two catalysts have an irregular shape. However, the particle size of Cat-1 is more uniform when compared to Cat-2. Compared to the particle size, Cat-1 has a smaller particle size than Cat-2. The particle size of Cat-1 of 2.5–12.5 μm. Meanwhile, Cat-2 has a particle size of 5–50 μm. Figure 2b, d shows the surface morphology of the particles of the two catalysts.

The elemental analysis results using EDX on Cat-1 are shown in Fig. 3, where Fig. 3a is the EDX mapping area of Cat-1. Elemental mapping results show that the particles consist of Zr (Fig. 3b), Ni (Fig. 3c), and Zn (Fig. 3d). Ni and Zn appear to

Table 1 Crystallite size of the ZrO_2-based catalysts prepared by various impregnations methods

Catalyst	*d* NiO (nm)	*d* ZnO (nm)
Cat-1	19.45	30.20
Cat-2	14.79	30.21

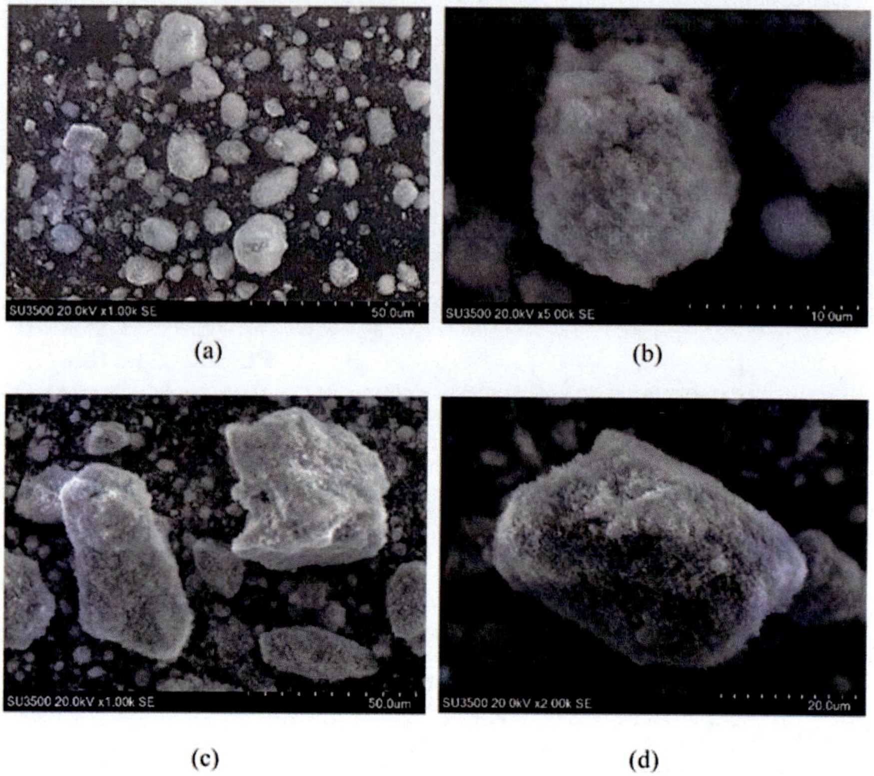

Fig. 2 SEM micrograph of Cat-1 with 1000x **a** and 5000x **b** magnification and Cat-2 with 1000x **c** and 2000x **d** magnification

be homogeneously dispersed on the surface of ZrO_2. In addition, Fig. 3e shows that the percentage of Ni and Zn particles homogeneously dispersed on the ZrO_2 surface is close to the theory (10%) namely 8.18 and 7.20%, respectively.

The elemental analysis results on Cat-2 using EDX are shown in Fig. 4, where Fig. 4a is the Cat-2 EDX mapping area. Elemental mapping results reveal that the particles consist of Zr (Fig. 4b), Ni (Fig. 4c), and Zn (Fig. 4d). According to the elemental mapping, the dispersion of Ni is more even than Zn. This is because sequential impregnation is used, with Ni being contacted first with the support and followed by Zn [27]. The percentages of Ni and Zn on the surface of ZrO_2 in Cat-2 are also close to the theoretical value of 10%, at 8.68 and 7.57%, respectively.

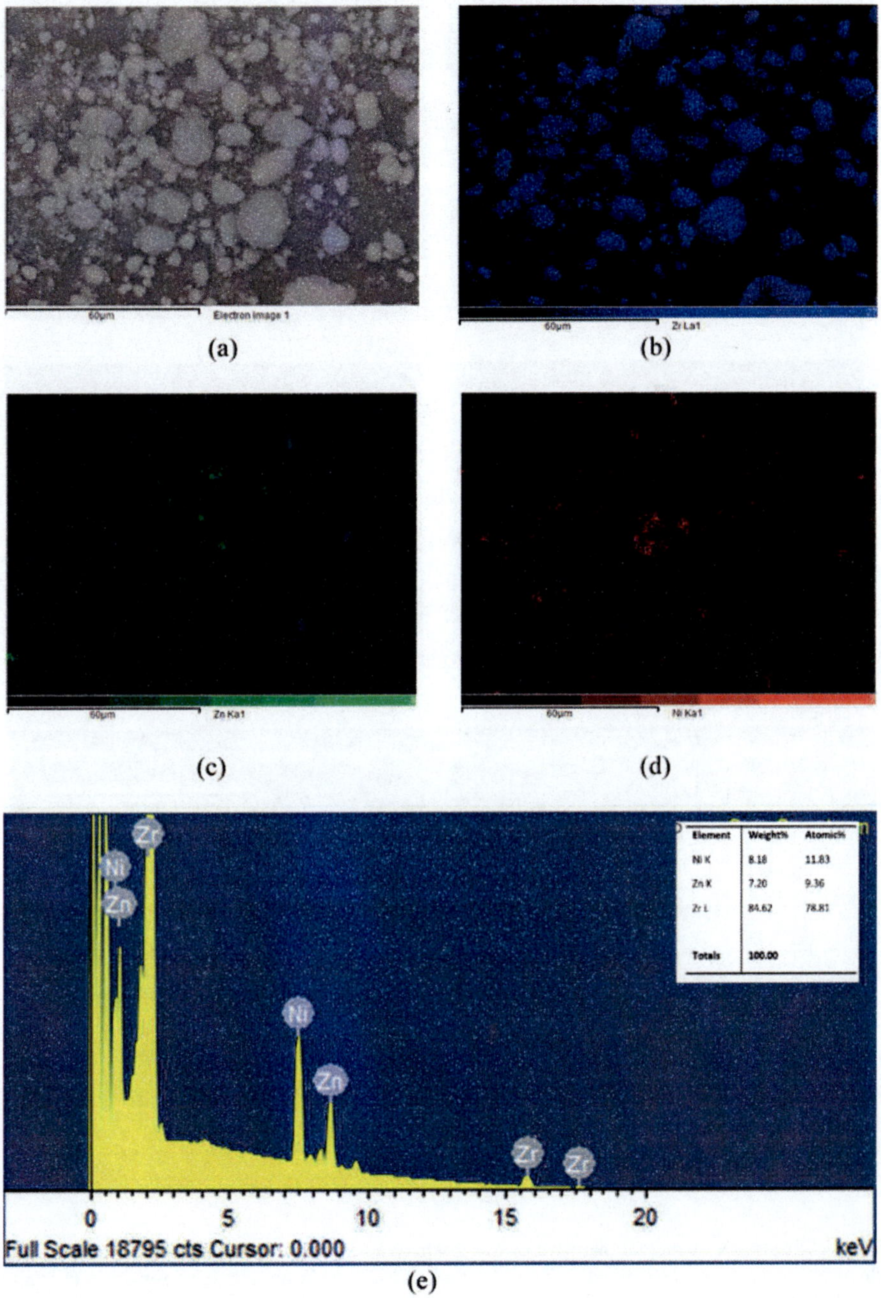

Fig. 3 SEM–EDX elemental mapping of Cat-1 where **a** secondary electron image and analogous elemental mapping of the element. **b** Zr, **c** Zn, **d** Ni, **e** EDX spectrum and table for the atomic and weight percentage of various elements

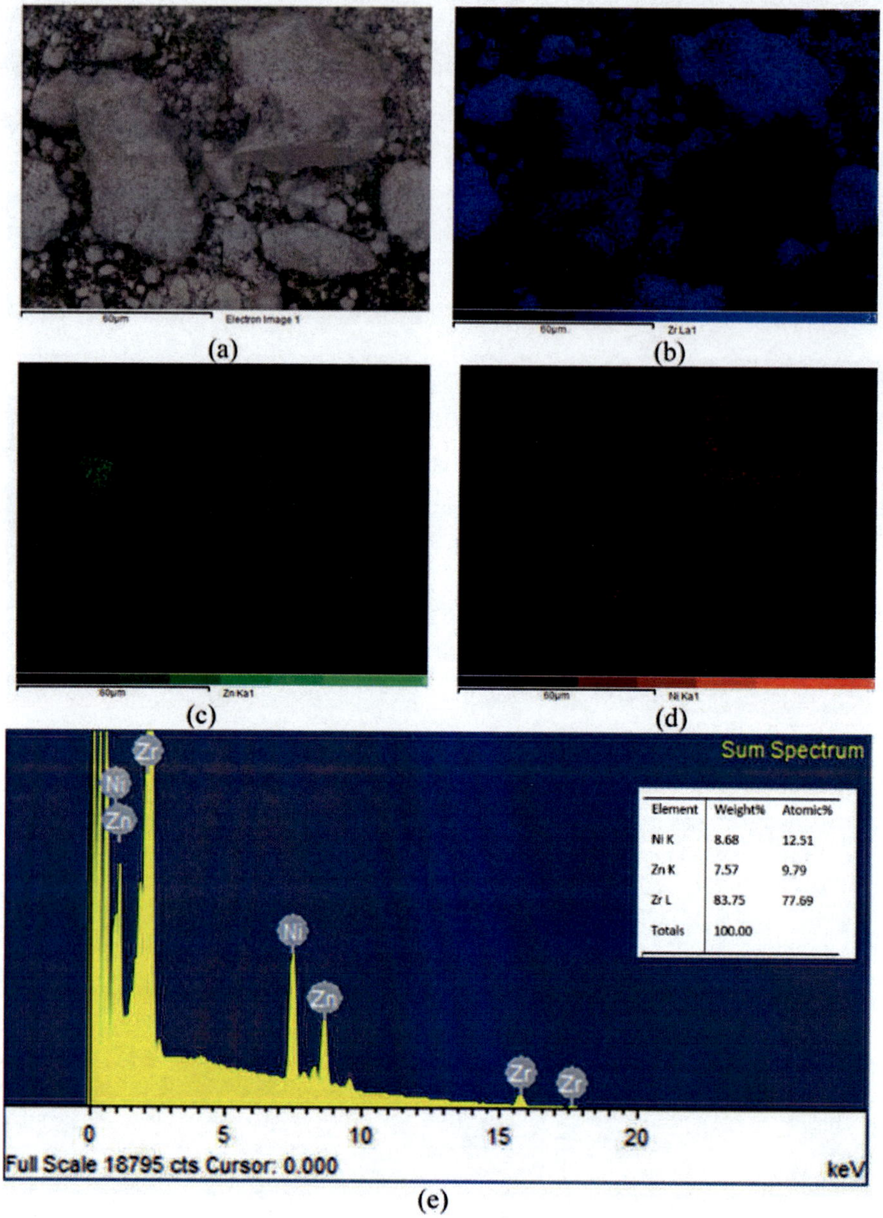

Fig. 4 SEM–EDX elemental mapping of Cat-2 where **a** secondary electron image and analogous elemental mapping of the element **b** Zr, **c** Zn, **d** Ni, **e** EDX spectrum and table for the atomic and weight percentage of various elements

4 Conclusion

Both catalysts were synthesized successively. XRD diffractogram shows that both catalysts consist of ZrO_2, NiO, and ZnO phases. The NiO crystallite sizes in Cat-2 (14.79 nm) are smaller than that of Cat-1 (19.45 nm), but the ZnO crystallite size in both catalysts is almost the same, 30.20 nm. The results of SEM–EDX show that both catalysts have irregular particle shapes, with the particle size of Cat-2 being larger than Cat-1. The EDX elemental analysis showed that the amount of Ni and Zn in both catalysts was close to the theoretical amount (10%). However, further research is needed with further characterization and application to a specific reaction to determine which of the two catalysts has better performance. This is because each chemical reaction has different catalyst requirements, and each catalyst will produce different results in each reaction despite having the same properties.

Acknowledgements This research is a part of the research Grant in the Research Organization of Nanotechnology and Materials-BRIN (No. 20/III.10/HK/2022). The authors also acknowledge the facilities, scientific and technical support from Laboratorium Bahan Maju Nuklir, and Advanced Characterization Laboratories Serpong National Research and Innovation Agency through E-Layanan Sains-BRIN.

References

1. Védrine, J.C.: Heterogeneous catalysis on metal oxides (2017)
2. Global climate strike. Nat Catal. **2**(10), 831 (2019)
3. Rizwanul Fattah, I.M., Ong, H.C., Mahlia, T.M.I., Mofijur, M., Silitonga, A.S., Rahman, S.M.A., et al.: State of the art of catalysts for biodiesel production. Front Energ. Res. **8**, 101 (2020)
4. Abdullah, S.H.Y.S., Hanapi, N.H.M., Azid, A., Umar, R., Juahir, H., Khatoon, H., et al.: A review of biomass-derived heterogeneous catalyst for a sustainable biodiesel production. Renew. Sustain. Energ. Rev. **70**, 1040–1051 (2017)
5. Fatimah, I., Yanti, I., Suharto, T.E., Sagadevan, S.: ZrO_2-based catalysts for biodiesel production: a review. Inorg. Chem. Commun. **143**, 109808 (2022)
6. Ghanbari, B., Kazemi, Z.F., Taheri, R.Z., Aghaei, E.: High-impact promotional effect of Mo impregnation on aluminum-rich and alkali-treated hierarchical zeolite catalysts on methanol aromatization. ACS Omega **5**(21), 11971–11986 (2020)
7. Mehrabadi, B.A.T., Eskandari, S., Khan, U., White, R.D., Regalbuto, J.R.: A review of preparation methods for supported metal catalysts. Adv. Catal. **61**, 1–35 (2017)
8. Erdogan, B., Arbag, H., Yasyerli, N.: SBA-15 supported mesoporous Ni and Co catalysts with high coke resistance for dry reforming of methane. Int. J. Hydrogen Energ. **43**(3), 1396–1405 (2018)
9. Li, Z., Huang, Z., Ding, S., Li, F., Wang, Z., Lin, H., et al.: Catalytic conversion of waste cooking oil to fuel oil: catalyst design and effect of solvent. Energy **157**, 270–277 (2018)
10. Koo, K.Y., Lee, J.H., Jung, U.H., Kim, S.H., Yoon, W.L.: Combined H_2O and CO_2 reforming of coke oven gas over Ca-promoted Ni/$MgAl_2O_4$ catalyst for direct reduced iron production. Fuel **153**, 303–309 (2015)
11. Lycourghiotis, S., Kordouli, E., Sygellou, L., Bourikas, K., Kordulis, C.: Nickel catalysts supported on palygorskite for transformation of waste cooking oils into green diesel. Appl. Catal. B Environ. **259**, 118059 (2019)

12. Takanabe, K., Nagaoka, K., Nariai, K., Aika, K.: Titania-supported cobalt and nickel bimetallic catalysts for carbon dioxide reforming of methane. J. Catal. **232**(2), 268–275 (2005)
13. Zhang, J., Wang, H., Dalai, A.K.: Development of stable bimetallic catalysts for carbon dioxide reforming of methane. J. Catal. **249**(2), 300–310 (2007)
14. Wu, H., Liu, J., Liu, H., He, D.: CO_2 reforming of methane to syngas at high pressure over bi-component Ni-Co catalyst: the anti-carbon deposition and stability of catalyst. Fuel **235**, 868–877 (2019)
15. Zhao, J., Zhou, W., Ma, J.: Pretreatment as the crucial step for biogas reforming over Ni–Co bimetallic catalyst—a mechanistic study of CO_2 pretreatment. Int. J. Hydrogen Energ. **39**(25), 13429–13436 (2014)
16. Wang, Z., Wang, C., Chen, S., Liu, Y.: Co–Ni bimetal catalyst supported on perovskite-type oxide for steam reforming of ethanol to produce hydrogen. Int. J. Hydrogen Energ. **39**(11), 5644–5652 (2014)
17. Ferrari, M., Delmon, B., Grange, P.: Influence of the impregnation order of molybdenum and cobalt in carbon-supported catalysts for hydrodeoxygenation reactions. Carbon N Y **40**(4), 497–511 (2002)
18. Bacariza, M.C., Graça, I., Lopes, J.M., Henriques, C.: Ni-Ce/zeolites for CO_2 hydrogenation to CH_4: effect of the metal incorporation order. ChemCatChem **10**(13), 2773–2781 (2018)
19. Salerno, P., Mendioroz, S., López, A.A.: Al-pillared montmorillonite-based NiMo catalysts for HDS and HDN of gas oil: influence of the method and order of Mo and Ni impregnation. Appl. Catal. A Gen. **259**(1), 17–28 (2004)
20. Abdul Razak, N.A., Mijan, N.-A., Taufiq-Yap, Y.H., Derawi, D.: Production of green diesel via hydrogen-free and solventless deoxygenation reaction of waste cooking oil. J. Clean Prod. **366**, 132971 (2022)
21. Naika, H.R., Lingaraju, K., Manjunath, K., Kumar, D., Nagaraju, G., Suresh, D., et al.: Green synthesis of CuO nanoparticles using *Gloriosa superba L.* extract and their antibacterial activity. J. Taibah Univ. Sci. **9**(1), 7–12 (2015)
22. Hongloi, N., Prapainainar, P., Seubsai, A., Sudsakorn, K., Prapainainar, C.: Nickel catalyst with different supports for green diesel production. Energy **182**, 306–320 (2019)
23. Romero-Sáez, M., Dongil, A.B., Benito, N., Espinoza-González, R., Escalona, N., Gracia, F.: CO_2 methanation over nickel-ZrO_2 catalyst supported on carbon nanotubes: a comparison between two impregnation strategies. Appl. Catal. B Environ. **237**, 817–825 (2018)
24. Bampos, G., Bebelis, S., Kondarides, D.I., Verykios, X.: Comparison of the activity of Pd–M (M: Ag Co, Cu, Fe, Ni, Zn) bimetallic electrocatalysts for oxygen reduction reaction. Top Catal. **60**(15), 1260–1273 (2017)
25. Yan, X., Zhang, Q., Zhu, M., Wang, Z.: Selective hydrogenation of benzene to cyclohexene over Ru–Zn/ZrO_2 catalysts prepared by a two-step impregnation method. J. Mol. Catal. A Chem. **413**, 85–93 (2016)
26. Souza, I.C.A., Manfro, R.L., Souza, M.M.V.M.: Hydrogen production from steam reforming of acetic acid over Pt–Ni bimetallic catalysts supported on ZrO_2. Biomass Bioenerg. **156**, 106317 (2022)
27. Jing, J., Yang, Z., Huo, J., Bai, H., Li, W.: Metal precursor impregnation sequence effect on the structure and performance of NiCo/MgO catalyst. Int. J. Hydrogen Energ. **44**(16), 8089–8098 (2019)

Assessment of Pollution and Sources of Metals in the Brantas River in East Java, Indonesia

Cicik Oktasari Handayani, Hidayatuz Zu'amah, and Sukarjo

Abstract Heavy metals in river water can significantly harm the river's environment and the surrounding community's health. This study examines the level and spread of trace element pollution in the Brantas River. Water samples were obtained from 25 distinct sampling points along the Brantas River. The water samples contained trace metals in lead, cadmium, chrome, nickel, copper, iron, and manganese. Using a contamination factor (CoFa), contamination degree (DC), and trace element pollution index (TEPI), the element contamination level in the Brantas River was assessed. In contrast, analysis of correlation, principal component (PCA), and cluster (CA) were utilized to evaluate the origins of trace metals. In succession, the lead, cadmium, and copper content have surpassed the quality level in 20 places, nine locations, and two locations, respectively. The average values of CoFa are as follows: Pb (2.63); Cd (1.09); Cu (0.74); Ni (0.40); Fe (0.26); Cr (0.10); and Mn (0.02). The DC result indicates that 76% of river water falls into moderate-to-heavy contamination. The TEPI value is greater than 100 (129.28). Hence, the water from the Brantas River cannot be used as a source of drinking water. Statistical investigations demonstrated that most trace elements originated from different anthropogenic sources.

1 Introduction

Due to the threat, it poses to aquatic ecosystems and human health, heavy metal poisoning of river water has become a global issue [1]. Heavy metals are poisonous, persistent, and accumulative [2]. Population growth, fast industrialization, and intensive agricultural practices are the primary drivers of heavy metal pollution in river water [3].

The Brantas River, at 320 km in length, is the longest in East Java Province. The town utilizes the Brantas River as a water supply for industry, domestic use, energy

C. O. Handayani (✉) · H. Zu'amah · Sukarjo
Research Center for Horticultural and Estate Crop, National Research and Innovation Agency, Cibinong Science Center, Jl. Raya Jakarta–Bogor Cibinong, Indonesia 16915
e-mail: cici002@brin.go.id

A. Basit et al. (eds.), *Proceedings of the International Conference on Radioscience, Equatorial Atmospheric Science and Environment and Humanosphere Science*, Springer Proceedings in Physics 290, https://doi.org/10.1007/978-981-19-9768-6_15

production, and agriculture [4]. In addition, up to 60% of the rice harvested in the province of East Java comes from fields near the Brantas River [5].

The Brantas River's water quality has deteriorated due to several excessive anthropogenic activities throughout its length. As a result, the Brantas River is suspected of being contaminated with heavy metals [6] from industrial, household, and agricultural waste discharge. The trace metal content in river water must be evaluated based on the description of the threat heavy metal pollution in river water poses to aquatic ecosystems and human health. Determining the level of contamination and distribution of heavy metal sources in the Brantas River is the objective of this investigation.

2 Method

2.1 Research Site

Brantas River is found in East Java Province at coordinates 111°31′00″–112°57′05″ east longitude and 7°11′43″–8°17′35″ south latitude. The Brantas River starts in Sumber Brantas Village, Batu City, and flows through Malang, Blitar, Tulungagung, Kediri, Jombang, and Mojokerto. The Brantas River is 320 km long and gets an average of 2000 mm of rain per year, of which about 85% falls during the rainy season.

There are 25 points where water samples can be taken. First, choose where to take water samples by looking at river water sources, such as clean water, polluted water, and water already used. The location where water samples were collected is shown in Fig. 1 (see Fig. 1). The technique of taking water samples at each location point is very dependent on the water discharge, so the sampling technique is adjusted to the standard rules in SNI 6989.57:2008 Section 57 regarding the method of taking surface water samples. River water samples were taken using a simple water sampling device with a ballast.

Water samples were prepared by filtering them using filter paper. Then, the clear water extract filtrate was measured using an (Atomic Absorption Spectrophotometer (AAS). Heavy metals analyzed include Pb, Cd, Cr, Cu, Mn, Ni, and Fe. Heavy metal analysis was carried out at the Laboratory of the IAERI, Ministry of Agriculture of the Republic of Indonesia.

2.2 Heavy Metal Contamination Assessments

Contamination Factor (CoFa)

The CoFa value is the ratio of metal concentration in water to metal concentration, which is the standard for river water quality that has been defined [7]. The formula

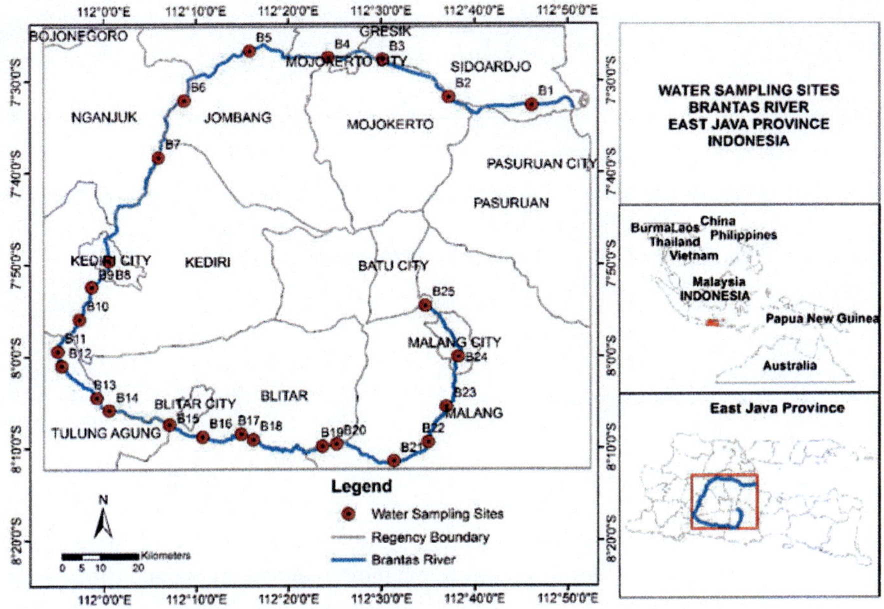

Fig. 1 Location of the Brantas River water sampling

computes the contamination factor in (1):

$$\text{CoFa} = \frac{C_{\text{si}}}{C_{\text{bi}}} \tag{1}$$

C_{si} is the metal concentration measured in the water. At the same time, C_{bi} is the metal concentration standard for river water quality, with the classification of contamination factors referred to by Gupta et al. [8].

Contamination Degree (DC)

DC is a thorough approach for measuring heavy metal contamination [9]. The formula computes the DC value in (2):

$$\text{DC} = \frac{1}{n}\sum_{i=1}^{n=8} \text{CoFa}_i \tag{2}$$

n is the amount of heavy metal observed.

Trace Element Pollution Index (TEPI)

The TEPI determines whether river water can be used to make drinking water. If the TEPI value of river water is less than 100, it is safe to drink. If the TEPI value is higher than 100, it will hurt the health of the people who drink it [10].

$$\text{TEPI} = \frac{\sum W_i Q_i}{\sum W_i} \tag{3}$$

$$Q_i = \frac{\sum (M_i - I_i)}{\sum (S_i - I_i)} \times 100 \tag{4}$$

where Q_i is a sub-index of the i-th parameter; W_i is the unit weight of the i-th parameter; n is the number of parameters; M_i is the monitored value of the i-th heavy metal parameter; I_i is the outstanding value of the i-th parameter; and S_i is the expected value of the i-th parameter.

2.3 Multivariate Analysis

Principal component analysis (PCA) was used to extract the principal components (PC) from the sampling point and evaluate the possible sources and variations of heavy metal in river water samples. Additionally, cluster analysis, or hierarchical cluster analysis (HCA), was carried out to examine the similarities and differences between the concentration levels of all identified heavy metals in river water sampling sites.

3 Result and Discussion

3.1 Heavy Metal Concentration in the Brantas River's Water

The Brantas River water samples had the highest Pb, Cd, Cr, Cu, Mn, Ni, and Fe (mg/L) at 0.178, 0.021, 0.012, 0.025, 0.066, 0.034, and 0.289, respectively. From highest to lowest, the metal concentrations were Pb > Cd > Cu > Fe > Ni > Cr > Mn, based on the highest value. Table 1 shows the statistical description of heavy metals in river water at each place where water samples were taken.

Concentrations of heavy metals Pb, Cd, and Cu in some samples have exceeded the quality standards, respectively, namely 80%, 36%, and 8% samples. In contrast, other metals such as Cr, Mn, Ni, and Fe did not exceed the quality standards.

3.2 Heavy Metal Contamination Assessments

Compared to other heavy metals, the average CoFa value of Pb metal is the highest, at 2.63. Based on the average CoFa values from the highest to the lowest, respectively, they are Pb > Cd > Cu > Ni > Fe > Cr > Mn. The projected heavy metal contamination

Table 1 The statistical description of heavy metal concentrations at the Brantas River water sampling location

Statistical	Pb	Cd	Cr	Cu	Mn	Ni	Fe
n	25	25	25	25	25	25	25
Average	0.079	0.011	0.006	0.015	0.007	0.020	0.078
Minimum	0	0.005	0	0.01	0	0.01	0.039
Maximum	0.178	0.021	0.012	0.025	0.066	0.034	0.289
SD	0.052	0.005	0.003	0.004	0.017	0.007	0.068
Water quality standards[a]	0.030	0.010	0.050	0.020	0.400	0.050	0.300

[a]IGR: 22/2021[11]

factor in river water is low, with a contamination factor value of less than one. The CoFa values for Pb metal at each sampling site vary, but the overall weight is more than one. Cd and Cu metals have CoFa values greater than 1 in many places, whereas Cr, Mn, Ni, and Fe do not. The percentage of all metal contamination factors observed in several categories can be seen in Table 2.

Degree of Contamination

The DC is also used to determine the overall concentration of heavy metals in river water. The DC values for the moderately contaminated, moderate to extremely contaminated, and heavily contaminated types are 20%, 76%, and 4%, respectively (see Fig. 2). Most of the water test regions along the Brantas River are contaminated with heavy metals, with moderate to severe pollution being the most prevalent.

Trace Element Pollution Index

The TEPI value calculated based on the average trace element from all water sampling locations can be seen in Table 3. The TEPI value is 129.28 (> 100), which shows that based on the average concentration of all metals in the water in the Brantas River, it is not feasible to be used as clean water and consumed by the community.

Table 2 Percentage of contamination factors in each category

Heavy metal	Mean	Low degree	Moderate degree	Considerable degree	Very high degree
		%			
Pb	2.63	20.00	44.00	36.00	0.00
Cd	1.09	64.00	36.00	0.00	0.00
Cr	0.12	100.00	0.00	0.00	0.00
Cu	0.74	92.00	8.00	0.00	0.00
Mn	0.02	100.00	0.00	0.00	0.00
Ni	0.40	100.00	0.00	0.00	0.00
Fe	0.26	100.00	0.00	0.00	0.00

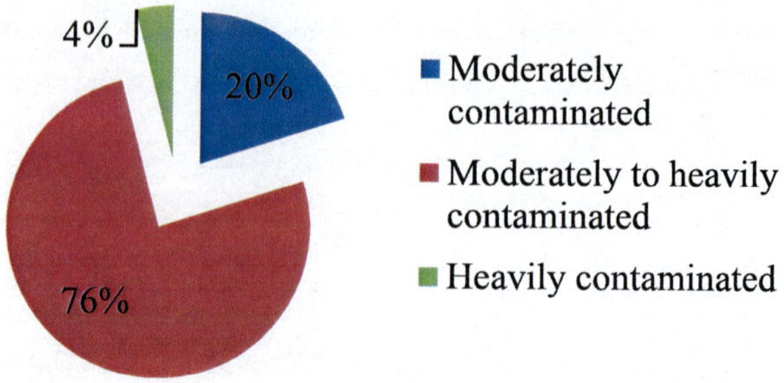

Fig. 2 DC percentage value in each category

Table 3 TEPI values of the average concentration of heavy metals at all water sampling locations

Para-meters	Mean value mgl/L (M_i)	Standard permissible value (S_i)	Highest desirable value (I_i)	Unit weighting factor (W_i)	Q_i	$Q_i * W_i$	TEPI
Pb	0.079	0.030	0	10	263.49	2634.86	129.28
Cd	0.011	0.010	0	33.33	108.57	3618.77	
Cr	0.006	0.050	0	2	11.60	23.21	
Cu	0.015	0.020	0	0.05	73.68	3.68	
Mn	0.007	0.400	0	0.2	1.73	0.35	
Ni	0.020	0.050	0	0.5	40.26	20.13	
Fe	0.078	0.300	0	3.33	26.14	87.11	

Multivariate Analysis

Principal component analysis was used to identify the main components associated with heavy metal sources in Brantas River water at different water sampling locations. There are two factors according to this criterion that can represent 64.71% of the total variance so that it can show information on most of the heavy metals in river water. The two main components proportions were 37.430% and 27.275%, respectively (Table 4).

The first principal component (PC1) has a cumulative variance value of 37.430%. Therefore, PC1 can represent a mixed source of lithogenic and anthropogenic activity. The evaluation of heavy metals shows that Mn, Ni, and Cr do not pollute river water; this indicates that these metals may have natural sources, such as rock erosion [12]. On the other hand, river water is contaminated with Pb and Cd metals from anthropogenic sources. Metals like Cd and Pb probably come from intensive agricultural activities [13]. However, Pb and Cd metals can also come from industrial activities [14], domestic waste [15], and motor vehicles [16].

The second principal component (PC2) has a cumulative variance value of 27.275%. PC2 can be considered as representing a natural source because the value of the majority of Cu and Fe metals is still below the quality standard. The presence of Cu and Fe metals in river water is probably related to the environment around the Brantas River, which is in a volcanic environment [6].

Table 4 PCA with varimax rotation for all heavy metals found in the studied water sample

Component	Element							Eigen-values	% of variance	Cumulative %
	Pb	Cd	Cr	Cu	Mn	Ni	Fe			
PC1	**−0.78**	**0.67**	**−0.8**	−0.03	**0.63**	**0.62**	0.34	2.620	37.430	37.430
PC2	−0.13	**−0.62**	−0.23	**0.89**	−0.37	0.04	**0.74**	1.909	27.275	64.705

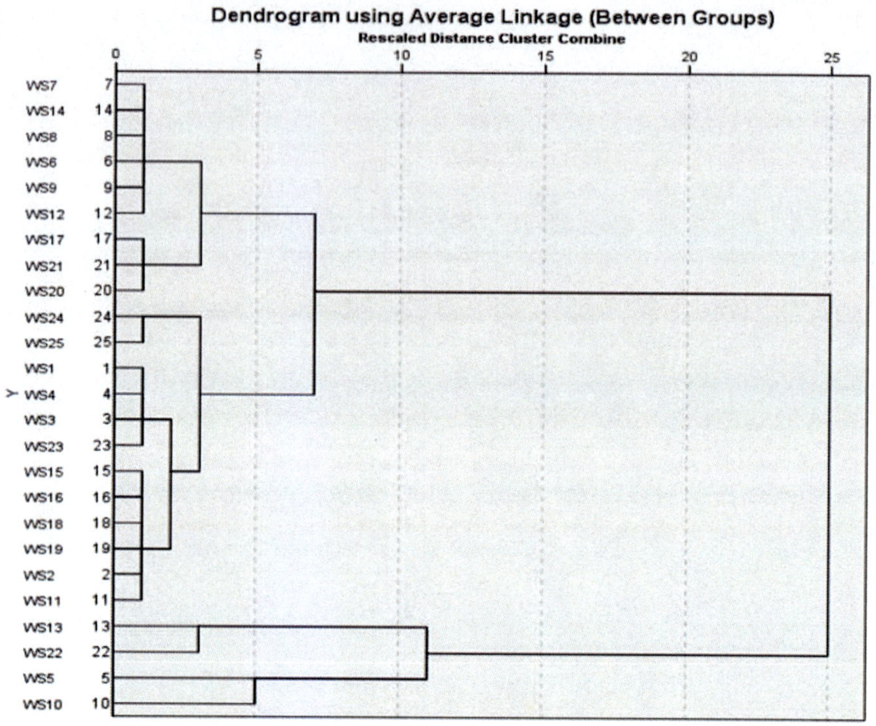

Fig. 3 Dendrogram shows the clustering of heavy metals

The similarities and differences in the characteristics of the concentration values of all heavy metals observed between water sampling locations can be seen in HCA (see Fig. 3). The dendrogram shows the process/stages in the formation of clusters, starting from the initial stage on the far left with many clusters to the right with two clusters formed. The character of the heavy metal values of Pb, Cd, Cr, Ni, Cu, Fe, and Mn at 21 water sampling locations was almost uniform. In comparison, for the other four locations, namely, WS5, WS10, WS13, and WS22, the characters were different from the other 21 locations.

4 Conclusion

There were twenty samples, nine samples, and two samples that exceeded the quality standard (Government Regulation of the Republic of Indonesia Number 22 of 2021) for the heavy metals Pb, Cd, and Cu, respectively. The average value of CoFa is as follows: Pb (2.63) > Cd (1.09) > Cu (0.74) > Ni (0.40) > Fe (0.26) > Cr (0.10) > Mn (0.02). The DC result indicates that 76% of river water falls into moderate-to-heavy contamination. The TEPI rating of 129.28 indicates that the water of the Brantas

River cannot be used as a source of drinking water for the local community. Multiple statistical tests, such as PCA, Pearson correlation matrix, and CA, showed that heavy metals originated from natural and anthropogenic sources.

References

1. Rajeshkumar, S., Liu, Y., Zhang, X., Ravikumar, B., Bai, G., Li, X.: Studies on seasonal pollution of heavy metals in water, sediment, fish and oyster from the Meiliang Bay of Taihu Lake in China. Chemosphere **191**, 626–638 (2018)
2. Parween, M., Ramanathan, A.L., Raju, N.J.: Assessment of toxicity and potential health risk from persistent pesticides and heavy metals along the Delhi stretch of river Yamuna. Environ. Res. **202**, 111780 (2021)
3. Luo, M., Yu, H., Liu, Q., et al.: Effect of river-lake connectivity on heavy metal diffusion and source identification of heavy metals in the middle and lower reaches of the Yangtze River. J. Hazard Mater. **416**, 125818 (2021)
4. Roestamy, M., Fulazzaky, M.A.: A review of the water resources management for the Brantas River basin: challenges in the transition to an integrated water resources management. Environ. Dev. Sustain. 1–16 (2021)
5. Sujono, I.: Restorasi Air Sungai Brantas (Water Restoration of Brantas River). Osf, Surabaya (2019)
6. Mariyanto, M., Amir, M.F., Utama, W., et al.: Heavy metal contents and magnetic properties of surface sediments in volcanic and tropical environment from Brantas River, Jawa Timur Province, Indonesia. Sci. Total Environ. **675**, 632–641 (2019)
7. Hoang, H.G., Lin, C., Tran, H.T., et al.: Heavy metal contamination trends in surface water and sediments of a river in a highly-industrialized region. Environ. Technol. Innov. **20**, 101043 (2020)
8. Gupta, N., Yadav, K.K., Kumar, V., et al.: Appraisal of contamination of heavy metals and health risk in agricultural soil of Jhansi city, India. Environ. Toxicol. Pharmacol. **88**, 103740 (2021)
9. Abrahim, G.M.S., Parker, R.J.: Assessment of heavy metal enrichment factors and the degree of contamination in marine sediments from Tamaki Estuary, Auckland, New Zealand. Environ. Monit. Assess. **136**(1), 227–238 (2008)
10. Jazza, S.H., Najim, S., Adnan, M.A.: Using Heavy Metals Pollution Index (HPI) for assessment quality of drinking water in Maysan Province in Southern East in Iraq. Egypt J. Chem. **65**(2), 1–2 (2022)
11. Peraturan Pemerintah No 22 Tahun 2021: Peraturan Pemerintah Nomor 22 Tahun 2021 tentang Pedoman Perlindungan dan Pengelolaan Lingkungan Hidup. Sekr Negara Republik Indones **1**(078487A), 483 (2021)
12. Li, W., Qian, H., Xu, P., et al.: Distribution characteristics, source identification and risk assessment of heavy metals in surface sediments of the Yellow River, China. CATENA **216**, 106376 (2022)
13. Fei, X., Lou, Z., Xiao, R., Ren, Z., Lv, X.: Source analysis and source-oriented risk assessment of heavy metal pollution in agricultural soils of different cultivated land qualities. J. Clean Prod. **341**, 130942 (2022)
14. Bi, X., Zhang, M., Wu, Y., et al.: Distribution patterns and sources of heavy metals in soils from an industry undeveloped city in Southern China. Ecotoxicol. Environ. Saf. **205**, 111115 (2020)

15. Yang, Q., Zhang, L., Wang, H., Martin, J.D.: Bioavailability and health risk of toxic heavy metals (As, Hg, Pb and Cd) in urban soils: a Monte Carlo simulation approach. Environ. Res. **214**, 113772 (2022)
16. Huang, C., Zhang, L., Meng, J., et al.: Characteristics, source apportionment and health risk assessment of heavy metals in urban road dust of the Pearl River Delta, South China. Ecotoxicol. Environ. Saf. **236**, 113490 (2022)

Utilization EDGAR Fifth Version as Input Emission of WRF-Chem Model for Simulating Ozone and $PM_{2.5}$ Over Jakarta and Its Surroundings Area

Prawira Yudha Kombara and Ninong Komala

Abstract The simulation of $PM_{2.5}$ and Ozone has been conducted in this study in Jakarta and the surrounding area. $PM_{2.5}$ and ozone at the surface level are one of a pollutant that can harm human health, especially the respiration system. Therefore, it is an important thing to learn the distribution pattern of the pollutants. The simulation was conducted using the WRF-Chem model utilizing Emission Database for Global Atmospheric Research (EDGAR) fifth version as anthropogenic emission input for the model. The simulation was done from 2nd until August 04, 2021. The result of the simulation was compared with observation data from The Ministry of Environment and Forestry (KLHK) and the U.S. Embassy. There are three points of observation data that be used for the comparison. Those points are Central Jakarta, Serang, and Bekasi point. Based on the comparison process, the suitability pattern between model results and observation data is still a tough thing to be reached for both pollutants. Furthermore, the daily pattern is still not similar yet too. In addition, based on Root Mean Square Error (RMSE) calculation, the $PM_{2.5}$ parameter gives better result than the ozone parameter for both observation points.

1 Introduction

As one of the big cities, Jakarta is still facing the problem of air pollution. The enhancement number of residents every year has resulted in massive community activities using motorized vehicles, causing the air quality in Jakarta increasingly declines. In addition, the increase in industrial and factory activities around the Jakarta area also has contributed to degradation of air quality. The problem of poor

P. Y. Kombara (✉) · N. Komala
Research Center for Climate and Atmosphere, National Research and Innovation Agency, Bandung, Indonesia
e-mail: praw005@brin.go.id

A. Basit et al. (eds.), *Proceedings of the International Conference on Radioscience, Equatorial Atmospheric Science and Environment and Humanosphere Science*, Springer Proceedings in Physics 290, https://doi.org/10.1007/978-981-19-9768-6_16

air quality has resulted in degradation in the quality of life of the citizens of DKI Jakarta due to unhealthy air. Based on IQAir [1], Jakarta is ranked as the ninth most polluted city in the world.

The problem of air pollution can be handled with the right policies. We can start from an academic study of how the pattern of air pollution that occurs in Jakarta. Understanding the pattern of air pollution is critical thing to arrange a right policy. Understanding the pattern of air pollution in Jakarta and its surroundings can be done by using several methods such as model simulation, spatial interpolation of ground observation data, or estimation with linear regression [2–5]. However, if we want to understand the overall pattern and mechanism of air pollution that occurs, the use of model simulation is the right choice. Simulation models for air pollution can be done using numerical models, such as the WRF-Chem model. The WRF-Chem model is a numerical model that combines the weather and chemical components of the atmosphere [6–9]. The WRF-Chem model requires meteorological and emission input data. Therefore, to produce a good simulation, the latest emission data is needed as input data for the model.

Earlier this year in February 2022, the fifth version of the EDGAR emission data was released to the public. The EDGAR is the Emissions Database for Global Atmospheric Research that contains global anthropogenic emission data [10]. This latest version of EDGAR data can be used as emission input data for the WRF-Chem model. However, the fifth version of the EDGAR data has not been tested for air pollution simulations in Indonesia. Therefore, in this study, we tried to use the fifth version of EDGAR data for emission input of the WRF-Chem model. The pollutant parameters that will be simulated in Jakarta and its surroundings for this trial are $PM_{2.5}$ and ozone gas (O_3). These two parameters are kinds of pollutants that are often encountered and become a problem in big cities. In addition, these two pollutants are very dangerous to human health, especially in the respiratory system so they can cause death [11, 12].

2 Data and Method

The data used in this study consists of

1. Data Final Reanalysis (FNL)

FNL data is reanalysis data produced by NCEP-NCAR. The FNL data used in this study has a temporal resolution of 6 h and a spatial resolution of 0.25°. This FNL data is used as meteorological input data for the WRF-Chem model. The web address to download this data is as follows: https://rda.ucar.edu.

2. The fifth version of The EDGAR Data

The EDGAR data is global anthropogenic emission data produced by Joint Research Centre. This data has a monthly temporal and spatial resolution of 0.1°. The version of the data used in this study is the fifth version. This data will be used as input data for anthropogenic emissions to run the WRF-Chem model. The fifth version of EDGAR data can be downloaded on the page: https://zenodo.org/record/6130621#.YzQOKzRBy5d.

3. KLHK air quality observation data

KLHK is the ministry that handles air pollution conducts continuous observations by installing an air quality measurement instrument called AQMS throughout Indonesia. This data is used to evaluate the simulation results of the WRF-Chem model. Two stations will be used in this study, namely Serang station, Banten (−6.1208 S, 106.1731 E), and Bekasi station, West Java (−6.237452 S, 106.992943 E).

4. $PM_{2.5}$ observation by U.S. Embassy

In the U.S. Embassy office, central Jakarta (−6.18108 S, 106.83015 E), there is an instrument that measures $PM_{2.5}$ operationally. This instrument is operated by AirNow, and they provide $PM_{2.5}$ data at hourly intervals. The data can be obtained from https://www.airnow.gov/international/us-embassies-and-consulates/#Indonesia$Jakarta_Central.

2.1 Model WRF-Chem Configuration

In this study, two domains were used to run the WRF-Chem model. The first domain has a spatial resolution of 5 km, and the second domain has a spatial resolution of 1 km. The domain design used in this study is shown in Fig 1. The simulation was run for three days from August 2–4, 2021.

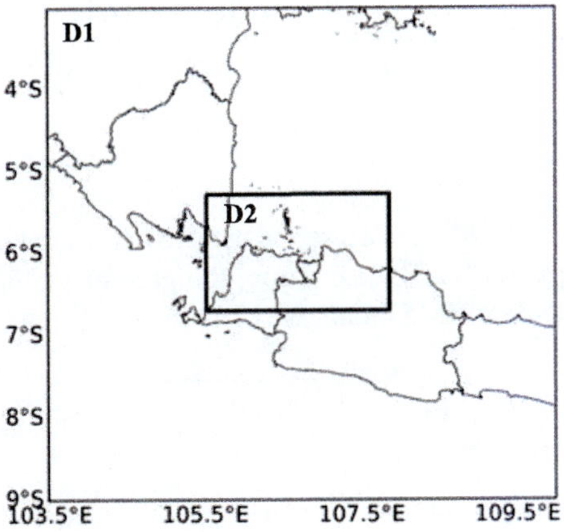

Fig. 1 Domain design for WRF-Chem simulation

Table 1 Configuration of parameterization schemes

Parameterization schemes	1st domain	2nd domain
Microphysics	WSM6	WSM6
Cumulus	BMJ	–
Boundary layer	YSU	YSU
Longwave radiation	RRTMG	RRTMG
Shortwave radiation	RRTMG	RRTMG
Chemical	MOSAIC using 8 sectional aerosol bins	MOSAIC using 8 sectional aerosol bins

In addition, to run the WRF-Chem model, a schema configuration of several parameterizations is required. Table 1 shows the list of the parameterization schemes are used.

2.2 *Model Evaluation*

The simulation results of the model were evaluated by two methods. The two methods are correlation and root mean square error (RMSE).

Correlation

Correlation is a value that expresses the relationship between two variables. The range of correlation value is from −1 to 1. To calculate the correlation, you can use (1).

$$r = \frac{\sum(F_i - \overline{F_i})(O_i - \overline{O_i})}{\sqrt{\sum(F_i - \overline{F_i})^2}\sqrt{\sum(O_i - \overline{O_i})^2}} \quad (1)$$

RMSE

Root mean square error (RMSE) is a method used to determine the difference between two variables. The greater the RMSE value, the greater the difference between the two variables. To calculate RMSE can use (2).

$$\text{RMSE} = \sqrt{\frac{1}{N}\sum\nolimits_{i=1}^{N}(F_i - O_i)^2} \quad (2)$$

Variables F and O represent model simulation variables and observations variables.

3 Result

3.1 $PM_{2.5}$ Parameter

The results of the comparison between model simulations and observations for the $PM_{2.5}$ parameter for three days at three stations are shown in Fig. 2. The simulation results of $PM_{2.5}$ at the Bekasi station is more similar to the observation data (Fig. 2b) than the Central Jakarta and Serang station points (Fig. 2a, c). Even so, for the $PM_{2.5}$ pattern at the Bekasi station point, there is still lag in several time windows. In general, the $PM_{2.5}$ simulation results at Bekasi station are slightly overestimated; in other words, the magnitude of the model simulation is slightly larger than the observation data.

Furthermore, the comparison results at the points of the Central Jakarta and Serang stations are not as good as the Bekasi station. The $PM_{2.5}$ simulation results for the Serang station are quite overestimate or the simulation magnitude is greater than the observation data, so it is difficult to find a similarity between the simulation results and the observation data. The opposite happened at the Central Jakarta station, where the simulation results were very underestimated compared to the observation data. Based on the simulation results, the maximum value for three days occurs on August 3, 2021, at 03–04 AM for Bekasi and Serang stations, while for Central Jakarta stations, it occurs on August 2 at around 10 PM.

The results for calculating the correlation and RMSE parameters for $PM_{2.5}$ at three stations are shown in Table 2. As discussed previously, the Bekasi station points have a smaller difference between simulation results and observations than Central Jakarta and Serang stations. This can be seen from the RMSE value for Bekasi station, which is only 4.3427, while Jakarta and Serang stations are larger. Then the correlation at three stations has not shown good results yet. As previously mentioned, the pattern between the simulation results and the observation data still does not show suitability. This is confirmed by the correlation value that does not exceed 0.5 from all stations.

If the simulation results and three-day observation data are composited daily, a diurnal pattern will be obtained as shown in Fig. 3. The daily pattern at Bekasi station (Fig. 3b) shows a pattern that is almost similar between simulation result and observation data. Based on the simulation result, the maximum magnitude at the Bekasi station occurred at 6 AM, while the observation data occurred three hours later. Then, the Serang station still shows a large enough difference between the simulation results and the observation data. The daily pattern from the simulation results shows two peaks that occur at 04 and 08 AM, while the observation data tends to be flatter even though there is a little fluctuation. Likewise, for the Central Jakarta station. In addition, the very large difference between the simulation results and the observation data, the suitability of the pattern has not been produced.

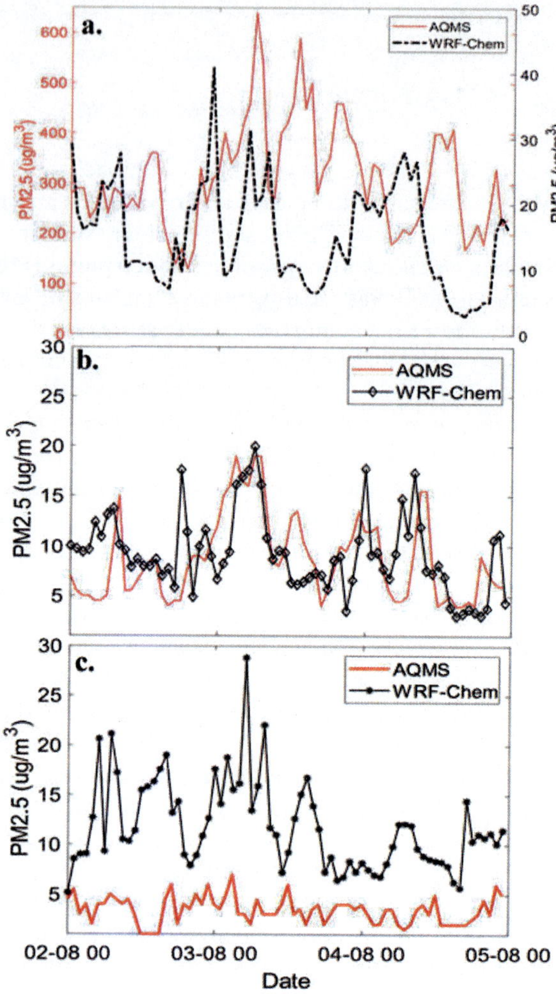

Fig. 2 Comparison of model result and ground station for $PM_{2.5}$ parameter in **a** Central Jakarta, **b** Bekasi station, and **c** Serang station

Table 2 Correlation and RMSE result for the $PM_{2.5}$ parameter in Bekasi and Serang station

Station	Correlation	RMSE
Centre Jakarta	−0.0755	314.2647
Bekasi	0.4247	4.3427
Serang	−0.0207	9.5634

The 24-h pattern for the $PM_{2.5}$ parameter has been mentioned in several previous studies. The $PM_{2.5}$ pattern in the Bandung Basin based on the model results can be said to have a diurnal pattern. The lowest value occurs at 10 AM and the maximum value is at 06 AM [4]. When compared to Jakarta and its surroundings area, based on the simulation results in August 2021, $PM_{2.5}$ in the surrounding Jakarta area

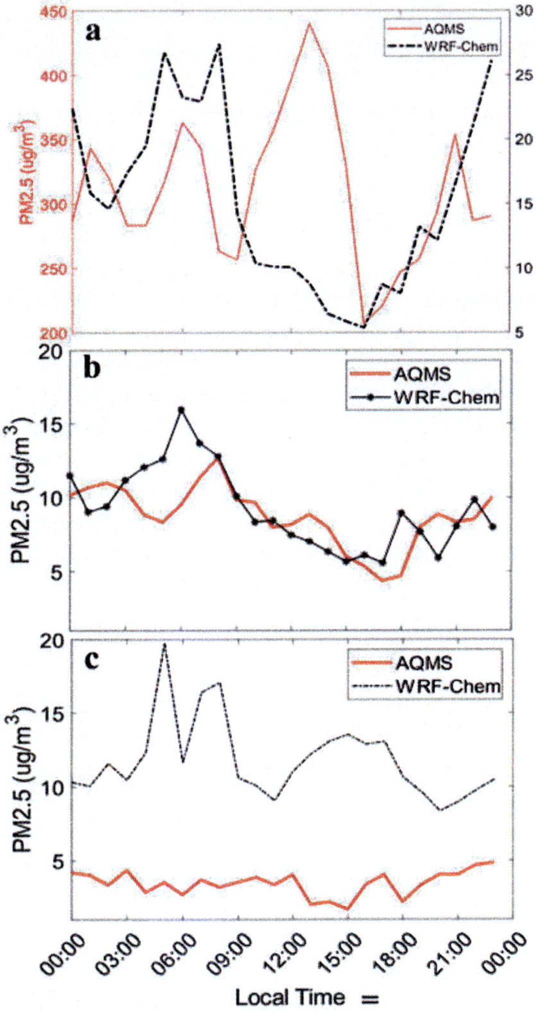

Fig. 3 Diurnal pattern of $PM_{2.5}$ parameter in **a** Central Jakarta, **b** Bekasi station, and **c** Serang station

has a semi-diurnal pattern. The difference in 24-h patterns could be due to differences in meteorological conditions, topography, and land use. Furthermore, based on Hutauruk et al. [13], the $PM_{2.5}$ pattern in Jakarta is similar to the diurnal pattern. This difference may be due to Hutauruk et al. [13] averaging the data for several years so that there is a possibility that the semi-diurnal pattern does not appear because of the averaging process.

The reason why there is still a large discrepancy at station points other than Bekasi and the different pattern between the simulation results and the observation data has not been confirmed. Many things affect the simulation results of the WRF-Chem model such as initial meteorological conditions, domain design, selection of physical and chemical parameterization schemes, and emission input data. Pratama and Sofyan [7] stated in their research, meteorological conditions greatly affect the dynamics of pollutants in the atmosphere, so they assimilate data for meteorological initial input to obtain more realistic initial conditions. However, in this study, we only update global anthropogenic emission data.

3.2 *Ozone Parameter*

The comparison of simulation results and observation data for ozone parameters is shown in Fig. 4. The magnitude of the simulation results looks greater than the observation data. This shows the simulation results are overestimate. The result of overestimation occurred in two stations, either at Bekasi station or Serang station. For the suitability of the pattern, the Bekasi station almost did not show a suitable pattern between the simulation results and the observation data. This is due to the observation data pattern which tends to be more sloping than the simulation results even though there is a slight fluctuation. In the case of Serang station, there is a slight pattern match between the simulation results and the observation data.

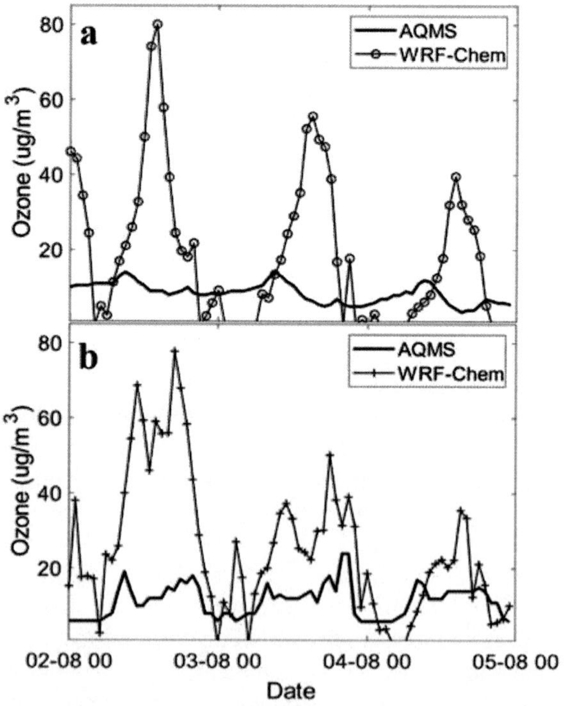

Fig. 4 Comparison of model result and ground station for ozone parameter in **a** Bekasi station, and **b** Serang station

Table 3 Same as Table 2, for the ozone parameter

Station	Correlation	RMSE
Bekasi	−0.0310	22.0868
Serang	0.4863	21.8090

The unsuitability of the pattern and the large difference is strengthened by the results of the calculation of correlation and RMSE which are shown in Table 3. Based on the previous explanation, the unsuitability of the data pattern has not been seen at the Bekasi station. This is confirmed by the correlation value which is only −0.031, while for Serang station, there is still a pattern match even though the correlation value is still below 0.5. The difference between the magnitude of the simulation results and the observation data at the two stations is still quite large. This is shown by the RMSE value which reached 22.0868 for Bekasi station and 21.8090 for Serang station.

Similar to the $PM_{2.5}$ parameter, a daily composite for ozone parameter is also carried out to be able to see the daily pattern as shown in Fig. 5. Comparison of daily patterns at Bekasi station still shows a pattern discrepancy between the simulation results and the observation data. The maximum value from the observation data occurs first, while from the simulation results the maximum value occurs only a few hours later. Then for the Serang station, the maximum value based on the simulation results occurred twice, namely at 10 AM and 05 PM. In addition, pattern conformity occurs after 09 AM. Even so, the conformity of the pattern that occurs cannot be said perfect because there is a large difference in magnitude between the simulation results and the observation data. Based on Kitada et al. [2], ozone in Jakarta is produced after sunrise, after 6 AM exactly. It means the result of ozone simulation in this study matches with Kitada et al. [2] study result. We can see in Fig. 5 that ozone concentration from the simulation started to increase from 6 am. In other words, while the sun begins to rise and the photochemical reaction was begun.

4 Conclusion

The simulation results for $PM_{2.5}$ and ozone after replacing the fifth version of EDGAR data to the WRF-Chem model have not shown satisfactory results. There is still a discrepancy in the pattern and a large difference between the magnitude of simulation results and the observation data, especially for the ozone parameter. In general, the simulation results for the $PM_{2.5}$ parameter are much better than the ozone parameter, especially at the Bekasi station. The results of the $PM_{2.5}$ parameter simulation at the Bekasi station show a pattern that resembles the pattern of observation data. In addition, the value of the difference between the simulation results and the observation

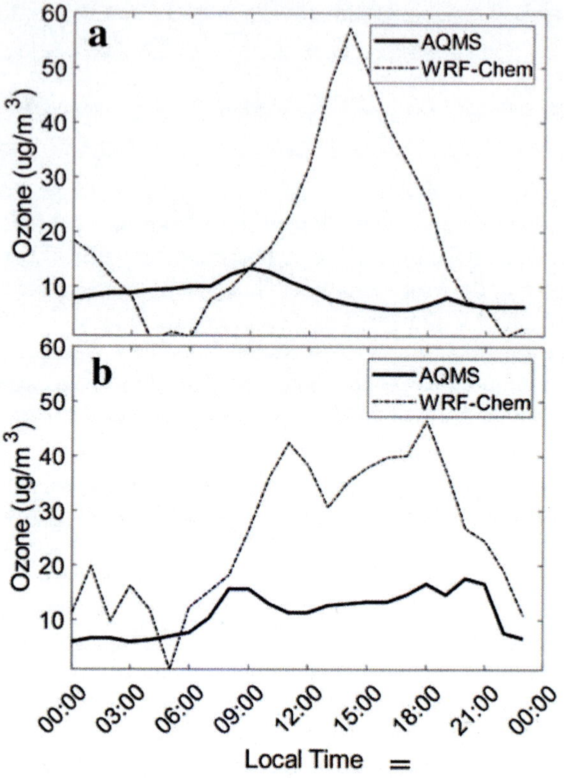

Fig. 5 Diurnal pattern of ozone parameter in **a** Bekasi station, and **b** Serang station

data at the Bekasi station can be said to be quite small so that the simulation results can follow the magnitude of the observation data. Generally, the simulation results for both parameters can be said to overestimate the observation data except at the Central Jakarta station which results in an underestimate of $PM_{2.5}$ simulation on the observation data.

Acknowledgements We would like to thank to the Minister of Environment and Forestry for sharing the data of the related stations.

References

1. IQAir: World Air Quality Report. 2020 World Air Qual. Rep. 1–41 (2020)
2. Kitada, T., Sofyan, A., Kurata, G.: Numerical simulation of air pollution transport under sea/land breeze situation in Jakarta, Indonesia in dry season. NATO Sci. Peace Secur. Ser. C Environ. Secur. 243–251 (2008). https://doi.org/10.1007/978-1-4020-8453-9_27

3. Solihah, K.I., Martono, D.N., Haryanto, B.: Identifying the best spatial interpolation method for estimating spatial distribution of PM2.5 in Jakarta. IOP Conf. Ser. Earth Environ. Sci. 893 (2021). https://doi.org/10.1088/1755-1315/893/1/012043
4. Salsabila, H., Turyanti, A., Nuryanto, D.E.: Estimation of the spatial distribution of maximum PM10 and PM2.5 concentration in Bandung City and surrounding countries using WRF-Chem Model (case study in July and October 2018). IOP Conf. Ser. Earth Environ. Sci. 893 (2021). https://doi.org/10.1088/1755-1315/893/1/012044
5. Cholianawati, N., Cahyono, W.E., Indrawati, A., Indrajad, A.: Linear regression model for predicting daily PM2.5 using VIIRS-SNPP and MODIS-Aqua AOT. IOP Conf. Ser. Earth Environ. Sci. 303 (2019). https://doi.org/10.1088/1755-1315/303/1/012039
6. Zhang, Q., Tong, P., Liu, M., Lin, H., Yun, X., Zhang, H., Tao, W., Liu, J., Wang, S., Tao, S., Wang, X.: A WRF-Chem model-based future vehicle emission control policy simulation and assessment for the Beijing-Tianjin-Hebei region, China. J. Environ. Manage. **253**, 109751 (2020). https://doi.org/10.1016/j.jenvman.2019.109751
7. Pratama, A., Sofyan, A.: BANDUNG MENGGUNAKAN WRFCHEM DATA ASIMILASI PM 10 AIR POLLUTION DISPERSION ANALYSIS IN BANDUNG CITY USING WRFCHEM DATA ASSIMILATION PENDAHULUAN pengaruh penting terhadap kehidupan manusia dan ekosistem. Pencemaran udara berasal pengamatan yang dilakukan o. J. Tek. Lingkung. **26**, 19–36 (2020)
8. Hong, J., Mao, F., Min, Q., Pan, Z., Wang, W., Zhang, T., Gong, W.: Improved PM2.5 predictions of WRF-Chem via the integration of Himawari-8 satellite data and ground observations. Environ. Pollut. **263**, 114451 (2020). https://doi.org/10.1016/j.envpol.2020.114451
9. Grell, G.A., Peckham, S.E., Schmitz, R., McKeen, S.A., Frost, G., Skamarock, W.C., Eder, B.: Fully coupled "online" chemistry within the WRF model. Atmos. Environ. **39**, 6957–6975 (2005). https://doi.org/10.1016/j.atmosenv.2005.04.027
10. Crippa, M., Solazzo, E., Huang, G., Guizzardi, D., Koffi, E., Muntean, M., Schieberle, C., Friedrich, R., Janssens-Maenhout, G.: High resolution temporal profiles in the Emissions Database for Global Atmospheric Research (2020)
11. Ismiyati, Marlita, D., Saidah, D.: Pencemaran Udara Akibat Emisi Gas Buang Kendaraan Bermotor. J. Manaj. Transp. Logistik. **1**, 241–248 (2014)
12. Fitri, D.W., Afifah, N., Anggarani, S.M.D., Chamidah, N.: Prediction concentration of PM2.5 in Surabaya using ordinary kriging method. AIP Conf. Proc. 2329 (2021). https://doi.org/10.1063/5.0042284
13. Hutauruk, R.C.H., Rahmanto, E., Pancawati, M.C.: Variasi Musiman dan Harian PM2.5 di Jakarta Periode 2016–2019. Bul. GAW Bariri. **1**, 20–28 (2020)

GEMS Satellite Identification on Volcanic Ash Distribution of Mount Dukono Eruption in April 2022

Amalia Nurlatifah, Emmanuel Adetya, Asri Indrawati, Risyanto, Sumaryati, Ninong Komala, Nani Cholianawati, and Prawira Yudha Kombara

Abstract The Geostationary Environment Monitoring Spectrometer (GEMS) is geostationary satellite that is used for air quality monitoring aims. Utilization of GEMS now becomes one of the alternative solution of air quality monitoring problem in Asia. In this research, GEMS is used for monitoring the volcanic ash distribution as impact of Mount Dukono Eruption in April 2022. GEMS is used with wind hourly data from ERA-5. HYSPLIT Forward Trajectory Model is used to determine the direction and the distribution of ash from the eruption of Mount Dukono. Wind data shows that in April 2022, the prevailing wind moves from the southeast and west. This causes the volcanic ash tend to move from southeast to the northwest and east. This is in line with the results of the HYSPLIT model plot at 500, 1335, and 3000 m AGL. All plots indicate that the volcanic ash tends to move toward to the northwest and east. The results of satellite imagery from GEMS show a significant concentration of sulfur dioxide and aerosol in the North Halmahera area at the time of the incident and move to the northwest and east over time.

1 Introduction

The Geostationary Environment Monitoring Spectrometer (GEMS) is the first Geostationary Earth orbit (GEO) satellite that is used for monitoring air quality [1]. GEMS was on boarding the Geostationary Korea Multi-Purpose Satellite 2 (GEO-KOMPSAT-2) satellite series, which also hosts the Advanced Meteorological Imager (AMI) and Geostationary Ocean Color Imager 2 (GOCI-2). GEMS is providing some products for better understanding about air quality monitoring such as air quality

A. Nurlatifah (✉) · E. Adetya · Sumaryati · N. Komala · N. Cholianawati · P. Y. Kombara
Research Center of Climate and Atmosphere, National Research and Innovation Agency, Central Jakarta, Indonesia
e-mail: amalia.nurlatifah@brin.go.id

A. Indrawati · Risyanto
Laboratory Management, Research Facilities and Science, and Technology Park, National Research and Innovation Agency, Central Jakarta, Indonesia

A. Basit et al. (eds.), *Proceedings of the International Conference on Radioscience, Equatorial Atmospheric Science and Environment and Humanosphere Science*, Springer Proceedings in Physics 290, https://doi.org/10.1007/978-981-19-9768-6_17

and meteorology condition monitoring, the long-range transport of air pollutants, emission source distributions, and chemical processes.

Air pollution is one of the most important problems in the world. There is 4.2 million deaths per year due to stroke, lung disease, lung cancer, acute, and chronic respiratory illnesses as impact of air pollution [2]. Source of air pollution can vary, and the sources are divided into two types, namely natural sources and sources from anthropogenic activities. The source of air pollution can be from household combustion devices, motor vehicles, and industrial facilities (anthropogenic activities), forest fires, and volcano eruptions (natural source).

Volcanic gases that can be the greatest potential hazards are sulfur dioxide, carbon dioxide, and hydrogen fluoride [3–7]. Volcanoes also can spew ash, a type of particulate matter air pollution, into the air for miles downwind of the eruption. Gases and aerosol that emit from volcano eruption can cause damage and hazard such as acid rain, climate change, long-range transport of ash/aerosol, decrease of visibility that can inhibit social and economic activity, and also a lot of toxic pollutants that can affect health [8–10]. So that, the volcano eruptions can cause air pollution problem that can be a source of damage in a lot of aspect such as health, social, and economy.

Air quality monitoring as impact from volcano can be beneficial for managing the volcano eruption disaster. Managing the volcano eruption is very important so that the impact of that can be mitigated. Indonesia Maritime Continent as one of the areas that located in the ring of fire area is surrounded by many active volcanoes which can erupt at any time. One of this active volcano is Mount Dukono.

Mount Dukono is an active and fairly remote volcano in eastern Indonesia. This mount is located in the northern end of Halmahera Island. Remote Dukono volcano on the island of Halmahera has been erupting continuously since 1933, with ash explosions, frequent ash plumes, and sulfur dioxide (SO_2) plumes.

Today, studies on monitoring volcanic activity, including eruptions, have developed a lot. Previous studies show that major volcanic eruptions emit large amounts of solid particles and volatile gases in the troposphere and the stratosphere [7, 11, 12]. Another study showed, in a statistical approach, the global distribution of volcanic SO_2 degassing during the last century (1900–2000) and further indicated that each eruption (even for non-monitored ones) could affect the stratosphere, based on empirical observations [13]. This study aims to identification the distribution of ash and sulfur dioxide as impact from Mount Dukono eruption in April 2022 with GEMS.

2 Data and Method

2.1 Research Location

The research location is Mount Dukono which is located in North Halmahera Province, precisely at 127.89 E and 1.693 N. The research conducted in April 24–29, 2022, 23:45 UTC/April 25–30, 2022 08:45 LT.

2.2 Data

The data that is used in this study is as follows:

1. GEMS Data

   The GEMS data that is used in this research has daily temporal resolution and 20 km × 20 km spatial resolution. In this study, the GEMS data used include:

   - The slant column and the total column of SO_2 were used to analyze the distribution of volcanic ash from the eruption of Mount Dukono. As we know that the volcano ash is major contained SO_2 so that the SO_2 concentration can reflect the activity of volcano. The SO_2 algorithm fits the spectrum over the 310–326 nm (310–340 nm in volcanic regions) window for planetary boundary layer (PBL) SO_2 SCD retrievals using a hybrid algorithm based on DOAS and PCA (Li et al. 2013).
   - Aerosol optical depth in the 550 nm spectrum which reflects the aerosol conditions in the study area during the eruption.

2. ERA-5 Wind Data

   ERA-5 is the fifth generation ECMWF reanalysis for the global climate and weather for the past 4–7 decades. The data assimilation system used to produce ERA-5 is the IFS Cycle 41r2 4D-Var. Wind data from ERA-5 that is used for this research has 0.25° spatial resolution and daily temporal resolution. The wind component that is used for this research is u-wind component and *v*-wind component.

2.3 Method

2.3.1 HYSPLIT Forward Trajectory Model

The Hybrid Single-Particle Lagrangian Integrated Trajectory (HYSPLIT) transport model is a complete system used to calculate and estimate air parcel trajectories such as transport, dispersion, chemical transformation, and deposition simulations [14]. HYSPLIT was run on the NOAA web (https://www.arl.noaa.gov/hysplit/). In this study, the HYSPLIT model is used with the output data from GDAS. This GDAS data has a spatial resolution of 0.25°. The model is run for an altitude of 500, 1335, and 3000 m AGL with a tracking time of 1 h ahead.

3 Results and Discussion

On April 24–29, 2022, PVMBG (Center of Volcanology and Geological Hazard Mitigation) reported an increase in the activity of Mount Dukono, located in North

Halmahera. This area is one of the areas covered by GEMS monitoring. Figure 1 shows the total value of the SO_2 column at the time of the incident (April 24–29, 2022). In Fig. 1a, it can be seen that on April 24, 2022, at 23:45 UTC (April 25, 2022, at 08:45 LT), GEMS can provide an illustration that there is an increase in the concentration of the total column SO_2 in the study area. This condition lasted for quite a long time, namely until the next day on April 25, 2022, at 23:45 UTC (April 26, 2022, at 08:45 LT) (Fig. 1b). Figure 1b shows the concentrated peak of the total column SO_2 value in the study area. This indicates that Mount Dukono's activity reached its maximum at the time of this incident. Mount Dukono's activity was still recorded high the next day, namely on April 26, 2022, at 23:45 UTC (April 27, 2022, at 08:45 LT) which is also supported by the GEMS depiction of the total value of the SO_2 column (Fig. 1c). Figure 1c shows that the total SO_2 column value in the study area at this time still looks high. Interestingly, at this time it was seen that the total concentration of the SO_2 column was spread lengthwise to the northwest of the study. This gives an indication that the SO_2 resulting from the increased activity (eruption) of Mount Dukono on the third day of the study has been dispersed to the northwest. To prove it, a trajectory/dispersion model plot and wind movement analysis are needed which will be discussed in the next subchapter.

The total value of the SO_2 column at Mount Dukono is seen decreasing on April 27, 2022, at 23:45 UTC (April 27, 2022, at 08:45 LT) (Fig. 1d), then increasing again on April 28, 2022, at 23:45 UTC (April 29, 2022, at 08:45 LT) and tends to move

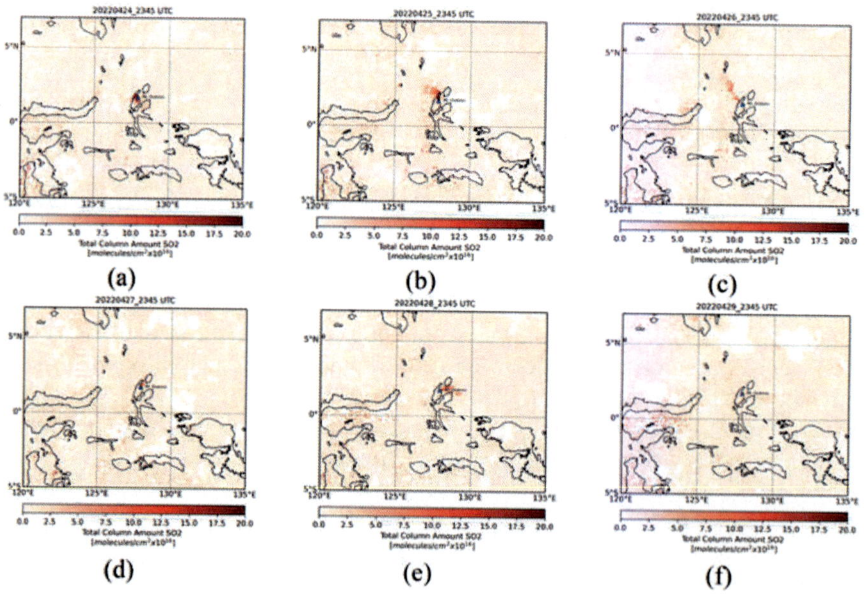

Fig. 1 GEMS image for total column amount of SO_2 (molecules/cm^2 $\times$ 10^6) in Mount Dukono on **a** April 24, 2022, **b** April 25, 2022, **c** April 26, 2022, **d** April 27, 2022, **e** April 28, 2022, and **f** April 29, 2022. All images are on 23:45 UTC

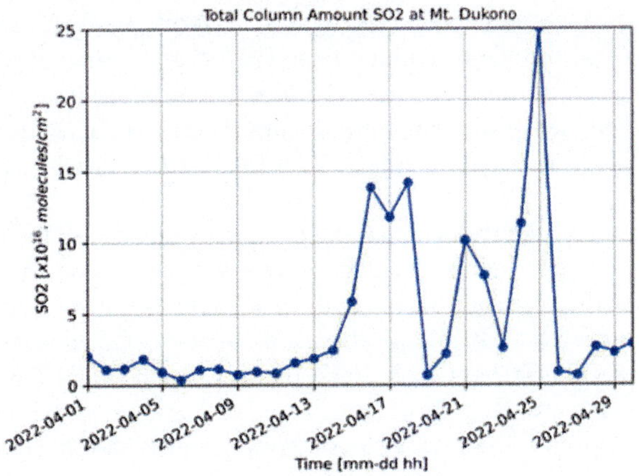

Fig. 2 Total column amount of SO_2 at mount Dukono in April 2022 23:45 UTC

toward the northeast (Fig. 1e), then decreases again the next day, namely on April 29, 2022, at 23:45 UTC (April 30, 2022, at 08:45 LT) (Fig. 1f).

To clarify the description of the total concentration value of the SO_2 column due to the increased activity of Mount Dukono, Fig. 2 shows the time series of SO2 concentration against time in April 2022 including the time of the incident. As has been found in Fig. 1, it can be seen that the total column SO_2 on April 25, 2022, at 23:45 UTC is the time with the highest concentration of slant column SO_2. At this time, MAGMA PVMBG ESDM reported that there had been a significant increase in activity at Mount Dukono. Figure 2 also shows that there has been a significant increase in the value of SO_2 concentration at Mount Dukono during the eruption (April 24–29, 2022 23:45 UTC/April 25–30, 2022, 08:45 LT). This supports the report from PVMBG that during this period there was a significant increase in activity at Mount Dukono.

From the explanation above, it can be concluded that GEMS is quite good in describing the total concentration value of the SO_2 column around Mount Dukono during the period of increased activity of Mount Dukono (April 24–29, 2022, 23:45 UTC/April 25–30, 2022, 08:45 LT).

In the presentation of Fig. 1c, it can be seen that the SO_2 emitted from Mount Dukono is seen spreading to the northwest on April 26, 2022, 23:45 UTC/April 27, 2022, 08:45 LT. Then in Fig. 1e, it can be seen that SO_2 is dispersed to the northeast on April 28, 2022, 23:45 UTC/April 27, 2022, 08:45 LT. From the two images, it can be seen that GEMS depicts the movement of SO_2 resulting from the Mount Dukono eruption. However, to be able to measure its performance in describing the direction of pollutant dispersion from volcano emissions, this must also be proven by depicting the dispersion model and wind direction plots.

Figure 3 shows a plot of the concentration value of the SO_2 slant column along with the trend of its movement described by GEMS. Figure 4 shows the results of the forward trajectory volcano ash modeling from the HYSPLIT model which illustrates the direction in which the pollutants move, while Fig. 5 shows a plot of wind direction at an altitude of 750 mb from ERA-5 at the time of the incident.

Figure 3a shows a plot of the concentration value of the SO_2 slant column at Mount Dukono at the beginning of the increase in activity, namely on April 25, 2022, at 23:45 UTC/April 26, 2022, at 08:45 LT. It can be seen that there is a tendency for volcano ash to move to the northwest, although the concentration is still dominant around the study area. This is in accordance with the results of the HYSPLIT model plot presented in Fig. 4a where this plot shows that at that time the volcano ash at the three altitudes (500, 1335, and 3000 m AGL) showed a movement to the northwest. The plot of the wind direction at an altitude of 750 mb on April 26, 2022, 00:00 UTC/April 27, 2022, 09:00 LT also shows the dominant wind moving to the northwest (Fig. 5a).

Figure 3b shows a plot of the concentration value of the slant column SO_2 at Mount Dukono during the middle period of increased activity, namely on April 26, 2022, at 23:45 UTC/April 27, 2022, at 08:45 LT. It can be seen that SO_2 or volcano ash has moved far to the northwest. This is in accordance with the results of the HYSPLIT model plot presented in Fig. 4b where this plot shows that at that time the volcano ash at all three altitudes (500, 1335, and 3000 m AGL) showed a movement to the northwest, further than the plot on the previous day, in Fig. 4a. The plot of the wind direction at an altitude of 750 mb on April 27, 2022, 00:00 UTC/April 28, 2022, 09:00 LT also shows the dominant wind moving to the northwest (Fig. 5b). Interestingly, the SO_2 GEMS plot and the HYSPLIT model on this date show a further movement of SO_2 compared to the previous day. This is supported and validated by the results of the plot of wind movement and speed on April 27, 2022, 00:00 UTC/April 28, 2022, 09:00 LT (Fig. 5b) where the plot results show that the wind movement is faster than before, on April 26, 2022, 00:00 UTC/27 April 2022 09:00 LT (Fig. 5a). Thus, there are indications that this faster wind speed is causing volcano ash to move faster and farther today (April 27, 2022, 00:00 UTC/April 28, 2022, 09:00 LT) compared to the previous day.

Figure 3c shows a plot of the concentration value of the slant column SO_2 at Mount Dukono at the end of the increase in activity, namely on April 28, 2022, at 23:45 UTC/April 29, 2022, at 08:45 LT. It can be seen that there is a tendency for volcano ash to change its movement compared to the previous two images, namely to the east. This is in accordance with the results of the HYSPLIT model plot presented in Fig. 4c where this plot shows that at that time volcano ash at two altitudes (500 and 1335 m AGL) also showed an eastward movement. The plot at these two altitudes shows the movement of volcano ash to the east although the movement is not too far away. This is supported by the plot of the wind direction at an altitude of 750 mb on April 29, 2022 00:00 UTC/April 30, 2022, 09:00 LT also shows that the wind at the study site is dominantly moving eastward at a low speed (Fig. 5c).

From the explanation above, it can be concluded that GEMS describes the movement of SO_2 from the volcano ash from the Mount Dukono eruption at the time of the incident which tends to move toward the northwest initially, then at the end of the volcanic ash eruption it moves to the east. This is confirmed by the results of the

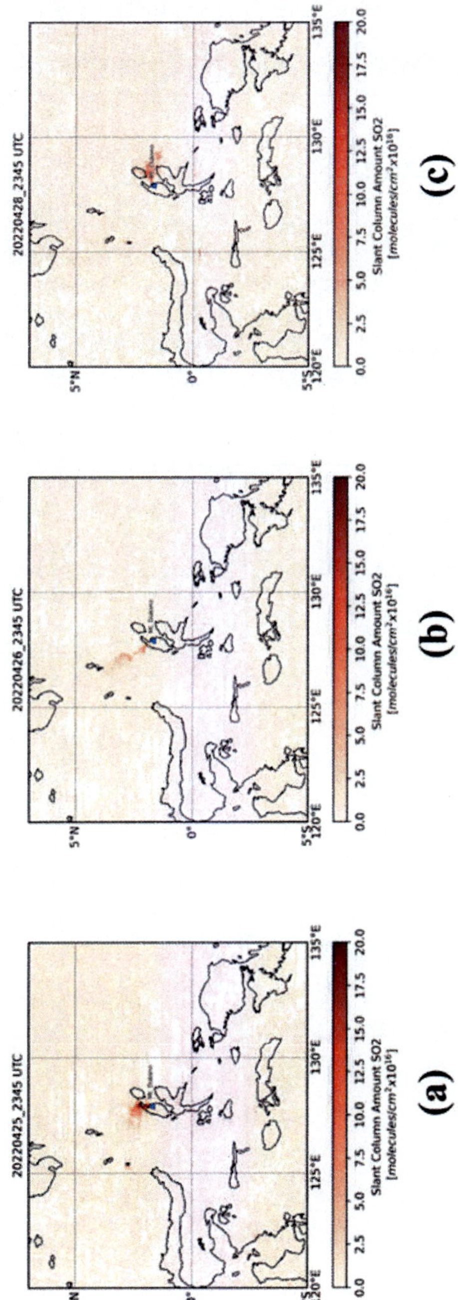

Fig. 3 Slant column concentration of SO_2 in mount Dukono in **a** April 25, 2022 23:45 UTC, **b** April 26, 2022 23:45 UTC, **c** April 28, 2022 23:45 UTC from GEMS

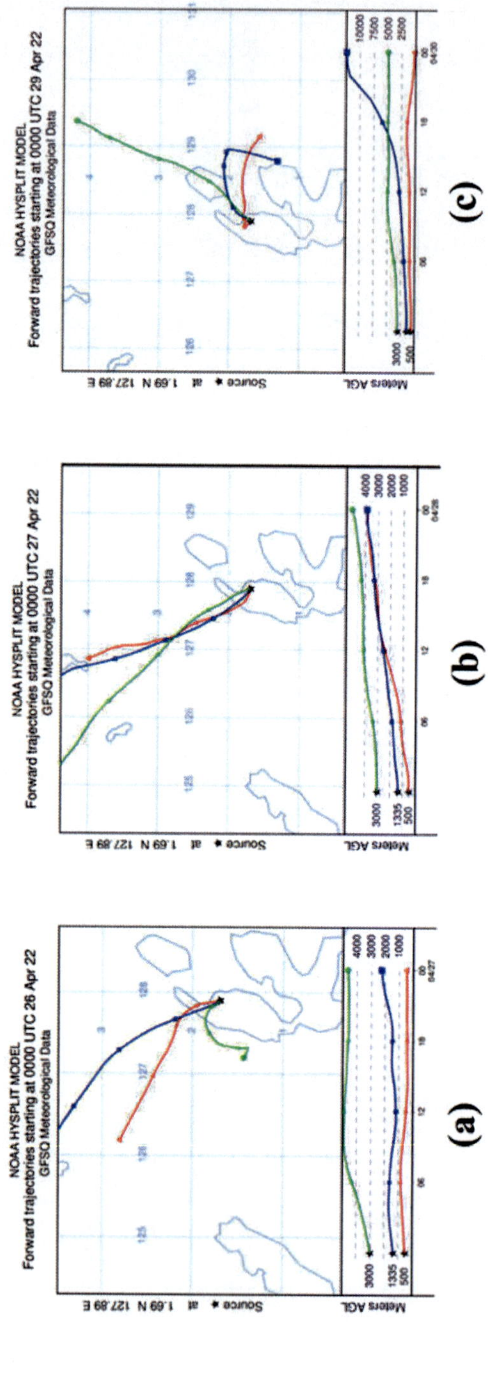

Fig. 4 HYSPLIT forward trajectory plot in mount Dukono on 500, 1335, and 3000 m AGL in **a** April 25, 2022 23:45 UTC, **b** April 26, 2022 23:45 UTC, and **c** April 28, 2022 23:45 UTC

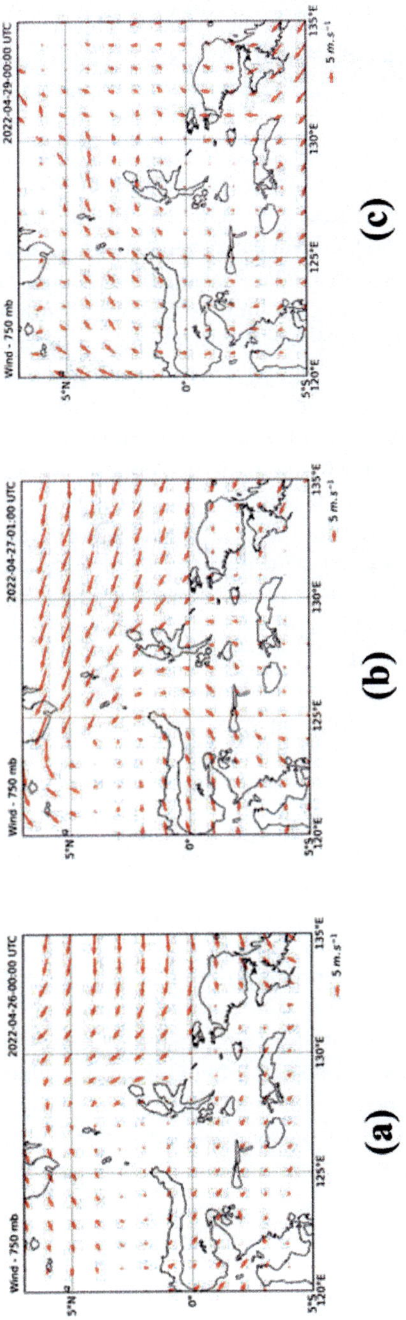

Fig. 5 Wind plot in mount Dukono on 750 mb in **a** April 26, 2022 00:00 UTC, **b** April 27, 2022 00:00 UTC, **c** April 29, 2022 00:00 UTC

HYSPLIT plot at 500, 1335, and 3000 m AGL and the wind plot at 750 mb from the ERA-5 which shows the same movement.

In addition to SO_2, aerosols are also one of the dominant components of volcanic eruptions. Aerosol optical depth (AOD) expresses the density of an area caused by aerosols. The higher the AOD value, the more concentrated the aerosol concentration in the area, and vice versa. Under normal conditions, the AOD value only ranges from 0.1 to 0.15.

Figure 6 shows the AOD value during the eruption of Mount Dukono (April 25, 2022–April 29, 2022, LT). In Fig. 6a, it can be seen that on April 24, 2022, at 23:45 UTC (April 25, 2022 08:45 LT), GEMS can provide an image that there is an increase in the AOD value in the study area. The AOD value in Mount Dukono is close to 1 even though the AOD value in the surrounding area is only around 0.2–0.4 (Fig. 6a). This continued until the next day where the AOD value at Mount Dukono had value close to 1 (Fig. 6b). The peak of activity from Mount Dukono occurred on April 26, 2022 23:45 UTC/April 27, 2022, 08:45 LT where the AOD at Mount Dukono and around North Halmahera has value close to 1, and some in the surrounding area are around 0.6 (Fig. 6c). This is also in line with the depiction of the total concentration value of the SO_2 column at this time (Fig. 1c) which also has a fairly high value. The depiction of the AOD value close to 1 which is quite widespread also illustrates that at this time volcano ash containing aerosols has spread in the area around the study area. AOD values close to 1 are also spread over the northwestern region of Mount Dukono. This confirms again that at this time the volcano ash from Mount Dukono tends to spread to the northwest.

AOD values that are getting denser and closer to 1 becoming more widespread at Mount Dukono and its surroundings on April 27, 2022, 23:45 UTC/April 28, 2022, 08:45 LT (Fig. 6d). This indicates that there is a wider distribution of volcano ash around the study area due to the eruption of Mount Dukono. A fairly high AOD value also still occurred the next day, on April 28, 2022, 23:45 UTC/April 29, 2022, 08:45 LT (Fig. 6e). Figure 6e shows that the AOD value close to 1 occurs in the area around Mount Dukono and the surrounding eastern area. This is in line with the results of the plot in the previous discussion which stated that at this time the volcano ash from the Mount Dukono eruption was moving eastward. The AOD value decreased as Mount Dukono's activity decreased on April 29, 2022, 23:45 UTC/April 30, 2022, 08:45 LT (Fig. 6f).

Overall, GEMS is able to describe the condition of SO2 emissions from volcanoes, especially in the case of the eruption of Mount Dukono in Indonesia in April 2022. This proves that GEMS can be used for monitoring Mount Dukono's activities. Case studies and similar results have also been reported in a study by [15]. Other study said that satellites play a significant and reliable role in monitoring volcanic activity (Ebmeier et al., 2018).

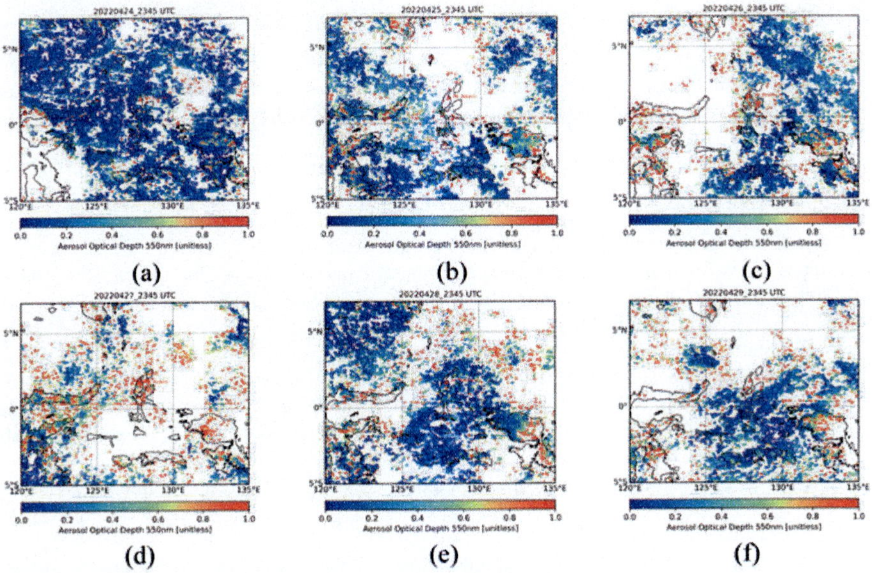

Fig. 6 GEMS image for aerosol optical depth (AOD) in mount Dukono on **a** April 24, 2022, **b** April 25, 2022, **c** April 26, 2022, **d** April 27, 2022, **e** April 28, 2022, **f** April 29, 2022. All images are on 23:45 UTC

4 Conclusion

GEMS data states that there has been an increase in the concentration of SO_2 and aerosols due to the increased activity of Mount Dukono on April 24–29, 2022, 23:45 UTC. On April 25–26, 2022, 23:45 UTC, volcano ash tends to move to the northwest. This is also confirmed by the results of the HYSPLIT plot which shows the movement of ash in the same direction. The wind movement at 750 mb of ERA-5 also shows consistent results where the prevailing wind in the study area tends to move to the northwest. On April 28, 2022, 23:45 UTC, volcano ash tends to move eastward. This is also confirmed by the results of the HYSPLIT plot which shows the movement of ash in the same direction. The wind movement at 750 mb of ERA-5 also shows consistent results where the prevailing wind in the study area tends to move to the east.

Acknowledgements Authors would like to thank to National Institute of Environmental Research (NIER), Republic of Korea and The United Nations Economic and Social Commission for Asia and the Pacific (UNESCAP) for providing GEMS dataset and other technical/material supports for conducting air quality studies in Indonesia.

References

1. GEMS Science Team: New era of air quality monitoring from space: geostationary environment monitoring spectrometer (GEMS). Bull. Am. Meteor. Soc. **101**(1), E1–E22 (2020). https://doi.org/10.1175/BAMS-D-18-0013.1
2. https://www.who.int/news-room/fact-sheets/detail/ambient-(outdoor)-air-quality-and-health Accessed 19 Sept 2022
3. Bhugwant, C., Siéja, B., Bessafi, M., Staudacher, T., Ecormier, J.: Atmospheric sulfur dioxide measurements during the 2005 and 2007 eruptions of the Piton de La Fournaise volcano: implications for human health and environmental changes. J. Volcanol. Geoth. Res. **184**(1), 208–224 (2009). https://doi.org/10.1016/j.jvolgeores.2009.04.012
4. Hobbs, P.V., Radke, L.F., Lyons, J.H., Ferek, R.J., Coffman, D.J., Casadevall, T.J.: Airborne measurements of particle and gas emissions from the 1990 volcanic eruptions of Mount Redoubt. J. Geophys. Res. 96(D10), 18735–18752 (1991). https://doi.org/10.1029/91JD01635
5. Gurwinder Sivia, S., Gheusi, F., Mari, C., Di Muro, A., Tulet, P.: Numerical simulations and parameterizations of volcanic plumes observed at Reunion Island. 10th EGU General Assembly, Apr 2013, Vienna, Austria. pp.EGU2013--10090. ⟨hal-00965476⟩
6. Tabazadeh, A., Turco, R.P.: Stratospheric chlorine injection by volcanic eruptions: HCl scavenging and implications for ozone. Science **260**(5111), 1082–1086 (1993). https://doi.org/10.1126/science.260.5111.1082
7. Evans, B.M., Staudacher, T.H.: In situ measurements of gas discharges across fissures associated with lava flows at reunion island. J. Volcanol. Geoth. Res. **106**(3–4), 255–263 (2001)
8. Re, M.: World Map of Natural Hazards, Munich Reinsurance Company (1998)
9. Brosnan, D.M.: Coral reefs: volcanic impacts, ecological impacts of the Montserrat volcano: a pictorial account of its effects on land and sea life. Sustainable Ecosystems Institute (2000). http://www.sei.org/impacts.html
10. Brantley, S., Myers, B.: Mount St. Helens—from the 1980 eruption to 2000 (Fact Sheet 036–00; Online Version 1.0). https://doi.org/10.3133/fs03600
11. Krueger, A.J., Walter, L.S., Bhartia, P.K., Schnetzler, C.C., Krotkov, N.A., Sprod, I., Bluth, G.J.S.: Volcanic sulfur dioxide measurements from the total ozone mapping spectrometer instruments. J. Geophys. Res. **100**(7), 14057–14076 (1995). https://doi.org/10.1029/95JD01222
12. Schneider, D.J., Rose, W.I., Coke, L.R., Bluth, G.J.S., Sprod, I.E., Krueger, A.J.: Early evolution of a stratospheric volcanic eruption cloud as observed with TOMS and AVHRR. J. Geophys. Res. **104**(4), 4037–4050 (1999). https://doi.org/10.1029/1998JD200073
13. Halmer, M.M., Schmincke, H.-U., Graf, H.-F.: The annual volcanic gas input into the atmosphere, in particular into the stratosphere: a global data set for the past 100 years. J. Volcanol. Geoth. Res. **115**(3), 511–528 (2002). https://doi.org/10.1016/S0377-0273(01)00318-3
14. Draxler, R.R., G.D. Rolph (2003) HYSPLIT (HYbrid single-particle Lagrangian integrated trajectory) Model access via NOAA ARL READY website: http://www.arl.noaa.gov/ready/hysplit4.html. NOAA Air Resources Laboratory, Silver Spring, MD
15. Gouhier, M., Pinel, V., Belart, J.M.C., et al.: CNES-ESA satellite contribution to the operational monitoring of volcanic activity: the 2021 Icelandic eruption of Mt Fagradalsfjall. J Appl. Volcanol. **11**, 10 (2022). https://doi.org/10.1186/s13617-022-00120-3
16. Ebmeier, S.K., Andrews, B.J., Araya, M.C., et al.: Synthesis of global satellite observations of magmatic and volcanic deformation: implications for volcano monitoring and the lateral extent of magmatic domains. J Appl. Volcanol. **7**, 2 (2018). https://doi.org/10.1186/s13617-018-0071-3

Emission Inventory, Investigation of EKC Presence, Handling of SO_2 Emission, and PBL SO_2 Column Problems in Jakarta

Toni Samiaji

Abstract Sulfur dioxide (SO_2) gas produced from burning objects (including fuel) containing sulfur harms the environment. SO_2 and NO_2 gases can cause acid rain. Based on the results of BMG/BMKG (Meteorological and Geophysical Agency/Meteorology, Climatology, and Geophysics Council) data processing, rainwater's power of hydrogen (pH) in Jakarta decreased from pH 5.47 in 1985 to 4.75 in 2020. In this regard, have SO_2 emissions in Jakarta increased? This study aims to analyze the trend of total SO_2 emission, investigate the presence of Environmental Kuznets Curve (EKC) and minimize the column and emission of SO_2 in Jakarta. The data are fuel consumption, sulfur content, population, SO_2 column in Planetary Boundary Layers (PBL), Gross Regional Domestic Product (GRDP), waste, and industrial production. The method used is the calculation of SO_2 emissions using tier 2, the relationship between the economy and the environment based on the investigation of the existence of the EKC in Jakarta and minimizing the column and emission of SO_2 with a Strength, Weakness, Opportunity, Threat (SWOT) analysis. The results found that the total SO_2 emissions in Jakarta from 1970–2021 tended to increase from 41 to 119 kt. There was no EKC in Jakarta, then based on SWOT analysis, the right strategy to minimize emissions and SO_2 column in Jakarta is diversification.

1 Introduction

As a basis for making policies related to dealing with air pollution problems, planning for controlling air pollution in urban areas, and determining the layout of urban spaces, it is necessary to carry out an inventory of pollutant emissions and their sources, one of which is the inventory of SO_2. From previous research, SO_2 inventory from fuel consumption in Jakarta in 2011 showed that SO_2 emissions were from

T. Samiaji (✉)
Research Center for Climate and Atmosphere, National Research and Innovation Agency, Jl. Sangkuriang, Cisitu, Bandung, West Java, Indonesia
e-mail: toni003@brin.go.id

A. Basit et al. (eds.), *Proceedings of the International Conference on Radioscience, Equatorial Atmospheric Science and Environment and Humanosphere Science*, Springer Proceedings in Physics 290, https://doi.org/10.1007/978-981-19-9768-6_18

industry and power plants at 78.22, followed by the transportation sector at 21.73 and the household sector at 0.05% [1]. Meanwhile, the trend of SO_2 emissions in Jakarta carried out by Lestari (2016) shows that SO_2 from 2010 to 2015 from power plants tended to decrease while those from transportation, industry, household, and commercial tended to increase [2]. This phenomenon raises the question of the trend from all these sectors when added up. International Institute for Applied Systems Analysis (IIASA), in the Asian model of Regional Air Pollution Information and Simulation (RAINS), conducts an inventory of SO_2 not only from fuels but also from industrial processes [3]. The three studies did not involve sources of SO_2 from burning waste, even based on the DKI (a particular area of the capital) Jakarta Provincial Sanitation Service, DKI Jakarta's waste production per day is up to thousands of tons, especially in a year. This study will inventory SO_2 emissions from fuel combustion, industrial processes, and household waste burning as a differentiator from previous research from 1970 to 2021 in Jakarta is a novelty in this study, and then, the trend is seen. So the research problem is how is the SO_2 emissions trend from anthropogenic activities (fuel and household waste burning plus industrial processes) in Jakarta from 1970 to 2021.

Increased fuel combustion causes economic growth but will emit SO_2 gas into the air and cause adverse effects on the environment, or the environment degrades. One of economic growth describes per capita income. When per capita income is associated with environmental degradation and forms an inverted U curve, it is called the EKC [4] found. In China, Wang et al. (2016) found the EKC from 1990 to 2012 when linking SO_2 emissions with Gross Domestic Product (GDP) [5], and also in 139 cities in India from 2001 to 2013 [6]. Next is in 12 European countries, when SO_2 emissions per capita are related to GDP per capita in 1870–2001, the EKC is found [7]. In EKC, not only SO_2 emissions but SO_2 concentrations also are associated with GDP [8]. For example, Listyarini (2008) associated Jakarta GRDP with SO_2 concentrations from 1993 to 2004, but the EKC was not found [9]. The concentration of SO_2 from the surface to the height of the PBL forms a column of PBL SO_2. Then the research problem is if the author connects GRDP per capita with the PBL SO_2 column, is the EKC found in Jakarta?

If SO_2 emissions and PBL SO_2 columns in Jakarta increase, what efforts can reduce emissions and column SO_2? In this regard, do not be careless in making decisions. For this reason, one way is to use a SWOT analysis [10]. People widely use SWOT analysis in strategic management [11], including in product marketing strategies [12] and dissecting business cases [13]. In addition, they use the SWOT analysis to assess the impact of a development project on the environment [14] or combine it with the cloud model for the sustainable water resources assessment [15] and others. People have not discussed the best way to reduce SO_2 emissions and its column based on SWOT analysis. In this study, using SWOT analysis, the writer will look for the right strategy to reduce SO_2 emissions and PBL SO_2 columns in Jakarta if they increase.

2 Methods and Data

The data used in this study are the sulfur content in fuel 1979–2025, fuel consumption in Jakarta 1970–2018, PBL SO_2 column in 2005–2021, waste production 1990–2014, the population of Jakarta and Indonesia 1970–2019, GRDP and GDP 1970–2019, industrial production 1990–2017. The author obtained fuel consumption data from the Limited Liability Company of Pertamina and BPS (Central Statistics Agency) Jakarta Province, SO_2 column National Aeronautics and Space Administration (NASA), the sulfur content the Directorate General of Oil and Gas, Ministry of Energy and Mineral Resources, and the Ministry of Environment and Forestry, population, industrial production, waste, and GRDP/GDP Jakarta and Central BPS.

To calculate SO_2 emissions in Jakarta, the writer use the equation

$$E = A \times \mathrm{EF} \tag{1}$$

where

E = emissions; A = activity rate; EF = emissions factor [16]. The activity rate can be in the form of the amount of fuel consumption each year.

The author determined SO_2 or SOx emissions from industrial processes by the number of particular goods produced or raw materials required and emission factors, as shown in 2.

$$E_{\mathrm{pollutant}} = \mathrm{AR}_{\mathrm{production}} \times \mathrm{EF}_{\mathrm{pollutant}} \tag{2}$$

Source [17, 18].

where $E_{pollutant}$ = SO_2 or SOx emissions from industrial processes, $AR_{production}$= the rate of activity of industrial processes, namely the number of particular goods produced or the number of raw materials needed, $\mathrm{EF}_{pollutant}$ = emission factor of SO_2 or SOx. The particular goods are goods that the process of their formation emits SO_2 gas. According to Kuenen et al. [17], Kuenen et al. [18], Maddox [19], and Kuenen and Dellaert [20], SO_2 or SOx emission factors from industrial processes depend on the type of goods produced.

The researcher may calculate SO_2 emissions from burning waste by the formula (2) above, only $E_{\mathrm{pollutant}}$ = SO_2 gas emission, $AR_{\mathrm{production}}$ = the rate of activity, namely the amount of garbage in the Tonnage burned annually.

To make EKC, Selden and Song (1994) used the formula

$$Y = \beta_0 + \beta_1 x + \beta_2 x^2 + \varepsilon \tag{3}$$

where Y is emission per capita, β_0 is a constant, β_1 and β_2 are the regression coefficient, x is GDP per capita, and ε is an error [21]. In another study at (3), Y is the ambient SO_2 concentration, x is GRDP [9], while in this study, Y is the PBL SO_2

column, and x is GRDP/capita. Considering that the SO_2 concentration that occurs in Jakarta is not solely due to SO_2 emissions from Jakarta, there is a contribution from the area around Jakarta caused by the wind that carries pollutants and clouds [22]. To overcome the problem of SO_2 gas in Jakarta, the goverment of Jakarta and its surroundings need the right strategy and action. To overcome this problem, here, the author uses a SWOT analysis.

Figure 1 shows a SWOT diagram; the x-axis extends from weaknesses to strengths, and the y-axis stretches from threats to opportunities. Deciding the exact strategy coordinates on the SWOT diagram, it is determined by

$$X = T_S - T_W \tag{4}$$

$$Y = T_O - T_T \tag{5}$$

Ingaldi and Skurkova [24], where X are the coordinates on the x-axis, Y y-axis, T_S is the total score of the internal aspects of strength, T_W is weakness, T_O is the external factor of opportunity,

and T_T is a threat. While the total score of each element is determined by

$$T_S = \sum_{i}^{n_s} R_{S_i} \times W_{S_i} \tag{6}$$

$$T_W = \sum_{i}^{n_w} R_{W_i} \times W_{W_i} \tag{7}$$

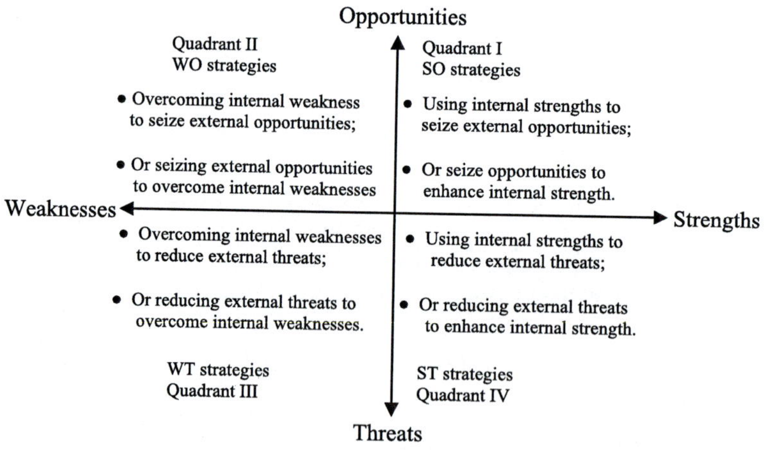

Fig. 1 SWOT diagram [23]

$$T_O = \sum_{i}^{n_o} R_{O_i} \times W_{O_i} \tag{8}$$

$$T_T = \sum_{i}^{n_t} R_{T_i} \times W_{T_i} \tag{9}$$

Primadona and Rafiqi [25], where R_{S_i} is the rating of each factor point of strength, R_{W_i} weakness, R_{O_i} opportunity, and R_{T_i} threat. Then W_{S_i} is the factor weight of strength, W_{W_i} weakness, W_{O_i} opportunity, and W_{T_i} threat. In addition, the sum of each weight of each factor is equal to 1, namely [26]

$$\sum_{i}^{n_s} W_{S_i} = 1 \tag{10}$$

$$\sum_{i}^{n_w} W_{w_i} = 1 \tag{11}$$

$$\sum_{i}^{n_o} W_{O_i} = 1 \tag{12}$$

$$\sum_{i}^{n_t} W_{T_i} = 1 \tag{13}$$

3 Results and Discussion

After analyzing the trend of total SO_2 emissions in Jakarta from 1970 to 2021, the author found that emissions tend to increase by natural logarithm, as shown in (14).

$$y = 19867 \ln(x) + 40943 + \varepsilon \tag{14}$$

y is the total SO_2 emission in Jakarta from 1970 to 2021, x is the year number, and ε is an error. Based on (14), the SO_2 in Jakarta tends to increase from 41 to 119 kt from 1970 to 2021. Then based on the analysis of variance (ANOVA), as given in Table 1, it is found that the regression significance is 0.02 (smaller than 0.05). It means that the model shown by (14) has a 98% confidence level that is correct.

Table 1 The significance and regression coefficient of the independent variable

R^2	Independent variable/model	Coefficient	Sig
0.104	Constant	40,943	0.120
	Regression (ANOVA)	–	0.020
	Ln(year number)	19,867	0.020

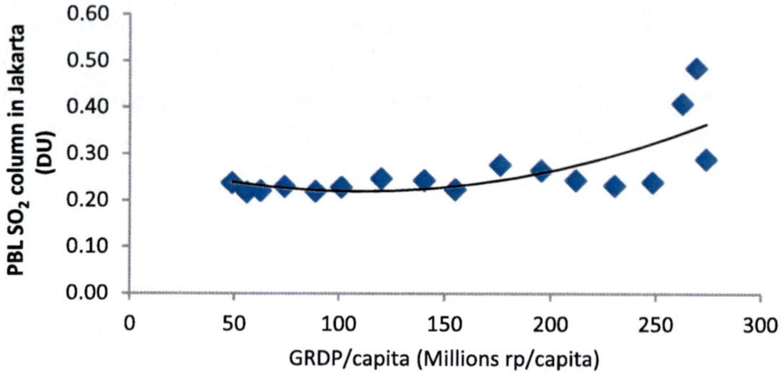

Fig. 2 Relationship curve of PBL SO_2 column with GRDP per capita for the period 2005–2021 in Jakarta

Table 1 shows that the significance of the constant is 0.12, meaning that the value of this constant is 88% true. The value of R^2 (determination factor) in Table 1 0.104 shows the importance of y in (14) 10.4% influenced by ln(x). It means the total SO_2 emissions in Jakarta, 10.4%, are affected by time. In comparison, the remaining 89.6% is influenced by other factors, for example, fuel consumption, industrial processes, waste burning, and sulfur content in fuel.

Figure 2 shows the relationship curve of the PBL SO_2 column with GRDP per Capita for 2005–2021 in Jakarta. Based on (3), the curve has the equation.

$$y = (5E - 06)x^2 - 0.0012x + 0.2838 + \varepsilon \tag{15}$$

However, y is the PBL SO_2 column, and x is the GRDP per capita. Because the shape of the curve is U, the writer did not find EKC. In addition, when relating the ambient SO_2 concentration with GRDP in Jakarta, EKC was not found [9]. Figure 2 means increasing community welfare lowers the SO_2 column. Still, with increasing community welfare, the community is increasingly consumptive, meaning that fuel use and waste production also increase, thus increasing the SO_2 column. Rising industrial production can cause increased waste production. Due to industrial production increases, SO_2 emissions also increase, thus increasing the PBL SO_2 column. The model significance shown in (15) is 0.008, so the U-shaped model is 99.2% correct. In this study, to analyze the existence of EKC in Jakarta, only GRDP/capita against the SO_2 PBL column, because of the ANOVA significance of GRDP and GRDP per capita on SO_2 emissions, GRDP per capita on SO_2 emissions per capita, GRDP and GRDP per capita on ambient SO_2 concentrations, GRDP and GRDP per capita to the sulfate ion concentration in rainwater is higher than 0.05. The maximum significance value of 0.05 is a general requirement for a model to be accepted in statistics. For GRDP against the PBL SO_2 column, although the significance of the ANOVA is less than 0.05, the regression coefficient significance for GRDP (or GRDP squared) is not defined (or cannot be determined). This model has a minimal regression coefficient, so the author considered the value of the PBL SO_2 column equal to constant because the product of the regression coefficient by GRDP (or GRDP squared) is very small and thought zero.

Judging from the column curve of SO_2 to GRDP per capita, with the increase in people's welfare, the environment in Jakarta, which initially experienced an improvement, became a slump. According to the equation model (15), in only six years, namely, from 2005 to 2011, the SO_2 column experienced a decrease and then increased. The large SO_2 column is due to the high SO_2 emission. The high source of SO_2 emissions is coal-fired steam power plants [27]. In addition, there are several industries not only from burning fuel but also from their industrial processes emitting SO_2 gas as cement factories [28], metals [29], glass [30], oil and gas refineries [30], and petrochemicals [18]. Some of these factories are in Jakarta, and others are around Jakarta, as shown in Fig. 3. Considering what contributes to the SO_2 concentration in Jakarta is not only SO_2 emissions from Jakarta but also from around Jakarta [31], as well those who contribute to the PBL SO_2 column Jakarta are not only SO_2 emissions from Jakarta but also from outside Jakarta. These phenomena are shown in Fig. 3 of the airflow by the wind entering Jakarta [22]. So, strategies and actions are needed to reduce the value of the PBL SO_2 column in Jakarta. Not only SO_2 emissions from Jakarta need to be considered but also SO_2 emissions outside Jakarta and the wind direction. Therefore by using SWOT analysis, the writer hopes to find the strategies to reduce emissions and the value of the PBL SO_2 column in Jakarta.

Table 2 shows the SWOT matrix to reduce the value of the PBL SO_2 column and SO_2 emissions in Jakarta and its surroundings. Based on the SWOT matrix and (4) and (5) equations, to determine the strategy for reducing the SO_2 column value and SO_2 emissions in Jakarta and its surroundings on the SWOT diagram, $X = 3.043 - 1.4 = 1.643$ is obtained, while $Y = 2.696 - 3 = -0.304$. The coordinates are = (1.643–0.304), so on the SWOT diagram (Fig. 1), it falls in quadrant IV or the Strength Threat (ST) strategy. The $\tan\theta = \frac{Y}{X} = \frac{-0.304}{1.643}$ is -0.187, so the angle (Θ) formed by the S axis and line from the center of the coordinate to point (x, y) is $-10.5°$ less than $-45°$. This point location is in a diversification strategy [26]. This strategy means that a suitable method to reduce the value of the SO_2 column and SO_2 emissions in Jakarta and its surroundings is to maximize internal strength while reducing external threats. In this case, the community of Jakarta and its vicinity must implement strength factors such as regulations, use measuring equipment and information media, and utilize human and financial resources as well as possible. Meanwhile, the people must reduce threat factors such as the use of fossil fuels, motor vehicle congestion, sectoral egos, and population surges.

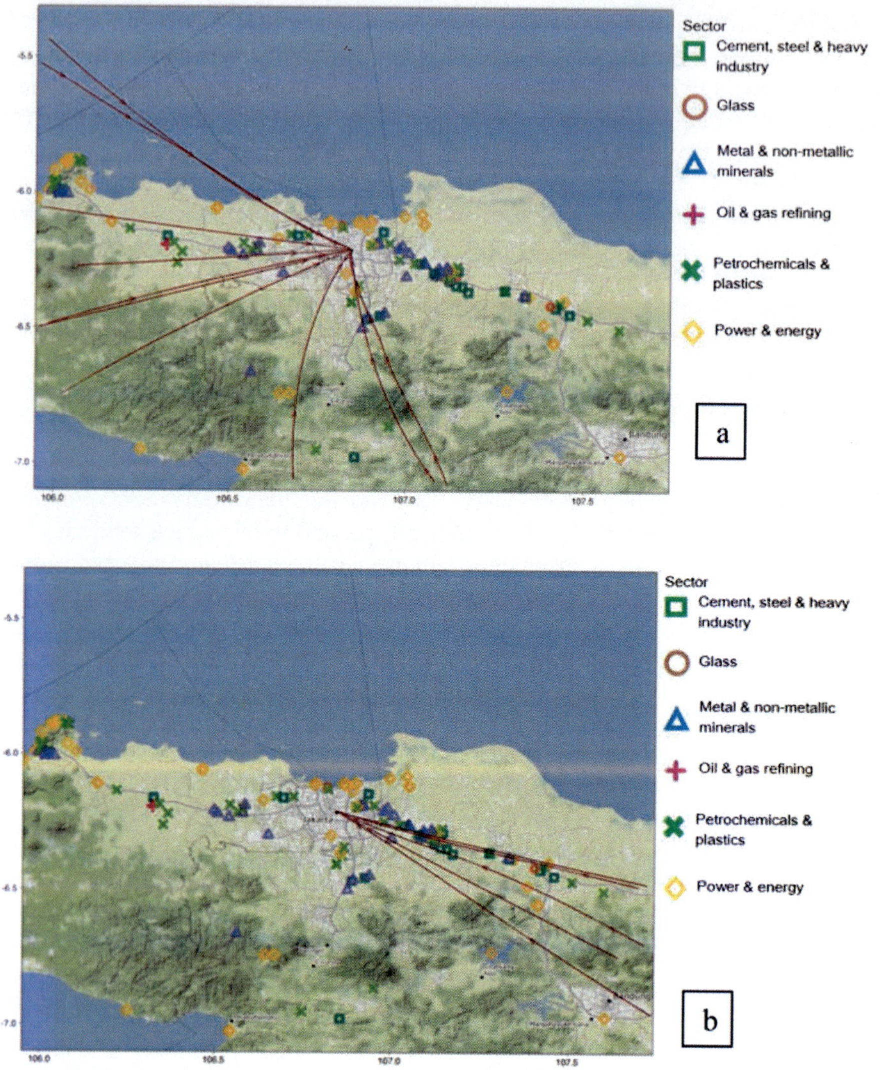

Fig. 3 Airflow (indicated by red lines) flowing into Jakarta in 2017–2020 as a result of the HYSPLIT and PROPER models during **a** the rainy season, **b** the dry season, including bringing SO_2 pollutants from outside Jakarta. *Source* [22]

Table 2 SWOT matrix for decreasing SO_2 column values and SO_2 emissions

Strength factors	Rating	Weight	Score
1. Having Air Quality Monitoring System (AQMS), measurement via satellite, High Volume air sampler, tools emission meter, anemometer	4	0.085	0.340
2. DKI Governor Regulation No. 31 of 2008 concerning emission thresholds for motor vehicle exhaust	4	0.085	0.340
3. Having human resources	3	0.064	0.191
4. Having financial resources	3	0.064	0.191
5. There is a Regional Environmental Laboratory at the Jakarta Environment Agency, BMKG	4	0.085	0.340
6. DKI Regional Regulation No. 2 of 2005 concerning Control of Air Pollution	4	0.085	0.340
7. Having information media to the public	2	0.043	0.085
8. Decree of the Governor of DKI No. 551 of 2001 concerning the determination of ambient air quality standard	3	0.064	0.191
9. DKI Governor Regulation No. 141 of 2007 concerning the use of CNG (Concentrate Natural Gas) for public transportation and local government operational vehicles	3	0.064	0.191
10. DKI Governor Regulation No. 88 of 2019 regarding traffic restrictions with an odd–even system	3	0.064	0.191
11. Decree of the Governor of Jakarta No. 545 of 2016 concerning the determination of the location and motor vehicle-free day schedule	1	0.021	0.021
12. Minister of Environment Regulation No. 12 of 2010 concerning the implementation of air pollution control in the area	1	0.021	0.021
13. Minister of Environment and Forestry Regulation No. 1 of 2021 concerning Company Performance Rating Program in Management Environment	2	0.043	0.085
14. Law No. 32 of 2009 About the Protection and Environmental Management	2	0.043	0.085
15. The company/factory must provide Marginal External Cost, namely include the cost of pollution or environmental damage in the production cost calculation	1	0.021	0.021
16. Governor of Jakarta, in supervising the air quality of Jakarta Province, appoint a Regional Environmental Supervisory Officer at the Office of Environment Jakarta Province	1	0.021	0.021
17. There is SO_2 transport from Jakarta to outside Jakarta	4	0.085	0.34
18. The existence of Intelligent Transport System-Area Traffic Control System (ITS-ATCS)	1	0.021	0.021
19. The Environment Minister Decree No. 45 of 1997 concerning the Air Pollution Standard Index	1	0.021	0.021
Total	47	1	3.043
Weakness factors	Rating	Weight	Score

(continued)

Table 2 (continued)

1. There has been no socialization of these policies to community/business entity	1	0.2	0.2
2. Human resources and funds are still lacking	1	0.2	0.2
3. Ineffective management and coordination	2	0.4	0.8
4. Dependence on fossil fuels is still large	1	0.2	0.2
Total	5	1	1.4
Opportunity factors	Rating	Weight	Score
1. Installing the SO_2 emission reduction device on the chimney, exhaust	4	0.087	0.348
2. Building a tall chimney	1	0.022	0.022
3. Inventory of emissions both in Jakarta and outside Jakarta	2	0.043	0.087
4. Modeling the distribution of SO_2 gas from outside Jakarta	2	0.043	0.087
5. Mixing limestone in coal burning	4	0.087	0.348
6. Controlling illegal parking	1	0.022	0.022
7. Relocating industries to areas far from Jakarta	4	0.087	0.348
8. Preparing Environmental Protection and Management Plan	3	0.065	0.196
9. Reduced fuel oil usage	3	0.065	0.196
10. Making a Strategic Environmental assessment	1	0.022	0.022
11. Analyzing environmental impacts	1	0.022	0.022
12. Switching to mass transportation	1	0.022	0.022
13. Testing SO_2 emissions in motor vehicles and factory chimneys periodically	1	0.022	0.022
14. Reducing sulfur content in fossil fuels	4	0.087	0.348
15. Planting a tree that absorbs SO_2 gas	1	0.022	0.022
16. Energy conservation	1	0.022	0.022
17. Supervision of compliance with emission quality standards, both mobile sources and sources are not moving	1	0.022	0.022
18. Conducting air pollution control education and training for a person or entity whose activities make the air polluted	1	0.022	0.022
19. Energy-saving culture	1	0.022	0.022
20. Using renewable energy	4	0.087	0.348
21. Modification of fuel injectors	1	0.022	0.022
22. Eco-driving	1	0.022	0.022
23. Development of emission permit trading system	1	0.022	0.022
24. Energy diversification to environmentally friendly energy	2	0.043	0.087
Total	46	1	2.696
Threat factors	Rating	Weight	Score
1. Lack of public/business entity awareness of the dangers of air pollution	1	0.036	0.036

(continued)

Table 2 (continued)

2. Fossil fuels as fuel	1	0.036	0.036
3. Motor vehicle congestion	4	0.143	0.571
4. There is SO_2 transport from outside Jakarta to Jakarta	1	0.036	0.036
5. Sectoral ego	2	0.071	0.143
6. Industry in Jakarta does not want to be relocated	3	0.107	0.321
7. Employees do not want to relocate	4	0.143	0.571
8. It costs money to install SO_2 emission reduction devices	1	0.036	0.036
9. The existence of government projects such as toll road construction	3	0.107	0.321
10. Construction of a waste power plant	3	0.107	0.321
11. Increasing population	4	0.143	0.571
12. There is a plan to increase coal consumption in RUED P (provincial region energy general plan) DKI	1	0.036	0.036
Total	28	1	3

4 Conclusion

The study found that SO_2 emissions in Jakarta from 1970 to 2021 tended to increase in a natural logarithm from 41 to 119 kt. Jakarta was no EKC where the y-axis is the PBL SO_2 column while the x-axis is GRDP per capita. Based on the SWOT analysis, the right strategy to minimize SO_2 emissions and SO_2 column in Jakarta, viz diversification. In this case, the people of Jakarta and its surroundings must implement strength factors such as regulations, use measuring equipment and information media, and utilize human and financial resources as well as possible. Meanwhile, they also must reduce threat factors such as the use of fossil fuels, motor vehicle congestion, sectoral egos, and population surges.

References

1. Darmanto, N.S., Dan Sofyan, A.: Analysis of the distribution of air pollutants NO_2, SO_2, CO, and O_2 in Jakarta using WRF-CHEM. J. Environ. Eng. **18**(1), 54–64 (2012)
2. Erou, A., Fadhillah, F.: Inventarisasi emisi Provinsi DKI Jakarta. Seri Lembar Informasi Pencemaran Udara. Indonesian Center for Environmental Law. Icel.or.id (2019)
3. Amann, M., Dhoondia, J.: Rains-Asia User's manual. International Institute for Applied Systems Analysis, A-2361 Laxenburg, Austria (1994)
4. Uchiyama, K.: Environmental Kuznets curve hypothesis. In: Uchiyama, K.(ed.). Environmental Kuznets Curve Hypothesis and Carbon Dioxide Emissions. Development Bank of Japan Research Series. https://doi.org/10.1007/978-4-431-55921-4_2 (2016)
5. Wang, Y., Han, R., Kubota, J.: Is there an environmental Kuznets curve for SO_2 emissions? A semi—parametric panel data analysis for China. Renew. Sustain. Energy Rev. **54**, 1182–1188 (2016)
6. Sinha, A., Bhattacharya, J.: Estimation of environmental Kuznets curve for SO_2 emission: a case of Indian cities. Ecol. Ind. **72**, 881–894 (2017)

7. Markandya, A., Golub, A., Pedroso-galinato, S.: Empirical analysis of national income and SO_2 emissions in selected European Countries. Environ. Res. Econ. **35**, 221–257 (2006)
8. Stern, D.I.: The rise and fall of the environmental Kuznets curve. World Dev. **32**(8), 1419–1439 (2004)
9. Listyarini, S., Tarumingkeng, R.C., Fauzi, A., Hutagaol, P.: Kurva Lingkungan Kuznet: Relasi Antara Pendapatan Penduduk Terhadap Polusi Udara. Jurnal Organisasi dan Manajemen **4**(1), 46–54 (2008)
10. Stacchini, A., Guizzardi, A., Mariotti, A.: Smoothing down arbitrariness in planning: From SWOT to participatory decision making. Land Use Policy **119**. https://doi.org/10.1016/j.landusepol.2022.106213 (2022)
11. Rahim, A.R., Radjab, E.: Manajemen strategi. Cetakan 1, Lembaga Perpustakaan dan Penerbitan Universitas Muhammadiyah Makassar (2017). ISBN: 978-602-8187-60-2
12. Prastiyo, A., Fathoni, A., Malik, D.: Red ginger product marketing strategy through SWOT analysis at UD Company. Ungaran blessing. J. Manag. **4**(4) (2018) ISSN: 2502–7689
13. Rangkuti, F.: Analisis SWOT: Teknik Membedah Kasus Bisnis. Gramedia, Jakarta. (1999) ISBN 979-605-718-2
14. Rathi, A.K.A.: Evaluation of project-level environmental impact assessment and SWOT analysis of the EIA process in India. Environ. Impact Assess. Rev. **67**, 31–39 (2017). https://doi.org/10.1016/j.eiar.2017.08.004
15. Gao, X., Chen, L., Sun, B., Liu, Y.: Employing SWOT analysis and normal cloud model for water resource sustainable utilization assessment and strategy development. Sustainability **9**, 1439 (2017). https://doi.org/10.3390/su9081439
16. Hatta, G.M., Asaad, I.: Regulation of the State Minister of the Environment No. 12 of 2010 concerning the implementation of air pollution control in the regions, Appendix II: Guidelines for the inventory of ambient air quality data and sources of air pollution (2010)
17. Kuenen, J., Berdowski, J., Most, P.V.D., Slager, J.M., Mulder, W., Hlawiczka, S., et al.: Iron and steel production. In: European Environment Agency. EMEP/EEA Air Pollutant Emission Inventory Guidebook, p. 21. Publications Office of the European Union, Luxembourg. ISBN: 978-92-9213-806-6 (2016a)
18. Kuenen, J., Appelman, W., Rentz, O., Oertel, D., Berdowski, J., Jonker, J., et al.: Chemical industry. In: European Environment Agency. EMEP/EEA air pollutant emission inventory guidebook, p. 26. Publications Office of the European Union, Luxembourg. ISBN: 978-92-9213-806-6 (2016b)
19. Maddox, R.R.: Gas and Liquid Sweetening, 2nd ed. Campbell Petroleum Series, Norman, OK (1974)
20. Kuenen, J., Dellaert, S.: Other metal production. In: European Environment Agency. EMEP/EEA air pollutant emission inventory guidebook, p. 5. Publications Office of the European Union, Luxembourg. ISBN: 978-92-9213-806-6 (2016)
21. Susandi, A.: The impact of international greenhouse gas emissions reduction on Indonesia. PhD thesis. Report on earth system science. Max Planck Institute for Meteorology. ISSN: 1614–1199. Hamburg (2004)
22. Myllyvirta, L., Suarez, I., Uusivuori, E., Thieriot, H., dan Anhauser, A.: Pencemaran udara lintas batas di Provinsi DKI Jakarta, Banten, dan Jawa Barat. Centre for Research on Energy and Clean Air (CREA) (2020)
23. Li, C., Negnevitsky, M., Wang, X.: Prospective assessment of methanol vehicles in China using FANP-SWOT analysis. Transp. Policy **96**, 60–75 (2020). https://doi.org/10.1016/j.tranpol.2020.06.010
24. Ingaldi, M., Skurkova, K.L.: Company strategy determination in food companies using the SWOT method. Acta Technol. Agric. **3**. https://doi.org/10.2478/ata-2014-0015 (2014)
25. Primadona, Y., Rafiqi, Y.: SWOT analysis on the competition strategy of the Madina Purbaratu minimarket, Tasikmalaya City. Sharia Econ. J. **2**(1) (2019). P-ISSN: 2548–5032
26. Zhou, J., He, P., Qin, Y., Ren, D.: A selection model based on SWOT analysis for determining a suitable strategy for prefabrication implementation in rural areas. Sustain. Cities Soc. **50**, 101715 (2019). https://doi.org/10.1016/j.scs.2019.101715

27. Song, H., Yang, M.: Analysis of Effectiveness of SO_2 Emission Reduction in Shanxi, China by Satellite Remote Sensing. Atmosphere **5**, 830–846. (2014)https://doi.org/10.3390/atmos5040830
28. Zhang, T., Wu, C., Li, B., Wang, J., et al.: Linking the emission of cement plants to the sulfur characteristics of their limestones: a study of 80 NSP cement lines in China. J. Clean. Prod. **220**, 200–211 (2019). https://doi.org/10.1016/j.jclepro.2019.02.133
29. Sun, X., Liu, Z., Huang, W., Ji, L., et al.: Looping upcycling SO_2 into value-added H_2S by fast-induced reduction process for heavy metals treatment in the nonferrous smelting industry. Fuel **331**, 125867 (2023). https://doi.org/10.1016/j.fuel.2022.125867
30. Kharol, S.K., Fioletov, V., McLinden, C.A., Shephard, M.W., et al.: The ceramic industry at Morbi is a large source of SO_2 emissions in India. Atmos. Environ. **223**, 117243 (2020). https://doi.org/10.1016/j.atmosenv.2019.117243
31. Samiaji, T.: The scenario of sulfur content reduction on coal to fulfil the quality standards of SO_2 concentration in the province of DKI Jakarta. Dissertation. Environmental Science Graduate School of Padjadjaran University. Bandung, Indonesia (2022)

Impact of Vortex on Rainfall as a Trigger for Tropical Cyclones (TC) in Maritime Continent of Southern Indonesia (Case Studies: Victoria, Ernie, and Seroja TC)

Luthfiyah Jannatunnisa, Nurjanna Joko Trilaksono, and Muhammad Ridho Syahputra

Abstract Tropical cyclones can have a destructive impact because they can directly affect weather conditions and wave height. In some cases, before becoming a tropical cyclone, it begins with a vortex that can produce weather disturbances such as rain, storms, and even tropical cyclones themselves. In this study, we investigated whether there are other possible vortex events, especially for the Maritime Continent of Southern Indonesia region. The data used are wind and relative vorticity from the reanalysis of the European Center for Medium-Range Weather Forecast (ECMWF) Reanalysis fifth Generation (ERA5) during the period of vortex events that grew into cyclones around the south of Indonesia from 2013 to 2021, then grouped for single, double, and triple vortex. A spatial rainfall intensity review was done using Global Satellite Mapping of Precipitation (GSMaP) satellite data when the vortex phenomenon occurred to assess the impact of the vortex event. Results showed that single vortex events were more common with seven events, double vortex with six events, and triple vortex with three events during 2013–2021. Single vortex formation area tends to be in the west of the Sumatra Region, double vortex formation tends to spread in the south of Indonesia, and the triple vortex occurs along the southern sea of Indonesia (west to east). The impact on rainfall spatially can be seen with the more vortex, the wider rainfall occurs in the south of Indonesia.

1 Introduction

Tropical cyclones (TC)s are one of the phenomena that can cause hydrometeorological disasters in the territory of Indonesia. Although rare, these tropical cyclones

L. Jannatunnisa
Master Program in Earth Sciences, Faculty of Earth Science and Technology, Institut Teknologi Bandung, Bandung, Indonesia

N. J. Trilaksono (✉) · M. R. Syahputra
Atmospheric Sciences Research Group, Faculty of Earth Science and Technology, Institut Teknologi Bandung, Bandung, Indonesia
e-mail: jpatiani@itb.ac.id

A. Basit et al. (eds.), *Proceedings of the International Conference on Radioscience, Equatorial Atmospheric Science and Environment and Humanosphere Science*, Springer Proceedings in Physics 290, https://doi.org/10.1007/978-981-19-9768-6_19

have a destructive impact because they can directly affect weather conditions and even storm waves [1, 2]. TC is an intense vortex event in the atmosphere which is classified as a synoptic event on the atmospheric motion scale [3] because it has a life span of days to weeks and a typical size of about 200–500 km but can reach 1000 km (World Meteorological Organization—WMO). In some cases, before becoming a tropical cyclone, it begins with a vortex that can produce weather disturbances such as rain, storms, and even tropical cyclones itself [4, 5]. Previous research has shown that the vortex in southern Indonesia that has a high frequency (in the December–January–February season) is a vortex in the East Indian Ocean region, and the impact is that there is significant rainfall for the southern part of Sumatra and Java Island [6]. There is also a finding of double vortex in the northern and southern parts of the East Indian Ocean which can result in an increase in rainfall in several locations in Indonesia [7].

Therefore, from previous studies, it will be investigated whether there are other possible vortex events. The distribution of vortex events in the southern region of Indonesia has various characteristics [6]. So that a vortex characterization process will be carried out which is reviewed based on the number of vortexes, especially those that initialize the occurrence of tropical cyclones. This study aims to review vortex events in the Maritime Continent of Southern Indonesia based on differences in the number of vortex events at one time, then review the differences in their impact on rainfall.

2 Data and Method

This study uses zonal and meridional wind data (u, v) as well as relative vorticity from the reanalysis of the European Center for Medium-Range Weather Forecast (ECMWF) Reanalysis fifth Generation (ERA5) during the period of vortex events that grew into tropical cyclones around the region of Southern Indonesia from 2013 to 2021. The data can be accessed at the link https://cds.climate.copernicus.eu/#!/home. The vortex identification method will be carried out following the research procedure conducted by Chang et al. [8], namely the vortex is identified when there is a clockwise wind circulation in the southern hemisphere (southern Indonesia) at an altitude of 925 hPa and wind speed of more than 2 m/s in the southern hemisphere. Four grid points 2.5° × 2.5° from the center of circulation are located [8]. Once identified, spatial rainfall intensity is reviewed using Global Satellite Mapping of Precipitation (GSMaP; available online at https://hokusai.eorc.jaxa.jp) satellite data, by calculating rainfall anomalies, which are the differences in rainfall values (in mm/hour) during cyclone maturation with its initial vortex. It aims to evaluate the difference in the impact of these phenomena.

3 Results and Discussion

The result will be divided into the following sections, the first section is about the vortex event in southern Indonesia, the second to fourth sections will discuss case study vortex analysis for the Victoria, Ernie, and Seroja TC, and the last section will present a discussion on the impact of all TC cases on the rainfall.

3.1 Vortex Events in Maritime Continent of Southern Indonesia

The single vortex formation area tends to be in the west of the Sumatra Region (Fig. 1a) except for the case of Lili and Marcus, the double vortex tends to spread in the southern part of Java and Papua (Fig. 1b), and the triple vortex occurs along the southern waters of Indonesia (Fig. 1c). From this vortex event, one case was taken from each single, double, and triple vortex classification (marked with bold letters). Furthermore, Table 1 shows the incidence of vortexes that developed into tropical cyclones during the period 2013–2021. From the statistical results obtained, there were more single vortex events, with seven events followed by a double vortex with six events; in addition, three triple vortex events were found during the period 2013–2021.

3.2 Single Vortex on Victoria Cyclone Event (2013)

Victoria TC developed in the Eastern Indian Ocean with fast and consistent cyclone movement toward the south–southeast. Figure 2 shows streamlines and vortices for initial vortex events (a and b), tropical cyclone (c and d), and cyclone dissipation (e and f). Seen on April 04 at 00 UTC, an initial vortex was formed which indicates a single vortex in the southwest of Sumatra. Then when entering the tropical cyclone stage on April 6 at 12 UTC, it was seen that there was an increase in wind speed and the relative vorticity value was more negative which tended to be circular and concentrated.

To evaluate the impact of this phenomenon, an anomaly of rainfall (in mm/hour) is shown at the time of tropical cyclone and vortex initialization stage. Figure 3 shows a positive anomaly that tends to occur in southern Sumatra to several areas on the island of Java, this condition indicates that during the Victoria TC, there was an increase in rainfall in the area compared to when the initial vortex was detected. While the negative anomaly value indicates when the initial vortex is formed, there is an increase in rainfall in the waters of the eastern Indian Ocean and some areas in central Sumatra.

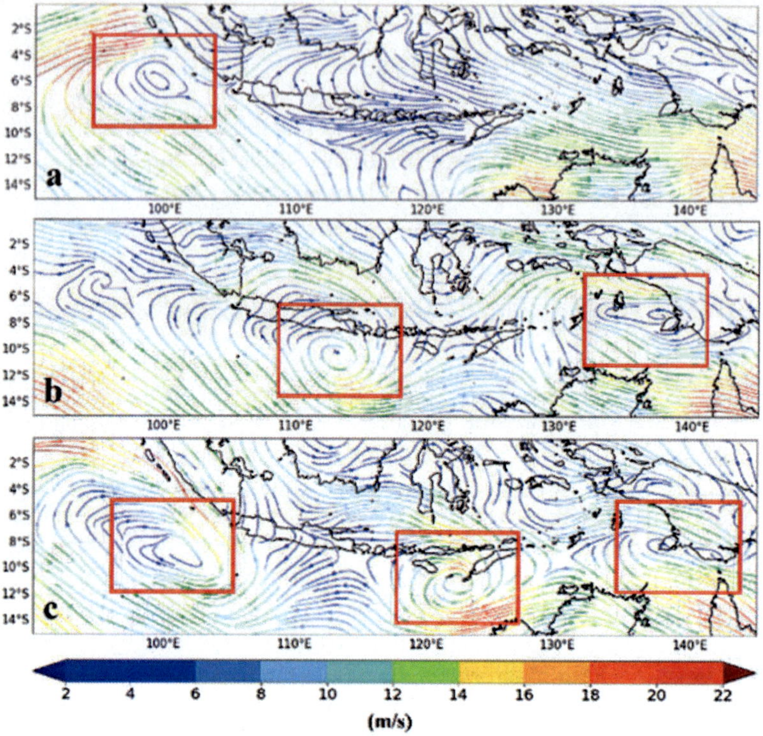

Fig. 1 Initial vortex in the southern Indonesian maritime continent in the period 2013–2021 for case studies of **a** single vortex on Victoria TC, **b** double vortex on Ernie TC, and **c** triple vortex on Seroja TC

Table 1 Classification of vortex events in Maritime Continent of Southern Indonesia for case studies of vortexes that developed into tropical cyclones for the period 2013–2021

Single	Double	Triple
Manga (2020)	Nora (2018)	**Seroja (2021)**
Lili (2019)	**Ernie (2017)**	Riley (2019)
Marcus (2018)	Frances (2017)	Narelle (2013)
Flamboyant (2018)	Cempaka and Dahlia (2017)	–
Memories (2018)	Yvette (2016)	–
Victoria (2013)	Daffodils (2014)	–
Alessia (2013)	–	–

The bold letters indicate each single, double, and triple vortex classification case that was evaluated further in this study

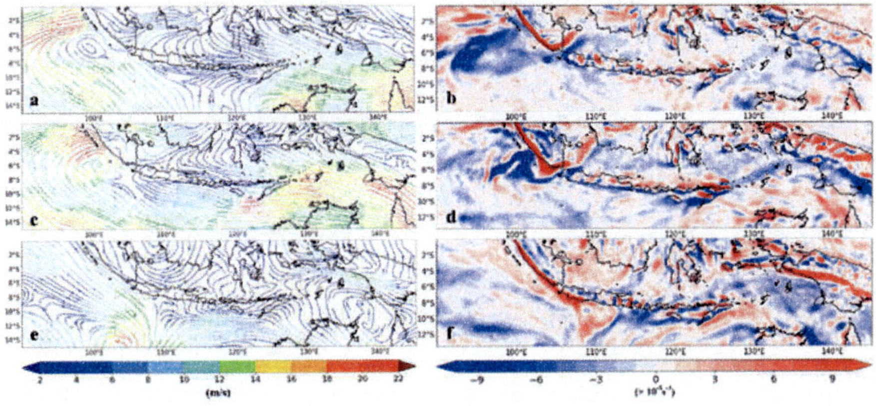

Fig. 2 Figures **a**, **c**, and **e** are streamlines plots and **b**, **d**, and **f** plots of relative vorticity at an altitude of 925 hPa at the identified single vortex stage on April 4, 2013, at 00 UTC (a and b), the Victoria TC on April 6, 2013, at 12 UTC (c and d) and the dissipation stage of the cyclone or cyclone away from the study area on April 10, 2013, at 06 UTC (e and f)

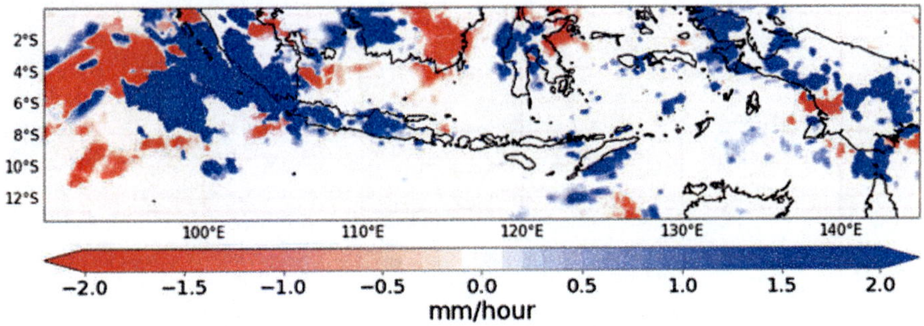

Fig. 3 Anomalies of rainfall during Victoria TC and the initialization of the formation of a single vortex. Positive values (blue shaded) indicate rainfall anomalies when TC occurs, and negative values (red shaded) indicate rainfall anomalies during initialization of a single vortex

3.3 Double Vortex on Ernie Cyclone Event (2017)

Cyclone Ernie grew in the southern waters of Indonesia, on April 4, southwest, and experienced a peak intensity of category five on April 7 at 12 UTC. Figure 4 shows streamlines and vorticity for initial vortex events (a and b), tropical cyclone (c and d), and cyclone dissipation (e and f) stage. On April 5, at 00 UTC, an initial vortex was formed, indicating the presence of a double vortex in the southern part of Java and around the Arafura Sea. Then when entering the tropical cyclone stage on April 7 at 00 UTC, it was seen that there was an increase in wind speed and the value of relative vorticity was more negative which tended to be circular and concentrated and the vortex in the Arafura Sea was strengthening.

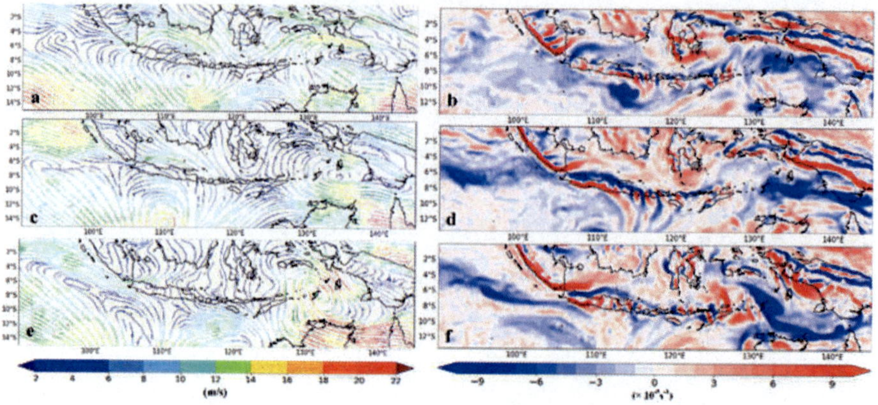

Fig. 4 Figures **a**, **c**, and **e** are streamlined plots and **b**, **d**, and **f** plots of relative vorticity at an altitude of 925 hPa at the identified double vortex stage on April 5, 2017, at 00 UTC (a and b), the Ernie TC on April 7, 2017, at 00 UTC (c and d) and the dissipation stage of the cyclone or cyclone away from the study area on April 9, 2017, at 00 UTC (**e** and **f**)

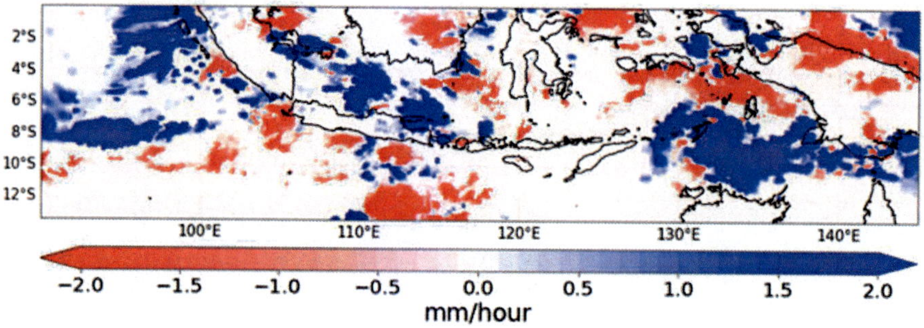

Fig. 5 Same as Fig. 3, but for Ernie TC

Figure 5 shows a positive anomaly occurred in the eastern part of Java, Java Sea, North Sea of Australia, and South of Papua during the strengthening of the cyclone. While the negative anomaly value indicates when the initial vortex is formed, there is an increase in rainfall in the waters of southern Java and the Arafura Sea. Meanwhile, for the Nusa Tenggara archipelago and its surroundings, there was a reduction in rainfall due to the double vortex.

3.4 Triple Vortex on Seroja Cyclone Event (2021)

Tropical Cyclone Seroja is quite well known because it has a destructive impact on East Nusa Tenggara and its surroundings. This cyclone moves slowly enough over several days in the region to cause continuous heavy rains, floods, landslides, and even fatalities. Cyclone Seroja is a category three cyclone and is a cyclone that experienced the Fujiwara effect [9] with Cyclone Odette and was able to survive far to the south. Figure 6 shows streamlines and vortices for initial vortex events (a and b), tropical cyclone (c and d), and cyclone dissipation (e and f) stage. Seen on April 2 at 18 UTC, an initial vortex was formed indicating the presence of a triple vortex along the southern part of the southern Indonesian maritime continent ocean (southwest, south, and southeast). Then when entering the tropical cyclone stage on April 3 at 18 UTC, it was seen that there was an increase in wind speed and the relative vorticity value was more negative which tended to be circular and concentrated. Meanwhile, when Cyclone Seroja moved south on April 7 at 06 UTC, it was seen that the vortex that was previously in the southeast was already outside the study area, and another vortex was formed in the eastern Indian Ocean.

Furthermore, Fig. 7 shows positive anomalies occurring in the southern region of Sumatra, western Java, southern Nusa Tenggara islands, and the Arafura Sea during the strengthening of the cyclone. While the negative anomaly value indicates that when the initial vortex is formed, there is an increase in rainfall in the waters of western Sumatra, the islands of Nusa Tenggara and its surroundings and the southern part of Papua. Meanwhile, for the central part of Java and Bali, there was a reduction in rainfall due to the triple vortex.

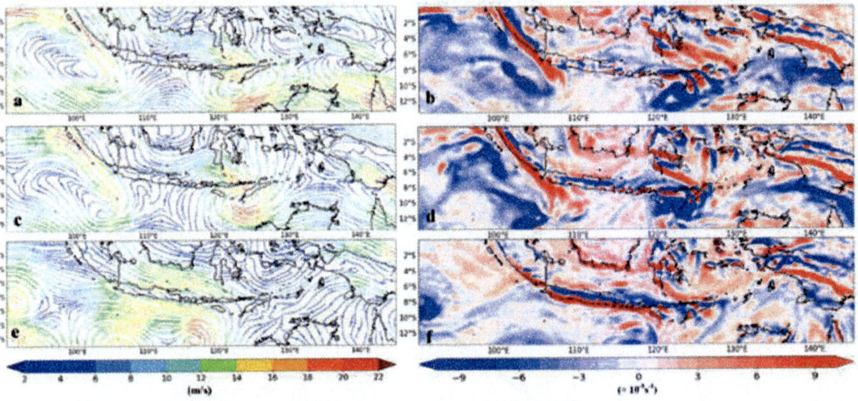

Fig. 6 Figures **a**, **c**, and **e** are streamlined plots and **b**, **d**, and **f** plots of relative vorticity at an altitude of 925 hPa at the single vortex identified stage on April 2, 2021, at 18 UTC (a and b), the Seroja TC on April 3, 2021, at 18 UTC (c and d) and the stage of dissipation of the cyclone or cyclone away from the study area on April 7, 2021, at 6 UTC (e and f)

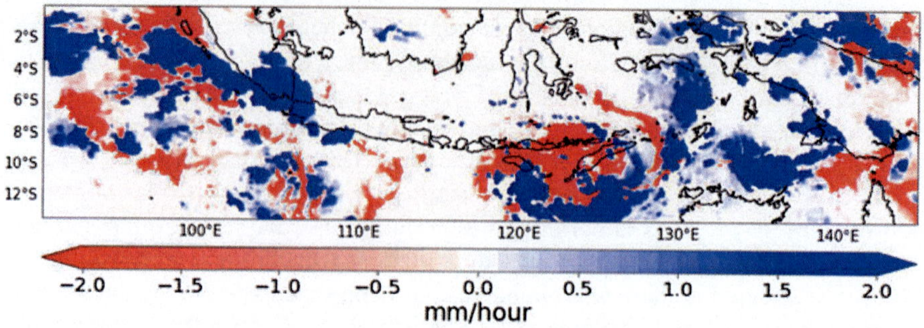

Fig. 7 Same as Fig. 3, but for Seroja TC

3.5 *Differences Impact of Rainfall on the Number of Vortex Events*

This section shows the impact of rainfall from vortex events quantitatively, by looking at rainfall during the initial vortex, tropical cyclone stage, and dissipation stage on five samples of weather observation stations in the study area. The locations are the Long Maritime Meteorological Station (Bandar Lampung City), Ahmad Yani Meteorological Station (Semarang, Central Java), David Constantijn Saudale Meteorological Station (Rote Ndao Regency, East Nusa Tenggara-NTT), Eltari Meteorological Station (Kupang City, NTT), and Merauke Climatology Station (Papua). In Fig. 8, the intensity of rainfall at Lampung station increased as the vortex formed until the tropical cyclone stage, then decreased after the cyclone weakened or moved away. However, in Central Java, rainfall tends to be constant and not too influenced by vortex events and tropical cyclones. Meanwhile, for other regions, observation data is not available.

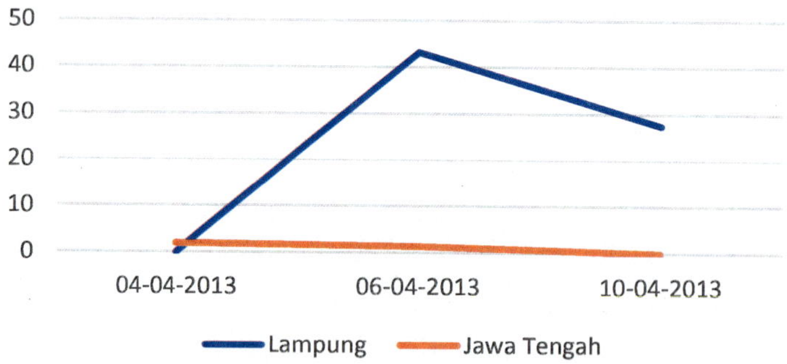

Fig. 8 Rainfall time series for the Victoria TC case study for Lampung (blue line) and Central Java (orange line) at the initial vortex formation (April 4, 2013), tropical cyclone stage (April 6, 2013), and cyclone dissipation stage (April 10, 2013)

Then for the double vortex in the case of Victoria TC is shown in Fig. 9. There is an increase in rainfall in the Central Java area until the cyclone grows, then decreases when the cyclone moves away. However, for the Lampung area, due to limited observation data, only an increase in rainfall is seen after the cyclone has moved away; this is due to the presence of strong winds of around 10–14 m/s (Fig. 4), and a new vortex system has emerged in the west of the area, Indonesia.

Figure 10 shows that the availability of data is much better than the previous case study, but for Central Java, it is not available. For the cities of Kupang, Rote Ndao, and Lampung, it was seen that there was an increase in rainfall from the initial vortex to the occurrence of tropical cyclone and weakened thereafter. However, in Merauke City, there was a decrease in rainfall intensity from the initial vortex to tropical cyclone stage due to the movement of the cyclone away from that location.

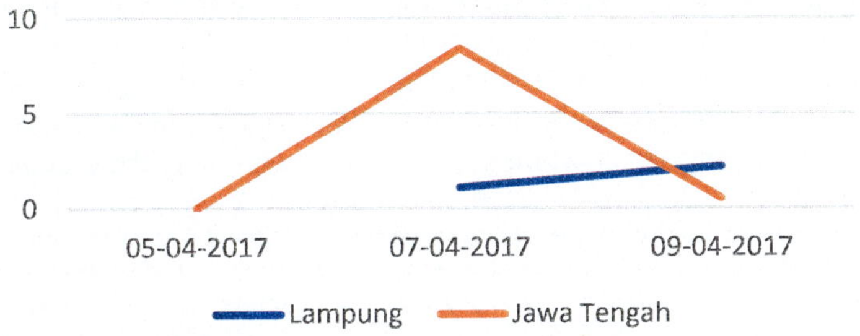

Fig. 9 Rainfall time series for the case study of Ernie TC for Lampung (blue line) and Central Java (orange line) at the initial vortex formed (April 5, 2017), tropical cyclone stage (April 7, 2017), and cyclone dissipation stage (April 9, 2017)

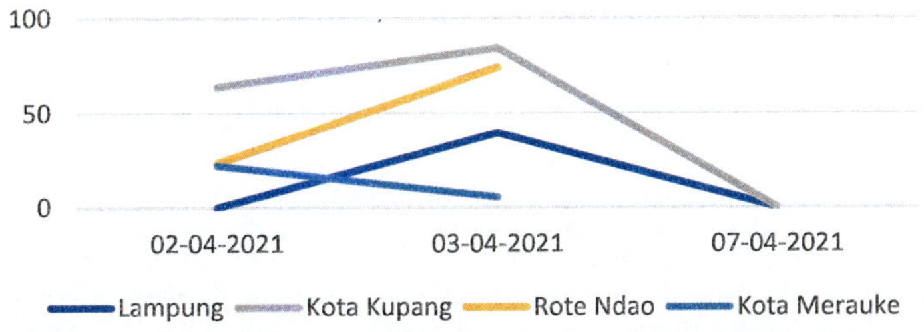

Fig. 10 Rainfall time series for the Seroja TC case study for Lampung (dark blue line), Central Java (orange line), Kupang City (gray), Rote Ndao (yellow), and Merauke City (light blue), at initial vortex forming (April 2, 2021), tropical cyclone stage (April 3, 2021), and cyclone weakening/removal (April 7, 2021)

4 Conclusions

From the observations of vortex events in the southern part of southern Indonesian maritime continent for each vortex case that preceded tropical cyclones during the period 2013–2021, the vortex events can be divided into single, double, and even triple vortexes. This difference in the number of vortexes causes differences in the resulting impact, by reviewing the rainfall anomaly. For single vortex events in the case of Cyclone Victoria, it appears that there is one initial vortex formed and the impact of rainfall only occurs around central Sumatra. In the case of Cyclone Ernie, there were double vortexes in the waters of southern Java and the Arafura Sea, then the impact of the rainfall that occurred was an increase in the area and a decrease in rainfall around the islands of Nusa Tenggara and its surroundings. Furthermore, in the case of Cyclone Seroja, it was seen that a triple vortex was formed, with the impact of an increase in rainfall in the waters of western Sumatra, the islands of Nusa Tenggara and its surroundings, as well as the southern part of Papua. So, with the difference in the number of vortexes that are increasing (single to triple), it will provide weather disturbances, one of which is in the form of more widespread rainfall in the southern Indonesian maritime continent area, especially the southern part.

Acknowledgements This work would not have been possible without the cooperation and guidance of the atmospheric sciences, Faculty of Earth Sciences and Technology, Institut Teknologi Bandung staff, who have supported this research and provided fruitful discussion, also Marine Technology Cooperation Research Center (MTCRC) which has funded the first author. Besides that, we thank ECMWF and JAXA GSMaP team for providing data for this research.

References

1. Suryantoro, A.: Tropical cyclones in south and southwest Indonesia from TRMM satellite monitoring and their possible links to high waves and tornadoes. Aerosp. Sci. Technol. Mag. 3(1) (2010)
2. Needham, H.F., Keim, B.D., Sathiaraj, D.: A review of tropical cyclone-generated storm surges: global data sources, observations, and impacts. Rev. Geophys. **53**(2), 545–591 (2015)
3. Chan, J.C.: The physics of tropical cyclone motion. Annu. Rev. Fluid. Mech. **37**, 99–128 (2005)
4. Hou, J., Wang, P., and Zhuang, S.: A new method of characterizing flow patterns of vortices and detecting the centers of vortices in a numerical wind field. J. Atmos. Oceanic Technol. (2017)
5. Emanuel, K.: Tropical cyclones. Annu. Rev. Earth Planet. Sci. **31**(1), 75–104 (2003)
6. Pratama, B.E.: Study of vortex activities in the Indonesian maritime continent. Master's Program Thesis, Institut Teknologi Bandung (2014)
7. Putra, I.K.Y.D., Trilaksono, N.J.: The impact of double vortex phenomena in eastern Indian Ocean on rainfall in western part of Indonesia. In AIP Conference Proceedings, vol. 1987, no. 1, p. 020045. AIP Publishing LLC (2018)
8. Chang, C.P., Harr, P.A., Chen, H.J.: Synoptic disturbances over the equatorial South China Sea and western maritime continent during boreal winter. Mon. Weather Rev. **133**(3), 489–503 (2005)
9. Diamond, H.J., Schreck, C.J., Allgood, A., Becker, E.J., Blake, E.S., Bringas, FG., Camargo, S.J., Chen, L., Coelho, C.A., Fauchereau, N., Goldenberg, S.B., Woolley, JM.: The Tropics. Bull. Am. Meteorol. Soc. 103(8), S193–S256 (2022)

Study of the Inter-Tropical Convergence Zone (ITCZ) Movement Over the Maritime Continent Region

Didi Satiadi, Ibnu Fathrio, and Anis Purwaningsih

Abstract Understanding the behavior of the ITCZ is crucial in order to understand the variability of weather and climate in the tropics. However, its behavior has not been fully understood, especially over the Maritime Continent region. This study aims to investigate the behavior of the ITCZ movement over the globe and the Maritime Continent region using a long-term monthly mean rainfall dataset from the fifth generation of ECMWF Atmospheric Reanalysis (ERA-5) from 2001 until 2020. In this investigation, the ITCZ positions were determined as the zonally averaged latitudinal position with maximum rainfall. In addition, the Australian monsoon indices (AUSMI) were used to investigate the relationship between seasonal migration of the ITCZ and the monsoon. The results of the study showed the bimodality of the ITCZ preferred locations, one in the north and another in the south of the equator. The seasonal meridional migration of the ITCZ generally followed that of the sun's apparent position with a 1–2 month delay. We found that the movements of the ITCZ were not always gradual, but sometimes exhibited cross-equatorial rapid movement or "jump". The seasonal migration of the sun's apparent position caused one ITCZ position to become more dominant than the other, such that the ITCZ appeared to jump from the less to the more dominant position. Further study also showed that the ITCZ jumps generally coincided with the onset or withdrawal of the Australian monsoon. The results of this study supported and provided further proof to numerical experiments by Chao (Chao in J Atmos Sci 57:641–651, 2000), Chao and Chen (J Atmos Sci 58:2820–2831, 2001, Clim Dyn 22:447–459, 2004) using a General Circulation Model (GCM) and previous studies by Satiadi (Pergerakan Meridional dan Zonal ITCZ dalam Kaitannya dengan Curah Hujan di Wilayah Benua Maritim Indonesia, Prosiding Seminar Nasional Sains Atmosfer dan Antariksa, pp. 44–56, 2011), Satiadi and Fathrio (Jurnal Sains Dirgantara 9:1–11, 2011, Bimodalitas dan Lompatan ITCZ: Teori, Pengamatan dan Simulasi, Prosiding Seminar Nasional Sains

D. Satiadi (✉) · I. Fathrio · A. Purwaningsih
Research Center for Climate and Atmosphere, National Research and Innovation Agency, Bandung, West Java, Indonesia
e-mail: satiadi@yahoo.co.uk

A. Basit et al. (eds.), *Proceedings of the International Conference on Radioscience, Equatorial Atmospheric Science and Environment and Humanosphere Science*, Springer Proceedings in Physics 290, https://doi.org/10.1007/978-981-19-9768-6_20

Atmosfer dan Antariksa, pp. 68–76, 2012) using the Multi-functional Transport Satellite (MTSAT) and the Tropical Rainfall Measuring Mission (TRMM) data that showed similar results.

1 Introduction

The Inter-tropical Convergence Zone (ITCZ) is a dominant phenomenon in the tropics that affects the weather and climate in this region, through modulation of wind, cloud, and rainfall. The ITCZ undergoes seasonal migration meridionally following the sun's relative position to the earth. The ITCZ is a component of the Hadley/Walker circulation, where surface air converges and rises upward. The latent heat released in the ITCZ drives and interacts with the Hadley/Walker circulation. The ITCZ is easily visible on satellite imagery as a series of tropical synoptic waves stretching west to east. Each wave is a cloud group oscillating between connected and split states for about two weeks, during which the split conditions can create a cyclone. The ITCZ is a continuous cloud band with varying latitudes in the longitudinal direction due to zonal variations of surface conditions such as the SST, land distribution, and land characteristics [1]. Waliser and Gautier [20] provided the climatological characteristics of the ITCZ using long-term satellite's Highly Reflective Cloud (HRC) dataset.

An early theory of the ITCZ came from Charney [5] was developed based on the theory of the Convective Instability of Second Kind (CISK), to explain the mechanism of tropical cyclone formation. This theory states that the convergence of water vapor at the boundary layer acts as fuel for the growth of convection. The stronger the convection will give a feedback, further strengthening the convergence at the boundary layer. According to Charney, a large Coriolis force plays an essential role in the formation of convection, so the ITCZ is very likely to be found in polar regions, with a large Coriolis force. However, because the supply of water vapor is more abundant at the equator, ITCZ also tends to form near the equator.

Subsequent research on ITCZ was carried out by Pike [13] and Sumi [19]. Pike and Sumi conducted experiments with a water planet model with uniform Sea Surface Temperature (SST) and showed the presence of ITCZ, which remained near the equator. This theory contradicts Charney's theory which predicts that uniform SST should form ITCZ at the poles. Furthermore, an explanation of the reasons why the ITCZ tends to form near the equator which also answers the experimental results of Sumi and Pike was given by Chao [1], Chao and Chen [4]. Chao and Chen also answered the problem of the occurrence of double ITCZ in Sumi's experiment, which was not caused by the horizontal resolution of the model used by Sumi.

The water planet model experiment conducted by Chao [1] using the Global Circulation Model (GCM) has shown how important the ITCZ phenomenon is and answered the previous ITCZ problems. According to the ITCZ theory proposed by Chao, the ITCZ tends to occupy certain stable latitude positions. At this position, an equilibrium is reached between the gravitational force or the effect of the earth's rotation and the location of the SST peak. The presence of preferred positions of the ITCZ was also found by Sikka and Gadgil [18] who found two favorable locations for a maximum cloud zone over the Indian longitudes (70° E–90° E) during April–October 1973–1977. Waliser and Somerville [21] also examined the preferred latitudinal position of the ITCZ based on observation, theory, and modeling analysis.

Chao explained that the movement of the ITCZ is closely related to the change of seasons or monsoons. Chao defines a monsoon as the ITCZ located at latitudes more than 10° from the equator. Chao also showed that the meridional movement of the ITCZ undergoes a sudden jump from the equator to higher latitudes, further defined as the onset of the monsoon. Lau and Yang [11] also studied seasonal variation, abrupt transition, and intra-seasonal variability associated with the Asian Summer Monsoon using a global circulation model. They found that the onset of the Asian Monsoon is followed by the sudden jump of the equatorial ITCZ. Such an abrupt seasonal variation of the ITCZ was also observed by Hu et al [10] using the Global Precipitation Climatology Project (GPCP) daily data. Hagos and Cook [7] also studied the mechanism behind the abrupt latitudinal shift of maximum precipitation known as the West African monsoon jump using a regional climate model. They concluded that the jump is an example of multiscale interaction in the climate system, in which an intra-seasonal scale event is triggered by the smooth seasonal evolution of SSTs and the solar forcing in the presence of land–sea contrast.

The monsoon theory based on the ITCZ is very different from the classical theory, which explains that the monsoon circulation is caused by the temperature difference between the land and the ocean. Chao's experiments have proven that atmospheric circulation associated with the monsoon persists even without the land. The temperature contrast between the land and the sea only modifies the formation of the monsoon circulation and affects the location of the ITCZ. Chao argues that this new definition of monsoon applies not only to the western Pacific region, where conditions resemble the conditions of the water planet model, but also to all monsoon circulations in the tropics. The movement of the ITCZ associated with this monsoon can be explained by understanding the equilibrium between the two attractive forces acting on the ITCZ caused by the earth's rotation (Coriolis force) and the location of the peak of the SST. The balance between these two pulling forces ultimately determines the positions favored by ITCZ. In line with the theory, [6] observed that the seasonal migration of the ITCZ was basically responsible for the variability of the Indian summer monsoon.

Research has been carried out to study the movement of the ITCZ especially in the Maritime Continent region. The aim of the research is to investigate the preferred locations and the seasonal migration of the ITCZ and its relationship with the Australian Monsoon. This research is an extension of previous studies by Satiadi [14], Satiadi and Fathrio [17], Satiadi and Adikusumah [15] using the Multi-functional Transport Satellite (MTSAT), the Tropical Rainfall Measuring Mission (TRMM), and the Commonwealth Scientific and Industrial Organization (CSIRO) Global Circulation Model (GCM) data. In the current investigation, longer term (20 years) ERA-5 reanalysis data were used, and more in depth analyses were presented.

2 Data and Methodology

The gridded data of monthly average rainfall from the European Center for Medium Range Weather Forecasting Reanalysis 5 (ERA-5) was used to determine the position and seasonal migration of the ITCZ. This data was downloaded from the Climate Data Store (CDS) Catalog ([8, 9]. The data was obtained for a particular domain (180° W–180° E, 30° S–30° N) with 0.25° × 0.25° latitude–longitude resolution, for the 2001–2020 period.

The position of the ITCZ was determined as the latitudinal position with the maximum rainfall. This was based on the assumption that the convergence zone of the Hadley circulation produces the tallest and thickest clouds and thus produces the most rainfall. Moreover, the mean latitudinal position of the ITCZ was calculated as the zonal mean (180° W–180° E). After the average position of ITCZ for each month was obtained, then the movement of ITCZ from month to month in a year was obtained and plotted.

Furthermore, we examined the relationship between the seasonal migration of the ITCZ and the Australian Monsoon over the year by comparing the monsoon indices with the ITCZ movement over the tropical area (180° W–180° E) and the Australian Monsoon area (110°–140° E). The Australian Monsoon Index (AUSMI) data for the 2001–2015 period were obtained from the Asia Pacific Data Research Center of the International Pacific Research Center (IPRC) (http://apdrc.soest.hawaii.edu/projects/monsoon/daily-data.html#mon). The AUSMI was defined as an area-averaged zonal wind at 850 hPa level within an area bounded by (5° S–15° S, 110° E–140° E). The index was calculated for its monthly mean, and then averaged over 15 years (2001–2015). The seasonal change of the average indices was then compared to the ITCZ movement to investigate the relationship between the ITCZ and the monsoon.

3 Results and Discussion

3.1 Bimodality of the ITCZ

The variability of the zonally averaged meridional profile of annual rainfall within 20 years (2001–2020), and its average were identified. A bimodal distribution of rainfall was indicated with two maxima, in the north of the equator (between 0° and 10° N), and in the south of the equator (between 10° S and 0°), separated by a minimum at the equator (see Fig. 1). The maxima corresponded to the average position of the ITCZ. However, the bimodal profile of the maxima was not symmetrical about the equator. The maximum in the north of the equator was generally higher (ranging from 6 to 8 mm/day) than that in the south of the equator (ranging from 4 to 6 mm/day). The asymmetry was presumably due to the difference in land/ocean composition between the northern and the southern hemispheres. The northern hemisphere (NH) has more land than ocean, while the southern hemisphere (SH) has more ocean than land. When the NH experiences summer, the air pressure will be lower in the NH than that in the SH, so that the wind will move from the SH to the NH. Conversely, when the SH experiences summer, the air pressure will be lower in the SH than that in the NH, so that the wind will move from the NH to the SH. However, because the NH is dominated by land while the SH is dominated by ocean, the summer pressure in the NH is lower than that in the SH, so that the ITCZ is pushed more strongly to the NH and spends more time in the NH. Such an asymmetry has also been found by Masunaga and L'ecuyer [12] in the East Pacific ITCZ. The zonally averaged meridional profile of annual rainfall is mainly determined by the variation of the Coriolis force and the Sea Surface Temperature (SST) with respect to latitude, so that the distribution is generally smooth as can be seen in Fig. 1.

The meridional profile of the zonally averaged monthly rainfall shows a similar pattern to that of the annual rainfall, but the monthly profile depicts more variability between the months. In general, the monthly profile also indicated a bimodal and asymmetric distribution (see Fig. 2). Distinct profile for each month was presumably caused by the different apparent position of the sun during its annual cycle, which influenced the monthly distribution of rainfall. The position of the maximum rainfall, and hence the ITCZ, alternated between the NH and the SH (Fig. 2). For example, in January, February, and March, the maximum was in the SH, while in other months, it was in the NH. In January, the maxima in the SH and the NS were almost balanced. The bimodality was more pronounced from December to April, when both maxima were clearly visible. However, it was less pronounced from May to November, when only the northern maximum was clearly visible. It can be seen from Fig. 2, the shifting of the profile during its annual cycle. In January, the profile was in the southernmost position. In the following months, the profile shifted gradually to the north, the southern maximum decreased, while the northern maximum increased,

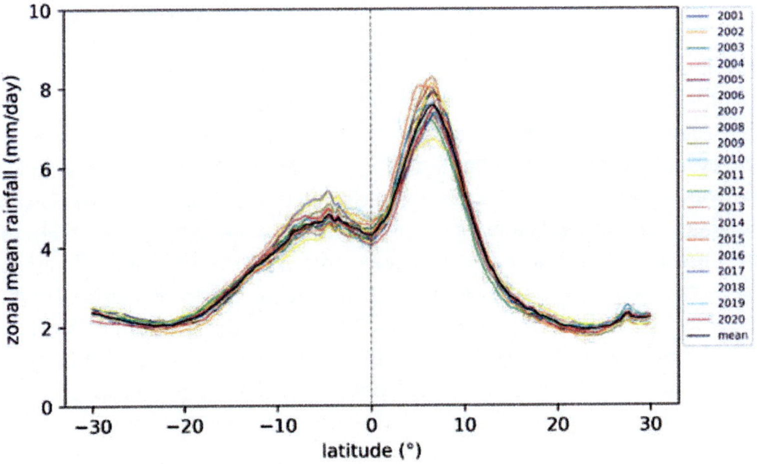

Fig. 1 Zonally averaged meridional profile of annual mean rainfall during 2001–2020

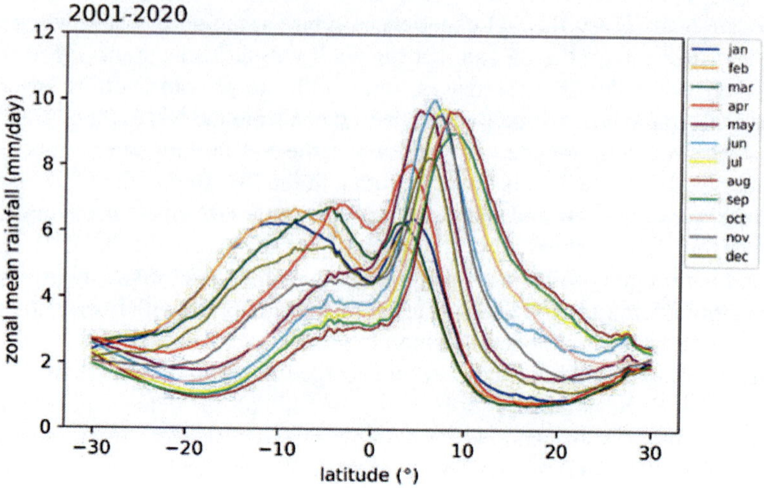

Fig. 2 Zonally averaged meridional profile of monthly mean rainfall (2001–2020)

so that the asymmetry increased. In August, the profile reached the northernmost position. Subsequently, in the following months, the profile receded gradually to the south, the northern maximum decreased, while the southern maximum increased, so that the symmetry increased. Figure 2 also shows that the rainfall distribution in the SH was more spread out in latitude than that in the NH, which was more localized in latitude.

3.2 *Seasonal Migration of the ITCZ*

The movement of the ITCZ during its annual cycle was not always gradual, but occasionally underwent a northward jump (between January and April) or southward jump (between December and January) (see Fig. 3). Moreover, it can be seen that the jump was actually the change of modality from the southern to the northern maximum and vice versa. The change of modality was related to the migration of the apparent position of the sun during its annual cycle. When the sun was in the SH (NH), then the southern (northern) maximum became more dominant than the other. It can be seen from Fig. 3 that the positions of the southern maximum have more variability than the northern one, which were more consistent. In general, the position of both maxima during its annual cycle followed the apparent position of the sun, which formed a sinusoidal pattern. However, the movement of the northern maximum seemed to experience a delay from that of the sun's apparent position for about 2 months. The delay was thought to be caused by the seasonal hysteresis phenomenon, where the output (maximum rainfall) over land, which dominated the NH, lagged behind the input (the annual cycle of the sun's apparent position). The hysteresis did not appear

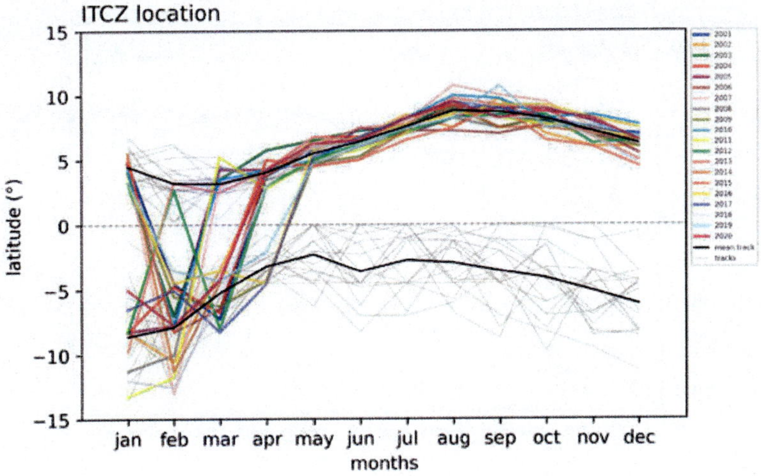

Fig. 3 Seasonal migration of ITCZ latitudinal position (2001–2020)

significant in the SH, which was dominated by the ocean, and hence responded very much slowly and weakly to the seasonal change of radiation. Zhang et al [22] investigated the ITCZ seasonal migration using GPCP data (1981–2010) over the Eastern Pacific, the Atlantic, the Indo-Western Pacific, and the global domain and found a similar track of the seasonal migration of the ITCZ. In February and March, the position of the northern maximum was the closest to the equator, while in August it was the farthest. In May, the southern maximum was the closest to the equator, while in January it was the farthest. In April, the positions of the maxima were the closest to each other.

3.3 *ITCZ Jump and Monsoon Onset*

Relationships between the ITCZ and the monsoon were analyzed using a combined plot of the latitudinal position of ITCZ with the Australian Monsoon Index (AUSMI) against the month (January–December) during 2001–2015 from the ERA-5 (Fig. 4). The AUSMI was defined as an area-averaged zonal wind at 850 hPa level within an area bounded by (5° S–15° S, 110° E–140° E). The average values of AUSMI were positive in December–January–February, when the westerly wind dominated, which corresponded to the wet season in the southern part of the IMC (Fig. 4). Conversely, they were negative in other months, when the easterly wind dominated, which corresponded to the dry season in the southern part of the IMC. The AUSMI change sign from positive to negative in around March and from negative to positive between November and December. It can be seen from Fig. 4 that the northward jump of the ITCZ occurred in March, coincident with the sign change (from positive to

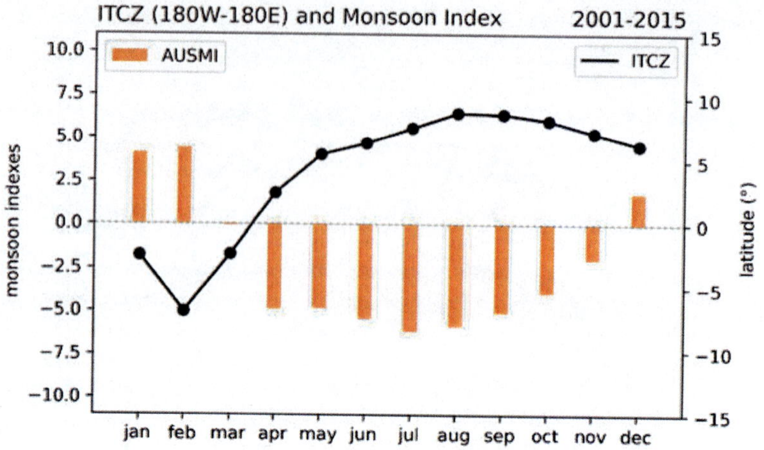

Fig. 4 Position of ITCZ (180° W–180° E) vs the Australian Monsoon Index (AUSMI)

negative) of the AUSMI, which also occurred in March. However, the southward jump of the ITCZ occurred in December, a bit late (about a month) compared to the sign change (from negative to positive) of the AUSMI, which occurred in November. The northward jump was quite rapid presumably due to the fact that the NH was dominated by landmass, which attracted the ITCZ. For the same reason, the southward jump was delayed due to the fact that the southward jump was withheld by the attraction of landmass in the NH.

The similar plot to Fig. 4 was employed for a different domain. This time, the zonal average was taken around the IMC, especially over the Australian Monsoon region (110° E–140° E). It can be seen from Fig. 5 that the northward jump of the ITCZ occurred in April, a bit late (about a month) compared to the sign change of the AUSMI, which occurred in March. However, the southward jump of the ITCZ in November coincided with the sign change of the AUSMI, which also occurred in November. The northward jump was delayed presumably due to the effect of the Australia Continent in this domain, which withheld the northward jump. For the same reason, the southward jump was rapid due to the attraction of the Australia Continent. Thus, the Australia continent may play an important role in determining the onset of the dry and wet season in this particular region.

4 Conclusions

Research has been conducted to study the meridional movement of the ITCZ using monthly rainfall data from ERA-5 during 2001–2020. In this study, the meridional position of the ITCZ was determined as the latitudinal position with the maximum rainfall in the zonally averaged meridional profile of monthly rainfall.

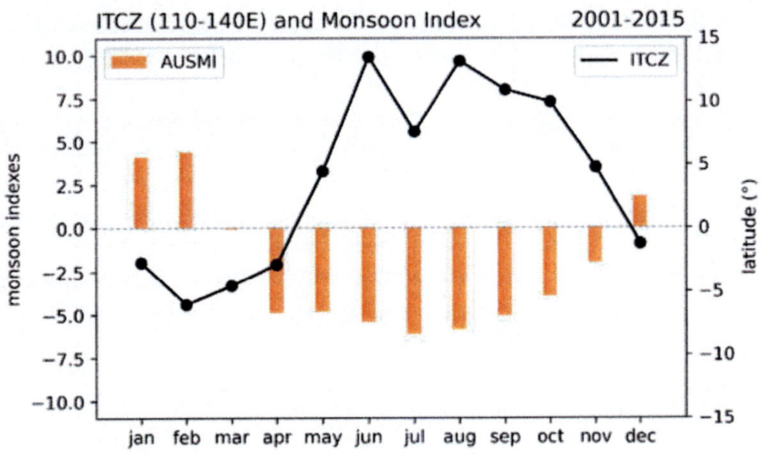

Fig. 5 Position of ITCZ (110°–140° E) vs the Australian Monsoon Index (AUSMI)

The zonally averaged meridional profile of annual rainfall exhibited a bimodal distribution with two maxima, one in the north of the equator and another one in the south of the equator, which corresponded to the two preferred locations of the ITCZ. The bimodal distribution was not symmetrical about the equator; the northern maximum was higher than the southern maximum. This asymmetry was presumably due to different land/sea composition between the northern and the southern hemisphere. More land in the NH and more ocean in the SH have caused the ITCZ to spend more time in the NH, which has the lowest pressure.

The zonally averaged meridional profile of monthly rainfall also exhibited a bimodal distribution. However, during the annual cycle, the position of the maximum alternated between the NH and the SH depending on the apparent position of the sun. In January–February–March, the maximum tended to be in the south, while in other months in the north.

The movement of the ITCZ generally followed the migration of the apparent position of the sun during its annual cycle, although it was delayed probably due to the seasonal hysteresis effect, especially in the NH. The movement of the ITCZ during its annual cycle did not always occur gradually, but occasionally experienced a northward or southward jump. The jump corresponded to the change in the modality from the southern maximum to the northern maximum and vice versa. We found that the poleward jumps occurred quite consistently every year during 2001–2020.

Comparison between the ITCZ migration and the Australian Monsoon Index (AUSMI) showed that in general, the ITCZ jump corresponded to the monsoon onset. In the global domain, the northward jump coincided with the withdrawal of the Australian Monsoon, although the southward jump occurred a bit late compared to the onset of the Australian Monsoon. In the AUSMI domain, the southward jump coincided with the onset of the Australian Monsoon, although the northward jump occurred a bit late compared to the withdrawal of the Australian Monsoon. Therefore the ITCZ southward jump in the AUSMI domain was better in detecting the onset of the Australian Monsoon (the start of the wet season over the IMC), whereas the ITCZ northward jump in the global domain was better in detecting the withdrawal of the Australian Monsoon (the start of the dry season over the IMC).

The results of this study supported and provided further proof to numerical experiments by Chao [1], Chao and Chen [2], Chao and Chen [4] using a General Circulation Model (GCM) and previous studies by Satiadi [14], Satiadi and Fathrio [17], Satiadi and Adikusumah [15] using the Multi-functional Transport Satellite (MTSAT) and the Tropical Rainfall Measuring Mission (TRMM) data that showed similar results.

References

1. Chao, W.C.: Multiple quasi equilibria of the ITCZ and the origin of monsoon onset. J. Atmos. Sci. **57**, 641–651 (2000)
2. Chao, W.C., Chen, B.: Multiple quasi-equilibria of the ITCZ and the origin of monsoon onset. Part II Rotational ITCZ attractors. J. Atmos. Sci. 58, 2820–2831 (2001)
3. Chao, W.C., Chen, B.: The origin of monsoons. J. Atmos. Sci. **58**, 3497–3507 (2001)
4. Chao, W.C., Chen, B.: Single and double ITCZ in an aqua-planet model with constant SST and solar angle. Clim. Dyn. **22**, 447–459 (2004)
5. Charney, J.G.: Tropical Cyclogenesis and the Formation of the ITCZ. Mathematical Problems of Geophysical Fluid Dynamics. In: W. H. Reid (ed.) Lectures in Applied Mathematics American Mathematical Society, vol. 13, pp. 355–368
6. Gadgil, S.: The monsoon system: land–sea breeze or the ITCZ? J. Earth Syst. Sci. **127**, 1 (2018). https://doi.org/10.1007/s12040-017-0916-x
7. Hagos, S.M., Cook, K.H.: Dynamics of the West African monsoon jump. J. Clim. **20**, 5264–5284 (2007)
8. Hersbach, H., Bell, B., Berrisford, P., Hirahara, S., Horanyi, A., Munoz-Sabater, J., Nicolas, J., Peubey, C., Radu, R., Schepers, D., Simmons, A., Soci, C., Abdalla, S., Abellan, X., Balsamo, G., Bechtold, P., Biavati, G., Bidlot, J., Bonavita, M., Chiara, G., Dahlgren, P., Dee, D., Diamantakis, M., Dragani, R., Flemming, J., Forbes, R., Fuentes, M., Geer, A., Haimberger, L., Healy, S., Hogan, R.J., Holm, E., Janiskova, M., Keeley, S., Laloyaux, P., Lopez, P., Lupu, C., Radnoti, G., Rosnay, P., Rozum, I., Vamborg, F., Villaume, S., Thepaut, J.-N.: The ERA5 global reanalysis. Q. J. R. Meteorol. Soc. **2020**, 1–51 (2020). https://doi.org/10.1002/qj.3803
9. Hersbach, H., Bell, B., Berrisford, P., Biavati, G., Horányi, A., Muñoz Sabater, J., Nicolas, J., Peubey, C., Radu, R., Rozum, I., Schepers, D., Simmons, A., Soci, C., Dee, D., Thépaut, J.N.: ERA5 monthly averaged data on single levels from 1959 to present. Copernicus Clim. Change Serv. (C3S) Clim. Data Store (CDS). (Accessed on 01-Sep-2022) (2019). https://doi.org/10.24381/cds.f17050d7
10. Hu Y et al.: Abrupt seasonal variation of the ITCZ and the Hadley circulation, Geophysical Research Letters, **34**, L18814. https://doi.org/10.1029/2007GL030950
11. Lau, K.M., Yang, S.: Seasonal variation, abrupt transition, and intraseasonal variability associated with the Asian summer monsoon in the GLA GCM. J. Clim. **9**, 965–985 (1995)
12. Masunaga, H., L'ecuyer, T.S.: Equatorial asymmetry of the east pacific ITCZ: observational constraints on the underlying processes. J. Clim. **24**, 1784–1800 (2011)
13. Pike, A.C.: Intertropical convergence zone studied with an interacting atmosphere and ocean model. Mon. Wea. Rev. **99**, 469–477 (1971)
14. Satiadi, D.: Pergerakan Meridional dan Zonal ITCZ dalam Kaitannya dengan Curah Hujan di Wilayah Benua Maritim Indonesia, Prosiding Seminar Nasional Sains Atmosfer dan Antariksa, pp. 44–56 (2011). ISBN 978-979-1458-53-5

15. Satiadi, D., Adikusumah, N.: Bimodalitas dan Lompatan ITCZ: Teori, Pengamatan dan Simulasi, Prosiding Seminar Nasional Sains Atmosfer dan Antariksa, pp. 68–76 (2012). ISBN 978-979-1458-64-1
16. Satiadi, D., Adikusumah, N.: ITCZ dan Monsun Indo-Australia, Prosiding Seminar Nasional Sains Atmosfer dan Antariksa, pp. 77–83 (2015). ISBN 978-979-1458-64-1
17. Satiadi, D., Fathrio, I.: Penentuan onset Monsun Di Wilayah indo-Australia Berdasarkan Lompatan ITCZ. Jurnal Sains Dirgantara **9**(1), 1–11 (2011)
18. Sikka, D.R., Gadgil, S.: On the maximum cloud zone and the ITCZ over Indian longitudes during southwest monsoon. Mon. Weather Rev. **108**, 1840–1853 (1980)
19. Sumi, A.: Pattern formation of convective activity over the aqua-planet with globally uniform sea surface temperature. J. Meteor. Soc. Japan **70**, 855–876 (1992)
20. Waliser, D.E., Gautier, C.: A satellite-derived climatology of the ITCZ. J. Clim. **6**, 2162–2173 (1993)
21. Waliser, D.E., Somerville, C.J.: Preferred latitude of the intertropical convergence zone. J. Atmos. Sci. **51**(12), 1619–1639 (1994)
22. Zhang et al.: Advances in research on the ITCZ: mean position, model bias, and anthropogenic aerosol influences. J. Meteorol. Res. **35,** 729–741 (2021)

Weather Condition Identification Using Edge Detection Method for Early Warning System

Aisya Nafiisyanti, Farid Lasmono, Ibnu Fathrio, Risyanto, Teguh Harjana, Didi Satiadi, and Acep Catur Nugraha

Abstract This study is a preliminary examination of edge detection methods for the given data set produced by a numerical weather prediction model. This research compares three edge detection methods: Laplacian, Sobel and Canny. Each method is applied to images that represent flight weather information, specifically low-level wind shear and low level. The area coverage is the Indonesian region with -10° to 10° latitude and 95° to 145° longitude range. Moreover, hue, saturation and value (HSV) colour space is performed in the pre-processing stage to give a better performance. The characteristics of the test results are obtained and then used as the basis for selecting the most effective method to be implemented in the system. The Laplacian and Canny methods are capable of fulfilling the requirement even though the three methods generally are capable of detecting edges in the given images.

1 Introduction

Image processing specifically in the atmospheric field and beyond has been used for various purposes and is implemented on different data sources. [1] applied a modified edge detection method which was developed from Canny methods on aerosol data from Lidar. Then, from this research, [2] exploited the detection results to identify Planetary Boundary Layer (PBL) height that plays a role to characterize the PBL itself. Another case was developed by [3], by extracting objects from Indian Remote Sensing Satellite (IRS) images that are derived from LISS-III, LISS-IV, and Cartosat-I sensors. Edge detection methods are also used by [4] to localize an Unmanned Aerial Vehicle (UAV) by detecting craters and calculating the craters' distance using the Mahalanobis method on the Martian surface. One sector that has the potential to use image processing is aviation by informing flight weather conditions. JATAYU as an air transportation network developed by the Center of Climate and Atmospheric Research, Indonesian National Research and Innovation Agency, is a system that uses

A. Nafiisyanti (✉) · F. Lasmono · I. Fathrio · Risyanto · T. Harjana · D. Satiadi · A. C. Nugraha
Indonesian National Research and Innovation Agency, Jakarta, Indonesia
e-mail: aisya.nafiisyanti@brin.go.id

A. Basit et al. (eds.), *Proceedings of the International Conference on Radioscience, Equatorial Atmospheric Science and Environment and Humanosphere Science*, Springer Proceedings in Physics 290, https://doi.org/10.1007/978-981-19-9768-6_21

images to convey information on its website [5] Each image describes the weather conditions of a parameter by classifying the values into several ranges. Each range is represented by a specific color. Thus, according to the classification, there might be objects in unregular shape in the image if a specific weather condition occurs.

Previously, the identification process, judgment, and conclusions of determining the occurrence of specific weather conditions were drawn using human vision. Then, the warning can be delivered through the system. However, this method requires human resources and consistency and also has inefficient performance. As the system moves toward automation, we attempt to shift the early warning process from human vision to computer vision.

This study uses edge detection as the approach to utilize computer vision. Detecting edges is one of the fundamental operations in image processing. It helps to reduce the amount of data (pixels) to process and maintains the necessary aspect of the image. Edges detection is a problem of fundamental importance in object extraction as it reduces image data and facilitates object detection. Edges identify object boundaries and are detected through abrupt changes in the gray level above a particular threshold. Edge detectors can exploit operators that are sensitive to the change in gray levels. The methods used in this study are traditional Sobel, Canny, and Laplacian operators. [6] stated that Canny has been the state-of-the-art method for edge detection until today and used the method for detecting abrupt changes in climate events by examining two dimensions: space and time. Also, [6] claimed that the method successfully managed to detect and quantify abrupt changes. However, other methods are widely used because they show good results, for example is Laplacian that becomes the basis for modified edge detection method in [7] by proposing the Laplacian Gradient algorithm for the detection of three-dimensional objects. Another example is [8] by proposing combination of Laplacian of Gaussian, Canny, and mathematical morphology operator for a timely and accurate ground fissures finding in coal mining area. [9] proposed modified Sobel method by generating new mask values resulting to the enhanced computation time. The purpose of this study is to find the best method for the given case.

2 Data and Method

2.1 Data

The data used in this research are obtained from Weather Research and Forecast model simulation results. The model produces Air Force Weather Aviation (AFWA) data as derived data from the main simulation results. Each parameter data is stored in an image in png format with a size of 1000 $\times$ 400 pixels. An image represents the condition of a parameter in an hour. In particular, this study used Low-Level Wind Shear (LLWS) and Low-Level Turbulence (LLT) data parameters on September 8 and 16, 2022. Since each day is represented by 24 images for each parameter, then

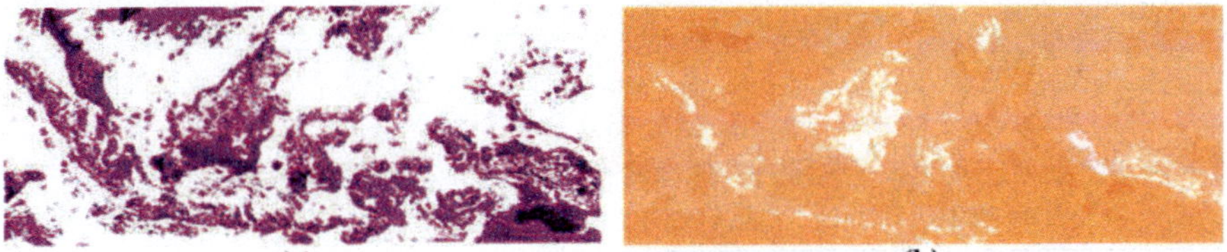

(a) (b)

Fig. 1 **a** Example of wind shear image and **b** example of turbulence image. Both represent information at one hour, and the white is seen as transparent

there are 96 images in total used in this study. These parameters at those specific times were chosen because of the continuous data availability and have a varying range of values. LLWS value is represented in purple color gradation. See Fig. 1, there are five color gradations and RGB coded to represent different range levels. Transparent (0, 0, 0), light purple (201, 0, 255), heavy purple (93, 0, 118), trans purple (51, 0, 64), and dark purple (137, 0, 174) represent value below 1.9, 2–3.9, 4–5.9, 6–7.9, and 8 and above, respectively. Meanwhile, the LLT value is represented in orange color gradation. There are five color gradations to represent different range levels. Transparent (0, 0, 0), peach orange (255, 199, 143), rajjah (255, 182, 104), yellow-orange (255, 165, 63), and dark orange (254, 141, 1) represent value below 0.15, 0.16–0.2, 0.21–0.4, 0.41–0.8, and 0.8 and above, respectively.

The idea of performing edge detection in this study is to find objects in form of blob(s) or line(s) in the color of trans purple and dark orange which represent range value 8 and above for LLWS, and 0.8 and above for LLT, respectively. The range necessarily is drawn with the lowest pixel intensity (a shape with the darkest color) within an image. The object with the lowest pixel intensity in this image represents an area whose value belongs to the highest range level. This means that the area is experiencing the most intense aviation weather events, in this case, turbulence events and wind shear events. Blob is a collection of pixels that share the same property that form irregular shapes and line is collection of interconnected pixels whose shape has two ends. This study excludes dot or single pixel because 1 pixel in the image is equivalent to 5 km^2 in actual size, so when compared to the actual conditions, this area is considered very small. See Fig. 2 and the following paragraph to understand the comparison.

Figure 2 is an example of severe turbulence event reported by National Weather Aviation Service Center (NWASC) on October 26, 2022. The incident occurred over southwestern Ohio near Indianapolis when the small aircraft fly at 3000 feet. The map shown has a scale of 2 cm:100 km. The black circle marked the estimated location of the turbulence and has a diameter of 1.5 cm or equals to 75 km. Thus, the circle has an area of 4,417.86 km^2 or 883 pixels in the image used in this study.

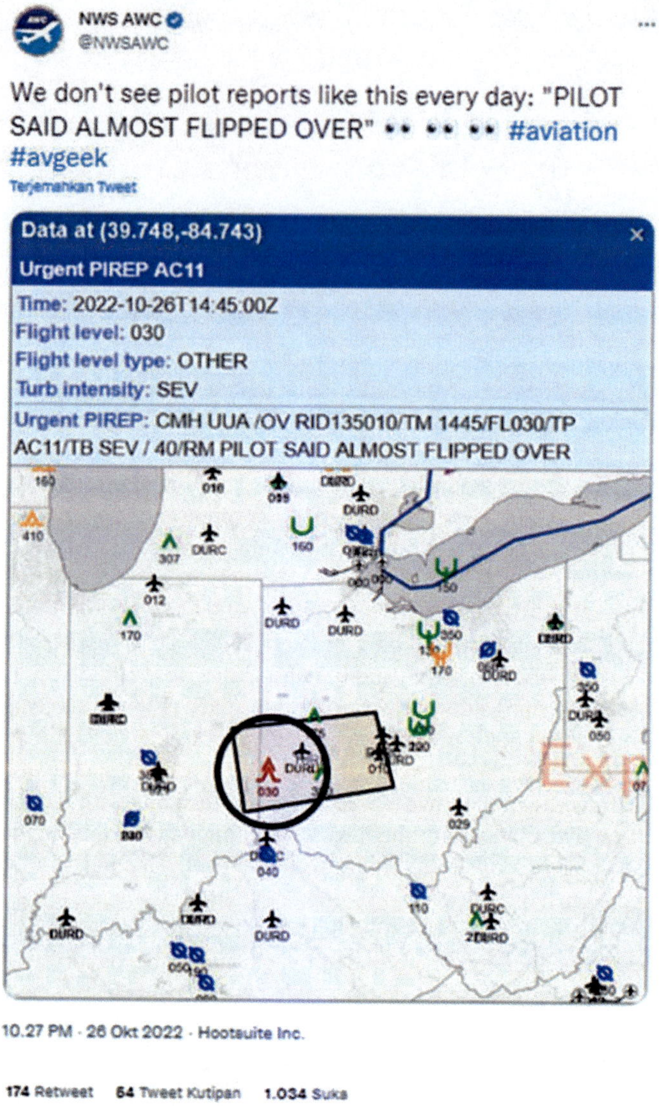

Fig. 2 Example of reported severe turbulence in aviation. *source*: https://twitter.com/NWSAWC/status/1585292028178534403?ref_src=twsrc%5Etfw%7Ctwcamp%5Etweetembed%7Ctwterm%5E1585292028178534403%7Ctwgr%5E2321ce5efa2d9c12b67f788d96649c1db78d747b%7Ctwcon%5Es1_&ref_url=https%3A%2F%2Fwww.newsweek.com%2Fsevere-turbulence-plane-almost-flips-over-1755237

2.2 Method

Sobel operator. Sobel is one of the popular methods among other operators for edge detection. It is based on the gradient/first-order derivative to perform convolution to find an edge by using a horizontal mask or vertical mask [10]. One of the advantages of this method is that it reduces noises in the image and gives brighter and clearer edges [11]. The vertical mask is simply the transposition of the horizontal mask as described in Fig. 3

The convolution process of an image starts from the top to bottom for the vertical mask and left to right for the horizontal mask. It is a multiplication of a pixel intensity value with its neighbors then weighted by the mask. As an illustration, Fig. 4 is the subset of pixels of an image. The gradient of P4 is determined by using Eqs. 1–7. Comparison as the weighing process is then performed to define whether Px_4 is an edge pixel or not by using Eq. 3. Px_4 is set as an edge pixel if the gradient is greater than the threshold, on the contrary, it is set as a non-edge pixel [12].

$$G_x = f_1 - f_2 \tag{1}$$

$$f_1 = (Px_6 + 2 \times Px_7 + Px_8) \tag{2}$$

$$f_2 = (Px_0 + 2 \times Px_1 + Px_2) \tag{3}$$

$$G_y = f_3 - f_4 \tag{4}$$

Fig. 3 **a** Horizontal mask and **b** vertical mask

(a)

-1	-2	-1
0	0	0
1	2	1

(b)

-1	0	1
-2	0	2
-1	0	1

Fig. 4 3*3 mask of image subset

Px_0	Px_1	Px_2
Px_3	Px_4	Px_5
Px_6	Px_7	Px_8

$$f_1 = (Px_2 + 2 \times Px_5 + Px_8) \tag{5}$$

$$f_1 = (Px_0 + 2 \times Px_3 + Px_6) \tag{6}$$

$$G = |G_x| + |G_y| \tag{7}$$

Canny Operator. The canny operator is another well-known method for edge detection. As well as Sobel, this method's advantage is that it can reduce noise. Traditionally, there are four stages of the Canny method [13]. The first is image smoothing using the Gaussian function, this step reduces the effect of the noises by using 3*3, 5*5, 7*7 mask, etc. The second step is calculating the gradient to obtain gradient magnitude and gradient direction. The calculation is tended to identify the drastic change in the gray area [14]. The third step is performing Non-Maxima Suppression (NMS). NMS uses 3*3 neighbors to perform interpolation of gradient magnitude along the gradient direction. If the gradient magnitude is greater than the interpolation of the two-gradient direction, then the pixel is selected as an edge point, otherwise, it is marked as a non-edge point. This selection supports the ideal final image where the edge's line is preferably thin to get a precise and unilateral edge. The last step is double thresholding to detect and connect edges. High and low thresholds are set and compared to the gradient magnitude of the pixel. If the gradient magnitude is greater than the high threshold, then the pixel is marked as an edge point, and if the gradient magnitude is less than the low threshold, the pixel is marked as the non-edge point. As the result of this step is discontinuous, the rest pixel which are connected to the edge point is marked as the edge point.

Laplacian operator. The Laplacian operator is based on zero-crossing/second-order derivative. As well as Canny, this method performs better in edge localization so that it gives an accurate location to the edge (Anas et al., 2019). However, it is sensitive to noise due to double derivation. The calculation is denoted in Eq. 8 [15].

$$\nabla^2 f = f(x+1, y) + f(x-1, y) + f(x, y+1) + f(x, y-1) - 4f(x, y) \tag{8}$$

where x and y are the coordinates of the pixel. Laplacian method uses a 3*3 mask as a filter as the following

HSV Color Space. Hue, Saturation, and Value (HSV) color space selection is implemented to perform edge detection better. Hue reflects the color that is dominantly observed by humans, saturation is the shade of the color, and value is the intensity or the tone of the color [16]. HSV is obtained from RGB using the following equation:

$$H = \arccos \frac{\frac{1}{2}(2R - G - B)}{\sqrt{(R-G)^2 - (R-B)(G-B)}} \tag{9}$$

$$S = \frac{\max(R, G, B) - \min(R, G, B)}{\max(R, G, B)} \tag{10}$$

$$V = \max(R, G, B) \tag{11}$$

This color space also has been used in various image recognition systems [17]. Thus, HSV color space is chosen to fit the idea of choosing the closest channel in which color space portrays shapes with the darkest color.

3 Result and Analysis

3.1 Preprocessing

Intuitively, channel V will be chosen since the study worked on pixel intensity, but that was not applicable for both parameters. LLT images are formed by a collection of pixels, the majority of which are orange gradations, while the rest is a few transparent areas (which are then converted to white). As seen in Fig. 1(b), there is less contrast compared to Fig. 1(a). In the channel V calculation result, the edges that define an object are not contrasting. In addition, the V channel captures all the edges in the image, including the edges of objects with non-low-intensity pixels. Meanwhile, the saturation channel provides better contrast for objects with low pixel intensity because it can differentiate the tone of each orange, even though it still captures another edge. Thus, channel S is selected for LLT, whereas channel V is selected for LLWS because the original image has stronger contrast since it has more transparent areas. The saturation channel is incapable to provide better differentiation since it is compared to white (maximum RGB).

3.2 Processing

According to Fig. 5(a–f), it is shown that the three methods are capable of detecting the edges of the objects in the images. They are also able to detect the smallest object in dots and lines form, which is supposedly identified as noise. Further explanation is methods performance paragraphs. However, there are some differences described in the following paragraphs.

Canny and Laplacian construct interconnected dots to form an edge for a blob. Meanwhile, Sobel constructs a blob with dotted lines as seen in Fig. 5(e) and (f). Besides, the interconnected dots in Canny and Laplacian form thinner lines compared to lines in Sobel. Although the results of Canny and Laplacian look similar, there is a difference that can be seen in Table 1.

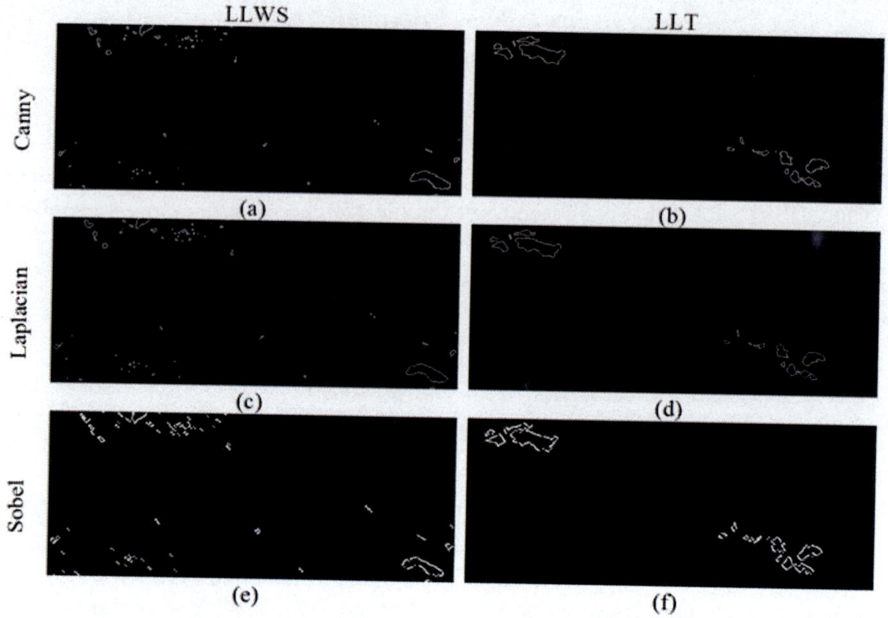

Fig. 5 Implementation results of Sobel, Canny, and Laplacian operator on LLWS and LLT images

Table 1 Number of objects counted on generated edge detection image results

Methods	LLWS	LLT
Sobel	74	40
Canny	54	8
Laplacian	43	7

To confirm the performance of the method, we implement object counting on each image result. The counting process results in a different number of objects in each method. As stated in Table 1, Laplacian collects the fewest number of objects, Canny identifies more objects than Laplacian, whilst Sobel collects the biggest number of objects on both datasets. Although it is stated in [18] that Laplacian is sensitive to noise, Canny appears to detect more objects than Laplacian. Since this study uses the traditional Canny method, the operator detects too detailed edges that causes the method is prone to noise [19], which similarly occurred in [20]. Different from the other two methods, Sobel does not connect the points marked as edges. Moreover, Sobel does not apply edge thinning so that the points are considered as one object. These reasons affect the calculation. Thus, the object counting process failed to recognize the original object.

The difference in the number of objects in the image of each method affects the determination of the object's central coordinate. Each method returns different locations of the objects. Take an example on Canny and Laplacian that give similar results, some coordinates shift slightly from one to another. This may affect the information conveyed on aviation.

4 Conclusion

Canny, Laplacian, and Sobel methods generally are capable of detecting edges in the given images. However, Sobel does not apply object thinning on each detected edge point/edge pixel, and each point next to each other is not interconnected, so that each point is considered as an object. Different from Sobel, Laplacian, and Canny form thinner point and create full blobs or lines which perfectly recognized as an object. The three methods are susceptible to noise since they detect detailed edges. In conclusion, traditional Laplacian and Canny methods are capable of fulfilling the requirement. Meanwhile, traditional Sobel needs improvement to meet the requirement.

References

1. Xiang, Y., Ye, Q., Liu, J., Zhang, T., Fan, G., Zhou, P., Lü, L., Liu, Y.: Retrieve of planetary boundary layer height based on image edge detection. Chin. J. Lasers **43** (2016)
2. Dang, R., Yang, Y., Hu, X.M., Wang, Z., Zhang, S.: A review of techniques for diagnosing the atmospheric boundary layer height (ABLH) using aerosol Lidar data, Remote Sens. **11** (2019)
3. Katiyar, S.K., Arun, P.V.: Comparative analysis of common edge detection techniques in context of object extraction, **50**, 68 (2014)
4. Higashino, S.I., Teruya, T., Yamada, K.: Position identification using image processing for uav flights in martian atmosphere. J. Robot. Mechatronics **33**, 254 (2021)
5. I. N. R. and I. Agency, JATAYU, http://jatayu.brin.go.id/new-jatayu/hello.php
6. Bathiany, S., Hidding, J., Scheffer, M.: Edge detection reveals abrupt and extreme climate events. Am. Meteorol. Soc. **33** (2020)
7. Al-Anssari, J., Naser, I., Ralescu, A.: Three-dimensional laplacian spatial filter of a field of vectors for geometrical edges magnitude and direction detection in point cloud surfaces. In: Proceeding 2019 IEEE international conference on humanized computing and communication HCC 2019, p. 83 (2019)
8. Xu, D., Zhao, Y., Jiang, Y., Zhang, C., Sun, B., X. He, Using improved edge detection method to detect mining-induced ground fissures identified by unmanned aerial vehicle remote sensing. Remote Sens. **13** (2021)
9. Vinista, P., Joe, M.M.: A novel modified sobel algorithm for better edge detection of various images. Int. J. Emerg. Technol. Eng. Res. **7**, 25 (2019)
10. Ahmed, A.S.: Comparative study among Sobel, Prewitt and canny edge detection operators used in image processing. J. Theor. Appl. Inf. Technol. **15** 19 (2018)
11. Setiadi, D.R.I.M., Jumanto, J.: An enhanced LSB-image steganography using the hybrid canny-Sobel edge detection. Cybern. Inf. Technol. **18**, 74 (2018)
12. Nausheen, N., Seal, A., Khanna, P., Halder, S.: A FPGA based implementation of Sobel edge detection. Microprocess. Microsyst. **56**, 84 (2018)

13. Song, R., Zhang, Z., Liu, H.: Edge connection based canny edge detection algorithm. Pattern Recogn. Image Anal. **27**, 740 (2017)
14. Sekehravani, E.A., Babulak, E., Masoodi, M.: Implementing canny edge detection algorithm for noisy image. Bull. Electr. Eng. Inf. **9**, 1404 (2020)
15. Bouganssa, I., Sbihi, M., Zaim, M.: Laplacian edge detection algorithm for road signal images and FPGA implementation. Int. J. Mach. Learn. Comput. **9**, 57 (2019)
16. Hema, D., Kannan, S.: Interactive color image segmentation using HSV color space. Sci. Technol. J. **7**, 37 (2019)
17. Qazanfari, H., Hassanpour, H., Qazanfari, K.: Content-based image retrieval using HSV color space features. Int. J. Comput. Inf. Eng. **13**, 537 (2019)
18. Anas, R., Elhadi, H.A., Ali, E.S.: Impact of edge detection algorithms in medical image processing. World Sci. News **118,** 130 (2019)
19. Liu, L., Liang, F., Zheng, J., He, D., Huang, J.: Ship infrared image edge detection based on an improved adaptive canny algorithm. Int. J. Distrib. Sens. Netw. **14** (2018)
20. Feng, Y., Zhang, J., Wang, S.: A new edge detection algorithm based on canny idea. In: AIP Conference Proceeding, vol. 1890 (2017)

Numerical Simulation of MCC Evolution Over Borneo Island Using a High-Resolution Model, Case Study: April 14–15, 2012

Ibnu Fathrio and Trismidianto

Abstract The evolution of the mesoscale convective complex (MCC), over Borneo on April 14–15, 2012, was successfully simulated using the Weather Research and Forecasting (WRF) model. This experiment uses ERA5 dataset as the initial and boundary condition on a 3 km spatial resolution. Meanwhile, the Betts–Miller–Janjic (BMJ), The WRF single moment 6 (WSM6), and the Yonsei University (YSU) were then selected as schemes for cumulus, microphysics, and planetary boundary layer, respectively. The model successfully captures the phase, duration, and lowest cloud top temperature of MCC. Despite the model's inability to simulate comparable spatial patterns of cloud distribution, this simulation shows promising results in simulating MCC evolution. A brief assessment of topography's role in MCC evolution is also discussed by contrasting normal topography with flat topography simulation.

1 Introduction

Since the 1980s, researchers have been studying mesoscale convective complexes (MCC), a long-lived and quasi-circular type of mesoscale convective systems (MCS) [1, 2]. The majority of MCC occurs in the northern hemisphere during the night, which lasts 10 h [3, 3]. The MCC is responsible for up to 18% of total precipitation in the Great Plains in the USA. According to a previous study [5], MCC contributes up to 60% of total rainfall in Subtropical South America. Given the importance of MCC in predicting hazardous weather, improving numerical model skills to predict MCS for weather forecasting and climate prediction is critical. As the climate warms, MCSs may become more frequent and intense [6].

I. Fathrio (✉) · Trismidianto
National Research and Innovation Agency, Bandung, Indonesia
e-mail: ibnu003@brin.go.id

A. Basit et al. (eds.), *Proceedings of the International Conference on Radioscience, Equatorial Atmospheric Science and Environment and Humanosphere Science*, Springer Proceedings in Physics 290, https://doi.org/10.1007/978-981-19-9768-6_22

[7] has documented MCC observation on the maritime continent. The MCC accounts for up to 20% of total rainfall in Indonesia. The MCC matures more on the continent than on the sea, particularly near mountainous areas. It has characteristics similar to the global population of MCCs. MCC populations are higher in the boreal spring, yet lower in the boreal summer. A previous study by Yulihastin et al. [8] has documented an effort to study the evolution of MCC using regional models on the maritime continent. They emphasized the importance of using up-to-date SST when simulating the long-lived MCC with persistent precipitation. Yulihastin et al. [9] successfully simulates the offshore propagation of MCC in southern Sumatra. They emphasized the well-simulated cold pool, which maintains favorable low-temperature conditions for the MCC.

Borneo, one of the largest islands in the maritime continents, has the highest contribution of the MCC to rainfall, up to 20% [10]. Borneo Island is home to five major cities as well as Indonesia's future capital city. The study of MCC that contributes to cause extreme rainfall is critical. Assessment of the regional model's abilities in simulating the evolution of the MCC over Borneo is required to improve the accuracy of the model prediction of the MCC. Unfortunately, only a few studies have discussed the evaluation of model performance to simulate MCC over Borneo. As a result, the goal of this study is to assess the performance of the regional model in simulating the evolution of MCC over Borneo Island. To evaluate the model, a study case of MCC on April 14–15, 2012, as discussed in [11], was chosen as a reference. In addition, this study emphasizes the role of mountainous regions in the northern part of Borneo in the initiation of the MCC.

2 Data and Method

This study focuses on a case of MCC event that occurred on the island of Borneo on April 14–15, 2012. Based on [1, 7] recognized this case as one of the convective systems that meets the MCC characteristics. Readers are encouraged to see [11] for more information on MCC characteristics in the maritime continent. To simulate the evolution of MCC, the Weather Research and Forecast (WRF) model [12] was utilized. The long-lived MCC is characterized using cloud top temperature from the MTSAT satellite's IR-1 channel [13]. This simulation consists of two nested domains with spatial resolutions of 9 km and 3 km, respectively (Fig. 1). The WRF Single Moment 6 (WSM6) microphysics scheme, the Yonsei University planetary boundary layer scheme, the Dudhia and RRTM schemes for shortwave and longwave radiation, respectively, the MM5 similarity for surface layer models, and the NOAH land surface model are all used in this simulation. The Betts–Miller–Janjic (BMJ) cumulus scheme is used for the first domain (9 km), but no cumulus scheme is used for the second domain (3 km) because convection is assumed to be resolved

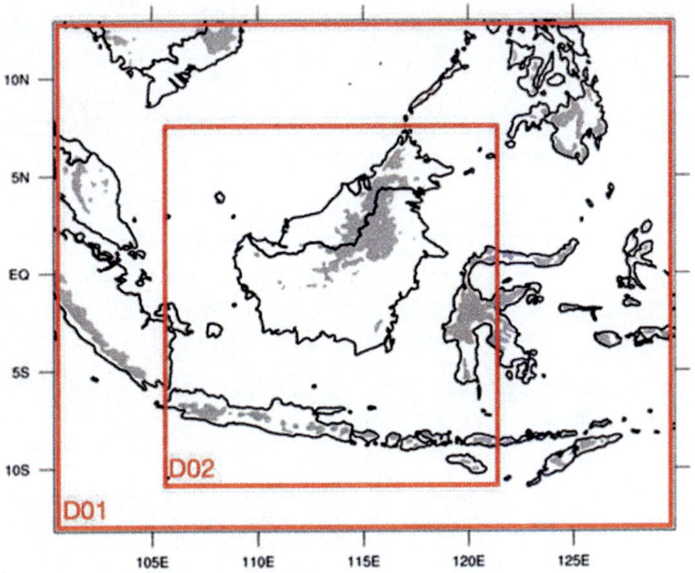

Fig. 1 Domain of simulation consists of the first domain (D01) and the second domain (D02) with a spatial resolution of 9 km and 3 km, respectively. The gray-shaded area denotes the topography greater than 500 m in height

explicitly. The European Center for Medium-Range Weather Forecasting (ECMWF) Re-Analysis 5 (ERA5) hourly reanalysis dataset [14] was used for the boundary and initial conditions. The time integration is two days, beginning at 1800 UTC on April 13, 2012, with a 12-h spin-up time.

To investigate the influence of topography on the initiation of MCC, the second simulation has been carried out by setting all topography over the continent in the first and second domains to one-meter height. Herein, the flat topography simulation is referred to as FLAT simulation, while the normal topography simulation is referred to as CTRL simulation.

3 Results and Discussion

The initiation stage of the MCC occurs on 15–19 LT (GMT + 7), April 14, 2012, when the convective cells are found in the northern, middle, and southern parts of the continent (Fig. 2a-d). A combination of mountain-valley breezes [11, 15] in the northern and middle parts of the continent, which are dominated by mountainous regions, and sea breezes contribute to the initiation of convective activity over the

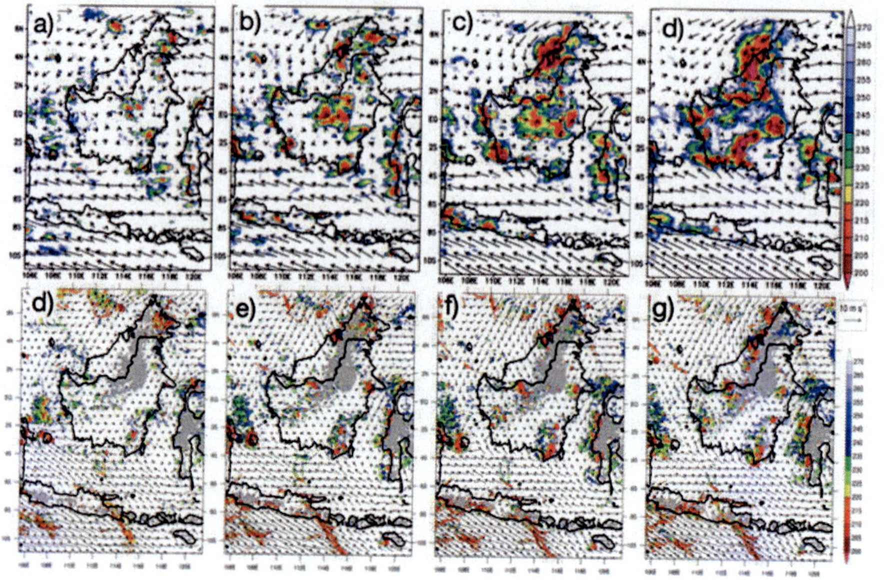

Fig. 2 Contour map of cloud top temperature derived from the observation **a–d** adapted from [10]. The initiation stage of MCC was on April 14, 2012 for **a** 1300 LT, **b** 1500 LT, **c** 1600 LT, and **d** 1700 LT. WRF output for cloud top temperature is shown at a similar time for **e** 1300 LT, **f** 1500 LT, **g** 1600 LT, and **h** 1700LT April 15, 2022, respectively. Cloud top temperature units are in Kelvin

middle and the coast of Borneo, respectively. The model simulates the initiation stage with a less convective cell found in the continent's middle and southern regions.

The MCC reaches a mature stage around midnight, when a broad convective area spreads across the middle, southern, and western parts of the continent (Fig. 3a-d). Convective cells tended to propagate landward to the middle of the continent. At the same time, expanding convective cells with cloud top temperatures of less than 200 K can be found in the southern part. The WRF model simulates the mature stages of MCC, showing the intensification and merging of convective cells in the continent's western and middle regions (Fig. 3f-h). In comparison with the observation, the model also reasonably simulates the minimum cloud top temperature. However, there was not a single major convective cell simulated by the model in the south.

The merging and weakening of convective cells in the western and middle parts of the continent are shown in the decaying stage (Fig. 4a-d), along with increased divergence of surface wind. Simultaneously, the land breeze promotes a squall line in the continent's south that propagates southward. This squall line takes six hours to reach the north Coast of Java Island. The model successfully simulates the decaying stages of MCC in the western part of the continent with surface wind divergence (Fig. 4e-h). The southern squall line has not yet been identified, but the land breeze is already present.

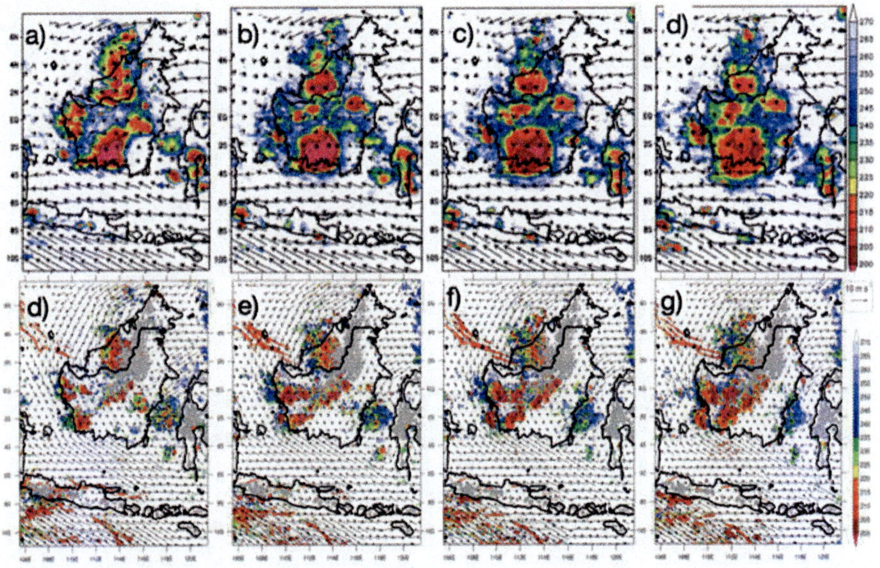

Fig. 3 Contour map of cloud top temperature derived from the observation **a–d** adapted from [10]. The growth stage of MCC is shown at **a** 2100 LT, while the mature stage is shown at **b** 2200 LT on April 14, 2022; **c** 0000 LT, and **d** 0100 LT on April 15, 2022. WRF output for cloud top temperature is shown at a similar time for **e** 2100 LT, and **f** 2200 LT on April 14, 2022; **g** 0000 LT, and **h** 0100 LT on April 15, 2022, respectively. Cloud top temperature units are in Kelvin

A few hours later, as the sea breeze returns, new convective cells form along the north and south coasts and in the middle of the continent (Fig. 5a-d). Meanwhile, the squall line in the south of the continent hits Java Island in the afternoon and lasts for 5–6 h before decaying. The model simulates the post-MCC stage and the squall-line propagation reasonably well (Fig. 5e-h). However, models simulate the presence of the squall line 4–5 h later than the actual squall line.

3.1 Influence of Topography in the Initial Stage of MCC

Figure 6 depicts the differences between CTRL and FLAT simulations, where topography enhances convective activity during the initiation and growth stages. Figure 6a–c shows developing convective cells to the west of the mountainous region in CTRL simulation (highlighted by the orange box in Fig. 6b), whereas less convective cells are visible in FLAT simulation. Figure 6c–e depicts how convective cells grow to the south of the mountainous region and then propagate south-westward (highlighted

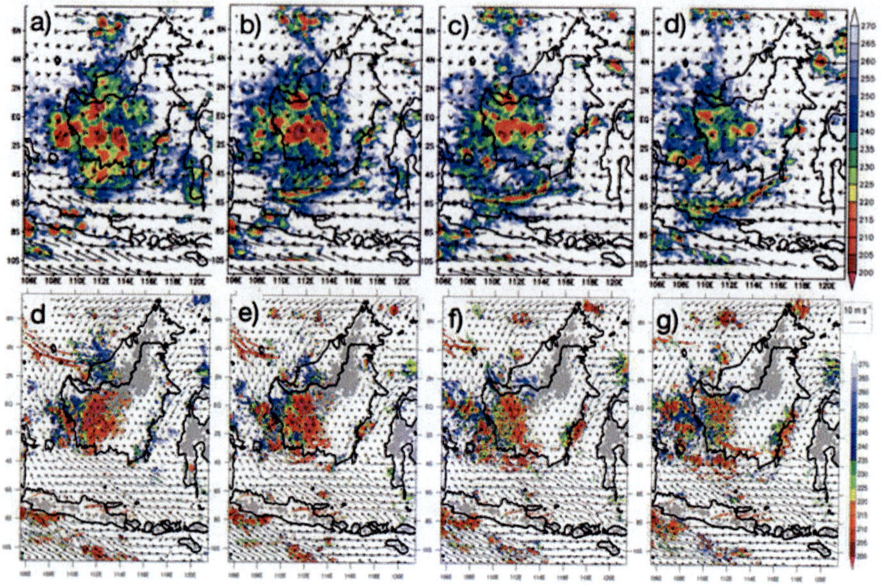

Fig. 4 Contour map of cloud top temperature derived from the observation **a–d** adapted from [10]. The decaying stage of MCC was shown at **a** 0300 LT, **b** 0500 LT, **c** 0700 LT, while the dissipation stage of MCC was shown at **d** 0900 LT on April 15, 2022. WRF output for cloud top temperature is shown at a similar time for **e** 0300 LT, **f** 0500 LT, **g** 0700 LT, and **h** 0900LT April 15, 2022, respectively. Cloud top temperature units are in Kelvin

by green boxes in Fig. 6). Following that, the convective cells merge with other convective cells growing in the continent's west to form larger convective cells. FLAT simulation, on the other hand, shows fewer convective cells growing in mountainous regions to the south. On the west coast of the continent, the FLAT simulation generates convective cells earlier than the CTRL simulation, as demonstrated by Tan et al. [16] in a no-topography experiment. This experiment supports previous findings that most continental MCC prefers high-elevation areas, which contribute to the generation of MCC in the middle of the continent [15].

4 Conclusion

Overall, the WRF model could simulate the evolution of MCC occurred on April 14–15, 2021, across the Borneo continent, from initiation to dissipation and post-MCC stages. Although there are differences in spatial distribution and the WRF model simulates smaller cell size than the observed MCC, the evolution phase and

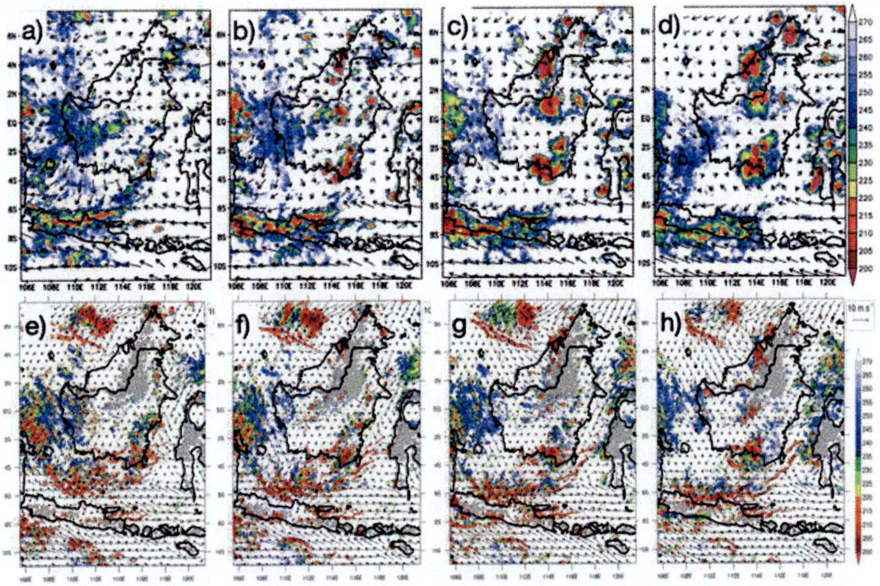

Fig. 5 Contour map of cloud top temperature derived from the observation **a**–**d** adapted from [10]. The post-MCC stages are shown at **a** 1300 LT, **b** 1500 LT, **c** 1700 LT, and **d** 1900 LT on April 15, 2022. WRF output for cloud top temperature is shown at a similar time for **e** 1300 LT, **f** 1500 LT, **g** 1700 LT, and **h** 1900 LT on April 15, 2022, respectively. Cloud top temperature units are in Kelvin

minimum cloud top temperature are successfully simulated. Some characteristics, such as the role of land-sea breeze, the influence of topography in the initiation stage, convective cell merging, and the decaying stage denoted by surface wind divergence, are simulated correctly. The model also simulates the post-MCC stage, such as an induced squall line that moves toward Java Island.

Furthermore, flat topography simulation was performed to confirm the role of the mountainous region, which gives rise to more convective cells, thereby increasing the strength and coverage of MCC over Borneo. This study allows researchers to delve deeper into the underlying mechanism of MCC life, which can last for several hours across Borneo. In the future, it will be interesting to see if this result is affected by the parameterization and initial-boundary condition used.

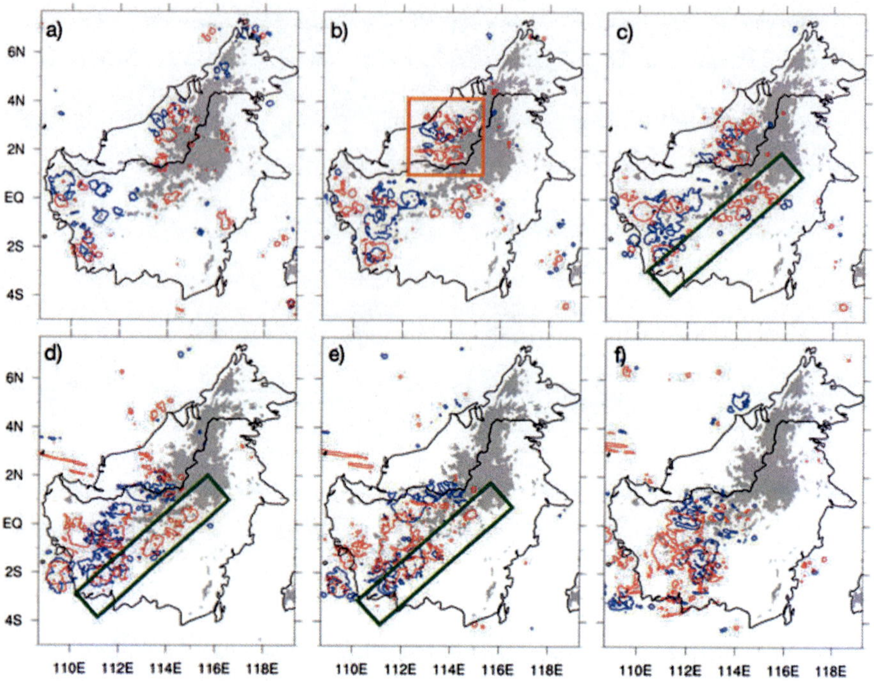

Fig. 6 Contour of simulated cloud top temperature of 210 K for CTRL simulation (red) and FLAT simulation (blue) at **a** 1900 LT, **b** 2000 LT, **c** 2100 LT, **d** 2200 LT on April 14, 2002, **e** 0000 and **f** 0200 LT on April 15, 2022. Orange circles highlight the presence of convective cells in CTRL. The gray-shaded area denotes the topography greater than 500 m in height. Orange and green boxes highlight the differences between CTRL and FLAT experiments

References

1. Maddox, R.A.: Mesoscale convective complexes Bull. Amer. Meteor. Soc. **61**, 1374–1387 (1980)
2. Houze, R.A.: Structure and dynamics of a tropical squall line system. Mon. Wea. Rev. **105**, 1540–1567 (1977)
3. Fritsch, J.M., Kane, R.J., Chelius, C.R.: The contribution of mesoscale convective weather system to the warm season precipitation in the United States. J. Clim. Appl. Meteorol. **25**, 1333–1345 (1986)
4. Laing, A.G.: Contribution of mesoscale convective complexes to rainfall in Sahelian Africa: estimates from Geostationary Infrared and Passive Microwave Data. J. Appl. Meteorol. **38**, 957–964 (1999)
5. Durkee J.D., Mote T.L., Shepherd, M.J.: The contribution of mesoscale convective complexes to rainfall across, subtropical South America. Int. J. Clim. **22** (2009)
6. Schumacher, R.S., Rasmussen, K.L.: The formation, character and changing nature of mesoscale convective systems. Nat. Rev. Earth Environ. **1**, 300–314 (2020)
7. Trismidianto.:The global population of mesoscale convective complexes (MCCs) over Indonesian maritime continent during 15 Years. IOP Conf. Ser.: Earth Environ. Sci. **166**, 012040 (2018)

8. Yulihastin, E., Nuryanto, D.E., Muharsyah, R.: Improvement of heavy rainfall simulated with SST adjustment associated with mesoscale convective complexes related to severe flash flood in Luwu, Sulawesi, Indonesia. Atmosphere **12**(11), 1445 (2021)
9. Yulihastin, E., Fathrio, I., Nauval, F., Saufina, E., Harjupa, W., Satiadi, D., Nuryanto, D.E.: Convective cold pool associated with offshore propagation of convection system over the East Coast of Southern Sumatra, Indonesia. Adv. Meteorol. (2021)
10. Trismidianto, Yulihastin, E., Satyawardhana, H., Nugroho, J.T., Ishida, S.: The contribution of the mesoscale convective complexes (MCCs) to total rainfall over Indonesian maritime continent. IOP Conf. Ser. Earth Environ. Sci. **54**, 012027, (2017)
11. Trismidianto.: Characteristics of the oceanic MCC, continental MCC, and coastal MCC over the Indonesian maritime continent. IOP Conf. Ser. Earth Environ. Sci. **149**, 012024 (2018)
12. Skamarock, W.C., Klemp, J.B., Dudhia, J., Gill, D.O., Liu, Z., Berner, J., Wang, W., et al.: A description of the advanced research WRF model version 4. National Center for Atmospheric Research: Boulder, CO, USA, 145, (2019)
13. Trismidianto, H.T.W., Ishida, S., Moteki, Q., Manda, A., Iizuka, S.: Development processes of oceanic convective systems inducing the heavy rainfall over the western coast of Sumatra on 28 October 2007. Sola **12**, 6–11 (2016)
14. Hersbach, H., Bell, B., Berrisford, P., Biavati, G., Horányi, A., Muñoz Sabater, J., Nicolas, J., Peubey, C., Radu, R., Rozum, I., Schepers, D., Simmons, A., Soci, C., Dee, D., Thépaut, J.-N.: ERA5 hourly data on single levels from 1959 to present. Copernicus Climate Change Service (C3S) Climate Data Store (CDS) (2018). Accessed on 01–07–2022
15. Trismidianto, Yulihastin, E., Satyawardhana, H., Ishida, S.: A composite analysis of the mesoscale convective complexes (MCCs) development over the Central Borneo and its relation with the propagation of the rainfall systems. IOP Conf. Ser. Earth Environ. Sci. **54**(1), 012036 (2017). IOP Publishing
16. Tan, H., Ray, P., Barrett, B., Dudhia, J., Moncrieff, M., Zhang, L., Zermeno-Diaz, D.: Understanding the role of topography on the diurnal cycle of precipitation in the maritime continent during MJO propagation. Clim. Dyn. **58**, 3003–3019 (2022)

Tropopause Height Variation Toward Different Land-Sea Convection Activities in Java Using GNSS-RO Data

Khanifah Afifi and Nurjanna Joko Trilaksono

Abstract The exchange of substances like ozone and water vapor is facilitated by the tropopause layer, which is the layer between the stratosphere and troposphere. Convective activity is one of the factors that affects the tropopause's height. The variation in tropopause height may be influenced by the distinct differences in diurnal convective activity between land and sea regions. The purpose of this research is to investigate diurnal variations in the height of the tropopause layer are affected by differences in convective activity between land and sea. The study area is Java Island and the sea that surrounds it (6°–9° S and 105°–115° E), and the study period runs from December 1, 2019, to November 30, 2020. The Constellation Observing System for Meteorology Ionosphere and Climate (COSMIC-2), Global Navigation Satellite System Radio Occultation (GNSS-RO) data, and Lapse Rate Tropopause methods are utilized to determine the height of the tropopause layer. In the meantime, the temperature black body gradient was used to estimate the convective activity. The findings demonstrated that, following the rise in tropopause height, the increase in convective activity over the ocean was more dominant in the late afternoon and early morning, while it was more dominant over land from noon until the afternoon in local time. The pattern of cloud formation in the ocean area is most closely matched by the tropopause height during March to May (MAM) season, whereas in the land area during December to February (DJF) season.

K. Afifi
Master Program in Earth Sciences, Faculty of Earth Sciences and Technology, Institut Teknologi Bandung, Bandung, Indonesia

N. J. Trilaksono (✉)
Atmospheric Sciences Research Group, Faculty of Earth Sciences and Technology, Institut Teknologi Bandung, Bandung, Indonesia
e-mail: jpatiani@itb.ac.id

A. Basit et al. (eds.), *Proceedings of the International Conference on Radioscience, Equatorial Atmospheric Science and Environment and Humanosphere Science*, Springer Proceedings in Physics 290, https://doi.org/10.1007/978-981-19-9768-6_23

1 Introduction

The Indonesian Maritime Continent (MC) has the main characteristic of diurnal variation of convective activity [1]. The existence of contrasting differences between diurnal convective activity in the land and sea areas is thought to influence the variation in tropopause height. Previous research showed that the diurnal variation in tropopause height was not significant enough in the Surabaya area [2]. However, that study used radiosonde data that had poor temporal resolution because radiosonde measurements were only performed at 00.00 and 12.00 UTC, so it was considered insufficient to study the diurnal variability of the tropopause. In addition, that study did not separate land and sea convection. The separation of land and sea convection is very important because there are contrasting differences in their diurnal variations [3].

Global Navigation Satellite System Radio Occultation (GNSS-RO) is a remote sensing technique that uses GNSS signals received by Low Earth Orbit (LEO) satellites by utilizing atmospheric refractivity. The GNSS signal will be deflected by the atmosphere before being captured by the LEO satellite and forming a deflection angle. The bending angle will produce an atmospheric refractivity profile. GNSS-RO provides data in global coverage and has a vertical resolution of 100 m [4]. RO data is of the highest quality at altitudes between 8 and 35 km and would be very useful for tropopause studies [5]. In addition, GNSS-RO data is also capable of providing data over the ocean area, which radiosondes and monitoring stations cannot provide.

Although many studies have investigated tropopause height variations in the tropics over long timescales such as annual [6], seasonal [7], and monthly [8], relatively few studies have identified tropopause height variations on diurnal timescales. Several studies have also been conducted to determine the effect of strong convection on tropical tropopause variations. However, the effect of differences in convective activity between land and sea on variations in tropopause height still needs to be studied further. In this study, the identification of variations in the height of the tropopause layer and its relation to differences in convective activity between land and sea will be developed, especially on the island of Java using diurnal GNSS-RO data.

2 Data and Method

In this study, GNSS-RO data provided by the Constellation Observing System for Meteorology Ionosphere and Climate (COSMIC-2) has a vertical resolution of 100 m and a spatial resolution of about 150 km [3]. COSMIC-2 is the result of cooperation between the USA and Taiwan which was launched on June 25, 2019. COSMIC-2 data can produce 5000 bending angle profiles and high-resolution refractivity per day in the tropics and subtropical [9]. The data taken is temperature and altitude from the wetPf2 atmospheric profile data. In addition, black body temperature (TBB) data from the IR1 channel of the Himawari-8 satellite is used to obtain the cloud top temperature which will then be used as a proxy in identifying convective activity. The Himawari-8 satellite is a geostationary satellite operated by the Japan Meteorological Agency (JMA). Himawari-8 was launched in October 2014 and is located at 140° E to observe eastern Asia and western Pacific regions. Himawari-8 satellite data was used in this research from Kochi University (http://weather.is.kochiu.ac.jp/archive-e.html). This data has a horizontal resolution of 0.05° × 0.05° and an hourly temporal with Portable Gray Map (PGM) format. The data period used is December 1, 2019–November 30, 2020, with the study area located at 6°–9° S and 105°–115° E.

The method used in this study includes the separation of data between land and sea using the Python global-land-mask module which has a resolution of 1 km. The latitude/longitude point of the data is checked whether it is on land or at sea and then grouped into land and ocean data before further processing. The height of the tropopause was identified using the minimum Lapse Rate Tropopause (LRT) method based on the definition from the World Meteorological Organization (WMO). WMO defined tropopause as the height at which the lapse rate decreases to 2° C/km or less, provided that within 2 km above this point, the average lapse rate does not exceed 2° C/km [10]. The convective activity was obtained from the average seasonal TBB gradient. The TBB gradient is the difference between the TBB value of a certain hour and the TBB value of the previous hour [11]. Both tropopause height variations and convective activity patterns were grouped into eight local time bins with 3-h intervals (7–10, 10–13, 13–16, 16–19, 19–22, 22–1, 1–4, and 4–7 local time (UTC + 7)).

3 Result

3.1 *Seasonal Tropopause Altitude Diurnal Pattern*

The diurnal pattern of tropopause height identified by the minimum LRT method for each season is presented in Fig. 1 (sea) and Fig. 2 (land). There is a diurnal variation of tropopause height in both sea and land areas. In general, the height of the tropopause in the sea area has a peak in the afternoon until the evening and early in the morning. Meanwhile, in the land, the peak is in the afternoon until late in the evening and late at night.

The maximum tropopause height above the ocean can be observed in the morning of each season. In December to February (DJF), March to May (MAM), and June to August (JJA) seasons, the maximum tropopause height occurs at 07.00–10.00 LT, while in September to November (SON) season the maximum height occurs at 04.00–07.00 LT. Toward noon the height of the tropopause above the ocean in all seasons decreases and begins to increase in the afternoon and into the evening. Meanwhile, the maximum tropopause height above the land can be observed at 10.00–13.00 LT during the DJF season, at 07.00–10.00 LT during the MAM and JJA seasons, and at 19.00–22.00 LT during the SON season. During the day in the land, the height of the

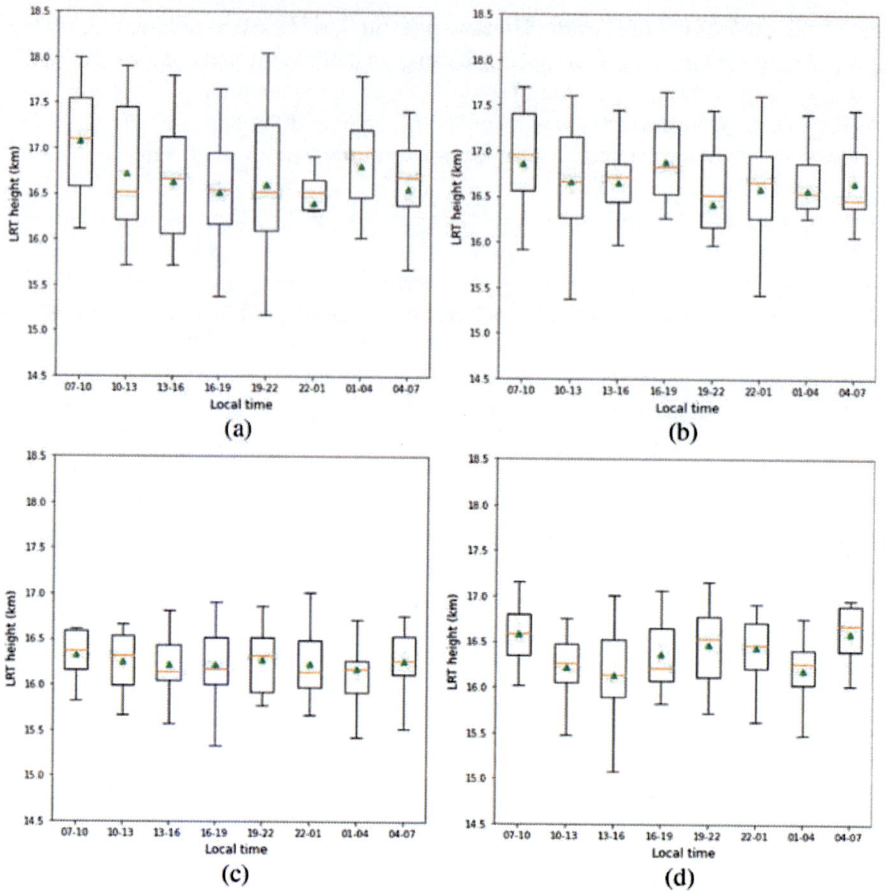

Fig. 1 Boxplot of the minimum LRT height above the ocean during the **a** DJF, **b** MAM, **c** JJA, and **d** SON seasons for the period December 1, 2019–November 30, 2020. The orange line is the median value, and the green triangle is the average value

tropopause tends to increase in almost all seasons except the JJA season. This is in accordance with research conducted by Suneeth [12].

Tables 1 and 2 show the variation in tropopause height in the ocean and land areas calculated using the Interquartile Range (IQR). The tropopause height in the ocean area ranges from 15 to 18 km with the highest variation of tropopause height occurring in the DJF season at 10.00–13.00 LT at 1.24 km and the lowest in the JJA season at 19.00–22.00 at 0.59 km. Meanwhile, in the land area, the tropopause altitude ranges from 14.7 to 18.1 km with the highest variation of tropopause height in the DJF season at 19.00–22.00 LT at 1.29 km and the lowest in the JJA season at 13.00–16.00 LT at 0.84 km. In general, the tropopause height variation is greater over land than over sea. This is in accordance with research conducted by Gettelman et al. [13].

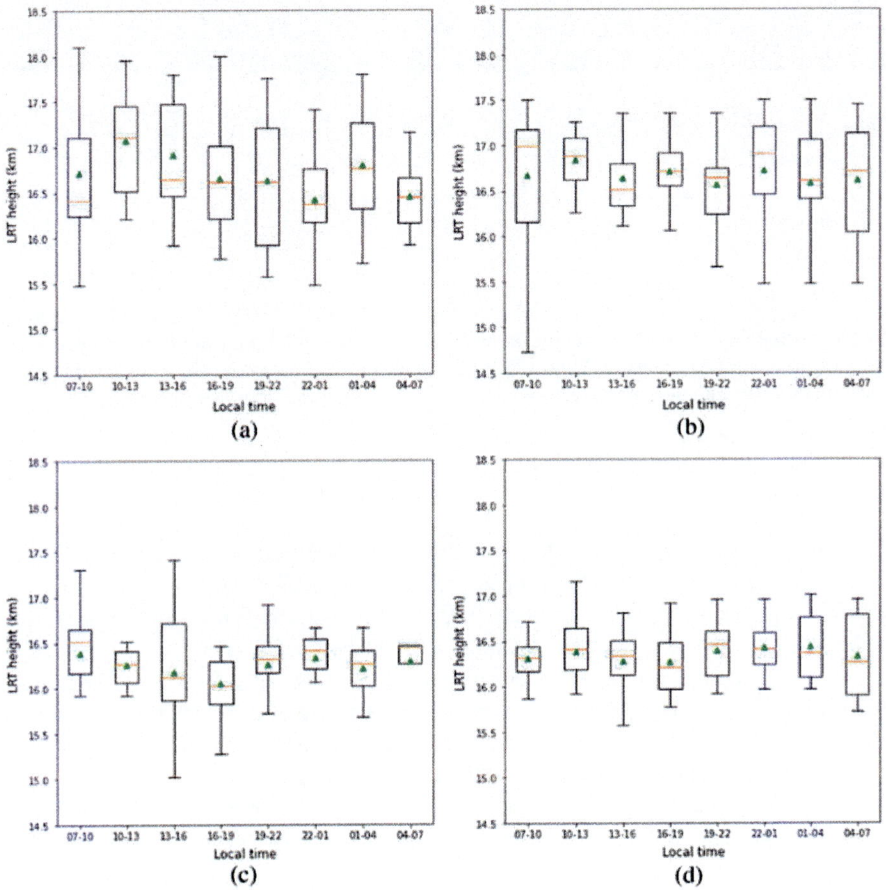

Fig. 2 Same as Fig. 1 but for minimum LRT height above the land

Table 1 IQR values of the average minimum LRT height in the sea area

Season	Local time							
	07-10	10-13	13-16	16-19	19-22	22-01	01-04	04-07
DJF	0.97	1.24	1.05	0.78	1.17	0.32	0.74	0.61
MAM	0.84	0.89	0.42	0.76	0.81	0.69	0.48	0.60
JJA	0.43	0.55	0.39	0.51	0.59	0.52	0.35	0.42
SON	0.45	0.42	0.63	0.57	0.66	0.50	0.38	0.50

Table 2 IQR values of the average minimum LRT height in the land areas

Season	Local time							
	07-10	10-13	13-16	16-19	19-22	22-01	01-04	04-07
DJF	0.87	0.94	1.00	0.79	1.29	0.60	0.94	0.50
MAM	1.03	0.46	0.46	0.35	0.51	0.74	0.64	1.09
JJA	0.48	0.35	0.84	0.47	0.30	0.32	0.40	0.20
SON	0.27	0.45	0.37	0.52	0.50	0.35	0.67	0.89

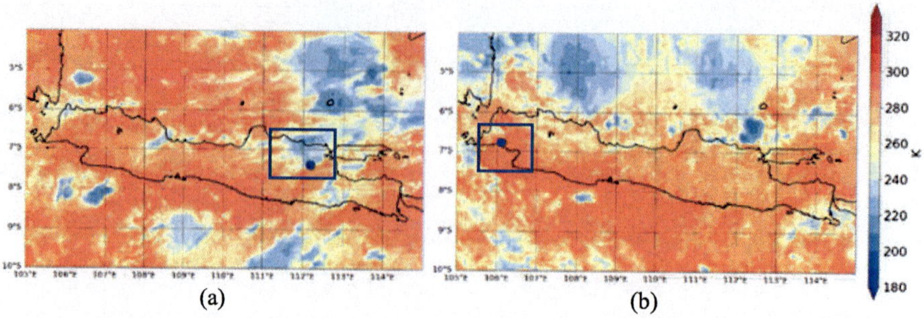

Fig. 3 Spatial plot of the TBB variable of the Himawari-8 satellite IR1 channel (cloud top temperature) at **a** January 27, 2020, at 09.00 LT and **b** March 21, 2020, at 09.30 LT. The blue dot is the location of the occultation profile

3.2 *Effect of Convective Activity on Tropopause Altitude*

The minimum tropopause height (14.7 km) was identified on March 21, 2020, at 09.30 LT, and the maximum height (18.1 km) was identified on January 27, 2020, at 09.00 LT in the land areas. Figure 3 shows the spatial pattern of cloud top temperatures at both times. In the land area on January 27 at 09.00 LT, there were convective clouds (deep convection events) characterized by TBB < 240 K [14]. While on March 21, 2020, at 09.30 LT, there were no convective clouds (clear sky conditions) characterized by TBB > 240 K. This indicates that at the time the tropopause was at maximum altitude, there are clouds with strong convection.

In this study, the average seasonal gradient of TBB for 24 h was used to identify the convective activity that occurred. The convective activity identified as a cloud growth process is characterized by a negative gradient. Cloud dissipation is indicated by a positive gradient value. The value of the TBB seasonal mean gradient for 24 h for land and sea areas has a clear difference between afternoon and evening. Based on Fig. 4, during the DJF season in the land area from the afternoon to evening the TBB gradient shows a negative value which indicates the occurrence of the cloud formation process. The peak of the negative TBB gradient occurred at 13.00–14.00 LT. At night, the TBB gradient shows a positive value associated with cloud dissipation which is thought to produce convective rain. Meanwhile, in the DJF season in the ocean area at 13.00–19.00, 20.00–21.00, and 23.00–00.00 LT, the process of cloud formation occurs, and in the early morning, cloud thinning occurs. In the MAM transition season in the land, it is 11.00–18. 00 LT occurring in the process of cloud formation with a peak at 13.00–14.00 LT, and at night until sunrise, there is cloud thinning. While in the ocean area at 14.00–18.00, 19.00–21.00, and 23.00–00.00 LT, the process of cloud formation occurs, and in the morning, there is cloud thinning. In the JJA season, the TBB gradient value is not very volatile and tends to be around the value 0, indicating the least convective activity, both growth and cloud dissipation. This is probably because the JJA season in Java is experiencing a dry season, so the atmospheric conditions tend to be cloudless, sunny, and rarely rainy. Meanwhile,

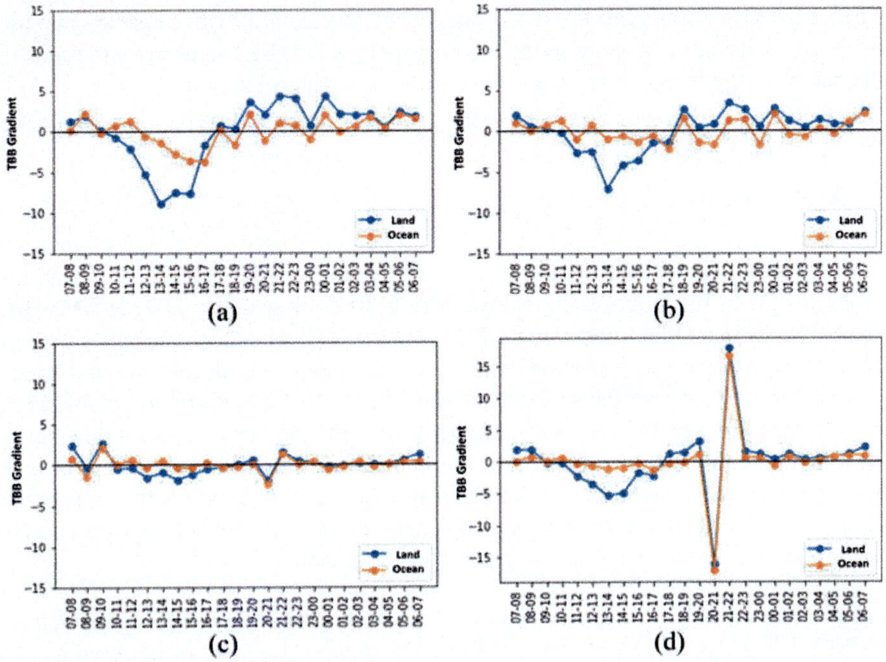

Fig. 4 Seasonal mean TBB gradient values between land and sea areas **a** DJF season, **b** MAM, **c** JJA, **d** SON. The orange line represents the ocean area, and the blue line represents the land area

during the SON transition season, both on land and in the ocean, there is a very high fluctuation of data that indicates the cloud growth process followed by cloud dissipation.

4 Discussion

In general, the diurnal variation of convective activity over the ocean in the form of cloud formation is more dominant in the late afternoon and early morning, while on land, cloud formation is more dominant in the noon until the afternoon local time, in sync with the diurnal variation in tropopause height. The diurnal pattern of TBB gradients in both DJF and MAM seasons shows that above the land, cloud formation is more dominant in the afternoon to evening local time. This can happen because land and sea have different responses to the reception of solar radiation. On the land, surface heating due to the reception of solar radiation takes place faster than in the ocean so that on land convective activity reaches its peak during the day, which in turn causes convective rain above the land to occur in the afternoon until evening [11]. Nitta and Sekine [1] stated that over the continents and large islands, convection reaches its maximum intensity in the late afternoon to evening, probably due to strong surface heating throughout the day. Local convective activity can cause an increase in the intensity of turbulence in the upper troposphere [15]; therefore, variations in the height of the tropopause may also be influenced by turbulence.

5 Conclusion

The effect of differences in convective activity between land and sea on variations in the height of the tropopause in Java during the four seasons: DJF, MAM, JJA, and SON has been identified using GNSS-RO data. There is a synchronous pattern of convective activity to variations in tropopause height in both sea and land areas. Convective activity over the seas around the island of Java is more dominant in the late afternoon and early in the morning, while over the land of Java, cloud formation is more dominant in the noon until the afternoon local time following the increase in tropopause height. In addition, it is known that seasonally, the height of the tropopause in the seas of Java Island that most closely matches the pattern of cloud formation is the MAM season, while on the land it is the DJF season.

Acknowledgements The authors wish to thank University for Corporation of Atmospheric Research (UCAR) for providing the GNSS-RO data and Kochi University for providing the Himawari-8 data. The first author was supported by the Marine Technology Cooperation Research Center (MTCRC).

References

1. Nitta, T., Sekine, S.: Diurnal variation of convective activity over the tropical western pacific. J. Meteorol. Soc. Jpn. **72**(5) (1994)
2. Prayuda, S.: Analysis of the Lapse Rate Tropopause (LRT) Characteristics and its relation to MJO activities in Surabaya. In: Proceedings of the STMKG Earth and Atmospheric (2018)
3. Johnston, B.R, Xie, F., Liu, C.: The effects of deep convection on regional temperature structure in the tropical upper troposphere and lower stratosphere. J. Geophys. Res. Atmos. **123**(3) (2018)
4. Kursinski, E.R, Hajj, G.A, Schofield, J.T, Linfield, R.P, Hardy, K.R.: Observing Earth's atmosphere with radio occultation measurements using the global positioning system. J. Geophys. Res. Atmos. **102**(19) (1997)
5. Foelsche, U., Borsche, M., Steiner, A.K, Gobiet, A., Pirscher, B., Kirchengast, G., Wickert, J., Schmidt, T.: Observing upper troposphere-lower stratosphere climate with radio occultation data from the CHAMP satellite. Clim. Dyn. **31**(1) (2008)
6. Seidel, D.J., Ross, R.J., Angell, J.K., Reid, G.C.: Climatological characteristics of the tropical tropopause as revealed by radiosondes. J. Geophys. Res. Atmos. **106**(D8) (2001)
7. Liu, Y., Xu, T., Liu, J.: Characteristics of the seasonal variation of the global tropopause revealed by COSMIC/GPS data. Adv. Space Res. 54(11). (2014)
8. Rieckh, T.: Tropopause characteristics from GPS radio ocultation data. Master Thesis. Wegener Center for Climate and Global Change. University of Graz (2013)
9. Schreiner, W. S., Weiss, J. P., Anthes, R. A., Braun, J., Chu, V., Fong, J., Hunt, D., Kuo, Y. H., Meehan, T., Serafino, W., Sjoberg, J., Sokolovskiy, S., Talaat, E., Wee, T. K., & Zeng, Z. COSMIC-2 Radio occultation constellation: first results. Geophys. Res. Lett. **47**(4) (2020)

10. WMO, M.: A three-dimensional science. WMO Bull. 6, 134–138 (1957)
11. Kusumawati, Y., Effendy, S., Aldrian, E. Spatial and temporal variations of convective rain on Java Island based on satellite imagery. Agromet J. **22**(1) (2008)
12. Suneeth, K.V., Das, S.S, Das, S.K.: Diurnal variability of the global tropical tropopause: results inferred from COSMIC observations. Clim. Dyn. **49**(9–10) (2017)
13. Gettelman, A., Salby, M.L, Sassi, F.: Distribution and influence of convection in the tropical tropopause region. J. Geophys. Res. Atmos. **107**(9–10) (2002)
14. Mehta, S. K., Ratnam, M. V., Murthy, B. V. K. Variability of the tropical tropopause over Indian monsoon region. J. Geophys. Res. Atmos. **115**(14) (2010)
15. Fujiwara, M., Yamamoto, M.K., Hashiguchi, H., Horinouchi, T., Fukao, S.: Turbulence at the tropopause due to breaking kelvin waves observed by the equatorial atmosphere radar. Geophys. Res. Lett. **30**(4) (2003)

Sensitivity Analysis of Constructed Analogue Statistical Downscaling Method for Extreme Rainfall Prediction

Trinah Wati, Tri Wahyu Hadi, Faiz Rohman Fajary, Ardhasena Sopaheluwakan, and Lambok M. Hutasoit

Abstract Sensitivity analysis of a statistical downscaling (SD) method is to assess a method's shortcomings, evaluate the reliability, compare the uncertainty, and fine-tune to see which parameter of the model can be changed. This study aims to analyse the sensitivity of constructed analogue (CA) SD method by evaluating the reliability of gaining extreme information and comparing the uncertainties of CA experiment schemes. There are four schemes, including multilinear regression (MLR), weight mean (WMEAN), MLR with principal component analysis/PCA (PC-MLR), and WMEAN with PCA (PC-WMEAN). The comparison of the schemes was based on predictors' uncertainty analysis, distribution, and extreme value using several metrics such as Verification Rank Histogram (VRH), Probability Density Function-Skill Score (PDF-SS), and Brier Score (BS). The predictor is the reanalysis dataset ERA-20C, while the predictand is rainfall from 11 observation rain gauge stations in Java Island, with leave one-year out cross-validation from 1970 to 2010. The results of VRH show the WMEAN and PC-WMEAN schemes have under-dispersed ensemble members. Meanwhile, the MLR shows positive bias, and PC-MLR shows negative bias. The MLR scheme has the highest score on PDF-SS. Although underestimated and overestimated, the PDF plot shapes of MLR and PC-MLR schemes are closer to observation than WMEAN and PC-WMEAN. The MLR and PC-MLR schemes are more capable of estimating daily and pentad rainfall of 50 and 100 mm than the WMEAN and PC-WMEAN schemes. This study implicates the development of the CA SD method for predicting/estimating extreme rain.

T. Wati (✉) · T. W. Hadi · F. R. Fajary · L. M. Hutasoit
Graduate Program of Earth Sciences, Faculty of Earth Sciences and Technology, Institut Teknologi Bandung, Bandung 40132, Indonesia
e-mail: trinah.wati19@students.itb.ac.id

T. Wati · A. Sopaheluwakan
Indonesia Agency for Meteorology, Climatology, and Geophysics, Jakarta 10610, Indonesia

A. Basit et al. (eds.), *Proceedings of the International Conference on Radioscience, Equatorial Atmospheric Science and Environment and Humanosphere Science*, Springer Proceedings in Physics 290, https://doi.org/10.1007/978-981-19-9768-6_24

1 Introduction

Statistical downscaling (SD) is one of the techniques to generate high-resolution regional climate models based on large-scale information from either reanalysis data or global climate models (GCM). The SD method with an analogue approach was previously developed for weather prediction [1, 2, 3]. Other studies [4, 5] used the method to gain local climate statistics consistent with synoptic-scale atmospheric conditions of GCM. The application of the analogue approach requires a long-term database to cover weather patterns or atmospheric conditions that may appear in future. The CA method has an advantage in the analogy of a local scale climate pattern with a synoptic-scale climate pattern, so that the spatial structure of the local climate is well presented in the simulation results. The method minimizes biases and uses the absolute value of the predictand in the process of prediction or estimation.

The capability of the CA SD method depends on the similarity between the patterns defined in the analogue selection [4]. Previous studies [6, 7] used the minimum Euclidean distance metric to calculate the similarity degree of weather patterns from predictors; another is cosine similarity [8, 9]. However, the shortcomings of the CA SD method were found, i.e. increasing spatial coherence in the result field, decreasing the final result's temporal variation caused by averaging process, and too much drizzle (light rain) production [10]. This study aims to analyse the sensitivity of the CA SD method by evaluating the reliability of gaining extreme information and comparing the uncertainties of CA experiment schemes.

2 Data and Method

The predictor data employ ERA-20C's 3-hourly averaged to daily zonal (U) and meridional (V) wind parameters at 850 hPa from 1970 to 2010. We derive U and V wind parameters as vectors to scalar values of stream function (ψ) or Psi and velocity potential (χ) or Phi using a decomposition of Helmholtz. The study uses a multi-window approach [8, 9], getting 14 ensemble members from seven monsoon windows to obtain the probabilistic rainfall prediction or estimation. Meanwhile, the predictand data employ daily rainfall of 11 stations in Java Island with the same periods. The location of monsoon windows and the stations are visualized in Fig. 1a and b.

There are two stages in CA: diagnosis and prognosis. Firstly, the diagnosis stage selects a subset of historical weather patterns of GCM predictors in training periods that are similar to the target periods. The selection of analogue uses the cosine similarity/$S(u)$ method [15] to calculate the degree of similarity between vectors $\vec{d}$ (u) in training periods and $\vec{d}$ (t) in target periods using the equation:

$$S(u) = \frac{\vec{d}(u).\vec{d}(t)}{\left|\vec{d}(u)\right|\left|\vec{d}(t)\right|} \quad (1)$$

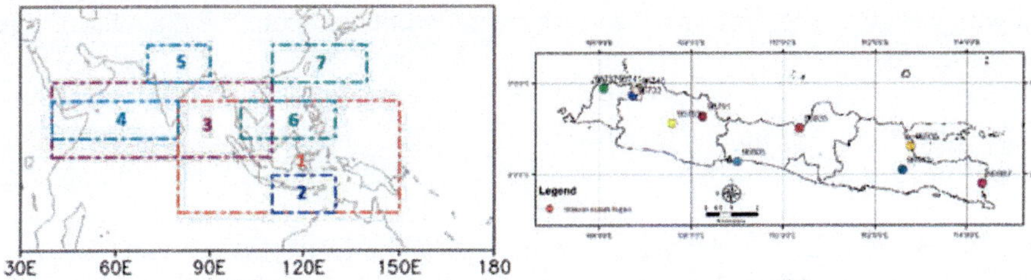

(a) (b)

Fig. 1 Location of monsoon windows [8] (**a**) and observation stations, (**b**) The monsoon windows are (1) Maritime Continent [11], (2) Australian Monsoon [12], (3) Webster and Young Monsoon [13], (4, 5) Indian Summer Monsoon [14], and (6, 7) Western North Pacific Summer Monsoon [14]

The subset number assigned is 30 subsets producing a constructed analogue and determining the most suitable multilinear combination [5, 8]. The application of EOF or PCA reduces the atmospheric circulation field's degree of freedom from predictors [4, 9]. The formulas for PCA are as follows:

$$F(u) = \sum_{k=1}^{M} c(u)_k a(u)_k \tag{2}$$

$$F(t) = \sum_{k=1}^{M} c(t)_k a(t)_k \tag{3}$$

where $F(u)$ and $F(t)$ are predictors and c and a are spatial and temporal components in training and target periods. M is the number of significant components (PCs), usually between 5 and 10. The PC scores are normalized to give equal weight to the variance described by all PCs.

The combination of 30 analogue subsets is applied in the prognosis stage using the predictand data to predict or estimate the predictand variable in the target time. The formula of analogue construction can be MLR [9], WMEAN [16], or other regression methods. The formulas of analogue construction in this study are as follows:

$$R_{\mathrm{CA}}(t) = \sum_{n=1}^{30} B_n R_n(t) + C \quad \text{for MLR} \tag{4}$$

$$R_{\mathrm{CA}}(t) = \sum_{n=1}^{30} W_n R_n \quad \text{for WMEAN} \tag{5}$$

where $R_{\mathrm{CA}}(t)$ is the CA simulation rainfall at the target periods (t) and n is the sum of the best analogue subsets at the diagnosis stage. B is the regression coefficient obtained between the predictors from the diagnosis stage for MLR, and W is the weighting between RMSE and the correlation coefficient from the diagnosis stage for WMEAN.

There are four CA schemes of the sensitivity analysis, i.e. MLR, WMEAN, MLR with principal component analysis (PC-MLR), and WMEAN with PCA (PC-WMEAN). We use cross-validation leave one-year out for the CA SD running. The comparison of the four schemes is based on predictor uncertainty, distribution, and extreme values analysis. Verification Rank Histogram (VRH) is a dispersion error measure of ensemble predictions representing observational uncertainty at a daily timescale. The shape of VRH that is uniform or flatter indicates the ensemble predictions close to the observations [17]. Distribution analysis uses the PDF Skill Score (*SS*) metric [18–20]; the formula is as follows:

$$SS = \sum_{k=1}^{nb} \min(f_m^k, f_o^k) \tag{6}$$

where f_m^k and f_o^k are relative frequencies of rain days (days with precipitation > 0.5 mm). A value in the *k*th bin belongs to the histograms of the CA SD result and observation, whereas Nb is the bin number of the empirical PDF calculation. The result of CA SD perfectly simulates the observed PDF with a maximum value of 1.

The verification for extreme values uses the Brier Score (BS) of 50 and 100 mm rainfall events daily and 100 mm on pentad data (five days accumulation). The formula is as follows:

$$BS = \frac{1}{N}\sum_{n=1}^{N}(f_n - O_n)^2 \tag{7}$$

where N is the number of sample data and f_n is the probability of CA SD results of daily and pentad rainfall events (50 and 100 mm). The formulation uses binary possibilities, and O_n is an observation (value 1 for events that occur and 0 for non-occurrence). BS has the best skill at zero and the worst at one.

3 Result and Discussion

The resulting period of the leave one-year out cross-validation approach is 1970 to 2010 at a daily timescale. Figure 2 shows VRH diagrams of four CA SD schemes at Banyuwangi Station in East Java. The VRH of the MLR schematic in Fig. 2a shows a positive bias of ensemble members [17]. Meanwhile, the VRH of WMEAN and PC-WMEAN schemes look similar and show under-dispersive or less spread of ensemble members. In contrast to the MLR scheme, the PC-MLR scheme exhibits a negative bias. All stations have a similar pattern of VRH, with the sample at Banyuwangi Station as a representation.

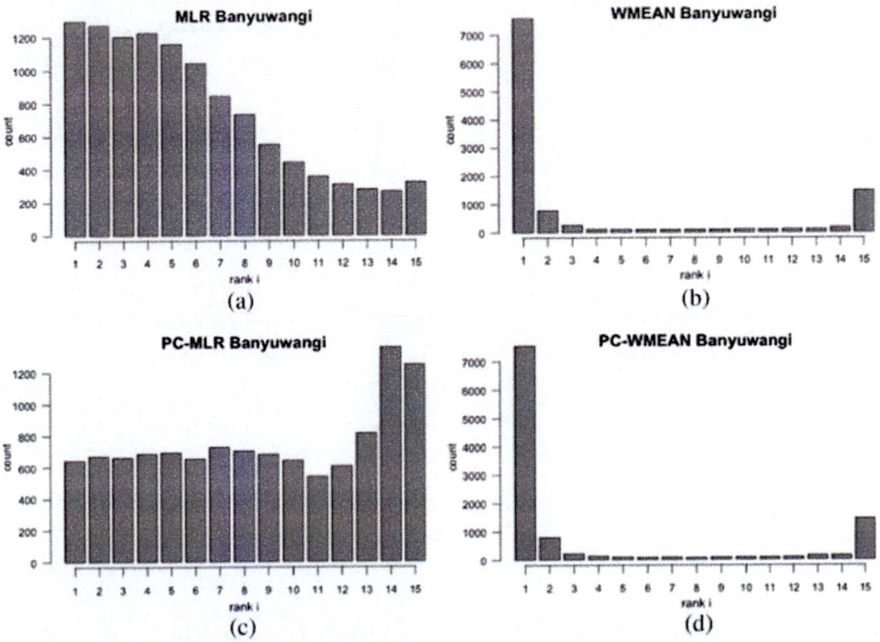

Fig. 2 Diagrams of VRH at Banyuwangi Station for CA SD scheme of MLR (**a**), WMEAN (**b**), PC-MLR (**c**), and PC-WMEAN (**d**)

The distribution analysis of the four schemes, as seen in Fig. 3a–d, compares the empirical PDF of rain days at Semarang Climate Station. It can be seen that the PDF graph of the WMEAN (b) and PC-WMEAN (d) schemes are different from the observations, while the PDF MLR (a) and PC-MLR (c) schemes are closer, but the PC-MLR scheme is very overestimated. The SS quantification measure for all stations presented in Table 1 shows the MLR scheme has the highest average of 0.80, ranging between 0.68 and 0.87. The lowest SS is the PC-MLR scheme with an average of 0.51, ranging from 0.31 to 0.83. Figure 3 is the representation for all stations that have similar PDF pattern.

The extreme value analysis uses the BS metric for 50 mm and 100 mm daily and 100 mm on pentad data presented in Figs. 4, 5 and 6. We consider analysing the pentad data due to the accumulation of rainfall for several days associated with flood events. The BS for 50 mm daily rainfall in Fig. 4 shows almost the same skills for all schemes, except for the MLR scheme, which has more representative stations with lower skills (blue). The BS for 100 mm daily rainfall in Fig. 5a–d shows a more significant number of stations in the MLR and PC-MLR schemes with better skills than in the WMEAN and PC-WMEAN schemes, while the BS for 100 mm pentad rain in Fig. 6 shows that the PC-WMEAN scheme has the lowest skill compared to other schemes.

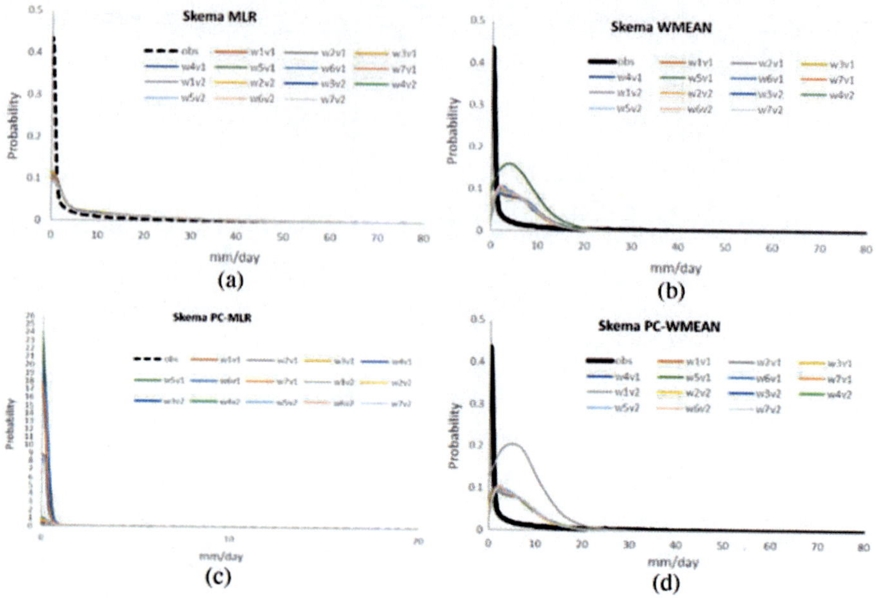

Fig. 3 Graphs of rain day empirical PDF at Semarang Climate Station for MLR (**a**), WMEAN (**b**), PC-MLR (**c**), and PC-WMEAN (**d**)

Table 1 SS of four CA SD schemes

CA SD Schemes	PDF skill score (SS)		
	Mean	Min	Max
MLR	0.80	0.68	0.87
WMEAN	0.65	0.55	0.73
PC-MLR	0.51	0.31	0.83
PC-WMEAN	0.65	0.56	0.74

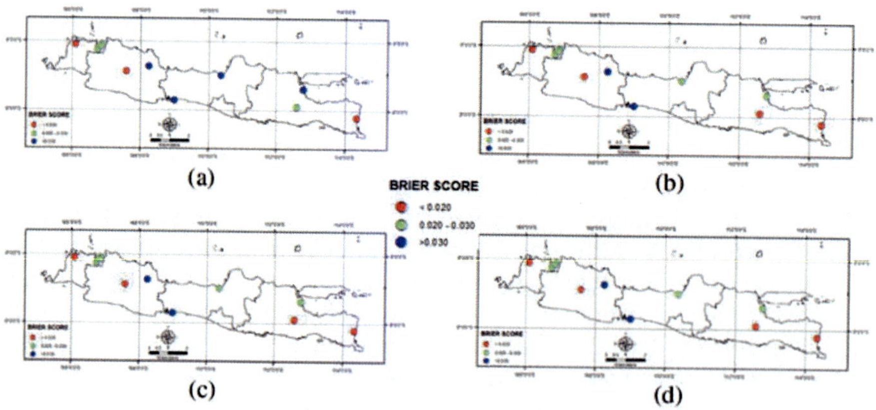

Fig. 4 BS of 50 mm daily rain for MLR (**a**), WMEAN (**b**), PC-MLR (**c**), and PC-WMEAN (**d**)

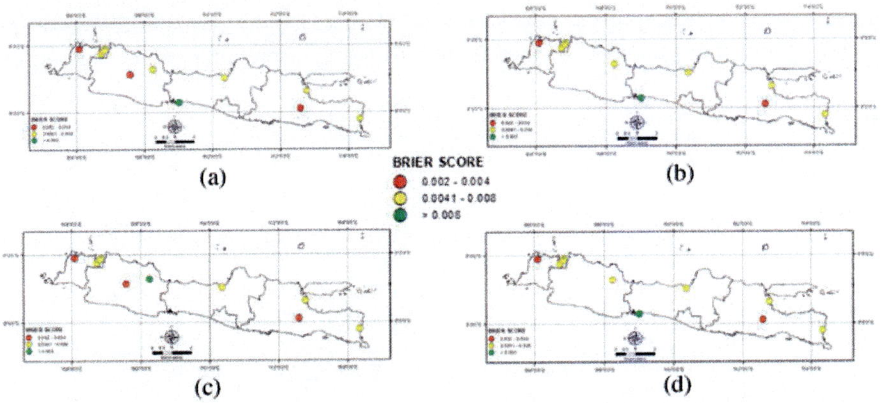

Fig. 5 BS of 100 mm daily rain for MLR (**a**), WMEAN (**b**), PC-MLR (**c**), and PC-WMEAN (**d**)

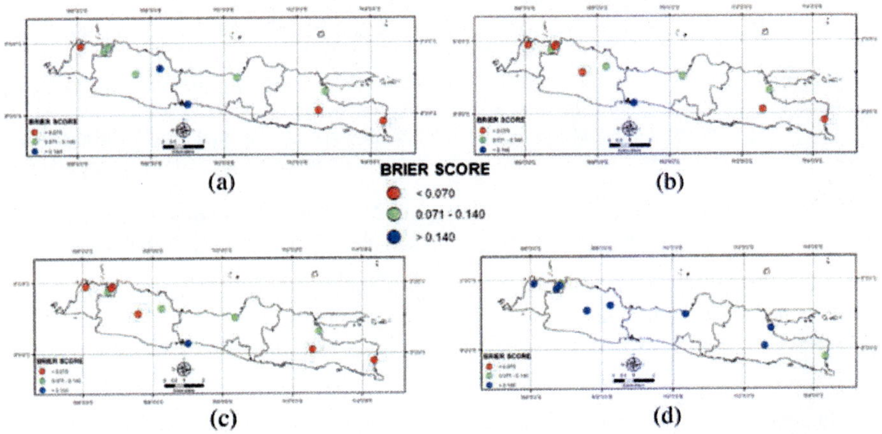

Fig. 6 BS of 100 mm pentad rain for MLR (**a**), WMEAN (**b**), PC-MLR (**c**), and PC-WMEAN (**d**)

4 Summary

Sensitivity analysis of four CA SD schemes in this study assesses uncertainty (VRH), distribution (SS), and extreme values (BS). The MLR and PC-MLR schemes have better distribution representation and a more significant number of stations with better extreme representatives than the WMEAN and PC-WMEAN schemes. The WMEAN and PC-WMEAN schemes based on the VRH show low dispersion, indicating less represented extreme values. PCA treatment did not affect the WMEAN scheme as seen in the same VRH pattern on WMEAN and PC-WMEAN schemes. On the contrary, in the MLR scheme, which has a positive bias VRH pattern, the PCA treatment gives more effect, resulting in the VRH pattern becoming negatively biased. It should be considered a calibration treatment to the result of CA SD for obtaining a more extreme value representation and closer to observed data. This study implicates the development of the CA SD method in estimating or predicting extreme rainfall.

References

1. Lorenz, E.N.: Atmospheric predictability as revealed by naturally occurring analogues. J. Atmos. Sci. **26**, 636–646 (1969)
2. Van den Dool, H.M.: Searching for analogues, how long must we wait? Tellus A **46**(3), 314–324 (1994)
3. Zorita, E., Hughes, J.P., Lettemaier, D.P., Von Storch, H.: Stochastic characterization of regional circulation patterns for climate model diagnosis and estimation of local precipitation. J. Clim. **8**(5), 1023–1042 (1995)
4. Zorita, E., Von Storch, H.: The analog method as a simple statistical downscaling technique: comparison with more complicated methods. J. Clim. **12**(8), 2474–2489 (1999)
5. Hidalgo, H.G., Dettinger, M.D., Cayan, D.R.: Downscaling with constructed analogues: daily precipitation and temperature fields over The United States (2008)
6. Timbal, B., McAvaney, B.J.: An analogue-based method to downscale surface air temperature: application for Australia. Clim. Dyn. **17**(12), 947–963 (2001)
7. Anuchaivong, P., Sukawat, D., Luadsong, A.: Statistical downscaling for rainfall forecasts using modified constructed analog method in Thailand. Sci. World J. (2017)
8. Syahputra, M.R.: Implementation on constructed analogue on seasonal rainfall forecast from CFS (Climate Forecast System) output; case study Island of Java and Sumatra. Master Thesis. Institut Teknologi Bandung. (2013). *In Bahasa*
9. Surmaini, E., Hadi, T.W., Subagyono, K., Puspito, N.T.: Prediction of drought impact on rice paddies in west java using analogue downscaling method. Indonesia J. Agric. Sci. **16**(1), 21–30 (2015)
10. Pierce, D.W., Cayan, D.R., Thrasher, B.L.: Statistical downscaling using localized constructed analogues (LOCA). J. Hydrometeorol. **15**(6), 2558–2585 (2014)
11. Robertson, A.W., Moron, V., Qian, J., Chang, C., Tangang, F., Aldrian, E., Koh, T.Y., Liew, J.: The maritime continent monsoon. The global monsoon system: research and forecast, 85–98 (2011)
12. Kajikawa, Y., Wang, B., Yang, J.: A multi-time scale Australian monsoon index. Int. J. Climatol. **30**(8), 1114–1120 (2010)
13. Webster, P.J., Yang, S.: Monsoon and ENSO: selectively interactive systems. Q. J. R. Meteorol. Soc. **118**, 877–926 (1992)
14. Wang, B., Wu, R., Lau, K.M.: Interannual variability of the Asian summer monsoon: contrasts between the Indian and the western North Pacific-East Asian monsoons. J. Clim. **14**(20), 4073–4090 (2001)
15. Garcia, E.: Cosine Similarity Tutorial. https://www.researchgate.net/publication/327248753 (2015)
16. Fernández, J., Sáenz, J.: Improved field reconstruction with the analog method: searching the CCA space. Climate Res. **24**, 199–213 (2003)
17. Hamill, T.M.: Interpretation of rank histograms for verifying ensemble forecasts. Mon. Weather Rev. **129**(3), 550–560 (2001)
18. Perkins, S.E., Pitman, A.J., Holbrook, N.J., McAneney, J.: Evaluation of the AR4 climate models'simulated daily maximum temperature, minimum temperature, and precipitation over Australia using probability density functions. J. Clim. **20**(17), 4356–4376 (2007)
19. Chhin, R., Yoden, S.: Ranking CMIP5 GCMs for model ensemble selection on regional scale: case study of the Indochina region. J. Geophys Res. Atmos. **123**(17), 8949–8974 (2018)
20. Wati, T., Hadi, T.W., Sopaheluwakan, A., Hutasoit, L.M.: Statistics of the performance of gridded precipitation datasets in Indonesia. Adv. Meteorol. (2022)

Meteorological Factors Influencing Coastal Flooding in Semarang, Central Java, Indonesia, on 23 May 2022

Teguh Harjana, Eddy Hermawan, Risyanto, Anis Purwaningsih, Dita Fatria Andarini, Ainur Ridho, Dian Nur Ratri, and Akas Pinaringan Sujalu

Abstract It has been well established that coastal flooding is caused by heavy rains, storm surges, and high tidal waves that potentially lead to tremendous damage. Using brightness temperature (TBB) data from the Himawari satellite and ocean surface wind data from ERA5, we have investigated meteorological factors influencing strong wind and increased seawater level over the northern coast of Semarang which caused the coastal flood on 23 May 2022. Our results indicated that there are two potential meteorological factors that contribute to the coastal flooding during that period. Firstly, the formation of MCS over the ocean on the northern coast of Semarang led to heavy rain over the coast and strong-surface wind speed that potentially enhances ocean tidal waves toward Semarang. Secondly, the unusually strong-surface easterly winds cause the increase in seawater level through the Ekman pumping mechanism. As for the latter, the strong easterly wind stress led to strong Ekman transport to the south of the flow (i.e., to the northern coast of Java), causing a large net transport of seawater toward this region as a result of a balance between Coriolis and turbulent (wind) drag forces. Our results provided a new perspective on the factors influencing the increased seawater level intruding into Semarang during the coastal flooding on 23 May 2022.

T. Harjana (✉) · E. Hermawan · Risyanto · A. Purwaningsih · D. F. Andarini
Research Center for Climate and Atmosphere, National Research and Innovation Agency (BRIN), Bandung 40173, Indonesia
e-mail: teguh.harjana@brin.go.id

A. Ridho
Cerdas Antisipasi Risiko Bencana Indonesia (CARI), Bandung 40293, Indonesia

D. N. Ratri
Meteorological, Climatological, and Geophysical Agency (BMKG), Jakarta 10720, Indonesia

Droevendaalsesteeg, Wageningen University and Research, 6708 Wageningen, The Netherlands

A. P. Sujalu
Universitas, 17 Agustus 1945 Samarinda, Samarinda 75123, Indonesia

A. Basit et al. (eds.), *Proceedings of the International Conference on Radioscience, Equatorial Atmospheric Science and Environment and Humanosphere Science*, Springer Proceedings in Physics 290, https://doi.org/10.1007/978-981-19-9768-6_25

1 Introduction

Semarang is one of the big cities in Indonesia and is located on the northern coast of Java Island. Semarang city is an urban area and has a specific problem related to the coastal environment, and it has been reported that the coastal area is a subside relative to the mean sea level [1]. The condition that the topography is under the sea level rise, due to the effects of land subsidence, makes Semarang vulnerable to coastal floods [2–8].

On 23 May 2022, Semarang experienced a tidal flood disaster that caused the breaking of the coastal embankment in the Tanjung Emas Area. The Semarang City Disaster Mitigation Agency (BPBD) on Tuesday announced that the breaking of the coastal embankment causes seawater to flood the land with inundations of 40 cm to 1.5 m. Moreover, Tanjung Emas Maritime Meteorological Station Semarang BMKG recorded the height of the tide before the embankment broke at 15.00 WIB (08Z) reaching 2 m [9]. It is also recorded by BPBD that the coastal flooding had affected 1255 households [10]. The Semarang tidal flood during this day occurred in the phase of the moon perbani (neap tides), so the influence on the tides was not as high as in the phase of the full moon (spring tides) [11]. The existence of a storm that occurred at 02.00 WIB early in the morning in the Java Sea and the presence of persistent winds since a few days earlier are suspected to be one of the triggers for the Semarang tidal flood [11].

In this paper, the factors that affect strong winds and sea level increases that caused coastal floods on 23 May 2022, on the northern coast of Semarang, will be investigated using satellite and reanalysis data. The detail of the data and method that are employed for this study is written in Sect. 2. Then, the results and the discussion are described in Sect. 3. Furthermore, this is followed by the conclusions and future works in Sect. 4.

2 Data and Methods

The fifth-generation reanalysis data from The European Center for Medium-Range Weather Forecasting (ECMWF ERA5) [12] was used in this study. Data used from ERA5 is Mean Sea Level Pressure (MSLP), u and v components of wind at 10 m, 925 hPa wind, and Sea Surface Temperature (SST). The Global Satellite Mapping of Precipitation (GSMaP) gauge-corrected version-7 standard precipitation product was used to investigate the presence of rain that occurs in the Java Sea [13]. The spatial resolution of the GSMaP data is 0.01×0.01 degrees, and the temporal resolution is 1 h (https://sharaku.eorc.jaxa.jp).

In addition, the presence of the Mesoscale Convective System (MCS) was identified using Temperature Black Body (TBB) from Band-13 (IR) of Himawari-8 satellite, using the "Grab 'em, Tag 'em, Graph 'em" (GTG) algorithm [14]. Criteria for the MCS phase are described in Table 1.

Table 1 Criteria for MCS phase

Physical characteristics	Criteria
BT(10.4)	243 K
Size	10,000 km^2
Duration	Size and temperature definition must be met for a period of 3 h
Initiation	Size and temperature definition are first satisfied
Termination	Size and temperature definition are no longer satisfied
Mature	Minimum mean of cloud temperature definition must be met

The GTG method has been widely used in analyzing the role of MCS in Indonesia, for example, in the case of heavy rains in Jakarta in January 2013 [15], Semarang [16], and New Capital City Nusantara of Indonesia [17].

In addition, Ekman transport and pumping velocity were also calculated which were derived from wind stress at an altitude of 10 m.

Ekman transport (EMT, $m^3 s^{-1} m^{-1}$) is calculated using the formula used by Dieng AL [18].

$$\mathrm{EMT}_x = \frac{\tau_y}{\rho_w f}, \quad \mathrm{EMT}_y = -\frac{\tau_x}{\rho_w f} \tag{1}$$

where ρ_w is the water density (1025 kg m^{-3}), f is the Coriolis factor (1/s), and the wind stress τ (Pa) is calculated using

$$\tau_x = \rho_a \cdot C_D \cdot \sqrt{u^2 + v^2} \cdot u, \quad \tau_y = \rho_a \cdot C_D \cdot \sqrt{u^2 + v^2} \cdot v \tag{2}$$

where ρ_a is the density of air (1.25 kg m^{-3}), C_D is the coefficient of drag, and *u, v* are components of wind at 10 m above sea level.

Ekman pumping velocity (EPV, ms^{-1}) is calculated using the curl of wind stress with the equation:

$$\mathrm{EPV} = \frac{1}{\rho_w f} \nabla \times \tau \tag{3}$$

3 Result and Discussion

A breaking seawater embankment of Tanjung Emas Semarang caused a coastal flood event of up to 1.5 m on 23 May 2022 at 08Z (15 WIB). Figure 1b denotes the evolution of sea level height at Tanjung Emas Semarang from 22 May 2022 at 00Z to 24 May

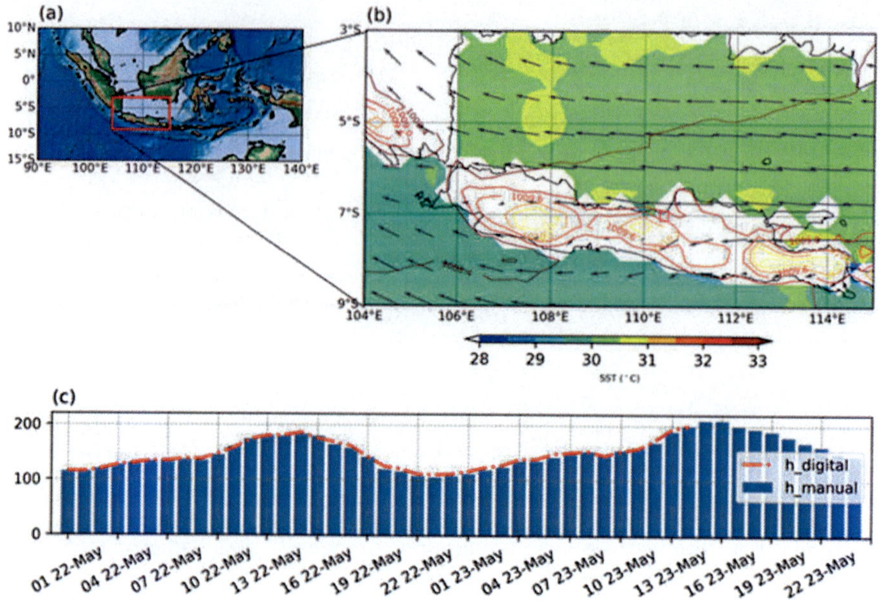

Fig. 1 Figures **a** and **b** show the geographical location of the northern coast of Java, and **c** shows a time series of sea wave heights at the time of the tidal wave on 23 May 2022

2022 at 00Z. It is worth noting that the high seawater level occurred at around 05Z–11Z on 22 and 23 May 2022 with a height between 180 and 210 cm. Specifically, the seawater level significantly increased since 23 May 2022 at 05Z and reached a peak of approximately 210 cm at 08Z and 09Z. This maximum sea level height affected the collapse of the seawater embankment and hence tidal flooding in many areas over Semarang and its surroundings. In addition, the hourly GSMaP precipitation data also shows heavy rainfall occurred in the Java Sea, next to the north of Central Java, from 22 May 2022 at 20Z till 23 May 2022 at 05Z (the GSMaP rain map is not shown here).

3.1 *The MCS Evolution Ahead of Coastal Flooding*

To understand the meteorological factors that modulate the increased seawater level during the coastal flooding in Semarang, we analyzed the evolution of MCS before and during the tidal flood event (see Fig. 2). A convective cloud began to develop on 22 May 2022 at 19Z in the coastal area of Central Java and was identified as the pre-MCS phase (Fig. 2a). This cloud grew rapidly (around 1 h), which showed the onset of the MCS initiation phase on 22 May 2022 at 20Z. At the same time, a rainfall of 5 mm/h occurred in the MCS areas. Then, the MCS developed to the growth stage, where the size of clouds increased further and became two cloud cells

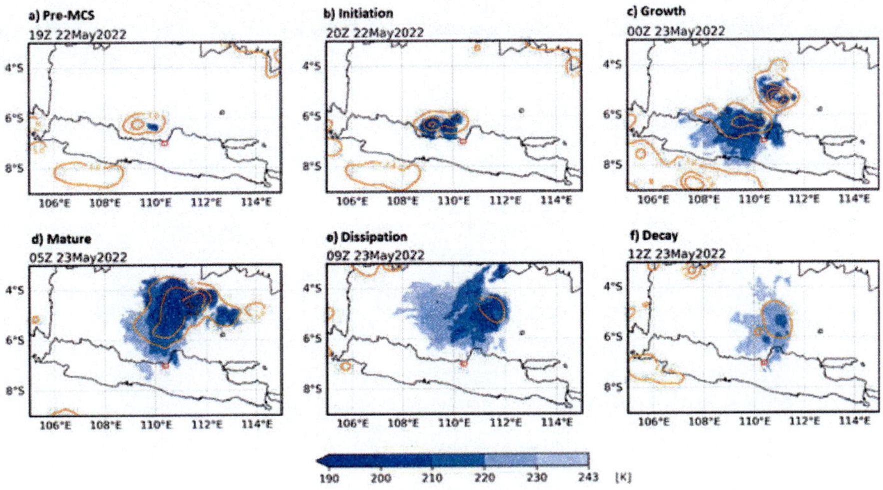

Fig. 2 Evolution of MCS (shaded) overlaid with GSMaP rain (yellow contour) in phases: **a** pre-MCS, **b** initiation, **c** growth, **d** mature, **e** dissipation, and **f** decay

to the north (Java Sea) and south area (the coastal and mainland) on 23 May 2022 at 00Z. Those MCSs propagated to the north and merged after five hours, which created a single deep MCS in the ocean. As a result, this mature stage of MCS produced heavy rainfall of 10 mm/hour over the coastal area of Central Java and the Java Sea. Thereafter, the MCS began to dissipate gradually until it reached the decay stage, which was followed by a decrease in rainfall intensity.

Overall, the evolution of MCS in the north of Semarang and surrounding areas beginning with the pre-MCS and ending with the decay stage during 12 h induced the increased storm surge over the northern coast of Java Island and hence enhanced the seawater mass due to the heavy precipitation.

3.2 Evolution of Convection and Near-Surface Easterly Wind

Figure 3 shows the evolution of convection (represented by cloud top temperature with TBB colder than 255 K), vertical velocity, and near-surface wind in each stage of MCS development as described in the previous section. A negative value of vertical velocity was identified in the coastal areas of Central Java since the pre-MCS stage. Interestingly, the highest magnitude of negative vertical velocity occurred over the ocean in the north of Semarang during the growth stage of MCS (23 May 2022 at 00Z), which is associated with the intense downward motion.

It also can be seen that persistent easterly wind was observed along the Java Sea before and during the tidal flooding event, with the magnitude varying between 7 ms^{-1} and 15 ms^{-1}. The near-surface easterly wind modulated the easterly wind

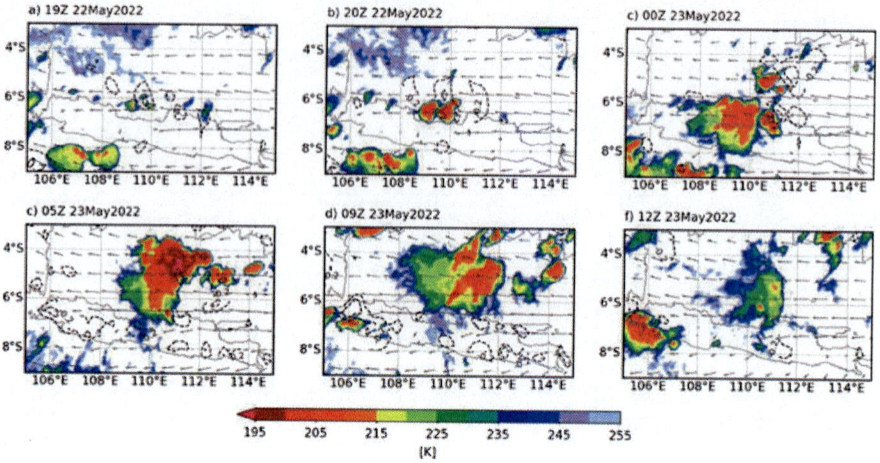

Fig. 3 Evolution of TBB (shaded), 925 hPa wind (vector), and vertical velocity (contour) associated for each phase of MCS

stress, and it induced the Ekman transport to the south areas of the northern coast of Java Island. The details of Ekman transport will be furtherly discussed in the next section.

3.3 *Ekman Mass Transport (EMT) and Ekman Pumping Velocity (EPV)*

Figure 4 shows the evolution of Ekman transport (EMT, vector) and Ekman pumping velocity (EPV, shaded) at 06Z (13 WIB) from 12 May 2022 to 23 May 2022. As seen in Fig. 4, on 12 May 2022, EMT is heading north, and on 14 May 2022, EMT is heading south. Several days before the tidal wave, this EMT consistently headed south toward the northern coast of Java Island which caused seawater to accumulate in the north of the island of Java. Ekman transport EMT on 23 May 2022 is much larger than the previous days.

Likewise, the EPV on 23 May 2022 had the lowest negative value compared to the previous days. A negative EPV indicates downwelling, and a positive EPV indicates upwelling. The EPV time series every 3 h from 10 May 2022 to 24 May 2022 is shown in Fig. 5.

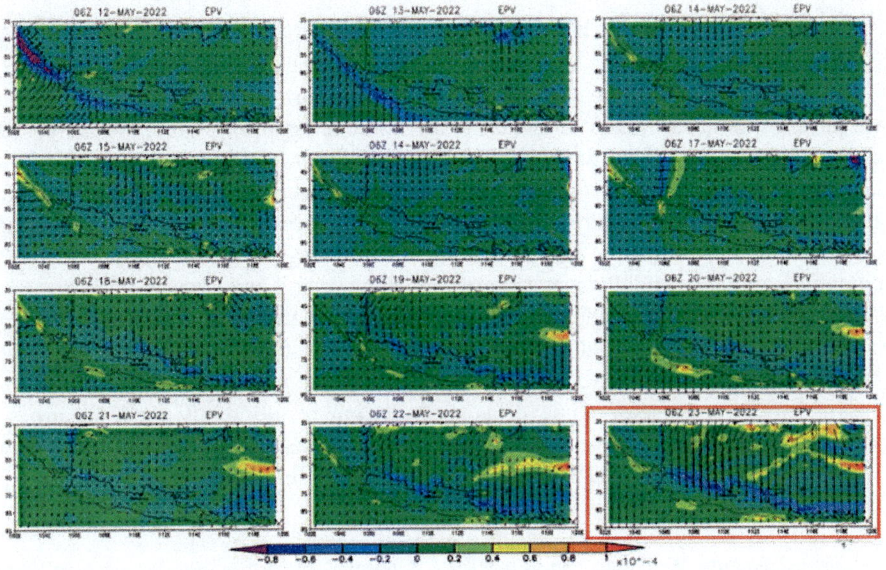

Fig. 4 Evolution of daily EPV (shaded) and EMT (vector) from 12 May 2022 at 06Z to 23 May 2022 at 06Z

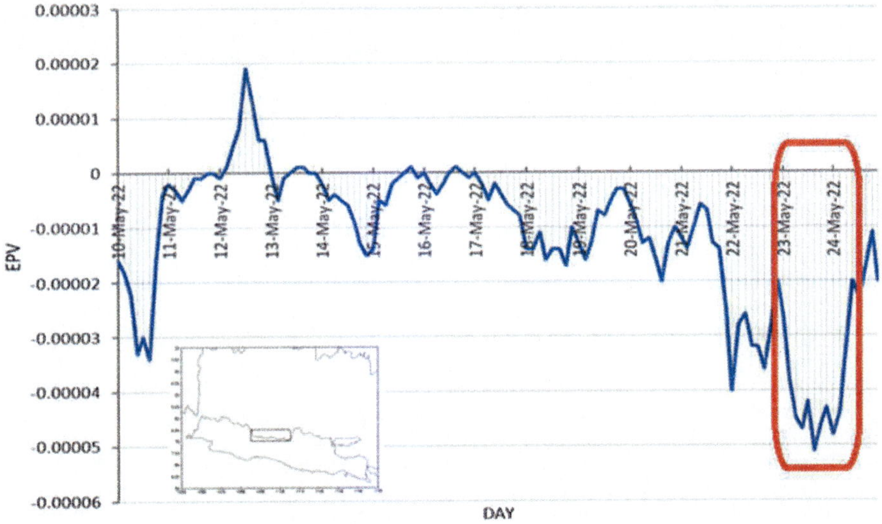

Fig. 5 Time series of EPV from 10 May 2022 to 24 May 2022 for the part of the northern coast area (108.5–110.5 E; 6.5–7 S)

4 Conclusion

Using TBB data from Himawari satellite and ocean surface wind data from ERA5, we have investigated meteorological factors influencing strong wind and increased seawater level over the northern coast of Semarang that caused the coastal flood on 23 May 2022. Our results suggested two potential meteorological factors that play an important role: (1) the formation of MCS over the ocean on the northern coast of Semarang that led to heavy rain over the coast and strong-surface wind speed that potentially enhances tidal waves toward Semarang Island and (2) unusually strong-surface easterly winds leading to increase seawater level through the Ekman pumping mechanism as a result of a balance between Coriolis and turbulent (wind) drag forces. Although our results did not yet quantify how much such factors contributed to the total increase in seawater level as a whole, our results really provided a new perspective on the factors influencing the increased seawater level intruding into Semarang during the coastal flooding on 23 May 2022, in addition to the previously known perspective of astronomical (tidal) forcing.

Acknowledgements We would like to thank Dr. Sandro W. Lubis from the Pacific Northwest National Laboratory (PNNL), who has given a lot of time for discussion, input, and improvement so that this paper can be completed. This paper was written with financial support from the DIPA of the Aeronautics and Space Organization, National Research and Innovation Agency (BRIN), in 2022.

Author Contribution All authors are the main contributors, have read the manuscript, and declare no conflict of interest.

References

1. Sutanta, H.: Spatial modeling of the impact of land subsidence and sea level rise in a coastal urban setting, case study: Semarang, Central Java, Indonesia. M.Sc. thesis, International Institute for Geo-Information and Earth Observation, ITC, Enschede, The Netherlands (2002)
2. Irawan, A.M., Marfai, M.A., Munawar, I.R., Nugraheni, S., Gustono T., Hasti., Rejeki, A., Widodo, A., Rikha, R., Mahmudiah., Faridatunnisa, M.: Comparison between averaged and localised subsidence measurements for coastal floods projection in 2050 Semarang, Indonesia. Urban Clim. **35**, 100760 (2021). https://doi.org/10.1016/j.uclim.2020.100760
3. Marfai, M.A., dan King, L.: Potential vulnerability implications of coastal inundation due to sea level rise for the coastal zone of Semarang City, Indonesia. Environ. Geol. (2007). https://doi.org/10.1007/s00254-007-0906-4
4. Al Dianty, M., Arbaningrum, R., Putuhena, F.J.: The linkage of effect climate change for determining design flood of Tenggang River. Geogr. Tech. (2020). https://doi.org/10.21163/gt_2020.151.17
5. Koch, M.: Intergrading earth & disaster science to enable sustainable adaptation & mitigation. In: Engineering, Information and Agricultural Technology in the Global Digital Revolution (2020). https://doi.org/10.1201/9780429322235-2
6. Kurniawati, W., Mussadun, Nugraha, M.F.: Spatial expression of Malay Kampung Semarang in facing flood disaster. In: IOP Conference Series: Earth and Environmental Science (2020). https://doi.org/10.1088/1755-1315/409/1/012049

7. van de Haterd, J., Budiyono, B., Darundiati, Y.H., Spaan, E.: Environmental change and health risks in coastal Semarang, Indonesia: importance of local indigenous knowledge for strengthening adaptation policies. Cities Heal. (2020). https://doi.org/10.1080/23748834.2020.1729451
8. Verrest, H., Groennebaek, L., Ghiselli, A., Berganton, M.: Keeping the business going: SMEs and urban floods in Asian megacities. Int. Dev. Plan. Rev. (2020). https://doi.org/10.3828/idpr.2020.3
9. BMKG Homepage, https://pasut.maritimsemarang.com. Accessed 2022/08/13
10. Tempo Homepage, https://en.tempo.co/read/1594634/coastal-flooding-hits-semarang-thousandsof-families-affected. Accessed 2022/05/24
11. CNN Homepage, https://www.cnnindonesia.com/teknologi/20220602182348-199-804098/4-penyebab-banjir-rob-semarang-siklus-bulan-hingga-badai. Accessed 2022/07/20
12. Hersbach, H., Bell, B., Berrisford, P., Hirahara, S., Horányi, A., Muñoz-Sabater, J., Nicolas, J., Peubey, C., Radu, R., Schepers, D., et al.: The ERA5 global reanalysis. Q. J. R. Meteorol. Soc. **146**, 1999–2049 (2020). https://doi.org/10.1002/qj.3803
13. Kubota, T., Shige, S., Hashizume, H., et al.: Global precipitation map using satellite-borne microwave radiometers by the GSMaP project: production and validation. IEEE Trans. Geosci. Remote Sens. **45**(7), 2259–2275 (2007). https://doi.org/10.1109/TGRS.2007.895337
14. Whitehall, K., Mattmann, C.A., Jenkins, G., Rwebangira, M., Demoz, B., Waliser, D., Kim, J., Goodale, C., Hart, A., Ramirez, P., et al.: Exploring a graph theory based algorithm for automated identification and characterization of large mesoscale convective systems in satellite datasets. Earth Sci. Inform. **8**, 663–675 (2015). https://doi.org/10.1007/s12145-014-0181-3
15. Nuryanto, D.E., Pawitan, H., Hidayat, R., Aldrian, E.: The occurrence of the typical mesoscale convective system with a flood-producing storm in the wet season over the Greater Jakarta area. Dyn. Atmos. Ocean. **96**, 101246 (2021). https://doi.org/10.1016/j.dynatmoce.2021.101246
16. Hermawan, E., Lubis, S.W., Harjana, T., Purwaningsih, A., Risyanto., Ridho, A., Andarini, D.F., Ratri, D.N., Widyaningsih.R.: Large-scale meteorological drivers of the extreme precipitation event and devastating floods of early-February 2021 in Semarang, Central Java, Indonesia. Atmosphere (2022). https://doi.org/10.3390/atmos13071092
17. Purwaningsih, A., Lubis, S.W., Hermawan, E., Andarini, D.F., Harjana, T., Ratri, D.N., Ridho, A., Risyanto, Sujalu, A.P.: Moisture origin and transport for extreme precipitation over Indonesia's New Capital City, Nusantara in August 2021. Atmosphere 13, 1391 (2022). https://doi.org/10.3390/atmos13091391
18. Dieng, A.L., Ndoye, S., Jenkin, G.S., Sail, S.M., Gaye, A.T.: Estimating zonal Ekman transport along coastal Senegal during passage of Hurricane Fred, 30–31 August 2015. Springer Nature Appl. Sci. **3**, 588 (2021). https://doi.org/10.1007/s42452-021-04578-5

Assessment of Predictability of Convective-Induced Turbulence Event Using High-Resolution Model, Case Study: Hong Kong Airlines Incident on 6 May 2016

Ibnu Fathrio, Aisya Nafiisyanti, Ina Juaeni, Muhammad Arif Munandar, Dita Fatria, Anis Purwaningsih, Fadli Nauval, Alfan Sukmana Praja, Elfira Saufina, Didi Satiadi, Teguh Harjana, Wendi Harjupa, and Risyanto

Abstract This study investigates the predictability of the mesoscale convective system (MCS) that caused severe turbulence on Hong Kong Airlines flight in the south of Banjarmasin on 6 May 2016 at 1830 UTC. This study is carried out by using The Weather Research and Forecasting (WRF) model by using high spatial resolution and the Global Forecast System (GFS) dataset as the primary forcing for boundary and initial conditions. The result shows that the turbulence could be well predicted by the model that is initiated on 6 May 2016 at 0600 UTC or 18 h before the incident happened. Despite differences in spatial distribution and timing of precipitation compared to the observation, this study also highlights the importance of using a high spatial resolution of 1 km to capture the convective-induced turbulence successfully.

1 Introduction

Hong Kong commercial aircraft was reported to experience severe turbulences on 6 May 2016 [1]. The plane departs from Denpasar Airport with an intended destination of Hong Kong Airport. KNKT reports [1] documented that severe turbulence hit the aircraft at 1830 UTC above the Java Sea, near Southern Kalimantan coastline. The weather radar displayed a magenta color in about 9 km distance indicating the severe intense area of rainfall and turbulence associated with the Cumulonimbus cloud. This incident injured 3 flight attendants and 11 passengers, which made the aircraft return to Bali Airport. According to KNKT reports, the aircraft used weather

I. Fathrio (✉) · A. Nafiisyanti · I. Juaeni · D. Fatria · A. Purwaningsih · F. Nauval · A. S. Praja · E. Saufina · D. Satiadi · T. Harjana · W. Harjupa · Risyanto
National Research and Innovation Agency, Bandung, Indonesia
e-mail: ibnu003@brin.go.id

M. A. Munandar
Indonesian Agency for Meteorological, Climatological, and Geophysics, Jakarta, Indonesia

A. Basit et al. (eds.), *Proceedings of the International Conference on Radioscience, Equatorial Atmospheric Science and Environment and Humanosphere Science*, Springer Proceedings in Physics 290, https://doi.org/10.1007/978-981-19-9768-6_26

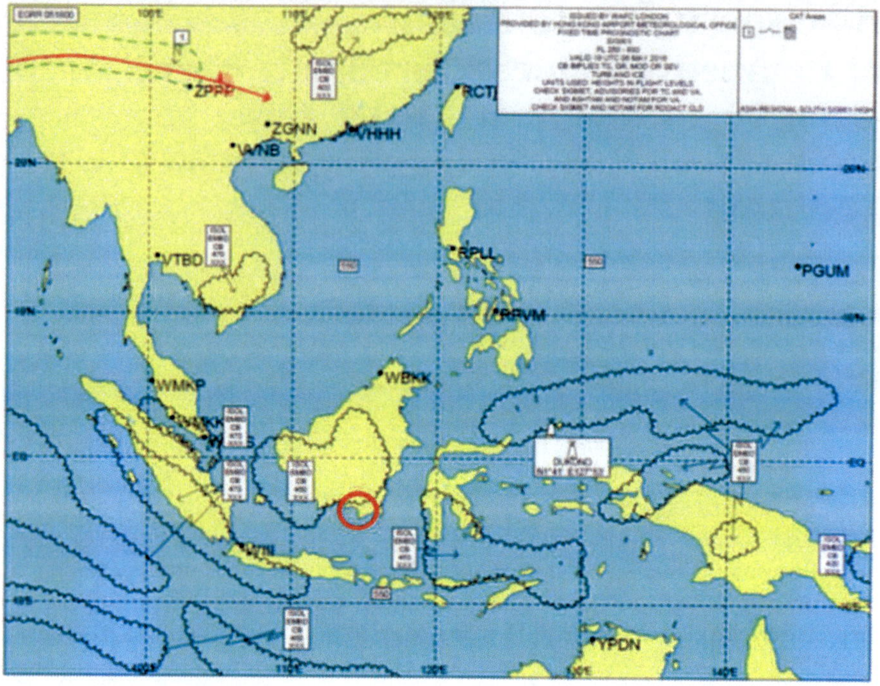

Fig. 1 Prognosis chart of flight level (FL) 250–630 valid for 6 May 2016 at 1800 UTC issued by WAFC London. This figure is obtained from Fig. 5 on KNKT report [1]

information issued by World Area Forecast Center (WAFC) London. The prognosis chart displayed a potential cumulonimbus cloud development area that unfortunately excludes the location where the aircraft experienced severe turbulence (Fig. 1).

This issue could be related to the model skill in predicting convective systems in the maritime continent. The spatial resolution of the model is an important factor in simulating convective systems in the tropics. A previous study [2] and [3] highlighted that 1 km horizontal resolution is required to properly affect the mesoscale convective system (MCS) in the maritime continent. Meanwhile, based on WAFC Management Report on March 2017–February 2018 obtained from [4], the WAFC still used global models with a spatial resolution of 17 km before 2017, which is too coarse to resolve the convective system. Therefore, higher spatial resolution models shown could complement the WAFC models [5]. Therefore, this study aims to evaluate the model's skill in predicting the Hong Kong turbulence incident by utilizing higher spatial resolution regional models up to 1 km. In addition, we also consider which initiation time of model predictions better simulates the incident in terms of timing and amplitude.

2 Data and Method

In this study, Weather Research Forecasting (WRF) [6] regional model was used to investigate the turbulence incident at 1830 UTC on 6 May 2016 in high spatial resolution. This study refers to [2] for selecting parameterization schemes, which has successfully simulates the evolution of MCS in the Southern Sumatra. In all domains, the model uses WRF Single Moment 6 (WSM6) as the microphysics scheme and Yonsei University as the planetary boundary layer scheme, and in all domains, Dudhia and RRTM schemes as shortwave and longwave radiation scheme, respectively, MM5 similarity as surface layer models, and NOAH land surface model as land surface scheme. The use of Betts-Miller-Janjic (BMJ) for the cumulus scheme in coarse resolution displays a promising result, as previously shown in [7]. Therefore, all domains utilize BMJ as cumulus scheme, except the innermost domain. The cumulus scheme on the innermost domains was set to the explicit (no cumulus scheme used).

The GFS products with a 0.25-degree spatial resolution are used as the initial and boundary conditions of simulation obtained from [8]. The product is available four times a day at 0000 UTC, 0600 UTC, 1200 UTC, and 18,000 UTC. All GFS input has a time lag of about 6 h to be available for download. Assuming the first 12 h is used as a spin-up time of simulation, and considering the incident happened at 1830 UTC, the latest GFS input that could be used is GFS initiated on 6 May 2016, at 0600 UTC. This data is available online around 1200 UTC, which is about six and a half hours before the turbulence happened. This study carried out a sensitivity test study by utilizing three different initial times of GFS input: 5 May 2016 at 1800 UTC, 6 May 2016 at 0000 UTC, and 6 May 2016 at 0600 UTC. Meanwhile, the sensitivity test on spatial resolution experiment was investigated for 1 km and 3 km (Fig. 2). To evaluate the 1 km domain (D03), the nesting domain consists of two parent domains with a spatial resolution of 9 km (D01) and 3 km (D02). In the case of 3 km, it only consists of one parent domain with a 9 km spatial resolution (D01). Table 1 displays the name andthe details of the six experiments. In addition, GSMaP precipitation data [9] and IR1 channel #13 (10.41 μm) of the Himawari-8 satellite were used as precipitation reference and proxy of cloud top temperature [1], respectively.

3 Results and Discussion

The evolution of the two mesoscale convective systems (MCS) is shown in Fig. 3. The first MCS, located to the east of the turbulence location, starts to grow at 1000 UTC (Fig. 3c), and the second MCS develops later at 1600 UTC to the west of turbulence location. The second MCS was primarily responsible for inducing turbulence hitting the aircraft at 1800 UTC (Fig. 3g) when all MCS reached mature stages. Eventually, the MCS dissipates after 1800 UTC (Figs. 3h and 3i). Figure 4 shows hourly precipitation derived from the GSMaP precipitation dataset, implying the turbulence

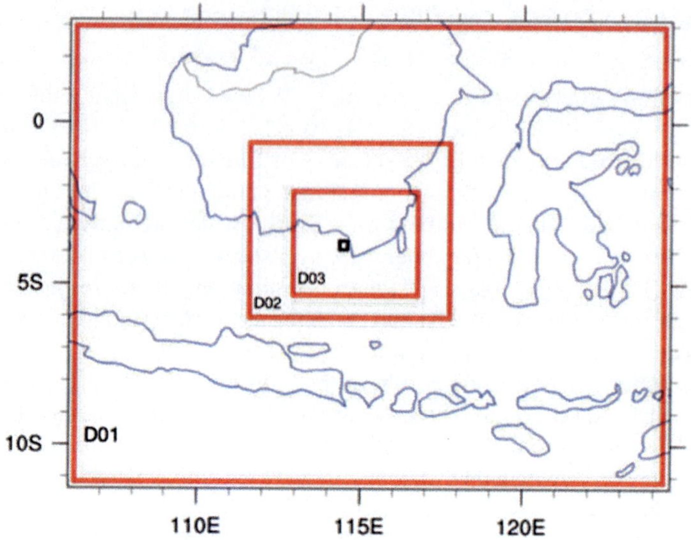

Fig. 2 WRF simulation domain. Details of the simulation scenario are explained in Table 1. For sensitivity test on spatial resolution, RUN 1, RUN 2, and RUN 3 simulation were run on three domains: D01, D02, and D03, while RUN 4, 5, and 6 simulations were run on two domains D01 and D02. The black square indicates the estimated location where the aircraft experiences turbulence based on KNKT reports

Table 1 WRF simulation scenario

Simulation name	GFS initial time	Spatial resolution of innermost domain and examined domain number
RUN I	1800 UTC, 5 May 2016	1km (D03)
RUN 2	0000 UTC, 6 May 2016	1km (D03)
RUN 3	0600 UTC, 6 May 2016	1km (D03)
RUN 4	1800 UTC, 5 May 2016	3km (D02)
RUN 5	0000 UTC, 6 May 2016	3km (D02)
RUN 6	0600 UTC, 6 May 2016	3km (D02)

incidence was related to the MCS that propagates southward. Furthermore, the incident happened near the area where the highest precipitation occurs (Fig. 4g, j) as the MCS reached the mature phase. It is expected that model that could properly predict the phase and spatial distribution of precipitation could also predict the turbulence accurately.

The skill of models in predicting the precipitation was examined, as shown in Fig. 5. Generally, all experiments show southward propagation of the rains. Yet, they simulate the early timing of precipitation peak, and the precipitation ends earlier than that of the observation. They also artificially generate the rainfall north of 3S, as

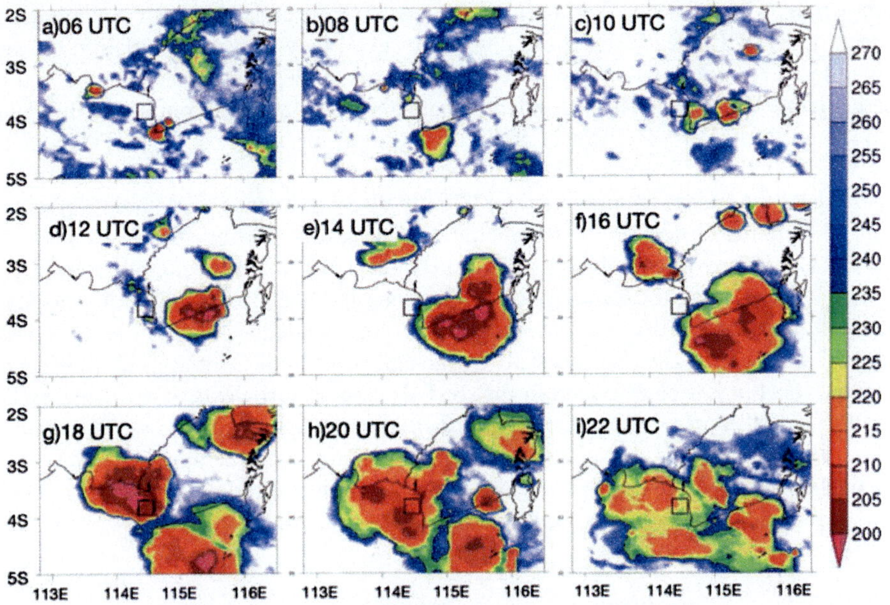

Fig. 3 Contour of black body temperature obtained from TBB-13 channel of Himawari-8 satellite on. The black box indicates the location of turbulence

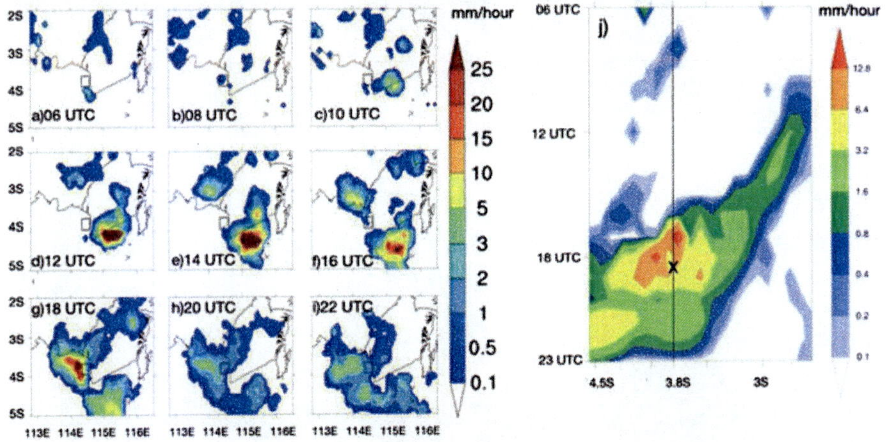

Fig. 4 Precipitation contour derived from GSMaP displayed in **a**–**i** 2D distribution and **j** latitude-time contour of precipitation averaged over 114° E to 115° E. The black square and X marker indicate the turbulence event location. Precipitation units are in mm hour^{-1}

shown in RUN 1, RUN 4, and RUN 5. The cause of the early timing of precipitation peak on RUN 1 and RUN 4 is unknown. However, they start the simulation earlier with less data assimilated than in RUN 2 which may degrade the accuracy. On the other hand, RUN 3 and RUN 6 underestimate the precipitation amplitude all the time in the north of 3.8S. The simulated precipitation also presents at 1500 UTC which is later than the other experiment. This could be related to shorter pre-conditioning periods of the model to simulate the MCS. Of all experiments, RUN 2 shows better results with the closest timing of precipitation peak to the observations.

Following previous analysis by Kim and Chun [10], this study examines the turbulence by estimating the gradient Richardson number and Brunt–Väisälä frequency. RUN 2 also successfully simulates the location of turbulence at a flight level of 410 (~12.5 km height) that is closest to the actual event both in timing, altitude, and spatial location, as can be seen in Fig. 6 (f-j). Greater turbulence kinetic energy is collocated with a strong gradient Richardson number less than 0.25 s^{-1} indicating severe turbulence and associated with the coolest top cloud temperature implying active convective activity. These features are not shown in RUN 1 and RUN 3. This suggests that the selection of GFS initiation time is crucial.

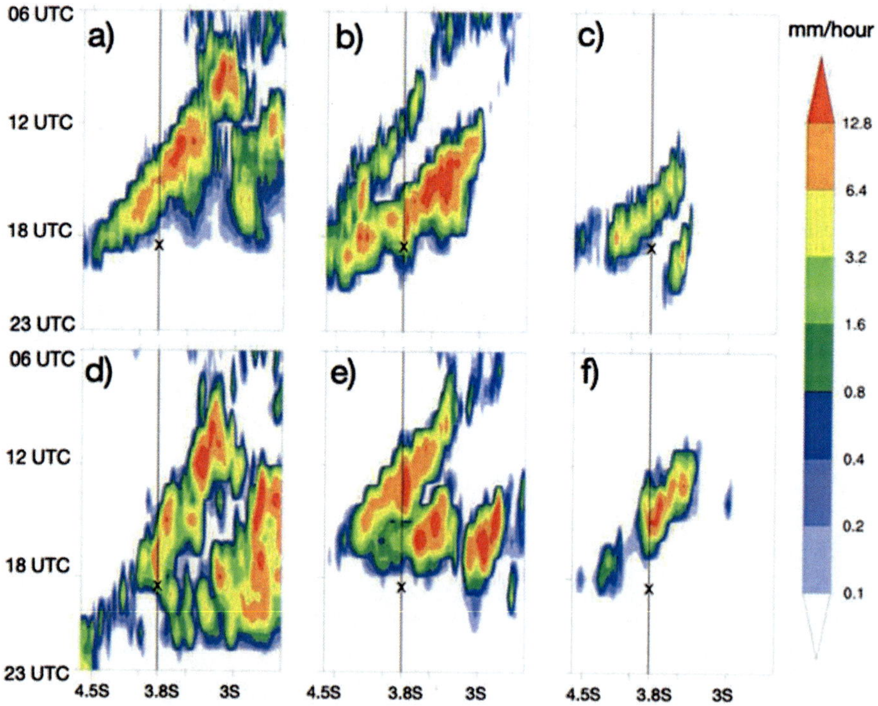

Fig. 5 Latitude-time contour of precipitation averaged over 114° E to 115° E for **a** RUN 1, **b** RUN 2, **c** RUN 3, **d** RUN 4, **e** RUN 5, and **f** RUN 6 experiments. The X marker indicates the turbulence incident

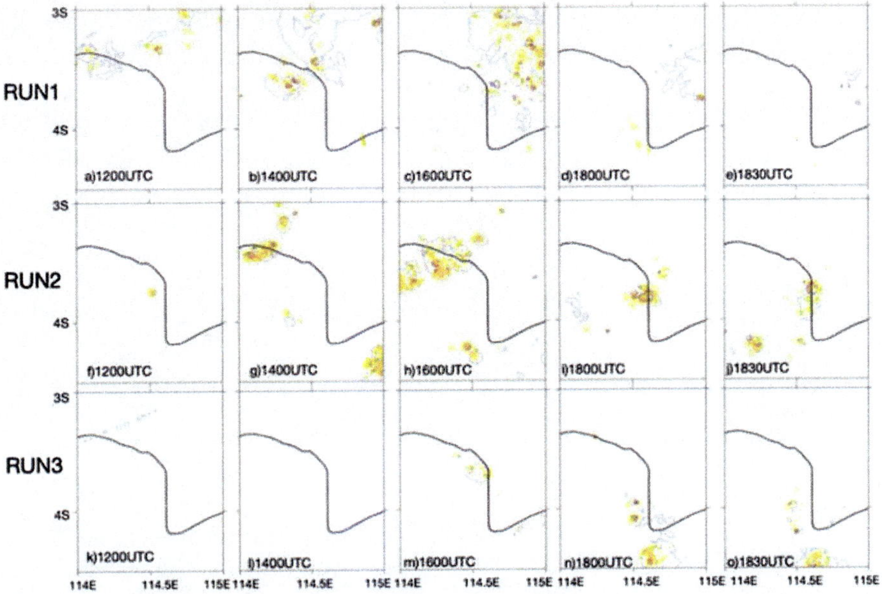

Fig. 6 Contour at ~ 12 km height or 410 flight level for cloud top temperature of 210 K (blue), Richardson number of 0.25 (purple), and turbulent kinetic energy (shaded) with yellow and orange colors that denote the value of 5 m^2s^{-2} and 10 m^2s^{-2}. The top panels **a–e**, middle panels **f–j**, and bottom panels **k–o** are for RUN 1, RUN 2, and RUN 3, respectively

Figure 7 shows that 3 km-simulation: RUN 4, RUN 5, and RUN 6 also fail to simulate the timing of turbulence. This implies that higher spatial resolution of 1 km could increase prediction skill in the precipitation phase. Although RUN 5 uses a similar GFS initiation time (on 6 May at 0000 UTC) as RUN 2, it predicts the turbulence event early on 1400 UTC. The turbulence happened at the top of the growing convective cloud, as shown in Fig. 8a–b, around 12 km height. The negative value of Brunt–Väisälä frequency indicates an unstable condition that contributes to a small gradient Richardson number. The cloud dissipates at 1830 UTC, yet the turbulence remains with comparable strength.

4 Conclusion

This study has shown that the WRF model could properly predict the incident of turbulence on Hong Kong Airlines at 1830 UTC on 6 May 2016 with an appropriate selection of GFS initiation time and running in higher spatial resolution of 1 km with three nested domain strategies. The simulated precipitation shows better timing and spatial distribution of rainfall related to the MCS activity. Moreover, this study recommends a pre-conditioning period of the model about 18 h before the peak of

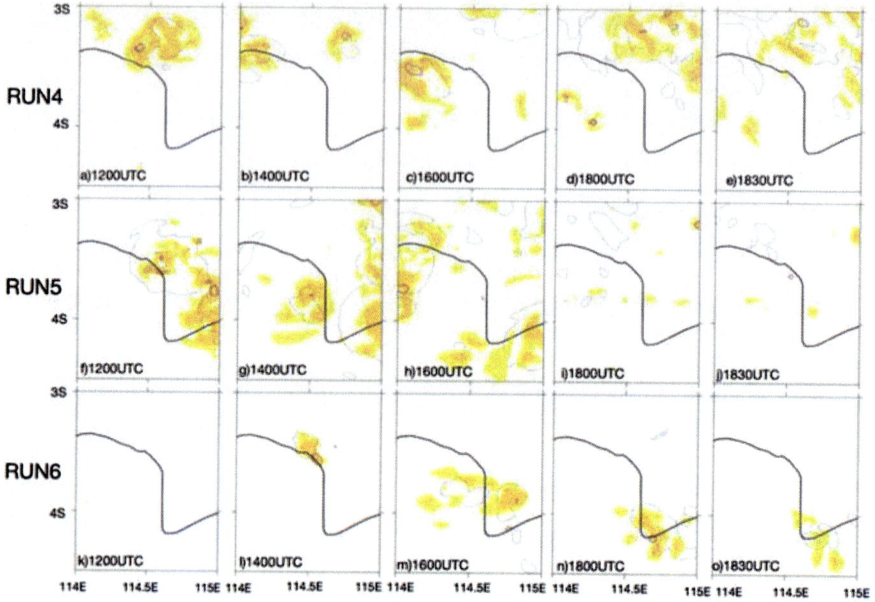

Fig. 7 Same as Fig. 6 but for RUN 4 **a–e**, RUN 5 **f–j**, and RUN 6 **k–o**

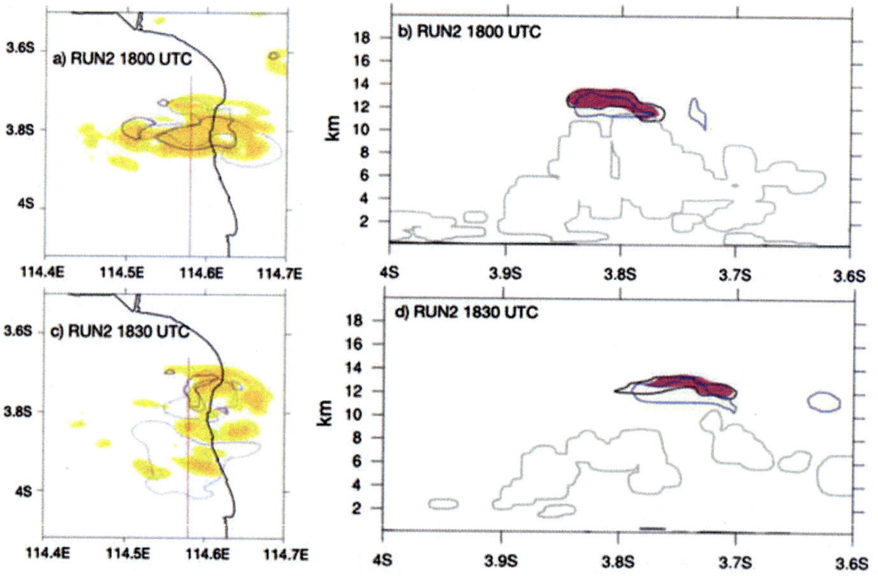

Fig. 8 Same as Fig. 6 but for RUN 2 on **a** 1800 UTC and **c** 1830 UTC. Latitude-height cross-section along the solid red line shown in **a** and **c** for cloud water mixing ratio of 0.05 g kg^{-1} (gray), cloud ice mixing ratio of 0.2 g kg^{-1} (blue), gradient Richardson number (black) of 0.25, and negative value of Brunt–Väisälä (s^{-1}; magenta-filled) at **b** 1800 UTC and **d** 1830 UTC

the MCS. The reason for the better prediction of GFS on 6 May 2016 at 0000 UTC (RUN 2) than GFS on 5 May 2016 at 1800 UTC (RUN 1 and RUN 4) could be related to more assimilated data being ingested by the former one.

References

1. KNKT: Aircraft Serious Accident Investigation Report. KNKT.14.05.14.04 Preliminary Report. 1–4 (2016)
2. Yulihastin, E., Fathrio, I., Nauval, F., Saufina, E., Harjupa, W., Satiadi, D., Nuryanto, D.E.: Convective cold pool associated with offshore propagation of convection system over the East Coast of Southern Sumatra, Indonesia. Adv. Meteorol. (2021)
3. Yulihastin, E., Nuryanto, D.E., Muharsyah, R.: Improvement of heavy rainfall simulated with SST adjustment associated with mesoscale convective complexes related to severe flash flood in Luwu, Sulawesi Indonesia. Atmosphere **12**(11), 1445 (2021)
4. www.icao.int, last accessed 2022/06/22
5. Rais, A.F., Putra, R.M., Fitrianto, M.A., Hermansyah, T.: A preliminary comparative study on the feasibility of a multipurpose numerical weather model for prediction of cumulonimbus Clouds in Indonesia. In: 2022 International Conference on Science and Technology, pp. 1–6. IEEE (2022)
6. Skamarock, W.C., Klemp, J.B., Dudhia, J., Gill, D.O., Liu, Z., Berner, J., Wang, W., et al.: A description of the advanced research WRF model version 4, National Center for Atmospheric Research: Boulder, CO, USA, 145, (2019)
7. Fonseca, R.M., Zhang, T., Yong, K.-T.: Improved simulation of precipitation in the tropics using a modified BMJ in the WRF model. Geosci. Model Develop. **8**(9), 2915–2928 (2015)
8. National Centers for Environmental Prediction,: NCEP GFS 0.25 degree global forecast grids historical archive. NCAR Research Data Archive, accessed 11 June 202, (2015)
9. Kubota, T., Shige, S., Hashizume, H., et al.: Global precipitation map using satellite-borne microwave radiometers by the GSMaP project: production and validation. IEEE Trans. Geosci. Remote Sens. **45**(7), 2259–2275 (2007)
10. Kim, J.H., Chun, H.Y.: A numerical simulation of convectively induced turbulence above deep convection. J. Appl. Meteorol. Climatol. **51**(6), 1180–1200 (2012)

The Effects of the Cross Equatorial Northerly Surge (CENS) on Atmospheric Convection and Convergence Over Jakarta and the Surrounding Area

Didi Satiadi, Anis Purwaningsih, Wendi Harjupa, Trismidianto, Dita Fatria Andarini, Fadli Nauval, Elfira Saufina, Fahmi Rahmatia, Ridho Pratama, Teguh Harjana, Risyanto, Ibnu Fathrio, Eddy Hermawan, Mutia Yollanda, and Dodi Devianto

Abstract The effect of the Cross Equatorial Northerly Surge (CENS) on the diurnal cycle of convection and moisture convergence over Jakarta and the surrounding area has been investigated. The data used in this study was the CENS indices, the Convective Available Potential Energy (CAPE), the Convective Inhibition Energy (CINH), the Vertically Integrated Moisture Fraction Convergence (VIMFC), and dew point temperature and rainfall from the fifth generation of the European Center for Medium-Range Weather Forecasting (ECMWF) Atmospheric Reanalysis (ERA5) during 2014. The data was depicted in a graph against the diurnal as well as the seasonal cycle, thus illustrating the multi-scale variability. The results showed a very strong influence of the CENS phenomena on the diurnal cycle of the CAPE, the CINH, the VIMFC, and rainfall in Jakarta and the surrounding area, over both the land and the sea areas. We found that in general, the CENS tended to suppress convection, but enhances moisture convergence. The important results of this study were that the increase of rainfall that occurred during active CENS was not actually caused by the increase of convection, but rather due to the increase in moisture convergence.

1 Introduction

Various atmospheric phenomena from synoptic to mesoscale significantly influence rainfall variability over the IMC. The Northerly Cold Surge (NCS) is one of the prominent synoptic disturbances that was responsible for the occurrence of heavy

D. Satiadi (✉) · A. Purwaningsih · W. Harjupa · Trismidianto · D. F. Andarini · F. Nauval · E. Saufina · F. Rahmatia · R. Pratama · T. Harjana · Risyanto · I. Fathrio · E. Hermawan
Research Center for Climate and Atmosphere, National Research and Innovation Agency, Bandung, West Java, Indonesia
e-mail: satiadi@yahoo.co.uk

M. Yollanda · D. Devianto
Mathematics Department, Andalas University, Padang, West Sumatera, Indonesia

A. Basit et al. (eds.), *Proceedings of the International Conference on Radioscience, Equatorial Atmospheric Science and Environment and Humanosphere Science*, Springer Proceedings in Physics 290, https://doi.org/10.1007/978-981-19-9768-6_27

rainfall, especially in the western part of Java Island. It often occurs during the peak of Boreal Winter (December–January–February; DJF) with the duration from days to a week [1]. Determination of the NCS has been defined and developed over time using different meteorological variables. Riehl classified the NCS as differences in pressure between 115° E and Hong Kong along 30° N that exceed 10 hPa [2]. Moreover, the NCS consists of two stages, a significant increase in pressure (P_s) and a steep drop in dew point (T_d) [3]. Lau and Chang concluded that based on the surface condition, the NCS can cause a sharp drop in surface temperature (4°–5°), and a decrease of humidity showed as a drastic drop of dew point (1°–2°), the enhancement of surface pressure (±3 hPa), and the strengthening of the northerly wind (5.5 m/s) through 12–24 LT [4]. Another study by Wu and Chan suggested that the surge is tracked when the pressure over Siberia reaches 1045 hPa. Then, the onset of the NCS in Hong Kong is identified by a rapid change in its synoptic condition around 12–14 LT [5]. A more recent study from Chang et al. defined the NCS as a transient gravitational wave-like motion caused by a wind-pressure imbalance, which allows the rapid propagation of energy from the mid-latitudes to the tropics [1]. Chang introduced a quantitative cold surge index by averaging the 925 hPa meridional wind over 110° E–117.5° E and 15° N and determining the average of more than 8 m/s as threshold for active cold surge [1]. Lim et al. provided an updated definition of the South China Sea cold surge based on low-level wind speed at 850 hPa over domain 5°–10° N and 107°–115° E and mean sea level pressure (MSLP) over domain 18°–22° N and 105°–122° E [6]. Furthermore, Abdillah et al. showed quantitative evidence linking the low-latitude CS to polar cold air outbreak in Siberia and defined multiple pathways of CS in the tropics [7].

The propagation of the NCS triggers the increasing rainfall intensity over some areas. The NCS is initially categorized as dry air masses. However, along its trajectory, from the mid-latitudes and sometimes extends to the South China Sea, air parcels are moistened by the ocean. This air mass propagation is strengthening the East Asia Hadley Cell associated with the enhanced outflow in the upper troposphere [8]. More specifically, the anomaly of cold air flow is weakened by an intensified surface heat flux. Thus, the cold characteristics of initial air mass are eliminated during the cold air masses propagation to the low latitude. Sometimes, this strong cold air outbreak can penetrate far to the south region and cross the equator and is called the Cross Equatorial Northerly Surge (CENS). The CENS as a dominant low-level circulation triggers deep convection and eventually leads to torrential rainfall over the equatorial region including the northern part of Java Island [9–14]. Over Jakarta, a strong NCS crossing the equator line played a critical role for several devastating flooding events, for example, in February 2013 [14], February 2020 [15], and February 2021 [16].

Furthermore, the complex local circulations (such as land–sea breeze and mountain valley wind) and their interaction with synoptic disturbances (such as the NCS and the CENS) play a vital role for the mechanisms of precipitation over the Java Island. There is a difference between rainfall patterns over the land and the ocean in the western part of Java Island. The increase of rainfall over the land (ocean) occurred in the afternoon (morning) [17], whereas in the coastal area of West Java, the maximum rainfall occurred frequently in the early morning (around 01 LT) during the wet season. Moreover, the interaction of local topography with the CENS might affect the variability of the amplitude and intensity of diurnal precipitation. The CENS modulated the early morning precipitation and extreme rainfall throughout the northern coast of West Java [9–20]. Nonetheless, the mechanism of the multi-scale interactions in modulating diurnal rainfall variability remains unclear [21]. Therefore, a thorough study addressing the mechanism is discussed in this paper.

This study aims to investigate the dynamic process of diurnal rainfall over the Jakarta area under the influence of CENS. We focus on analyzing the influence of the CENS on the convective and moisture convergence activities over the land and the sea areas utilizing several parameters, namely the Convective Available Potential Energy (CAPE), the Convective Inhibition (CINH), and the Vertically Integrated Moisture Flux Convergence (VIMFC). Convective processes were represented by the CAPE and the CINH. The CAPE is an indication of the instability or stability of the atmosphere that is used to assess the potential for convection development. Potential energy is represented by the total excess buoyancy, whereas the CINH measures the amount of energy that will inhibit an air parcel from rising from the surface to the level of free convection [22]. The convergence activities were represented by the VIMFC, which indicates the potential of moisture to converge over a certain location. Overall, this study gives understanding about how convective and convergence activities undergo over the land and the sea and thus influence the diurnal rainfall variation and how the CENS influences these activities.

2 Data and Methods

2.1 Data

This study used the hourly gridded dataset (0.25° × 0.25°) from the fifth-generation European Center for Medium-Range Weather Forecasts (ECMWF) Atmospheric Reanalysis of the global climate (ERA5) [23]. We obtained several parameters on a single level such as the CAPE, the CINH, the VIMFC, and precipitation from http://www.ecmwf.int/. We retrieved all the mentioned data for two locations (Onshore Jakarta: 6.4° S, 106.8° E; and Offshore Jakarta: 5.5° S, 106.8° E) representing the land and the sea areas (see Fig. 1). These parameters were employed to investigate the convective and moisture convergence activities associated with the CENS in 2014. The year 2014 was selected for this study, since the CENS phenomena were relatively strong and quite persistent, occurring intermittently for more than two weeks.

To identify the CENS, the daily mean wind and wind stress obtained from the Advanced Scatterometer (ASCAT) and Daily ASCAT (DASCAT) with field gridded data (0.25° × 0.25°) were used. The data was accessed for the same period as the ERA5 data, from ftp://ftp.ifremer.fr/ifremer/cersat/products/gridded/. The ASCAT is a real aperture radar with vertically polarized antennas operating at 5.255 GHz frequency (C-band) [24]. It transmits a long pulse with linear frequency modulation ("chirp").

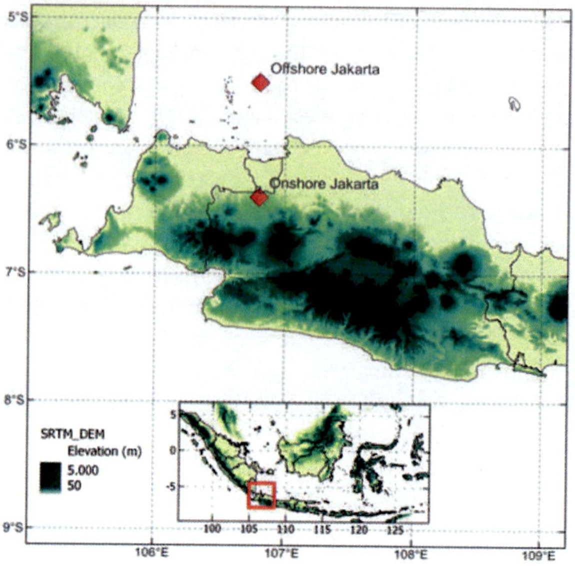

Fig. 1 Map of the study area in the western part of Java Island. The red marks denote the locations under study (Onshore Jakarta: 6.4° S, 106.8° E and Offshore Jakarta: 5.5° S, 106.8° E)

2.2 Methods

The Identification of the CENS. To determine the occurrences of the CENS, we applied a method from Hattori et al. (2011) by calculating its daily indices. The CENS index is identified by an area average of 925 hPa meridional wind over the area of 0°–5° S and 105°–115° E. Then, a threshold of 5 m/s was used to define an active phase of CENS [19].

Analysis of Convective and Convergence Processes. Convective processes were analyzed by investigating the diurnal and seasonal cycles of the CAPE and the CINH. The CAPE is the term used for energy available in the atmosphere for convection to occur. The CAPE is basically the type of energy available to lift an air parcel from the level of free convection to the level of neutral buoyancy [25]. Positive CAPE values point to conditions that are not stable, and there are growing chances of hail and thunderstorms. It is one of the most frequently used indicators of meteorological conditions that are convenient for the incidence of energetic precipitation events along with lightning, wind shear, and hailstorms [26, 27] . According to a study conducted by Chou and Neelin (2004), in convective zones, positive CAPE is maintained by increasing moisture content [28]. Conversely, the CINH measures the amount of energy that will prevent an air parcel from rising from the surface to the level of free convection [22].

Convergence processes were analyzed by investigating the diurnal and seasonal cycles of the VIMFC. The VIMFC is the horizontal rate of flow of moisture (water vapor, cloud liquid, and cloud ice), per square meter across the flow, for a column of air extending from the surface of the earth to the top of the atmosphere. Its horizontal divergence is the rate of moisture spreading outward from a point, per square meter.

This parameter is positive for moisture that is spreading out, or diverging, and negative for the opposite, for moisture that is concentrating, or converging. The VIMFC has a strong correlation with the rain occurrence [29, 30].

3 Results and Discussion

3.1 *The Effects of CENS Over the Land Area*

The hourly data from ERA5 and the CENS indices during 2014 was plotted to examine diurnal rainfall in relation to the conditions of the CAPE, the CINH, and the VIMFC under the influence of the CENS phenomena. Figure 2 shows rainfall (Fig. 2a), the CAPE (Fig. 2b), the CINH (Fig. 2c), the VIMFC (Fig. 2d), dew point temperature (Fig. 2e), and the CENS indices (Fig. 2f) against hour (y-axis) and day (x-axis) during 2014 over the land area (Onshore Jakarta) (6.4° S, 106.8° E). Although the CENS usually occurs only in DJF, one-year data was plotted to provide comparison with the normal conditions.

Results indicate that there was a change in temporal distribution of diurnal rainfall during the CENS phenomenon compared to that without the phenomenon (see Fig. 2a). Rainfall over the land area generally increased in the late afternoon around 12–18 local time. This result was expected presumably due to the intensification of atmospheric convection caused by the heating of the land surface by solar radiation during the day. However, during active CENS (in January and also February), rainfall occurrences seemed to be extended from the early morning to the evening.

The change of the magnitude of the CAPE during the CENS phenomenon was also identified (see Fig. 2b). Over the land area, the CAPE generally increased in the afternoon at around 10–16 local time, at an earlier time than that of rainfall. This result was expected presumably due to the atmospheric destabilization caused by the heating of the land surface due to solar radiation during the day. However, the growth of the CAPE tended to be suppressed during active CENS.

The behavior of CINH under the influence of the CENS was analyzed (see Fig. 2c). Over the land area, the CINH generally increased in the morning and in the evening and relatively reduced in the afternoon. The increase of the CINH in the morning was important for allowing the accumulation of the CAPE in the afternoon. Subsequently, the decrease of CINH in the afternoon was important for allowing the release of the CAPE in the form of convection and rainfall in the late afternoon. After the process, the CINH increased again in the evening until early morning, which suppressed convection. Similar to that of the CAPE, the CINH development seemed to be suppressed during active CENS. The CINH decreased possibly due to the wind convergence between the CENS and the land breeze, which provided a mechanical lifting through the inhibition layer [31].

Furthermore, to investigate the moisture convergence process, the VIMFC was analyzed during and without the CENS phenomenon (see Fig. 2d). Over the land

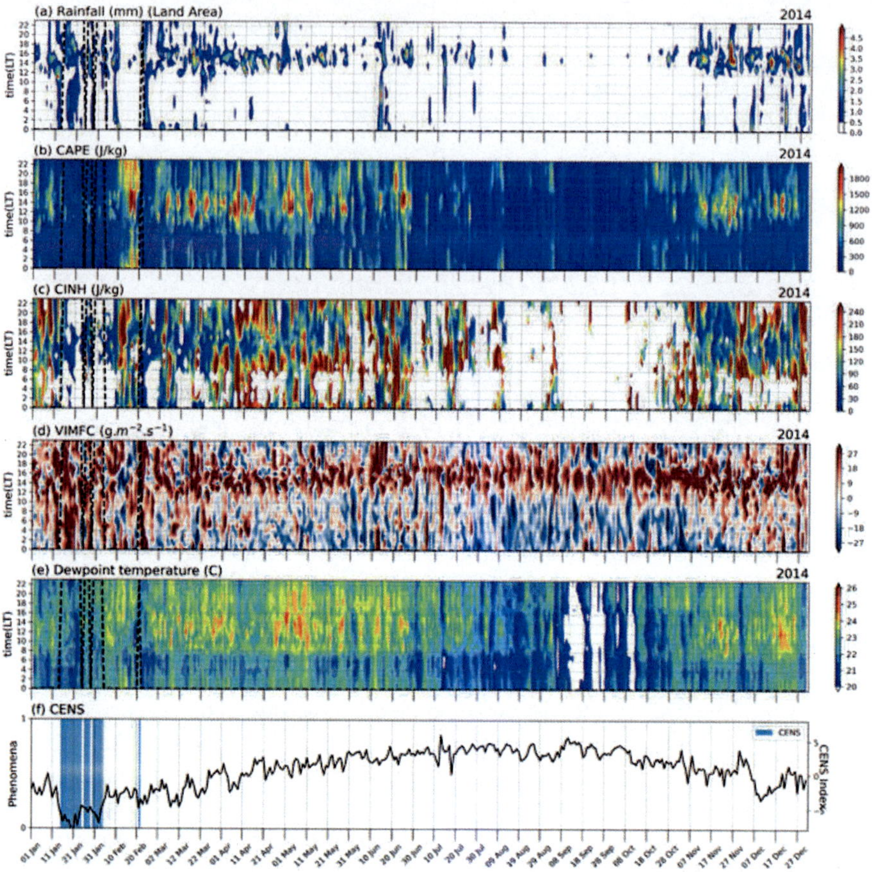

Fig. 2 **a** Rainfall, **b** CAPE, **c** CINH, **d** VIMFC, **e** dew point temperature, and **f** CENS indices over the land area (Onshore Jakarta) in 2014

area, the VIMFC generally increased in the late afternoon. This result was expected due to the intensification of atmospheric convection and also convergence caused by the heating of the land surface due to solar radiation during the day. Note that the onset of the VIMFC was more in phase with the rainfall than that of the CAPE, whereas during the active CENS (in January and also February), the VIMFC seemed to be extended from the early morning to the evening. The temporal distribution of the VIMFC was also more spread out during negative CENS indices.

Figure 2f shows the daily CENS indices during 2014. Positive (negative) indices mean that the wind propagated northward (southward). The magnitude of the indices indicated the wind strength. The CENS was active when the southerly wind speed was greater than 5.5 m/s, indicated by the blue lines. It can be seen from Fig. 2e that a relatively strong and persistent CENS occurred in January and also a brief one in February 2014.

Overall, the interaction between the CENS and convection or convergence processes over the land areas can explain which factor, whether convective or convergence, was responsible for the change in rainfall in terms of its timing and intensity. When the CENS was active, the CAPE and the CINH tended to be significantly suppressed, but the rainfall tended to be increased (See Fig. 2). The decrease of the CAPE development during active CENS was presumably due to the advection of the cold and dry wind gusts from the north, which reduced the dew point temperature (See Fig. 2f) and hence reduced the CAPE. Koseki (2013) has also shown that stronger monsoon does suppress convection due to lower the equivalent potential temperature. In contrast, the VIMFC tended to be enhanced and more extended in time from the early morning until the evening during active CENS. The results indicated that the increase of rainfall over the land during active CENS was not related to the increase of convection, which was actually suppressed, but was more related to the increase of the moisture convergence. The decrease of CINH was also an important factor in this mechanism, since it allowed the forced convection to take place easily, despite the lower CAPE.

3.2 *The Effects of CENS Over the Sea Area*

The same analysis as in Sect. 3.1 was conducted for the sea area to examine the diurnal rainfall in relation to the conditions of the CAPE, the CINH, and the VIMFC under the influence of the CENS and to compare the results between the land and the sea areas. Figure 3 shows rainfall (Fig. 3a), the CAPE (Fig. 3b), the CINH (Fig. 3c), the VIMFC (Fig. 3d), dew point temperature (Fig. 3e), and the CENS indices (Fig. 3f) against hour (y-axis) and days (x-axis) during 2014 over the sea area (Offshore Jakarta) (5.5° S, 106.8° E).

In general, there was a contrast between the phase of diurnal rainfall profile over the land and that over the sea areas. It can be seen from Fig. 3a that rainfall over the sea area generally increased during the night. This result was expected presumably due to the intensification of atmospheric convection caused by the heat and water vapor released by the sea surface during the night. Over the sea area, the CENS also influenced the intensity of rainfall. During the active CENS (in January and also February), there was an increase of rainfall during the night.

Moreover, the effects of CENS on the CAPE and the CINH profiles over the sea area were identified (Fig. 3b, c). Over the sea area, the CAPE was relatively high during the day and the night, but there was enhancement of CAPE particularly during night (see Fig. 3b). This result was also expected presumably due to the enhancement of atmospheric destabilization caused by the heating of the sea surface during the night. It can be seen from Fig. 3b that the CAPE development tended to be significantly suppressed during active CENS. Moreover, the CINH over the sea generally increased in the late afternoon and relatively reduced during the night (see Fig. 3c). The increase of the CINH in the late afternoon was important for allowing the accumulation of the CAPE in the evening. Subsequently, the decrease of CINH

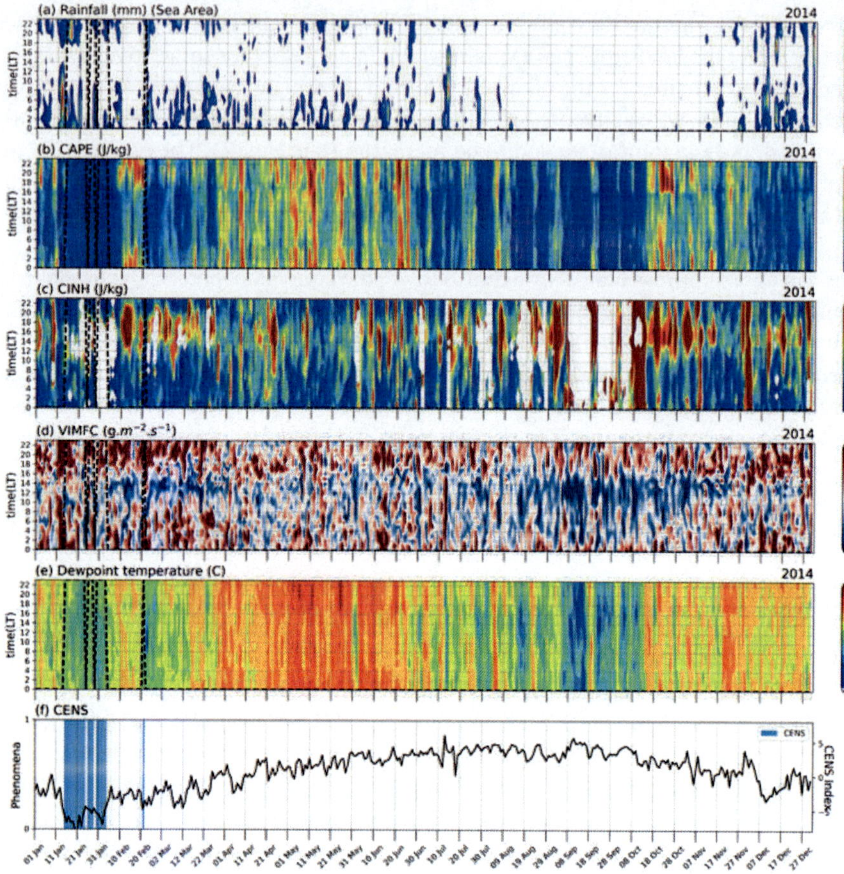

Fig. 3 **a** Rainfall, **b** CAPE, **c** CINH, **d** VIMFC, **e** dew point temperature, and **f** CENS indices over the sea area (Offshore Jakarta) in 2014

was important for allowing the release of the CAPE in the form of convection and rainfall during the night. The CINH seemed to be suppressed during active CENS, possibly due to the wind convergence, which provided a mechanical lifting through the inhibition layer.

It can be seen from Fig. 2d that over the sea area, the VIMFC was generally increased during the night. This result was expected presumably due to the intensification of atmospheric convection and also convergence caused by the heat released by the sea surface during the night. Figure 2d also shows that during negative CENS indices, the VIMFC was more intensified, especially when the CENS phenomena occurred (in January and also February).

Overall, the results over the sea were similar to that over the land area regarding the effects of CENS on convection, moisture convergence, and rainfall. When CENS was active, the CAPE as well as the CINH tended to be significantly suppressed, but the

VIMFC and rainfall tended to be increased during the night and early morning (see Fig. 3). The decrease of the CAPE development during active CENS was presumably due to the advection of cold and dry wind gusts from the north, which reduced the dew point temperature (See Fig. 2f) and hence reduced the CAPE.

Therefore, the increase of rainfall over the sea area during active CENS over the sea area was not related to the increase of convection, which was actually suppressed, but was more related to the increase of moisture convergence. The elimination of CINH was also an important factor in this mechanism, since it allowed the forced convection to take place easily, despite the lower CAPE.

4 Conclusions

Based on the analysis described in the previous sections, we found a very strong influence of the CENS on the diurnal cycle of convection, moisture convergence, and rainfall over Jakarta that was found both over the land and the sea areas. In general, the CENS was found to suppress the growth of the CAPE during the day and the night, over both the land and the sea. The decrease of the CAPE development during active CENS was presumably due to the advection of cold and dry wind gusts from the north, which reduced the dew point temperature and stabilized the atmosphere. The effect of the CENS on the CINH was similar to that on the CAPE. The CENS generally reduced the formation of the CINH significantly during the day and the night, over both the land and the sea, possibly due to the wind convergence between the CENS and the land breeze, which provided a mechanical lifting through the inhibition layer. In contrast, the CENS was found to enhance the VIMFC over both the land and the sea. Therefore, one of the important results from this study was that the increase of rainfall that occurred during active CENS was not related to the increase of convection, which was actually suppressed, but was more related to the increase of moisture convergence, which was more dominant when the CENS was active. The elimination of CINH was also an important factor in this mechanism, since it allowed the forced convection to take place easily, despite the lower CAPE.

Acknowledgements This research is supported by the National Research and Innovation Agency (BRIN) of the Republic of Indonesia. We are grateful for the availability of the fifth-generation European Center for Medium-Range Weather Forecasts (ECMWF) Atmospheric Reanalysis of the global climate (ERA5) data downloaded from http://www.ecmwf.int and also the ASCAT and DASCAT data downloaded from ftp://ftp.ifremer.fr/ifremer/cersat/products/gridded/.

References

1. Chang, C.P., Harr, P.A., Chen, H.J.: Synoptic disturbances over the equatorial South China Sea and western Maritime Continent during boreal winter. Mon. Weather Rev. **133**, 489–503 (2005)
2. Riehl, H.: Surface wind over the South China Sea during the northeast monsoon season. Navy Weather Res. Facility Tech. **24**, 22–68 (1968)
3. Chang, C.P., Millard, J.E., Chen, G.T.J.: Gravitational character of cold surges during winter MONEX. Mon. Wea. Rev. **111**, 293–307 (1983)
4. Lau, K.M., Chang, C.P.: Planetary scale aspects of the winter monsoon and atmospheric teleconnection. In: Chang, C.P., Krishnamurti, T.N., (eds.) Monsoon Meteorology, pp. 161–202. Oxford University Press (1987)
5. Wu, M.C., Chan, J.C.L.: Surface features of winter monsoon surges over South China. Mon. Weather Rev. **123**, 662–680 (1995)
6. Lim et al.: Impacts of Boreal winter monsoon cold surges and the interaction with MJO on Southeast Asia Rainfall. J. Clim. **30**(11), 4267–4281 (2017). https://doi.org/10.1175/JCLI-D-16-0546.1
7. Abdillah et al.: Cold surge pathways in East Asia and their tropical impacts. J. Clim. **34**(1), 157–170 (2021). https://doi.org/10.1175/JCLI-D-20-0552.1
8. Huang, P. et al.: Interannual variability of regional Hadley circulation intensity over western pacific during Boreal winter and its climatic impact over Asia-Australia Region. JGR Atmosphere **123**(1), 344–366. https://doi.org/10.1002/2017JD027919
9. Wu, P., Hara, M., Fudeyasu, H., et al.: The impact of trans-equatorial monsoon flow on the formation of repeated torrential rains over Java Island. SOLA **3**, 093–096 (2007)
10. Tangang, F.T., Juneng, L., Salimun, E., et al.: On the roles of the northeast cold surge, the Borneo vortex, the Madden-Julian Oscillation, and the Indian Ocean Dipole during the extreme 2006/2007 flood in southern Peninsular Malaysia. Geophys. Res. Lett. **35**(14), 1–6 (2008)
11. Yokoi, S., Matsumoto, J.: Collaborative effects of cold surge and tropical depression-type disturbance on heavy rainfall in Central Vietnam. Mon. Weather Rev. **136**, 3275–3287 (2008)
12. Chen, T.C., Tsay, J.D., Matsumoto, J., Alpert, J.: Development and formation mechanism of the Southeast Asian Winter Heavy Rainfall Events around the South China Sea. Part I: Formation and Propagation of Cold Surge Vortex. J. Clim. **28**, 1417–1443 (2015)
13. Hermawan, E., Lubis, S.W., Harjana, T., Purwaningsih, A., Risyanto, Ridho, A., Andarini, D.F., Ratri, D.N., Widyaningsih, R.: Large-scale meteorological drivers of the extreme precipitation event and devastating floods of early-February 2021 in Semarang, Central Java, Indonesia. Atmosphere **13**(7), 1092 (2022)
14. Yulihastin, E., Trilaksono, N.J.: Evolution of heavy rainfall in Jakarta flood case 2013 based on COSMO Model. Asia Pacific Center for Ecohydrology-UNESCO Research Center for Limnology-Indonesian Institute of Sciences, 167 (2014)
15. Saufina, E., Trismidianto, Risyanto, Fathrio, I., Harjupa, W.: Impact of cross equatorial northerly surge (CENS) on Jakarta heavy rainfall and its interaction with tropical cyclones (Case study: 18–25 February 2020). AIP Conf. Proc. **2366**, 050002 (2021)
16. Purwaningsih, A., Tank, A.K., Vila, J.: Atmospheric conditions associated with northerly surge, Borneo vortex and madden Julian oscillation during the extreme rainfall cases in early 2021 over the western part of the maritime continent. In: Yulihastin, E., Abadi, P., Sitompul, P., Harjupa, W. (eds.) Proceedings of the International Conference on Radioscience, Equatorial Atmospheric Science and Environment and Humanosphere Science, 2021. Springer Proceedings in Physics, vol. 275. Springer, Singapore (2022)
17. Yulihastin, E., Fathrio, I., Trismidianto, Nauval, F., Saufina, E., Harjupa, W., Satiadi, D., Nuryanto, D.E.: Convective cold pool associated with offshore propagation of convection system over the East Coast of Southern Sumatra, Indonesia. Adv. Meteorol. (2021)
18. Yulihastin, E., Hadi, T.W., Ningsih, N.S., Syahputra, M.R.: Early morning peaks in the diurnal cycle of precipitation over the northern coast of West Java and possible influencing factors. Ann. Geophys **38**, 231–242 (2020)

19. Hattori, M., Mori, S., Matsumoto, J.: The cross-equatorial northerly surge over the maritime continent and its relationship to precipitation patterns. J. Meteorol. Soc. Japan Ser. **II**(89), 27–47 (2011)
20. Mori, S., Hamada, J.I., Hattori, M., et al.: Meridional March of Diurnal Rainfall over Jakarta, Indonesia, Observed with a C-band Doppler Radar: an overview of the HARIMAU 2010 Campaign. Progress Earth Planetary Sci. **5**(47) (2018)
21. Yulihastin, E.: Propagation of convective systems associated with early morning precipitation and different northerly background winds over Western Java. J. Meteorol. Soc. Japan. Ser. II **100**(1), 99–113 (2022). https://doi.org/10.2151/jmsj.2022-005
22. DeAngelis, A.M., Qu, X., Zelinka, M.D., Hall, A.: An observational radiative constraint on hydrologic cycle intensification. Nature **528**, 249–253 (2015)
23. Hersbach, H., Bell, B., Berrisford, P., Biavati, G., Horányi, A., Muñoz Sabater, J., Nicolas, J., Peubey, C., Radu, R., Rozum, I., Schepers, D., Simmons, A., Soci, C., Dee, D., Thépaut, J.-N.: ERA5 hourly data on single levels from 1959 to present. Copernicus Climate Change Service (C3S) Climate Data Store (CDS) (2018). Accessed 2022/05/23
24. Klaes, K.D.: The EUMETSAT Polar System. Comprehensive Remote Sensing, 192–219 (2018)
25. Moncrieff, M.W., Miller, M.J.: The dynamics and simulation of tropical cumulonimbus and squall lines. Q. J. R. Meteorol. Soc. **102**, 373–394 (1976)
26. Zawadzki, I., Torlaschi, E., Sauvageau, R.: The relationship between mesoscale thermodynamic variables and convective precipitation. J. Atmos. Sci **38**, 1535–1540 (1981)
27. Singh, C., Ganguly, D., Dash, S.K.: Dust load and rainfall characteristics and their relationship over the South Asian monsoon region under various warming scenarios. J. Geophys. Res. Atmos. **122**, 7896–7921 (2017)
28. Chou, C., Neelin, J.D.: Mechanisms of global warming impacts on regional tropical precipitation. J. Clim. **17**, 2688–2701 (2004)
29. Kosum, C., Luadsong, A., Aschariyaphotha, N.: Vertically integrated moisture flux convergence over Southeast Asia and its relation to rainfall over Thailand. Pertanika J. Sci. Technol. **26**(1), 235–246 (2018)
30. Darand, M., Pazhoh, F.: Vertically integrated moisture flux convergence over Iran. Clim. Dyn. **53**, 3561–3582 (2019). Author, F.: Article title. Journal **2**(5), 99–110 (2016)
31. Colby, F.P., Jr.: Convective inhibition as a predictor of convection during AVE-SESAME II. Mon. Weather Rev. **112**(11), 2239–2252 (1984)

Comparison Influencing of El-Nino Southern Oscillation and Indian Ocean Dipole on Rainfall Variability During the Asian Winter and Summer Monsoon Over Indonesian Maritime Continent

Trisni Hadiningrum, Deni Okta Lestari, and Trismidianto

Abstract The Indian Ocean Dipole (IOD) and El-Nino Southern Oscillation (ENSO) are part of the atmospheric phenomena that considerably influence rainfall variability in Indonesia. This study has reported the impact of ENSO and IOD on Indonesian rainfall variability when they interact with each other. This analysis uses daily rainfall data from Aphrodite, sea surface temperature (SST), horizontal wind, Ocean Nino Index (ONI) index, and IOD index from 1961 to 2010. The results of this study indicate that the effect of LN and NIOD that co-occurs on rainfall variability in Indonesia is more robust in the AMJJASO than in the NDJFM. The impact of LN and PIOD simultaneously on rainfall is more visible in the NDJFM than in the AMJJASO. The effect of NIOD on rainfall is more dominant than EN when it co-occurs in the NDJFM and the AMJJASO. The impact of NIOD + EN on rain in Java is more significant during AMJJASO than during NDJFM. Events of PIOD and EN that co-occur will cause drought in almost all of Indonesia, and a high decrease in rainfall will occur on the island of Java during AMJJASO. The influence of LN on rainfall is more dominant than NetIOD during NDJFM and AMJJASO. The effect of LN + NetIOD on increasing rainfall in Indonesia is higher during AMJJASO than during NDJFM. The impact of EN, when it co-occurs with NetIOD, on the increase in rainfall in Sumatra is higher during NDJFM than during AMJJASO. The influence of PIOD and NetENSO on drought in Indonesia is quite strong if the two phenomena co-occur during AMJJASO.

T. Hadiningrum (✉) · D. O. Lestari
Department of Atmospheric and Planetary Science, Sumatera Institute of Technology, Lampung, Indonesia
e-mail: hadiningrumtrisnii@gmail.com

Trismidianto
Research Center for Climate and Atmospheric, National Research and Innovation Agency (BRIN), Bandung 40173, Indonesia

A. Basit et al. (eds.), *Proceedings of the International Conference on Radioscience, Equatorial Atmospheric Science and Environment and Humanosphere Science*, Springer Proceedings in Physics 290, https://doi.org/10.1007/978-981-19-9768-6_28

1 Introduction

Indonesian Maritime Continent (IMC) lies at the heart of the tropical region's warm pool. It is one of the world's most convectively active areas, affecting the global climate system. It is located between two continents (Asia and Australia) and between two oceans (the Indian and the Pacific Oceans) [1]. The IMC is a unique geographical region composed of a complex system of mountainous islands and consists of five large (Papua, Kalimantan, Sumatera, Sulawesi, and Java) and more than 13,000 smaller islands. It has about 5000 km or 1/8 of the equatorial circumference in longitude [2]. This individual IMC situation recognized the central importance of this region of heating as one of the primary energy sources for the entire global circulation system [1]. As it is, the air is primarily humid, and the enhanced cloudiness indicates massive exchanges of energy that are fundamentally important in the general circulation of the global atmosphere. Of this region, none is more important in global climate dynamics than the maritime continent region because of its role in providing energy for the operation of the north–south tropical Hadley cell and the east–west Walker circulation, both essential components of the global circulation [3]. McBride [4] and recently, Slingo et al. [5] have shown that the IMC heat source is a significant driver of global circulation.

The unique IMC conditions also make IMC weather and climate variability influenced by various weather systems and atmospheric phenomena occurring at different temporal and spatial scales, with time range ranging from intra-diurnal (e.g., sea breezes) to inter-annual (e.g., El-Nino Southern Oscillation) and spatial scales ranging from a few kilometers (e.g., individual cloud cells), to thousands of kilometers (e.g., the Madden–Julian Oscillation), to planetary-scale ENSO oscillations. The effect of these atmospheric phenomena occurs when these phenomena occur singly or interact with each other.

One of the interactions that is part of the atmospheric phenomenon that has a considerable influence on the rainfall variability in Indonesia is the interaction between the El-Nino Southern Oscillation (ENSO) and the Indian Ocean Dipole (IOD). ENSO is an irregularly periodical variation in winds and sea surface temperatures over the tropical Eastern Pacific Ocean, affecting much of the tropics and subtropics. The warming phase is known as El Niño, and the cooling phase is known as La Niña. IOD, also known as the Indian Niño, is an irregular oscillation of sea surface temperatures in which the western Indian Ocean becomes alternately warmer and colder than the eastern part of the ocean.

The previous research has shown that ENSO and IOD impact rainfall in Indonesia, where most of Indonesia have a positive rainfall anomaly ranging from 100 to 300 mm/day, except for Sumatra Island, with a rainfall anomaly of −200 mm/day. The same condition occurred in September–October–November and December–January–February, where only a few areas in Sumatra, Java, Kalimantan, and Sulawesi had negative rainfall anomalies [6]. Seasonal analysis shows that rainfall is low during June, July, and August, while high precipitation occurs in almost all parts of Indonesia during the DJF season [7]. The effect of ENSO and IOD has been reported by the previous studies using data analysis for 3-month periods (i.e., DJF (December, January, February), MAM (March, April, May), JJA (June, July, August), SON (September, October, November)). From several previous studies, there has never been a detailed explanation of the effect of ENSO and IOD interactions that co-occur in two different seasonal conditions. Therefore, this study aims to compare how much influence the interaction between ENSO and IOD has on the two seasons in Indonesia. However, this study uses a season division based on the Asian monsoon division, namely Winter Asian Monsoon as the rainy season, which is often referred to as November to March (NDJFM), and Boreal Summer Monsoon as the dry season occurs from April to October (AMJJASO) [8].

2 Data and Method

We use the ENSO and IOD index to analyze the condition of ENSO and IOD. The ENSO index is seen from the SST anomaly, Nino 3.4, and IOS Tahiti-Darwin, defined as the difference in the Sea Level Pressure (SLP) anomaly in the East Pacific, Tahiti (17.6° S, 149.6° W) with the SLP in the West Pacific, Darwin (12.4° S, 130° E). The Ocean Nino Index is obtained by representing the 3-month average of the ERSST.v5 SST anomaly in the Niño 3.4 region (5° N–5° S, 120° –170° W). The index data can be downloaded via http://www.cpc.ncep.noaa.gov/data/indices/oni.ascii.txt. The IOD index is seen from the difference in the western equatorial Indian Ocean SST anomaly (50° E–70° E/10° S–10° N) with the Indian Ocean SST anomaly off the coast of Sumatra (90° E–110° E/10°S—Equator). This IOD index can be downloaded free from the following link: https://psl.noaa/gov/data/timeseries/DMI. We classified the incidence of ENSO and IOD during the study period from January 1961 to December 2010.

In this study, rainfall forecast data use data from APHRODITE. Asian Precipitation-Highly-Resolved Observational Data Integration Toward Evaluation's (APHRODITE) gridded precipitation is a set of long-term (1951 onward) continental-scale daily products based on a dense network of rain-gage data for Asia, including the Himalayas, South, and Southeast Asia and mountainous areas in the Middle East. The gridded products are available for four sub-domains—Monsoon Asia, Middle East, Russia, and Japan—and a combined domain. The time-varying data have a 0.25×0.25 or 0.5×0.5 degree horizontal resolution in each part, except for Japan, which has a 0.05×0.05 degree horizontal resolution. Climatological daily mean precipitation and temperature data are available for Monsoon Asia at 0.05×0.05 degree resolution. This study used this data from January 1961 to December 2010. At this link, the data can be downloaded http://aphrodite.st.hirosaki-u.ac.jp/download/.

To analyze wind and sea surface temperature, we used product data from the National Centers for Environmental Prediction (NCEP) and the National Center for Atmospheric Research (NCAR) with a resolution of $2.5° \times 2.5°$ in the period January 1961 to December 2010. NCEP/NCAR Reanalysis 1 is an analysis/estimation of a complex project system to perform data assimilation using past data from 1948 to the present [9]. This data can be downloaded for free from the following link: https://psl.noaa.gov/data/gridded/data.ncep.reanalysis.html.

In this study, a composite spatial analysis was carried out for each predetermined event condition. By grouping the data into two seasons, such as AMJJASO and NDJFM, then it composite data when the ENSO and IOD phenomena occur. After that, we composite data on rainfall, SST, and wind. Then, we look for anomalies from the two data by subtracting the composite data at the time of the incident from the overall composite data for each season. To make it easier to do the analysis, we do some abbreviations, namely LaNina = LN, ElNino = EN, Negative IOD = NIOD, Positive IOD = PIOD, Neutral ENSO = NetENSO, Neutral IOD = NetIOD.

3 Result and Discussion

3.1 Statistical Data of ENSO and IOD

Figure 1 shows the frequent interactions between ENSO and IOD in the AMJJASO season, indicated by the dashed and standard red lines, and in the NDJFM season,

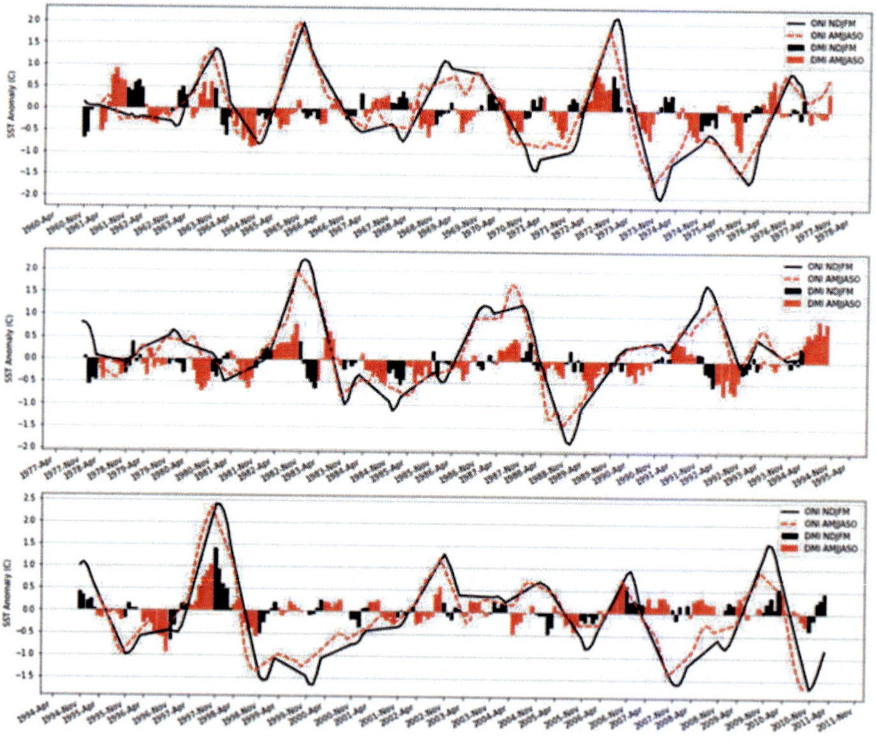

Fig. 1 Time series of monthly ENSO and IOD during AMJJASO and NDJFM over events in 1961–2010

indicated by the black lines. Figure 1 shows several LN occurrences simultaneously as NIOD, even with a high index, and the event of EN along with NIOD or PIOD in the AMJJASO and NDJFM seasons.

3.2 Interaction ENSO with IOD

When the LN coincides with NIOD during the NDJFM season, there is an increase in rainfall around eastern Kalimantan, East Java, northern Sumatra, and almost all of Sulawesi, with a slight increase in intensity (see Fig. 2a). About 20 mm/month, other areas experienced a decrease in rainfall. The increase in precipitation is very significant in almost all of Indonesia when LN and NIOD co-occur during the AMJJASO season, as shown in Fig. 2b. In fact, an increase in rainfall with a reasonably high intensity reaching 60–80 mm/month has occurred in almost all areas in Java and Kalimantan. This event shows that LN and NIOD in rainfall variability in Indonesia are more robust in the AMJJASO season than in the NDJFM season.

The effect of LN and PIOD co-occurs on rainfall during the NDJFM season, it can be seen that the influence of LN is quite dominant compared to PIOD, where the increase in rainfall only occurs in Java and eastern Indonesia, while in Sumatra, almost all experience a decrease in rainfall as shown in Fig. 2c [10]. Compared to the AMJJASO season, during the NDJFM season, rainfall increases only occur in

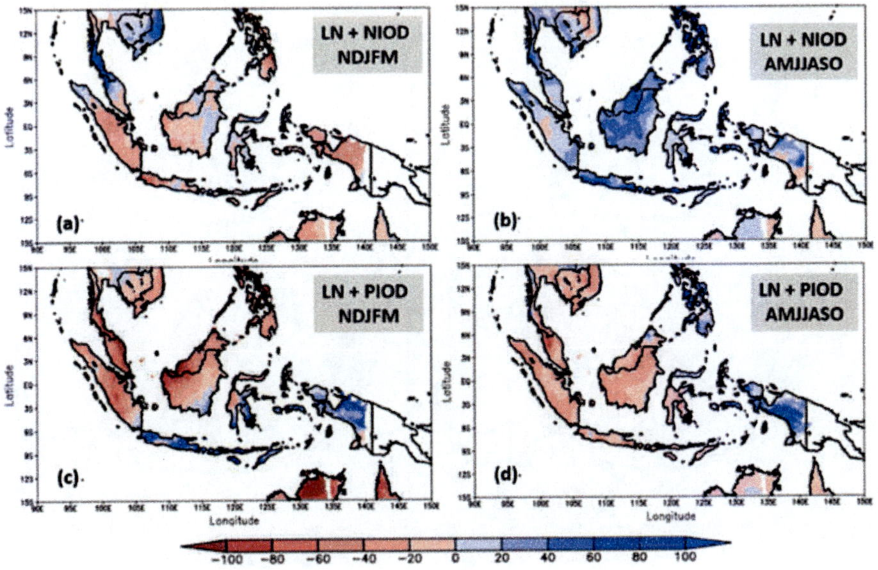

Fig. 2 Rainfall anomaly during NDJFM on **a** LN + NIOD and **c** LN + PIOD and during AMJJASO on **b** LN + NIOD and **d** LN + PIOD. Unit in mm/month

Papua (see Fig. 2d). Meanwhile, almost all other areas experienced a decrease in precipitation. Even the island of Java, which previously shared a significant increase in rainfall during NDJFM, was not seen during the AMJJASO season. This event shows that the simultaneous influence of LN and PIOD on rainfall is more visible in the NDJFM season than in the AMJJASO season. The impact of LN also seems to be more dominant than PIOD because, during PIOD, Indonesian territory should experience dryness [11].

When EN and NIOD co-occur, the dominant influence of NIOD compared to EN on rainfall in Indonesia both in the NDJFM and AMJJASO seasons, with a significant increase in rainfall in several areas in western Indonesia, especially in Sumatra and Java (See Fig. 3a, b). Regions that experience increased rainfall are more comprehensive during the AMJJASO season than during the NDJFM season. The increase in precipitation with an intensity of about 40–60 mm/month is seen evenly in Java during the AMJJASO season (see Fig. 3a) compared to NDJFM, only seen around East Java with a small intensity. About 20 mm/month (see Fig. 3b), therefore, EN + NIOD on rainfall in Indonesia is more significant in the AMJJASO season than in the NDJFM season.

EN and PIOD made Indonesia dry when this phenomenon occurred [12]. In Fig. 3c, d, almost all of Indonesia experience drought or decreased rainfall when EN and PIOD co-occur in the NDJFM or AMJJASO season. There was increased precipitation around northern Sumatra in both seasons, and other phenomena may need to be studied further. The decrease in rainfall during NDJFM ranges from 60–80 mm/month in almost all of Indonesia (Fig. 3c). The decline in rainfall even reaches

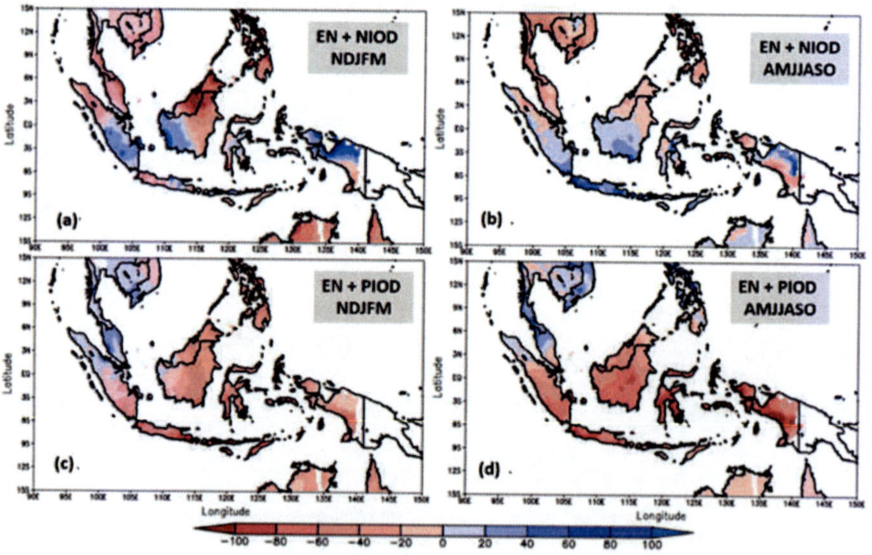

Fig. 3 Rainfall anomaly during NDJFM on **a** EN + NIOD and **c** EN + PIOD and during AMJJASO on **b** EN + NIOD and **d** EN + PIOD. Unit in mm/month

80–100 mm/month around the island of Java when the two phenomena co-occur in the AMJJASO season (Fig. 3d).

3.3 Interaction of ENSO and IOD If One is Neutral

Figure 4a, b shows the effect of LN on NetIOD during the NDJFM and AMJJASO seasons. The influence of LN is more dominant than NetIOD during NDJFM and AMJJASO, with an increase in rainfall around central and eastern Indonesia. Sumatra looks very dry; there is no increase in rainfall. The increase in rainfall broadly more during the AMJJASO season than NDJFM, where there was a fairly even increase in the islands of Kalimantan, Sulawesi, and Papua with small intensities of around 0–20 mm/month (see Fig. 4b). However, the increase in rainfall in northern Kalimantan is more significant during NDJFM than during AMJJASO (see Fig. 4a). This shows the LN effect at the same time as NetIOD on rainfall in northern Kalimantan is more significant during the NDJFM season than during AMJJASO.

The effect of EN during NetIOD is also clearly seen in Fig. 4c, d, which causes drought in almost all of central and eastern Indonesia during the NDJFM and AMJJASO Seasons. NetIOD conditions indicate an increase in rainfall in Sumatra and parts of Kalimantan. The increase in rainfall in Sumatra is higher during NDJFM than during AMJJASO. However, the distribution of areas experiencing increased

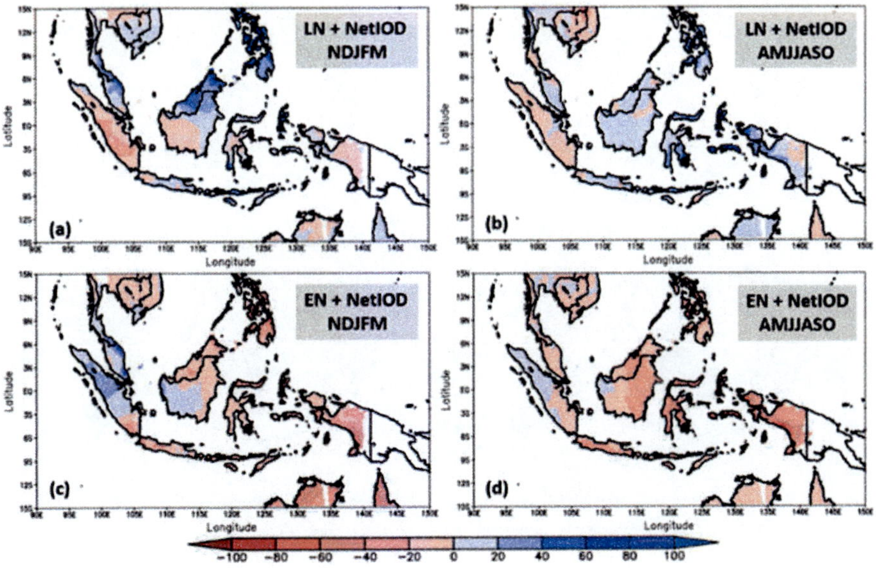

Fig. 4 Rainfall anomaly during NDJFM on **a** LN + NetIOD and **c** EN + NetIOD and during AMJJASO on **b** LN + NetIOD and **d** EN + NetIOD. Unit in mm/month

rainfall in Kalimantan is wider during AMJJASO than during NDJFM, with almost the same intensity. Then, the effect of EN during Net IOD on rainfall in Sumatra is higher during NDJFM than during AMJJASO. However, in Kalimantan, the impact is more significant during AMJJASO than during NDJFM [13].

NetENSO conditions will make it dry in central and eastern Indonesia when they co-occur with NIOD during the NDJFM season (see Fig. 5a). However, this did not happen during the AMJJASO season because there was an increase in rainfall in some areas of eastern Kalimantan and almost all of Papua although with a low intensity of around 0–20 mm/month (see Fig. 5b). The effect of NIOD that co-occurs with NetENSO on rainfall is more visible during AMJJASO than NDJFM; this is indicated by an increase in precipitation with low intensity in almost all of Java and southern Sumatra. However, it is seen in parts of Sumatra during NDJFM. This shows that the effect of NIOD on increasing rainfall in Sumatra and Java is more dominant than NetENSO, which acts more as a decrease in rainfall intensity when the two phenomena co-occur.

PIOD conditions should make western Indonesia dry [14]. This can only be seen during the AMJJASO season (see Fig. 5c) but not during NDJFM (See Fig. 5d). The significant increase in rainfall on the islands of Sumatra, Kalimantan, Papua, and Java when PIOD co-occurs with NetENSO during the NDJFM season is thought to be due to the influence of NetENSO followed by other phenomena that need further follow-up. During AMJJASO, the increase in rainfall was only visible over the island of Papua, while other areas experienced drought. This shows that the influence of

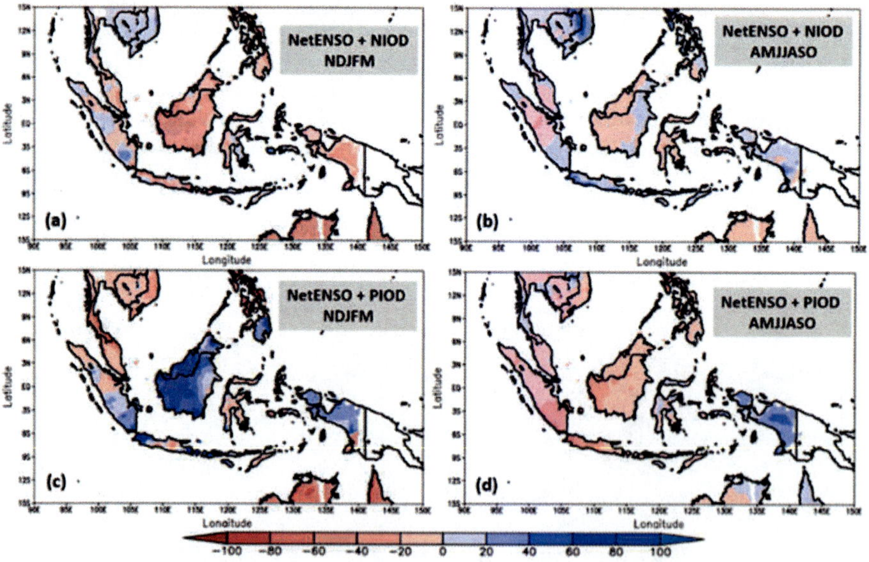

Fig. 5 Rainfall anomaly during NDJFM on **a** NetENSO + NIOD and **c** NetENSO + PIOD and during AMJJASO on **b** NetENSO + NIOD and **d** NetENSO + PIOD. Unit in mm/month

PIOD and NetENSO on drought in Indonesia is quite strong if the two phenomena co-occur during AMJJASO.

3.4 SST and Wind Anomaly Analysis

Figure 6a, b shows LN conditions that co-occur with NIOD, where there is a decrease in sea surface temperature around the Pacific Ocean and also in the Indian Ocean, while in Indonesia, there is an increase in sea surface temperature. This resulted in the Indonesian region experiencing wet conditions that received a supply of water vapor from the west and east. Figure 6a, b strengthens the analysis of Fig. 2a, b, where the influence of LN and NIOD that co-occurs is more substantial during AMJJASO than during NDJFM due to the decrease in sea surface temperature in the Pacific and Indian Oceans that is greater in intensity during the AMJJASO season than during NDJFM.

The decrease in sea surface temperature is very high in the Indonesian region when the PIOD co-occurs with the LN in the NDJFM season (see Fig. 6c) [15]. This strengthens the analysis of Fig. 2c, namely the decrease in rainfall in almost all Indonesian regions. This decrease in sea surface temperature also occurred in Indonesia during AMJJASO when these two phenomena occurred (See Fig. 6d). This makes parts of Indonesia dry, as shown in Fig. 2d.

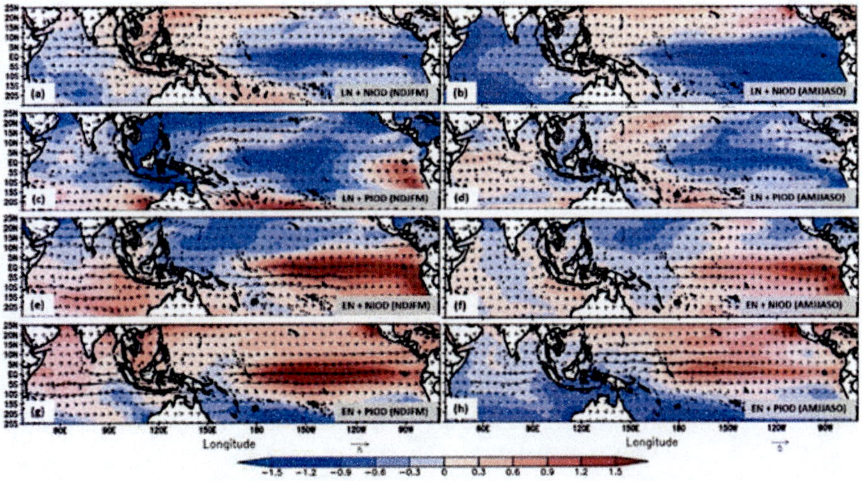

Fig. 6 Composite of SSTs anomaly (Shaded; °C) and 850 hPa wind anomaly (vector; m/s) during **a** LN + NIOD (NDJFM), **b** LN + NIOD (AMJJASO), **c** LN + PIOD (NDJFM), **d** LN + PIOD (AMJJASO), **e** EN + NIOD (NDJFM), f EN + NIOD (AMJJASO), **g** EN + PIOD (NDJFM) and **h** EN + PIOD (AMJJASO)

Figure 6c, d shows the occurrence of EN that coincided with NIOD in the NDJFM and AMJJASO seasons. It is clear that there is a decrease in sea surface temperature in eastern Indonesia caused by EN, while in western Indonesia, there is an increase in sea surface temperature due to NIOD (See Fig. 6c). This is what causes an increase in rainfall in west Indonesia, while the eastern part is experiencing drought as shown in Fig. 3a, b. Lack in Indonesia extends to almost all of Indonesia, when EN co-occurs with PIOD due to an increase in sea surface temperature in the Pacific and Indian oceans, causing the Indonesian region to experience a decrease in sea surface temperature, which causes drought (see Fig. 6e, f). This strengthens the analysis of Fig. 3c, d.

Figure 7a, b strengthens the analysis of Fig. 4a, b, where when the LN and NetIOD co-occur, the decrease in sea surface temperature in the Pacific Ocean causes an increase in sea surface temperature in Indonesia so that it becomes a hot pool which causes an increase in rainfall in Indonesia. However, the rise in the rain only occurs in parts of central and eastern Indonesia with low intensity; this is due to the influence of uncertain NetIOD conditions, which can make Indonesian areas wet or dry. Figure 7c, d clearly shows a decrease in sea surface temperature in the Pacific Ocean, indicating the presence of EN. This makes Indonesia dry, and even if there is an increase in rainfall, as shown in Fig. 4c, 4d, it is due to the NetIOD condition, which causes an increase in sea surface temperature in the western part of Indonesia.

When NetENSO conditions, the sea surface temperature in the Pacific Ocean is also uncertain, as is what happened in Indonesia. However, the concurrent occurrence of NIOD has increased sea surface temperature, which is visible in Indonesia, as shown in Fig. 7c, d. This causes an increase in rainfall in the western part of

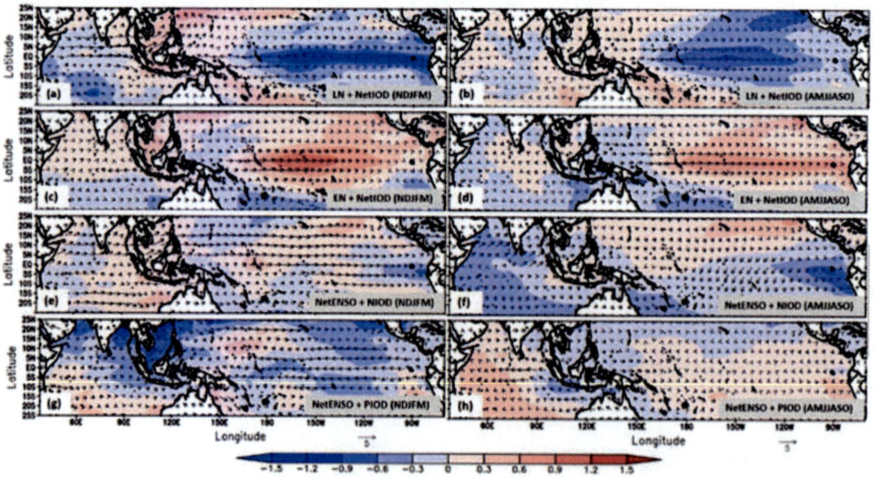

Fig. 7 Composite of SSTs anomaly (Shaded; °C) and 850 hPa wind anomaly (vector; m/s) during **a** LN + NetIOD (NDJFM), **b** LN + NetIOD (AMJJASO), **c** EN + NetIOD (NDJFM), **d** EN + NetIOD (AMJJASO), **e** NetENSO + NIOD (NDJFM), **f** NetENSO + NIOD (AMJJASO), **g** NetENSO + PIOD (NDJFM), and **h** NetENSO + PIOD (AMJJASO)

Indonesia, as shown in Fig. 5a, 5d. Meanwhile, when NetENSO co-occurs with PIOD, it should also cause drought in Indonesia, as shown in Fig. 7e, 7f, where there is a decrease in sea surface temperature in Indonesia. However, Fig. 7e is not in sync with Fig. 5c, where many areas experience an increase in rainfall; this is possible because of the influence of other phenomena that co-occur as the PIOD or NetENSO. Figure 7h supports the statement of Fig. 5d, which shows the drought in several parts of Indonesia due to a decrease in sea surface temperature in Indonesia.

4 Conclusion

From this study, it can be concluded that the effect of LN and NIOD that co-occurs on rainfall variability in Indonesia is more robust in the AMJJASO season than in the NDJFM season, with higher and more evenly distributed intensity in Kalimantan and Java. The effect of LN and PIOD simultaneously on rainfall is more visible in the NDJFM season than in the AMJJASO season. The influence of the LN also seems to be more dominant than the PIOD because, during the PIOD, parts of Indonesia must experience drought. The effect of NIOD on rainfall is more prevalent than EN when it co-occurs both in the NDJFM season and the AMJJASO season, with a significant increase in precipitation in some areas in western Indonesia, especially in Sumatra and Java. The effect of NIOD + EN on rainfall in Java is more critical during AMJJASO than during NDJFM. Events of PIOD and EN that co-occur will cause drought in almost all of Indonesia, and a high decrease in rainfall will occur on the island of Java during AMJJASO.

The influence of LN on rainfall is more dominant than NetIOD during NDJFM and AMJJASO, with an increase in precipitation around central and eastern Indonesia. The effect of LN + NetIOD on increasing rainfall in Indonesia is higher during AMJJASO than during NDJFM. The impact of EN, when it co-occurs with NetIOD, on the increase in rainfall in Sumatra is higher during NDJFM than during AMJJASO. However, the effect was more significant in Kalimantan during AMJJASO than during NDJFM. NetENSO conditions will dry in central and eastern Indonesia simultaneously with NIOD during the NDJFM season. However, this did not happen during the AMJJASO season because there was an increase in rainfall in some areas of eastern Kalimantan and almost all of Papua, although with low intensity. The influence of PIOD and NetENSO on drought in Indonesia is quite strong if the two phenomena co-occur during AMJJASO. The patterns of increasing and decreasing sea surface temperature anomalies and wind movements described in Figs. 6 and 7 are in sync and strengthen the analysis of the previous images regarding the influence of ENSO and IOD on rainfall in Indonesia.

Acknowledgements We are grateful for the availability of APHRODITE data which can be accessed at http://aphrodite.st.hirosaki-u.ac.jp/download/ and the ENSO index and IOD index sourced from this link http://www.cpc.ncep.noaa.gov/data/indices/oni.ascii.txt and https://psl.noaa/gov/data/timeseries/DMI. The availability of NCEP/NCAR Reanalysis data can be downloaded for free from the following link: https://psl.noaa.gov/data/gridded/data.ncep.reanalysis.html.

Declaration of Competing Interest

The authors report that they have no conflict of interest. All the authors of this paper are the main contributors. They have assisted in processing, analyzing, and interpreting the data and preparing the report as a whole.

References

1. Ramage, C.S.: Role of a tropical "maritime continent" in the atmospheric circulation. Mon. Weather Rev. **96**, 365–369 (1968)
2. Yamanaka, M.D.: Physical climatology of Indonesian maritime continent: an outline to comprehend observational studies. Atmos. Res. Elsevier. https://doi.org/10.1016/j.atmosres.2016.03.017
3. Sturman, A.P., Tapper, N.J.: The weather and climate of Australia and New Zealand Oxford University Press, Melbourne 476 (1996)
4. McBride, J.L.: Indonesia, Papua New Guinea, and Tropical Australia: The Southern Hemisphere summer monsoon. Meteorology of the Southern Hemisphere, Meteor. Monogr. (49), Amer. Meteor. Soc., 89–99 (1999)
5. Slingo, J., Inness, P., Neale, R., Woolnough, S., Yang, G.Y.: Scale interactions on diurnal to seasonal timescales and their relevance to model systematic errors. Ann. Geophys. **46**, 139–155 (2003)
6. Yulihastin, E., Febrianti, N.: Impacts of El Nino and IOD on the Indonesian Climate mechanism of air-sea interaction to change of diurnal rainfall over java View project Ais-sea interaction an it's impact on anomalously-wet dry season view project impacts of El Nino and IOD on the Indonesian Climate. Available: https://www.researchgate.net/publication/323783989 (2009)
7. Lestari, D.O., Sutriyono, E., Sabaruddin, S., Iskandar, I.: Respective influences of Indian Ocean Dipole and El Niño-Southern Oscillation on Indonesian Precipitation. J. Math. Fundam. Sci. **50**(3), 257–272 (2018). https://doi.org/10.5614/j.math.fund.sci.2018.50.3.3
8. Luh Putu Trisnawati, N.: Analisis Dampak El-Nino Terhadap Curah Hujan Pada Bulan Basah Dan Kering Di Kintamani (2018)
9. NCEP-NCAR Reanalysis 1, Physical Sciences Laboratory
10. Aditya, F.: Analisis Variabilitas Curah Hujan Di Kalimantan Barat Tahun 1991–2020, *Jurnal Meteorologi Klimatologi dan Geofisika* **2**(3) (2021)
11. As-syakur R.A.: Pola Spasial Hubungan Curah Hujan Dengan ENSO Dan IOD Di Indonesia - Observasi Menggunakan Data TRMM 3B43 (2012)
12. Nur'utami, M., Hidayat, R.: Influences of IOD and ENSO to Indonesian rainfall variability: role of Atmosphere-ocean Interaction in the Indo-pacific Sector Procedia Environ. Sci. (2016)
13. Safril, A.: Rainfall variability study in Kalimantan as an impact of climate change and El Nino. AIP Publishing. 978-0-7354-4064-7 (2020)
14. Kurniadi, A., Wllwr, E., Min, K.S., Seong, M.G.: Independent ENSO and IOD impacts on rainfall extremes over Indonesia. Int. J. Climatol. **41**, 3640–3656 (2021)
15. Kusuma D.W., Murdimanto, A., Aden, L.Y., Sukresno, B., Jatisworo, D., Hanintyo, R.: Sea surface temperature dynamics in Indonesia. IOP Conf. Ser. Earth Environ. Sci. **98**(1). https://doi.org/10.1088/1755-1315/98/1/012038

The Distribution and Characteristics of Mesoscale Convective Complex (MCC) and Its Relation with Rainfall During Madden-Julian Oscillation (MJO) Conditions Over Indonesia

Trismidianto

Abstract One of the atmospheric phenomena that significantly affects Indonesian rainfall is MJO, and the other phenomenon with a different time scale with MJO is MCC which rainfall until causing storms. This study was conducted to the analysis of MCC during MJO. MCCs were identified and tracked for 15 years (2001–2015) over IMC by infrared satellite imagery using an algorithm that combines criteria of cloud coverage, eccentricity, and cloud lifetime. The identification of the MJO phase is carried out using the MJO index from RMM1 and RMM2. The results showed that the existence of MCC can strengthen the influence of the MJO on increasing rainfall in Indonesia. MCC is more common in areas that have a high convective activity, which indicates the presence of MJO. The existence of MCC which occurs concurrently with MJO will increase its influence on the distribution of rainfall and does not even rule out the possibility of causing extreme rainfall in Indonesia. The effect of active MJO followed by the presence of MCC on rainfall in Indonesia is higher when compared to when MJO is not active. Monthly, in phases 2, 3, and 4, MCC mostly appeared in November, while in phases 1 and 5, MCC mainly occurred in March. MCC often appears in January when the MJO is in phases 6 and 7, and MCC is more commonly found in February when the MJO is in phase 8. The largest MCC cloud core size during the mature phase is found in phase 5. MCC with large average size is found in Region A in almost all phases. The average MCC size found has a cloud core size ranging from 100,000 to 200,000 km^2, but the largest sizes above 800,000 km^2 are found during phases 3 and 4 of the MJO. The most extended average duration of about 12.1 was found in MCC, which occurred during phase 2 of the MJO. Most of the MCCs that occurred were found on the continents for each MJO phase.

Trismidianto (✉)
Research Center for Climate and Atmospheric, National Research and Innovation Agency (BRIN), BRIN Kawasan Sains dan Teknologi Samadikun, Cisitu Sangkuriang, Bandung 40135, Indonesia
e-mail: trismidianto@brin.go.id

A. Basit et al. (eds.), *Proceedings of the International Conference on Radioscience, Equatorial Atmospheric Science and Environment and Humanosphere Science*, Springer Proceedings in Physics 290, https://doi.org/10.1007/978-981-19-9768-6_29

1 Introduction

Indonesia is one of the most active convective regions in the world that affects the global climate system [1]. One of the convective systems that dominate weather and climate in Indonesia is the mesoscale convective system (MCSs). In general, there are three classifications of MCS, namely squall line, which is a type of MCS in the form of liner cluster clouds that usually form along cold fronts. Bow echoes are a type of MCS that is shaped like a bow. Bow echo size is relatively smaller than other types. A special type of MCS is mesoscale convective complexes (MCCs), the most widely observed MCS with the largest size and long life span in the world [2]. The phenomenon of MCC has been studied by many researchers for more than a few decades in the world [3–9]. MCC does not only occur in middle latitudes but also occur in tropical areas, including Indonesia [3]. Trismidianto [10] found that 1028 MCCs occurred in Indonesia during the 15 years from 2001 until 2015, the average duration of the MCCs over IMC was approximate ~ 9.5 h, and the maximum duration was 10 h. Those observed in this study were usually nocturnal and reached a maximum at midnight. Trismidianto et al. [11] stated that MCCs contributed to total rainfall for 15 years over the Indonesian Maritime Continent by up to 20%. The most significant contribution was concentrated over Central Kalimantan, the South China Sea, the Indian Ocean, and Papua Island. MCC can result in bad weather and continuous rain [3, 11, 12]. They also concluded that MCC does not only affect rainfall in the location where the MCC occurs but also affects rainfall in the area around the MCC [11, 12].

The relationship between MCC and other phenomena that have a significant influence on the distribution of rainfall in Indonesia has been explained by several previous researchers. The link between ENSO and MCC has been reported by some earlier researchers, for example, Velasco and Fritsch [13] suggested that El Niño might play a role in MCC activities. Laing and Fritsch [14] also told the possibility of a relationship between El Niño and MCCs. Trismidianto and Satyawardhana [15] stated that the existence of the MCC was also accompanied by increased rainfall intensity at the locations of the MCC occurrences for all ENSO events. Rustiana et al. [16] stated that rainfall distribution during MCC was higher when the events of negative IOD compared to other events, especially in the western maritime of Sumatra until Kalimantan. In addition to the relationship between MCC and ENSO and IOD, several studies have explained the relationship between MCC and MJO, one of which is Aiqiu et al. [17], which states that the influence of MCC on significant rainfall is also supported by the existence of MJO phase 4. The substantial increase in rainfall when phase 4 active MJO is followed by the MCC. However, Aiqiu et al. [17] only took a case study of one MCC that occurred during MJO phase 4. As how the characteristics of MCC during various MJO conditions had not been explained in detail by the previous research; therefore, using MCC data that had been found by Trismidianto [10], this study will explain how the impact of MCC during several MJO conditions against rainfall distribution in Indonesia, due to both of atmospheric phenomena usually cause of floods in Indonesia.

2 Data and Method

MCC, in this study, was identified by inputting the temperature, latitude, and longitude values for each cloud shield pixel obtained from IR to a modified version of a computerized MCC program using MATLAB based on characteristics of MCCs [3]. Maddox [1] stated that MCC has a cloud shield with continuous brightness temperature (TBB)—32° C (in this study converted to 241 K) and must have an area of 100,000 km^2. The interior cold cloud has TBB −52° C (in this study converted to 221 K) and must have an area of 50,000 km^2 for over six hours, although the cloud shield does not have to maintain an eccentricity of 0.7 for its entire life cycle. TBB data obtained from Himawari generation infrared satellite imagery consisting of Geostationary Meteorological Satellite (GMS-5), Multi-functional Satellite Imagery Transport Satellite (MTSAT 1R and MTSAT 2), Himawari 8 and Geostationary Operational Environmental Satellites (GOES-9) with a spatial resolution of 0.05° × 0.05°. This data can be downloaded at http://weather.is.kochiu.ac.jp/sat/GAME.

The analysis was conducted classifying MCC events in the Asian Cold Monsoon, November to March (NDJFM), in several MJO phases. The interpolated outgoing longwave radiation (OLR) as a proxy for convection obtained from NOAA/NESDIS. The identification of the MJO phase is carried out using the MJO index, which is the daily time series effectively without the need for conventional time filtering and is called real-time multivariate MJO series 1 (RMM1) and 2 (RMM2). RMM1 and RMM2 provide information about the state of the MJO throughout the tropics and have been shown by Wheller and Hendon [18] to be true in all seasons. As long as MJO activity was identified, RMM1 led RMM2 around the fourth cycle, indicating eastward propagation of the MJO along the equator. MJO index data were obtained from the Climate Prediction Center (CPC) Website, NOAA (https://www.cpc.ncep.noaa.gov/products/precip/CWlink/MJO/). The MJO is considered strong (weak) if the amplitude is higher (less) than 1. Furthermore, the MJO event is divided into eight phases, where each phase of the MJO indicates the location of the MJO convective center. The convective center of the MJO propagates eastward from west Africa (phase 1) to the east, passing over the Indian Ocean (phases 2 and 3), MC (phases 4 and 5), migrating to the western Pacific (phases 6 and 7), and decaying at phase 8. However, in this study, we divided active MJO and inactive MJO by defining active and inactive MJO based on Lim et al. [19] and Xavier et al. [20], but there was a slight difference in active MJO by adding phase 5 during strong conditions. Active MJO is defined during phases 2, 3, 4, and 5, which have an amplitude value greater than or equal to 1, while inactive MJO is all days with MJO phases 2, 3, 4, and 5 with an amplitude value less than 1 added all the days when MJO in phases 1 and 5–8.

The MCC information is used to explain the MCC characteristics over the three longitude regions (90–105E as Region A, 106–125E as Region B, and 126–140E as Region C) consisting of the area interior size, duration of the life cycle, and interior eccentricity of MCC. In addition to analyzing the characteristics and distribution of MCC during these MJO phases, an analysis of the relationship between MCC and rainfall was also carried out using estimated rainfall data from GSMaP data [21] from JAXA (http://sharaku.eorc.jaxa.jp/GSMaP/). The spatial distribution of rainfall from GSMaP has spatial and time resolutions of $0.1^\circ \times 0.1^\circ$ and has time observation near real time. 850 hPa wind data obtained from ERA5, which provides hourly estimates of a large number of atmospheric, land, and oceanic climate variables. The data cover the Earth on a 30 km grid and resolve the atmosphere using 137 levels from the surface up to 80 km. ERA5 includes uncertainties for all variables at reduced spatial and temporal resolutions.

3 Result and Discussion

3.1 MJO Climatologies

The movement of the MJO that propagates from the Indian Ocean to the Pacific Ocean via Indonesia can be seen in Figs. 1a, b. MJO propagation is identified by the presence of convective activity as indicated by the negative OLR anomaly as shown in Fig. 1a. In phase 1, the MJO begins to occur in the Indian Ocean near India. In phases 2 and 3, MJO is already in the Indian Ocean near Sumatra Island. MJO was already above Indonesia when MJO entered phases 4 and 5. MJO was still in parts of eastern Indonesia, especially on Papua Island when MJO was in phases 6 and 7. MJO had already left Indonesia for the Pacific Ocean during phase 8 of the MJO. This convective activity is followed by an increase in rainfall as shown in Fig. 1b. During phase 1, increased rainfall was seen in the Indian Ocean around India where the MJO started. However, there is an interesting thing that shows an increase in rainfall over the island of Sumatra, even though there is no convective activity in the area. An increase in high rainfall occurred in the Indian Ocean, Sumatra Island, and some parts of Java seen when MJO was in phases 2 and 3. During phases 4 and 5, MJO was above Indonesia which caused an increase in rainfall over Indonesia. Increased rainfall on parts of the island of Papua was seen when the MJO was in phase 6. During phases 7 and 8, there was no visible increase in rainfall over Indonesia because the MJO had moved into the Pacific Ocean leaving Indonesia.

3.2 MCC Distribution During MJO Condition

3.2.1 Distribution of MCC in Each Phase of MCC

There are 508 MCCs that occur in the Asian winter in the month of NDJFM. Figure 2a shows the distribution of MCC events during the phases of the MJO. Increased rainfall occurred around the MCC event. In phase 1 to phase 4, increased rainfall is concentrated in the Indian Ocean and the island of Sumatra where there are many

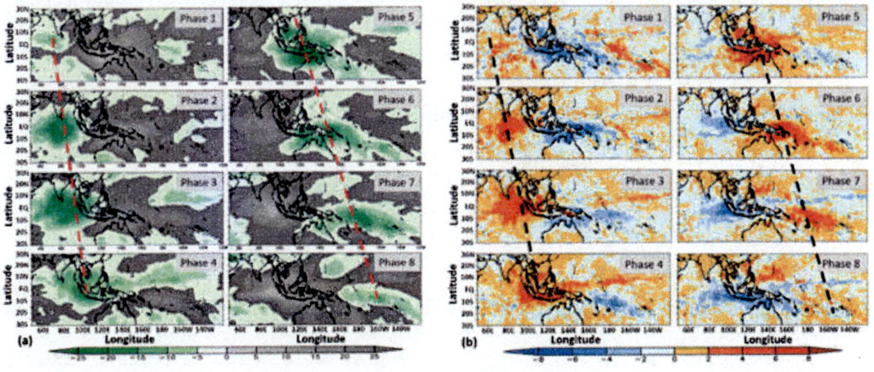

Fig. 1 Composites of **a** OLR anomaly (Wm^{-1}) and **b** rainfall anomaly (mmday^{-1}) for eight phases of the MJO cycle for November through March in the period of 2001–2015

MCCs. Increasing rainfall is moving to central and eastern Indonesia as the MJO system moves which are followed by many MCC events around the region. In phases 6 to 8, MJO has left Indonesian territory but there is still an increase in rainfall in Indonesian areas around the location of the MCC. It is estimated that the MCC that occurs during phases 6 to 8 is not caused by the presence of MJO but is possible due to local factors because the MCC that occurs is more dispersed and not concentrated in one place. In MJO phase 1, there were 46 MCCs that occurred in Indonesia, which were concentrated in Region A, around the Indian Ocean near Sumatra, and this was also followed by an increase in rainfall. MCC events in MJO phase 1 were also seen around Regions B and C but on a small scale so the effect on rainfall was not that significant. There were 64 MCCs that occurred in phase 2 of the MJO which were concentrated in the Indian Ocean and Kalimantan Island, but the average MCC size was medium. Similar to phase 1, the MCC in phase 2 of the MJO also affects the rainfall increase in Region A. In phase 2, the distribution of MCC is more on the Sumatra Island and also on the Borneo Island.

The impact of MCC on increasing rainfall in Region A is very significant, as seen in MJO phase 3. This is also in accordance with the results of Yulihastin et al. [22] that there is MJO modulation related to the formation of the MCS which is suspected of being MCC in Jakarta. There were 89 MCC events that occurred during phase 3 of the MJO focused on Region A and Region B, around the Indian Ocean near Sumatra Island and above Borneo Island with a large-scale average size. The presence of MCC that occurs in phase 3 of the MJO is believed to be able to increase the influence of MJO on increasing rainfall in Regions A and B, including in Java. Even though there are no MCC events on the island of Java, there are several MCC events that occur around the Java Island. Almost the same as in phase 3, a total of 70 MCC that occurred in phase 4 MJO also affected an increase in rainfall in Regions A and B which had the same pattern in phase 3, but the intensity of the increase in rainfall was smaller than in phase 3. MJO phase 4 is also seen to have an effect on

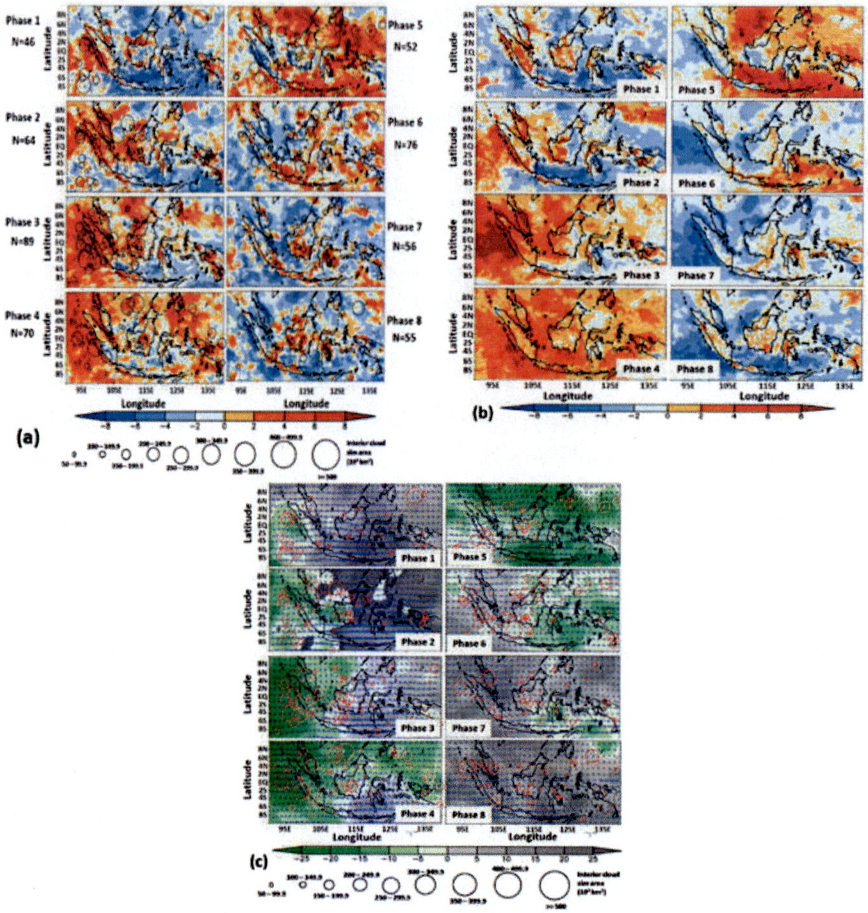

Fig. 2 Composites of rainfall anomaly in mmday-1 (shaded) during eight phases of the MJO cycle for November through March in 2001–2015 in condition **a** MCC occur, **b** no MCC, and **c** composites of OLR anomaly in Wm-1 (shaded) and 850 hPa wind in ms-1 (vector). The circle refers to the distribution of MCC occur, where the size of the circle indicates the size of the MCC when it occurs in the mature phase

increasing rainfall in parts of Region C, and this is also seen when MCC occurs in phase 4 of the MJO. The effect of MCC and MJO on increasing rainfall in Regions B and C can be seen in phase 5, where there were 52 MCC events which were more widely spread in Regions B and C although there were not many large ones. The growth of several MCCs in the Indian Ocean region can also be affected by MJO, or local scale influences from atmospheric phenomena but this is not discussed in the results of this study. The number of MCCs in MJO phases 6, 7, and 8, respectively, were 76, 56, and 55 events, including a fairly large number, but the average was on a small scale. In phases 6, 7, and 8, MCC is more widely spread in Regions B and C but on a small scale. The effect of MCC on increasing rainfall in Indonesia in phases 6, 7, and 8 is not too significant compared to other phases.

When compared with Fig. 2b, it can be seen that the increase in rainfall during the MJO followed by the presence of an MCC event is higher than when there is no MCC. This shows that the presence of MCC increases the influence of MJO on the distribution of rainfall in Indonesia. A large number of MCC distributions at the MJO location also indicates that there is MJO modulation of the development of the MCC. In phases 6, 7, and 8, when the MJO is without MCC, there is no significant increase in rainfall in the Indonesian. However, when compared to the MJO event followed by MCC, it can be seen that there was an increase in rainfall in the area around the MCC occurrence in phases 6, 7, and 8. This shows that the MCC that occurs in phases 6, 7, and 8 is not modulated by MJO but is possible due to local factors or diurnal convection, Fig. 2c reinforces the statement in Fig. 2a that the pattern of convective activity around the MCC event in each phase of the MJO is the same as the pattern of increasing rainfall, especially in phases 1–5, this reinforces the statement that the presence of MJO modulates the development of the MCC.

The percentage of extreme rainfall events in each MJO phase followed by the presence of MCC and without MCC using the 95th percentile threshold is shown in Fig. 3. From Fig. 3a, it can be seen that extreme rainfall events with a high percentage reaching more than 20% almost all occur around the location of the occurrence of MCC in each phase of MJO. When compared with Fig. 3b, it can be seen that the percentage of extreme rainfall events is not very significant when MJO events are not followed by MCC events that occur in almost all MJO phases. This shows that the presence of MCC that occurs simultaneously with MJO will increase its influence on the distribution of rainfall, and it is even possible that it can cause extreme rainfall in Indonesia.

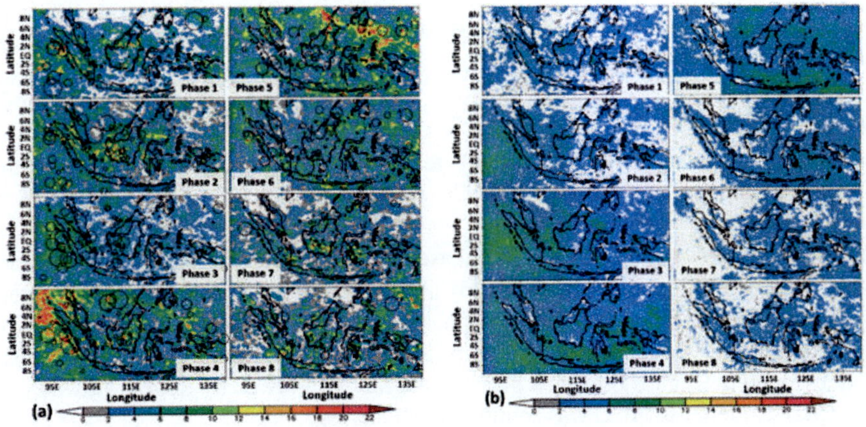

Fig. 3 Composites of the percentage of extreme rainfall based on percentile 95 during eight phases of the MJO cycle for November through March in 2001–2015 in condition **a** MCC occur and **b** no MCC

3.2.2 Distribution of MCC During MJO Active and Inactive

Figure 4 shows the distribution of MCC events when the MJO is active (Fig. 4a), and the MJO is inactive (Fig. 4b). It can be seen that the distribution of MCC events when the MJO is active is indeed less than when the MJO is inactive, but the effect on increasing rainfall intensity is higher when the MJO is active. During the active MJO, the area where the MCC occurs is always followed by a distribution of high rainfall increases, and this is different from when the MJO is inactive. The distribution of the increase in high rainfall is not seen in all areas of the MCC occurrence. This strengthens the analysis result that the presence of an active MJO initiates the occurrence of MCC, but this needs to be continued in a more detailed study related to the MJO mechanism in initiating the occurrence of this MCC. The existence of MCC when the MJO is active strengthens the influence of the MJO in increasing rainfall in the Indonesian Territory. However, MCC is not only initiated by the MJO, and it can also be influenced by other atmospheric phenomena, including local-scale phenomena, which are not discussed in the results of this study. The strengthening of the analysis in Fig. 4c, d is shown in Fig. 4c, d, which shows the same pattern of convective activity as the pattern of increasing rainfall when the MJO is active, and the MJO is inactive. The formation of convective activity when the MJO is active is also shown in Fig. 4c due to the strengthening of the wind. The percentage of extreme rainfall events is very significant, especially around MCC locations when the MJO is active (see Fig. 4e). This high percentage of extreme rainfall is not seen when the MJO is not active as shown in Fig. 4f even though there are many MCC occurrences. This shows that there is an interaction between MJO and MCC in influencing rainfall variability in Indonesia, and it is even possible to cause extreme rainfall.

3.3 Characteristics of MCC During MJO Condition

Statistically, as shown in Fig. 5a, it can be seen that most MCC events were found in phase 3, which is one of the active phases of the MJO. Phases 2, 4, and 5 also occur quite a lot of MCC but are still below the number of MCC events during phase 6. There is a large number of MCC events in phase 6, due to the influence of other atmospheric phenomena or local scale phenomena because the MCC formed is primarily small scale. Monthly, in phases 2, 3, and 4, MCC mostly appeared in November, while in phases 1 and 5, MCC mainly occurred in March. MCC often appears in January when the MJO is in phases 6 and 7. MCC also is more commonly found in February when the MJO is in phase 8. Figure 5b shows that MCC is most widely found in Region B in each phase of the MJO, followed by Region A, whereas in region C, the least MCC was found in each phase.

Figure 5c shows that, on average, the largest MCC cloud core size during the mature phase is found in phase 5, followed by phases 3 and 4, where the average value is above 150 thousand km^2. MCC with a large average size is found in Region A in almost all phases of the MJO except for phase 6, where the average large size

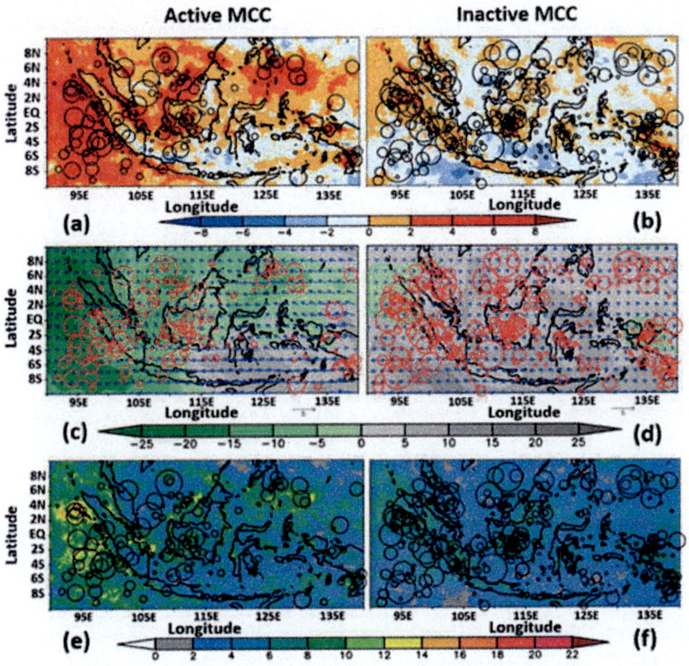

Fig. 4 Composites of rainfall anomaly in mmday^{-1} (shaded) during MCC occur in **a** MJO active and **b** MJO inactive. Composites of OLR anomaly in Wm-1 (shaded) and 850 hPa wind in ms-1 (vector) during **c** MJO active and **d** MJO inactive. The percentage of extreme rainfall based on percentile 95 during **e** MJO active and **f** MJO inactive, for November through March in the period 2001–2015. The circle refers to the distribution of MCC occur, where the size of the circle indicates the size of the MCC magnitude when it occurs mature phase

MCC is found in Region C. Figure 5d shows the distribution of the number of MCC based on size, which is divided into three categories, namely small scale (50,000–149,999 km^2), medium scale (150,000–299,999 km^2), and large scale (> = 300,000 km^2). The average MCC that occurs in each MJO phase is small scale. Medium-scale MCCs were mostly found in phases 2 and 3, while large MCCs were found during phases 3 and 4 of the MJO.

Figure 6a shows that the average MCC size found has a cloud core size ranging from 100,000–200,000 km^2, but the largest sizes above 800,000 km^2 are found during phases 3 and 4 of the MJO, and some large sizes above 500,000 km^2 are also found during phases 1, 6, and 8 MJO. We define the MCC duration as the duration between the time of the initial phase and the time of the dissipation phase based on the previous research [10]. The frequency and mean of the distribution of the duration of all MCC during every phase of MJO are shown in Fig. 6b. The average duration of MCC events in each phase was about 11 h, but the most extended average duration of about 12.1 was found in MCC during phase 2 of the MJO. Short-lived MCC often occurs in phase 6 of the MJO, and medium-lived MCC is usually found during phase 3 of the

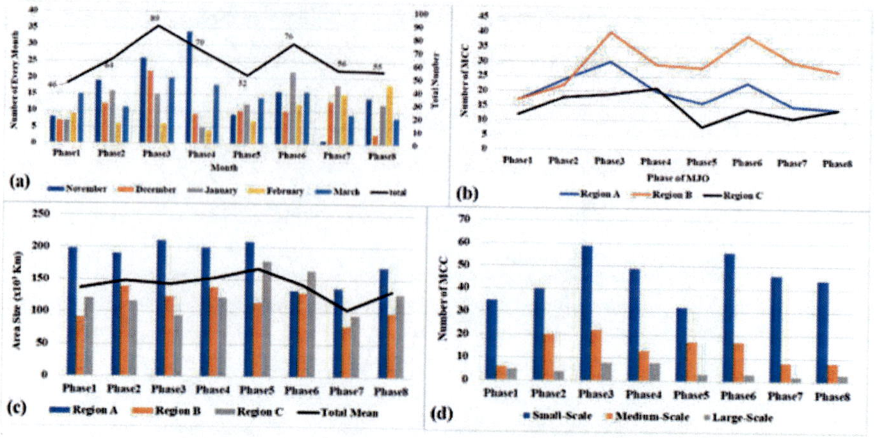

Fig. 5 Number of MCC events in **a** every month and **b** every region during MJO in its eight phases for November through March in 2001–2015. Area size of interior cloud of MCC events during the mature phase in **c** every region and **d** based on size category during MJO in its eight phases for November through March in 2001–2015

MJO, while long-lived MCC often occurs when the MJO is in phase 2. The average duration of MCC events is longer than MCC overall. Trismidianto [10] found that the mean duration was approximate ~ 9 0.5 h, and the maximum duration is 10 h. The distribution and average duration are significantly larger than the distribution and average for MCCs in southern Africa (~9.5 h; [23]). However, it is slightly shorter than the global average (~10 h; [14]) and much less than those found in subtropical South America (14 h; [24]). In most of the MCCs, around 686 events occurred during 8–12 h. Only 20 events are the long-lived MCCs that occurred over 20 h.

In this study, we defined the oceanic (continental) MCCs as a system that reaches the maximum extent while positioned over the ocean (land) as the following method by Morel and Senesi [25] and Blamey and Reason [23]. This research follows the method by Ogino [26] and Mori [27] to define the coastal region. The coastal MCC is defined as the MCCs that occur near the coast between the ocean and the continent during MCC reaching maximum extent or mature stage. The frequency of MCC occurrences on the coast, continent, and ocean in this study is shown in Fig. 6c. Most of the MCCs that occurred were found on the continents for each MJO phase, with the highest number in phase 3 and phase 6. The number of coastal MCC and ocean MCC events mostly occurred during phase 3 MJO.

4 Conclusion

The increase in rainfall during the MJO followed by the presence of an MCC event is higher than when there is no MCC. The increase in rainfall is also higher in the

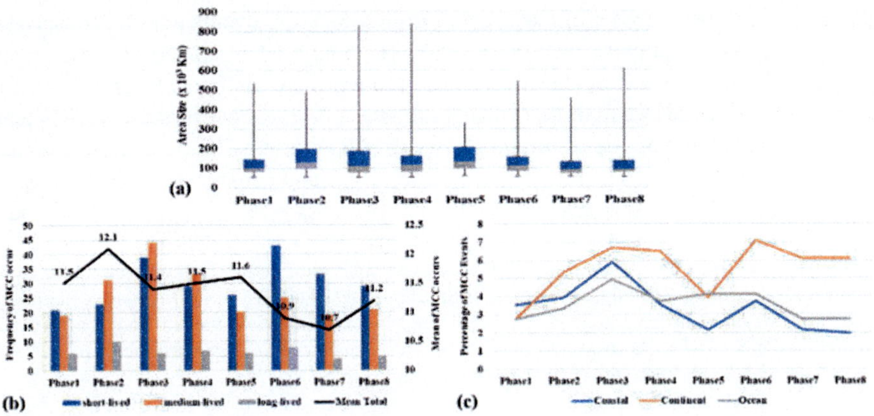

Fig. 6 **a** Box and plot whisker for area size of interior cloud of MCC events during the mature phase in in every phase of MJO for November through March in 2001–2015. Frequency distribution of duration time of MCCs, **b** based on duration category and **c** over the coastal, continent, and ocean during MJO in its eight phases for November through March in 2001–2015

presence of MCC during active MJO than when MJO is not active. The percentage of extreme rainfall events is not very significant when MJO events are not followed by MCC events that occur in almost all MJO phases. However, extreme rainfall events often occur around the MCC event location in each phase of the MJO. This shows that the existence of MCC which occurs concurrently with MJO will increase its influence on the distribution of rainfall and does not even rule out the possibility of causing extreme rainfall in Indonesia. The effect of active MJO followed by the presence of MCC on rainfall in Indonesia is higher when compared to when MJO is not active. This shows that the presence of MCC increases the influence of MJO on the distribution of rainfall in Indonesia. A large number of MCC distributions at the MJO location also indicates that there is MJO modulation of the development of the MCC. In phases 6, 7, and 8, when MJO is without MCC, there is no significant increase in rainfall in Indonesia. However, when compared to the MJO event followed by MCC, it can be seen that there is an increase in rainfall in the area around the MCC event in phases 6, 7, and 8. This indicates that the MCC that occurs in phases 6, 7, and 8 is not modulated by MJO but is possible due to local factors or diurnal convection.

Monthly, in phases 2, 3, and 4, MCC mostly appeared in November, while in phases 1 and 5, MCC mostly occurred in March. MCC often appears in January when the MJO is in phases 6 and 7, and MCC is more commonly found in February when the MJO is in phase 8. The largest MCC cloud core size during the mature phase is found in phase 5, followed by phases 3 and 4, where the average value is above 150 thousand km^2. MCC with a large average size is found in Region A in almost all phases of the MJO except for phase 6 where the average large size MCC is found in Region C. the average MCC size found has a cloud core size ranging from 100,000–200,000 km^2, but the largest sizes above 800,000 km2 are found during phases 3 and 4 of the MJO. The average duration of MCC events in each phase was about 11 h, but the longest average duration of about 12.1 was found in MCC that occurred during phase 2 of the MJO. The frequency of MCC occurrences on the coast, continent, and ocean in this study is shown in Fig. 6c. Most of the MCCs that occurred were found on the continents for each MJO phase, with the highest number in phase 3 and phase 6. The number of coastal MCC and ocean MCC events mostly occurred during phase 3 MJO.

Acknowledgements Thank you so much for the three sources of data, which we can get for free, and the data are beneficial to this research. This research uses data TBB from Himawari generation data (http://weather.is.kochiu.ac.jp/sat/GAME). MJO index data were obtained from the Climate Prediction Center (CPC) Website, NOAA (https://www.cpc.ncep.noaa.gov/products/precip/CWlink/MJO/), and GSMaP satellite (http://sharaku.eorc.jaxa.jp/GSMaP/).

References

1. Ramage, C.S.: Role of a tropical "maritime continent" in the atmospheric circulation. Mon. Weather Rev. **96**, 365–369 (1968)
2. Houze, R.A.Jr. Mesoscale convective system. Rev. Geophys, American Geophisical Union, 43 (2004)
3. Maddox, R.A.: Mesoscale convective complexes. Bull. Amer. Meteor. Soc. **61**, 1374–1387 (1980)
4. Miller, D., dan Fritsch, J.M.: Mesoscale convective complexes in western pacific region. Amer. Meteor. Soc. **119**, 2978–2992 (1991)
5. Laing, A.G, dan Fritsch, J.M.: Mesoscale convective complexes over the Indian monsoon region. Am. Meteorol. Soc. **121**, 2254–2263 (1993)
6. Velasco, I., dan Fritsch, J.M.: Mesoscale convective complexes in the America. J. Geophys.. Res. **92**, 9591–9613 (1987)
7. Durkee, J.D., dan Mote, T.L.: A climatology of warm season mesoscale convective complexes in subtropical South America. Int. J. Clim., 12 (2009)
8. Yuan, J., Houze, R.A.: Global variability of mesoscale convective system anvil structure from A-Train satellite data. J. Clim. **23**, 5864–5888 (2010). https://doi.org/10.1175/2010JCLI3671.1
9. Trismidianto.: A study of mesoscale convective complexes (MCCs) activities in the western Indian Ocean and their effects on convection over Sumatra Island. Tesis Magister, FITB, ITB (2012)
10. Trismidianto.: The global population of mesoscale convective complexes (MCCs) over Indonesian Maritime Continent during 15 Years IOP Conf. Ser.: Earth Environ. Sci. **166**, 012040, 1–17 (2018). https://doi.org/10.1088/1755-1315/166/1/012040
11. Trismidianto, Yulihastin, E., Satyawardhana, H., Nugroho, J.T., Ishida, S.: The contribution of the mesoscale convective complexes (MCCs) to total rainfall over Indonesian Maritime Continent. IOP Conf. Ser. Earth Environ. Sci. **54** (2017). https://doi.org/10.1088/1755-1315/54/1/012027
12. Trismidianto, Hadi, T.W., Ishida, S., Moteki, Q., Manda, A., Iizuka, S.: Development processes of oceanic convective systems inducing the heavy rainfall over the western coast of Sumatra on 28 October 2007. Sci. Online Lett. Atmos. (SOLA). **12**, 6—11 (2016). https://doi.org/10.2151/sola.2016-002
13. Velasco, I., Fritsch, J.M.: Mesoscale convective complexes in the America. J. Geophys. Res. **92**, 9591–9613 (1987)

14. Laing, A.G., Fritsch, J.M.: The global population of mesoscale convective complexes. Q. J. R. Meteorol. Soc. **123**, 389–405 (1997)
15. Trismidianto, Satyawardhana, H.: Mesoscale convective complexes (MCCs) over the Indonesian maritime continent during the ENSO events IOP Conf. Ser. Earth Environ. Sci. **149** 012025, 1–10. https://doi.org/10.1088/1755-1315/149/1/01202
16. Rustiana, S., Trismidianto, Satyawardhana, H.: The influence of ENSO and IOD during mesoscale convective complex (MCC) to rainfall in Indonesia. IOP Conf. Ser. Earth Environ. Sci. **303** 012006 (2019). https://doi.org/10.1088/1755-1315/303/1/012006
17. Aiqiu, L.F., Zakir, A., Trismidianto.: Analysis of mesoscale convective complex during madden Julian oscillation phase 4 (case study: heavy rain in Cilacap on Sept 16–17, 2016) IOP Conf. Ser. Earth Environ. Sci. **303**, 012063 (2019). https://doi.org/10.1088/1755-1315/303/1/012063
18. Wheeler, M.C., Hendon, H.H.: An all-season real-time multivariate MJO index: development of an index for monitoring and prediction. Mon. Weather Rev. **132**(8), 1917–1932 (2004)
19. Lim, S.Y., Marzin, C., Xavier, P., et al.: Impacts of boreal winter monsoon cold surges and the interaction with MJO on Southeast Asia rainfall. J. Clim. **30**, 4267–4281 (2017)
20. Xavier, P., Rahmat, R., Cheong, W.K., Wallace, E.: Influence of Madden-Julian Oscillation on South East Asia rainfall extremes—observations and predictability. Geophys. Res. Lett. **41**, 4406–4412 (2014)
21. Setiawati, M.D., Miura, F.: Evaluation of GSMaP daily rainfall satellite data for flood monitoring: case study—Kyushu Japan. J. Geosci. Environ. Prot. **4**, 101–117. ISSN Online: 2327-4344 (2016)
22. Yulihastin1, E., Trismidianto, Satyawardhana, H., Nugroho, G.A.: MJO modulation on diurnal rainfall over West Java during pre-monsoon and strong El Niño periods. IOP Conf. Ser. Earth Environ. Sci. **54** (2017). https://doi.org/10.1088/1755-1315/54/1/012029
23. Blamey, R.C., Reason, C.J.C.: The role of mesoscale convective complexes in southern Africa summer rainfall. J. Clim. **26** (2012)
24. Durkee, J. D., Mote, T.L., Shepherd, M.J.: The contribution of mesoscale convective complexes to rainfall across, subtropical South America. Int. J. Clim. **22** (2009)
25. Morel, C., Senesi, S.: A climatology of mesoscale convective systems over Europe using satellite infrared imagery II: characteristics of European mesoscale convective systems Q. J. R. Meteorol. Soc. **128**, 1973–1995 (2002)
26. Ogino, S.Y., Yamanaka, M.D., Mori, S., Matsumoto, J.: How much is the precipitation amount over the tropical coastal region? J. Clim. **29**, 1231–1236 (2016)
27. Mori, S., Hamada, J.I., Tauhid, Y.I., Yamanaka, M.D., Okamoto, N., Murata, F., Sakurai, N., Hashiguchi, H., Sribimawati, T.: Diurnal land-sea rainfall peak migration over Sumatra Island, Indonesian Maritime Continent, observed by TRMM satellite and intensive rawinsonde soundings Mon. Wea. Rev **132**, 2021–2039 (2004)

Diurnal Rainfall Pattern in Riau Islands as Observed by Rain Gauge and IMERG Data

Ravidho Ramadhan, Helmi Yusnaini, Marzuki Marzuki, Zahwa Vieny Adha, Mutya Vonnisa, and Robi Muharsyah

Abstract Diurnal rainfall is a dominant local phenomenon in the maritime continent due to the land–sea interaction. Factors controlling rain on small tropical islands may differ from large islands, resulting in the difference in the diurnal rainfall pattern. In this work, we investigated the diurnal rainfall in small islands of Riau Islands using automatic rain gauge (ARG) and Integrated Multi-satellite Retrieval for GPM (IMERG) data from December 2015 to October 2019. Diurnal rainfall in Riau Islands was observed in terms of precipitation amount (PA), precipitation frequency (PF), and precipitation intensity (PI). Mean PA, PF, and PI values from gauge observations show variation values, that is, 0.16–0.35 mm/h (PA), 1.67–3.20% (PF), and 7.04–19.11 mm/h (PI). Furthermore, the values of PA, PF, and PI also show different peaks time. The peak time of PA by rain gauge and IMERG observation occurs at 1100–1300 LST (Batam Island), 0300–0500 LST (Singkep Island), and 1100–1400 LST (Bintan Island), the PF peak occurs at 1100–1300 LST (Batam and Bintan Island) and 0500–0700 LST (Singkep Island), and the peak of PI tends to be fluctuated. This precipitation system's peak time is also dependent on rain duration. IMERG data also observed this pattern of PA, PF, and PI. IMERG data captured rainfall peaks migration from the ocean to the prominent islands in the Riau Islands, such as Batam, Bintan, and Lingga Islands.

1 Introduction

The Riau Islands is a province in Indonesia that is located in the northeast of Sumatra Island. It consists of more than 2400 islands with a total area of around 8200 km^2 [1].

R. Ramadhan · H. Yusnaini · M. Marzuki (✉) · Z. V. Adha · M. Vonnisa
Department of Physics, Universitas Andalas, Padang 25163, Indonesia
e-mail: marzuki@sci.unand.ac.id

R. Muharsyah
Agency for Meteorology, Climatology and Geophysics of Republic Indonesia, Jakarta, Indonesia

A. Basit et al. (eds.), *Proceedings of the International Conference on Radioscience, Equatorial Atmospheric Science and Environment and Humanosphere Science*, Springer Proceedings in Physics 290, https://doi.org/10.1007/978-981-19-9768-6_30

The number of islands is the largest among all provinces in Indonesia. Eventually, most of the population is focused on a few main islands, such as Batam and Bintan Islands. The population density in the Riau Islands is ranked 10th in Indonesia, with a population density of 258 people/km^2 [2].

The geographical condition of the Riau Islands, which is surrounded by oceans and located in the tropical region, causes hydrometeorological disasters that are very potential to occur over there. For example, hydrometeorological disasters are the only cause of natural disasters in the Riau Islands in 2021 [3]. It is composed of floods and cyclones that impacted almost 10 thousand residents. Thus, adequate government planning is needed to minimize the impact of the hydrometeorological disaster. Adequate planning can only be realized if detailed information about rainfall patterns is discovered.

One of the most local dominant phenomena of rainfall on tropical islands is diurnal cycles. Diurnal cycles are the result of strong land–sea interactions related to the land–sea breeze cycle [4]. A previous study was carried out to investigate the diurnal pattern of rainfall over the Indonesian Maritime Continent (IMC) [5–10]. However, most of these studies focus on the diurnal pattern on the big islands (>2000 km^2) in Indonesia, such as Sumatra, Java, Borneo, and Papua. Study on diurnal rainfall patterns on small islands such as the Riau Islands is still minimal. Thus, this study aims to investigate the pattern of diurnal rainfall in the Riau Islands region to analyze the diurnal characteristics of rainfall on small islands in the IMC.

2 Data and Methods

This study focuses on the most densely populated area of the Riau Islands Province. The regencies included in this study are Karimun, Bintan, and Lingga regencies. These regencies also included two municipalities that are Batam and Tanjung Pinang. In the observation area, there are five prominent islands with the largest size, namely Batam, Bintan, Lingga, Singkep, and Kundur Islands. Although these islands are relatively more extensive than others, these islands are not more than 1000 km^2 in size. Details of the arrangement of the islands can be seen in Fig. 1.

Diurnal rainfall pattern observations in Riau Islands were analyzed using four automatic rain gauges (ARG) from the Meteorology, Climatology and Geophysics Agency (BMKG) station from December 2015 to October 2019. The distribution of the stations used can be seen in Fig. 1. One of the stations is located on Singkep Island (ARG 01), one station is on Batam Island (ARG03), and two stations are located on Bintan Island (ARG02 and ARG04). ARG records rain observations every 10 min with an observation threshold of 0.2 mm, so that rainfall below 0.2 mm is not counted. Furthermore, the spatial pattern of diurnal rainfall is analyzed using grid rainfall data from satellite precipitation products (SPPs). SPPs data used in this study is Integrated Multi-Satellite Retrieval for GPM (IMERG) V06 data with a resolution of 0.1–30 min. IMERG data was used from 2015 until 2020. The accuracy and capability of the IMERG V06 data have been evaluated for the IMC region [11–13].

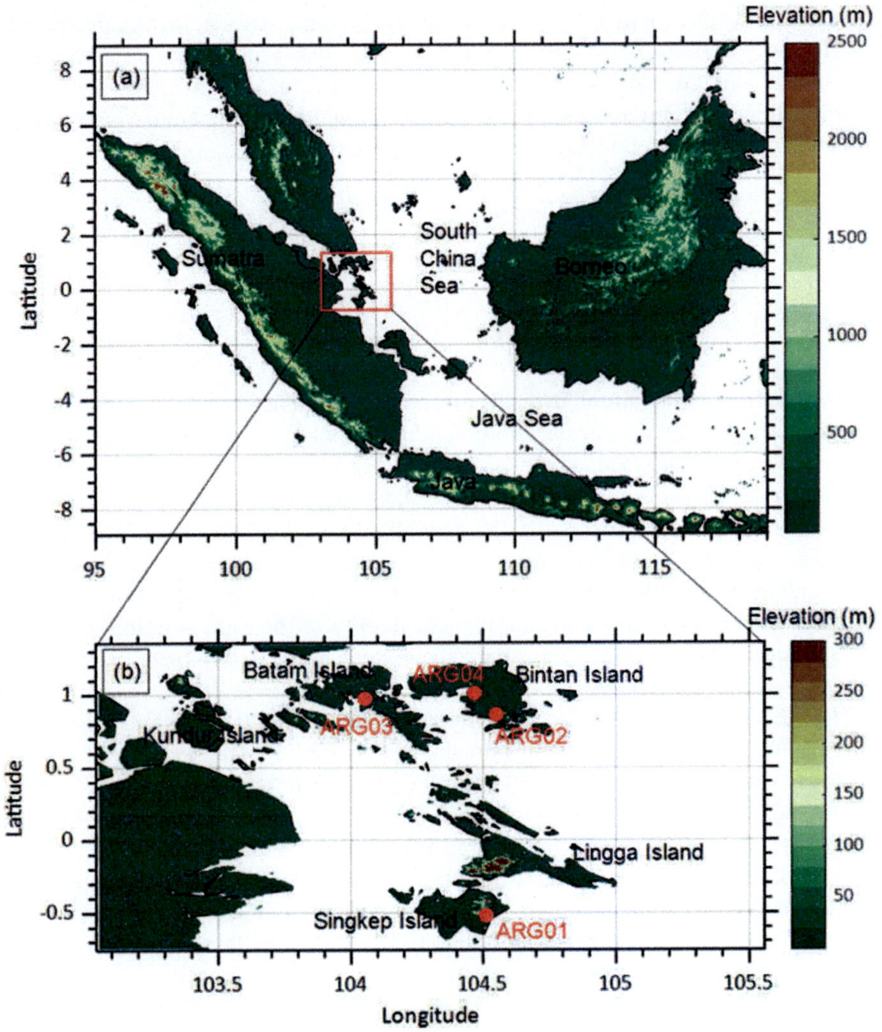

Fig. 1 Position and topography of Riau Islands (**a**) and location of ARG (**b**)

The diurnal rainfall pattern in Riau Islands is characterized by hourly precipitation amount (PA), precipitation frequency (PF), and precipitation intensity (PI) [5, 6]. First, PA is defined as the rate of accumulation of rainfall during the entire observation period to the total observation (mm/h). Second, PF is defined as the ratio of the number of counted rainfall events to total epoch data (%). Third, PI is defined as the ratio of accumulated rainfall to number of hourly data that counted as rainfall events (mm/h). A more detailed definition of the method of calculating PA, PF, and PI can be seen in Marzuki et al. [6].

The PA, PF, and PI values will be considered for each ARG point and the IMERG grid for every hour (0–23) in the Local Time Standard/LST unit (UTC + 7). Moreover, this hourly value of PA, PF, and PI can be used to determine the peak time of the diurnal rainfall pattern in the Riau Islands. Furthermore, this study also observed the effect of rain duration on the diurnal rainfall pattern in the Riau Islands, Indonesia. The rainfall duration is divided into short-duration rainfall (<3 h), medium-duration rainfall (3–6 h), and long-duration rainfall (>6 h).

3 Result and Discussions

The amount of annual rainfall in the Riau Islands is influenced by the duration rainfall category. The percentage of each duration rainfall category for the total rainfall events from ARG and IMERG data can be seen in Fig. 2. All ARGs show a dominant percentage of short-duration rainfall (>80%). On the other hand, the percentage of long-duration rainfall is very rare (<5%) for all ARGs (Fig. 2a). A consistent pattern was also found from IMERG observations for the entire Riau Islands (Fig. 2b,d). This pattern is consistent with the percentage for each rainfall duration found on Sumatra Island [6]. It indicates that the distribution of the total duration of rain on large islands such as Sumatra and small islands such as the Riau Islands has the same pattern.

IMERG shows the land–sea contrast of the rainfall duration percentage (Fig. 2b–d). A high percentage of short-duration rainfall is observed over the largest islands in Riau Islands, while long-duration rainfall shows a low percentage. On the other hand, a contrasting pattern appears over the ocean. It shows a lower (higher) percentage of short-duration (long-duration) rainfall over the ocean. The contrast patterns of the percentage of short-duration and long-duration rainfall events were also found on several islands in the IMC [6, 14, 15]. In addition, the contrast pattern of rainfall duration percentage is not observed over the smaller island. The pattern of smaller islands is the most similar to the ocean pattern.

Before observing the peak times for PA, PF, and PI, we compared the hourly PA, PF and PI for the five ARGs with the nearest IMERG grid point from the ARG location (Fig. 3). The Riau Islands' average PA, PF, and PI ARGs show values of 0.01–0.16–0.35 mm/h, 1.67–3.20% and 7.04–19.11 mm/h. On the other hand, IMERG obtained the average PA, PF, and PI from ARGs in Riau Islands, showing values in ranges of 0.21–0.30 mm/h, 5.98–7.32% and 3.31–4.15 mm/h. Thus, ARG and IMERG shown matched the average of PA, PF, and PI values.

Regarding the peak time identified, ARG and IMERG showed more suitable values for PA and PF and did not by PI. Although the PA and PF peaks from ARG and IMERG showed a suitable pattern, there was little difference between the peak times. This difference in the diurnal peak of IMERG versus the rain gauge was also observed by previous studies [6, 16]. IMERGs tend to observe slightly late peaks of PA and PF compared to ARGs. The PA peaks of ARG and IMERG tend to be observed in the morning to afternoon (0600–1800 LST). Furthermore, the PF peak

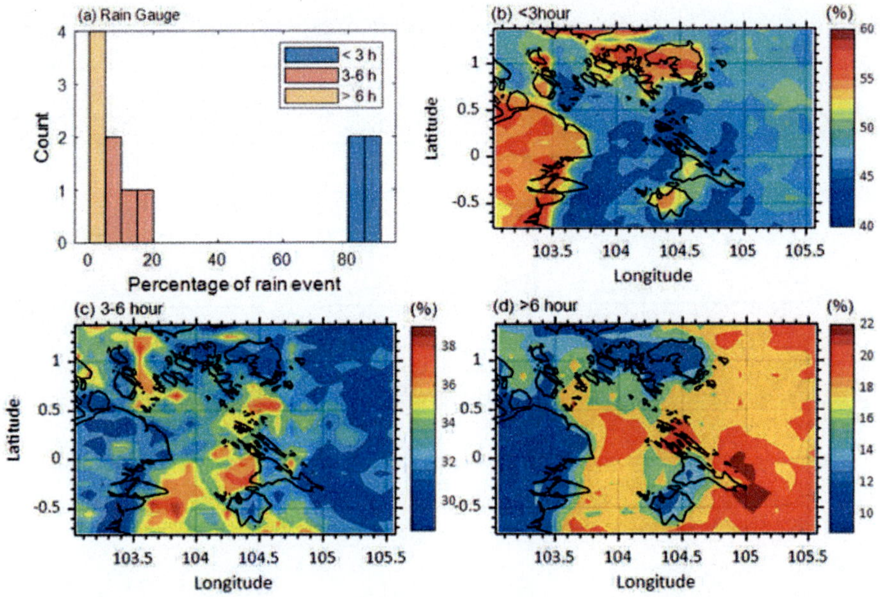

Fig. 2 Distribution of percentage of rain events from ARG observation (**a**) and IMERG data for (**b**) short-duration rainfall (<3 h), (**c**) medium-duration rainfall (3–6 h), and (**d**) long-duration rainfall (>6 h)

occurred later than the PA peak, ranging from 0800 to 1300 LST. An interesting thing was observed in ARG01, which showed an earlier peak of PA and PF in the early morning (Fig. 3a, b). This morning peak of the rainfall was probably caused by the location of ARG01 in a highland area (Fig. 1).

The spatial distribution of peak time and mean of PA, PF and PI from IMERG is shown in Fig. 4. In general, the mean values of PA, PF and PI show higher values over the ocean than over the islands, especially in the South China Sea (SCS). This high amount of rainfall in the ocean area is related to the tremendous amount of water vapor over the ocean than over the land [17]. A higher PA, PF and PI values in the ocean are consistent with a high percentage of long-duration rainfall events (Fig. 2).

IMERG shows the peak time of PA and PF from afternoon to evening (1500–1900 LST) over the largest islands in Riau Islands (Fig. 4). Its peak time of PA and PF over the largest islands does not show a clear spatial pattern between land and coast areas. Furthermore, the earlier peak time (1200–1500 LST) was observed at SCS. Contrary to SCS, the early morning peak time (0000–0600 LST) of PA and PF was observed in the strait between the Riau Islands and Sumatra Island. This difference pattern indicates a different mechanism of diurnal rainfall migration over those seas. In addition, the same pattern as the seas is also observed on smaller islands. This pattern indicates that the land–sea interaction is not dominant over the smaller islands in the Riau Islands.

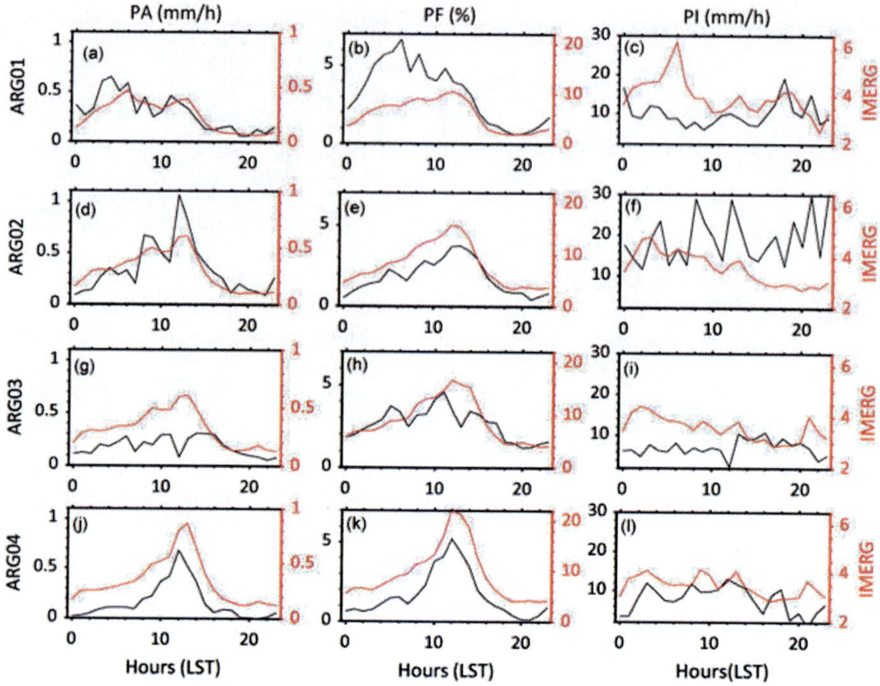

Fig. 3 Time series diurnal PA (first column), PF (second column), and PI (third column) from ARG01 (**a–c**), ARG02 (**d–f**), ARG03 (**g–i**), and ARG04 (**j–l**) with IMERG (the red line) in Riau Island. The x-axis shows the time in hours, and the y-axis shows the value of PA (mm/h), PF (%), and PI (mm/h) from ARG (right side) and IMERG (left side)

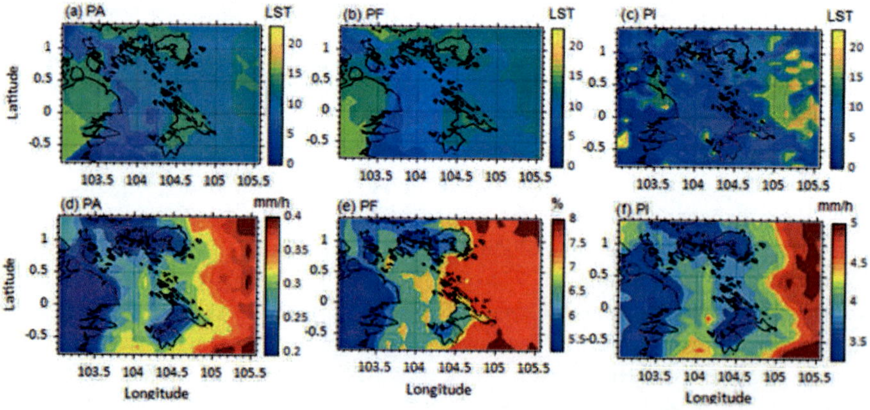

Fig. 4 Spatial distribution of peak time and mean of PA (**a** and **d**), PF (**b** and **e**), and PI (**c** and **f**) from IMERG observation during 2015–2020

4 Conclusions

The diurnal pattern of rainfall in the Riau Islands can be distinguished from ARG observation and IMERG data. The consistent observation from ARG observation and IMERG data indicates that IMERG data is the potential for diurnal analysis on small tropical islands. ARG observation and IMERG data showed that the peak of PA and PF occurred in the morning to evening (0600–1800 LST) and PI occurred in the morning (IMERG) and evening (ARG). The earlier peak time is dominated over the very small island in Riau Islands, while the later peak time is dominated over prominent islands such as Bintan, Batam, and Lingga Islands. The difference in peak rainfall between larger and smaller islands in Riau Islands indicates a different mechanism in the diurnal characteristics over Riau Islands. Thus, a more comprehensive evaluation is needed to represent a mechanism of diurnal cycles over the Riau Islands.

Acknowledgements This study was supported by Riset Publikasi Terindeks (RPT) grant from Universitas Andalas (Contract number: T/114/UN.16.17/PT.01.03/IS-RPT/2022). The author would like to thank the Meteorology, Climatology and Geophysics Agency (BMKG) for providing the data for research purposes. We also thank National Aeronautics and Space Administration (NASA) and the Japan Aerospace Exploration Agency (JAXA) which have provided IMERG final-run product version 06 (V06).

References

1. https://www.bps.go.id/statictable/2014/09/05/1366/luas-daerah-dan-jumlah-pulau-menurut-provinsi-2002-2016.html/. Accessed 25 Sept 2022
2. https://www.bps.go.id/indicator/12/141/1/kepadatan-penduduk-menurut-provinsi.html. Accessed 25 Sept 2022
3. https://dibi.bnpb.go.id/kwilayah/index. Accessed 25 Sept 2022
4. Yamanaka, M.D.: Physical climatology of Indonesian maritime continent: an outline to comprehend observational studies. Atmos. Res. **178**, 231–259 (2016)
5. Marzuki, M., Suryanti, K., Yusnaini, H., Tangang, F., Muharsyah, R., Vonnisa, M., Devianto, D.: Diurnal variation of precipitation from the perspectives of precipitation amount, intensity and duration over Sumatra from rain gauge observations. Int. J. Climatol. **41**(8), 4386–4397 (2021)
6. Marzuki, M., Yusnaini, H., Tangang, F., Muharsyah, R., Vonnisa, M., Harmadi, H.: Land–sea contrast of diurnal cycle characteristics and rain event propagations over Sumatra according to different rain duration and seasons. Atmos. Res. **270**, 106051 (2022)
7. Qian, J.H., Robertson, A.W., Moron, V.: Interactions among ENSO, the monsoon, and diurnal cycle in rainfall variability over Java Indonesia. J. Atmosp. Sci. **67**(11), 3509–3524 (2010)
8. Yulihastin, E., WahyuHadi, T., Sari Ningsih, N., RidhoSyahputra, M.: Early morning peaks in the diurnal cycle of precipitation over the Northern coast of West Java and possible influencing factors. In: AnnalesGeophysicae, vol. 38(1), pp. 231–242. Copernicus GmbH (2020, February)
9. Qian, J.H., Robertson, A.W., Moron, V.: Diurnal cycle in different weather regimes and rainfall variability over Borneo associated with ENSO. J. Clim. **26**(5), 1772–1790 (2013)
10. Hassim, M.E.E., Lane, T.P., Grabowski, W.W.: The diurnal cycle of rainfall over New Guinea in convection-permitting WRF simulations. Atmos. Chem. Phys. **16**(1), 161–175 (2016)

11. Ramadhan, R., Yusnaini, H., Marzuki, M., Muharsyah, R., Suryanto, W., Sholihun, S., ... Tokay, A.: Evaluation of GPM IMERG performance using gauge data over Indonesian Maritime continent at different time scales. Remote Sens **14**(5), 1172 (2022)
12. Ramadhan, R., Marzuki, M., Yusnaini, H., Muharsyah, R., Suryanto, W., Sholihun, S., ... Hashiguchi, H.: Capability of GPM IMERG products for extreme precipitation analysis over the indonesian Maritime Continent. Remote Sens. **14**(2), 412 (2022)
13. Yusnaini, H., Ramadhan, R., Marzuki, M., Ningsih, A.P., Hashiguchi, H., Shimomai, T., ... Sholihun, S.: Statistical comparison of IMERG precipitation products with optical rain gauge observations over Kototabang, Indonesia. Jurnal Ilmu Fisika| Universitas Andalas **14**(1), 10–20 (2022)
14. Ramadhan, R., Marzuki, M., Suryanto, W., Sholihun, S., Yusnaini, H., Muharsyah, R., Hanif, M.: Trends in rainfall and hydrometeorological disasters in new capital city of Indonesia from long-term satellite-based precipitation products. Remote Sensing Appl.: Soc. Environ. 100827 (2022)
15. Marzuki, M., Yusnaini, H., Ramadhan, R., Tangang, F., Amirudin, A.A.B., Hashiguchi, H., ... Vonnisa, M.: Characteristics of precipitation diurnal cycle over a mountainous area of Sumatra Island including MJO and seasonal signatures based on the 15-year optical rain gauge data, WRF model and IMERG. Atmosphere **13**(1), 63 (2021)
16. Tang, S., Li, R., He, J., Wang, H., Fan, X., Yao, S.: Comparative evaluation of the GPM IMERG early, late, and fnal hourly precipitation products using the CMPA data over Sichuan Basin of China. Water **12** (2020)
17. Wu, P., Hara, M., Hamada, J.I., Yamanaka, M.D., Kimura, F.: Why a large amount of rain falls over the sea in the vicinity of western Sumatra Island during nighttime. J. Appl. Meteorol. Climatol. **48**(7), 1345–1361 (2009)

Comparison of Statistical Properties of Rainfall Extremes Between Megacity Jakarta and New Capital City Nusantara

Sopia Lestari, Fadli Syamsudin, Teguh A. Pianto, Reni Sulistyowati, Erma Yulihastin, Dwiyoga Nugroho, Rahaden B. Hatmaja, Dava Amrina, Muhammad N. Habibi, and Namira N. Perdani

Abstract This research investigates the statistical properties of rainfall extremes (REs) over Jakarta and Kalimantan using daily rainfall gauges during 1980–2008 (28 years). The results show that in the Megacity Jakarta, the maximum intensity of 95th and 99th percentile of rainfall is 65 and 120 mm, respectively, and is larger compared to Kalimantan (62 and 100 mm, respectively). However, the intense REs over Jakarta occur less frequently compared to Kalimantan. In contrast, there is a lower intensity of REs over Kalimantan, but they are more frequent than Jakarta. Both over Jakarta and Kalimantan, frequency of REs varies and is influenced by season. In the dry season, the frequency of REs over Kalimantan is about two times higher than frequency of REs over Jakarta. By using the ERA-5 reanalysis datasets, patterns of wind speed and directions are different with the strong westerly wind prevailing around Jakarta. Meanwhile, the strong westerly wind and southwesterly winds exist in the western and northeastern parts of Kalimantan, respectively. Different wind features might be one of the causes of the different nature of REs over both regions.

S. Lestari (✉) · E. Yulihastin · R. B. Hatmaja
Research Center for Climate and Atmosphere, National Research and Innovation Agency, Jl. Dr. Djunjunan 133, Bandung, West Java, Indonesia
e-mail: sopi003@brin.go.id

T. A. Pianto · R. Sulistyowati
Research Center for Geospatial, National Research and Innovation Agency, Geostech Building, PUSPIPTEK, Serpong, South Tangerang, Banten, Indonesia

D. Nugroho
Ministry of Marine Affairs and Fisheries Republic of Indonesia, Merdeka Timur No.16, Jakarta Pusat, Indonesia

D. Amrina · M. N. Habibi
Meteorology Climatology and Geophysics Agency, Jl. Angkasa No .2 Kemayoran, Jakarta Pusat, Indonesia

F. Syamsudin · N. N. Perdani
Faculty of Fishery and Marine Science, Universitas Padjadjaran, Jl. Raya Bandung-Sumedang KM.21, West Java, Indonesia

A. Basit et al. (eds.), *Proceedings of the International Conference on Radioscience, Equatorial Atmospheric Science and Environment and Humanosphere Science*, Springer Proceedings in Physics 290, https://doi.org/10.1007/978-981-19-9768-6_31

1 Introduction

Mean rainfall is getting more intense and frequent worldwide, but it highly varies depending upon the topography [1]. Yet, the upper tail end of mean rainfall distribution has the most immense impacts on humans [2]. The heavy end-tail of distribution, or in this case, rainfall extremes (REs), has been observed and projected to increase as global climate warms [3]. A warmer climate pertaining to temperature rises has a relationship with an increase in atmospheric water vapor. These environments might relate to events associated with deep convection and REs as reported by current research [4].

To gain a high-level of confidence, numerous studies have attempted to understand the impact of climate change on mean rainfall changes; however, there have still been constraints in properly simulating these variations, particularly over the tropical regions [5], that is, because of the various factors affecting the changes and a lack of understanding of the interplay of large- and local-scale climate variability regulating REs development, mainly in the tropics.

Other than atmospheric moisture, changes in vertical motion are closely linked to REs [6]. Despite the warmer world occurs coincident with an increase of global REs, the regional REs are likely to occur and to be more complex due to varying topography and vertical change motion [6], mainly in areas with deep convection, such as in the tropics.

Indonesia, located in the center of the Maritime Continent (MC), is one of the most important regions in the tropics controlling global climate [7]. Improving knowledge of fundamental features and distributions of REs and aspects influencing them shall help global climate models resolving the deficiency in simulating the models. As in the current state, over the MC, understanding nature of mean rainfall, moreover, REs have been challenging due to the complex atmospheric system in the region [8].

Over the Indonesian Maritime Continent (IMC), former research has examined the influence of large- and local-scale drivers on mean rainfall and REs over the Jakarta region and their seasonal variations using 13 indices [9]. The analysis includes frequency, intensity, and contribution relative to the climatology. The previous studies also reported that REs vary and are confined by the season and slopes of the terrain [10]. Although this study has been extensively conducted to understand rainfall characteristics, the distribution of heavy tail end of rainfall has not been explored. In this study, we will examine general large-scale feature that likely induces REs in the two major regions over the IMC, that is, Jakarta and its surrounding regions and Kalimantan. Jakarta and Balikpapan located on Kalimantan Island are the respective current and future capital cities of Indonesia. The underlying large-scale conditions between the two domains will be investigated during the occurrence of REs events.

2 Data and Method

Although recent global studies have heavily relied on reanalysis and satellite datasets to measure rainfall, large uncertainties of analysis in rainfall extremes (REs) using these datasets still exist [11]. Stations or ground-based rainfall datasets have been still considered a trustworthy source to observe rainfall falling on the ground, particularly over areas having a highly localized nature of rainfall [12]. Despite the limited number of stations, we selected to utilize rainfall data from gauges to examine features of REs.

We focused on two regions over the IMC, Jakarta and its surroundings, and Kalimantan (Fig. 1) for the study area. Jakarta located in the north of Java Island in the western part of the IMC (Fig. 1a) is the current capital city of Indonesia with an area of 660 km square [13]. The greater Jakarta is locally known as Jabodetabek (Jakarta, Bogor, Depok, Tangerang, and Bekasi) and occupies the coastal areas in the north, lowland, and mountains in the south. This region has varying topography ranging from 20 to 2000 m [9].

Kalimantan or Borneo Island lies in the center of the IMC. Balikpapan, situated in the southeastern part of Kalimantan, will be the future capital city of Indonesia. This area has nearly 750.000-m square and is known as the third largest island in the world [14]. Areas in the northern part of Kalimantan are dominated by mountains with the height between 1000 and 2000 m (Fig. 1b).

Daily rainfall datasets were obtained from the National Agency for Meteorology, Climatology, and Geophysics (BMKG) during 1980–2008 (28 years) with seven and five stations located over the Jakarta and Kalimantan regions, respectively. Over Jakarta, the seven stations cover areas representing coastal, lowland, and mountainous regions extending from the north to the south meanwhile over Kalimantan, due to the limited datasets, the stations mostly cover lowland areas confined to three stations

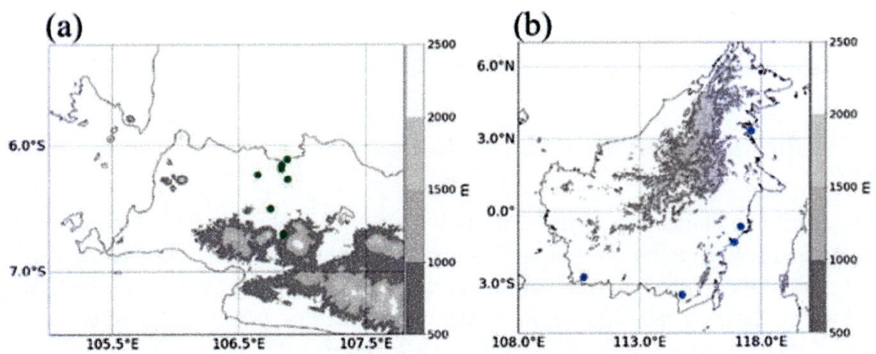

Fig. 1 Map of **a** Jakarta and its surrounding regions and **b** Kalimantan islands with green and purple dots show location of stations at Jakarta and Kalimantan, respectively. Gray contours show the surface topography with a height of between 500 and 2500 m

in the southeast, one station in the southwest, and one station in the northeast near the Makassar Strait.

Rainfall data treatment was applied, for example, checking spurious values of datasets and not including years with a number of missing days which are greater than 40 days. Hence, within 28 years of datasets, only 21 years of data were finally used.

Analysis of daily rainfall focused on examining the three indices consisting of (1) rainy days, a number of days with the rainfall intensity $\geq$ 1 mm/day (wday), (2) the 95th percentile of rainfall intensity (R95p), and (3) the 99th percentile of rainfall intensity (R99p). The analysis focused on looking at the rainy days (wday), rainfall due to very wet days (R95p), and rainfall due to extremely wet days (R99p) defined by World Meteorological Organization [15] and the previous study [9].

For each domain, the distribution of annual R95p and R99p was compared. We also calculated distributions of frequency of wday, R95p, R99p (frequency, %) and relative contribution of wday, R95p, R99p to total rainfall (contribution, %). Further, we also identified seasonal variations of frequency of wday, R95p, R99p.

Gridded datasets of ERA-5 ECMWF (the European Center for Medium-Range Weather Forecasts) were used with the period like station datasets (1980–2008). The data include vertical velocity on 500-hPa (omega 500), zonal (u), and meridional (v) winds at 850-hPa. The reanalysis datasets of ERA-5 have 0.1×0.1 degrees of horizontal and one-hour temporal resolutions and are widely used in the previous studies [10]. We used the ERA-5 data to understand the atmospheric conditions during 21 years of analysis that might be conducive to the formation of REs. Grids of vertical velocity were extracted nearest to the rainfall stations to examine the speed of air motion during the occurrence of REs. In this analysis, we included the median, R95p, and R99p of rainfall. Mean in vertical velocity, speed and wind direction selected only on the same day with R95p was explored to learn synoptic conditions during the REs events.

3 Results

3.1 Characteristics of Rainfall Events

This section explains distributions of rainy days (wday) and rainfall extremes (REs) (R95p and R99p) in the two major regions in Indonesia, Jakarta, and Kalimantan.

Figure 2 shows the whisker and box plots of thresholds of intensity of REs (R95p and R99p) during 1980–2008. It generally represents that Jakarta has a higher intensity of median compared to Kalimantan. Maximum R95p over Jakarta and Kalimantan is 65 and 62 mm, while maximum R99p is 120 mm and 100 mm, respectively. Between Jakarta and Kalimantan, there is a wide different median of R99p of around 15 mm. In contrast, median of R95p is somewhat similar for both regions, and the

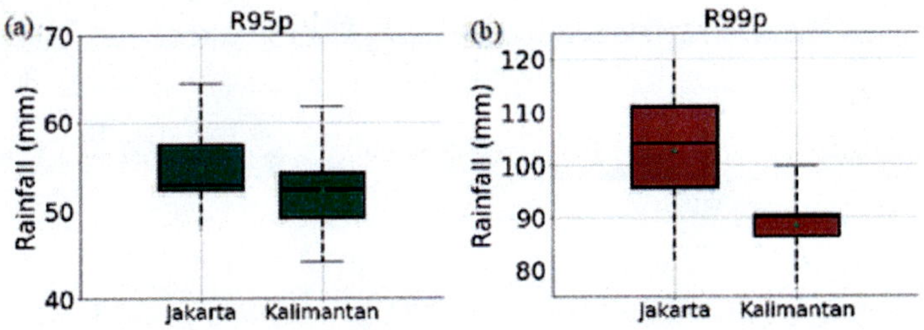

Fig. 2 Whisker and box plots representing distributions of REs of **a** R95p and **b** R99p over Jakarta and Kalimantan Regions during 1980–2008. Box displays range of interquartile with the minimum and maximum are shown in whiskers. Horizontal black line shows the median, and green triangle shows the mean

difference is only around 0.5 mm. This result indicates that Jakarta has more intense REs relative to Kalimantan.

We show whisker and box plots of the frequency and contribution of rainy days (wday), heavy (R95p), and extreme (R99p) rainfall relative to all days (Fig. 3). The aim is to seek the largest percentage of frequency and intensity of wday and REs over Jakarta and Kalimantan. Focusing on median, Fig. 3 shows that over Kalimantan, the three indices show more frequent rainfall rather than over Jakarta; however, the relative contribution to total rainfall is smaller than Jakarta, particularly for wday (Fig. 3b) implying that Kalimantan has more frequent but lighter intensity. Same as Kalimantan, a higher occurrence of rainfall but less intensity is also found in Melbourne compared to other major cities in Australia [16].

Even though one station in Kalimantan is in the northern part, most of stations are situated in the southern part and are categorized as a region with the peak of rainy season happening in Dec–Feb, like Jakarta [17]. However, interestingly, the characteristics of rainfall and REs are still relatively different between Jakarta and Kalimantan indicating that rainfall over the region is highly localized.

Despite a low frequency of R95p (~2,3% of total days) exists in Jakarta (Fig. 3c), relative contribution of R95p to total rainfall is high (25%) (Fig. 3d). This suggested that a low frequency, but a high intensity of REs happens over Jakarta while there is a high occurrence but low intensity in REs over Kalimantan. This result is comparable with [16] who found that R95p contributes to > 20% to total annual rainfall in Melbourne, Sydney, and Brisbane and global areas [16] when taking into account the median.

Consistent with the previous study [16], Jakarta and Kalimantan have a typical peak of rainy season occurring in Dec–Feb and the highest wday and REs also take place in this season. Figure 4 shows the variations of frequency in wday, R95p, R99p seasonally, and it is found that for the Jakarta region, there is *a clearly distinguishable frequency* between rainy (Dec–Feb) and dry (Jun–Aug) seasons of about 30% (wday), 2% (R95p), and 0.5% (R99p). Meanwhile, for the Kalimantan, the difference frequency is nearly 15% (wday), 0.7% (R95p), and 0.2% (R99p). This result illustrates that frequency for both wdays and REs over Jakarta is more strongly affected

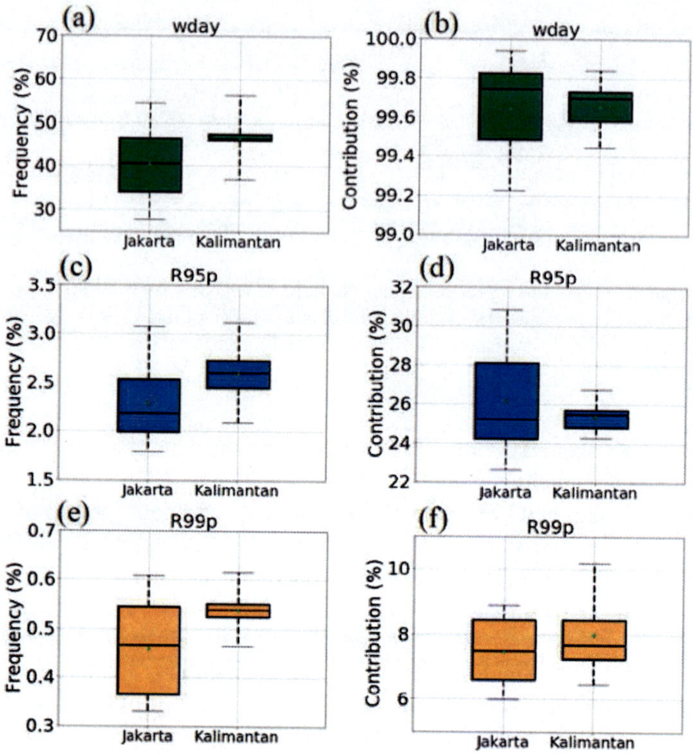

Fig. 3 Whisker and box plots representing distributions of **a**, **b** wday, **c**, **d** R95p, and **e**, **f** R99p over Jakarta and Kalimantan during 1980–2008. Box displays range of interquartile with the minimum and maximum are shown in whiskers. Horizontal black line shows the median, and green triangle shows the mean. **a**, **c**, **e** shows frequency of wday, R95p, R99p (frequency, %), while **b**, **d**, **f** presents relative contribution of wday, R95p, R99p to total rainfall (contribution, %)

by the monsoonal system compared to Kalimantan. The monsoonal system has also regulated the pattern in REs with the highest peak which occurs in Jan–Feb (Jakarta) and Nov–Dec (Kalimantan), while the lowest peak exists in Jun–Aug (Jakarta) and Sep (Kalimantan). The most striking result is that during the dry seasons (Jun-Sep), the REs over Kalimantan are almost doubled in frequency compared to REs over Jakarta in the same season.

3.2 Large-Scale Conditions

Vertical velocity is analyzed to understand synoptic environment that might be favorable to initiate REs development. Figure 5 illustrates the 500-hPa vertical velocity (hereinafter referred to as 500-omega) selected from the nearest grids of locations of rainfall stations at the same time as the occurrence of rainfall event (median, R95,

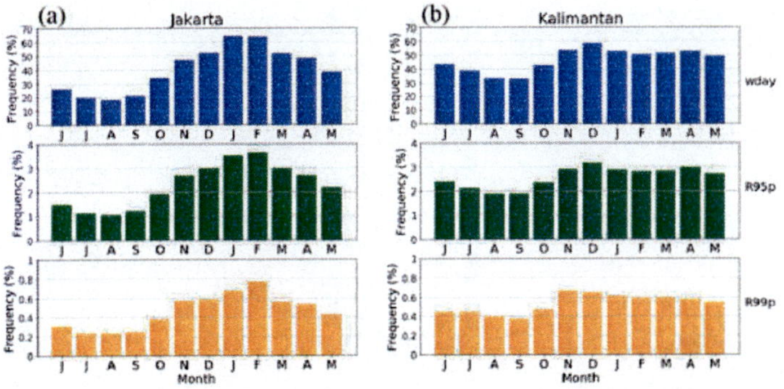

Fig. 4 Seasonal variations of frequency (%) in wday, R95p, and R99p over **a** Jakarta and **b** Kalimantan regions during 1980–2008

R99p). The negative values of 500-omega indicate the upward motion of the air, while positive 500-omega shows the downward motion of the air.

Median rainfall, R95p, and R99p generally occur corresponding to local condition of raising air at around 0 and -0.4 Pa S^{-1} at Jakarta and Kalimantan (Fig. 5). Spatial distributions of 500-omega show that the strong ascent associated with R95p also exists on Jakarta and Kalimantan, with a stronger ascent found over the mountainous region (Figures not shown).

In addition to 500-omega, speed and direction of 850-hPa winds indicate that strong westerly wind (2 ms^{-1}) is predominant over Jakarta (Fig. 6a). But, over Kalimantan, wind direction varies with the westerly wind occurs in the west, southeasterly wind in the north and southeast, while southwesterly in the northeast with the highest wind speed exists in the northeast (2 ms^{-1}) (Fig. 6b). Strong westerly and southwesterly winds might induce convergence over the central and eastern part of Kalimantan responsible for the development REs. A different wind direction between Jakarta and

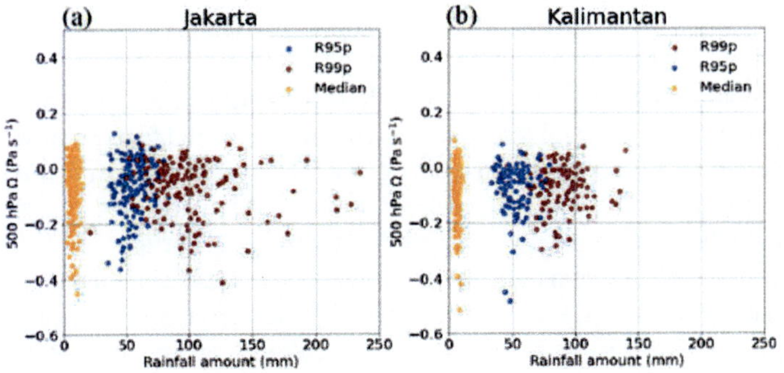

Fig. 5 Scatter plots between median rainfall, R95p, and R99p and 500-omega from station datasets and ERA-5 analysis, respectively, over **a** Jakarta **b** and Kalimantan during 1980–2008. The negative values of 500-omega indicate ascent, while positive values show descent

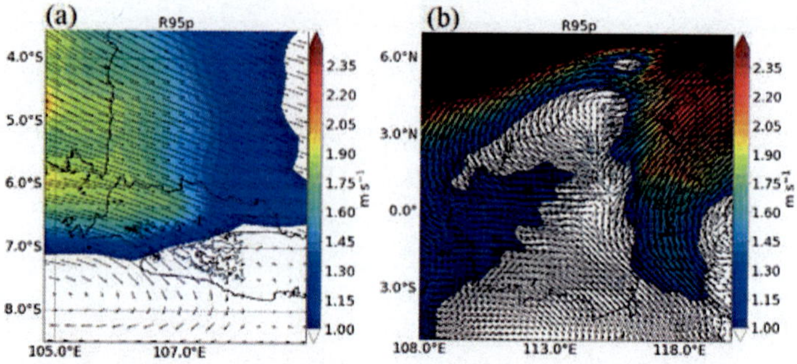

Fig. 6 Spatial maps of speed and direction of the 850-hPa wind from ERA-5 reanalysis, over (**a**) Jakarta (**b**) and Kalimantan during 1980–2008. The magnitude of wind speed is shown in the filled contour (ms^{-1}) while the wind direction is indicated by the vector

Kalimantan is likely related to "dipole pattern" induced by complex local–regional physical processes as reported by the previous studies [18, 19]. However, incorporating a comprehensive analysis of physical background responsible for the different features of REs over both regions shall be done in future studies.

4 Conclusion

This study has shown characteristics of distribution in rainy days (wday) and rainfall extremes of the 95th percentile (R95p) and the 99th percentile (R99p) analyzed from observational datasets during 1980–2008. Results indicate that in the Megacity Jakarta, maximum intensity of 95th and 99th percentile of rainfall is 65 and 120 mm, respectively, and is larger compared to Kalimantan (62 and 100 mm, respectively). This indicates more intense rainfall happens in Jakarta and its surrounding regions (Fig. 2). However, compared to Kalimantan, over Jakarta, the frequency of wday, R95p, and R99p shows a lower percentage implying that Jakarta has less frequent but higher intensity of rainfall and REs rather than Kalimantan (Fig. 3). Yet, although Jakarta has a lower frequency of REs, relative contribution of R95p to total rainfall is high (25%) (Fig. 3d). Over Jakarta, there is a distinct seasonal feature of wday shown by the different frequency between wet and dry seasons of about 30% (Fig. 4a). The most notable result is that during the dry seasons (Jun-Sep), the REs over Kalimantan are almost doubled in frequency compared to REs over Jakarta.

Rainfall extremes are apparent to be initiated by a large-scale pattern of upward motion of air in the vicinity area in Jakarta and Kalimantan (Fig. 5). However, features of wind speed and directions are different between those two regions, with the strong westerly wind prevailing over Jakarta and its surrounding regions. Meanwhile, the strong westerly wind and southwesterly winds exist in the western and northeastern parts of Kalimantan, respectively. The convergence area of wind seems to play a role in generating REs over the Kalimantan region. However, this is a preliminary study. Further research shall explore a more detailed explanation of the underlying mechanism which causes the different features of REs over the Megacity Jakarta and Kalimantan.

Acknowledgements This research has been supported by the National Research and Innovation Agency under the Cruise Day Facilitation Program with grant number 376/II/FR/3/2022 and RIIM program Batch 1 with grant number 65/II.7/HK/2022. We thank the Indonesian Agency for Meteorology, Climatology, and Geophysics (BMKG) for providing rainfall datasets.

References

1. Trenberth, K.: Changes in precipitation with climate change. Climate Res. **47**(1), 123–138 (2011)
2. Merz, B., Basso, S., Fischer, S., Lun, D., Blöschl, G., Merz, R., ... Schumann, A.: Understanding heavy tails of flood peak distributions. Water Resour. Res. **58**(6), (2022)
3. Allan, R.P., Soden, B.J.: Atmospheric warming and the amplification of precipitation extremes. Science **321**(5895), 1481–1484 (2008)
4. Groisman, P.Y., Knight, R.W., Easterling, D.R., Karl, T.R., Hegerl, G.C., Razuvaev, V.N.: Trends in intense precipitation in the climate record. J. Clim. **18**(9), 1326–1350 (2005)
5. O'Gorman, P.A., Schneider, T.: The physical basis for increases in precipitation extremes in simulations of 21st-century climate change. Proc. Natl. Acad. Sci. **106**(35), 14773–14777 (2009)
6. White, B.A., Jakob, C., Reeder, M.J.: Fundamental ingredients of Australian rainfall extremes. J. Geophys. Res.: Atmosph. **127**(17) (2002)
7. Ramage, C.S.: Role of a tropical "maritime continent" in the atmospheric circulation. Mon. Weather Rev. **96**(6), 365–370 (1968)
8. O'Gorman, P.A., Schneider, T.: The physical basis for increases in precipitation extremes in simulations of 21st-century climate change. Proc. Natl. Acad. Sci. **106**, 14773–14777 (2009)
9. Lestari, S., King, A., Vincent, C., Karoly, D., Protat, A.: Seasonal dependence of rainfall extremes in and around Jakarta, Indonesia. Weather and Climate Extremes **24** (2019)
10. Lestari, S., Protat, A., Louf, V., King, A., Vincent, C., Mori, S.: Subdaily rain-rate properties in western Java analyzed using C-band Doppler radar. J. Appl. Meteorol. Climatol. **61**(9), 1199–1219 (2022)
11. Lockhoff, M., Zolina, O., Simmer, C., Schulz, J.: Representation of precipitation characteristics and extremes in regional reanalyses and satellite-and gauge-based estimates over western and central Europe. J. Hydrometeorol. **20**(6), 1123–1145 (2019)
12. Alexander, L.V., Fowler, H.J., Bador, M., Behrangi, A., Donat, M.G., Dunn, R., Funk, C., Goldie, J., Lewis, E., Rogé, M.: On the use of indices to study extreme precipitation on sub-daily and daily timescales. Environ. Res. Lett. **14**, 125008 (2019)
13. Abidin, H.Z., Andreas, H., Gumilar, I., Fukuda, Y., Pohan, Y.E., Deguchi, T.: Land subsidence of Jakarta (Indonesia) and its relation with urban development. Nat. Hazards **59**(3), 1753–1771 (2019)
14. Zhang, C.: Borneo–a future gem of Asia-Pacific. In: The 33rd Annual Pan-Pacific Conference PPBA (2016)
15. WMO.: Guidelines on analysis of extremes in a changing climate in support of informed decisions for adaptation. World Meteorological Organization (2009). Accessed 19 Sept 2020. https://library.wmo.int/doc_num.php?explnum_id=9419
16. Warren, R.A., Jakob, C., Hitchcock, S.M., White, B.A.: Heavy versus extreme rainfall events in southeast Australia. Q. J. R. Meteorol. Soc. **147**(739), 3201–3226 (2021)

17. Aldrian, E., Dwi Susanto, R.: Identification of three dominant rainfall regions within Indonesia and their relationship to sea surface temperature. Int. J. Climatol. **23**, 1435–1452 (2003)
18. Qian, J.H., Robertson, A.W., Moron, V.: Diurnal cycle indifferent weather regimes and rainfall variability over Borneoassociated with ENSO. J. Clim. **26**(5), 1772–1790 (2013)
19. Kurniadi, A., Weller, E., Min, S.K.: Seong, and Min-Gyu: independent ENSO and IOD impacts on rainfall extremes over Indonesia. Int. J. Climatol. **41**(6), 3640–3656 (2021)

Analysis of Mesoscale Convective Complex (MCC) that Often Appears on the Northern Coast of the Borneo Island

Trismidianto

Abstract Mesoscale convective complex (MCC) was portrayed as an organized ensemble of convective elements, whose life cycle is longer than that of the individual convective parts and the largest of the convective storms. MCC was identified by using the program in MATLAB by inputting the temperature, latitude, and longitude values for each cloud shield pixel obtained from brightness temperature (TBB) data based on characteristics of MCC. The data used are a combination of satellite data and reanalysis data. This study analyzes 29 MCC events frequently found in the same location on the Northern Coast of Borneo Island. The study results show that MCCs that occur on the Northern Coast of Borneo Island on average have a long-life span and tend to be nocturnal MCC. The development of the MCC on the Northern Coast of Borneo Island was supported by the presence of surface convergent winds and land breezes. The new convective system began to appear during which the MCC decayed and migrated westward after the MCC dissipated. The MCC on the Northern Coast of Borneo Island has an impact on rainfall, and the MCCs in this region are possibly one of the triggers to the development of the other convective system over the South China Sea. From this study result, if MCC occurs on the Northern Coast of Borneo Island, we can predict its movement and influence.

1 Introduction

One of the convective systems that dominates weather and climate over the Indonesian Maritime Continent (IMC) is the mesoscale convective systems (MCS). MCS is a complex of organized storm clouds on a larger scale than ordinary/individual storms but smaller thanextratropical cyclones and usually lasts for a few hours or

Trismidianto (✉)
Research Center for Climate and Atmospheric, National Research and Innovation Agency (BRIN), Bandung 40173, Indonesia
e-mail: trismidianto@brin.go.id

A. Basit et al. (eds.), *Proceedings of the International Conference on Radioscience, Equatorial Atmospheric Science and Environment and Humanosphere Science*, Springer Proceedings in Physics 290, https://doi.org/10.1007/978-981-19-9768-6_32

more [1]. A particular type of MCS is mesoscale convective complexes (MCCs), and the most widely observed MCS with the largest size and long-life span globally [2]. The phenomenon of MCC has been studied by the previous researchers for more than a few decades in the world [2–7]. MCC over IMC has been also studied by the previous studies [8–13]. MCC plays a very active role in the distribution of energy, momentum, and mass in the atmosphere and occurs over a broad spectrum of time and space scales [12, 14]. MCCs have their organization; they are not passive components in atmospheric flow but interact with large-scale dynamics [15–17]. MCC triggers were strongly correlated with orography, and maximums of MCC triggers were all observed near ridges [8–10, 11].

Trismidianto [8] found that there were 1028 MCCs that occurred during the period 2001–2015 at IMC. Of the1028 MCCs found, there were nine areas as favorite locations occurring of MCCs. Some are above the oceans and coastal areas, and others are above the inland areas and highlands [8, 10, 11]. The nine areas are the Indian Ocean near Sumatra Island, along the West Coast of Sumatra Island, South China Sea on the Northern Coast of Borneo Island, Central Kalimantan around 3°S–1°N; 111°–115°E, East Kalimantan, around 1°–4°N; 116°–120°E, Makassar Strait near the coast of Kalimantan Island, Central Sulawesi about 4°–2°LS; 120°–123°E, Merauke above Papua Island at locations around 8°–5°S and 135.5°–140°E, and Cendrawasih Bay near the coast of Papua Island at around 132.5°–136.5°E and 4°–1°S [8, 10–11].

Of the nine favorite areas for MCC occurrence, several have been discussed in the previous paper by Trismidianto [8, 11–13]. For example, Trismidianto et al. [12] have discussed the development of MCC in Central Kalimantan, but the discussion has focused more on analyzing the atmospheric conditions when MCC occurred in the region. Trismidianto [13] has described the characteristics and mechanisms of MCC events in the Indian Ocean around Sumatra. While the characteristics of MCC that occur in the South China Sea on the North Coast of Borneo Island have never been discussed, the characteristics and movement mechanisms of MCC in each location are often different. So, it is interesting to know the characteristics and development and movement of MCC in the South China Sea on the North Coast of Borneo Island using a composite analysis. This is expected to be able to find out how the influence of the MCC on the rainfall distribution around it so that it can be an analysis of extreme rainfall predictions when the MCC occurs at the same location.

2 Data and Method

MCC was identified by inputting the temperature, latitude, and longitude values for each cloud shield pixel obtained from the equivalent black body temperature (TBB), derived from the hourly infrared data to a modified version of a computerized MCC program using MATLAB based on characteristics of MCCs [2]. The MCC information is used to explain the characteristic of MCC by analyzing the number of MCC occurrences, the duration time of the MCC life cycle, the area size of the cloud, and eccentricity and analyzing the evolution characteristics of MCC. TBB data obtained from Himawari generation infrared satellite imagery consisting of Geostationary Meteorological Satellite (GMS-5), Multi-functional Satellite Imagery Transport Satellite (MTSAT 1R and MTSAT 2), and Himawari 8 and Geostationary Operational Environmental Satellites (GOES-9) with a spatial resolution of 0.05° × 0.05°. This data can be downloaded at https://www.data.jma.go.jp/mscweb/en/himawari89/cloud_service/cloud_service.htm.

The estimated rainfall data, corresponding to the MCCs in this research, were obtained from the real-time TRMM TMPA (TMPA-RT) 3B41RT v7 dataset, which has hourly temporal resolution and 0.25° × 0.25° spatial resolution. (Huffman, J.G. 2013). These data could be downloaded from ftp://disc2.nascom.nasa.gov/data/TRMM/Gridded/3B41RT/. Using the spatial analysis method, we analyze surface wind and atmospheric conditions using ECMWF (European Center for Medium-Range Weather Forecasts) ERA5 and CCMP data. ERA5 provides hourly estimates of a large number of atmospheric, land, and oceanic climate variables. The data cover the Earth on a 30 km grid and resolve the atmosphere using 137 levels from the surface up to 80 km. ERA5 includes uncertainties for all variables at reduced spatial and temporal resolutions [18]. The surface wind data were obtained from the CCMP, which covers the global ocean for 20 years with 6-hourly temporal resolution and 25-km spatial resolution. The dataset is produced using a variational analysis method to combine extensive cross-calibrated multiple satellite datasets with in situ data analyzes [19].

To explain the similarities and consistency of the evolution and propagation of the MCCs in the other several regions, this research is utilized composite analysis adapted from McAnelly and Cotton [20–21]. We create a composite region by taking the region that is most frequently concentrated of MCCs that occurred in the South China Sea on the Northern Coast of Borneo Island, as shown in Fig. 1a in symbol SCS. The region which was chosen also has the grid locations of the MCC center, and the duration of the time life cycle is almost the same by taking the point of the life cycle of the nearest hour. The example composite analysis approach to the evolution of the MCC development is illustrated in Fig. 1b, where the timing of the TBB-defined life cycle characteristics is indicated along an hourly time axis for each MCC. In this study, analysis was also carried out on four important stages, which are also referred to as critical stages in the growth of MCC, i.e., initial, mature, decay, and dissipation or post-MCC stage, following the previous research, Cotton et al. [22], which have defined eight stages in the life cycle. Of an MCC: MCC-12 h, pre-MCC, initial, growth, mature, decay, dissipation, and post-MCC; however, the most important period for an MCC is four of the critical stages.

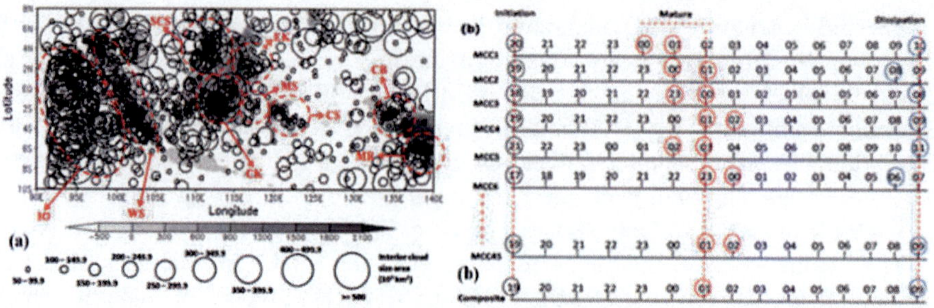

Fig. 1 **a** Geographical distribution of the MCCs in the IMC during 15 years (2001–2015). Locations are for the MCC at the time of the maximum extent of the interior cloud size area when the mature stage, source of figure from [8, 10, 11]. **b** Illustrated of the stratification the MCC composite. The time shows in LT

3 Result and Discussion

3.1 MCC Characteristics

A total of 29 MCCs were found to occur frequently in the same area in the South China Sea on the Northern Coast of Borneo Island. Variations in actual MCC totals per month, as shown in Fig. 2a, are reported with a prominent peak of MCC activity in October and followed by January, April, and June. From the results of Trismidianto [8], it was found that many MCCs occurred in April and October. In this study, we define the MCC duration as the duration between the time of the initial stage and the time of the dissipation stage. The frequency average distribution of the duration of all MCC systems in each month is shown in Fig. 2a in the line graph. The average duration was approximately ~11.53 h. The average duration of MCC life is greater than the overall MCC results globally at IMC, which averages around 9.5 h [8]. However, it is slightly longer than the global average (~10 h; [4]) and much less than those found in subtropical South America (14 h; [6]). The long-lived MCCs occurred in June, October, and November. Most of the short-lived MCCs occurred in May, February, and December. Figure 2b shows that the MCC that frequently occurs on the Northern Coast of Borneo Island has an average area size of the interior cloud of MCC during the mature stage of around 101,000–150,000 km^2. The MCC eccentricity that often occurs on the Northern Coast of Borneo Island generally ranges from 0.90–0.95 (see Fig. 2c).

In general, MCC in IMC usually develops from late afternoon to nearly midnight (see Fig. 3). The most significant frequency of initiation events occurred between the time of 1700–2300 LT. LT is local time (UTC + 7). The MCC system reaches maximum size levels after midnight or early morning at mostly between 0100–0500 LT, and then, MCC begins to decay in the early morning hours after sunrise, especially around 0700–1100 LT. The MCC system dissipated from noon to late afternoon, around 1200–1400 LT. This indicates that the life cycle characteristics of

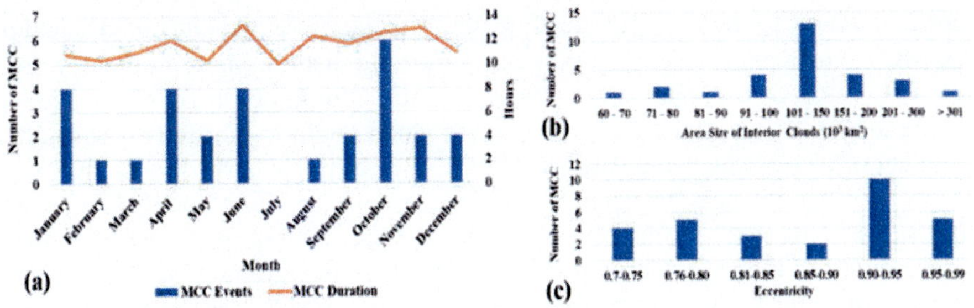

Fig. 2 **a** Number of MCCs that occur and the duration of the MCCs each month, **b** area size of the interior cloud of MCC during the mature stage of MCC, and **c** number of MCC eccentricity at the time of maximum extent

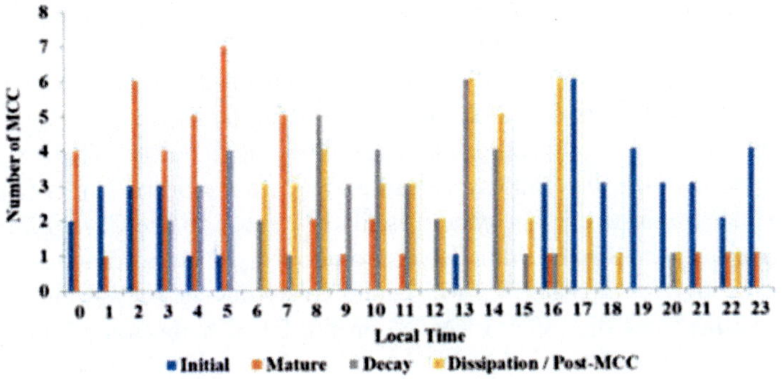

Fig. 3 Frequency distribution of all MCCs during the critical stage (initial, mature, decay, and dissipation or post-MCC stage)

MCC tend to be nocturnal, which is following the characteristics of MCC in globally that occur in IMC [8]. This result is similar to the previous studies that documented regional populations of MCCs were found to be nocturnal, among others; MCC in the Americas and the Western Pacific region, Southern Africa [2, 3, 5, 6, 23]. It follows that the global population is also predominantly nocturnal [24].

3.2 *MCC Evolution Using Composite Method*

Figure 4 illustrates the average time of TBB in the hours of initiation, maturation, decay, and dissipation stages for the 29 MCC cases using the composite method. The MCC that occurs on the Northern Coast of Borneo Island begins to develop from small-scale clouds on the Northern Coast of Borneo Island (see Fig. 4a) and begins to develop after one day from the initial phase even the interior cloud or cloud core has begun to be seen (see Fig. 4b). The interior cloud and cloud shield of the MCC are clearly visible from the day before the mature phase to the day after (see Fig. 4c-e). During the mature phase, the MCC looks very large, even covering the mainland and coastal areas. The interior cloud begin to shrink two days after the mature phase which indicates that MCC will begin to enter the decay phase (Fig. 4f). When it enters the decay phase (Fig. 4g), the interior cloud of MCC starts to disappear followed by the cloud shield from MCC starts to decrease in scale. The position of the MCC during the decay phase is also different compared to the location when the MCC is newly formed and mature, this shows that there has been a movement of the MCC toward the ocean. Two days after the decay phase, the MCC begins to move toward the ocean, leaving the North Coast of Borneo Island as the initial location of MCC formation (see Fig. 4h). Figure 4i and j shows that MCC is starting to dissipate spreading to the South China Sea, and it is also seen that there is distribution around the mainland from the Borneo Island. The development of MCC in this region is likely generated by the diurnal cycle of cloud systems from the South China Sea [25]. They describe that the diurnal cycle over the South China Sea disappears during the day and migrates to Borneo Island. The system reaches its maximum size around morning. The system begins to decay and dissipate in the morning to noon when the cloud system migrates into the South China Sea after the MCC dissipates. However, this system is likely related to the development of the diurnal cycle over the South China Sea as a cloud trigger. According to Houze [25], the diurnal cycle over the South China Sea begins to develop at midnight.

Figure 5 illustrates the effect of MCC on rainfall distribution which shows the same pattern as the evolution of MCC development from TBB data. It can be seen that the peak of maximum rainfall does occur when the MCC reaches its maximum

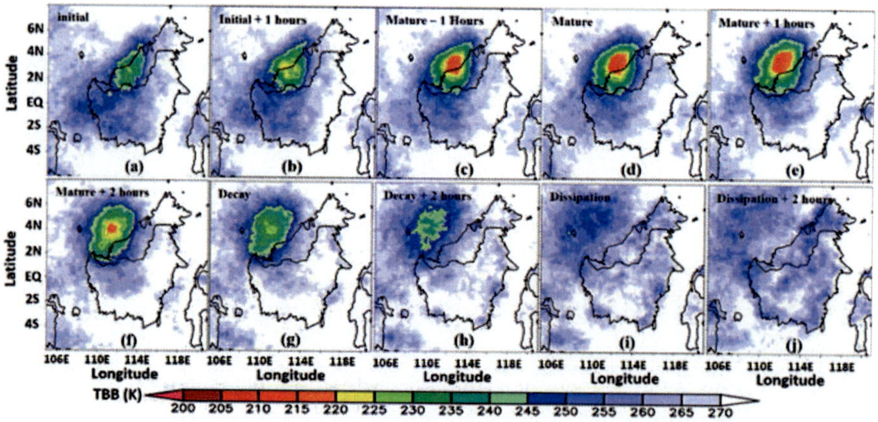

Fig. 4 Horizontal distribution of composite of the TBB for 29 MCC composite in this study area. The time describes the average timing of the hours of initiation, mature, decay, and dissipation stage of MCC in terms of the normalized life cycle using composite method. The unit for TBB is Kelvin

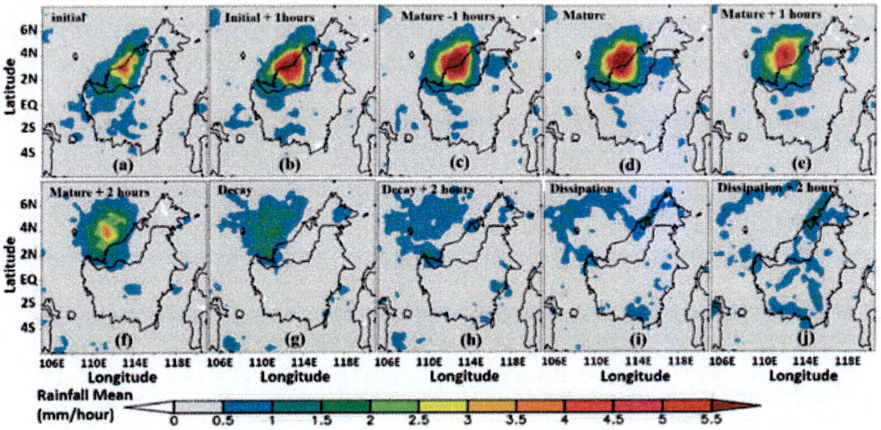

Fig. 5 Same with Fig. 4 but for the horizontal distribution of rainfall composite for 29 MCC composite in this study area

peak. When the MCC decayed, it was also seen that there was a distribution of rainfall toward the South China Sea and also some around Kalimantan. This shows that the MCC that occurred on the Northern Coast of Borneo Island affected the distribution of rainfall in several areas in Kalimantan.

3.3 Cold Pool and Convergent Wind Support the MCC Development

Figure 6a shows that the development of MCC on the North Coast of Borneo Island is triggered by the presence of surface convergent winds that interact with land breezes to form a small-scale convective system. The existence of surface convergent wind support from all directions, especially from the South China Sea, which is getting stronger, makes the convective system that was small in scale become an MCC system, as shown in Fig. 6b. During the mature MCC phase, the cold pools also began to be indicated by the surface potential temperature (theta) which is relatively smaller than the surrounding area, even though the land breeze was not seen clearly. According to Engerer et al. [26], a cold pool is an area of downdraft air cooled by evaporation that spreads out horizontally beneath a precipitating cloud. The difference in surface theta could have acted as a trigger for the development of new convective systems to form along the leading edge of the cold pool that began to be seen in the decay stage of the MCC, as shown in Fig. 6c. The divergent outflow due to the cold pool and in conjunction with the evening sea breeze makes the new convective system migrate widely throughout the surrounding area of the MCC system area, predominantly along the South China Sea and some of over Kalimantan Region, as shown in Fig. 6c and d.

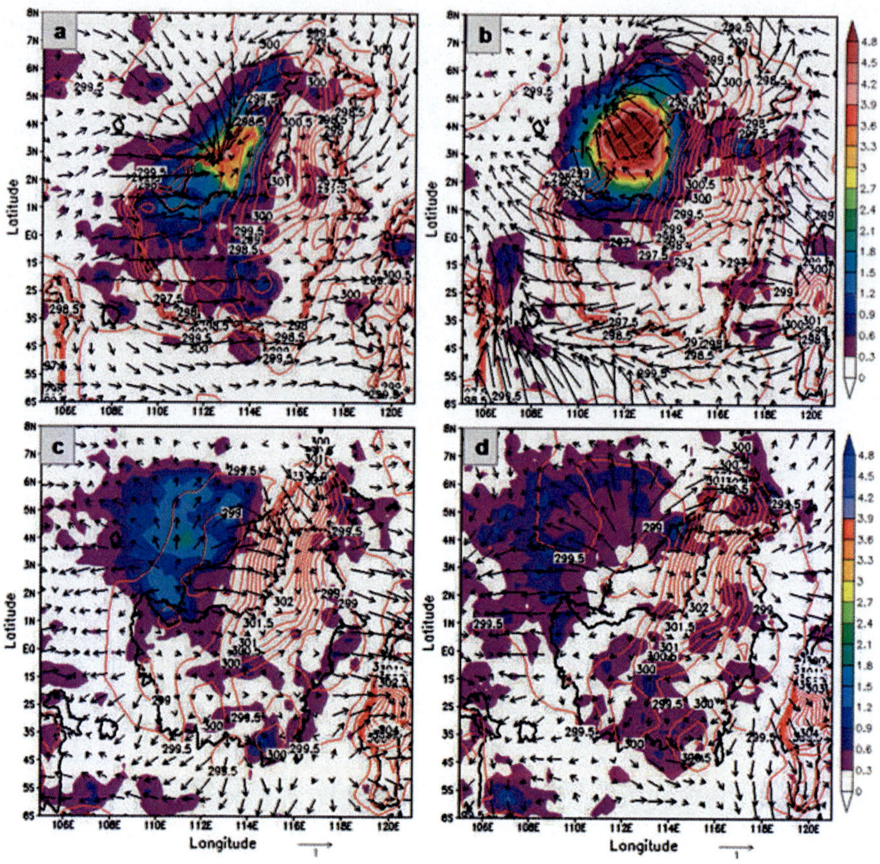

Fig. 6 Horizontal distribution of the composite analysis of the average rainfall for 29 MCC cases in this study (shaded), wind surface vector anomaly (vector), and surface theta (contour) for **a** initial stage, **b** mature stage, **c** decay stage, **d** dissipation stage

Figure 7a and b shows that during the initial and mature phases, there is a surface convergent wind in the vicinity of the MCC occurrence. This reinforces the statement in Fig. 6a and b that there is support from the surface convergent wind in the development of MCC. During the decay and dissipation phases (Fig. 7c and d), there is a surface divergent wind which also strengthens the analysis from Fig. 6c and d which shows that there is a divergent outflow when the MCC decays and dissipates. At the upper level, there is a divergent area around the MCC location during the initial and mature phases (see Fig. 7e and f). The area of upper layer divergence is also visible during the decay and dissipation phases but appears to be starting to shift from the location where the MCC system was formed (see Fig. 7g and h).

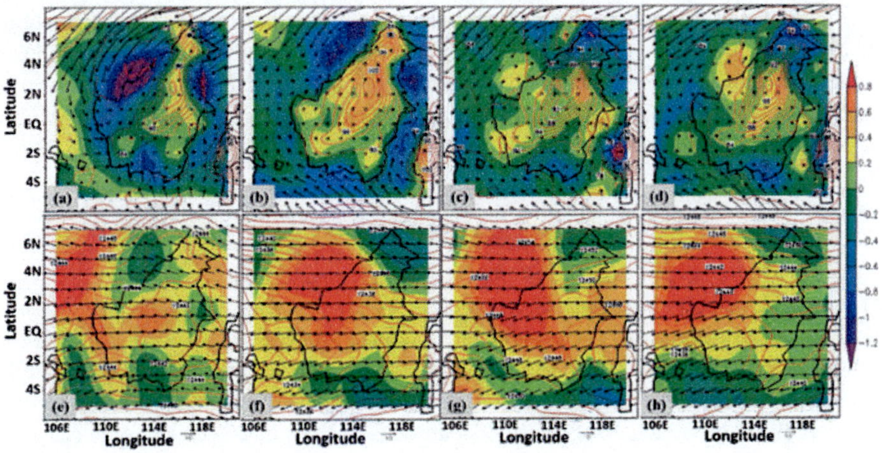

Fig. 7 Composite analysis of the geopotential height (m, black contour) and divergence (10–5 s-1, shaded) in 1000 hPa (**a**–**d**) and 200 hPa (**e**–**h**) during initial stage (**a** and **e**), mature stage (**b** and **f**), decay stage (**c** and **g**), and dissipation stage (**d** and **h**) using ECMWF ERA5 data

Figure 8a shows that during the initial and mature phases, there is a very strong convergence in the surface layer. The presence of this convergence wind is also seen in Fig. 8b, with the easterly wind in the surface layer toward the MCC system. Meridional winds on the surface layer also show movement from north to south toward the MCC formation system (see Fig. 8c). Surface convergence winds during the initial and mature phases shown in Fig. 8 support the statements in Fig. 6a and b. While the surface divergence winds shown in Fig. 6c and d cannot be shown in Fig. 8a where there is no wind divergence in the surface layer during the decay and dissipation phases, weak divergence winds are seen in the intermediate layer. However, surface divergence winds can be demonstrated by the presence of westerly winds from the MCC system (see Fig. 8b) and northerly winds from the MCC system (see Fig. 8c).

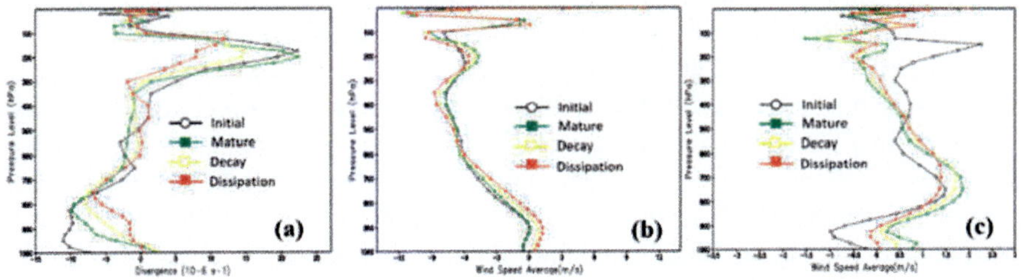

Fig. 8 Composite analysis of the **a** divergence (10–6 s-1), **b** zonal wind (m/s), and **c** meridional wind (m/s) for 29 MCC case with average latitude 2–5 N and longitude 112–115 E during MCC phase of initial, mature, decay, and dissipation using ECMWF ERA5 data

Figure 9 also strengthens the analysis of Fig. 6 regarding the presence of a cold pool when the MCC reaches its maximum or mature stage and sometime after the mature stage, which causes a cold pool outflow that triggers the occurrence of new convective cells. This is indicated by an increase in pressure during the mature phase of MCC, followed by a decrease in the value of equivalent potential temperature (theta-E) and theta as well as a decrease in humidity and an increase in surface temperature. The cold pools were associated with theta and theta-E decreases [27] and generated by these individual convective cells in an MCS typically spread out over the surface and combine to form a large mesoscale cold pool covering a contiguous area on the scale of the entire MCS [1].

The formation of a cold pool when MCC occurs will trigger the formation of new convective cells which will cause continuous rain because the new convective cells that are formed induce convective surrounding them to form other new convective systems [9]. This is what makes MCC can cause heavy and even extreme rainfall. Figure 10 shows the percentage of extreme rainfall events during the MCC in this case study. It can be seen that extreme rainfall occurred during the development of MCC and reached its peak. However, during MCC decay and dissipation, several areas also experienced extreme rainfall due to the MCC.

4 Conclusion

The results of the study show that as many as 29 MCCs that frequently occur on the North Coast of Borneo Island have a long-life span with an average duration of around ~11.53 h, which is longer than the average duration of all MCCs that occur in Indonesia, which is around 9.5 h [8]. The characteristics of the MCC life cycle tend to be nocturnal which are the same as the global MCC characteristics that occur in Indonesia [8]. The MCC that occurs on the North Coast of Borneo Island develops triggered by surface convergent winds which are indicated as westerly and northerly winds from the South China Sea which are supported by the presence of land breeze, as illustrated in Fig. 11. The system reaches its maximum limit due to interaction with the mainland winds the night that follows the rainfall is also getting the maximum. Cold pools begin to form when MCC reaches the mature phase. This cold pool triggers the formation of new convective cells which will cause continuous rain because the new convective cells that are formed induce convective surrounding them to form other new convective systems. New convective systems begin to emerge during the MCC in the decay and dissipation phases. New convective cells migrate westward after the MCC disappears. The MCC on the Northern Coast of Borneo Island has an impact on rainfall, and the MCCs in this region are possible as one of the triggers to the development of the other convective system over the South China Sea by Houze [25]. Houze [25] showed that the morning precipitation over the sea near the Northwest of Borneo Island is due to the convergence of the onshore and monsoon winds.

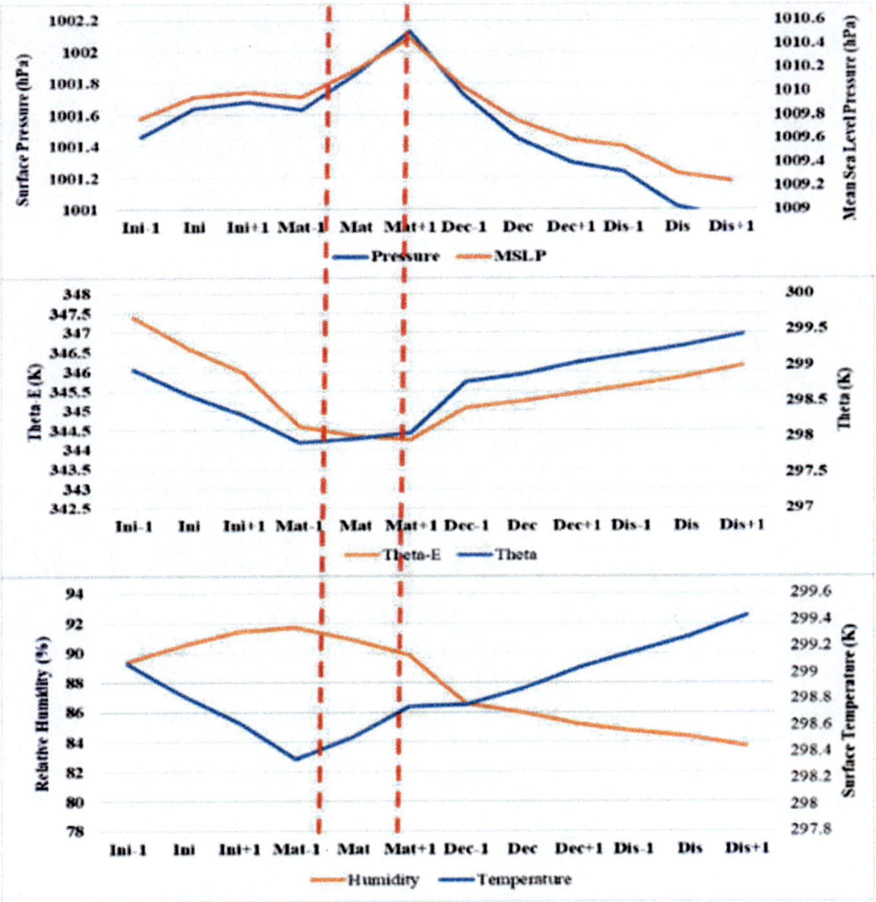

Fig. 9 Composite of surface pressure (hPa) and mean sea level pressure (hPa) in the above figure, equivalent potential temperature (Theta-E), and potential temperature (Theta) in unit Kelvin (middle figure), and the below figure shows the relative humidity (%) and surface temperature (K) during initial-1, initial, initial + 1, mature-1, mature, mature + 1, decay-1, decay, decay + 1, dissipation-1, dissipation, and dissipation + 1. All of this data is from ERA5. The area between the two red lines is an indication of the time of the cold pool

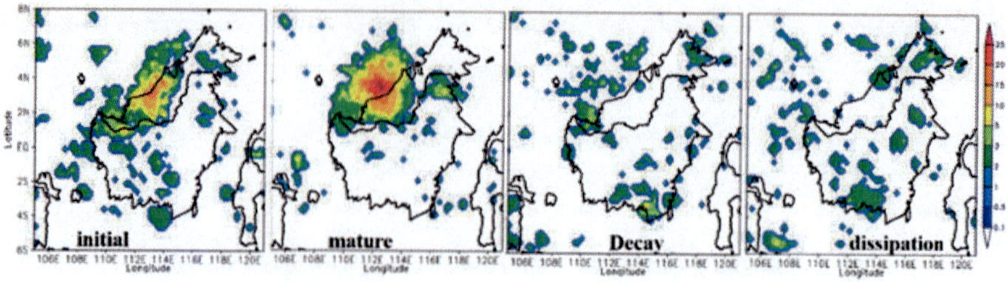

Fig. 10 Percentage of extreme rainfall events during MCC in the stage of initial, mature, decay, and dissipation using the composite method

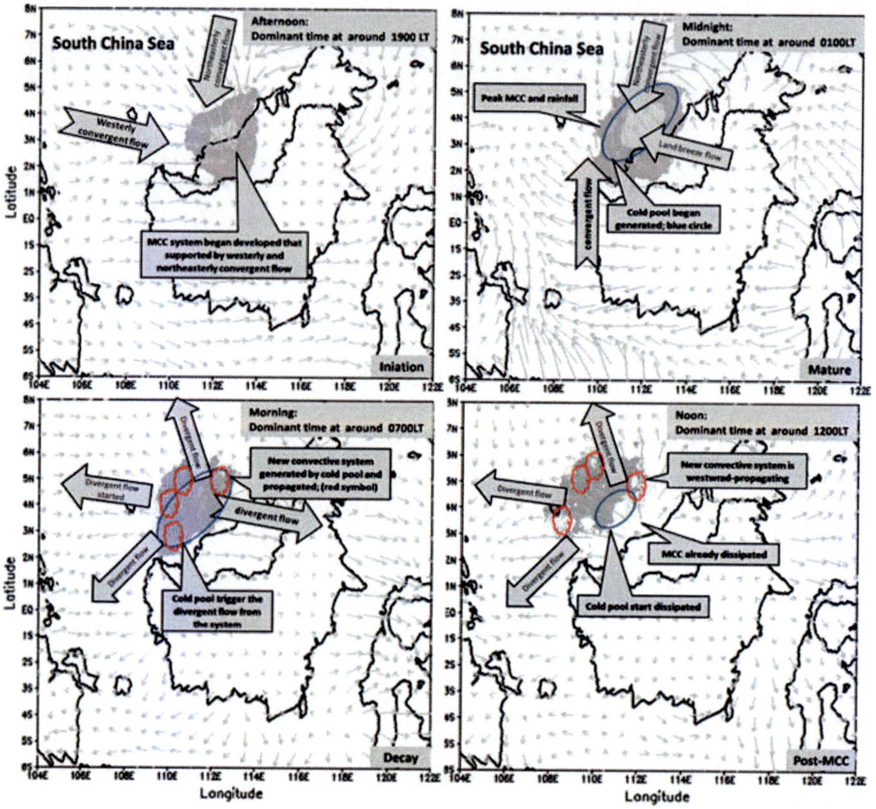

Fig. 11 Schematic representations of the evolution and propagation of the MCC over this study related to the development of the convective system over the South China Sea. Light and dark gray areas illustrated the MCC cloud shield and convective clouds, respectively

References

1. Houze, R.A. Jr.: Mesoscale convective system. In: Review of Geophisics, American Geophisical Union, vol. 43 (2004)
2. Maddox, R.A.: Mesoscale convective complexes. Bull. Amer. Meteor. Soc. **61**, 1374–1387 (1980)
3. Miller, D., dan Fritsch, D., J.M.: Mesoscale convective complex in Western Pasific Region. Amer. Meteor. Soc. **119**, 2978–2992 (1991)
4. Laing, A.G., dan Fritsch, J.M.: Mesoscale convective complexes over the Indian Monsoon Region. American Meteorol. Soc. **121**, 2254–2263 (1993)
5. Velasco, I., dan Fritsch, J.M.: Mesoscale convective complexes in the America. J. Geophysic. Res. **92**, 9591–9613 (1987)
6. Durkee, J.D., dan Mote, T.L.: A climatology of warm season Mesoscale convective complexes in subtropical South America. Int J. Clim, **12** (2009)
7. Yuan, J., Houze Jr, R.A.: Global variability of mesoscale convective system anvil structure from A-Train satellite data. J. Climate **23**, 5864–5888 (2010). https://doi.org/10.1175/2010JCLI3671.1
8. Trismidianto.: The global population of Mesoscale convective complexes (MCCs) over Indonesian maritime continent during 15 Years. In: IOP Conference Series: Earth and Environmental Science, vol.166 (2018). https://doi.org/10.1088/1755-1315/166/1/012040
9. Trismidianto, Hadi, T.W., Ishida, S., Moteki, Q., Manda, A., Iizuka, S.: Development processes of Oceanic convective systems inducing the heavy rainfall over the Western Coast of Sumatra on 28 October 2007. Scient. Online Lett. on the Atmos. (SOLA). **12**, 6–11 (2016). https://doi.org/10.2151/sola.2016-002
10. Trismidianto, Yulihastin, E., Satyawardhana, H., Nugroho, J.T., Ishida, S.: The contribution of the Mesoscale convective complexes (MCCs) to total rainfall over Indonesian Maritime Continent. In: IOP Conference Series: Earth and Environmental Science, vol. 54, (2017a) https://doi.org/10.1088/1755-1315/54/1/012027
11. Trismidianto.: Characteristics of the oceanic MCC, continental MCC, and coastal MCC over the Indonesian maritime continent. In: IOP Conference Series: Earth and Environmental Science, vol. 149, (2018). https://doi.org/10.1088/1755-1315/149/1/012024
12. Trismidianto, Yulihastin, E., Satyawardhana, H., Ishida, S.: A composite analysis of the Mesoscale convective complexes (MCCs) development over the Central Kalimantan and its relation with the propagation of the rainfall systems. In: IOP Conference Series: Earth and Environmental Science, vol. 54, (2017). https://doi.org/10.1088/1755-1315/54/1/012036
13. Trismidianto.: The large-scale meteorological condition during the critical stage of Mesoscale convective complexes (MCCs) over Indian Ocean. In: IOP Conference Series: Earth and Environmental Science, vol. 374, (2019). https://doi.org/10.1088/1755-1315/374/1/012039
14. Ashley, W.S.: A distribution of mesoscale convective complexe rainfall in the United State. American Meteorol. Soc. **131**, 3003–3301 (2003)
15. Maddox, R.A.: Large scale meteorological conditions associated with midlatitude mesoscale convective complexes. Mon. Wea. Rev. **111**, 1475–1493 (1983)
16. Hartmann, D.L., Hendon, H.H., Houze Jr, R.A.: Some implications of the mesoscale circulations in tropical cloud clusters for large-scale dynamics and climate. J. Atmos. Sci. **41**, 113–121 (1984). https://doi.org/10.1175/1520-0469(1984)041,0113: SIOTMC.2.0.CO;2
17. Chen, S.S., Houze, R.A., Jr.: Diurnal variation and life-cycle of deep convective systems over the tropical Pacific warm pool. Q. J. R. Met. Soc. **123**, 357–388 (1997)
18. Dee, D.P., Uppala, S.M., Simmons, A.J., Berrisford, P., Poli, S., Kobayashi, S., Andrae, U., Balmaseda, M.A., Balsamo, G., Bauer, D., Bechtold, P., Beljaars, A.C.M., Van de Berg, L., Bidlot, J., Bormann, N., Delsol, C., Dragani, R., Fuentes, M., Geer, A.J., Haimberger, L., Healy, S.B., Hersbach, H., Holm, E.V., Isaksen, L., Kallberg, K., Kohler, M., Matricardi, M., McNally, A.P., Monge-Sanz, B.M., Morcrette, J.J., Park, B.K., Peubey, C., De Rosnay, P., Tavolato, C., Thepaut, J.N., Vitart, F.: The ERA-Interim reanalysis: configuration and performance of the data assimilation system. Quart. J. Roy. Meteor. Soc. **137**, 553–597 (2011)

19. Atlas, R., Hoffman, R.N., Ardizzone, J., Leidner, S.M., Jusem, J.C., Smith, D.K., Gombos, D.: A cross-calibrated, multiplatform ocean surface wind velocity product for meteorological and oceanographic applications. Bull. Amer. Meteor. Soc. **92**, 157–174 (2011)
20. McAnelly, R.L., Cotton, W.R.: Meso-β-scale characteristics of an episode of meso-α-scale convective complexes. Mon Wea Rev. **114**, 1740–1770 (1986)
21. McAnelly, R.L., Cotton, W.R.: The precipitation life cycle of mesoscale convective complexes over the central United States. Mon. Wea. Rev. **117**, 784–808 (1989)
22. Cotton, W.R., George, R.L., Wetzel, P.J., McAnelly, R.L.: A long-lived mesoscale convective complex. Part I: The mountain-generated component. Mon. Wea. Rev. **111**, 1983—1918 (1983)
23. Blamey, R.C., Reason, C.J.C.: The role of mesoscale convective complexes in Southern Africa summer rainfall. J. Clim **26** (2012)
24. Laing, A.G. dan Fritsch, J.M, The global population of mesoscale convective complexes. Q. J. R. Meteorol. Soc. **123**, 389–405 (1997)
25. Houze, R.A., Jr.: Cloud clusters and large-scale vertical motions in the tropics. J. Meteorol. Soc. Jpn. **60**, 396–410 (1982)
26. Engerer, N.A., Stensrud, D.J., dan Coniglio, M.C.: surface characteristics of observed cold pools. Monthly Weather Rev. **136**, 4839–4849 (2008)
27. Tompkins, A.M.: Organization of tropical convection in low vertical wind shears: the role of cold pools. J. Atmos. Sci. **58**, 1650–1672 2001

Impact of Cold Surge Based on Its Strength on Rainfall Distribution in Western Indonesia

Alfan Sukmana Praja and Trismidianto

Abstract Cold surge (CS) is a mass of cold air carried by the north–south (meridional) wind circulation due to high pressure disturbances in the Siberian region that flows to the equator and south through the north coast of Java and is one of the synoptic scale weather phenomena that has a significant influence during Asian winter monsoon (NDJFM). This study was conducted to analyze the impact of CS based on the category of strength to the distribution of rainfall in Indonesia. Analysis of CS events was carried out by calculating the CS index which is the average meridional wind speed at the level of 850 hPa in the region of 110−115 °E and 7.5−12.5 °N. Based on its strength, CS is divided into three, namely weak CS (8–10 ms-2), moderate CS (10–12 ms-2), and strong CS (>= 12 ms-2). Rainfall data are obtained from 20 years GSMaP data and wind data and other parameters from ERA 5 data. The results showed the area that experiences an increase in rainfall decreases when the strength of the CS increases. However, rainfall intensity increased in the South China Sea, Karimata Strait, and the Java Sea. When CS is weak, there is generally an increase in rainfall from day-1 to day+ 1. While for moderate CS, the increase in rainfall starts when a cold surge occurs until day+ 2. Lastly, when CS is strong, an increase in rainfall and potential for cold surges has occurred from day-2 to day+ 2. Strong CS can increase rainfall more in the Java Sea area than Java Island, whereas weak and moderate CS increase rainfall more in Java Island than in the Java Sea from day-1 to day+1.

1 Introduction

Indonesia's strategic geographical location has attracted world researchers to observe the dynamics of climate and weather in Indonesia. As a tropical area, Indonesia has high rainfall and changes yearly. Changes in rainfall on a seasonal scale in

A. S. Praja (✉) · Trismidianto
Research Center for Climate and Atmosphere, National Research and Innovation Agency, BRIN KST Samadikun, Jl. Cisitu Sangkuriang, Bandung, West Java, Indonesia
e-mail: alfa004@brin.go.id

A. Basit et al. (eds.), *Proceedings of the International Conference on Radioscience, Equatorial Atmospheric Science and Environment and Humanosphere Science*, Springer Proceedings in Physics 290, https://doi.org/10.1007/978-981-19-9768-6_33

Indonesia are heavily influenced by Monsoon, ENSO, MJO, and so on. Changes in the daily scale can occur by disturbances from the north called a cold surge. The value of fluctuations in air pressure and wind on a large scale can trigger a cold surge phenomenon, namely a large mass flow of air from mainland Siberia in the meridional east season accompanied by a decrease in temperature and air pressure in Hong Kong at a surface layer of 1000–850 hPa.

The cold surge will flow through the South China Sea, Karimata strait, the Bangka Islands to the Java Sea. Increased rainfall in the Java Sea, which can reach the mainland, can trigger flooding in areas near the coast, such as Jakarta. Therefore, cold surge analysis is crucial to increase awareness in dealing with floods due to high rainfall, especially from November to March. In these months, the rainfall on the island of Java tends to be high, along with frequent cold surges.

The relationship between MJO and cold surge has been studied in some parts of Indonesia [1–3]. Generally, when MJO is active, cold surges can increase rainfall even up to 150% [1]. However, the type of cold surge that often causes extreme rain in western Indonesia has not been further elaborated. Thus, this information is very important to anticipate the occurrence of high rainfall.

The impact of cold surges on the tropics is that it can trigger cumulus convection so that cold surges can increase extreme rainfall and precipitation in several parts of western Indonesia [1, 2]. Cold surges can occur once or twice a month for up to 2 weeks [4]. The high potential intensity of rainfall in an area can be seen from water vapor transport [5, 6].

The previous research on cold surges has not specifically divided the influence of cold surge strength on rainfall in western Indonesia. So in this study, the authors divide the cold surge into three categories, namely weak, moderate, and strong, and explain its influence on rainfall distribution in western Indonesia. A brief explanation of the MJO and extreme rain is provided along with the case study.

2 Data and Method

The data used in this study include daily rainfall data in the month of NDJFM (November, December, January, February, and March) as months that have high rainfall rates [7] and correspond to the time of frequent occurrence of cold surges [8]. Rainfall data from 2000 to 2019 were obtained from GSMap, downloaded from sharaku.eorc.jaxa.jp/GSMaP, with a spatial resolution of 0.1×0.1 degrees.

This study also used zonal wind data and meridional winds with the same period used in rainfall data. Wind data are obtained from ECMWF which can be downloaded at apps.ecmwf.int/datasets/data/interimfull-daily/ with a spatial resolution of 0.25 $\times$ 0.25 degrees. We also use outgoing long-wave radiation(OLR) [9] data from NCEP/NCAR.

Analysis of CS events was carried out by calculating the CS index, which is the average meridional wind speed at the level of 850 hPa in the region of 110–115 °E and 7.5–12.5 °N [10]. Based on its strength, CS is divided into three, namely, weak

CS (8–10 ms-2), moderate CS (10–12 ms-2), and strong CS (>= 12 ms-2) [4]. All of 193 CS events consisting of 140 weak CS (73%), 37 moderate CS (19%), and 16 strong CS events (8%).

In this study, we defined active MJO as when MJO is active in the Indian Ocean (phases 2 and 3) and active in the maritime continent (phases 4 and 5) with an amplitude greater than or equal to 1, otherwise called inactive MJO. The calculation of extreme rainfall uses the 95th percentile of the cumulative daily precipitation distribution [11]. The measure of moisture transport is obtained from the total water vapor carried according to the movement of the wind [1, 5].

3 Result and Discussion

3.1 Rainfall Distribution During CS Event

In western Indonesia, when CS occurs, rainfall in the month of NJDFM increases to 21 mm/day, especially along the South China Sea route, the Karimata Strait to the Java Sea, and parts of the Indian Ocean to southern Java (Fig. 1). Meanwhile, when there is no CS, there is relatively no rainfall anomaly in western Indonesia.

Based on Fig. 2, the wind direction in the weak, moderate, and strong CS events is relatively the same, mostly blowing from the South China Sea, the Karimata Strait to turning toward the Java Sea. However, in strong CS, a vortex around Kalimantan triggers the Borneo vortex [12]. The highest increase in rainfall intensity in the South China Sea area along the CS route, the Karimata Strait to the Java Sea, occurs when the CS is strong. However, the highest increase in rainfall intensity in the western part of Java and the Indian Ocean near Sumatra was caused by moderate CS. Besides the increase in rainfall intensity, there is a decrease in rainfall intensity in parts of the Indian Ocean near Sumatra when the CS is strong.

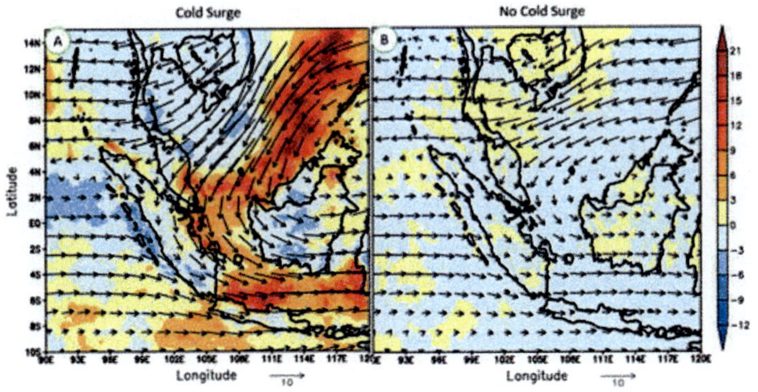

Fig. 1 Rainfall Distribution during CS dan No CS

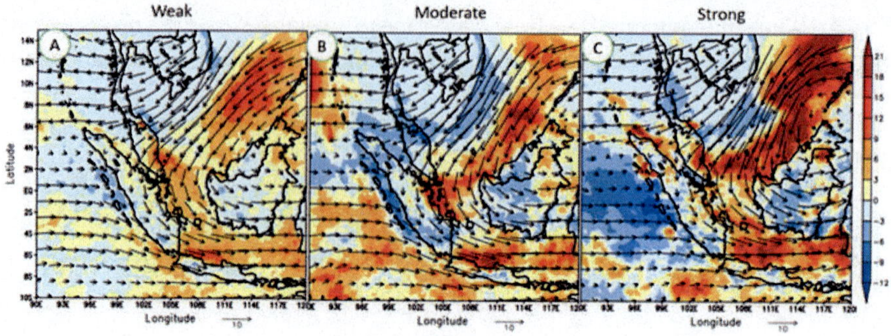

Fig. 2 Rainfall anomaly when CS is weak, moderate, and strong

A comparison of extreme rainfall patterns when CS is weak, moderate, and strong is presented in Fig. 3 following. When CS is weak, the highest extreme rainfall occurs around the South China Sea and parts of the Java Sea, Batam, and northern Bangka waters. At moderate CS, extreme rainfall events with an average of up to 30% occur in the South China Sea, North Brunei, and North Java to the Java Sea. Meanwhile, when CS is strong, extreme rainfall occurs up to 35% in the South China Sea, north of the island of Kalimantan, to the Java Sea. Generally, CS strong produces higher extreme rainfall in western Indonesia than when CS is weak, and CS is moderate.

The previous studies have shown that 1–2 days of lag time are required for cold surges to increase rainfall [1]. This lag time analysis aims to see the movement of each type of CS and rainfall before and after the occurrence of CS. In Fig. 4 lag time analysis, the increase in rainfall occurs when CS is weak and moderate in the Java Sea area. As for CS strong, from day-1 to day+2, there was a fairly consistent increase in rainfall in the Java Sea area.

From day-2 until day+2 in weak CS, there is no significant change on the island of Sumatra and Kalimantan island with anomaly changes ranging from 0 to 4 mm. Around the island of Java, the most significant increase in rainfall anomaly occurs

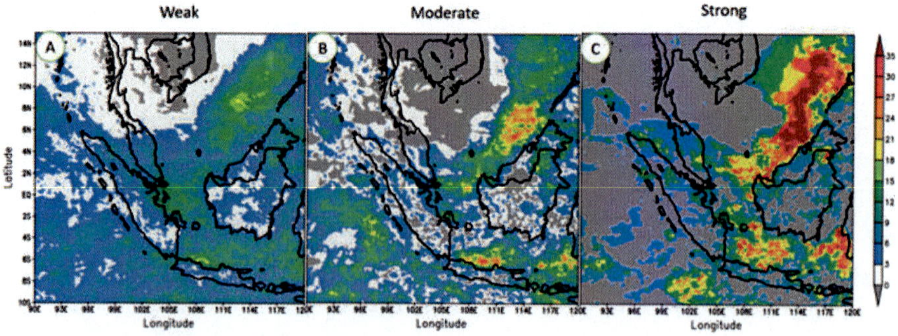

Fig. 3 Percentage of extreme rainfall events

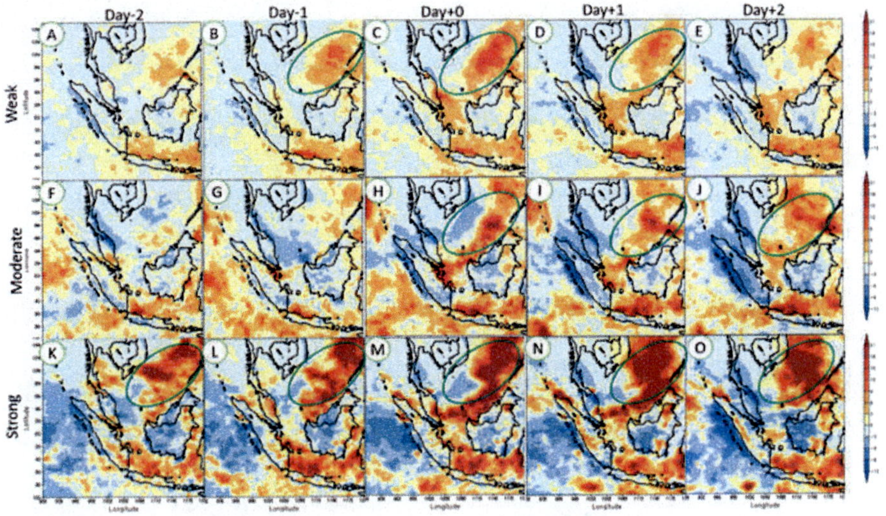

Fig. 4 Lag time analysis day-2 to day+2 for weak, moderate, and strong CS

when day-1 in weak CS. When CS is moderate, there is a decrease in rainfall of up to 12 mm from day-2 to day +2 on the islands of Sumatra and Kalimantan. However, in the western part of Java Island, the highest rainfall is from day-2 to day+2.

When CS strong occurs, the series of days of the occurrence of CS have different intensities between West Kalimantan and East Kalimantan. The western part of Kalimantan generally has a decrease in rainfall, while the eastern part experiences an increase in rainfall from day-2 to day +2. High rainfall occurs on day-2 and day-1, while on CS day up to day+ 2, very high rainfall occurs in the South China Sea when CS is strong.

When CS is weak, there is generally potential for cold surges from day-1 to day+1, when CS is moderate, CS day to day+2 still occurs CS. Meanwhile, when CS is strong, from day-2 to day +2 cold surge, an increase in rainfall still occurs in the South China Sea.

3.2 *Impact of CS During MJO*

Figure 5 shows the difference between events when MJO is active and inactive for weak, moderate, and strong CS. The influence of MJO on rainfall is relatively strong in the Indian Ocean, West Sumatra, especially when CS is weak and CS is moderate. When CS is weak with MJO takes place, rainfall increases up to 15 mm, while when inactive MJO only reaches 6 mm. Even when the CS is moderate, the active MJO generally increases in rainfall by up to 18 mm compared to the inactive MJO, which typically decreases in Sumatra's West Indian Ocean region.

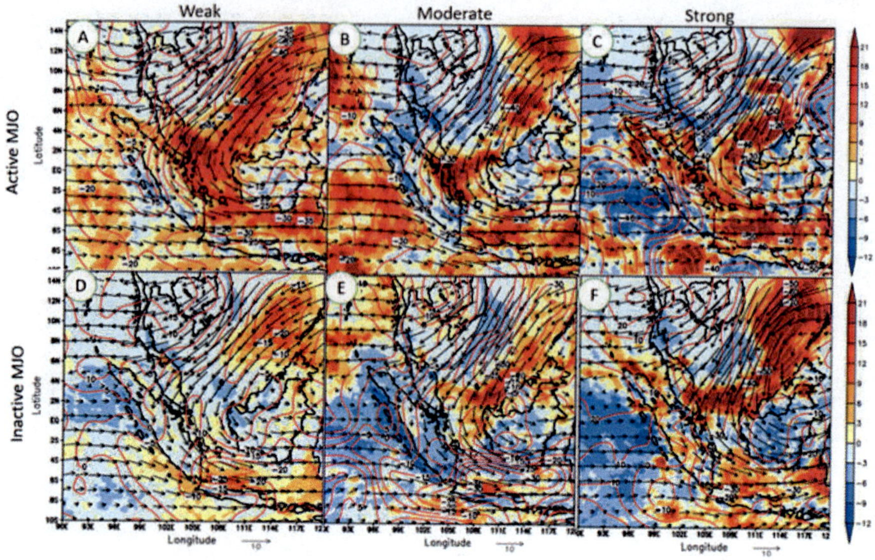

Fig. 5 CS based on its strength when MJO is active and inactive

When MJO is active, the distribution of the increase in rainfall gets narrower in contrast to the increasing strength of the CS. When CS is weak, almost all areas experience increased rainfall from the Indian Ocean to Kalimantan. However, when the CS is moderate, parts of northern Sumatra and the Riau Islands decrease in rainfall intensity. When the CS was strong, the areas that experienced a decrease in rainfall were more expansive: the Indian Ocean, West Sumatra, Riau Islands, North Sumatra, and parts of southern Java.

In contrast to when the MJO is inactive, the distribution of the increase in rainfall when the CS is weak occurs in the South China Sea, most of the Java Sea, and parts of the Indian Ocean near Sumatra. The increase distribution becomes narrower when CS is moderate, which only occurs in the South China Sea, the Karimata Strait, and a small part of the Java Sea. However, when CS is strong, the distribution of the increase in rainfall extends to cover the South China Sea, Java Sea, southern Java Island, and parts of Sumatra.

The OLR value reaches −70 when MJO is inactive for strong CS in the South China Sea, even lower than when MJO is active, which only reaches −50. However, this is the opposite in the Java Sea area, which has a lower OLR value when MJO is active, namely −50, and when MJO is inactive, −30. In general, the OLR value when MJO is active has a short range of −40 to −50 in the South China Sea area, both when CS is weak, moderate, or strong. But when MJO is inactive, there is a significant difference when CS is weak and CS is moderate, which only reaches −20 to −30 against strong CS, which can reach −70.

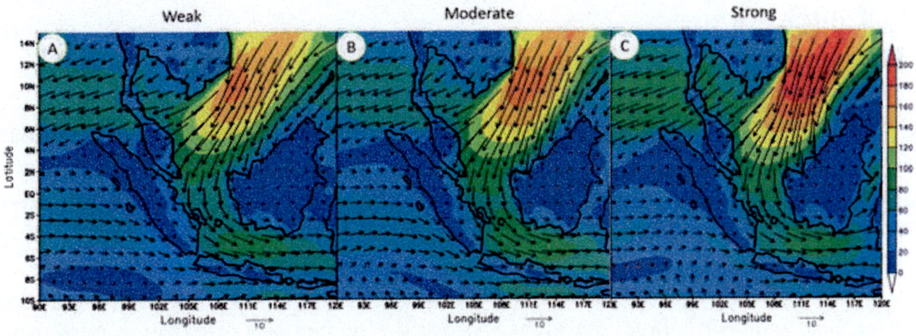

Fig. 6 Moisture transport when weak, moderate, and strong CS

3.3 *Moisture Transport and Case Study Analysis*

Moisture transport for weak and moderate CS tends to be influenced by flows from the Indian Ocean. Meanwhile, moisture transport in pure strong CS is obtained only from the north, not from the Indian Ocean (Fig. 6).

Observation data in the Kemayoran area represent the occurrence of high-intensity rainfall and flood events in the Jakarta area during weak, moderate, and strong cold surges. Rainfall in Fig. 7 represents weak, moderate, and strong CS on 17 January 2014, 3 February 2018, and 13 January 2009, with 147.9, 50.7 and 102 mm/day, respectively [13]. Based on the Hovmoller diagram in Fig. 7, day-1 to day+ 1, strong CS can increase rainfall more in the Java Sea area than Java Island, whereas weak and moderate CS increase rainfall more in Java Island than the Java Sea. This increase is also triggered by the moisture transport in Fig. 6, which explains that the flow from the Indian Ocean to the island of Java is not significant when the CS is strong.

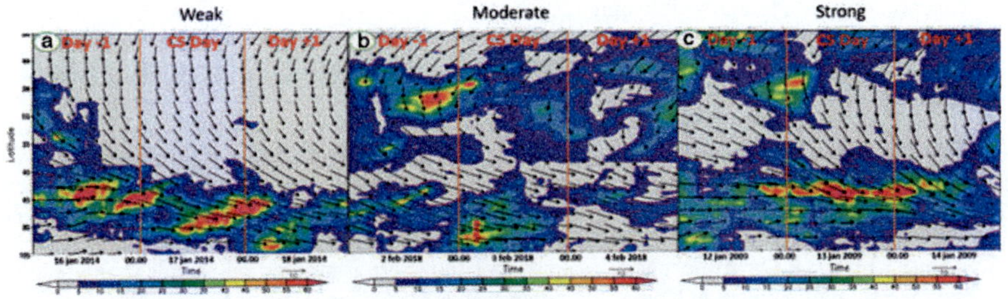

Fig. 7 Time-latitude cross sections of meridional wind and precipitation average of 100°–120°E when **a** weak, **b** moderate, and **c** strong cold surge from day-1 to day+1

4 Conclusion

The area that experiences an increase in rainfall decreases when the strength of the CS increases. However, rainfall intensity increased in the South China Sea, Karimata Strait, and the Java Sea. The increase in rainfall intensity and the narrowing of the distribution area also occur when CS and MJO are active. In contrast to inactive MJO, although the intensity of rainfall increases, the distribution area narrows during moderate CS and expands again during strong CS. Therefore, the percentage of extreme rainfall events in weak and moderate CS is relatively high for the South China Sea and North Java Sea areas. As for the percentage of extreme rainfall in Strong CS, the coverage is wider distributed in the South China Sea, the Karimata Strait to the Java Sea, and the southwest of the island of Java.

The lag time analysis of each CS has differences related to the increase in rainfall and the potential for CS two days before to 2 days after. When CS is weak, there is generally an increase in rainfall from day-1 to day+1. While for moderate CS, the increase in rainfall starts when a cold surge occurs until day+2. Lastly, when CS is strong, an increase in rainfall and potential for cold surges has occurred from day-2 to day+2. Moisture transport for weak and moderate CS is influenced by flow from the Indian Ocean, while it is not for strong CS. Based on the case study, strong CS can increase rainfall more in the Java Sea area than Java Island, whereas weak and moderate CS increase rainfall more in Java Island than in the Java Sea from day-1 to day+1.

References

1. Fauzi, R.R., Hidayat, R.: Role of cold surge and MJO on rainfall enhancement over Indonesia during East Asian winter monsoon. In: IOP Conference Series: Earth and Environmental Science. vol. 149(1), IOP Publishing (2018)
2. Lim, S.Y., Marzin, C., Xavier, P., Chang, C.P., Timbal, B.: Impacts of boreal winter monsoon cold surges and the interaction with MJO on Southeast Asia rainfall. J. Clim. **30**(11), 4267–4281 (2017)
3. Pang, B., Lu, R., Ling, J.: Impact of cold surges on the Madden-Julian oscillation propagation over the Maritime Continent. Atmosp. Sci. Lett. **19**(10), e854 (2018)
4. Chang, C.P., Harr, P.A., Chen, H.-J.: Synoptic disturbances over the equatorial South China Sea and western maritime continent during boreal winter. Monthly Weather Rev. **133**(3), 489–503 (2005)
5. Zhou, T.J., Yu, R.C.: Atmospheric water vapor transport associated with typical anomalous summer rainfall patterns in China. J. Geophys. Res. D: Atmos. **110**(8), 1–10 (2005)
6. Wicaksono, G.B., Hidayat, R.: Extreme rainfall in Katulampa associated with the atmospheric circulation. Proc. Environ. Sci. **33**, 155–166 (2016)
7. Xavier, P. et al.: Influence of Madden-Julian oscillation on Southeast Asia rainfall extremes: observations and predictability. Geophys. Res. Lett. **41**(12), 4406–4412 (2014)
8. Yulihastin, E., Hadi, T.W., Ningsih, N.S.: Diurnal rainfall propagation relate to cold surge-cold tongue interaction over the northern coast of West Java. In: IOP Conference Series: Earth and Environmental Science. vol. 303(1), IOP Publishing (2019)
9. Liebmann, B., Smith, C.A.: Description of a complete (interpolated) outgoing longwave radiation dataset. Bull. Am. Meteor. Soc. **77**(6), 1275–2127 (1996)

10. Koseki, S., Koh, T.Y. ,Teo, C.K.: Borneo vortex and mesoscale convective rainfall. Atmosp. Chem. Phys. **14**(9), 4539–4562 (2014)
11. Reed, A.T., Stansfield, A.M., Reed, K.A.: Characterizing long Island's extreme precipitation and its relationship to tropical cyclones. Atmosphere **13**(7), (2022)
12. Saragih, R.M., Fajarianti, R., Winarso, P.A.: Atmospheric study of the impact of Borneo vortex and Madden-Julian oscillation over Western Indonesian maritime area. J. Phys.: Conf. Ser. **997**(1), IOP Publishing (2018)
13. BMKG Homepage. https://dataonline.bmkg.go.id. Last Accessed 08 Oct 2022

Numerical Simulation of Low-Level Wind Shear at Soekarno-Hatta Airport Associated with Landward Propagation of Mesoscale Convective System

Dita Fatria Andarini, Muhammad Arif Munandar, and Ibnu Fathrio

Abstract Low-level wind shear (LLWS) is one of meteorological phenomena that can produce major risks for aircraft. This study aims to investigate the occurrence of low-level wind shear in Soekarno-Hatta Airport, Jakarta (12 December 2019), using a high-resolution numerical weather prediction of the weather research forecasting (WRF) model. Moreover, the hourly rainfall data from Global Satellite Mapping of Precipitation (GSMaP) was utilized to analyze the convective system. According to pilot reports (PIREPS), strong LLWS was recorded at around 10:00–11:00LT. The rainfall distribution of GSMaP confirmed that the LLWS event occurred with the presence of a convective system that propagated from the North of Java Island. The maximum rainfall was also observed around the airport, with an intensity of about 30 mm. The WRF model successfully simulated the phase and amplitude of precipitation to explain this convective activity. In addition, there were two LLWS events observed from the model results; 09:50–10:00LT and 10:30LT in the east and west airport, respectively. The surface wind speed of more than 10 m/s associated with strong downdraft, divergence, and high magnitude of radar reflectivity (>40 dB) induced those LLWS processes.

1 Introduction

Vertical wind shear, defined as the variation of wind direction and speed by height, is a crucial parameter in the atmosphere that could be associated with small to large-scale atmospheric phenomena. At low-level altitudes, the occurrence of significant

D. F. Andarini (✉) · I. Fathrio
National Research and Innovation Agency, Jakarta, Indonesia
e-mail: dita005@brin.go.id

I. Fathrio
e-mail: ibnu.fathrio@brin.go.id

M. A. Munandar
Indonesian Agency for Meteorological, Climatological and Geophysics, Jakarta, Indonesia

A. Basit et al. (eds.), *Proceedings of the International Conference on Radioscience, Equatorial Atmospheric Science and Environment and Humanosphere Science*, Springer Proceedings in Physics 290, https://doi.org/10.1007/978-981-19-9768-6_34

vertical wind shear could be dangerous for aircraft. It could happen when the airplane experiences a microburst produced below the base of a thunderstorm. The microburst causes the plane to encounter strong headwinds and tailwinds and displaces the aircraft distance from the ideal take-off and landing path [1].

The previous studies have shown how improvement of numerical weather prediction (NWP) skill results in better low-level wind shear (LLWS) prediction. Storm and Basu (2010) employed weather research forecasting (WRF) simulation to forecast low-level wind shear in the United States Great Plains as an exchange for costly observation and better spatial–temporal coverage [2, 3]. Their results revealed the dependency of the model on utilizing a planetary boundary layer scheme in simulating the wind shear. Moreover, [4] and [5] also used a high-resolution model up to 200 m to predict airflow variation at the Hong Kong International airport as a possible cause of low-level wind shear occurrence. Similarly, [6] evaluated the output of high spatio-temporal resolution of NWP to predict LLWS in the runway corridor of Hon Kong International Airport. It depicted positive skills for short-term prediction.

In terms of the implementation of data assimilation to improve the prediction results, [7] successfully applied LIDAR to improve WRF's ability to detect LLWS events. They highlighted that LIDAR data assimilation could improve the forecast for a short duration after the events, while increased vertical level had less contribution to improving the forecast result. A more recent study by Choi et al. [8] also implemented statistical analysis of ensemble model output to simulate the low-level wind shear at several airports in South Korea, where the results showed a better prediction than the raw ensemble method.

Indonesia is an archipelago country, consisting of five major islands and thousands of small islands. Thus, aircraft is a vital transportation means to connect all areas of the nation. More than 250 airports are available in Indonesia for the domestic and international flights. In this case, information on flight weather disturbances, especially LLWS, is crucial in supporting the safety of Indonesian aviation. However, there are only a few studies that utilize the use of NWP and observations in predicting LLWS in Indonesia. One of them is the research conducted by Sasmito et al. [9], which analyzed the Lion Air plane crash in 2013 and showed the potential impact of LLWS. Therefore, this study re-simulated one of the LLWS events related to mesoscale convection activity on 12 December 2019, around 10:00–11:00 local time (LT) at Soekarno-Hatta International Airport, Jakarta. This study aims to evaluate the skill of WRF models run in high spatial resolution for convective-induced LLWS cases in Jakarta airport that has not been investigated in the previous studies.

2 Data and Method

WRF experiments were applied to simulate LLWS events on 12 December 2019, around 10:00–11:00LT, recorded by a pilot report. Unfortunately, no maximum wind data was obtained when an LLWS disturbance occurred from the LLWS warning system at the airport. In detail, the WRF model configuration consists of 3 nested domains with a spatial resolution of 9 km, 3 km, and 1 km, respectively (see Fig. 1). The WRF model configuration applied in this study was the WRF Single Moment 6 (WSM6) microphysics scheme and Yonsei University planetary boundary layer scheme. Regarding the cumulus scheme, the Betts-Miller Janjic scheme was implemented in Domain 1 (D01) and Domain 2 (D01). Meanwhile, Domain 3 (D03) used no cumulus scheme to simulate convection explicitly. These configurations had successfully implemented to simulate the propagation of convective systems that induced heavy rainfall over southern Sumatera [10] and diurnal precipitation over the Maritime Continent [11].

Furthermore, the evaluation was conducted only on the D03, which is assumed to have better skill in elucidating the convective system as the cause of the LLWS. In addition, the Air Force Weather Agency (AFWA) diagnostic tools in WRF were also applied for further analysis. It allows us to analyze the weather disturbances on the aircraft from specific variables such as low-level wind shear, wind gust, and turbulence [12].

In order to evaluate the convective system during the LLWS event, gridded rainfall data from the Global Satellite Mapping of Precipitation gage-corrected (GSMaP_gage) standard version 7 was employed for the period of 12 December 2019, 04:00LT to 14:00LT [13]. This data has 0.1×0.1 degrees of horizontal resolution and a 1-h interval. The GSMaP precipitation was used as a reference to evaluate the convective system that occurred during the incidence. In addition, this study also utilized the horizontal wind component data in surface layer from the European

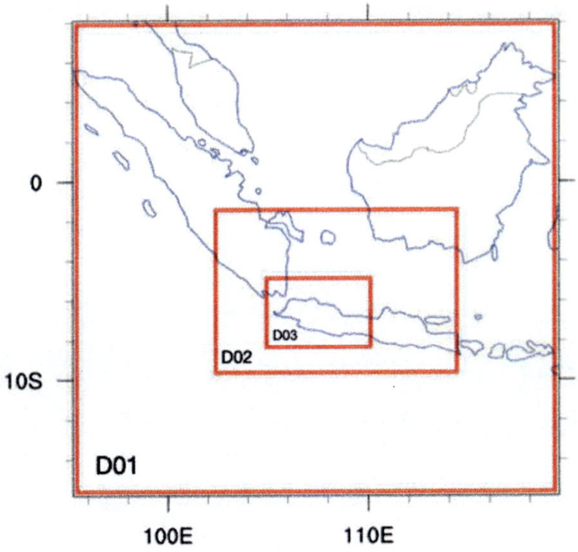

Fig. 1 Simulation domain of WRF model. The red boxes represent domain 1 (D01), domain 2 (D02), and domain 3 (D03) for 9, 3 and 1 km resolution, respectively

Center for Medium-Range Weather Forecast reanalysis 5 (ERA-5) [14]. The spatial resolution of this data is 0.25 × 0.25 degree, while the temporal resolution is 1 h.

3 Results and Discussion

3.1 Mesoscale Convective System

A westward propagation of the mesoscale convective system (MCS) was detected originating from the Java Sea, north of central Java, from the night to the morning, as shown from the spatial distribution of hourly rainfall in Fig. 2. The convective system initiated at around 04:00LT. It then reached its mature stage next to the north of western Java at 07:00LT. Three hours later, the southern part of this mesoscale convective system propagated southwestward and hit the Northern part of Java Island, including Soekarno-Hatta airport. The maximum precipitation was observed at about 30 mm per hour when it reached the airport. Eventually, the MCS decayed at 14:00 LT. Generally, the MCS activity induced heavy rainfall over the airport at around 10:00LT.

In 09:50–10:30 LT, pilot reports informed that the strong low-level wind shear was identified. Unfortunately, there were no actual wind speed data at the airport during that period. According to the horizontal wind components from ERA-5 data, a strong northerly surface wind occurred over the study area since 04:00LT. This condition was persistent and it increased at about 10:00LT and 12:00LT with a magnitude of more than 5 m/s. Moreover, the peak of precipitation was also found in the same period. As a result, it confirmed that low-level wind shear was generated by the convective systems. This is also in agreement with the previous study that the unstable

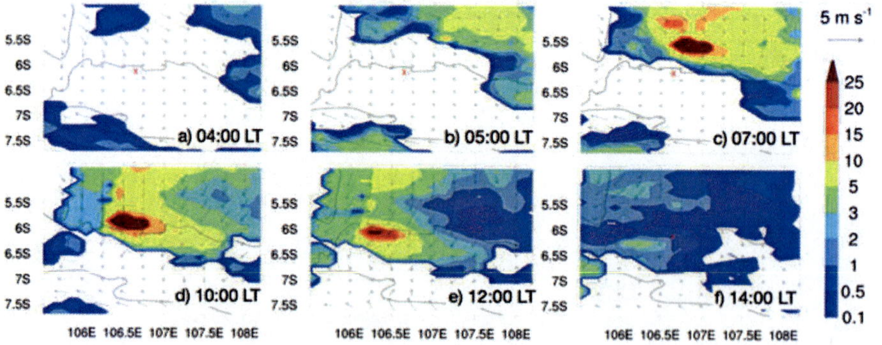

Fig. 2 Spatial distribution of precipitation from GSMaP superimposed with ERA-5 surface wind on 12 December 2019 at **a** 04:00 LT and **b** 05:00, and 12 December 2019 at **c** 07:00 LT, **d** 10:00 LT, **e** 12:00 LT, and **f** 14:00 LT. X marker denotes the location of the airport

atmospheric conditions are conducive to convective cloud development, and hence, are favorable for the LLWS process [15].

3.2 WRF Simulation

The propagation of the convective system was clearly seen from the 1 km rainfall simulated by the WRF model (Fig. 3). Firstly, the convective system concentrated around the north of Central Java at around 05:00LT. Then, it propagated to the southwest (land area) and reached a peak in the airport at about 10:00LT. In this stage, the rainfall reached its maximum value of around 30 mm, which is consistent with the observation data, as shown in Fig. 2. Eventually, the convection system began to decay in the next two hours. From this result, it is worth noting that the WRF model successfully simulated the evolution of the convective system both in amplitude and phase.

To investigate a more detailed analysis of the LLWS event induced by the convection system, this study also explored other parameters simulated by a high-resolution WRF model, such as surface wind and derived radar reflectivity. Figure 4 shows the spatial distribution of derived radar reflectivity superimposed with the divergence and wind vector. The surface wind speed reached a value of greater than 10 m/s twice associated with strong downdraft, divergence pattern of surface wind, and high magnitude (> 40 dBz) of simulated radar reflectivity as the index of heavy precipitation. This pattern could be associated with a microburst that spreads out at the surface, generating a strong surface wind speed or LLWS. However, the recorded downdraft speed seemed weaker than the previous study [16], and models simulated the location of LLWS events 5–10 km from the airport.

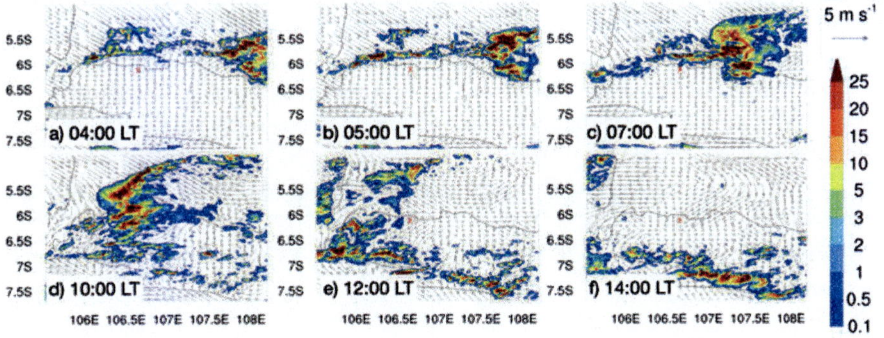

Fig. 3 Evolution of rainfall (filled contour) and surface wind (vector) simulated by the WRF model on 12 December 2019 at **a** 04:00 LT and **b** 05:00 LR, **c** 07:00 LT, **d** 10:00 LT, **e** 12:00 LT, and **f** 14:00 LT. X marker denotes the location of the airport

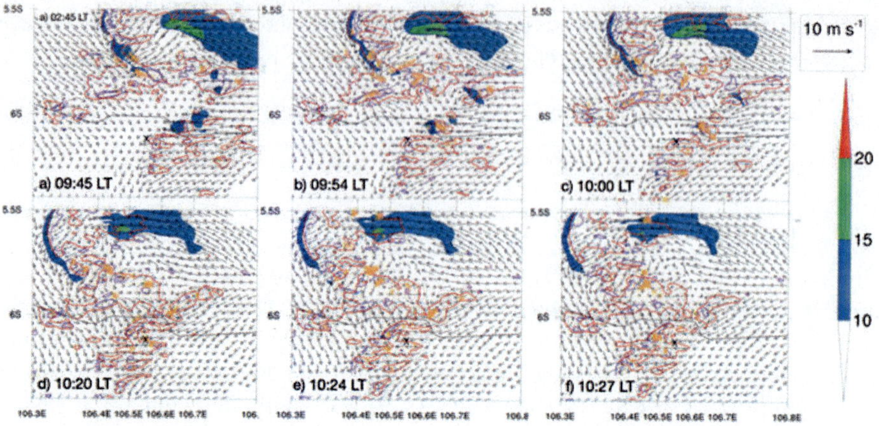

Fig. 4 Spatial distribution of wind speed (filled contour) and surface wind divergence (solid purple line) on 12 December 2021 at **a** 09:45 LT, **b** 09:54 LT, **c** 10:00 LT, **d** 10:20 LT, **e** 10:24 LT, and **f** 10:27 LT. Only values (multiplied by 1000 for clarity) of 1.5, 3 and 6 (units in m^{-1}) are shown for surface wind divergence. Location of vertical velocity less than 2 m/s in 0–2 km height is denoted by the orange cross marker. Derived radar reflectivity greater than 40 dBz at 2 km height is denoted by the red line contour. The first LLWS event and second LLWS are represented in top panels (**a**, **b**, **c**) and lower panels (**d**, **e**, **f**), respectively. All variables are estimated by using AFWA diagnostic tool. X marker denotes the location of the airport

There were two LLWS events that occurred in the period of 09:45LT–10:30LT. The first LLWS was identified around the east of the airport at about 09:50–10:00 LT, indicated by the strong easterlies with the magnitude greater than 15 m/s (Fig. 4a–c). The high amplitude of radar reflectivity also dominated the western area of the airport. Moreover, the second LLWS was observed next to the west of the airport at around 10:24LT, as shown in Fig. 4d–f. This consistent high magnitude of radar reflectivity and divergence favored the environment to produce the second LLWS event.

Furthermore, Fig. 5 displays the distribution of LLWS and wind gusts during the first and second LLWS events. A strong LLWS and maximum wind gust near the airport with values greater than 10 m/s (~20knots) occurred during the first event (Fig. 5a-c). Similarly, the high value of LLWS was also found nearby the airport (next to the west) in the second event, as depicted in Fig. 5d-f. Those results are consistent with the high intensity of reflectivity and strong easterly surface wind during the two LLWS events, as previously explained in Fig. 4.

4 Conclusion

WRF simulation on high spatial resolution (1 km) has successfully simulated the evolution of mesoscale convective systems that triggered strong low-level wind

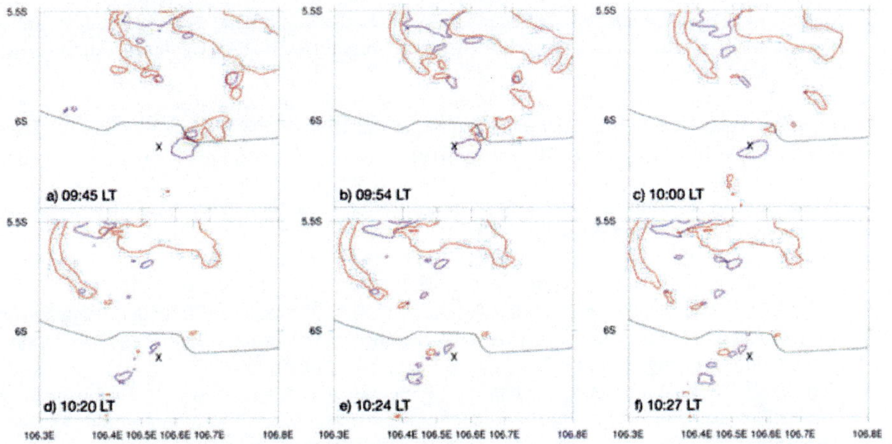

Fig. 5 Contour of wind gust (red) and low-level wind shear (purple) that represent wind speed of 10 m/s for the first event at **a** 09:45 LT, **b** 09:54 LT, **c** 10:00 LT, and the second event at **d** 10:20 LT, **e** 10:24 LT, and **f** 10:27 LT on 12 December 2019. All variables are estimated by using AFWA diagnostic tool. X marker denotes the location of the airport

shear in the vicinity of Jakarta International Airport around 09:50–10:30LT on 12 December 2019. The phase and maximum precipitation were simulated comparably with the observation from the GSMaP data. Further analysis of LLWS and wind gust was carried out by utilizing AFW diagnostic tool to reveal the presence of significant LLWS and wind gust. The model results identified two LLWS events; 09:50–10:00LT at the east and around 10:24LT at the west of the airport. Those events were modulated by more than 10 m/s of the surface wind associated with strong downdraft, divergence, and high magnitude of radar reflectivity (>40 dB).

The location of the LLWS event cannot be determined since there was no available data. Nevertheless, WRF models could predict the possible location of LLWS events 5–10 km away from the airport, which should be considered as an alert for the take-off and landing of the aircraft. This could also be partly caused by the selection of parameterization schemes in the parent domains, especially the cumulus scheme as a crucial factor in determining the strength, timing, and location of convection. Therefore, sensitivities studies of the model should be further investigated in future studies.

References

1. HKO, I.: Windshear and Turbulence in Hong Kong–information for pilots. Hong Kong Observatory and International Federation of Air Line Pilots' Associations (2005)
2. Storm, W., Basu, S.: The WRF model forecast-derived low-level wind shear climatology over the United States Great Plains. Energies **2010**(3), 258–276 (2010). https://doi.org/10.3390/en3020258

3. Skamarock, W.C., Klemp, J.B., Dudhia, J., Gill, D.O., Liu, Z., Berner, J., Wang, W. et al.: In: A description of the Advanced Research WRF Model Version 4, National Center for Atmospheric Research, Boulder, CO, USA, pp. 145. (2019)
4. Chan, P.W., Hon, K.K.: Performance of super high resolution numerical weather prediction model in forecasting terrain-disrupted airflow at the Hong Kong international airport: case studies. Meteorol. Appl. **23**(1), 101–114 (2016)
5. Chan, P.W., Hon, K.K.: Observation and numerical simulation of terrain-Induced windshear at the Hong Kong International Airport in a planetary boundary layer without temperature inversions. Adv Meteorol Artic ID (2016)
6. Hon, K.: Predicting low-level wind shear using 200-m-resolution NWP at the Hong Kong international airport. J. Appl. Meteorol. Climatol. **59**(2), 193–206 (2020)
7. Li, L., Xie, N., Fu, L., Zhang, K., Shao, A., Yang, Y., Ren, X.: Impact of lidar data assimilation on low-level wind shear simulation at Lanzhou Zhongchuan international airport, China: a case study. Atmosphere **11**, 1342 (2020). https://doi.org/10.3390/atmos11121342
8. Choi, H.-W., Kim, Y.-H., Han, K., Kim, C.: Probabilistic forecast of low level wind shear of Gimpo, Gimhae, Incheon and Jeju international airports using ensemble model output statistics. Atmosphere **12**, 1643 (2021). https://doi.org/10.3390/atmos12121643
9. Sasmito, A., Permana, D.S., Praja, A.S., Haryoko, U.: Pengaruh Microburst dan Low-Level Wind Shear (LLWS) Pada Kasus Kecelakaan Pendaratan Pesawat Lion Air Tanggal 13 April 2013 Di Bali. Jurnal Meteorologi dan Geofisika **21**(1), 1–8 (2020)
10. Yulihastin, E., Fathrio, I., Nauval, F., Saufina, E., Harjupa, W., Satiadi, D., Nuryanto, D.E.: Convective cold pool associated with offshore propagation of convection system over the East Coast of Southern Sumatra, Indonesia. Adv. Meteorol. (2021)
11. Fonseca, R.M., Zhang, T., Yong, K.-T.: Improved simulation of precipitation in the tropics using a modified BMJ in the WRF model. Geoscientif. Model Developm. **8**(9), 2915–2928 (2015)
12. Creighton, G., Kuchera, E., Adams-Selin, R., McCormick, J., Rentschler, S., Wickard, B.: AFWA diagnostics in WRF (2014)
13. Kubota, T., Shige, S., Hashizume, H., et al.: Global precipitation map using satellite-borne microwave radiometers by the GSMaP project: production and validation. IEEE Trans. Geosci. Remote Sens. **45**(7), 2259–2275 (2007)
14. Hersbach, H., Bell, B., Berrisford, P., Hirahara, S., Horányi, A., Muñoz-Sabater, J., Nicolas, J., Peubey, C., Radu, R., Schepers, D., et al.: The ERA5 global reanalysis. Q. J. R. Meteorol. Soc. **146**, 1999–2049 (2020). https://doi.org/10.1002/qj.3803
15. Lin, C., Zhang, K., Chen, X., Liang, S., Wu, J., Zhang, W.: Overview of low-level wind shear characteristics over Chinese mainland. Atmosphere **12**(5), 628 (2021)
16. Bolgiani, P., Fernández-González, S., Valero, F., Merino, A., García-Ortega, E., Sánchez, J.L., Martín, M.L.: Simulation of atmospheric microbursts using a numerical mesoscale model at high spatiotemporal resolution. J. Geophys. Res.: Atmosp. **125**(4), e2019JD031791 (2020)

Prediction of CENS, MJO, and Extreme Rainfall Events in Indonesia Using the VECM Model

Mutia Yollanda, Wendi Harjupa, Dodi Devianto, Dita Fatria Andarini, Fadli Nauval, Elfira Saufina, Anis Purwaningsih, Wendi Harjupa, Trismidianto, Teguh Harjana, Risyanto, Fahmi Rahmatia, Ridho Pratama, and Didi Satiadi

Abstract A study of two global circulations crossing Indonesia, the Cross-Equatorial Northerly Surge (CENS) and the Madden–Julian Oscillation (MJO), can trigger extreme rainfall occurrences over Indonesia, has been carried out. The purpose of this study is to predict the presence of these circulations and the impact of these circulations on the distribution of rain throughout Indonesia. The territory of Indonesia was divided into 14 clusters to determine the effect of CENS and MJO and observe the relation of which with extreme rainfall patterns in each area. The method used to predict these three phenomena is a statistical approach called vector autoregressive (VAR), which indicates the occurrence of global circulation and extreme rain in Indonesia. In the studied area, there is a cointegration or long-term relationship between variables, which can be anticipated using the vector error correction model (VECM). It is also found that the presence of CENS and MJO circulation can be predicted with the model error rate measured based on the MAPE values of 1.8518 and 0.0619. When CENS is active (inactive), rainfall occurrences occur more (less) in the southern part of Indonesia, respectively. Moreover, the VECM model provides a pretty good prediction for the variables CENS, MJO, and rainfall intensity in 14 clusters of Indonesia.

1 Introduction

Hydrometeorological disaster data recorded by the National Disaster Management Agency (BNPB) from 2000 to 2022 show that hydrometeorological disasters such as floods and landslides are the most frequent disasters that have a significant effect on

M. Yollanda (✉) · D. Devianto
Department of Mathematics and Data Science, Andalas University, Padang, Indonesia
e-mail: mutiayollanda@gmail.com

W. Harjupa · D. F. Andarini · F. Nauval · E. Saufina · A. Purwaningsih · W. Harjupa · Trismidianto · T. Harjana · Risyanto · F. Rahmatia · R. Pratama · D. Satiadi
Research Center for Climate and Atmosphere, National Research and Innovation Agency (BRIN), Bandung, Indonesia

A. Basit et al. (eds.), *Proceedings of the International Conference on Radioscience, Equatorial Atmospheric Science and Environment and Humanosphere Science*, Springer Proceedings in Physics 290, https://doi.org/10.1007/978-981-19-9768-6_35

life (https://gis.bnpb.go.id/ [1]). For example, based on the BNPB statement, from January to September 2020 alone, 99% of disasters were recorded as hydrometeorological disasters. The highest number of incidents was the flood as many as 791 times and followed by other disasters, including 573 tornadoes, 387 landslides, 314 forests and land fires, 26 tidal waves or abrasion, 22 droughts, 13 earthquakes, and volcanic eruptions. 5. The total number of natural disasters is 2131 incidents. A number of these incidents have a consequence of lost life and property. The BNPB recorded 322 fatalities, 454 were injured, and 4,481,641 were displaced and affected by the disaster. While infrastructure damage, the disaster impacted the housing sector, the 31,749 housing units, 627 educational facilities, 653 worship facilities, and 128 health facilities were damaged (cnbcindonesia.com 01/10/22) [2].

One of the causes of hydrometeorological disasters is the high intensity of rainfall that occurs. Rain is the result of a long process. The rainfall process results in the transformation and retransformation of mechanical energies. The influence of weightlessness keeps the clouds hanging over the sky, resulting in freefall until their growing weight is counterbalanced by gravity. Consequently, huge clouds are broken into water molecules that hit the earth as rain after traveling under gravity through the troposphere, acquiring heat via air friction, and rising in tropospheric temperature with height [3]. The previous research found that the high rainfall in the Indonesian cluster was caused by the large number of atmospheric circulations that occurred in both the Cross-Equatorial Northerly Surge and the Madden–Julian Oscillation [4–10].

The Madden–Julian Oscillation (MJO) is associated with enhanced low-level convergence and ascent in the synoptic scale, while winds in the upper troposphere diverge from the updraft. This circulation is accompanied by enhanced deep convection. Its cycle is split up into eight phases, each of which corresponds to 1/8 of the phases. An individual MJO event can last anywhere from 30 and 60 days. The increase and decrease in rainfall in the Indonesia Maritime Continent (IMC) also depend on whether the MJO convection center stops over the eastern Indian Ocean, passes through the IMC, or moves to the Western Pacific [11]. Teleconnection associated with the MJO is strongest when the MJO convection passes through Indonesia [12]. Indonesia is a known barrier to the eastward propagation of the MJO [13].

The CENS is a dominant low-level circulation that triggers deep convection and eventually leads to heavy rainfall and flood event over the equator cluster, including the northern part of Java Island [14–17]. For example, the study investigated that the strongest northerly wind played a critical role in the devastating flood event in Jakarta [14]. Similarly, a strong CENS crossing the equator line induced extreme rainfall, and a major flood occurred over Jakarta in February 2013 [18].

Many studies have been carried out to see the effect of oscillations that pass through Indonesia on the occurrence of rain in Indonesia, including analyzing the combined impact of several of these oscillations, which aims to better understand the distribution of rain throughout Indonesia and of course can predict the rainfall that occurs. Rain prediction has been developed using many methods, including a numerical weather prediction (NWP) [19]. Prediction with NWP requires a highly capable computer to carry out the prediction process for a large area and considerable funding. For this reason, other methods are needed to predict extreme rain events with resources that are not too large. Besides predicting rain events, the predictions of global events such as MJO and CENS are also needed to provide early warning that the circulation can produce rain with high intensity.

The VAR model is a development of the autoregressive (AR) time series model with more than one variable. All variables in the VAR model are assumed to be endogenous variables (variables whose value is determined by other variables in the model) and are interrelated. The application of VAR models to climate and weather changes attracts researchers to study past behavior, present, and future of the variables used and then used the VAR model to predict the value of the climate and weather variables for the future period [20–23]. With the case of cointegration between variables, the VAR model has not been able to model these variables, so a correction model is needed, namely the vector error correction model (VECM) for error correction in the previous VAR model. The VECM model is starting to attract attention from various fields because it can determine the relationship between one variable and another. In the case of COVID-19, the VECM technique is approached to see the relationship between air pollution and COVID-19 hospitalizations in the state of Kuwait [24].

This paper will use the rainfall data, CENS, and MJO as endogenous and interrelated variables. Based on the crucial things above, this paper aims to use the VAR method to predict rain events, CENS, and MJO to provide alerts to related parties. This paper is divided into several parts, namely Sect. 2, describing the data and methods. Section 3 contains the results and discussion, and Sect. 4 includes the conclusions of this study.

2 Data and Methodology

2.1 CENS, MJO, and Rainfall Data

The rainfall data used in this study were obtained from the Global Satellite Mapping and Precipitation (GSMaP) gage-corrected (GSMaP_gage) version-7 from 2009 to 2019. GSMaP gage is a standard product of GSMaP that has been corrected by daily rain gages from NOAA CPC Gage-Based Analysis [25]. This data have a horizontal resolution of 0.1 $\times$ 0.1 degrees and can be accessed from https://sharaku.eorc.jaxa.jp/GSMaP was employed to deploy the vector autoregressive (VAR) model.

To identify CENS indices, we used the daily wind component data (mean wind and wind stress field) from advanced scatterometer (ASCAT) and daily ASCAT (DASCAT). ASCAT is an actual aperture radar, operating at 5.255 GHz frequency (C-band) and using vertically polarized antennas [26]. It transmits a long pulse with linear frequency modulation ("chirp"). These data have a spatial resolution of 0.25° $\times$

0.25° degrees and can be accessed from ftp://ftp.ifremer.fr/ifremer/cersat/products/gridded/. Moreover, MJO indices data were obtained from the Climate Prediction Center (CPC) of the National Oceanic and Atmospheric Administration (NOAA) at https://www.cpc.ncep.noaa.gov/products/precip/CWlink/MJO/. The MJO Index-2 was selected over 100°E over 18 years from 2001–2018.

2.2 *Building of Vector Error Correction Model*

2.2.1 Stationarity Data

Before forming the vector autoregressive (VAR) model, the time series data first perform some tests on the initial data, whether the data are stationary or not. Therefore, the stationarity of time series data concerning variance or mean value can be carried out successively using the Box-Cox transformation [27] and the augmented Dickey–Fuller (ADF) test [28].

2.2.2 Model Vector Autoregressive (VAR)

The VAR model is an extension of the autoregressive (AR) time series model with more than one variable. All variables in the VAR model are assumed to be endogenous variables (variables whose value is determined by other variables in the model) and are interrelated. The general form of the VAR model with a lag p, VAR(p), can be calculated as follows [28]:

$$X_{p,t} = \phi_0 + \sum_{i=1}^{p} \phi_i X_{i,t-i} + e_t, i = 1, 2, \ldots, p, t = 1, 2, \ldots \tag{1}$$

where

$X_{p,t}$ is the dependent variable $(X_{1,t}, X_{2,t}, \ldots, X_{p,t})$ sized $p \times 1$,
$X_{i,t-1}$ is the independent variable $(X_{1,t-i}, X_{2,t-i}, \ldots, X_{p,t-i})$, sized $p \times 1$,
ϕ_0 is vector intercept size $p \times 1$,
ϕ_i is a coefficient matrix of size $p \times p$ for each $i = 1, 2, \ldots, p$,
e_t is the remainder vector $(e_{1,t}, e_{2,t}, \ldots, e_{p,t})$ sized $p \times 1$.

2.2.3 Cointegration Test

If data are cointegrated, then the modeling for that data cannot use the VAR model. The basis of cointegration is that many time series data can deviate from their average in the short term but move together. The hypothesis in Johansen's test is as follows:

H_0 : There are as many as r, where $r = 0, 1, \ldots, k - 1$ cointegration equation (no cointegration or long-term relationship between variables).

H_1 : There are cointegration equations (cointegration occurs between variables).

2.2.4 Vector Error Correction Model (VECM)

VECM is a restricted form of VAR due to the existence of non-stationary but cointegrated data forms. The VECM specification specifies the long-term relationship of endogenous variables to converge to their cointegration relationship but still allows the presence of short-term dynamics. The VECM has the following equation:

$$\Delta X_{p,t} = \phi_{p,0} + \Pi X_{p,t-1} + \sum_{i=1}^{p-1} \Phi_i^* \Delta X_{p,t-i} + e_t \tag{2}$$

where $X_{p,t} = X_{p,t-1} + \Delta X_{p,t}$, $\Phi_j^* = -\sum_{i=j+i}^{p} \Phi_i$, $j = 1, 2, \ldots, p - 1$, and $\Pi = -(1 - \Phi_1 - \Phi_2 - \ldots - \Phi_p)$.

2.2.5 Optimum Lag Determination

VAR estimation is very sensitive to the lag length used. Therefore, it is necessary to know the optimal lag before estimating the VAR. Determination of the optimal amount of lag can be determined using Akaike information criterion (AIC), Schwarz information criterion (SC), final prediction error (FPE), or Hannan Quinnon (HQ). Determination of the optimal lag using the information criteria by selecting the measures with the smallest value [29]. This optimal lag is applied to eliminate the autocorrelation problems in the VAR model so that they will no longer arise.

2.2.6 Impulse Response Functions (IRFs)

After modeling using the VAR model, it is necessary to have a method that can characterize the dynamic structure generated by VAR. IRF shows the response of each endogenous variable over time to the shock of the variable itself and other endogenous variables. Thus, IRF is used to see the contemporary effect of a dependent variable if it gets a shock or innovation from the independent variable by one standard deviation.

2.2.7 Mean Absolute Percentage Error (MAPE)

MAPE is a useful measure to measure the error of the estimated value of the model. The error in the estimated value of the model expresses an average absolute percentage of residual, which is written as follows [30]:

$$MAPE = \frac{1}{N}\sum_{t=1}^{N}\left|\frac{Y_t - \hat{Y}}{Y_t}\right| \times 100\% \tag{3}$$

where N is the number of predictions made, Y_t is the actual data, and $\hat{Y}_t$ is the data forecast using the VAR model. The MAPE value as the modeling accuracy is indicated as "excellent" if the MAPE value < 10%, "good" if the MAPE value is 10–20%, "sufficient" if the MAPE value is 20–50%, and "inaccurate" if the MAPE value > 50%.

2.3 *Rain Cluster Division*

In finding out the rainfall distribution in Indonesia, the Indonesian cluster is divided into 14 clusters. Because we use global-scale atmospheric circulation, the area of the cluster division is quite large. Some parts of Indonesia are not included in the territorial division due to many ocean areas, such as Maluku and East Nusa Tenggara (Fig. 1 and Table 1).

The research area is divided to be 14 clusters. The division is just to simply analysis for the method that we used. It is very important to make the cluster size smaller, which can impact our prediction result. The prediction will be more precise. However, it will generate many clusters, and of course, it will cause the analysis to become more difficult since we will discuss the impact of circulations on the many clusters.

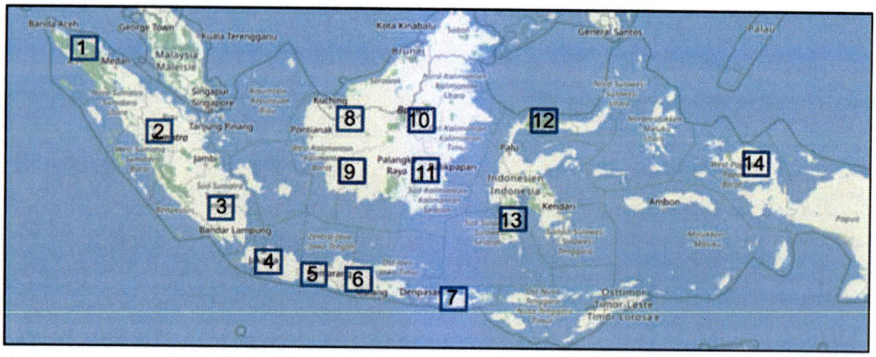

Fig. 1 Division of research area

Table 1 Study area coordinate

Locations	Longitude 1	Longitude 2	Latitude 1	Latitude 2
1	96.3	97.3	3.8	4.8
2	100	101	0	−1
3	104	105	−3.8	−4.8
4	107	108	− 6.3	−7.3
5	109	110	−6.7	−7.7
6	111	112	−7	−8
7	117	118	−8	− 9
8	110.1	111.1	1.2	0.2
9	110.1	111.1	−1.2	−2.2
10	114	115	1.2	0.2
11	114	115	−1.2	−2.2
12	121	122	1.2	0.2
13	119.6	120.6	−4	−5
14	132	133	−0.7	−1.7

3 Result and Discussion

In this chapter, we will discuss vector autoregressive modeling on MJO amplitude, CENS, and rainfall time series data in Indonesia which is partitioned into 14 clusters.

3.1 Time Series Data Plot

The data used are daily amplitude, CENS, and rainfall data from 14 clusters in Indonesia starting from January 2001 to December 2019 with 6939 data. The initial step is to plot daily data on amplitude, CENS, and rainfall from 14 clusters in Indonesia which can be shown in Fig. 2.

Figure 2 shows Indonesia's amplitude, CENS, and rainfall plots have an up-and-down trend over time. It means that the time series data for amplitude, CENS, and rainfall in Indonesia are not stationary or do not fluctuate at a certain number. It shows a transformation step so that the time series data to be analyzed becomes stationary in terms of the mean and variance. The pattern of peak rainfall events in Fig. 2 shows a clear relationship between the CENS index and peak rainfall that clustered rainfall can be seen in 4, 5, 6 and 7.

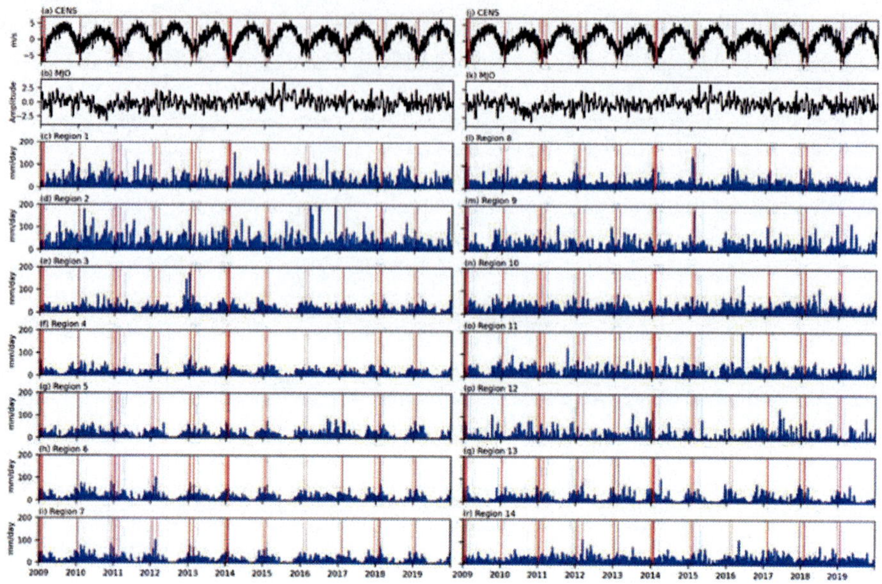

Fig. 2 Time series data plot

3.2 Data Transformation

The next step is to perform the Box-Cox transformation and augmented Dickey–Fuller test that will be displayed as follows:

Based on Table 2, because the statistical value of the ADF test is less than the critical point value, namely −1.96, the reject H_0. It means that all variables do not have a unit root, or in other words, the data are stationary with respect to the middle value. While the Box-Cox parameters contained in Table 2 can be used to transform the variance values.

Table 2 Box-Cox parameters and Dickey–Fuller augmented test statistical values

Variable	Parameter Box-Cox	ADF test	Variable	Parameter Box-Cox	ADF test
CENS	1.902137	−4.644505	Cluster 7	−0.8781	−13.28698
Amplitude	0.907747	−2.031191	Cluster 8	0.016774	−16.15948
Cluster 1	−0.40141	−2.645304	Cluster 9	−0.22655	−17.04338
Cluster 2	−0.23157	−14.82022	Cluster 10	−0.03125	−14.56749
Cluster 3	−0.30515	−4.958705	Cluster 11	−0.16825	−15.77734
Cluster 4	− 0.11661	−15.20612	Cluster 12	−0.58326	−13.1753

Table 3 Results of the AIC, HQ, SC, and FPE criteria

Lag	Criteria			
	AIC	HQ	SC	FPE
1	−5.5221E + 01	−5.5069E + 01	−5.4793E + 01*	1.0422E-24
2	−5.5365E + 01	−5.5070E + 01*	−5.4535E + 01	9.0249E-25
3	−5.5385E + 01	−5.4948E + 01	−5.4152E + 01	8.8474E-25*
4	−5.5373E + 01	−5.4794E + 01	− 5.3739E + 01	8.9460E-25
5	−5.5336E + 01	−5.4614E + 01	−5.3299E + 01	9.2870E-25
6	−5.5314E + 01	−5.4450E + 01	−5.2875E + 01	9.4905E-25
7	−5.5277E + 01	−5.4270E + 01	−5.2435E + 01	9.8509E-25

3.3 Determining Optimum Lag

The following are the scores for the Akaike information criterion (AIC), Schwarz information criterion (SC), Hannan-Quinn information criterion (HQ), and Akaike's final prediction error (FPE):

Table 3 shows that the lowest AIC values occur in lag 3, HQ in lag 2, SC in lag 1, and FPE in lag 3. With the diversity of the results of the optimal criteria, it is necessary to rank between criteria. After the average ranking is obtained for the four criteria, the best ranking result is lag 3. It means that the VAR model will be optimal at lag 3.

3.4 Cointegration Test

The cointegration test indicates the possibility of a long-term relationship between variables. A cointegrated pair of variables indicates a long-term relationship. The result of the cointegration test shows the trace statistic of 107.9833 which is greater than the critical values of 9.24. It means that the null hypothesis H_0 is rejected, or in other words, there are cointegration or long-term relationships between variables.

3.5 VECM

Due to the indication of cointegration, the VECM model will be formed as a model built after correcting the previous VAR model error. Let X_1 as CENS, X_2 as amplitude, and $X_i, i = 3, 4, \ldots, 16$ as the rainfall for each cluster in Indonesia, respectively. The following are the parameters of the VECM model.

Table 4 shows the parameter estimation of CENS, MJO, and rainfall of 14 clusters in Indonesia. The obtained parameter values are implemented to create the formula of VECM.

3.6 Test of Impulse Response Functions (IRFs)

Due to the indication of cointegration, the VECM model will be formed as a model that was built after correcting the previous VAR model error. The following are the parameters of the VECM model.

Figure 3 shows how the movement occurred due to the shock from the CENS phenomenon on rainfall in 14 clusters of Indonesia. It can be seen that before the movement of CENS, all clusters in Indonesia were at their equilibrium point. Then in the next period, several clusters began to show their response to the CENS phenomenon, both positive responses that occurred in clusters 1, 2, 3 and 4. In contrast, the rest gave a negative response to the CENS phenomenon. Then in the next period, it was seen that the movement of the amount of rainfall in 14 clusters of Indonesia began to reach the equilibrium point, which means that the response given to all 14 clusters began to reach stability by showing that the standard deviation was convergent toward a value of 0.

Figure 4 shows how the movement occurred due to the shock of the amplitude phenomenon on rainfall in 14 clusters of Indonesia. It can be seen that before the movement of amplitude, all clusters in Indonesia were at their equilibrium point. Then in the next period, several clusters began to show their response to the CENS phenomenon, both negative responses that occurred in the 6, 7, 8, 9, 10 and 13 clusters to the amplitude phenomenon. In the second period, it can be seen that there are all 14 cluster's rainfall responses to the amplitude phenomenon started to converge to point 0, except for clusters 10 and 11. In the movement given the amplitude phenomenon to rainfall in 10 clusters, rainfall in the cluster begins to show a negative response to rainfall. In contrast to the rainfall in cluster 11, the amplitude phenomenon began to respond positively. Then in the next period, it was seen that the movement of the amount of rainfall in 14 clusters of Indonesia began to reach the equilibrium point, which means that the response given to all 14 clusters began to reach stability by showing that the standard deviation was convergent toward a value of 0.

3.7 Comparison of the Original Data and the Estimated Data Using the VAR Model

Figure 5 shows how the data on the estimated CENS variables, amplitude, and intensity of rainfall in 14 clusters of Indonesia compare with the observed data. Based on Fig. 5, it can be seen that the data from the VECM modeling (blue) are getting close

Table 4 Parameter estimate of VECM for CENS, amplitude, and rainfall in Indonesia

Model	Parameter																
	X1	X2	X3	X4	X5	X6	X7	X8	X9	X10	X11	X12	X13	X14	X15	X16	ECT
X1	−2.52	–	–	0.002	–	0.004	–	–	−0.005	−0.005	−0.004	−0.007	-	−0.003	–	–	−0.006
X2	3.6e-04	0.002	–	–	–	–	–	–	–	–	–	–	–	–	–	9.8e-04	−4.5e-04
X3	–	–	−0.474	−0.031	–	–	−0.070	-	–	−0.061	–	0.008	-0.048	–	–	–	0.164
X4	–	–	0.043	−0.657	0.108	−0.099	–	0.137	–	−0.093	–	–	−0.118	–	–	−0.092	0.765
X5	–	–	0.027	0.036	-0.465	–	0.038	–	–	–	−0.026	-0.044	–	–	–	–	−0.094
X6	−0.615	–	–	0.022	–	−0.460	–	–	–	–	–	–	–	–	–	–	–
X7	−0.349	–	–	0.031	–	–	−0.499	–	–	–	–	–	–	–	–	–	−0.061
X8	−0.544	–	–	-0.042	–	–	–	−0.513	–	0.023	–	−0.042	–	–	–	–	−0.150
X9	–	–	–	–	–	–	–	–	−0.453	–	–	−0.036	0.045	–	0.047	–	−0.052
X10	0.737	–	0.023	0.057	0.059	–	–	−0.049	–	−0.459	0.031	−0.104	0.037	–	−0.049	0.033	−0.086
X11	0.716	–	–	0.070	0.049	–	–	−0.048	−0.058	–	−0.466	−0.109	–	–	–	0.039	−0.184
X12	–	–	–	0.184	0.033	0.054	0.075	−0.152	−0.059	0.079	–	−0.785	0.050	−0.046	−0.057	0.131	−0.592
X13	–	–	–	0.086	–	–	0.071	-0.069	−0.071	–	0.04	−0.068	-0.483	–	−0.057	0.053	−0.188
X14	–	–	−0.031	–	–	–	–	–	–	–	–	–	–	−0.480	–	–	–
X15	−0.447	–	−0.025	–	–	–	–	–	–	0.037	0.04	–	0.037	–	−0.441	−0.029	–
X16	0.172	–	–	−0.054	–	–	−0.040	0.049	–	−0.045	–	0.047	–	0.035	0.043	−0.498	–

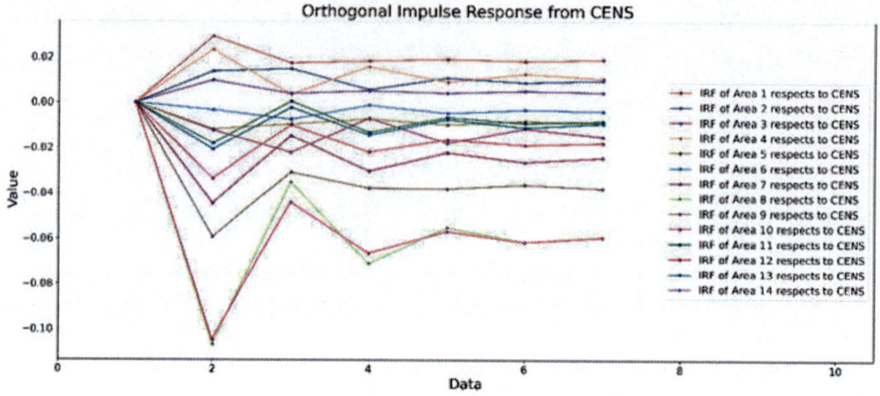

Fig. 3 Effect of CENS on rainfall intensity in 14 clusters

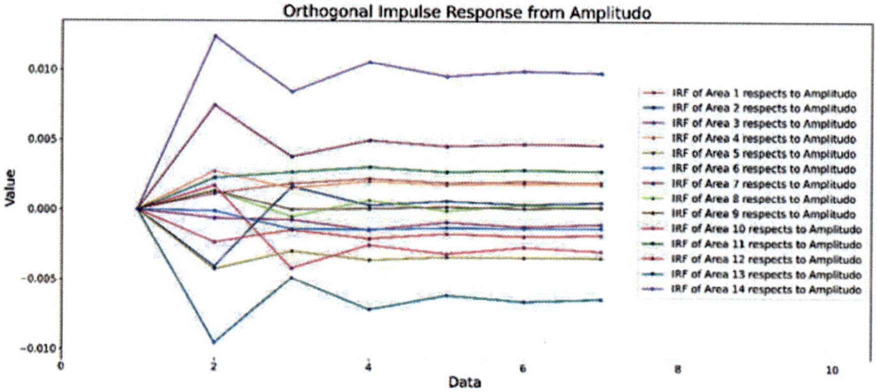

Fig. 4 Effect of amplitude on rainfall intensity in 14 clusters

to the observed value (red). It indicates that the VECM modeling results obtained are pretty good for predicting the values of CENS, amplitude, and intensity of rainfall for 14 clusters in Indonesia for the next period.

3.8 VECM Modeling Accuracy Values

To determine the accuracy of the VECM model, the following is the MAPE value for each variable in Table 5.

In explaining the model's accuracy for each variable, the MAPE value is required. The MAPE value that is lower than 10 indicates that the model for each variable has better performance to be applied in forecasting for the next period. Based on Table 5, all MAPE values are approaching zero or lower than 10% so that all the VECM

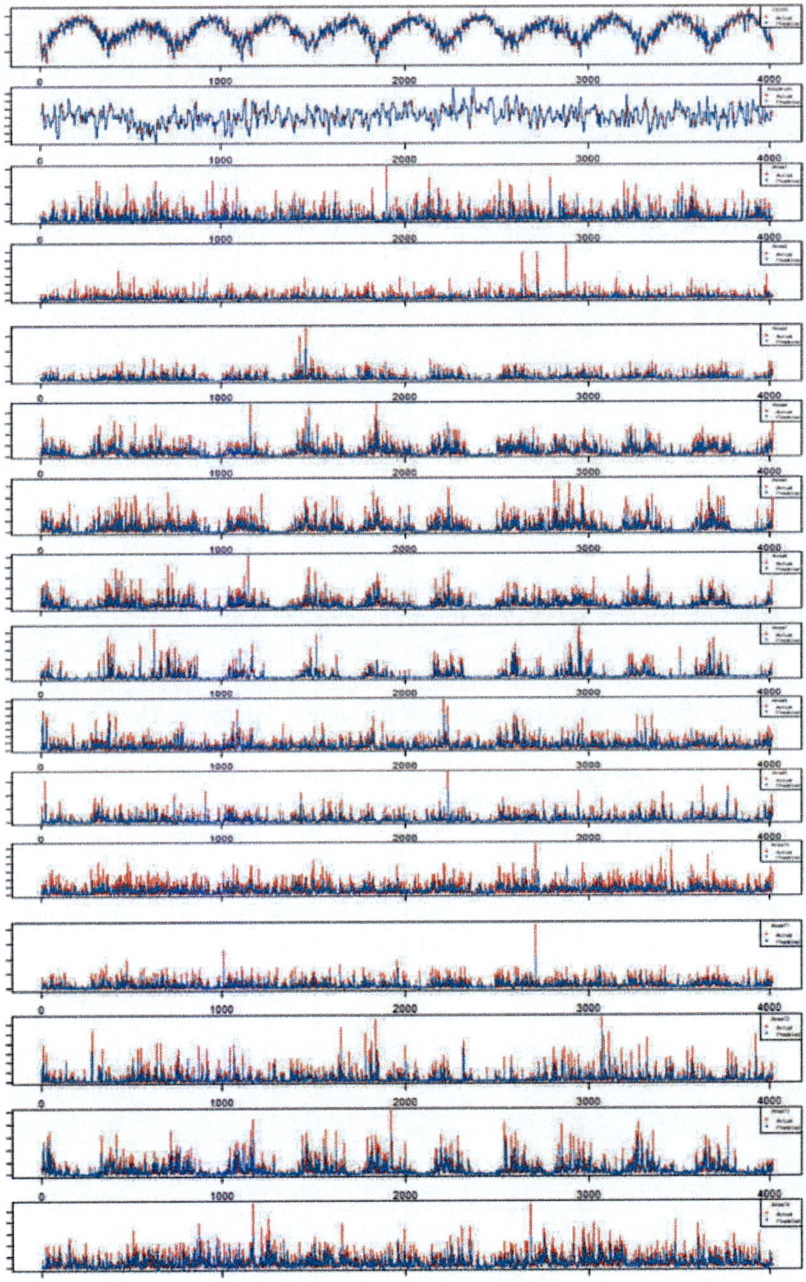

Fig. 5 Grafik Perbandingan data Aktual dan Pemodelan VECM

Table 5 MAPE values

Variable	MAPE(%)	Variable	MAPE(%)	Variable	MAPE(%)
CENS	1.852	Cluster 5	0.489	Cluster 10	0.967
Amplitude	0.062	Cluster 6	0.578	Cluster 11	0.821
Cluster 1	0.945	Cluster 7	0.438	Cluster 12	0.611
Cluster 2	1.327	Cluster 8	0.730	Cluster 13	0.548
Cluster 3	0.626	Cluster 9	0.741	Cluster 14	0.683
Cluster 4	0.478	–	–	–	–

models obtained are very good at predicting the value of each variable for the future period.

3.9 *Forecasting Results for the Following Period*

The forecasting value will be shown along with the forecasting results' lower and upper limits on January 1st 2019 in this following Table 6.

Table 6 shows the forecasting value for Indonesia's CENS, amplitude, and rainfall. Besides, there are also the lower and upper values as the forecasting ranges for one period ahead. Based on Table 6, most of the actual values are already within the

Table 6 Forecasting value in one day ahead

Variable	Forecast	Lower	Upper	The actual value	Conclusion
CENS	−5.034	−6.854	−3.214	−4.068	Within range
Amplitude	0.333	−0.292	0.959	0.370	Within range
Cluster 1	−1.290	−29.969	27.388	0.000	Within range
Cluster 2	15.679	−19.224	50.582	0.004	Within range
Cluster 3	24.637	4.943	44.331	0.731	Out of range
Cluster 4	37.144	21.088	53.201	25.171	Within range
Cluster 5	47.288	29.676	64.901	12.357	Out of range
Cluster 6	10.425	−6.285	27.135	12.059	Within range
Cluster 7	6.451	−9.339	22.241	11.508	Within range
Cluster 8	29.091	5.167	53.016	18.055	Within range
Cluster 9	29.368	4.077	54.660	9.339	Within range
Cluster 10	32.813	10.066	55.560	61.631	Out of range
Cluster 11	28.009	3.267	52.751	29.537	Within range
Cluster 12	4.695	−16.260	25.650	21.674	Within range
Cluster 13	20.549	3.035	38.063	20.369	Within range
Cluster 14	−0.454	−21.555	20.646	0.153	Within range

forecasting range except for clusters 3, 5 and 10. They are still out of the forecasting range because daily forecasts are still relatively unstable and fluctuate freely.

4 Conclusions

Rainfall in the 14 clusters over Indonesia is forecasted to be stable and within the prediction range in the upcoming periods. Similar trends are also observed for the CENS and amplitude of the MJO where the predicted data are not beyond the lower and upper forecasted value. In other words, this condition is the same as in the previous periods fluctuating at the same point, which neither overestimates nor underestimates.

Furthermore, the comparison of rainfall prediction results in those clusters with VECM, and actual data can be seen by the mean absolute percentage error (MAPE) test values which were barely under 1 (good performance) for the whole study areas. In addition, the amplitude of the MJO data also shows a fairly enough performance in which the MAPE values are close to zero. Unfortunately, the MAPE value of forecasting CENS is clearly observed at more than 1%, precisely at 1.852. Somehow, this value remains considerably categorized as a good performance.

To sum up, the statements above clearly indicate that the VECM performs well in forecasting the CENS, amplitude of the MJO, and rainfall over Indonesia. Therefore, this VECM is recommended to be implemented in developing the early warning system to mitigate and manage the upcoming hazards in Indonesia. For getting better results and more supporting evidence of this performance, however, it is suggested to carry out further this study in predicting other global phenomena such as Cold Surge (CS), Southerly Surge (SS), and Indian Oscillation Dipole (IOD) and including their relationships with extreme rainfall in Indonesia.

References

1. BNPB.: Geoportal Data Bencana Indonesia. https://gis.bnpb.go.id/. Last Accessed 04 Oct 2022
2. Dob.: RI Supermarket Bencana, Sampai September Ada 2000 Bencana—CNBC Indonesia. https://www.cnbcindonesia.com/news/20201001104931-4-190809/ri-supermarket-bencana-sampai-september-ada-2000-bencana. Last Accesed 01 Oct 2022
3. Amajama, J.: Physics of rainfall. J. Scientif. Eng. Res. **3**(1), 51–54 (2016)
4. As-syakur, A.R., Osawa, T., Miura, F., Nuarsa, I.W., Ekayanti, N.W., Dharma, I.G.B.S., Adnyana, I.W.S., Arthana, I.W., Tanaka, T.: Maritime continent rainfall variability during the TRMMera: the role of monsoon, topography and El Niño Modoki. Dyn. Atmos. Oceans **75**, 58–77 (2016)
5. Peatman, S.C., Matthews, A.J., Stevens, D.P.: Propagation Of the Madden–Julian oscillation through the Maritime Continent and scale interaction with the diurnal cycle of precipitation. Q. J. R. Meteorol. Soc. **140**, 814–825 (2014)
6. Saufina, E., Trismidianto, Risyanto, Fathrio, I., Harjupa, W.: Impact of cross-equatorial northerly surge (CENS) on Jakarta heavy rainfall and its interaction with a tropical cyclone

(Case study: 18–25 February 2020). In: AIP Conference Proceedings, vol. 2366, pp. 050002. (2021). https://doi.org/10.1063/5.0059995

7. Purwaningsih, A., Tank, A.K., Vila, J.: Atmospheric conditions associated with northerly Surge, Borneo Vortex and Madden Julian Oscillation during the extreme rainfall cases in early 2021 over the Western Part of the Maritime Continent. In: Yulihastin, E., Abadi, P., Sitompul, P., Harjupa, W. (eds.) Proceedings of the International Conference on Radioscience, Equatorial Atmospheric Science and Environment and Humanosphere Science, 2021. Springer Proceedings in Physics, vol. 275, Springer, Singapore (2022). https://doi.org/10.1007/978-981-19-0308-3_57
8. Lestari, D.O., Sutriyono, E., Sabaruddin, Iskhaq Iskandar.: Respective influences of Indian Ocean Dipole and El NiñoSouthern Oscillation on Indonesian precipitation. J. Mathem. Fundamental Sci. **50**(3), 257–272 (2018)
9. Kurniadi, A., Weller E., Ki Ming, S., Gyu Seong, M.: Independent ENSO and IOD impacts on rainfall extremes over Indonesia. Int. J. Climatol. **41**(2) (2021). https://doi.org/10.1002/joc.7040
10. Hadi, T., Horinouchi, T., Tsuda, T., Hashiguchi, H., Fukao, S.: Sea-breeze circulation over Jakarta, Indonesia: a climatology based on boundary layer radar observations. Mon. Weather Rev. **130**(9), 2153–2166 (2001)
11. Zhang, C.: Madden–Julian oscillation: bridging weather and climate. Bulletin of the American Meteorol. Soc. **94**, 1849–1870 (2013). https://doi.org/10.1175/BAMS-D-12-00026.1
12. Adames, Á.F., Wallace, J.M.: Three-dimensional structure and evolution of the MJO and its relation to the mean flow. J. Atmos. Sci. **71**(6), 2007–2026 (2014). https://doi.org/10.1175/JAS-D-13-0254.1
13. Nitta, T., Mizuno, T., Takahashi, K.: Multi-scale convective systems during the initial phase of the 1986/87 El Niño. J. Meteorol. Soc. Japan Ser. II **70**(1B) 447–466 (1992). https://doi.org/10.2151/jmsj1965.70.1B_447
14. Wu, P., Hara, M., Fudeyasu, H., Yamanaka, M.D., Matsumoto, J., Syamsudin, F., Sulistyowati, R., Djajadihardja, Y.S.: The impact of trans-equatorial monsoon flow on the formation of repeated torrential rains over Java Island. Sola **3**, 93–96 (2007)
15. Tangang, F.T., Juneng, L., Salimun, E., Vinayachandran, P.N., Seng, Y.K., Reason, C.J.C., Behera, S.K., Yasunari, T.: On the roles of the northeast cold surge, the Borneo vortex, the Madden-Julian oscillation, and the Indian Ocean Dipole during the extreme 2006/2007 flood in southern Peninsular Malaysia. Geophys. Res. Lett. **35**, L14S07 (2008). https://doi.org/10.1029/2008GL033429
16. Yokoi, S., Matsumoto, J.: Collaborative effects of cold surge and tropical depression-type disturbance on heavy rainfall in central Vietnam. Mon. Weather Rev. **136**(9), 3275–3287 (2008)
17. Chen, T.-C., Tsay, J.-D., Matsumoto, J., Alpert, J.: Development and formation mechanism of the Southeast Asian winter heavy rainfall events around the South China Sea. Part I: formation and propagation of cold surge vortex. J. Climate **28**(4), 1417–1443 (2015). http://www.jstor.org/stable/26194409
18. Yulihastin, E, Trilaksono, N.J.: Evolution of heavy rainfall in Jakarta flood case 2013 based on COSMO model. In: Asia Pacific Centre for Ecohydrology-UNESCO Research Center for Limnology-Indonesian Institute of Sciences, **167** (2014). https://doi.org/10.13140/2.1.3913.9525
19. Pu Z., Kalnay E.: Numerical weather prediction basics: models, numerical methods, and data assimilation. In: Duan, Q., Pappenberger, F., Thielen, J., Wood, A., Cloke, H., Schaake J. (eds.) Handbook of Hydrometeorological Ensemble Forecasting. Springer, Berlin, Heidelberg (2018)
20. Shahin, M.A., Ali, M.A., Ali, A.B.M.S.: Vector autoregression (VAR) modeling and forecasting of temperature, humidity, and cloud coverage. In: Computational Intelligence Techniques in Earth and Environmental Sciences. Springer, Dordrecht (2014)
21. Adenomon, M.O., Oyejola, B.A.: Forecasting meteorological time series data with a reduced form vector autoregressive (Var) model and three univariate time series techniques: a comparative study. Soc. Basic Sci. Res. Rev. **2**, 139–152 (2014)

22. Ramli, I., Rusdiana, S., Basri, H., Munawar, A.A., Azelia, V.: Predicted rainfall and discharge using vector autoregressive models in water resources management in the high hill Takengon. In: IOP Conference Series: Earth and Environmental Science, vol. 271 (2019)
23. Abdallah, W., Abdallah, N., Marion, J., Oueidat, M., Chauvet, P.: A vector autoregressive methodology for short-term weather forecasting: tests for Lebanon. SN Appl. Sci. **2**, 1–9 (2020)
24. Alsaber, A.R., Setiya, P., Al-Sultan, A.T., Pan, J.: Exploring the impact of air pollution on COVID-19 admitted cases. Japanese J. Statist. and Data Sci. **5**, 379–406 (2022)
25. Chen, M., Shi, W., Xie, P., Silvia, V.B.S., Kousky, V.E., Higgins, R.W., Janowiak, J.E.: Assessing objective techniques for gauge-based analyses of global daily precipitation. J. Geophys. Res. **113**, D04110 (2008). https://doi.org/10.1029/2007JD009132
26. Klaes, K.D.: The EUMETSAT polar system. Comprehensive Remote Sensing 192–219 (2018)
27. Brockwell, P.J.: In: Introduction to Time Series and Forecasting, 3rd edn. Springer Nature, Switzerland (2016)
28. Wei, W.W.S.: Time Series Analysis. Addison Wesley, New York (2006)
29. Lutkepohl, H.: New Introduction to Multiple Time Series Analysis. Springer, Berlin (2005)
30. Gentle, J.: Statistical Analysis of Financial Data: With Examples In R. CRC Press, Boca Raton (2020)

Implementation of Zero Runoff to Reduce Runoff Discharges in Timbang Langsa Village, Langsa City

Eka Mutia, Ellida Novita Lydia, Wan Alamsyah, and Danil Rahmad Priatna

Abstract Factors causing flooding include high rainfall, reduced water catchment areas, blockages due to indiscriminate garbage disposal, tree cutting and reduced forest area. The reduction in the catchment area can lead to an increase in the amount of rainwater runoff that occurs on the surface. Flooding can also occur due to the dimensions of the channel that are not able to accommodate the flood discharge or the collector channel that is not able to accommodate the water discharge from the Interceptor Channel. Langsa City is currently carrying out relocation activities for residents living on the banks of the Krueng Langsa river. The relocation site is located in Timbang Langsa village. This community relocation area was originally an oil palm plantation owned by PT. Timbang Langsa who is no longer productive. Changes in land use are seen as one of the causes of flooding in the area. Based on this, it is necessary to develop a way to reduce Runoff discharge, by making an alternative land cover. The purpose of this study is to reduce the flood discharge that occurs with a change from the value of runoff coefficient (C). The research method used is a survey method. The results obtained that the runoff discharge (Q) value was significantly reduced on the land cover as a mixed garden with a runoff discharge (Q) value ranging from 0.0077 to 0.0718 m^3/sec from the previous runoff discharge (Q) value ranging from 0.0187 to 0.0979 m^3/sec.

E. Mutia · E. N. Lydia (✉) · W. Alamsyah · D. R. Priatna
Universitas Samudra, Jalan Prof. Dr. Syarif Thayeb, Meurandeh, Langsa-Aceh, Indonesia
e-mail: ellidanovita@unsam.ac.id

E. Mutia
e-mail: ekamutia@unsam.ac.id

W. Alamsyah
e-mail: wanalamsyah@unsam.ac.id

D. R. Priatna
e-mail: danil.unsam19@gmail.com

A. Basit et al. (eds.), *Proceedings of the International Conference on Radioscience, Equatorial Atmospheric Science and Environment and Humanosphere Science*, Springer Proceedings in Physics 290, https://doi.org/10.1007/978-981-19-9768-6_36

1 Introduction

Rapid population growth can cause serious environmental problems in the world [1–3]. The increase in population will certainly be followed by development developments, causing reduced infiltration functions and the construction of artificial drainage networks can also change the natural path of water/runoff [4–6]. Surface runoff is an important component in the hydrological cycle that occurs in urban areas. This is related to changes in land use which result in changes in land cover in urban areas [7]. The consequences that can occur from population growth include increased land cover, soil compaction, vegetative cover, development of rainwater drainage systems, infiltration, groundwater recharge and reduced evapotranspiration. Surface runoff and peak flow will increase in volume and magnitude because the runoff will be quickly discharged into water bodies [8, 9].

Langsa City is currently making improvements to the riverbank area which was originally a densely populated area to be used as a waterfront tourist destination. The community previously lived on the banks of the river and then relocated to a new place. The relocation site was originally an oil palm plantation. Changes in land use at the two locations are an effort to reduce the flood discharge that occurs, but currently in the relocation area, flooding is still happening and it is difficultly to have enough clean water. This phenomenon is interesting to study to obtain an appropriate method to overcome the problem of flooding in coastal areas. The purpose of this study is to obtain an appropriate runoff coefficient based on the planned land cover to reduce runoff discharge. The reduction of runoff discharge aims to make the infiltration that occurs greater than runoff; this aims to overcome floods in Timbang Langsa village.

2 Methods

2.1 Study Area

The research will be conducted in the relocation area of Timbang Langsa village, Langsa Baro sub-district, Langsa City, Aceh, Indonesia (see Fig. 1). Geographically, the research location is located between latitude 4°32′4.35″N–4°32′7.24"N and longitude 97°55′28.22″ E–97°55′59.29"E in Langsa City, Aceh, with an area of 0.097 km^2 [10]. There is a lot of rainfall throughout the year in Langsa. The most rainfall in Langsa city occurs in Mei, with an average rainfall of 170 mm. January has the least rainfall, with an average rainfall of 1010 mm [11].

The main types of vegetation found at the research site were weeds, ornamental plants, fruit plants and tubers. The livelihoods of the majority of the population are entrepreneurship, employees and traders.

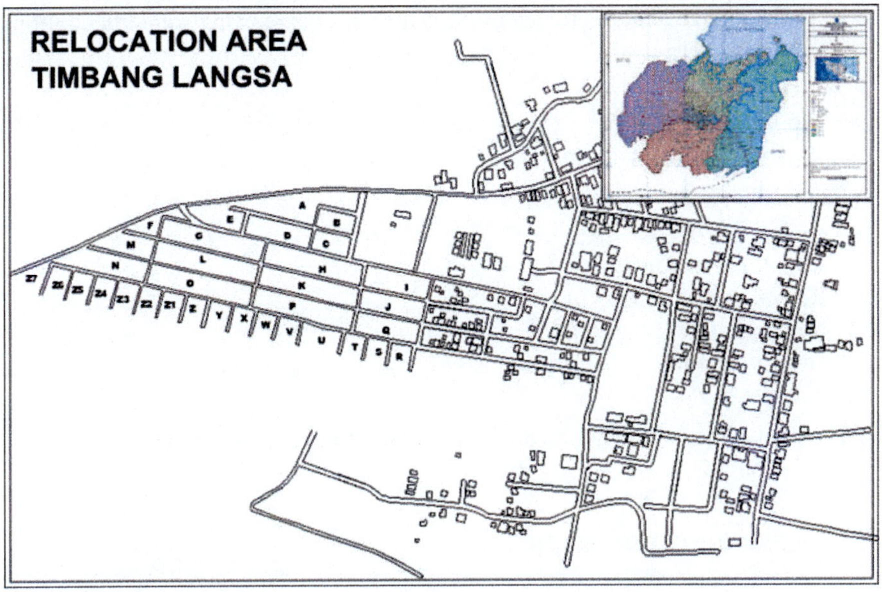

Fig. 1 Research site map

2.2 *Data Collection and Processing*

The data needed in this study are measurement data of the research area, elevation data of research area, rainwater catchment delineation data and Qanun Kota Langsa number 12 of 2013 concerning the Langsa City Spatial Plan for 2012–2032.

This type of research is quantitative research with a descriptive survey method. The research stage consists of three stages, namely the preparation stage, the survey stage and the data processing and analysis stage.

- Stages of preparation

At this stage, the activities carried out are secondary data collection consisting of maximum daily rainfall data for the city of Langsa in 2009 – 2020.

- Survey stage in the field

At this stage, the activities carried out are primary data collection consisting of data on the area of the flow area through declination and elevation measurements. The primary data were obtained by surveying the field.

- The data processing and analysis stage consist of:
  - The stages of research implementation begin with the delineation of rainwater catchment areas
  - Survey data calculation analysis

– Hydrological analysis to determine runoff discharge using a modified rational formula [12]

$$Q = 0.278C.Cs.I.A \tag{1}$$

Description:

Q: Runoff discharge
C: Runoff coefficient
Cs: Retention coefficient
I: Rain intensity
A: Area of stream.

- The final stage is to recommend the zero runoff concept to be used

3 Results and Discussion

3.1 Runoff Coefficient

The studied conducted in a residential area which is relocation location for people who previously lived on the banks of the Krueng Langsa river. The allotment of land at this location is a coupling housing, a yard that functions as a green open space and an asphalt road. By Fig. 1, it can be seen that the study area is divided into 33 watersheds. The existing watersheds have almost the same shape and number of houses as one. So what will be shown in Tables 1, 2, 3 and 4 are several watersheds, namely A, C, N, O, Q, Z6 and Z7 watersheds. The runoff coefficient (C) will be calculated for the C value for housing and yard, while the C value for asphalt is not taken into account. This is because the runoff discharge that will be accommodated is runoff that is in the yard of each house.

Table 1 Runoff Area

Runoff area	Number of houses (couple)	The yard area (km^2)	Residential area (km^2)	Area of runoff (km^2)	Value of C
A	6	0.0063	0.0004	0.0067	0.3638
C	12	0.0010	0.0011	0.0021	0.4807
N	22	0.0033	0.0018	0.0051	0.4374
O	32	0.0032	0.0026	0.0058	0.4612
Q	12	0.0026	0.0009	0.0035	0.4159
Z6	2	0.0013	0.0005	0.0018	0.4226
Z7	2	0.0015	0.0002	0.0017	0.3787

Table 2 Runoff coefficient value

Runoff area	Area of runoff (km^2)	C value (land)	C value (grass)	C value (mixed garden)	C value (paving)
A	0.0067	0.3638	0.3260	0.1276	0.6945
C	0.0021	0.4807	0.3136	0.3136	0.3136
N	0.0051	0.4374	0.1480	0.0477	0.3342
O	0.0058	0.4612	0.4616	0.3613	0.6477
Q	0.0035	0.4159	0.2099	0.2099	0.2099
Z6	0.0018	0.4226	0,2016	0.0650	0.4552
Z7	0.0017	0.3787	0.4114	0.2749	0.6650

Table 3 Runoff discharge (m^3/det)

Runoff area	Area of runoff (km^2)	Q value (land)	Q value (grass)	Q value (mixed garden)	Q value (paving)
A	0.0067	0.0337	0.0297	0.0100	0.0711
C	0.0021	0.0979	0.0937	0.0718	0.1353
N	0.0051	0.0746	0.0700	0.0459	0.1153
O	0.0058	0.0317	0.0299	0.0208	0.0482
Q	0.0035	0.0260	0.0238	0.0130	0.0460
Z6	0.0018	0.0785	0.0729	0.0437	0.1289
Z7	0.0017	0.0187	0.0169	0.0077	0.0341

Table 4 Clean water needs

Stream area	Number of houses (couple)	Q value—m3/s (soil)	Q value—m3/s (mixed garden)	Water needs (m3/day)		Reservoir (3 days)—m3	
				Home	Flow area	Home	Flow area
A	6	0.0337	0.0100	0.652	3912	1956	11,736
C	12	0.0979	0.0718	0.652	7824	1956	23,472
N	22	0.0746	0.0459	0.652	14,344	1956	43,032
O	32	0.0317	0.0208	0.652	20.864	1956	62.592
Q	12	0.0260	0.0130	0.652	7824	1956	23,472
Z6	2	0.0785	0.0437	0.652	1304	1956	3912
Z7	2	0.0187	0.0077	0.652	1.304	1956	3912

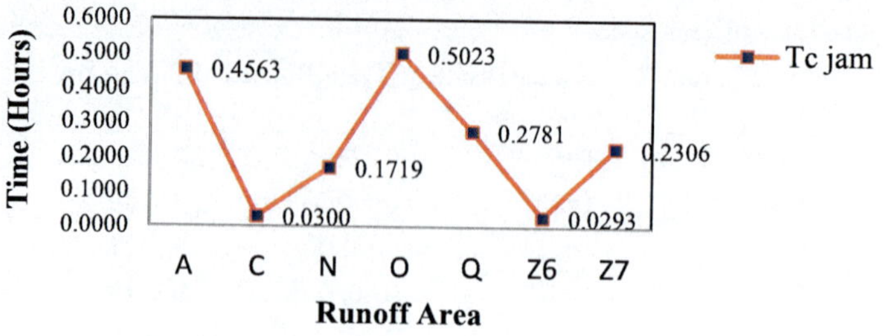

Fig. 2 Time concentration

The area of the study area is 0.097 km^2, with the value of C in each flow area in Table 1 following (the value of C is calculated for only a few different typical flow areas). The value of C is taken with the assumption that all open land is clay.

The value of C in each flow area ranges from 0.3 to 0.5. The lowest value of C found in runoff area A. In that area, the house area is only 5.97% of the area of runoff A. The highest value of C found in flow area C with a Residential area of 52.38%. This shows that the more green open space, the greater the infiltration of flood discharge.

3.2 *Time Concentration (Tc)*

The time concentration (Tc) will determine the length of time the water flows. This calculation is to estimate the time needed to accommodate the flow in a container for further storage as a source of clean water. The calculation of the concentration time can be in Fig. 2.

In Fig. 2, it can be seen that the lowest concentration time is in the Z6 flow area with a Tc value of 0.0293 and the highest Tc is in the O flow area with a value of 0.5023. The value of Tc depends on the shape of the flow area. The shape of the short flow area will result in a shorter time compared to the shape of the elongated flow area [13].

3.3 *Runoff Discharge (Qt)*

Runoff discharges the research site to calculate how much discharge is generated at the site. Runoff discharge will then be reduced by harvesting rainwater so the flood problem can be resolved (Fig. 3).

Runoff discharge ranges from 0.02 to 0.10 m^3, depending on the overall flow area outside.

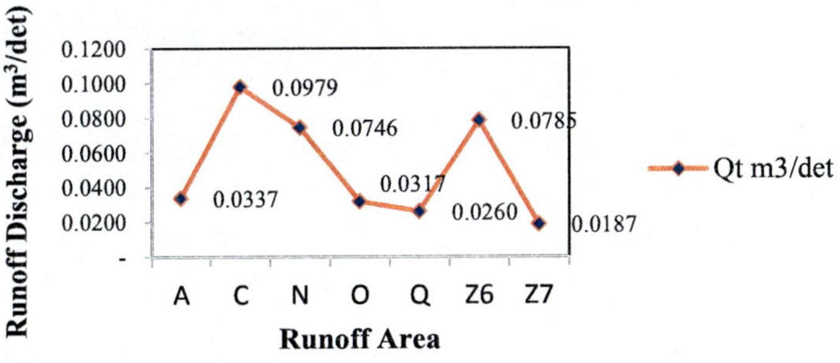

Fig. 3 Runoff discharge

3.4 *Clean Water Needs*

Everyone's need for clean water is different. In Indonesia, the need is based on SNI 6728. 1: 2015 concerning the preparation of the spatial balance of natural resources, the water demand for medium-sized cities is 100–125 L/per person/day [14]. Clean water at the research site is sourced from the Regional Drinking Water Company (PDAM) Tirta Kemuning, Langsa City.

The community is still experiencing difficulties with clean water because its distribution is not yet smooth. At the research location, it is assumed that if in 1 house there are 5 people, then the need for clean water is 625 L/person/day. We will try to fulfil this need for clean water by harvesting rainwater from runoff.

3.5 *Reduction of Flood Discharge*

Reduction of runoff that occurs at the research site will be carried out by changing the C value. Several alternative land covers will be tried to make recommendations on which one has a lower C value so that it will reduce the resulting Q (Fig. 4).

Based on Tables 2 and 3, the smallest C value and Q value occur in land cover used as mixed gardens. So if the open space in each housing is mostly used as mixed gardens, it can reduce runoff discharge by 25–70%.

The next step is to reduce runoff by harvesting rainwater. The amount of clean water needed in each watershed will be different according to the number of houses built.

The largest water demand is in the O runoff area, because it has more houses than other runoff area, but has a low Q value. Further research is needed to calculate the adequacy of clean water when compared to the existing plan.

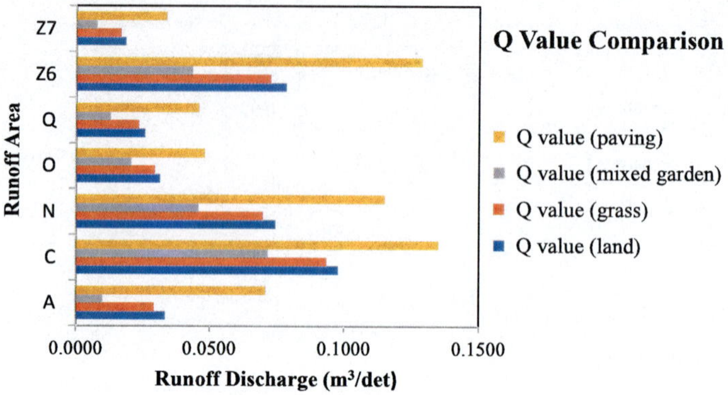

Fig. 4 Comparison of runoff discharge value

4 Conclusion

The appropriate runoff coefficient value is a C value of 0.1276–0.2749 with mixed garden land cover types. Land cover with mixed gardens reduces runoff discharge by 25–70%, so that the community using mixed garden land cover types will reduce the flood discharge that occurs.

Acknowledgements The author would like to thank LPPM and PM of Samudra University who have provided research funds and all related parties who have helped so that this research can run smoothly. This research was funded by the University of Samudra's DIPA in 2022.

References

1. Liu, W. et al.: Identifying city-scale potential and priority areas for retrofitting green roofs and assessing their runoff reduction effectiveness in urban functional zones. J. Clean Prod. **332** (2022). https://doi.org/10.1016/j.jclepro.2021.130064
2. Chan, F.K.S., et al.: 'Sponge City' in China—a breakthrough of planning and flood risk management in the urban context. Land Use Policy **76**, 772–778 (2018). https://doi.org/10.1016/j.landusepol.2018.03.005
3. Fang, C., et al.: Modeling regional sustainable development scenarios using the urbanization and eco-environment coupler: case study of Beijing-Tianjin-Hebei urban agglomeration, China. Sci. Total Environ. **689**, 820–830 (2019). https://doi.org/10.1016/j.scitotenv.2019.06.430
4. Petroselli, A., Wałęga, A., Młyński, D., Radecki-Pawlik, A., Cupak, A., Hathaway, J.: Rainfall-runoff modeling: a modification of the EBA4SUB framework for ungauged and highly impervious urban catchments. J. Hydrol. (Amst) **606** (2022). https://doi.org/10.1016/j.jhydrol.2021.127371
5. Lizárraga-Mendiola, L. et al.: Hydrological design of two low-impact development techniques in a semi-arid climate zone of central Mexico. Water (Switzerland) **9**(8) (2017). https://doi.org/10.3390/w9080561

6. Walega, A., Salata, T.: Influence of land cover data sources on estimation of direct runoff according to SCS-CN and modified SME methods. Catena (Amst) **172**, 232–242 (2019). https://doi.org/10.1016/j.catena.2018.08.032
7. Chang, B., Wherley, B., Aitkenhead-Peterson, J.A., McInnes, K.J.: Effects of urban residential landscape composition on surface runoff generation. Sci. Total Environ. **783** (2021). https://doi.org/10.1016/j.scitotenv.2021.146977
8. Hong, W., Guo, R., Tang, H.: Potential assessment and implementation strategy for roof greening in highly urbanized areas: a case study in Shenzhen, China. Cities **95** (2019). https://doi.org/10.1016/j.cities.2019.102468
9. Walsh, C.J., et al.: Principles for urban stormwater management to protect stream ecosystems. Freshwater Sci. **35**(1), 398–411 (2016). https://doi.org/10.1086/685284
10. Badan Pusat Statisik Kota Langsa,: Kecamatan Langsa Barő Dalam Angka 2022. Langsa, Badan Pusat Statistik Kota Langsa (2022)
11. Badan Pusat Statistik Kota Langsa.: Kota Langsa dalam Angka (Langsa Municipality in Figures) 2022. Langsa, Badan Pusat Statistik Kota Langsa (2022).
12. Mutia, E., Alamsyah, W.: Penataan Jaringan Drainase Berdasarkan Tata Ruang Kota Langsa. Jurutera **2**(1), 38–47 (2015)
13. Suripin.: Sistem Drainase Perkotaan yang BerkelanjutanSistem Drainase Perkotaan yang Berkelanjutan. Jakarta, Penerbit Andi (2004)
14. Afriyanda, R., et al.: Analisis Kebutuhan Air Bersih Domestik Di Desa Penjajap Kecamatan Pemangkat Kabupaten Sambas. PWK, Laut, Sipil, Tambang **6**(2), 1–11 (2018)

Monthly Rainfall Prediction Using Vector Autoregressive Models Based on ENSO and IOD Phenomena in Cilacap

Fadli Nauval, Mutia Yollanda, Dodi Devianto, Wendi Harjupa, Dita Fatria Andarini, Elfira Saufina, Anis Purwaningsih, Fahmi Rahmatia, Ridho Pratama, Trismidianto, Teguh Harjana, Risyanto, and Didi Satiadi

Abstract A study of El Nino-Southern Oscillation (ENSO) and Indian Oscillation Dipole (IOD) as climate elements to predict extreme rainfall is critical to be carried out. This study aims to forecast monthly rainfall by using Vector Autoregressive (VAR) model in the Cilacap. Rainfall data were obtained by the Global Satellite Mapping of Precipitation (GSMaP) and ENSO and IOD by the National Oceanic and Atmospheric Administration (NOAA) from March 2000 to December 2018. The data was used to predict monthly rainfall for the next twelve months in 2019. The VAR model with a minimum lag of length 2 was considered to be the best lag and selected to model the rainfall in the study area. Response function reveals that changes in rainfall significantly affect changes in rainfall after some time lags. The results of the actual data accuracy test with predictive data from the Vector Autoregressive (VAR) model were conducted with the *root mean square error* (*RMSE*) and the *mean absolute percentage error* (*MAPE*). The RMSE value was 0.3291, 0.047, 0.061 and the MAPE value was 87.0671, 0.1845, 0.217. Rainfall in the Cilacap is predicted to be stable in the future. This condition is the same as the rainfall in the previous periods, which fluctuated at the same point.

1 Introduction

Indonesia, which is well-known as Indonesia Maritime Continent (IMC) is a county with a tropical climate that is vulnerable to extreme rainfall [1]. Moreover, the IMC experiences the impact of climate change in which the air temperature in Indonesia has increased by 0.30 °C since 1900. This increase in temperature occurred

F. Nauval (✉) · W. Harjupa · D. F. Andarini · E. Saufina · A. Purwaningsih · F. Rahmatia · R. Pratama · Trismidianto · T. Harjana · Risyanto · D. Satiadi
Research Center for Climate and Atmosphere, National Research and Innovation Agency (BRIN), Bandung, Indonesia
e-mail: fadli.nauval@brin.go.id

M. Yollanda · D. Devianto
Department of Mathematics, UNAND, West Sumatera, Indonesia

A. Basit et al. (eds.), *Proceedings of the International Conference on Radioscience, Equatorial Atmospheric Science and Environment and Humanosphere Science*, Springer Proceedings in Physics 290, https://doi.org/10.1007/978-981-19-9768-6_37

throughout the season. Changes in weather and seasons are marked by an increase in rainfall (La Nina) that occurs in an area, while in other regions, there is a decrease in rainfall (El Nino) of 2–3% [1].

Variability in the IM is known to be modulated by two ocean–atmosphere phenomena, El Nino–Southern Oscillation (ENSO) and the more recently described Indian Ocean Dipole (IOD; [2, 3]). The possible linkages between ENSO and IOD are also an active area of investigation [4]. Furthermore, although climate teleconnections are of a global scale, local modulations—forced, for example, by fine-scale topography of terrain or land-sea contrasts—may play an important role. Therefore, understanding the role of multiscale interaction in regional climate is critical for climate prediction and change studies and applications in weather- and climate-related risk management. Java Island is Indonesia's most populated island and most important industrial and agricultural region. It is located in the deep tropics of the Southern Hemisphere, at the center of the Asian–Australian monsoon region [5]. It is zonally elongated and slightly tilted in the northwest-southeast direction, with a central mountain range that runs the length of the island.

Based on data on disaster events in Indonesia from 2001 to 2019 released by the National Disaster Management Agency (BNPB), it is known that Java Island, especially Central Java Province, is the area with the highest hydrometeorological disasters, both floods, landslides, and strong winds (BNPB disaster plot). The spatial distribution of disaster events, especially floods in Central Java, shows that the most dominant area experiencing flooding is Cilacap (spatial plot of the distribution of flood events in Central Java), so it is necessary to predict rain in the Cilacap area as an effort to mitigate disasters in the area.

Figure 1 depicts Indonesia's number of hydrometeorological disasters from 2000 to 2019. Overall, the hazard taking place the most in the given period was flood, and the area often experiencing the catastrophic was Jawa Tengah.

Figure 2 illustrates how many events of flood disasters across the Central Java province were observed in each district from 2000 to 2019. In general trend, Cilacap was an area that dealt with flood disasters the most. Moreover, the number of flood hazards in Central Java dominantly occurred in the southern and northern parts in the given period. As for southern Central Java, the rate of flood events was higher than the middle areas such as Cilacap, Banyumas, Kebumen, and Wonogiri at 155, 66, 56, and 62, respectively, while the areas of Purbalingga, Banjarnegara, and Wonosobo only experienced floods at less than 20 events in the last two decades. Furthermore, a similar trend in the southern part also took place in the northern areas.

Changes in rain patterns every day and every month are very important to take action. More accurate predictions are needed to avoid large losses of property and lives. Rain prediction using a numerical model has been carried out. One of the models used is Numerical Weather Prediction (Pu and Kalnay 2018). There are still shortcomings where the accuracy of using numerical models is still small [6]. For this reason, another approach is needed so that it can produce more accurate rain predictions.

One of the common prediction methods is the Vector Autoregressive (VAR) method. VAR was initially developed by an Econometric expert, Christopher A.

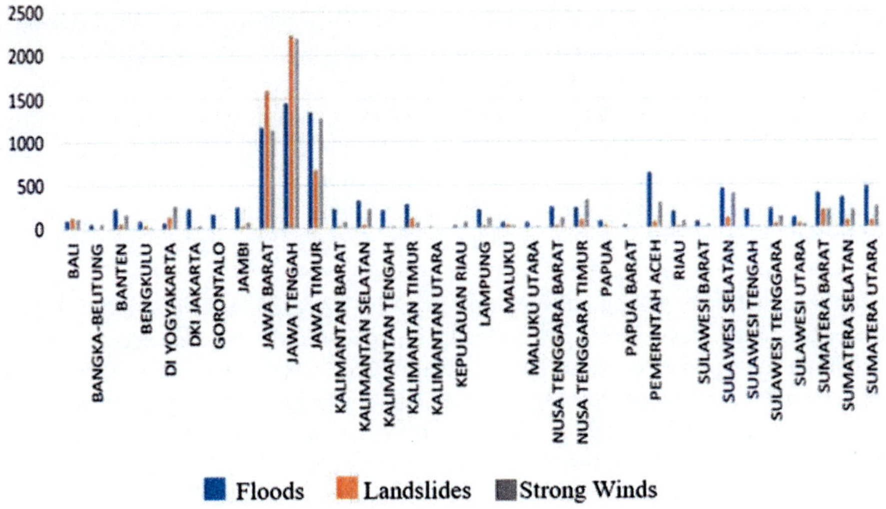

Fig. 1 The number of occurrences in Indonesia

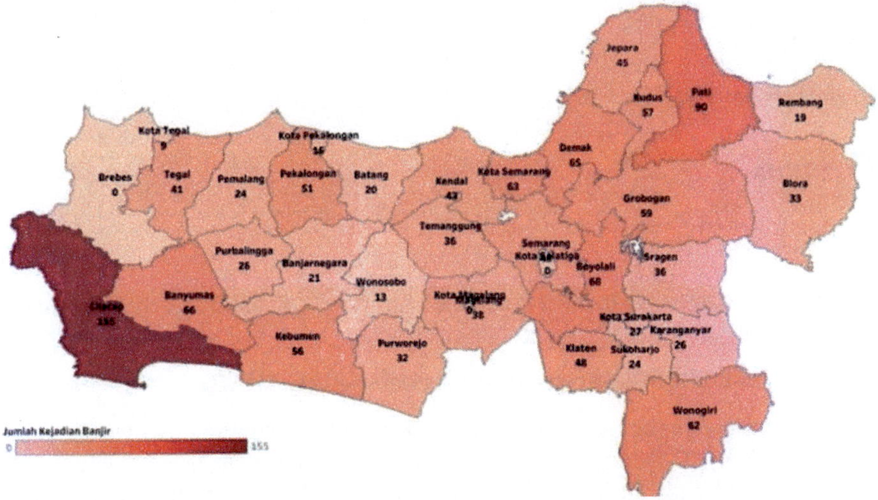

Fig. 2 Distribution of flood disasters in Central Java

Sims, as an alternative model approach to the multiple equation models with consideration of a theoretical approach that aims to be able to capture economic phenomena well [7]. VAR is a dynamic equation system. The Vector Autoregressive (VAR) model can be used to predict time series data for more than one variable, whereas in this model, we do not need to distinguish between independent variables and variables [7]. In the VAR model, all variables are endogenous (bound). With a high level of accuracy, the VAR model can be used for operational purposes so that it can be a reference for those who need it.

In the meteorology field, the VAR model has been used to improve weather forecast system. In Lebanon, the parameters of weather such as temperature, humidity, and precipitation are implemented to evaluate the built model [8]. The results were very promising (average of about 96.67% of precision between real values and three predicted parameters: temperature minimum, maximum humidity and precipitation) in the field of short-term weather parameter forecasting. Moreover, the similar parameters were also trained to carry out the establishment of the VAR model in Bangladesh to enhance Climate Change observation [9]. The VAR model was found to be the best. Structural analyses were performed using forecast error variance decomposition and impulse response function. These structural analyses divulged that the temperature, humidity, and cloud coverage would be interrelated and endogenous in future. Finally, temperature, humidity, and cloud coverage were forecasted from January 2011 to December 2016 using the best selected model VAR. Therefore, the utilization of precipitation data is necessary to deepen further in order to take a preventive action toward upcoming hazards so that the high rate of floods occurrences in Cilacap is critical to conduct the research of the use of the VAR model by using the ENSO and IOD data.

2 Methodology

2.1 Data

1. **Flood Data BNPB**

Data on disaster events in Indonesia obtained from the National Disaster Management Agency (BNPB)

2. **Global Satellite Measurement of Precipitation (GSMAP)**

Rainfall data were obtained from Global Satellite Mapping of Precipitation (GSMaP_MVK). GSMaP_MVK is a product that integrates passive microwave radiometer data with infrared radiometer data in order to have high temporal (1 hour) and spatial (0.1 degrees) resolution global precipitation estimates. The product (GSMaP_MVK) is produced based on a Kalman filter model that refines the precipitation rate propagated based on the atmospheric moving vector derived from two successive IR images. The detail of the algorithm can be found in [10]. Data dapat diunduh di https://sharaku.eorc.jaxa.jp/GSMaP.

3. **Sea Surface Temperature (SST) Data**

ENSO and IOD Index data from SST data downloaded from NOAA.

2.2 Method

1. **Stationarity Data**

Before forming the Vector Autoregressive (VAR) model, the time series data first perform some tests on the initial data, whether the data is stationary or not. Therefore, the stationarity of time series data concerning variance or mean value can be carried out successively using the Box-Cox transformation (Brockwell, 2016) and the Augmented Dickey-Fuller (ADF) test [11].

2. **Model Vector Autoregressive (VAR)**

The VAR model is an extension of the Autoregressive (AR) time series model with more than one variable. All variables in the VAR model are assumed to be endogenous variables (variables whose value is determined by other variables in the model) and are interrelated. The general form of the VAR model with a lag p, VAR(p), can be calculated as follows [12]:

$$X_{p,t} = \phi_0 + \sum_{i=1}^{p} \phi_i X_{i,t-i} + e_t, i = 1, 2, \ldots, p, t = 1, 2, \ldots \tag{1}$$

where;

$X_{p,t}$ is the dependent variable $(X_{1,t}, X_{2,t}, \ldots, X_{p,t})$ sized $p \times 1$,
$X_{i,t-1}$ is the independent variable $(X_{1,t-i}, X_{2,t-i}, \ldots, X_{p,t-i})$, sized $p \times 1$,
ϕ_0 is vector intercept size $p \times 1$,
ϕ_i is a coefficient matrix of size $p \times p$ for each $i = 1, 2, \ldots, p$,
e_t is the remainder vector $(e_{1,t}, e_{2,t}, \ldots, e_{p,t})$ sized $p \times 1$.

3. **Cointegration Test**

If data is cointegrated, then the modeling for that data cannot use the VAR model. The basis of cointegration is that many time series data can deviate from their average in the short-term but move together. The hypothesis in Johansen's test is as follows:

H_0 : there are as many as r, where $r = 0, 1, \ldots, k-1$ cointegration equation (no cointegration or long-term relationship between variables).

H_1 : there are cointegration equations (cointegration occurs between variables).

4. **Vector Error Correction Model (VECM)**

VECM is a restricted form of VAR due to the existence of non-stationary but cointegrated data forms. The VECM specification specifies the long-term relationship of endogenous variables to converge to their cointegration relationship but still allows the presence of short-term dynamics. The VECM has the following equation:

$$\Delta X_{p,t} = \phi_{p,0} + \Pi X_{p,t-1} + \sum_{i=1}^{p-1} \Phi_i^* \Delta X_{p,t-i} + e_t \tag{2}$$

where

$$X_{p,t} = X_{p,t-1} + \Delta X_{p,t},\ \Phi_j^* = -\sum_{i=j+i}^{p} \Phi_i,\ j = 1, 2, \ldots, p-1, \text{ and}$$
$$\Pi = -(1 - \Phi_1 - \Phi_2 - \ldots - \Phi_p \tag{3}$$

5. **Optimum Lag Determination**

VAR estimation is very sensitive to the lag length used. Therefore, it is necessary to know the optimal lag before estimating the VAR. Determination of the optimal amount of lag can be determined using Akaike Information Criterion (AIC), Schwarz Information Criterion (SC), Final Prediction Error (FPE), or Hannan Quinnon (HQ). Determination of the optimal lag using the information criteria by selecting the measures with the smallest value [12]. This optimal lag is applied to eliminate the autocorrelation problems in the VAR model so that they will no longer arise.

6. **Impulse response functions (IRF)**

After modeling using the VAR model, it is necessary to have a method that can characterize the dynamic structure generated by VAR. IRF shows the response of each endogenous variable over time to the shock of the variable itself and other endogenous variables. Thus, IRF is used to see the contemporary effect of a dependent variable if it gets a shock or innovation from the independent variable by one standard deviation.

7. **Mean Absolute Percentage Error (MAPE)**

MAPE is a useful measure to measure the error of the estimated value of the model. The error in the estimated value of the model expresses an average absolute percentage of residual, which is written as follows:

$$MAPE = \frac{1}{N}\sum_{t=1}^{N}\left|\frac{Y_t - \hat{Y}}{Y_t}\right| \times 100\% \tag{4}$$

where N is the number of predictions made, Y_t is the actual data, and $\hat{Y}_t$ is the data forecast using the VAR model.

Figure 3 indicates the topography of the Cilacap area. It is clearly seen that Cilacap is located on Java Island and close to the Indian Ocean. Interaction between land and sea affects the uniqueness of rainfall patterns as studied in this research.

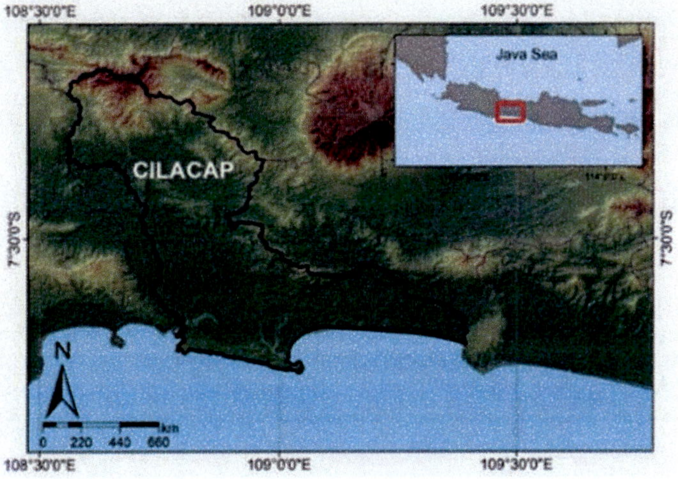

Fig. 3 Topography of Cilacap

3 Result and Discussion

3.1 Time Series of Rainfall Data and the ENSO Index, and the IOD

Figure 4a depicts a graph of rainfall time series data, IOD anomaly, and ENSO anomaly for the monthly period starting from March 2000 to December 2018. Moreover, it can be seen that the three time series data still fluctuate randomly and do not move around a fixed value. That is, these three data must be stationary to the variance and the mean so that the fluctuations in the data are more directed at a certain interval and the movement is centered on a fixed value. Therefore, a Box-Cox transformation is needed to overcome the problem of stationary variance and differencing methods to make the time series data move at a middle value.

Figure 4b shows that there is a clear relationship between the ENSO, IOD, and peak rainfall. Those three data indicate stationarity one another. Thus, the time series data can be easily analyzed in terms of the mean and variance.

3.2 Determining Optimum Lag

Determination of the optimal amount of lag can be determined using Akaike Information Criterion (AIC), Schwarz Information Criterion (SC), Final Prediction Error

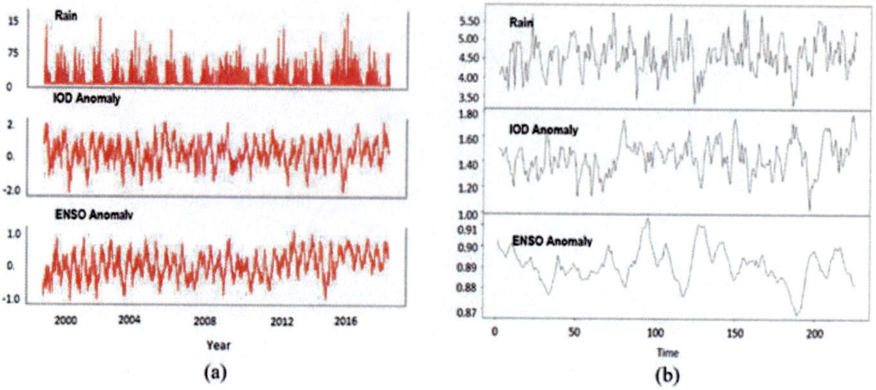

Fig. 4 Time series of anomaly data of mohly rainfall **a** before differencing, **b** after differencing

(FPE) or Hannan Quinnon (HQ). Determination of the optimal lag using the information criteria is obtained by selecting the criteria that have the smallest value (Table 1).

VAR Estimation Results:

Estimated coefficients for equation Rain:

$$RAIN(t) = 134.5828899 + 0.2835074RAIN(t-1) + 26.6413343IOD(t-1) + 26.4434951ENSO(t-1) + 0.1647146RAIN(t-2) - 27.4772601IOD(t-2) - 24.5771315ENSO(t-2) \quad (5)$$

Table 1 Criteria of optimum lag

Lag	Criteria			
	AIC	HQ	SC	FPE
1	−17.98138	−17.9051	−17.79263	1.551650e-08
2	−18.33049	−18.19702	**−18.00018***	**1.094459 e-08***
3	−18.30355	−18.11287	−17.83168	1.124497e-08
4	−18.30727	−18.05939	−17.69384	1.121851e-08
5	**−18.30654***	**−18.00146***	−17.55155	1.111766e-08
6	−18.31614	−17.95385	−17.41959	1.15E-08

* represents the criteria of optimum lag given by the data

Estimated coefficients for equation IOD:

$$\begin{aligned} IOD(t) &= 0.0553247229 - 0.0001777911 RAIN(t-1) \\ &+ 0.7539071521 IOD(t-1) + 0.1444971616 ENSO(t-1) \\ &+ 0.0000895992 RAIN(t-2) - 0.1165418043 IOD(t-2) \\ &- 0.1303052468 ENSO(t-2) \end{aligned} \tag{6}$$

Estimated coefficients for equation ENSO:

$$\begin{aligned} ENSO(t) &= -0.0622970607 - 0.0002176436 RAIN(t-1) \\ &+ 0.2096981972 IOD(t-1) + 1.4310148716 ENSO(t-1) \\ &+ 0.0005191951 RAIN(t-2) - 0.2639312138 IOD(t-2) \\ &- 0.5001126226 ENSO(t-2) \end{aligned} \tag{7}$$

It can be seen in Fig. 5 that there are fluctuations in rainfall intensity in the first month of stability, and then fluctuations occur until the 5th month, which then stabilizes in the 5th month and so on. While the IRF from ENSO for Rainfall, it can be seen that there is a fluctuation on the 2nd day, and in the following month, there is stability at the equilibrium point.

Table 2 shows that the performance of the VAR model predicts the rain, IOD, and ENSO data in the following month. It can be clearly seen that those three data are

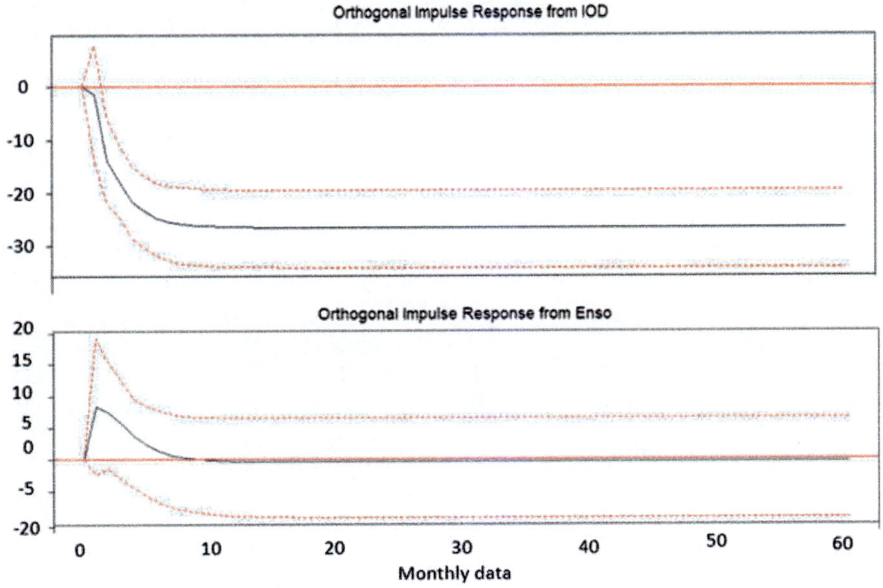

Fig. 5 Orthogonal impulse response

Table 2 Forecasting interval of VAR model

Rain						
	Forecast	Lower	Upper	CI	Actual	
Jan 2019	201.6998	27.27569	376.1238	174.4241	326.038	Within predicted range
IOD						
Jan 2019	0.194283	−0.18436	0.572929	0.378646	0.452	Within predicted range
ENSO						
Jan 2019	0.192421	−0.24527	0.630111	0.43769	0.520	Within predicted range

in the predicted range. It also indicates that the VAR model successfully predicts the upcoming index or value in the forecasted data. In order to evaluate the performance, statistical evaluation was carried out.

Figure 6 indicates that VAR model estimation is able to evaluate the performance of those three data. Moreover, the relationship between VAR model estimation and observation is significantly strong for ENSO and IOD data in particular. Somehow, VAR model only predicts a narrow range in terms of middle value only. In other words, the VAR model is not able to predict both the highest and the lowest peak of rainfall compared to observation data (Table 3).

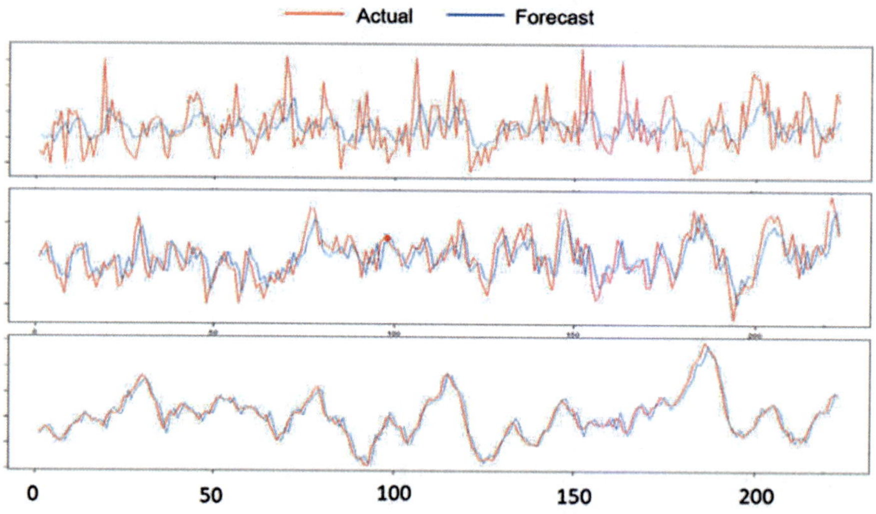

Fig. 6 Model VAR estimation

Table 3 VAR model accuration

	Root mean squared error (RMSE)	Mean absolute percentage error (MAPE)
Rainfall	87.06771	0.3291528
IOD	0.1845223	0.04796626
ENSO	0.2178605	0.06163001

4 Conclusion

Rainfall in Cilacap is predicted to be stable and within the prediction range in the upcoming periods. Similar trends are also observed for the IOD and ENSO, where the predicted data is not beyond the lower and upper forecasted values. In other words, this condition is the same as in the previous periods fluctuating at the same point, which neither overestimates nor underestimates.

Furthermore, the comparison of rainfall prediction results in such a study area with the VAR model and actual data can be seen by the *mean absolute percentage error* (*MAPE*) test values which were barely under 1 (good performance) in the study area. In addition, the IOD and ENSO data also indicate a better performance, in which the MAPE values are completely close to zero. Unfortunately, the RMSE value of forecasting rainfall is clearly observed more than one, which is precisely at 87.06771. Somehow, the value of RMSE for the IOD and the ENSO remains considerably categorized as a good performance where the value of which is less than one.

To sum up, the aforementioned statements indicate crystal clear that the VAR model performs well in forecasting the IOD and ENSO and is fairly moderate in predicting rainfall in Cilacap. Therefore, this VAR model is recommended to be implemented in developing the early warning system to mitigate and manage the upcoming hazards in Indonesia. For getting better results and more supporting evidence of this performance, however, it is suggested to further carry out this study in predicting rainfall, the IOD, and the ENSO in other areas in Indonesia.

References

1. Chang, C.P., Wang, Z., McBride, J., Liu, C.-H.: Annual cycle of Southeast Asia– Maritime Continent rainfall and asymmetric monsoon transition. J. Climate **18**, 287–301 (2005)
2. Saji, N.H., Goswami, B.N., Vinayachandran, P.N., Yamagata, T.: A dipole mode in the tropical Indian Ocean. Nature **401**, 360–363 (1999)
3. Ashok, K., Guan, Z., Saji, N.H., Yamagata, T.: Individual and combined influences of the ENSO and Indian Ocean Dipole on the Indian summer monsoon. J. Climate **17**, 3141–3155 (2004b)
4. Wang, X., Wan, C.: Different impacts of various El Nin˜o events on the Indian Ocean Dipole. Clim. Dyn. **2014**(42), 991–1005 (2014)

5. Ramage, C.S.: Role of a tropical "maritime continent" in the atmospheric circulation. Mon. Weather Rev. **96**, 365–369 (1968). https://doi.org/10.1175/1520-0493(1968)096%3c0365:ROATMC%3e2.0.CO;2
6. Li, J.: Assessing the accuracy of predictive models for numerical data: Not $\boldsymbol{r}$ nor $\boldsymbol{r}^2$, why not? Then what? (2007)
7. Widarjono, A.: Ekonometrika Teori dan Aplikasi untuk Ekonomi dan Bisnis. Edisi kedua. Yogyakarta, Ekonisia (2007)
8. Abdallah, W., Abdallah, N., Marion, J., Oueidat, M., Chauvet, P.: A vector autoregressive methodology for short-term weather forecasting: tests for Lebanon. SN Appl. Sci. **2**, 1–9 (2020)
9. Shahin, M.A., Ali, M.A., Ali, A.B.M.S.: Vector Autoregression (VAR) Modeling and Forecasting of Temperature, Humidity, and Cloud Coverage. Computational Intelligence Techniques in Earth and Environmental Sciences. Springer, Dordrecht (2014)
10. Ushio, T., Sasashige, K., Kubota, T., Shige, S., Okamoto, K., Aonashi, K.: A kalman filter approach to the global satellite mapping of rainfall (GSMaP) from combined passive microwave and infrared radiometric data. J. Meteorol. Soc. Japan **87A**, 137–151 (2009)
11. Lutkepohl, H.: New Introduction to Multiple Time Series Analysis. Springer, Berlin (2005)
12. Wei, W.W.S.: Time Series Analysis: Univariate and Multivariate Methods. Addison-Wesley Publishing Co., USA, United State of America (2006)

Extreme Rainfall Clusters in Borneo and Their Synoptic Climate Causes

Narizka Nanda Purwadani, Mohamad Rahman Djuwansah, Muhammad Rais Abdillah, Faiz Rohman Fajary, and Ida Narulita

Abstract A detailed study of the relationship between synoptic climate phenomena and high rainfall is essential to predict the occurrence of hazardous rainfall. We use Global Satellite Mapping of Precipitation (GSMaP) data and ERA-5 Reanalysis data to identify extreme rainfall events and the triggering anomalies. Extreme rainfall was identified as a 97th percentile maximum threshold. Then, k-means clustering applied to cluster spatial patterns of extreme rainfall. The influencing climatic factors anomaly reveals the synoptic climate phenomena in the Western Maritime Continent that provoke extreme rainfall on Borneo Island. Strong northerly moisture transport in December-January–February and March–April-May seasons may be related to cold surge events and lead to heavy rainfall in northern or western Borneo. Westerly wind anomalies toward the Philippines that can occur in all seasons may be associated with Madden Julian Oscillation and cause heavy rainfall around North Kalimantan. Also, positive sea surface temperature anomalies in the Java Sea may trigger wet anomalies around Central Kalimantan.

N. N. Purwadani
Master Program in Earth Sciences, Faculty of Earth Sciences and Technology, Institut Teknologi Bandung, Bandung, Indonesia

M. R. Djuwansah
Research Centre for Environment and Clean Technology, National Agency for Research and Innovation, Bandung, Indonesia

M. R. Abdillah (✉) · F. R. Fajary
Atmospheric Sciences Research Group, Faculty of Earth Sciences and Technology, Institut Teknologi Bandung, Bandung, Indonesia
e-mail: m.rais@itb.ac.id

I. Narulita
Research Centre for Climate and Atmospheric Science, National Agency for Research and Innovation, Bandung, Indonesia

Faculty of Engineering, University of Indonesia, Depok, Indonesia

A. Basit et al. (eds.), *Proceedings of the International Conference on Radioscience, Equatorial Atmospheric Science and Environment and Humanosphere Science*, Springer Proceedings in Physics 290, https://doi.org/10.1007/978-981-19-9768-6_38

1 Introduction

Flood is one of Indonesia's most frequent hydrometeorological disasters, including on Borneo Island. Social media-based data in 2015–2021 from a global database of flood events record floods in Borneo can occur in all parts of the island in all seasons [1]. Still, the highest flood events occur in the December-January–February (DJF) season. One of the phenomena that trigger flooding is extreme rainfall. Therefore, a study on extreme rainfall in Borneo is needed so flood disaster mitigation can run more effectively.

Previous climatological studies have examined the phenomena affecting rainfall variability around the Borneo region and noted the importance of synoptic disturbances [2, 3]. However, the spatial pattern of extreme rainfall that commonly occurs on the island of Borneo due to those synoptic disturbances is still not well understood. Extreme rainfall events seem to recur from time to time according to repeated atmospheric circulation conditions [4]. Thus, it is possible to classify the spatial patterns of extreme rainfall [5]. Our study focuses on finding the spatial pattern of extreme rainfall that commonly occur on Borneo Island and its relationship with synoptic phenomena to increase the predictability of extreme rainfall, as one of the flood triggers.

2 Methodology

2.1 *Data*

Daily rainfall data is taken from near-real-time gauge-adjusted rainfall of the Global Satellite Mapping of Precipitation (GSMaP) at $0.1° \times 0.1°$ grids [6], which could be effective for extreme rainfall monitoring in Asia–Pacific [7]. This data can be downloaded using the FTP address hokusai.eorc.jaxa.jp. Specific humidity, wind fields, and sea surface temperature (SST) are obtained from the European Center for Medium-Range Weather Forecasts Reanalysis v5 (ERA5) and are used for a composite analysis of atmospheric conditions during extreme rainfall events. ERA5 has a 0.25° horizontal resolution, provided at hourly intervals [8], and is free for download from the climate data store website https://cds.climate.copernicus.eu/. All the datasets are retrieved for the study period of 2001–2021.

2.2 *Methods*

GSMaP rainfall data are masked on Borneo Island to exclude extreme rainfall in the ocean. An extreme rainfall day is identified if one or more grid points of daily rainfall in Borneo are greater or equal to its 97th percentile of the cumulative distribution function of daily rainfall maxima for the study period. The identified date is classified as the date of the extreme rainfall event.

To reveal dominant spatial patterns of extreme rainfall events in Borneo, k-means clustering is applied to the correlation matrix of extreme rainfall events in each season (DJF, March–April-May (MAM), June-July–August (JJA), and September–October-November (SON)). The number of clusters in each season is chosen using the trial-and-error method. For each season, k-means clustering is applied for 2, 3, 4, or 5 clusters. Then, the number of clusters with the most representative spatial patterns (no similar pattern between clusters) is chosen. A composite analysis of moisture flux, SST, and wind anomalies is used to determine synoptic conditions that trigger extreme rainfall events in each cluster. Moisture flux (MF) is obtained from integrating specific humidity (q) and wind field (V) from the surface to 700 hPa:

$$MF = \frac{1}{g} \int_{\text{surface}}^{700\,\text{hPa}} qV\, dp \tag{1}$$

where g is the gravitational constant and p is pressure. Then the divergence of moisture flux (MFD) can be calculated using

$$MFD = \nabla \cdot MF \tag{2}$$

Positive (negative) values of MFD denote divergence (convergence) areas.

3 Results

3.1 Temporal Distribution of Extreme Rainfall

The value of extreme rainfall refers to the above-mentioned threshold from 2001–2021 (191 mm/day). During the study period, there were 231 total extreme rainfall events, which vary per year (Fig. 1a). The most frequent occurrences were in 2019, with twenty-six events. In the monthly distribution (Fig. 1b), extreme rainfall mostly occurs in December, January, February, and May, or DJF for the quarterly seasons.

3.2 Spatial Distribution

Borneo extends around the equator for which heavy rainfall can occur throughout the year. However, each season has a different circulation pattern so it may trigger different spatial patterns of extreme rainfall. The clustering result shows four clusters

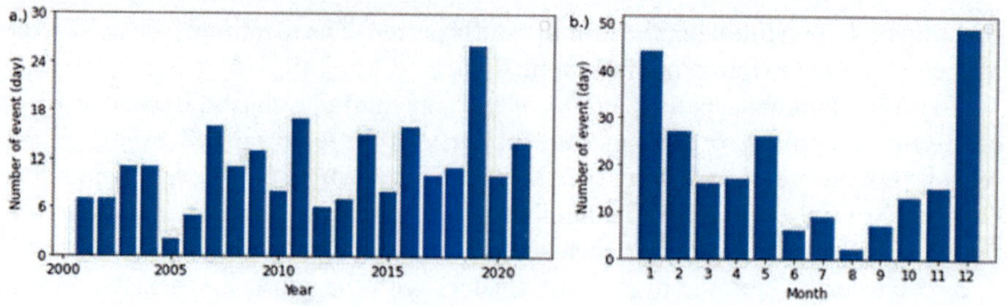

Fig. 1 Yearly (**a**) and monthly (**b**) number of extreme rainfall events during the period 2001–2021 in Borneo

(C1, C2, C3, and C4) of extreme rainfall in DJF and MAM and three clusters (C1, C2, and C3) in JJA and SON. Note that C1 in DJF is not related to C1 in other seasons. C1 in DJF means the first spatial pattern of extreme rainfall in the DJF season. Clusters of extreme rainfall in each season show a certain spatial pattern that is different from other clusters. Each cluster may be caused by similar triggers but may also be by different triggers.

3.3 *The Synoptic Patterns of Extreme Rainfall Events*

DJF (Fig. 2). Extreme rainfall events in C1 occur when negative SST anomalies cover the entire South China Sea (SCS) with northerly moisture transport anomalies from the Philippines. This moisture transport converges in northern Borneo and causes extreme rainfall in the Sabah region. In C2, there are positive SST and easterly moisture transport anomalies in the Java Sea. These positive SST anomalies moisten southern Borneo and trigger extreme rainfall in Central Kalimantan and South Kalimantan. SST and moisture transport anomalies at C3 are slightly like C1. The difference lies in the moisture transport at C3 originating from SCS and converging in western Borneo, thus causing extreme rainfall in West Kalimantan and Sarawak region. In C4, there are positive SST anomalies around northern Borneo and westerly moisture transport anomalies to the Philippines. These conditions cause wet anomalies in North Kalimantan and SCS.

MAM (Fig. 3). In C1, extreme rainfall events occur when there are westerly moisture transports anomalies in Borneo Island that converge around western Sulawesi. This moisture convergence causes heavy rainfall in East Kalimantan and South Kalimantan. C2 in MAM is similar to C1 in DJF with negative SST anomalies in SCS and northerly moisture transport from the Philippines. This moisture transport causes moisture convergence in northern Borneo and extreme rainfall around the Sabah region. C3 in MAM is slightly similar to C2 in DJF with positive SST anomalies and easterly moisture transport in the Java Sea, thus causing rainfall anomalies around

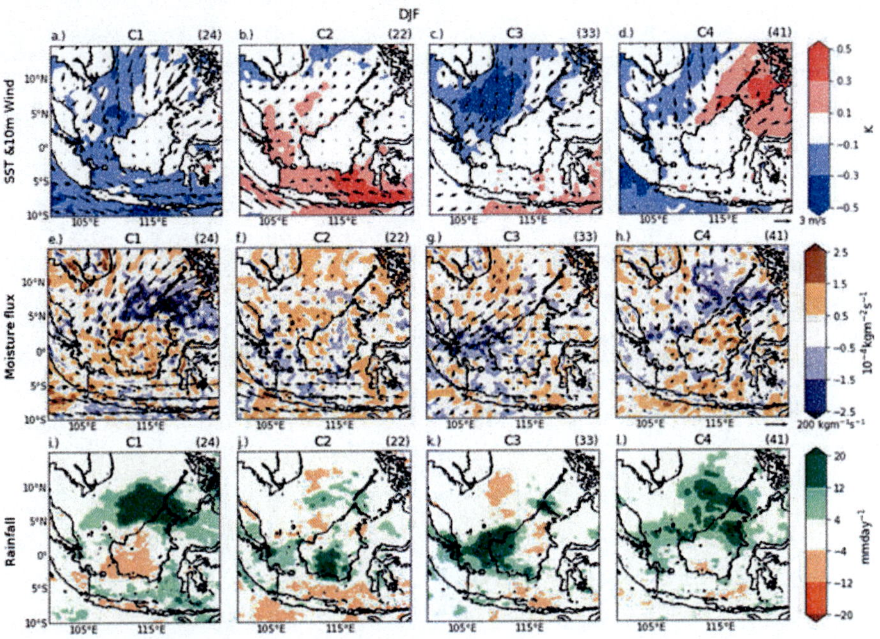

Fig. 2 Clusters composite anomalies of [top] SST (shaded) and 10 m wind (arrows); [middle] vertically integrated 1000–700 hPa moisture divergence (shaded) and moisture transport (arrow); and [bottom] rainfall (shaded) in DJF season. The number in brackets indicates the number of members in each cluster

Central Kalimantan. C4 in MAM is quite similar to C4 in DJF with westerly moisture transport anomalies to the Philippines and heavy rainfall around North Kalimantan and SCS.

JJA (Fig. 4). Extreme rainfall events in JJA occur when the SST anomaly is positive in almost all of the Western Maritime Continent (WMC). The difference lies in the direction of moisture transport. In C1, there are northeasterly moisture transport anomalies from the Philippines and SCS. This moisture transport converges on the SCS, the Karimata Strait, and the west of the Java Sea, causing rainfall anomalies on the coast of this part of Borneo. C2 in JJA is slightly similar to C4 in DJF and C4 in MAM with westerly moisture transport anomalies to the Philippines and heavy rainfall around North Kalimantan and SCS. C3 shows a similar case to C1 but with weaker northeasterly moisture transport anomalies in SCS and stronger anomalies in the Philippines and the Celebes Sea. This moisture transport converges in Makassar Strait and causes extreme rainfall in East Kalimantan and South Kalimantan.

SON (Fig. 5). Extreme rainfall events in SON occur when there is northeasterly moisture transport from SCS. This moisture transport turns west at C1 and converges around the southern part of SCS. This moisture convergence causes extreme rainfall around the Sarawak region. In C2, the moisture transport turns south to the Karimata Strait. Moreover, there are positive SST anomalies around the Java Sea in C2. This

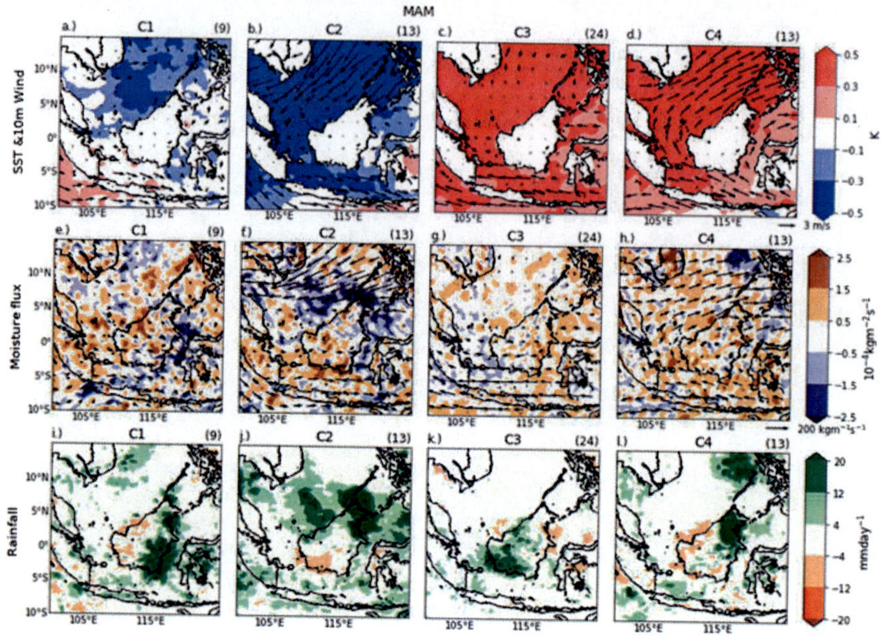

Fig. 3 Same with Fig. 2 but in MAM season

warm SST moistens the southern part of Borneo, thus causing heavy rainfall in Central Kalimantan. In C3, the moisture transport turns east resulting in westerly moisture transport anomalies to the Philippines. This pattern is analogous with C4 in DJF, C4 in MAM, and C2 in JJA with wet anomalies in North Kalimantan.

4 Discussion

The Asian monsoon has the most significant effect on rainfall patterns in WMC [9, 10]. The complex terrain of the archipelago strongly influences the local wind circulation and rain distribution in WMC [10]. Interannual climate phenomena, such as ENSO and IOD, affect the amount of annual rainfall [11, 12]. In the intra-seasonal time range, synoptic phenomena such as strong northerly wind known as Cold Surge (CS) [13], vortices around the waters and mainland of Borneo known as Borneo Vortex (BV) [2], and Madden Julian Oscillation (MJO) [14] interfere with the variation of monsoonal rainfall in the region.

Referring to the average climate regime pattern in WMC [15], we hypothesized several synoptic climate phenomena causing rainfall anomalies in Borneo. Based on previous analysis, this study can identify the intensified northerly wind known as CS as a possible specific synoptic climate affecting the heavy rainfall in northern Borneo at DJF C1 and MAM C2 and in western Borneo at DJF C3. The difference in

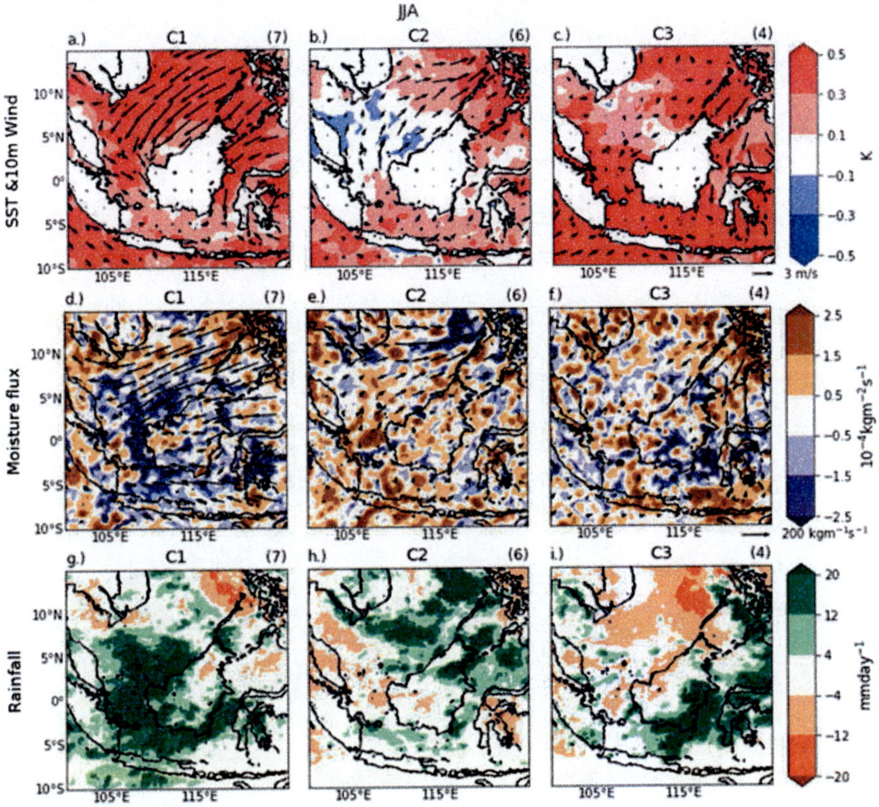

Fig. 4 Same with Fig. 2 but in JJA season

the location of the wet anomalies may occur due to the difference in the direction of moisture transport in the CS, from the Philippines in DJF C1 and MAM C2, and from SCS in DJF C3. The wet anomaly around North Kalimantan at DJF C4, MAM C4, JJA C2, and SON C3 may be related to the MJO phenomenon with convective center around the Philippines because the wind patterns in the four clusters show westerly wind anomalies to the Philippines. This is in line with the statement that westerly winds exist in and to the west of the center of MJO large-scale convection [14]. Positive SST anomalies in the Java Sea may also have an important role in extreme rainfall events around Central Kalimantan such as in DJF C2, MAM C3, and SON C2. This preliminary conclusion still needs confirmation by observing more in detail the relationship between rainfall intensity and the indexes of CS, MJO, and vorticity particularly that related to the BV. Meanwhile, knowledge about the influence of other climate variations, both synoptic and interannual, on extreme rainfall in Borneo still requires further investigation and will be the focus of further studies.

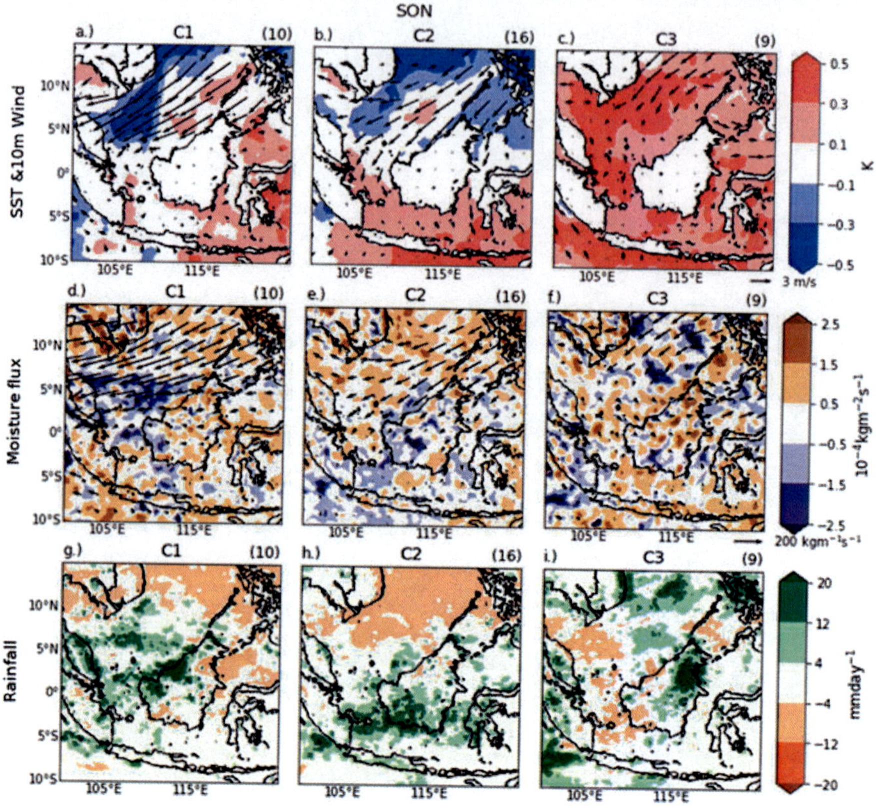

Fig. 5 Same with Fig. 2 but in SON season

5 Conclusion

This study has clustered the spatial patterns of extreme rainfall in Borneo during 2001–2021 to investigate the synoptic conditions associated with each cluster. There are several synoptic patterns associated with extreme rainfall in Borneo. Strong northerly moisture transport with negative SST anomalies occurring in the DJF and MAM seasons may be associated with CS events and cause heavy rainfall around northern or western Borneo. Westerly wind anomalies with positive SST around northern Borneo that can occur in all seasons may be related to the MJO with convective centers around the Philippines and causing heavy rainfall around North Kalimantan. In addition, positive SST anomalies around the Java Sea may also have an important role in triggering extreme rainfall events around Central Kalimantan.

Acknowledgements We would like to thank the Earth Sciences Research Organization BRIN for its financial support through the Disaster Home Financing Program (WBS2-15) for the fiscal year 2022. This study was also partially supported by the FITB-ITB Collaborative Research PPMI.

NNP was also partly supported by the project titled "Marine Science & Technology Cooperation (MTCRC) between Korea and Indonesia" funded by the Ministry of Oceans and Fisheries, Korea.

References

1. de Bruijn, J.A., de Moel, H., Jongman, B., et al.: A global database of historic and real-time flood events based on social media. Sci Data **6**, 311 (2019). https://doi.org/10.1038/s41597-019-0326-9
2. Chang, C.-P., Harr, P.A., Chen, H.-J.: Synoptic disturbances over the equatorial South China Sea and Western Maritime Continent during boreal winter. Mon. Weather Rev. **133**, 489–503 (2005). https://doi.org/10.1175/MWR-2868.1
3. Lim, S.Y., Marzin, C., Xavier, P., et al.: Impacts of boreal winter monsoon cold surges and the interaction with MJO on Southeast Asia rainfall. J. Clim. **30**, 4267–4281 (2017). https://doi.org/10.1175/JCLI-D-16-0546.1
4. Jacobeit, J., Homann, M., Philipp, A., Beck, C.: Atmospheric circulation types and extreme areal precipitation in Southern Central Europe. Adv. Sci. Res. **14**, 71–75 (2017)
5. Gvoždíková, B., Müller, M., Kašpar, M.: Spatial patterns and time distribution of central European extreme precipitation events between 1961 and 2013. Int. J. Climatol. **39**, 3282–3297 (2019). https://doi.org/10.1002/joc.6019
6. Kubota, T., Aonashi, K., Ushio, T., et al.: Global satellite mapping of precipitation (GSMaP) products in the GPM era. In: Satellite Precipitation Measurement. pp. 355–373. Springer (2020)
7. Tashima, T., Kubota, T., Mega, T., et al.: Precipitation extremes monitoring using the near-real-time GSMaP product. IEEE J. Selected Topics in Appl. Earth Observ. Remote Sens. **13**, 5640–5651 (2020)
8. Hersbach, H., Bell, B., Berrisford, P., et al.: The ERA5 global reanalysis. Q. J. R. Meteorol. Soc. **146**, 1999–2049 (2020). https://doi.org/10.1002/qj.3803
9. Robertson, A.W., Moron, V., Qian, J.-H., et al.: The maritime continent monsoon. In: The Global Monsoon System: Research and Forecast, pp. 85–98. (2011)
10. Chang, C.-P., Lu, M.-M., Lim, H.: Monsoon convection in the Maritime Continent: interaction of large-scale motion and complex terrain. Meteorol. Monographs **56**, 6.1–6.29 (2016). https://doi.org/10.1175/AMSMONOGRAPHS-D-15-0011.1
11. Aldrian, E., Dümenil Gates, L., Widodo, F.H.: Seasonal variability of Indonesian rainfall in ECHAM4 simulations and in the reanalyses: the role of ENSO. Theoret. Appl. Climatol. **87**, 41–59 (2007)
12. Narulita, I., Fajary, F.R., Syahputra, M.R., et al.: Spatio-temporal rainfall variability of equatorial small island: case study Bintan Island, Indonesia. Theor. Appl. Climatol. **144**, 625–641 (2021). https://doi.org/10.1007/s00704-021-03527-x
13. Abdillah, M.R., Kanno, Y., Iwasaki, T., Matsumoto, J.: Cold surge pathways in East Asia and their tropical impacts. J. Clim. **34**, 157–170 (2021). https://doi.org/10.1175/JCLI-D-20-0552.1
14. Zhang, C.: Madden-Julian oscillation. Rev. Geophys. **43** (2005). https://doi.org/10.1029/2004RG000158
15. Chang, C.-P., Wang, Z., McBride, J., Liu, C.-H.: Annual cycle of Southeast Asia—Maritime Continent rainfall and the asymmetric monsoon transition. J. Clim. **18**, 287–301 (2005). https://doi.org/10.1175/JCLI-3257.1

Analysis of Wind Variations and Differences on the Airport Runways: a Case Study at I Gusti Ngurah Rai Airport, Bali

Kadek Sumaja and Amanda Pasa Kencana

Abstract Surface wind conditions are strongly influenced by several factors, such as geographical location, seasons, and temperature differences. I Gusti Ngurah Rai Airport is located on the narrow land on Bali Island and surrounded by the ocean. Thus, its runway is prone to significant wind variation due to the changing season as well as the influence of land and sea breeze. Moreover, there is an essential issue in determining the runway in use for the takeoff and landing process when there are significant differences in wind direction and speed between two runways. Therefore, an analysis was carried out to determine the variation of wind direction and speed on runways 09 and 27 in I Gusti Ngurah Rai, Bali. The 10 years of observation data were retrieved from AWOS, then processed to define the wind data in both runways. The wind rose method was utilized to analyze the wind pattern on both runways in each season. Then, the delta calculation was utilized to determine the significant difference in wind direction and speed on the two runways. The result showed that the wind variation on both runways were generally similar and the significant wind direction differences mainly occurred during the transition season. Moreover, the highest number of significant wind direction differences generally emerged before noon and at night while the lowest number was in the evening. Lastly, wind speeds above 10 KT mostly occurred on runway 09 when the wind direction in both runways was significantly different.

1 Introduction

Wind information on the airport runway is crucial to support efficiency and safety during the takeoff and landing process of aircraft [1–3]. The takeoff and landing

K. Sumaja (✉)
I Gusti Ngurah Rai Meteorological Office, Indonesian Agency for Meteorology Climatology and Geophysics, Bali, Indonesia
e-mail: kadek.sumaja@bmkg.go.id

A. P. Kencana
Bandung Institute of Technology, Bandung, Indonesia

A. Basit et al. (eds.), *Proceedings of the International Conference on Radioscience, Equatorial Atmospheric Science and Environment and Humanosphere Science*, Springer Proceedings in Physics 290, https://doi.org/10.1007/978-981-19-9768-6_39

activities are strongly affected by the wind pattern on the runway. The aircraft started the process from one end point of the runway which is known as the runway in use. I Gusti Ngurah Rai Airport has a single runway, which stretches from east to west, where the eastern touch point is called runway 27 and the touch point at the western end of the runway is called runway 09. In performing safe and efficient takeoff and landing process, the aircraft must go downwind [2]. Thus, reliable wind information is crucial at each end of the runway. To provide this information, along the runway at the I Gusti Ngurah Rai Airport were installed three wind sensors as part of the Automatic Weather Observation System (AWOS) at the eastern (runway 27), the middle, and the western point (runway 09).

The wind direction in Bali Island, where I Gusti Ngurah Rai Airport is located, changes every season due to the influence of the monsoon [4]. Moreover, the runway at I Gusti Ngurah rai airport is flanked by the ocean in the west and east. Therefore, wind variations are most likely to arise due to local influences such as land and sea breezes as well as heat capacity and surface roughness between sea and water [5]. This geographical condition not only leads to the occurrence of crosswind, wind gust, and wind shear but also significant differences in wind direction between both runways [6]. Therefore, this study aims to analyze the wind direction variation and its differences on both I Gusti Ngurah Rai Airport's runways to determine the time occurrence and frequency of these significant differences.

2 Data and Methods

This study utilized wind observation data on runway 09 and runway 27 from wind sensors of the Automatic Weather Observation System (AWOS) for 10 years period (2012–2021), as shown in Fig. 1. The time used is in Coordinated Universal Time (UTC) as recorded in AWOS. The wind direction and speed were retrieved from 30 min' weather data known as MET REPORT, and weather data when significant weather conditions occurred known as SPECIAL REPORT. Wind direction is the direction from which the wind blows and wind speed defines how strong the wind blows. The wind data was shortened based on the completeness of the wind data on the two runways, so around 97% of total data were passed pre-processing and utilized in this research.

In addition, this study also employed the 10 years general wind data at I Gusti Ngurah Rai Airport to classify the runway wind data analysis into a group with similar patterns. Then, the data was divided based on the direction and wind speed on runway 09 and 27 using Microsoft Excel. Furthermore, it was categorized into certain periods based on the normal wind [7, 8]. Thus, four wind periods were obtained, namely December-January–February (DJF) was a westerly wind, May–June-July–August-September (MJJAS) was an easterly wind, March–April (MA) was a transition period from westerly to easterly wind, and October–November (ON) was a transition period from easterly to westerly wind, as shown in Fig. 2.

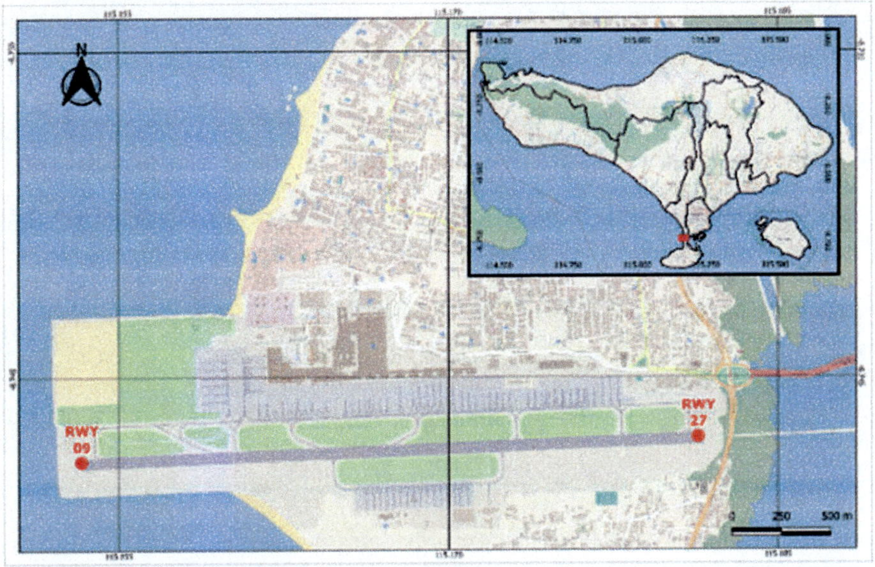

Fig. 1 Runways 09 and 27 at I Gusti Ngurah Rai Airport

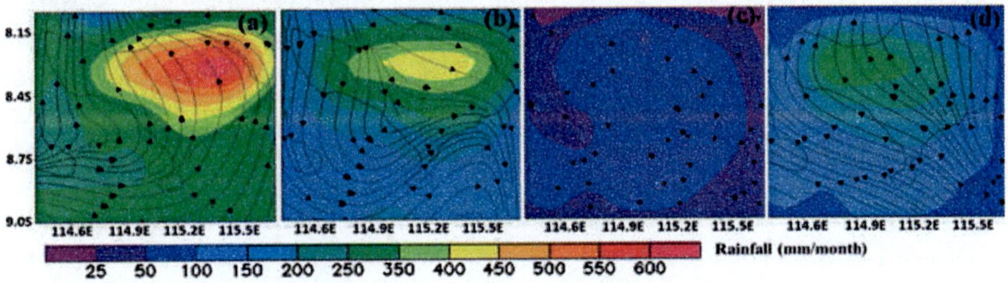

Fig. 2 Wind and rainfall in Bali period **a** DJF, **b** MA, **c** MJJAS, and **d** ON

The analysis is carried out by looking for the difference in wind direction between runways 09 and 27. The wind direction is measured between 1 and 360° (in form of a cycle from the north, east, south, west, and back to the north) and the highest difference would be 180°. Therefore, when the difference value is more than 180°, the value that would be used in the analysis is the result of subtraction between 360° and the initial value. For instance, if the difference in wind direction between both runways is 200°, then the value that will be used is $360^{\circ}–200^{\circ} = 160^{\circ}$. Then, referring to the BMKG regulations regarding significant changes or differences in wind direction, the results of the difference in wind direction to be analyzed are those whose values are above 60° [9, 10]. Then from the difference in the significant wind direction, it will be selected for direction with a speed above 10 Knots (KT), and time of occurrence for each season. The wind difference was considered to interfere with the takeoff and landing process because it complicates the selection of the runway in use [1].

3 Result and Analysis

3.1 Wind Pattern in Runways 09 and 27

The wind patterns on runway 09 were generally similar to those of runway 27, as shown in Figs. 3 and 4. For the DJF, the wind direction on runway 09 varied from east to southwest, in contrast on Runway 27 it only differed from north to west. Furthermore, in the MA and ON periods, it had more variation. The wind on runway 09 was blown almost from every direction, slightly different from the wind direction conditions on runway 27 with no wind blowing from the southwest. A unique condition occurs in the MJJAS period, where the wind was blown only from the east and southeast. This indicates that the transition season, MA and ON, have a high potential for significant wind direction differences between runways 09 and 27. It is due to the uncertain wind direction during the transition season caused a high fluctuation in wind movement on the runway and lead to an increase in the wind direction difference on runway 09 and 27 [11]. Moreover, some wind variation occurred during the DJF period, especially in runway 09. This period is the peak of the rainy season in Bali, so it has high air instability and a lot of convective clouds [8, 12]. This convective cloud during its mature state produced numerous wind gusts which most likely affect the wind on runway 09 where located beside the sea that has fewer obstacles [13]. Although the possibility of its occurrence is uncertain, wind gusts usually occur due to the influence of storms, convective clouds, and air instability [14, 15]. The MJJAS period had the smallest wind variation. It is because this period is the eastern monsoon season. This easterly wind carried less water vapor and has a more stable atmospheric condition that led to a lack of cloud growth and stable air movement [4].

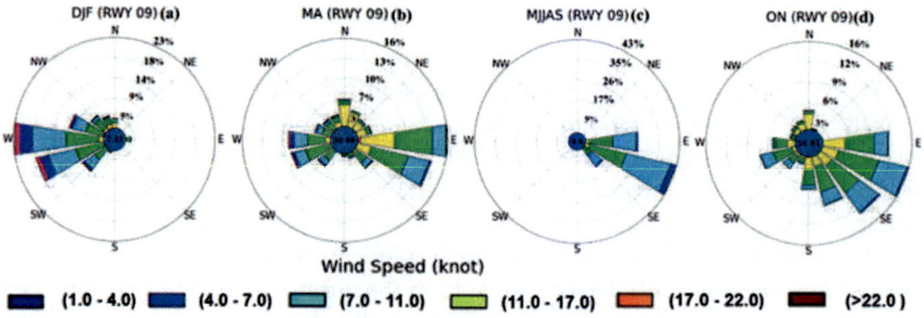

Fig. 3 Wind pattern in runway 09 period **a** DJF, **b** MA, **c** MJJAS, and **d** ON

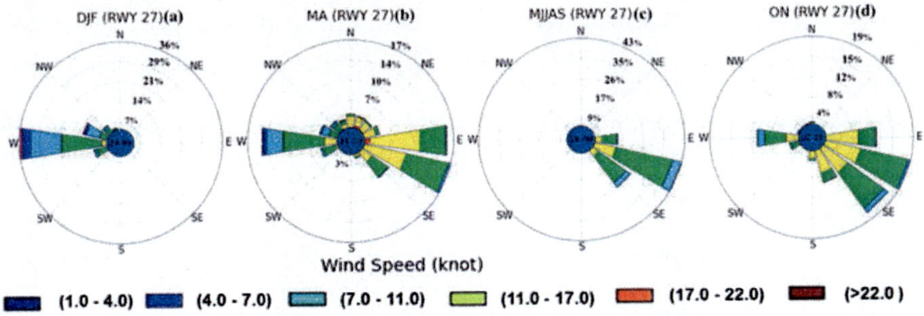

Fig. 4 Wind pattern runway 27 period **a** DJF, **b** MA, **c** MJJAS, and **d** ON

3.2 *Significant Wind Difference Between Runway 09 and 27*

Significant wind direction difference

The direction discrepancy in wind data between runways 09 and 27 is then analyzed for significant events or when the difference is above 60^0. The number of occurrences of significant differences was categorized according to periods of normal wind data (DJF, MA, MJJAS, and ON) and compared with the total incidence in each season to determine the percentage of the occurrence. The results were shown in Fig. 5a, where the ON period has the highest percentage of occurrence (18%) of significant wind direction difference between runway 09 and 27. Then followed by months MA, MJJAS, and DJF with 14%, 13%, and 10%, respectively. Thus, it is necessary to put more attention out for the high incidence of significant wind differences in October–November. Since this period is the beginning of the rainy season in Bali, it would cause the potential for disturbance to increase [8, 12].

Furthermore, wind speed data, when the difference in wind direction on runway 09 and 27 was significant, was selected as the maximum wind speed. This aimed to determine the potential occurrence of maximum wind disturbance on this significant

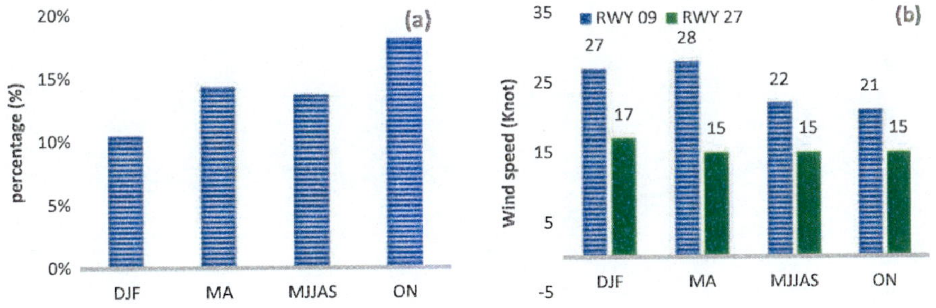

Fig. 5 **a** Occurrence of significant wind direction difference between runway 09 and 27, **b** Maximum wind speed during significant difference in wind direction between runway 09 and 27

wind variation. The maximum wind speed on run-way 09 is always higher than that of runway 27, as shown in Fig. 5b. Although the MA Period has the highest maximum wind speed (28 KT) that occurs on runway 09, the DJF period has a high maximum wind speed on both runways (above 17 KT). The MJJAS and ON periods have almost the same maximum wind speed, namely 21–22 KT on runway 09 and 15 KT on runway 27. These results indicate that airlines and navigation parties should pay more attention to the DJF month because of the significant difference in wind direction and speed. Moreover, the maximum winds on both runways were considerably high and this period is the peak of the rainy season in Bali [8, 12, 13].

In addition, the significant wind direction difference between the two runways was displayed in the hourly range to find out at what time the significant wind direction difference on the two runways occurred. The results showed that the number of significant differences in wind direction generally occurred in the morning, then increased during the day, decreased at night, rosed in the middle of the night, and slowly dropped to the morning. This was caused by the location of the runway which is located between two oceans. Thus, the wind variation most likely occurred due to the differences in heat capacity and surface roughness between sea and land. The boundary area between land and ocean has a characteristic of local wind that changes periodically in form of the sea and the land breeze that happens diurnally [11]. These breezes tend to influence the monsoonal wind on this runway to depend on its speed [5].

The most significant difference in direction during DJF and ON periods mostly occurred at 14 UTC and 1 UTC, respectively. The ON month also shows the number of significant wind direction differences in the afternoon (18 to 19 UTC) and night (13–14 UTC). Furthermore, significant differences in MA and MJJAS months occurred in the morning at 2 UTC and in the evening at 12 and 13 UTC. Generally, the lowest difference in wind direction across all seasons was in the evening from 9 to 10 UTC (Fig. 6).

Wind Speed when the difference in wind direction is significant

Wind speed above 10 knots on runway 09 and 27 during significant wind direction difference was displayed as the frequency of occurrence, as shown in Fig. 7a–d.

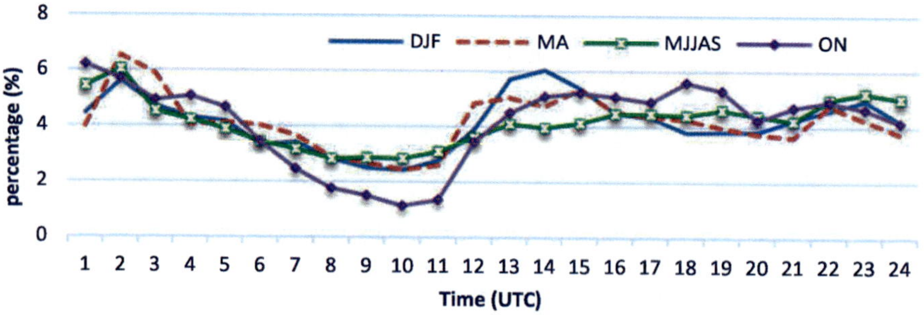

Fig. 6 The occurrence (%) of significant wind direction difference between runway 09 and 27

Generally, when the wind direction in both runways was significantly different, the frequency of wind speeds above 10 KT on runway 09 was always higher than on runway 27. This is due to runway 09 is directly adjacent to the open ocean compared to runway 27 in the east, which has bays and narrow straits as its borders. This would generate a factor where the direct heating of the sun during the day is greater than the heating by the earth's surface [16]. Thus, the influence of sea breezes on runway 09 was mostly higher since the sea breezes have a greater speed than land breezes [17]. A wind speed of 11 KT was also the most frequent in each period. Wind speeds above 10 KT on runway 27 mainly occurred in the DJF and MA periods, followed by ON and the least occurred in the MJJAS periods. Therefore, the DJF and MA periods could potentially disrupt takeoff and landing activities.

In addition, the significant wind data which has wind speed above 10 KT was classified into the normal wind periods. It can be seen in Fig. 8 that the DJF period has the most events, followed by MA, then the MJJAS and ON periods. Although there were many incidents of significant wind direction differences, only a few have velocities above 10 KT that occur simultaneously on both runways. As a result, there would be a rare circumstance for the Air Traffic Controller (ATC) to face the challenging situation of determining the runway in use. Because when the wind direction on both runways is significantly different, but the speed on one runway is low, and the other is high, the runway chosen for takeoff and landing would be the runway with the highest wind speed. However, if both runways have high wind

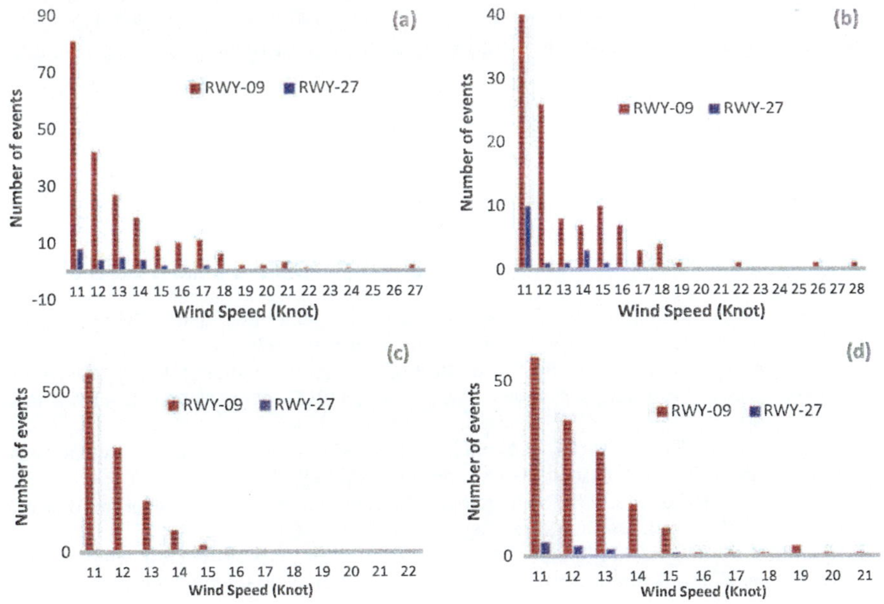

Fig. 7 Wind speed when the difference of wind direction was above 60° for period **a** DJF, **b** MA, **c** MJJAS, and **d** ON

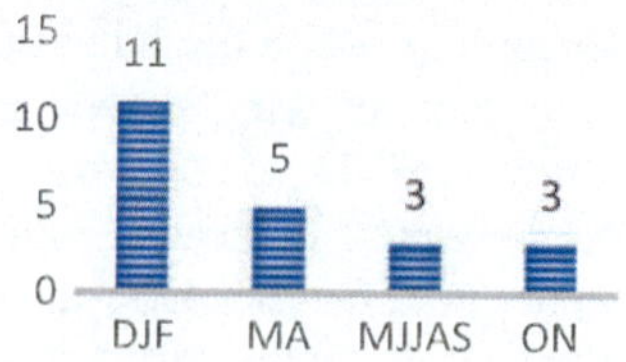

Fig. 8 Frequency of wind speed above 10 knots on both runways 09 and 27 for wind direction difference of more than 60°

speeds, the ATC will find it difficult to determine the runway in use for the takeoff and landing process. Such conditions rarely occurred at I Gusti Ngurah Rai airport, because from 186,631 observation data, there were only 22 incidents. Therefore, it can be said that the wind conditions at I Gusti Ngurah Rai airport are very supportive for flight operations services.

4 Conclusion

The wind patterns on runway 09 were generally similar to those of runway 27 and the significant differences of wind direction mostly occur during the transition season MA and ON. Moreover, the highest number of significant wind direction differences mostly occurred before noon (2–3 UTC) and at night (13–14 UTC) while the lowest number mostly occur in the evening (10–11 UTC). Lastly, the maximum wind speed as well as the frequency of wind speeds above 10 KT on run-way 09 was always higher than that of runway 27 when the wind direction in each runway was significantly different.

References

1. Federal Aviation Administration.: Takeoffs and departure climbs. Airpl. Fly. Handb.1–14. (2013)
2. van Es, G.W.H., Karwal, A.K.: Safety aspects of tailwind operations. Flight Saf. Found. (2001). [Online]. Available: https://www.skybrary.aero/bookshelf/books/1148.pdf
3. Connor, A.O., Kearney, D.: Evaluate the effect of turbulence on aircraft during landing and take-off phases. Int. J. Aviat. Aeronaut. Aerosp. **5**(4) (2018). https://doi.org/10.15394/ijaaa.2018.1284
4. Tjasyono, H.B.: The character of rainfall in the Indonesian monsoon. In: International Symposium Equatorial Monsoon Systems, pp. 1–11. (2008)
5. Tjasyono, B., Harijono, S.W.B.: Meteorologi Indonesia : Awan dan Hujan Monsun.Volume 2, 1st edn, vol. 2, Jakarta, Badan Meteorologi dan Geofisika (2007)
6. Domenico Cimini, F.S.M., Visconti, G.: In: Integrated Ground-Based Observing Systems. Berlin, Heidelberg: Springer (2011). https://doi.org/10.1007/978-3-642-12968-1
7. Pusparini, N., Winardi, Irmawan, D.: Analisa angin zonal dalam menentukan awal musim hujan di Bali bagian selatan. Bul. Fis. **16**(2), 1–10 (2015)
8. BMKG.: Informasi Hujan Bulanan Wilayah Bali. (2022). http://balai3.denpasar.bmkg.go.id/info-hujan-bulanan Accessed 28 Sep 2022

9. ICAO.: ICAO Annex 3—Meteorological Service for International Air Navigation. 17th edn, no. July 2010. International Civil Aviation Organization (2001)
10. BMKG.: Pembuatan dan Penyampaian METAR dan SPECI dalam Pelayanan Informasi Cuaca untuk Penerbangan. Indonesia (2017)
11. Rifai, A., Rochaddi, B., Fadika, U., Marwoto, J., Setiyono, H.: Kajian Pengaruh Angin Musim Terhadap Sebaran Suhu Permukaan Laut (Studi Kasus : Perairan Pangandaran Jawa Barat). Indones. J. Oceanogr. **2**(1), 98–104 (2020). https://doi.org/10.14710/ijoce.v2i1.7499
12. BMKG-Staklim Jembrana.: Buletin Prakiraan Musim Hujan 2020/2021, Jembrana (2021)
13. Risuana, I.G.S., Hendrawan, I.G., Suteja, Y.: Distribusi Spasial Total Padatan Tersuspensi Puncak Musim Hujan Di Permukaan Perairan Teluk Benoa, Bali. J. Mar. Aquat. Sci. **3**(2), 223 (2017). https://doi.org/10.24843/jmas.2017.v3.i02.223-232
14. Seregina, L.S., Haas, R., Born, K., Pinto, J.G.: Development of a wind gust model to estimate gust speeds and their return periods. Tellus A Dyn. Meteorol. Oceanogr. **66**(1), 22905 (2014). https://doi.org/10.3402/tellusa.v66.22905
15. Jungo, P., Goyette, S., Beniston, M.: Daily wind gust speed probabilities over Switzerland according to three types of synoptic circulation. Int. J. Climatol. **22**(4), 485–499 (2002). https://doi.org/10.1002/joc.741
16. Kurnia Anzhar, Y.S.B.S.: Pola Angin Laut dan Angin Darat di Daerah Ujung Lemahabang, Semenanjung Muria. J. Pengemb. Energi Nukl. **2**(4), 199–206 (2000). https://doi.org/10.17146/jpen.2000.2.4.2024
17. Pahlevi, A.R.: Simulasi Interaksi Angin Laut dan Bukit Barisan dalam Pembentukan Pola Cuaca di Wilayah Sumatera Barat Menggunakan Model WRF-ARW. Pros. Semin. Nas. Metod. Kuantitatif **1**(1), 7–17 (2017)

Diurnal Variation of Rainfall Over Bangka Belitung Islands Determined from Rain Gauge and IMERG Observations

Helmi Yusnaini, Zahwa Vieny Adha, Ravidho Ramadhan, Marzuki Marzuki, and Robi Muharsyah

Abstract The diurnal cycle of rainfall is the primary circulation of the atmosphere, which is highly dependent on an area's topographic conditions, land surface, and land-sea contrast. Understanding diurnal characteristics are beneficial in providing an overview of regional weather and the influence of topography on the diurnal cycle of rainfall. From previous research, it can be seen that there are differences in the diurnal cycle between the western and eastern parts of Sumatra. However, research in the eastern part of Sumatra is very limited, especially on small islands. Therefore, in this study, the diurnal rainfall characteristic on small islands of east Sumatra, i.e., Bangka Belitung Islands, from December 2015 to October 2019, was investigated using rain gauge and Integrated Multi-satellite Retrieval for GPM (IMERG) data. The diurnal characteristics of rainfall are defined as precipitation amount (PA), precipitation frequency (PF), and precipitation intensity (PI). The average PA, PF, and PI on the island by rain gauge (IMERG) observation, respectively, are 0.15–0.27 mm/h (0.22–0.35 mm/h), 2.69–4.88% (6.15–8.59%), and 5.14–7.87 mm/h (3.35–3.84 mm/h). Rain peaks from PA, PF, and PI dominated the early afternoon (1200–1500 LST). Meanwhile, the PI peaks by IMERG show different peaks between the southern and northern parts of the island, i.e., early morning (0200 LST).

1 Introduction

The diurnal cycle of rainfall is a local pattern of rainfall distribution caused by an intense convection process (interaction between the surface and the atmosphere). These cycles tend to be driven by local circulations such as land, sea, mountain, and valley winds [1]. As an area that receives high solar radiation yearly, this diurnal cycle is significant in the tropics, especially the Indonesian maritime continent (IMC) [2].

H. Yusnaini · Z. V. Adha · R. Ramadhan · M. Marzuki (✉)
Department of Physics, Universitas Andalas, Padang 25163, Indonesia
e-mail: marzuki@sci.unand.ac.id

R. Muharsyah
Agency for Meteorology, Climatology and Geophysics of Republic Indonesia, Jakarta, Indonesia

A. Basit et al. (eds.), *Proceedings of the International Conference on Radioscience, Equatorial Atmospheric Science and Environment and Humanosphere Science*, Springer Proceedings in Physics 290, https://doi.org/10.1007/978-981-19-9768-6_40

IMC condition with complex topography, variation in shape and size of the island, and unique shoreline shape causes inhomogeneous diurnal cycles [3]. In addition, the diurnal circulation is also strongly influenced by global atmospheric interactions such as monsoons, Madden–Julian Oscillation (MJO), ENSO, and other possible factor.

The characteristics and mechanisms of the diurnal cycle of rainfall for the Sumatra region have been carried out by many researchers [4, 5, 6, 7]. As-Syakur et al. [5] found that the peak of diurnal rainfall was observed in the early afternoon in the Barisan Mountains and the small islands on the east coast, such as Bangka and Belitung. Its study used data from Tropical Rainfall Measuring Mission multi-satellite analysis (TRMM 3B42) version 07 for 17 years. The latest research on the diurnal characteristics of rainfall in the Sumatra region was carried out by Marzuki et al. [6 & 7] by utilizing the Automatic Rain Gauge (ARG) and Integrated Multi-satellite Retrievals for the Global Precipitation Measurement (IMERG) final-run product version-6 for a period of 5 years (2015–2019). They found that the peak accumulation (PA), frequency (PF), and intensity (PI) of rainfall on Bangka Island were dominant during the day (1400 Local Standard Time/ LST (UTC + 7)).

The average PA and PF are slightly low in the area compared to areas in the Barisan Mountains, both ARG and IMERG observations. Observations of the diurnal characteristics of rainfall in the Bangka and Belitung areas are still very limited [8, 9]. Prasetya et al. [9] found that topography greatly influences and controls local circulation, such as land and sea breeze, in forming convective clouds around the hills on the Bangka and Belitung Islands. They also found that the most robust sea breeze activity was observed during the March to May (MAM) period. Supari and Setiawan [8] found that rainfall patterns in the Bangka and Belitung areas are strongly influenced by the movement of the Inter-Tropical Convergence Zone (ITCZ) and monsoons. Due to limited research on the diurnal characteristics of rainfall for the Bangka and Belitung areas, here we will describe the climatology of diurnal characteristics by utilizing data with good quality, namely ARG and IMERG, following research that has been done previously by Marzuki et al. [6, 7, 10].

2 Data and Methods

This study uses surface rainfall data from 8 stations of an automatic rain gauge (ARG) which spread across Bangka Island (7 stations) and Belitung Island (1 station). The data has an observation period from August 2015 to October 2019 and is provided by the Meteorology, Climatology, and Geophysics Agency (BMKG). ARG data has a temporal resolution of 10 min with a minimum sensitivity of 0.2 mm/10 min. Furthermore, the satellite data used is IMERG V06 data with an observation period of 2015–2020. The IMERG data is grid data with a spatial resolution of 0.1 × 0.1 and a temporal 30 min. More detailed information about these data can be read in the report of Tan et al. [11]. This data is provided free to the public by NASA which can be accessed on the web https://disc.gsfc.nasa.gov/datasets/. Several studies have carried out IMERG data validation with surface rainfall data. Validation of hourly IMERG data for the Sumatra area has been carried out by Ramadhan et al. [12, 13], and Yusnaini et al. [14].

Bangka Belitung Islands is divided into two main islands, namely Bangka Island and Belitung Island, which are small islands to the east of Sumatra Island with an area of 81.725 km^2 [8]. The topographical conditions of the Bangka and Belitung Islands can be seen in Fig. 1. These two islands generally have a distinctive topography consisting of hilly areas with the highest peak of 705 m above sea level and directly adjacent to the sea [9]. The distribution of ARG station coordinates can be seen in Fig. 1b.

The diurnal characteristics of rainfall were analyzed according to Marzuki et al. [6, 7]. Precipitation amount (PA) is rainfall accumulation from all available observation data (mm/h). Precipitation frequency (PF) is the ratio of the total data that detects the occurrence of rain to the total of all data under observation (%). Precipitation Intensity (PI) was calculated from the total rainfall at the time of the rainfall (mm/h). In addition, this study also groups rain events into three different durations, i.e., <3 h, 3–6 h, and > 6 h. Rain events are defined as events within a specific period without data gaps for more than one hour [6, 7, 15, 16]. Rain that occurs after a significant data gap of one hour will be classified as a new rain event.

3 Result and Discussion

Figure 2 shows the annual mean of annual rainfall from IMERG observations for the 2015–2020 observation period. The average annual rainfall is found to be relatively high on the Bangka and Belitung Island, namely 2500 to 3400 mm/year. Uniquely, the average rainfall on Belitung Island is higher (3400 mm/year) than on Bangka Island (2500 mm/year). Although lower than Belitung Island, rainfall on Bangka Island shows a different distribution between West Bangka (2653 mm/year) and South Bangka (2525 mm/year).

Differences in the spatial distribution of annual rainfall on the Bangka and Belitung Islands were also observed from the TRMM Multi-satellite Precipitation Analysis 3B42 Real-Time product at daily scale (TMPA 3B42RT) for ten years by Supari and Setiawan [8]. They found that rainfall was smaller in the northern part of Bangka Belitung and higher in the central part of West Bangka and North of South Bangka (1800 and 3000 mm/year, respectively). The spatial distribution of observations from TMPA 3B42RT differs from IMERG observation. This condition is due to the difference in the observation data used of the two satellite data and the difference in spatial resolution between the two satellites, i.e., 0.25° (TMPA 3B42RT) and 0.1° (IMERG). Based on the grouping of rain types carried out by Aldrian et al. [17], Bangka and Belitung Islands are included in type B. The peak of rain is strongly influenced by the phenomenon of the ITCZ and monsoon, with two peaks of rain (April and December) [8, 17].

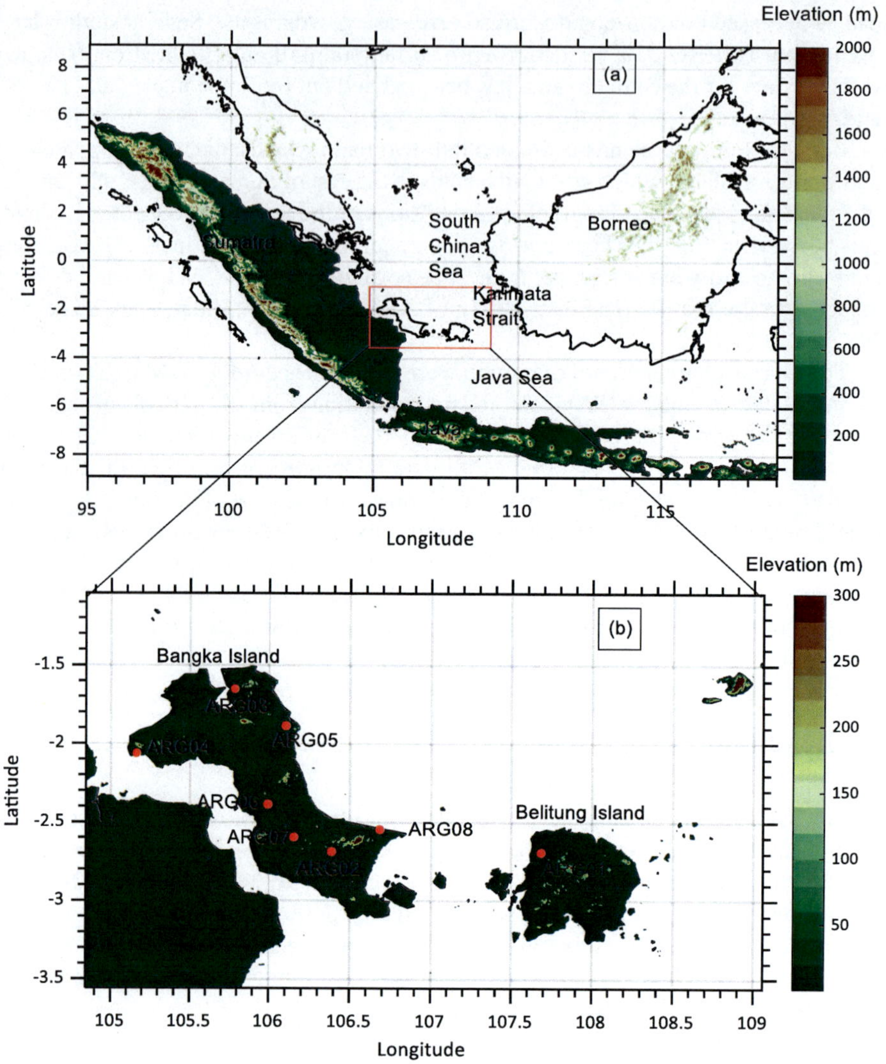

Fig. 1 Position and topography of Bangka and Belitung Islands **a** and location of ARG **b**

In this work, rainfall will be defined if the rainfall is 0.2/10 min from ARG and 0.6 /30 min from IMERG. Figure 3 shows the percentage of rain events with three different durations from the ARG and IMERG observations. ARG observations show that more than 80% of rain events on Bangka and Belitung Islands originate from short-duration rain events (< 3 h), and rain events which more than 6 h are rare (smaller than 5%) (Fig. 2a). Similar to ARG observations, IMERG also shows the Bangka and Belitung are dominated by rain with a small duration of 3 h (61.67 and 51.61%) (Fig. 2b).

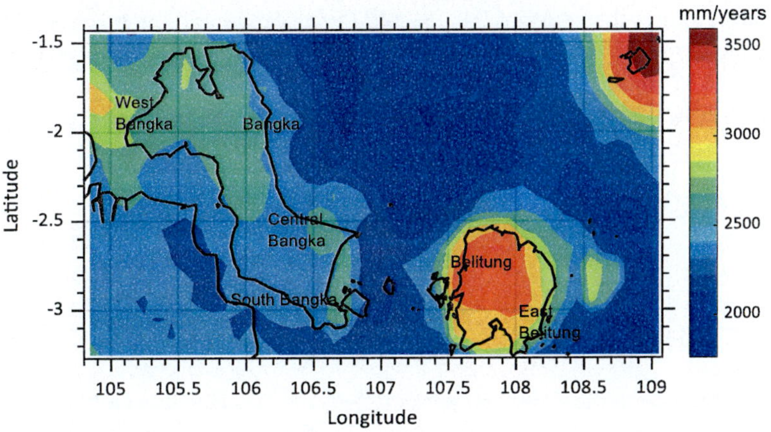

Fig. 2 Spatial distribution of annual rainfall by IMERG observation during 2015–2020

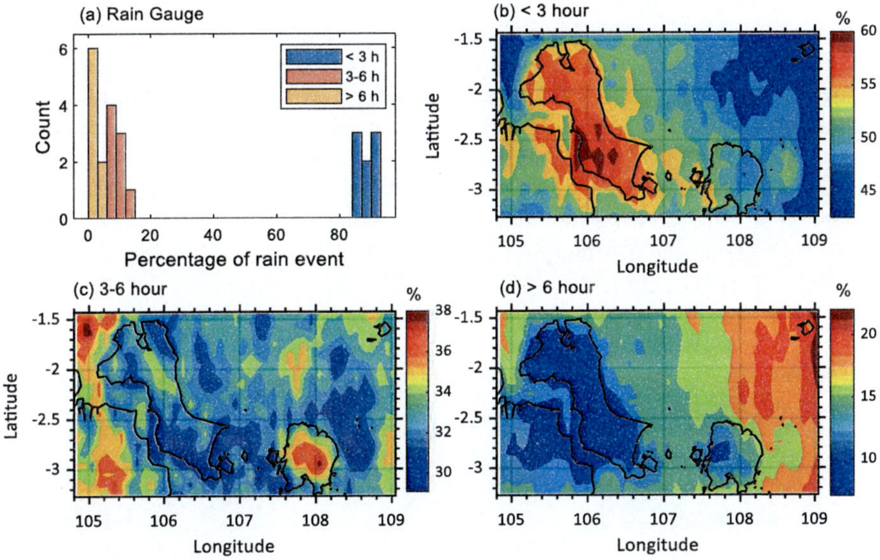

Fig. 3 Distribution of percentage of rain events for duration of < 3 h, 3–6 h, > 6 h under observation of ARG **a** dan IMERG **b**, **c**, **d**

Rainfall events with a 3–6 h duration from IMERG observations showed a higher percentage on Bangka Island (38.56%) and lowered on Belitung Island (35.31%). Meanwhile, the incidence of rain with a duration of >6 h was found to be very low over the two islands (7.50–15.00%) compared to the surrounding seas, especially around the Karimata Strait. The dominance of rain with a duration of <3 h was not only found on the Bangka and Belitung Islands but also on the Sumatra Island by Marzuki et al. [7].

Figure 4 shows the peaks of PA, PF, and PI (first row) and the mean PA, PF, and PI (second row) results from the IMERG observations. In general, the peak times of PA, PF, and PI on the Bangka and Belitung Island tend to be constant in the early afternoon (1400–1500 LST). Based on the analysis of temperature data conducted by Prasetya and Adzani [9], the peak of the rain that occurred in the afternoon came from changes in air temperature that occurred at 0700–1000 LST. Furthermore, the PI peaks were found to be quite different in areas close to the coast, especially in West Bangka and the southern side of Belitung Island. The peak of PI in this area was found in the early morning (0200 LST). More broadly, this IMERG observation shows a fairly dominant peak difference between land and sea areas, where the peaks of PA, PF, and PI in the ocean area are dominant from night to morning (2200–0800 LST).

The difference in the dominant peaks of PA, PF, and PI between land and sea is supported by the larger average PA, PF, and PI in the ocean area than in the land area. The average of PA and PF were found to be higher on Belitung Island (0.36 mm/hour and 8.64%) than on Bangka Island (0.24 mm/hour and 7.28%). Meanwhile, the average PI did not significantly differ between the two islands (3.46–3.73 mm/hour). This condition shows a significant influence from the migration of clouds formed in the ocean to land. Very high PA and PF values were observed in the sea area east of the island around the South China Sea (SCS). SCS has a very large role in the amount of water vapor formed in the east of the Sumatra Island [21]. Some of the water vapor will fall as rain on the sea, while some will be carried by the wind to the land. The closer proximity of Belitung to SCS causes it to get higher water vapor transport compared to Bangka.

Figure 5 shows the distribution of the PA, PF, and PI diurnal peaks of the ARG and IMERG. ARG only makes observations for one location point, while IMERG

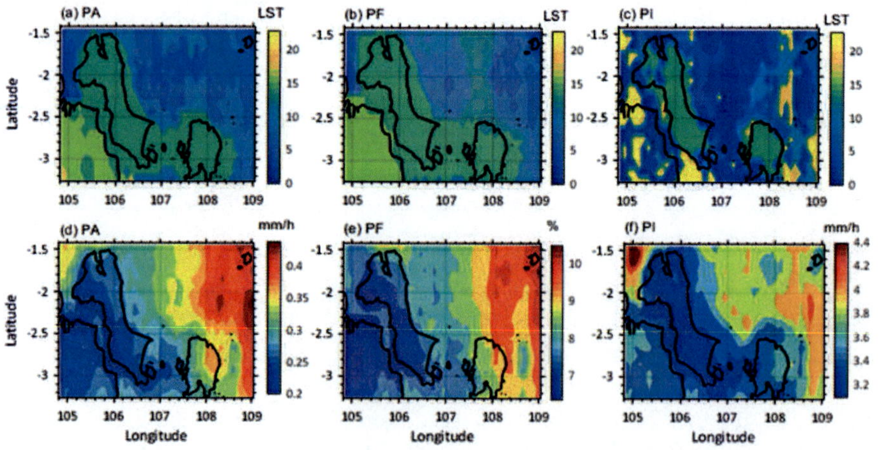

Fig. 4 Spatial distribution of peak PA **a**, PF **b**, PI **c** and mean PA **d**, PF **e**, PI **f** by IMERG observation during 2015–2020

estimates rainfall in one grid area with an area of 123.9 km^2. This difference allows for a difference between the observed diurnal peaks of ARG and IMERG. The PA peaks of the 5 ARG and IMERG stations (Fig. a, b, m, p, s) showed the same peak at 1300 LST. IMERGs also had a peak PA that was one hour slower than ARG (Fig. g) and one hour faster than ARG (Fig. 5). The peak difference not too far is also observed from the PF peak. The PF value between ARG and IMERG was observed to be very different, where peak value IMERG had a higher (interval up to 20%) than ARG (interval up to 12%). The difference in the PF value interval is caused by the difference in data length and the sensitivity of rain detection from each instrument [10]. Likewise, the observed mean PI of IMERG was found to be lower than that of ARG observations. This condition is caused by the IMERG Algorithm, using passive microwave (PMW), which tends to underestimate shallow and warm rain [7, 18].

The peak time of the rain that occurred in the two most likely also experienced a shift to global scale atmospheric phenomena, so more specific discussion was needed. This difference is also caused by the sensitivity of the sensors involved in the formation of the satellite data. Satellite sensors show better performance (more sensitive) in detecting convective rainfall, whose appearance is one hour different from other rains, especially stratiform rain [19]. This diurnal peak delay has also been reported by previous researchers [20].

4 Conclusions

The diurnal variation of rainfall in the Bangka and Belitung areas has been observed with good performance from the rain gauge and IMERG observations, especially for PA, PF, and PI. The peak times of PA, PF, and PI show differences that are not contradictory, namely in the early afternoon (1200–1500 LST). Even so, the three observation stations showed very different PI peaks, namely at the ARG01, ARG03, and ARG04 stations located in Belitung, North, and West Bangka. The observations of IMERG are dominant during the day (1300 LST) while ARG has a peak at night time (1900–2300 LST). Meanwhile, the average PA, PF, and PI show a higher distribution on the Belitung than Bangka Island, supported by the average annual rainfall, which is also high on Belitung Island (3000 mm/year). In addition, the duration of rain events was also found to be different between the islands of Bangka and Belitung, and the dominant event duration rainfall was < 3 h in Bangka and 3–6 h in Belitung.

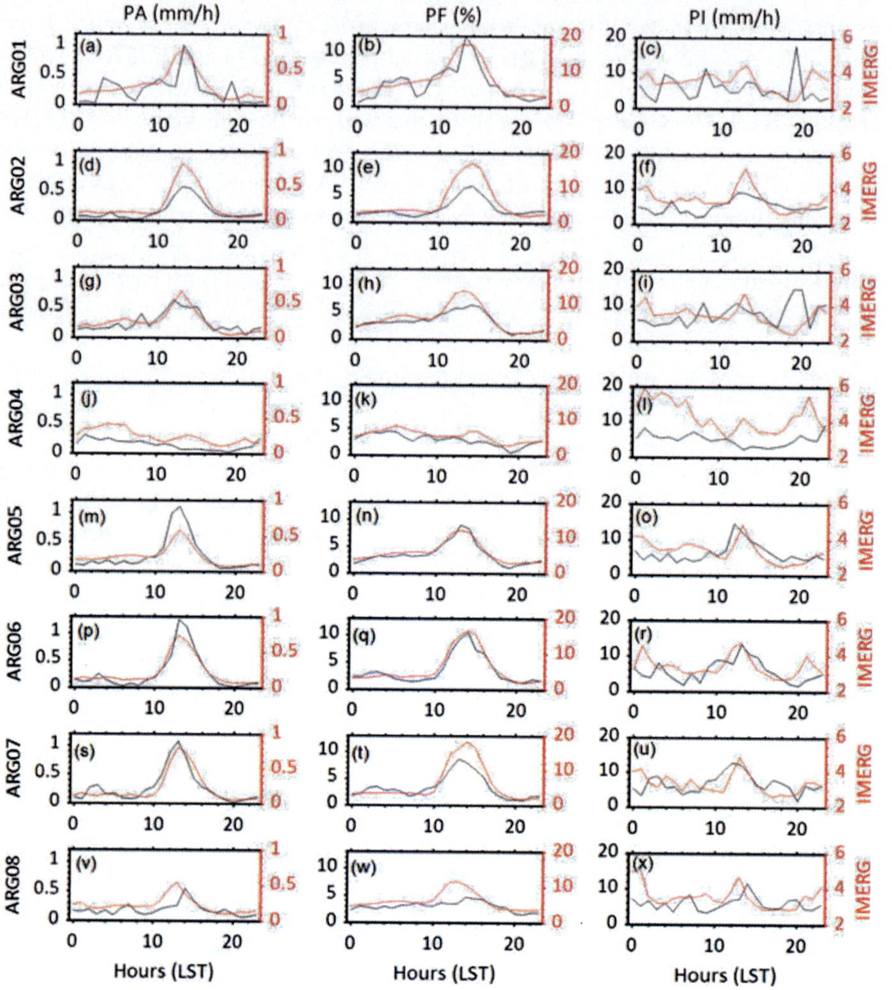

Fig. 5 Time series diurnal PA (first column), PF (second column), and PI (third column), from 8 station of ARG (the black line) and IMERG (the red line) over Bangka and Belitung Island

Acknowledgements This study was supported by Riset Publikasi Terindeks (RPT) grant from Universitas Andalas (Number: T/114/UN.16.17/PT.01.03/IS-RPT/2022). The author would like to thank the Meteorology, Climatology and Geophysics Agency (BMKG) for providing the data for research purposes. We also thank National Aeronautics and Space Administration (NASA) and the Japan Aerospace Exploration Agency (JAXA) which has provided IMERG final run product version 06 (V06).

References

1. Qian, J.H.: Why precipitation is mostly concentrated over islands in the Maritime Continent. J. Atmos. Sci. **65**(4), 1428–1441 (2008)
2. Yamanaka, M.D.: Physical climatology of Indonesian maritime continent: an outline to comprehend observational studies. Atmos. Res. **178**, 231–259 (2016)
3. Ohsawa, T., Ueda, H., Hayashi, T., Watanabe, A., Matsumoto, J.: Diurnal variations of convective activity and rainfall in tropical Asia. J. Meteorol. Soc. Japan Ser. II **79**(1B), 333–352 (2001)
4. Mori, S., Jun-Ichi, H., Tauhid, Y.I., Yamanaka, M.D., Okamoto, N., Murata, F., Sakurai, N., Hashiguchi, H., Sribimawati, T.: Diurnal land–sea rainfall peak migration over Sumatera Island, Indonesian Maritime Continent, observed by TRMM satellite and intensive rawinsonde soundings. Mon. Weather Rev. **132**(8), 2021–2039 (2004)
5. As-syakur, A.R., Imaoka, K., Ogawara, K., Yamanaka, M.D., Tanaka, T., Kashino, Y., Osawa, T.: Analysis of spatial and seasonal differences in the diurnal rainfall cycle over Sumatera revealed by 17-year TRMM 3B42 dataset. SOLA **15**, 216–221 (2019)
6. Marzuki, M., Yusnaini, H., Ramadhan, R., Tangang, F., Amirudin, A.A.B., Hashiguchi, H., Vonnisa, M.: Characteristics of precipitation diurnal cycle over a mountainous area of Sumatra Island including MJO and seasonal signatures based on the 15-year optical rain Gauge data WRF model and IMERG. Atmosphere **13**(1), 63 (2021)
7. Marzuki, M., Yusnaini, H., Tangang, F., Muharsyah, R., Vonnisa, M., Harmadi, H.: Land–sea contrast of diurnal cycle characteristics and rain event propagations over Sumatra according to different rain duration and seasons. Atmos. Res. **270**, 106051 (2022)
8. Supari, S., Setiawan, N.: Rainfall variability over Bangka Belitung island based on validated TRMM product. Jurnal Meteorologi dan Geofisika **14**(1), 9–17 (2013)
9. Prasetya, R., Adzani, R.: Analysis of small scale topography and local precipitation in Bangka Belitung Island. JFA (Jurnal Fisika dan Aplikasinya) **16**(2), 102–109 (2020)
10. Marzuki, M., Suryanti, K., Yusnaini, H., Tangang, F., Muharsyah, R., Vonnisa, M., Devianto, D.: Diurnal variation of precipitation from the perspectives of precipitation amount, intensity and duration over Sumatra from rain gauge observations. Int. J. Climatol. **41**(8), 4386–4397 (2021)
11. Tan, J., Huffman, G.J., Bolvin, D.T., Nelkin, E.J.: IMERG V06: changes to the morphing algorithm. J. Atmos. Oceanic Tech. **36**(12), 2471–2482 (2019)
12. Ramadhan, R., Yusnaini, H., Marzuki, M., Muharsyah, R., Suryanto, W., Sholihun, S., Tokay, A.: Evaluation of GPM IMERG performance using Gauge data over indonesian Maritime Continent at different time scales. Remote Sens. **14**(5), 1172 (2022)
13. Ramadhan, R., Muharsyah, R., Yusnaini, H., Vonnisa, M., Hashiguchi, H., Suryanto, W.: Evaluation of GPM IMERG products for extreme precipitation over Indonesia. J. Phys.: Conf. Ser. **2309**(1), 012008. (2022) IOP Publishing
14. Yusnaini, H., Ramadhan, R., Marzuki, M., Ningsih, A.P., Hashiguchi, H., Shimomai, T., Sholihun, S.: Statistical comparison of IMERG precipitation products with optical rain gauge observations over Kototabang, Indonesia. J. Ilmu Fisikal Universitas Andalas **14**(1), 10–20 (2022)
15. Li, L., Li, J., Chen, H., Yu, R.: Diurnal variations of summer precipitation over the Qilian mountains in Northwest China. J. Meteorol. Res. **33**(1), 18–30 (2019)
16. Yu, R., Xu, Y., Zhou, T., Li, J.: Relation between rainfall duration and diurnal variation in the warm season precipitation over central eastern China. Geophys. Res. Lett. **34**(13), (2007)
17. Aldrian, E., Dwi Susanto, R.: Identification of three dominant rainfall regions within Indonesia and their relationship to sea surface temperature. Int. J. Climatol.: A J. Royal Meteorol. Soc. **23**(12), 1435–1452 (2003)
18. Shige, S., Kida, S., Ashiwake, H., Kubota, T., Aonashi, K.: Improvement of TMI rain retrievals in mountainous areas. J. Appl. Meteorol. Climatol. **52**(1), 242–254 (2013)
19. Alfahmi, F., Boer, R., Hidayat, R., Perdinan, P., Sopaheluwakan, A.: The impact of concave coastline on rainfall offshore distribution over Indonesian. Sci. World J. **2019**, 1–11 (2019)

20. Tan, J., Huffman, G.J., Bolvin, D.T., Nelkin, E.J.: Diurnal cycle of IMERG V06 precipitation. Geophys. Res. Lett. **46**, 13584–13592 (2019)
21. Xu, T.F., Wei, Z.X., Susanto, R.D., Li, S.J., Wang, Y.G., Wang, Y., Fang, G.H.: Observed water exchange between the South China Sea and Java Sea through Karimata strait. J. Geophys. Res.: Oceans **126**(2), e2020JC016608 (2021)

Analysis of Upwelling Variations Caused by ENSO Intensification in the Southern Makassar Strait

Gandhi Napitupulu, Naffisa Adyan Fekranie, Susanna Nurdjaman, Totok Suprijo, and Luki Subehi

Abstract The occurrence of upwelling in the area south of Makassar Strait is relatively rich in organic materials. This study aimed to examine upwelling areas and distribution patterns in the southern Makassar Strait under different climate circumstances, particularly strong and weak ENSO. This study utilized sea surface temperature (SST) and chlorophyll-a data from 2013 and 2015 from the Level 3 Modis picture. The result shows that the upwelling began in the southern Makassar Strait and spread to the southwest, influenced by the east monsoon wind pattern. Strong El Niño also affected the duration of the upwelling in south Makassar Strait to be longer (two months, August–September) than during the weak ENSO (a month, August).

1 Introduction

Indonesia's sea area exceeds its land area. Over two-thirds of the land area is the sea. Sea is a moving medium, above and below the surface, causing water circulation on a small and large scale. Climate is a global factor that influences the movement and distribution of water masses, as well as the ocean's physical, chemical, and biological properties. Climate is primarily determined by variations in solar energy received by

G. Napitupulu
Department of Earth Science, Faculty of Earth Sciences and Technology, Institut Teknologi Bandung, Bandung, Indonesia

N. A. Fekranie
Department of Oceanography, Faculty of Earth Sciences and Technology, Institut Teknologi Bandung, Bandung, Indonesia

S. Nurdjaman (✉) · T. Suprijo
Oceanography Research Group, Faculty of Earth Sciences and Technology, Institut Teknologi Bandung, Bandung, Indonesia
e-mail: susanna@itb.ac.id

L. Subehi
Research Center for Limnology and Water Resources, National Research and Innovation Agency, Bogor, Indonesia

A. Basit et al. (eds.), *Proceedings of the International Conference on Radioscience, Equatorial Atmospheric Science and Environment and Humanosphere Science*, Springer Proceedings in Physics 290, https://doi.org/10.1007/978-981-19-9768-6_41

the earth's surface, which result in temperature and air pressure variations. Because of the difference in air pressure, the wind may move from high-pressure to low-pressure areas. Differences in air pressure also affect humidity and precipitation, causing characteristic variations between regions with high and low rainfall [1].

Upwelling is an increase in deep-sea water mass (between 100 and 200 m deep) at the surface (up to the euphotic layer) due to a variety of causes. One of the causes is current divergence, where the wind blows parallel to the shore or at the equator, pushing the water mass away from the shore [2]. In addition, the Coriolis force drives the surface water mass away from the shore, resulting in coastal upwelling. The water mass beneath the surface will fill the gap left by the surface water mass as it moves away from the shore.

Upwelling was observed in several Indonesian seas, including one south of the Makassar Strait. Therefore, the study on south Makassar Strait's upwelling area will benefit the fisheries sector. Moreover, in fisheries, the sea surface temperature (SST) variability can be used to determine the upwelling location, and the variability of chlorophyll-a concentration can be used to determine the fertility level of the waters.

A previous study proved that the interannual variability of cooler SST along the Java and Sumatra Islands is related to a dipole in the Indian Ocean, which is independent of ENSO and results from air–sea interaction inherent to the Indian Ocean [3]. It was also concluded that the 1997/1998 anomaly, despite coincidence with the ENSO, may primarily be due to Indian Ocean internal dynamics rather than a direct response to external influences [4]. Meanwhile, some studies argue that interannual variability in the Indian Ocean region is unrelated to ENSO [5]. However, most of the interannual SST variability in that region can be attributed to ENSO-related fluctuations over the Pacific Ocean. Another study also showed that interannual variability in the region, while coincident with ENSO, is not controlled entirely by ENSO [6].

The upwelling area can be identified through the variability of sea surface temperature and chlorophyll-a concentrations. It can be done using remote sensing data because it has a broader area coverage than direct checks in the field and can also be done regularly. Although the wavelength used in remote sensing imagery is sensitive to changes and differences in sea surface temperature and chlorophyll-a concentrations, it cannot penetrate clouds. The unavailability of continuous and complete data can be impacted because the research location is in the tropics, where cloud cover is very high.

This study aims to identify the upwelling area and its distribution through visualization of chlorophyll-a and sea surface temperature (SST) distribution under different climate circumstances, particularly under the co-influence of ENSO and IOD. We observed the condition of chlorophyll-a, SST, and upwelling in south Makassar Strait during the weak El Niño and normal IOD, and during the strong El Niño and normal IOD.

2 Study Area and Methods

Figure 1 shows the Southern Makassar Strait, the area of interest in this study. The study area is located on the latitude of 0–10°S and the longitude of 114–124°E. This study utilizes sea surface temperature (SST) and chlorophyll-a data in the area south of the Makassar Strait from Aqua Moderate Resolution Imaging Spectroradiometer (MODIS) satellite imagery level 3 with a monthly composite of 4 km resolution. The research time frame is one year, from January 2015 to December 2015, corresponding to four seasons. SeaDAS 7.2 is used to evaluate the downloaded NetCDF formatted picture data.

The Aqua MODIS satellite image is occasionally obscured by clouds, depending on the circumstances. As a result, the downloaded satellite image is first processed through cloud masking using SeaDAS to enter a formula that prevents the cloud value from being read as the sea surface temperature value. If the cloud masking process is not used, the resulting data is irrelevant and does not match the expected sea surface temperature value. Processing at SeaDAS continues with image cropping based on the intended research area, then exporting the cropped data into a readable image format.

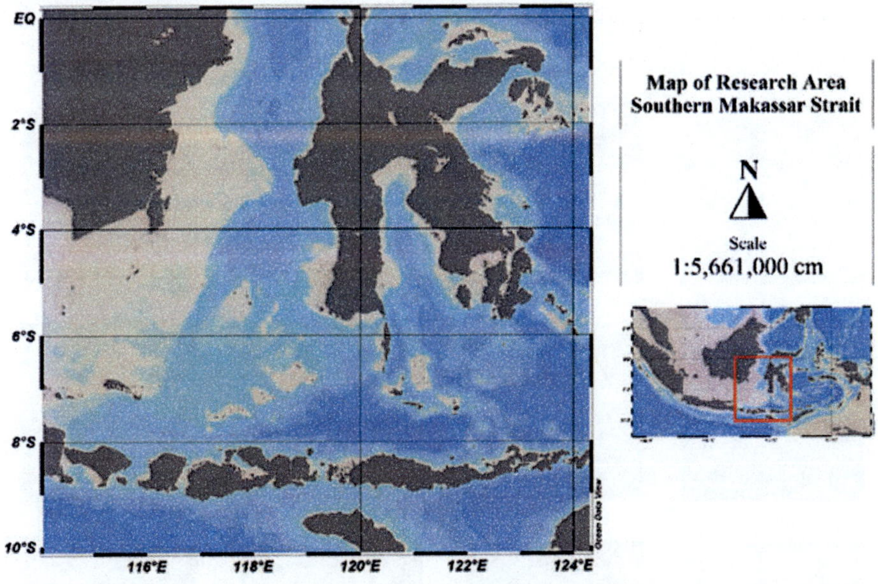

Fig. 1 Map of research area

3 Results and Discussion

3.1 Distribution Pattern of Chlorophyll-A During Weak El Niño and Normal IOD

Variations in chlorophyll-a concentrations were collected from AQUA-Modis imaging across the Makassar Strait's southern waters to explore the changes in chlorophyll-a from January–December 2013 (Fig. 2). The data in 2013 was chosen to examine the distribution of chlorophyll-a under normal conditions because 2013 is a normal year (weak ENSO and IOD). In the west season (January–February), chlorophyll concentration in the Makassar Strait is lower on average than in other seasons. However, relatively high concentrations are found in coastal areas, presumably due to the influence of nutrient input from the land caused by the relatively high level of rainfall in the west season. The regional distribution pattern of chlorophyll-a concentrations is the same as the west monsoon from April to May (transition season I).

Furthermore, it can be shown that at the beginning of the transition to the east monsoon in May, there were no noticeable indications of an increase in chlorophyll-a concentrations in the southern part of the Makassar Strait. The concentration of chlorophyll-a increased, reaching a relatively high-level beginning in June and

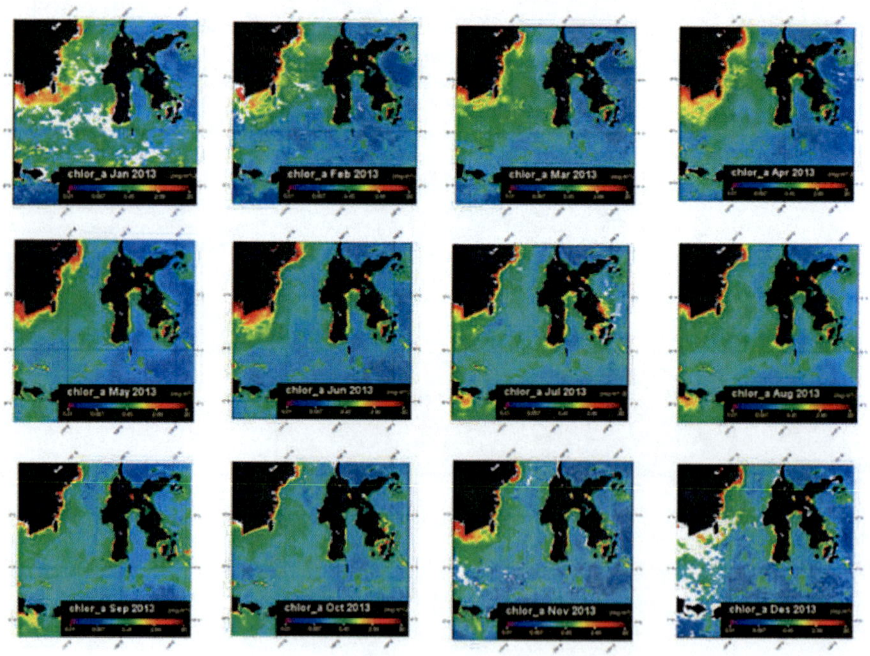

Fig. 2 Spatial distribution of Chlorophyll-a in the southern waters of the Makassar Strait in 2013

peaking in August. The high chlorophyll-a concentration during the east monsoon season, preceded by low SST, indicated upwelling. It is consistent with a prior study that discovered upwelling occurred in this location during the east monsoon (June–August) [7].

In June, it was seen that the distribution pattern of the relatively high concentration of chlorophyll-a was still around the coastal areas, especially the southern part of the Makassar Strait. Meanwhile, from July–August, the distribution pattern began to distribute to the southwest of the island of Sulawesi toward the Flores Sea, with the level of chlorophyll concentration ranging from 0.8–1.2 mg/m^3. In addition, the widespread pattern sees an increase in the concentration of chlorophyll-a of distribution in the southern part of the Makassar Strait.

At the beginning of transitional season II (September), the spatial pattern of upwelling distribution was still clearly visible. The spatial pattern of upwelling distribution was still visible at the start of transitional season II (September). The end of transitional season II (November) is assumed to mark the end of the upwelling phenomena, as indicated by the appearance of chlorophyll concentrations, which began to decline again at the end of transitional season II.

It was reported that surface chlorophyll in the Makassar Strait fluctuates, increases closer to the coast, and decreases further offshore. The high quantity of chlorophyll is one of the reasons that can cause chlorophyll blooms to emerge on the surface as a result of the high intensity of sunlight [8]. It is also evidenced by previous studies which explained that the oceanographic conditions of the waters substantially impact the level of chlorophyll concentration on the surface [9]. Therefore, the chlorophyll distribution is more abundant on the sea surface along the offshore South Sulawesi from west to south of the South Sulawesi mainland coast.

3.2 *Distribution Pattern of Sea Surface Temperature During Weak El Niño and Normal IOD*

Monthly variations of sea surface temperature (SST) around the southern waters of the Makassar Strait from January–December 2013 are shown in Fig. 3. It shows that the SST distributions range between 28–30.5 °C.

SST from December to February (West Season) tends to be warm. However, Fig. 3 shows an anomaly of lower SST in January with a value between 27.5–29.5 °C. SST increased in March and reached the highest temperature in April, which caused warmer water conditions. On the other hand, the chlorophyll-a concentration was low from December–April. High SST and low chlorophyll-a concentration in December–April in the Makassar Strait in the southern part of Sulawesi Island are influenced by west monsoon wind patterns. Longshore winds generated during the west monsoon can cause a downwelling process. Meanwhile, longshore winds during the east monsoon can cause an upwelling process. Surface winds that blow during

Fig. 3 Spatial distribution of Sea Surface Temperature in the southern waters of the Makassar Strait in 2013

December–April caused the Ekman Transport to flow northeast to the southern coast of Sulawesi Island and drove the downwelling phenomenon [10].

The value of SST during May shows an even temperature distribution pattern within the range of 29.5–30.5 °C, which then increases to 30.0–31.5 °C in June. Entering July, the sea surface temperature experienced a significant change. In the southern part of the mainland of South Sulawesi, it tended to fall in the range of 28.5–29.5 °C. While in the north, the SST ranges from 30.0–30.5 °C. The SST fluctuations were possibly caused by the southeast trade winds, which had been blowing strongly from the southeast. SST variations are also conceivable due to river discharges and decreased precipitations.

However, during the transition to the east monsoon period (May), symptoms of a decline in temperature began to appear in the southern part of the Makassar Strait. It was even more visible in June and July, indicating the upwelling onset. In August, it appeared with a stratified temperature distribution pattern across the South Makassar Strait. The image processing results in July revealed a decrease in SST in South Sulawesi Province that did not match with an increase in chlorophyll content, indicating a time lag between SST and chlorophyll changes.

In transitional season II (September–October), the distribution of SST shows that upwelling was getting weaker, indicated by a decrease of chlorophyll-a concentration and an increase in SST values in the southern part of the Makassar Strait. The upwelling phenomenon in the east monsoon and transition II show a spatial pattern of SST distribution starting from the south of Sulawesi Island and extending to the Flores Sea.

It is concluded from the SST distribution that upwelling begins in June, occurs in the southern part of the Makassar Strait, and reaches its peak in August. The August SST distribution showed widespread sea surface temperatures with low values indicating the peaking and widening of the upwelling distribution area. The decline in SST and an increase in chlorophyll-a concentration in June–August (East Monsoon Period) indicated an upwelling process assumed to be influenced by longshore winds.

Upwelling Fluctuations Comparison of Literature Study Results

A prior study examined the variability of sea surface temperature and seasonal chlorophyll-a, as well as the intensity and type of upwelling in the Makassar Strait [11]. The data used include sea surface temperature, chlorophyll-a, and wind data derived from Aqua MODIS Level 3 images from 2007–2010. The results showed that monsoon wind influences the variability of chlorophyll-a and sea surface temperature in the Makassar Strait. The movement pattern and high wind speed, followed by a decrease in SST and an increase in chlorophyll-a concentration, indicate an upwelling phenomenon in the Makassar Strait in the southern part of Sulawesi Island.

The comparison of SST and chlorophyll-a distribution patterns for 2010 and 2013 shows that SST formation begins in June on average. An increase followed this decrease in SST in chlorophyll-a concentration, which spreads in the waters of the Makassar Strait's southern end. SST formation in 2010 started in the second week of June, peaked in the second week of August, and finished in the second week of October. SST formation occurred in the second week of June, followed by an increase in chlorophyll-a concentrations in the fourth week of June, which peaked in the fourth week of August and ended in the fourth week of September. The 2010 phenomenon was very similar to the 2013 phenomenon. SST formation in 2013 began in June, peaked in August, and ended in October. SST formation in June was followed by an increase in chlorophyll-a concentrations in June, which then increased in August and ended in September. Based on the analysis of the temperature and chlorophyll-a distribution patterns, it is seen that the peak of the upwelling phenomenon for 2013 occurred in August, as indicated by the decrease in SST in this result, which is not significantly different from the distribution pattern of SST and chlorophyll-a for 2010.

When the upwelling peaked in August, the pattern of upwelling spread could be seen toward the southwest of Sulawesi Island. A drop in SST and an increase in chlorophyll-a concentration coincided with the expansion of the upwelling distribution area.

A prior study examined the spatial distribution of SST and the vertical temperature distribution in Makassar Strait [12]. As a result, the peak of the southeast monsoon conditions in the observation area occurred in August 2015, and upwelling occurred in the area. The temperature difference of nearly 3.0 °C became a strong indicator of water mass transfer from the water column to the surface. The southeast monsoon wind, which passed through the observation surface area and caused mass transport of water away from the coast, was the most influential factor in this event.

Furthermore, an upwelling simulation in the south Makassar Strait was conducted [13]. The model that was developed by MAJAFLOX between June and August revealed that the upwelling mechanism was primarily related to the Southeast Monsoon. Strong winds caused vertical diffusivity during this season. The intensity of the wind pressure increased heat loss through heat flux, resulting in strong vertical mixing. The ITF in the Makassar Strait contributed to the mechanism of upwelling. The ITF caused recirculation and formation of a vortex, which caused convergence on the ITF pathway and divergence in coastal areas. This recirculation is routed through the Selayar Strait in the south of Sulawesi Island, with some current redirected to the Flores Sea. According to the study, minors played a role in reducing the intensity of upwelling. Meanwhile, the topographical complexity caused ITF recirculation.

3.3 Distribution Pattern of Chlorophyll-A During Strong El Niño and Normal IOD

Figure 4 shows the monthly distribution of chlorophyll-a from January–December 2015. The 2015 data was chosen to examine the spread of chlorophyll-a under extreme conditions (Strong ENSO and IOD). Chlorophyll-a concentrations in the Makassar Strait were lower during the west monsoon (January–February) than during other seasons. Coastal areas, on the other hand, showed higher concentrations. It is most likely due to the high precipitation's influence on nutrient input from the land. The chlorophyll-a concentration increased by up to 0.2 mg/m^3 compared to the western season of 2013. Chlorophyll-a concentration and distribution in the transitional season I have a similar pattern to the west monsoon.

During the east monsoon period (June–August), there were constant increases in chlorophyll-a concentrations in the southern part of the Makassar Strait. In June, the distribution pattern of relatively high chlorophyll-a concentrations was seen around the coastal areas, particularly the southern part of the Makassar Strait. In contrast, in July–August, the distribution pattern began to spread to the southwest of the island of Sulawesi toward the Flores Sea with high levels of chlorophyll concentration ranging from 1–1.4 mg/m^3.

The temporal distribution of chlorophyll-a in the southern Makassar Strait in 2015 revealed that the increase in chlorophyll concentration began in June when the chlorophyll-a concentration started to rise in the range of 0.9–1 mg/m^3 and continued to increase until it peaked at the beginning of transitional season II (September). Strong El Niño and IOD + influence affect the highest chlorophyll-a concentration occurrence and its period. In 2013, the peak of chlorophyll-a concentrations occurred in August, while in 2015, the high concentration of chlorophyll-a was modulated in August and peaked in September.

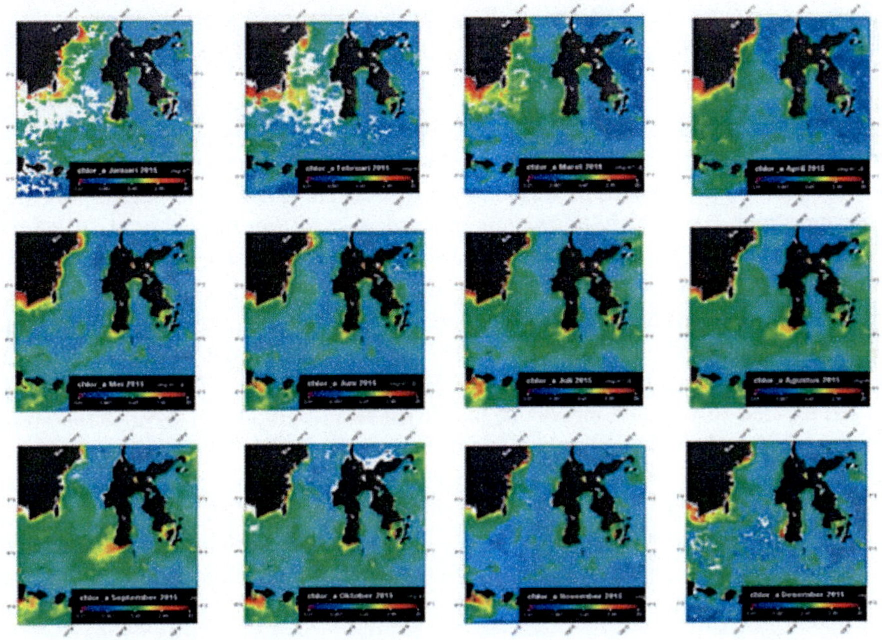

Fig. 4 Spatial distribution pattern of Chlorophyll-a in the southern part of the Makassar Strait in 2015

The increased chlorophyll concentration corresponds to the upwelling phenomenon during the eastern monsoon period. During the east monsoon in a typical year, high chlorophyll only extends to the southern coast of Sulawesi. Easterly wind anomalies caused significant upwelling and cooler water. It also brought the thermocline to a shallower depth along the shores of Java and Sumatra during the 2015 El Niño peak [14].

El Niño and IOD + occurrences increased chlorophyll-a concentrations along the shores of Sulawesi and Kalimantan, extending south to the Makassar Strait (Fig. 4). The southeastern half of the region has more chlorophyll anomalies than the further south. The seasonal variability of the monsoon and the interannual variability of El Niño and Indian Ocean Dipoles influenced the spatial distributions of chlorophyll-a.

Higher chlorophyll-a concentrations along the southern shores of Sulawesi and Kalimantan were related to a strong El Niño in 2015. During the east monsoon period, chlorophyll-a levels were higher than in the west monsoon period.

3.4 *Distribution Pattern of Sea Surface Temperature During Strong El Niño and Normal IOD*

Monthly variations of sea surface temperature (SST) around the southern waters of the Makassar Strait from January–December 2015 are shown in Fig. 5. SST distribution pattern in the south part of the Makassar Strait in December–February (west season) shows a relatively high-temperature distribution in the range of 29–32 °C. The high range of SST is still visible in March and April (transition season I). As the east monsoon season began in May, a temperature drop started to appear in the southern part of the Makassar Strait. The decline of SST became more noticeable in July, indicating the beginnings of upwelling, and the SST had the lowest value in September, showing the peak of upwelling.

Compared to the east monsoon, the distribution of SST values in transitional season II (October–November) shows that the indications of upwelling are weakening, as indicated by an increase in SST values in the southern part of the Makassar Strait. In general, the east monsoon upwelling phenomenon and transition II exhibit a spatial pattern of SST distribution beginning in the southern part of Sulawesi Island and extending to the Flores Sea. The upwelling in 2015 started in July in the

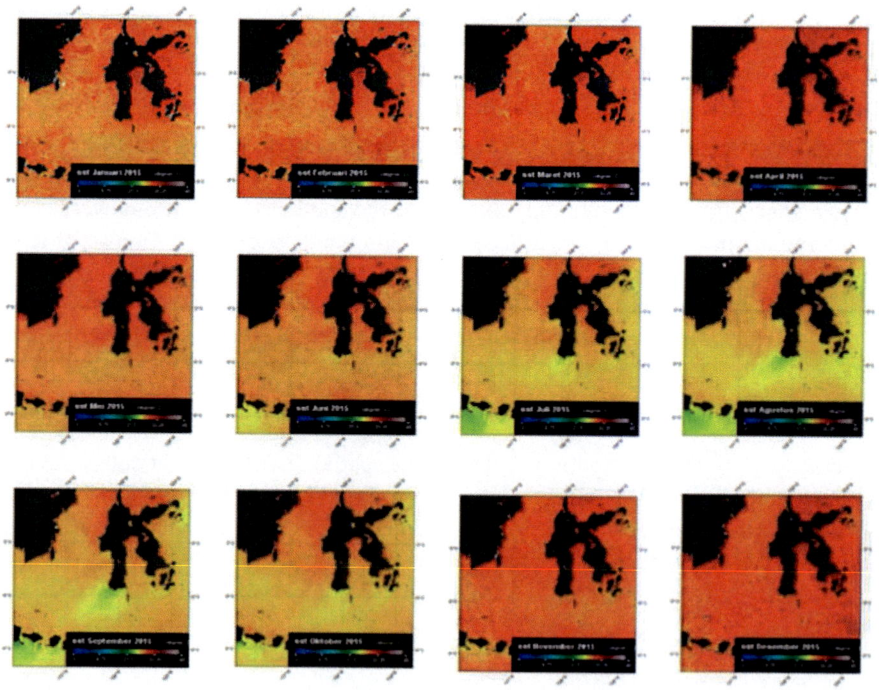

Fig. 5 Spatial distribution of Sea Surface Temperature in the southern waters of the Makassar Strait in 2015

southern part of the Makassar Strait and peaked in August. In August, there was a widespread increase in sea surface temperatures with low values, indicating that the upwelling distribution area peaked and expanded. Strong El Niño and IOD increase sea surface temperature in the MAM period by about 0.2–1 °C and decrease sea surface temperature by 0.2–0.8 °C in the JJA period.

In 2013, the lowest SST occurred in August, while in 2015, the lowest SST was last in August–September. The distribution and duration of upwelling between the strong and weak El Niño and normal IOD show that the influence of strong El Niño and IOD + affects the lowest SST occurrence and its period. Therefore, we can conclude that a strong El Niño also affected the duration of the upwelling in south Makassar Strait to be longer (two months) than during the weak ENSO (a month). This result supports a prior study that found an extension on the Java-Sumatra upwelling during El-Niño episodes, both time and space [14].

4 Conclusion

Upwelling south of the Makassar Strait is generated locally by the alongshore winds associated with the monsoon and ENSO. In normal years, upwelling occurs from early June to August. The distribution of SST values in the south of the Makassar Strait during the east season is higher than during the west season. The low SST value in the east monsoon period (May–August) was followed by an increase in chlorophyll-a concentration, indicating upwelling to the south of the Makassar Strait. The pattern of spread of upwelling in the east monsoon begins in June, reaches its peak in August, and ends in October, with the peak of upwelling occurring in August. The spatial distribution pattern of SST and chlorophyll-a concentrations in the southern waters of the Makassar Strait during the east monsoon also shows that the distribution pattern shifts to the southwest. The strong El Niño effect affecting the monsoon transition causes extended upwelling in both time and space. The period of peak upwelling that occurs is getting longer to two months.

Acknowledgements The author would like to thank the Korea-Indonesia Marine Technology Cooperation Research Center (MTCRC) Scholarship for funding during the research. Thanks also go to the Aqua Moderate Resolution Imaging Spectroradiometer (MODIS) for providing and accessing the data used in this research.

References

1. Nontji, A.: Archipelago Sea. Anugerah Nontji. Jakarta, Djambatan, Thurman (2005)
2. Setiawan, R.Y., Kawamura, H.: Summertime phytoplankton bloom in the South Sulawesi Sea. IEEE J. Selected Topics in Appl. Earth Observ. Remote Sens. **4**(1), 241–244 (2010)
3. Saji, N.H., Goswami, B.N., Vinayachandran, P.N., Yamagata, T.: A dipole mode in the tropical Indian Ocean. Nature **401**(6751), 360–363 (1999)

4. Webster, P.J., Moore, A.M., Loschnigg, J.P., Leben, R.R.: Coupled ocean–atmosphere dynamics in the Indian Ocean during 1997–98. Nature **401**(6751), 356–360 (1999)
5. Venzke, S., Latif, M., Villwock, A.: The coupled GCM ECHO-2. Part II: Indian ocean response to ENSO. J. Climate **13**(8), 1371–1383 (2000)
6. Murtugudde, R., McCreary, J.P., Jr., Busalacchi, A.J.: Oceanic processes associated with anomalous events in the Indian Ocean with relevance to 1997–1998. J. Geophys. Res.: Oceans **105**(C2), 3295–3306 (2000)
7. Sverdrup, H.U., Munk, W.H.: Empirical and theoretical relations between wind, sea, and swell. EOS Trans. Am. Geophys. Union **27**(6), 823–827 (1946)
8. Syahdan, M., Atmadipoera, A.S., Susilo, S.B., Gaol, J.L.: Variability of surface chlorophyll-a in the Makassar Strait–Java Sea, Indonesia. Int. J. Sci.: Basic and Appl. Res. (IJSBAR) **14**(2), 103–116 (2014)
9. Le Bouteiller, A., Blanchot, J., Rodier, M.: Size distribution patterns of phytoplankton in the western Pacific: towards a generalization for the tropical open ocean. Deep Sea Res. Part A. Oceanograph. Res. Papers **39**(5), 805–823 (1992)
10. Mackey, D.J., Parslow, J., Higgins, H.W., Griffiths, F.B., O'Sullivan, J.E.: Plankton productivity and biomass in the western equatorial Pacific: biological and physical controls. Deep Sea Res. Part II: Topical Stud. Oceagraph. **42**(2–3), 499–533 (1995)
11. Sarianto, D.: Analisis Daerah Penangkapan Ikan Cakalang (Katsuwonus pelamis) di Sekitar Bacan dan Obi, Halmahera Selatan. SEMAH Jurnal Pengelolaan Sumberdaya Perairan **2**(3) (2018)
12. Utama, F.G., Atmadipoera, A.S., Purba, M., Sudjono, E.H., Zuraida, R.: Analysis of upwelling event in Southern Makassar Strait. In: IOP Conference Series: Earth and Environmental Science, vol. 54(1), pp. 012085. IOP Publishing (2017)
13. Atmadipoera, A.S., Widyastuti, P.: A numerical modeling study on upwelling mechanism in Southern Makassar Strait. J. Ilmu dan Teknologi Kelautan Tropis **6**(2) (2014)
14. Susanto, R.D., Gordon, A.L., Zheng, Q.: Upwelling along the coasts of Java and Sumatra and its relation to ENSO. Geophys. Res. Lett. **28**(8), 1599–1602 (2001)

The Influence of Climatic-Oceanographic Factors on Triggering Sea Level Variations in the Equatorial Malacca Strait, Indonesia

Ulung Jantama Wisha, Yusuf Jati Wijaya, and Yukiharu Hisaki

Abstract The Strait of Malacca is an area of significance. Climate anomaly may affect the surrounding Strait of Malacca, where the sea level variation is supposed to be controlled by several climatic factors, such as the Indian Ocean Dipole (IOD) and El-Nino Southern Oscillation (ENSO). This study will elucidate the correlation between sea level and several climatic-oceanographic factors that potentially play a significant role in shaping the fluctuation of sea level in the Malacca Strait. The approximately 27-year sea level data (1992–2019) measured from altimetry satellite and tide gauge was employed. A linear regression method was used to define the velocity estimation of sea level at designated times in the form of time series. Both altimetry and tide gauge data showed an upward trend over 27 years, with 0.24 cm/year and 0.39 cm/year, respectively. The sea level trend declined over the last ten years (2009–2019). The decline trends were -0.18 cm/year and -0.03 cm/year for altimetry and tide gauge data, respectively. The seasonal comparison between sea level vs. DMI and sea level vs. SOI was weakly correlated with a low correlation value ($R^2 < 0.1$). Of particular concern, the zonal current is strongly correlated with sea level variations with R^2 of about 19–36%. Moreover, the Madden Julian Oscillation (MJO) and zonal current showed a strong possible influence, particularly during the northeast monsoon.

U. J. Wisha (✉) · Y. Hisaki
Department of Physics and Earth Sciences, Graduate School of Engineering and Science, University of the Ryukyus, Nishihara 903-0213, Japan
e-mail: ulun002@brin.go.id

U. J. Wisha
Research Center for Oceanography, National Research and Innovation Agency (BRIN), 14430 Jakarta, Indonesia

Y. J. Wijaya
Department of Oceanography, Faculty of Fisheries and Marine Science, Diponegoro University, 50275 Semarang, Indonesia

A. Basit et al. (eds.), *Proceedings of the International Conference on Radioscience, Equatorial Atmospheric Science and Environment and Humanosphere Science*, Springer Proceedings in Physics 290, https://doi.org/10.1007/978-981-19-9768-6_42

1 Introduction

The arbitrarily erratic climate variability impacts all aspect of life on earth. The most observable tendency is sea level and temperature changes. Sea level change modeling is widely used to evaluate coastline and climate changes [1]. Seasonal variability of the sea mainly relates to meteorological factors (atmospheric pressure, wind, evaporation, and precipitation), water balance (river runoff), evaporation, and topography and water density alterations [2]. Sea level change is not geographically uniform. Therefore, a precise estimation of sea level change is imperative for controlling its impact on coastal regions [3].

Several cases of coastal damage (coastal erosion and instability) that may be led by climate change and sea level rise have been reported to occur in the Malacca Strait [4, 5]. The Malacca Strait is an area of significance. It is a narrow stretch of water bordered by the Peninsular Malaysia and the east coast of Sumatra, linking the Indian and Pacific Oceans via the South China and Java seas [6, 7]. Climate anomaly may affect the area in the surrounding Malacca Strait, where the sea level anomaly (SLA) is supposed to be controlled by several climatic factors, such as the Indian Ocean Dipole (IOD) and El-Nino Southern Oscillation (ENSO). Moreover, several scholars have also reported the tendency of sea level rise in the equatorial Malacca Strait [8–11]. Concerning sea level monitoring, several tide gauge stations are scattered in surrounding Malaysia and Singapore, and these recorded sea level data will be assessed in this study. However, the altimetry data are also considered.

On the other hand, ocean current influence on inducing sea level anomaly in the Malacca Strait has never been assessed comprehensively and should be investigated. Therefore, it is essential to determine the combined effect of climatic-oceanographic factors on SLA detected in the equatorial Malacca Strait since previous studies on sea level rarely consider the implication of climatic-oceanographic factor influences [9, 10]. This study will elucidate the correlation between sea level variations and several climatic and oceanographic factors that potentially play a significant role in shaping the fluctuation of sea level in the equatorial Malacca Strait.

2 Materials and Methods

2.1 *Study Site*

This study will examine the sea level variation in the equatorial Malacca Strait, located between the western Peninsular Malaysia and the eastern coast of Sumatra, Indonesia. This area is significant, connecting the Andaman Sea (Indian Ocean) and South China Sea (Pacific Ocean) with a funnel-shaped strait formation (around 65 km in the south and broadens northward of about 250 km) [12].

Geologically, the Malacca Strait is a part of the Sunda Shelf, a large, low-relief land surface at the beginning of Quaternary time. The strait current configuration is formed by the inundation of postglacial sea level rise due to the higher latitude ice melting [13]. However, the updated review of sea level trends in the Malacca Strait is imperative since many developments in the coastal area are expected to get massive in the future, where sea level monitoring becomes the basis consideration of it.

2.2 *Data Acquisition and Analysis*

The approximately 27-year sea level data (1992–2019) measured from altimetry satellite and tide gauge were retrieved from The Sea Level Explorer established by University of Colorado Boulder, University of Hawaii Sea Level Center and NASA Jet Propulsion Lab., web page: https://ccar.colorado.edu/altimetry/. While to compare the annual record of sea level, another source of long-term tide gauge measurement was employed, namely Permanent Service for Mean Sea Level (PSMSL), web page: https://www.psmsl.org/data/. The location of tide gauge survey used in this study is shown in Table 1.

To evaluate the influence of climatic factors on sea level variations, we used the Dipole Mode Index (DMI) and Southern Oscillation Index (SOI) data retrieved from https://psl.noaa.gov/data. On the other hand, as an equatorial region, Madden Julian Oscillation (MJO) might play a role in determining anomaly in sea level. The real-time multivariate (RMM) index was retrieved from http://www.bom.gov.au/climate/mjo/.

As we considered the influence of the Indian Ocean's zonal current on sea level, Global Observed Ocean Physics 3D Quasi-Geostrophic Currents (OMEGA3D) data were used to prove the hypothesis of zonal currents influence in the equatorial Malacca Strait. The marine Copernicus provides the weekly data with a grid resolution of 1/4° [14], retrieved from a webpage: https://data.marine.copernicus.eu/.

A linear regression method was employed to define the velocity estimation of sea level at designated times in the form of time series, so that sea level projection is possible for other periods [15]. Furthermore, a simple correlation analysis was also used to examine how strongly two variables are related to each other [16]. In this case, a correlation analysis between SLA, DMI-SOI, sea surface temperature, and zonal currents was performed.

Table 1 Tide gauge survey coordinates

Location	Longitude	Latitude	Coverage year
Tanjong Pagar (Singapore)	103.9167	1.2500	1992–2019
Raffles Light House (Singapore)	103.7500	1.1667	1973–2020
Kukup (Malaysia)	103.4428	1.3253	1986–2018

3 Results and Discussion

3.1 Sea Level Trends Over 27 Years of Observation

Figure 1 shows the sea level profile over 27 years in the equatorial Malacca Strait. Overall, the sea level value surveyed by altimetry and tide gauge showed a similar fluctuation over times. However, the arbitrarily erratic profile of sea level detected from altimetry data was observed since in the middle of 2012 up to 2018. Both altimetry and tide gauge data showed an upward trend over 27 years with 0.24 and 0.39 cm/year, respectively. The annual amplitude detected was not too different with a deviation of 0.5 cm. By contrast, the semi-annual amplitude detected by tide gauge measurement was 0.66 cm higher than altimetry data (Table 2).

Of particular concern, the sea level trend tended to decline over the latest 10 years (2009–2019), even though the trend was not too significant. The decline trends were 0.18 cm/year and -0.03 cm/year for altimetry and tide gauge data, respectively (Table 2). However, the annual and semi-annual amplitudes were similar. Despite the problematic comparison between altimetry and tide gauge in estimating the sea level due to the difference vertical reference system, many review papers have explained the efforts to validate sea level derived from satellite altimetry with tide gauge data [3].

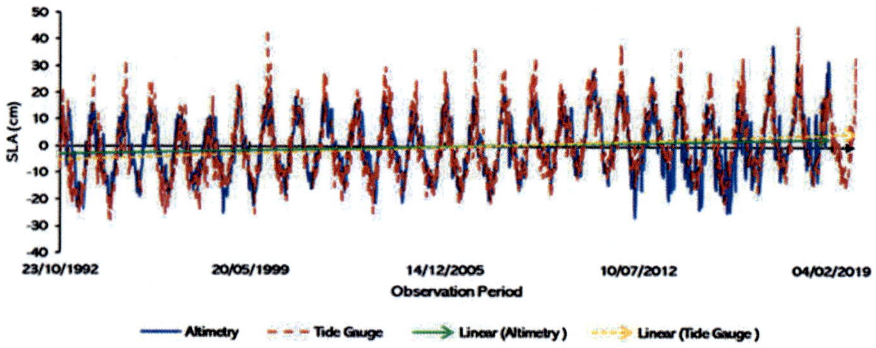

Fig. 1 Sea level data over 27 years derived from altimetry versus tide gauge measurement (RMSE = 0.788 cm)

Table 2 Sea level trend over the period of study

Observation Period	1992–2019		2009–2019	
Source	Altimetry	Tide gauge	Altimetry	Tide gauge
Trend	0.24 cm/year	0.39 cm/year	−0.18 cm/year	−0.03 cm/year
Annual amplitude	12.96 cm	13.46 cm	12.53 cm	13.73 cm
Semi-annual amplitude	1.82 cm	cm	2.29 cm	1.38 m

3.2 *Correlation Analysis of Sea Level Variations Versus Climatic-Oceanographic Parameters*

Monthly comparison between sea level and DMI-SOI is shown in Fig. 2. Overall, the response of sea level in the equatorial Malacca Strait caused by IOD and ENSO varied considerably. The strongest symptoms of sea level anomalies (SLAs) were commonly observed during the northeast monsoon over the period of 1992–2019. Higher sea level generally correlated with negative dipole mode and positive SOI value. Positive mode of IOD induces higher air pressure centered in the Indian Ocean and vice versa for the negative phase of IOD, triggering the westerly wind to flow eastward and higher sea level in the eastern Indian Ocean [17]. These mechanisms possibly impact the Malacca strait channel via the Bay of Bengal even though the influence is not too significant. Another study [18] described that the ENSO-related SLAs predomination in the South China Sea was stonger during certain periods. However, the influence becomes less significant at 8-month lag time. This statement triggers a consideration that the ENSO-driven SLAs is probably insignificant.

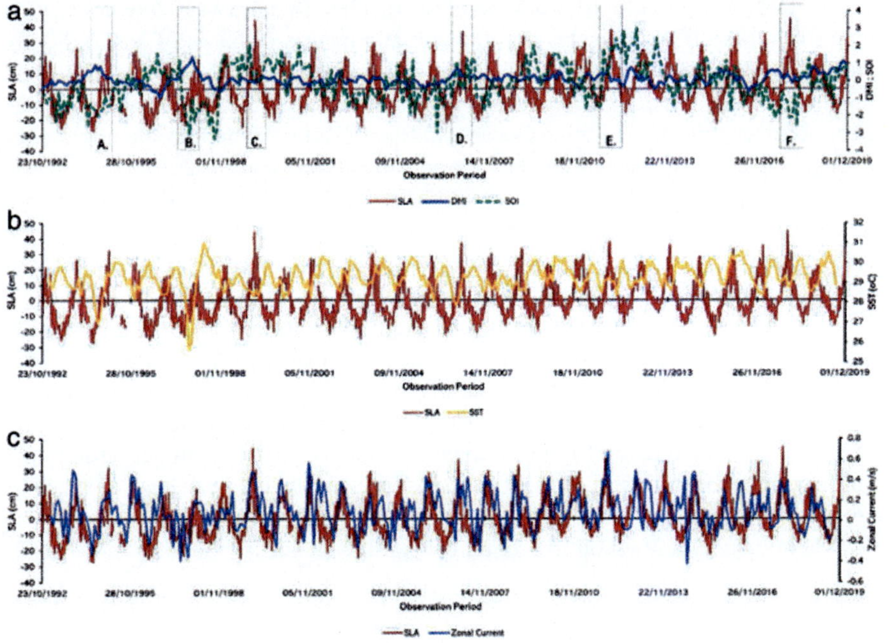

Fig. 2 Monthly comparison of sea level profile vs. DMI and SOI **a**, sea level vs. SST **b**, and sea level vs. Zonal Current **c**; dashed black squares denote the sampling of SLA data that will be compared to RMM index of MJO

Concerning the sea surface temperature (SST), many scholars observed the substantial assimilation variability between SST and SLA [19]. However, this correlation is not detected in the equatorial Malacca Strait. The insignificant phase of IOD–ENSO reflecting the fluctuation of sea level anomaly is also proven by the comparison between sea level and SST, where the opposite pattern of SST and sea level modulations was observed during 27 years of observation (Fig. 2b). More interestingly, during the northeast monsoon in 1998, the lowest temperature reaching 26 °C was detected and peaked at 31 °C in the following period. However, after that period, the SST fluctuation seemed more uniform and stable (ranging from 29–30.5 °C), relying on the influence of IOD–ENSO.

In contrast, we observed a strong modulation pattern between zonal currents and sea level in the study area (Fig. 2c). Overall, the zonal current positively flowing into the Malacca Strait is followed by the increase in sea level and vice versa for the negative zonal current profiles (flowing toward the Andaman Sea), ranging from −0.3 to 0.76 m/s. However, several inconsistent patterns were generally identified during the second transitional and the northeast monsoon.

To prove the previous hypotheses, we initiated a correlation analysis between SLA vs. DMI and SLA vs. SOI (Fig. 3a and b). Overall, the seasonal comparison between SLA vs DMI and SLA vs SOI was weakly correlated with a very low correlation value ($R^2 < 0.1$). This indicates that the climatic factors (IOD and ENSO) did not solely control the variability of sea level in the equatorial Malacca Strait. However, any other factor may influence the anomaly in sea level in the Malacca Strait, such as zonal currents from the Indian Ocean and Bay of Bengal which frequently induce a “tongue” pattern of sea surface temperature and chlorophyll-a [20].

Concerning the correlation between SLAs vs. SST (Fig. 3c), it seems more correlated than the SLA vs. DMI-SOI correlation. However, it is still weak in all monsoon periods, with R^2 ranging from 5 to 10%. A stronger correlation over 27 years of observation was observed during the second transitional season.

We found that zonal current regimes strongly correlated with SLAs during the northeast monsoon and the second transitional period with the R^2 of about 19–36% (Fig. 3d). However, in the other period of monsoons, the correlation tended to be weak, whereby the other triggering factors may be more influencing the anomaly in sea levels, such as IOD and ENSO. Since the study area is positioned around the equatorial line, the influence of MJO may be strong during the peak level of sea level anomaly over 27 years. Thus, in the following subsection, we will be addressed the possible influence of MJO on triggering the significant sea level anomaly.

3.3 Possible Influence of Madden Julian Oscillation on Triggering Sea Level Anomaly

Several sampling periods were chosen to depict the possible influence of MJO on shaping the SLA profile (Fig. 4). The possibility of MJO-determined the significant

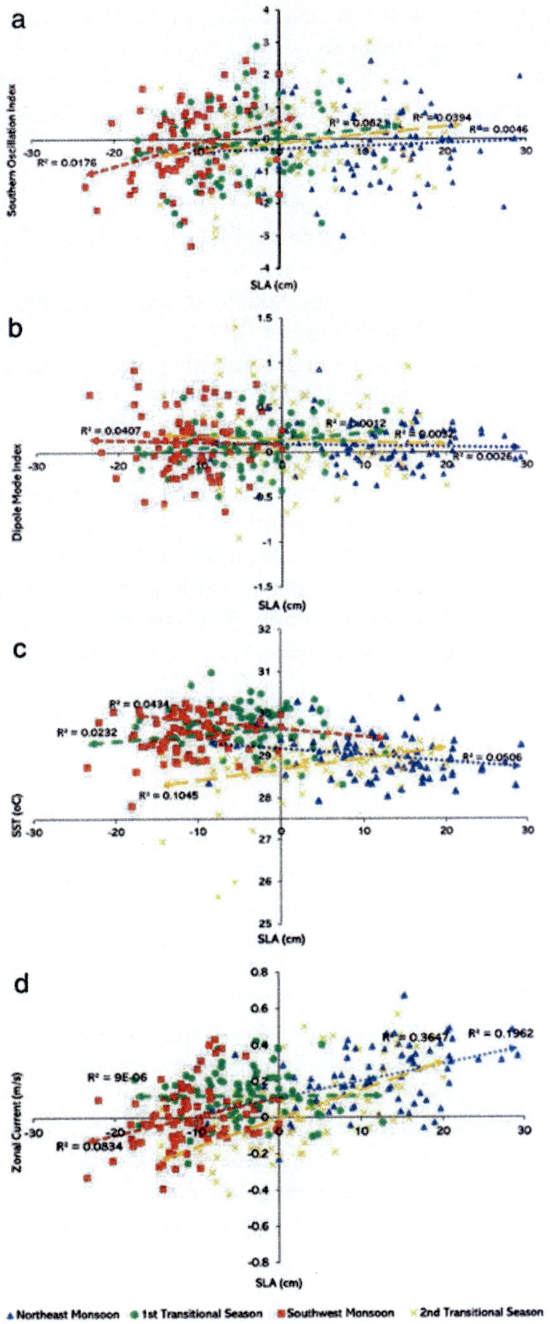

Fig. 3 Correlation analysis of SLA vs. DMI **a**, SLA vs. SOI **b**, SLA vs. SST **c**, and SLA vs. zonal current **d** over the period of study

phase of SLAs denoted in Fig. 2a was sufficiently high. Except for June 1994 and September 1997 (Fig. 4a and b), the strong influence on SLAs during the northeast monsoon (mainly in December) where the positive value of RMM1 showed the robust influence of MJO on triggering evaporation in the maritime continent [21], affecting SLAs in the Malacca Strait.

Most of the highest upward sea level fluctuations were observed during the northeast monsoon. This state considers that MJO-induced evaporation plays a significant role in determining the anomaly in sea level. The more substantial MJO frequencies were observed in the Indian Ocean before it was finally moving toward the maritime continent of Southeast Asian countries, such as in October–November 1999, October–December 2006, and October–December 2011 (Fig. 4c—e) with the RMM1 value of about 2. According to [22], after inducing evaporation in the Indian Ocean region, the MJO will usually move in a clockwise direction (from west to east).

By contrast, in the middle of October 2017, when the sea level increased by about 10 cm from the previous high phase, a robust MJO frequency in the maritime continent was detected with an RMM1 value of 3.5. In November 2017, the RMM index showed a weak frequency of MJO where the RMM value was within the center circle, meaning that it is arduous to discern MJO using the RMM method [23]. However, in early December 2017, the RMM index showed sufficient predomination in the maritime continent with a value of about 1.5 (Fig. 4f).

4 Conclusion

An upward sea level trend was observed in the Malacca Strait over 27 years of observation (1992–2019). However, a little decline trend was identified during the latest ten years. The IOD and ENSO did not solely control the variability of SLA, shown by a weak correlation between SLA vs. DMI and SLA vs. SOI. By contrast, the SST fluctuation is sufficiently correlated with SLA though the phases showed opposite patterns. We observed a strong correlation between SLA vs. zonal current, with a 19–36% of correlation during the northeast monsoon.

The MJO strongly influenced evaporation in the maritime continent during the second transitional season up to the early northeast monsoon. However, other factors may have a more substantial influence on triggering sea level anomalies in different monsoon periods. We only assessed the data using a regression and correlation analysis in this study. It is recommended to conduct other advanced statistical research to understand better the variation effect of every climatic-oceanographic factor on sea level.

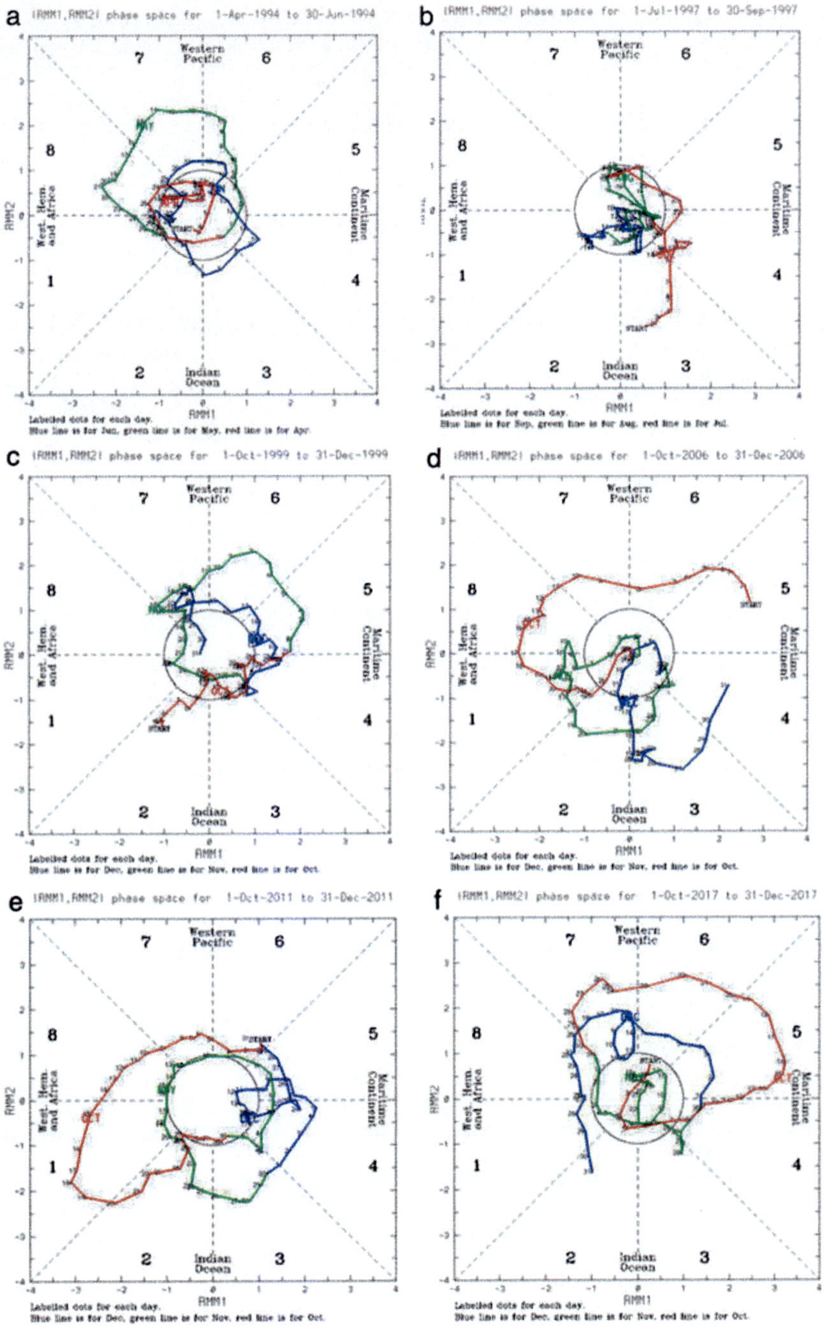

Fig. 4 RMM Index during the highest phase of SLAs in the equatorial Malacca Strait denoted in Fig. 4. *Source* http://www.bom.gov.au/climate/mjo/

Acknowledgements We would like to thank University of Colorado Boulder, University of Hawaii Sea Level Center, NASA Jet Propulsion Lab., and Marine Copernicus for providing the primary data used in this study. gratitude is also given to Department of Physics and Earth Sciences, University of the Ryukyus, and MEXT Scholarship. This study is funded by the Grant-Aid for Scientific Research (C-2)—Ministry of Education, Culture, Science, and Technology of Japan (20k04708).

References

1. Wang, Y., Marsooli, R.: Dynamic modeling of sea-level rise impact on coastal flood hazard and vulnerability in New York City's built environment. Coast. Eng. **169**, 103980 (2021)
2. Woodworth, P.L., Melet, A., Marcos, M., Ray, R.D., Wöppelmann, G., Sasaki, Y.N., Cirano, M., Hibbert, A., Huthnance, J.M., Monserrat, S., Merrifield, M.A.: Forcing factors affecting sea level changes at the coast. Surv. Geophys. **40**(6), 1351–1397 (2019)
3. Pajak, K., Kowalczyk, K.: A comparison of seasonal variations of sea level in the southern Baltic Sea from altimetry and tide gauge data. Adv. Space Res. **63**(5), 1768–1780 (2019)
4. Awang, N.A., Jusoh, W.H.W., Hamid, M.R.A.: Coastal Erosion at Tanjong Piai, Johor Malaysia. J. Coastal Res. **71**(10071), 122–130 (2014)
5. Ghazali, N.H.M.: Coastal erosion and reclamation in Malaysia. Aquat. Ecosyst. Health Manage. **9**(2), 237–247 (2006)
6. Ibrahim, Z.Z., Yanagi, T.: The influence of the Andaman Sea and the South China Sea on water mass in the Malacca Strait. La mer **43**, 33–42 (2006)
7. Chuan, G.K., Cleary, M.: In: Environment and Development in the Straits of Malacca. Routledge (2005)
8. Tay, S.H.X., Kurniawan, A., Ooi, S.K., Babovic, V.: Sea level anomalies in straits of Malacca and Singapore. Appl. Ocean Res. **58**, 104–117 (2016)
9. Luu, Q.H., Tkalich, P., Tay, T.W.: Sea level trend and variability around Peninsular Malaysia. Ocean Sci. **11**(4), 617–628 (2015)
10. Tkalich, P., Vethamony, P., Luu, Q.H., Babu, M.T.: Sea level trend and variability in the Singapore Strait. Ocean Sci. **9**(2), 293–300 (2013)
11. Kurniawan, A., Tay, S.H.X., Ooi, S.K., Babovic, V., Gerritsen, H.: Analyzing the physics of non-tidal barotropic sea level anomaly events using multi-scale numerical modelling in Singapore regional waters. J. Hydro-Environ. Res. **9**(3), 404–419 (2015)
12. Thia-Eng, C., Gorre, I.R., Ross, S.A., Bernad, S.R., Gervacio, B., Ebarvia, M.C.: The Malacca straits. Mar. Pollut. Bull. **41**(1–6), 160–178 (2000)
13. Gee, T., Partners, L., Ashford, K.: Geological and geotechnical considerations for seismic hazards for Penang Island, Malaysia. In: Proceedings of the XVII ECSMGE-2019, Reykjavik, Iceland, pp. 1–6. (2019)
14. Buongiorno Nardelli, B.: A multi-year time series of observation-based 3D horizontal and vertical quasi-geostrophic global ocean currents. Earth Syst. Sci. Data **12**(3), 1711–1723 (2020)
15. Park, J., Sweet, W.: Accelerated Sea level rise and florida current transport. Ocean Sci. **11**(4), 607–615 (2015)
16. Heyen, H., Zorita, E., Von Storch, H.: Statistical downscaling of monthly mean North Atlantic air-pressure to sea level anomalies in the Baltic Sea. Tellus A **48**(2), 312–323 (1996)
17. Rahmawan, G.A., Wisha, U.J.: Tendency for climate-variability-driven rise in Sea level detected in the Altimeter Era in The Marine Waters of Aceh, Indonesia. Int. J. Remote Sens. Earth Sci. (IJReSES) **16**(2), 165–178 (2020)
18. Liu, Q., Feng, M., Wang, D.: ENSO-induced interannual variability in the southeastern South China Sea. J. Oceanogr. **67**(1), 127–133 (2011)
19. Athie, G., Marin, F.: Cross-equatorial structure and temporal modulation of intraseasonal variability at the surface of the tropical Atlantic Ocean. J. Geophys. Res.: Oceans **113**(C8) (2008)

20. Mandal, S., Behera, N., Gangopadhyay, A., Susanto, R.D., Pandey, P.C.: Evidence of a chlorophyll "tongue" in the Malacca Strait from satellite observations. J. Mar. Syst. **223**, 103610 (2021)
21. Virts, K.S., Wallace, J.M., Hutchins, M.L., Holzworth, R.H.: Diurnal lightning variability over the Maritime Continent: impact of low-level winds, cloudiness, and the MJO. J. Atmos. Sci. **70**(10), 3128–3146 (2013)
22. Moum, J.N., Pujiana, K., Lien, R.C., Smyth, W.D.: Ocean feedback to pulses of the Madden–Julian oscillation in the equatorial Indian Ocean. Nat. Commun. **7**(1), 1–7 (2016)
23. Kim, H.M., Kim, D., Vitart, F., Toma, V.E., Kug, J.S., Webster, P.J.: MJO propagation across the Maritime Continent in the ECMWF ensemble prediction system. J. Clim. **29**(11), 3973–3988 (2016)

Zooplankton Response to Harmful Algae Blooms (HABs) Species Phytoplankton

Hanung Agus Mulyadi, Arief Rachman, and Nurul Fitriya

Abstract Zooplankton is essential as a grazer of harmful algae blooms (HABs) phytoplankton species. This review will outline the HABs species of phytoplankton and the interaction and response of zooplankton to HABs species. HABs species of phytoplankton produce toxins, e.g., paralytic shellfish poisoning (PSP), diarrheic shellfish poisoning (DSP), amnestic shellfish poisoning (ASP), and ciguatera fish poisoning (CFP) that potentially harm zooplankton taxa. Zooplankton has various responses to HABs species. Some zooplankton taxa, mainly copepods, impacted the adverse toxicity produced by HABs species of phytoplankton. It can be seen by decreasing the egg productivity and egg viability and injurious swimming behavior of copepods. On the contrary, some zooplankton taxa are more tolerant of responding to HABs species. The HABs occurrence altered zooplankton structure communities. Before and after the blooms, the zooplankton was dominated by herbivorous and free spawner zooplankton, e.g., copepods and tunicates. While during the blooms, the zooplankton was dominated by carnivorous groups, e.g., jellyfish, chaetognath, and copepods. Copepod *Acartia hudsonica* has an adaptation strategy to respond to a noxious toxin produced by *Alexandrium fundyense*. This species is more tolerant to PSP toxin if they are historically exposed *A. fundyense*. PSP toxin tolerance can be measured through a toxic feeding assay.

1 Introduction

Phytoplankton is microalgae that produces essential nutrients through photosynthesis. Thus, as a primary consumers, zooplankton feeds on phytoplankton and is

H. A. Mulyadi (✉) · A. Rachman · N. Fitriya
Research Center for Oceanography, National Research and Innovation Agency, Jl. Pasir Putih 01, Ancol, Jakarta Utara 14430, Indonesia
e-mail: hanu001@brin.go.id

H. A. Mulyadi
Department of Marine Science and Technology, Faculty of Fisheries and Marine Science, IPB University, Jl. Aghatis 01, Bogor, Indonesia

A. Basit et al. (eds.), *Proceedings of the International Conference on Radioscience, Equatorial Atmospheric Science and Environment and Humanosphere Science*, Springer Proceedings in Physics 290, https://doi.org/10.1007/978-981-19-9768-6_43

linked to planktivorous organisms in the upper tropic level. However, some phytoplankton species produce a toxin potentially harmful to marine organisms. Lists of harmful microalgae in the world [1] including specific toxic algae in the Western Pacific [2] was reported.

The term blooms of algae are defined as increased populations and species of algae that potentially cause harm for marine organisms and humans, known as harmful algae blooms (HABs). Toxin-producing phytoplankton from HABs species includes those that cause red tide, paralytic shellfish poisoning, diarrheic shellfish poisoning, amnestic shellfish poisoning, and ciguatera. [2].

Toxins produced by HABs species cause mass mortalities of fishes [3] and are potentially fatal to humans [4]. Although the effect of toxins in fishes and bivalves has been studied, the interaction and response of zooplankton are less clear. Some zooplankton taxa impacted the adverse toxicity produced by HABs. However, the other taxa of zooplankton are more tolerant of responding to HABs occurrence. This review outlined the zooplankton response to HABs species of phytoplankton. The structure of this article is divided into two sections. The first provides the HABs species of phytoplankton, and the second section outlines the response of zooplankton to HABs.

2 Materials and Methods

A literature reference on the HABs species and the response of zooplankton to HABs collected from journal, proceeding, and book using Google Search (GS) database by August 24, 2022. The other repositories were chosen, e.g., Scopus, Web of Science, and ResearchGate. Data analysis was conducted by identifying the topics. Finally, a summary of zooplankton response and interactions to HABs species is given.

3 HABs Species Phytoplankton

HABs species taxonomic list identified more than 180 species [1] Table 1. These consist of some groups, e.g., diatom, dinoflagellates, and cyanobacteria. The HABs species, mostly dinoflagellates, constitute more than half of the total [1] (Fig. 1).

An influential group of HABs species of phytoplankton and dinoflagellates consists of seven orders and one family. Order Gonyaulacales first, followed by Gymnodiniales, Prorocentrales, Dinophysiales, and Amphidinales, family Amphidomataceae, and order Thoracosphaerales, respectively. Saxitoxin (STX) is produced by dinoflagellates [5]. Tetrahydropurine, which causes paralytic shellfish poisoning, is related to saxitoxin (PSP). The main adverse consequence of STX-group toxins is neurotoxicity. It is believed that their binding to voltage-gated sodium channels and the subsequent blocking of ion conductance via these channels is the main molecular mechanism for their neurotoxicity [5].

Table 1 HABs species of phytoplankton in the world [1]

No.	Species	Group
1	*Nitzschia bizertensis*	Diatom
2	*Nitzschia navis-varingica*	Diatom
3	*Pseudo-nitzschia abrensis*	Diatom
4	*Pseudo-nitzschia australis*	Diatom
5	*Pseudo-nitzschia batesiana*	Diatom
6	*Pseudo-nitzschia bipertita*	Diatom
7	*Pseudo-nitzschia brasiliana*	Diatom
8	*Pseudo-nitzschia caciantha*	Diatom
9	*Pseudo-nitzschia calliantha*	Diatom
10	*Pseudo-nitzschia cuspidata*	Diatom
11	*Pseudo-nitzschia delicatissima*	Diatom
12	*Pseudo-nitzschia fraudulenta*	Diatom
13	*Pseudo-nitzschia fukuyoi*	Diatom
14	*Pseudo-nitzschia galaxiae*	Diatom
15	*Pseudo-nitzschia granii*	Diatom
16	*Pseudo-nitzschia hasleana*	Diatom
17	*Pseudo-nitzschia kodamae*	Diatom
18	*Pseudo-nitzschia lundholmiae*	Diatom
19	*Pseudo-nitzschia multiseries*	Diatom
20	*Pseudo-nitzschia multistriata*	Diatom
21	*Pseudo-nitzschia obtusa*	Diatom
22	*Pseudo-nitzschia plurisecta*	Diatom
23	*Pseudo-nitzschia pseudodelicatissima*	Diatom
24	*Pseudo-nitzschia pungens*	Diatom
25	*Pseudo-nitzschia seriata*	Diatom
26	*Pseudo-nitzschia simulans*	Diatom
27	*Pseudo-nitzschia subcurvata*	Diatom
28	*Pseudo-nitzschia subfraudulenta*	Diatom
29	*Pseudo-nitzschia subpacifica*	Diatom
30	*Pseudo-nitzschia turgidula*	Diatom
31	*Amphidinium carterae*	Dinoflagellates
32	*Amphidinium gibbosum*	Dinoflagellates
33	*Amphidinium klebsii*	Dinoflagellates
34	*Amphidinium operculatum*	Dinoflagellates
35	*Dinophysis acuminata*	Dinoflagellates
36	*Dinophysis acuta*	Dinoflagellates
37	*Dinophysis caudata*	Dinoflagellates

(continued)

Table 1 (continued)

No.	Species	Group
38	*Dinophysis fortii*	Dinoflagellates
39	*Dinophysis infundibulum*	Dinoflagellates
40	*Dinophysis miles*	Dinoflagellates
41	*Dinophysis norvegica*	Dinoflagellates
42	*Dinophysis ovum*	Dinoflagellates
43	*Dinophysis sacculus*	Dinoflagellates
44	*Dinophysis tripos*	Dinoflagellates
45	*Phalacroma mitra*	Dinoflagellates
46	*Phalacroma rotundatum*	Dinoflagellates
47	*Alexandrium affine*	Dinoflagellates
48	*Alexandrium andersonii*	Dinoflagellates
49	*Alexandrium australiense*	Dinoflagellates
50	*Alexandrium balech*	Dinoflagellates
51	*Alexandrium catenella*	Dinoflagellates
52	*Alexandrium fragae*	Dinoflagellates
53	*Alexandrium hiranoi*	Dinoflagellates
54	*Alexandrium leei*	Dinoflagellates
55	*Alexandrium minutum*	Dinoflagellates
56	*Alexandrium monilatum*	Dinoflagellates
57	*Alexandrium ostenfeldii*	Dinoflagellates
58	*Alexandrium pacificum*	Dinoflagellates
59	*Alexandrium pseudogonyaulax*	Dinoflagellates
60	*Alexandrium tamarense*	Dinoflagellates
61	*Alexandrium tamiyavanichii*	Dinoflagellates
62	*Alexandrium taylorii*	Dinoflagellates
63	*Centrodinium punctatum*	Dinoflagellates
64	*Coolia tropicalis*	Dinoflagellates
65	*Fukuyoa paulensis*	Dinoflagellates
66	*Fukuyoa ruetzleri*	Dinoflagellates
67	*Fukuyoa yasumotoi*	Dinoflagellates
68	*Gambierdiscus australes*	Dinoflagellates
69	*Gambierdiscus balechii*	Dinoflagellates
70	*Gambierdiscus belizeanus*	Dinoflagellates
71	*Gambierdiscus caribaeus*	Dinoflagellates
72	*Gambierdiscus carolinianus*	Dinoflagellates
73	*Gambierdiscus carpenteri*	Dinoflagellates
74	*Gambierdiscus cheloniae*	Dinoflagellates

(continued)

Table 1 (continued)

No.	Species	Group
75	*Gambierdiscus excentricus*	Dinoflagellates
76	*Gambierdiscus lapillus*	Dinoflagellates
77	*Gambierdiscus pacificus*	Dinoflagellates
78	*Gambierdiscus polynesiensis*	Dinoflagellates
79	*Gambierdiscus scabrosus*	Dinoflagellates
80	*Gambierdiscus silvae*	Dinoflagellates
81	*Gambierdiscus toxicus*	Dinoflagellates
82	*Gonyaulax spinifera*	Dinoflagellates
83	*Gonyaulax taylorii*	Dinoflagellates
84	*Lingulodinium polyedra*	Dinoflagellates
85	*Ostreopsis fattorussoi*	Dinoflagellates
86	*Ostreopsis heptagon*	Dinoflagellates
87	*Ostreopsis lenticularis*	Dinoflagellates
88	*Ostreopsis mascarenensis*	Dinoflagellates
89	*Ostreopsis ovata*	Dinoflagellates
90	*Ostreopsis rhodesiae*	Dinoflagellates
91	*Ostreopsis siamensis*	Dinoflagellates
92	*Protoceratium reticulatum*	Dinoflagellates
93	*Pyrodinium bahamense*	Dinoflagellates
94	*Gymnodinium catenatum*	Dinoflagellates
95	*Karenia bicuneiformis*	Dinoflagellates
96	*Karenia brevis*	Dinoflagellates
97	*Karenia brevisulcata*	Dinoflagellates
98	*Karenia concordia*	Dinoflagellates
99	*Karenia cristata*	Dinoflagellates
100	*Karenia longicanalis*	Dinoflagellates
101	*Karenia mikimotoi*	Dinoflagellates
102	*Karenia papilionacea*	Dinoflagellates
103	*Karenia selliformis*	Dinoflagellates
104	*Karlodinium armiger*	Dinoflagellates
105	*Karlodinium conicum*	Dinoflagellates
106	*Karlodinium corsicum*	Dinoflagellates
107	*Karlodinium digitatum*	Dinoflagellates
108	*Karlodinium gentienii*	Dinoflagellates
109	*Karlodinium veneficum*	Dinoflagellates
110	*Margalefidinium fulvescens*	Dinoflagellates
111	*Margalefidinium polykrikoides*	Dinoflagellates

(continued)

Table 1 (continued)

No.	Species	Group
112	*Polykrikos hartmannii*	Dinoflagellates
113	*Takayama cladochroma*	Dinoflagellates
114	*Heterocapsa circularisquama*	Dinoflagellates
115	*Vulcanodinium rugosum*	Dinoflagellates
116	*Prorocentrum borbonicum*	Dinoflagellates
117	*Prorocentrum caipirignum*	Dinoflagellates
118	*Prorocentrum cassubicum*	Dinoflagellates
119	*Prorocentrum concavum*	Dinoflagellates
120	*Prorocentrum cordatum*	Dinoflagellates
121	*Prorocentrum emarginatum*	Dinoflagellates
122	*Prorocentrum hoffmannianum*	Dinoflagellates
123	*Prorocentrum leve*	Dinoflagellates
124	*Prorocentrum lima*	Dinoflagellates
125	*Prorocentrum mexicanum*	Dinoflagellates
126	*Prorocentrum rhathymum*	Dinoflagellates
127	*Prorocentrum steidingerae*	Dinoflagellates
128	*Prorocentrum texanum*	Dinoflagellates
129	*Luciella masanensis*	Dinoflagellates
130	*Pfiesteria piscicida*	Dinoflagellates
131	*Pfiesteria shumwayae*	Dinoflagellates
132	*Amphidoma languida*	Dinoflagellates
133	*Azadinium dexteroporum*	Dinoflagellates
134	*Azadinium poporum*	Dinoflagellates
135	*Azadinium spinosum*	Dinoflagellates
136	*Anabaena cylindrica*	Cyanobacteria
137	*Anabaena lapponica*	Cyanobacteria
138	*Anabaenopsis arnoldii*	Cyanobacteria
139	*Anabaenopsis milleri*	Cyanobacteria
140	*Anagnostidinema amphibium*	Cyanobacteria
141	*Anagnostidinema lemmermannii*	Cyanobacteria
142	*Annamia toxica*	Cyanobacteria
143	*Aphanizomenon flosaquae*	Cyanobacteria
144	*Aphanizomenon gracile*	Cyanobacteria
145	*Blennothrix lyngbyacea*	Cyanobacteria
146	*Calothrix parietina*	Cyanobacteria
147	*Chrysosporum bergii*	Cyanobacteria
148	*Chrysosporum ovalisporum*	Cyanobacteria

(continued)

Table 1 (continued)

No.	Species	Group
149	*Cuspidothrix issatschenkoi*	Cyanobacteria
150	*Cyanobium bacillare*	Cyanobacteria
151	*Cylindrospermum stagnale*	Cyanobacteria
152	*Dolichospermum flosaquae*	Cyanobacteria
153	*Dolichospermum lemmermannii*	Cyanobacteria
154	*Dolichospermum macrosporum*	Cyanobacteria
155	*Dolichospermum mendotae*	Cyanobacteria
156	*Dolichospermum planctonicum*	Cyanobacteria
157	*Dolichospermum sigmoideum*	Cyanobacteria
158	*Dolichospermum spiroides*	Cyanobacteria
159	*Geitlerinema carotinosum*	Cyanobacteria
160	*Geitlerinema splendidum*	Cyanobacteria
161	*Gloeotrichia echinulata*	Cyanobacteria
162	*Kamptonema formosum*	Cyanobacteria
163	*Limnospira fusiformis*	Cyanobacteria
164	*Limnothrix redekei*	Cyanobacteria
165	*Lyngbya majuscula*	Cyanobacteria
166	*Microcystis aeruginosa*	Cyanobacteria
167	*Microcystis botrys*	Cyanobacteria
168	*Microcystis flosaquae*	Cyanobacteria
169	*Microcystis ichthyoblabe*	Cyanobacteria
170	*Microcystis panniformis*	Cyanobacteria
171	*Microcystis viridis*	Cyanobacteria
172	*Microcystis wesenbergii*	Cyanobacteria
173	*Nodularia spumigena*	Cyanobacteria
174	*Phormidium uncinatum*	Cyanobacteria
175	*Planktothrix agardhii*	Cyanobacteria
176	*Planktothrix prolifica*	Cyanobacteria
177	*Planktothrix rubescens*	Cyanobacteria
178	*Raphidiopsis curvata*	Cyanobacteria
179	*Raphidiopsis mediterranea*	Cyanobacteria
180	*Raphidiopsis raciborskii*	Cyanobacteria
181	*Snowella lacustris*	Cyanobacteria
182	*Trichormus variabilis*	Cyanobacteria

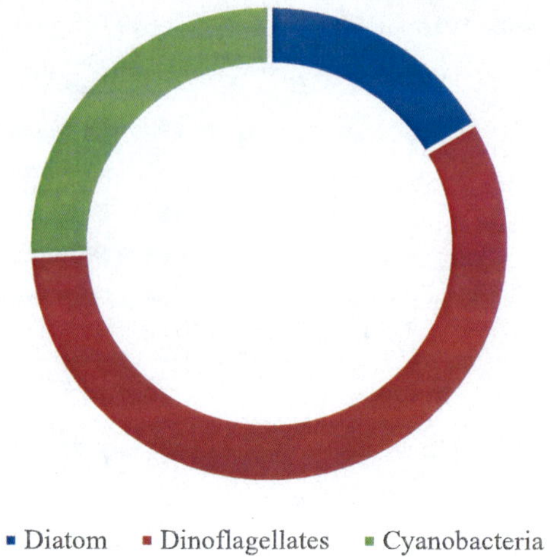

Fig. 1 Comparison of HABs species phytoplankton diatom, dinoflagellates, and cyanobacteria

HABs species phytoplankton in the Western Pacific has been reported [2]. Detailed lists of HABs species of paralytic shellfish poisoning (PSP), diarrheic shellfish poisoning (DSP), amnestic shellfish poisoning (ASP), ciguatera fish poisoning (CFP), and red tide causative species are given in Table 2.

The PSP-causing species *Alexandrium catanella*from the Chilean Coast produces the carbamoyl gonyautoxins (GTX1/GTX4) and N-sulfocarbamoyl derivatives (C1/C2, B1)[6]. Neosaxitoxin is the primary toxin produced by *Alexandrium minutum* from The Bay of Plenty (New Zealand), with modest amounts of saxitoxin and gonyautoxin (GTX1-4) [7]. In contrast, Japan strain of *Alexandrium tamarense* produces C2, gonyautoxin 4, neosaxitoxin, and gonyautoxin [8].*Gymnodinium catenatum*, another PSP-causing species, also produces *N*-sulfocarbamoyl analog (C1–4), gonyautoxin 5 (GTX5), gonyautoxin 6 (GTX6), and decarbamoyl derivatives, decarbamoyl saxitoxin (dcSTX), decarbamoyl neosaxitoxin (dcNeo), and decarbamoyl gonyautoxin 3 (dcGTX3) [9].

DSP is caused by species of dinoflagellates in the genus *Dinophysis*. These taxa can produce pectenotoxin (PTX) and okadaic acid (OA) groups [10]. Pectenotoxin-2 (PTX-2) and dinophysistoxin-1 (DTX-1) are dinoflagellates toxin produced by *Dinophysis acuminata* from New Zealand [11]. Meanwhile, diatom groups provide the causal of ASP. Domoic acid (DA) is produced by the diatom genus *Pseudo-nitzchia* that relevant to ASP [12].

Ciguatoxins (CTXs) are toxic fish poisoning agents produced by the *Gambierdiscus* species *G. australes, G. excentricus, G. silvae, G. carolinianus,* and *G. caribeaus* [13]. While *Cochlodinium polykrikoides*, the species that causes red tide, produces O_2 and H_2O_2, as well as reactive oxygen species (ROS)—a mediated ichthyotoxic mechanism that is detrimental to fish [14].

Table 2 HABs species of phytoplankton in the Western Pacific [2]

No.	Species	Toxin
1	*Alexandrium catenella*	PSP causative species
2	*A. minutum*	PSP causative species
3	*A. tamarense*	PSP causative species
4	*A. tamiyavanichii*	PSP causative species
5	*Gymnodinium catenatum*	PSP causative species
6	*Pyrodinium bahamense*	PSP causative species
7	*Dinophysis acuminata*	DSP causative species
8	*D. caudata*	DSP causative species
9	*D. fortii*	DSP causative species
10	*D. miles*	DSP causative species
11	*D. mitra*	DSP causative species
12	*D. norvegica*	DSP causative species
13	*Pseudo-nitzchia delicatissima*	ASP causative species
14	*P. multiseries*	ASP causative species
15	*P. pseudodelicatissima*	ASP causative species
16	*Gambierdiscus toxicus*	CFP causative species
17	*Ostreopsis lentricularis*	CFP causative species
18	*O. ovata*	CFP causative species
19	*O. siamensis*	CFP causative species
20	*Prorocentrum cancavum*	CFP causative species
21	*P. lima*	CFP causative species
22	*Cochlodinium polykrikoides*	Red tide causative species
23	*Heterocapsa circularisquama*	Red tide causative species
24	*Karenia mikimotoi*	Red tide causative species
25	*Noctiluca scintillans*	Red tide causative species
26	*Prorocentrum minimum*	Red tide causative species
27	*Chattonella antiqua*	Red tide causative species
28	*Heterosigma akashiwo*	Red tide causative species

4 Zooplankton Response to HABs Species

HABs occurrences have been reported in waters around the world, including those in Indonesia [15, 16], Japan [17], Korean [18], and Chinese waters [19, 20]. These events harm on aquaculture activities by causing fish mortality and a loss of economic value. Understanding the interaction and response of zooplankton to HABs is essential.

Zooplankton serves a crucial ecological role in regulating ecosystem functioning. HABs have a deleterious impact on zooplankton and can quickly cause drastic environmental changes. The interaction or response of zooplankton to HABs species is given in Table 3.

Table 3 Zooplankton response/interaction to HABs

No	HABs species	Response/interaction	References
1	*Alexandrium catanella*	Significant changes in the grazing or survival of copepod *A. tonsa* are unlikely to result from transient exposure (24 h) to poisonous *A. catenella*	[21]
2	*Alexandrium fundyense*	In copepod *Calanus finmarchicus*, three cytosolic Glutathione S-Transferase (GSTs) reacted short and fast to the *A. fundyense* diet. Copepod *C. finmarchicus* to be tolerant to *A. fundyense*, specific GSTs do not need to be long-term up regulated	[22]
		Copepod C. *finmarchicus* resistant to saxitoxin of *A. fundyense*. Detoxification did not play the significant role in *C. finmarchicus* response. Increasing the frequency of *A. fundyense* blooms may have an impact on *C. finmarchicus* populations by altering individual energy budgets and limiting their capacity to store lipids	[23]
		Copepod *Temora longicornis* easily digested *A. fundyense*, which drastically changed the swimming behavior of copepods without impacting their fitness or survival	[24]
		Paralytic shellfish toxins (PST) tolerance copepods using feeding assay shows that the frequency distribution of ingestion rates was polymodal when copepods consumed *A. fundyense*. The unhealthy diet always resulted in much higher ingestion rates than the benign diet. Tolerance to PST is connected to grazer intake of toxic *A. fundyense*	[25]

(continued)

(continued)

No	HABs species	Response/interaction	References
3	*Alexandrium minutum*	Copepodamides, a group of eight tiny compounds generated by copepods, are polar lipids that connect taurine to isoprenoid fatty acids conjugates of various compositions. The bloom-forming dinoflagellate Alexandrium minutum produces up to a 20-fold increase in paralytic shellfish toxins in response to pico—to nanomolar concentrations of copepodamides. Different species of copepod exhale unique mixtures of copepodamides, which help explain the species-specific defense reactions were seen in phytoplankton	[26]
		Paralytic shellfish toxin (PST) synthesis in the toxic algal bloom-forming dinoflagellate *Alexandrium minutum* is triggered by watery cues from the copepod *Acartia tonsa*. Induced *A. minutum* was more resistant to additional copepod grazing and contained up to 2.5 times more toxins than controls	[27]
4	*Alexandrium minutum* and *Gyrodinium corsicum*	Dinoflagellates that are toxic to copepods interact with them differently depending on the species. *Gyrodinium corsicum* can be grazed on by the copepod *Euterpina acutifrons* without having any negative impact on their ability to consume or grow. Although both copepod species ingested *A. minutum* in the short term, the toxic effects of these flagellates would limit their ability to affect other detrimental marine dinoflagellates in the long term	[28]
5	*Dinophysis* spp.	Copepod grazing can greatly slow down or halt the expansion of *Dinophysis* spp. populations due to the vast quantity of grazers and the unfavorable prey population growth conditions	[29]
6	*Gymnodinium catenatum*	Swimming speed and distance of copepods *Tigriopus japonicus* decline, detoxifying mechanism	[21]

(continued)

(continued)

No	HABs species	Response/interaction	References
7	*Karenia brevicornis*	Copepods *Acartia tonsa* on diets high in *K. brevis* had lower survival rates and less egg output per female, but the proportion of hatched eggs was unchanged. Copepods tried to compensate for poor food quality by consuming more of it	[30]
8	*K. brevicornis* and *Karlodinium veneficum*	Toxic algae species may change top-down control exercised, as well as how they may affect the copepod's eating behavior	[31]
9	*Karenia brevis and Peridinium foliaceum*	*Acartia tonsa* is not significantly impacted by the presence of	[32]
10	*Ostreopsis* cf. *ovata*	The copepod turned out to be the *Ostreopsis* cf. *ovata*-tolerant organism. However, when fed with the poisonous dinoflagellate, its fecundity and fertility ratios were decreased, indicating a potential reprotoxic effect. The existence of the toxic cells inside the digestive tract of the copepod when it was fed a mono-diet of *O.* cf. *ovata* suggested that these main grazers might act as a route for toxins through the marine food web, even if fecal pellet production is reduced dramatically	[33]
11	*Prorocentrum donghaiense*	Zooplankton population and community change during three phases (growth, maintenance, dissipation) of HABs	[20]
		Copepods and tunicates, which are herbivorous and free-spawners, dominated the zooplankton community before and after the bloom, whereas small jellyfish, chaetognaths, and copepods, which are carnivores and egg-brooding lineages prevailed during the bloom	[19]

(continued)

(continued)

No	HABs species	Response/interaction	References
12	*Prorocentrum lima*	Nonspecific detoxification responses to oxidative stress after exposure to *Prorocentrum lima*. The drastically enhanced mRNA expression of the proteasome, spliceosome, and ribosome as well as the increased fatty acid oxidation and oxidative phosphorylation all pointed to the proteolysis of damaged proteins and activation of energy expenditure. Exposure to okadaic acid (OA) accelerated the hydrolysis of chitin, which may further interfere with *A. salina*molting and reproductive processes	[34]
13	*Pseudo-nitzschia*	After consuming a domoic acid (DA), the arctic copepods *Calanus hyperboreus* and *C. glacialis* lowered their escape response	[35]

The variability in studies of zooplankton feeding on hazardous HABs species is mainly caused by variations in the concentrations of harmful algae toxins in dietary treatments. The average duration of studies on copepods feeding, including those on toxic algal species, has been 24 h [36]. To determine the effects of various species of harmful algae on various grazers, more times (a few days or a week) are required.

According to the species, toxic algae have varying effects on grazers. More intriguingly, the impacts also vary across and even between populations of a species. The population-specific reaction of grazers to toxic algae is significantly driven by local adaptation, which refers to the fact that grazer populations that have previously been exposed to toxic algal blooms perform better on average when faced with toxic algae than populations that are ignorant of these blooms [35]. It is crucial to determine the grazer's abundance before, during, and after blooms.

It is essential to comprehend how grazers and harmful algae interact. Toxic algae have a deleterious effect on grazers in various ways, such as decreased metabolism, fertility, and consumption. Through both direct (toxic) and indirect (e.g., food limitation) implications, these negative impacts on primary consumers eventually propagate to higher trophic levels of the food chain. However, grazing is an approach that could prevent harmful algal blooms.

5 Conclusion

Zooplankton has various responses to HABs species. Some of zooplankton taxa impacted the adverse toxicity produced by HABs species of phytoplankton. However, the other taxa of zooplankton are more tolerant of responding HABs. Zooplankton population which has previously been exposed to toxic algal blooms performs better on average when faced with toxic algae than the population grazers that are ignorant of HABs.

Acknowledgements This study is a part of the research revealing biodiversity of coastal zooplankton through eDNA, DNA barcoding, and morphological features, funded by the LIPI's COREMAP-CTI to HAM under agreement number 17/A/DK/2021. HAM is the main contributor, while AR and NF contribute equally as associate contributors.

References

1. Lundholm, N., Churro, C., Fraga, S., Hoppenrath, M., Iwataki, M., Larsen, J., Mertens, K., Moestrup, Ø., Zingone, A. (eds.) (2009 onwards). IOC-UNESCO Taxonomic Reference List of Harmful Micro Algae. Accessed at https://www.marinespecies.org/hab on 22 Aug 2022. https://doi.org/10.14284/362
2. Omura, T., Iwataki, M., Borja, V.M., Takayama, H., Fukuyo, Y.: Marine Phytoplankton of the Western Pacific. Kouseisha Kouseikaku, Tokyo (2012)
3. Sakamoto, S., Lim, W.A., Lu, D., Dai, X., Orlova, T., Iwataki, M.: Harmful algal blooms and associated fisheries damage in East Asia: current status and trends in China, Japan, Korea and Russia. Harmful Algae **102**, 101787 (2021)
4. Cusick, K.D., Sayler, G.S.: An overview on the Marine Neurotoxin, Saxitoxin: genetics, molecular targets, methods of detection and ecological functions. Mar. Drugs **11**, 991–1018 (2013)
5. EFSA.: Scientific opinion of the panel on contaminants in the food chain on a request from the European Commission on Marine biotoxins in shellfish—saxitoxin group. The EFSA J. **1019**, 1–76 (2009)
6. Krock, B., Seguel, C.G., Cembella, A.D.: Toxin profile of Alexandrium catenella from the Chilean coast as determined by liquid chromatography with fluorescence detection and liquid chromatography coupled with tandem mass spectrometry. Harmful Algae **6**, 734–744 (2007)
7. Chang, F.B., Anderson, D.M., Kulis, D.M., Till, D.G.: Toxin production of Alexandrium minutum (Dinophyceae) from the Bay of Plenty New Zealand. Toxicon **35**(3), 393–409 (1998)
8. Shimada, H., Motylkova, I.V., Mogilnikova, T.A., Mikami, K., Kimura, M.: Toxin profile of Alexandrium tamarense (Dinophyceae) from Hokkaido, Northern Japan and Southern Sakhalin, Eastern Russia. Plankton Benthos Res **6**(1), 35–41 (2011)
9. Costa, P.R., Robertson, A., Quilliam, M.A.: Toxin profile of Gymnodinium catenatum (Dinophyceae) from the Portuguese coast, as determined by liquid chromatography tandem mass spectrometry. Mar. Drugs **13**, 2046–2062 (2016)
10. Fux, E., Smith, J.L., Tong, M., Guzman, L., Anderson, D.M.: Toxin profiles of five geographical isolates of Dinophysis spp. from North and South America. Toxicon **57**, 275–287 (2011)
11. Mackenzie, L.A.: A long-term time series of Dinophysis acuminata blooms and associated shellfish toxin contamination in port underwood, Marlborough sounds New Zealand. Toxins **11**, 74 (2019)

12. Turk Dermastia, T., Dall'Ara, S., Dolenc, J., Mozetiˇc, P.: Toxicity of the Diatom Genus Pseudo-nitzschia (Bacillariophyceae): insights from toxicity tests and genetic screening in the Northern Adriatic Sea. Toxins **14**, 60 (2022)
13. Rossignoli, A.E., Tudó, A., Bravo, I., Díaz, P.A., Diogène, J., Riobó, P.: Toxicity Characterisation of *Gambierdiscus* Species from the Canary Islands. Toxins (Basel). **12**(2), 134 (2020)
14. Kim, D., Oda, T., Muramatsu, T., Kim, D., Matsuyama, Y., Honjo, T.: Possible factors responsible for the toxicity of Cochlodinium polykrikoides, a red tide phytoplankton. Comparative Biochem. Physiol. Part C **132**, 415–423 (2002)
15. Likumahua, S.: The recent bloom of Pyrodinium bahamense var compressum in Ambon Bay Eastern Indonesia. Mar. Res. Indonesia **38**(1), 31–37 (2013)
16. Sidabutar, T., Cappenberg, H., Srimariana, E.S., Muawanah, A., Wouthuyzen, S.: Hamrful algal blooms and their impact on fish mortalities in Lampung Bay: an overview. In: IOP Conference Series: Earth and Environmental Science, vol. 944, pp. 012027. IOP Publishing (2021)
17. Imai, I., Inaba, N., Yamamoto, K.: Harmful algal blooms and environmentally friendly control strategies in Japan. Fish. Sci. **87**, 437–464 (2021)
18. Shin, Khim, B.K., Jang, L.H., Lim, J., Jo, Y.H.: Convolutional neural network model for discrimination of harmful algal blooms (HAB) from non-HABs using Sentinel 3 OLCI imagery. ISPRS J. Photogrammetry and Remote Sens **191**, 250–262 (2022)
19. Zhang, Z., Zhuang, Y., Chen, H., Lu, S., Li, Y., Ge, R., Chen, C., Liu, G.: Effects of Prorocentrum donghaiense bloom on zooplankton functional groups in the coastal waters of the East China Sea. Marine Pollut. Bulletin **172**, 112878 (2021)
20. Shen, A., Chen, W., Xu, Y., Ho, K.C.: Zooplankton population and community structure changes in response to a harmful Algal Bloom caused by *Prorocentrum donghaiense* in the East China Sea. J. Marine Sci. Eng. **10**, 291 (2022)
21. Han, J., Park, J.S., Park, Y., Lee, J., Shin, H.H., Lee, K.W.: Effects of paralytic shellfish poisoning toxin-producing dinoflagellate *Gymnodinium catenatum* on the marine copepod *Tigriopus japonicus*. Marine Pollut. Bulletin **163**, 111937 (2021)
22. Roncalli, V., Jungbluth, M.J., Lenz, PH.: Glutathione S-transferase regulation in *Calanus finmarchicus* feeding on the toxic *Dinoflagellate Alexandrium* fundyense. PLoS ONE **11**(7), e0159563 (2016). https://doi.org/10.1371/journal.pone.0159563
23. Roncalli, V., Cieslak, M.C., Lenz, P.H.: Transcriptomic responses of the calanoid copepod Calanus finmarchicus to the saxitoxin producing dinoflagellate Alexandrium fundyense. Sci. Rep. **6**, 25708 (2016). https://doi.org/10.1038/srep25708
24. Lasley-Rasher, R.S., Nagel, K., Angra, A., Yen, J.: Intoxicated copepods: ingesting toxic Phytoplankton leads to risky behaviour. Proc. R. Soc. B **283**, 20160176 (2016). https://doi.org/10.1098/rspb.2016.0176
25. Senft, C., Avery, D.E., Dam, H.G.: A novel approach to identifying PST tolerant copepods: an individual ingestion assay. Harmful Algae **10**, 804–810 (2011)
26. Selander, E., Kubanek, J., Hamberg, M., Andersson, M.X., Cervin, G., Pavia, H.: Predator lipids induce paralytic shellfish toxins in bloom-forming algae. PNAS **112**(20), 6395–6400 (2015)
27. Selander, E., Thor, P., Toth, G., Pavia, H.: Copepods induce paralytic shellfish toxin production in marine dinoflagellates. Proc. R. Soc. B. **273**, 1673–1680 (2006)
28. Marinho da Costa, R., Fernandez, F.: Feeding and survival rates of the copepods Euterpina acutifrons Dana and Acartia grani Sars on the dinoflagellates Alexandrium minutum Balech and Gyrodinium corsicum Paulmier and the Chryptophyta Rhodomonas baltica Karsten. J. Experim. Marine Biol. Ecol. **273**,131–142 (2002)
29. Kozlowsky-Suzuki, B., Carlsson, P., Ruhl, A., Graneli, E.: Food selectivity and grazing impact on toxic Dinophysis spp. by Copepods feeding on natural plankton assemblages. Harmful Algae **5**, 57–68 (2006)
30. Prince, E.K., Lettieri, L., McCurdy, K.J., Kubanek, J.: Fitness consequences for copepods feeding on a red tide dinoflagellate: deciphering the effects of nutritional value, toxicity, and feeding behavior. Oecologia **147**, 479–488 (2006)

31. Hong, J., Talapatra, S., Katz, J., Tester, P.A., Waggett, R.J., Place, A.R.: Algal toxins alter copepod feeding behavior. PLoS ONE **7**(5), e36845 (2012). https://doi.org/10.1371/journal.pone.0036845
32. Speekmann, C.L., Hyatt, C.J., Buskey, E.J.: Effects of Karenia brevis diet on RNA:DNA ratios and egg production of Acartia tonsa. Harmful Algae **5**, 693–704 (2006)
33. Anne-Shopie, P., Julie, R., Laurence, G.G., Shopie, M., Eva, T., Thomas, OP., Rodolphe, L., Stephane, G.: Effects of the toxic dinoflagellate Ostreopsis cf. ovata on survival, feeding and reproduction of a phytal harpacticoid copepod. J. Experim. Marine Biol. Ecol. **516**, 103–113 (2019)
34. Gong, Y., Zhang, K., Geng, N., Wu, M., Yi, X., Liu, R., Challis, J.K., Codling, G., Xu, EG., Giesy, J.P.: Molecular mechanisms of Zooplanktonic toxicity in the Okadaic acid producing dinoflagellate Prorocentrum lima. Environm. Pollut. **279**, 116942 (2021)
35. Haroardottir, S., Krock, B., Wohlrab, S., John, U., Nielsen, T.G., Lundholm, N.: Can domoic acid affect escape response in copepods? Harmful Algae **79**, 50–52 (2018)
36. Turner, J.F.: Planktonic marine copepods and harmful algae. Harmful Algae **32**, 81–93 (2014)

Verification of Significant Wave Height of the INA-WAVES Against Sentinel 3 Altimetry Data: Case Study for Bali Waters

Timothy Kenoly, Tirtha Wijaya, Andri Ramdhani, Mahardika Jalu Pradana, and Subekti Mujiasih

Abstract INA-WAVES is a wave model application developed by the Agency for Meteorology, Climatology, and Geophysics of the Republic of Indonesia (BMKG) and integrated into an Ocean Forecast System (OFS). It was derived from the wave spectrum model of Wave Watch III and utilized to provide many essential seas dynamics information. Because of the strategic role, it would be wise to verify the performance regularly. This study used altimetry data of Sentinel 3A and B to verify significant wave height (SWH) analysis data of INA-WAVES. The period of data was from October 2020 to August 2021. The authors demonstrated the data by a range category (qualitatively) and the root mean square error (RMSE) (quantitatively). The RMSE score assessed the model accuracy in six locations around Bali Island: two points in the Bali Sea, one in the southern Bali strait, and three in the south of Bali. Generally, a visual verification revealed that the percentage of equality between the model and the altimetry is promising, such as the Bali Sea (78.95%), the

T. Kenoly
Marine Science Program, Faculty of Marine Sciences and Fisheries, Udayana University, Bali, Indonesia

T. Wijaya
Agency for Meteorology, Climatology, and Geophysics of the Republic of Indonesia (BMKG), Regional Office III, Jl. Raya Tuban, Kuta, 80361 Badung, Bali, Indonesia

A. Ramdhani · M. J. Pradana
Agency for Meteorology, Climatology, and Geophysics of the Republic of Indonesia (BMKG), Jl. Angkasa I No.2 Kemayoran, Jakarta Pusat, DKI, Jakarta 10610, Indonesia

S. Mujiasih (✉)
Research Center for Climate and Atmosphere, National Research and Innovation Agency (BRIN), Jl. Cisitu Sangkuriang, 40135 Bandung, West Java, Indonesia
e-mail: sube002@brin.go.id

GeoHydrodynamics and Environment Research (GHER), Department of Astrophysics, Geophysics and Oceanography (AGO), University of Liège, Sart-Tilman B5a, 4000 Liège, Belgium

A. Basit et al. (eds.), *Proceedings of the International Conference on Radioscience, Equatorial Atmospheric Science and Environment and Humanosphere Science*, Springer Proceedings in Physics 290, https://doi.org/10.1007/978-981-19-9768-6_44

northern Bali Strait (90.91%), and the southern Lombok Strait (82.35%). In addition, the quantitative result confirmed that the Bali Sea has low RMSE values, 0.505 and 0.687 m. Unfortunately, the performance in the south of Bali (Indian Ocean), the Badung Strait, and the northern Lombok Strait should be evaluated further. In particular, in the Indian Ocean, the RMSE shows 1.267 m and 1.12 m.

1 Introduction

Waves are one of the natural phenomena in the waters. Winds in the middle of the sea can cause water surface disturbances, later becoming waves. The duration and strength of this wind and weather conditions, such as storms, could affect the size of the waves formed [1]. Long-term data on wave variations, such as significant wave height data, in certain areas, are essential because they can be used to determine the variability and to predict weather conditions in the region [2, 3]. Furthermore, it can also be used to understand the potential changes that can occur in the coastal area where the waves are located [4]. Wave conditions in the waters around Bali vary throughout the year. In the north of Bali, significant wave heights ranged from 0.5 to 1.25 m in December–January–February (DJF) and June–July–August (JJA) but decreased to 0–0.75 m in March–April–May (MAM) and September–October–November (SON). In the east of Bali, significant wave heights tended to be the same throughout the year ranging from 0.5 to 1.5 m. In the western part of Bali, in the month of DJF, the significant wave height ranged from 0.5 to 0.75 m, increased in the month of MAM to 0.75–1.5 m, and remained in the same range until the end of the year. In comparison, wave heights in Bali's southern waters were significantly higher than in other areas, 0.75−2 m throughout the year [5].

BMKG developed a wave model from Wave Watch III coupled with Simulated Waves Nearshore. The model was integrated into the Ocean Forecast System (OFS). INA-WAVES used the nesting technique on three domains and considered bathymetry variation and forcing from NOAA's GFS. INA-WAVES first performed calculations for the Indonesian territorial waters domain by running the model on the global field so that the boundary condition values outside the part will be used to calculate a predictive value in the region [6]. Significant wave height of OFS was used for maritime weather information for various Indonesian water and shipping activities [7]. Therefore, verifying the data from the INA-WAVES model is essential to ensure that the delivered information can reduce the risk at sea. Similar research had been carried out previously by Mujiasih and Nuryanto [8], but with the object of verification, namely the Wind waves 5.0 model, which BMKG also used before using the INA-WAVES model. A performance assessment should accompany the improvement of the wave model from Wind Wave 5.0 to the INA-WAVES.

Meanwhile, the technology of remote sensing on the ocean is progressively enhanced, indicated by the presence of Sentinel 3A and B, which has a better resolution than their predecessor. Sentinel 3A and B are two satellites with radar altimetry sensors capable of recording changes in sea level and significant wave heights. Sentinel 3 is equipped with a high-resolution Synthetic Aperture Radar (SAR) to provide complete coverage of the earth [9]. Due to its high resolution and quality, Sentinel 3 altimetry data have the potential to be used as a data verifier from the results of the INA-WAVES significant wave height analysis. The presence of Sentinel 3 led to a research question of whether the INA-WAVES model output conforms to the satellite or not. Therefore, the authors aimed to evaluate the INA-WAVES model output against the Sentinel altimetry data to understand the suitability of the INA-WAVES spatially and temporally.

2 Data and Methods

2.1 Data Collection

Three different data were used in this study: INA-WAVES output visual data, INA-WAVES output quantitative data, and Sentinel 3 altimetry data. The INA-WAVES model was verified against the Sentinel altimetry to know the performance of the INA-WAVES model output. The verification area was divided into seven water areas around Bali Island as in Fig. 1: the Bali Sea, the northern and southern Bali Straits, the northern and southern Lombok Straits, the Badung Straits, and the Indian Ocean south of the Bali region. The INA-WAVES model output can be accessed via https://peta-maritim.bmkg.go.id/ofs-static, which we focused on significant wave height data. However, only 57 days of 77 days of Sentinel data availability were processed and analyzed due to the unavailability of INA-WAVES data. The total available data were 57 days from January 1–September 24, 2021. The INA-WAVES model used analytical data at 00UTC and 12UTC hours, closest to the available Sentinel 3 altimetry data. We analyzed them visually and quantitatively. The data availability for visual analysis is 57 days at 1:44–1:48 UTC and 14:03–14:06 UTC.

In general, there are 77 days in which Sentinel satellites 3A and B pass through 1 or more than seven water areas of BMKG regional III, which are the object of this research. Some regions did not appear on specific dates. It is due to Sentinel 3 satellite does not pass through the area concerned. Sentinel 3 data can be accessed via Copernicus's website, coda.eumetsat.int/#/home [9]. The altimetry data used was in the form of radar imagery in the waters (SR_2_WAT_) from the Synthetic Aperture Radar Altimeter (SRAL) instrument [10]. Level 2 (L2) and Non-time Critical data were chosen to ensure the product has gone through the geophysical correction process and has been consolidated by the data provider first. The across-track resolution of this data is ± 300 m with a temporal resolution of 27 days. To facilitate a more accessible quantitative analysis, we used six points. The 6 points were based on the path of Sentinel 3 that intersect each other to provide better spatio-temporal data. The 6 points were located in the Indian Ocean south of Bali (points 1, 2 and 3), the Bali Sea (points 4 and 5), and the Southern Bali Strait (point 6), which can be seen on the map in Fig. 1.

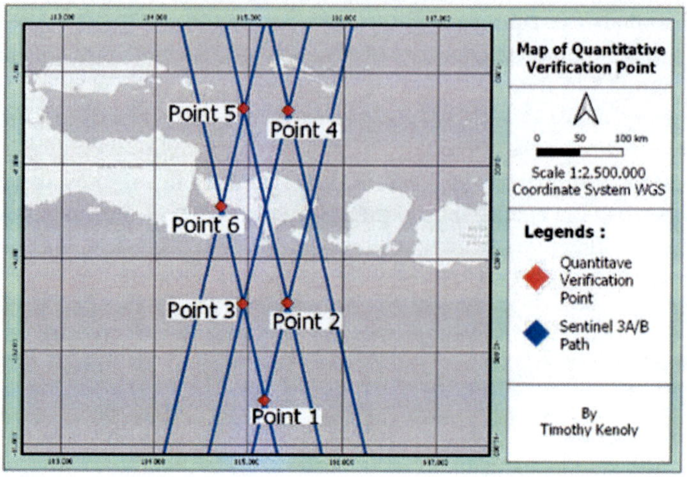

Fig. 1 Sentinel 3A and B path around Bali [9]

We displayed two samples of significant wave maps, both Sentinel and INA-WAVE, as in Figs. 2 and 3. Every single track of Sentinel 3 can pass through some areas. For instance, scanning on January 1, 2021, covered the Bali Sea, Lombok Strait, and the Indian Ocean. The other date, such as January 21, 2021, covered the Bali Sea, the Bali Strait, and the Indian Ocean. So, we used Sentinel data depending on the time of scanning over the Bali area.

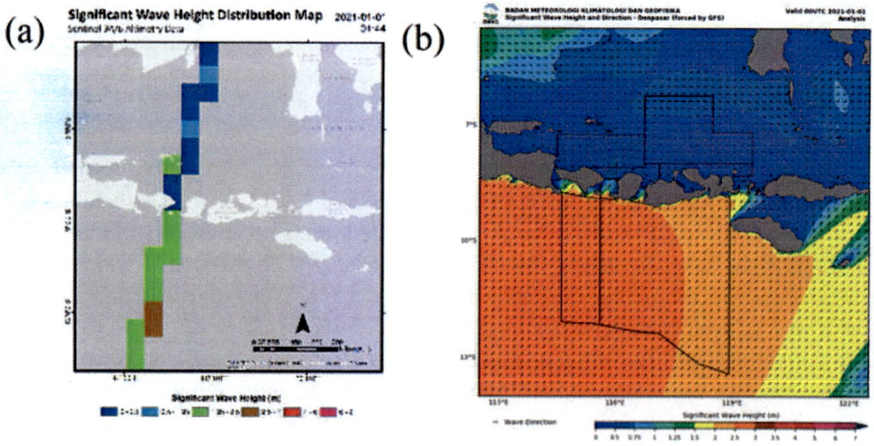

Fig. 2 Sample SWH distribution map on January 1, 2021. **a** Sentinel 3 **b** INA-WAVES

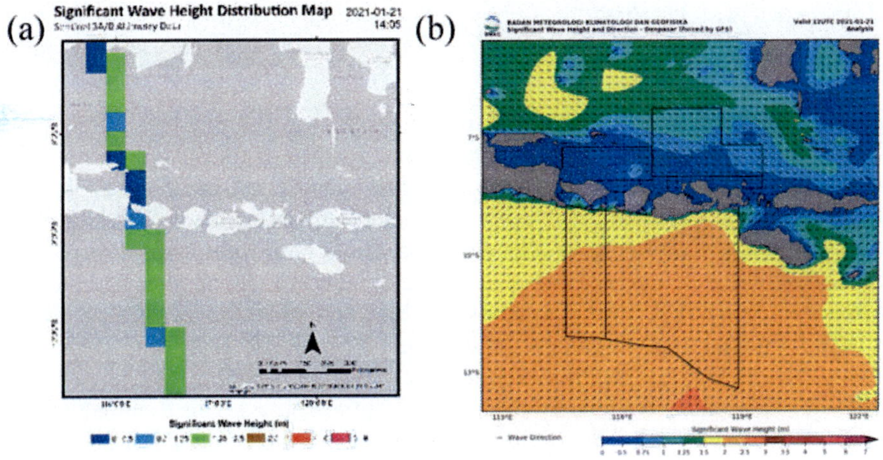

Fig. 3 Sample SWH distribution map on January 21, 2021. **a** Sentinel 3 **b** INA-WAVES

2.2 *Data Processing*

Visual data processing was focused on significant wave height data from the Ku channel and frequency of 20 Hz (swh_ocean_20_ku) of the Sentinel 3 standard measurement data products and their respective coordinate variables (lon_20_ku, lat_20_ku). This variable was used because it is the actual significant wave height measurement by the satellite, with as many as 20 measurements in 1 s. These data were then interpolated on the specific grid size. The grid size was not the actual measurement of the Sentinel three measurement area but the result of a visualization to facilitate analysis. The significant wave height values were then categorized into six categories, namely calm (0–0.5 m), low (0.5–1.25 m), moderate (1.25–2.5 m), high (2.5–4 m), very high (4–6 m), and extreme (6–9 m). Through the point in Fig. 1, an area of approximately 500 m will be drawn to the respective point's north, south, west, and east. This area will be used as a filter to determine the value of the significant wave height from the Sentinel 3 measurement results, which will later be analyzed. As for the INA-WAVES quantitative data to be explored, it was obtained by finding the closest point from the same 6 points that have been determined on the day and time that was closest to the time of the previous Sentinel 3 data availability.

2.3 Analysis Method

2.3.1 Visual Analysis

The analysis was carried out by a simple visual method. The data being compared are in the form of significant wave height distribution maps from the INA-WAVES analysis against significant wave height distribution maps from Sentinel 3 altimetry data. The suitability score was obtained based on color observations on the two maps describing the range of significant wave heights at each location. The suitable category, scored by 1, was obtained when the SWH value range of INA-WAVES analysis is equal to, or the upper limit is within, the Sentinel 3 value range. The unsuitable category, scored by 0, was obtained when the SWH value range of INA-WAVES analysis is not equal to, or the upper limit is outside, the Sentinel 3 value range. However, note that the significant wave height values from the INA-WAVES model and Sentinel 3 altimetry data were compared based on the range category. The authors compared with a simple visual method, so it did not involve exact values at a specific coordinate point from Sentinel 3 and INA-WAVES data, instead based on seven different water areas mentioned before.

2.3.2 Quantitative Analysis

The quantitative analysis method used was root mean square error (RMSE). This method has been widely used in various meteorological and climatic studies [11]. The result value of the RMSE calculation does not have a category, but the smaller the RMSE value, the better or more accurate the existing modeling data is. There is no maximum limit for the value of the RMSE, but the best RMSE value that can be achieved is 0. The equation in the RSME method [12] is as follows:

$$RMSE = \sqrt{\frac{1}{N}\sum_{i=1}^{N}(F_i - O_i)^2} \quad (1)$$

With F_i being the INA-WAVES model's wave height output data, O_i is the wave height value measured by Sentinel 3 [12]. In comparison, n is the total amount of data. The value of the validation data was obtained using the measurement points from Sentinel 3, closest to the modeling points from INA-WAVES. Determination of the nearest point relied on the Pythagorean theorem approach to get the distance between both points, which was obtained by taking the root of the sum of squares of their respective latitudes and longitudes.

3 Results and Discussion

3.1 *Visual Analysis*

Table 1 shows that the suitability of the INA-WAVES significant wave height model against Sentinel 3 data, in general, reached 61.36%. In addition, there were 5 out of 7 observed water areas with more than 50% of suitability. The authors found that the northern part of the Bali Strait has the highest suitability percentage, exceeding 90% (10:1). A comparison of 10:1 means ten suitable events and one unsuitable occurrence. Then, the southern part of the Lombok Strait and the Bali Sea was around 80% (14:3 and 45:12). Unfortunately, in the Indian Ocean, the south of the Bali region and the northern part of the Lombok Strait only achieved suitability in the range of 40.35% (23:34) and 40% (6:9). Furthermore, it was found that in the Indian Ocean south of the Bali region, the discrepancy tends to occur in the range of the moderate wave category (1.25–2.5 m). The discrepancy in this range reached 82%, or 28 out of 34, and was dominated by overestimating INA-WAVES modeling results. The Northern Lombok Strait region has the most discrepancies in the calm wave category, with 6 out of 9 differences. These discrepancies occur when the INA-WAVES model overestimates the significant wave height. At the same time, the other three discrepancies arise in the high wave category when the INA-WAVES model performs an underestimation.

Due to the limited data obtained, the temporal analysis of the northern Lombok Strait cannot be carried out intensively. Although the Northern Lombok Strait and the Indian Ocean south of the Bali Region have a low number of conformities, it does not mean that the results of the INA-WAVES model in these two areas are always inaccurate. Generally, when reviewing a specific location based on the path traversed by Sentinel 3, the suitability of the INA-WAVES model is quite good.

3.2 *Quantitative Analysis*

The quantitative data analysis obtained six RMSE values at six ordered different points, namely 1.267 m (point 1), 1.129 m (point 2), 0.618 m (point 3), 0.505 m (point 4), 0.989 m (point 5), and 0.687 m (point 6). The best RMSE lies at point 4 (the Bali Sea). The lowest RMSE value is scored by point 4 or in the Bali Sea area. Likewise, the RMSE value of point 5, which is also below 1, can strengthen the argument that the INA-WAVES significant wave height model performs best in the Bali Sea area. Then, the RMSE value at point 6, namely the southern part of the Bali Strait, is also relatively low. It indicates that the accuracy of the INA-WAVES model is quite good, regardless of the suitability value of the visual analysis, which is not so high. Meanwhile, the RMSE value at points 1 and 2 in the Indian Ocean south of the Bali region is relatively high. It supports the argument that the performance of

Table 1 Suitability level of INA-WAVES and Sentinel 3 data in each region

Regions	Total events	Total suitability	Total unsuitability	Suitability percentage (%)	Description
Bali Sea	57	45	12	78.95	Highest suitability in March and June
Northern Lombok Strait	15	6	9	40.00	Highest suitability in June and September
Southern Lombok Strait	17	14	3	82.35	Highest suitability in every month except March, April, and June
Badung Strait	26	17	9	65.38	Highest suitability in January, February, and September
Northern Bali Strait	11	10	1	90.91	Highest suitability in every month except June
Southern Bali Strait	37	20	17	54.05	Highest suitability in September
Indian Ocean South of Bali	57	23	34	40.35	Highest suitability in May and June
Grand Total	220	135	85	61.36	Highest suitability in June

the INA-WAVES model for the Indian Ocean south of the Bali region is not good compared to other areas in this study.

The scatterplot of the six parts in 2021 (Fig. 5) depicts a similar pattern to the scatterplot at the same points in 2020 (Fig. 4). At points 1 and 2, it is seen that the INA-WAVES model performs an overestimation. While at point 3, the INA-WAVES value tends to be between the range of Sentinel 3 measurement values or even below. At points 4, 5, and 6, a pattern appears where the Sentinel 3 value tends to be in the range of the Sentinel 3 measurement value, but it has sometimes exceeded the measurement value. However, when analyzed further, the RMSE score of point 3 is far lower when compared to the RMSE value of points 1 and 2. According to Figs. 4c and 5c, an overestimation of significant wave height value by the INA-WAVES model in point 3 does not occur. The results of the INA-WAVES model are in the range of Sentinel 3 measurement values. It also strengthens the previous estimate, which states that most non-conformities generated by INA-WAVES are in the form of overestimating significant wave height estimates. It refers to a wave climatology in this area which the southeast side tends to be affected by disturbance from a remote location (swell).

The INA-WAVES model's tendency to overestimate could occur due to 2 different things. The first is the level of accuracy of the INA-WAVES model performance. It might be due to the model performing nesting on three different domains. Areas with low suitability get significant wave height values from the domain that does not consider bathymetry. While, the second is that Sentinel 3A and B have limited area of coverage.

Given that the across-track resolution of Sentinel 3 is only 300 m wide, it is estimated that when the INA-WAVES model can model all the ranges of significant

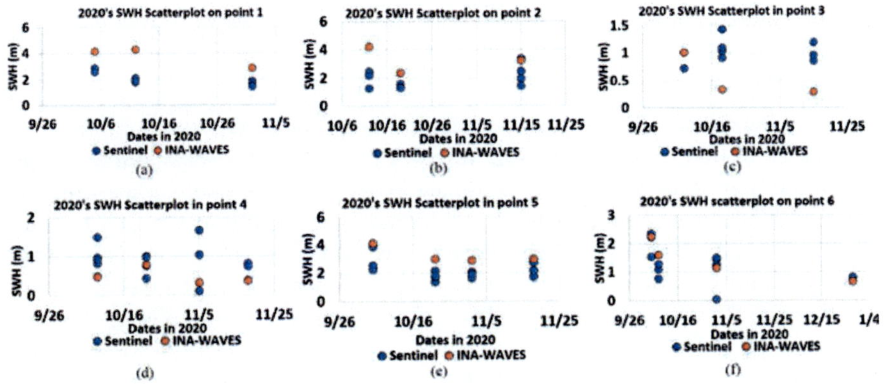

Fig. 4 Scatterplot graph of Significant wave height results from Sentinel 3 measurements and INA-WAVES model in 2020

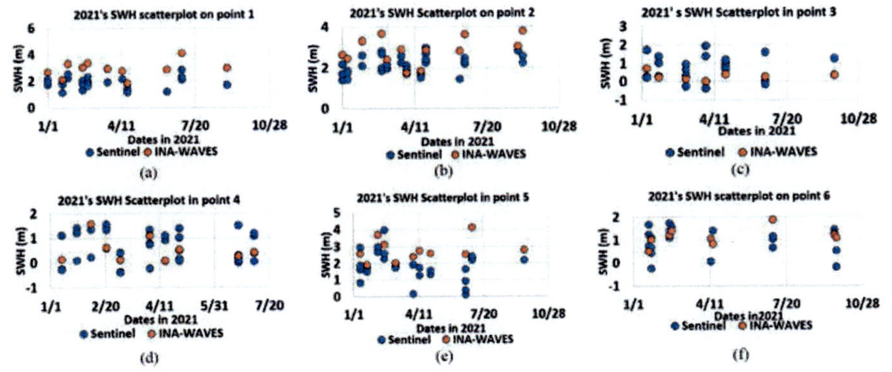

Fig. 5 Scatterplot graph of Significant wave height results from Sentinel 3 measurements and INA-WAVES model in 2021

wave heights in all parts of the observed area, Sentinel 3 only records in certain regions of the area concerned. As a result, Sentinel 3 altimetry data have the potential to read parts that have only low significant wave heights. It will significantly affect large areas of water, especially the Bali Sea or the Indian Ocean south of the Bali region. The suitability fluctuations that occur temporally are believed to be affected by the changing seasons. It is known that the climate in Indonesia can be divided into 4 phases, namely the rainy season in December, January, and February (DJA), the first transition season in March, April, and May (MAM), and the dry season in June, July, and August (JJA), and the September, October, and November (SON) second transition season. The fluctuation pattern generally follows the changing seasons, where a decrease in suitability tends to occur from January to February, March to May, or June to August. In contrast, the increase in conformity can occur from February to March, May to June, and August to September.

4 Conclusion

Based on the analysis in the previous section, two conclusions are as follows: The authors found that the lowest RMSE values were in the Bali Sea and the southern part of the Bali Strait, which were 0.505 and 0.687 m, respectively. Meanwhile, the highest RMSE values were found in the Indian Ocean south of the Bali region at 1.267 and 1.12 m. Besides, most discrepancies occur in the calm and moderate waves categories due to the INA-WAVES overestimated. Several reasons are considered to be the cause of this. The INA-WAVES model's performance conditions are suspected of contributing to the discrepancy, followed by the lack of accuracy of Sentinel 3 altimetry data in a large area of water due to the small across-track resolution.

Acknowledgements The authors thank the Agency for Meteorology, Climatology, and Geophysics of the Republic Indonesia Regional III, Denpasar, and the Center of Marine Meteorology (BMKG), who have facilitated this research. Sentinel Altimetry data were downloaded from https://coda.eumetsat.int/#/ accessed on 6 October 2021 with an update link: https://navigator.eumetsat.int/start.

Author Contributions All authors contributed as the main contributor to this paper. All authors read and approved the final paper.

References

1. Komar, P.D.: Beach processes and erosion an introduction. In: CRC Handbook of Coastal Processes and Erosion. Florida, CRC Press (2018)
2. Wirjohamidjojo, S., Sugarin, S.: Praktek Meteorologi Kelautan. Badan Meteorologi dan Geofisika (BMG). Jakarta (2008)
3. World Meteorological Organization: Guide to Marine Meteorological Services, 3rd edn. Secretariat of The World Meteorological Organization, Switzerland (2001)

4. Miller, J.K., Dean, R.G.: A simple new shoreline change model. Coast. Eng. **51**(7), 531–556 (2004)
5. Kurniawan, R., Khotimah, M.K.: Ocean wave characteristics in Indonesian Waters for sea transportation safety and planning. IPTEK The J. Technol. Sci. **26**(1) (2016)
6. Pusat Meteorologi Maritim.: BMKG Ocean Forecast System. Diklat Teknis Meteorologi Maritim Tahun 2020. Pusat Pendidikan dan Pelatihan BMKG Jakarta, Februari (2020)
7. Supriyadi, E.: Pemanfaatan Satelit Altimetri Untuk Verifikasi Tinggi Gelombang Signifikan Ocean Forecast System (OFS)-Model BMKG. J. Meteorologi dan Geofisika **19**(2), 93–102 (2019)
8. Mujiasih, S., Nuryanto, D.E.: Utilization of categorical and continuous combination for wave model verification. Mausam **66**(3), 617–624 (2015)
9. Copernicus Online Data Access (CODA).: Product synthetic aperture radar altimetry non-time critical (2020). Accessed on October 6th of 2021, from https://coda.eumetsat.int/#/home
10. European Organization for the Exploitation of Meteorological Satellites (EUMETSAT): Sentinel-3 SRAL Marine User Handbook. EUMETSAT, Germany (2017)
11. Chai, T., Draxler, R.R.: Root mean square error (RMSE) or mean absolute error (MAE)? Arguments against avoiding RMSE in the literature. Geoscient. Model Developm. **7**(3), 1247–1250 (2014)
12. Collaboration for Australian Weather and Climate Research (CAWCR).: Forecast Verification Methods Across Time and Space Scales (2017). Accessed on October 3rd of 2021, from https://www.cawcr.gov.au/projects/verification/

Warming of the Upper Ocean in the Indonesian Maritime Continent

Mochamad Furqon Azis Ismail, Asep Sandra Budiman, Abdul Basit, Erma Yulihastin, Herlina Ika Ratnawati, Dewi Surinati, Adi Purwandana, Widodo Setiyo Pranowo, Subekti Mujiasih, Rahaden Bagas Hatmaja, and Praditya Avianto

Abstract The Indonesian Maritime Continent (IMC) is the biggest tropical archipelago on Earth lying in the middle of the warm pool of the eastern Indian and western Pacific. The variability of ocean temperatures in the IMC plays a significant role in the global Earth's climate system. There is compelling evidence that the global oceans have warmed over the past 50 years, increasing their upper-ocean heat content (OHC). Here, we present new estimates of the long-term variability of the OHC in the IMC based on the OCEAN5 reanalysis outputs. We show that the upper-ocean heat of the IMC has not warmed uniformly. In addition, notable seasonal, interannual to interdecadal variations in the OHC were observed in the basin. This study provides the first insight into the region's OHC trend and spatial dynamic.

1 Introduction

An increase in the concentration of anthropogenic greenhouse gases in the atmosphere has been shown to be significantly linked to heat being trapped in the Earth's climate system [1], resulting in global warming [2]. More than 90% of the excess energy at the top of the atmosphere is preserved in the ocean [3], raising the ocean's temperature, while the residual heat is present in warming the atmosphere and land surface [2]. Observational assessments over the last few decades indicate that the heat is mainly stored in the upper ocean [4]. The importance of the ocean as a heat reservoir for Earth's climate system has been highlighted by recent evidence that the world's oceans have warmed over the past 50 years [3, 5]. Changes in the upper-ocean heat content (OHC) contribute to variations in the global energy budget [6].

M. F. A. Ismail (✉) · A. S. Budiman · A. Basit · E. Yulihastin · H. I. Ratnawati · D. Surinati · A. Purwandana · W. S. Pranowo · S. Mujiasih · R. B. Hatmaja · P. Avianto
National Research and Innovation Agency of Indonesia (BRIN), Jakarta 10343, Indonesia
e-mail: mfur001@brin.go.id

M. F. A. Ismail
Humboldt Fellow, GEOMAR Helmholtz Centre for Ocean Research Kiel, 24148 Kiel, Germany

A. Basit et al. (eds.), *Proceedings of the International Conference on Radioscience, Equatorial Atmospheric Science and Environment and Humanosphere Science*, Springer Proceedings in Physics 290, https://doi.org/10.1007/978-981-19-9768-6_45

Upper OHC is one of the climate indicators that provide reliable evidence of global warming. The upper OHC also serves as the ocean's memory for the potential predictability of climate variability on interannual to decadal time scales [7]. Analysis of upper OHC variations is vital for understanding interannual and decadal global climate variability such as El Niño–Southern Oscillation (ENSO), sea level rise, tropical cyclones, and marine heatwaves. There is significant decadal variability in the upper OHC as opposed to a steady warming trend [5]. On interannual to decadal time scales, recent observational and modeling work has revealed that internal climatic variability such as ENSO may affect the OHC [8]. Additionally, spatial distributions of OHC show that warming is not constant throughout space [9]. Therefore, it is essential to quantify the OHC trend and its variability in the marginal seas from a regional perspective.

The Indonesian Maritime Continent (IMC) is a major landmass in Southeast Asia that includes the Indonesian Archipelago, straits, and marginal seas (Fig. 1a). The IMC is a junction of the seasonally reversing Indo-Australian monsoonal winds. These winds determine the mesoscale circulation and may provide heat storage and transport of OHC in the IMC. The region is also critical for regional air-sea interactions and climate because it comprises a significant portion of the Indo-Pacific warm pool, the focal point of the atmospheric large-scale upward motion. This convection is associated with the ascending branch of the Walker circulation, whose interannual zonal migration is a key component of the El Niño–Southern Oscillation (ENSO) variability [10]. In terms of ocean dynamics, the IMC is an important conduit of an interbasin flow of mass and heat transport from the tropical western Pacific to the eastern Indian known as the Indonesian Throughflow (ITF).

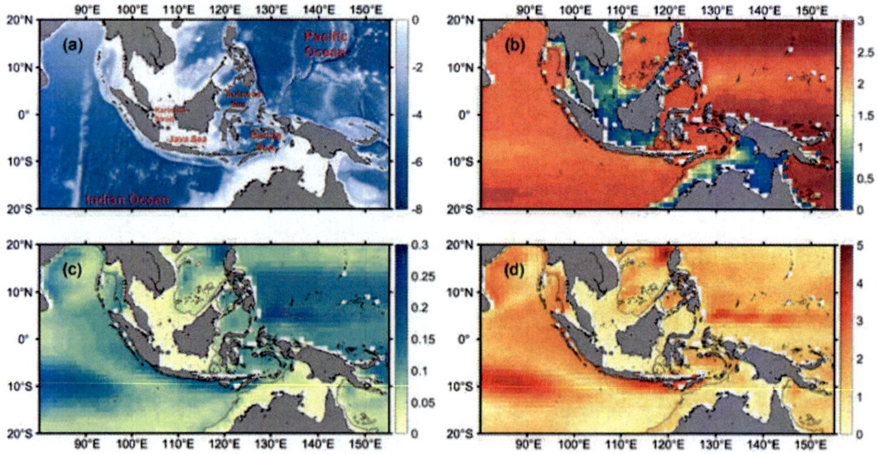

Fig. 1 **a** Bathymetry of the IMC and its surrounding waters, including the Banda Sea, Sulawesi Sea, Java Sea, and Karimata Strait. **b** mean ($\times\ 10^{10}$ J m^{-2}), **c** standard deviation ($\times\ 10^{10}$ J m^{-2}), and **d** magnitude of seasonal cycle ($\times\ 10^{9}$ J m^{-2}) of OHC for the 0–300 m in the study region from 1990 to 2021. Black lines exhibit 500 m isobath

Ocean observations are crucial to our understanding of upper OHC. Although there has been a tremendous improvement in global OHC observations, the IMC still lacks OHC data. Previous studies in the IMC have focused on marine heatwave identifications [11–13], eddies [14], and upwelling [15]. The upper OHC receives much less attention despite being an integral part of the regional ocean atmosphere and climate variability. This study investigates that the upper OHC variability, long-term trend, and anomaly in the IMC are investigated using ocean reanalysis output. The data description and the analytical tool used are explained in Sect. 2. The seasonal variability of OHC, along with its increasing trend and anomaly, is presented and discussed in Sect. 3. Finally, Sect. 4 lists the conclusions.

2 Material and Methods

The near-global gridded OHC data in this study is based on the OCEAN5 reanalysis system [16], which is provided by Copernicus Climate Change Service (C3S) Climate Data Store (https://doi.org/10.24381/cds.67e8eeb7). The OCEAN5 system uses the Nucleus for European Modelling of the Ocean and the NEMOVAR ocean assimilation system [17]. The OHC data is a combined product of the model with observations from various global platform (e.g., satellite, ships, buoys), resulting in ocean information that is continuous in time without spatial gaps. This data has sufficient length, consistency, and continuity to investigate the long-term OHC variability in the IMC. Here, we use the OHC for the upper 300 m from January 1990 to December 2021. Several major basins within the IMC have been selected as representative of the eastern region consisting of the Banda Sea (125°-131° E, 4°-7° S) and the Sulawesi Sea (119–125 °E; 2–5 °N), and the western region: The Java Sea (109–115 °E; 4–7 °S) and the Karimata Strait (105–109 °E; 2 °S–2 °N).

The formula of the OHC can be written:

$$OHC = \rho C \int_{0}^{z} T(z)dz \quad (1)$$

where ρ is the density of sea water (1026 kg m^{-3}), C is the specific heat of sea water (3990 J kg^{-1} K^{-1}), z is the depth limit of the calculation (300 m), and T is the temperature at each depth (°K).

The multivariate ENSO index version 2 (MEIv2) is obtained from the NOAA PSL, Boulder, Colorado, USA, from their website at https://psl.noaa.gov/enso/mei/. The MEIv2 is a new version of the MEI that utilizes 5 variables, viz. sea level pressure, sea surface temperature, surface zonal and meridional winds, and outgoing longwave radiation [18].

3 Results and Discussion

Figure 1b shows the long-term mean of the upper 300 m OHC in the IMC from 1990 to 2021. A pair of OHC maximum of about 3×10^{10} Jm^{-2} exists in the western Pacific, coinciding with the location of the western Pacific warm pool. These OHC maxima observe along approximately 10–20° N and 130–155° E and offshore across northern and eastern Papua and northeastern Australia (Fig. 1b). In contrast, the OHC minima ($> 0.5 \times 10^{10}$ Jm^{-2}) mostly contained in the shallow basin region of the Karimata Strait, the Natuna Sea, the Java Sea, and the Arafura Sea. The OHC standard deviations are larger ($> 0.2 \times 10^{10}$ Jm^{-2}) in the western Pacific, along the main ITF route including the Sulawesi Sea and the Banda Sea, offshore of the west (south) Sumatra (Java) region, and the Indian Ocean along 5° S–13° N and 80–90° E. The ITF is likely responsible for these significant standard deviation values. In contrast, the OHC standard deviations are lower ($< 0.05 \times 10^{10}$ Jm^{-2}) in the Karimata strait and the Java Sea (Fig. 1c).

In order to understand the seasonal variability of upper OHC in the IMC, we first assessed the seasonal cycle spatial patterns in the IMC region. The magnitude of the seasonal cycle of OHC for the 300 m from 1990 to 2021 is shown in Fig. 1d. In general, the seasonal cycle of OHC displays a similar pattern to the OHC standard deviations. It is observed that the magnitude of the seasonal cycle within the marginal sea of the IMC, viz. the Sulawesi Sea, the Java Sea, and the Karimata Strait, is lower ($< 1 \times 10^{10}$ Jm^{-2}) than the Banda Sea, which is reaching about 2×10^{10} Jm^{-2}. A comparison of the temporal means of the seasonal cycle of the marginal seas in the IMC over the period 1990 to 2021 is presented in Fig. 2. The mean seasonal cycle in the Banda Sea and Sulawesi is higher than in the Java Sea and Karimata Strait. The mean seasonal cycle of OHC in the Banda Sea exhibits a maximum OHC value of about 25×10^9 Jm^{-2} during the northwest monsoon season in February and March (Fig. 2a). Minimum OHC value of about 23×10^9 Jm^{-2} is observed during the southeast monsoon season from July to September.

The seasonal cycles of OHC in the Banda Sea suggest that maximum and minimum are likely influenced by the seasonally reversing monsoon (Fig. 2a). Further comparison of mean seasonal cycles of the OHC among the Sulawesi Sea, the Java Sea, and the Karimata Strait identifies clearer bimodal variations (Fig. 2b–d). Predominantly higher OHC in those basins was observed during the southeast monsoon season (May to August) and pre-northwest monsoon (October to November). It appears that the pre-northwest monsoon wind has a clear influence on the seasonal cycle pattern in the Sulawesi Sea, the Java Sea, and the Karimata Strait. In the Sulawesi Sea, the peak and bottom values of OHC reach about 24.20×10^9 Jm^{-2} and 23.50×10^9 Jm^{-2} in June and February, respectively. The highest seasonal cycle of OHC in the Java Sea and the Karimata Strait attains a maximum of about 54.50×10^8 Jm^{-2} and 40.50×10^8 Jm^{-2}, respectively, in the month of May/June, and secondary high values in November (Fig. 2c and d).

To explore the increasing rate of upper OHC at diverse spatial locations, the linear tendency in the OHC for the 0–300 m layers is quantified for the entire observation

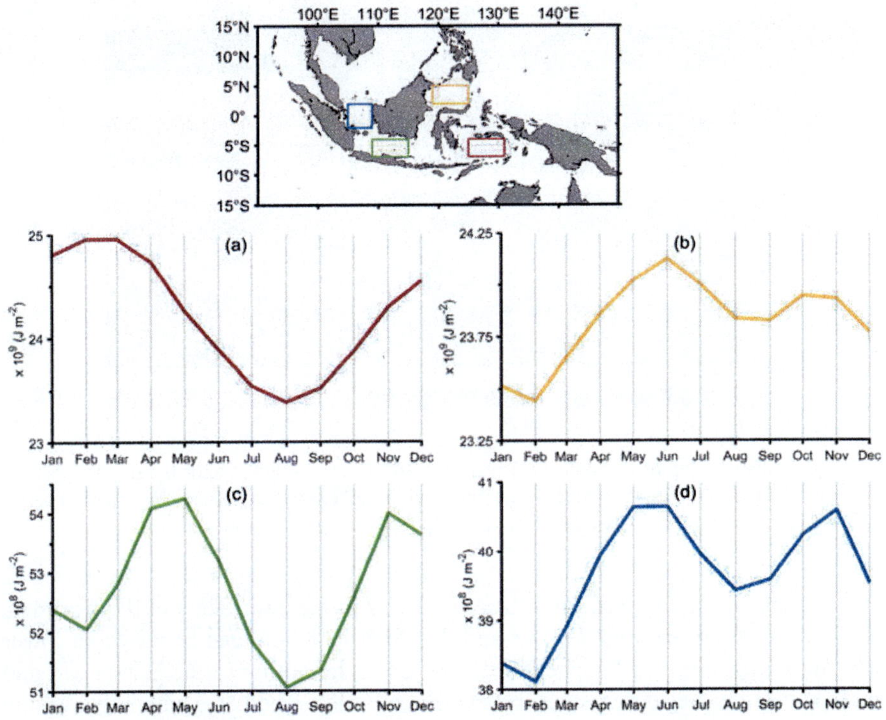

Fig. 2 Mean seasonal cycle of OHC over the period 1990 to 2021 in the domain-averaged over: **a** the Banda Sea; **b** the Sulawesi Sea; **c** the Java Sea; **d** the Karimata Strait

period. In Fig. 3, we show the linear decadal trend of OHC in the IMC computed for the 0–300 m layers from 1990 to 2021. The OHC linear trend is positively significant for most regions, ranging from 0.2×10^8 Jm^{-2} $decade^{-1}$ to 7.0×10^8 Jm^{-2} $decade^{-1}$. The trend with larger values is found in the vicinity of the Halmahera eddy reaching up to 7.0×10^8 Jm^{-2} $decade^{-1}$. The Halmahera eddy is associated with the convergence of warm sea surface temperatures. Another maximum of OHC trends (~6.8×10^8 Jm^{-2} $decade^{-1}$) is observed along the western Pacific. This is a region where the sea surface temperature resides very warm (> 28 °C) the whole year (Fig. 3). In general, the eastern basins in the IMC such as the Banda Sea and Sulawesi Sea have higher positive trends than the western basin as the Java Sea and Karimata Strait. The Banda Sea and Sulawesi Sea trends denote positive increases ranging from 3×10^8 Jm^{-2} $decade^{-1}$ to 6×10^8 Jm^{-2} $decade^{-1}$.

We evaluate the relationship of OHC variability in the IMC to the ENSO. Figure 4 shows the time series of deseasonalized OHC anomaly averaged over the Banda Sea, the Sulawesi Sea, the Java Sea, and the Karimata Strait in the upper 300 m in reference to the 1990 to 2020 baseline period, along with their linear decadal trend. The interannual and interdecadal variability of upper OHC in the Banda Sea and the Sulawesi Sea is quite similar in terms of their magnitude and their phase as well

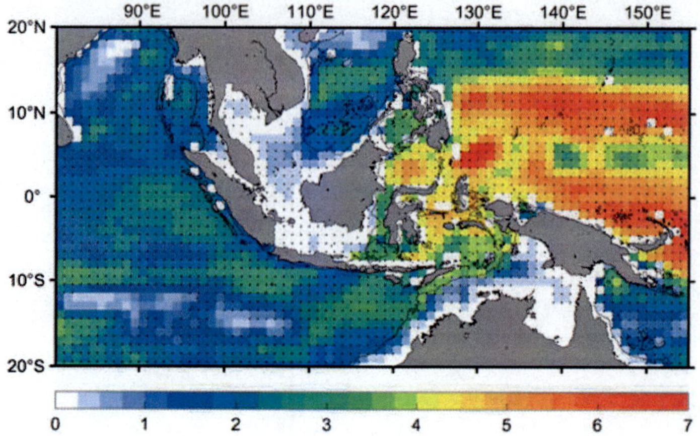

Fig. 3 Geo-distribution of OHC decadal trend ($\times$ 10^8 J m^{-2} $decade^{-1}$) for the 0–300 m in the IMC over the period of 1990 to 2021. Black lines exhibit 500 m isobath. The black stippled points indicate a significant trend where p-values less than 0.05

(Fig. 4a and b). In contrast, the OHC variability in the Java Sea and the Karimata Straits has similar patterns (Fig. 4c and d), albeit its magnitude is one order lower than the Banda Sea and Sulawesi Sea. The OHC variability in the Banda Sea and the Sulawesi Sea negatively correlates with the ENSO. The years with negative and positive anomalies of OHC in the Banda Sea and the Sulawesi Sea coincide with the strong El Niño and La Niña events, respectively (Fig. 4a and b). In contrast, the variability of OHC in the Java Sea and the Karimata Strait is such that the anomaly is positive (negative) during the El Niño (La Niña) events with a lag of about 3–6 months (Fig. 4c and d).

It is clear from Fig. 4 that there is an increasing trend of OHC anomaly in our study region. Using the Mann–Kendall rank statistic test, these trends are evaluated for significance and determined at a 99.9% significant level. The Sulawesi Sea has the highest decadal warming trend of OHC of about $4.2 * 10^8$ J m^{-1} $decade^1$, followed by the Banda Sea reaching up to $4.2 * 10^8$ J m^{-1} $decade^1$. The OHC trend in the Java Sea and the Karimata Strait is about $2.0 * 10^7$ J m^{-1} $decade^1$ and $1.7 * 10^7$ J m^{-1} $decade^1$, respectively.

4 Conclusions

In this study, the variability of upper OHC over the period of 1990 to 2021 in the IMC is investigated based on the OCEAN5 reanalysis outputs. We show that the upper OHC in the internal seas of the IMC viz. the Banda Sea, the Sulawesi Sea, the Java Sea, and the Karimata Strait tend to increase, though it is spatially distinct. In addition, we have documented the seasonal, interannual, and interdecadal evolution

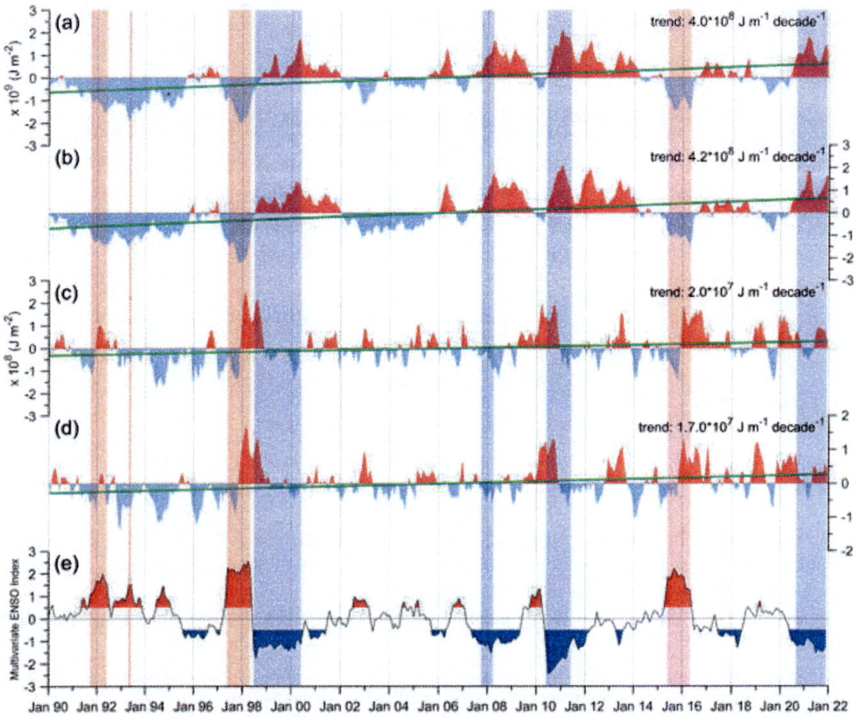

Fig. 4 Time series of deseasonalized OHC anomaly (J m^{-2}) of 0–300 m layer relative to 1990 to 2020 baseline climatology spatially averaged over: **a** the Banda Sea, **b** the Sulawesi Sea, **c** the Java Sea, and **d** the Karimata Strait with the climatological seasonal cycle removed. **e** Temporal distribution of the multivariate ENSO index over the same period in **a** to **d**. Blue (red) shadings denote the two strong El Niño (La Niña) periods identified by the multivariate ENSO index exceeding ± 1. Straight green line indicates the linear trend of OHC year^{-1}

of upper OHC in the Banda Sea, the Sulawesi Sea, the Java Sea, and the Karimata Strait and quantified their long-term trend. For most regions, the OHC linear trend is positive, ranging from 0.2 × 108 Jm^{-2} decade^{-1} to 7.0 × 108 Jm^{-2} decade^{-1}. Compared to the western basins, including the Java Sea and Karimata Strait, the eastern basins of the IMC, viz. the Banda Sea and Sulawesi Sea, exhibit stronger positive tendencies. The ENSO and the OHC fluctuation in the Banda and Sulawesi Seas are inversely correlated. Strong El Niño and La Niña episodes both occur during the years with negative and positive OHC anomalies, particularly in the Banda Sea and the Sulawesi Sea, respectively.

Acknowledgements The OHC data was downloaded from the Copernicus Climate Change Service (C3S) Climate Data Store (https://doi.org/10.24381/cds.67e8eeb7). Mochamad Furqon Azis Ismail was supported by the Alexander von Humboldt-Stiftung and contributed as the main contributor. Asep Sandra Budiman, Abdul Basit, Erma Yulihastin, Herlina Ika Ratnawati, Dewi Surinati, Adi Purwandana, Widodo Setiyo Pranowo, Subekti Mujiasih, Rahaden Bagas Hatmaja, and Praditya Avianto contributed as associate contributors to this paper.

References

1. IPCC.: The Physical Science Basis: Working Group I Contribution to the Fifth Assessment Report of the Intergovernmental Panel on Climate Change (2014)
2. Trenberth, K.E., Fasullo, J.T., Balmaseda, M.A.: Earth's energy imbalance. J. Climate **27**, 3129–44 (2014)
3. Levitus, S., Antonov, J., Boyer, T.: Warming of the world ocean, 1955–2003. Geophys. Res. Lett. **32** (2005)
4. Levitus, S., Antonov, J.I., Boyer, T.P., Baranova, O.K., Garcia, H.E., Locarnini, R.A., Mishonov, A.V., Reagan, J.R., Seidov, D., Yarosh, E.S.: World ocean heat content and thermosteric sea level change (0–2000 m), 1955–2010. Geophys. Res. Lett. **39** (2012)
5. Levitus, S., Antonov, J.I., Boyer, T.P., Stephens, C.: Warming of the World Ocean. Science **80**(287), 2225–2229 (2000)
6. Trenberth, K.E., Fasullo, J.T.: Tracking earth's energy: from El Niño to global warming. Surv. Geophys. **33**, 413–426 (2012)
7. Xue, Y., Balmaseda, M.A., Boyer, T., Ferry, N., Good, S., Ishikawa, I., Kumar, A., Rienecker, M., Rosati, A.J., Yin, Y.: A comparative analysis of upper-ocean heat content variability from an ensemble of operational ocean reanalyses. J. Clim. **25**, 6905–6929 (2012)
8. Cheng, L., Trenberth, K.E., Palmer, M.D., Zhu, J., Abraham, J.P.: Observed and simulated full-depth ocean heat-content changes for 1970–2005. Ocean Sci. **12**, 925–935 (2016)
9. Palmer, M.D., Good, S.A., Haines, K., Rayner, N.A., Stott, P.A.: A new perspective on warming of the global oceans. Geophys. Res. Lett. **36** (2009)
10. Yoneyama, K., Zhang, C.: Years of the Maritime Continent. Geophys. Res. Lett. **47**, e2020GL087182 (2020)
11. Iskandar, M.R., Ismail, M.F.A., Arifin, T., Chandra, H.: Marine heatwaves of sea surface temperature off south Java. Heliyon **7**, e08618 (2021)
12. Ismail, M.F.A.: Characteristics of marine heatwaves off West Sumatra derived from high-resolution satellite data. Hunan Univer. Natural Sci. **48**, 130–136 (2021)
13. Ismail, M.F.A., Gerhaneu, N.Y., Yulihastin, E., Ratnawati, H.I., Purwandana, A.: Assessment of marine warming in Indonesia: a case study off the coast of West Sumatra. IOP Conf. Ser. Earth Environ. Sci. **718**, 012006 (2021)
14. Ismail, M.F.A., Ribbe, J., Arifin, T., Taofiqurohman, A., Anggoro, D.: A census of eddies in the tropical Eastern boundary of the Indian Ocean. J. Geophys. Res. Oceans **126**, 1–9 (2021)
15. Budiman, A.S., Bengen, D.G., Nurjaya, I.W., Arifin, Z., Ismail, M.F.A.: The spatio-temporal variability of chlorophyll-A and its physical variables in the South Java Sea shelf. Hunan University Natural Sci. **48** (2021)
16. Hersbach, H., Bell, B., Berrisford, P., Biavati, G., Horányi, A., Muñoz Sabater, J., Nicolas, J., Peubey, C., Radu, R., Rozum, I.: ERA5 hourly data on single levels from 1979 to present. Copernicus Climate Change Service (C3S) Climate Data Store (CDS) (2018)
17. Zuo, H., Alonso-Balmaseda, M., Mogensen, K., Tietsche, S.: OCEAN5: the ECMWF ocean reanalysis system and its real-time analysis component. ECMWF Tech. Memo (2018)
18. Zhang, T., Hoell, A., Perlwitz, J., Eischeid, J., Murray, D., Hoerling, M., Hamill, T.M.: Towards probabilistic multivariate ENSO monitoring. Geophys. Res. Lett. **46**, 10532–10540 (2019)

Air-Sea Interaction Over Southeast Tropical Indian Ocean (SETIO) During Storm Intensification Episodes in the Early Dry Season Period

Namira Nasywa Perdani, Ankiq Taofiqorahman, Erma Yulihastin, Rahaden Bagas Hatmaja, Gammamerdianti, Eka Putri Wulandari, Noersomadi, and Haries Satyawardhana

Abstract Previous studies have mentioned that ocean–atmosphere interaction over the Maritime Continent played a prominent role in determining climate and weather variability on the intraseasonal variation scale in Indonesia. One of them could be proven by the anomalously wet-dry season over Java Island, which was generated by the warming of sea surface temperatures (SST) in the Indonesian Sea. However, the mechanism of the ocean–atmosphere interaction in these local seas in generating weather disturbances such as thunderstorms was still poorly understood. This study investigates the ocean–atmosphere interaction in the southern seas of Indonesia in inducing storms during the anomalously wet-dry season episodes. Here, we explored satellite and reanalysis datasets, including ocean and atmosphere variables in sub-daily time resolution and various spatial resolutions, i.e., 0.25°, 0.083° and 0.1° derived from The European Centre for Medium-Range Weather Forecast (ECMWF) reanalysis 5th Generation (ERA5), Copernicus Marine Environment Monitoring Service (CMEMS), and Global Satellite Mapping of Precipitation (GSMaP), respectively. Those variables were investigated over the Southeast Tropical Indian Ocean (SETIO) during the early dry season of May–June 2021. Our finding described that during these episodes, the increase of rainfall over southwestern Indonesia did not correspond to the position of the Intertropical Convergence Zone (ITCZ) and might have been influenced by air-sea interaction over a regional scale. We also confirmed that storm activities over southwestern Indonesia were associated with strengthening sea surface current (SSC) along the west coast of Sumatra and forming eddies in the southern Java Sea. We also noticed that variability in both SSC and SST in the southern Java Sea, which occurred on a sub-daily scale, may influence a dynamic weather pattern over southwestern Indonesia.

N. N. Perdani (✉) · A. Taofiqorahman
Marine Sciences Department, Faculty of Fisheries and Marine Sciences, Padjajaran University, Bandung, Indonesia
e-mail: namiranasywa@gmail.com

E. Yulihastin · R. B. Hatmaja · Gammamerdianti · E. P. Wulandari · Noersomadi · H. Satyawardhana
National Research and Innovation Agency, Jakarta, Indonesia

A. Basit et al. (eds.), *Proceedings of the International Conference on Radioscience, Equatorial Atmospheric Science and Environment and Humanosphere Science*, Springer Proceedings in Physics 290, https://doi.org/10.1007/978-981-19-9768-6_46

1 Introduction

The ocean–atmosphere interaction is a coupling system that can influence weather and climate cycles in the world. Both are exchanged heat energy and momentum to create a dynamic between the ocean and the atmosphere. The ocean–atmosphere dynamics play an important role in triggering convective activity in Indonesia Maritime Continent (IMC) from local to large scale [1]. On a local scale, land-sea breezes interaction can increase convective activity that causes local rainfall due to the diurnal SST variability [2–4]. In global scales, phenomena such as ENSO, IOD, and MJO influenced convective activity in various time and spatial scales. The complex dynamics of the ocean–atmosphere interaction in the IMC may produce seasonal anomalies in Indonesia, particularly on Java Island.

A previous study stated that the anomalously wet in dry season over Java was most influenced by local ocean 37% compared to remote forcings [5]. Therefore, the warming SST in southern Java affects convective activities on the island of Java during the dry season. Furthermore, previous studies have suggested that the slightest change in SST can cause large-scale changes in atmospheric circulation [6–9]. However, the dynamical processes over surface level between air-sea components over southern waters of IMC in increasing convective activity associated with storms during the anomalously wet in dry season were still less of knowledge. Therefore, this study aims to investigate the ocean–atmosphere interaction in the southern waters of Indonesia in triggering storms during the anomalously wet in dry season by exploring three storm episodes during May to June 2021.

2 Methods

We use satellite and reanalysis data of ocean and atmospheric parameters to investigate the air-sea interaction in developing convective activity associated with the convective storm in southern IMC. Hourly oceanic parameter data such as sea surface temperature (SST) and sea surface currents (SSC) are taken from the Copernicus Marine Environment Monitoring Service (CMEMS) (https://marine.copernicus.eu/), with a $0.083° \times 0.083°$ spatial resolution. The hourly atmospheric parameter data such as Outgoing Longwave Radiation (OLR), surface temperature (2 m), latent heat, and divergence wind (850 mb) were obtained from the European Center for Medium-Range Weather Forecast (ECMWF) reanalysis 5th Generation (ERA5) with a spatial resolution of $0.25° \times 0.25°$ [10]. Furthermore, hourly rainfall data are obtained from the Global Satellite Mapping of Precipitation (GSMaP) with rain gauge data validation [11]. The GSMaP_gauge data have been developed to fill the data gaps in the rainfall forecast between satellite and rain gauge data due to differences in spatial and temporal resolution [12]. Therefore, there is an increasing spatial resolution of GSMaP from $0.25° \times 0.25°$ to $0.1° \times 0.1°$ for the GSMaP_gauge used in this study.

The area study is located between 20–3°S; 95–120°E, so-called Southeast Tropical Indian Ocean (SETIO). To describe the mean state of the atmosphere–ocean condition, we used a synoptical analysis of the average May–June 2021 and defined the results as a background condition during the early period of the dry season. Furthermore, we perform a composite analysis to investigate the diurnal variation in every six-hourly data. For SST anomaly, we calculate the anomaly by subtracting diurnal data from the data average for two months (May–June).

3 Results

3.1 *Background Conditions*

Globally, convective activity in the IMC during the early dry season 2021 does not correspond to the Intertropical Convergence Zone (ITCZ) position, as seen in Fig. 1a and b. However, the condition of convective activity in May showed normal conditions, but for June, it did not show a strong signal of the dry season. In this condition, the sun's position should move closer to the north latitude, and the Australian monsoon winds begin to strengthen, causing cooling of SST in southern Indonesia. Still, in this case, it will not occur in June 2021. This month shows the SST warming south of Indonesia, and SETIO is shown in Fig. 1c. The increasing of SST in the SETIO area causes a vortex to form, which is indicated by a wind divergence of 850 mb (Fig. 1d). This condition describes globally that the beginning of the dry season in Indonesia in 2021 shows a wet-dry season signal based on OLR, SST Climatological Anomaly, and 850 mb wind divergence data.

3.2 *Ocean–Atmosphere Interaction in Sub-Daily Cases*

We did an average of May—June 2021 every six hours, which shows that convective activity is still concentrated in the SETIO region during the night-to-early morning time, early morning-to-morning time, morning-to-middle day, afternoon-to-night-time (Fig. 2a), and its OLR (Fig. 2b). This group of convective clouds in the SETIO area is centered above the eddies as shown in Fig. 2d. On the other hand, we can detect diurnal SST variability between night-to-early morning time, early morning-to-morning time, morning-to-middle day time, and afternoon-to-night-time. In the early morning-to-morning time, the SST anomaly (Fig. 2c) decreased in the coldest conditions. In contrast, in the afternoon-to-night-time, the SST anomaly increased in the warmest conditions among other periods. We can conclude that the influence of the ocean dominates to maintain convective activity in the SETIO over the atmosphere, such as temperature at 2 m (Fig. 2e) and latent heat (Fig. 2f) during the early dry season.

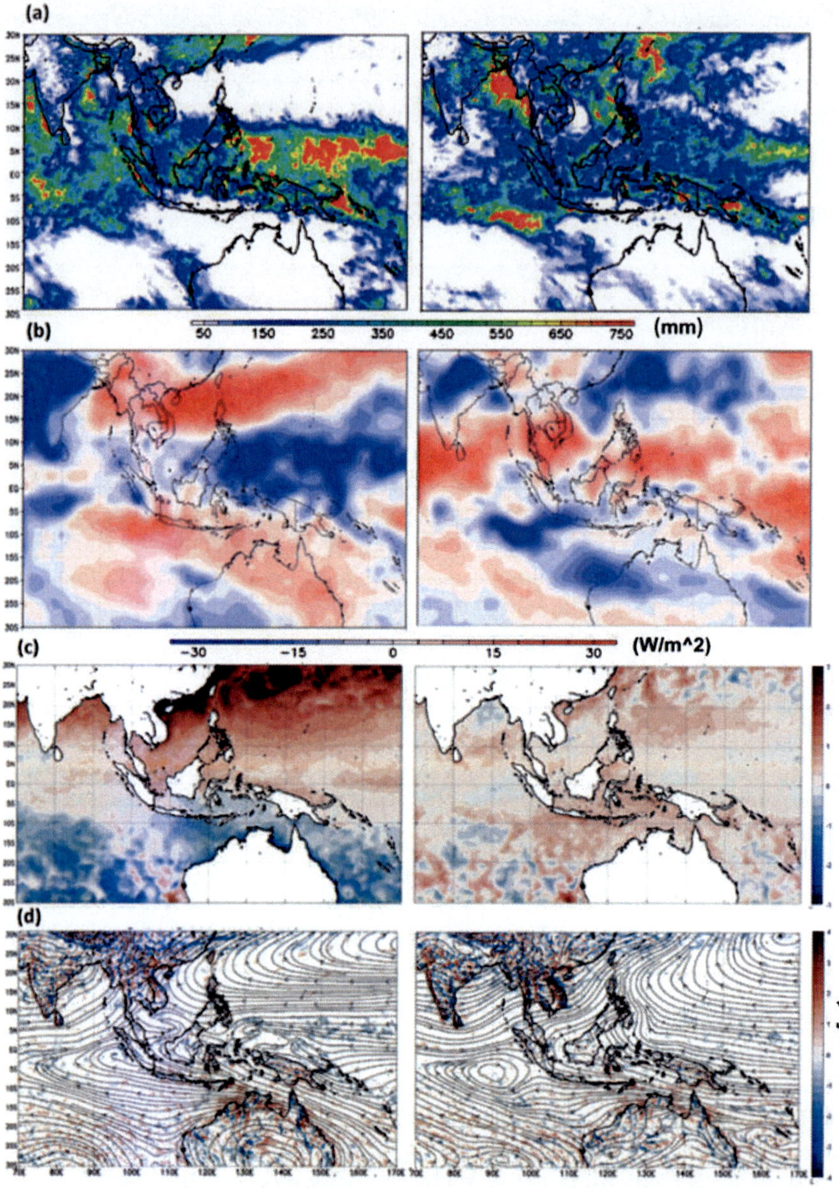

Fig. 1 Averages for May (left) and June (right) of 2021 for: **a** rainfall derived from GSMaP satellite data, **b** OLR carried from NCEP/NCAR reanalysis, c SST anomaly, and **d** 850 mb wind divergence from CMEMS and ERA5 reanalysis data, respectively

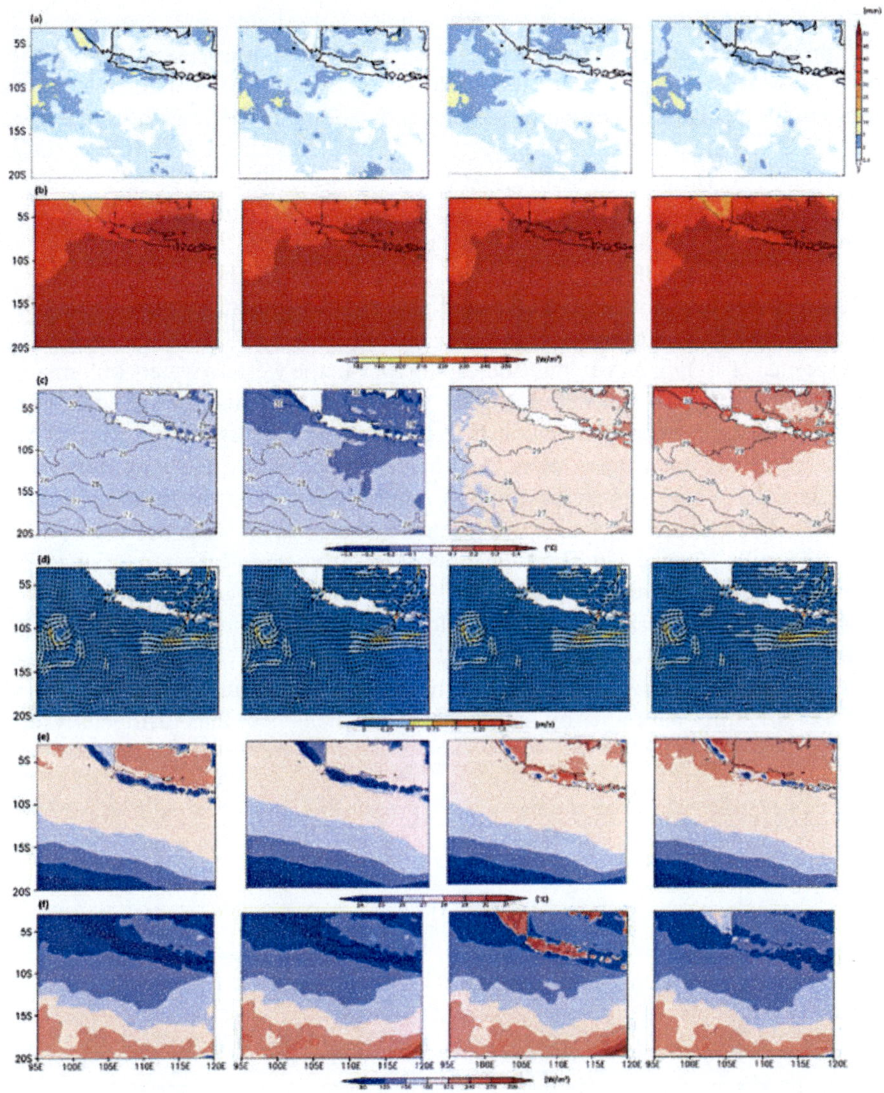

Fig. 2 Average May–June 2021 of six-hourly data from left to right: night-to-early morning time (20:00–01:00 LST); early morning-to-morning time (02:00–07:00 LST); morning-to-middle day time (08:00–13:00 LST); and afternoon-to-night-time (14:00–19:00 LST) for: **a** rainfall, **b** OLR, **c** SST anomaly, **d** SSC, **e** temperature at 2 m, and **f** latent heat

We did breakdown scales on convective activity daily, divided into 3 cases with the high precipitation affecting Java Island (Fig. 3). The first case was on May 27–28, 2021. On May 27 at night, around the southwest of the island of Sumatra, rain clouds began to form (Fig. 4a). The storm got bigger spreading to the coast of Sumatra on May 28, 2021, in the early morning-to-morning time; this can be seen from its OLR cover (Fig. 4b). The storm is getting bigger caused of the increasing of SST (Fig. 4c) in the sea near Sumatra due to the Kelvin waves signed by the increasing of SSC near Sumatera Island (Fig. 4d). Rain causes a drop in temperature at 2 m (Fig. 4e) in the early morning-to-middle day time periods. Then, there is another convective activity at 20S formed due to the increase of latent heat near Australian waters (Fig. 4f). We can conclude that the forming of convective clouds in May 2021 was influenced by the presence of Kelvin waves, thus increasing SST in the western waters of Sumatra, while further south the convective activity was caused by an increase in latent heat.

The next case is concentrated convective clouds over the island of Java (Fig. 5b) and the detection line-shaped storm (Fig. 5a). This convective activity is formed in the south of Java due to an increase in SST in the area as seen from its SST anomaly (Fig. 5c). The increasing of SST occurs due to the mass of water carried by SSC (Fig. 5d) called Indonesian Through Flow (ITF) from the Makassar Strait to the Indian Ocean through the Lombok strait. Then, the rain clouds propagated to the north of Java in the early morning-to-middle day time, but the movement of these rain clouds was caused by temperature at 2 m in the Java Sea. Same as the first case, there is convective activity at the southern Latitude. This rain cloud is not caused by SST variability but rather by high latent heat values near the Australian waters (Fig. 5f).

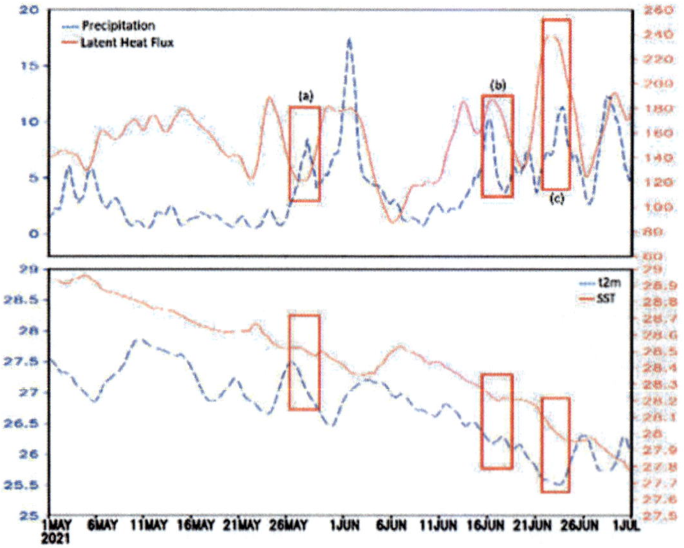

Fig. 3 Time series of averages precipitation, latent heat flux, temperature at 2 m, and SST for: **a** 27–28 May 2021 (case 1), **b** 16–17 June 2021 (case 2), and **c** 23–24 June 2021 in SETIO area

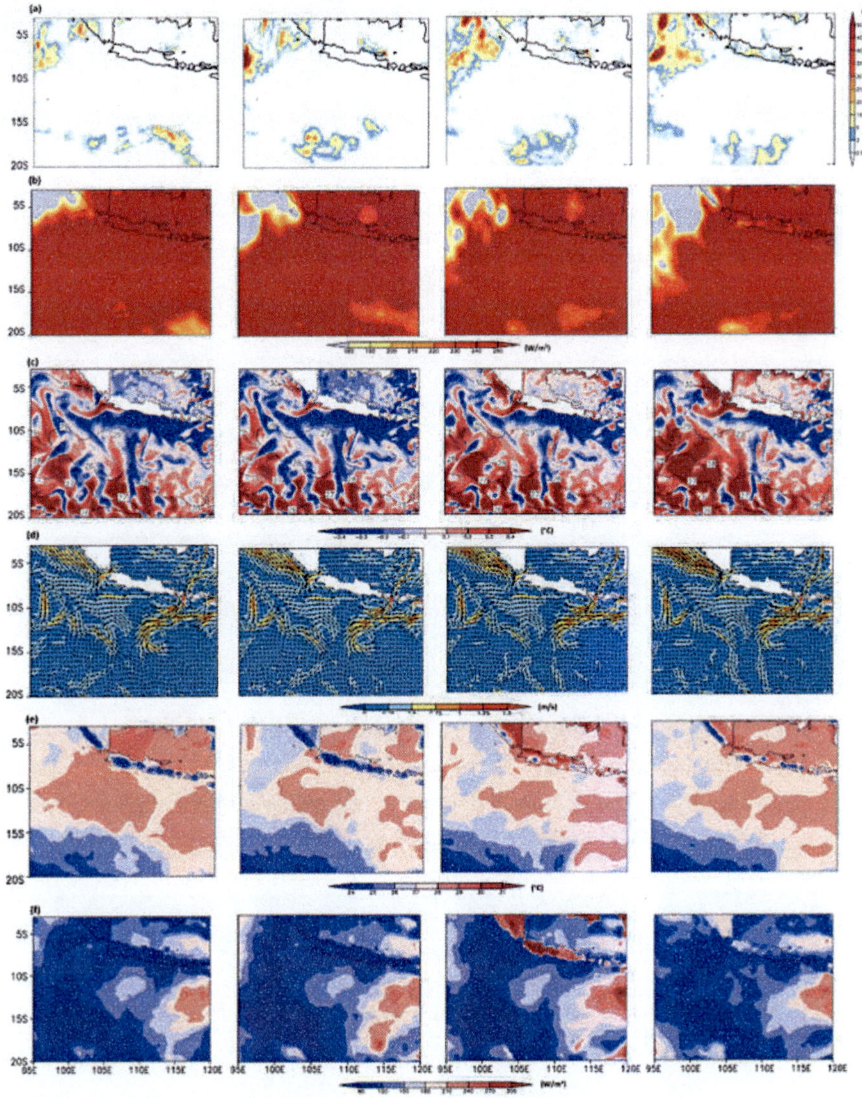

Fig. 4 Same as Fig. 2, but for case 1

The following week there was another line-shaped storm. Initially, a storm had begun to form during the night-to-early morning on June 23 (Fig. 6a). Then the next day, on June 24, the storm pattern became clearer to form a line, and there were multi-cells of rain clouds in the north of the island of Java. It can be seen from the OLR (Fig. 6b) that convective activity in the north is moving toward the south, which makes convective storms even more amplified in the morning-to-middle daytime on June 24. The factor that triggered the concentration of convective clouds was the

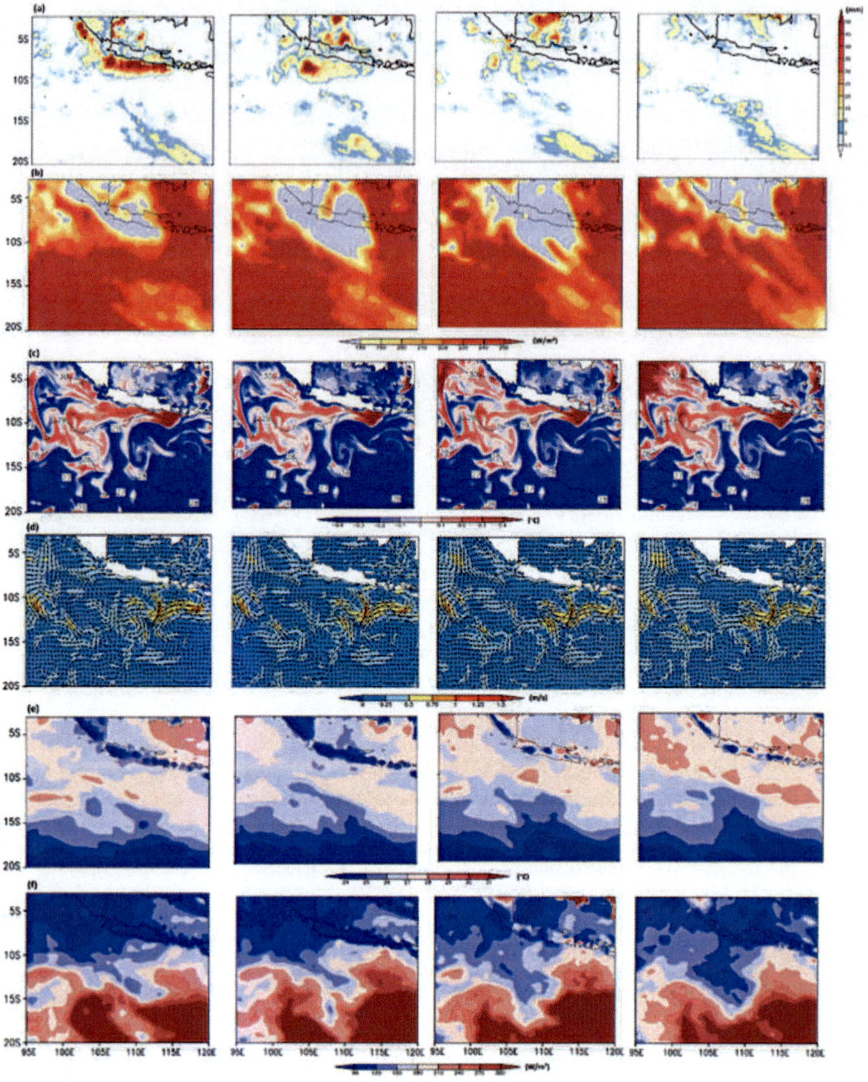

Fig. 5 Same as Fig. 2, but for case 2

increase in SST (Fig. 6c) due to ITF (Fig. 6d) and latent heat (Fig. 6f). Rain clouds make the affected area experience a temperature at 2 m drops with a value reaching 26 °C shown in Fig. 6e.

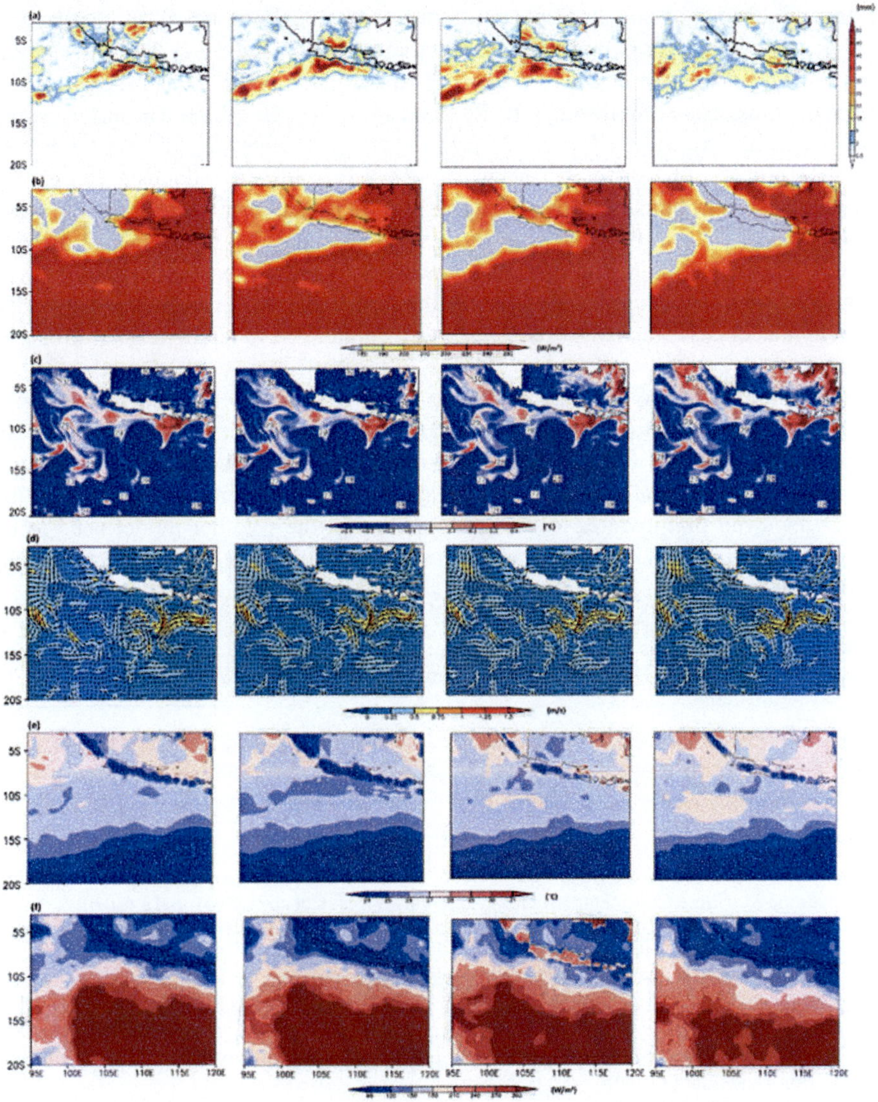

Fig. 6 Same as Fig. 2, but for case 3

4 Conclusion

We investigated Air-Sea Interaction Over Southeast Tropical Indian Ocean (SETIO) in The Early Dry Season Period causing a lot of convective activity to occur above the Maritime Continent; it is proven that in June 2021, convective activity is not strongly associated with the position of the ITCZ. The factor that maintains convective activity is the influence of ocean–atmosphere interactions in local waters. We found that on a six-hourly average for two months, convective activity formed just above the eddies in the SETIO area. In addition, the ocean factor, such as SST carried by the ITF from the Makassar strait through the Lombok strait, supplies the warm surface water mass, causing increased SST variability in southern Indonesia. This can be proven by several cases of storms that occur sub-daily in southern Indonesia, dominated by ocean factors compared to atmospheric factors if the storm occurs near the coast of the island of Java. Previous studies have investigated that the influence of local waters dominates the occurrence of the anomaly wet-dry season on the island of Java compared to the impact of global disturbances [5]. It differs from the convective activity that occurs close to Australian waters; an increase in latent heat value causes this.

This study still has limitations on the other parameters to observe the effect of the air-sea interaction in the SETIO area, causing the dry season anomaly in Indonesia. Therefore, for further studies, it is necessary to consider other ocean–atmosphere parameters to improve the weather model in Indonesia. In addition, this study reviewed that updates to the data inputted in the weather model need to be carried out on a sub-daily to capture convective activity on the IMC with higher resolution.

Acknowledgements This research has been supported by the National Research and Innovation Agency under the Cruise Day Facilitation Program with grant number 376/II/FR/3/2022.

References

1. Xue, P., Malanotte-Rizzoli, P., Wei, J., Eltahir, E.A.B.: Coupled Ocean-atmosphere modeling over the Maritime Continent: a review. J. Geophys. Res. Ocean. **125**(6), 1–18 (2020)
2. Birch, C.E., et al.: Scale interactions between the MJO and the western Maritime Continent. J. Clim. **29**(7), 2471–2492 (2016)
3. Im, E.S., Eltahir, E.A.B.: Simulation of the diurnal variation of rainfall over the western Maritime Continent using a regional climate model. Clim. Dyn. **51**(1–2), 73–88 (2018)
4. Thompson, B. et al.: A high-resolution atmosphere–ocean coupled model for the western Maritime Continent: development and preliminary assessment, vol. 52, pp. 7–8. Springer, Berlin, Heidelberg (2019)
5. Suaydhi, Y.E.P.M.F., Sofiati, I.: Oceanic effect on precipitation development in teh Maritime Continent during anomalously-wet dry seasons in Java Indones. J. Geogr. **53**(3), 328–339 (2021)
6. Bjerknes, J.: Monthly weather reyiew atmospheric teleconnections from the equatorial Pacific Mon. Weather Rev. **97**(3), 163–172 (1969)
7. Hendon, H.H.: Indonesian rainfall variability: impacts of ENSO and local air-sea interaction. J. Clim. **16**(11), 1775–1790 (2003)
8. Zhang, C., Anderson, S.P.: Sensitivity of intraseasonal perturbations in SST to the structure of the MJO. J. Atmos. Sci. **60**(17), 2196–2207 (2003)
9. Li, Y., et al.: Assessing the role of the ocean–atmosphere coupling frequency in the western Maritime Continent rainfall Clim. Dyn. **54**(11–12), 4935–4952 (2020)
10. Hersbach, H., et al.: The ERA5 global reanalysis Q. J. R. Meteorol. Soc. **146**(730), 1999–2049 (2020)

11. Kubota, T., et al.: Global satellite mapping of precipitation (GSMaP) products in the GPM Era Adv. Glob. Chang. Res. **67**, 355–373 (2020)
12. Bui, H.T., Ishidaira, H., Shaowei, N.: Evaluation of the use of global satellite–gauge and satellite-only precipitation products in stream flow simulations. Appl. Water Sci. **9**(3), 1–15 (2019)

Anomalous Sea Surface Temperature and Chlorophyll-a Induced by Mesoscale Cyclonic Eddies in the Southeastern Tropical Indian Ocean During the 2019 Extreme Positive Indian Ocean Dipole

Rahaden Bagas Hatmaja, M. Rizqi Ramadhan, Sigit Kurniawan Jati Wicaksana, and Suaydhi

Abstract The positive Indian Ocean Dipole (pIOD) event in 2019 was notable as the strongest event in the last two decades. This event is characterized by cold sea surface temperature anomaly (SSTA), high chlorophyll-a concentrations (CHL), and intense mesoscale eddy activity off the coast of Java in the southeast tropical Indian Ocean (SETIO), coincident with upwelling season in June to November. In this study, the modulation effects of mesoscale eddies on the local SSTA and CHL are investigated by analyzing daily reanalysis and remote-sensed data from January to December 2019. The variations of SSTA and CHL are dominated by upwelling-induced SSTA and CHL up to ~ 80%. Nevertheless, the eddy-induced SSTA and CHL have remarkable contribution about ~ 25% to their monthly variation, especially during upwelling season that coincident with the peak of pIOD from June to September. In this period, higher number and stronger cyclonic eddies (CEs) are generated, which significantly contributes to the colder SSTA and higher CHL. Moreover, stronger CEs induce colder SSTA and higher CHL by both horizontal advection, which transports cold, high-CHL water from the coast to 400 km offshore, and vertical upwelling, which raises nutrient-rich cold water from the subsurface to the surface.

R. B. Hatmaja (✉) · Suaydhi
Research Centre for Climate and Atmosphere, National Research and Innovation Agency, Bandung, Indonesia
e-mail: rahadenbagas@gmail.com

M. R. Ramadhan
Oceanography Department, Faculty of Earth Sciences and Technology, Bandung Institute of Technology, Bandung, Indonesia

S. K. J. Wicaksana
Research Centre for Oceanography, National Research and Innovation Agency, Jakarta, Indonesia

A. Basit et al. (eds.), *Proceedings of the International Conference on Radioscience, Equatorial Atmospheric Science and Environment and Humanosphere Science*, Springer Proceedings in Physics 290, https://doi.org/10.1007/978-981-19-9768-6_47

1 Introduction

In 2019, the Indian Ocean experienced the strongest positive Indian Ocean Dipole (IOD) event in the last two decades [1]. This event was characterized by low sea surface temperature (SST) anomalies and high chlorophyll-a (CHL) concentrations off the coast of Java in the southeastern tropical Indian Ocean (SETIO), while reversed characteristics in the western pole near Somalian coast [1, 2]. In Fig. 1a, the daily Dipole Mode Index (DMI) from 1 January to 31 December 2019, which is calculated as the difference of the SST anomaly (SSTA) in the eastern equatorial Indian Ocean (90–110 °E, 10 °S–0 °N) and the SSTA in the western equatorial Indian Ocean (50–70 °E, 10 °S–10 °N), shows that the pIOD event initiated in May and reached its peak in October up to ~ 2.5 °C, then weakens in the following months. The average SSTA and CHL concentrations in October reached over ~ –3 °C and ~ 0.5 mg/m^3 on the east IOD zone, respectively. Moreover, these anomalous ocean activities were attributed to the enhanced upwelling during positive IOD (pIOD) event that advected higher-nutrient cold deep water to the surface [1, 3]. However, this significant effect of wind-driven coastal upwelling on SETIO was merely limited to 100 km from the coast [3].

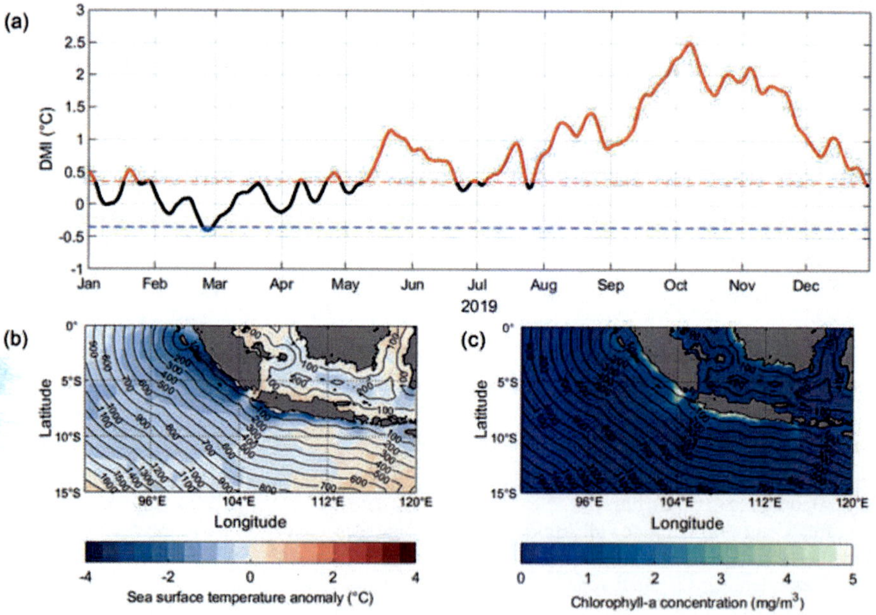

Fig. 1 **a** Time series of Dipole Mode Index (DMI) from 1 January to 31 December 2019 and distribution of mean **b** sea surface temperature anomaly (SSTA) as well as **c** chlorophyll-a concentration (CHL) on the eastern pole of the Indian Ocean Dipole (IOD) during positive IOD's peak in October 2019

In addition, the widespread of SSTA and CHL for more than 200 km offshore during pIOD event are also controlled by high cyclonic eddies (CE) activity [4], as the result of baroclinic instabilities that are associated with the South Java Current (SJC), South Equatorial Current (SEC), and Indonesian Throughflow (ITF) on this region [5–8]. Additionally, the wind forcing of pIOD event can enhance the vertical shear of the background currents and favor energetic CEs off Java coast [4]. Previous study has investigated the effects of mesoscale eddies on the CHL bloom off the Java coast during the 2006 pIOD event and showed that the peak CHL occurred around the peak phase of the IOD, and the active CEs contributed to eddy upwelling that increased the offshore CHL [9]. On the other hand, another study demonstrated that the anti-cyclonic eddy (AE) activities in the SETIO were stronger during the 1994 IOD event, which countered the anomalous cooling [10]. These results imply a probable linkage between mesoscale eddies and SSTA, as well as CHL, on the interannual timescale. Therefore, the investigation of relative effects of mesoscale eddies on the SSTA and CHL distribution off the coast of Java during 2019 IOD, as the recent event, is needed.

2 Data and Methods

2.1 Data

The assessment of SSTA and CHL of this study is conducted by using daily SSTA and CHL data from 1 January to 31 December 2019, which were retrieved from NOAA Optimum Interpolation SST V2 High-Resolution Dataset on a 0.25° global grid (https://psl.noaa.gov/) and Copernicus Global Ocean Colour (Copernicus-GlobColour) Satellite Observations product on a 4 km global grid (https://marine.copernicus.eu/), respectively. The daily SSTA data were calculated relative to a 1971–2000 climatology. Moreover, the mesoscale eddies were identified based on daily merged satellite altimetry multimission data, which includes sea level anomaly (SLA) and surface geostrophic currents data with a spatial resolution of 0.25° for the same period. SLA is defined as the height of water over the mean sea surface, that is computed with respect to a twenty-year mean reference period (1993–2012). These data were collated during the following satellite altimetry missions: Jason-1/2/3, Sentinel-3A, HY-2A, SARAL/AltiKa, Cryosat-2, ERS1/2, TOPEX/Poseidon, ENVISAT, GFO, and ERS1/2, which are provided by the E.U. Copernicus Marine Environment Monitoring Service (https://marine.copernicus.eu/).

2.2 Eddy Detection and Tracking Methods

The analysis of this study is conducted for eddies with a lifetime more than ten days [11]. In addition, only eddies generated between 105–117 °E and 6–12 °S are tracked. The eddies were identified and tracked using the hybrid eddy census algorithm. The hybrid algorithms combine a geometric criterion with an Okubo-Weiss parameter resulting in a more reliable detection from SLA compared to only using an Okubo-Weiss parameter or a geometric criterion [12]. In this hybrid method, the presence of a geostrophic eddy is defined as a coherent region detected within a closed SLA contour with minimum 2 cm and an Okubo-Weiss parameter below $-2 \times 10^{-12}\ s^{-2}$. Moreover, a Hanning filter is applied twice on the Okubo-Weiss parameter [12, 15].

The Okubo-Weiss parameter (W) is computed as follows [13, 14]:

$$W = S_n^2 + S_s^2 + \xi^2 \quad (1)$$

$$S_n = \frac{\partial u}{\partial x} - \frac{\partial v}{\partial y}, S_s = \frac{\partial v}{\partial x} + \frac{\partial u}{\partial y}, \xi = \frac{\partial v}{\partial x} + \frac{\partial u}{\partial y} \quad (2)$$

where u and v are the horizontal velocity components in the x and y directions.

The eddy tracking method basically assumes that an eddy detected in one SLA map is the same eddy identified in the following SLA map if the distance in a nondimensional property space is minimal. This methods follows the following equation [16]:

$$X_{e1,e2} = \sqrt{\left(\left(\frac{\Delta X}{X_0}\right)^2 + \left(\frac{\Delta R}{R_0}\right)^2 + \left(\frac{\Delta \xi}{\xi_0}\right)^2\right)} \quad (3)$$

where ΔX is the spatial range between the eddy cores X_{e1} and X_{e2}, ΔR is the variation of diameter, $\Delta\xi$ is the variation of vorticity, X_0 is a characteristic length scale (25 km), R_0 is a typical eddy radius scale, and ξ_0 is a characteristic vorticity (10 − 5 s − 1).

3 Results and Discussion

3.1 The Characteristics of Mesoscale Eddies in the Southeastern Tropical Indian Ocean

In 2019, a total of 96 CEs and 80 AEs were detected, which distributed more than 100 km offshore (Fig. 2a). Areas of a higher total number of CEs and AEs are dominated in the vicinity of the SEC, across the eastern edge of the SJC near 115 °E

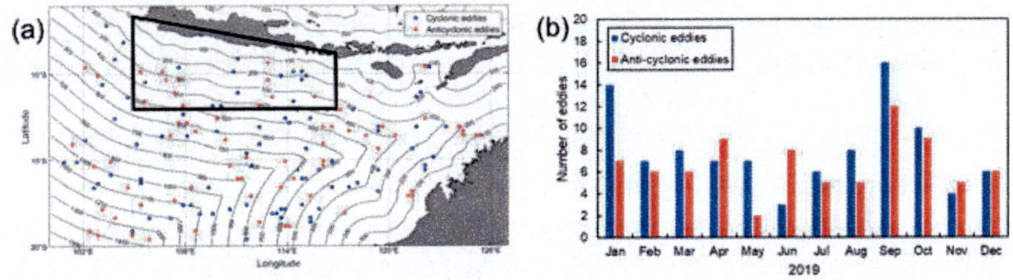

Fig. 2 **a** The spatial distribution and **b** monthly number of detected mesoscales eddies in the Southeastern Tropical Indian Ocean (SETIO) in 2019. Black contour denotes offshore distance with interval of 100 km. Black box denotes the area of interest

and 12 °S, within the Savu Sea, and the ITF region [6, 8, 17]. Therefore, the analysis of mesoscale eddies' influence on SSTA and CHL in SETIO will be focused on 6° to 12 °S and 105° to 117 °E (area of interest).

In addition, temporal variations of mesoscale eddies in Fig. 2b show fluctuation of CEs' and AEs' formation. Number of AEs' formation dropped dramatically in May and escalated in June, as the number of CEs showed reversed dynamics. However, the number of CEs multiplied in the following months until its peak in September. This rapid growth, as well as more significant number of CEs than AEs, is suggested as the influence of strengthened ITF outflow transport in Lombok Strait that modulates the formation of CE during the peak of pIOD event in this region [4, 18].

As the CE is more dominant than AE in 2019, the mesoscale eddies' characteristics focus on CE properties (Table 1). Along with significant number of CEs' formation from July to October, the size of CEs was also increasing. From this period, the mean amplitude (radii) of the CEs ranges from 3.97 cm (89.1 km) to 6.52 cm (91.6 km). Besides its most significant number, the CEs in September also live longest, with mean lifetime extended than 50 days or almost two months later.

Table 1 Monthly cyclonic eddy properties derived from tracked eddies, for mean lifetime (τ), mean amplitude (η), and mean radii (R)

Month	Jan	Feb	Mar	Apr	May	Jun	Jul	Aug	Sep	Oct	Nov	Dec
η (cm)	4.86	4.29	6.16	3.88	4.22	5.03	3.97	4.76	6.52	5.66	4.71	4.43
R (km)	91.0	93.1	88.6	87.7	92.5	94.7	89.1	77.7	89.3	91.6	88.3	88.0
τ (days)	26.0	26.9	35.1	31.3	22.3	22.0	20.5	25.6	51.5	25.3	20.3	15.3

3.2 *The Variations of Eddy-Induced Sea Surface Temperature Anomalies and Chlorophyll-a Concentration*

Based on spatial distribution SSTA, CHL, and CE during pIOD event, the analysis of eddy-induced SSTA and CHL is focused on the area of interest that discussed on the previous section (see Fig. 2a). To examine the impact of CE on SSTA and CHL variation, the regional-averaged SSTA and CHL are divided into two categories based on offshore distance: upwelling region (0–100 km offshore) and mesoscale eddies region (100–400 km offshore). Based on monthly variation of mean SSTA and CHL in the SETIO (Fig. 3), both parameters show similar characteristics that are dominated by upwelling-induced SSTA and CHL up to ~ 80%, especially during the upwelling season coincident with pIOD's generation from June to October [19]. Nevertheless, the eddy-induced SSTA and CHL have remarkable contribution (~25%) to its monthly variation from June to September, which is associated to rapid growth of CEs' number (Fig. 2b). It can be concluded that CE has significant impact in modulating SSTA and CHL in SETIO during the 2019 pIOD event, especially from June to October, which peaked up to –3.3 °C and 1.7 mg/m^3. This impact can be explained through vertical upwelling that raises nutrient-rich cold water from the subsurface to the surface, as well as horizontal advection that transports cold water with high CHL near the coast to the offshore area.

In addition to inspect the influence of CE on the SSTA and CHL's variations, the relationship between the eddy's amplitude that characterized by the SLA of the eddy's center and the mean SSTA, as well as CHL inside the eddies, is discussed furthermore. As shown in Fig. 4, stronger CEs (with larger SLA in the center) induce colder SSTA and higher CHL. Nonetheless, the strong correlation between CE's strength, SSTA, and CHL is significant to the eddies with amplitude larger than 6 cm. It is suggested because several newborn eddies, with an amplitude of less than 6 cm, were formed near to the upwelling region, thus resulting in a bias in the eddy-induced SSTA and CHL variations on these ranges.

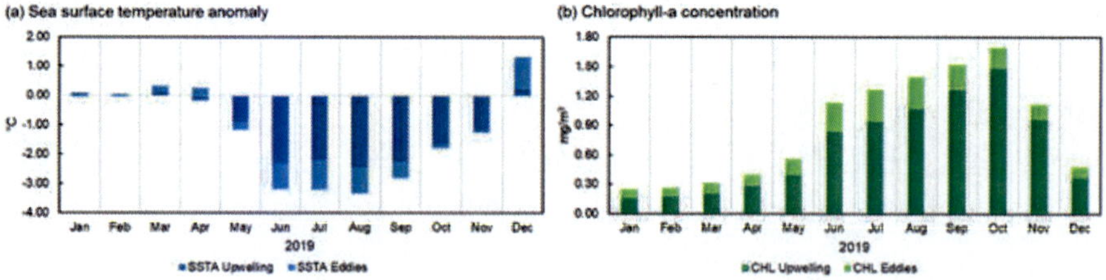

Fig. 3 Monthly mean of area-averaged **a** SSTA and **b** CHL on upwelling's region (0–100 km offshore) and mesoscale eddies' region (100–400 km offshore) over the area of interest

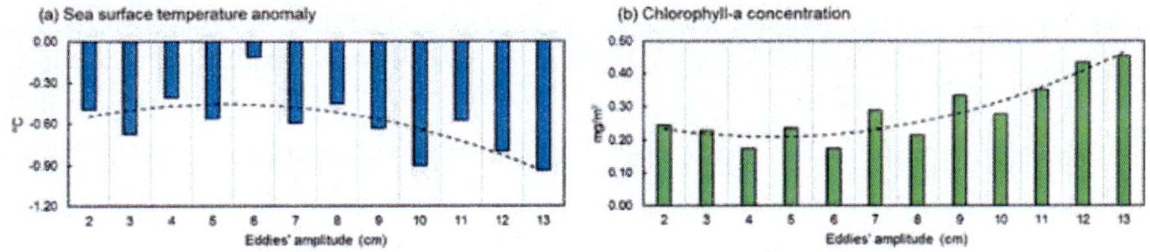

Fig. 4 Eddy-induced **a** sea surface temperature anomaly and **b** chlorophyll-a concentration as a function of cyclonic eddies' amplitude. The curves are second-order polynomial fits

4 Conclusion

The pIOD event in 2019 is characterized by cold SSTA, high CHL, and intense CE in SETIO, which peak in October. The modulation effects of CE on the offshore SSTA and CHL distribution within 400 km in the SETIO were investigated using satellite data from January to December 2019. Based on eddies detection and tracking methods, 96 CEs and 80 AEs are detected and tracked. These eddies dominated in the vicinity of the SEC, across the eastern edge of the SJC near 115 °E and 12 °S, within the Savu Sea, and the ITF region. In addition, temporal variations of mesoscale eddies show that the number of CEs grows rapidly from June until its peak in September. It is suggested as the influence of strengthened ITF outflow transport in Lombok Strait modulates the formation of long-lived CE. Moreover, the size of CEs was growing along with its swift growth, with the mean amplitude (radii) of the CEs ranging from 3.97 cm (89.1 km) to 6.52 cm (91.6 km).

Furthermore, monthly variation of mean SSTA and CHL in the SETIO is dominated by upwelling-induced SSTA and CHL up to ~80%, especially during the upwelling season coincident with pIOD. However, the eddy-induced SSTA and CHL have remarkable contribution by ~25%, which is associated to rapid growth of CEs' number. It can be concluded that CE has significant impact on modulating SSTA and CHL in SETIO during the 2019 pIOD event from June to October. This impact can be explained through both vertical upwelling process of nutrient-rich cold water that rises from the subsurface to the surface, as well as horizontal advection that transport cold water with high CHL near the coast to the offshore area.

Moreover, the eddy-induced SSTA and CHL are strongly correlated with the eddy amplitude, as characterized by SLA at the eddy center, with strong CEs inducing colder SSTA and higher CHL anomalies. It is interesting to note that this close relation is merely significant to the eddies with amplitude larger than 6 cm, due to its bias as the result of the formation of newborn eddies, with an amplitude of less than 6 cm, that firstly generated close to the upwelling region. Nevertheless, this study results in a new insight for understanding the physical-biogeochemical oceanic processes in SETIO as the eastern pole of IOD, which led for fisheries management in this region.

Acknowledgements The authors gratefully acknowledge the E.U. Copernicus Marine Service Information and Giovanni for providing satellite data. This research was supported by the Indonesian National Research and Innovation Agency under Aeronautics and Space Research Fund Program [No. 6/III/HK/2022].

References

1. Shi, W., Wang, M.: A biological Indian Ocean Dipole event in 2019. Sci Rep **11**, 2452 (2021). https://doi.org/10.1038/s41598-021-81410-5
2. Saji, N.H., Goswami, B.N., Vinayachandran, P.N., Yamagata, T.: A dipole mode in the tropical Indian Ocean. Nature **401**, 360–363 (1999). https://doi.org/10.1038/43854
3. Hatmaja, R.B., Munthe, C.C., Yulihastin, E., Pramudia, K.E.: Atmospheric response to the Southern Java upwelling variability associated with positive Indian Ocean Dipole Event. pp. 25–37. (2022)
4. Yang, G., Zhao, X., Li, Y., et al.: Chlorophyll variability induced by mesoscale eddies in the Southeastern tropical Indian Ocean. J. Mar. Syst. **199**, 103209 (2019). https://doi.org/10.1016/j.jmarsys.2019.103209
5. Azis Ismail, M.F., Ribbe, J., Arifin, T., et al.: A census of Eddies in the tropical Eastern boundary of the Indian Ocean. J Geophys Res Ocean **126**, 1–9 (2021). https://doi.org/10.1029/2021JC017204
6. Feng, M., Wijffels, S.: Intraseasonal Variability in the south equatorial current of the East Indian Ocean. J. Phys. Oceanogr. **32**, 265–277 (2002). https://doi.org/10.1175/1520-0485(2002)032%3c0265:IVITSE%3e2.0.CO;2
7. Jia, F., Wu, L., Qiu, B.: Seasonal modulation of eddy kinetic energy and its formation mechanism in the Southeast Indian Ocean. J. Phys. Oceanogr. **41**, 657–665 (2011). https://doi.org/10.1175/2010JPO4436.1
8. Yu, Z., Potemra, J.: Generation mechanism for the intraseasonal variability in the Indo-Australian basin. J. Geophys. Res. **111**, C01013 (2006). https://doi.org/10.1029/2005JC003023
9. Iskandar, I., Sasaki, H., Sasai, Y., et al.: A numerical investigation of eddy-induced chlorophyll bloom in the southeastern tropical Indian Ocean during Indian Ocean Dipole—2006. Ocean Dyn. **60**, 731–742 (2010). https://doi.org/10.1007/s10236-010-0290-6
10. Ogata, T., Masumoto, Y.: Interactions between mesoscale eddy variability and Indian Ocean dipole events in the Southeastern tropical Indian Ocean—case studies for 1994 and 1997/1998. Ocean Dyn. **60**, 717–730 (2010). https://doi.org/10.1007/s10236-010-0304-4
11. Faghmous, J.H., Frenger, I., Yao, Y., et al.: A daily global mesoscale ocean eddy dataset from satellite altimetry. Sci. Data **2**, 150028 (2015). https://doi.org/10.1038/sdata.2015.28
12. Halo, I., Backeberg, B., Penven, P., et al.: Eddy properties in the Mozambique Channel: A comparison between observations and two numerical ocean circulation models. Deep Res. Part II Top Stud. Oceanogr. **100**, 38–53 (2014). https://doi.org/10.1016/j.dsr2.2013.10.015
13. Okubo, A.: Horizontal dispersion of floatable particles in the vicinity of velocity singularities such as convergences. Deep Sea Res. Oceanogr. Abstr. **17**, 445–454 (1970). https://doi.org/10.1016/0011-7471(70)90059-8
14. Weiss, J.: The dynamics of enstrophy transfer in two-dimensional hydrodynamics. Phys. D Nonlinear Phenom. **48**, 273–294 (1991). https://doi.org/10.1016/0167-2789(91)90088-Q
15. Chelton, D.B., Schlax, M.G., Samelson, R.M.: Global observations of nonlinear mesoscale eddies. Prog. Oceanogr. **91**, 167–216 (2011). https://doi.org/10.1016/j.pocean.2011.01.002
16. Penven, P.: Average circulation, seasonal cycle, and mesoscale dynamics of the Peru current system: a modeling approach. J. Geophys. Res. **110**, C10021 (2005). https://doi.org/10.1029/2005JC002945

17. Qiu, B., Mao, M., Kashino, Y.: Intraseasonal variability in the Indo-Pacific throughflow and the regions surrounding the Indonesian Seas. J. Phys. Oceanogr. **29**, 1599–1618 (1999). https://doi.org/10.1175/1520-0485(1999)029%3c1599:IVITIP%3e2.0.CO;2
18. Sprintall, J., Révelard, A.: The Indonesian throughflow response to Indo-Pacific climate variability. J. Geophys. Res. Ocean **119**, 1161–1175 (2014). https://doi.org/10.1002/2013JC009533
19. Saji, N.H., Yamagata, T.: Structure of SST and surface wind variability during Indian Ocean Dipole mode events: COADS observations. J. Clim. **16**, 2735–2751 (2003). https://doi.org/10.1175/1520-0442(2003)016%3c2735:SOSASW%3e2.0.CO;2

Modeling of Surface Current-Driven Water Pollution and ASEAN Countries Index Water Quality Assessment in the Rupat Strait, Riau Province, Indonesia

Koko Ondara and Ulung Jantama Wisha

Abstract Seawater degradation has become a concern in the strategic location of Rupat Strait, where various industries exist. This study aims to determine the current-driven water pollution in the Rupat Strait. We directly surveyed water quality parameters (pH, salinity, temperature, and DO). The other environmental parameters, such as BOD5, ammonia, phosphate, nitrate, TSS, and heavy metals, were analyzed in the laboratory. Two-dimensional hydrodynamic of MIKE21 flow model FM was simulated to understand the sea current surface patterns whereby the modeled tidal current data was superimposed with water quality to predict the current-driven pollutant. Water quality and environmental parameters showed a different response to the tidal current pattern. The assessed parameters were above (below) the quality standard, indicating water pollution and degradation, particularly for nutrients and heavy metals, with a deviation of more than 0.03 mg/L compared to the quality standards for marine waters. The highest pollutant accumulation was detected in the middle of the strait, where the current magnitude ranged from 0.02 to 0.21 m/s. In contrast, in the north and south gates of the waterway, the pollutant concentration was not too significant, even though the current profiles tend to get more robust. Aside from hydro-oceanographic features, the sources and intake of industrial waste-induced pollution play a substantial role in determining the pollution level within the Rupat Strait.

Koko Ondara and Ulung Jantama Wisha are the main contributor of this manuscript.

K. Ondara (✉) · U. J. Wisha
Research Center for Oceanography, National Research and Innovation Agency, 14430 Jakarta, Indonesia
e-mail: koko003@brin.go.id

U. J. Wisha
Department of Physics and Earth Sciences, University of the Ryukyus, Nishihara 903-0213, Japan

A. Basit et al. (eds.), *Proceedings of the International Conference on Radioscience, Equatorial Atmospheric Science and Environment and Humanosphere Science*, Springer Proceedings in Physics 290, https://doi.org/10.1007/978-981-19-9768-6_48

1 Introduction

Rupat Strait is a small channel positioned between the Malacca Strait and Rupat Island, Riau Province, with a channel length of approximately 72.4 km and a width ranging from 3.8 to 8 km. Rupat Strait is a significant area, becoming the center of oil and gas industries with a dense shipping activity. Rupat Island is composed of peatland, resulting in a vulnerable coastal area when an extreme oceanographical state occurs at certain times [1]. This condition has become a significant concern for a long time because the Rupat Strait supports tourism, fishery, and industry [2]. On the other hand, the rapid development in the surrounding Rupat Island leads to a downgrade in the water quality. It is in line with the increase of industrial wastes and oil spills that jeopardize the marine biota and even the local people.

The primary pollutant in the Rupat Strait is derived from offshore oil spilling, ballast water wastes from tanker ships and other water transportation modes [3–5], and heavy metal from any coastal activities [6–9]. The primary source of heavy metals is from industry, mining, household, and agricultural wastes [10–12]. The increased heavy metal and other pollutant concentrations will directly affect the biota's survival ability [13]. It can accumulate within the body of marine biota and in the sediment [6, 13, 14, 16–20]. An integrated environmental study is crucial because water pollution potency is very alarming in the Rupat Strait and may be controlled by tidal currents-induced transport mechanisms. Furthermore, the influence of oceanographic factors on triggering pollutant distribution within the strait is not apparent yet, and this aspect should be investigated. Therefore, this study aims to determine the water mass transport mechanisms and their influence on pollutant distribution within the Rupat Strait based on hydrodynamic modeling, spatial analysis approaches and the water quality results will be compared to the ASEAN countries index water quality assessment.

2 Materials and Methods

2.1 Study Site

Rupat Strait is a semi-enclosed channel formed by Rupat Island and the eastern coast of Sumatra (Dumai City). Administratively, it belongs to Riau Province, Indonesia (Fig. 1a). Geographically, it is positioned between 101.2–101.8° East and 1.6–2.15° North. Malacca Strait borders this water area in the north and south, Dumai City in the west, and Rupat Island in the East.

The most significant economic activity in the surrounding Dumai City is industrial sectors, such as the crude oil industry and oil and gas product distribution. Moreover, the port, fishery, and anthropogenic activities increase water pollution as the wastes

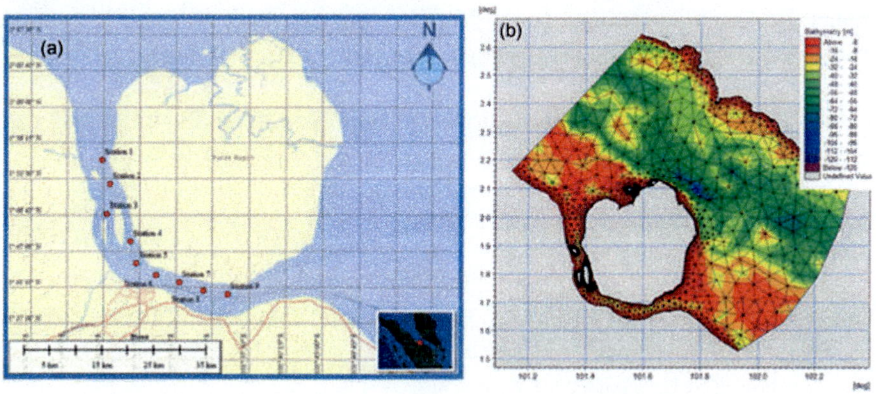

Fig. 1 Research location map, water quality sampling site (**a**); model mesh (**b**)

flow into the strait in the form of hydrocarbon (sulfide, ammonia, phenol, mercaptan, and other organic and acidic compounds) [13]. Loading and unloading activities in the port also frequently leave the liquid waste into the ocean, where oil spilling is undoubtedly avoided [21].

2.2 Hydrodynamic Model Simulation

MIKE21 flow model FM with unstructured triangle mesh [22] was employed to determine the current surface patterns within the Rupat Strait. It was simulated from June 1, 2022 to December 28, 2022, representing two significant states of moon phase-induced tidal elevation (spring and neap tidal conditions) [23]. In the meshing stage (Fig. 1b), bathymetry and coastline data were provided by the bathymetry map of the Indonesian Navy, calibrated using in situ bathymetry survey result using single beam echo sounder Teledyne Odom collected in September 2020. A tidal model predicting the surface elevation fluctuation was used as boundary conditions of the model developed. The model setup is given in Table 1.

2.3 Water Sampling and Analysis

The field sampling was conducted from August 29, 2022 to September 5, 2020. Nine sampling sites were chosen to represent the water condition of Rupat Strait (Fig. 1a). All areas were surveyed during the same tidal conditions (ebb tides), considering the land-sourced waste intake to the water. It commenced from the Northern gate of Rupat Strait and ended at the Southern waterway (Fig. 1). We considered the

Table 1 Flow model flexible mesh setup

Parameter	Applied in the simulation
Time	Number time of steps 210 Time steps interval 3600 Start date 1/6/2020 01:00 to 28/12/2020 23:00
Mesh boundary	Indonesian navy bathymetry map calibrated by field measurement in 2020 with six open boundaries
Solution technique	Time integration: low order, fast algorithm Space discretization: low order, fast algorithm Minimum time step: 0.01 s Maximum time step: 3600 s Critical CFL number: 0.8
Flood and dry	Drying depth: 0.005 m Flooding depth: 0.05 m Wetting depth: 0.1 m
Initial water level	-0.37 m
Density	Density type: Barotropic Reference temperature: 30 °C Reference salinity: 28 PSU
Bed resistance	Manning number: 0.22 $m^{1/3}/s$

nature of Rupat Strait and the location of industry and settlement areas to choose the sampling site (Fig. 2).

Water bodies were sampled at a depth of 1 m underneath the surface level using a Nansen bottle. The water sampled was then stored in a plastic bottle. It was then analyzed in the marine laboratory of Riau University. The analyzed parameters consisted of BOD5, ammonia, phosphate, nitrate, and heavy metals (nickel, zinc, lead, copper, and cadmium). The result was then compared to the standard quality established by ASEAN Marine Water Quality criteria (AMWQC), Indonesia, Thailand, the Philippines, and Malaysia [24, 25].

An in situ survey (Fig. 1a) using the TOA-DKK water quality checker was also done at the same site. All environmental data were then spatially analyzed using an inverse distance weighted spatial analysis tool to generate the distribution map of the pollutant within the study area. The distribution map was then overlaid with the hydrodynamic model result sampled with the same tidal conditions. In this case, overlaying the low spring tidal current pattern is feasible since the water quality survey was done during ebb tides.

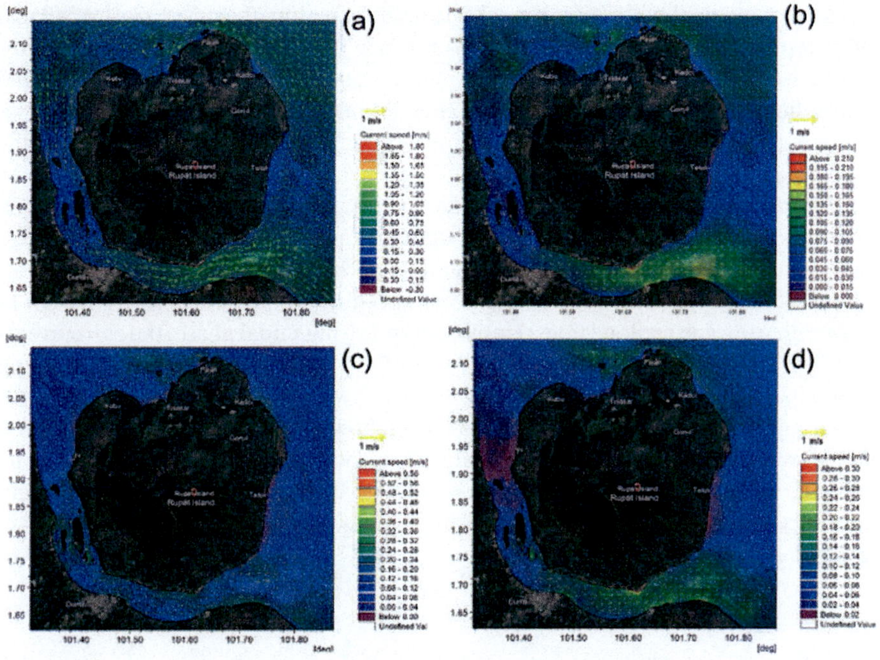

Fig. 2 Tidal current patterns in the Rupat Strait during spring high tidal conditions (**a**); spring low tidal conditions (**b**); neap high tidal conditions (**c**); neap low tidal conditions (**d**)

3 Results and Discussion

3.1 Tidal Current Patterns in the Rupat Strait

Based on tidal harmonic analysis, the semi-diurnal co-tidal showed a predomination in the study area where the M2 and S2 amplitudes were the highest among the significant tidal constituents (Table 2). The Formzahl was 0.34, showing that the mixed tide predominates the study area with prevailing semi-diurnal. Moreover, according to the least-square-based analysis, within 18.6 years of prediction, the highest elevation reaches 1.53 m above the mean sea level, and the lowest level is 1.46 m beneath the mean sea level. The average tidal range was 2.99 m.

Table 2 Tidal harmonic analysis results

Constituents	M2	S2	N2	K2	K1	O1	P1	M4	MS4
Amplitude (m)	0.77	0.36	0.11	0.11	0.14	0.25	0.05	0.00	0.00
Phase lag (°)	258.8	82.8	245.7	112.4	−12.3	−37.3	155.1	15.1	118.6

Within a channel with semi-diurnal tide predomination, the phase lags of M2 and M4 determine the flood or ebb current dominance [26]. The value of $2gM2 - gM4$ was between 90° and 270°, indicating that the study area has more extreme and shorter flood tides than the ebb currents (flood dominant) [27].

During the spring high tidal conditions (Fig. 2a), the current velocity in the Northern gate of Rupat Strait reached 0.82 m/s, where the northerly current flowed Southward, passing the Rupat channel. In contrast, the more robust southerly flow ranging from 0.5 to 1.05 m/s entered the strait via the Southern gate. The water mass was confluence in Mampu Island in the middle of the waterway, decreasing velocity and directional alteration [28]. The current speed in this area ranged from 0.15 to 0.49 m/s during spring flood tides. Unlike the spring high tidal phase, the ebb currents flowed in the opposite direction, getting out of the strait with a magnitude ranging from 0.07 to 0.15 m/s (Fig. 2b).

During the neap phases, the current direction showed the same pattern as the spring one. However, the velocity decreased significantly, ranging from 0.04 to 0.32 m/s and 0.02 to 0.18 m/s during flood and ebb tides. Overall, the current motions are more robust during the spring phase of tides, influencing the transport mechanism throughout the strait [30].

3.2 *The Spatial Distribution of Environmental Parameters*

The influence of hydrodynamics within the Rupat Strait on distributing the pollutant intake is shown in Fig. 3. Overall, the current tidal regimes did not solely control the distribution and accumulation of pollutants. Except for the total suspended solid (TSS) parameter, the highest concentration was identified in the middle of the strait (around Mampu Island).

Water temperature ranged from 30.6 to 31.4 °C with the highest value observed in the surrounding Mampu Island as the southerly and northerly tidal currents flowed toward the middle of the strait, bringing the higher (lower) temperature derived from the Northern (Southern) Malacca Strait (Fig. 4a). Higher TSS concentration was found in the Southern Rupat Strait, where the current motions were more robust in this area (>1 m/s). The TSS ranged from 90.3 to 145.9 mg/L (Fig. 3b). On the other hand, the biological oxygen demand (BOD5) ranged from 8.4 to 9.4 mg/L. Even though the deviation was not too significant, a higher concentration was observed in the surrounding Mampu Island as the current regimes transported the water mass toward the middle of the strait.

The nutrient parameter distribution showed the same patterns for ammonia, nitrate, and phosphate (Fig. 3d–f), where the higher concentration was identified in the middle of the strait. The average N:P ratio was too high (197.6:1), showing that the nutrient pollution was alarming, resulting in imbalanced water conditions. The super high ammonia concentration could cause this state to be detected throughout the study area. According to [29], ammonia can be derived from the solution of organic nitrogen generated by microbes and fungi (ammonification) and from minerals entering water

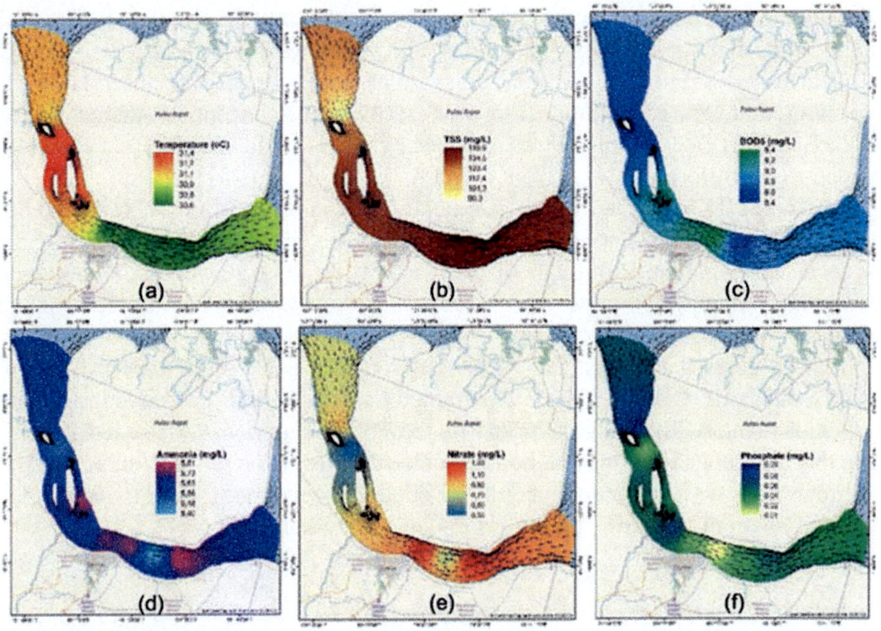

Fig. 3 Spatial distribution of water quality parameters superimposed with tidal current pattern during sampling time; temperature **a**; TSS **b**; BOD5 **c**; ammonia **d**; nitrate **e**; and phosphate **f**

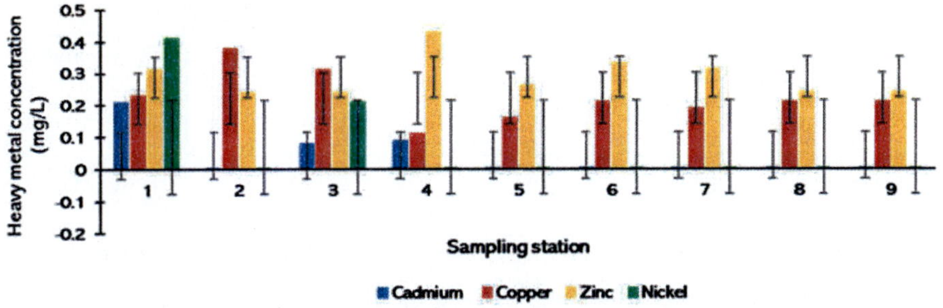

Fig. 4 Heavy metal concentration in the Rupat Strait (Cd and Ni $\times 10^{-1}$). The black lines denote the standard deviation of every data

bodies through soil erosion. However, industrial and anthropogenic wastes also increase the intake of nutrients within the Rupat Strait.

Concerning marine pollutants, the most dangerous pollutant is heavy metals. Since the concentration of heavy metals is difficult to determine in the water bodies, we only identified several dominant heavy metals detected in the study area, such as cadmium, copper, zinc, and Nickel (Fig. 4). All sampling stations identified the other heavy metal parameters (copper and zinc) except nickel and cadmium. Zinc was detected at stations 1 and 3, while cadmium was only detected at stations 1, 3 and 4, situated in the Northern gate of Rupat Strait, ranging from 0.008 to 0.2 mg/L.

Generally, a significant concentration of nickel and copper was observed in the Northern strait, with a magnitude from 0.04 to 0.39 mg/L. The remnant station located in the Southern gate of Rupat Strait, the lower concentration of nickel and cadmium tended to be uniform, ranging from 0.032 to 0.2 mg/L. Due to the nature of heavy metal tending to be settled in the sediment, its quantification in the water bodies could not be used for determining the influence of water motions in transporting the heavy metal compounds throughout the Rupat Strait [31]. However, the industrial wastes and shipping activities are more contributed to elevating heavy metal content in the semi-enclosed strait of Rupat.

Table 3 gives the assessment of the environmental parameters toward several quality standards. Overall, almost all parameter concentrations exceeded the quality standards in the ASEAN region, primarily the TSS and nutrients. These results indicate that the Rupat Strait may be polluted since the elevated industrial, shipping, and anthropogenic wastes are undoubtedly avoided. However, long-term monitoring is necessary to control marine pollutions in the Rupat Strait.

Table 3 Comparison of the environmental parameters to several quality standards

Parameter	Average value	Quality standard	Pollution status
TSS (mg/L)	126.11	I = 80, M = 50, T = nd, P = 30, A = nd	Highly polluted
BOD5 (mg/L)	8.84	I = 20, M = nd, T = nd, P = nd, A = nd	Unpolluted
Ammonia (mg/L)	5.56	I = 0.3, M = 0.05, T = 0.07, P = nd, A = 0.07	Highly polluted
Phosphate (mg/L)	5.64	I = 0.015, M = 0.075, T = 0.045, P = nd, A = 0.045	Highly polluted
Nitrate (mg/L)	0.86	I = 0.08, M = 0.06, T = 0.02, P = nd, A = 0.06	Highly polluted
Nickel (mg/L)	0.01	I = 0.05, M = 0.06, T = 0.02, P = nd, A = 0.06	Unpolluted
Zinc (mg/L)	0.29	I = 0.05, M = 0.05, T = 0.05, P = nd, A = 0.05	Highly polluted
Copper (mg/L)	0.22	I = 0.008, M = 0.003, T = 0.008, P = 0.02, A = 0.008	Highly polluted
Cadmium (mg/L)	0.004	I = 0.001, M = 0.002, T = 0.005, P = 0.01, A = 0.01	Slightly polluted

(Water quality standard for marine biota: I = in Indonesia, M = in Malaysia, T = in Thailand, P = in The Philippines, A = ASEAN Marine Water Quality Criteria (AMWQC), nd = no data)

4 Conclusion

To conclude, the water motions within the Rupat channel are more dynamic during spring tidal phases, primarily at the Southern gate of the strait. It is categorized as a dominant flood channel with more intense flood currents. The hydrodynamic variability within the Rupat Strait does not solely control the distribution of the environmental parameters assessed in this study. The most impactful parameters were temperature, TSS, BOD5, and nutrients. Copper is the most found heavy metal in the Rupat Strait, where the concentration exceeds the standard quality. Based on the average concentrations of environmental parameters, Rupat Strait is categorized as a polluted water area caused by intensive wastewater from industrial and anthropogenic processes. It is recommended that regular monitoring is necessary to prevent the long-term impact of marine pollution.

Acknowledgements We would like to thank The Research Institute for Coastal Resources (RICRV) for research funding in 2020 and those who have helped in completing the field survey and this article.

References

1. Putra, S: Analisis Laju Perubahan Garis Pantai Mengunakan Metode Linier Regression Rate (LRR) dan Metode End-Point Rate (EPR) (Studi Kasus di Pulau Rupat Kabupaten Bengkalis). Riau University (2014)
2. Sandhyavitri, A., Fatnanta, F., Ramadhan, H.R., Suprayogi, I.: Combination of a Coastal Vulnerability Index (CVI) and social economic approaches in prioritizing the development of Riau Coastlines. Indonesia. MATEC Web Conf. **276**, 02006 (2019). https://doi.org/10.1051/matecconf/201927602006
3. Thamrin, K., Anwar, A., Mubarak, M.: Pola Penyebaran Minyak di Perairan Selat Rupat Dumai, Riau. J. Ilmu Lingkung. **1**(1) (2007)
4. Arabova, Z.M., Arabov, M.S., Prochorov, E.M., Saadati, P.A.: Problems and ways to reduce anthropogenic impact on waters of The World Ocean. In: Vestnik of Astrakhan State Technical University. Series: Marine Engineering and Technologies. pp. 41–47. (2019). https://doi.org/10.24143/2073-1574-2019-3-41-47
5. Tian, Y., Chen, J., Huang, J., Lu, S.: Research of ship antifouling paint pollutants analysis and treating technique. Chinese J. Environ. Eng. **11**(2), 1164–1170 (2017). https://doi.org/10.12030/j.cjee.201601127
6. Riani, E., Sudarso, Y., Cordova, M.R.: Heavy metals effect on unviable larvae of dicrotendipes simpsoni (diptera: Chironomidae), a case study from Saguling Dam Indonesia. AACL Bioflux **7**(2), 76–84 (2014)
7. Riani, E.: Perubahan Iklim dan Kehidupan Akuatik (Dampak pada Bioakumulasi Bahan Berbahaya dan Beracun & Reproduksi). IPB Press, Bogor (2012)
8. Gavhane, S.K., Sapkale, J.B., Susware, N.K., Sapkale, S.J.: Impact of heavy metals in riverine and estuarine environment: a review. Res. J. Chem. Environ **25**(5), 226–233 (2021)
9. Gao, L.: Microplastics aged in various environmental media exhibited strong sorption to heavy metals in seawater. Mar. Pollut. Bull. **169** (2021). https://doi.org/10.1016/j.marpolbul.2021.112480

10. Safitri, N.A., Hamidy, R.: Konsentrasi Logam Berat (Cd dan Pb) pada Sedimen Permukaan Perairan Teluk Bayur Provinsi Sumatera Barat Indonesia. J. Environ. Sci. **2**(3), 85–94 (2009)
11. Bartzas, G., Tsakiridis, P.E., Komnitsas, K.: Nickel industry: Heavy metal(loid)s contamination—sources, environmental impacts, and recent advances on waste valorization. Current Opinion in Environ. Sci. Health **21** (2021). https://doi.org/10.1016/j.coesh.2021.100253
12. Luo, P.: Heavy metals in water and surface sediments of the Fenghe River Basin, China: assessment and source analysis. Water Sci. Technol. **84**(10–11), 3072–3090 (2021). https://doi.org/10.2166/wst.2021.335
13. Syahminan, E., Riani, S., Rifardi, A.: Telaahan Logam Berat Pb dan Cd pada Sedimen di Perairan Barat Laut Dumai—Riau. J. Pengelolaan Sumberd. Alam dan Lingkung **5**(2), 133–140 (2015)
14. Mulyani, S. Lani T.I.G., Sujana, A.E.: Identifikasi Cemaran Logam Pb dan Cd Pada Kangkung Yang Ditanam di Daerah Kota Denpasar. Bumi Lestari J. Environ. **12**(2) (2012)
15. Çelik, A.: Heavy metal monitoring around the nesting environment of green sea turtles in Turkey. Water. Air. Soil Pollut. **169**(1–4), 67–79 (2006). https://doi.org/10.1007/s11270-006-1562-0
16. Stanković, S., Jović, M., Mihajlović, M.L.D., Joksimović, J., Tanaskovski, B.: Metal pollution determined by pollution indices for sea grass *P. oceanica* and surface sediments. Arch. Biol. Sci. **67**(1), 91–101 (2015). https://doi.org/10.2298/ABS140410010S
17. Ali, R., Bream A.: The effects of sewage discharge on the marine Gastropod Gibbula sp., collected from the coast of Al-Hanyaa, Libya. Egypt. Acad. J. Biol. Sci. B. Zool. **2**(2), 47–52, (2010). https://doi.org/10.21608/eajbsz.2010.15889
18. Marcovecchio, J.E., Moreno, V.J., Perez, A.: Determination of heavy metal concentrations in biota of Bahia Blanca. Argentina. Sci. Total Environ. **75**(2–3), 181–190 (1988). https://doi.org/10.1016/0048-9697(88)90031-9
19. Gómez-Parra, A., Forja, J., DelValls, M.T., Sáenz, A.I., Riba, I.: Early contamination by heavy metals of the Guadalquivir estuary after the Aznalcollar mining spill (SW Spain). Mar. Pollut. Bull. **40**(12), 1115–1123 (2000). https://doi.org/10.1016/S0025-326X(00)00065-5
20. Tack, F., Verloo, M.: Chemical speciation and fractionation in soil and sediment heavy metal analysis: a review. Int. J. Environ. Anal. Chem. **59**(2–4), 225–238 (1995). https://doi.org/10.1080/03067319508041330
21. Syahminan, S., Rifardi, R., Ruli, E.: Vertical content an analysis crude oil at core in Dumai coastal waters (2011)
22. Sutherland, G., Foreman, M., Garrett, C.: Tidal current energy assessment for Johnstone Strait, Vancouver Island. Proc. Inst. Mech. Eng. Part A J. Power Energy **221**(2), 147–157 (2007). https://doi.org/10.1243/09576509JPE338
23. Brooks, D., Townsend, A.D.W.: Variability of the coastal current and nutrient pathways in the eastern Gulf of Maine. J. Mar. Res. **47**(2), 303–321 (1989). https://doi.org/10.1357/002224089785076299
24. ASEAN Secretariat: ASEAN Marine Water Quality Management Guidelines and Monitoring Manual. 1st ed. Jakarta, New Millenium Pty Ltd (2008)
25. Ministry of Natural Resources and Environment.: Malaysia Marine Water Quality Criteria and Standard. pp. 1. (2016), [Online]. Available: http://www.doe.gov.my/
26. Wisha, U.J., Wijaya, Y.J., Hisaki, Y.: Real-time properties of hydraulic jump off a tidal bore, its generation and transport mechanisms: a case study of the Kampar River Estuary, Indonesia. Water **14** (2561), (2022). https://doi.org/10.3390/w14162561
27. Wisha, U.J., Wijaya, Y.J., Hisaki, Y.: Tidal bore generation and transport mechanism in the Rokan River Estuary, Indonesia: Hydro-oceanographic perspectives. Reg. Study. Mar. Sci. **52** (102309), (2022). https://doi.org/10.1016/j/rsma.2022.102309
28. Sutherland, G., Foreman, M., Garrett, C.: Tidal current energy assessment for Johnstone Strait, Vancouver Island. Proc. Inst. Mech. Eng. Part A J. Power Energy **221**(2), 147–157, (2007). https://doi.org/10.1243/09576509JPE338
29. Peirong, Z., Wei, L.: Use of fluidized bed biofilter and immobilized *Rhodopseudomonas palustris* for ammonia removal and fish health maintenance in a recirculation aquaculture system. Aquac. Res. **44**(3), 327–334 (2013). https://doi.org/10.1111/j.1365-2109.2011.03038.x

30. Ondara, K., Husrin, S.: Characteristic of breaking waves and analysis of sediment transport in Teluk Kendari. Jurnal Ilmu dan Teknologi Kelautan Tropis, (2018). https://doi.org/10.2944/jitkt.v9i2.19293
31. Chakraborty, S., Chakraborty, P., Nath, B.N.: Lead distribution in coastal and estuarine sediments around India. Mar. Pollut. Bull. **97**(1–2), 36–46 (2015). https://doi.org/10.1016/j.marpolbul.2015.05.056

Identification of Seasonal Water Mass Characteristics in West Sumatra Waters

Gandhi Napitupulu, Ivonne M. Radjawane, Nabila Afifah Azuga, Khafid Rizki Pratama, Naffisa Adyan Fekranie, and Hansan Park

Abstract The eastern Indian Ocean is western tropical waters with very complex dynamic and physical processes affecting the area along the line of the West Sumatra coast. This study aims to analyze the water masses in the West Sumatra waters adjacent to the equatorial latitude based on the seasonality of December–February (DJF), March–May (MAM), June–August (JJA), and September–November (SON) from 2003 to 2018 which uses World Ocean Database (WOD) data. The tendency of the observed water mass characteristics to be dominant in the surface layer is more concentrated by the influence of water masses from the area around West Sumatra and intrusion from the surrounding waters. Meanwhile, in the middle to deep layers, more water mass patterns are observed from the western Indian Ocean region with a temperature range of −10 °C and salinity of 34.6–36.7 PSU. Characteristics based on seasonality are more indicative of a water mass mixing process. The DJF season tends to have a more uniform sea surface temperature than the JJA season. In addition, transitional seasons I and II (MAM and SON) tend to approach the main seasonal conditions. This is evidenced by the different Mixed Layer Depth (MLD) contours in the DJF season, which are thinner than in the JJA season. The temperature uniformity in the horizontal profile seems to vary in the JJA season compared to the DJF, so the thermocline observed in the JJA season is deeper. The identified water column density conditions show the same pattern seasonally.

G. Napitupulu · N. A. Azuga · K. R. Pratama
Department of Earth Sciences, Faculty of Earth Sciences and Technology, Institut Teknologi Bandung, Bandung, Indonesia

I. M. Radjawane (✉)
Oceanography Research Group, Faculty of Earth Sciences and Technology, Institut Teknologi Bandung, Bandung, Indonesia
e-mail: ivonnemr@itb.ac.id

N. A. Fekranie
Department of Oceanography, Faculty of Earth Sciences and Technology, Institut Teknologi Bandung, Bandung, Indonesia

I. M. Radjawane · H. Park
Korea-Indonesia Marine Technology Cooperation Research Center, Bandung Institute of Technology, Bandung, Indonesia

A. Basit et al. (eds.), *Proceedings of the International Conference on Radioscience, Equatorial Atmospheric Science and Environment and Humanosphere Science*, Springer Proceedings in Physics 290, https://doi.org/10.1007/978-981-19-9768-6_49

1 Introduction

The western waters of Sumatra are a unique and complex area because their geographical location is crossed by the equator and flanked by the Asian continents–the Australian continent and the Indian Ocean–the Pacific Ocean [1], making these waters dynamic [2]. It is due to the significant influence of the monsoon system, Sumatra's west coast, and Indonesian Crossflow (*ARLINDO/Arus Lintas Indonesia*), the Indian Ocean Dipole (IOD), and El-Nino Southern Oscillation (ENSO) phenomena on the dynamics of the water mass characteristics of Sumatra's western waters [3].

The mass of water is formed from the composition of temperature, salinity, and density at a certain water depth [4]. Temperature, salinity, density, and pressure will develop stratification in the waters vertically, affecting the stability of the water mass [5]. Water mass identification is accomplished by analyzing temperature and salinity changes [6]. The water temperature is divided vertically into three layers: the mixed layer, the thermocline layer, and the inner layer. The mixed layer is formed by the stirred water mass caused by the movement of winds, currents, and tides on the sea surface, resulting in a homogeneous temperature value in the layer [7].

Seasonally, the east monsoon temperature is lower than the west monsoon temperature. It is inversely proportional to the salinity value, higher during the east monsoon and lowers during the west monsoon [6]. Seasonal winds, also known as monsoons, are seasonal winds that regularly blow due to differences in the intensity of land heating by the solar radiation system, causing differences in air mass pressure [8]. Winds carry masses of cold and moist air during the west monsoon, which occurs from October to March, causing rain in various affected locations. During the east monsoon, which arises from April to September, the winds carry dry air masses, resulting in a dry season in the affected areas [9].

Wind plays a vital role in moving and stirring the surface water mass [5]. However, wind movements do not affect the mixing and movement of water masses in deep waters (500–2000 m depth) because the speed of the current generated by the wind becomes zero at this depth, which is known as the depth without movement or the Ekman depth [10].

The eastern Indian Ocean water mass is dominated by two major streams that originate in the Subtropical Indian Ocean (SIC) and the Australasian Mediterranean Sea [4]. Another study identified five types of water masses in the Indian Ocean's northeastern waters, including Bengal Bay Water (BBW), South Indian Central Water (SICW), Indian Equatorial Water (IEW), Subtropical Lower Water (SLW), and Northern Salinity Minimum Water (NSMW) [11]. The water mass in the Indian Ocean around western Australia consists of Antarctic Bottom Water (ABW), Lower Circumpolar Deep Water (LCDW), Upper Circumpolar Deep Water (UCDW), Antarctic Intermediate Water (AIW), Australasian Mediterranean Water (AMW), Subantarctic Mode Water (SMW), and South Indian Central Water (SCW) [12].

Another study suggested that the Equatorial Undercurrent (EUC) moves high salinity water masses to Indonesian waters [13]. The Arabian Sea High Salinity

Water (ASHSW) moves from the western Indian Ocean to the upper thermocline in the eastern Indian Ocean.

Many studies have been conducted on water masse's characteristics in the eastern Indian Ocean waters, but few have focused solely on the waters of western Sumatra. Therefore, this study aims to investigate the seasonal characteristics of water mass in the western waters of Sumatra and to identify the factors that influence mixing in the western waters of Sumatra when viewed from vertical and horizontal cross-sections of these waters.

2 Study Area and Methods

The study area includes the western waters of Sumatra at coordinates of 0–10° south latitude and 90–100° east longitude (Fig. 1). The data used in this study was obtained from the World Ocean Database (WOD) and processed using Ocean Data View (ODV) software from 2003 to 2018. The data is available for download at https://www.ncei.noaa.gov/products/world-ocean-database.

The salinity, temperature, and pressure (depth) data from 2003 to 2018 were utilized to determine the density and mass properties of the water. The water mass in four seasons: the western season (DJF), the transitional season I (MAM), the east season (JJA), and the transitional season II (SON), were analyzed separately by displaying a T-S diagram, vertical distribution, and horizontal distribution at 0–100 m and 0–1000 m (Table 1).

Fig. 1 Map of the study area

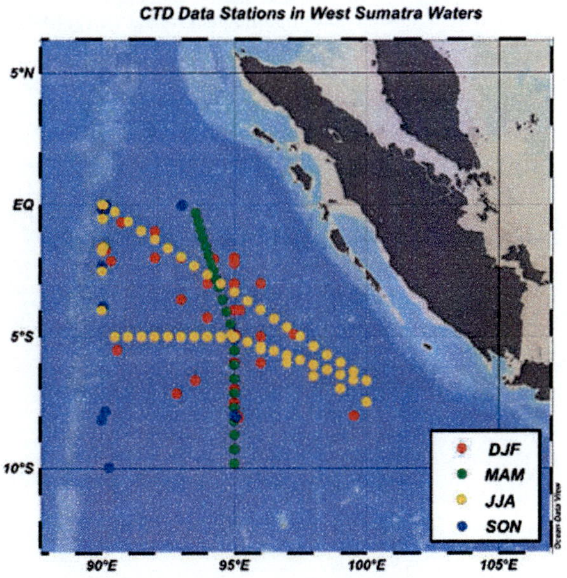

Table 1 Characteristics of global water mass temperature and salinity (*Source* Emery, 2001)

Layer	Indian Ocean	Pacific Ocean
Upper waters (0–500 m)	Bengal Bay Water (BBW) 25.0–29 °C, 28.0–35.0‰) Arabian Sea Water (ASW) (24.0–30.0 °C, 35.5–36.8‰) Indian Equatorial Water (IEW) (8.0–23.0 °C, 34.6–35.0‰) Indonesian Upper Water (IUW) (8.0–23.0 °C, 34.4–35.0‰) South Indian Central Water (SICW) (8.0–25.0 °C, 34.6–35.8‰)	Pacific Subarctic Upper Water (PSUW) (3.0–15.0 °C, 32.6–33.6‰) Western North Pacific Central Water (WNPCW) (10.0–22.0 °C, 34.2–35.2‰) Eastern North Pacific Central Water (ENPCW) (12.0–20.0 °C, 34.2–35.0‰) Eastern North Pacific Transition Water (ENPTW) (11.0–20.0 °C, 33.8–34.3‰) Pacific Equatorial Water (PEW) (7.0–23.0 °C, 34.5–36.0‰) Western South Pacific Central Water (WSPCW) (6.0–22.0 °C, 34.5–35.8‰) Eastern South Pacific Central Water (ESPCW) (8.0–24.0 °C, 34.4–36.4‰) Eastern South Pacific Transition Water (ESPTW) (14.0–20.0 °C, 34.6–35.2‰)
Intermediate waters (500-1500 m)	Antarctic Intermediate Water (AAIW) (2-10°C, 33.8–34.8‰) Indonesian Intermediate Water (IIW) (3.5–5.5°C, 34.6–34.7‰) Red Sea-Persian Gulf Intermediate Water (RSPGIW) (5-14^{0}C, 34.8–35.4‰)	Pacific Subarctic Intermediate Water (PSIW) (5.0–12.0 °C, 33.8–34.3‰) California Intermediate Water (CIW) (10.0–12.0 °C, 33.9–34.4‰) Eastern South Pacific Intermediate Water (ESPIW) (10.0–12.0 °C, 34.0–34.4‰) Antarctic Intermediate Water (AAIW) (2–10 °C, 33.8–34.5‰)
Deep and abyssal waters (1500 m-bottom)	Circumpolar Deep Water (CDW) (1.0–2.0°C, 34.62–34.73‰)	Circumpolar Deep Water (CDW) (0.1–2.0 °C, 34.62–34.73‰) Subantarctic Surface Water (SASW) (3.2–15.0 °C, 34.0–35.5‰)

3 Results and Discussion

3.1 Water Mass Characteristics

The temperature-salinity (T-S) diagram shows a decrease in temperature from the surface (0 m) to a depth of about 5000 m. The characteristics of the water mass are arranged by the composition of temperature, salinity, and density [5]. The surface layer temperature in Sumatra's western waters is 32 °C, with salinity values ranging from 32.5 to 32.8 PSU, and the density value is 22.5 kg/m^3. The temperature value decreases significantly as depth increases, with a minimum value of 3 °C at 5000 m. It is due to differences in the penetration of sunlight into the water, which causes the surface layer to be warmer than the deep layer [15]. Temperatures are higher at shallow depths and lower at deeper water layers [16].

Meanwhile, the salinity value increases as the depth increases, with a maximum salinity value of 34.8 PSU at a depth of 5000 m. At the surface layer, salinity tends to be lower due to the inclusion of freshwater sources in the area [17]. The density of water also increases along with depth. The maximum density value is 27.5 kg/m^3. The density of seawater varies depending on salinity and temperature. The density of water is lighter in the surface layer due to the freshwater influx. It gets heavier in deeper waters and becomes constant at 2000 m [10]. According to the T-S diagram analysis (Fig. 2), there are six types of water masses found in Sumatra's western waters at coordinate positions 0–10° south latitude and 90°–100° east longitude. The water masses are Bengal Bay Water (BBW), Indonesian Upper Water (IUW), South Indian Central Water (SICW), Antarctic Intermediate Water (AAIW), Red Sea-Persian Gulf Intermediate Water (RSPGIW), and Circumpolar Deep Water (CDW).

BBW water masses dominate the surface layer (depth 0–120 m) with temperatures ranging from 25 to 29 °C and salinities ranging from 28 to 35 PSU. At a depth of 150–650 m, the mass of water is dominated by SICW, which has a temperature range of 8–25 °C and a salinity range of 34.6–35.8 PSU, followed by IUW, which has a temperature range of 8–23 °C and a salinity range of 34.4–35 PSU.

At a depth of 650–1000 m, a mass of water is identified as the RSPGIW, with temperature values ranging from 5 to 14 C and salinity values ranging from 34.8 to 35.4 PSU. Another water mass that dominates at this layer is AAIW, which has a temperature range of 1–10 °C and a salinity range of 33.8–34.8 PSU. While at depths of 3000–5000 m, the dominating water is CDW, which has a temperature of 1–2 °C and a salinity of 34.62–34.73 PSU.

The characteristics of the water mass in the western waters of Sumatra indicate intense mixing. According to the T-S diagram, the mixing is speculated to occur due to the influence of internal waves from tides and strong global currents. The Indian Ocean plays a significant role in the global current system, causing continuous mixing of water masses in the western area of Sumatra. One of the areas for the generation and propagation of internal solitary waves (ISW) is the southern part of the Andaman and Nicobar Islands, which border the northwest waters of Sumatra. It is caused by

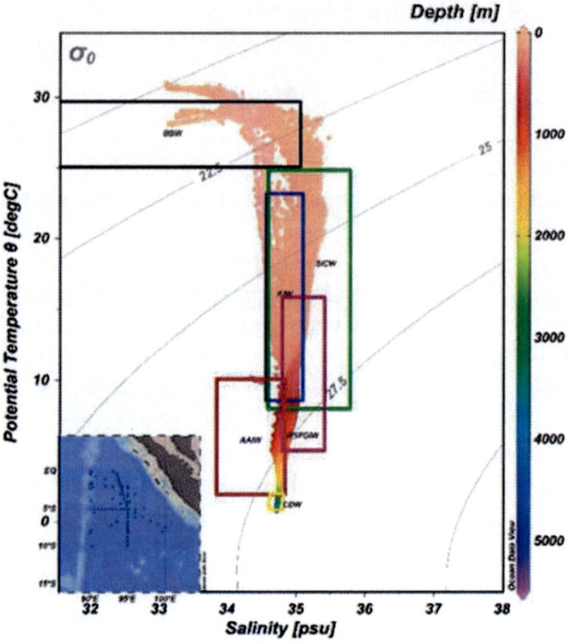

Fig. 2 Characteristics of water masses

a strong barotropic semidiurnal tidal current from the Indian Ocean hitting the ridge off the coast of Breueh Island to the northwest.

The internal wave formation along the ridge is primarily caused by barotropic tidal currents of the semidiurnal component (M_2). The major axis of the M2 ellipse is oriented northeast (flood) and southwest (ebb) with a strong sea level elevation amplitude and magnitude that reach about 0.5 m and 5.0 m/s, respectively. During high tide, the magnitude of the barotropic tidal current from the Indian Ocean that cuts along the ridge dramatically increases, and the current direction is divided into several directions. Some of the current flows to the southeast as it enters the passage between Pulau Weh, Pulau Breueh, and Sumatra and then flows partly across the north side of Pulau Weh to the east before turning east or southeast. The strait between Nasi Island and mainland Sumatra also contributes to the tidal currents (Prasetya 2020).

3.2 Seasonal Variability of Water Mass

Temperature Variations

Figure 3 shows the variation of water mass for the four seasons. The temperature values at the surface and before the thermocline layer at low-latitude stations are higher than at high-latitude stations during the western season (Fig. 3a), as shown by an isothermal line rising by 28 °C. The SST distribution is clear in the minimum range at higher latitudes during the western season. The high value of SST at low-latitude stations is caused by the intensity of solar radiation, which tends to be below the equator during this period. The westerly monsoon winds bring cooler air from the continents of Asia and Europe that moves toward mainland Australia through the Northeast Indian Ocean. Strong mixing has also been observed to bring SICW water mass from India [11].

Similar to the pattern observed during the west monsoon period, intense mixing occurred, shown by mixing occurrence at a shallower depth and wider range of water mass column. It happened because the surface winds that blew during this period

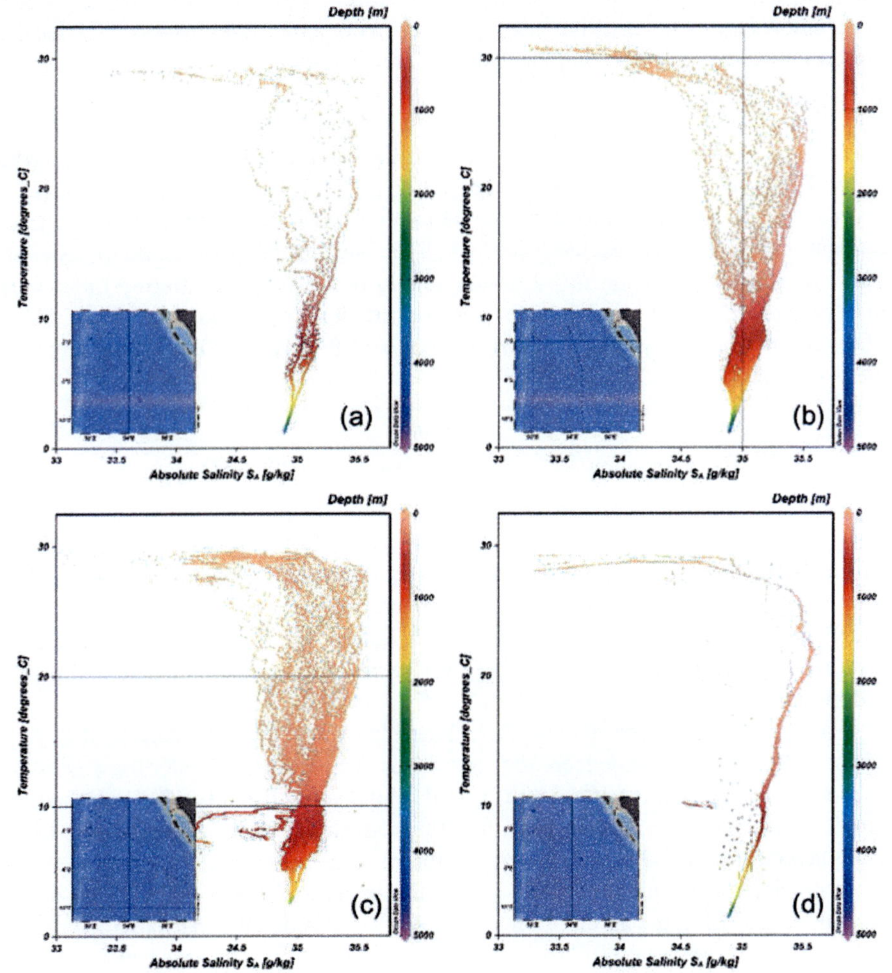

Fig. 3 Variation of Water Mass Characteristics during **a** December–January–February (DJF) **b** March–April–May (MAM) **c** June–July–August (JJA) **d** September–October–November (SON)

were not as strong, and the direction was more random. As a result, the mixing strength of the surface layer is not as homogeneous.

In contrast to previous periods, the SST value is generally higher toward stations with higher latitudes during the east monsoon period of June–August (JJA). Furthermore, during the transitional season II (SON), wind direction and currents are unstable and weak. The temperature difference between low and high-latitude stations is not particularly noticeable. The water mass is generally unstable beneath the thermocline layer (200–250 m).

The influence of currents and topography causes water mass parcels to be distorted. Compared to other seasons, the east monsoon has the most instability due to strong mixing caused by tides in the western part of Sumatra [18]. Tides potentially generate mixing when an internal wave occurs [23]. In western Sumatra waters, turbulent flow is common and results in turbulent mixing that distorts the water mass (lower density water masses can be above higher density water masses).

Salinity Variations

The diagrams show that seasonal variation of the salinity profile is insignificant (the difference between seasons is only about 0.2–0.3 PSU). Salinity values in the surface layer are typically lower, ranging from 33 to 34.5 PSU. The highest surface salinity is identified during the east monsoon, when the western region of Sumatra receives the least rainfall, resulting in a reduced supply of fresh water entering the waters and, presumably, high salinity values. The salinity increases from 35 to 35.5 PSU at 200–500 m. Salinity values range from 34.5 to 34.8 PSU at 1000–3000 m.

3.3 Seasonal Variability of Horizontal Temperature and Salinity

The temperature distribution in the longitude horizontal section (Fig. 4a) shows that the thermocline layer is deeper in the east than in the west. It indicates that the water mass was transported from the West Indian Ocean to the East Indian Ocean. According to Yoga and Karmen (2013), the surface temperature distribution in the Indonesian region ranges from 27.8 to 32 °C.

Water temperatures, particularly in Indonesian waters, are influenced by the cycle of changing seasons. Aside from the season, the temperature of water mass is influenced by the sun's intensity, depth, and the land around it [3]. The temperature profile in a seawater column reveals that the seawater column is divided into three main layers: the mixed surface layer, the thermocline layer, and the deep layer. To more precisely determine the three layers, the upper and lower thermocline limits are calculated using the assumption that a layer is considered a thermocline layer if the temperature gradient (100 m) is equal to or larger than 2 °C. However, it is not the case for West Sumatran waters. As shown in Fig. 4a, the decrease of thermocline depth in the eastern area indicates an upwelling process. The thermocline in longitude

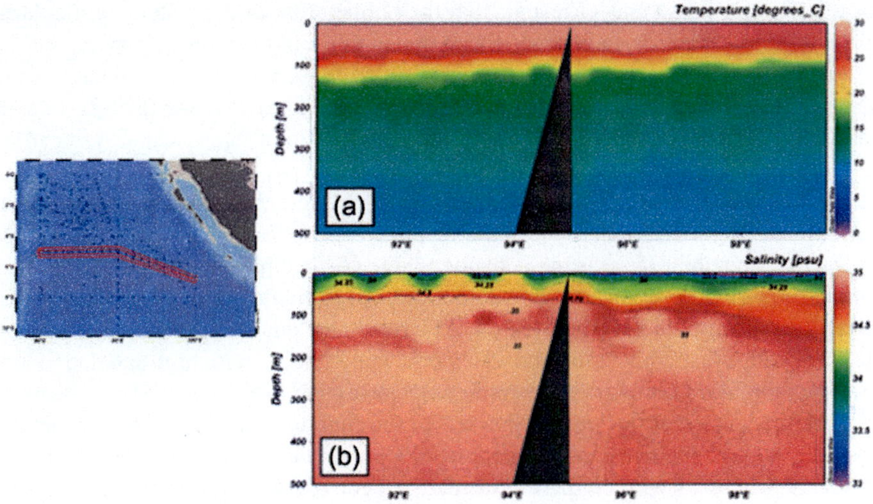

Fig. 4 Horizontal cross-section of **a** temperature and **b** salinity in longitude for the depth of 500 m in West Sumatra waters

92 E occurred at 0–120 m, while it occurred at shallower than 100 m in longitude 98 E. Besides the upwelling caused by underwater currents, the bathymetric conditions also play a role in determining the thermocline depth in the region.

Figure 4b shows salinity contours that vary with longitude. It shows that the salinity values are higher at longitudes 92–94 E than at longitudes 98 E, and therefore, the salinity pattern seems to move from west to east. Interestingly, salinity is almost uniformly in the 35 PSU range at depths of more than 100 m. It happened due to a mass transfer of water from the West Indian Ocean, an intrusion from the more saline western water mass (saline). The dominant water mass flow toward the east is at a depth of 80–150 m from latitude 1.2° south latitude to 1.5° north latitude. Meanwhile, the current direction is westward in the surface layer and at depths of 250–500 m. There is a current reversal between the Equatorial Under Current (EUC) and the current in the surface layer [13] (Fig 6).

The wind is stronger during the east and transition seasons and lasts longer. As a result, the thermocline layer will be deeper. Strong winds will cause stronger currents and waves, stirring the water mass in the marine water column, causing the mixed layer to become deeper and the thermocline layer to descend [19]. Generally, as the wind speed above the water increases, the thermocline column becomes thinner, and the thermocline gets deeper.

The results show that the temperature in the west is higher than in the east, indicating an upwelling process that occurs in the waters of West Sumatra. The characteristics of the water mass carried from the West Indian Ocean allow it to remain in West Sumatra's waters. The West Indian Ocean has a broader thermal process than the eastern region. This concept reveals that the temperature of the waters, particularly those in Indonesia, is strongly influenced by the cycle of changing seasons.

Aside from the season, the temperature is also influenced by the intensity of the sun, depth, and land. The uniform temperature tends to exist at the 0–100 m layer for the western region and the 80 m for the eastern region. Meanwhile, water mass underneath the thermocline layer tends to vary toward depth.

During the DJF season, the water mass distributed across the eastern Indian Ocean has low salinity. Meanwhile, the mass of water distributed in this area has higher salinity during the JJA season. It happens due to precipitation processes formed during the DJF period, such as the Madden Julian Oscillation (MJO) that causes rain to fall across Sumatra's western region. As a result, the depth of the mixed layer may be shallower. Meanwhile, the wind transports water masses with high salt due to low precipitation along Sumatra's western coast during the JJA season. The movement of water masses from the west to the east of the Indian Ocean becomes slow as the easterly wind strengthens and heads toward China's mainland. It can be compared to Fig. 5, where varying temperatures tend to decrease at lower latitudes (8° south latitude) compared to higher latitudes (2° south latitude). Furthermore, the visible depth tends to decrease at higher latitudes with a thermocline depth range of 0–50 m.

Figure 7a shows that in the west monsoon, the vertical contour of density varies from 22 to 24 kg/m^3 and tends to occur at the 0–150 m thermocline. Density values are higher at depths of 200–500 m, ranging from 26 to 30 kg/m^3. Meanwhile, during the transitional season I (MAM), the density contours of 22–24 kg/m^3 occurred at a depth of shallower than 100 m, and the density contours of 26–27 kg/m^3 occurred at a depth of 250–500 m. During the east monsoon (JJA), the density ranging from 22 to 23 kg/m^3 appeared at 0–120 m and the density of 26–27 kg/m^3 at a depth of more than 200 m.

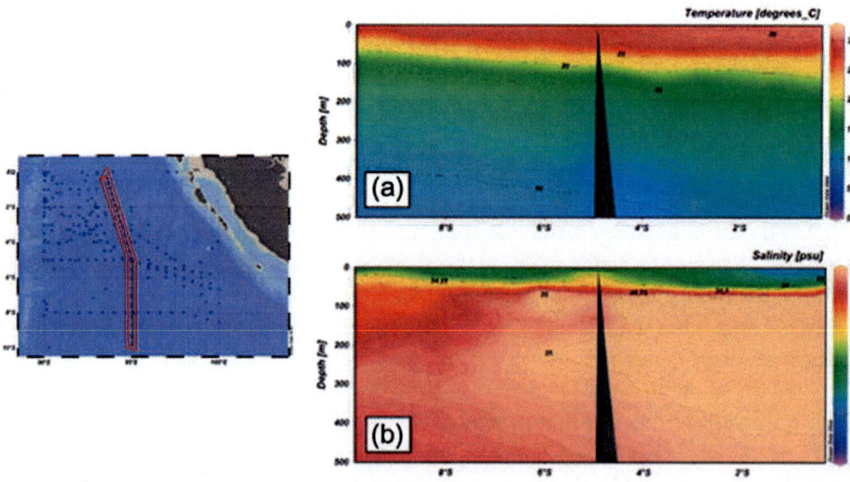

Fig. 5 Vertical cross-section of **a** temperature and **b** salinity in latitude for the depth of 500 m in West Sumatra waters

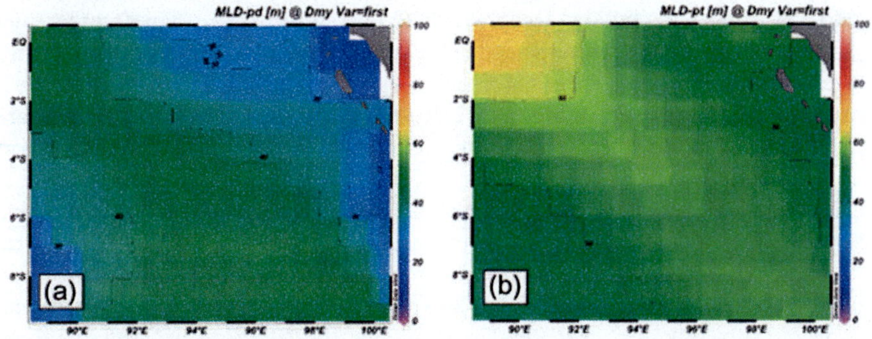

Fig. 6 Spatial plot for Mixed Layer Depth in **a** December–January–February (DJF) 2016 and **b** June–July–August (JJA) 2016 in West Sumatra waters

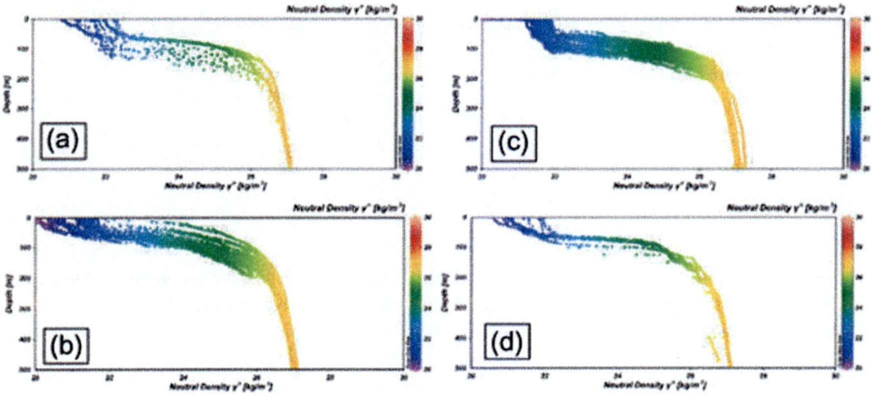

Fig. 7 Vertical cross-section of density (kg/m3) **a** December–January–February (DJF) **b** March–April–May (MAM) **c** June–July–August (JJA) **d** September–October–November (SON)

4 Conclusion

The temperature declined from the surface layer to a depth of 5000 m, whereas the salinity increased with depth. The western waters of Sumatra have a maximum temperature of 32 °C and a minimum temperature of 3 °C, while the highest salinity value is 34.8 PSU, and the lowest value is 32.5 PSU. In general, the distribution pattern of SST during the west monsoon and transitional season I tend to be higher at low latitudes. Meanwhile, the SST value at high latitudes rises during the east monsoon. During transitional season II, however, the temperature difference between low and high-latitude locations is not very significant. The T-S diagram analysis reveals six different types of water masses in Sumatra's western waters, including Bengal Bay Water (BBW), Indonesian Upper Water (IUW), South Indian Central Water (SICW), Antarctic Intermediate Water (AAIW), Red Sea-Persian Gulf Intermediate Water

(RSPGIW), and Circumpolar Deep Water (CDW). Horizontally, the temperature distribution in Sumatra's western and the horizontal temperature profile in Sumatra's western waters indicate the presence of an upwelling phenomenon, defined by a temperature differential of more than 2 °C/100 m of water depth. In the western half of the waterways, the salinity profile is higher. During the east monsoon and the transition season, the thermocline layer becomes deeper due to strong winds that trigger the mixing of water masses until the mixing layer. Meanwhile, the thermocline layer tends to be shallower during the west monsoon. The salinity value of the waters in the west monsoon tends to be lower than in the east monsoon. The density profile at a depth of 0–150 m ranges from 22 to 24 kg/m^3 in the west season and transitional season I, 22 to 23 kg/m^3 in the east season, and 21–22 kg/m^3 in the transitional season II. Meanwhile, at depths more than 200 m, density levels during the west season range from 26 to 30 kg/m^3, whereas density values during transitional seasons I, east season, and transitional season II range from 26 to 27 kg/m^3.

Acknowledgements The author would like to thank the Korea-Indonesia MTCRC (Marine Technology Cooperation Research Center) Scholarship for funding during the research. Thanks also go to the World Ocean Database (WOD) for providing and accessing the data used in this research.

References

1. Supriyadi, E., Hidayat, R.: Identification of upwelling area of the Western territorial waters of Indonesia from 2000 to 2017. Indonesian J. Geograp. **52**(1), 105–111 (2020)
2. Napitu, R., Surbakti, H., Diansyah, G.: Identifikasi Karakteristik Massa Air Perairan Selat Bangka Bagian Selatan. Maspari J.: Marine Sci. Res. **8**(2), 91–100 (2016)
3. Purba, M.: Dinamika perairan selatan Pulau Jawa-Pulau Sumbawa saat Muson Tenggara. Torani **17**(2), 140–150 (2007)
4. Purba, N.P., Pranowo, W.S., Faizal, I., Adiwira, H.: Temperature-Salinity stratification in the Eastern Indian Ocean using argo float. In: IOP Conference Series: Earth and Environmental Science, June, vol. 162(1), pp. 012010. IOP Publishing (2018)
5. Al Tanto, T., Hartanto, T., Purba, M., Pranowo, W.S.: Karakteristik Massa Air di Perairan Barat Daya Pulau Sumba, Provinsi Nusa Tenggara Timur. Jurnal Kelautan Nasional **15**(1), 23–36 (2020)
6. Siregar, S.N., Sari, L.P., Purba, N.P., Pranowo, W.S., Syamsuddin, M.L.: Pertukaran massa air di Laut Jawa terhadap periodisitas monsun dan Arlindo pada tahun 2015. Depik **6**(1), 44–59 (2017)
7. Nofiyanti, K., Kunarso, K., DK.: A. R. D. T. Kajian Kedalaman Mixed Layer Dan Termoklin Kaitannya Dengan Monsun Di Perairan Selatan Pulau Jawa (Doctoral dissertation, Diponegoro University) (2020)
8. Pentury, R., Pietersz, J.H., Tuapattinaja, M.A., Pello, F.S., Huliselan, N.V., Hulopi, M., Tupan, C.I.: Potensi Komunitas Mangrove Pantai Tial Kabupaten Maluku Tengah. TRITON: Jurnal Manajemen Sumberdaya Perairan **16**(2), 68–76 (2020)
9. Simanjuntak, P.P., Safril, A.: Analisa angin zonal dan meridional dalam menentukan awal musim hujan di Kota Jambi. Jurnal Teori dan Aplikasi Fisika **8**(1), 43–50 (2020)
10. Rofiqa, Z., Muliadi, M., Risko, R.: Estimasi Potensi Tenaga Arus Laut Permukaan sebagai Pembangkit Listrik di Perairan Selatan Selat Makassar. Prisma Fisika **6**(3) (2005)
11. Al Ayubi, M.A., Surbakti, H., Mbay, L.O.N.: Identifikasi Massa Air Di Perairan Timur Laut Samudera Hindia (Doctoral dissertation, Sriwijaya University) (2012)

12. Woo, M., Pattiaratchi, C., Feng, M.: Hydrography and water masses in the south-eastern Indian Ocean. SFRME Collaborative Research Project. School of Environmental Systems Engineering, Australia (2006)
13. Kusmanto, E., Siswanto, S.: Analisis Masa Air Dan Estimasi Transport Arus Bawah Ekuator Pada Bujur 90° E Selama Indonesia Prima 2017. Jurnal Meteorologi dan Geofisika **19**(2), 59–69 (2019)
14. Emery, W.J.: Water types and water masses. Ocean circulation. Science 1556–1567 (2003)
15. Santoso, A.D.: Pemantauan hidrografi dan kualitas air di Teluk Hurun Lampung dan Teluk Jakarta. Jurnal Teknologi Lingkungan **6**(3) (2011)
16. Sidabutar, E.A., Sartimbul, A., Handayani, M.: Distribusi suhu, salinitas dan oksigen terlarut terhadap kedalaman di Perairan Teluk Prigi Kabupaten Trenggalek. JFMR (Journal of Fisheries and Marine Research) **3**(1), 46–52 (2019)
17. Kalangi, P.N., Mandagi, A., Masengi, K.W., Luasunaung, A., Pangalila, F.P., Iwata, M.: Sebaran suhu dan salinitas di Teluk Manado. Jurnal Perikanan dan Kelautan Tropis **9**(2), 70–75 (2013)
18. Purwandana, A., Purba, M., Atmadipoera, A.S.: Distribusi Percampuran Turbulen di Perairan Selat Alor (Distribution of Turbulence Mixing in Alor Strait). ILMU KELAUTAN: Indonesian J. Marine Sci. **19**(1), 43–54 (2014)
19. Hutabarat, M.F., Purba, N.P., Astuty, S., Syamsudin, M.L., Kuswardani, A.R.: Variabilitas lapisan termoklin terhadap kenaikan mixed layer depth (MLD) di Selat Makassar. Jurnal Perikanan Kelautan **9**(1) (2018)
20. Prasetya, I.A.: Studi Gelombang Internal di Kawasan Perairan Pulau Weh, Aceh (2020)
21. Wyrtki, K.: Physical oceanography of the Southeast Asian waters (Vol. 2). University of California, Scripps Institution of Oceanography (1961)
22. Sulaiman, A.: Turbulensi Laut Banda. Jakarta, Badan Pengkajian dan Penerapan Teknologi (2000)

The Climate Comfort and Risk Assessment for Tourism in Bali, Indonesia

Kadek Sumaja, I Kadek Mas Satriyabawa, Sindy Maharani, and Weny Anggi Mustika

Abstract Tourism is one of the main economic resources in Bali that is mostly affected by climate condition which tends to change. However, it is still unclear how climate change may affect tourism in tropical archipelagos like Bali Island. Moreover, climate comfort is one of the most important factors that must be considered to increase the number of tourists visiting because it may alter the pattern of tourist visits and preferences. Therefore, this study will analyze the comfort level of the tourism climate using the Holiday Climate Index (HCI) and humidity index (humidex) methods. Moreover, the climate comfort level in observation points is mapped to analyze the comfortable level for each month in Bali. In addition, HCI and humidex values will be combined with the number of foreign visitors to Bali to analyze the relationship between climate comfort indicators and tourist visits. The results of this study indicated that there were six quite comfortable months in the majority area of Bali, while the most comfortable month for traveling in Bali was in August and the least comfortable month was in December and January. The number of tourist arrivals to Bali had a relationship with the HCI and humidex values. The results of this study are expected to assist tourists, the tourism sector, and local governments in making pre-travel decisions based on climatic comfort information and tourism policy to enhance the Bali tourism business.

1 Introduction

Tourism is one of the main economic resources in Bali. However, unfortunately, it is one of the economic sectors that is mostly affected by climate conditions which tends

K. Sumaja (✉) · I. K. M. Satriyabawa · S. Maharani
I Gusti Ngurah Rai Meteorological Office, Indonesian Agency for Meteorology Climatology and Geophysics, Bali, Indonesia
e-mail: kadek.sumaja@bmkg.go.id

W. A. Mustika
Regional III Office, Indonesian Agency for Meteorology Climatology and Geophysics, Bali, Indonesia

A. Basit et al. (eds.), *Proceedings of the International Conference on Radioscience, Equatorial Atmospheric Science and Environment and Humanosphere Science*, Springer Proceedings in Physics 290, https://doi.org/10.1007/978-981-19-9768-6_50

to change. The most of tourist attractions in Bali are related to outdoor and social activities, such as sightseeing, adventure (trekking, rafting, surfing), and cultural tourism [1, 2]. Therefore, the condition of the climate parameters will affect most of the tourist preferences in Bali. However, the potential impact of climate change on tourism in tropical archipelagos like Bali Island is largely unknown.

Climate conditions could affect the number of tourists visits as well as tourists alternatives and it is one of the key elements that must be considered to increase the number of tourists visiting [3, 4]. Moreover, climate comfort is one of the most important factors that must be considered into increasing the number of tourists visiting because it may alter the pattern of tourist visits and tourist preferences [5]. Thus, information about the climate conditions in a tourist destination is considered substantial. However, the abundance of information related to local climate will only confuse travelers. Consequently, simpler information is imperative to represent the climate comfort of a region. Some of the methods are the Holiday Climate Index (HCI) and the humidity index (humidex).

The Holiday Climate Index (HCI) highlights the significance of climate as a factor in sustaining the competitiveness of coastal and urban destinations. HCI is affected by various climate parameters such as maximum and average temperature, relative humidity, cloud cover, precipitation, and wind speed. Also, it measures how hot we feel by combining the effects of perceived warm temperatures and humidity. HCI has a strong influence on tourist visits especially international tourists, because they need to adapt to climate conditions in destination area [6]. Several studies related to climate comfort that had been carried out in Bali and other regions of Indonesia consist of the Tourism Climate Index (TCI), Temperature Humidity Index (THI), and Holiday Climate Index (HCI). However, this study only used a few observation points and did not display monthly information neither numerically nor spatially, so it was less representative. The information on these two convenience indexes would aid tourists and related stakeholders to design their trips and to develop tourist areas. Thus, this study aims to analyze and map the comfort level of the tourism climate using the Holiday Climate Index (HCI) and humidity index (humidex). Moreover, finding the most comfortable month and the relationship between climate comfort indicators and tourist visits.

2 Data and Methods

Data of this research consists of BMKG's land observation data and ERA5 reanalysis data (https://cds.climate.copernicus.eu/) from 2001 to 2020 in 24 observation points (Fig. 1). This data comprises maximum and average temperature (°C), relative humidity (%), cloud cover (%), precipitation (mm/day), and wind speed (km/h). Furthermore, the number of tourists arriving in Bali is obtained from Bali Statistical Agency (https://bali.bps.go.id) from 2008 to 2019. The latest data was from 2019 because this research would like to analyze tourist visits in normal occurrences before the COVID-19 pandemic [8]. Climate parameters will be calculated and rated according to the following formula and table [7, 9].

Firstly, thermal comfort (TC) was calculated using (1). Then TC as well as other climate parameters were rated according to Table 1. Furthermore, HCI urban and beach were calculated using the rated value. HCI beach comprised of observation points categorized as coastal with an elevation below 10 m, and the other observation points were calculated using HCI urban as shown in (2) and (3). Humidex was calculated using (4). Then, the HCI and humidex values were converted into HCI criteria and humidex criteria according to Table 2. Finally, those criteria were plotted into monthly climate comfort maps using Q-GIS software.

Table 1 Parameter rating

Rating	Thermal comfort (°C)	Precipitation (mm/day)	Wind speed (km/hr)	Cloud cover (%)
10	23–25	0,00	1–9	11–20
9	20–22 or 26	< 3,00	10–19	1–10 or 21–30
8	27–28	3,00–5,99	0 or 20–29	0 or 31–40
7	18–19 or 29–30	–	–	41–50
6	15–17 or 31–32	–	30–39	51–60
5	11–14 or 33–34	6,00–8,99	–	61–70
4	7–10 or 35–36	–	–	71–80
3	0–6	–	40–49	81–90
2	37–39 or (−1)–(−5)	9,00–12,00	–	91–99
1	≤ -6	–	–	100
0	≥ 39	> 12,00	50–57	–
-1	–	> 25,00	–	–
-10	–	–	> 70	–

Table 2 HCI and humidex

Index	HCI	Index	Humidex
0–19	Dangerous	< 20	No discomfort
20–39	Unacceptable	20–29	Little discomfort
40–49	Marginal	30–39	Some discomfort
50–59	Acceptable	40–45	Great discomfort
60–69	Good	> 45	Dangerous
70–79	Very good	–	–
80–89	Excellent	–	–
90–100	Ideal	–	–

Fig. 1 Research location

$$TC = 0.8T + \frac{RH + T_x}{500} \quad (1)$$

$$\text{HCI Urban} = 2(TC) + 4(A) + 3(P) + (W) \quad (2)$$

$$\text{HCI Beach} = 4(TC) + 2(A) + 3(P) + (W) \quad (3)$$

$$\text{Humidex} = \text{T} + \frac{5}{9}*\left(\left(6.112*10^{\left(\frac{7.5*T}{237.7+T}\right)}*\frac{RH}{100}\right) - 10\right) \quad (4)$$

TC is thermal comfort (°C), Tx is the maximum temperature (°C), T is the average temperature (°C), RH is relative humidity (%), A is cloud cover (%), P is precipitation (mm/day), and W is wind speed (km/h).

The association between the HCI or humidex and the number of visitors would be measured using the correlation coefficient (r) as shown in (5), where S_i is the HCI or humidex, G_i is the tourist visit, σ_S and σ_G are their standard deviations, and n is the number of data pairs.

$$r = \frac{\sum_{i=1}^{n}(S_i - \overline{S})\left(G_i - \overline{G}\right)}{(n = 1)\sigma_s\sigma_G} \quad (5)$$

3 Result and Analysis

3.1 Climate Parameters in Bali

The results showed the variation of each climate parameter in Bali (Fig. 2). The average annual temperature ranged from 24.2 to 26.6 °C and the average maximum temperature in Bali ranged from 27.6 to 30.2 °C. The annual average temperature and the highest maximum temperature occurred in November and hit the lowest in August. Furthermore, the lowest and highest wind speed was blown in March and August, respectively, from 4.0 to 10.8 km/hr. The average daily rainfall in Bali was distributed from 1.2 to 11.3 mm/day, where January and February were the wettest months and August was the driest month in Bali. Humidity was classified from 80 to 86%, the highest was in January and the lowest was in September (Fig. 2).

The average cloud cover in Bali varied between 54 and 88%, where the most of clouds covered the sky in December, January, and February. Therefore, December, January, and February were the most humid months, with the highest rainfall, the hottest, and the most cloud cover, whereas in July and August, the opposite condition occurred, with fewer clouds, the lowest temperature, lower humidity, less rainfall, and stronger wind speed. This result was consistent with earlier studies that showed temperature, rain, and cloud cover were the main climate factors that affect how comfortable the climate was at the destination [6, 7, 10–12].

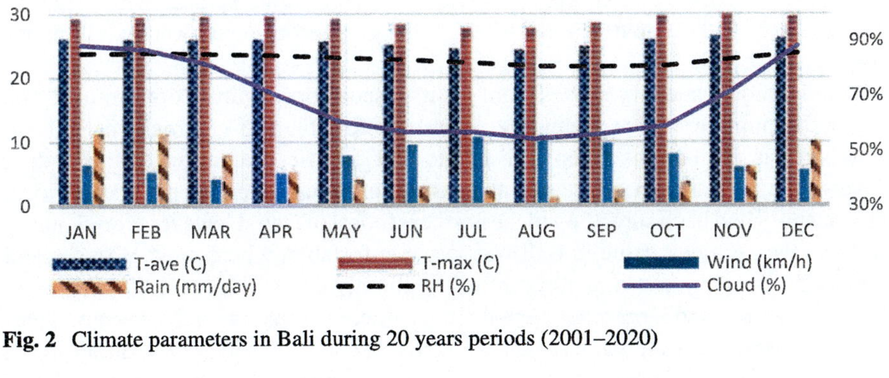

Fig. 2 Climate parameters in Bali during 20 years periods (2001–2020)

3.2 Holiday Climate Index (HCI) and Humidity Index (Humidex)

The Holiday Climate Index (HCI) in Bali differed from an 'unacceptable' to an 'excellent' level as shown in Fig. 3. May to September was the month with the highest HCI value in Bali, where the range of HCI values in all regions was 'very good' and 'excellent'. Furthermore, the comfort level increased in July and August, when the most areas of Bali were in the 'excellent' category. Then, in June and September, the Bali area was dominated by the 'very good' HCI category, with a small percentage of 'excellent' and 'good' ratings appearing in May. The period with the next highest HCI category was October, November, and April. In October, most areas of Bali were in the very 'good' category, and there were still 'excellent' ones. Still, in some areas, they began to become 'marginal' and 'acceptable' categories in central Bali and the southwestern coast of Bali. Then in November, most of region in Bali was categorized as HCI 'good', with some 'marginal' and 'acceptable' and only slightly 'excellent' in the highlands area. In April, the Bali region was divided into HCI 'good' and 'very good' ratings with almost equal areas.

December to March was the months with the lowest HCI values. In March, Bali province was dominated by 'good' and 'acceptable' criteria, followed by a few 'marginal' areas, and began to appear 'unacceptable' on the north coast of Bali. Then, December, January, and February were the most uncomfortable months in Bali because most areas of Bali were ranked as 'marginal'. In February, almost a third of the Bali area was categorized as marginal. Although in December, there were several areas at the tip of West Bali and Nusa Penida Island which was measured as 'good' category, the 'marginal' category areas also began to expand. Even, the 'unacceptable' class started to appear in the southern Bali region. In January, the marginal areas were almost the same as in December. Still, the 'unacceptable' areas had expanded and covered the main tourist areas on the island of Bali, such as Denpasar, Kuta, Jimbaran, and Nusa Dua. This outcome was following earlier studies, which showed that these monthly climate comfort levels changed and peaked throughout the dry season [6, 10, 13].

The humidex category ranges from great 'discomfort' to little 'discomfort'. The lower the humidex value means the more comfortable the environment. July to October was the months when Bali Island had a 'little discomfort' category that covered the highlands in northeast Bali. During July, August, and September periods, this category had been spreading to the northeast coast of Bali. However, from January to June, the humidex value in Bali was less comfortable, where most of this island was in the category of 'some discomfort', with only a few areas of a 'little discomfort'. November and December were the most uncomfortable months, because apart from most areas of Bali being ranked as 'some discomfort' and a 'little discomfort', several 'discomfort' areas had emerged. 'Great discomfort' areas were widely spread on the southern and northern coasts of the western part of Bali.

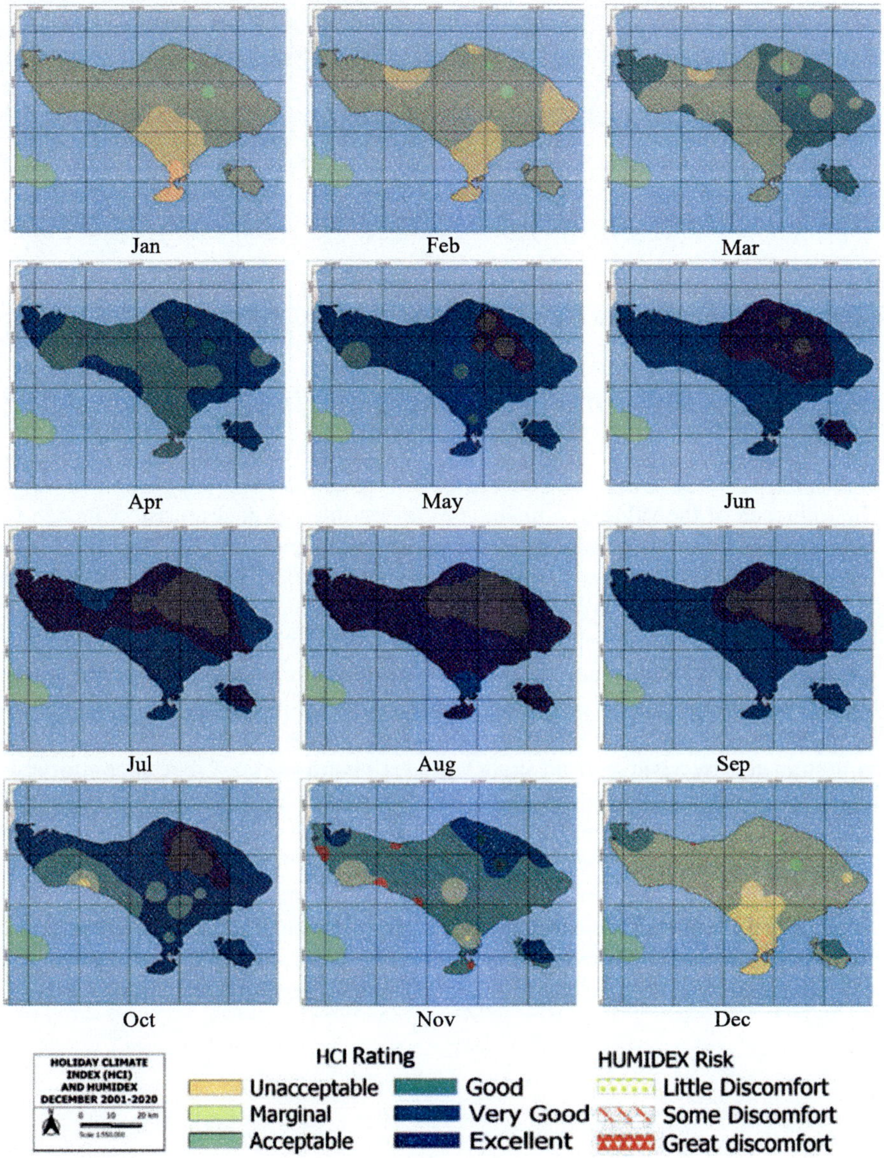

Fig. 3 Holiday climate index (HCI) and humidity index (humidex)

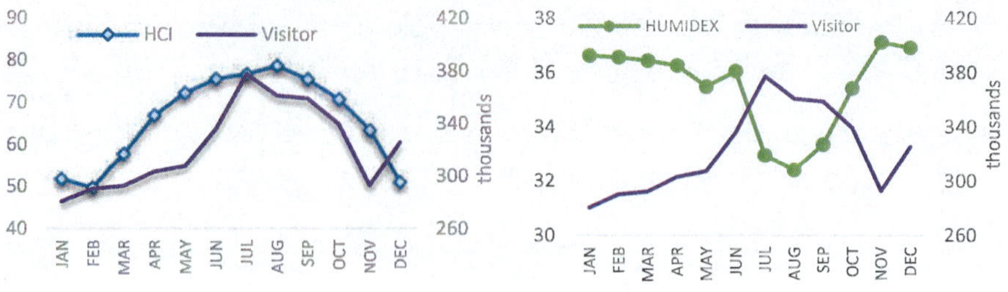

(a) (b)

Fig. 4 Relationship between HCI and humidex with international visitors

3.3 Relation Between HCI, Humidex, and International Visitor

The influence of the value of climate comfort on international tourist visits can be seen from the correlation value between the two elements (Fig. 4). The correlation between HCI and Humidex with international visitors is 0.76 and −0.86. It shows that HCI had a very strong positive relationship and humidex had a very strong negative relationship with international tourist arrivals. Thus, when the HCI value is low, tourist visits were also low and contrarily. As regards humidex, the number of tourist visits increased as the value of the humidex decreased. Since the higher the humidex value, the more uncomfortable the conditions in the area. Thus, foreign tourists who are dominated by visitors from high latitudes country would feel uncomfortable [14]. This result was relevant to the previous research that states climate comfort information such as HCI and humidex was one of the crucial factors that should take into account to boost the number of visitors [7, 9, 15]. For future consideration, this result will help the tourism sector and local governments in making pre-travel decisions based on climatic comfort information and tourism policy to enhance the Bali tourism business. Moreover, this can be the basis for further research on the mitigation plan to overcome extreme weather or climate impact on tourism.

4 Conclusion

Based on the results and analysis, the HCI rating in Bali was 'good' to 'excellent' from June to September, it was 'unacceptable' to 'good' from December to March and the other months indicate a transitional HCI condition. Moreover, the humidex category of 'little' to 'some discomfort' occurred for most months, except November and December which have 'great discomfort' classes. The HCI rating and humidex showed that the most comfortable months for traveling in Bali were June, July, and August. The number of tourist arrivals to Bali was directly proportional to HCI and inversely proportional to humidex, where the correlation value of these two indices was very strong.

References

1. Bali, D.P.P.: Daya Tarik Wisata Candi. Dinas Pariwisata Provinsi Bali (2018). https://disparda.baliprov.go.id/ Accessed 27 Sep 2022
2. Bali, D.P.P.: Analisa Pasar dan Indeks Kepuasan Wisatawan Mancanegara. Dinas Pariwisata Provinsi Bali, Denpasar (2019)
3. Susanto, J., Zheng, X., Liu, Y., Wang, C.: The impacts of climate variables and climate-related extreme events on island country's tourism: evidence from Indonesia. J. Clean. Prod. **276**, 124204 (2020). https://doi.org/10.1016/j.jclepro.2020.124204
4. Craig, C.A., Feng, S.: A temporal and spatial analysis of climate change, weather events, and tourism businesses. Tour. Manag. **67**, 351–361 (2018). https://doi.org/10.1016/j.tourman.2018.02.013
5. Scott, D., Rutty, M., Amelung, B., Tang, M.: An inter-comparison of the holiday climate index (HCI) and the tourism climate index (TCI) in Europe. Atmosphere (Basel) **7**(6), 23–28 (2016). https://doi.org/10.3390/atmos7060080
6. Dini, L.R.Z., Sobirin.:Tingkat Kenyamanan Iklim Di Pulau Bali Berdasarkan tourism climate index. In: 8th Industrial Research Work National Seminar, pp. 678–684. (2017)
7. Hasanah, N.A.I., Maryetnowati, D., Edelweis, F.N., Indriyani, F., Nugrahayu, Q.: The climate comfort assessment for tourism purposes in Borobudur Temple Indonesia. Heliyon **6**(12), e05828 (2020). https://doi.org/10.1016/j.heliyon.2020.e05828
8. Provinsi Bali, B.P.S.: Banyaknya Wisatawan Mancanegara Bulanan ke Bali Menurut Pintu Masuk (Orang) (2022). https://bali.bps.go.id Accessed 25 Sep 2022
9. Demiroglu, O.C., Saygili-Araci, F.S., Pacal, A., Hall, C.M., Kurnaz, M.L.: Future holiday climate index (HCI) performance of urban and beach destinations in the Mediterranean. Atmosphere (Basel) **11**(9), 1–30 (2020). https://doi.org/10.3390/ATMOS11090911
10. Yazdanpanah, H., Barghi, H., Esmaili, A.: Effect of climate change impact on tourism: a study on climate comfort of Zayandehroud River route from 2014 to 2039. Tour. Manag. Perspect. **17**, 82–89 (2016). https://doi.org/10.1016/j.tmp.2015.12.002
11. Scott, D., Gössling, S., De Freitas, C.R.: Preferred climates for tourism: case studies from Canada, New Zealand and Sweden. Clim. Res. **38**(1), 61–73 (2008). https://doi.org/10.3354/cr00774
12. Amelung, B., Nicholls, S.: Implications of climate change for tourism in Australia. Tour. Manag. **41**, 228–244 (2014). https://doi.org/10.1016/j.tourman.2013.10.002
13. Scott, D., Mcboyle, G.: Using a 'tourism climate index'to examine the implications of climate change for climate as a tourism resource. Tourism and Recreation. International Society of Biometeorolog, Porto Carras, pp. 69–88. (2002)
14. Provinsi Bali, B.P.S.: Banyaknya Wisatawan Mancanegara yang Datang ke Bali Menurut Kebangsaan, 2014–2020 (2022). https://bali.bps.go.id Accessed 25 Sept 2022
15. He, P., Qiu, Y., Wang, Y.D., Cobanoglu, C., Ciftci, O., Liu, Z.: Loss of profit in the hotel industry of the United States due to climate change. Environ. Res. Lett. **14**(8) (2019). https://doi.org/10.1088/1748-9326/ab2dce

Wind Effect on Spectral Ratio Analyses of Acoustic Waves Excited by Volcanic Explosion: Preliminary Result of Application at Sakurajima Volcano, Japan

Mohammad Hasib, Takeshi Nishimura, Albertus Sulaiman, Titi Anggono, Syuhada, Febty Febriani, Cinantya Nirmala Dewi, Aditya Dwi Prasetio, and Trinugroho

Abstract We investigate the wind effect on spectral ratio analyses of acoustic waves (infrasound) at Sakurajima volcano, Japan. Our preliminary result shows that spectral amplitude ratios of acoustic waves are different for different stations. This result implies that the propagation and site effects are not effectively removed. To clarify why the spectral ratio method does not work, we suppose that temporal variations of the wind in the atmosphere around the crater and volcano change the propagation paths of acoustic waves. We use assimilation wind data from European Centre for Medium-Range Weather Forecasts (ECMWF) to quantify the wind speed. Time resolution is 6 h, and spatial resolution is $0.125^\circ \times 0.125^\circ$. The acoustic signals are recorded by Japan Meteorological Agency (JMA). In the present study, we examine the spectral ratio of acoustic waves at Sakurajima volcano from January 2012 to December 2013 without and within data selection during low wind speed (< 3.3 m/s). Our results show that the spectral ratios of acoustic waves at different stations are similar when we select acoustic data during low wind speeds (< 3.3 m/s). This result indicates that the propagation paths of acoustic waves change due to wind conditions around the volcano.

M. Hasib (✉) · T. Anggono · Syuhada · F. Febriani · C. N. Dewi · A. D. Prasetio · Trinugroho
Research Center for Geological Disasters, National Agency of Research and Innovation (BRIN), Bandung, Indonesia
e-mail: moha050@brin.go.id

T. Nishimura
Department of Geophysics, Tohoku University, Sendai, Japan

A. Sulaiman
Research Center for Climate and Atmosphere, BRIN, Bandung, Indonesia

A. Basit et al. (eds.), *Proceedings of the International Conference on Radioscience, Equatorial Atmospheric Science and Environment and Humanosphere Science*, Springer Proceedings in Physics 290, https://doi.org/10.1007/978-981-19-9768-6_51

1 Introduction

Sakurajima volcano is one of the most active volcanoes in Japan and it is located on the island of Kyushu in southern Japan at the position of 31.58 N, 130.65 E. Sakurajima volcano generates explosion earthquakes every year with Vulcanian-type eruption [1]. Since the volcanic explosion occurs, the volcano effectively evokes not only seismic waves, but also the acoustic waves. In this study, we are only focused on the investigation of acoustic waves excited by volcanic explosion. Sakurajima volcano frequently exploded from January 2012 to December 2013 (see Fig. 1). The availability of the acoustic signal is satisfied.

Application of the spectral ratio method on the volcanoes was first applied by Hasib et al. in [2]. They used the spectral ratio method on the seismic waves of explosion earthquakes and explained that the process can retrieve the source information of explosion earthquakes. The spectral ratio method retrieves the source information by eliminating the geometrical spreading, instrument response, and site effect [3, 4]. In this study, we apply the spectral ratio to acoustic waves. We examine the spectral ratio of the acoustic waves to clarify the source information of acoustic waves excited by the volcanic explosions. When a volcanic explosion occurs, the seismic and acoustic signals are evoked efficiently. Hence, the concept of spectral ratio on acoustic waves may be similar to the seismic waves. The difference between seismic and acoustic signals of the volcanic explosion is in the radiation medium.

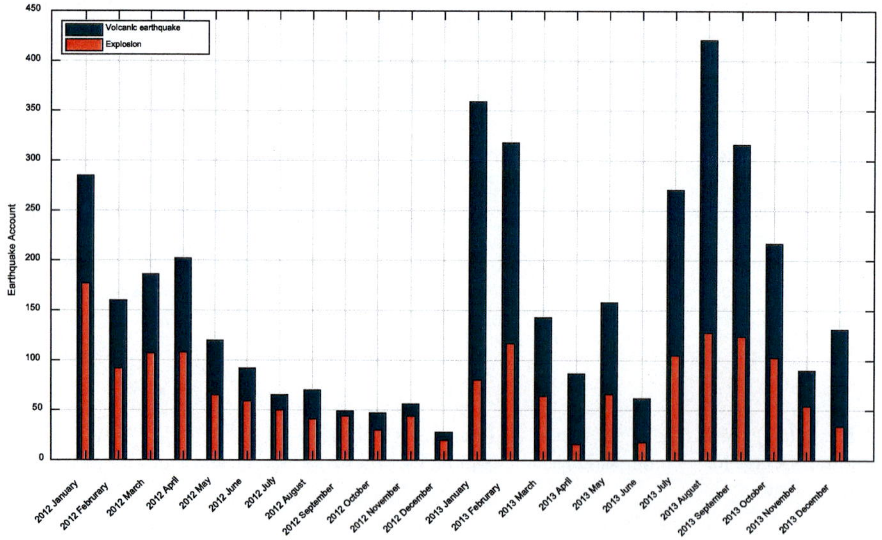

Fig. 1 Seismic activity at Sakurajima volcano from January 2012 to December 2013 recorded by Japan Meteorological Agency (JMA). The dark green bar indicates the volcanic earthquake. The red bar indicates the volcanic explosion

Acoustic wave radiates via the atmosphere around the volcano. Since the investigation of the spectral ratio of acoustic waves in our analysis is about two years. We should carefully consider the wind condition around the Sakurajima volcano. The contamination of atmospheric condition may be affected the acoustic waves [5–7]. Therefore, we examine the results of the spectral ratio of acoustic waves within and without wind condition selection to clarify the wind effect on acoustic wave propagation. We marked that we only clarified the wind effect on the spectral ratio of acoustic waves based on observation data in this study.

2 Method

We use the acoustic station is operated by Japan Meteorological Agency (JMA). The acoustic waves are measured by infrasonic microphones (ACO Ltd, Type 7144, 0.1–100 Hz). All the acoustic station is distributed at about 2.91–4.97 km away from the active crater (see Fig. 2). We apply the spectral ratio method in acoustic signal to understand the source information of volcanic explosion in more detail. However, the important remark is the applicability of the spectral ratio method on acoustic waves evoked by volcanic explosion since this condition is the first application on acoustic waves. First, we apply the spectral ratio without any data selection. We analyse 1164 acoustic signals excited by the volcanic explosion at Sakurajima volcano that occurred from January 2012 to December 2013. Second, we classify the acoustic event based on their amplitude into four classes (Class I, II, III, and IV) to examine the spectral ratio (see Fig. 3a). The smallest class (Class I) is the reference in our analysis. Third, we calculate the average of spectra in each class. Fourth, we take the ratio Class IV, III, and II to Class I. At last, we calculate the spectral ratio for different stations.

The propagates of acoustic signal evoked by volcanic explosion may be disturbed by wind conditions around the crater (see Fig. 4). Hence, we calculate the spectral ratio into two states which are the calculation of acoustic signal without and within wind selection to understand the effect of wind on spectral ratio. For the analysis of acoustic signal during low wind speed (within wind selection), we select the acoustic event by assimilation wind data from European Centre for Medium-Range Weather Forecasts (ECMWF) to understand the temporal variation of wind conditions around the Sakurajima volcano during January 2012 to December 2013 [8, 9]. The assimilated wind data is obtained from reanalysis of the global atmosphere since 1979 with the time resolution of 6 h, and spatial resolution of $0.125^{\circ} \times 0.125^{\circ}$. The procedure of assimilated wind data extraction from the database can be explained as follows: first, we select and extract the wind data at different elevations corresponding to atmospheric pressure level (see Fig. 5).

Second, we select the area calculation (around Sakurajima volcano) at a position longitude of 130, 62 E-130, 74 E, and latitude of 31, 54 N-31, 62 N, then extract the wind data. Third, we select the wind data starting from 31 m above

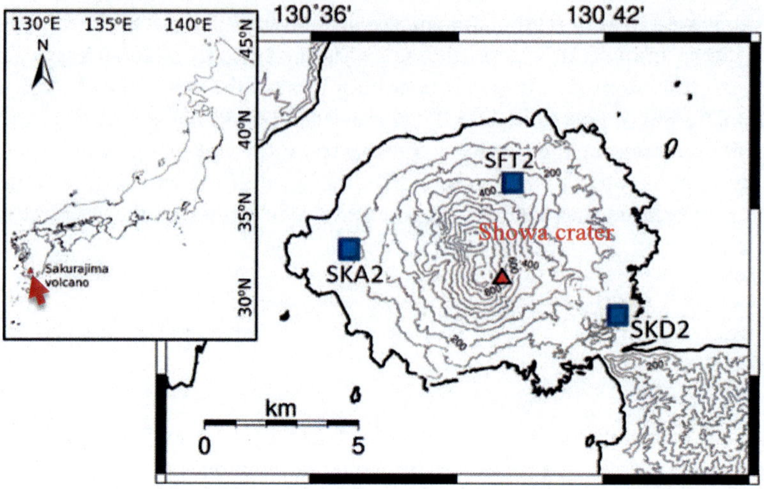

Fig. 2 Distribution of acoustic stations at Sakurajima volcano. The red triangle represents the active crater (Showa crater). The blue squares represent the acoustic station. The arrow indicates the Sakurajima Volcano location

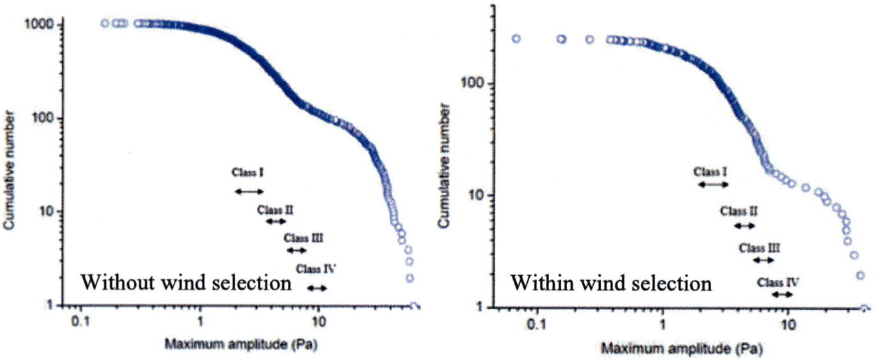

Fig. 3 Acoustic event classification in our analysis from January 2012 to December 2013, **a** without wind selection, and **b** within wind selection (low wind speed). Double-sided arrow indicates the range of class in our analysis

sea level (minimum elevation) up to 800 m above sea level (Showa summit elevation). One example of the result of simulated wind extraction can be seen in Fig. 6. We characterize the wind condition based on the Beaufort wind scale [11].

At last, we select the acoustic events that occur during the low wind speed condition is characterized as a light breeze, in which wind speed is less than 3.3 m/s (<3.3 m/s). However, due to acoustic data selection, we examine 102 acoustic signals in conditions of low wind speeds from January 2012 to December 2013. An example of the acoustic waveform evoked by a volcanic explosion during low wind speed can

Fig. 4 Schematic figure of acoustic wave propagation evoked by a volcanic explosion around the volcano (Modified from Shimatami et al. in [10])

be seen in Fig. 7. We do the same procedure as spectral ratio calculation without wind selection for spectral ratio calculation during low wind speed. The classification of the acoustic events during low wind speed for spectral ratio calculation is shown in Fig. 3b.

3 Result

The result of the spectral ratio calculation of acoustic signal without wind selection at the different stations is shown in Fig. 8. The amplitude ratios of acoustic waves at different stations are different, especially at the SFT2 station. The characteristics are observed in other classes. This result implies that the hypothesis used in the spectral ratio is not satisfied. The result means the spectral ratio method is not well applicable for acoustic waves evoked by the volcanic explosions when we do not consider the wind condition in atmosphere around the volcano.

On the other hand, the result of the spectral ratio calculation of acoustic waves at different stations during low wind speeds is well described in Fig. 9. The amplitude ratio is similar to each other for different stations. The characteristics are observed for different classes. The result is identical to the study spectral ratio on the seismic waves by Hasib et al. in [2]. The results clarify that the spectral ratio method is applicable for the acoustic signals when we select the acoustic data during low wind speed (<3.3 m/s). This means the wind condition around the volcano affects the acoustic signal propagation. The result reflects that the propagation of acoustic path is more sensitive to the wind condition that is affected the recording of acoustic waveform. We discuss the consequence of wind conditions to the observed acoustic waveform in the discussion section.

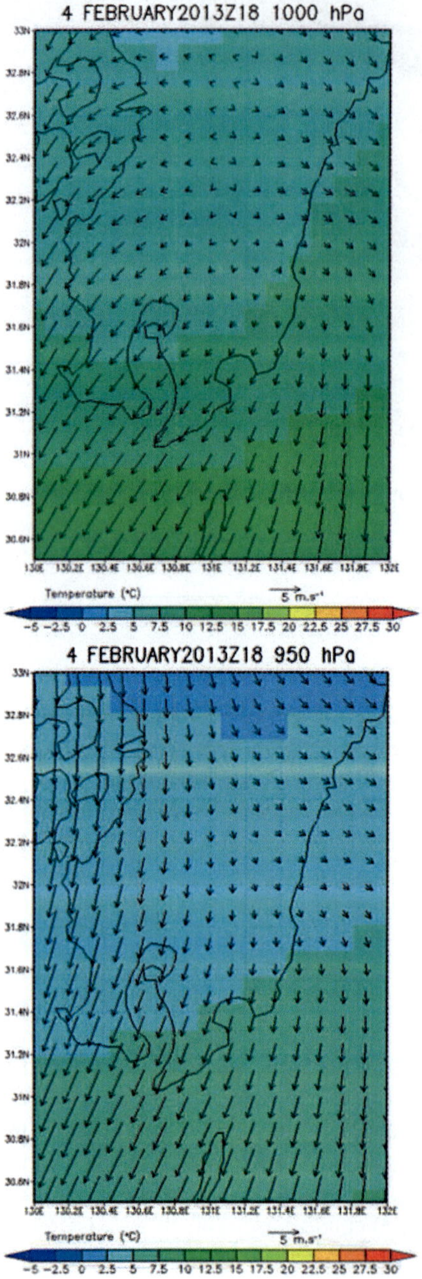

Fig. 5 Snapshots of wind data simulation at Kyushu Island on 04 February 2013, at a range of 18.00–24.00 JST for four elevations

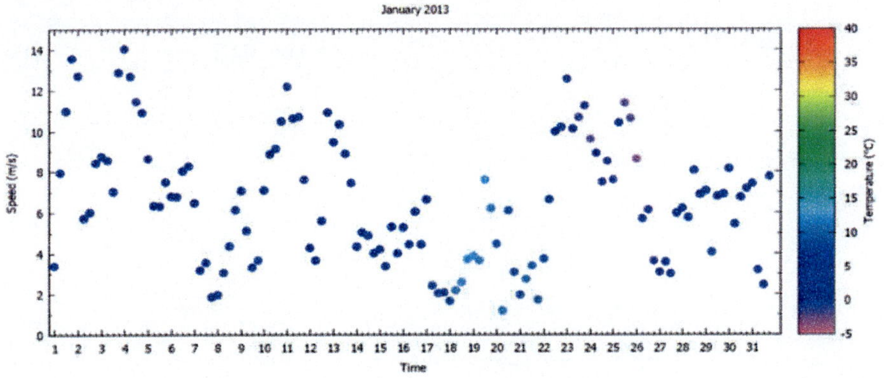

Fig. 6 Example of temporal wind speed around Sakurajima volcano during January 2013. Colours correspond to the temperature of the atmosphere

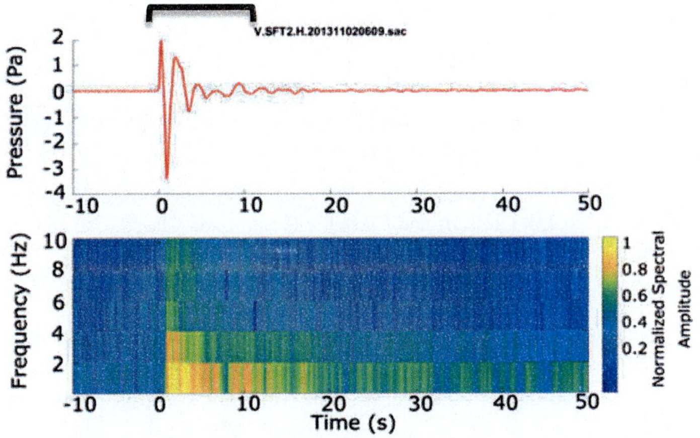

Fig. 7 Example of acoustic signal at Sakurajima volcano during low wind speed on 02 November 2013 at station SFT2. Colours correspond to the temperature of the atmosphere. Black upper brackets indicate the time window for spectral ratio calculation

4 Discussion

We observed that the spectral ratio method does not work when we do not select the acoustic data due to wind conditions. The spectral ratio result for different stations should be similar because the source of the volcanic explosion is the same as that recorded for all stations. This condition is well explained in seismic studies by the previous study [2, 12–14]. We clarify the problem by investigating on observed waveform of the acoustic waves to understand wind's effect on the acoustic signals. Figure 10 shows an example of acoustic signals generated during high and low wind speeds at the Sakurajima volcano. We see that the observed acoustic signal during

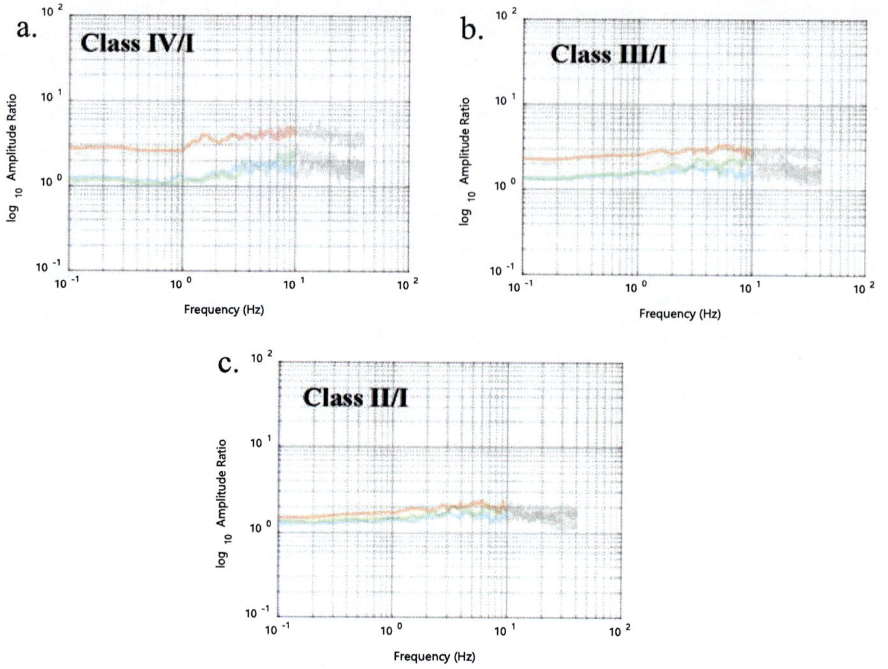

Fig. 8 Observed spectral ratio of the acoustic wave without wind selection. Spectral ratio calculation **a** Class IV and Class I, **b** Class III and Class I, **c** Class II and Class I. Colours correspond to the different acoustic stations. The grey colour indicates the data that are not analysed in this study

high wind speed conditions around the volcano is more scattered and a longer time occurrence compared to the acoustic signal that is generated during low wind speed (<3.3 m/s). The characteristic is observed for different stations. This suggests that the wind affects the record of the acoustic signal. The wind condition around the volcano effectively changes the propagation of acoustic wave excited by a volcanic explosion. Our interpretation based on observation data is supported by Lacanna et al. in [15] using simulating acoustic wave propagation using a two-dimensional finite difference time domain (2D FDTD) method to understand the wind effect on the acoustic waves. They acted the acoustic waves propagating for about < 60 km from Sakurajima volcano to Kirishima volcano within wind conditions and no-wind conditions. The simulation of the acoustic waves without wind parameters shows no disturbance of acoustic wave propagation. However, the simulation of acoustic wave within wind parameter clearly shows the perturbance of acoustic wave propagation when considering wind parameters in their study. We suggest that selecting the acoustic data is necessary and crucial for applying the spectral ratio analysis on acoustic waves evoked by volcanic explosion.

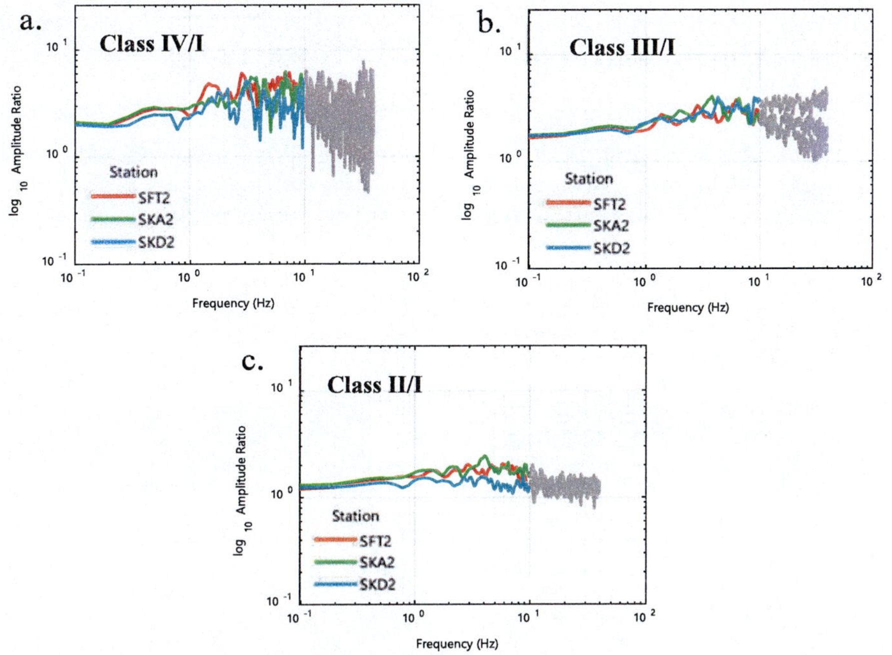

Fig. 9 Observed spectral ratio of the acoustic wave during low wind speed. Spectral ratio calculation **a** Class IV and Class I, **b** Class III and Class I, **c** Class II and Class I. Colours correspond to the different acoustic stations. The grey colour indicates the data that are not analysed in this study

5 Conclusion

We have investigated the spectral ratio of the acoustic waves from January 2012 to December 2013 without and within considering wind conditions. We observe that the spectral ratio is different for the different stations when we do not select the acoustic data. This means that the spectral ratio method on acoustic signals excited by volcanic explosion does not work when we do not determine the acoustic data by considering wind conditions around the volcano. On the other hand, the result of the spectral ratio is similar for the different stations by considering wind conditions, which means that the spectral ratio is applicable to the acoustic signals when we select the acoustic signal during low wind speed (<3.3 m/s). Our result shows that propagation paths of acoustic waves are affected by wind conditions around the volcano. Our result suggests that the selection of acoustic signal evoked by a volcanic explosion during low wind speed is an important process to retrieve the source information of volcanic explosion by taking spectral ratio calculation.

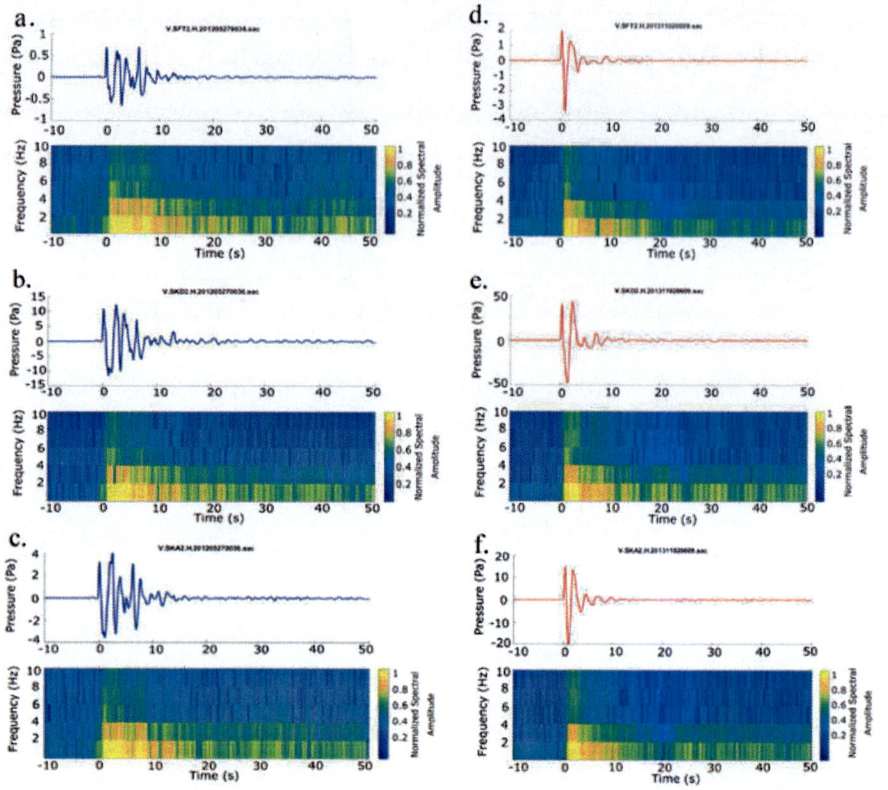

Fig. 10 Examples of the observed acoustic signal during high wind speed on 27 May 2012, recorded at **a** SFT2, **b** SKD2, and **c** SKA2 stations. Examples of the observed acoustic signals during low wind speed on 02 November 2013 were recorded at **d** SFT2, **e** SKD2, and **f** SKA2 stations

Acknowledgements We thank Japan Meteorological Agency (JMA) for providing the acoustic data. We thank European Centre for Medium-Range Weather Forecasts (ECMWF) for providing the wind data. This study was supported by National Research and Innovation Agency (BRIN) No. SP DIPA-124.01.1.690501/2022.

References

1. Iguchi, M., Takeshi, T., Yusaku, O., Sadato, U., Shigeru, N.: Characteristics of volcanic activity at sakurajima volcano's showa crater during the period 2006 to 2011. Bull. Volcanol. Soc. Jpn **58**, 115–135 (2013)
2. Hasib, M., Nishimura, T, Nakahara, H.: Spectral ratio analyses of explosion earthquakes at Sakurajima volcano, Japan. Volcanol. Geothermal Res. (2019)
3. Mayeda, K., Malagnini, L., Walter, W.R.: A new spectral ratio method using narrow band coda envelopes: evidence for non-self-similarity in the Hector Mine sequence. Geophys. Res. Lett. **34**, L11303 (2007)

4. Mayeda, K., Malagnini, L.: Source radiation invariant property of local and near regional shear wave coda: Application to source scaling for the Mw 5.9 Wells, Nevada sequence. Geophys. Res. Lett. **37**(7) (2010)
5. Arrowsmith, S.J., Hedlin, M.A.H.: Observations of infrasound from surf in southern California. Geophys. Res. Lett. **32**, L09810 (2005)
6. Fee, D., Garcés, M.: Infrasonic tremor in the diffraction zone. Geophys. Res. Lett. **34**, L16826 (2007)
7. Fee, D., Matoza, R.S.: An overview of volcano infrasound: from Hawaiian to plinian, local to global. J. Volcanol. Geotherm. Res. **249**, 123–139 (2013)
8. Bao, X., Zhang, F.: Evaluation of NCEP–CFSR, NCEP–NCAR, ERA-Interim, and ERA-40 reanalysis datasets against independent sounding observations over the Tibetan Plateau. J. Clim. **26**(1), 206–214 (2013)
9. Song, L., Liu, Z., Wang, F.: Comparison of wind data from ERA-Interim and buoys in the Yellow and East China Seas. Chin. J. Oceanol. Limnol. **33**(1), 282–288 (2015)
10. Shimatami, J., Takahashi, H., Ichihara, M., Takahata, T., Shimayoma, I.: Monitoring volcanic activity with high sensitivity infrasound sensor using a piezoresistive cantilever. IEEE (2019)
11. Mather, J.R.: Beaufort wind scale. In: Oliver, J.E. (ed) Encyclopedia of World Climatology. Encyclopedia of Earth Sciences Series. Springer, Dordrecht (2005)
12. Izutani, Y., Kanamori, H.: Scale-dependence of seismic energy-to-moment ratio for strike-slip earthquakes in Japan. Geophys. Res. Lett. **28**(20), 4007–4010 (2001)
13. Izutani, Y.: Radiated energy from the mid Niigata, Japan, earthquake of October 23, 2004, and its aftershocks. Geophys. Res. Lett. **32**(21) (2005)
14. Somei, K., Asano, K., Tomotaka, I., Ken, M.: Source scaling of inland crustal earthquake sequences in Japan using the S-wave coda spectral ratio method. Pure Appl. Geophys. **171**(2014), 2747–2766 (2014)
15. Lacanna, G., Ichihara, M., Iwakuni, M., Takeo, M., Iguchi, M., Ripepe, M.: Influence of atmospheric structure and topography on infrasonic wave propagation. J. Geophys. Res.: Solid Earth **119**(4), 2988–3005 (2014)

Spatial Assessment Impact of Tsunami Hazard on the Transportation Infrastructure in Phuket South of Thailand

Sirapop Rukhajee, Thanadon Brahmasakha Na Sakolnagara, Jakkrapong Srisuwan, Wasana Putklang, and Muhammad Hanif

Abstract Thailand is the country close to the ring of fire and in 2004 faced a Tsunami disaster which caused a lot of damage to coastal cities, people and property. This study aims to assess the impact of the tsunami on transportation infrastructure in several wave height scenarios in Phuket, Thailand. Method in this research uses spatial analysis using data such as Shuttle Radar Topography Mission (SRTM) and land cover for making cost raster surface roughness and overlay to take information about infrastructure and social facilities. From the modeling of tsunami hazard in Phuket will determine the wave height at the coast for 5, 10 and 15 m. In the scenarios of tsunami wave as 5 m high, around 2,639.1 hectares of beach and urban areas around the beach will be damaged, along with about 168.96 km of highway damage. If the scenarios of tsunami wave as 10 m high, around 10.047,15 hectares of beach and urban areas around the beach will be damaged, along with approximately 885,41 km of highway damage. While the scenarios of a tsunami wave as 15 m high, an area of approximately 15,559.23 hectares and a highway of approximately 1561.12 km will be damaged. There are many transportation infrastructures on the coast that could potentially be affected by the tsunami, such as airports, ship ports and bus stations.

1 Introduction

A tsunami is a disaster; the occurrence of large volume water waves caused by a very high-energy impulse so that water from the ocean can reach the land surface beyond the normal limits of the coastline. A tsunami of such magnitude caused devastation to villages, towns, significant damage to buildings and infrastructure [1]. The southern region of Thailand is a peninsula land area surrounded by water. Southern Thailand is in a position not far from the ring of fire that crosses the Indonesian island of Sumatra. On December 25, 2004, when there was a large earthquake in the ring of fire in the

S. Rukhajee (✉) · T. B. N. Sakolnagara · J. Srisuwan · W. Putklang · M. Hanif
Geoinformatics, Department of Computer Science, College of Computing, Khon Kaen University, Khon Kaen, Thailand
e-mail: sirapop.ru@kkumail.com

A. Basit et al. (eds.), *Proceedings of the International Conference on Radioscience, Equatorial Atmospheric Science and Environment and Humanosphere Science*, Springer Proceedings in Physics 290, https://doi.org/10.1007/978-981-19-9768-6_52

province of Aceh, Indonesia, the earthquake produced a tsunami that affected the Phuket area, and it killed more than 8000 people. At the time of the incident, although the epicenter of the earthquake was far from the territory of Thailand, the tsunami waves reached the coast of Phuket and southern Thailand at a high of Tsunami wave around of 10 to 11 m [2, 8]. The tsunami disaster destroyed a building, house, infrastructure, many foreign tourists died because of the tsunami, because Phuket city has many popular resorts [3].

Recently, around of the ring of fire north of Sumatra Island, earthquakes often occur on a large scale, the last earthquake that occurred in Aceh at a scale of 4.7 m, September 24, 2022. This earthquake occurred in the Meulaboh City area of 6.2 [5, 9]. This condition shows that tectonic activity with quite large earthquakes continues to occur around the earthquake location which caused the Aceh to Thailand Tsunami in 2004. The current condition of Phuket has also recovered and developed around a coastal area as tourism destinations. With the dynamics of socio-economic development, the researcher considers it important to conduct a tsunami disaster assessment for the Phuket area. This study aims to assess the impact of the tsunami on road infrastructure in several wave height scenarios in Phuket, Thailand (Fig. 1).

This research was conducted on one of the islands, namely Phuket, which is located in the coastal area of the southern part of the Kingdom of Thailand. This island was once affected by a tsunami that originated from a tectonic earthquake that triggered the Tsunami in 2004 by a fault in Indonesia.

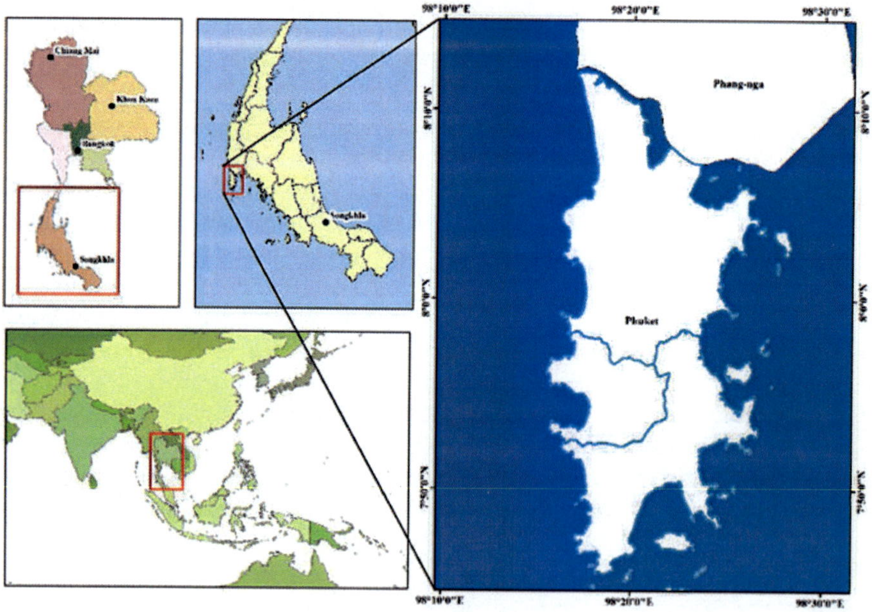

Fig. 1 Study area Phuket Island South of Kingdom of Thailand

2 Material and Method

2.1 Material

For mapping the tsunami inundation area, we used several open access data accessed from various sites, including radar data such as Shuttle Radar Topography Mission (SRTM) from USGS access on the website https://eathexplorer.usgs.gov/. In this research, we collect supporting data such as administrative boundary and land use maps which is obtained from the Center for Geoinformatics and Space Northeast Thailand, Khon Kaen University. Far there, more about road infrastructure data is obtained from the open street map http://openstreetmap.org/. The tool for analysis data is ArcGIS software where provided of the Geoinformatics Technology laboratory of Khon Kaen University.

2.2 Method

2.2.1 Analysis Surface Roughness

In the process of mapping the tsunami inundation, we will go through several process, such as preprocessing which make a land surface roughness and tsunami inundation [10]. The land surface roughness or H_{loss} is cost value for tsunami inundation. Here, surface roughness is the ability of the land surface to reduce the transmission of tsunami wave energy coming from the coast [11, 12]. In the analysis of surface roughness. We use two input data, namely land use and slope. The value of the land use coefficient is based on the ability of each object to reduce or dampen tsunami waves. Meanwhile, the use of slope indicators where the more inclined a slope is, the easier it will be to break up wave energy coming from the sea, while flat slopes do not have a big influence on breaking waves, where land surface roughness (n) is land use, this will be calculated by the mathematical equation H_{loss} (Table 1).

In a report issued by Berryman [4], it has been explained in detail about a mathematical equation. This equation is used to assess the land surface roughness index to the tsunami wave transmission of the tsunami disaster with the following Eq. 1:

$$H_{\text{loss}} = \left(\frac{167n^2}{H_0^{\frac{1}{3}}}\right) + 5\sin S \tag{1}$$

where

H_{loss}: the land surface roughness,
n: the roughness coefficient of various types of land use,
H_0 : the maximum height scenario of tsunami waves on the coastline,

Table 1 Coefficient of land surface roughness

Code	Land cover	Coefficient value (n)
R1	Water body (river, lake)	0.007
R2	Shrub	0.04
R3	Forest, jungle, tree	0.07
R4	Built-up areas, residential	0.035
R5	Bare land, default area, open field	0.015
R6	Agricultural area	0.025
R7	Built up	0.045
R8	Mangroves	0.025
R9	City center	0.01
R10	Fishpond	0.01

Sources [4, 6]

s : the morphometric condition or slope of the slope.

The result of H_{loss} will become grid value for cost distance analysis for tsunami inundation. In this process about *S* where used the data from the data of SRTM which was extract into slope map in raster format.

2.2.2 Tsunami Inundation

The next stage where results of the H_{loss} analysis are obtained, these data are used as costs for the cost distance function to be executed. The cost (land surface roughness) does not depend on the direction of the wave, while the cost in terms of slope depends on the direction of travel [4]. In this case, the cost distance analysis, where (*cd s*) is the coastline as the source of the arrival of the tsunami waves, and the raster cost source (*cr s*) is the raster resource value at a given location. The raster cost (*cr a*) is the cell H_{loss} value at a given location. Indarto [7] describes (2) of cost distance:

$$\text{Cost Distance} = Cds\frac{Crs + ra}{2} \tag{2}$$

where

Cds : is a distance source cost,
Crs : is raster source cost, a raster cost.

In 2004, the tsunami disaster was hit the coast area in Phuket, where the height of tsunami wave around 10 to 11 m. Meanwhile, in this study, we will assess the tsunami inundation with several scenarios wave high such as 5, 10 and 15 m.

2.2.3 Overlay

The last stage is analysis highway damage as the impact of Tsunami inundation. Where we using overlay maps, which are carried out to extract information on the number of infrastructures affected by the tsunami that hit Phuket in a predetermined scenario. With the use of the overlay technique, the length or area of infrastructure affected by the tsunami in the Phuket coastal area can be known.

3 Result and Discussion

Modeling of tsunami inundation in the coastal area of Phuket can show the distribution of large affected areas. Overall, from Fig. 2, we can observe the tsunami inundation area can be seen in three scenarios of tsunami wave heights. The area that had a major impact from the tsunami was the eastern coastal area of Phuket Island due to the relatively flat slope conditions and the river traversed. This condition accelerates the transmission of water movement from the ocean to the land because the surface roughness factor in the form of slopes and land use cannot reduce the height of the tsunami waves.

Thalang sub-district is the sub-district most widely affected by tsunami inundation in various modeling scenarios. It happens because this administrative area is also relatively wider than Kathu and Mueang Phuket sub-districts. Nevertheless, the Thalang sub-district has a relatively flat morphological condition of slopes in the eastern region and has a relatively larger river estuary than other areas. The condition of the river estuary, a body of water, accelerates the transmission of the movement of tsunami waves on land. This phenomenon differs from the conditions in the eastern part of the island of Phuket. This area tends to have a rough morphology in the form of hills and slopes that are relatively more sloping than the eastern part of the island of Phuket, look at Fig. 2.

From the results of the extraction of the tsunami inundation area based on the sub-district area on Phuket Island, we then overlay it with a road map. This result is done to see the details of the tsunami's impact on road infrastructure. Figure 2 reveals detailed information about the tsunami inundation in each sub-district on the island of Phuket and large area inundation from the first scenario; the maximum wave of tsunami as coastal is 5 m high, Thalang district will be affected by 1.238,05 hectares of flooding, Kathu district will be damaged 101,34 hectares and Muang Phuket district will be damaged 1.299,71 hectares. Most of the damaged areas were in Muang Phuket district in the eastern part of the city.

For the second scenario, the maximum wave of tsunami as coastal is 10 m high, Thalang district will be affected by 5.273,12 hectares of flooding, Kathu district will be damaged 428,05 hectares and Muang Phuket district will be damaged 4.345,98 hectares. The eastern area along the beach and the coast from Thalang to Mueang Phuket district will be severely damaged by tsunami because the area near coast is flat, while third scenario, the maximum wave of tsunami as coastal is 15 m high,

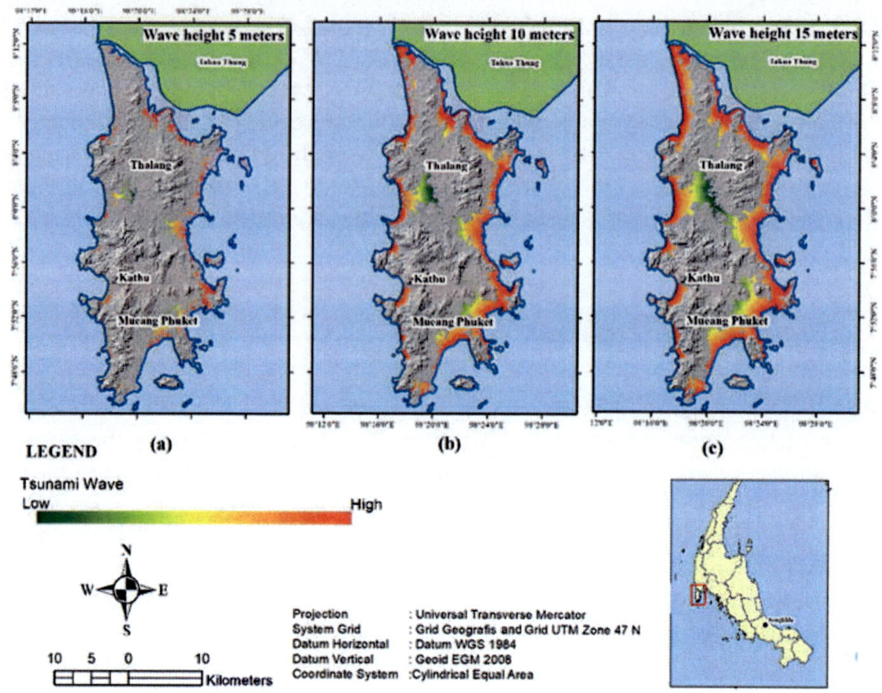

Fig. 2 Following figure is a model of the tsunami simulation results on Phuket Beach, **a** a tsunami with a wave height of 5 m, **b** wave height of 10 m and **c** wave height of 15 m

Thalang district will be affected by 8.567,82 hectares of flood damage, Kathu district will be damaged 678,91 hectares and Muang Phuket district will be damaged 63125 hectares. The entire coastal and beach area in Phuket will be severely damaged. Most of the urban areas in Mueang Phuket and Thalang district will be affected by tsunami.

Figure 3 reveals the distribution of the appearance of the sample areas affected by the tsunami. Changes in the tsunami wave height scenario show significant differences in the area affected and the length of road infrastructure that will be damaged by this disaster. A lot of transportation infrastructure located in coastal areas affected by the tsunami. Areas that experience this impact are mainly located in relatively flat topography and are located on the coast and close to river mouths. However, from Fig. 4, it shows a comparison of the inundation area and length of infrastructure affected by the tsunami disaster. For the highway damage in Phuket, the first tsunami waves of 5 m, Thalang district will have roads damaged by floods for 5063 km, Kathu district will be damaged 1556 km and Mueang Phuket district will be damaged 102,77 km. Most of the highway damage was in the urban area along the coast area in the east of Phuket. The second tsunami waves of 10 m, Thalang district will have roads damaged by floods for 292,56 km, Kathu district will be damaged 7515 km and Mueang Phuket district will be damaged 517,70 km. Most of the highways in urban areas in Thalang district and Muang Phuket district were severely

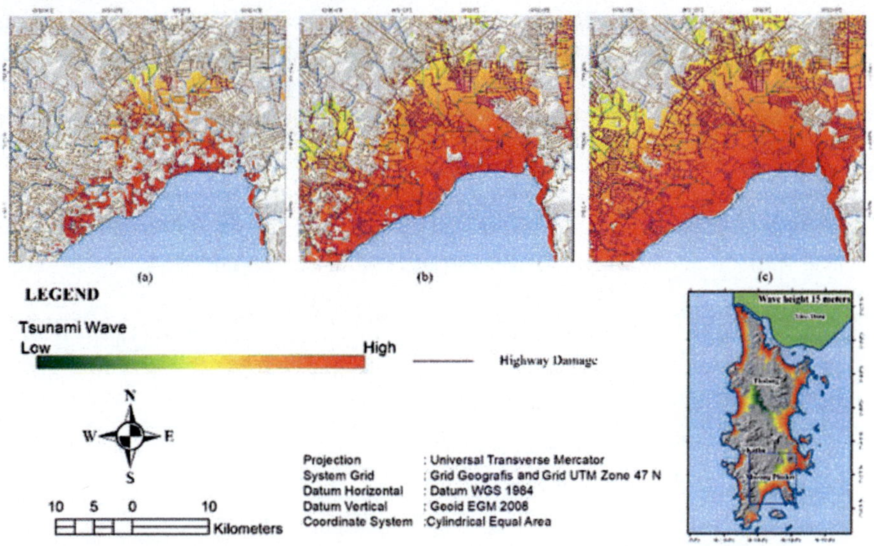

Fig. 3 The appearance of one of the locations that is zoomed in shows the transportation infrastructure that was affected by the tsunami **a** a tsunami with a wave height of 5 m, **b** wave 10 m and **c** wave 15 m

damaged including the area around the beach. While the tsunami waves of 15 m will be more damaging to the highway than the 5 and 10 m wave high, Thalang district will have highway damaged by floods for 612,06 km, Kathu district will be damaged 113,95 km and Mueang Phuket district will be damaged 835,11 km. Damage will be in the urban area of all districts except the highway in the city of Kathu district that has not been damaged.

From this research, there are several transportation locations on the coast that are likely to be affected by the tsunami disaster. Especially airports located on the coast, there are 14 port ships, and approximately 10 bus stations. This shows that in a disaster emergency response situation, if a tsunami disaster occurs, in the worst scenario, namely 15 m, access to evacuation efforts and distribution of disaster relief

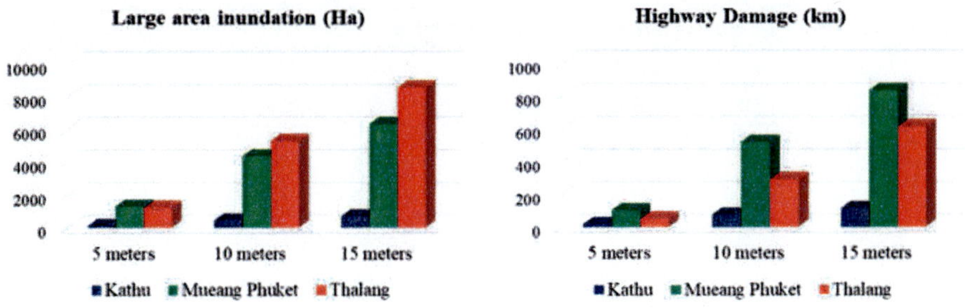

Fig. 4 Chart of large of area inundation (left) and highway damage (right)

is quite difficult because many transportation networks are affected, and the locations of the most important transportation facilities are also affected, such as airports, look at Table 2. The impact of the tsunami disaster on transportation infrastructure caused huge losses. This condition will become a problem of access to enter the Phuket Island in an effort to deal with the response and evacuation after natural disaster is happened.

Table 2 Infrastructure around coastal area will affect because of tsunami

Type		Name
Airport	1	Phuket International Airport (Located in Thalang district)
	1	Phuket Yacht Haven Marina harbor
	2	Ao Por Pier
	3	Bang Rong Pier
	4	Laem Hin Pier
	5	Port to Phi Phi Island
	6	Tian Sin Pier
	7	Jian wanich Pier
Port ships	8	Island Hopper Pier
	9	Rashada Harbor Shipyard
	10	Ao Makham Pier
	11	Cruise Terminal Phuket
	12	Maiton Pier
	13	Chalong Bay Port
	14	Rawai Pier
	1	Kamala sub-district administrative office
	2	Kata Beach
	3	Nai Harn Beach
	4	Rawai Beach
	5	Chalong Bay
Bus stations	6	Makham Bay
	7	Saphan Hin
	8	The New Thai community
	9	Sarasin bridge
	10	Bang Rong Pier

4 Conclusion

From the modeling of tsunami hazard on transportation infrastructure in Phuket will determine the wave height at the coast for 5, 10 and 15 m. In the scenarios of tsunami wave as 5 m high, around 2639.1 hectares of urban areas near the coast will be moderate damaged, along with approximately 168.96 km of highway damage, some areas and highways in urban areas near the coast in the eastern of Phuket will be damaged. When tsunami wave as 10 m, its covering area around 10.047,15 hectares of beach and urban areas around the coastline. It will be severely damaged, along with approximately 885,41 km of highway damage, especially in the city of Thalang and Mueang Phuket district. While the scenarios of tsunami wave as 15 m, the area around covered 15,559.23 hectares and a highway damage of approximately 1561.12 km. In addition, the tsunami can reach most areas and make severely affected to the city in Thalang district and Mueang Phuket district, because the urban area of Thalang district located in the coastal area that looks like a gulf and the urban area. Mueang Phuket district is located in coastal plains that tsunami can destroy toward the shore and continue to the land easily, while the urban area in Kathu district is surrounded by mountains, so this area is not affected by the tsunami wave. We hope that in the future research, data will be updated to improve information so that it becomes a strong idea, and increasing the quality information for tsunami disaster mitigation efforts.

Acknowledgements We want to thank the entire research team of the department of Geoinformatics, College of Computing, Khon Kaen University. Furthermore, we express our great appreciation to the organizers of the international seminar in Indonesia INCREASE BRIN 2022. They have provided an opportunity for us to join this conference so that we can share space to convey our scientific views on the topic of disaster studies in coastal areas related to oceanography.

References

1. Moon, W.C., Sidek, L.M., Lau, T.L., Puay, H.T., Majid, T.A., Wahab, A.K.A., Teo, F.Y.: A shared vision on the 2004 Indian Ocean Tsunami in Malaysia: hazard assessments, post-disaster measures and research. J. Mar. Sci. Eng. **10**, 1088 (2022). https://doi.org/10.3390/jmse10081088
2. Bhanupong, N.: Adjustment and recovery in Thailand two years after the Tsunami. Asian Development bank Institute. Kasumigaseki Building 8F 3–2–5 Kasumigaseki, Chiyoda-ku, Tokyo 100–6008, Japan (2007)
3. Kaiser, G., et al.: Mapping tsunami impacts on land cover and related ecosystem service supply in Phang Nga, Thailand. J. Nat. Hazards Earth Syst. Sci. **13**(3095–3111), 2013 (2013)
4. Kelvin, B.: Review of Tsunami hazard and risk in New Zealand. Institute of Geological and Nuclear Sciences client report. (2006)
5. BMKG.: Badan Meteorologi Klimatologi Indonesia. (2022). https://data.bmkg.go.id/gempabumi/
6. Mohd, A.R. et al.: Risiko Bencana Indonesia. Badan Penanggulangan Bencana Indonesia. (2016)

7. Dan Faisol Arif, I.: Konsep Dasar Analisis Spasial. Andi Publisher. Jakarta (2013). ISBN: 978-979-29-3354-3
8. Ratih, P.F. et.al.: Investigating a conservation area based on Tsunami hazard mapping in landuse planning of sand Dune Parangtritis area, Yogyakarta, Indonesia. In: Proceedings of the Pakistan Academy of Sciences: Life and Environmental Sciences, vol. 56(2) (2019)
9. Edy, P.K.I.M., Hanna, P.C.: Mapping of Tsunami disaster evacuation pathways based on Tsunami altitude scenario using Network Analyst Method (case study: Palu City, Central Sulawesi). J. Appl. Geospatial Info. (2020)
10. Fitria, P.R., Tetuko Sri, S.J.: Tsunami inundation hazard mapping as monitoring and conservation assessment in Parangtritis Coastal Area, Indonesia. Atsumi International Scholarship Foundation (2022)
11. Nofrizal, A.Y.: Prediction of sea water flooding hazard on settlement area in Padang City as climate changes impact using GIS and remote sensing. In: IOP Conference Series: Earth and Environmental Science. vol. 303, (2019). https://doi.org/10.1088/1755-1315/303/1/012025
12. Hanif, M., et al.: Impact of coastal flood on building, infrastructure, and community adaptation in Bukit Bestari Tanjungpinang. J. Gea **21**(2), 102–111 (2021). https://doi.org/10.17509/gea.v21i2.38911

Potential Application of Machine Learning on Agriculture and Capture Fisheries

Bernadetta Rina Hastiestari and Dewi Syahidah

Abstract Disease and pests have been reported to deteriorate the yield of crops and aquatic farming. These deteriorating effects should be alleviated because these sectors are important, not only for supporting the national economy but also providing enough food for all. Some methods have been implemented to increase the agricultural output such as the application of insecticide, fungicide and some chemical medicine, but fewer approaches based on digitalization have been applied. The previous naked eye diagnostics are not error-free; therefore, a more advanced technique is required to feel the gap. Adapting a machine learning process for the detection of pests and diseases on crop plants and aquatic animals should be initiated. Artificial intelligence (AI) has helped many farmers to detect diseases and pests in the field. Therefore, this paper analyzes the current and future effects of climate change dealing with the prevention of disease and pest outbreaks in agricultural systems and fisheries, various practical applications of machine learning (ML) tools in agriculture and capture fisheries domains as well as a perspective of approaching climate-smart system for agriculture and fishery farming.

1 Introduction

In agriculture, plant disturbing organisms can be divided into pests, diseases and weeds. Pest is an organism injuring part of plants such as leaf, stem, fruit and root, for example, insects, mollusk, vertebrates and mites [1]. Plant disease induces physiological disturbance in plants; therefore, the affected plants are not vigorous compared to the healthy plants [2]. These phenomena deteriorate yield and affect food security.

B. R. Hastiestari (✉)
Research Centre for Genetic Engineering, National Research and Innovation Agency, Indonesia (BRIN), Jakarta, Indonesia
e-mail: bern003@brin.go.id

D. Syahidah
Research Centre for Veterinary, National Research and Innovation Agency, Indonesia (BRIN), Jakarta, Indonesia

A. Basit et al. (eds.), *Proceedings of the International Conference on Radioscience, Equatorial Atmospheric Science and Environment and Humanosphere Science*, Springer Proceedings in Physics 290, https://doi.org/10.1007/978-981-19-9768-6_53

Similarly to agriculture, fisheries can make a country developed and prosperous, where one of the fisheries business fields is the capture fisheries. These sectors are one of the most important food production sectors in addition to aquaculture systems. Both systems reportedly contributed 401 billion USD or approximately 179 million metric tons of aquatic animals in 2018. These two sectors of aquatic food production provide a global source of livelihood for an estimated 59.5 million people [3].

In order to improve economic, social and environmental sustainability, the adoption of smart technology based on innovative data and analytical systems along the fish supply chain is becoming more evident [4–6]. For example, tracking the movement of fishing vessels, the use of fishing gear and fish stocks through machine learning (ML) algorithms can predict and estimate fishing effort with unprecedented temporal and spatial resolution [7].

In subsequent sub-sections of this paper, several basic information on the various practical applications of ML tools in agriculture and capture fisheries domains from a representative selection of scientific pieces of previous literature will be cited and discussed. This includes an overview of some of the research and applications in agriculture and capture fisheries management aspects. Some of the future perspectives related to the large-scale adoption of advanced technologies will be pointed out at the end of this paper. This paper is written to survey how to detect pests and diseases in agriculture and fishery sectors which are increasingly using machine learning approaches and provide some recommendations for increasing yield and productivity.

2 Theoretical Framework

Machine learning is a method to make automatization possible. The automation system is possible to be applied in many applications including agriculture and fishery. Factors contributing to yield deterioration are pest, disease and lack of management and monitoring. Pest and disease are almost undetectable to the naked eye and must be carried out by an expert. Monitoring of plants and fish are also problematic if the area is large. Several methods for disease detection have been used based on laboratory-based approaches, for example, genome-based detection using polymerase chain reaction (PCR) [8] and immunological and serological disease detection [9]. However, this method metered plant infection, which was time-consuming and costly for machines and chemicals. Therefore, in this technological era, ML brings new insights to solve these problems with automatic recognition. Several studies on the detection of plant pests and diseases have been implemented in Indonesia using remote sensing technology.

Depiction of a learning problem can be defined for improving some performances in executing some tasks (Fig. 1), starting with preliminary observation of the objects, such as the pest and disease of the objects, then the potential problem is identified with consultation to literatures. ML analysis could be conducted by recording and processing the data, such as spectral data of the objects. System could be designed

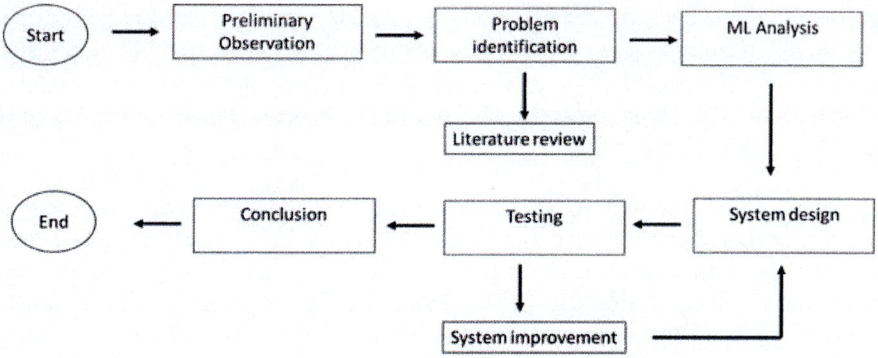

Fig. 1 Research flowchart

by prototyping the displays of early detection. ML algorithms could be presented in many candidate programs such as mathematical functions and certain programming languages. Then, the function could be tested in order to obtain programs such as search process and identification.

3 Application of ML

3.1 Application in Agriculture

The appearance of disease symptoms in the plant can be detected through discoloration of plant parts such as leaves, stems and fruit. These symptoms are hardly detected by the naked eye and should be done by an expert. This will be problematic if the planting area is large. Some methods for disease detection have been employed based on laboratory-based approaches [10–12]. However, these methods are plant destructive, time-consuming and high cost for machines and chemicals. Therefore, some research in plant pest and disease detection has been applied in Indonesia by using "Pakar" and "Image Processing" to detect disease and pest at early stages.

"Pakar" is a system for solving problems based on database gathered from textbooks from resembling an expert [13]. "Pakar" system contains a created program that can analyze information [14, 15]. According to these authors, components of the Pakar system comprise (1) a Knowledge base: the core of the system consisting of facts and rules. (2) User Interface: part of the system that accommodates the communication between user and the computer (3) Inference Engine: system part composed of thinking and system pattern which analyze and searching the answer or conclusion, (4) Development Engine: part of a system to facilitate inference engine development, whereas searching method is employed by using methods, for example, forward chaining (fc) which is searching method from data and facts to find solutions [16], certainty factor (cf) that is used to describe the level of confidence of the problem[17] and backward chaining (bc) is a searching method with applying opposite direction from fc, in which the searching process is started from the conclusion to find the problem [18]. This system has been applied to detect pests and diseases of red onion [19]. Honggowibowo [20] has reported the application this system to detect rice tungro virus using fc and bc.

"Image processing" is a part of the machine learning approach employing a deep learning algorithm to improve the accuracy of the input data from the image. Some traditional methods have been employed such as random forest that can be used to solve the drawback of overfitting of image training dataset as well as manage arithmetical and categorical data [21]. In Indonesia, these methods have been applied for some plants in Indonesia, i.e., Naive Bayes method for corn [22], Convolutional Neural Network (CNN) for potato disease detection [23] and Vector Machine method for tea [24].

3.2 Application in Fishery Management of Fisheries Resources

All activities are undertaken to harvest natural fish, and other aquatic living water resources in marine and freshwater environments are referred to as capture fisheries. The two objectives of applying computational methods (DM and ML) in capture fisheries and related activities (Table 1.) are briefly discussed in the following subsections.

Management of fisheries resources

Research on fisheries resource management has been conducted in the Niger delta fisheries of Central Mali is the starting point for AI development [25]. The authors used object-oriented modeling and multi-agent simulations to artificially simulate multifarious environments representing fishermen, fish communities, biotope, fishing gear and many other interacting entities, and tested different fishing scenarios and hypotheses to derive qualitative fisheries management options based on ecological, economic and sociological knowledge; functional relationships in fisheries such as environmental patterns associated with fisheries abundance fluctuations were identified using time series decomposition, principal components analysis and k-means clustering; the spatiotemporal relationship between environmental factors and fish distribution (temperature and assemblage) was determined using spatiotemporal assignment mining model (STAMM) [25]; fish community based on ecologically relevant hydrologic indicators was analyzed by using genetic programming algorithm [26]; biological associations in marine capture fisheries data were identified by using apriori algorithm for categorical data analysis [27]; automated age estimation system from otolith images for age-based fish stock assessment studies was developed using CNN and CT scanning [28].

Table 1 Methods used in previous studies for capture fisheries

Objectives	Methods	Uses	References
Management of fisheries resources	Spatiotemporal assignment mining model (STAMM)	Identification of fisheries abundance fluctuations	25
	Self-organizing feature maps	Measuring fish community diversity index	26
	Apriori algorithm	Investigation of the relationship between fish communities and water quality	27
	CNN and CT scanning	Assessing age-based fish stock	28
Monitoring fishing and fish catch	DBSCAN and k-nearest neighbor	Modeling implicit species exchanges	29
	SFFS and k-nearest neighbor	Prediction of the potential fishing areas	30
	AGRID +	Analyzing potential fishing zones	31, 32
	OPENCV 3.0	Comparing tools by automated counting of fish escapements video captures	33
	Mask R-CNN	Developing fishery management	34
	Potential Fishing Zone (PFZ)	Analyzing potential fishing areas based on datamining approach	35

3.3 Monitoring Fishing and Fish Catch

In fishing technology and fish catch monitoring, some studies use different methods for different objectives. For instance, potential fishing zones by integrating the mean of fish catches for each area with sea surface temperature and chlorophyll concentration were determined using density-based spatial clustering of applications with noise (DBSCAN) and k-nearest neighbor [29]; potential tuna fishing zones by combining oceanographic features with fish biology and habitat preferences were predicted using sequential forward floating selection (SFFS) and k-nearest neighbor [30]; potential fishing zones by computing the mean of fish catches for each area were determined using advanced grid-based isodensity line clustering (AGRID +) [31]; dimensional position via time and depth-referenced images of marine organisms for improved interpretation of acoustic echograms, fish distribution and behavior was analyzed using in-trawl stereo imaging and dimensional analysis [32]; the escapement rate of unwanted fish through a selective device was calculated using open source computer vision library (OPENCV 3.0) [33]; fish size by automatic segmentation to reduce catches of undersized fish in trawling was measured using mask region–Convolutional Neural Network (R-CNN) [34].

In Indonesia, where tuna is the main exported commodity to the European Union, many studies aimed to assist fishermen in finding the fishing zone of the fish (tuna). The previous investigation addressed the fishing management [35], tuna potential fishing zone (PFZ) prediction and oceanographic parameters affecting the prediction of tuna PFZ [36].

4 Conclusion

It is considered indispensable to apply unconventional systems for data pool, organization and brainpower in almost scopes of agriculture and capture fisheries nowadays. The unprecedented bulk of agriculture and capture fisheries is now promising with rapid improvements in information/communication knowledge, sensor networks and independent boards. For future work, the most promising to be applied in the case of agriculture and capture fisheries in Indonesia are image processing for early disease detection and potential fishing zone for fishing management, respectively.

References

1. Marer, P.J.: In: Landscape Maintenance Pest Control, vol. 7, UCANR Publications (2006)
2. Katan, J.: Diseases caused by soilborne pathogens: biology, management and challenges. JPP **1**, 305–315 (2017)
3. Arneson, P.A.: Plant disease epidemiology: temporal aspects. Plant Health Instr. (2011)
4. Grulke, N.E: The nexus of host and pathogen phenology: understanding the disease triangle with climate change. The New Phytol. **189**(1), 8–11 (2011)
5. Forrest, J.R.K.: Complex responses of insect phenology to climate change. Curr. Opin. Insect. Sci. **17**, 49—54 (2016)
6. FAO: The State of World Fisheries and Aquaculture 2020. Sustainability in action, Rome (2020)
7. Beyan, C., Browman, H.I: Setting the stage for the machine intelligence era in marine science. ICES J. Mar. Sci. **77**(4), 1267–1273 (2020)
8. Martinelli, F., Scalenghe, R., Davino, S., Panno, S., Scuderi, G., Ruisi, P., Villa, P., Stroppiana, D., Boschetti, M., Goulart, L.R., Davis, C.E.: Advanced methods of plant disease detection a review. Agronomy for Sustain. Developm. **35**(1), 1–25 (2015)
9. Pazos, A.J., Român, G., Acosta, C.P., Abad, M., Sánchez, J.: Stereological studies on the gametogenic cycle of the scallop, Pecten maximus, in suspended culture in Ria de Arousa (Galicia, NW Spain). Aquaculture **142**(1–2), 119–135 (1996)
10. O'Donncha, F., Grant, J.: Precision aquaculture. IEEE Internet Thing Mag. **2**(4), 26–30 (2019)
11. Baldi, P., La Porta, N.: Molecular approaches for low-cost point-of-care pathogen detection in agriculture and forestry. Front. Plant Sci. **11**, 570862 (2020)
12. Kliot, A., Kontsedalov, S., Lebedev, G., …, Ghanim, M.: Fluorescence in situ hybridizations (FISH) for the localization of viruses and endosymbiotic bacteria in plant and insect tissues. J. Vis Ex. **84**, e51030. 24 (2014)
13. Kusrini, S.: Sistem Pakar Teori dan Aplikasi, Andi offset, Yogyakarta (2006)

14. Sumpala, A.T., Sutoyo, M.N.: Sistem Pakar Untuk Mendiagnosa Hama dan Penyakit Tanaman Kakao Menggunakan Metode Forward Chaining dan Certainty Factor. In Seminar Nasional Teknologi Informasi Komunikasi dan Industri, pp. 261–267. (2018)
15. Fernando, F., Fauzi, A.: Sistem Pakar Diagnosa Hama Penyakit Tanaman Padi dan Holtikultura Menggunakan Metode Forward Chaining Berbasis Android. J. Teknik Komputer **5**(2), 84–91 (2019)
16. Farizi, A.: Sistem Pakar Untuk Mendiagnosa Kerusakan Komputer Dengan Menggunakan Metode Forward Chaining. Edu Komputika J. **1**(2) (2014)
17. Rohajawati, S., Supriyati, R.: Sistem Pakar: Diagnosis Penyakit Unggas Dengan Metode Certainty Factor. CommIT **4**(1), 41–46 (2010)
18. Tarigan, F.A.: Sistem Pakar Untuk Mendiagnosa Penyakit Ginjal dengan Metode Backward Chaining. Jurnal Times **3**(2), 25–29 (2014)
19. Prambudi, D.A., Widodo, C.E., Widodo, A.P. : Expert system application of forward chaining and certainty factors method for the decision of contraception tools. In: E3S Web of Conferences, vol. 31, pp. 10009. EDP Sciences (2018)
20. Honggowibowo, A.S.: Sistem pakar diagnosa penyakit tanaman padi berbasis web dengan forward dan backward chaining. Telkomnika **7**(3), 187 (2009)
21. Bharate, A.A., Shirdhonkar, M.S.. A review on plant disease detection using image processing. In: 2017 International Conference on Intelligent Sustainable Systems (ICISS), pp. 103–109. IEEE (2017)
22. Sihotang, H.T.: Sistem pakar untuk mendiagnosa penyakit pada tanaman jagung dengan metode bayes (2019)
23. Rozaqi, A.J., Sunyoto A., Arief, M.R.: Deteksi Penyakit Pada Daun Kentang Menggunakan Pengolahan Citra dengan Metode convolutional neural network. Creative Info. Technol. J. **8**(1), 22 (2021). https://doi.org/10.24076/citec.2021v8il.263
24. Chen, J., Liu, Q., Gao, L.: Visual tea leaf disease recognition using a convolutional neural network model. Symmetry **11**(3) (2020)
25. Bousquet, F., Cambier, C., Morand, P.: Distributed artificial intelligence and object- oriented modelling of a fishery, Math. Comput. Model. **20**(8) (1994) 97–107 (1994)
26. Plaza, F., Salas, R., Yáñez, E.: Identifying ecosystem patterns from time series of anchovy (*Engraulis ringens*) and sardine (*Sardinops sagax*) landings in Northern Chile. J. Stat. Comput. Simul. **88**(10), 1863–1881 (2018)
27. Tsai, W.P., Huang, S.P., Cheng, S.T., Shao, K.T., Chang, F.J.: A data-mining framework for exploring the multi-relation between fish species and water quality through self-organizing map. Sci. Total Environ. **579**, 474–483 (2017)
28. Moore, B.R., Maclaren, J., Peat, C., Anjomrouz, M., Horn, P.L., Hoyle, S.: Feasibility of automating otolith ageing using CT scanning and machine learning. New Zealand Fisheries Assessment Report **58**(23) (2019)
29. Xu, J., Wickramarathne, T.L., Chawla, N.V., Grey, E.K., Steinhaeuser, K., Keller, R.P., ... Lodge, D.M.: Improving management of aquatic invasions by integrating shipping network, ecological, and environmental data: data mining for social good. In: Proceedings of the 20th ACM SIGKDD International Conference on Knowledge Discovery and Data Mining, pp. 1699–1708. (2014)
30. Hidayanto, A.N., Fahmi, H., Fitrianah, D., Arymurthy, A.M.: Oceanographic features selection to predict the tuna potential fishing zones using SFFS method. Int. Mathem. Forum **11**(24), 1157–1166) (2016)
31. Fitrianah, D., Fahmi, H., Hidayanto, A.N., Arymurthy, A.M.: A data mining based approach for determining the potential fishing zones. Int. J. Inf. Educ. Technol. **6**(3), 187–191 (2016)
32. Rosen, S., Holst, J.C.: DeepVision in-trawl imaging: sampling the water column in four dimensions. Fish. Res. **148**, 64–73 (2013)
33. Simon, J., Kopp, D., Larnaud, P., Vacherot, J.P., Morandeau, F., Lavialle, G., Morfin, M.: Using automated video analysis to study fish escapement through escape panels in active fishing gears: application to the effect of net colour. Mar. Policy **116**, 103785 (2020)
34. Garcia, R., Prados, R., Quintana, J., Tempelaar, A., Gracias, N., Rosen, S., ... Løvall, K.: Automatic segmentation of fish using deep learning with application to fish size measurement. ICES JMS **77**(4), 1354–1366 (2020)

35. Fitrianah, D., Hidayanto, A.N., Fahmi, H., Gaol, J.L., Arymurthy, A.M.: ST-AGRID: a spatio temporal grid density based clustering and its application for determining the potential fishing zones. IJSEA **9**(1), 13–26 (2015)
36. Fitrianah, D., Praptono, N.H., Hidayanto, A.N., Arymurthy, A.M.: Feature exploration for prediction of potential tuna fishing zones. IJIEE **5**(4), 270 (2015)

Climate Change Risk Assessment Toward Agriculture and Food Security in Sumedang Regency

Hadi Ferdiansyah, Nugrahana Fitria Ruhyana, and Erti Nurfindarti

Abstract Agriculture is one of the economic activities vulnerable to climate change. One of the areas in West Java Province where the economy of the community still depends on the agricultural sector and is quite sensitive to climate change is Sumedang Regency. The el Nino phenomenon in 2015 caused a decline in farm production and food security in Sumedang Regency. This study aimed to conduct a risk assessment of climate change in the agricultural sector in Sumedang Regency. The mixed method with a sequential explanatory design approach was chosen to map the results of risk analysis which was followed up with focus group discussions with stakeholders to determine location priorities and activities needed to reduce climate change risks in the agricultural sector. The analysis results show nine villages in Sumedang Regency that are prioritized in lowering the risk of the agricultural sector and food security from the threat of landslides.

1 Introduction

Climate change is the mother of all externalities whose impact is very large, complicated, and uncertain compared to other environmental problems. Most human activities, such as industry, agriculture, and household, produce greenhouse gases that trigger climate change. The impact of climate change also affects all human activities. Weather anomalies impact agriculture, energy use, health, and other aspects [1, 2]. Ironically, low-income countries that contribute the least to emitting greenhouse gases are more vulnerable to the impacts of climate change [3–5].

H. Ferdiansyah (✉) · N. F. Ruhyana
Regional Research and Development Planning Agency of Sumedang Regency, Sumedang, Indonesia
e-mail: daanhadi@gmail.com

E. Nurfindarti
Regional Development Planning Agency of Serang City, Serang, Indonesia

N. F. Ruhyana
Universitas Sebelas April, Sumedang, Indonesia

A. Basit et al. (eds.), *Proceedings of the International Conference on Radioscience, Equatorial Atmospheric Science and Environment and Humanosphere Science*, Springer Proceedings in Physics 290, https://doi.org/10.1007/978-981-19-9768-6_54

Climate change is a climatic condition from the average change and variability of its characteristics. These condition changes can persist for a long time, usually decades or more. Therefore, climate change is a process that can occur naturally. Nevertheless, there are other external causes, the emissions from continuous human activities in land use and other activities that affect the atmosphere's composition [6]. The climate change happening in the world is also being experienced in Indonesia, although the changes are not too significant. Indonesia's annual average air temperature has increased by almost 0.1 °C per year. Annual rainfall also has decreased by 2 to 3% per year across Indonesia over the past century [7].

Based on previous research predict that extreme climate events in Indonesia will increase [8]. The increase in extreme climate events can cause hydrometeorological disasters. Hydrometeorological disasters such as floods and landslides are closely related to extreme weather in the form of high-intensity rain [9, 10]. In addition, soil conditions, physical properties of the soil surface, and drainage systems can also affect the occurrence of floods and landslides [11].

The impact of global climate change on the national economy, especially in the agricultural sector [12–15]. It will also affect national food security. The challenge of climate change on food security is considered the stability of production on agricultural land [6]. The agricultural sector is most threatened, suffering, and vulnerable to climate change [16]. Food crops are the most vulnerable commodities to changes in rainfall, extreme weather events, and increases in air temperature. Generally, this is because food crops are seasonal crops that are relatively sensitive to weather stress, especially excess and lack of water [17].

Agriculture is the main contributor to the regional economy of the Sumedang Regency. From 2011 to 2020, the average Gross Regional Domestic Product (GRDP) distribution in the agriculture, forestry, and fisheries sectors reached 21.27% of the total GRDP in Sumedang Regency. On the other hand, agricultural production in Sumedang Regency is affected by annual rainfall changes. The high rainfall in Sumedang Regency causes floods, flash floods, and landslides (Fig. 1).

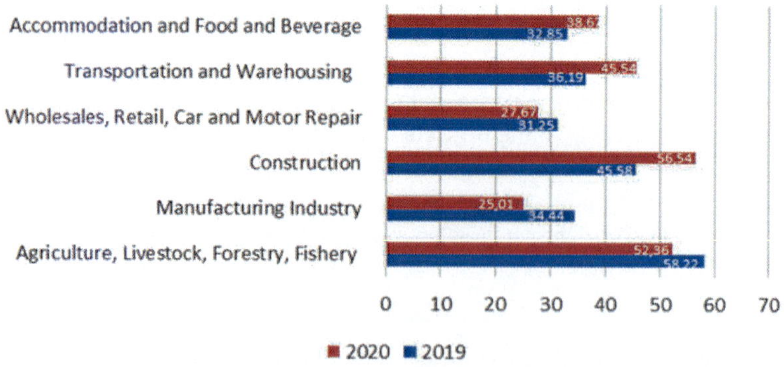

Fig. 1 Proportion of 40% of community groups with the lowest welfare level in every business field in Sumedang Regency. *Source* Susenas, Central Bureau of Statistics (2019; 2020), processed

Landslides are the higher-risk disaster in Sumedang. Landslides were more frequent than other disasters; they occurred in 2020 155 times [18]. Landslides caused economic losses, including some agricultural areas buried by landslide material. The landslide disaster in Sumedang Regency caused more casualties than other hydrometeorological disasters [19, 20]. The impact of landslides is very expensive for households, in addition to having to change or leave farmland, reconstructing their homes, and causing migration outside the area [6]. Climate change causes crop failure, decreases food production, and affects Indonesia's food security [21]. The poor in rural and urban areas is a vulnerable group affected by climate change [22]. Further consequences of this phenomenon are food access and malnutrition problems, especially among children and people living in remote areas.

The government's regulation seeks to mainstream climate change issues in synergy with national and regional development planning. Climate change risk assessment is needed as a basis for adaptation planning so that communities have resilience and the ability to overcome the dangers of climate change [6]. However, the problem in regional development planning is not merely disaster and climate change issues. Development planning also considers regional financial capacity. So the development planning process is necessary to formulate development priorities related to regional development targets, the level of urgency, and leverage for improving regional development performance.

Based on the description above, the Regional Government of Sumedang Regency seeks to integrate the climate change issue and hydrometeorological disasters into the Regional Medium-Term Development Plan (RPJMD). Through a risk assessment, expected that activities and locations that are prioritized can be determined to reduce the impact of climate change on the agricultural sector and food security in the Sumedang Regency.

2 Method

A risk assessment is a method for determining the nature and level of risk by analyzing the level of hazard and conditions of vulnerability that potentially harm people, property, public services, livelihoods, and the environment exposed to a disaster [23]. The risk assessment can provide information on risk profiles, patterns, and changes to determine priorities, select alternative strategies, and formulate sensitive public policies for disaster risk and climate change adaptation [24, 25]. Risk assessment is a tool to assess, visualize, and communicate various levels of exposure, vulnerability, and disaster risk in an area [26]. The leveling of risk from a disaster and the impact of climate change is obtained from the results of the intersection between hazard (H) and vulnerability (V) and divided by adaptive capacity (C) [27]. In a mathematical function, the level of risk is formulated as follows:

$$R = H \times V/C \tag{1}$$

The indicators determination of hazard and vulnerability can identify from dominant factors that affect risk leveling. The indicators determination depends on the intended target for reducing risks and vulnerabilities [28–30].

The data used as dangers of climate change indicators in Sumedang Regency are foretime and prediction rainfall data. The foretime rainfall data is secondary data accessed from the CHIRPS grid-type database site (data.chc.ucsb.edu), while the prediction rainfall data (years 2020–2045) has resulted from the modeling results of the Center for Climate Change Studies, Bandung Institute of Technology. The intended use of rainfall data is so that it can be processed spatially.

The hazard indicators for the frequency of landslides gained from the 2016–2020 Disaster Event Report released by the Regional Disaster Management Agency of Sumedang Regency (2020). The parameters in distribution areas prone to landslides are based on the Regional Regulation of 2018 concerning the Spatial Planning of the Sumedang Regency for 2018–2038.

The vulnerability and adaptive capacity indicator data use Village Potential data (Podes) released by the Central Statistics Agency and other supporting data obtained from several Regional Apparatuses in Sumedang Regency. Determining vulnerability indicators from Podes data is based on the Focus Discussion Group (FGD) results. Each hazard indicator data, vulnerability, and adaptive capacity was then analyzed using the t-score method to standardize the value of each variable with different units. The score results show the level of risk in the agricultural sector and food security—in each village—when facing landslides.

To reduce the risk of landslides as the impact of climate change on the agricultural sector and food security in Sumedang Regency, the indicators of hazard, vulnerability, and adaptive capacity agreed upon from the FGD results are as follows (Table 1).

The making of a landslide risk map is processed through the calculation of scores and risk classifications resulting from the mapping of vulnerability and adaptation capacity, as well as the danger of landslides, using the VCA matrix multiplication method, which shows in Table 2 [31].

Table 1 Hazard and vulnerability indicators in risk analysis of landslides and on agriculture sector and food security in Sumedang Regency

Hazard	Vulnerability	Adaptive capacity
• Maximum rainfall data (foretime and prediction) • Landslide occurrences data • Data on areas prone to landslides	• The ratio of rice field area to a village area • Main livelihood • Population density • The proportion of poor people • Road accessibility quality	• Source of irrigation water • Agricultural infrastructure development • Community empowerment in environmental conservation and disaster management • Early warning system • Markets and their distance • Health facilities and their distance

Table 2 Determination of risk category with VCA matrix

Risk (H*V/C)		Hazard (H)				
		Higher	High	Moderate	Low	Lower
Vulnerability/adaptive capacity (V/C)	Higher	Highest	Highest	Higher	High	Moderate
	High	Highest	Higher	High	Moderate	Low
	Moderate	Higher	High	Moderate	Low	Lower
	Low	High	Moderate	Low	Lower	Lowest
	Lower	Moderate	Low	Lower	Lowest	Lowest

3 Results and Discussion

Analysis of vulnerability and adaptive capacity is used to understand how communities are exposed to long-term hazards or disturbances. Understanding vulnerability will result in efforts to build resilience (resilience) from hazards/disruptions. From this analysis, it can be distinguished between vulnerable and non-vulnerable villages that deal with the disaster. From the results of the vulnerability analysis and adaptive capacity visualized in Fig. 2, known that the level of vulnerability of the agricultural sector and food security in most areas in the Sumedang Regency is at a medium–high level.

The agricultural sector's high to higher vulnerability level and food insecurity tend to be concentrated in the outermost areas of the Sumedang Regency. This condition is in line with the findings of several studies, which revealed that economic

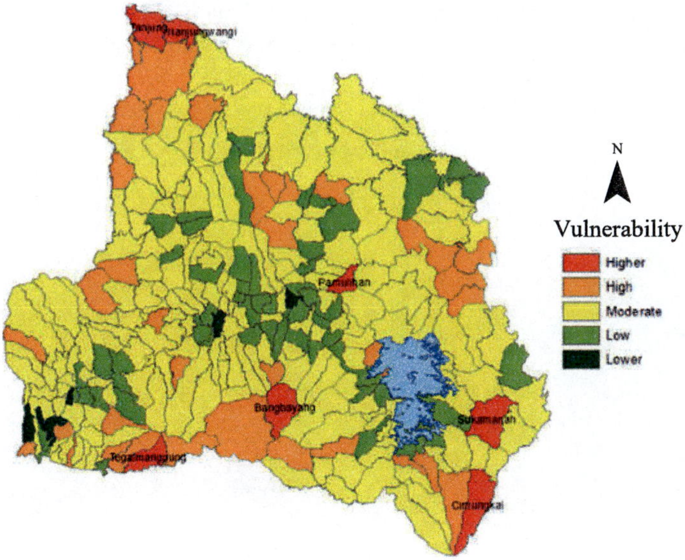

Fig. 2 Level of the vulnerability of agriculture and food security to landslides

vulnerability to food insecurity increased in people who lived in rural and the periphery of the region [32, 33]. Seven of the 277 villages in Sumedang Regency have higher levels of agricultural vulnerability and food insecurity. The seven villages are Tanjung, Nanjungwangi, Tegalmanggung, Bangbayang, Pamulihan, Sukamanah, and Cimungkal.

Analysis of landslide hazards in each village includes indicators of maximum rainfall based on foretime data and future predictions, the intensity of landslides in the last five years, and overlaid with maps of landslide-prone areas. From the landslide hazard analysis (using foretime rainfall data), it is known that most of the Sumedang Regency area faces a moderate level of landslide hazard. Of 277 villages, only one with a higher level of landslide hazard is Cimanintin.

Meanwhile, predictive rainfall data increases potential landslide hazards in the Sumedang Regency. The distribution of the danger of landslides with higher levels is also expanding, covering 38 villages. Predictions of increased rainfall in Sumedang Regency in the future will increase the potential for landslides, especially in villages that are in areas prone to landslides. A comparison map of landslide hazards between current conditions and future predictions is presented in Fig. 3.

From the overlay between vulnerability analysis and landslide hazard analysis, a map of agricultural sector risk and food security is obtained when landslide hazard occurs. Landslides as attributed to climate change influencing agricultural production, income, and household food security [34, 35]. This risk map consists of current risk (baseline) and prediction of future risk. In the baseline conditions, it can be seen that the agricultural sector and food security in most areas of the Sumedang Regency are facing a moderate level of landslide risk. However, there are four villages with a very high level of risk from the agricultural sector and food security against landslides, namely Pamulihan, Cimungkal, Sukamanah, and Cimanintin.

Meanwhile, in the future, it is predicted that the level of risk in the agricultural sector and food security will increase and become more widespread. It is predicted that there will be eight villages with the highest level of risk in agricultural and food

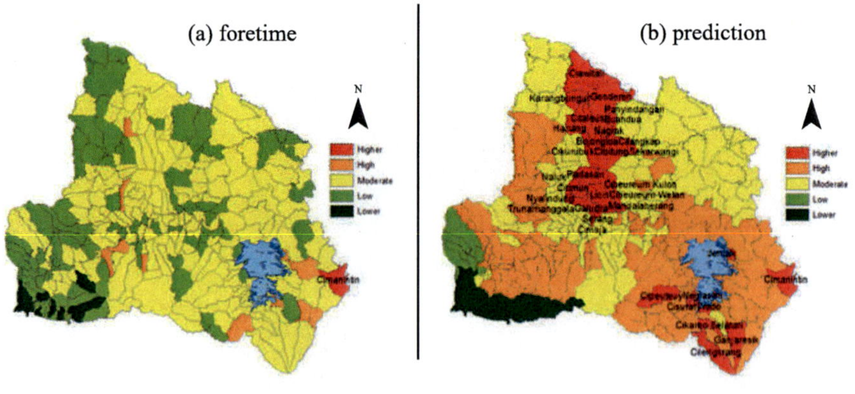

Fig. 3 Landslide hazard map

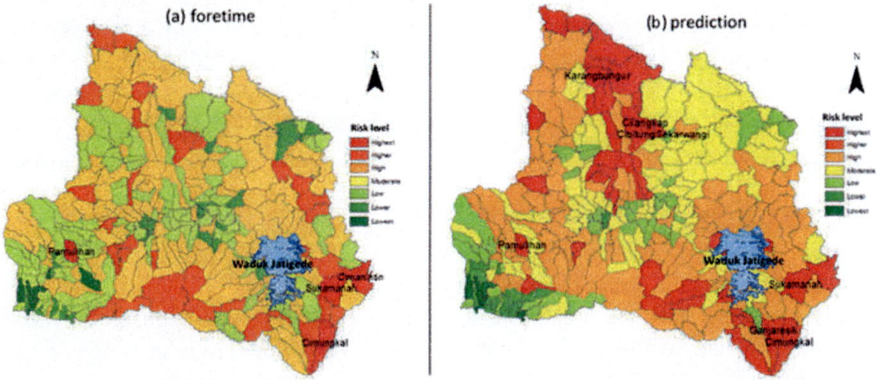

Fig. 4 Map of agricultural sector risks and food security against landslides

security in the future compared to other villages, namely Pamulihan, Karangbungur, Cilangkap, Cibitung, Sekarwangi, Ganjaresik, Cimungkal, and Sukamanah Village. Villages with low to lower levels in dealing with landslides are located in the downtown area of Sumedang Regency and close to the Greater Bandung Metropolitan Area, both in current conditions and predictions in the future. A comparison map of the risks of the agricultural sector and food security against landslides between current conditions and future predictions is presented in Fig. 4.

From the results of this risk analysis, it is known that villages at high risk of facing landslides have a high level of vulnerability, and low adaptive capacity, located in landslide-prone areas. It is because of frequent landslides and the influence of high rainfall in the last five years. Villages with high vulnerability are villages with high population density and poverty rates; most of the population works in the agricultural sector, and most areas are rice fields and poor quality road access. Meanwhile, villages with low adaptive capacity are regions with limited agricultural infrastructure, low community empowerment, no early warning system, few trade facilities, and inadequate health facilities.

These findings align with the previous studies that people living in landslide-prone areas are vulnerable to a lack of connectivity for food, water, transportation, communication, access to difficult health and education facilities, and poor hygienic conditions [34, 36, 37]. Efforts are needed to reduce the risks to the agricultural sector and food security due to landslides, especially land use changes, adaptive agricultural land use planning and diversification, and land and water conservation in landslide-prone areas [36, 38–40].

4 Conclusion

Risk analysis is one of the tools that can be applied to determine priority locations with the highest risk of climate change impacts. This analysis can also determine activities that can reduce these risks. This analysis is very beneficial in the development planning process, which is confronted with various other development problems and limited funds.

The agricultural sector in Sumedang Regency is at risk of landslides due to extreme weather. As a result, it can threaten regional food security due to decreased availability. Likewise, the poor, mostly working as farmers, will be more miserable if crop failure occurs due to catastrophic weather such as a landslide. Therefore, nine priority villages are to reduce the risk to the agricultural sector and food security against landslides, namely Pamulihan, Cimungkal, Sukamanah, Cimanintin, Karangbungur, Cilangkap, Cibitung, Sekarwangi, and Ganjaresik.

5 Recommendation

The climate change risk analysis in Sumedang Regency will be more accountable and implementable if supported by quality data. Valid and reliable data from all agencies and other stakeholders in Sumedang Regency is needed. The process of determining the location, development sector, or type of priority hazard needs to be iterated repeatedly to make it more targeted, including by re-verifying the findings of this risk assessment.

The steps that need to be taken by the Sumedang Regency Government from the results of this study are to prioritize villages with high to highest risk in landslides risk reduction. In addition, the strategy that needs to be taken is to increase adaptive capacity and reduce vulnerability in these villages.

References

1. Kogan, F., Guo, W.: Strong 2015–2016 El Niño and implication to global ecosystems from space data. Int. J. Remote Sens. **38**(1), 161–178 (2017). https://doi.org/10.1080/01431161.2016.1259679
2. Berhane, S.A., Bu, L.: Aerosol-cloud interaction with summer precipitation over major cities in Eritrea. Remote Sens. **13**(4), 1–21 (2021). https://doi.org/10.3390/rs13040677
3. De Silva, M.M.G.T., Kawasaki, A.: Socioeconomic vulnerability to disaster risk: a case study of flood and drought impact in a rural Sri Lankan community. Ecol. Econ. **152**(April), 131–140 (2018). https://doi.org/10.1016/j.ecolecon.2018.05.010
4. Rossati, A.: Global warming and its health impact. Int. J. Occup. Environ. Med. **8**(1), 7–20 (2017). https://doi.org/10.15171/ijoem.2017.963
5. Ward, P.S., Shively, G.E.: Disaster risk, social vulnerability, and economic development. Disasters **41**(2), 324–351 (2017). https://doi.org/10.1111/disa.12199

6. Sudmeier-Rieux, K. et al.: A neglected disaster: landslides and livelihoods in Central-Eastern Nepal. In: Landslide Science and Practice, vol. 4, Berlin, Heidelberg: Springer, pp. 169–176. (2013)
7. Febrianti, N.: Hubungan Pemanasan Global dengan Kondisi Suhu Udara. In: Seminar Nasional Matematika UNPAR 4, pp. 299–305. (2009)
8. Baettig, M.B., Wild, M., Imboden, D.M.: A climate change index: Where climate change may be most prominent in the 21st century. Geophys. Res. Lett. **34**(1), L01705 (2007). https://doi.org/10.1029/2006GL028159
9. Kurniawan, P.M.R.: Analisis Cuaca Ekstrem Terkait Bencana Hidrometeorologi Di Jayapura (Studi Kasus Hujan Lebat Tanggal 22 Februari 2014). J. Meteorol. Klimatologi dan Geofis. **5**(3), 1–13 (2018)
10. Ferdiansyah, A.: Potensi parameter keluaran raob (rawinsonde observation programs) sebagai indikator kunci dalam analisis curah hujan (2012)
11. Rahmadana, A.D.W., Rosyidi, S.A.P., Setiawan, D.M., Nuraini, R.: Geomorphological spatial model of risk analysis for natural disasters in heritage railway tunnel. IOP Conf. Ser. Earth Environ. Sci. **971**(1), 012024 (2022). https://doi.org/10.1088/1755-1315/971/1/012024
12. Beach, R.H. et al.: Climate change impacts on US agriculture and forestry: benefits of global climate stabilization. Environ. Res. Lett. **10**(9) (2015). https://doi.org/10.1088/1748-9326/10/9/095004
13. Manuel, L., Chiziane, O., Mandhlate, G., Hartley, F., Tostão, E.: Impact of climate change on the agriculture sector and household welfare in Mozambique: an analysis based on a dynamic computable general equilibrium model. Clim. Change **167**(1–2) (2021). https://doi.org/10.1007/s10584-021-03139-4
14. Ozer, O.O., Ozden, A., Kan, M.: Impact of global climate change on agricultural production: Balkan countries in 2050. J. Environ. Prot. Ecol. **21**(1), 151–158 (2020). [Online]. Available https://www.scopus.com/inward/record.uri?eid=2-s2.0-85082116643&partnerID=40&md5=4a2c309ff0b5389290ac1aeff02fef94
15. Baker, J.S. et al.: Evaluating the effects of climate change on US agricultural systems: sensitivity to regional impact and trade expansion scenarios. Environ. Res. Lett. **13**(6) (2018). https://doi.org/10.1088/1748-9326/aac1c2
16. Adib, M.: Pemanasan Global, Perubahan Iklim, Dampak dan Solusinya di Sektor Pertanian. J. Biokultur **III**(2), 420–429 (2014)
17. Surmaini, E., Runtunuwu, E.: "Upaya sektor Pertanian dalam Menghadapi Perubahan Iklim", *Upaya Sekt.* Pertan. dalam Menghadapi Perubahan Iklim **30**(1), 1–7 (2015)
18. Regional Disaster Management Agency of Sumedang Regency.: Rekapitulasi Bencana Daerah 2015–2020, Sumedang (2020)
19. cnnindonesia.com.: longsor-sumedang-40-orang-tewas-1126-jiwa-terdampak @ www.cnnindonesia.com. (2021)
20. kompas.com.: 4 Orang Tewas Tertimbun Longsor di Sumedang (2016). https://regional.kompas.com/read/2016/09/21/11433411/4.orang.tewas.tertimbun.longsor.di.sumedang
21. Asnawi, R.: Perubahan iklim dan kedaulatan pangan di Indonesia. Tinjauan produksi dan kemiskinan. Sosio Inf. **1**(03), 52857 (2015)
22. Srinivasa Rao, C., Gopinath, K.A., Prasad, J.V.N.S., Prasannakumar, Singh, A.K.: Climate resilient villages for sustainable food security in tropical India: concept, process, technologies, institutions, and impacts. Advances in Agronomy, vol. 140. Academic Press Inc., ICAR-Central Research Institute for Dryland Agriculture, Hyderabad, Telangana, India, pp. 101–214. (2016). https://doi.org/10.1016/bs.agron.2016.06.003
23. Murphy, R.: Disasters. Essentials Concepts Global Environment Government, pp. 66–68. (2020)
24. Yusuf, A.A., Francisco, H.: Climate change vulnerability mapping for Southeast Asia EEPSEA special and technical paper. EEPSEA Special Technical Paper, pp 32. (2009)
25. Ramli, M.W.A., Alias, N.E., Yusop, Z., Taib, S.M.: Disaster risk index: a review of local scale concept and methodologies. IOP Conf. Ser. Earth Environ. Sci. **479**(1) (2020). https://doi.org/10.1088/1755-1315/479/1/012023

26. de Almeida, L.Q., Welle, T., Birkmann, J.: Disaster risk indicators in Brazil: a proposal based on the world risk index. Int. J. Disaster Risk Reduct. **17**, 251–272 (2016)
27. International Strategy for Disaster Reduction (ISDR), Living with Risk: A Global Review of Disaster Reduction. Geneva (2002)
28. Cardona, O.D., Carreño, M.L.: Updating the indicators of disaster risk and risk management for the Americas. J. Integr. Disaster Risk Manag. **1**(1), 27–47 (2011). https://doi.org/10.5595/idrim.2011.0014
29. Cardona, O.D. et al.: Determinants of risk: exposure and vulnerability. In: Managing the Risks of Extreme Events and Disasters to Advance Climate Change Adaptation: Special Report of the Intergovernmental Panel on Climate Change, Cambridge University Press, pp. 65–108. (2012)
30. IDB.: Indicators of disaster risk and risk management: program for Latin America and the Caribbean: summary report. Inter-American Dev. Bank (2010)
31. Bayuaji, D., Nugraha, A., Sukmono, A.: Analisis Penentuan Zonasi Risiko Bencana Tanah Longsor Berbasis Sistem Informasi Geografis (Studi Kasus : Kabupaten Banjarnegara). J. Geod. Undip **5**(1), 326–335 (2016)
32. Tröger, S.: Interpretations of 'work' in the context of social transformation–consequences for food-security in the rural periphery of Tanzania. Geogr. Rundsch. **55**(7–8), 24–29 (2003). [Online]. Available https://www.scopus.com/inward/record.uri?eid=2-s2.0-0041698176&partnerID=40&md5=08114412f322d24f5a296c23df92effa
33. Mitu, M.M.P., Islam, K., Sarwar, S., Ali, M., Amin, M.R.: Spatial differences in diet quality and economic vulnerability to food insecurity in Bangladesh: results from the 2016 household income and expenditure survey. Sustain. **14**(9) (2022). https://doi.org/10.3390/su14095643
34. Thanh Thi Pham, N., Nong, D., Raghavan Sathyan, A., Garschagen, M.: Vulnerability assessment of households to flash floods and landslides in the poor upland regions of Vietnam. Clim. Risk Manag. **28** (2020). https://doi.org/10.1016/j.crm.2020.100215
35. Deng, X., Xu, D., Zeng, M., Qi, Y.: Landslides and cropland abandonment in China's mountainous areas: Spatial distribution, empirical analysis and policy implications. Sustain. **10**(11) (2018). https://doi.org/10.3390/su10113909
36. Sattenpalli, S., Parkash, S.: Landslides, land use systems and food security. **4**, 151–161 (2013). https://doi.org/10.1007/978-3-642-31337-0_20
37. Humayain, K.M., Nahid, M.: Vulnerabilities to landslides in Chittagong metropolitan area, Bangladesh. Disaster Adv. **14**(1), 37–47 (2021). [Online]. Available: https://www.scopus.com/inward/record.uri?eid=2-s2.0-85098883576&partnerID=40&md5=acd2b7ff670657806ae68fd9a50b580e
38. Jumiyati, S., Hadid, A., Toknok, B., Nurdin, R., Paramitha, T.A.: Climate-smart agriculture: Mitigation of landslides and increasing of farmer's household food security. IOP Conf. Ser. Earth Environ. Sci. **708**(1), 012073 (2021). https://doi.org/10.1088/1755-1315/708/1/012073
39. Tiwari, P.C., Joshi, B.: Rainfall variability, landslides and food security in Himalaya. In: Landslide Science and Practice, vol. 4, Berlin, Heidelberg, Springer, pp. 183–189. (2013)
40. Rodgers, J. et al.: Resilience challenges in South and Southeast Asian hill cities. In: 1th National Conference on Earthquake Engineering 2018: Integrating Science, Engineering, and Policy, NCEE 2018, vol. 1, pp. 630–639. (2018). [Online]. Available: https://www.scopus.com/inward/record.uri?eid=2-s2.0-85085487178&partnerID=40&md5=ec20e3f5cdce3210ba37a88ae6286846.

Measuring the Impact of Climate Resilience Actions in Agriculture: A Preliminary Study

Elza Surmaini, Woro Estiningtyas, and Yayan Apriyana

Abstract The increasing scale of potential climate change impacts gives urgency to address agricultural adaptation more coherently to effectively manage potential climate risks impact. Since 2020, the Indonesian government has had to report the climate resilience actions (CRA) in the National Determined Contribution (NDC). For achieving the CRA goals and targets, the Ministry of National Development Planning/Bappenas has developed a mechanism for monitoring, evaluating, and reporting the CRA in Aksara, an application for Planning and Monitoring Indonesia's Low Carbon Development and CRA. The objectives of the study are to measure the impact of CRA in the agricultural sector and propose correction factors (CF) for intervention areas. South Sumatra, West Java, DI Yogyakarta, and West Nusa Tenggara Provinces were selected for the study areas. The activities evaluated are tertiary irrigation networks rehabilitation, pump and pipe irrigation, small water reservoir, and organic fertilizing processing unit (OFPU). The impact of these activities was carried out through in-depth interviews with representative farmer groups, extension workers, and staff of the agricultural office. The results conclude the intervention area of the rehabilitation of tertiary irrigation networks was found nearly to the survey, while the rest are much lower than the value used in Aksara. The result concludes that the intervention area of pumping irrigation is close to the default value used in Aksara. We recommend using CF to increase the accuracy of area interventions as a parameter to calculate the reduction in GDP loss from CRA. CF for tertiary irrigation network rehabilitation range from 0.86–0.95, pipe irrigation 0.67–0.75, small water reservoir 0.12–0.57, and OFPU 0.53–0.72.

E. Surmaini (✉) · W. Estiningtyas · Y. Apriyana
Research Centre for Climate and Atmosphere, National Research and Innovation Agency, Jl Dr. Djunjunan 133, Bandung, Indonesia
e-mail: elza001@brin.go.id

A. Basit et al. (eds.), *Proceedings of the International Conference on Radioscience, Equatorial Atmospheric Science and Environment and Humanosphere Science*, Springer Proceedings in Physics 290, https://doi.org/10.1007/978-981-19-9768-6_55

1 Introduction

Changes in climate and extreme weather have already occurred and are increasing challenges for the agriculture sector. The impacts are expected to continue or intensify in the future. Farmers in vulnerable areas are particularly affected by climate change impacts, such as floods, droughts, and other extreme weather events [1–3]. Effective management of change-induced challenges requires an adaptation strategy that may vary from any location.

The 23rd Conference of Parties of the United Nations Framework Convention on Climate Change (COP-UNFCCC) in Bonn released the Koronivia Joint Work on Agriculture (KJWA). KJWA emphasizes the importance of adaptation, adaptation co-benefits, and resilience of agricultural systems in the face of climate change [4]. There is a growing body of literature on the specific adaptation responses and capacities that farmers rely on when facing climate variability and change [5–7]. According to [8] IPCC (2000), adaptation is an adjustment of natural or human systems in response to actual or expected climatic stimuli or their effects, for reducing harm or taking advantage of opportunities by making appropriate adjustments and changes.

The Government of Indonesia issued the 2020–2024 Climate Resilience Development Policy, a reference for climate resilience actions (CRA) for various ministries and sub-national governments. CRA Monitoring, Evaluation, and Reporting (MER) process implemented in the application of Aksara. The Aksara platform only focused on the programs and activities of Ministries and National Governmental Institutions [9]. Aksara calculates the impact of the programs tagged in the climate budget. Currently, the impact measurement calculated in Aksara is based on data on potential impacts. Different field conditions such as climate, topography, soil, and farmer characteristics cause these impacts not always meet their potential. Thus, conducting a field study is necessary to validate its impact.

The study aims to measure the impact of CRA and propose the correction factor of the area of intervention of CRA. Its scope is therefore limited to the ministry agriculture activities, the local context and private sector are not part of the study.

2 Study Area

The study areas were selected based on the location priority of the climate resilience development (Fig. 1). There are three classifications of priority locations: Super Priority/SP (the location which has a high potential hazard, high vulnerability, and high risk); Top Priority/TP (the location which has high potential hazard and high vulnerability or and high risk); and Priority/P (the location in which has high potential hazard). SP and TP locations were selected as study areas as shown in Table 1. Study areas were selected based on recommendations from the Directorate General of Agricultural Infrastructure and Facilities (DGAIF), Ministry of Agriculture.

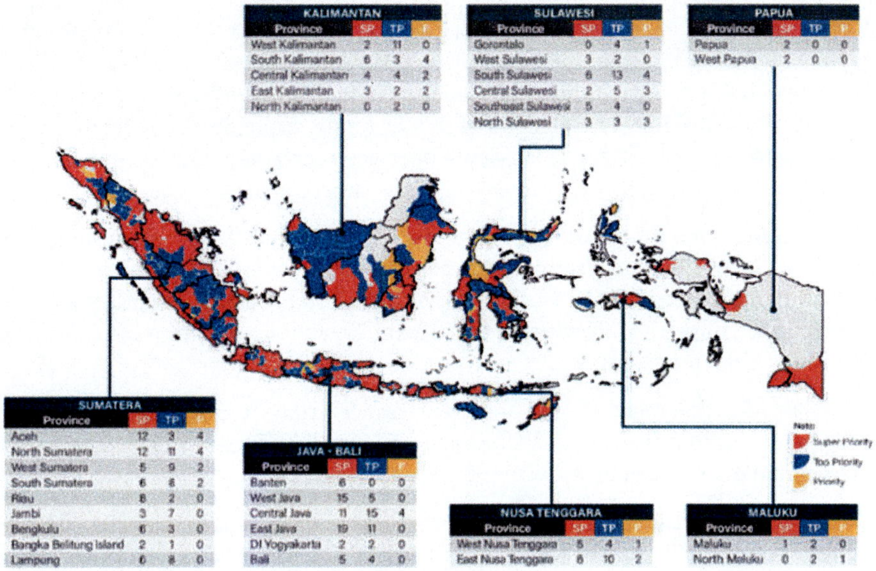

Fig. 1 Location priority for the CRA (Bappenas, 2021)

Table 1 Study areas for measuring the impact of CRA

Province	District
South Sumatera	Muara Enim (SP), Ogan Komering Ulu (SP), Palembang (TP)
West Jawa	Subang (SP), West Bandung (TP), Purwakarta (TP)
DI Yogyakarta	Kulonprogo (SP), Bantul (TP), Gunung Kidul (SP)
West Nusa Tenggara	East Lombok (SP), Central Lombok (SP), West Lombok (SP)

3 Data and Method

3.1 Data

The data used are a list of climate budget tagging activities at the DGAIF, location priority for CRA, a list of prospective farmers and potential locations (CPCL) for climate-resilient activities in 2020, and data activities of Aksara

3.2 Method

The research stages are as follows: (1) identification of climate budget tagging (CBT) activities at the DGAIF, (2) selection of farmer group respondents, (3) identification of parameter input in Aksara, (3) in-depth interviews, and (4) analysis of intervention area and correction factor. The flowchart of the step of analysis is detailed in Fig. 2

Identification of climate budget tagging activities. CBT is a government-led process of identifying, measuring, and monitoring climate-relevant public expenditures in the agriculture sector (World Bank 2021). The tagged climate budget program was identified through FGD with the DGIF team. Selection adaptation activities were referred to CBT programs.

Selection of farmer group respondents. Farmer group respondents in the study areas were selected from the CPCL documented by the DGIF by considering the number and distribution of the activities.

Identification parameter input in Aksara. Aksara has several parameters used as input to assess the impact of activities. The parameter values are determined based on activity planning data and have not been validated in the field.

In-depth interview. This study used specific qualitative methodologies to gather information, as described by [10, 11]. In-depth interview was organized with key local participants namely, member representatives of farmer groups, extension workers, and agricultural officers to gather information on the impact of adaptation activities. A questionnaire was prepared to collect information related to the intervention area, planting index, productivity, the impact of activities, etc.

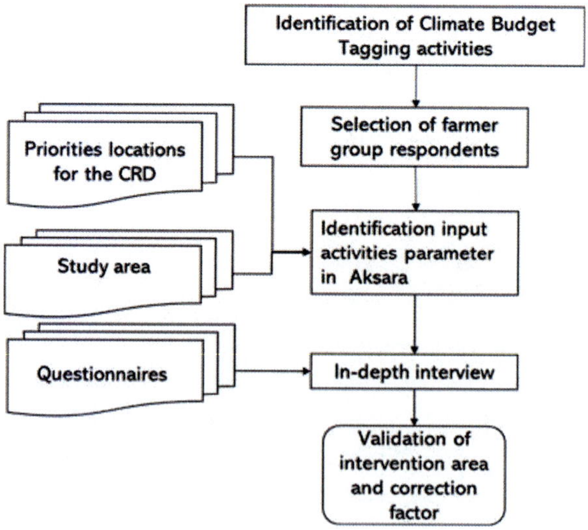

Fig. 2 Flowchart of measuring the climate resilience action

Table 2 Tagged climate resilience actions in DGAIF

No	Activity	Total	Location
1	Tertiary irrigation network rehabilitation	186,264 Ha	31 provinces, 290 districts
2	Pump irrigation	1000 unit	32 provinces, 293 districts
3	Pipe irrigation	138 unit	24 provinces, 67 districts
4	Small water reservoir	400 unit	29 provinces, 215 districts
5	Organic fertilizer processing unit	940 unit	30 provinces, 219 districts

Correction factor. Correction factors for each parameter were calculated by dividing the value used in Aksara and the intervention area observed.

4 Result and Discussion

4.1 Climate Resilient Actions

Activities included in CBT of the fiscal year 2020 in DGAIF are detailed in Table 2. The activities were spread across most provinces and implemented in several districts. Generally, activities in DGAIF are related to increasing irrigation water. Enabling irrigation during the dry season has therefore been crucial in assuring improved productivity and cropping index [12]

The parameters used in these activities are described in Table 3. All activities use the same input parameters, namely the area of rice fields intervened (Ha). The area receiving organic fertilizer is used as a parameter of the OFPU. Crop water coefficient (ratio agriculture yield per total irrigation (ton/m^3), rice price (Rp/kg), and land productivity (Ha) using official data from the Statistics Bureau.

Farmer group respondents were selected based on the CPCL list. In the CPCL list, location information is available at the village level, the name and contact number of the chair of the farmer group. Because the number of farmer groups was not evenly distributed, respondent's numbers vary among locations. For example, farmer groups received pipe irrigation only in Gunung Kidul district, DI Yogyakarta Province and Purwakarta districts, West Jawa Province, while OFPU activities were found in three districts. A total of 34 farmer groups who received the program were selected for in-depth interviews as detailed in Table 4.

A comparison of data on the intervention area in Aksara and from in-depth interviews in each study area is detailed in Table 5. The correction factor is then calculated based on the comparison of the two values. This correction factor can be used to calculate the area of intervention more accurately.

Tertiary irrigation rehabilitation activities are generally carried out on semi-technical rice fields. Based on the interview, this activity has increased the intervention area by about 20–40 ha, varying among districts. The values are lower than

Table 3 Input parameter used in Aksara

Parameter	Tertiary irrigation network rehabilitation	Pump irrigation	Pipe irrigation	Small water reservoir	OFPU
Increased water discharge for irrigation (m3/year)	√	√	√	√	
Area of rice fields intervened (Ha)—filled by the user	√	√	√	√	√
Crop water coefficient (ton/m3)-default	√	√	√	√	
Rice price (Rp/kg)-default	√	√	√	√	√
Land productivity (Ha)-default	√	√	√		√
The area receiving organic fertilizer (Ha)-filled by the user					√

the default used in Aksara in all provinces range 21–48 ha. The default area intervention of the pump irrigation range 6.7—12.9 ha was slightly lower than the default value by 8–15 Ha. Conversely, the interview concludes that the irrigation intervention area from the pipe irrigation range of 15–20 ha is far below its potential value range of 20–40 Ha. According to respondents, this is caused by low river discharge during the dry season.

Water reservoir activities consist of two types, namely small reservoirs and channel reservoirs. The intervention area of the channel reservoir ranges from 20–60 ha, while small reservoirs ranging from 1–5 ha. In Aksara, these two reservoir types are combined with the default intervention area of 42 ha. In South Sumatra, there are only small reservoirs with an intervention area range of 2—5 ha, much lower than the potential value used in Aksara. Based on the interview, farmers stated that rainfall and runoff collected in the reservoirs are not sufficient to meet the high-water demand for rice crops in the dry season. With an annual precipitation of less than 2000 mm/year, and only 3–4 wet months, rainwater is available for just one growing season [13]. In West Java, there are small reservoirs and channel reservoirs whose average value of the intervention area is 20 Ha, far below the default used in Aksara. The difference in the area of interventions between the two types of reservoirs should be distinguished in Aksara, for the calculation of the GDP value to be more accurate.

Table 4 Selected farmers group respondents for an in-depth interview

Activity	Province	District	Sub-district (Farmers group)
Tertiary irrigation network rehabilitation	South Sumatera	Ogan Komering Ulu	Sindang sari (Sri Utama)
	West Jawa	Purwakarta	Darandang (Warga Tani II), Kasomalang (Sukatani 2),
	DI Yogyakarta	Kulon Progo	Panjatan (Anagung, Muda mudi, Tirtawening, Ginanjar Mulyo)
		Bantul	Imogiri (Ngupoyo Tirto), Bambang Lipuro (Tirto Lumintu),
		Gunung Kidul	Ponjong (Sido Guyub)
	West Nusa Tenggara	East Lombok	Sakra Barat (Tampi Asih III, Tani Maju)
		West Lombok	Kediri (Otakde), Gerung (Tunas Budi)
Pump irrigation	South Sumatera	Ogan Komering Ulu	Ulu Ogan (Siang Bulan, Pioner Jaya)
		Ogan Komering Ilir	Lempuing (Sidodadi II)
		Palembang	Gandus (Kertawijaya)
	West Jawa	Purwakarta	Darandang (Lingga Jaya I)
		Subang	Pegaden Barat (Subur Jaya)
		Bandung Barat	Padalarang (Cimerang 1)
	DI Yogyakarta	Gunung Kidul	Karangmojo (Gatakrejo)
	West Nusa Tenggara	Central Lombok	Jonggat (Taek Gunung), Janapria (Terentem 1)
Pipe irrigation	West Jawa	Purwakarta	Darandang (Sugih Mukti),
	DI Yogyakarta	Gunung Kidul	Karang Mojo (Mudirahayu)
Small water reservoir	South Sumatera	Ogan Komering Ulu	Tanjung Lubuk (Suka Makmur)

(continued)

Table 4 (continued)

Activity	Province	District	Sub-district (Farmers group)
	West Jawa	Purwakarta	Darandang (Tirta Surya Pertiwi)
	DI Yogyakarta	Bantul	Imogiri (Bumi Mukti)
OFPU	South Sumatera	Palembang	Sematang Barang (Tumbuh Murni)
	West Jawa	Purwakarta	Darandang (Tirta Surya Pertiwi)
	DI Yogyakarta	Kulon Progo	Wates (Sinar Mataram II)
		Bantul	Bambang Lipuro (Mandiri)
		Gunung Kidul	Sewanu (Marsudi Rejeki)
	West Nusa Tenggara	West Lombok	Mariji (Pade Jujur Pendem)

Table 5 Comparison of the intervention area used in the Aksara and observation

Activity	Province	Intervention (Ha)		Correction factor
		Aksara	Survey	
Tertiary irrigation network rehabilitation	South Sumatera	48	40	0.94
	West Jawa	35	30	0.86
	DI Yogyakarta	21	20	0.95
	West Nusa Tenggara	43	40	0.93
Pump irrigation	South Sumatera	12.9	15	1.17
	West Jawa	9.5	10	1.06
	DI Yogyakarta	10.0	10	1.00
	West Nusa Tenggara	6.7	8	1.19
Pipe irrigation	West Jawa	40.0	20	0.67
	DI Yogyakarta	20.0	15	0.75
Small water reservoir	South Sumatera	42.0	5	0.12
	West Jawa	35.0	20	0.57
OFPU	South Sumatera	7.1	5	0.70
	West Jawa	5.6	3	0.53
	DI Yogyakarta	5.5	4	0.72
	West Nusa Tenggara	5.5	3	0.54

The intervention area from OFPU activity varies among locations. During 2021, some farmer groups have produced organic fertilizer, but others have not mastered the process. The intervention area of 5.5–7 ha was used in Aksara. However, a lower area of intervention ranging from 3–5 ha was found in the field survey. The obstacle faced by farmers is the management of organic fertilizer production through OFPU. The technical guidance of the OFPU application is highly recommended for improving the quality of the fertilizer produced.

The research shows that the intervention area for all activities is lower than the default used in Aksara, except for the pump irrigation which is slightly valued above the default with a range of 1.00−1.19. The CF for irrigation network rehabilitation range from 0.86−0.95, pipe irrigation 0.67−0.75, small water reservoir 0.12−0.57, and OFPU 0.53−0.72. The CF are recommended to apply in Aksara for increasing the accuracy of the intervention area as an indicator for calculating the reduction in GDP loss from CRA.

5 Conclusion

To achieve the CRA goals and targets, the Bappenas have developed a mechanism for monitoring, evaluating, and reporting the climate resilience activities called Aksara. Currently, the impact measurement calculated in Aksara is based on data on potential areas of intervention of the activities. Corrections need to validate the size of the intervention area as an indicator to calculate the decrease in GDP loss due to the CRA.

The research shows that the intervention area for all CRA is lower than the default value in Aksara, except for the pump irrigation which is slightly above the default value with a range of 1.00−1.19. The CF are recommended for increasing the accuracy of the intervention area as input for calculating the reduction in GDP loss from CRA. Correction factors for tertiary irrigation network rehabilitation range from 0.86−0.95, pipe irrigation 0.67−0.75, small water reservoir 0.12−0.57, and OFPU 0.53−0.72.

Given the very important accuracy of the area interventions for calculating the decrease in GDP loss of CRA, this study needs to be expanded to other locations to attain more comprehensive recommendations.

Acknowledgements This study was financially supported by Indonesian Center for Agricultural Land Resources Research and Development Grant Number 4584.SDA.512.056A/2021. The authors sincerely thank the Directorate General of Agriculture Infrastructure and Facilities for supporting data service

References

1. Surmaini, E., Hadi, T.W., Subagyono, K., Puspito, N.T.: Early detection of drought impact on rice paddies in Indonesia by means of Niño 3.4 index. Theor. Appl. Climatol. **121** (3–4) 6692015 (2015)
2. Arora, N.K.: Impact of climate change on agriculture production and its sustainable solution. Environ. Sustain. **2**, 95–96 (2019)
3. Boer, R., Surmaini, E.: Economic benefits of ENSO information in crop management decisions: case study of rice farming in West Java Indonesia. Theoret. Appl. Climatol. **139**(3–4), 14352020 (2016)
4. St-Louis, M., Schlickenrieder, J., Bernoux, M.: The Koronivia joint work on agriculture and the convention bodies: an overview. Rome, Food and Agriculture Organization of the United Nations (2018)
5. Belliveau, S., Smit, B., Bradshaw, B.: Multiple exposures and dynamic vulnerability: evidence from the grape industry in the Okanagan Valley Canada. Glob. Environ. Chang. **16**(4), 364–378 (2006)
6. Nelson, R., Kokic, P., Crimp, S., et al.: The vulnerability of Australian rural communities to climate variability and change: part II—integrating impacts with adaptive capacity. Environ. Sci. Pol. **13**, 18–27 (2010)
7. Vermeulen, S.J., Dinesh, D., Measures for climate change adaptation in agriculture opportunities for climate action in the agricultural system. CCAFS Info note. Copenhagen. Denmark. CGIAR Research Program on Climate Change, Agriculture and Food Security (CCAFS) (2016)
8. Internationa Panel on Climate Change. Summary for policymakers. In: Parry, M., et al (eds.) Climate Change 2007: Impacts, Adaptation, and Vulnerability. Contribution of Working Group II to the Fourth Assessment Report of the Intergovernmental Panel on Climate Change. Cambridge University Press, Cambridge (2007)
9. Kementerian PPN/Bappenas.: Executive summary Climate resilience development policy. Ministry of National Development Planning/Bappenas (2021)
10. Van Huynh, C., Le, Q.N.P., Nguyen, M.T.H., Tran, P.T., Nguyen, T.Q., Pham, T.G., Nguyen, L.H.K., Nguyen, L.T.D., Trinh, H.N.: Indigenous knowledge in relation to climate change: adaptation practices used by the Xo Dang people of central Vietnam. Heliyon **6**, e05656 (2020)
11. Limpo, S.L., Fahmid, I.M., Fattah, A., Rauf, A.W., Elza Surmaini, E., Saptana, M., Syahbuddin, H., Andri, B.K.: Integrating indigenous and scientific knowledge for decision making of rice farming in South Sulawesi, Indonesia. Sustainability **14**, 2952 (2022)
12. Su, Y., Colton, J.S., Matin, M.D.A., Krupnik, T: .J18. A simplified field irrigation pump testing method for developing countries: a case study in Bangladesh. Irrig. and Drain. **67**, 559–571 (2018)
13. Mulyani, A., Sarwani, M.: Characteristics and potential of sub-optimal land for agricultural development in Indonesia. J. Sumberdaya. Lahan **7**, 47–55 (2013). (In Indonesian)

Soil CO_2 Emissions Through Peat Decomposition in a Strong El Niño Year Were Higher Than in a Normal Year

Hidayatuz Zu'amah, Cicik Oktasari Handayani, and Nur Wakhid

Abstract The long dry season during the El Niño event usually causes the groundwater level to drop significantly. As a result, soil CO_2 emissions through peat decomposition are potentially increased compared to a normal year. Even so, studies on the effect of the El Niño years on peat decomposition are rare in tropical peatlands. The study objective was to compare soil CO_2 emissions through peat decomposition during an El Niño and a normal year. A field study on soil CO_2 emissions through peat decomposition measurement was undertaken in a smallholder rubber plantation on peat soil from July to October 2015, utilizing a trenching plot during the dry season. Also, as a comparison, soil CO_2 emissions from peat decomposition were recorded during the same period in 2014. In 2015, a strong El Niño year, soil CO_2 emissions from peat decomposition were significantly larger than in 2014, a normal year. Also, soil CO_2 emissions through peat decomposition in 2015 showed a significant negative correlation with groundwater level. Our results showed that different climatic conditions affect the peat decomposition process on tropical peat soil. This result is expected to be used as important information regarding Indonesia's peatland management in dealing with future climate change.

1 Introduction

Tropical peatlands that are usually associated with swamp forests under waterlogged conditions represent a huge store of carbon (C) as a result of organic matter accumulation over decades. Only roughly 17% of the world's peatlands are tropical, but they store up to 104.7 petagrams of carbon, including the peatlands of the Congo Basin

H. Zu'amah (✉) · C. O. Handayani
Research Center for Horticultural and Estate Crops, National Research and Innovation Agency, Cibinong Science Center, Jl. Raya Jakarta–Bogor, Cibinong, Bogor 16915, Indonesia
e-mail: aya.zuzun@gmail.com

N. Wakhid
Research Center for Ecology and Ethnobiology, National Research and Innovation Agency, Cibinong, Bogor 16911, Indonesia

A. Basit et al. (eds.), *Proceedings of the International Conference on Radioscience, Equatorial Atmospheric Science and Environment and Humanosphere Science*, Springer Proceedings in Physics 290, https://doi.org/10.1007/978-981-19-9768-6_56

[1]. Therefore, this ecosystem is crucial to the global C cycle and the mitigation of greenhouse gas emissions [2]. Indonesia has the biggest tropical peatland region in Southeast Asia, which is spread throughout the islands of Sumatra, Kalimantan, Sulawesi, and Papua [3]. Central Kalimantan has the most peatland in Kalimantan, with approximately 2.6 million ha of peat and a C deposit of up to 1220 Mg C ha^{-1} [3, 4].

However, this huge C deposit is now vulnerable to emitting a large proportion of C emissions and increasing fire risks due to land conversion. The land conversion on peatland is usually equipped with drainage construction, which lowers the groundwater level and disturbs the soil condition [5]. The mega rice project's failure in the late 1990s degraded Central Kalimantan's peat.

Several studies have quantified soil CO_2 emissions from tropical peatland plantations, but few have separated peat decomposition and heterotrophic respiration from total soil respiration [6–9]. Most of the publications have typically measured the soil CO_2 efflux (emission rates) using closed chamber methods [6]. Thus, it is critical to exclude root respiration during carbon dioxide (CO_2) emissions' measurement in order to estimate CO_2 emissions from peat decomposition only using closed chamber methods. In tropical peatlands, particularly during El Niño years, field measurements of CO_2 emissions through peat decomposition are limited. The El Niño events cause most areas in Indonesia to experience drought, including peatlands [10]. As a consequence, precipitation has dropped, the dry season is prolonged, and the groundwater level has decreased more. Also, fire risk is increased [11], and CO_2 emissions through peat decomposition are accelerated. Therefore, direct measurement of CO_2 emissions through peat decomposition during the El Niño event is critical to understand the impact of climate conditions. The study's objective was to compare CO_2 emissions through peat decomposition in El Niño and normal years.

2 Methods

2.1 Study Site

The study was conducted in peat soil at Central Kalimantan, Indonesia (02°29′50″S, 114°11′20″E) in a smallholder rubber plantation. In most areas, the peat was about 5 m deep. The research site was a peat swamp forest before it became a component of the failed mega rice project. Before rubber trees were planted there in 2007, peat fires also frequently impacted the site. The trees were planted at a spacing of 3 × 6.5 m, with pineapple intercropped between the tree rows. Due to rubber trees being defoliated during the dry season, a high amount of leaf litter was found on the soil surface during the measurement.

2.2 Groundwater Level Measurement

To understand the effect of environmental factor, groundwater level was measured during peat decomposition measurement. Every hour, the groundwater level was measured using an automatic water logger (HOBO U20, Onset, Massachusetts, USA) placed vertically through a perforated PVC pipe introduced into the soil [12]. Also, data on precipitation from Palangkaraya were used to show the connection between groundwater level and precipitation [12].

2.3 Soil CO_2 Emissions Through Peat Decomposition Measurement

Soil CO_2 emissions through peat decomposition were measured during the dry seasons in 2014 and 2015, normal and strong El Niño years [13], respectively. The dry season in 2014 and 2015 was estimated to last four months, from July, August, September, and October, based on the criterion of monthly precipitation of less than 100 mm [14]. Therefore, in this study, the peat decomposition was measured monthly from July through October in 2014 and 2015, respectively. The measurement was conducted on the chamber base using a closed chamber system. The chamber was opaque to avoid the acceleration of temperature due to the sunshine on the day of measurement. The chamber was made from PVC and had a diameter of 30 cm and a height of 20 cm. Because there is a groove at the top of the chamber base, the chamber and base were water sealed to prevent the gas from escaping.

A portable infrared gas analyzer (GMP343, Vaisala, Helsinki, Finland) was mounted inside the chamber to monitor the soil CO_2 emissions from peat decomposition. Along with the gas analyzer, a temperature data logger (LR5011, HIOKI) and a DC data logger (LR5042, HIOKI, Nagano, Japan) were installed within the chamber. Every six months, the gas analyzer would be calibrated using standard gases and run in open path mode. Each base measured CO_2 emissions and air temperature every 10 s for three minutes. An automated logger measured of soil temperature around the bases at 5 cm depth. As soil temperature varies throughout the day, measurements were taken in the morning, during the day, and in the afternoon. The soil CO_2 efflux (μmol m^{-2} s^{-1}) was computed from air temperature (Ta, °C) and the rate of increase in CO_2 concentration (dC/dt, μmol m^{-2} s^{-1}) using Eq. (1) [15]:

$$\text{Soil}\,CO_2\,\text{efflux} = \frac{dC}{dt}\cdot\frac{V}{V'\left(\frac{273.15+Ta}{273.15}\right)}\cdot\frac{1}{A} \quad (1)$$

where V is the volume of the chamber (0.0144 m^3), V' is the molar volume of air at 0 °C (0.0224 m^3 mol^{-1}), and A is the chamber's ground area (0.0707 m^2). CO_2 concentrations from the final two minutes of monitoring were used to determine *dC/dt* using the least-squares method.

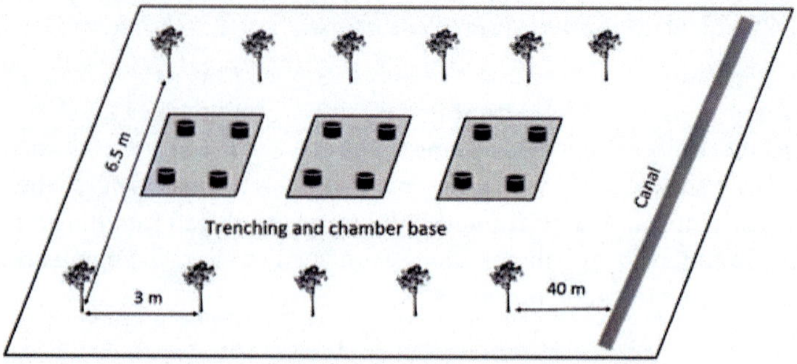

Fig. 1 Design of trenching plots and distribution of chamber bases

To exclude root respiration, soil CO_2 emissions from peat decomposition were measured using a trenching plot. At a distance of around 40 m from a drainage canal, three 1 × 1 m^2 trenching plots were established (see Fig. 1). In order to prevent root invasion within the plots, the roots surrounding each plot were severed and trenched into the peat at a depth of 1 m by inserting plastic boards on their four sides. Four chamber bases were put 2 cm into the peat surface to prevent disturbing it during measurements (see Fig. 1). Also, to avoid CO_2 emissions through litter decomposition [16, 17], rubber leaf that accumulated in the trenching plot was cleaned before measurement.

2.4 Data Analysis

Analysis of variance (ANOVA) and Tukey's multiple comparison test were utilized in order to make comparisons between the datasets. The linear regression and the Pearson correlation tests were used to investigate the association between the rate of peat decomposition and the groundwater level. The analysis of the data was created with the help of the software packages Microsoft Excel and Real Statistics Resource Pack (Charles Zaiontz, www.real-statistics.com).

3 Results and Discussion

3.1 Groundwater Level

Monthly precipitation between July and October was measured at less than 100 mm both in 2014 and 2015 (see Fig. 2). There was little difference between the four month precipitation averages of 2014 (40 mm $month^{-1}$) and 2015 (39.2 mm $month^{-1}$). In contrast to 2014, the lowest rainfall registered at 16 mm in July 2015 had only

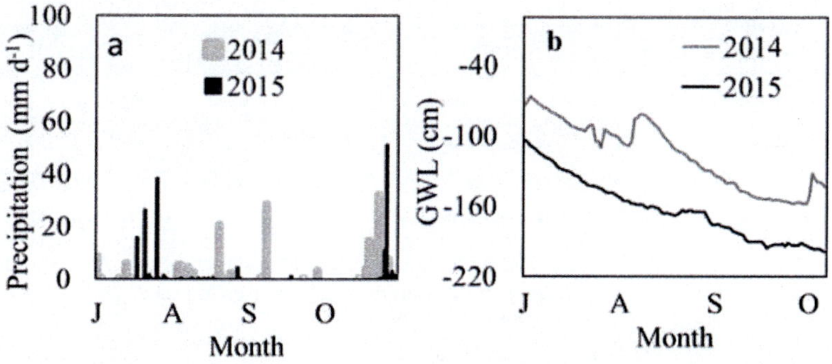

Fig. 2 Daily variations of **a** precipitation and **b** groundwater level from July to October in 2014 (gray) and 2015 (black line), respectively. The precipitation and groundwater level data were adapted from Wakhid and Nurzakiah [12]

0.9 mm month^{-1} in September (see Fig. 2). During 4 months, the average of groundwater level was -1.17 and -1.63 m, in 2014 and 2015, respectively. Groundwater levels were 0.46 m lower in 2015 (El Niño year) than in 2014. Groundwater levels in 2015 were continuously lower than one meter and varied from -1.04 to -1.99 m, while in 2014, the groundwater level varied from -0.67 to -1.58. The lowest groundwater level was measured in October, and the highest level near the soil surface was recorded in July, the same for 2014 and 2015.

3.2 Soil CO_2 Emissions Through Peat Decomposition

Soil CO_2 emissions through peat decomposition between three replications (morning, daytime, and afternoon) in 2014 and 2015 were not significantly different ($p > 0.05$). Thus, daily averages were calculated for each trenching's four chamber bases based on data from three replicates collected in 2014 and 2015 (see Table 1). The soil CO_2 emissions between the three trenching plots in 2014 were significantly different. However, in 2015, data from three trenching plots were not significantly different (see Table 1).

Data of CO_2 emissions in 2014 and 2015 were averaged using all three trenching plots for each month to see the seasonal variation of soil CO_2 emissions through peat decomposition (see Fig. 2). Even though small, the CO_2 emissions' variations in 2014 and 2015 followed the variation of groundwater level (see Figs. 2 and 3). In 2014 and 2015, soil CO_2 emissions from peat decomposition ranged from 1.95 to 2.73 and 5.02 to 8.58 μ mol m^{-2} s^{-1}, respectively (see Fig. 3). The highest CO_2 emissions through peat decomposition were measured in October both in 2014 and 2015.

The highest soil CO_2 emissions through peat decomposition were recorded to coincide with the lowest groundwater level (see Figs. 2 and 3). In 2015, soil CO_2

Table 1 Mean soil CO_2 efflux from three trenching plots (mean ± 1SD)

Plot	Soil CO_2 efflux (μ mol m^{-2} s^{-1})	
	2014	2015
Trenching 1	3.06 ± 0.51a	7.76 ± 1.09
Trenching 2	1.86 ± 0.33b	6.00 ± 1.20
Trenching 3	1.89 ± 0.58b	6.53 ± 2.29

According to the Tukey's HSD test that follows ANOVA, different letter suffixes denote significant differences ($p < 0.05$) between trenching

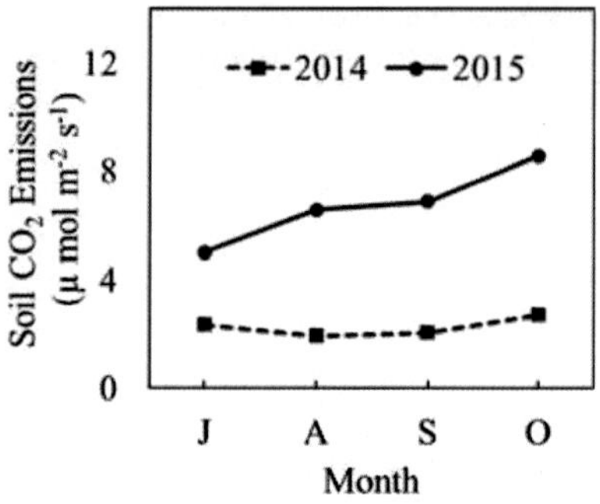

Fig. 3 Monthly variations of soil CO_2 emissions through peat decomposition, from July to October, in 2014 (dashed line) and 2015 (solid line), respectively

emissions showed a significant negative correlation with groundwater level ($p < 0.001$, $r^2 = 0.97$). However, in 2014, the relationship between soil CO_2 emissions through peat decomposition and groundwater level was not significant (see Fig. 4).

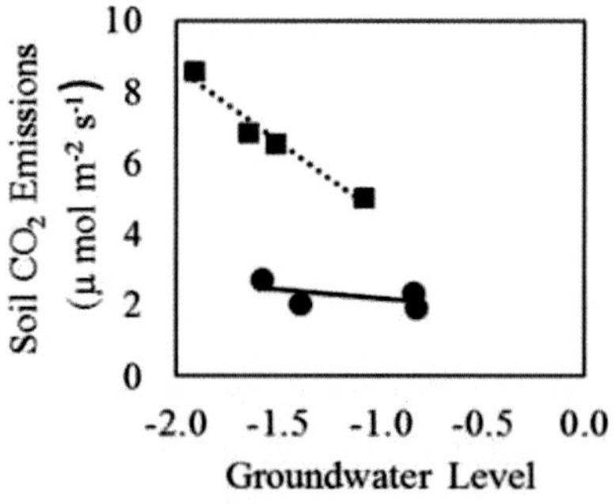

Fig. 4 Relationship between soil CO_2 emissions through peat decomposition and groundwater level, from July to October, in 2014 (black circle) and 2015 (black square), respectively. Each symbol denotes an average of three trenching plots ($n = 3$). A line is fitted for 2014 (solid line, $p = 0.45$, $y = -0.50x + 1.69$, $r^2 = 0.30$) and 2015 (dotted line, $p < 0.05$. 2014: and 2015: $y = -4.12x + 0.41$ $r^2 = 0.97$)

Table 2 Soil CO_2 emissions through peat decomposition during four months derived from mean efflux (mean ± 1SD)

Peat decomposition	Soil CO_2 emissions during 4 months (g C m^{-2})	
	2014	2015
Trenching 1	390 ± 65	990 ± 139
Trenching 2	238 ± 42	765 ± 153
Trenching 3	241 ± 74	834 ± 293
Mean (n = 3)	290 ± 87a	863 ± 115b

Different suffixed letters denote a significant difference ($p < 0.05$) between normal and El Niño according to the Tukey's HSD test following ANOVA

The results of measurement were simply converted into four months (123 days) of soil CO_2 emissions through peat decomposition using the mean of three trenching plots in 2014 and 2015 (see Table 2). In 2015, a year with a severe El Niño event, the average rate of soil CO_2 emissions through peat decomposition in the dry season was nearly three times higher than in the typical year of 2014 (see Table 2). The soil CO_2 emissions through peat decomposition during four months in 2015 significantly differed in 2014 ($p < 0.001$).

Soil CO_2 emissions through peat decomposition in 2015 showed significant negative linearity with groundwater level because peat carbon in plantations is vulnerable to drought events such as a strong El Niño event [18]. Even though only 4 months, soil CO_2 emissions through peat decomposition in 2014 were only slightly lower than annual soil CO_2 emissions through peat decomposition in a burnt forest on peat soil (372 g C m^{-2} $year^{-1}$) [19]. Also, soil CO_2 emissions through peat decomposition in 2015 were slightly lower than annual soil CO_2 emissions through peat decomposition from an oil palm plantation established on peat soil (930 g C m^{-2} $yearr^{-1}$) [20]. The dry season, particularly in an El Niño event, is likely to be stated as a peak season of soil CO_2 emissions through peat decomposition on tropical peat soil.

4 Conclusion

Soil CO_2 emissions through peat decomposition were compared during the dry season between a normal year in 2014 and a strong El Niño year in 2015 on a smallholder rubber plantation in Central Kalimantan, Indonesia. The soil CO_2 emissions through peat decomposition in a strong El Niño year were almost three times larger than that in a normal year. The result showed that climatic conditions should be considered carefully during carbon emissions measurement on tropical peatlands. Also, the effort for climate change mitigation on tropical peatlands should be greater during an El Niño year.

References

1. Dargie, G.C., Lewis, S.L., Lawson, I.T., Mitchard, E.T.A., Page, S.E., Bocko, Y.E., Ifo, S.A.: Age, extent and carbon storage of the central Congo Basin peatland complex. Nature **542**, 86–90 (2017)
2. Strack, M., Davidson, S.J., Hirano, T.: The potential of peatlands as nature-based climate solutions. Current Climate Change Reports **8**, 71–82 (2022)
3. Anda, M., Ritung, S., Suryani, E., Sukarman, S., Hikmat, M., Yatno, E., Mulyani, A., Subandiono, R.E., Suratman, S., Husnain, H.: Revisiting tropical peatlands in Indonesia: Semi-detailed mapping, extent and depth distribution assessment. Geoderma **402**(115235), 1–14 (2021)
4. Murdiyarso, D., Hergoualc'h, K., Verchot, L.V.: Opportunities for reducing greenhouse gas emissions in tropical peatlands. PNAS 107 19655–19660 (2010)
5. Furukawa, Y., Inubushi, K., Ali, M., Itang, A.M., Tsuruta, H.: Effect of changing groundwater levels caused by land-use changes on greenhouse gas fluxes from tropical peatlands. Nutr. Cycl. Agroecosyst. **71**, 81–91 (2005)
6. Wakhid, N., Hirano, T., Okimoto, Y., Nurzakiah, S., Nursyamsi, D.: Soil carbon dioxide emissions from a rubber plantation on tropical peat. Sci. Total Environ. **581–582**, 857–865 (2017)
7. Hergoualc'h, K., Hendry, D.T., Murdiyarso, D., Verchot, L.V.: Total and heterotrophic soil respiration in a swamp forest and oil palm plantations on peat in Central Kalimantan, Indonesia. Biogeochemistry **135**, 203–220 (2017)
8. Ishikura, K., Hirano, T., Okimoto, Y., Hirata, R., Kiew, F., Melling, L., Aeries, E.B., Lo, K.S., Musin, K.K., Waili, J.W., Wong, G.X., Ishii, Y.: Soil carbon dioxide emissions due to oxidative peat decomposition in an oil palm plantation on tropical peat. Agricul. Ecosyst. Environ. **254**, 202–212 (2018)
9. Manning, F.C., Kho, L.K., Hill, T.C., Cornulier, T., Teh, Y.A.: Carbon emissions from oil palm plantations on peat soil. Frontiers in Forests and Global Change **2**(37), 21 (2019)
10. Dadap, N.C., Cobb, A.R., Hoyt, A.M., Harvey, C.F., Feldman, A.F., Im, E.S., Konings, A.G.: Climate change-induced peatland drying in Southeast Asia. Environ. Res. Lett. **17**, 7 (2022)
11. Nurdiati, S., Bukhari, F., Julianto, M.T., Sopaheluwakan, A., Aprilia, M., Fajar, I., Septiawan, P., Najib, M.K.: The impact of El Niño southern oscillation and Indian Ocean Dipole on the burned area in Indonesia. Terrestrial, Atmosp. Oceanic Sci. **33**, 16 (2022)
12. Wakhid, N., Nurzakiah, S.: Soil CO_2 emissions from a rubber plantation on tropical peat during a strong El Niño year. In: IOP Conference Series: Earth and Environmental Science, 1st International Conference on Sustainable Tropical Land Management 16−18 September 2020, Bogor, Indonesia, vol. 648, pp. 012098. (2021)
13. Schiermeier, Q.: Hunting the Godzilla El Niño. Nature **526**, 490–491 (2015)
14. Malhi, Y., Pegoraro, E., Nobre, A.D., Pereira, M.G.P., Grace, J., Culf, A.D., Clement, R.: Energy and water dynamics of a central Amazonian rain forest. J. Geophys. Res. **107**(D20), 8061 (2002)
15. Sano, T., Hirano, T., Liang, N., Hirata, R., Fujinuma, Y.: Carbon dioxide exchange of a larch forest after a typhoon disturbance. For. Ecol. Manage. **260**(12), 2214–2223 (2010)
16. Wakhid, N., Hirano, T.: Soil CO_2 emissions and net primary production of an oil palm plantation established on tropical peat. Mires and Peat **27**, 13, 11 (2021)
17. Wakhid, N., Hirano, T.: Carbon dioxide emissions from frond Decomposition in oil palm plantations on tropical peat. J. Oil Palm Res. in press (2022)
18. Kwon, M.J., Haraguchi, A., Kang, H.: Long-term water regime differentiates changes in decomposition and microbial properties in tropical peat soils exposed to the short-term drought. Soil Biol. Biochem. **60**, 33–44 (2013)
19. Hirano, T., Kusin, K., Limin, S., Osaki, M.: Carbon dioxide emissions through oxidative peat decomposition on a burnt tropical peatland. Glob. Change Biol. **20**, 555–565 (2014)
20. Dariah, A., Marwanto, S., Agus, F.: Root- and peat-based CO_2 emissions from oil palm plantations. Mitig. Adapt. Strat. Glob. Change **19**, 831–843 (2014)

The Level of Paddy Rice Farmer's Adoption on Sustainable Agricultural Practices in Ciamis Regency, West Java

Abdul Mutolib, Candra Nuraini, Indah Listiana, Helvi Yanfika, Raden Ajeng Diana Widyastuti, Ali Rahmat, and Rinaldi Bursan

Abstract Intensive rice cultivation has been shown to increase rice production. However, in the long term intensive cultivation causes a decrease in soil and water quality and reduced production. Sustainable rice cultivation is an effort to apply environmental-friendly agricultural methods and techniques to avoid environmental damage and saturation of land resources. The study was conducted in Pamarican District, Ciamis Regency, West Java Province, from July to September 2021. The number of samples was 40 farmer households selected by simple random sampling. The study used a descriptive–qualitative method. The level of farmer adoption was measured using a Likert scale. Sustainability indicators used in this study include: (1) land cultivation, (2) use of organic fertilizers, (3) use of organic pesticides and herbicides, (4) utilization of straw waste (post-harvest), and (5) utilization of refugia plants. The results showed that the indicators of the use of organic fertilizers were classified in high category, land cultivation and utilization of straw waste (post-harvest) were classified in the medium category, and the use of organic pesticides and herbicides and the use of refugia plants as natural enemies of plant pest organisms were in a low category. Collaboration between stakeholders is needed to improve sustainable agricultural practices in Pamarican District.

A. Mutolib (✉)
Graduate Program, University of Siliwangi, Tasikmalaya, Indonesia
e-mail: abdul.mutolib@unsil.ac.id

C. Nuraini
Faculty of Agriculture, University of Siliwangi, Tasikmalaya, Indonesia

I. Listiana · H. Yanfika · R. A. D. Widyastuti
Faculty of Agriculture, University of Lampung, Bandar Lampung, Indonesia

A. Rahmat
Research Center for Limnology and Water Resources, National Research and Innovation Agency, Bogor, Indonesia

R. Bursan
Faculty of Economics, University of Lampung, Bandar Lampung, Indonesia

A. Basit et al. (eds.), *Proceedings of the International Conference on Radioscience, Equatorial Atmospheric Science and Environment and Humanosphere Science*, Springer Proceedings in Physics 290, https://doi.org/10.1007/978-981-19-9768-6_57

1 Introduction

Rice cultivation is an essential commodity for the people of Indonesia. Most Indonesian people consume rice as a staple food, and this causes the need for rice as a primary staple to increase yearly. In 2020, rice production in Indonesia is 31.51 million tons, and in 2021, it will decrease to 31.36 million tons [1]. The availability of rice is closely related to the country's food security; the decline in production and the scarcity of rice can cause hunger and other potential economic problems. The high demand for rice encourages the government and other related parties to increase rice productivity at a higher limit.

The government has attempted to increase rice production in several ways, one of which is through the intensification of rice farming. This intensification is an effort to maximize the productivity of agricultural land in order to obtain more extraordinary results by implementing five farming businesses. Five farming businesses include various techniques, namely modern soil management, use of superior varieties, application of good irrigation techniques, use of agricultural fertilizers, and control of plant pests using chemical pesticides and herbicides [2]. In Indonesia, the intensification program was implemented through the Mass Guidance Program (BIMAS) in 1965 [3].

The agricultural intensification program, especially in rice crops, succeeded in leading Indonesia to be self-sufficient in rice in 1984 [4]. Unfortunately, after self-sufficiency, rice production and productivity in the 1990s tended to stagnate while the population increased. Demand for rice also continued to increase, so rice needs were forced to be met by imports. This illustrates that farming with an intensive system does not guarantee sustainable production. Intensive land management, such as the continuous use of high doses of inorganic fertilizers, stimulates the acidification process of the soil so that the area of acid land increases. Farming systems with intensive methods also encourage soil saturation due to high external chemical inputs, so they cannot support increased crop productivity.

Furthermore, using non-organic (chemical) pesticides and herbicides has encouraged pest and disease resistance and the extinction of natural enemies, thereby increasing the potential for crop failure. The decline in crop production is one indicator that the intensive method of agriculture is unsustainable and leads to a decrease in crop production in the long term. Therefore, it is necessary to develop agriculture to maintain stable and sustainable agricultural production.

Agricultural cultivation currently leads to a sustainable cultivation system, one of which is through organic farming systems. The organic farming system is one of the sustainable agricultural innovations that can answer the demands and challenges of today's agricultural world. Some research results show that by implementing organic farming, farmers will get benefits, including reducing production costs, getting higher prices than conventional products, increasing income, avoiding dependence on input suppliers, increasing the economic sustainability of farmers, and improving the health of farmers and their families [5]. However, the adoption rate of organic rice farming is still relatively low [6, 7].

This study will not examine the implementation of organic rice farming systems but will focus on sustainable agricultural practices that farmers have adopted. The use of chemical inputs and intensive cultivation techniques can still be applied. However, some sustainable farming practices must also be encouraged to create a balance in the agricultural system, especially in rice plants. Agriculture should be carried out environmentally friendly, making the best use of natural resources and available inputs. This system is known as sustainable agriculture, namely, an agricultural system that does not damage, does not change, and is harmonious, in harmony with the environment [8, 9].

Pamarican District in Ciamis Regency, West Java Province, is the center of rice cultivation and has been designated as the location of the Farmers' Corporation by the Ministry of Agriculture of the Republic of Indonesia. As the location of the Farmer's Corporation, the focus of the development of Pamarican District is not only on the marketing aspect of rice but also on sustainable production. Therefore, this study aims to determine the level of adoption of rice farmers in sustainable rice farming in Pamarican District, Ciamis Regency.

2 Method

The research location was determined purposively in Pamarican District, Ciamis Regency, West Java Province. The location was chosen deliberately because Pamarican District is the center of lowland rice in the Ciamis Regency. Pamarican District is one of the areas designated by the Ministry of State-Owned Enterprises (BUMN) as a center for developing and empowering farmers to support food self-sufficiency and the welfare of farmers.

The research was carried out from July to September 2021 with 40 rice farmers in Pamarican, Sukahurip, and Sukajaya villages. The three villages were chosen intentionally as the center of lowland rice in Pamarican District. The number of research respondents was selected using a simple random method to ensure that the data represented the farmer population. The data collected in this study are primary data and secondary data. Primary data were obtained from direct interviews with farmers. The secondary data were obtained from relevant agencies such as the Central Statistics Agency and Agricultural Extension Center; reports from relevant agencies, journals, and other sources support research on Farmers' Adoption of Sustainable Agricultural Practices.

The research used descriptive–qualitative. According to Sukmadinata, descriptive–qualitative research intends to describe existing phenomena, both natural and human-engineered, which pay more attention to the characteristics, quality, and interrelationships between activities [10]. Data collection techniques used are field observations and interviews with research questionnaires. The level of adoption of rice farmers toward sustainable agricultural practices uses a Likert scale with low, medium, and high categories. The low, medium, and high categories were obtained through the calculation of scores in the questionnaire. Each sustainability indicator has 12 questions with an answer score range of 1–5. So, the value obtained from each indicator is 12–60. Categories were obtained by dividing the score into three quartiles (low category scores between 12 and 28, medium category scores between 28.1 and 44, and high category scores between 44.1 and 60). Data were analyzed by an interactive analysis model, which includes data reduction, data presentation, and conclusion drawing/verification [11].

3 Results and Discussion

3.1 *Sustainable Agricultural Practices*

The results of the study have identified five sustainable rice farming practices in Pamarican District, Ciamis Regency, which include (1) land cultivation, (2) use of organic fertilizers, (3) use of organic pesticides and herbicides, (4) utilization of straw waste (post-harvest), and (5) utilization of refugia plants. Each indicator has a different level of adoption in society. Some indicators are adopted with a high level of adoption. However, some are adopted at a low rate, and the number of adopters (farmers who apply sustainable agricultural practices) is low. Nevertheless, sustainable agricultural practices that already exist in the community are social capital that can be developed more massively and collectively. An explanation of the practice of sustainable lowland rice cultivation in the research location is shown in Table 1.

3.2 *The Level of Paddy Rice Farmer's Adoption on Sustainable Agricultural Practices*

The level of farmer adoption of sustainable agricultural practices was analyzed using a Likert scale with a low, medium, and high scale. Although farmers understand that sustainable agricultural practices positively impact rice farming, the level of application in the field is still low. Factors that influence the low adoption of sustainable agricultural practices are habits and convenience factors. Most communities still apply less sustainable agricultural practices, so other farmers are reluctant to switch to sustainable farming practices. Furthermore, from the convenience aspect, it is easier for farmers to obtain chemical fertilizers and pesticides than organic fertilizers and natural pesticides. The level of adoption of rice farmers toward sustainable agricultural practices in Pamarican District, Ciamis Regency, is shown in Table 2.

Table 2 shows that sustainable agricultural practices are dominated in the medium and low categories. From all indicators, only the use of organic fertilizers is categorized as high. Land cultivation and utilizing straw waste (post-harvest) were categorized in the medium group. In contrast, the utilization of organic pesticides–herbicides and utilization of refugia plants were categorized as low. In general, the factor

Table 1 Sustainable agriculture practices in rice farming in Pamarican District, Ciamis Regency

No	Practice	Explanation
1	Land cultivation	Tillage is done to loosen the soil. Soil tillage is carried out before the transfer of rice from the nursery. The function of tillage is to support the growth of soil roots and control weeds. However, the adopted tillage is dominated by intensive tillage. Several studies have explained that minimum tillage affects the availability of higher humic acid than intensive tillage [12]
2	Utilization of organic fertilizers	Most people have adopted the practice of using organic fertilizers. Usually, organic fertilizer comes from goat, chicken, or cow manure. Organic fertilizers are implemented on the land after processing (piracy). Farmers obtain organic fertilizer from purchases and partly from independently raised livestock. Organic fertilizers have increased rice plant's productivity [13]
3	Utilization of organic pesticides and herbicide	It is a sustainable agricultural practice using natural ingredients such as pesticides and herbicides. Natural pesticides and herbicides or organic pesticides have the advantage that they do not cause pest resistance and the death of natural enemies of pests and diseases. For example, several farmers in Pamarican District have used organic pesticides to overcome pests and diseases in rice plants
4	Utilization of straw waste (post-harvest)	Rice straw is beneficial for the fertility of rice fields. Research by Ansari and Mukhlis (2014) stated that straw could affect organic C, total N, available P, soil K-dd, and rice plant height [14]. Therefore, some respondent farmers have adopted the use of a straw to increase paddy fields' fertility
5	Utilization of refugia plants	Refugia is a microhabitat that provides temporary shelter for natural enemies of pests, such as predators and parasitoids, and provides an advantage for biotic ecosystem interactions, such as pollinators [15]. Types of refugia planted by rice farmers include Kenikir flowers, (*Cosmos*) corn (*Zea mays*), and long beans (*Vigna cylindrical (L.)*). Long bean and corn refugia in paddy fields have been proven to balance preventing pest attacks that can cause rice crop failure [16]

Table 2 Level of adoption of rice farmers on sustainable agricultural practices in Pamarican District, Ciamis Regency

No	Practice	Score	Category
1	Land cultivation	39.18	Medium
2	Utilization of organic fertilizers	53.62	High
3	Utilization of organic pesticides and herbicide	25.33	Low
4	Utilization of straw waste (post-harvest)	32.84	Medium
5	Utilization of refugia plants	18.76	Low

Information: Category Scores (Low: 12–28, Medium: 28.1–44, High: 44.1–60)

that drives the low practice of sustainable agriculture, especially the utilization of organic pesticides–herbicides and the utilization of refugia plants, is that farmers prefer the use of chemicals to control pests and weeds. Using chemical pesticides and herbicides is considered more efficient and has a more tangible impact (a faster process). Natural pesticides and herbicides are cheaper and can be made from local materials, but they are considered less effective. In addition, farmers plant refugia to satisfy food needs (vegetables), not to overcome pests and diseases. Therefore, few farmers use refugia to overcome pest and disease problems in rice plants.

4 Conclusion

Sustainable agricultural practices in rice farming in Pamarican District, Ciamis Regency, consist of five types, namely (1) land cultivation, (2) use of organic fertilizers, (3) use of organic pesticides and herbicides, (4) utilization of straw waste (post-harvest), and (5) utilization of refugia plants. Indicators of organic fertilizer are categorized as high in sustainable farming practices, while land cultivation and utilization of straw waste (post-harvest) are categorized in the medium group. Two indicators that fall into the low category are the utilization of organic pesticides–herbicides and the utilization of refugia plants categorized as low. The factor that drives the low practice of sustainable agriculture in the utilization of organic pesticides–herbicides and utilization of refugia plants is that farmers prefer the use of chemicals that are considered more effective in controlling pests than natural pesticides and herbicides.

The adoption of sustainable agriculture in general in Pamarican District is still low; therefore, increasing understanding and practice of sustainable agriculture needs to be encouraged massively, continuously and multi-stakeholder. Farmer's perceptions of sustainable agricultural practices are still low. Farmers think that it is not important to practice environmental-friendly agriculture because it does not provide economic benefits, and this causes farmers to only focus on the production aspect. The government needs to cooperate with extension workers and other stakeholders in an effort to encourage sustainable agricultural practices in order to encourage healthier and more sustainable rice production in the future.

References

1. BPS.: Luas Panen Dan Produksi Padi di Indonesia 2021(Harvest Area and Rice Production in Indonesia 2021). Jakarta, BPS (2022)
2. Nurvianto, Trilaksana, A.: Intensifikasi Pertanian Tanaman Padi Dalam Rangka Swasembada Beras Provinsi Jawa Timur Tahun 1969–1984 (Intensification of Rice Crops Agriculture in the Context of Rice Self-Sufficiency in East Java Province 1969–1984). Avatara. e-Journal Pendidikan Sejarah **2**(3), 598–608 (2014)
3. Wahyuni, S., Indraningsih, K.S.: Dinamika Program Dan Kebijakan Peningkatan Produksi Padi (Dynamics of Rice Production Improvement Programs and Policies). Forum Penelitian Agro Ekonomi **21**(2), 143–156 (2003)
4. Riawanti, W.: Kajian Peran Elit Politik Dalam Kebijakan Pangan: Jebakan Impor Pangan Pasca Reformasi (The Study of the Role of Political Elites in Food Policy: The Post-Reform Food Import Trap). Natapraja: Jurnal Kajian Ilmu Administrasi Negara **3**(1), 25–40 (2015)
5. Listiana, I., Mutolib, A., Bursan, R., Yanfika, H., Widyastuti, R.A.D., Rahmat, A.: Institutional strengthening of farmer group to support sustainable agriculture and food security in Pesawaran regency. J. Phys.: Conf. Ser. **1796**(1), 012028 (2021)
6. Widiarta, A., Adiwibowo, S.: Widodo.: Analisis keberlanjutan pertanian organik di kalangan petani (Analysis of the sustainability of organic farming among farmers). J. Sodality **5**(1), 79–100 (2011)
7. Herawati.: Kapasitas petani mengelola padi sawah ramah lingkungan di Sulawesi Tengah (Farmer's capacity to manage environmentally friendly lowland rice in Central Sulawesi) [disertasi]. Bogor, Institut Pertanian Bogor (2018)
8. Salikin, K.A.: In: Sistem pertanian berkelanjutan (Sustainable farming system). Yogyakarta, Kanisius (2003)
9. Leeuwis, C.: In: Communication for Rural Innovation, Rethinking Agricultural Extension. Yogyakarta, Kanisius (2009)
10. Sukmadinata.: In: Metode Penelitian Pendidikan (Educational Research Methods). Bandung. Remaja Rosdakarya (2011)
11. Miles, M.B., Huberman, A.M., Saldana, J.: In: Qualitative Data Analysis A Methods Sourcebook, 4th edn. Thousand Oaks, Califorinia, SAGE Publications (2014)
12. Putra, R.Y.A., Sarno, S., Wiharso, D., Niswati, A.: Pengaruh Pengolahan Tanah dan Aplikasi Herbisida Terhadap Kandungan Asam Humat Pada Tanah Ultisol Gedung Meneng Bandar Lampung (Effect of Soil Processing and Herbicide Applications on Humic Acid Content in Ultisol Soil Meneng Building, Bandar Lampung). J. Agrotek Tropika **5**(1), 51–56 (2017)
13. Adi, I.G.P.R., Puja, I.N.: Increased rice crop productivity through compost and NPK fertilization. Agrotrop **9**(2), 178–187 (2019)
14. Ansari, H., Mukhlis, J.: Effect of fertilizer dosage and rice straw to the contents of soil nutrient and rice production in the SRI (System of Rice Intensification) cropping system. J. Online Agroekoteknologi **2**(3), 1048–1055 (2014)
15. Keppel, G.V., Niel, K., Wardell-Johnson, G., Yates, C., Byrne, M., Mucina, L., Schut, A.G.T., Hopper, S., Franklin, S.: Refugia: Identifying and understanding safe havens for biodiversity under climate change. Glob. Ecol. Biogeogr. **21**(4), 393–404 (2012)
16. Setyadin, Y., Abida, S.H., Azzamuddin, H., Rahmah, S.F., Leksono, A.S.: Efek Refugia Tanaman Jagung (*Zea mays*) dan Tanaman Kacang Panjang (*Vigna cylindrica*) pada Pola Kunjungan Serangga di Sawah Padi (Oryza sativa) Dusun Balong, Karanglo, Malang (Effect of Refugia on Corn (*Zea mays*) and Long Bean (*Vigna cylindrica*) on Insect Visitation Patterns in Rice (*Oryza sativa*) Hamlet of Balong, Karanglo, Malang). Jurnal Biotropika **5**(2), 54–58 (2017)

Index-Based Insurance for Climate Risk Management in Indonesia Agriculture

Woro Estiningtyas, Kiki Kartikasari, Perdinan, and Saktyanu Kristyantoadi Dermoredjo

Abstract Agriculture in Indonesia, especially food crops, is significantly affected by climate change. Agricultural activities are highly dependent on climate, especially rainfall. Farmers as food farming actors receive the most significant impacts and risks due to climate uncertainty. Therefore, we need protection through agricultural insurance. Existing agricultural insurance needs to be complemented by another option that has the potential to be developed in Indonesia, namely Weather Index Insurance (WII). In this system, the insured is not plants, but a weather index which is compiled and determined based on the relationship between crop failures and changes in climatic conditions due to floods or droughts. This paper is a preliminary study aimed at analyzing a climate index for rice drought to support the development of agricultural insurance in Indonesia. The study was conducted in Indramayu Regency as one of the rice centers in West Java Province. The climate index is determined based on long-term historical rainfall data (1966–2022) associated with climate-related disasters' droughts using the Historical Burn method with a case study in Sliyeg Sub-District. The results of the study showed that based on 56 years of rainfall data and drought events, obtained an Exit value of 26 mm and a Trigger value of 34.5 mm. This index value becomes a reference in insurance claims regardless of crop failure in the field. Intensive socialization, surveys, and interviews are needed to explore willingness to accept and farmers' responses in the context of implementing WII.

W. Estiningtyas (✉)
Research Center for Climate and Atmosphere, BRIN, Jakarta, Indonesia
e-mail: Woro004@brin.go.id

K. Kartikasari
Carbon and Environmental Research Indonesia, Bogor, Indonesia

Perdinan
IPB University, Bogor, Indonesia

S. K. Dermoredjo
Research Centre for Economics of Industry, Services, and Trade, BRIN, Jakarta, Indonesia

A. Basit et al. (eds.), *Proceedings of the International Conference on Radioscience, Equatorial Atmospheric Science and Environment and Humanosphere Science*, Springer Proceedings in Physics 290, https://doi.org/10.1007/978-981-19-9768-6_58

1 Introduction

Global warming due to increasing Greenhouse Gas (GHG) concentration in the atmosphere has been widely known to change the global and regional climate. The intensity and frequency of extreme climate events are also increasing in many parts of the globe. IPCC [1] confirmed that several physical and biological changes are connected with global warming. In Indonesia, extreme climate events have been consistently linked to drought and flood [2]. The prolonged dry season during the period of 1844–1960 caused about 30 severe droughts with a return period of 4 years, while in the following decades of 1961–2006, the frequency of same intensity disasters was increasing to 2–3 years [3]. More than 75% of the droughts are also associated with El Niño [4].

Agriculture, especially crops, is one of the vulnerable sectors to climate change [5]. Rice, the main crop as well as a staple food in Indonesia, is very vulnerable to extreme climate events [6, 7]. Extreme climate has long been the main cause of crop failure in [8] and yield loss [9, 10]. Farmers often face multiple burdens of financial loss and frustration when crop failure occurs simultaneously due to the increasing frequency of climate extremes.

In parallel to increasing climate-related challenges to agriculture production, there is an increasing demand for food from multiplying populations worldwide. In 2050, the world population is projected to increase by 34% or equal to 9.1 billion people, predominantly in developing countries [11]. Rapid population growth threats the sustainable use of natural resources including food, energy, and water while the shortage has already occurred due to climate anomalies and a declining number of productive agricultural land. Climate has long been identified to significantly affect food availability, quality, and access. Uncertainty of rainfall pattern and intensity that lead to drought, flood, or pest invasion are challenges that farmers constantly have to deal with. An instrument for protection from climate risks like Weather Index Insurance (WII) is necessary.

WII is a financial instrument that covers the risk of weather in agriculture [12] or mitigates the risk of negative impacts from extreme climate [13, 14]. Weather Index Insurance has a more practical mechanism and lower operational cost compared to traditional insurance because it does not need field verification [15]. Payouts are based on the index, not field measures or real loss [16]. WII can be applied to cover extreme climate impacts of floods, droughts, and pest and disease attacks. In this paper, WII is focused on drought in rice plants. WII has the advantage of not having to do field checks for insurance claims, thus saving more costs. This study specifically selected rice as the commodity to cover by the insurance due to drought, though application to another commodity with high economic value is possible with an adjustment of the methodology to define the index. This paper presents the results of the analysis of the drought climate index for rice plants which is an initial study in order to support the development of agricultural insurance in Indonesia.

2 Methodology

Research and study were conducted in the Sliyeg Sub-District, Indramayu, West Java Province (Fig. 1). Sliyeg was located in the tail end of the irrigation network, making it vulnerable to drought and flood [17]. Indramayu was among 27 key rice production centers in West Java, with 15% contribution to total production or 14% of harvested areas in the province [18].

This study was mainly based on decadal (10 days) rainfall observation data from Cikedung station from 1966 to 2022 (56 years) which is the closest station to Sliyeg. The weather index is often used in agriculture analysis considering the significant correlation between weather and crop loss [19]. WII does not necessarily insure the crop but the index of climate instead. This index is developed and defined based on an in-depth assessment of crop loss due to drought and climate variabilities in a specific location. An in-depth study based on long-term drought data shows that the impact of drought is yield loss due to changes in rainfall characteristics that affect the availability of water for plants. The index is calculated using Historical Burn Analysis (HBA) developed by IRI Columbia University. This approach is preferable for many locations with limited climate data (only rainfall data is available in the full set). In principle, HBA learns historical patterns for future projection, assuming the same pattern applies. Therefore, it needs long historical data to identify reliable Triggers and Exits for the index in the future.

The initial stage is the identification of the insured period (index window) and the calculation of decadal rainfall for the same period. The selected index window for Sliyeg is May–September, considering that droughts frequently occurred during those months. Further, identified CAP represents maximum rainfall in one decade. The level of CAP is related to potential daily evapotranspiration in respective locations; for tropical areas, it is 5 mm/day or 50 mm in a decade [20]. If the rainfall in 10 days is less than the cap, the actual rainfall data will be used for calculation. Else, the CAP value will be used in the calculation if the rainfall is higher. This adjusted

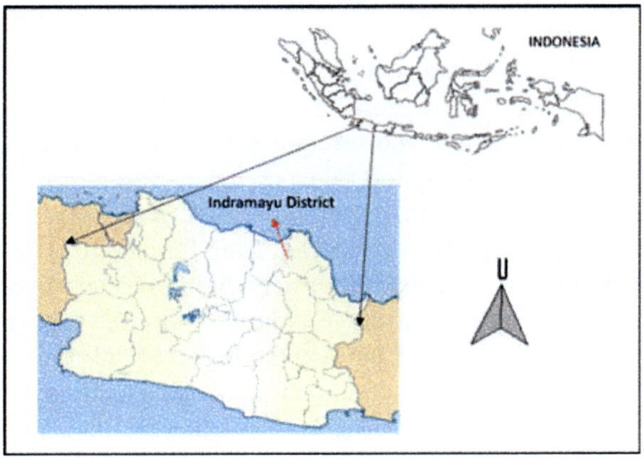

Fig. 1 Study location

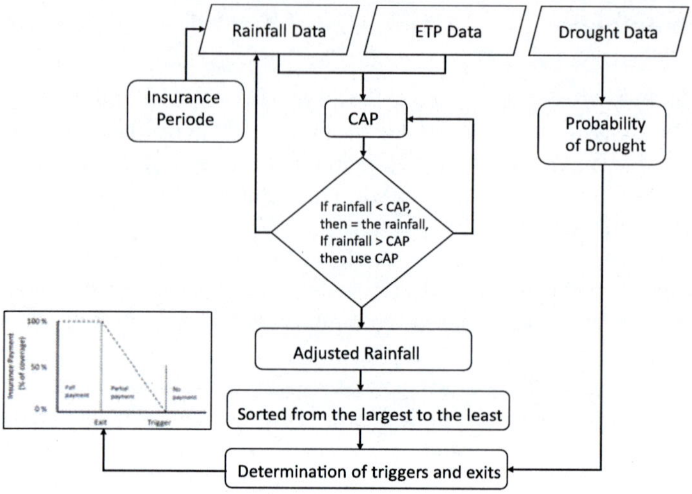

Fig. 2 Climate index analysis with Historical Burn method

rainfall data are prepared for each year and then sorted from the highest to the lowest value. A drought occurrence record is used to estimate the probability of drought in a certain period and to determine the Trigger and Exit for the insurance scheme. Full payouts will be disbursed if the total rainfall during the insured period is lower than the Exit value. Partial disbursement applies if total rainfall is between Trigger and Exit. Based on the WII concept, full payment will be disbursed if the total rainfall during the coverage period is lower than the Exit value. Partial disbursement applies if the total rainfall is between the Trigger and Exit, whereas if the total rainfall is higher than the Trigger, farmers cannot claim compensation (Fig. 2).

3 Result and Discussion

Sliyeg is a vulnerable area to extreme climate events, drought in particular. Harvest loss is prominently caused by drought (79%), followed by pest and disease (14.5%), and flood (5.6%), and drought generally occurs for 1–9 months [21]. This result verified the assumption to experience severe drought frequently due to its location in the tail end of the irrigation network. Risk management is necessary to minimize the impact on farmers' income. Therefore, there is a need for a solution to minimize the risks experienced by farmers due to drought events. Based on data from 1966 to 2021 (55 years) average monthly rainfall, August is the month with the lowest rainfall (27 mm) and January the highest (273 mm) with average 129 mm (Fig. 3). This becomes a consideration for choosing an index window. In this study, the index window used is May–August where this period is in the dry season.

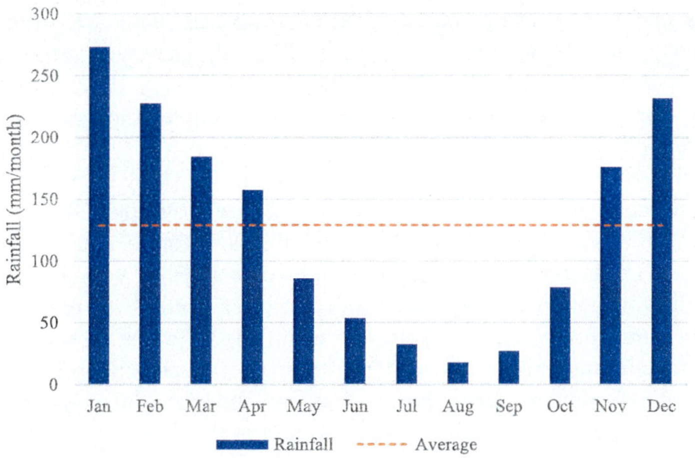

Fig. 3 Average monthly rainfall for the period 1966–2021

The adjusted rainfall analysis during the index window (May–August) is 26–600 mm. This lowest rainfall value is used as the Exit value in the index calculation. Furthermore, the return period of drought events in the study location is obtained every 4 years, so in a period of 56 years, it is estimated that there will be 14 drought's event. This value is used as the basis for determining Triggers from adjusted rainfall data that has been sorted from large to small. The 14th data from the bottom obtained a value of 115 mm which is the Trigger value (Fig. 4). Based on this result, there are three possibilities to occur with an insured period of May–September: (i) full payouts are disbursed when total rainfall is less than 26 mm; (ii) partial payment when the rainfall is ranging between 26 and 115 mm, proportionally to the rainfall; and (iii) no payouts when rainfall is more than 115 mm.

The climate index that we have identified will further determine Premium design and insurance claim mechanism, along with crop production and socio-economic data. The climate index insurance concept requires farmers to renew the insurance contract annually in adjustment to the revision of the climate index, to ensure that the index is representative and up to date. This year's contract is made to cover or

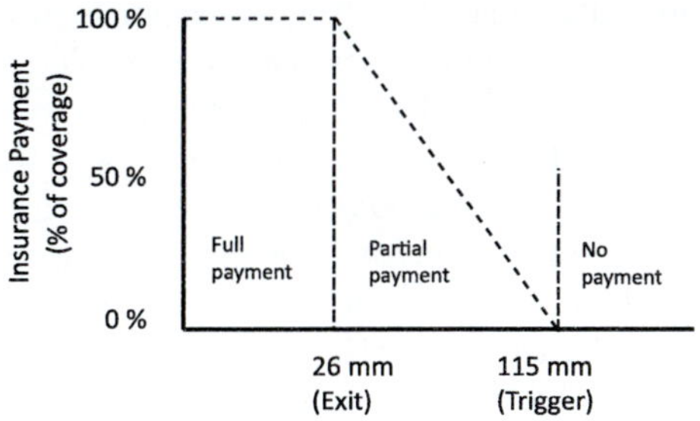

Fig. 4 Climate index in Sliyeg for insurance period of May–August

insure the upcoming year's plantation. Premium is based on the risk level; if drought risk for next year is high, then the Premium will be high and vice versa. Premium is intensively discussed among relevant stakeholders, on average 10–15% of production costs or even lower. Martirez [22] reported that an example of multi-risks' insurance during the high-risked year is 12.27%.

Implementation of WII highly depends on response [23] and the adaptive capacity of farmers to climate risks. Estiningtyas et al. [24] reported that 34% of farmers in Sliyeg are middle-aged (41–50 years old) with the latest formal education at primary level (49%). They own 1.8 ha of agricultural land on average, with irrigation (61%). Though their irrigation is categorized into last class or the least priority, they can only plant maximum two times of rice consecutively + fallow each year. There are several strategies to minimize the loss from climate disasters, but options for shifting the planting calendar are rarely chosen. Farmers tend to do the usual practices they learned from the older generation. Also, planting at the same time helps avoid new pests or disease invasions.

Another important aspect of the implementation of agriculture insurance is a willingness to pay (WTP) from farmers. WTP is often used as an indicator to offer additional agriculture extension services [25]. WTP indicates the maximum costs paid by customer for specific good or service, typically a ratio between the subjective values of the buyer to the good or service [26]. WTP is important during the discussion on insurance premium. A study in Sliyeg Sub-District reveals that 82.5% of farmers are willing to pay various levels of premium and the highest percentage is willing to pay 200–300 K IDR per season per hectare [27]. This WTP needs to be equipped with willingness to accept (WTA), namely the willingness of farmers to accept the index and payment.

The main regulation to support agriculture insurance is Law No.19/2013 about farmer's protection and empowerment. Farmer's protection is defined as any attempt to support farmers from challenging access to production units, business certainty, price fluctuation, harvest loss, high-cost distribution, and climate change. Farmers empowerment to cover any initiatives to build the capacity of farmers in agriculture business via training and education, coaching, development of system and infrastructure for agriculture products distribution, consolidation and guarantee on land ownership, access to relevant knowledge, technology, or information. A wide range of instruments has been developed and implemented for farmer's protection such as an early warning system, harvest loss compensations, and relevant climate change adaptation measures. Yet, an insurance mechanism is necessary for farmer's protection [28]. Farmers may see a clearer picture of how WII is derricking climate risks on their crops, while insurers can learn about the climate risks in agriculture and use it for further evaluation and economic assessment.

Potential development of WII in Indramayu District is: (1) Indramayu District is one of the rice production centers in West Java or even Indonesia, and successful implementation of agriculture programs in Indramayu will have positive impacts at the national level; (2) Indramayu District is vulnerable to climate anomalies, which is potential for WII; (3) agriculture has long been the main livelihood for majority of the population in Indramayu, and these farmers pay significant attention to any initiatives to further develop agriculture sector; (4) farmers' need for money prior to the onset of seasons (usually to buy production units) is a window of opportunity for insurance mechanism and relevant institutions, e.g., cooperatives, farmers group, etc.; (5) high willingness to pay is a very promising start. On the other hand, challenges on WII are: (1) Institutional supports at the national, regional, as well as local levels need to be prepared; (2) competent human resources are needed to assess if index insurance is effective and efficient for risk management; (3) market of insurance—especially in developing countries, just emerged. Start-up can be costly and play an important role in further growth [29]; (4) as a recently introduced program, payment for the premium might still be limited and need more stimuli. According to [22], the implementation of WII requires several stages starting from the preparation process until implementation: (i) farmer interviews, stakeholder interviews, data procurement, index design, policy trials, learning documentation (6 months), (ii) policy deployment, policy explanation, consumer feedback, policy purchase (3 months), (iii) weather index monitoring (4 months), and (iv) claim calculation, claim payment (1 month).

Implementation of WII is ideally tailored to local conditions (physical and socio-economic characteristics) in each rice production center, making the scheme location-specific. Identification of potential locations is key in the planning and development of WII. Large agriculture areas in Indonesia with various levels of vulnerability to climate change need supportive policies and options for farmers' protection and empowerment to minimize the risks. These potential and challenges of WII are important inputs for further development of agriculture insurance. WII is one of the risk management's options to minimize the impact received by farmers due to extreme climate events such as drought. Intensive socialization needs to be done so that the concepts and mechanisms of WII are understood for smooth implementation.

4 Conclusion

The key basis for determining the weather index is the detection of rainfall variability and its close relationship with the occurrence of rice drought. A study in the Sliyeg Sub-District discovered an initial value of the weather index for rice drought of Exit 26 mm and Trigger 115 mm. The window index must be chosen carefully because it is related to the index value. Processes and preparations are required for implementation of WII. In-depth surveys and interviews, as well as intensive socialization, are required in the planning and implementation of WII to reduce climate risk, especially in the agricultural sector. The determination of the right weather index makes WII one of the alternative forms of agricultural insurance that has the potential to be developed and implemented.

References

1. IPCC.: Climate Change 2007: The Physical Science Basis. Summary for Policymakers. Intergovernmental Panel on Climate Change, Geneva (2007)
2. Boer, R., Perdinan.: Adaptation to climate variability and climate change: its socio-economic aspect. In: Proceeding of Workshop on „Climate Change: Impacts, Adaptation, and Policy in South East Asia. Economy and Environmental Program for Southeast Asia. Bali (2008)
3. Boer, R.: Indonesian Country report: climate variability and climate change and their implications. Government of Indonesia, Jakarta (2007)
4. Boer, R., Subbiah.: Agriculture drought in Indonesia. In: Boken, V.J., Cracknell, A.P., Heathcote, R.L. (eds.) Monitoring and Predicting Agriculture Drought: A Global Study. Oxford University Press, New York. pp. 330–344. (2005)
5. Estiningtyas, W., Surmaini, E., Susanti, E.: Vulnerability of the food crop to climate change (in Indonesian). J. Sumberd. Lahan 2016, Special Edition, 85–96 (2016). ISSN 1907-0799
6. Boer, R., Surmaini, E.: Economic benefits of ENSO information in crop management decisions: case study of rice farming in West Java Indonesia. Theor. Appl. Climatol. **2020**(139), 1435–1446 (2020). https://doi.org/10.1007/s00704-019-03055-9
7. Surmaini, E., Hadi, T.W., Subagyono, K., Puspito, N.T.: Early detection of drought impact on rice paddies in Indonesia by means of Niño 3.4 index. Theor. Appl. Climatol. **121**, 669–684 (2015). https://doi.org/10.1007/s00704-014-1258-0
8. Hadi, P.U., Saleh, C., Bagyo, A.S., Hendayana, R., Marisa, Y., dan I. Sadikin.: Studi Kebutuhan Asuransi Pertanian Pada Pertanian Rakyat. Laporan Hasil Penelitian. Pusat Penelitian Sosial Ekonomi Pertanian, Badan Litbang Pertanian, Departemen Pertanian (2000)
9. Jiahua, Z., Yao, F., Hao, C., Boken, V.: Impacts of temperature on rice yields of different rice cultivation systems in southern China over the past 40 years. Physics and Chemistry of the Earth 87–88, pp. 153–159. Elsevier (2016)
10. van der Kooi, C.J., Reich, M., Löwc, M., De Koka, L.J., Tausza, M.: Growth and yield stimulation under elevated CO_2 and drought: a meta-analysis on crops. Environ. Experim. Botany **122**, 150–157. Elsevier (2016)
11. FAO.: How to Feed the World in 2050 Synthesis_Report (2022). https://www.fao.org/fileadmin/templates/wsfs/docs/expert_paper/How_to_Feed_the_World_in_2050.pdf
12. Prokopchuk, O., Prokopchuk, I., Mentel, G.: Index insurance as an innovative tool for managing weather risks in the Agrarian sector of economics. J. Competitiveness **10**(3), 119–131 (2018). https://doi.org/10.7441/joc.2018.03.08
13. Norton, M.T., Turvey, C., Osgood, D.: Quantifying spatial basis risk for weather index insurance. The J. Risk Finance **14**(1), 20–34 (2013); Emerald Group Publishing Limited, pp. 1526–5943. (2012). https://doi.org/10.1108/15265941311288086.
14. Singh, P., Agrawal, G.: Efficacy of weather index insurance for mitigation of weather risks in agriculture an integrative review. Int. J. Ethics and Syst. (IJOES) **35**(4), 584–616 (2019); Emerald Publishing Limited, pp. 2514–9369. (2019). https://doi.org/10.1108/IJOES-09-2018-0132
15. Osgood, D.: Index insurance and climate risk management in Malawi. World Bank, CRMG, IRI, CU-CRED (2007)
16. Manuamorn, O.P.: A feasibility study on weather index insurance for agriculture in Indonesia—weather index insurance in the context of agricultural risk management and relevant international experiences. Bahan Presentasi dalam Weather Index Insurance Seminar 2010, Finding of the Feasibility Study for Covering Weather Risk on Maize Production. International Finance Corporation, Jakarta 15 April (2010)
17. Estiningtyas, W., dan Hamdani, A.: Respon Perilaku Usahatani Padi Terhadap Risiko Iklim Ekstrem dan Serangan OPT. Jurnal Meteorologi dan Geofisika **16**(1) tahun 2015 1–10. Badan Meteorologi, Klimatologi dan Geofisika. Jakarta (2015)
18. BPS. Jawa Barat dalam angka (2022). https://jabar.bps.go.id/publication/2022/02/25/0d261f828b581d8082bbc6c1/provinsi-jawa-barat-dalam-angka

19. Colliera, B., Skeesa, J., Barnett, B.: Weather index insurance and climate change: opportunities and challenges in lower income Countries. The Geneva Papers, 2009, 34, (401–424). The Geneva Papers on Risk and Insurance—Issues and Practice (2009)
20. IRI.: Index insurance capacity building exercises for Indonesia release 2.0. IRI Earth Institute Columbia University (2012)
21. Estiningtyas, W., Buono, A., Boer dan, R., Las, I.: Penggunaan Metode Fuzzy Similarity dalam Penentuan Cakupan Wilayah Indeks Curah Hujan. Jurnal Meteorologi dan Geofisika **14**(2). Badan Meteorologi, Klimatologi dan Geofisika. Jakarta (2013)
22. Martirez.: Microensure, Helping the poor weather life's storm. Bahan Tayang (2009)
23. Jin, J., Wang, W., Wang, X.: Farmer's Risk Preferences and Agricultural Weather Index (2016)
24. Estiningtyas, W.: Pengembangan Model Asuransi Indeks Iklim Untuk Meningkatkan Ketahanan Petani Padi Menghadapi Perubahan Iklim. Disertasi. Program Studi Klimatologi Terapan. Fakultas Pasca Sarjana. Institut Pertanian Bogor (2012a)
25. Aydogdu, M.H.: Evaluation of farmer's willingness to pay for agricultural extension services in GAP-Harran Plain Turkley. J. Agr. Sci. Tech. **2017**(19), 785–796 (2017)
26. Klaus, W., Skiera, B.: Measuring consumer's willingness to pay at the point of purchase. J. Market. Res. **XXXIX**, 228–241 (2002)
27. Estiningtyas, W.: Analisis Usahatani Padi Untuk Mendukung Pengembangan Asuransi Indeks Iklim (Weather Index Insurance) : Studi Kasus di Kabupaten Indramayu. Jurnal Pengkajian dan Pengembangan Teknologi Pertanian. **15**(2) (2012). Balai Besar Pengkajian dan Pengembangan Teknologi Pertanian Badan Penelitian dan Pengembangan Pertanian Kementerian Pertanian (2012b)
28. Pasaribu, S.M.,. Setiajie I.A., Agustin, N.K., Lokollo, E.M., Tarigan, H., Hestina dan, J., Supriyatna, Y.: Pengembangan Asuransi UsahaTani Padi Untuk Menanggulangi Risiko Kerugian 75% Akibat Banjir, Kekeringan dan Hama Penyakit. Pusat Analisis Sosial Ekonomi dan Kebijakan Pertanian. Badan Litbang Pertanian (2010)
29. IFC.: Weather index insurance for Maize production in Eastern Indonesia. A Feasibility Study. Report. International Finance Corporation and Australia Indonesia Partnership (2009)

Near-Future Projections of Rainfall, Temperature, and Solar Radiation in Sumatra Island Under Climate Change Scenarios

Misnawati, E. Susanti, E. Surmaini, Y. R. Fanggidae, E. R. Dewi, Suciantini, M. R. Syahputra, U. A. Linarka, and A. Sopaheluwakan

Abstract The global climate has been changing noticeably in recent decades, and this trend is expected to continue. Climate change can be identified by changes in the mean or variability of climatic properties and is defined as changes in climate that persist over long periods, usually decades or longer. We present an assessment of projected climate change over Sumatra Island, focusing on minimum temperature, maximum temperature, rainfall, and solar radiation. We select the outputs from three global climate models (GCMs) from the Coupled Model Intercomparison Project Phase 5 with Representative Concentration Pathways 4.5 and 8.5 which are CSIRO, HadGEM, and CNRM covering the historical period 1996–2005 and the near-future period 2026–2045 slice to the rainy season and the dry season. The results show that the rate of increasing/decreasing rainfall, minimum temperature, maximum temperature, and solar radiation is higher in RCP8.5 than in RCP4.5 for both seasons. The box plot of rainfall and solar radiation shows insignificant change to historical under RCP4.5 and RCP8.5 emission scenarios for all models, but the minimum temperature and maximum temperature are projected to increase in an average range of 0.7–1.1 °C. The spatial distribution change of rainfall, solar radiation, minimum temperature, and maximum temperature is projected to increase and decrease over different areas in Sumatra for both seasons under two RCPs, and HadGEM indicates the highest change in rainfall of 1–500 mm/month over southern Sumatra for the dry season under both RCPs. The minimum temperature and maximum temperature increase and decrease of 0.1–1.0 °C. The solar radiation is projected to homogeneously decrease in an average range of 0.1–2 MJ/m^2/day under both emission

Misnawati (✉) · E. Susanti · E. Surmaini · Y. R. Fanggidae · E. R. Dewi · Suciantini
National Research and Innovation Agency, Jakarta, Indonesia
e-mail: misnawati.msaleh@gmail.com; misn003@brin.go.id

M. R. Syahputra
Bandung Technology Institute, Bandung, Indonesia

U. A. Linarka · A. Sopaheluwakan
Meteorology, Climatology, and Geophysical Agency, Jakarta, Indonesia

A. Basit et al. (eds.), *Proceedings of the International Conference on Radioscience, Equatorial Atmospheric Science and Environment and Humanosphere Science*, Springer Proceedings in Physics 290, https://doi.org/10.1007/978-981-19-9768-6_59

scenarios for the rainy season and dry season. The findings of this study suggest the importance of developing climate change adaptation and mitigation strategies over Sumatra.

1 Introduction

Climate change is a crucial issue that is currently happening and is predicted to continue even if current emissions are managed to reduce [1–3]. Indicators of climate change can be determined by looking at changes in trends from various climate parameters. Global research shows a trend in average air temperature and changes in rainfall patterns compared to historical climate conditions. Changes in the thermodynamics of atmospheric conditions as a result of the increase in global air temperature affect the spatial and temporal changes in the quantity of heat and humidity, resulting in increasingly intense extreme climate events [4–6]. These changes have a direct impact on climate-related disaster events [7].

The most common climate-related disasters in Indonesia are hydrometeorological disasters [8]. The agricultural sector is one of the sectors affected by climate-related disasters, and this can directly threaten agricultural production, especially food crops [9]. Adaptation and mitigation measures in the agricultural sector need to be taken to anticipate the impacts of climate change [10, 11]. Future projections of rainfall, air temperature, and solar radiation are three climatic parameters that are directly related to plant growth that needs to be carried out to know the possible conditions that will be faced in the agricultural sector in the future. An understanding of future conditions is essential for planning mitigation strategies and the types of adaptations carried out in each region, considering the spatial and temporal variations [5, 12].

Sumatra is an island in Indonesia with the second largest rice field area [13] that has an important role in maintaining food security. Therefore, studies on climate projections in Sumatra need to be carried out to get an overview of the future of rainfall conditions, temperature, and solar radiation. A previous study in Java showed that the climate change scenarios RCP 4.5 and 8.5 of the two models used had an impact on reducing rice production in aggregate, but there are certain districts whose productivity increases in 2045 [14]. This research is a preliminary study to prepare future climate projection data for analyzing rice productivity over Sumatra and to identify areas that are potentially vulnerable to food security due to decreased rice as affected by reduced rainfall and also areas that have the opportunity to increase productivity due to increased rainfall that can be achieved by proper adaptation measures.

To find out future climate change projections for Sumatra, the GCM model is used as it is the most advanced tool currently available. General circulation models (GCMs) have been widely used in studies of future climate change [15, 16]. However, the outputs of GCMs must be downscaled before being used at regional or local scales. In this study, the outputs from three global climate models (GCMs) from the Coupled Model Intercomparison Project Phase 5 with Representative Concentration

Pathways (RCP) 4.5 and 8.5 are used by CSIRO, HadGEM, and CNRM. RCP 4.5 is the most realistic scenario to achieve, while RCP 8.5 is the worst scenario if mitigation measures to reduce emissions are not carried out [17]. The CSIRO, HadGEM, and CNRM in this study are the result of dynamic downscaling using the RegCM4 regional model for the Southeast Asian region conducted by BMKG together with countries in the Southeast Asia region through the Southeast Asia Regional Climate Downscaling (SEACLID) collaboration which have carried out the process of downscaling the output of Coupled Model Intercomparison Project 5 (CMIP5) for the Southeast Asia region. The results of the downscaling in the form of high-resolution climate change scenario data (25 × 25 km) for the Southeast Asia region can be used to study the impact of climate change in various sectors including the agricultural sector [14]. This study aims to provide an overview of future projections of rainfall, temperature, and solar radiation in Sumatra Island under climate change scenarios RCP 4.5 and 8.5, which then can be used as a basis for projecting rice production in Sumatra.

2 Data and Methods

2.1 Study Area

Sumatra is located at 95−105°E and 6°N to 6°S covered by various topographies from plain areas to the plateau with elevations ranging from 0 to more than 3500 m above sea level. On the western side, Sumatra is covered by Bukit Barisan Mountain which extends from northern Aceh to Southern Lampung (Fig. 1). Sumatra has two types of rainfall patterns: monsoonal rainfall has a wet season from November to March and a dry season from May to September and equatorial rainfall has two peaks of rainfall that generally occurs from October to November and from March to May [18]. Sumatra is characterized by rainforest and humid climate with a total annual rainfall of about 1500–3500 mm/year, where the temperature ranges between 22 °C and 26 °C [19].

2.2 Data

In this study, three CMIP5 models from the Southeast Asia Regional Climate Downscaling/Coordinated Regional Climate Downscaling Experiment-Southeast Asia (SEACLID/CORDEX-SEA) project [20] were selected to represent future climate change over Sumatra. Two Representative Concentration Pathways' (RCPs) scenarios, RCP 4.5 and RCP85 over the period 2026–2045, are examined in this study. The RCPs represent the pathways of greenhouse gas emissions throughout the twenty-first century, as they relate to different radiative forcing stabilization levels

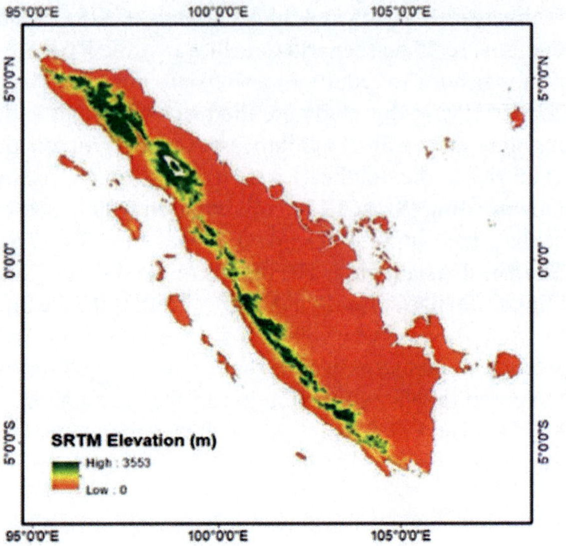

Fig. 1 Map of the study area

Table 1 List of CORDEX-SEA output considers in this study

Contribution by country	GCM models	RCP	GCM resolution
Indonesia	CSIRO MK3.6 (CSIRO, Australia)	RCP4.5, RCP8.5	1.8653° × 1.875°
Philippines	HadGEM2-ES (Hadley Center, UK)	RCP4.5, RCP8.5	1.25° × 1.875°
Vietnam	CNRM- CM5 (CNRM, France)	RCP4.5, RCP8.5	1.4008° × 1.40625°

by 2100. RCP4.5 indicates a medium stabilization scenario without overshooting pathways to 4.5 W/m2 of radiative forcing and RCP8.5 refers to a "business as usual" scenario without policy-driven mitigation, where a rising pathway is leading to 8.5 W/m2 by the end of twenty-first century [21]. The chosen models in this study are summarized in Table 1.

2.3 *Methodology*

The CORDEX-SEA dataset was used in this work at a daily scale consisting of rainfall, minimum temperature, maximum temperature, and solar radiation for the historical (1996–2005), RCP4.5, and 8.5 projected periods (2026–2045). The climatology historical period data should be at least 30 years [22], but as explained in the introduction that this research is a preliminary study to prepare climate data for crop modeling to project rice productivity in Sumatra, so we use ten years' data period (1996–2005) that has the same length as crop yield data to validate the result of crop modeling.

The projected changes in rainfall, minimum temperature, maximum temperature, and solar radiation are described on a seasonal basis, rainy season (RS) from October to March, and the dry season (DS) from April to September for each model and each period under RCP4.5 and RCP8.5 compared to the historical periods by a boxplot. The changes are calculated for each grid cell, each RCP, and each model separately, and the results are presented by a map.

3 Results

3.1 Projection of Rainfall, Minimum Temperature, Maximum Temperature, and Solar Radiation Under RCP4.5 and RCP8.5 Scenarios

Box plot of the mean monthly rainfall and mean daily solar radiation over Sumatra for historical (1996–2005) and near-future projection (2026–2035 and 2036–2045) under RCP4.5 and RCP8.5 scenarios for three models CSIRO, CNRM, and HadGEM is presented in Fig. 2. The CNRM shows that the projected rainfall under both emission scenarios is relatively stable to the historical period with an average value of up to 190 mm/month. Similarly, projected rainfall by CSIRO and HadGEM tends to be consistent without showing a significant change to the historical period with the approximate value of 340 mm/month and 420 mm/month, respectively. Moreover, the median, 25th percentiles, and 75th percentiles of projected rainfall are considered at the same level. However, the maximum value of each model is quite different, CSIRO has the highest maximum value, and the lowest maximum value is shown by CNRM. A similar situation was also projected for solar radiation, and the near-future projection did not show a significant value to the historical period. All three models have the median, 25th percentiles, and 75th percentiles at almost the same value. Under both RCP and period time, CNRM has an average value, minimum value, and maximum value of about 8.5 MJ/m^2/day, 4.5 MJ/m^2/day, and 10.5 MJ/m^2/day, respectively. CSIRO has an average value, minimum value, and maximum value of about 9.5 MJ/m^2/day, 6 MJ/m^2/day, and 12 MJ/m^2/day, respectively, while HadGEM has an average value, minimum value, and maximum value of about 9.4, 5.5 and 11.5 MJ/m^2/day.

Figure 3 shows the box plot of the mean minimum and maximum temperature projection under two emission scenarios for historical (1996–2005) and near-future projections (2026–2035 and 2036–2045). This figure indicates that the minimum temperature and maximum temperature which are projected will increase for all models under both RCPs. Under RCP4.5, the minimum temperature projected by CNRM, CSIRO, and HagGEM has an average increase of 0.8 °C, 0.9 °C, and 1.0 °C,

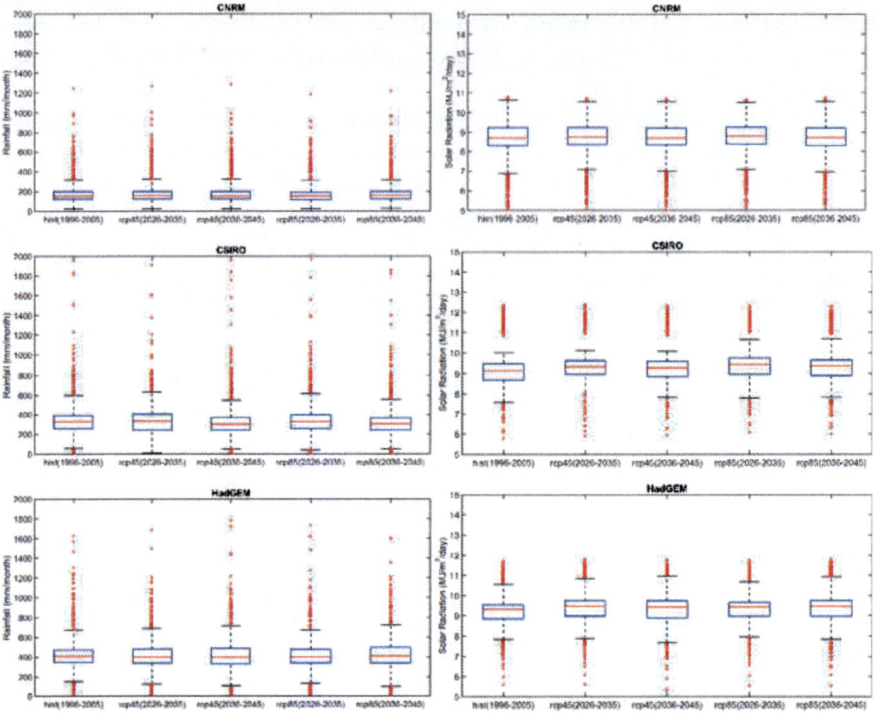

Fig. 2 Projections of rainfall (left panel) and solar radiation (right panel) over Sumatra in historical (1996–2005) and near-future projections (2026–2035 and 2036–2045) under RCP4.5 and RCP8.5 scenarios

respectively. Under RCP8.5, a minimum temperature increases on average 0.9 °C, 1.0 °C, and 1.1 °C, respectively. In the cast of maximum temperature, both RCPs will experience an increase in maximum temperature except the maximum temperature projected by HadGEM in 2036–2045 under RCP8.5. An average increase in maximum temperature is about 0.7 °C for RCP4.5 and 0.9 °C for RCP8.5 for CNRM and CSIRO. Under RCP4.5, HadGEM projects a higher increase in maximum temperature at value 0.9 °C, while under RCP8.5, the maximum temperature is seen to increase by 0.9 °C for the 2026–2035 period, but the maximum temperature in the 2036–2045 period will experience a decrease of 0.9 °C.

3.2 *Spatial Distribution Projected Change of Rainfall*

Figure 4 presents the spatial distribution of projected monthly rainfall changes under RCP4.5 and 8.5 for two time slices (2026–2035 and 2036–2045). In the rainy season

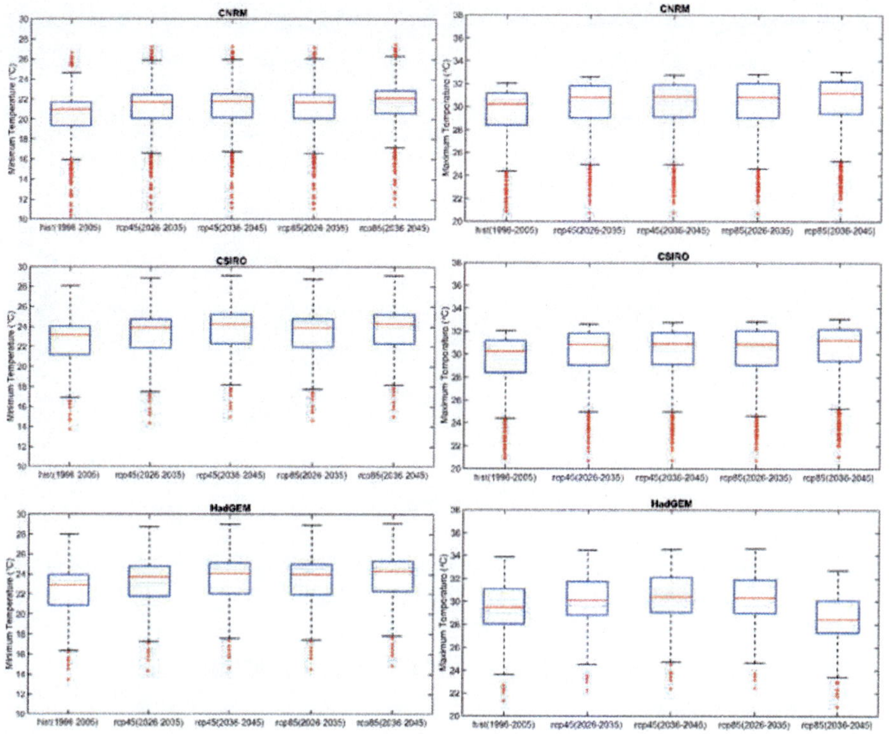

Fig. 3 Same as Fig. 2 but for minimum temperature (left panel) and maximum temperature (right panel)

(RS), CNRM shows a decrease of 1–200 mm/month over the whole of Sumatra relative to the historical period (1996–2005) under both scenarios. Similarly, CSIRO indicates a homogeneously decrease over Sumatra under RCP4.5 and 8.5 reaching values up to 1–200 mm/month; however, an increase of 1–100 mm/month is recorded over Lampung area under RCP4.5 at 2026–2035 time periods. Unlike CNRM and CSIRO, the projected rainfall of HadGEM2 differs from one part of Sumatra to the other. HadGEM projected an increase of about 1–300 mm/month over southern Sumatra (Lampung, South Sumatra, Jambi, and Bengkulu) under RCP4.5 and RCP8.5 for all periods unless RCP4.5 scenario, the rainfall tends to decrease 1–200 mm/month over the continental areas at 2036–2045. In the case of the dry season, CNRM and CSIRO show a decreasing amount of monthly rainfall in the range of 1–200 mm/month and reaching 300 mm/month in a small part of Sumatra, while HadGEM2 indicates a significant reduction of 1–500 mm/month for both the examined two future periods over southern Sumatra. Thus, an increase of up to 1–200 mm/month is projected over high-altitude areas of Bukit Barisan Mountain. The results agree with the previous study [20] which showed that rainfall will experience to decrease over the southern part of Sumatra during the dry season.

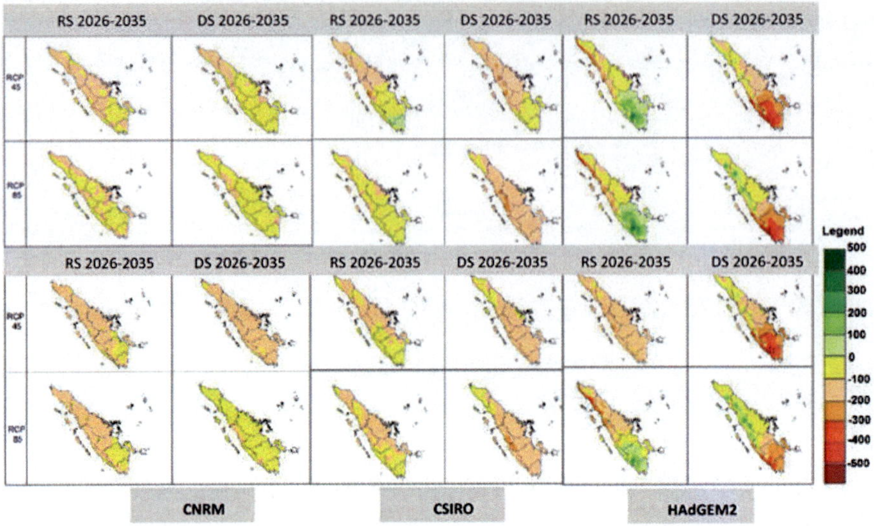

Fig. 4 Spatial distribution of projected change in monthly rainfall under RCP4.5 and RCP8.5 for the periods 2026–2035 and 2036–3045 in the rainy season and dry season

3.3 Spatial Distribution Projected Change of Minimum Temperature

Figure 5 shows the spatial distribution projected change of minimum temperature in two time periods under both RCP4.5 and RCP8.5 emission scenarios. This figure reveals a decrease in minimum temperature in the range of 0.1–1 °C under RCP 4.5 in both seasons at two time periods for CNRM. Under RCP8.5, a decrease of 0.1–1 °C is projected over Sumatra at 2026–2035 for both seasons, while at 2036–2045, an increase of 0.1–1 °C is seen over the area of Sumatra for both seasons. For CSIRO, a decrease of 0.1–1 °C is projected under the rainy season for both emission scenarios at the 2026–2035 time period, while an increase is seen under the 2036–3045 time period for both scenarios in the rainy season. For the dry season, CSIRO shows a decrease of 0.1–1 °C under RCP4.5 and an increase of 0.1–1 °C under RCP8.5 for both periods. HadGEM2 shows an increase in the larger area of Sumatra than CSIRO and CNRM. In the rainy season, the minimum temperature change is higher and larger than in the dry season reaching a value of 0.1–2 °C in most of part Sumatra for the first period and seeing an increase all over Sumatra for the second period under both emission scenarios, while in the dry season, RCP8.5 shows a higher change in minimum temperature than RCP4.5.

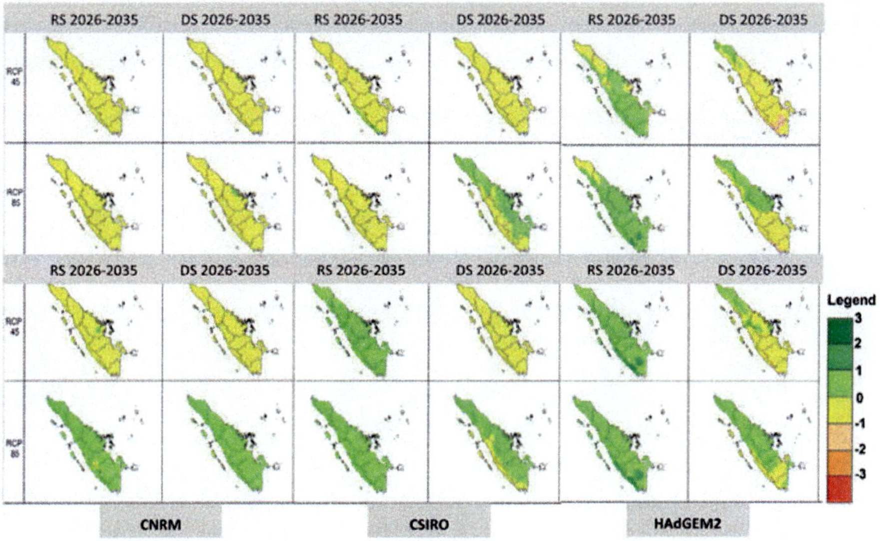

Fig. 5 Spatial distribution of projected change in minimum temperature under RCP4.5 and RCP8.5 for the periods 2026–2035 and 2036–2045 in the rainy season and the dry season

3.4 Spatial Distribution Projected Change of Maximum Temperature

The spatial distribution of the difference in maximum temperature between the projected and historical periods over Sumatra is illustrated in Fig. 6. The maximum temperature increases and decreases in different parts of Sumatra. CNRM shows a consistent pattern of the decrease of 0.1–2 °C under both emission scenarios in 2026–2035 for both examined seasons. At the 2036–2045 time period, a decrease of 0.1–2 °C is projected over Sumatra under RCP4.5 for both seasons unless a small part of central Sumatra shows an increase ranging from 0.1 to 1 °C, while under RCP8.5, an increase of the projected maximum temperature is seen in a larger area in both the rainy and dry seasons.

A similar distribution was also projected for CSIRO; however, CSIRO shows a much larger increase in value than CNRM in most of Sumatra, especially under RCP8.5 in both seasons. In the case of HadGEM2, the projected change of maximum temperature is similar to CNRM and CSIRO under RCP4.5 in both seasons, where an increase and decrease in the value of about 0.1–1 °C, respectively, are seen in a different part of Sumatra; under RCP8.5, HadGEM2 projects a higher decrease in maximum temperature than CNRM and CSIRO reaching value 3 °C. This condition may occur due to increased rainfall in the same area as the maximum temperature (Fig. 4), an increase in rainfall will increase the air humidity, and temperature will decrease because more energy is used to evaporate than to warm the air [23].

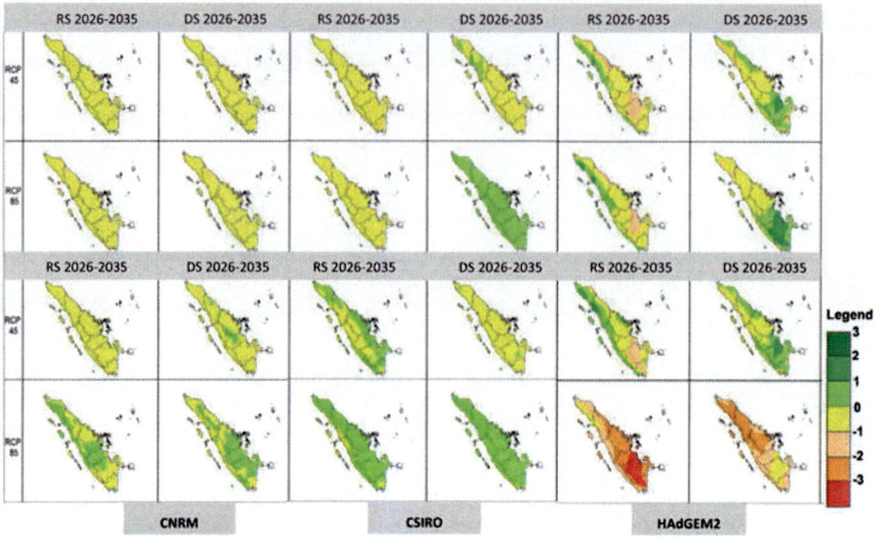

Fig. 6 Spatial distribution of projected change in maximum temperature under RCP4.5 and RCP8.5 for the periods 2026–2035 and 2036–3045 in the rainy season and dry season

3.5 Spatial Distribution Projected Change of Solar Radiation

In Fig. 7, the spatial distribution projected change in solar radiation is shown for two RCPs in two distinct periods and seasons. Generally, solar radiation is projected to homogeneously decrease in an average range of 0.1–2 MJ/m^2/day. At the 2026–2035 period under both RCPs, CNRM shows that solar radiation projected decreases of 0.1–1 MJ/m^2/day over most of Sumatra for the rainy season, while in the dry season, a decrease is higher in range value of 0.1–2 MJ/m^2/day. Similarly, HadGEM2 also indicated a decrease of 0.1–1 MJ/m^2/day over most of Sumatra for the rainy season, and a higher decrease up to 2 MJ/m^2/day is seen in the dry season unless in a very small part of Sumatra which shows an increase up to 1 MJ/m^2/day. For the CSIRO under both RCPs in two different periods and seasons, solar radiation change is indicated to decrease up to 1–2 MJ/m^2/day over most Sumatra, while the higher decrease of 3 MJ/m^2/day is seen over South Aceh, Nagan Raya, and West Aceh. An increase up to 0.1–1 MJ/m^2/day is indicated in a small part of Sumatra over the east coast of Sumatra.

Based on above results, statistically and spatially each model shows a different change in rainfall, minimum temperature, maximum temperature, and solar radiation. This is possibly occurred because each model has a different performance in simulating each parameter of the climate [24, 25]. Changes in rainfall, temperature, and solar radiation will affect the increasing extreme climate events: flooding, drought, pest and disease outbreak, and change in the planting time of crops [5, 6].

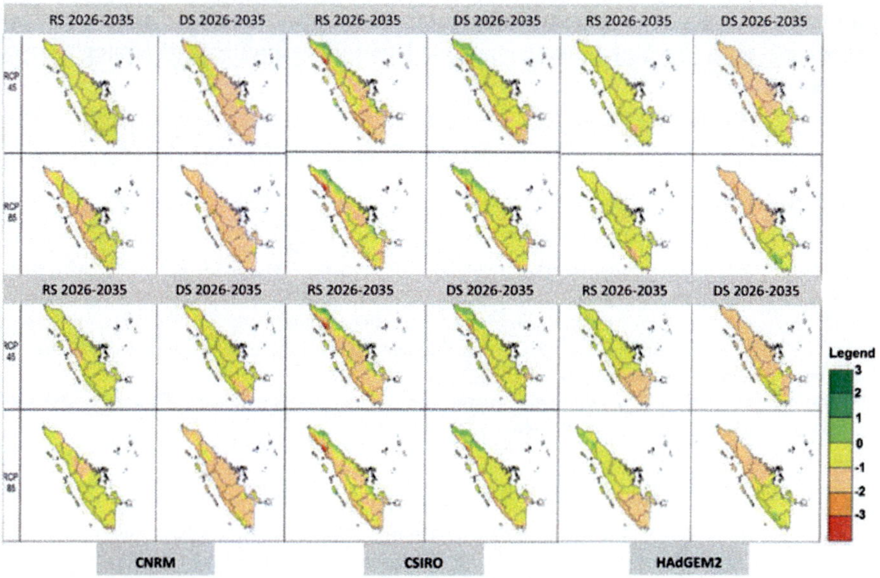

Fig. 7 Spatial distribution of projected change in solar radiation (MJ/m^2/day) under RCP4.5 and RCP8.5 for the periods 2026–2035 and 2036–3045 in the rainy season and dry season

4 Conclusion

The general aim of this study is to provide an assessment of projected changes in climate parameters (rainfall, solar radiation, and minimum and maximum temperature) over Sumatra for the near-future (2026–2045) period relative to the historical (1996–2006) climate condition. The following conclusions are summarized that the box plot analysis describes an insignificant change in rainfall and solar radiation under RCP 4.5 and RCP8.5 emission scenarios for all models (CNRM, CSIRO, and HadGEM), but the minimum temperature and maximum temperature are projected to increase in an average range of 0.7–1.1 °C. Under RCP4.5 and RCP8.5, rainfall is projected to increase and decrease over different areas in Sumatra for both seasons (rainy season and dry season). A higher decrease in rainfall is projected by HadGEM over southern Sumatra. Spatial distribution projected change of minimum and maximum temperature in both seasons reveals an increase and decrease of 0.1–1.0 °C over a different part of Sumatra under both RCPs, but the HadGEM projects a higher decrease in maximum temperature under RCP8.5 reaching a value of 3.0 °C over southern Sumatra. Solar radiation is projected to homogeneously decrease in an average range of 0.1–2 MJ/m^2/day unless over the east coast of Sumatra, solar radiation seems to increase about 0.1–1 MJ/m^2/day under both emission scenarios for the rainy season and dry season. This study provides information on Sumatra areas that will be vulnerable to climate change in terms of increase/decrease in rainfall, minimum temperature, maximum temperature, and solar radiation. Thereby, policy-makers can act to develop climate change adaptation and mitigation strategies over Sumatra.

References

1. IPCC.: Summary for policymakers. In: Stocker, T.F., Qin, D., Plattner, G.- K., Tignor, M., Allen, S.K., Boschung, J., Nauels, A., Xia, Y., Bex, V., Midgley, P.M. (eds.) Climate Change 2013: The Physical Science Basis. Contribution of Working Group to the Fifth Assessment Report of the Intergovernmental Panel on Climate Change. Cambridge University Press, Cambridge, United Kingdom and New York, NY, USA, pp. 3–29. (2013)
2. Kellogg, W., Schware, R.: Climate Change and Society, Consequences of Increasing Atmospheric Carbon Dioxide, 1st edn. Routledge, New York (2019)
3. Sippel, S., Meinshausen, N., Fischer, E.M., Székely, E., Knutti, R.: Climate change now detectable from any single day of weather at global scale. Nat. Clim. Chang. **10**(1), 35–41 (2020)
4. Chen, D., Rojas, M., Samset, B.H., Cobb, K., Diongue Niang, A., Edwards, P., Emori, S., Faria, S.H., Hawkins, E., Hope, P., Huybrechts, P., Meinshausen, M., Mustafa, S.K., Plattner, G.-K., Tréguier, A.-M.: Framing, context, and methods. In: Masson-Delmotte, V., Zhai, P., Pirani, A., Connors, S.L., Péan, C., Ber-ger, S., Caud, N., Chen, Y., Goldfarb, L., Gomis, M.I., Huang, M., Leitzell, K., Lonnoy, E., Matthews, J.B.R., Maycock, T.K., Waterfield, T., Yelekçi, O., Yu, R., Zhou, B. (eds.) Climate Change 2021: The Physical Science Basis. Contribution of Working Group I to the Sixth Assessment Report of the Intergovernmental Panel on Climate Change (IPCC). Cambridge University Press, Cambridge, United Kingdom and New York, NY, USA, pp. 147–286. (2021)
5. Luo, M., Liu, T., Meng, F., Duan, Y., Bao, A., Frankl, A., De Maeyer, P.: Spatiotemporal characteristics of future changes in precipitation and temperature in Central Asia. Int. J. Climatol. **39**(3), 1571–1588 (2019)
6. Kitoh, A., Hosaka, M., Adachi, Y., Kamiguchi, K.: Future projections of precipitation characteristics in East Asia simulated by the MRI CGCM2. Adv. Atmos. Sci. **22**, 467–478 (2005)
7. ADB.: Asian development outlook 2015. Asian Development Bank, Manila (2015)
8. Azizah, M., Subiyanto, A., Triutomo, S., Wahyuni, D.: The effect of climate change on hydrometeorological disasters in Cisarua District—Bogor. PENDIPA J. Sci. Educ. **6**(2), 541–546 (2022)
9. WFP.: How climate change affect hunger, Word Food Program (2022). Available at http://www.wfp.org/climate-change. Last Accessed 22 Sep 2022
10. Hasegawa, T., Sakurai, G., Fujimori, S., et al.: Extreme climate events increase risk of global food insecurity and adaptation needs. Nat. Food **2**, 587–595 (2021)
11. Hussain, S., Huang, J., Ahmad, S., Nanda, S. et al.: Rice production under climate change: adaptations and mitigating strategies. In: Fahad, S., Hasanuzzaman, M., Alam, M. et al. (eds.) Environment, Climate, Plant and Vegetation Growth. Springer, Cham (2020)
12. Al-Ghussain, L.: Global warming: review on driving forces and mitigation. Environ. Prog. Sustain. Energy **38**(1), 13–21 (2019)
13. BBSDLP.: Luas Baku Lahan Sawah Indonesia. Balai Besar Penelitian dan Pengembangan Sumber Daya Lahan Pertanian. BBSDLP, Bogor (2019)
14. Susanti, E., Dewi, E.R., Surmaini, E., Sopaheluwakan, A., Linarko, A., Syahputra, M.R.: The projection of rice production in Java Island to support Indonesia as the world food granary. In: E3S Web of Conferences, vol. 306, EDP Sciences (2021)

15. Sorribas, M.V., Paiva, R.C., Melack, J.M., Bravo, J.M., Jones, C., Carvalho, L., Beighley, E., Forsberg, B., Costa, M.H.: Projections of climate change effects on discharge and inundation in the Amazon basin. Clim. Change **136**, 555–570 (2016)
16. Reshmidevi, T., Kumar, D.N., Mehrotra, R., Sharma, A.: Estimation of the climate change impact on a catchment water balance using an ensemble of GCMs. J. Hydrol. **556**, 1192–1204 (2017)
17. Pérez, R.S.J.J.L., Pérez, L., Barras, R.M.G., Pecci, J., Palacios, M.: Health impact economic cost due to climate change over three European Cities. Int'l J. Adv. Agricul. Environ. Eng. **4**(1), 43–49 (2017)
18. Edvin, A., Susanto, R.D.: Identification of three dominant rainfall regions and their relationship to sea surface Int. J. Climatol. **23**, 1435–1452 (2003)
19. Susanti, E., Surmaini., Heryani, N., Estiningtyas, E., Suciantini., Apriyana, Y.: Updating of The agro-climate resources map of Indonesia to support agricultural planning. Jurnal Tanah dan Iklim **45**(1), 47–58 (2021)
20. Supari, Tangang, F., Juneng, L., Cruz, F., Chung, J.X., Ngai, S.T., Salimun, E., Mohd, M.S.F., Santisirisomboon, J., Singhruck, P., PhanVan, T., Ngo-Duc, T.., Narisma, G., Aldrian, E., Gunawan, D., Sopaheluwakan, A.: Multi-model projections of precipitation extremes in Southeast Asia based on CORDEX-Southeast Asia simulations. Environ. Res. **184**, 109350 (2020)
21. Moss, R.H., Edmonds, J.A., Hibbard, K.A,, Manning, M.R., Rose, S.K., van Vuuren, D.P., Carter, T.R., Emori, S, Kainuma, M., Kram, T., Meehl, G.A., Mitchell, J.F., Nakicenovic, N., Riahi, K., Smith, S.J., Stouffer, R.J., Thomson, A.M., Weyant, J.P., Wilbanks, T.J.: The next generation of scenarios for climate change research and assessment. Nature **463**(7282), 747–56 (2010)
22. Robinson, P.J., Henderson-Sellers, A.: In: Contemporary Climatology. 2nd edn. Routledge, London, Pearson Prentice Hall, pp. 4–5. (1999) ISBN 987-0582276314
23. Pearlmutter, D., Krüger, E.L., Berliner, P.: The role of evaporation in the energy balance of an open-air scaled urban surface. Int. J. Climatol. **29**, 911–920 (2019). https://doi.org/10.1002/joc.1752
24. Tangang, F., Chung, J.X., Juneng, L., et al.: Projected future changes in rainfall in Southeast Asia based on CORDEX–SEA multi-model simulations. Clim. Dyn. **55**, 1247–1267 (2020)
25. Handoko, U., Boer, R., Aldrian, E., Latifah, A.L., Dasanto, B.D., Apip., M..: Comparison performance of the multi-regional climate model (RCM) in simulating rainfall and air temperature in Batanghari Watershed. Inter. J. Sci. Technol. **8**, 52–67 (2019)

An Assessment of Groundwater Quality in the Southeastern Part of Rajasthan, India

Sushmi Nimje, Niruti Gupta, and Yash Kumar Mittal

Abstract Globally, India is positioned 120th among 122 countries in WaterAid's water quality index. According to reports of the World Resources Institute, Rajasthan is one of the states facing acute water scarcity. Amid water scarcity, water quality issues add to the problem and impact people's health. The study evaluates the groundwater quality status in parts of Kota, Chittorgarh, and Bundi of Rajasthan. The research assesses the quality of water and evaluates the reason behind it. Due to rapid development, agriculture and power plants, the region is facing quality issues. The water quality assessment is conducted using GIS and the water quality index. The result indicates that many considered parameters were above the permissible limit, and many villages are affected by more than one parameter of bad water quality. The WQI calculated for 2010, 2015 and 2020 indicates how the water quality has degraded over the years. The calculated index plays an important role in sustainable water development as it provides the best assessment tool to evaluate water quality. This study on aspects of groundwater quality can help engineers/planners make decisions and implement the best management practices for the groundwater.

1 Introduction

Groundwater is the largest source of potable water in major cities. Domestic, agriculture, and industrial sectors majorly depend on groundwater [1]. In India, annual utilizable groundwater resources are measured as 433 bcm [2]. Groundwater recharge occurs through rainfall. According to literature, in 2015, India received about 4000

S. Nimje (✉) · N. Gupta · Y. K. Mittal
Malaviya National Institute of Technology, Jaipur, India
e-mail: 2019rar9032@mnit.ac.in

N. Gupta
e-mail: ngupta.arch@mnit.ac.in

Y. K. Mittal
e-mail: yash.arch@mnit.ac.in

A. Basit et al. (eds.), *Proceedings of the International Conference on Radioscience, Equatorial Atmospheric Science and Environment and Humanosphere Science*, Springer Proceedings in Physics 290, https://doi.org/10.1007/978-981-19-9768-6_60

bcm of average annual rainfall along with snowfall, out of which 3000 bcm is received in the monsoon season (June–September), and water usable is only 1123 bcm.

The need for water also increases with increased urban growth, industrialization, and development. The stress on these groundwater resources can result in an alteration of flow direction and a change in nature with other water bodies [3]. There is a constant increase in demand for water; on the other side, supply chain of water either remains constant or decreases because of climate change and global warming. This water demand and supply gap can cause a water crisis that has already begun in some countries, including India. Limited research is available regarding the linkage between climate change and groundwater interactions. Only recently, the impact on groundwater quality due to climate change has been studied. Through modifications in hydrological processes such as recharge, storage, and discharge, climate change can affect the quality of ground water [4]. Additionally, the idea of global warming, changes in land use, deforestation, urbanization, and area waterproofing may all contribute to the deterioration of water quality [5].

There is excessive pressure on the country's water resources because of India's growing population and economic growth. An Indian adult requires about 135 L of water every day [6]. However, this refers to the direct and visible water used. In addition to this, we constantly use water virtually.

Water demand is gradually rising and will continue to increase across all sectors. Balancing the demand and available supply of water will be vital for India's economic growth and development to meet its growing water demands. The planning report indicates that India's water demand will surpass supply by two times in the year 2030, which will result in severe water scarcity for hundreds of millions of people. Major key challenge in this crisis is the unavailability of water data. Data systems in relation to water in India are limited in their efficiency, coverage, and robustness [7]. When groundwater has pollutants, it is challenging to ensure the groundwater draft and adequate water quality by preventing the pollutants from reaching the groundwater source. Therefore, regular groundwater quality monitoring should be conducted, and measures should be found to protect it from pollution [8]. This study is an approach to assess groundwater quality using remote sensing and a water quality index to ensure a sustainable groundwater approach. The assessment will help identify quality-affected areas that need urgent attention to ensure the safe quality of water.

2 Material and Methods

2.1 The Study Area

In Northwestern India, Rajasthan is India's largest state by area [9]. The climate in Rajasthan varies from arid to semiarid. Southeastern part of the state is more humid. Rajasthan stands at a critical juncture due to its alarming decrease in groundwater level [10, 11]. All rivers in Rajasthan are rainfed and transient, except for the Chambal and Mahi. They run through a small part of the southeastern and southern parts of the state, respectively [9]. Rajasthan is a water-stressed area and also has to deal with the serious issue of widespread water contamination. Household sewage and industrial effluents are draining into natural waters, disturbing the area's water quality, and creating eutrophication. Various reports submit that in the entire country, Rajasthan has 51% of the Fluoride-affected regions and 42% of the saline-affected areas [12]. Over 75% of the towns and villages are affected by bad water quality, influencing 20 million individuals in the state. Climate change will further lead to more groundwater extraction due to the limited availability of surface water, and an increase in water demand will further deteriorate the groundwater quality, putting serious risks to people's health [13]. As per the study report of the Wildlife Institute of India (2010), in last 17–18 years, it is observed that the average amount of water used for irrigation by the state Rajasthan and Madhya Pradesh, through the construction of Gandhi Sagar Dam and water abstraction via Kota Barrage, decreased by 22.6% and 41.4%, respectively. At the same time, water use for non-irrigation (industrial and drinking water purposes) has increased three-fold, resulting in a water shortage downstream. By 2002–03, the net water use for non-irrigation purposes was almost 41%.

The study area comprises blocks of Kota, Chittorgarh, and Bundi of Rajasthan. Kota is an education hub famous for competition coaching classes, Chittorgarh is well known for its tourism and history, and Bundi is known for forts and baolis (step well). There are no major industries in this region, but Kota has four power plants within its 50 km radius. First is the Rajasthan atomic power plant, with a capacity of 995 MW. It is a nuclear power plant in Rawatbhata and uses Chambal water for cooling purposes. Next in line is the Rana Pratap Sagar Hydro Dam, situated in the Rawatbhata Block of Chittorgarh district. The third is Kota Super Thermal Power Plant which generates power from coal. It is situated on the bank of the Chambal River and is within Kota city, and the fourth is the Jawahar Sagar Power Plant, a hydropower plant in Bundi Block. There is a Kota Barrage in Kota city, which is used for irrigation and diverting the waters through the state. About 60%–70% of the area is under agriculture and forest, and agriculture is the main occupation around these areas. The study area also has the Chambal River, and power plants are located along the river, as shown in Fig. 1.

2.2 *Methodology*

For research, the secondary data were collected from various government departments like the PHE laboratory and the water resource department of the districts. The data were also collected through the JJM portal of Jal Shakti Ministry, GoI. The chemical and biological parameters both were considered while collecting data. The parameters considered were pH, TDS, Arsenic, Nitrate, Fluoride, Chloride, Hardness, BOD, COD, and alkalinity. The final seven parameters are pH, TDS, Arsenic, Nitrate, Fluoride, Chloride, and alkalinity. The maps are prepared individually on GIS of the villages affected by these parameters for assessment. The final score map

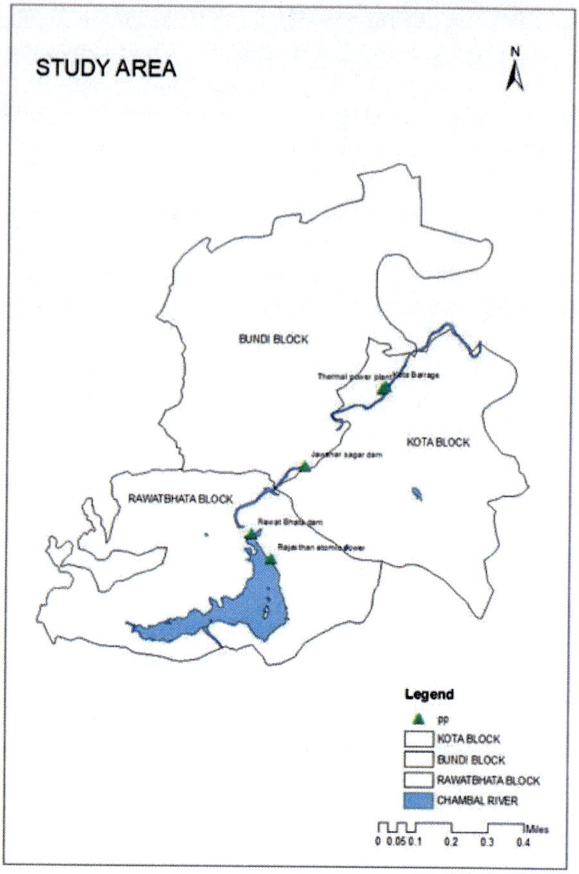

Fig. 1 Study area

is prepared using these individual maps. Villages affected with only one parameter are given a score of one. Villages affected with seven parameters are given a score of seven. Hence, the final score map is prepared with scoring one to seven. Severely affected areas scored seven, and the least affected area scored one. The data were compiled for water quality index values, and the mode was taken for each parameter for calculation; hence, the assessment was done.

Multiple indicators can be used to assess water quality, but water quality index (WQI) is one of the most reliable and easier methods for calculating water quality. Multiple authors have used this method for the assessment of water quality [14–16]. The data from 2020, 2015 and 2010 are used for assessment. The parameters considered for predicting water quality are pH, TDS, Nitrate, Iron, Fluoride, Chloride, Hardness, and alkalinity.

The average values of each water quality parameter for each year, along with the limits of the Bureau of Indian Standards [6], are considered. The WQI is computed using the permissible values of drinking water suggested by the Bureau of Indian Standards.

The unit weight (W_i) of the various water quality parameters is inversely proportional to the recommended standards for the corresponding parameters.

$$W_i = K/S_i \tag{1}$$

where W_i = unit weight for the ith parameter, S_i = standard value for ith parameters, and K = proportional constant.

The value of K has been considered 1 here and is calculated using the mentioned equation below:

$$K = 1/\sum(1/S_n) \tag{2}$$

On summation, all the Wi values should be equal 1 (unity).

According to [14], the value of quality rating or sub-index (Q_i) is calculated using the equation as given below:

$$Q_i = 100\,[(V_o - V_i)/(S_n - V_i)] \tag{3}$$

where V_o = observed value of ith parameter at a given sampling site, V_i = ideal value of ith parameter in pure water, and S_n = standard permissible value of ith parameter.

Final or overall WQI is calculated using the following formula:

$$\text{WQI} = \sum W_i Q_i / \sum W_i \tag{4}$$

3 Results and Discussion

3.1 Water Quality Assessment Using GIS

GIS is an efficient and powerful assessment tool in water management [15, 16]. Water quality assessment is done using ArcGIS software 10.1, as shown in the figures below. The villages in the study area affected by pH, TDS, Arsenic, Nitrate, Fluoride, Chloride, Hardness, and alkalinity are marked on the map. The parameters are discussed below, along with the maps.

pH

pH is a measure of how acidic or basic the water is. The pH of water in its purest form is 7, but according to most agencies [6, 17], the recommended pH of drinking water is between 6.5 and 8.5. The pH of water is important as it helps maintain alkalinity in the cells to function in the body. In this study, the pH ranges from 6 to 9.

TDS

The total dissolved solid represents the total concentration of the dissolved substance in water. The permissible limit, according to BIS, for TDS is up to 500 mg/l. The TDS in the study area is quite high and is found to be ranging from 1000 to 9000 mg/l.

Arsenic

A few traces of Arsenic are found in the study area but are not evident; hence, its distribution is not shown on the map. Arsenic is bad for health; no such traces should be found in drinking water.

Nitrate

Nitrate can be found in groundwater due to contaminated runoff, chemicals, or fertilizers in the soil. The permissible limit for Nitrate is 45 [18]. The Nitrate in the study area is found to be high as 900 mg/l.

Fluoride

Fluoride is naturally found in water, and when it is found in high quantity, can cause tooth decay. The Fluoride in the study is found to be as high as 4 mg/l. The permissible limit for Fluoride in drinking water is up to 1 mg/l.

Chloride

Chloride concentration can corrode pipelines and can harm appliances. The standard level of Chloride is not to exceed 250 mg/l. In context to the study area, the levels are up to 3500 mg/l.

Alkalinity

A recommended amount of alkalinity is good in drinking water as it helps neutralize the acid changes in water. The permissible limit for alkalinity is 20–200 mg/l. The study area has alkalinity up to 1500 mg/l.

Hardness

Hardness is generally caused by carbonates of magnesium and calcium. The permissible limit of Hardness is 600 mg/l [18]. In the study area, the Hardness is up to 1500 mg/l (Fig. 2).

A score matrix is made using these maps (Fig. 3). Villages affected by bad-quality parameters are scored from one to seven. The most affected village scored seven, and the least affected village scored one. The reason behind the degraded water quality of the study area can be power plants and agriculture. According to land-use land cover maps, from 1985 to 2005, the area under cropland and water bodies decreased significantly. Because of increased built-up land and mixed forest, urbanization and commercialization happened in the area. Built-up land increased four times from 1985 to 2005 with an increase in population due to power plants. Urbanization happened near the power plants and rivers during the same period. Also, the major portion is under agriculture and forest in the study area. For analysis,

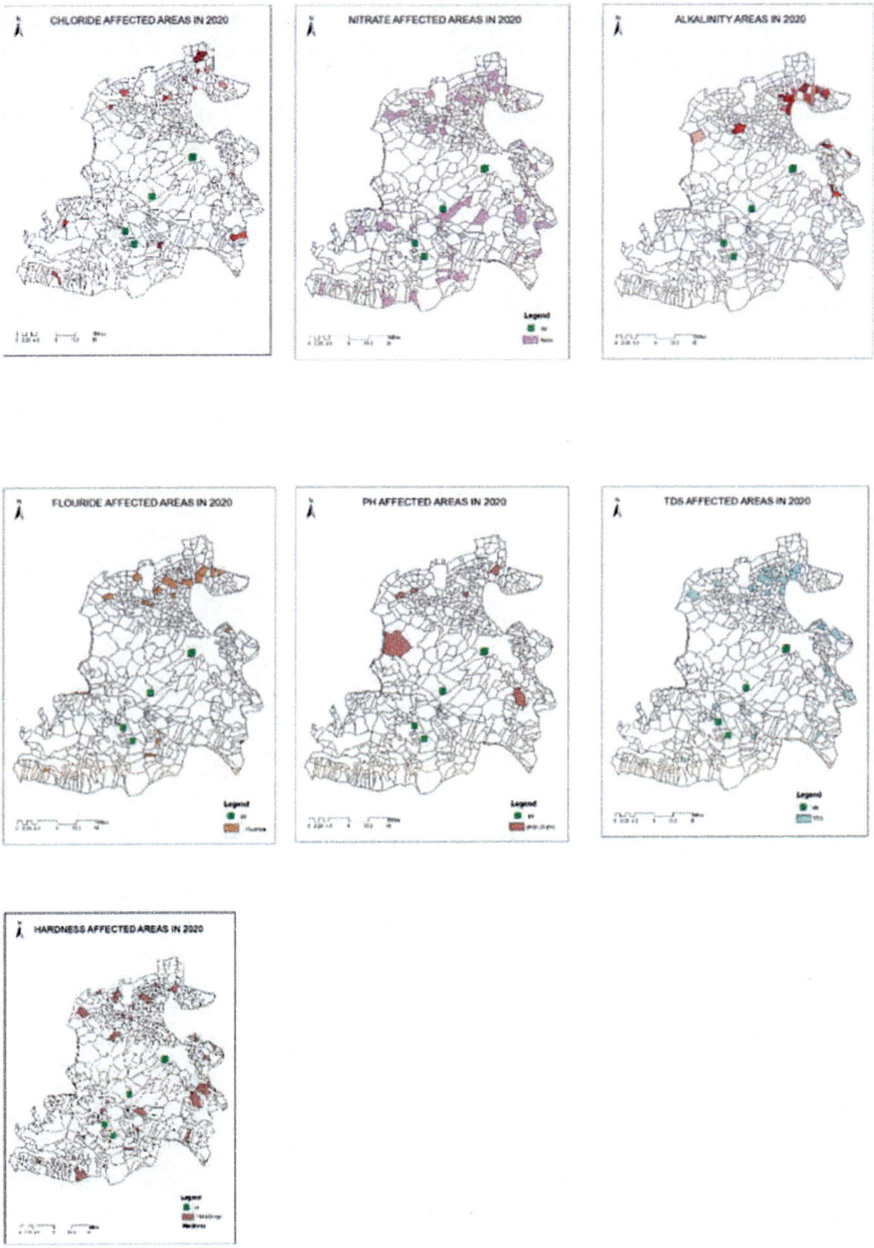

Fig. 2 Quality maps

a buffer of 15 km is created around all the power plants to see the areas affected by pollution due to power plants (for most of the power plants, EIA is done in the buffer area of 10–20 km). Some of the quality-affected villages are in the buffer area. The effluents from the power plant can be one of the responsible reasons for the degradation of groundwater quality [19]. The high temperature of discharged water from the power plant can harm the ecosystem. Air pollution caused by these power plants is also responsible for acid rain, which increases the Nitrate content in the groundwater. Another reason for degraded water quality can be chemical fertilizers used in agriculture. Other quality-affected areas are the ones where agriculture is the prime occupation. The fertilizers penetrate the soil and contribute to degrading the groundwater quality.

Fig. 3 Quality score map

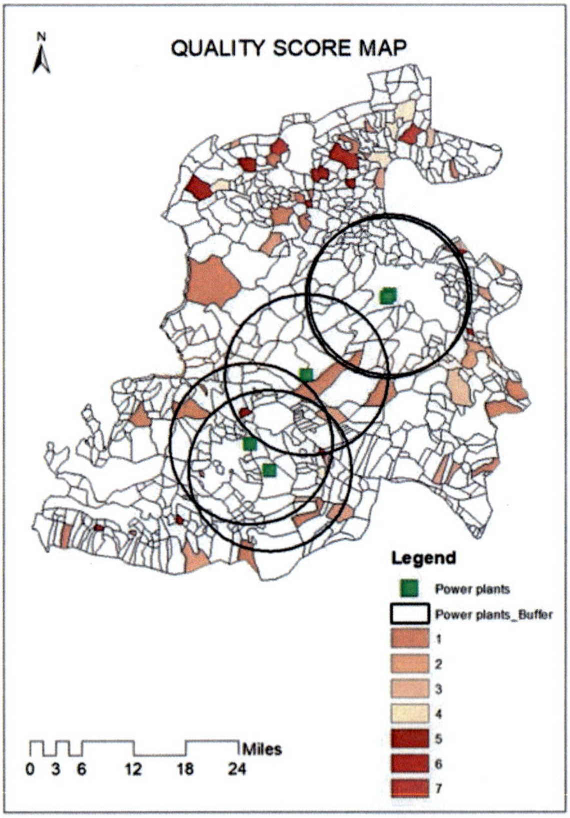

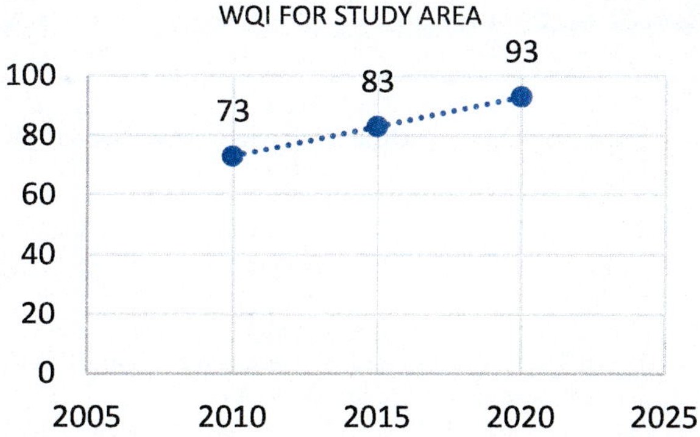

Fig. 4 Water quality index for study area

4 Water Quality Index Analysis

The calculated WQI values for the years 2010, 2015, and 2020 are shown in Fig. 4. It is observed from the graph that water quality falls from the poor to very poor category. Water quality for most of the parameters is not within the permissible limits in the study area. Parameters above permissible limits are pH, TDS, Nitrate, Fluoride, Chloride, and alkalinity. The figure also helps to understand how the water quality has decreased over ten years and quality will continue to degrade if necessary steps are not taken.

5 Conclusion

The study's outcome in the area of Bundi, Kota, and Chittorgarh reveals that the groundwater quality is bad in the area, which raises serious concerns for people's health. The bad quality of water in the area is due to the presence of pH, TDS, Nitrate, Fluoride, Chloride, Hardness, and alkalinity. Based on the high value of WQI, groundwater is not safe for domestic purposes. The WQI results indicate that the groundwater is deteriorating every year and will continue to do so if necessary steps are not taken. According to spatial analysis, the study area is mostly affected by Hardness, Nitrate, and Fluoride. The score map concludes that the most affected areas with the above parameters are Ajeta, Kanjri Silor, Amarpura, Gadegal, Matoonda, and Ramganj. The study also concludes that the two possible reasons for bad water quality in the region can be power plants and agriculture. With climate change and global warming, the deterioration of water quality is possible. Further study can be conducted to find out the reason behind the quality issue in the study area.

References

1. Vasant Madhav Wagh, D.B.: Major ion chemistry and quality assessment of the groundwater resources of Nanded tehsil, a part of Southeast Deccan Volcanic Province, Maharashtra, India. Environ. Earth Sci. Volume (2016)
2. Report of the Ground Water Resource Estimation Committee.: New Delhi, Ministry of Water Resources, River Development & Ganga Rejuvenation (2017)
3. Howard, K.W.: Urban groundwater issues—an introduction. In: Current Problems of Hydrogeology in Urban Areas. Urban Agglomerates and Industrial Centres, pp. 1–15. Netherlands, Kluwer Academic Publishers (2002)
4. DiogoCosta, H.: Impacts of climate change on groundwater in the Great Lakes Basin: a review. J. Great Lakes Res. 1613–1625 (2021)
5. Maurizio Barbieri, M.D.: Climate change and its effect on groundwater quality. Environ. Geochem. Health 1–12 (2021)
6. BIS.: National Building Code of India (2016)
7. Aayog, N.: Composite water management index. New Delhi: Ministry of Jal Shakti and Ministry of Rural Development (2019)
8. Arjun Ram, S.K.: Groundwater quality assessment using water quality index (WQI). Appl. Water Sci. (2021)
9. Rachel, M., Coyte, A.S.: Co-occurrence of geogenic and anthropogenic contaminants in groundwater from Rajasthan, India. Sci. Total Environ. 1216–1227 (2019)
10. Prashant Bhakar, A.P.: Groundwater quality assessment in a hyper-arid region of Rajasthan, India. Natural Resour. Res. **28** (2018)
11. Abhas Jain, S.K.: Prevalence of fluoride in ground water in Rajasthan State: extent, contamination levels and mitigation. Open J. Water Pollut. Treatm. (2014)
12. Reddy, V.R.: Water sector performance under scarcity conditions: a case study of Rajasthan, India. Water Policy Uncorrected Proof 1–18 (2010)
13. Rajasthan State Action Plan.: Government of Rajasthan (2014)
14. Brown, R.M., M.N.: A water quality index—do we dare? Water Sew Works pp. 339–343 (1972)
15. Huzefa Shaikh, H.G.: Hydrogeochemical characterization of groundwater from semiarid region of WESTERN India for drinking and agricultural purposes with special reference to water quality index and potential health risks assessment. Appl. Water Sci. (2020)
16. Sadat-Noori, S.M., Ebrahim, E.: Groundwater quality assessment using the water quality index and GIS in Saveh-Nobaran aquifer, Iran. Environ. Earth Sci. 3827–3843 (2014)
17. Agency, U.E. (n.d.). Epa. Retrieved from U.S. Environmental Protection Agency: https://www.epa.gov/caddis-vol2/ph
18. Standards, B.O.: Indian Standard drinking Water—Specifications (2012)
19. Nazish Sarang, L.S.: Impact of sewage discharge on water quality and benthic diversity of Kota Barrage, Kota, Rajasthan India. J. Indus. Pollut. Control **30**(2), 199–204 (2014)

Application of Remote Sensing Data to Assess Tidal Inundation and Its Potential for Food Crop Cultivation

Lilik Slamet Supriatin, Indah Susanti, Sinta Berliana Sipayung, and Martono

Abstract One of the causes of the global food crisis is climate change. To overcome the global food crisis, it is done by opening up new lands for the cultivation of food crops. The diversity of locations for food crop cultivation is very necessary. One of the locations that has not been widely developed for food crop cultivation is tidal land. One of the obstacles to the development of tidal land is the inundation of tidal land. Periodic monitoring of tidal inundation is able to determine the temporal pattern of inundation that can be used for the preparation of types of food crops that can be cultivated. Monitoring locations in South Kalimantan and Central Kalimantan. Monitoring the height of tidal inundation is carried out using remote sensing technology. Data source from the ERA5-Land Satellite (Land Monthly Averaged by Hour of Day) ECMWF. Periodic data 1981–2020. Analysis of tidal inundation heights was carried out during the periods December–February (DJF), March–May (MAM), June–August (JJA), and September–November (SON) at three depths (1, 2 and 3 m). The results show that in a year, there are two patterns of inundation which is the DJF and MAM pattern and JJA and SON one. DJF and MAM pattern has more (flooded) than stagnant water conditions in the JJA and SON months' groups. Flooded conditions affect the type of rice variety to be planted. Characteristics of rice plants related to waterlogging or tidal land are plant height. In the DJF and MAM month groups, rice varieties with higher plant height can be planted than those planted in the JJA and SON month groups (variety paddy Seratus Malam and Kartuna).

L. S. Supriatin (✉) · S. B. Sipayung · Martono
Research Center for Climate and Atmosphere, National Research and Innovation Agency, West Java, Indonesia
e-mail: lilik_lapan@yahoo.com

I. Susanti
Directorate of Repository, Multi Media, and Scientific Publishing, National Research and Innovation Agency, West Java, Indonesia

A. Basit et al. (eds.), *Proceedings of the International Conference on Radioscience, Equatorial Atmospheric Science and Environment and Humanosphere Science*, Springer Proceedings in Physics 290, https://doi.org/10.1007/978-981-19-9768-6_61

1 Introduction

Agriculture is a sector of life that is affected and vulnerable to climate change. The climate that is always changing today will affect the agricultural sector, especially on cropping patterns, planting season (planting time), production, and yield quality [8]. There is a strong inverse relationship between climate change and agricultural production of food crops. This is because the food plants consumed by the Indonesian people are derived from seasonal plant species that are very sensitive to water availability stress [17]. Added by Suberjo [13] states that changes in weather and climate will reduce agricultural production by 5–20% so that the yield that can be harvested is only about 80–95%. The fall of high-intensity rain with a short duration results in flooding of rice fields and root rot or prolonged drought which results in drought. Climate change causes rainfall that falls disproportionately to the water needs of food crops [9]. It is different with Angles et al. [1], which states that the decline in agricultural production in India is due to changes and decreases in rainfall intensity.

Climate change causes an increase in rainfall in certain places as well as drought in different places. In addition to the direct impact of climate change on agricultural crops, the impact of climate change will be more widespread in farming communities. Solihin [11] states that the long-term impact of climate change is a decrease in the area of agricultural land caused by many farmers switching to other jobs. Previously, Ayunwuy [3] had conducted research related to the impact of climate change on agricultural human resources in Nigeria which yielded similar results to Solihin [11]. Therefore, adaptation actions are needed in the food crop agricultural sector to adjust weather patterns to climate change. One of the adaptations in the food crop agriculture sector is to develop tidal swamp land as the location of new land. This type of land does not have limited availability and water requirements for rice plants, although the threat of sea water ingress due to sea level rise and global warming haunts this tidal land.

Tidal land is a land resource that has not been used optimally in Indonesia. Indonesia has tidal areas on four large islands, namely Sumatra (along the east coast of mainland Riau Province, South Sumatra, Jambi, and a little bit in North Sumatra and Lampung, the west coast of Aceh, Bengkulu and West Sumatra), Kalimantan (the west coast of West Kalimantan, the west coast of South Central Kalimantan and South Kalimantan, as well as the east and northeast coasts of East Kalimantan), and Papua (south coast of Fak-Fak, southeast coast of Merauke, Bintuni, Manokwari, Sorong, north coast of Nabire, Sarmi, and the Mamberamo river valley), Sulawesi (the southwest coast of Palu, Mamuju, Bone Bay, the northeast coast of Palopo, and the south coast of Toli-Toli) [12]. Indonesia's tidal land area ranges from 20.13 to 25.82 million ha, with the largest being Sumatra (8.41 million ha), Papua (7.49 million ha), and Kalimantan (6.99 million ha) [2]. Figure 1 shows the location of tidal swamp land in Indonesia (yellow color represents tidal land).

Identification and inventory of tidal land can be quickly and cheaply knowing using remote sensing technology. Advances in remote sensing technology have been

Fig. 1 Location of tidal land in Indonesia. *Source* Ministry of Public Works and Public Housing of the Republic of Indonesia

able to monitor the height of puddles on tidal land. Liliesand et al. [6] stated that remote sensing technology is superior in determining the boundaries of objects in an area or landscape more accurately when compared to the field survey method. With remote sensing technology, the earth's surface in the form of land, oceans, coasts, beaches, swamps, tidal lands, lakes, hills, and mountains with all kinds of land cover forms can be seen in printed images or digital data. Remote sensing data analysis can identify and inventory topographic conditions, levels of erosion and environmental degradation, land cover, hydrological and drainage conditions.

Therefore, the research will conduct an inventory of temporal tidal inundation in the locations of South Kalimantan and Central Kalimantan as well as its potential for the development of food crop agriculture, especially rice plants. The purpose of this study was to determine temporally the height of standing water on tidal land and its potential for the development of rice plants.

2 Data and Method

2.1 Data

The main data used in this study are the height of the tidal inundation. Supporting data are rainfall (1961–2020) and types of rice varieties. The data of rice varieties used are plant height and plan age. Locations for monitoring the height of tidal inundation are in South Kalimantan and Central Kalimantan (1.5 U–4 S and 109 E–118 E). Monitoring the height of tidal inundation is carried out using remote sensing technology. Data come from ECMWF's ERA5-Land (Land Monthly Averaged by Hour of Day) satellite. ERA5-Land data are used because ERA5-Land is a reanalysis dataset that provides a consistent view of changes in soil variables over decades with improved resolution compared to ERA5. ERA5-Land has been produced by replaying the soil component of the ECMWF ERA5 climate reanalysis. Reanalysis combines model data with observations from around the world into a complete and globally consistent dataset using the laws of physics. Data period is 1981–2020.

2.2 Method

The analysis of tidal inundation heights was carried out during the DJF wet season period (December–February), the transitional season from the rainy season to the MAM dry season (March–May), the JJA dry season (June–August), and the transitional season from dry season to rainy season SON (September–November) at three depths, namely 1, 2 and 3 m. The results of data analysis are in the form of volume ratio between water and soil. To get the inundation height, the following formulation is used:

$$\text{Inundation height} = \text{volume ratio of water to soil} \times \text{depth}$$

The height of the inundation will be compared with the height of the rice plant to obtain rice varieties that have the potential to be planted on tidal land. For the month group, it will be analyzed to determine the type of rice variety based on the age of the rice plant.

3 Results and Discussion

The results of spatial processing of remote sensing data for the volume of inundation at a depth of 1, 2 and 3 m at the tidal land research location are presented in Figs. 2, 3 and 4. Figures 2, 3 and 4 show the volume of water in the soil for the wet month (DJF), dry month (JJA), transitional months (MAM and JJA) at three depths (1, 2 and 3 m). The volume of water in the soil is the ratio between water (m^3) and soil (m^3). The greater the ratio between water and soil, the greater the water that is inundated in the tidal land. Volume ratio between water with soil ever has been used by Sabater [10].

The use of water-to-soil ratio because tidal land is an organic soil. Water on land in tidal areas comes from rivers and ocean flows. When the sea is high tide, some of the sea water will enter the tidal land so that this type of land never experiences limitations in water availability. Tidal land conditions like this are in accordance with the development of rice plants. Rice plants will have a lot of weeds, when planted with in the fields (dry land). Rice plants cultivated in paddy fields have limited water requirements, if the climate is always changing and is not in accordance with the rice growth phase, but in tidal land, paddy plants continue to grow, without experiencing stress on water scarcity.

Based on Fig. 2, it can be seen that in the DJF and MAM moon groups, the ratio between water and soil at a depth of 1 m is dominated by the ratio value of 0.36–0.39.

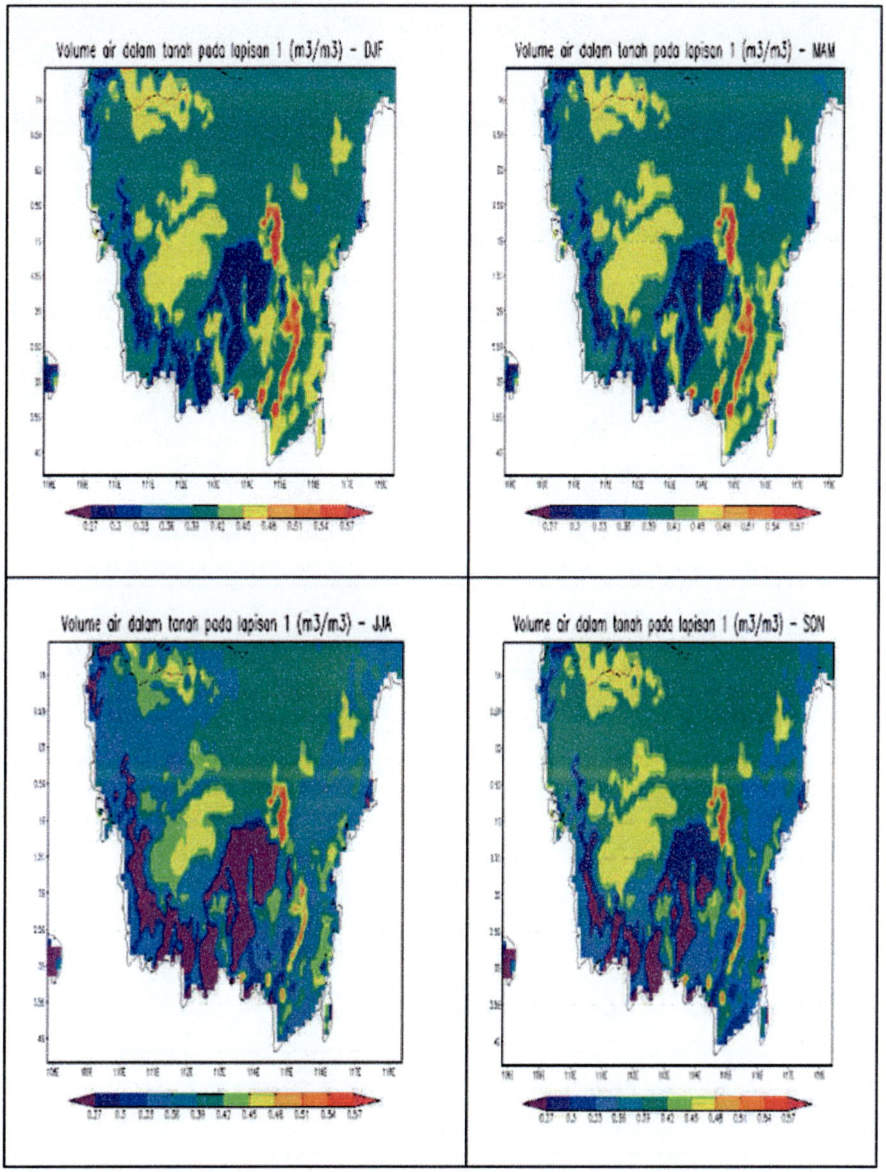

Fig. 2 Volume ratio between water and soil at a depth of 1 m in four groups of months (DJF, MAM, JJA, and SON)

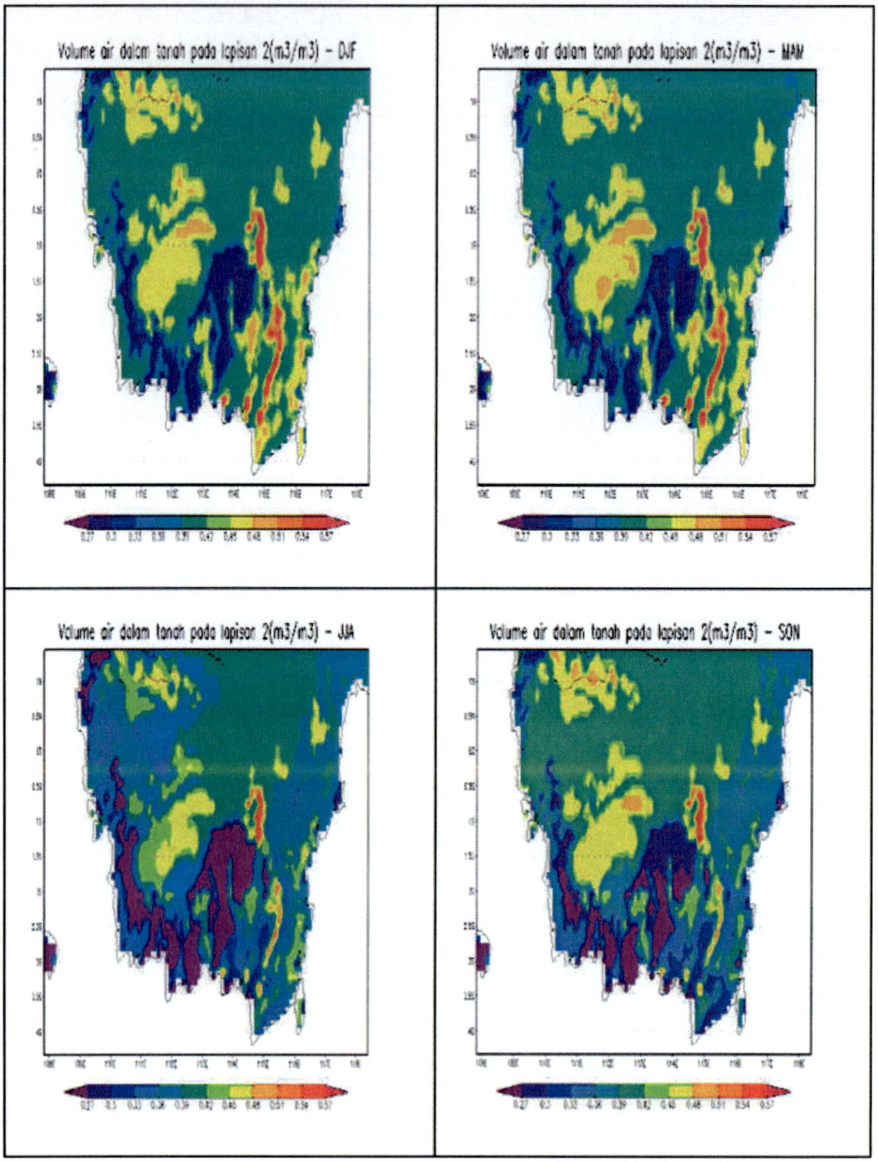

Fig. 3 Volume ratio of water to soil at a depth of 2 m in four groups of months

This means that it can be equated with a height of 36–39 cm. Even in DJF, there is a ratio of 0.27–0.30 (27–30 cm) which is not found in MAM. In the MAM ratio of 0.27–0.30, the water increases so that it becomes 0.3–0.33 (30–33 cm). The MAM group also has the largest ratio among other month groups, which is 0.54–0.57 (54–57 cm). This is because the rainfall at the location at DJF is relatively smaller than

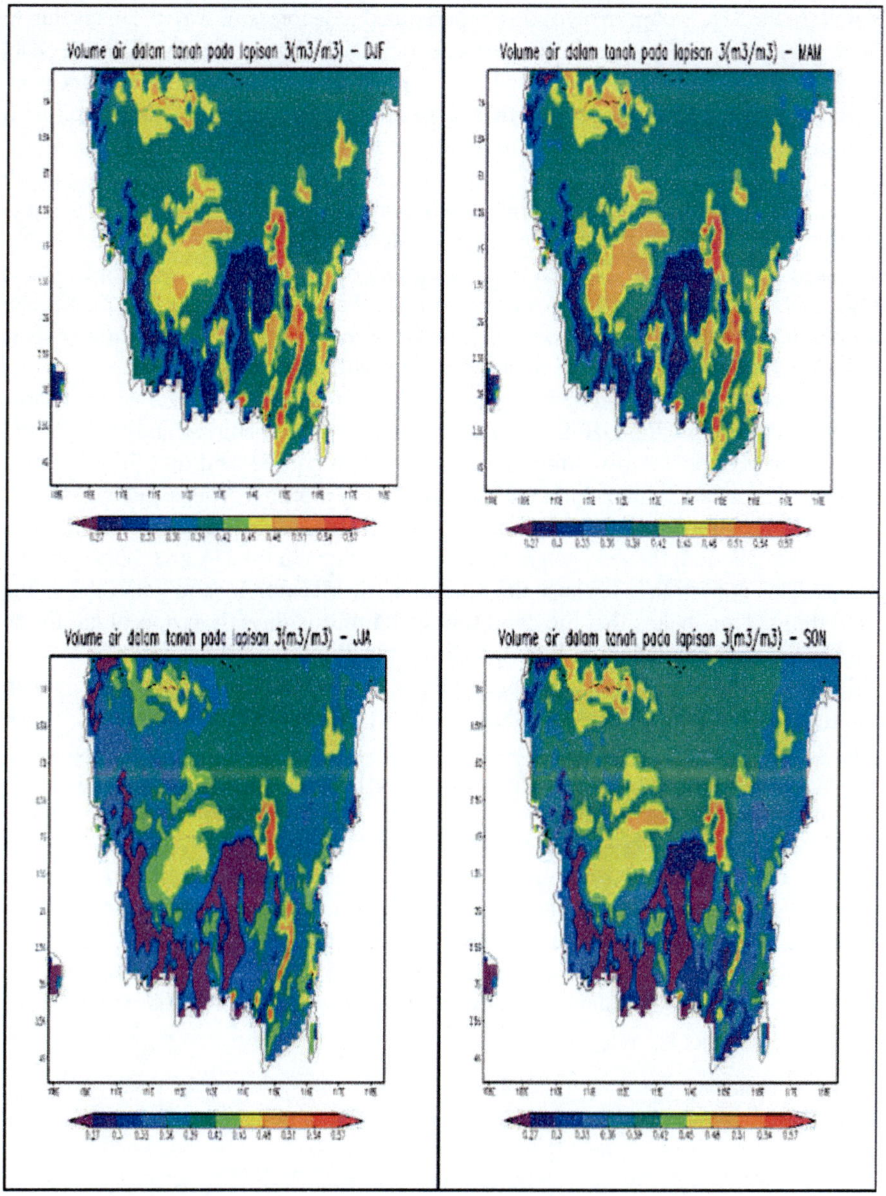

Fig. 4 Volume ratio of water to soil at a depth of 3 m for four groups of months

MAM. In the JJA group, many ratios experienced a decrease in water. This is due to the dry season starting to enter the research location. On the other hand, in the SON group, many of the ratios experienced an increase due to the start of the rainy season.

Based on Fig. 3, it can be seen that at a depth of 2 m, it is still dominated by a ratio of 0.36–0.39 (height 72–78 cm), but what is interesting here is the presence of a water-to-soil ratio of 0.48–0.51 (96–102 cm). This ratio does not exist at a depth of 1 m. The dynamics of the volume ratio of water to soil at a depth of 2 m is equal to a depth of 1 m, when JJA has decreased than MAM and will rise again on SON.

Based on Fig. 4, it can be seen that for a depth of 3 m, the ratio of water to soil is 0.54–0.57 (162–171 cm) in a greater percentage than the depth of 1 and 2 m. It can be concluded that the deeper the tidal land, the higher the water-to-soil ratio. This is because the deeper you dig, the more water you will find.

Based on Figs. 2, 3 and 4, it can be seen that for depths of 1, 2 and 3 m, the water conditions for the DJF and MAM moon groups are similar and the JJA and SON moon groups are also similar. So, in a year, it can be said that there are two patterns of inundation, namely in the DJF and MAM months' groups and the JJA and SON month's groups. In the DJF and MAM months' groups, the water conditions were more (flooded) than the stagnant water conditions in the JJA and SON months' groups. This condition is because the rainfall at the research location also shows the same thing. This means that the condition of rainfall is directly proportional to the height of the tidal inundation. Figure 5 shows the climatological rainfall.

Based on Fig. 5, it can be seen that the climatological rainfall of South Kalimantan has a peak of rainfall in March (300 mm) and December (275 mm) for each year (period 1961–2020). Meanwhile, the monthly rainfall from June to November experienced a downward trend. So that between monthly rainfall with tidal inundation has the same pattern. This condition can be used as information for types of rice varieties that have the potential to be planted in tidal land. This is supported by the statement of Suriadikarta [16] and Haryono [5], which states that tidal land is one form of agroecology that can be developed for agricultural land, especially food crops. The depth of the tidal land at the study site is between 1 and 3 m.

Rice plants based on their metabolism are divided into C3, C4, and Crassulaceae acid metabolism (CAM) plants. Included in the C4 group of plants is rice. C4-type

Fig. 5 Climatological rainfall in South Kalimantan

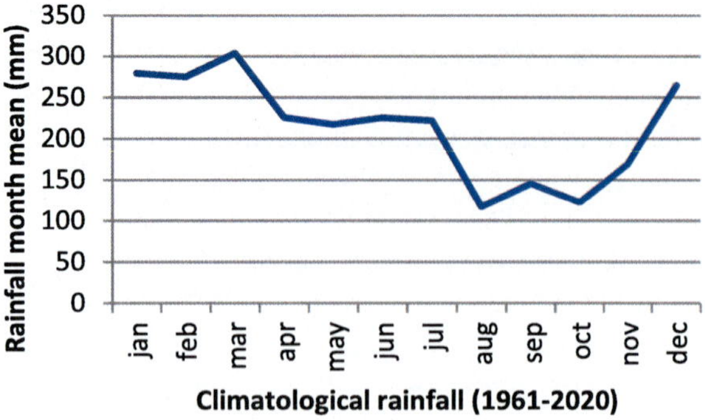

Table 1 Age and height of rice varieties that have the potential to be planted on tidal land

Paddy varieties	Age (days)	Height (cm)
Bengawan	155–160	145–185
Si Gadis	140–145	145–150
Remaja, Jelita, Dara, Syintha, Arimbi, Bathara	140–160	145–165
Gajah Lampung	145	160
Seratus Malam, Kartuna	105–120	140–150
Inpara (inhibrida padi rawa)	115–135	92–111

Source Balai Besar Penelitian Padi, Subang, West Java

plants are larger in water use so that rice plants are cultivated on tidal land where the availability of water is always maintained [4]. Table 1 presents the height and age of rice varieties that have the potential to be planted in tidal land.

Based on Table 1, it can be seen that the range of height of rice plants that can be planted is 92–185 cm with a plant age ranging from 105 to 160 days (3–5 months). This is because when the tide rises, the rice plant does not sink which can lay down and rot the plant. This is in line with the statement from Sudana [14] which divides tidal land into three, namely shallow tidal land (water puddles of less than 50 cm for less than 3 months), middle tidal land (50–100 cm puddles of water for 3–6 months), and deep tidal land (waterlogging of more than 100 cm for more than 6 months).

When compared with Figs. 2, 3 and 4, the tidal areas that are suitable for the height of the rice plant are those with a depth of 2 and 3 m. This is because the inundation height at a depth of 2 m ranges from 96 to 102 cm (Fig. 3), and at a depth of 3 m, it ranges from 162 to 171 cm (Fig. 4). Rice varieties that have the potential to be planted at a depth of 3 m are Bengawan (plant height 145–183 cm) and other varieties can be planted at a depth of 2 m.

Based on Table 1, it can also be seen that apart from the Inpara variety, all rice varieties in the table above are local varieties. Local varieties are more tolerant of pests and diseases, although yields are low. Supriatin [15] from his research which correlated the age of rice varieties with the number of pests and diseases resulted in a correlation coefficient of $r = -0.24$, and r for the relationship between plant height and the number of pests and diseases was −0.38. This shows that the higher and longer the age of the rice variety, the less the number of pests and diseases that attack the variety. In addition to the advantages of rice varieties planted on tidal land, they are attacked by a small number of pests and also rivers that carry mud (sediment) and flow into tidal land will also drown pyrite (a type of sulfur), so it will be safe for rice plants [7].

4 Conclusion

Remote sensing technology has been able to monitor water conditions in tidal areas. Most of the tidal areas in South Kalimantan and Central Kalimantan have a volume ratio between water and soil of 0.27–0.3 (height 27–30 cm). In a year, there are two patterns of inundation, namely in the DJF and MAM months is one groups and the JJA and SON is other month groups. In the DJF and MAM month's groups, the water conditions were more (flooded) than the stagnant water conditions in the JJA and SON month's groups. Flooded conditions affect the type of rice varieties to be planted. The characteristic of rice plants related to waterlogging or tidal land is plant height. Rice varieties that can be planted in the DJF and MAM month's groups may be planted with rice varieties that have a higher plant height than if planted in the JJA and SON month's groups.

References

1. Angles, S., Chinnadurai, M., Sundar, A.: Awa reness on impact of climate change on dryland agriculture and coping mecha- nisms of dryland farmers. Indian J. Agricul. Econ. **66**, hlm 365−372 (2011)
2. Darman, M., Arsyad, D.M., Saidi, B.B., dan Enrizal.: Pengenbangan Inovasi Pertanian di Lahan Rawa Pasang Surut Mendukung Kedaulatan Pangan. Jurnal Pengembangan Inovasi Pertanian **7**(4), 169−176 (2014)
3. Ayunwuy, K., Ogunlade, Oyetoro.: Farmers perception of impact of climate changes on food crop production in Ogbomoso agricultural zone of Oyo State, Nigeria. Continental J. Agricul. Econ. **4**, 19−25 (2010)
4. Graves, J., Reavey, D.: In: Global Environmental Changes Plants, Animals and Communities, London, Longman (1996)
5. Haryono.: In: Lahan Rawa: Lumbung Pangan Masa Depan Indonesia. pp. 141. IAARD Press, Jakarta (2013)
6. Lillesand, T.M., Keifer, R.W., Chipman, J.W.: Remote Sensing and Image Interpretation, 5th edn. Willey, New York (2004)
7. Mawardi, Sudira, P., Sunarminto, B.H., Gunawan, T., Purwanto, B.H.: Pengaruh Pasang Surut terhadap Pengendapan Lumpur di Lahan Sawah Rawa Kawasan Sungai Barito Kalimantan Selatan. Jurnal Agritech **38**(3), 273−281 (2018)
8. Nurdin.: Antisipasi Perubahan Iklim untuk keberlanjutan ketahanan pangan di Sulawesi Utara, Universitas Negeri Gorontalo (2011)
9. Ruminta, R.: Analisis penurunan produksi padi akibat perubahan iklim di Kabupaten Bandung Jawa Barat. Jurnal Kultivasi **15**(1), 37–45 (2016). https://doi.org/10.24198/kultivasi.v.15il.12006
10. Sabater, M.J.: ERA5-Land monthly averaged data from 1981 to present. Copernicus Climate Change Service (C3S) Climate Data Store (CDS) (2019). https://doi.org/10.24381/cds.68d2bb30
11. Solihin, A., dan Sukartini, N.M.: Respon Petani terhadap perkembangan teknologi dan perubahan iklim: Studi kasus di Desa Gadungan, Tabanan, Bali. Jurnal Ekonomi Kuantitatif Terapan **6**(hlm), 128–139 (2013)
12. Subagyo, H.: Karakteristik dan Pemgelolaan Lahan Rawa Pasang Surut. Balai Besar Penelitian dan Pengembangan Sumberdaya Lahan Pertanian. Badan Penelitian dan Pengembangan Pertanian. Departemen Pertanian. Jakarta (2006). ISBN 979-9474-52-3

13. Suberjo.: Adaptasi Pertanian dalam Pemanasan Global, (Dosen Faperta UGM dan Mahasiswa S3 University of Tokyo) (2009). http://subejo.staff.ugm.ac.id/?p=108
14. Sudana, W.: Potensi dan prospek lahan rawa sebagai sumber produksi pertanian. Jurnal Analisis Kebijakan Pertanian **3**, 141–151 (2005)
15. Supriatin, L.S.: Sistem Pertanian Padi Sawah pada Iklim Tropika Basah Indoensia, Disertasi Program Pasca Sarjana Ilmu Lingkungan Universitas Indonesia (2014)
16. Suriadikarta, D.A.: Teknologi Pengelolaan Lahan Rawa Berkelanjutan : Studi kasus Kawasan EX PLG Kalimantan Tengah. (Technology for Sustainable Management of Tidal Swampy Areas: Case Study in former PLG in Central Kalimantan Povince) Jurnal Sumberdaya Lahan **6**(1) (2012)
17. Surmaini, E., Lahan Pertanian, B.B.L.S., Runtunuwu, E., Las, I.: Upaya Sektor Pertanian dalam Menghadapi Perubahan Iklim (2016)

Rice Projection in Sumatra by 2045 Regarding Climate Projection and Crop Model

E. R. Dewi, E. Susanti, E. Surmaini, Suciantini, Misnawati, U. A. Linarka, F. Ramadhani, A. Dariah, A. Sopaheluwakan, and M. R. Syahputra

Abstract The sustainable increase of rice productivity faces challenges due to the impact of climate change. The climate projection such as CORDEX-SEA with Agricultural Production System Simulator (APSIM) can provide an overview of how the projection of rice productivity in Sumatra Island in the future is not yet known. This study aimed to assess the projection of rice productivity in Sumatra Island in 2026–2035 and 2036–2045 for the RCP4.5 and RCP8.5 scenarios under irrigated and rainfed fields at two planting dates. The simulation results of the CSIROMK3.6 model in the rainy (on September 30 and October 20 planting date) and dry seasons (on April 20 and May 10 planting date) indicated that for RCP4.5 and RCP8.5 in 2026–2035 and 2036–2045 under both irrigated and rainfed fields have increased by more than 0% to more than 50%. Meanwhile, the projection of rice productivity in the dry season at the planting date on April 20 and May 10 in 2026–2035 and 2036–2045 under both rice fields showed that most of Sumatra Island experienced an increase of more than 0–30% for RCP4.5 and 0–20% for RCP8.5. In closing, for the 2026–2035 and 2036–2045 projections both growing seasons and planting dates, the CSIROMK3.6 model shows an increase in productivity. Adaptation options to maintain rice productivity by setting the planting time show differences in productivity. It is important to determine the right time to plant for optimum productivity.

E. R. Dewi (✉) · E. Susanti · E. Surmaini · Suciantini · Misnawati · U. A. Linarka · F. Ramadhani · A. Dariah
National Research and Innovation Agency, Jakarta, Indonesia
e-mail: elsarakhmi@gmail.com

A. Sopaheluwakan
Meteorological, Climatological and Geophysical Agency, Jakarta, Indonesia

M. R. Syahputra
Bandung Institute of Technology, Bandung, Indonesia

A. Basit et al. (eds.), *Proceedings of the International Conference on Radioscience, Equatorial Atmospheric Science and Environment and Humanosphere Science*, Springer Proceedings in Physics 290, https://doi.org/10.1007/978-981-19-9768-6_62

1 Introduction

Indonesia's population is projected to reach 305 million people in 2035 (www.bps.go.id/linkTabelStatis/view/id/1274), and this will increase the need for food supply. To meet the demands of a growing population, food production should be increased by up to 50% by 2030 [1]. Concurrently, world food production is projected to decrease by 1–7% until 2060 due to the impact of climate change. Around 20% of the world's population will be at risk of starvation if the impact of climate change is not counted on.

Paying attention to strategic factors and potential resources that can be utilized optimally, Indonesia has ambitions to become one of the world's food granaries in the future [2]. Indonesia persists in confidence to immediately obtain self-sufficiency and food sovereignty.

The projected rice production in the future may support Indonesia as the world food granary by 2045. It is important to analyze a spatial projection of rice production, mainly in food production centers. The impact of climate change needs to be analyzed as adaptation strategies. Furthermore, these strategies have to be evolved such as developing more resistant areas to the impacts of climate change, building reliable infrastructure, and introducing more tolerant varieties of extreme conditions [3].

The process of downscaling the output of the Coupled Model Intercomparison Project 5 (CMIP5) for the Southeast Asia region has been carried out by the Meteorological, Climatological, and Geophysical Agency/Badan Meteorologi, Klimatologi dan Geofisika (BMKG) as Indonesia representative in collaboration with the Southeast Asia Regional Climate Downscaling (SEACLID/CORDEX-SEA) [4]. The results of the downscaling have been developed in the form of high-resolution (25 $\times$ 25 km and 5 $\times$ 5 km) climate change scenario data for the Southeast Asia region. These data can be used for analyses of the impact of climate change in agricultural and other sectors. The baseline and projection data for this present study use climate data from CORDEX-SEA, with the GCM model of CSIROMK3.6.

To examine long-term climate variability and change, the World Meteorology Organization (WMO) suggests employing a baseline of 30 years (1961–1990) as an attribution [5, 6]. Nowadays, an updated baseline of 30-year period (1981–2010) is used to provide weather and climate extremes and forecasts' information [6, 7]. However, the Intergovernmental Panel on Climate Change (IPCC) employs a baseline of 20 years (1986–2005) in the fifth assessment report [8]. The baseline period selection is usually set by the accessibility of the required climate data. In this study, a baseline of 15-year period (1991–2005) was used because of the adequacy of climate data and the projection period used was also 15 years (2026–2035 and 2036–2045).

The impact assessment of climate projections can be integrated with a crop simulation model, namely Agricultural Production System Simulator (APSIM) that can simulate agricultural production systems in response to environmental and climate risks [9]. Rice production in 2045 (the target time for Indonesia as the world food granary) can be projected through the integration of climate projections and crop simulation models. Additionally, rice production anomalies due to climate change until the period 2045 can be assessed, then can be identified which areas will be affected by increasing temperatures, and will have the potential as new centers of rice production. This information may support the achievement of rice production targets.

A study on the projection of rice production in Java Island was conducted by Susanti et al. [10] to sustain Indonesia as the world food granary. This study uses two climate projection models CSIROMK3.6 and HadGEM2 for the period 2021–2035 and 2036–2050, then uses the APSIM plant model simulation. The simulation results based on the CSIROMK3.6 climate projection show that the impact of production is high during the rainy season (RS) compared to the dry season (DS). The area can be maintained or developed for rice crops in districts with increased production. For the RCP4.5 (RS) scenario on irrigated land, 41 districts out of 191 districts on the island of Java show a change in high-end production. Changes in production during 2036–2050 indicate that fewer districts are experiencing high production changes. APSIM simulation results based on HadGEM2 climate projections show that more districts have decreased production in both year periods for both scenarios. Management strategies related to planting time, application of fertilization, and irrigation need to be analyzed further to obtain adaptation options in the field. This study aimed to identify the projection of rice productivity and compile a map of projected changes at the district level in Sumatra Island in 2021–2035 and 2036–2045 using the Representative Concentration Pathway (RCP) 4.5 and 8.5 scenarios.

2 Materials and Methods

2.1 Scope of Activities

The basis of this research was to integrate the system of climate projections and crop simulations. Global models were downscaled to regional models as climate projections, and the output then was used to run the crop simulations. The daily climate projection datasets, i.e., rainfall, maximum temperature, minimum temperature, and solar radiation were developed for each grid for Sumatra Island. Soil datasets were grouped from the land survey results at the Indonesian Centre for Agricultural Land Resources Research and Development (ICALRRD). Rice data were collected through farmer cultivation practices and focus group discussions (FGDs) with the local agricultural agencies and extension workers.

The APSIM model was then parameterized so that the simulation values were close to the observation values. The next step was to run crop simulations for the baseline (1991–2005) and projection (2021–2035 and 2036–2045) periods for the RCP4.5 and RCP8.5 scenarios during the rainy and dry seasons with recommended applications on irrigated rice fields. The baseline and projection productivity's difference was the value of change in rice productivity to existing productivity. Furthermore, technology and regional development recommendation can be approached through the projected map of changes in rice productivity for the periods 2021–2035 and 2036–2045.

2.2 Research Data

The research was carried out from May to December 2021 in eight provinces on Sumatra Island (Aceh, North Sumatra, West Sumatra, Riau, Jambi, Bengkulu, South Sumatra, and Lampung) (Fig. 1). The data used were: (1) daily climate data, namely rainfall, minimum temperature, maximum temperature, and radiation throughout Sumatra. Sources of data used are CORDEX-SEA baseline data (1996–2005) and projections (2026–2035 and 2036–2045) from the Global Climate Model (GCM) obtained from BMKG as well as daily climate data from observations from climate/rain stations BMKG, PU irrigation, as well as the Ministry of Agriculture; (2) soil data, namely: soil physics data (texture) and soil chemistry data (organic matter: C, N, C/N, pH) collected from the ICALRRD, Indonesian Soil Research Institute (ISRI), Pustaka, and Harmonized World Soil Database (HWSD); (3) rice crop data include varieties, plant age, plant phenology phase, rice production, and plant cultivation technology carried out by farmers; (4) crop management data such as planting time, type of rice field (irrigated/rainfed), and fertilization; and (5) daily soil and climate data at the parameterized location.

2.3 APSIM Model Setup

The APSIM model version 7.9 was configured with modules for rice, groundwater (SOILWAT), nitrogen in the soil (SOILN), crop residues (surface OM), and crop management practices [9]. These modules were then connected to the APSIM central engine to simulate rice crops in a rice-based cropping system on Sumatra Island.

Parameterization of APSIM Model. The soil modules, namely SOILWAT and SOILN, were parameterized according to standard practice in APSIM. For the selected sites, the parameters for the SOILWAT module include bulk density, field capacity, permanent wilting point, pH, and soil texture. Parameters for the SOILN module include organic carbon (OC), pH, Finert (inert C fraction), and Fbiom (microbial biomass fraction). Finert and Fbiom, different collections of organic matter based on specific default values, represent a portion of the total organic carbon at a given location [11, 12].

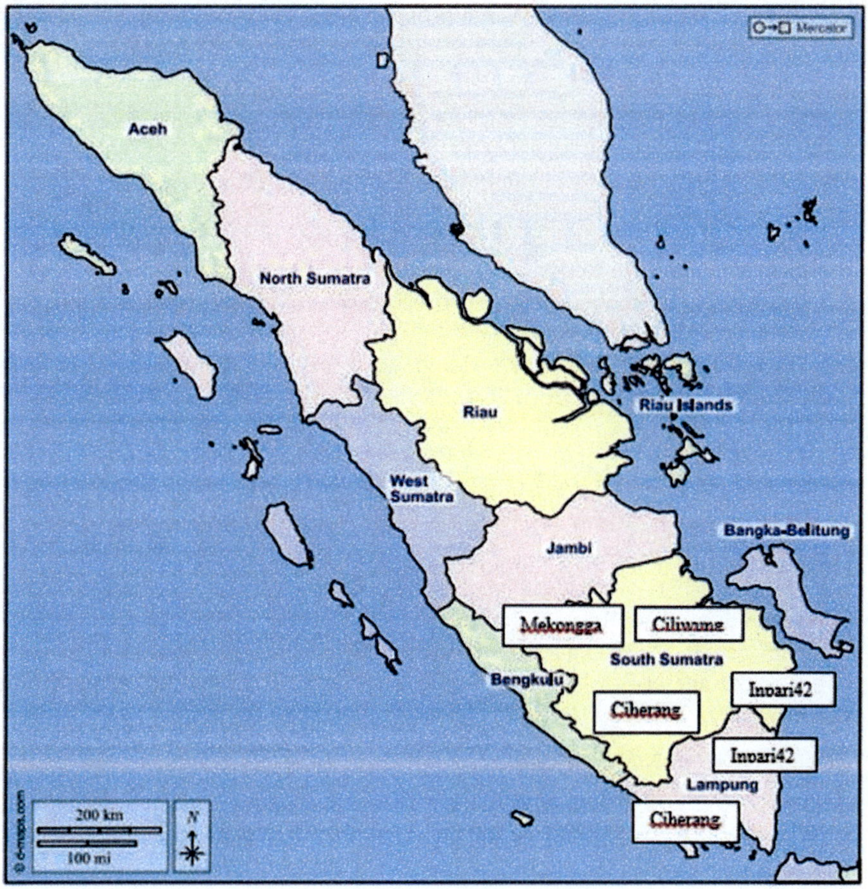

Fig. 1 Map of eight provinces (Aceh, North Sumatra, Riau, West Sumatra, Jambi, Bengkulu, South Sumatra, and Lampung) in Sumatra Island and location of variety parameterization (South Sumatra and Lampung)

For rice crops, rice varieties were parameterized using the dominant rice varieties from the farmer's field according to the procedures described in APSIM. The phenological parameters included the early growth phase, the active tillering phase, the panicle formation phase, and the flowering phase. In the next stage, the parameter values of each phase of phenological development were adjusted until the value of the simulation phase matches the value of the observation phase.

Analysis of Crop Simulation. The crop model simulation was developed to identify and analyze Sumatra rice productivity projections for 2026–2035 and 2036–2045 using downscaling CORDEX-SEA climate projections scenarios RCP4.5 and RCP8.5. The recommended cultivation technology applied to the rice planting system was described as follows:

a. Planting time includes two growing seasons, namely RS and DS:

- The dominant planting time for RS according to the integrated cropping calendar (RS 2019/2020; RS 2020–2021; and 2021/in22) is September III–October I and October I–III, so the recommended planting time is September 30 and October 20.
- For DS is dominant in May I–II and April II–III, so the planting time to be used for simulation is April 20 and May 10.

b. Irrigation:

- Irrigated rice fields: 5 mm intervals of irrigation 2 days apart until rice crop was 95 days old.
- Rainfed rice fields: irrigation only came from rainfall.

c. Fertilizer application:
NPK 15-10-12 (300 kg ha^{-1}) plus urea (200 kg ha^{-1}), NPK fertilizer was given once at the age of 7–14 days after planting (DAP) and urea 2 times when the plant was 21–25 DAP (100 kg ha^{-1}) and 35–40 DAP (100 kg ha^{-1}).

The next system was compiled using APSIM's Manager and simulated starting on January 1, 2026, and running until December 31, 2045, for each RCP scenario. Soil water, soil mineral N, and soil organic matter were reset to initial conditions.

2.4 *Plant Phenology Data*

Rice crop data include varieties, plant age, plant phenology phase, and rice productivity. The variety used in this research was determined by the availability of data for parameterization. Data for parameterizations were collected from the yield of the farmer's field. The required data were planting date, harvest date, 50% flowering age (days) and grain yield (kg ha^{-1}), type of irrigation, time of application, and dose of fertilizer. The compiled data were conducted from April to October 2021. The research locations were in Pringsewu Regency, Lampung, and Ogan Komering Ilir (OKI) Regency, South Sumatra (Fig. 1).

2.5 Parameterization

The APSIM model was parameterized using data field experimental data conducted by farmers in Pringsewu Regency, Lampung Province, and OKI Regency, South Sumatra Province in 2021. The rice varieties used were Mekongga, Ciherang, Ciliwung, and Inpari42. Daily climate data for the period April 1–September 30, 2021, in Pringsewu Regency, Lampung, and OKI Regency, South Sumatra, were used as input for climate data.

The Ciherang rice variety was chosen because it is the dominant variety used in the two provinces; then, the Ciherang variety was calibrated by adjusting the phenological parameters of the rice plant in ORYZA2000. The parameters used are the results of the best fitting between simulation and observation data. The phenological parameters were set based on the record time of the planting phase, formation, and physiological maturity in the farmer's field. Parameter values are adjusted until the phenological phase simulation matches the observed values. Planting dates used were September 30 and October 20 for RS and April 20 and May 10 for DS with irrigation of 5 mm per 2 days. Fertilization was applied three times at 7–14 DAP (300 kg ha^{-1} of NPK 15-10-12 and 200 kg ha^{-1} of urea), 21–25, and 35–40 DAP (100 kg ha^{-1} of urea), respectively. The HWSD database and the defaults in the SoilWat and SoilN modules on APSIM were used as input for the soil data.

Figure 2 shows the comparison of rice productivity between observation and simulation. The simulation value shows the same productivity for all varieties compared to the observation value, but for the Ciherang variety (irrigated rice field), the observation value was lower than the simulation value. This condition may occur because the water set during the simulation was continuously available while irrigation was not continuously available in the field. The APSIM model predicted rice productivity evenly according to water and fertilizer application.

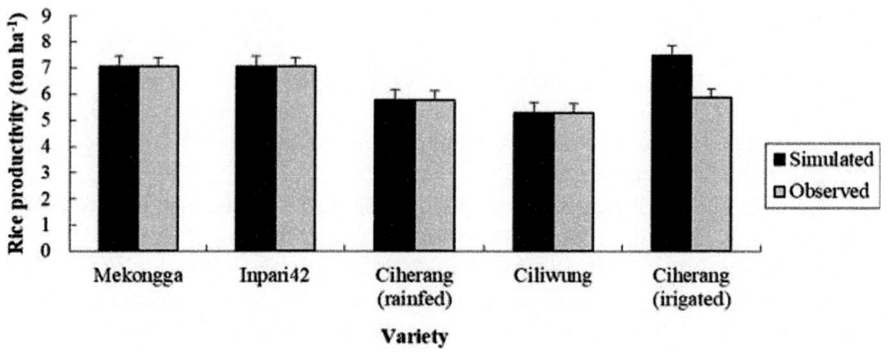

Fig. 2 Simulated and observed values of rice productivity (ton ha^{-1}) based on farmers' fields in Pringsewu Regency, Lampung, and OKI Regency, South Sumatra in 2021

3 Results and Discussion

3.1 Projected Rice Productivity in Sumatra

Rice productivity anomalies in Sumatra were projected by a crop simulation model. To run the APSIM model, a CORDEX-SEA project model specifically CSIROMK3.6 was used as the input of climate. Thereafter, the model was simulated for the 1991–2005 baseline period and two projection periods of 2026–2035 and 2036–2050. APSIM simulations were run on irrigated and rainfed rice fields during RS and DS 2021–2035 and 2036–2050 for the scenario of RCP4.5 and RCP8.5 at two planting times.

Projection of Rice Productivity in 2026–2035 on Irrigated and Rainfed Rice Fields in RS at Planting Time on September 30. The projection results show an increase in productivity for the RCP4.5 and 8.5 scenarios on irrigated and rainfed fields throughout Sumatra Island (Fig. 3). Rice productivity is projected to increase from 10–20 to > 50%. On the irrigated rice fields for the RCP4.5 scenario, a 10–20% increase in productivity will be only found in coastal areas, especially a small area on the west coast of Aceh, Lampung, and southern Bangka. This is possibly due to a decrease of rainfall up to 100 mm month^{-1} less than in other areas (a decrease of rainfall of 200 mm month^{-1}). Increased productivity up to > 50% will be found in the eastern part of Aceh, the central part of North Sumatra, most of West Sumatra except the northwest coast, almost all of Jambi except the eastern part, the northern part, and a few areas in the western part of South Sumatra, while for other areas are in the range of 20–50%. For the RCP8.5, the area that will experience an increase in productivity of > 50% on irrigated rice fields extends to almost most of Aceh (except the western part), North Sumatra, and West Sumatra. The area that will experience an increase of between 10 and 20% is wider on rainfed rice fields, especially in the province of Riau, Lampung, and small islands in the western part of Sumatra Island, especially for the RCP4.5 scenario. Rice productivity will increase by > 50% almost dominating, especially at RCP8.5. Different projection results will be found in the Riau Islands Province and a small area in the Bangka Belitung Province. At this location, most areas will be projected a decline in productivity of up to 50%, mainly because of a change toward a minimum temperature increase of 1 °C and a decrease of rainfall up to 200 mm month^{-1}.

Projection of Rice Productivity in 2026–2035 on Irrigated and Rainfed Rice Fields in RS at Planting Time on October 20. The projection of rice productivity for the planting time on October 20, 2026–2035, will increase for the RCP4.5 and RCP8.5 scenarios. The increase in productivity with the RCP8.5 scenario will be lower than RCP4.5. The decrease in productivity on irrigated rice fields only will occur on Simeulue Island with a decrease of 0–10%. Rice productivity on rainfed rice fields will be projected to increase for RCP4.5 and RCP8.5. The increase in productivity for the RCP8.5 will be smaller than the change in productivity for the RCP4.5. In general, the increase in productivity for the RCP4.5 and RCP8.5 will reach more than 50% for most of Sumatra (Fig. 4).

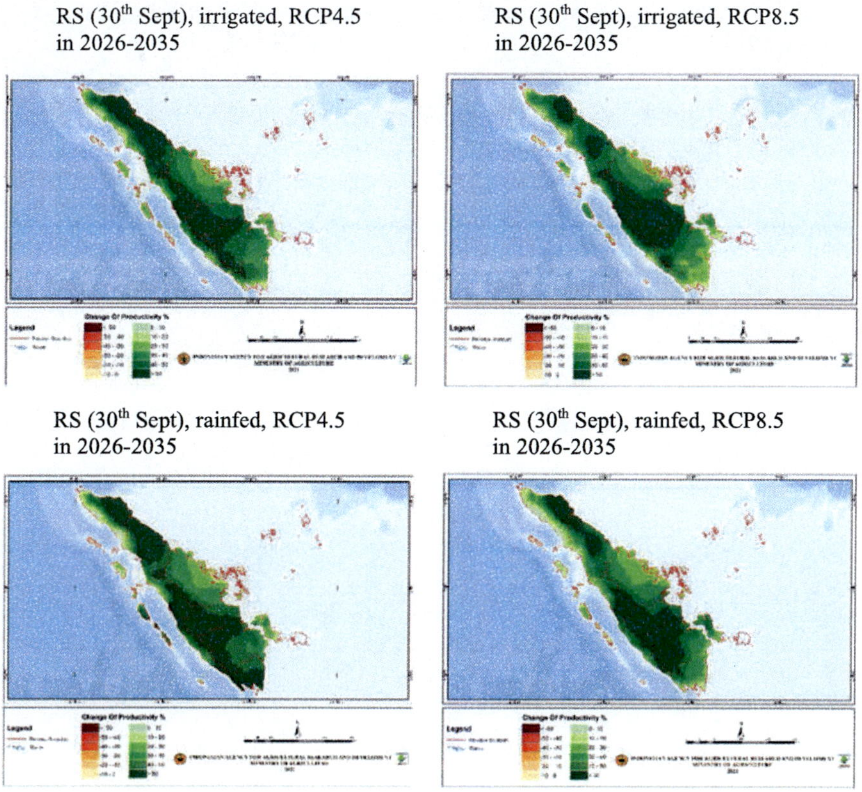

Fig. 3 Projection of changes in rice productivity on irrigated and rainfed rice fields in RS for the RCP4.5 and RCP8.5 scenarios at planting time on September 30 in the period of 2026–2035

Projection of Rice Productivity in 2026–2035 on Irrigated and Rainfed Rice Fields in DS at Planting Time on April 20. The projection of rice productivity in the period of 2026–2035 in DS at planting time on April 20 with the RCP4.5 scenario shows that most areas of Sumatra will experience an increase in productivity of 0–30%, while for the RCP8.5 scenario, increase in productivity ranges from 0 to 20% (Fig. 5). The areas that experience a decline in yields of 0–20% will be wider, namely the Aceh region, the west coast (including Simeulue, Nias, Mentawai Islands), and the east coast (including Riau Islands, Bangka Belitung Islands) Sumatra Island. The highest productivity increase for the planting date of April 20 about 30–40% will be found in the scenario of RCP8.5 on rainfed rice fields located in the western part of South Sumatra and Lampung. Overall, the decline in production in rainfed rice fields will be more extensive than in irrigated rice fields, because of less rainfall and water availability.

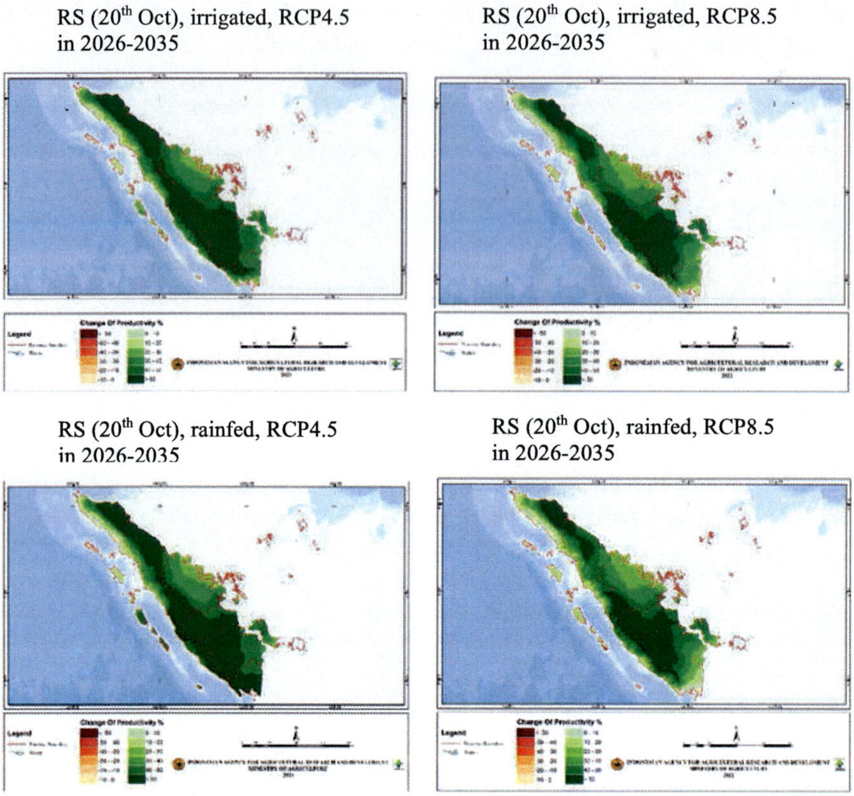

Fig. 4 Projection of changes in rice productivity on irrigated and rainfed rice fields in RS for the RCP4.5 and RCP8.5 scenarios at planting time on October 20 in the period of 2026–2035

Projection of Rice Productivity in 2026–2035 on Irrigated and Rainfed Rice Fields in DS at Planting Time on May 10. Figure 6 shows that most of the west coast of Sumatra to the middle part is projected to increase in productivity on irrigated and rainfed rice fields in DS at planting time on May 10 for both RCP scenarios. Concurrently, rice productivity will be projected to decrease in the North of Aceh Province and the east coast. Areas experiencing decreased productivity will be wider at RCP8.5 compared to RCP4.5.

Projection of Rice Productivity in 2036–2045 on Irrigated and Rainfed Rice Fields in RS at Planting Time on September 30. The increase in productivity will be found throughout Sumatra Island for both RCP4.5 and 8.5, on irrigated and rainfed rice fields, except in some areas in the Riau Islands Province which are projected to decline (Fig. 7). Overall changes in productivity will increase in 2036–2045 which are the same as in 2026–2035, starting from 10–20% to > 50%. Most changes from 30–40% to > 50%. For RCP4.5 irrigated rice fields, productivity will increase to > 50%. It will be found in the eastern part of Aceh, the central part of North Sumatra,

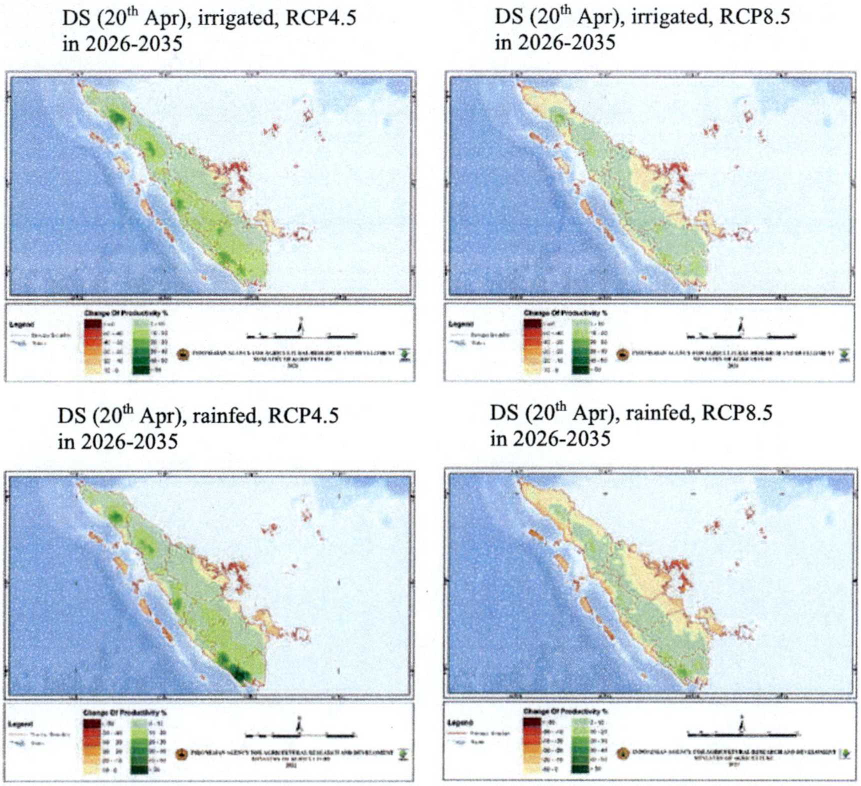

Fig. 5 Projection of changes in rice productivity on irrigated and rainfed rice fields in DS for the RCP4.5 and RCP8.5 scenarios at planting time on April 20 in the period of 2026–2035

most of West Sumatra except the northwest coast, almost all of Jambi except the eastern part, the northern part, and a few areas in the western part of South Sumatra. For RCP8.5, the increase in productivity will be broader in scope, spanning most of Aceh, North Sumatra, almost all of West Sumatra, the western part of Riau, almost all of Jambi and northern parts, and a few areas in the western part of South Sumatra. On rainfed rice, productivity increase will be almost the same as in irrigated rice fields. Productivity will increase by > 50% almost dominating Sumatra, especially for RCP8.5. A decline in productivity by up to 50% is projected to occur in the Riau Islands Province and a small part of the Bangka Belitung Province.

Projection of Rice Productivity in 2036–2045 on Irrigated and Rainfed Rice Fields in RS at Planting Time on October 20. Changes in rice productivity with the planting date of October 20 for the period 2036–2045 will experience a significant increase in productivity reaching more than 50% for most areas of Sumatra both on irrigated and rainfed rice fields for RCP4.5 and RCP8.5. The decline in production

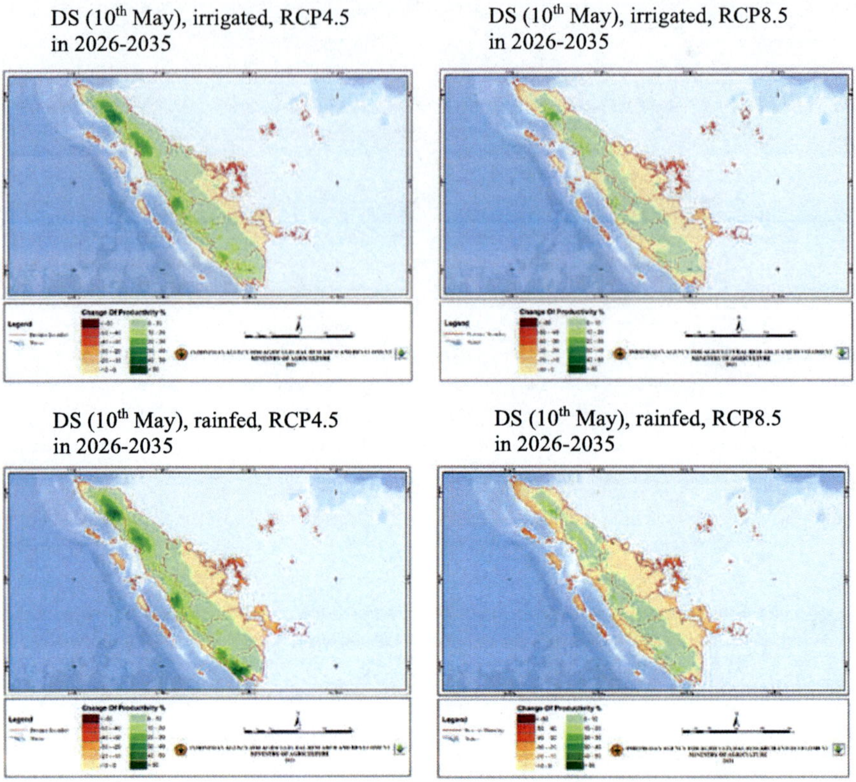

Fig. 6 Projection of changes in rice productivity on irrigated and rainfed rice fields in DS for the RCP4.5 and RCP8.5 scenarios at planting time on May 10 in the period of 2026–2035

will be only seen on Simeulue Island for RCP4.5 on both irrigated and rainfed rice fields (Fig. 8).

Projection of Rice Productivity in 2036–2045 on Irrigated and Rainfed Rice Fields in DS at Planting Time on April 20. Rice productivity in 2036–2045 at planting time on April 20 will experience an increase for both RCP scenarios on irrigated and rainfed rice fields (Fig. 9). A productivity increase of 0–20% is projected to occur in the central part of Sumatra Island from North Sumatra to South Sumatra. Meanwhile, a decrease in productivity between 0 and 20% will occur on the west coast of Sumatra, Aceh, and the east coast of Sumatra (Riau, Riau Islands, Bangka Belitung). On irrigated rice fields, the pattern formed will be not so different between scenarios RCP4.5 and RCP8.5, but on rainfed rice fields, RCP4.5 shows that the area experiencing a yield decrease of 0−10% will be wider than RCP8.5. The highest increase in productivity will occur in the RCP8.5 scenario of rainfed rice fields which is projected to increase by 30–40% located in the western part of South Sumatra and Lampung, but this increase only will occur in a small part of Sumatra. Overall, the

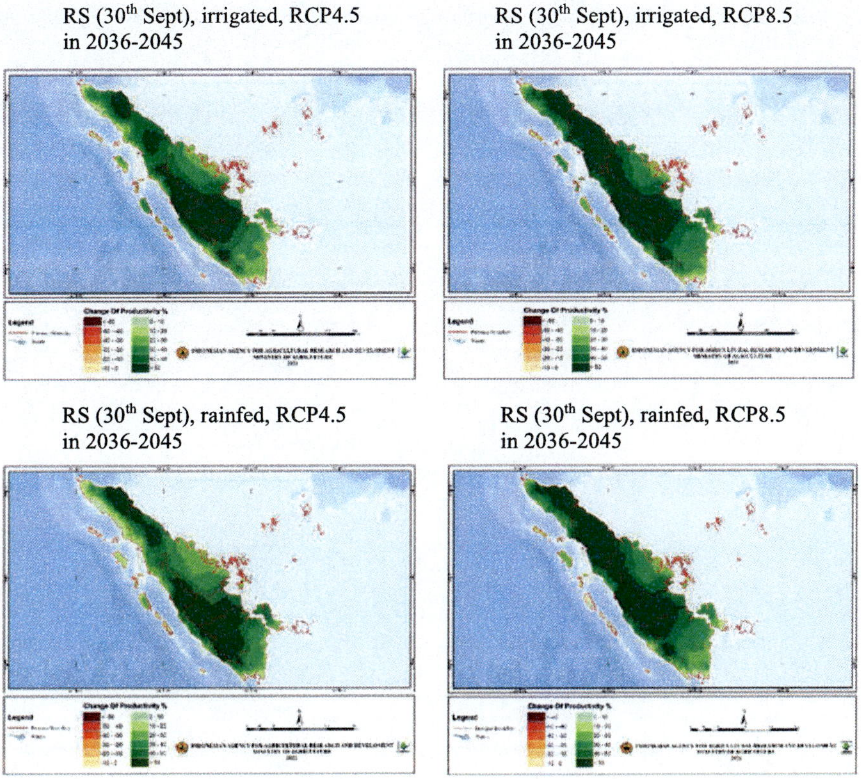

Fig. 7 Projection of changes in rice productivity on irrigated and rainfed rice fields in RS for the RCP4.5 and RCP8.5 scenarios at planting time on September 30 in the period of 2036–2045

area that will experience a decrease in yield will be more rainfed than irrigated rice fields.

Projection of Rice Productivity in 2036–2045 on Irrigated and Rainfed Rice Fields in DS at Planting Time on May 10. Figure 10 shows the projection of rice productivity on irrigated and rainfed rice fields in DS with a planting time of May 10. For RCP4.5 in 2036–2045, rice productivity will be almost the same as RCP4.5 in 2026–2035. The increase in productivity will occur in most of the west coast of Sumatra to the central part, while in the North of Aceh Province and the east coast will show a decrease. Areas experiencing decreased productivity will increase under the RCP8.5 scenario.

In DS, the results of the CSIROMK3.6 projection on both planting times show an increase in productivity, except for the planting date of May 10, both on irrigated and rainfed rice fields for the projections of 2036–2045, especially for RCP8.5. Generally, the increase in productivity in the 2026–2035 projection will be greater for the RCP8.5 scenario (Fig. 11). This condition is expected to occur because the planting time conditions are quite suitable for each phase of plant growth and the availability

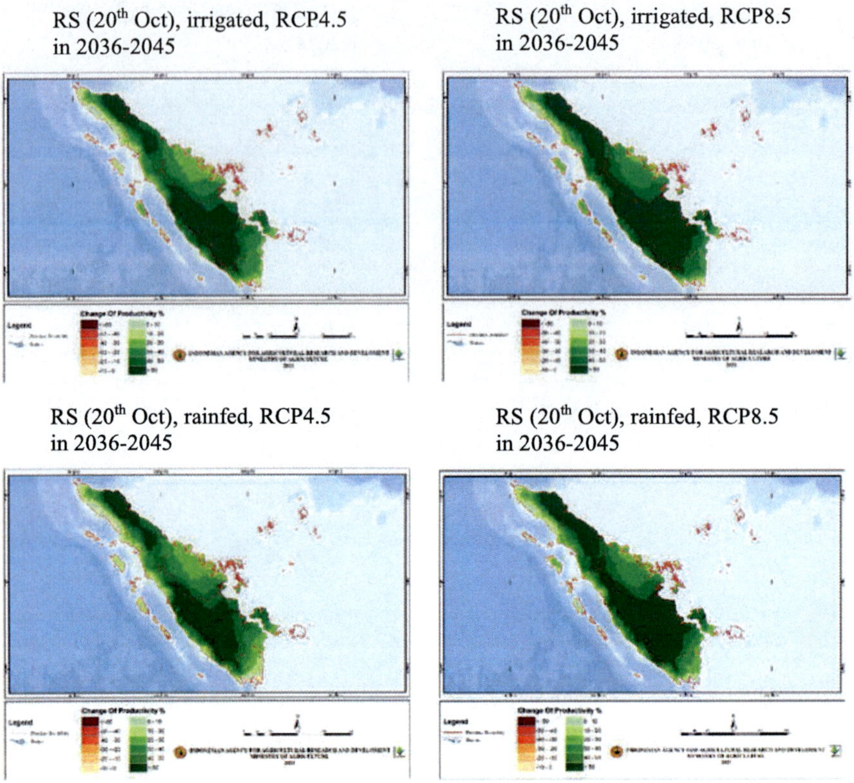

Fig. 8 Projection of changes in rice productivity on irrigated and rainfed rice fields in RS for the RCP4.5 and RCP8.5 scenarios at planting time on October 20 in the period of 2036–2045

of water and other climatic elements, such as radiation intensity, temperature, and air humidity support optimal plant growth. Plant metabolism is directed by temperature, and optimum temperature can increase growth and prevent crop damage. Solar radiation drives crop production, while relative humidity regulates crop transpiration and water balance [13, 14]. On the other hand, in the years 2036–2045, the decline in productivity occurred due to the dynamics of climate elements that were not by plant growth and development.

In RS, the results of the analysis using the CSIROMK3.6 model show that average productivity will increase, with the RCP8.5 scenario having an impact on increasing yields that will be greater than RCP4.5 (Fig. 12). Different scenarios and planting times have different impacts on average productivity depending on the climate projections of the model; this shows that the adaptation options made must be adapted to the climatic conditions in each region of Sumatra considering the spatially different impacts of climate change such as previously described.

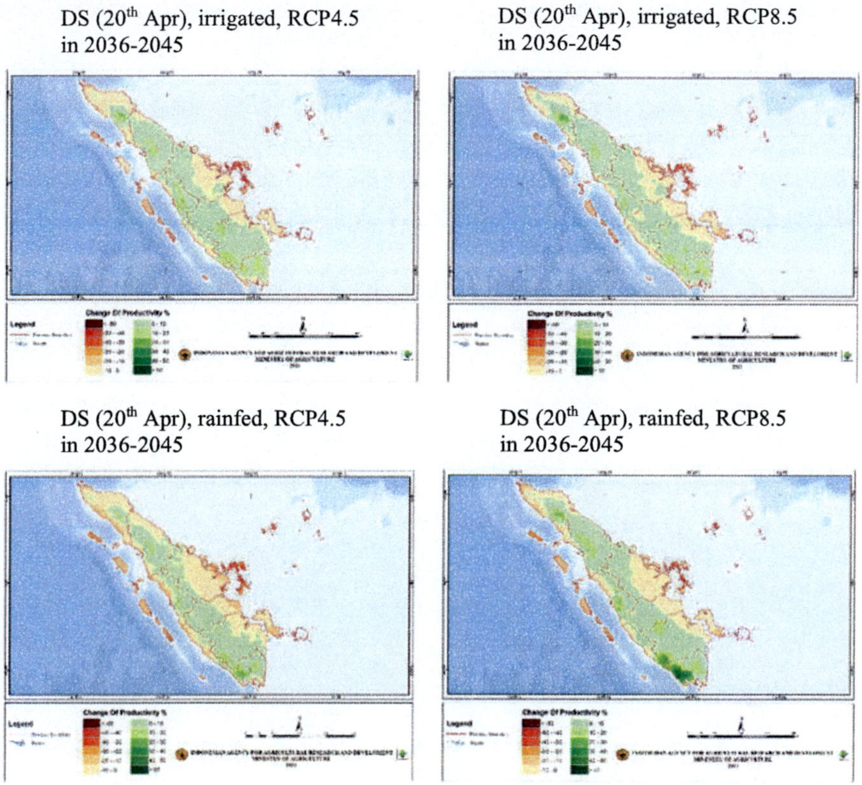

Fig. 9 Projection of changes in rice productivity on irrigated and rainfed rice fields in DS for the RCP4.5 and RCP8.5 scenarios at planting time on April 20 in the period of 2036–2045

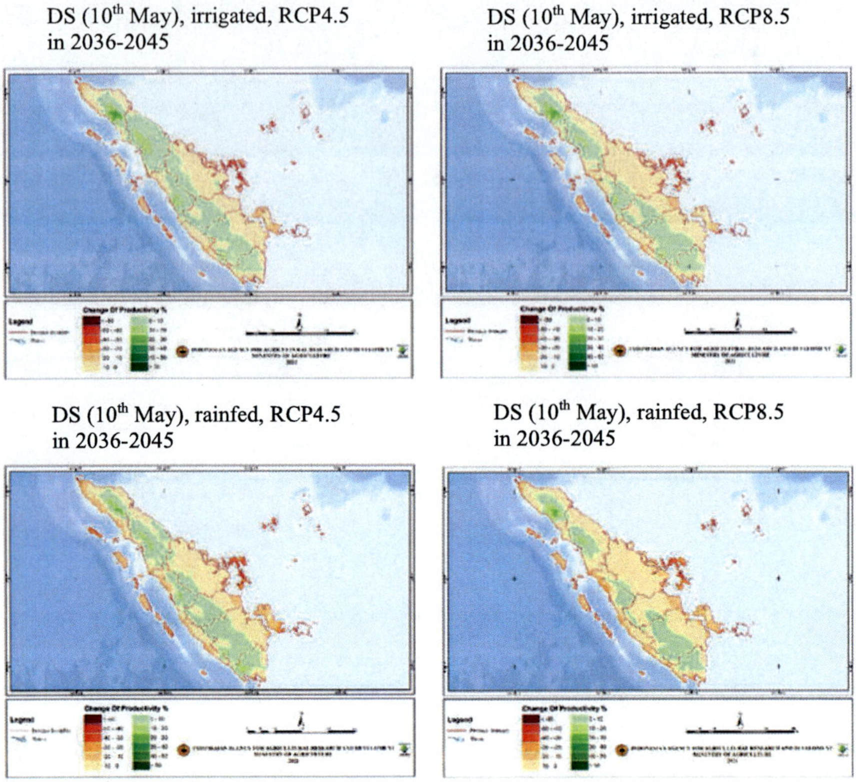

Fig. 10 Projection of changes in rice productivity on irrigated and rainfed rice fields in DS for the RCP4.5 and RCP8.5 scenarios at planting time on May 10 in the period of 2036–2045

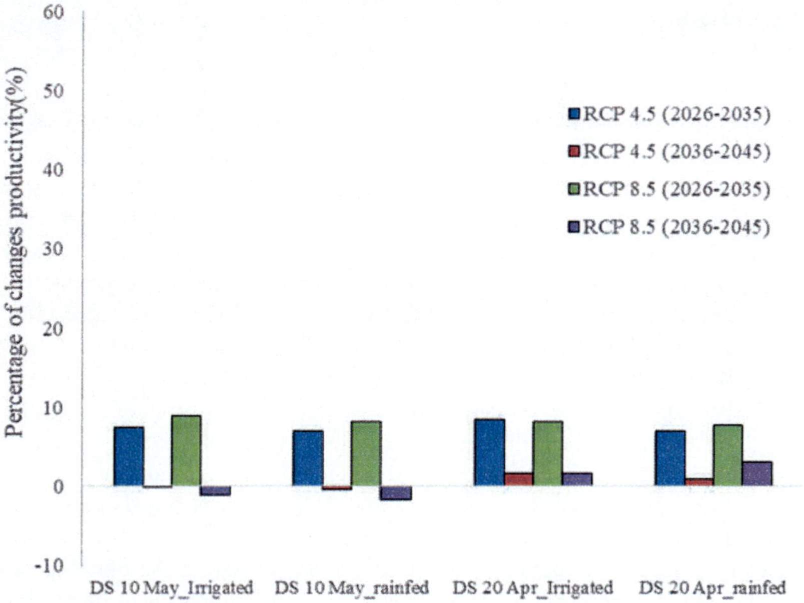

Fig. 11 Changes in the average rice productivity in Sumatra Island on irrigated and rainfed rice fields in DS

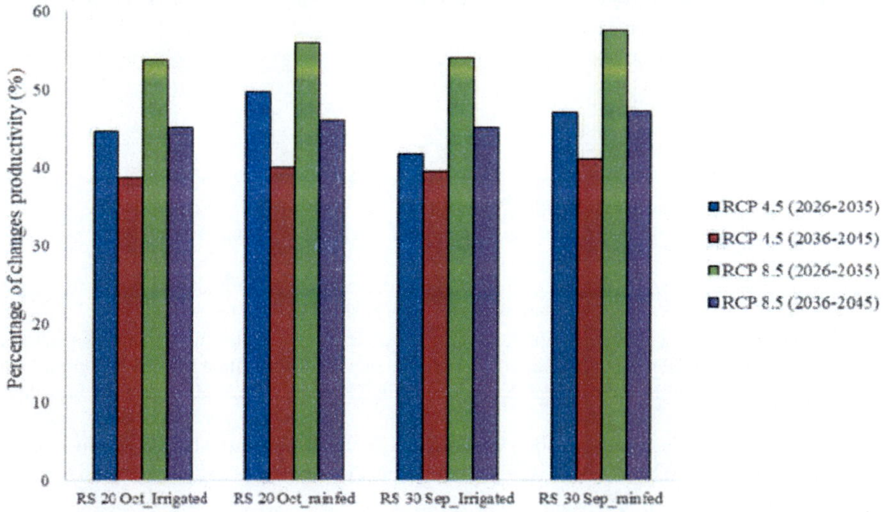

Fig. 12 Changes in the average rice productivity in Sumatra Island on irrigated and rainfed rice fields in RS

4 Conclusions

Different scenarios and planting times have different impacts on the average rice productivity depending on the climate projections of the model, and this shows that adaptation options must be adapted to the climatic conditions in each region.

The projection of rice productivity with the CSIROMK3.6 model in RS at planting time on September 30 and October 20 shows that in RCP4.5 and RCP8.5 in 2026–2035 and 2036–2045, both on rainfed and irrigated rice fields have increased by more than 0% to more than 50%. Meanwhile, the projection of rice productivity in DS with initial planting on April 20 and May 10 in 2026–2035 and 2036–2045 on rainfed and irrigated rice fields shows that most of Sumatra Island has increased by more than 0–30% at RCP4.5 and increased 0–20% on RCP8.5.

The CSIROMK3.6 model has a high increase in productivity, and this is indicated by a large difference in the radiation intensity data, where the average value of the radiation intensity of the model is sufficiently high. The average value of the radiation intensity of the CSIROMK3.6 for RCP4.5 and RCP8.5 scenarios also looks higher than the historical average value.

Adaptation options to maintain rice productivity by setting the planting time in RS and DS show differences in productivity. So, it is important to determine the right planting start for optimum productivity. To get the right adaptation options, it is necessary to run the model by changing the factors that affect productivity including the amount and irrigation system, fertilization, planting time, and rice variety.

References

1. Word Food Program.: How Climate Change Affects Hunger (2016). http://www.wfp.org/climate-change. Last Accessed 21 July 2016
2. Ministry of Agriculture.: Grand design pengembangan lumbung pangan beroreintasi eskpor di wilayah perbatasan, Indonesia Ministry of Agriculture, Jakarta (2017)
3. Surmaini, E., Faqih, A.: Kejadian Iklim Ekstrem dan Dampaknya Terhadap Pertanian Tanaman Pangan di Indonesia. Junal Sumberdaya Lahan **10**(2), 115–128 (2016)
4. Sopaheluwakan, A., Satyaningsih, R., Linarka, A., Rizal, J., Mulsandi, A., dan Kadarsah.: Analisa Proyeksi Perubahan Iklim di Indonesia dengan Menggunakan Data Coordinated Regional Downscaling Experiment South East Asia (CORDEX-SEA). Cordex Meeting, short paper (2016)
5. World Meteorological Organization.: The Role of Climatological normal in a changing climate Technical Report WMO/TD-No. 1377; WCDMP-No.61 (2007). https://library.wmo.int/doc_num.php?explnum_id=4546
6. World Meteorological Organization.: https://public.wmo.int/en/media/pressrelase/no-997-scientist-urge-more-fequent-updates-of-30-year-climate-baselines-keep (2014)
7. World Meteorological Organization.: WMO Guidelines on the Calculation of Climate Normals Technical Report WMO-No.1203 (2017). https://library.wmo.int/doc_num.php?explnum_id=4166
8. IPCC.: Climate Change 2014; Synthesis Report. Contribution of Working Groups I, II, and III to the Fifth Assessment Report of the Intergovernmental Panel on Climate Change Technical

Report. Intergovernmental Panel on Climate Change (2014). https://www.ipcc.ch.report.ar5.syr/
9. Keating, B.A., Carberry, P.S., Hammer, G.L., Probert, M.E., Robertson, M.J., Holzworth, D., Huth, N.I., Hargreaves, J.N.G., Meinke, H., Hochman, Z., McLean, G., Verburg, K., Snow, V., Dimes, J.P., Silburn, M., Wang, E., Brown, S., Bristow, K.L., Asseng, S., Chapman, S., McCown, R.L., Freebairn, D.M., Smith, C.J.: An overview of APSIM, a model designed for farming system simulation. Europ. J. Agronomy **18**, 267–288 (2003)
10. Susanti, E., Dewi, E.R., Surmaini, E., Sopaheluwakan, A., Linarko, A., Syahputra, M.R.: The projection of rice production in Java Island to support Indonesia as the world food granary. In: E3S Web of Conferences, vol. 306, pp. 01011. (2021)
11. Probert, M.E., Carberry, P.S., McCown, R.L., Turpin, J.E.: Simulation of legume-cereal systems using APSIM Australian. J. Agricul. Res. **49**, 317–327 (1998)
12. Luo, Z., Wang, E., Fillery, I.R.P., Macdonald, L.M., Huth, N., Baldock, J.: Modelling soil carbon and nitrogen dynamics using measurable and conceptual soil organic matter pools in APSIM. Agric. Ecosyst. Environ. **186**, 94–104 (2014)
13. Hatfield, J.L., Prueger, J.H.: Temperature extremes: effect on plant growth and development. Weather and Climate Extremes **10**, 4–10 (2015)
14. Ferrante, A., Manani, L.: Agronomic management for enhancing plant tolerance to abiotic stresses: high and low values of temperature, light intensity, and relative humidity. Horticulturae **4**, 21 (2018)

Multiyear La Niña Events and Poor Harvest of Sea Salt in Madura Island

Rikha Bramawanto and Suaydhi

Abstract Previous research stated that simultaneous La Niña and negative IOD caused sea salt harvest failure in Indonesia. In the last two decades, these happened in 2010 and 2016. It caused a scarcity of sea salt and a significant increase in salt prices. Currently, Indonesian salt farmers are upset because La Niña, which began in 2020, continues until now (mid-year 2022), becoming a multiyear La Niña. Multiyear La Niña events also occurred in 1973–1975 and 1998–2000. This article compares sea salt production and microclimate conditions in the current multiyear La Niña with the past multiyear La Niña. The results show that the three multiyear La Niña events affect the poor harvest of sea salt production in the Madura District. Understanding the character of multiyear La Niña can be an anticipatory effort for salt stakeholders to face multiyear La Niña in the future.

1 Introduction

Indonesia's position is between two oceans, the Indian and the Pacific Oceans. It makes Indonesia get global phenomena effect in the form of irregular oscillations of sea surface temperature in the Indian and Pacific Oceans, known as Indian Ocean Dipole (IOD), and El Niño Southern Oscillation (ENSO) [1, 2]. These two phenomena are closely related to the formation of rainfall in Indonesia [3–5]. The anomaly of negative sea surface temperatures in the Indian Ocean (negative IOD) and the Pacific Ocean (La Niña) makes the surface temperature of the waters in Indonesia warmer, triggering the formation of more rain than its normal conditions [6, 7]. Otherwise, the positive sea surface temperature anomaly in the Indian Ocean (positive IOD) and the Pacific Ocean (El Niño) triggers high rainfall in both oceans. This condition makes the weather in Indonesia drier than usual and can result wider drought [8, 9]. In addition, Indonesia occupies the equatorial region with a tropical

R. Bramawanto (✉) · Suaydhi
Research Center for Climate and Atmosphere, National Research and Innovation Agency (BRIN), Bandung, Indonesia
e-mail: rikha.bramawanto@brin.go.id

A. Basit et al. (eds.), *Proceedings of the International Conference on Radioscience, Equatorial Atmospheric Science and Environment and Humanosphere Science*, Springer Proceedings in Physics 290, https://doi.org/10.1007/978-981-19-9768-6_63

climate and only has two seasons, the rainy and the dry season. Indonesia is also affected by the Asian–Australian monsoon system [10, 11].

Climate variability in Indonesia associated with El Niño and La Niña events [12]. The climate variability affects the community's livelihoods and many sectors in Indonesia, including health, agriculture, marine and fisheries, forestry, etc. IOD and ENSO affect inter-annual variability in the incidence of dengue hemorrhagic fever and malaria in all provinces in Indonesia [13]. The short rainy season during the 2015 El Niño reduced the main crops (rice, corn, and soybean) in Langkat, North Sumatera, Indonesia [14]. Analysis of historical data on El Niño events and rice harvests in Java and Bali shows the need for water storage, preparation of drought-resistant varieties, crop diversification, and early warning systems as adaptation strategies in rice farming in Indonesia [15]. There are differences in the characteristics of seasonal upwelling in the waters south of Java during the 2010 strong La Niña and super El Niño 2015 [16]. Coral geochemical records in Indonesia show the effects of ENSO and IOD on salinity and SST [17]. The intensity of El Niño has an impact on the severity of land and forest fires in fire-prone areas of provinces in Indonesia [18].

The Australian Bureau of Meteorology (BoM) and the World Meteorological Organization (WMO) have confirmed that the century's first "triple dip" La Niña is ongoing [19, 20]. The closest "triple dip" La Niña events in the past occurred in 1973–1975 and 1998–2000 [21] (Fig. 1). "Triple dip" La Niña is a multiyear La Niña phenomenon that occurs in three consecutive years [22]. Experts claim that the "triple dip" La Niña has affected temperatures and rainfall patterns, triggering floods and droughts in various parts of the world [23]. The ongoing drought in southern South America and the Horn of Africa suggests a "triple dip" La Niña impact [24]. Several meteorological agencies in Asia and Australia warned of the flood risk caused by this phenomenon [25]. Currently, salt farmers in Indonesia are upset because the last La Niña sequentially continues from late 2020 to the present (mid-2022), which inhibits salt production [26].

The rain that persists throughout the dry season disrupts sea salt production. This abnormal condition is one of climate variability impacts on the availability of salt stocks in Indonesia. Previous studies have stated that strong La Niña coupled with

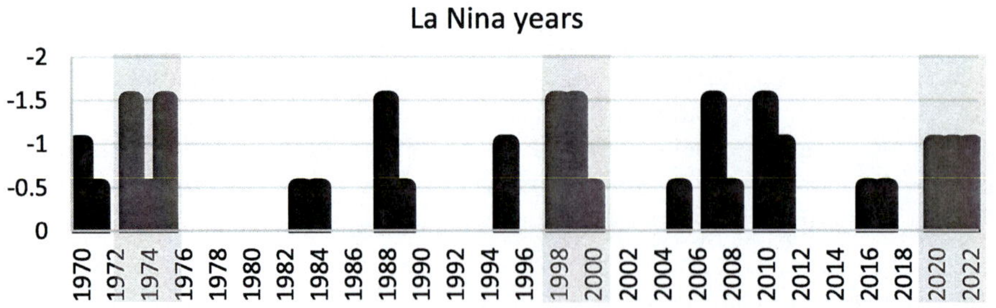

Fig. 1 La Niña years (−0.5 = weak, −1 = moderate, −1.5 = strong) and the appearance of the "triple" dip La Nina from 1970 to present. *Source* https://ggweather.com/enso/oni.htm (processed data)

negative IOD can cause salt harvest failure in Indonesia [27]. These extreme conditions occurred in 2010 and 2016. The salt harvest failure in 2016 triggered a scarcity of salt [28]. It caused the price of salt increases abnormally [29]. Unfortunately, the salt policy in Indonesia did not consider the forecast of climatic variabilities in the following years to anticipate that kind of condition.

This article compares sea salt production and microclimate conditions in the current multiyear La Niña with the past multiyear La Niña in Madura. This information can enrich the data to predict the arrival of multiyear La Niña in the next period, especially in Madura and its surroundings.

2 Data and Methods

In this study, the monthly average total precipitation data is from the fifth-generation atmosphere reanalysis data of the European Centre for Medium-Range Weather Forecasts (ERA5 ECMWF) [30] with a horizontal resolution of $0.25° \times 0.25°$. The study area is the island of Madura and the surrounding waters at latitudes −6 to −8 ^{0}S and longitudes 112 to 115 ^{0}E. The climatology is taken as the period 1970–2021. The Nino 3.4 index and dipole mode index are from NOAA ESRL Physical Sciences Laboratory. The Nino 3.4 index and dipole mode index are based on SST dataset from the Hadley Centre Sea Ice and Sea Surface Temperature dataset (HadISST) spanning from 1990 to 2021 with a horizontal resolution of 1° latitude × 1° longitude [31]. Old salt production data (1990–2018) were obtained from PT Garam and Ministry of Marine Affairs and Fisheries (KKP). The latest salt production data (2019–2021) were taken from the annual report of PT Garam. The salt referred to in this study is salt made from sea water which is processed in ponds by the community and PT Garam. The data are processed using MS excel and ocean data view ODV 5.6.2.

The monthly precipitation patterns for triple dip La Niña events in 1973–1975, 1998–2000, and 2020–2022, at three salt farms in Madura Island (Sumenep, Pamekasan, and Sampang districts), were compared to each other. In this study, we illustrate monthly average total precipitation spatial distribution to determine the meteorological conditions during the triple dip La Niña events, the La Niña events combined with negative IOD and normal ENSO. Then an analysis of the characteristics of salt production was carried out in several La Niña events, whether they only occurred for one year, two years, or three years successively.

June and October are critical points in the salt-making process. The salt harvest period will come late when the June rainfall is still high. The salt harvest period will be short when it starts to rain in October. To compare the critical point when occurs of multiyear La Niña, La Niña + negative IOD simultaneously, and ENSO normal, then June and October precipitation data are taken in 1993, 1999, 2010, 2014, 2016 and 2021 as a representation of the three conditions above. The rainfall in 1993 and 2014 associated with normal ENSO led to normal salt production. The 2010 and 2016 rainfall related to simultaneous La Niña and negative IOD events caused salt harvest failure. However, the rain in 1999 and 2021 associated with the La Niña Multiyear event resulted in low salt production. To simplify the analysis, the monthly mean data of total precipitation is plotted in a spatial distribution.

3 Result and Discussion

The high rainfall over the last three years, especially during the dry season, has been causing a decline in salt production in Indonesia. The salt production in 2020 and 2021 is 1.36 and 1.09 million tons, while the salt harvest is estimated to be around 0.6 million tons or even lower in 2022. Indonesia's salt production in the last three years was below the average from 2010–2022 (about 1.6 million tons). If the calculation does not include salt harvest failures in 2010 and 2016, then the salt production for the last three years will have a wider deviation (Fig. 2). Low salt production also occurred in 2011, 2013, and 2017. The low salt harvest in 2011 and 2017 occurred after crop failures in previous years. The salt production ratio between below average to above average was 8:5 in the last 12 years.

To find out condition of salt production in the past during the triple dip La Niña, we need salt production data from 30 to 50 years back. It was hard to find valid data on salt production in Indonesia before 2010. Therefore, this study focuses on salt production in Madura Island, especially in salt ponds owned by PT Garam, in Sumenep, Pamekasan, and Sampang districts. We managed to obtain the best available data on salt production since 1990.

The results of combining data on the occurrence of ENSO events, represented by a Nino3.4 index, with salt production in Madura mostly show concomitant patterns between La Nina events with decreased production and El Nino with increased salt production in the last 32 years. (Fig. 3). Of the fourteen La Niña (weak, moderate,

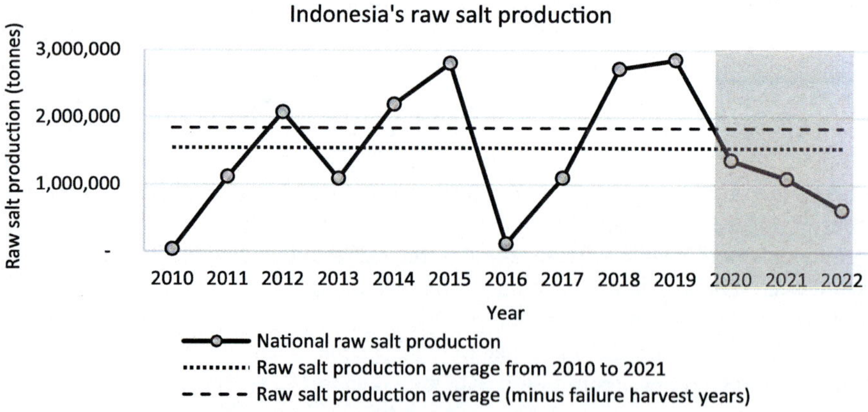

Fig. 2 Indonesia's raw salt production from 2010 to 2021. Salt production in 2022 is the prediction by the Ministry of Marine Affairs and Fisheries [32]

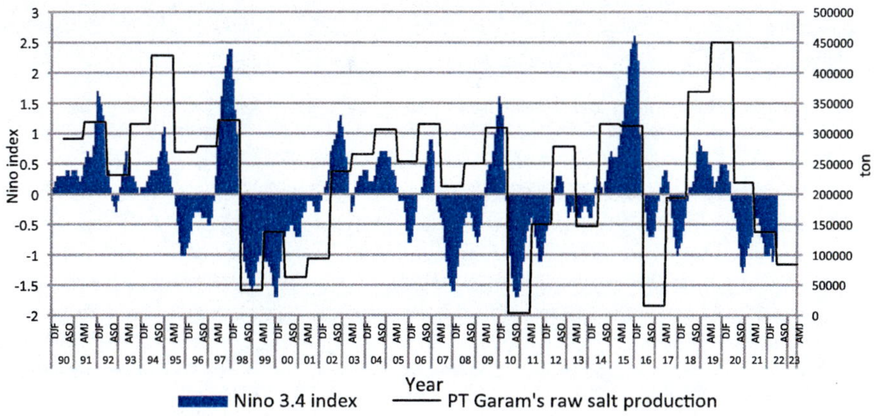

Fig. 3 PT Garam's raw salt production from 1990 to 2022 in various ENSO conditions. A full-fledged El Niño or La Niña episode was classified after exceeding the threshold for at least five consecutive overlapping three-month seasons

strong) events since 1990, there were six times (42.9%) moderate salt harvests in 1995, 2005, 2007, 2008, 2017, and 2020, and four times (28.6%) little harvests in 1999, 2011, 2013 and 2021, two times (14.25%) very little harvest in 1998 and 2000, and two times (14.25%) crop failure in 2010 and 2016. The strong La Niña in 1999 and 2007 did not mean that salt production had dropped drastically. Crop failure and very little salt harvest can occur when the previous year there was a strong El Niño as happened in 1997–1998, 2009–2010, and 2015–2016.

Strong La Niña both occurred at the start of the triple dip La Niña in 2020 and 1998, and there was a high salt production in the previous year (2019 and 1997). However, the salt production character of the triple dip La Niña 2020–2022 is different from that of the triple dip La Niña 1998–2000. The difference is strongly suspected to be caused by a phenomenon that occurred in the previous year. A weak El Niño in 2019 followed by a strong La Niña did not drastically reduce salt production in 2020. Meanwhile, a strong El Niño in 1997 followed by a strong La Niña made salt production decrease drastically in 1998.

In Madura, the triple dip La Niña rainfall in 2020–2022 generally has a similar pattern to the "triple dip" La Niña in 1998–2000 and 1973–1975 (Fig. 4). The drought in the "triple dip" La Niña event lasts only about 2–3 months, so it can be said to be a wet dry season. Generally, the dry season in Madura is about 5–6 months. The driest conditions only occur in the months of June–July–August. The shortest drought in the "triple dip" La Niña event occurred in June 1998. The dry season that was not completely dry occurred in 1973. The highest rainfall occurred in 2020–2022. The pattern of rainfall fluctuations in the Sumenep, Pamekasan, and Sampang districts is relatively the same, although Sampang rainfall is slightly higher than that in Pamekasan and Sumenep districts.

June monthly average rainfall in the Madura waters, especially the northern part of the island, was still high when La Niña and negative IOD occurred in 2010 and 2016,

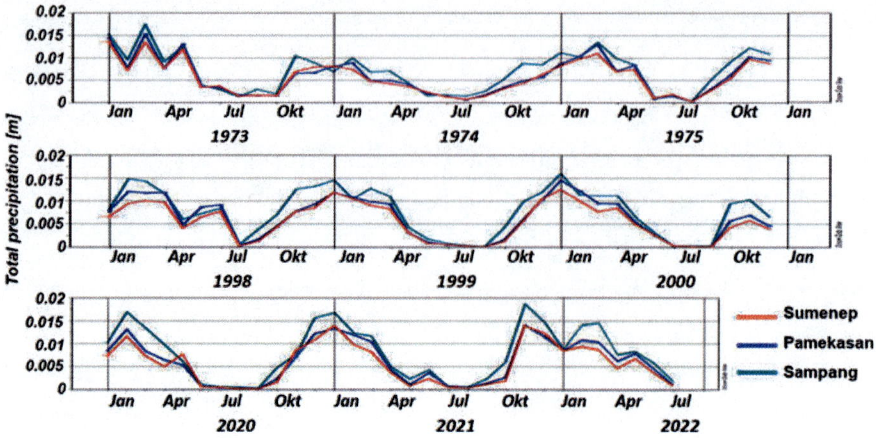

Fig. 4 Comparison of monthly average total precipitation in Sumenep, Pamekasan, and Sampang on three "triple dip" La Niña events in 1973–1975, 1998–2000, and 2020–2021

around 5–7.5 mm. During the multiyear La Niña (1999 and 2021) and normal ENSO (1993 and 2014), the monthly average rainfall in June is relatively the same, around 0–2.5 mm. The weather conditions in June 2021 and 1993 were wetter than in June 2021 and 2014 (Fig. 5a). The monthly average rainfall in October shows a significant difference. The island of Madura, especially the Sampang District, entered the rainy season with monthly average rainfall of 10–12.5 mm in October 2010 and 2016. The monthly average total precipitation on Madura Island was relatively small, 0–2.5 mm, in October 1999 and 2021. Meanwhile, the general condition of the Madura Island was still dry in October 1993 and 2014 (Fig. 5b).

This study focused on several La Niña events as factors that affect the decline of salt production. We get some interesting information after combining PT Garam's salt production data with past La Niña events. Although not all La Niña events caused a decrease in production or failure to harvest salt in PT Garam's ponds, salt stakeholders should be wary of any La Nina events, both single and multiyear La Nina events. There was 57.1% poor salt harvest during La Nina events in the last 32 years. We need to do further analysis involving other climatic variables such as IOD, Asian–Australian monsoon, and so on to identify La Nina species that can significantly reduce salt production, including whether it is within the Modoki criteria or not.

Comprehensive knowledge of the influence of climate on raw salt production is very strategic, considering the high demand for salt, especially for industrial purposes, which reaches 3.7 million tons per year. In the future, the government must be able to provide the latest valid weather-climate information and accurate forecasts for salt farmers and import policymakers. On the other hand, the opportunity to utilize the supporting resources for salt production is still very potential, both extensively and intensively.

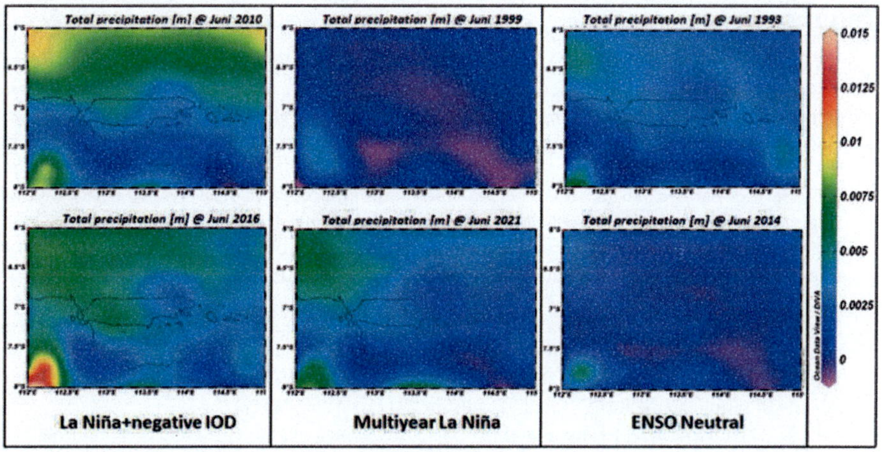

(a)

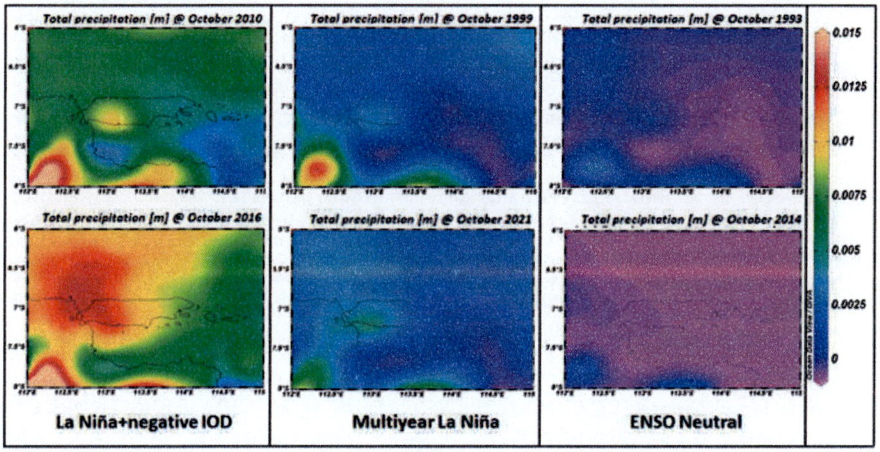

(b)

Fig. 5 Comparison of monthly average rainfall spatial distribution in June and October at the time of La Niña + negative IOD, multiyear La Niña, and ENSO neutral events

4 Conclusion

The multiyear La Niña has had an impact on the decline in salt production in Indonesia. Multiyear La Niña did not result in salt harvest failure as in 2010 and 2016, but the harvest was below the normal average. Experience and understanding of the character of multiyear La Niña can be an anticipatory step for salt stakeholders to predict the arrival of multiyear La Niña in the next period and face multiyear La Niña in the future.

References

1. Wieners, C.E., Dijkstra, H.A., de Ruijter, W.P.M.: The interaction between the Western Indian Ocean and ENSO in CESM. Clim. Dyn. **52**, 5153–5172 (2019)
2. Dewi, Y.W., Wirasatriya, A., Sugianto, D.N., Helmi, M., Marwoto, J., Maslukah, L.: Effect of ENSO and IOD on the variability of Sea surface temperature (SST) in Java Sea. In: IOP Conference Series: Earth Environ 530. The 5th International Conference on Tropical and Coastal Region Eco Development 17–18 September 2019, Semarang, Indonesia (2019)
3. Nur'utami, M.N., Hidayat, R.: Influences of IOD and ENSO to Indonesian rainfall variability: role of atmosphere-ocean interaction in the Indo-Pacific sector. Proc. Environ. Sci. **33**, 196–203 (2016)
4. Saputra, A., Hermawan, E., Darmawan, D.: Model development of Niño 3.4 and Indian Ocean Dipole (IOD) Anomalies teleconnection. J. Sains Dasar **7**(2), 111–116 (2018)
5. Narulita, I.: ENSO and IOD impact to rainfall variability on Cerucuk watershed, Belitung Island (In Indonesian). Jurnal Tanah dan Iklim **41**(1), 45–60 (2017)
6. Lim, E.P., Hendon, H.H.: Causes and predictability of the negative Indian Ocean Dipole and its impact on La Niña during 2016. Sci. Rep. **7**, 12619 (2017)
7. Lestari, D., Sutriyono, E., Kadir, S., Iskandar, I.: Impact of 2016 Weak La Niña Modoki event over the Indonesian Region. Int. J. GEOMATE. **17**, 156–162 (2019)
8. Avia, L., Sofiati, I.: Analysis of El Niño and IOD phenomenon 2015/2016 and their impact on rainfall variability in Indonesia. In: IOP Conference Series: Earth and Environmental Science, vol. 166, pp. 012034. (2018)
9. Lestari, D.O., Sutriyono, E., Sabaruddin, Iskandar, I. : Respective influences of Indian Ocean Dipole and El Niño-Southern oscillation on Indonesian precipitation. J. Math. Fund. Sci. **50**(3), 257–272 (2018)
10. Wheeler, M.C., McBride, J.L.: Australian-Indonesian monsoon. In: Intraseasonal Variability in the Atmosphere-Ocean Climate System. Springer Praxis Books. Springer, Berlin, Heidelberg (2005)
11. Duan, Y., Liu, H., Yu, W., Liu, L., Yang, G., Liu, B.: The onset of the Indonesian-Australian summer monsoon triggered by the first-branch Eastward-propagating Madden–Julian Oscillation. J. Clim. **32**(17), 5453–5470 (2019)
12. Xie, S., Annamalai, H., Schott, F.A., McCreary, J.P., Jr.: Structure and mechanisms of South Indian Ocean climate variability. J. Clim. **15**(8), 864–878 (2002)
13. Kurnianingsih, Wirasatriya, A., Lazuardi, L., Kubota N., Ng, N.: IOD and ENSO-related time series variability and forecasting of Dengue and Malaria incidence in Indonesia. In: 2020 International Symposium on Community-centric Systems (CcS), pp. 1–8. IEEE (2020)
14. Ramija, K.E., Sudrajat, A., Irwandi, H., Ariantono, J.Y.: Influence of El Niño 2015/2016 on climate variability and production of main crops in Langkat regency. Agromet **35**(2), 98–107 (2021)
15. Naylor, R.L., Battisti, D.S., Vimont, D.J., Falcon, W.P., Burke, M.B.: Assessing risks of climate variability and climate change for Indonesian rice agriculture. PNAS **14**(19), 7752–7757 (2007)
16. Atmadipoera, A.S., Jasmine, A.S., Mulia, P., Kuswardani, A.R.T.D.: Upwelling characteristics in the Southern Java waters during strong La Niña 2010 and super El Niño 2015. Jurnal Ilmu Dan Teknologi Kelautan Tropis **12**(1), 257–276 (2020)
17. Cahyarini, S.Y., Henrizan, M.: Coral based-ENSO/IOD related climate variability in Indonesia: a review. In: IOP Conference Series: Earth and Environmental Science, Global Colloquium on GeoSciences and Engineering, 18–19 October 2017, Bandung, Indonesia, vol. 118 (2017)
18. Nurdiati, S., Sopaheluwakan, A., Septiawan, P.: Spatial and temporal analysis of El Niño impact on land and forest fire in Kalimantan and Sumatra. Agromet **35**(1), 1–10 (2021)
19. BBC Homepage: https://www.bbc.com/weather/features/62890361. Last Accessed 21 Sep 2022
20. WMO Homepage: https://public.wmo.int/en/media/press-release/wmo-predicts-first-%E2%80%9Ctriple-dip%E2%80%9D-la-ni%C3%B1a-of-century. Last Accessed 21 Sep 2022

21. Climateextremes Homepage: https://climateextremes.org.au/wp-content/uploads/Multi-year-La-Nina-Events-ARC-Centre-of-Excellence-for-Climate-Extremes.pdf. Last Accessed 21 Sep 2022
22. WMO Homepage: https://public.wmo.int/en/our-mandate/climate/el-ni%C3%B1ola-ni%C3%B1a-update. Last Accessed 21 Sep 2022
23. Nature Homepage: https://www.nature.com/articles/d41586-022-01668-1. Last Accessed 21 Sep 2022
24. VOA Homepage: https://www.voanews.com/a/prolonged-la-ni%C3%B1a-likely-to-worsen-drought-in-horn-of-africa/6614144.html. Last Accessed 21 Sep 2022
25. ABC Homepage: https://www.abc.net.au/news/2022-09-13/bom-declares-third-la-nina-summer-weather-flood-risk/101424100. Last Accessed 21 Sep 2022
26. RRI Homepage: https://rri.co.id/sumenep/daerah/1470732/cuaca-tidak-menentu-produksi-garam-di-pamekasan-terhambat?utm_source=news_main&utm_medium=internal_link&utm_campaign=General%20Campaign. Last Accessed 21 Sep 2022
27. Bramawanto, R., Abida, R.F.: The climatology aspect review (ENSO and IOD) against the production of salt in Indonesia (In Indonesian). Jurnal Kelautan Nasional **12**(2), 91–99 (2017)
28. Helmi, A., Sasaoka, M.: Dealing with socioeconomic and climate-related uncertainty in small-scale salt producers in rural Sampang Indonesia. J. Rural Stud. **59**, 88–97 (2018)
29. Harini, S., Mulyanto, A.D., Iriany, A., Astuti, C.C.: Interpolation methods of spatial distribution salt price during scarcity period in Malang Raya using geographic information system. Int. J. Adv. Trends in Comput. Sci. Eng. **8**(6), 3120–3129 (2019)
30. Hersbach, H., Bell, B., Berrisford, P., Hirahara, S., Horányi, A., Muñoz-Sabater, J., ……. Thépaut, J.: The ERA5 global reanalysis. Quart. J. Royal Meteorol. Soc. Part A **146**(730), 1999–2049 (2020)
31. Rayner, N.A., Parker, D.E., Horton, E.B., Folland, C.K., Alexander, L.V., Rowell, D.P., Kent, E.C., Kaplan, A.: Global analyses of sea surface temperature, sea ice, and night marine air temperature since the late nineteenth century. J. Geophys. Res. Atmos.**108**(D14) (2003)
32. Pusat Riset Kelautan Homepage: http://pusriskel.litbang.kkp.go.id/index.php/en/publikasi/prediksi-produksi-garam. Last Accessed 24 Sep 2022

Climate Indicators Triggering Attacks of Rice Stem Borer as Early Detection Information

Suciantini, Erni Susanti, Elza Surmaini, Misnawati, and Yudi Riadi Fanggidae

Abstract In rice cultivation, one of the climate-related disasters that often occurs is the attack of plant pests and diseases. In Indonesia, especially on Java Island, the primary pest of rice plants with the most widespread attack is the rice stem borer (RSB). The purpose of this paper is to compile the relationship between RSB attacks and climate parameters. In the future, the relationship obtained will be used as input for preparing early detection model, as one of the adaptations technology efforts. The relationship between RSB attack and climate parameters was analyzed using multiple regressions equation to get the best model. The climate parameters used include; minimum air temperature, maximum air temperature, average air temperature, humidity, and rainfall. The attack area data is used from seven districts in Central Java. The analysis results show that the distribution of additional monthly rice stem borer attacks in seven districts in Central Java which mainly occurs in the rainy season (October–March) and the highest in March. Climatic parameters that have a high and significant correlation to the added area of RSB attack are maximum air temperature and humidity.

1 Introduction

Agricultural production, especially food crops, is greatly affected by variability and climate change [1–4]. Climate change can have a negative impact by increasing crop damage due to the disasters it causes. Climate-related disasters can trigger a decline in food production, especially rice crops. One of the climate-related disasters to that can disrupt production is the attack of pests and plant diseases [5]. Increased exposure, increased status, decreased respiration, and the rate of pests and diseases in Indonesia are related to climate change [6].

Suciantini (✉) · E. Susanti · E. Surmaini · Misnawati · Y. R. Fanggidae
The Research Center for Climate and Atmosphere, National Research and Innovation Agency, Bandung, Indonesia
e-mail: suci022@brin.go.id

A. Basit et al. (eds.), *Proceedings of the International Conference on Radioscience, Equatorial Atmospheric Science and Environment and Humanosphere Science*, Springer Proceedings in Physics 290, https://doi.org/10.1007/978-981-19-9768-6_64

Rice stem borer (RSB) is one of the most significant pests in Indonesia, apart from rats and brown planthoppers [7, 8]. The growth of rice plants can be attacked by RSB at all stages [9], from the nursery to the mature stage [10]. In Indonesian terms, they are known as *sundep* (dead hearts) and *beluk* (white ear heads), which represent RSB in the vegetative and generative phases [11–14]. At the beginning of planting, the intensity of the RSB attack is high, while in the productive phase, it decreases [10]. In addition, [15] stated that the dynamic nature of the pest population is influenced by the environmental constraints.

Climate is very influences on the abundance of RSB [16]. Climate parameters are thought to be the trigger for pest and disease explosions, including the RSB. Population distribution, development level, and pest phenology are primarily determined by air temperature [12]. However, the study of climate elements for early detection of pest and disease outbreaks is still minimal.

The purpose of this paper is to compile the relationship between RSB attacks and climate parameters. In the future, the relationship obtained will be used as input for the preparation of early detection models, as one of the adaptation technology efforts on agricultural production to avoid the adverse effects of plant attacks of pests and diseases. Prevention of development pests and plant diseases can reduce loss of crop yields, resulting in economic losses can be reduced.

2 Material and Methods

The study was distributed from January to December 2021. Developing an early detection model of RSB attack data in Central Java Province (Regencies of Batang, Brebes, Tegal, Pekalongan, Pemalang, City of Tegal, and City of Pekalongan) with a data period of 2008–2019. The data was obtained from The Center for Protection of Food Crops, Horticulture, and Plantations (BPTPHP) in Central Java Province. The materials needed are long series of pest attack data and climate data in food crop production centers. Based on the availability of climate data, the observed climate data include; minimum air temperature, maximum air temperature, average air temperature, humidity, and rainfall. The climate data used in this study is observation data from the Cacaban station, Tegal Regency, Central Java Province, and Climate data derived from NASA POWER. The climate data used for analysis is data from the BMKG's Tegal climate station, which was obtained from BMKG's data online. The period of the data obtained was 2005–2020 year. The BMKG's Tegal station daily climate data consists of rainfall data, T maximum, T minimum, T mean, and air humidity (RH).

NASA POWER data was used to fill in blank data from observational data. NASA POWER data was accessed via the link https://power.larc.nasa.gov/data-access-viewer/ by entering the coordinates of the Cacaban Station, Tegal (Latitude − 6.8672, Longitude 109.1372). The filling in blank data was preceded by analyzing the bias correction of NASA POWER data on observational climate data using regression equations.

$$\omega = \alpha + \beta\lambda \quad (1)$$

ω = observational data, α: regression constant, β: regression coefficient, and λ: NASA POWER data. After the regression equation was obtained, it was used to calculate the corrected NASA POWER data. This corrected NASA POWER data was used to fill in the unavailable observation data [17, 18].

The data analysis method used is correlation analysis followed by multiple linear regression analysis with a stepwise procedure. Analysis of the relationship between the extent of pest attack and climate parameters in several studies [19] used multiple linear regression equations. Climatic factor data was used as an independent variable (x), and data on the extent of rice stem borer attacks as a response variable (Y). The regression equation is as follows:

$$Y = a + b_1x_1 + b_2x_2 + b_3x_3 + b_4x_4 + b_5x_5 \quad (2)$$

where: Y = area of attack of RSB, x = climatic elements, a = intercept, and b= regression coefficient.

3 Result and Discussion

3.1 Result Complement Data with NASA POWER

NASA POWER data was chosen because the data is available, easy to use, and user-friendly [20]. Figures 1 and 2 show a comparison of observational data with data from NASA POWER; the two types of data show almost similar patterns in RH and temperature, but there are variations in rainfall. The difference in the highest value was obtained on October I. The result of NASA POWER is more significant than that recorded from the observation station.

3.2 The Distribution of Additional Monthly Rice Stem Borer (RSB) Attack

The rice stem borer (RSB) is the primary pest of rice plants with the most widespread attack in Indonesia, especially on the island of Java [21]. In this paper, the data used is the added area of RSB attacks in Batang, Brebes, Tegal, Pekalongan, Pemalang regencies, and the City of Tegal and Pekalongan. The highest attack area of the seven regencies/cities was Pemalang Regency, followed by Pekalongan and Batang regencies. Stem borer attacks fluctuate from year to year. An extreme event of additional RSB attacks occurred in 2012 in Pemalang Regency. Figure 3 shows that

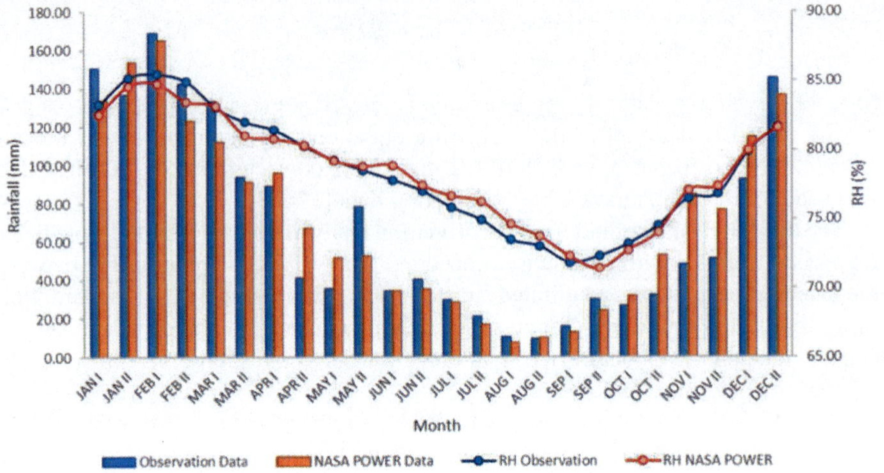

Fig. 1 Plotting of 2-week average rainfall and RH observations with rainfall and RH data from POWER data for the 2008–2019 data period

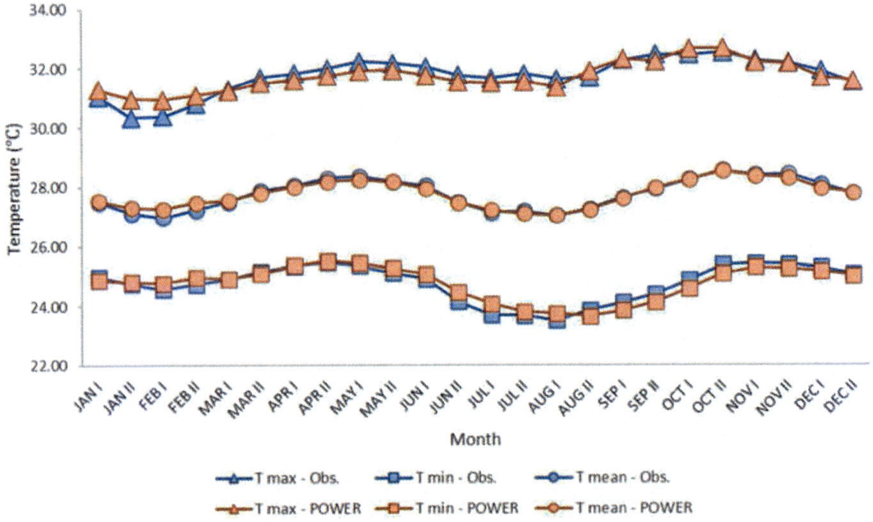

Fig. 2 Plotting of 2-week average data T_{max}, T_{min}, and T_{mean} (Tegal station) with POWER data

the average added area of RSB attack is 300 ha per 2 weeks, and in the last two years, it is seen that the added area of RSB attack has decreased below the average value. The decrease of the attack can occur due to several things, including caused of simultaneous planting in a large area (1000 ha) with a span of 15 days [22]. In general, many factors influence the dynamics of pest development, including: crop

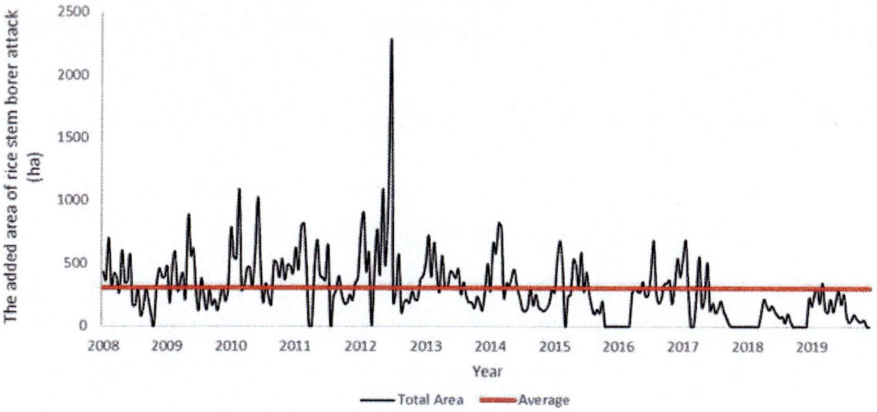

Fig. 3 Dynamics of RSB added area for the period 2008–2019 total of seven districts/cities

cultivation, cropping patterns, the presence of natural enemies, control methods, and climate change [23].

The distribution of additional monthly RSB attacks in seven regencies/cities of Central Java (Fig. 4) shows that RSB attacks occur mainly in the rainy season (October–March) and the highest in March. This statement is in line with the opinion of [16], which stated that the explosion of RSB attacks occurs when the intensity of rainfall is high. The area added attacks are increase in January, February, March, May, and June. However, in July, it was found that data stating that the area added to the attack had reached almost 2500 ha. This data is outlier or extreme observation. Knowing the magnitude of the spike in the area added to the attack, it is necessary to study at what phase the pest is present and the factors that may interfere with its ecological balance because this magnitude is very useful for controlling the pest [24].

From May to September, widespread incidents of RSB attacks continue. This result is shown by the relatively high minimum observation value, around 300 ha on May and June. The median monthly attack area added from January to July was in the range between 500 and 1000 ha, except for February, where it was > 1000 ha, the highest average monthly area added attack area. The median area added by RSB from August to December was lower at < 500 ha. In December, the median additional attack area increased from the previous months. This result indicates that in the rainy season, the size of attack is higher than in the dry season [16, 25]. Based on the illustration in the boxplot, the monthly maximum attacks occurred on March (> 1500 ha). Also in March, the variation in the area of addition of RSB was most varied, with the lowest quartile of ± 100 ha, a median of 600–700 ha, the highest quartile of 1200 ha, and a maximum value of around 1600 ha (Fig. 4).

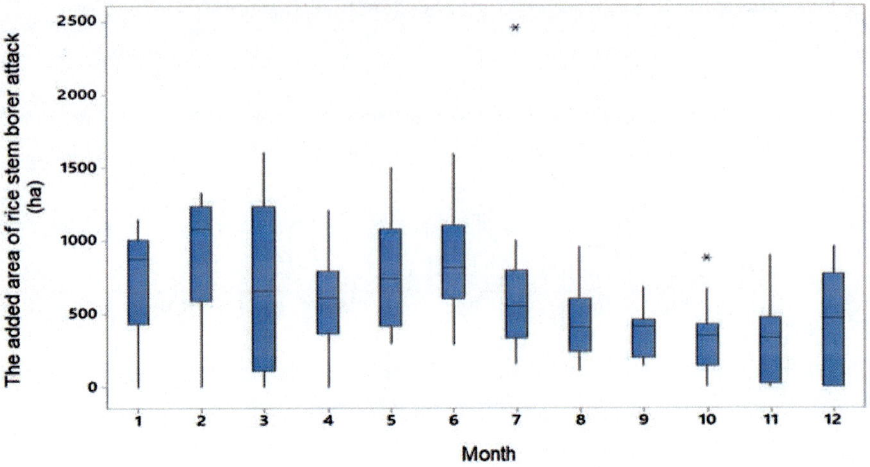

Fig. 4 Distribution of additional monthly RSB attacks in seven regencies/cities of Central Java

3.3 The Relationship of Area Affected by Rice Stem Borer (RSB) Attack with Climatic Parameters

To see the relationship between climate parameters and the attack of the RSB, a regression analysis was performed between the climate parameters lag1 to lag3 with the attack area of RSB. Table 1 shows that the five climate parameters with a relatively high and significant correlation are T_{max} and RH. The maximum T correlation was highest at lag1 (2 weeks before the attack), which was −0.474, and humidity at lag3 (1.5 months (6 weeks) before the attack), which was 0.418. From the results of the correlation analysis, it can be seen that there is an inverse relationship between temperature and the area of attack of the RSB. The lower the air temperature, the greater the attack area. This result is in line with the statement of [26], which stated that the increase in RSB (yellow stem borer) attacks was caused by low temperatures and high humidity. The difference in RH is directly proportional to the extent of the attack. RSB is very important to watch out for, especially for areas that have experienced two rice plantings.

Table 1 Correlation of area affected by RSB attack with climatic parameters

The added area of attack (ha)	T_{max} (°C)	T_{min} (°C)	T_{mean} (°C)	Rainfall (mm)	RH (%)
Lag0	– 0.398	– 0.032	– 0.220	0.250	0.313
Lag1	**– 0.474**	0.024	– 0.227	0.287	0.394
Lag2	– 0.405	0.105	– 0.126	0.326	0.397
Lag3	– 0.406	0.185	– 0.075	0.317	**0.418**

The bold letters represents the two highest correlations between RSB attacks and climate parameters

Table 2 Correlation of log area affected by stem borer attack with climatic parameters

The added area of attack (ha)	T_{max} (°C)	T_{min} (°C)	T_{mean} (°C)	Rainfall (mm)	RH (%)
Lag0	– 0.345	– 0.142	– 0.250	0.189	0.193
Lag1	**– 0.350**	– 0.099	– 0.236	0.160	0.240
Lag2	– 0.338	– 0.015	– 0.167	0.214	0.290
Lag3	– 0.331	0.093	– 0.085	0.207	**0.347**

The bold letters represents the two highest correlations between RSB attacks and climate parameters

The additional area of the RSB attack was also tried to be logarithmic, then seeing the correlation, and it turned out that the correlation value was lower than without logging, as shown in Table 2. However, the climatic elements that had an effect are the same, namely RH at lag3 (1.5 months before the attack) which was 0.347 and maximum temperature at lag1 (2 weeks before the attack) which was -0.350.

The dominant climatic factors that are the key to the spread of RSB attacks were analyzed using stepwise regression. The results of the stepwise regression between RSB attack area and climate parameters obtained five dominant climate parameters in step 7, namely the maximum temperature at Lag1 and Lag3, average temperature, and humidity at Lag1 and Lag3. The basis for regression decision-making stepwise used is the lowest Mallows Cp, the lowest S, and R-sq (coefficient of determination) is 26,66%, and R-sq Adjusted (coefficient of determination corrected) is 25,35% (the highest). A low R-sq shows that only 26,66% make up the model, while the rest is determined by unknown factors. Based on stepwise regression, the equation obtained is:

$$\begin{aligned} \text{RSB}_{\text{attack area}} = {} & 3697 - 119T_{\max}(\text{Lag1}) - 125T_{\max}(\text{Lag3}) \\ & + 170T_{\text{mean}} - 12.7\text{RH}(\text{Lag1}) + 7.7\text{RH}(\text{Lag3}) \end{aligned}$$

To know the triggering factors for the high prevalence of RSB attacks, it is necessary to look at the data starting from 6 weeks before the attack. The maximum temperature data to watch out for ranges from 28 to 33 °C, especially between 30.5 and 33 °C, the average temperature is 26–29 °C, and the RH is between 70 and 90%. According to [27], for the growth of yellow stem borer eggs, the optimum temperature is around 24–29 °C. If the temperature and humidity conditions are simultaneously met or are in the temperature and humidity range, then it can be initial information to anticipate attacks. For this reason, the data provider needs to pay attention to the data obtained from the field periodically. By studying the correlation and climatic parameters that trigger RSB attacks, it is expected to predict crop loss due to RSB attacks. Furthermore, the relationship obtained can be used as an early warning, part of the adaptation technology to reduce the risk of agricultural losses.

4 Conclusions

The rice stem borer (RSB) attack was significantly triggered by maximum temperature and humidity. In seven districts/cities in Central Java, RSB attacks mostly occur in the rainy season and are highest in March. Based on stepwise regression analysis, the determinant predictors were obtained, namely maximum temperature at Lag1 and Lag3, average temperature, and RH at Lag1 and Lag3. The R-sq received is relatively low at 26.66%. The maximum temperature data to watch out for ranges from 28 to 33 °C, especially between 30.5 and 33 °C, the average temperature is 26–29 °C, and the RH is between 70 and 90%.

There are still many processes to go through in developing a tool for an early warning system. The correlation between RSB attacks and climate parameters is only the initial stage for the preparation of an early warning system, so this activity should be continued so that it can be used as a reference for climate information for agriculture.

References

1. Apriyana, Y., Susanti, E., Suciantini, Ramadhani, F., Surmaini, E.: Analysis of climate change impacts on food crops production in dry land and design of information system. Informatika Pertanian **25**(1), 69–80 (2016)
2. Surmaini, E., Faqih, A.: Climate events and their impacts on food crop in Indonesia. Jurnal Sumberdaya Lahan **10**(2), 115–128 (2016)
3. Ruminta.: Analysis of decreasing production of paddy due to climate change in Bandung district West Java. Jurnal Kultivasi **15**(1), 37–45 (2016)
4. Kotir, J.H.: Climate change and variability in Sub-Saharan Africa: a review of current and future trends and impacts on agriculture and food security. Environ. Dev. Sustain **13**, 587–605 (2011)
5. Maulana, W., Suharto, W.: Response of some varieties of rice (Oryza Sativa L.) to Pest Borer and "Walang Sangit" (Leptocorisa acuta Thubn.) attack. Agrovigor **10**(1), 21–27 (2017)
6. Wiyono, S.: Perubahan iklim dan ledakan hama penyakit tanaman. Keanekaragaman Hayati Di Tengah Perubahan Iklim : Tantangan Masa Depan Indonesia. KEHATI, Jakarta (2007)
7. Susanti, E., Surmaini, E., Estiningtyas, W.: Climate parameters as indicators of early warning attack on pest and diseases of plant. Jurnal Sumberdaya Lahan **12**(1), 59–70 (2018)
8. Nurhayati, E., Koesmaryono, Y., Impron.: Predictive modeling of rice yellow stem borer population dynamics under climate change scenarios in Indramayu. In: The 3rd International Seminar on Sciences "Sciences On Precision And Sustainable Agriculture" (ISS-2016). IOP Conf. Series: Earth and Environmental Science vol. 58. Bogor, Indonesia (2017).
9. Cahyono, G.R., Nurmahaludin.: Rancang bangun alat perangkap hama tanaman padi menggunakan arduino mega 2560. Jurnal Poros Teknik **7**(2), 54–105 (2015)
10. Pertiwi, E.N., Mudjiono, G., Rachmawati, R.: Hubungan populasi ngengat penggerek batang padi yang tertangkap perangkap lampu dengan intensitas serangan penggerek batang padi di sekitarnya. Jurnal HPT **1**(2), 88–95 (2013)
11. Aryantini, L.T., Supartha., Wijaya, I.N.: Population abundance and rice stem borer attack on rice in Tabanan regency. Jurnal Agroekoteknologi Tropika **4**(3), 203–212 (2015)
12. Baehaki.: Hama penggerek batang padi dan teknologi pengendalian. Iptek Tanaman Pangan **8**(1), 1–14 (2013)

13. Rahaman, M.M., Stout, M.J.: Comparative efficacies of next-generation insecticides against yellow stem borer and their effects on natural enemies in rice ecosystem. Rice Sci. **26**(3), 157–166 (2019)
14. Jasrotia, P., Khippal, A., Yadav, J., Kashyap, P.L., Kumar, S., Singh, GP.: Effect of weather variables on the incidence of yellow stem borer (Scirpophaga incertulas W.) and leaf folder (Cnaphalocrocis medinalis G.) in rice. J. Cereal Res. **11**(3), 247–251 (2019)
15. Junaedi, E., Yunus, M., Hasriyanty.: Parasitoids and it is parasitisme on white rice stem borer (Scirpophaga innotata WALKER) in two different altitudes of rice fields (Oryza sativaL.) in District of Sigi. e-J.Agrotekbis **4**(3), 280–287 (2016)
16. Koem, S., Koesmaryono, Y., Impron.: Population phenology modeling of rice yellow stem borer Scirpophaga incertulas (Walker) based on climate effect. Jurnal Entomologi Indonesia **11**(1), 1–10 (2014)
17. Rodrigues, G.C., Braga, R.P.: Evaluation of NASA POWER reanalysis products to estimate daily weather variables in a hot summer mediterranean climate. Agronomy **11**, 1207 (2021)
18. Marzouk, O.A.: Assessment of global warming in Al Buraimi, sultanate of Oman based on statistical analysis of NASA POWER data over 39 years, and testing the reliability of NASA POWER against meteorological measurements. Heliyon **7**, e06625 (2021)
19. Susanti, E., Ramadhani, F., June, T., Amien, I.: The use of climate information for developing early warning system to brown plant hopper attack on paddies. Jurnal Tanah dan Iklim **30**, 47–58 (2009)
20. Rodrigues, G.C., Braga, R.P.: Estimation of daily reference evapotranspiration from NASA POWER reanalysis products in a hot summer Mediterranean climate. Agronomy **11**, 2077 (2021). https://doi.org/10.3390/agronomy11102077
21. Misnawati, M., Boer, R., June, T., Faqih, A.: Perbandingan metodologi koreksi bias data curah hujan Chirps. Limnotek **25**(1), 18–29 (2018)
22. Sinaga, A., Rohaeni, W.R., Marbun, O.: Studi perkembangan populasi hama penggerek batang padi hasil tangkapan light trap di Kecamatan Cilamaya Wetan – Karawang MK 2013. Seminar Nasional Agro Inovasi berbasis Sumberdaya Lokal dalam Meningkatkan Kesejahteraan Masyarakat dan Petani
23. Ilyas, A., Djufry, F.: Correlation and regression analysis pest population and natural enemy dynamic of certain varieties rice after implementation of superior IPM in Bone South Sulawesi. Informatika Pertanian **22**(1), 29–36 (2013)
24. Makarim, A.K., Widiarta, I.N., Hendarsih, S., Abdulrachman, S.: Petunjuk Teknis Pengelolaan Hara dan Pengendalian Hama Penyakit Tanaman Padi Secara Terpadu. Departemen Pertanian, Jakarta (2003)
25. Misnaheti, B., Aisyah, D.: Tren perkembangan penggerek batang pada tanaman di Sulawesi Selatan. http://www.peipfi-komdasulsel.org/wpcontent/uploads/2011/06/410-415-TREN-PERKEMBANGANPENGGEREK-Misnaheti.pdf. Last accessed 10 Mar 2021
26. Patel, S., Singh, C.P.: Seasonal incidence of rice stem borer, Scirpophaga incertulas (Walker) on different varieties of rice in relation to weather parameters. J. Entomol. Zool. Stud. **5**(3), 80–83 (2017)
27. Saleh, T.W., Buri, N., Saragih, A.A.: Keragaan hama, penyakit dan musuh alami pada budidaya beberapa varietas padi gogo di lahan sawah. In Prosiding Temu Aplikasi Teknologi & Seminar Nasional Pertanian dan Peternakan: Akselerasi Inovasi Pertanian Era Industri 4.0 Mendukung Sapira. Hal, pp. 163–170 (2020)

Improvement of the Cropping Index and Farmers' Resilience in Rainfed Fields Through the Application of Climate Smart Agriculture

Aris Pramudia, Abriani Fensionita, Yunita Fauziah Rahim, Asis Purwoko, Andriarti Kusumawardani, and Muhammad Takdir Mulyadi

Abstract One of the causes of low agricultural production on rainfed land is cultivation techniques that are highly dependent on rainfall conditions, with a high risk of crop failure in the second planting season. This condition is further exacerbated by farmers' limited understanding and skills in adapting to rainfall variability. This paper presents an experience applying climate smart agriculture to rainfed agriculture in Lawang Subdistrict, Malang Regency, East Java Province, Indonesia. Several technics were implemented, including setting a planting schedule according to rainfall conditions, cropping pattern technique with the optimum composition of commodity types on water availability, conducting simple observations of rainfall and soil water content levels, making biopores, and application of biopesticides and organic fertilizers. The learning has some positive impacts, including early planting can be accelerated by 20–30 days from October III–November I to September III–October I, the cropping index increases from 2 times to 4–5 times a year, improving the cropping patterns from rice-paddy or rice-fallow into maize-rice-maize-peanuts-long beans, seedling age is shorter than 20–30 days after sowing to 12–15 days after sowing. Before the implementation of CSA, farmers experienced harvests 1–2 times a year, with a 60% risk of crop failure in the second planting season. After implementing CSA, farmers have harvested 4–5 times with several various commodities. Rice production also increased from 4.0 to 4.2 t/ha to 6.5–7.0 t/ha. However, CSA is a specific location technology; therefore, the application of CSA in rainfed fields still needs to be developed in other areas or regions.

A. Pramudia (✉)
Research Center for Climate and Atmospheric, National Research and Innovation Agency, Kota, Bandung, West Java 40173, Indonesia
e-mail: aris029@brin.go.id

A. Fensionita · Y. F. Rahim · A. Purwoko · A. Kusumawardani · M. Takdir Mulyadi
Directorate of Food Crops Protection, Directorate General of Food Crops, Indonesia Ministry of Agriculture, Jakarta 12520, Indonesia

A. Pramudia
Innovation Center for Tropical Science, Bogor 16113, Indonesia

A. Basit et al. (eds.), *Proceedings of the International Conference on Radioscience, Equatorial Atmospheric Science and Environment and Humanosphere Science*, Springer Proceedings in Physics 290, https://doi.org/10.1007/978-981-19-9768-6_65

1 Introduction

Rainfed paddy fields are fields for rice cultivation that utilize water sources only from rainfall [1, 2]. This field is usually in a flat area or a relatively wide basin but has no water source other than rainfall. Plants cultivated in rainfed rice fields are generally seasonal food crops, such as rice and palawija. Food crop production on rainfed land is generally lower than on irrigated fields because it has limited planting time, which only depends on rainfall conditions [3, 4] and has low soil fertility [5]. In agricultural areas of Indonesia, rainfed areas generally can only be planted two times a year [6]. The second planting usually coincides with the end of the rainy season or the beginning of the dry season, so it has a high risk of crop failure. This condition is further exacerbated by the limited understanding and skills of the farmer in anticipating and adapting to rainfall fluctuation and climate variability.

To overcome the problem of food crop agriculture on rainfed fields, one solution is to applicate some climate smart agriculture concepts [7]. Climate smart agriculture (CSA) is an approach that transforms and reorients agricultural production systems and food value chains so that they both support sustainable agriculture and ensure food security in conditions of climate change. Based on the three main pillars, the CSA concept aims to: (1) Increase agricultural productivity and income sustainably, (2) implement adaptation and build resilience to climate change, and (3) reduce and, or eliminate greenhouse gas emissions, where possible [8].

This paper presents an experience of applying climate smart agriculture to food agriculture on rainfed land in Lawang Subdistrict, Malang Regency, East Java Province, Indonesia. Several things were implemented, including setting a planting schedule according to rainfall conditions, designing intercropping cropping patterns with the composition of adaptive types of commodities to water availability conditions, conducting simple rainfall observations, observing simple soil water content levels, making biopores, using biological pesticides, as well as the provision of organic fertilizers on the farmers' fields.

2 Methodology

2.1 Study Location

The study was conducted in rainfed rice fields managed by the Farmers' Group Aman Makmur in Srigading Village, Lawang Subdistrict, Malang Regency. The rainfall data used to represent the study location is rainfall data from the Karangploso Climatology Station, obtained from the BMKG online, accessed via the link http://dataonline.bmkg.go.id/data_iklim [9]. Other cultivation technical information is obtained through the Srigading Village Farmers Group, Lawang Subdistrict, Malang Regency. This study was carried out in 2020–2021.

2.2 *Rainfall Analysis for Improving the Cropping Calendar Recommendations and the Cropping Patterns*

In determining the planting schedule on rainfed land, farmers referred to the rules, namely (1) planted rice as the first commodity, (2) thus, the first planting schedule waited for the start of the rainy season, and (3) the next planting was applied after the previous plant has harvested. For the first planting, farmers estimated the start of the rainy season by the frequency of heavy rain days. Referring to the criteria for determining the initial rainy season from the BMKG, the first planting approach was when the rainfall intensity during three successive decades is > 50 mm/decade [10]. Rainfall prediction information was obtained from BMKG.

The innovation of the study introduced in mentoring farmers in this study was to modify the rules in determining the planting schedule and crops management, as follows: (1) the first commodity to be cultivated does not have to be rice, but plants that have fewer water requirements, (2) the planting time can be faster than the beginning of the rainy season, (3) the next planting is applied without having to wait for the harvest of the previous crop, (4) land space is prepared for planting the next crop, and (5) other modifications adapted in the field.

Modifications to the planting schedule and plant composition settings applied in the field were as follows:

(1) The first crop was maize.
(2) The second crop was rice with the *jajar legowo 2:1* technique.
(3) The third crop was maize, planted when the rice was two months old, in an empty row in the *jajar legowo* technique.
(4) The fourth crop was peanuts, planted on former rice fields after the rice was harvested.
(5) The fifth crop was long beans planted in rows of maize that had been harvested.

The approach taken in planting maize is when the intensity of rainfall at the beginning of planting and during the corn growing season is > 25 mm/decade (Table 1).

2.3 *Capacity Building of Farmers and Improvement of Land Quality*

To improve land quality and farmer capacity, and to increase land productivity [11], the following steps were taken: (1) use biological pesticides to control the pest, (2) applicate biological fertilizers on the land, (3) design biopores on the land, (4) teach farmers how to measure rainfall, and (5) teach farmers how to monitor groundwater level conditions. A simple rainfall gauge was made from a milk canister, and the

Table 1 Application of CSA in the field and its purpose

CSA application	Goals
1. Rainfall analysis	Determining the optimum planting time and adjusting the water availability
2. Management of cropping patterns and crop rotation	Implement the effective combination of crop commodities
3. Application of an effective cultivation system	Combining the *jajar legowo 2:1* technique, inserting planting time, intercropping
4. Improvement of land quality	
4.1. Use of biological pesticides	Application of healthy and environmentally friendly agricultural cultivation
4.2. Use of biofertilizer	Utilization of agricultural waste for green manure
4.3. Designing biopores	Provision of pore space to increase water availability
5. Increasing the farmer's capacity	
5.1. Rainfall measurement	Farmers can measure rainfall simply using a milk canister and a ruler
5.2. Groundwater monitoring	Farmers can monitor the soil water level simply using a paralon and a ruler

rainfall that was collected in the canister was measured using a ruler. Monitoring the soil water level was carried out by making a control hole from a paralon as deep as 40 cm and measuring the water level using a ruler (Table 1).

3 Results

3.1 Rainfall Pattern

The research location has a monsoon rainfall pattern, which is a monomodal pattern that is closely related to Indonesia or Southeast Asia monsoon wind circulation [12]. With this pattern, the research location alternates between one wet period and one dry period. In the agricultural calendar, the wet period is usually called the rainy season, while the dry period is usually called the dry season. The annual rainfall in the study area is 2041 mm/year, spreading 1732 mm in the rainy season and 309 mm in the dry season. With this annual rainfall, the study area is categorized as a rainfed agricultural area [1] with a moderate climate [13]. There are six wet months with an intensity of more than or equal to 200 mm/month, and six dry months with a power of less than or equal to 100 mm/month (Fig. 1). The Agroclimate Zone is C3, meaning it has a potential planting period of six months, especially for rice plants [14].

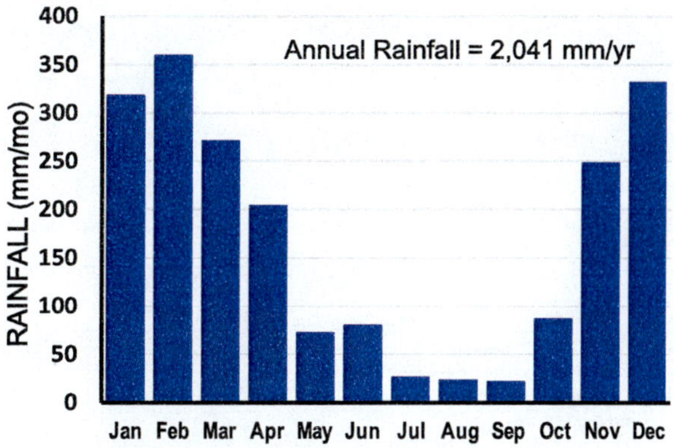

Fig. 1 Figure of monthly precipitation fluctuations and annual rainfall in Malang

3.2 *Improvement of the Crop Planting and Cropping Pattern Management*

The learning process on farmers' cultivation illustrates that the first planting follows the start of the rainy season, at October II–III or November I–II [15, 16]. The commodity applied in the first growing season is rice, which is grown conventionally using the tiled spacing technique. In the second planting season, several commodities applied by farmers include rice, maize, and other crops. The selection of commodities is adjusted to the conditions of rainfall experienced by farmers (Fig. 2).

To introduce the implementation of CSA for improving the cropping schedules and the cropping patterns, farmers were suggested to do not to apply the continuously flooded rice. Farmers could apply maize which requires less water than rice for the first crop. By applying maize as the first crop, the planting time can be carried out 20–30 days earlier, from September III–October I (Fig. 2). After harvesting during the first planting season, the season is at the peak of the rainy season. Maize harvest is done by cutting the base of the stem.

The second planting season is carried out around December II–III or January I–II. In this second planting season, rice plants were applied with the *jajar legowo* 2:1 planting technique. The planting *jajar legowo* 2:1 is a rice planting technique by emptying one row of rice plants and tucking the plants in that row into two rows on either side. Thus, there are two consecutive rows of rice plants with closer spacing between plants and interspersed with the third row of empty rows [17]. The technique of irrigating water for rice crops is adjusted to the rainfall conditions during the growing season. If water through rainfall is considered sufficient, then irrigation water can be applied through continuous inundation. However, if the water through rainfall is estimated to be limited, the provision of water can be carried out continuously for 3–5 days, following drying conditions also for 3–5 days, alternately (intermittent technique).

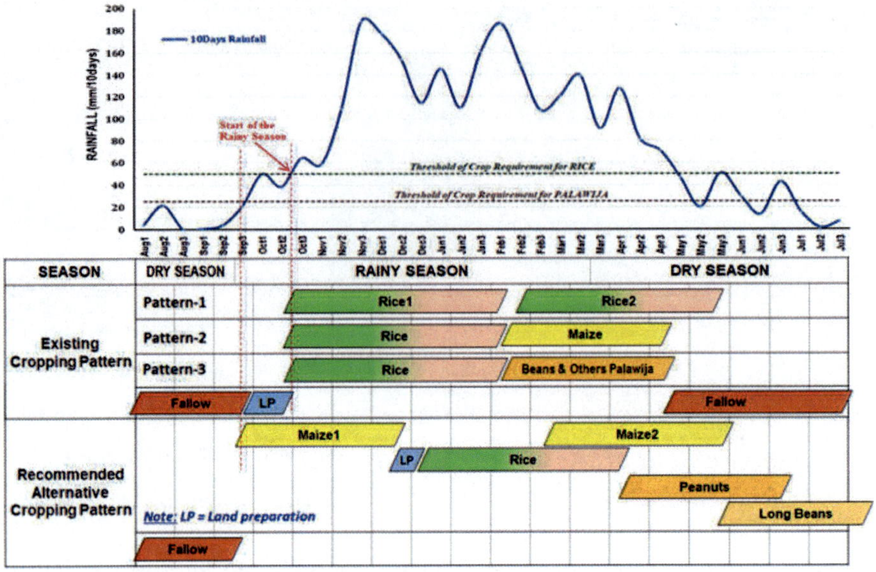

Fig. 2 Average rainfall conditions, the scheme of existing cropping patterns, and the recommendations for alternative cropping patterns that apply the CSA concept to rainfed land in Lawang Subdistrict, Malang Regency

When the rice plants are two months old or enter the generative phase and do not experience flooding again, the third planting is carried out around March I–II, applying maize. Planting is carried out on empty rows between 2-row rice plants that enter the generative phase. So simultaneously on the land, there are generative phase rice plants and vegetative phase maize plants.

When the maize plant enters the generative phase, namely in March III–April I, the rice harvest is carried out by slashing, but the remaining forage plants are left on the land. After the rice was harvested in April II–III, the fourth planting was carried out on the former rice fields. The fourth crop applied is peanuts. Thus, there is an intercropping system between corn plants in the generative phase and peanut plants.

In May II–III, maize was harvested by leaving the maize stalks on the land. After the maize harvest, if it is predicted that there will be sufficient rainfall, the fifth planting will be carried out in May III–June I by applying long beans. Long bean seeds are planted in rows between maize stalks still left in the field. When the long bean plant grows taller, the remaining maize stalks serve as vines for the long bean plants. Thus at this stage, there is an intercropping system between the peanut and the long bean.

In June II–III, peanuts were harvested. The last is the harvest of long beans can be done in August II–III. Using that cropping pattern, farmers experience 4–5 planting times in a year using various food crop commodities.

Fig. 3 Installation of biopores, simple rainfall measurement using milk canister, and monitoring of soil water levels in the fields

3.3 Improvement of Land Quality and Farmer Capacity

To prepare better fields, biopores were made in the field and the application of biofertilizers. The biopores were built with a size of about 50 cm × 50 cm with a depth of 100–150 cm. Biopore holes are filled with organic litter to facilitate absorption and the soil water-holding capacity. Improvement of farmer capacity is carried out through technology transfer and assistance in implementing technology carried out by farmers on the land. After going through the process of assisting with the application of CSA on rainfed rice fields, farmers can measure rainfall using canister tubes and can monitor the depth of the soil water table on the field using pipes and rulers (Fig. 3).

3.4 The Impact of CSA Implementation

The learning and mentoring process of CSA implementation for farmers has several positive impacts, including:

- The planting time of the first cropping season can be accelerated up to 20–30 days.
- The cropping index increased from 2 times a year to 4–5 times a year.
- The seedling duration can be shortened from 20 to 30 days after sowing to 12–15 days after sowing.
- The harvest index with 1–2 times a year and a 60% risk of crop failure in the second planting season, can be increased to 4–5 times a year with minimum risk of crop failure.
- Rice production increased from 4.0 to 4.2 t/ha to 6.5–7.0 t/ha.

Table 2 Added value was obtained after the application of CSA to the field

Treatment/parameter	Before CSA implementation	After mentoring and CSA implementation
1. Cropping calendar	October II–III/November I–II	September III–October I
2. Cropping pattern	Rice-palawija (maize/peanut/beans) Rice-rice	Maize-rice-maize-peanut-long beans
3. Cropping seasons	2 times a year	4–5 times a year
4. Crop rotations	Sequential, conventional	Combination of sequential, rotational, intercropping, inserting planting time
5. Rice seed	Not certified, 75–100 kg/ha	Certified, 30 kg/ha
6. Seedling duration	20–30 days after sowing	12–15 days after sowing
7. Organic fertilizer	–	1–2 t/ha
8. Chemical fertilizer		
– Urea	400 kg/ha	150 kg/ha
– Phonska	–	150 kg/ha
– SP36	100 kg/ha	–
– ZA	100 kg/ha	–
9. Pest management	Chemical pesticide	POC, biological pestiside, MOL
10. Harvest	1–2 times In the second harvest, 60% failed due to a lack of water	4–5 times Supported by the management of cropping patterns, biopores, and intermittent irrigation
11. Rice productivity	4.0–4.2 t/ha	6.5–7.0 t/ha

- Farmer's knowledge and skills in weather observation, weather data utilization, and other CSA applications are increased.
- Soil fertility has been increasing, indicated by the application of fertilizer which is getting lower from year to year (Table 2).

As with other research, they are engineering the implementation of various technologies in rainfed agriculture that can increase productivity, soil fertility, and farmers' income and capacity [18, 19].

4 Conclusion

Farming systems, by applying the principles of climate smart agriculture, such as engineering cropping schedules and crop patterns, increasing groundwater infiltration, and monitoring rainfall and groundwater levels, can increase the cropping index, harvest index, and agricultural production.

Applying climate smart agriculture through engineering schedules and cropping patterns could change farmers' habits. However, because this is not easy, this process will be an 'exciting' challenge in the development of CSA-based agriculture. However, it is believed that the application of CSA is very site-specific, where one land or area can be different from another land or area. For this reason, the application of CSA on rainfed or dry land still needs to be developed or replicated in other regions or areas.

Acknowledgements Acknowledgments are conveyed to the Directorate of Food Crop Protection, Directorate General of Food Crops, Ministry of Agriculture and Rainfed Farmers' Group Aman Makmur in Srigading Village, Lawang Subdistrict, Malang Regency, East Java Province, Indonesia, for sharing information and field experiences so that this paper is compiled.

References

1. Singh, M., Tiwari, N.K., Kumar, N., Dabur, K.R., Dehinwal, A.K.: Dry and rainfed agriculture-characteristics and issues to enhance the prosperity of Indian farming community. Pharmacol. Life Sci. Bull. Env. Pharmacol. Life Sci. **6**(10), 32–38 (2017) [Online]. Available: http://bepls.com/OCT_2017/6.pdf
2. Mevada, K.D., Poonia, T.C., Saras, P., Deshmukh, S.P.: Rainfed agriculture and watershed management 6th semester B.Sc. (Tech) Agri. Course No. : Agron 6. In: 10 Title of Course: Rainfed Agriculture and Watershed Management. S.K. Kataria & Sons (2021)
3. Sulaiman, A.A., Candradijaya, A., Syakir, M.: Technological advancement and the economic benefit of Indonesian rain-fed farming development. Adv. Agric. (2019). https://doi.org/10.1155/2019/9689037
4. Baig, M.B., Shahid, S.A., Straquadine, G.S.: Making rainfed agriculture sustainable through environmentally friendly technologies in Pakistan: a review. Int. Soil Water Conserv. Res. **1**(2), 36–52 (2013). https://doi.org/10.1016/S2095-6339(15)30038-1
5. Benauli, A.: Kajian Status Hara N, P, K Tanah Pada Sawah Tadah Hujan (Studi Kasus Tiga Desa di Kecamatan Beringin). Agrosains J. Penelit. Agron. **23**(1), 55 (2021). https://doi.org/10.20961/agsjpa.v23i1.49239
6. Arifin, S.A.A., Sadat, M.A.: J. Agribisnis **9**(2), 130–138 (2021). [Online]. Available: https://ejournals.umma.ac.id/index.php/agribis/article/view/1080
7. Pramudia, A., Apriyana, Y., Susanti, E., Suciantini, Harmanto.: Pertanian cerdas iklim indonesia: konsep dan teknologi. In: Sulaiman Y, Listianingsih W (eds) Pertanian Cerdas Iklim Indonesia: Konsep dan Teknologi. Bekasi, pp. 175–192 (2022)
8. FAO: Climate Smart Agriculture, Sourcebook. Rome (2013)
9. BMKG: Data-Online BMKG (2022). http://dataonline.bmkg.go.id/data_iklim
10. BMKG: Prakiraan Musim Kemarau 2020 di Indonesia. BMKG (2020)
11. Erythrina, E., et al.: Assessing opportunities to increase yield and profit in rainfed lowland rice systems in Indonesia. Agronomy **11**(4), 1–15 (2021). https://doi.org/10.3390/agronomy11040777
12. Tjasyono, H.B.: The character of rainfall in the Indonesian monsoon. In: International Symposium on Equatorial Monsoon System, pp. 1–11 (2008). Available: http://file.upi.edu/Direktori/SPS/PRODI.PENDIDIKAN_IPA/BAYONG_TJASYONO/Kumpulan_Makalah/The_Character_of_Rainfall.pdf
13. Susanti, E., et al.: Pemutakhiran Peta Sumberdaya Agroklimat Indonesia untuk Mendukung Perencanaan Pertanian. In: Updating of the Agro-climate Resources Map of Indonesia to

Support Agricultural Planning, vol. 45, no. 1, pp. 47–58 (2021) [Online]. Available: http://ejurnal.litbang.pertanian.go.id/index.php/jti/article/view/12370
14. Oldeman, L., Las, I., Darwis, S.N.: An Agroclimatic Map of Sumatra, vol. 52. Central Research Institute for Agriculture, Bogor (1979)
15. Pramudia, A., et al.: Cropping calendar analysis for dry season 2020 in Indonesia. IOP Conf. Ser. Earth Environ. Sci. **648**(1) (2021). https://doi.org/10.1088/1755-1315/648/1/012117
16. Apriyana, Y. et al.: The integrated cropping calendar information system: a coping mechanism to climate variability for sustainable agriculture in Indonesia. Sustainability **13**(11):1–22 (2021) [Online]. Available: https://www.mdpi.com/2071-1050/13/11/6495
17. Nengah Surata Adnyana, I.: Efektifitas Sistem Tanam Jajar Legowo 2:1 Dengan Sistem Tegel Terhadap Produktivitas Padi Sawah Di Subak Babakan Cangi, Desa Batuan Kaler, Kecamatan Sukawati, Kabupaten Gianyar. dwijenAGRO **10**(2):127–133 (2020). https://doi.org/10.46650/dwijenagro.10.2.1037.127-133
18. Rahmadiyah, R., Tanjung, F., Hariance, R.: Analisis Perbandingan Usahatani Padi Sawah Irigasi Dengan Padi Sawah Tadah Hujan Di Kecamatan Koto Tangah Kota Padang. JOSETA J. Socio Econ. Tripical Agric. **1**, 9–23 (2019) [Online]. Available: http://joseta.faperta.unand.ac.id/index.php/joseta/article/view/177
19. Widyantoro, Toha, H.M.: Optimalisasi Pengelolaan Padi Sawah Tadah Hujan Melalui Pendekatan Pengelolaan Tanaman Terpadu. Pros. Semin. Nas. Serealia Meningkat. Peran Penelit. Serealia. Menuju Swasembada Pangan Berkelanjutan, pp. 648–657 (2010)

Applications of Soil Conditioner Polyacrylamide to Suppress Runoff and P (Phosphorus) Nutrients Loss at the Sweet Corn Cultivation Under Climate Change Issue

Niken Ida Lovita, Ali Rahmat, Yuliana Eva Agasi, Sendi Purnama Hidayat, Yogina Lestari Ayu Situmorang, and Dwi Rustam Kendarto

Abstract Climate change is a phenomenon that treats the environment. Climate change will trigger extreme weather conditions such as high rainfall, which promotes runoff, flooding, and landslide. High rainfall will disintegrate soil particles that can cause surface runoff. The surface runoff will cause the transport of top soil, which contains a lot of nutrients for the plants, and this will make the land infertile and decrease its productivity. To control surface runoff and soil productivity can be done by applying a soil amendment, such as polyacrylamide (PAM). Polyacrylamide is an organic-anionic polymer that could stabilize the soil structure by maintaining a water-permeable pore structure while forming the soil surface. This study aims to determine the effect of polyacrylamide on the reduction of surface runoff and P nutrient loss in sweet corn cultivation. The results showed that polyacrylamide had an effect in reducing surface runoff and P nutrient loss. Polyacrylamide with a dose of 60 kg/ha is the optimal dose that can reduce surface runoff by 14% and reduce the P nutrients loss by 12.16% during eight rainstorms. Meanwhile, for yield and sweet corn's weights, the dose of polyacrylamide 15 kg/ha was the most optimal.

1 Introduction

Global warming and climate change are two of the most extensively researched and debated environmental issues [1]. Climate change is defined as a change in the state of the climate that can be identified (e.g., using statistical tests) by changes in the mean and variability of its properties over an extended period of time, typically decades

N. I. Lovita · Y. E. Agasi · S. P. Hidayat · Y. L. A. Situmorang · D. R. Kendarto
Department Agriculture Engineering, Faculty of Agricultural Industrial Technology, Padjadjaran University, Sumedang, Indonesia

A. Rahmat (✉)
Research Center for Limnology and Water Resources, National Research and Innovation Agency, Jakarta Pusat, Indonesia
e-mail: alir001@brin.go.id

A. Basit et al. (eds.), *Proceedings of the International Conference on Radioscience, Equatorial Atmospheric Science and Environment and Humanosphere Science*, Springer Proceedings in Physics 290, https://doi.org/10.1007/978-981-19-9768-6_66

or longer [2, 3]. Changes in global climate and the Earth's hydrological cycle [4–6] have increased heavy rainfall, increasing the risk of surface runoff and flooding [7, 8]. Surface runoff is one of the causes of high critical land in Bogor Regency. High rainfall is one of the main causes of high surface runoff, especially in Indonesia [9]. The average rainfall in Bogor Regency is in the range of 100–150 mm/day which is included in the very high category [10]. The high surface runoff will cause the loss of topsoil which contains a lot of nutrients transported by surface runoff. Transport of nutrients by surface runoff will cause eutrophication of surface water bodies (lakes, reservoirs, and rivers).

Due to the high rainfall and steep slopes, the soil is prone to erosion and surface runoff. The efforts can be made to mitigate the high surface runoff by improving the soil's physical properties. The physical properties of the soil can be improved by using soil amendments, one of which is polyacrylamide (PAM). PAM is water-soluble and high molecular weight. PAM is frequently used as a soil amendment to improve infiltration, reduce surface runoff, prevent soil erosion, and prevent nutrient loss [11]. Previous research in China by Jiang et al. [12] found that PAM at 20 kg/ha is the optimal dose for reducing surface runoff and can reduce phosphorus transport in surface runoff by 27%. According to Kabede et al. [13], PAM at up to 40 kg/ha is the optimal dose for reducing surface runoff.

According to the above explanation, using PAM to control runoff and nutrient loss is one strategy for dealing with climate change. However, the effectiveness of PAM must be investigated because different locations and times will produce different results. The purpose of this study was to determine the effectiveness of PAM at five different dosage levels in suppressing surface runoff and P nutrient loss in tropical soils in Indonesia under sweet corn cultivation.

2 Methods

The research was carried out at the Research Center for Limnology and Water Resources, National Research and Innovation Agency, Cibinong, Bogor, West Java, in January – May 2022. The materials used in this study were sweet corn seeds, manure, KCl fertilizer, TSP fertilizer, urea fertilizer, and polyacrylamide (PAM). This study was arranged using a completely randomized design method (CRD) consisting of five treatment doses and repeated as many as rain events that occurred. The doses of polyacrylamide applied in this study were: P0 (0 kg/ha), P1 (15 kg/ha), P2 (30 kg/ha), P3 (45 kg/ha), P4 (60 kg/ha) using plants sweet corn trial with a plot size of 2 × 6 m totaling 5 plots. The data analysis used in this study was the ANOVA statistical test. If there was a significant difference, then the LSD test was carried out at an error level of 5%. There are three components analyzed in this study, namely surface runoff (runoff), nutrients (P), and components of sweet corn production (cob length, cob diameter, weight of cob with cob, and weight of cob without cob). Surface runoff (runoff) is calculated using the following formula [14]:

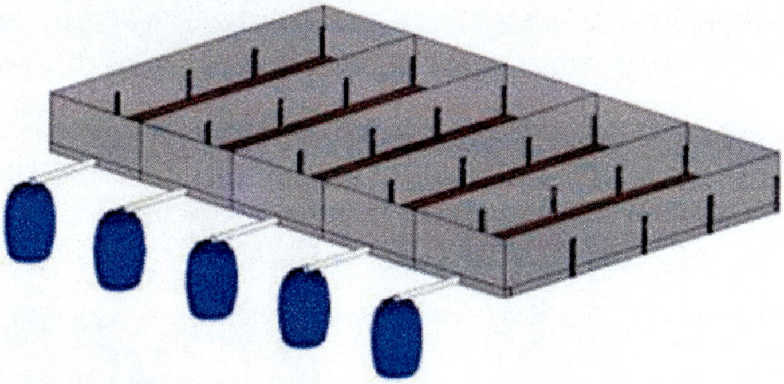

Fig. 1 Plot illustration

$$\mathrm{Ro} = V/A \tag{1}$$

Information:

Ro = Runoff (mm)

V = Volume of runoff (ml)

A = Cross-sectional area of the plot (m^2).

The P nutrients were analyzed using the 25% HCL extraction method at the ICBB Laboratory - PT. Indonesia's Biotechnology Biodiversity. Sweet corn production components were measured using a ruler, caliper, and scales.

Tillage is processed to speed up the planting process. The tools used for processing the soil are hoes, machetes, and shovels. Tillage is carried out by clearing the land of weeds. The land is divided into five plots, with each plot measuring 2 m × 6 m. Each plot was separated using a zinc plate (Fig. 1).

The fertilizer used is 10 tons/ha for manure, 65.49 kg/ha for TSP fertilizer, 200 kg/ha for urea, and 124.46 kg/ha for KCl. Before sowing on the surface of the land, the fertilizers are mixed until homogeneous with a predetermined dose for each plot. After loosening the soil, basic fertilizer is sprinkled over the entire surface of the soil and then distributed the fertilizer equally. After a few hours, PAM will be applied with a predetermined dose for each plot and the plot will be left for 7–10 days before planting.

Sweet corn seeds were planted with a spacing of 75 cm × 40 cm and 2 seeds per planting spot. The irrigation is carried out by depending on the condition of the land and on the rainfall that occurs. It will be harvested when the sweet corns are 60–80 days old.

Sampling will be carried out on the day after the rain, taken every morning at 9.00 WIB. Surface runoff samples were taken from the conductor, which was used as a reservoir for runoff and erosion on the land. Surface runoff and erosion water samples were separated using a sieve. Surface runoff samples were calculated using

Fig. 2 Surface runoff sample

a measuring cup. In addition, surface runoff is also calculated from the evaporation of water during the erosion sample oven process (Fig. 2).

3 Methods

3.1 General Condition of Research Land

The research land has a latosol soil type with characteristics of red–yellow soil, slow permeability, and clay texture. In addition, the slope is 44%, which is in the steep category. The condition research land has a steep slope and also has very slow soil permeability, which can increase the possibility of runoff (Table 1).

Table 1 Physical and chemical properties of the soil in the research area

Soil properties	Value	Category
Texture: % Silt	8	Clay
% Sand	9	Clay
% Clay	83	
Permeability (cm/h)	0.93	Very slow
P nutrients (mg/100 g)	69	–

3.2 *Effect of Polyacrylamide on Surface Runoff*

The runoff will occur when rain falls at a high enough intensity with a relatively long duration. The kinetic energy of rain will destroy soil aggregates, so the crushed soil particles will clog the soil pores. Blockage of soil pores by finer particles causes surface runoff because the soil cannot absorb water.

The results of the ANOVA analysis showed that the PAM (polyacrylamide) treatment was not significantly different in all treatments ($P > 0.05$) (Table 2). PAM-treated plots generally produced lower surface runoff than those without PAM treatment (P0). For six rainfall events, namely the first rain event to the sixth rain event, P1 always produces the lowest runoff while P4 produces the largest runoff. In addition to P4, P0 also generated large surface runoff for six rainfall events. The cause of the large surface runoff at P0 is the absence of PAM granules scattered on the soil surface. Successive rain events caused the destruction of soil aggregates and the formation of seals on the soil which caused an increase in surface runoff at P0. The high runoff at P4 is caused by the higher the PAM dose level, the smaller the soil pores will make it difficult for rainwater to seep into the soil. The results of this analysis are in accordance with research conducted by Kabede et al. [13] that higher PAM (polyacrylamide) dose levels resulted in greater surface runoff at the beginning of the rain event.

P1 resulted in the lowest runoff for the six rainfall events. The low runoff at P1 was due to the 15 kg/ha PAM dose that could produce optimal aggregate stability and low soil viscosity compared to higher PAM doses to suppress runoff that occurred.

In the last two rain events, the runoff at P4 decreased significantly and became the lowest of the other treatments. The drastic decrease in surface runoff at P4 was caused by the soil drying up in the before two last rains, which caused the soil viscosity to decrease. When it rained, the PAM ware still in the soil absorbed some of the water, reducing runoff at higher PAM doses.

For eight rainfall events, P3 and P0 produced the most surface runoff when compared to other plots. Many non-growing or dead crops contributed to the significant increase in surface runoff volume at P3. In this case, vegetation serves to intercept and control the amount of rainwater that falls to the ground surface; due to many plants was dead, the function of plant to intercept and control the water reach the ground surface was decreasing or does not work, with the consequence will promote high runoff.

Table 2 Results of ANOVA analysis between treatments and surface runoff

Component	Treatments				
	P0	P1	P2	P3	P4
Runoff (mm)	1252.3a	1259.264a	1459.753a	1691.305a	1085.361a

Description: similar letter notation means there is no significant difference at the Duncan test level which has a value of 5%

Higher PAM doses can increase surface runoff at the beginning of a rain event but can reduce erosion. This follows the research conducted by Agasi [15], which stated that the 60 kg/ha PAM dose was the optimal dose for reducing erosion due to surface runoff. PAM with a dose of 15 kg/ha can suppress surface runoff at the beginning of the rain event, but its effectiveness decreases in the last rain event. With the highest dose of 60 kg/ha, PAM increased the surface runoff at the beginning of the rain event but could reduce the surface runoff in the last few rain events. The use of PAM at a dose of 30 kg/ha did not produce a large surface runoff during eight rain events. The use of PAM doses that are too high is not good for the environment, and also, from an economic point of view, the costs incurred are quite large, so their use is not effective as well as the use of PAM doses that are too low in effectiveness cannot last until the last rain event. The dose of PAM used in suppressing surface runoff was not significantly different between treatments, so it can be concluded that the PAM dose of 30 kg/ha was the optimal dose in reducing surface runoff during the eight rain events in this study, both for environmental safety and also for the economy.

3.3 Effect of Polyacrylamide on P Nutrients Loss

One of the important pathways for the P nutrients loss from agricultural land is surface runoff. Surface runoff carries important soil particles, one of which is P nutrients loss.

The results of ANOVA analysis showed that PAM treatment was significantly different in the P nutrients loss ($P > 0.05$). Table 3 shows that the largest P nutrient loss due to runoff and erosion is in the P0. P4 resulted in the smallest P nutrient loss during the eight rainfall events. Figure 3 shows that during the six rains, from the first rain to the sixth rain, the amount of P nutrients loss decreased according to the increasing dose of PAM in the soil. P4 always resulted in the smallest P nutrient loss, while the P0 always resulted in the largest P nutrient loss for six rainfall events. The plot with PAM treatment had lower P nutrient loss compared to P0 without PAM treatment. The high P nutrients loss in P0 was due to the absence of PAM in the soil, which caused the soil to be easily carried away by surface runoff which contained nutrients. P4, with a higher PAM dose level compared to other treatments, produced soil aggregates that were very binding to one another. During rain, the soil and nutrients were not easily carried away by surface runoff.

Table 3 Results of ANOVA analysis between treatments and P nutrients loss

Component	Treatments				
	P0	P1	P2	P3	P4
P nutrients loss (mg/100 g)	82.2300c	78.7250b	74.2813a	74.4475a	71.8550a

Description: similar letter notation means there is no significant difference at the Duncan test level which has a value of 5%

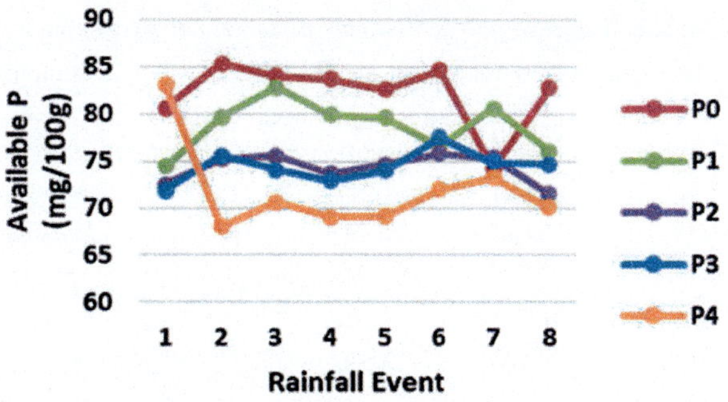

Fig. 3 P nutrients loss during eight rainstorm

Table 4 Comparison of the effect of PAM on the loss of nutrient N with control plot

Treatment	P Nutrients loss (%)			
	Treatment plots	Control plot (P0)	Difference	Percentage drop (%)
P1	78.725		3.505	4.26
P2	74.28125	82.23	7.94875	9.66
P3	74.4475		7.7825	9.46
P4	71.855		10.375	12.16

There was a drastic increase in P nutrient loss in P1 and P3 during the sixth and seventh rain events. The increase in P nutrient loss in P1 could be due to the insufficient dose level of PAM in P1 to retain the P nutrients loss by surface runoff. This increase in P nutrients loss could be caused because after seven rain events the effectiveness of PAM decreased with increasing time and intensity of rain. The increase in P nutrients loss in P3 could be due to the death of the vegetation in P3 toward the end of the rain event (Table 4).

The PAM dose rate of 60 kg/ha reduced P nutrients loss by 12.16% during eight rainy events. This result follows the research conducted by Agasi [15], which stated that the PAM dose of 60 kg/ha was the optimal dose for reducing erosion so that the nutrient loss would be less in this plot. P0 always yielded the greatest P nutrients loss for the eight rainfall events. The P nutrients loss in the P0 without PAM application was 3 times greater than that in P4.

3.4 *Effect of Polyacrylamide on Sweet Corn Production*

Sweet corn was planted on February 18, 2022, but after ten days after planting, some sweet corn plants did not grow so replanting was carried out. Replanting is done at the planting point that did not grow on March 1, 2022 (Table 5).

The results of the analysis showed that polyacrylamide (PAM) had an effect on the diameter of the cobs ($P < 0.05$), but it was not significantly different for the weight of

Table 5 Effect of polyacrylamide production of sweet corns

Treatment	With corn husks leaves			Without corn husks leaves		
	Corn weight (g)	Corn length (cm)	Corn diameter (cm)	Corn weight (g)	Corn length (cm)	Corn diame ter (cm)
P0	231.36a	5.4a	25.62a	162.29a	4.71a	17.18a
P1	296.13a	5.91a	29.00ab	203.21a	4.82a	19.76b
P2	266.08a	5.65a	28.31ab	18,826a	4.8a	1812ab
P3	246.91a	5.53a	27.37ab	17,677a	4.73a	1893ab
P4	297.35a	5.86a	31.06b	18,207a	4.62a	1956ab

Description: similar letter notation means there is no significant difference at the Duncan test level which has a value of 5%

the cobs and the length of the fruit, whether with husk leaves or not. The yields in plot 4 produced the largest fruit weight with husk leaves and diameter of fruit with husk leaves compared to other plots, while the highest fruit length with husk leaves was in plot 1. The results of the largest fruit weight without husk leaves were found in plot 1 as well as fruit length and fruit diameter without husk leaves. Yields increased with the PAM application when compared to control plots without PAM application. Plots with PAM application can increase by 25% of fruit weight without husk leaves, 2% of fruit length without husk leaves, and 15% of fruit diameter without husk leaves than the control plot. This is because PAM can increase the availability of nutrients for plants. Increased crop yields also can be caused by PAM where expected can increase water infiltration in the soil so that the availability of water for plants is met. PAM increases the movement of water on the soil surface so that more water flows into the planting row compared to the control plot. The dose level of 15 kg/ha contained in plot 1 is the most optimal dose in increasing crop yields because in plot 1, infiltration is always high from the beginning of the rain event to the last rain event during the study. This can also be seen in a surface runoff where plot 1 always produces low surface runoff when compared to plots treated with other PAMs.

4 Conclusion

The conclusion of the research is that the 30 kg/ha dose is optimal in reducing runoff by14%. The PAM dose of 60 kg/ha is the optimal dose in suppressing the P nutrients loss by 12.16%. A dose of 15 kg/ha is the optimal dose to increase sweet corn yields.

References

1. Rahmat, A., Zaki, M.K., Effendi, I., Mutolib, A., Yanfika, H., Listiana, I.: Effect of global climate change on air temperature and precipitation in six cities in Gifu Prefecture, Japan. J. Phys.: Conf. Ser. **1155**, 012070 (2019)
2. Sturrock, R.N., Frankel, S.J., Brown, A.V., Hennon, P.E., Kliejunas, J.T., Lewis, K.J., Worrall, J.J., Woods, A.J.: Climate change and forest diseases. Plant. Pathol. **60**(1), 133–149 (2011)
3. Rahmat, A., Mutolib, A.: Comparison air temperature under global climate change issue in Gifu city and Ogaki City, Japan. Indones. J. Sci. Technol. **1**, 37–46 (2016)
4. Allen, M.R., Ingram, W.J.: Constraints on future changes in climate and the hydrologic cycle. Nature **419**, 224–232 (2002)
5. Held, I.M., Soden, B.J.: Robust responses of the hydrological cycle to global warming. J. Climate. **19**, 5686–5699 (2006)
6. Wentz, F.J., Ricciardulli, L., Hilburn, K., Mears, C.: How much more rain will global warming bring? Science **317**, 233–235 (2007)
7. Trenberth, K.E.: Conceptual framework for changes of extremes of the hydrological cycle with climate change. Clim. Change **42**, 327–339 (1999)
8. Trenberth, K.E.: Changes in precipitation with climate change. Clim. Res. **47**, 123–138 (2011)
9. Kartasapoetra, A.G.: Teknologi Konservasi Tanah dan Air. Rineka Cipta. Jakarta (2005)
10. Badan Pusat Statistik Kabupaten Bogor. Jumlah Curah Hujan Menurut Bulan dan Pos Hujan (mm/hari). Badan Meteorologi Klimatologi dan Geofisika Stasiun Klimatologi Bogor. https://bogorkab.bps.go.id/indicator/151/80/1/jumlah-curah-hujan-menurut-bulan-dan-stasiun-pos-hujan.html (2022).
11. Entry, J.A., Sojka, R.E.: The efficacy of polyacrylamide to reduce nutrient movement from an irrigated field. Trans. Am. Soc. Agric. Eng. **46**(1), 75–83 (2003)
12. Jiang, T., Teng, L., Wei, S., Deng, L., Luo, Z., Chen, Y., Flanagan, D.C.: Application of polyacrylamide to reduce phosphorus losses from a Chinese purple soil: a laboratory and field investigation. J. Environ. Manage. **91**(7), 1437–1445 (2010)
13. Kebede, B., Tsunekawa, A., Haregeweyn, N., Mamedov, A.I., Tsubo, M., Fenta, A.A. Meshesha, D.T., Masunaga, T., Adgo, E., Abebe, G., Berihun, M.L.: Effectiveness of polyacrylamide in reducing runoff and soil loss under consecutive rainfall storms. Sustainability **12**(4), 1–18 (2020)
14. Ziliwu, Y.: Pengaruh Beberapa Macam Tanaman Terhadap Aliran Permukaan dan Erosi. Universitas Diponegoro. http://eprints.undip.ac.id/11329/1/2002MTS1772.pdf (2002)
15. Agasi, Y.E.: Kajian Aplikasi Soil Conditioner Polyacrylamide Untuk Menekan Erosi dan Kehilangan Unsur Hara N (Nitrogen) Di Lahan Budidaya Jagung Manis. Under Graduate Thesis. Padjajaran University (2022)

Elemental Analysis of Breadnut Seed Biochar and Its Potential Application as a Soil Amendment

Sukamto and Ali Rahmat

Abstract Soil acidification can be promoted by long-term cultivation and the excess use of fertilizer. Moreover, increased precipitation due to climate change, can exacerbate leaching and contribute to soil acidification which correlates to the decrease in soil fertility. It has significant adverse impacts not only on agricultural production but also on economic aspects. To overcome this problem, various ways have been applied to improve soil fertility by reducing soil acidity. Recently, waste-based biochar products were used as a promising soil amendment. In this study, breadnut seed-based biochar has been successfully produced by using thermal treatment at 250 and 350 °C. The elemental composition has been evaluated by using X-ray fluorescence spectrometer (XRF). According to the elemental analysis, there is no significant difference in the amount of elemental composition for the different thermal treatments. Generally, breadnut seed-based biochar obtained at 250 °C shows higher nutrient content such as Mg, P, S, K, Ca, and Fe which are 0.716%, 5.476%, 1.757%, 71.063%, 19.685%, and 0.583%, respectively. The presence of these nutrients on the breadnut seed-based biochar increases its potential application for soil amendment to reduce soil acidity and enhance soil fertility. Increasing soil fertility and increasing crop productivity was one strategy for climate change mitigation indirectly.

1 Introduction

The 2022 Global Report on Food Crises (GRFC) reported the highest food insecurity which is nearly 25%. Due to the COVID-19 pandemic, conflict, and extreme climate change, more than 193 million people in 53 countries are suffering from food crisis [1]. Moreover, the increasing of population around the world leads this problem turn to worse. On the other hand, various efforts have been conducted to

Sukamto
Hokkaido University, Sapporo, Japan

A. Rahmat (✉)
National Research and Innovation Agency, Jakarta Pusat, Indonesia
e-mail: alir001@brin.go.id

A. Basit et al. (eds.), *Proceedings of the International Conference on Radioscience, Equatorial Atmospheric Science and Environment and Humanosphere Science*, Springer Proceedings in Physics 290, https://doi.org/10.1007/978-981-19-9768-6_67

face this issue in the last decade such as food diversification and agriculture intensification. However, aiming to increase agriculture production for fulfilling food needs, agriculture intensification promotes the excessive use of fertilizers that can reduce soil quality such as soil acidification. Moreover, because of rising temperatures and changes in precipitation (amount and frequency), climate change is expected to have a significant impact on soils and ecosystems. Increased precipitation, can exacerbate leaching and contribute to soil acidification [2]. Soil acidification changes the soil quality by decreasing soil pH < 5.5 which can be caused by the deposition of acid [3] and excessive use of ammonium-based fertilizers [4]. The decrease in soil pH causes an increase in aluminum and manganese toxicities toward crops as well as a decrease in the content of phosphorus and nutrient ions such as calcium, potassium, magnesium, and molybdenum [5]. This condition has a negative impact on soil quality which results in a decrease in crop yields. Therefore, appropriate, and efficient steps must be taken to maintain or improve soil acidity so that global food availability can be achieved, and the world food crisis can be overcome.

Recently, various materials have been applied for recovering soil acidification such as lime, gypsum, red mud, organic waste, and residual crops. These materials show good abilities to increase soil pH and show cation exchange. However, the high heavy metals contents on these materials can cause soil contamination and promotes dangerous side effect for environmental [6]. Owing by this limitation, previously, some researchers have focused to use biochar as the soil amendment. Bicohar was obtained from pyrolyzed feedstocks at high temperatures ranging from 200 to 700 °C, biochar has shown excellent ability to improve soil quality compared to other organic materials [7]. Generally, biochar is alkaline in nature due to the presence of Mg, K, Ca, and Fe ions. It promotes liming effects when biochar is applied into the acid soil which contribute to the ameliorate soil acidity and increase of crop production [8].

With the respect of biochar application for agronomic purposes, the content of nutrient ions on the biochar is one of the most important aspects that determine its ability to improve soil quality. The concentration of nutrient ions depends on the used feedstock and pyrolysis temperature. Moreover, due to different pyrolysis temperature, toxic elements may be presented on the biochar that also can promote dangerous impact for plant growth, environment, and human health [9]. Therefore, the elemental analysis is a crucial way to ensure that biochar contains sufficient and safe ions for plant growth and production.

In this study, breadnut seed biochar was created by using pyrolysis method, where the temperature of pyrolysis at 250 and 350 °C. To evaluate the elemental content of biochar, non-destructive and environmentally friendly analysis method was applied, where X-ray fluorescence spectrometry (XRF) has been used as the main technique. This study not only offers environmentally friendly method to analysis total elemental composition of biochar, but also promote the use of organic waste by product such as breadnut seed to be functional materials as a soil amendment for agronomic application.

Fig. 1 Dried raw breadnut seed (**a**) and biochar product (**b**)

2 Methods

2.1 *Material*

Breadnut seed was obtained from market. Breadnut seed waste was collected and then washed by using clean water. After wash, breadnut seed waste was air-dried for 5 h after that coarsely blended. The blended breadnut seed then was dried in an oven at 105 °C for 24 h.

2.2 *Biochar Preparation and Characterization*

Dried breadnut seed was placed and burned in a furnace at 250 and 350 °C for 4 h [10, 11]. The obtained biochar then was labeled with BBc250 and BBc350 for biochar that was obtained at 250 °C and 350 °C, respectively. The raw material and biochar product is shown in Fig. 1. Both BBc250 and BBc350 were milled and sieved to 355 μm in size. The biochar powder was analyzed by using X-ray fluorescence spectrometer (XRF) (Omnian ED-XRF Panalytical Epsilon 3 XLE) at Lampung Advanced Characterization-BRIN. The data was analyzed descriptively by comparing the total elemental composition of biochar obtained at 250 and 350 °C.

3 Results and Discussions

3.1 *Elemental Analysis of Biochar Product*

For several thousand years, biochar which is a solid product obtained from pyrolyzed biomass has been well known and produced for various application such as heat production, metallurgical application, building material, and agriculture. Generally,

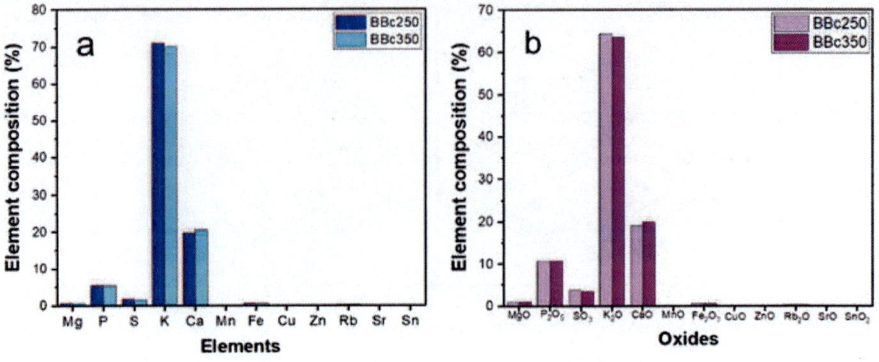

Fig. 2 Element (**a**) and oxide (**b**) composition of BBc250 and BBc350

biochar can be obtained from decomposition of biomass during pyrolysis process. The conversion process is depending on the temperature and characteristic of feedstock that use as the starting material. In this study, pyrolysis process has been conducted in the temperature 250 and 350 °C, and it is referred to as mild pyrolysis (torrefaction). Torrefaction is used to retain and concentrate most of the energy content in the solid product and increase mechanical properties of the products significantly [12]. During the torrefaction process, there are three temperature zones: (1) non-reactive zone (from 50 to 150 °C), (2) reactive zone (from 150 to 200 °C), and (3) destructive zone (from 200 to 300 °C) which includes some biomass reactions such as dehydration, devolatilization and carbonization of hemicellulose, depolymerization and devolatilization/softening of lignin, and depolymerization devolatilization of cellulose [13]. In this study, after pyrolyzed for 4 h under 250 and 350 °C, the obtained biochar was analyzed by using X-ray fluorescence spectrometer (XRF), and the results are depicted in Fig. 2.

According to the graph in Fig. 2, it can be shown that biochar produces from mild pyrolysis at 250 and 350 °C contain various mineral compositions. The highest element composition and oxide composition is achieved by potassium. Where in elements form the percentage are 71.06% and 70.229% for BBc250 and BBc350, respectively. Moreover, the percentage potassium oxide form on the biochar which reached up to 60% for both BBc250 and BBc350. Beside potassium, calcium in form element and oxides lied on the second place after potassium. Another mineral that also presented on the BBc250 and BBc350 are Mg, P, S, Mn, Fe, Cu, Zn, Rb, Sr, and Sn. However, for BBc350 there is no Sn content detected on the element as well as its oxide. Naturally, wet breadnut seed contains these minerals including K, Ca, P, Fe, Cu, Mn, Na, and Mg [14]. Generally, during the biochar production, the carbonaceous solid is expected to form. However, in this study, carbon element is not formed and detected. It might be caused by the low temperature that used in the pyrolysis process. In terms of very high carbon content (up to 95%) on biochar, high temperature close to 1000 °C is required.

3.2 Potential Application of Biochar as Soil Amendment

The increase of world population across the globe promotes significant effect not only on the social aspect but also economics. Recently, this condition turned to worse due to the conflict, COVID-19 pandemic, and extreme climate change. Apart from all the most crucial needs, food availability is the vital thing that should be fulfilled. According to the 2022 Global Report on Food Crises (GRFC), more than 193 million people in 53 countries are suffering from food crisis. Owing to that problem, agriculture sector plays important role to face food crisis. Alternative programs have been enrolled to overcome the problem including agriculture intensification which then promotes the excessive use of fertilizers. Instead of reducing food crisis, the use of excessive fertilizer gives dangerous impact due to soil acidification and quality degradation.

Generally, soil acidification is a natural process which can be occurred almost in tropical and subtropical regions. It is caused by several processes such as H^+ ions precipitation, presence of acidic gasses including SO_2 and NO_3, and the application of ammonium-based fertilizers (NH_4^+) [15, 16]. Soil acidification is indicated by the decrease of soil pH (<5.5) which can be caused by the replacement of alkaline cations such as Ca^{2+}, Mg^{2+}, and K^+ by H^+ and Al^{3+} ions. Moreover, soil acidification can be promoted by the dissolution of Al and Mn minerals. This condition increases the metal toxicities of Al, Mn, and Fe and reduces important mineral such as phosphorus which reduce the soil fertility and crop production in consequence [17].

According to the elemental analysis, both BBc250 and BBc350 biochar contain alkaline cations such as Ca^{2+}, Mg^{2+}, and K^+. When biochar is applied to the acid soil, alkaline cations which presented on the biochar will displace exchangeable H^+ and Al^{3+} ions on acid soil. Free H^+ ion will combine with OH^- ion and from neutral water or CO_3^{2-} or HCO_3^- ions to form unstable H_2CO_3, which will be readily dissociated to form CO_2 and H_2O as illustrated on Fig. 3. These liming effects can increase the soil pH and improve soil fertility especially for the acid soil that contains high toxic aluminum to increase crop productivity.

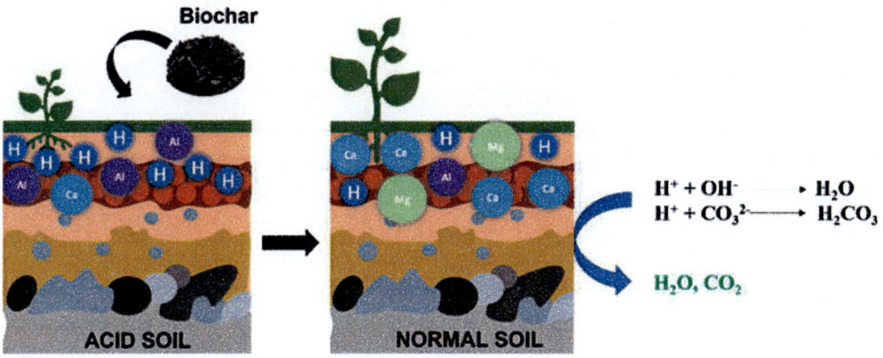

Fig. 3 Liming effects of acid soil by biochar illustration

3.3 Potential Application of Biochar as Climate Change Mitigation

Climate change caused by rising levels of greenhouse gases (GHG) in the atmosphere is expected to have disastrous consequences for our planet [18]. As a result, this should spur action to reduce emissions and increase GHG removal from the atmosphere.

Biochar contributes to climate change mitigation in a variety of ways. First, slow pyrolysis can convert unstable carbon from biological materials into stable carbon (biochar), which can bind C in the soil for hundreds of years [19]. Second, biochar in soil has the potential to reduce greenhouse gas emissions other than CO_2. Soil is a major source and absorber of greenhouse gases like carbon dioxide (CO_2), methane (CH_4), and nitrous oxide (N_2O). The incorporation of biochar into soil has been shown to influence the C and N transformation and retention processes. This process, along with other biochar-influenced mechanisms, has the potential to significantly reduce soil GHG emissions. Recent research has found that biochar increases CH_4 uptake [20] and reduces N_2O emissions from soil [21]. This could significantly increase the benefit of greenhouse gas mitigation.

Biochar has been shown to increase soil fertility by raising pH, retaining nutrients, and retaining soil moisture, resulting in increased biomass production by plant species. As a result, more C is released from the atmosphere and stored in plant tissues. Apart from the obvious economic benefits of increasing crop yields, it also increases the amount of waste biomass available for slow pyrolysis and bioenergy production, as well as the amount of biochar available for long-term carbon sequestration. According to the above, biochar has a direct effect on mitigating climate change by absorbing greenhouse gases and storing carbon, and it indirectly helps in the absorption of carbon in the air (carbon sequestration) into plant biomass by improving and increasing soil fertility.

4 Conclusion

In this study, breadnut seed waste has been successfully converted into a biochar and showed high content of minerals such as Ca, K, Mg, P, S, Mn, Fe, Cu, Zn, Rb, and Sr. Mild pyrolysis under 250 and 350 °C showed unsignificant difference on the mineral compositions. The high content of alkaline cations presented on the obtained biochar indicates that the biochar obtained from breadnut seed using mild pyrolysis has a high potential to be applied as soil amendment to ameliorate soil acidity and increase crop productivity. Increasing soil fertility and increasing crop productivity was one strategy for climate change mitigation indirectly.

Acknowledgements The authors acknowledge the facilities, scientific and technical support from Advanced Characterization Laboratories Lampung, National Research, and Innovation Agency through E- Layanan Sains, Badan Riset dan Inovasi Nasional.

Authorship Sukamto and Ali Rahmat Setiawan are main contributor/author in this article.

References

1. World Food Programme: 2022 Global Report on Food Crises: Joint Analysis for Better Decision. Retrieved 28 Sept 2022, from https://www.wfp.org/publications/global-report-food-crises-2022
2. Brinkman, R., Brammer, H.: The influence of a changing climate in soil properties. In: Transactions 14th ISSS Congress, Kyoto, vol. 5, pp. 283–287 (1990)
3. Duan, L., Chen, X., Ma, X., Zhao, B., Larssen, T., Wang, S., Ye, Z.: Atmospheric S and N deposition relates to increasing riverine transport of S and N in southwest China: implications for soil acidification. Environ. Pollut. **218**, 1191–1199 (2016)
4. Cai, Z., Wang, B., Xu, M., Zhang, H., Zhang, L., Gao, S.: Nitrification and acidification from urea application in red soil (Ferralic Cambisol) after different long-term fertilization treatments. J. Soils Sediments **14**, 1526–1536 (2014)
5. Fenn, M.E., Huntington, T., McLaughlin, S., Eagar, C., Gomez, A., Cook, R.: Status of soil acidification in North America. J. For. Sci. **52**, 3–13 (2006)
6. Li, J.Y., Wang, N., Xu, R.K., Tiwari, D.: Potential of industrial byproducts in ameliorating acidity and aluminum toxicity of soils under tea plantation. Pedosphere **20**, 645–654 (2010)
7. Lehmann, J., Joseph, S.: Biochar for Environmental Management: Science, Technology, and Implementation. Routledge (2015)
8. Jeffery, S., Verheijen, F., Van Der Velde, M., Bastos, A.: A quantitative review of the effects of biochar application to soils on crop productivity using meta-analysis. Agric. Ecosyst. Environ. **144**, 175–187 (2011)
9. Enders, A., Hanley, K., Whitman, T., Joseph, S., Lehmann, J.: Characterization of biochars to evaluate recalcitrance and agronomic performance. Bioresour. Technol. **114**, 644–653 (2012)
10. Rahmat, A. Chemical properties of biochar from Date Palm Seed (*Phoenix dactylifera* L.) under low temperature pyrolysis as soil amendment candidate. Appl. Res. Sci. Technol. **1**(2), 116–120 (2021)
11. Rahmat, A., Ramadhani, W.S., Nur, M., Sutiharni, Mutolib, A.: Chemical properties of salacca seed biochar under low temperature of pyrolysis. Appl. Res. Sci. Technol. **2**(1), 43–46 (2022)
12. Weber, K., Quicker, P.: Properties of biochar. Fuel **217**, 240–261 (2018)
13. Tumuluru, J.S., Sokhansanj, S., Wright, C.T., Boardman, R.D., Yancey, N.A.: A review on biomass classification and composition, co-firing issues and pretreatment methods. In: Proceedings of the American Society of Agricultural and Biological Engineers Annual International Meeting, pp. 2053–2083 (2011)
14. Adeleke, R.O., Abiodun, O.A.: Nutritional composition of breadnut seeds (*Artocarpus camansi*). Afr. J. Agric. Res. **5**(11), 1273–1276 (2010)
15. Dashuan, T., Shuli, N.: A global analysis of soil acidification caused by nitrogen addition. Environ. Res. Lett. **10**, 024019 (2015)
16. Singh, S., Tripathi, D.K., Singh, S., Sharma, S., Dubey, N.K., Chauhan, D.K., Vaculík, M.: Toxicity of aluminium on various levels of plant cells and organism: a review. Environ. Exp. Bot. **137**, 177–193 (2017)
17. Goulding, K.W.T.: Soil acidification and the importance of liming agricultural soils with reference to the United Kingdom. Soil Use Manage. **32**, 390–399 (2016)

18. IPCC (Intergovernmental Panel on Climate Change): In: Eggleston, H.S., Buendia, L., Miwa, K., Ngara, T., Tanabe, K. (eds.) 2006 IPCC Guidelines for National Greenhouse Gas Inventories, Prepared by the National Greenhouse Gas Inventories Programme. Published for the IPCC by IGES, Japan (2006)
19. Lehmann, J., Gaunt, J., Rondon, M.: Biol-char sequestration in terrestrial ecosystems—A review. Mitig. Adapt. Strat. Glob. Change **11**, 403–427 (2006)
20. Rondon, M.A., Molina, D., Hurtado, M., Ramirez, J., Lehmann, J., Major, J., Amezquita, E.: Enhancing the productivity of crops and grasses while reducing greenhouse gas emissions through bio-char amendments to unfertile tropical soils. In: 18th World Congress of Soil Science July 9–15, Philadelphia, Pennsylvania, USA (2006)
21. Yanai, I.Y., Toyota, K., Okazaki, M.: Effects of charcoal addition on N_2O emissions from soil resulting from rewetting air-dried soil in short-term laboratory experiments. Soil Sci. Plant Nutr. **53**, 181–188 (2007)

The Impact of Climate Change on Meteorological Drought in Yogyakarta

Sinta Berliana Sipayung, Indah Susanti, Edy Maryadi, Amalia Nurlatifah, Adi Witono, Lilik Slamet, and Laras Tursilowati

Abstract Drought is a regional phenomenon that is naturally affected by climate conditions. Climate change, which changes the pattern of rainfall and temperature, is very influential on the drought conditions in an area. Yogyakarta Province is one of the regions in the south of Indonesia with a distinctive regional character. By using rainfall data and the Palmer Drought Severity Index (PDSI) from 1958 to 2019, it can be shown that there is a change in the potential for two drought periods of 30 years for five districts in Yogyakarta Province. Changes are distinguished for the December–January–February period (DJF) which represents the wet season, and the June–July–August period (JJA), which represents the dry season. The results of the data analysis have shown an increase in rainfall for the wet season and a decrease in rainfall in the two periods. However, an increase in rainfall during the wet season does not mean that there has been a decrease in the level of drought. PDSI data for the first period showed an opportunity for an index of more than four by 21.55% for Bantul, 38.87% for Gunung Kidul, and 39.85% for Yogyakarta City, 39.63% for Kulon Progo, and 39.49% for Sleman. This figure has increased for Bantul in the second period, but for other districts, it has increased. This indicates that the data increase in rainfall that occurred, accompanied by an increase in temperature and evapotranspiration, causes the level of drought continues to increase. If this trend persists, there will be indications that rainfall in DIY will increase during the DJF season, but the potential for drought will continue.

S. B. Sipayung (✉) · A. Nurlatifah · A. Witono · L. Slamet · L. Tursilowati
Research Center for Climate and Atmosphere, National Research and Innovation Agency, Jakarta Pusat, Indonesia
e-mail: s_berlianasipayung@yahoo.com

I. Susanti
Directorate of Repository, Multimedia, and Scientific Publishing, National Research and Innovation Agency, Jakarta Pusat, Indonesia

E. Maryadi
Research Center for Data Science and Information, National Research and Innovation Agency, Jakarta Pusat, Indonesia

A. Basit et al. (eds.), *Proceedings of the International Conference on Radioscience, Equatorial Atmospheric Science and Environment and Humanosphere Science*, Springer Proceedings in Physics 290, https://doi.org/10.1007/978-981-19-9768-6_68

1 Introduction

Climate change is currently still a global issue that is quite important to continue to be studied. One indicator of climate change is the increase in the average temperature of the earth's surface, causing changes in other elements such as rising seawater temperatures, increasing evaporation, and changing rainfall patterns which ultimately affect various aspects of life. Climate change will also affect the hydrological cycle, especially in Indonesia, creating new challenges in water management. Indonesia is an archipelagic country that is very vulnerable to the impacts of climate change [42], which causes the carrying capacity of the environment to decline. The results of the study [19] showed that during the twentieth century, the intensity of rainfall in some parts of Indonesia decreased, while in some other areas, it increased by around 2–3%. From the projection results using model scenario data, in Indonesia, there has been an increase in rainfall except in southern Indonesia [28].

Several indicators and impacts of climate change in Indonesia include changes in rainfall patterns, increases in air temperature, shifts in seasons, sea level rise, increased frequency of natural disasters, and threats to biodiversity [26]. In addition, the volume, pattern, intensity, and frequency of rainfall have also changed, which tends to result in more extreme events [5]. In extreme wet conditions, when there is excessive rainfall, there is a surge in runoff due to a decrease in the ability of the soil to absorb rainwater, which in turn causes flooding. In extremely dry conditions, when rainfall is very minimal, it causes land drought and water supply deficits for the needs of the population.

Analysis of long-term trends in atmospheric variables such as rainfall, temperature, and evapotranspiration has been used extensively, for example, rainfall by Warburton and Schulze [12, 40], temperature by Kruger and Shongwe 21, Warburton Toucher et al. [39], while evaporation and soil moisture have been carried out by Malisawa and Rautenbach [24], as proxies for detecting climate change. They used temperature data and ran the method developed by Thornthwaite [33], the potential evapotranspiration (PET) of water runoff was calculated. In addition, the factor of changes in local groundwater holding capacity is also considered to simulate the impact of soil degradation patterns due to anthropogenic activities. The results show that degradation occurring over 30 years period (1950–1980) can hurt local water resources in wet and dry climate areas.

Drought is a disaster that causes significant losses to economic, ecological, and environmental sectors because drought causes a decrease in water supply, crop losses, environmental degradation, desertification, and forest fires [11, 18, 27, 29]. The occurrence of climate change has affected the patterns of temperature, evapotranspiration, and precipitation in various parts of the world. Several climate models indicate the potential for droughts to occur more frequently with increasing severity [22, 23, 38], especially in the middle latitudes [31, 41]. Droughts can be classified into four types: meteorological, hydrological, agricultural, and socioeconomic. Meteorological drought is a condition of lack of water caused by an imbalance of rainfall and evaporation, so these two parameters are often used for meteorological drought analysis [4, 13].

To measure drought and monitor its development, many drought indices have been developed and applied [8, 15, 20, 37]. Among them, the Palmer Drought Severity Index [30] is the most prominent meteorological drought index used in the United States for drought monitoring and research. PDSI and its variants have been used to measure long-term changes in above-ground aridity in the twentieth century [9, 10, 34–36] and the twenty-first century [2, 3]. PDSI has also been widely used in tree ring-based reconstructions of past droughts in North America and other regions, for example, Cook et al. [6, 7]

Yogyakarta (Daerah Istimewa Yogyakarta/DIY) is one of the regions in Indonesia located on the island of Java and has 3.66 million inhabitants and 3,185.80 km^2. As a province with a large population, Yogyakarta is faced with many challenges, including providing clean water as a source of human life. In addition, Yogyakarta is also an area of monsoon rainfall with considerable variability, so it can experience a relatively significant impact from extreme weather due to climate change. This study aims to identify and analyze the impact of climate change on meteorological drought in Yogyakarta from 1958 to 2019 using the PDSI method.

2 Data and Method

2.1 Data

Rainfall Data from TerraClimate

TerraClimate is a collection of monthly climate data and climate water balance available on global terrestrial surfaces. TerraClimate uses an interpolation principle that combines high spatial resolution climate data from the WorldClim dataset, with coarser spatial resolution, but more detailed temporal resolution than CRU TS4.0 and 55-year Japan Reanalysis (JRA55). Conceptually, this procedure applies the interpolated time variation anomaly from the CRU TS4.0/JRA55 to the high spatial resolution climatological data from WorldClim to create a high spatial resolution dataset with a broader range of temporal data (Earth Engine Data Catalog 2021). The rainfall data from TerraClimate was obtained using Google Earth Engine.

PDSI

PDSI is a drought index data that uses available temperature and rainfall data to estimate relative drought. This is a standard index whose values range from -10 (dry) to $+10$ (wet). The PDSI has been quite successful in measuring long-term drought. This is because PDSI uses temperature data and a physical water balance model, it can capture the primary effect of global warming on drought through changes in potential evapotranspiration (NCAR-UCAR 2022). PDSI variations have been used to measure long-term changes in land surface dryness [14].

Table 1 Classification of dry and wet conditions as defined by Palmer [30] for the Palmer Drought Severity Index (PDSI)

Value PDSI	Categorized
4.0 or more	Extreme wet
3.0–3.99	Severe wet
2.0–2.99	Moderate wet
1.0–1.99	Slight wet
0.5–0.99	Incipient wet spell
0.49 to −0.49	Near normal
−0.5 to −0.99	Incipient dry spell
−1.0 to −1.99	Slight drought
−2.0 to −2.99	Moderate drought
−3.0 to −3.99	Severe drought
−4.0 or less	Extreme drought

2.2 *Method*

Classification of PDSI
PDSI was originally developed initially by Palmer [30]. It is a meteorological drought index that presents a suitable measurement of humidity conditions in a location concerning time and is calculated using precipitation and evapotranspiration potential. To get the soil moisture anomaly index, the Palmer method is used, which starts from the calculation of temperature, potential evapotranspiration, and water availability. Palmer drought index is determined from several constants to get the climatically appropriate for existing conditions (CAFEC) value until the Palmer method drought index is obtained as follows:

$$X = \left(\frac{z}{3}j - 1 + \Delta X\right) \tag{1}$$

$$\Delta X = \left(\frac{z}{3}\right)j - 0,103\left(\frac{Z}{3}\right)j - 1 \tag{2}$$

Which: X = Palmer's Drought Index, Z = Moisture deviation index (anomaly). The PDSI value classification is presented in Table 1.

3 Results and Discussion

One of the impacts that may arise due to climate change, according to NASA, is a shift in rainfall patterns (NASA 2022). This shift in rainfall patterns affects many aspects, one of which is the possibility of drought and water balance conditions in an area. Shifts in rainfall patterns occurred in several places in Yogyakarta. In the Bantul area, during the DJF season in 1989–2019, the amount of rainfall tends to be higher than in the same season in 1958–1988. However, different things happen in the dry season, namely, in the JAS season. In the JAS season in 1989–2019, there was less rainfall compared to the same season in 1958–1988 (Fig. 1a). The same condition also occurs in the Gunung Kidul area. In the 1989–2019 DJF season, rainfall was more than in the same season in 1958–1988 (Fig. 1b). However, in the JJA season, rainfall is minor (Fig. 1b). Conditions in the Bantul and Gunung Kidul areas are also similar to those conditions in the cities of Yogyakarta, Kulon Progo, and Sleman (Fig. 1c–e). This change in rainfall patterns also affects changes in the potential for drought events in DI Yogyakarta.

Figure 2 shows a boxplot of PDSI data for 60 years in DI Yogyakarta. The data is grouped into two periods, namely, 1958–1988 and 1989–2019. In the PDSI boxplot in all regions, it can be seen that there is a significant change in the distribution of the data. It can be seen that in the period 1958–1988, PDSI values were more distributed at the bottom of the boxplot (small values). However, from 1989 to 2019, the PDSI value was widely distributed at the top of the boxplot (the value increased). This indicates that for the entire DI Yogyakarta area, there is an indication of an increase in rainfall in the DI Yogyakarta area which is associated with a decrease in the probability of extreme dry events.

Changes in the value of seasonal rainfall are more clearly seen when the maximum value is identified. In the first dry season, all areas of Yogyakarta Province showed a maximum rainfall value of above 300 mm, while in the second dry season, it was only around 200 mm. The most significant decrease from the maximum dry season rainfall occurred in Kulon Progo, which was 257.756 mm. For the wet season, there is an increase in the maximum value of rainfall. However, the changes in value are not as significant as the change in the wet season. The maximum wet season rainfall increased in Sleman by 69.535 mm (Table 2).

Like rainfall, the comparison of the first and second periods of the overall PDSI probability distribution for the seasons in five districts in Yogyakarta, does not show a significant shift. However, in general, it shows the condition of the second period, which is wetter than the first period. This can be seen from the probability value which tends to shift to a more significant value (Table 3).

The general hydrological conditions in Yogyakarta, which are wetter, can also be seen by comparing the maximum PDSI values between the first and second periods. For the wet season, there has been a shift in the maximum PDSI value of around 1.4–2. Although in the dry season, there has been a decrease in the PDSI value, both maximum and minimum, the shift in the PDSI value, is not as significant as the shift in the wet season. Extreme dry conditions (minimum PDSI in the dry season), only shifted around 0.002–0.2. However, an increase in rainfall during the wet season does not mean that there has been a decrease in the level of drought. PDSI data from the first period showed an opportunity for an index of more than four by 21.55% for Bantul, 38.87% for Gunung Kidul, 39.85% for Yogyakarta City, 39.63% for Kulon Progo, and 39.49% for Sleman.

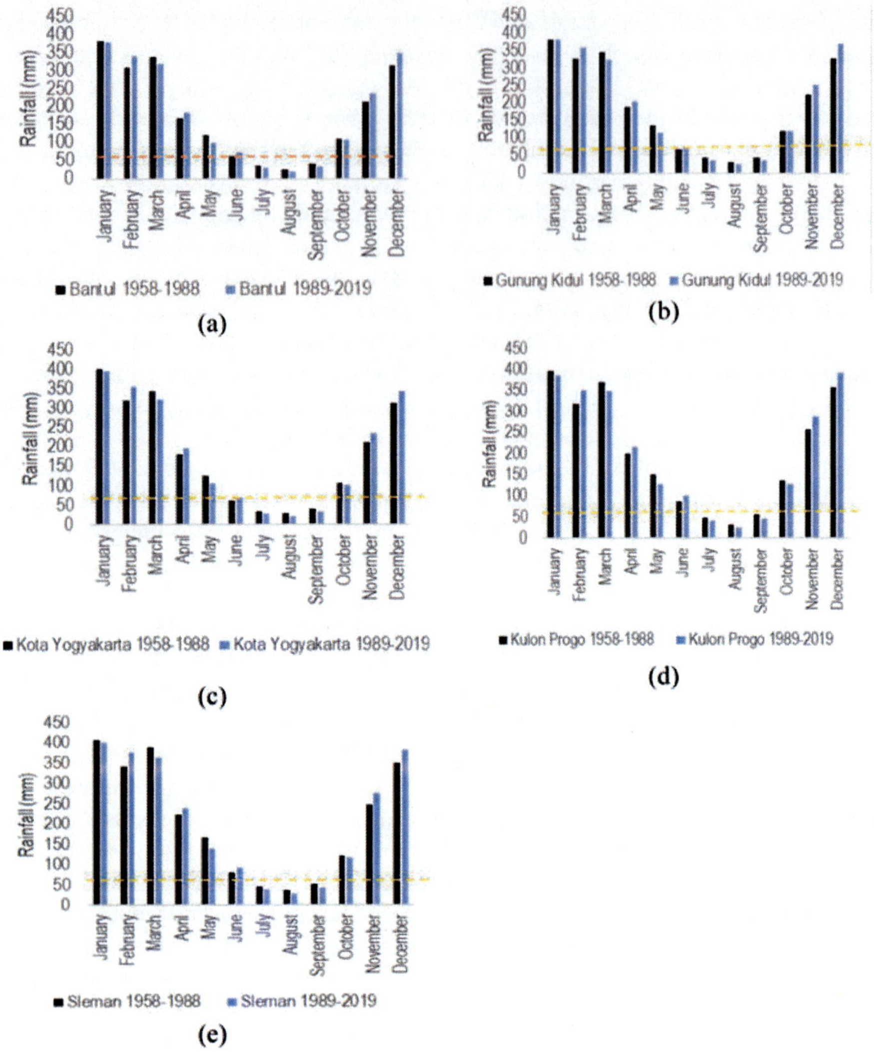

Fig. 1 Graph of rainfall composite for three decades in each district in DI Yogyakarta, **a** Bantul, **b** Gunung Kidul, **c** Yogyakarta City, **d** Kulon Progo, **e** Sleman

4 Conclusion

TerraClimate data monthly temporal and spatial resolution of around 4 km by using composite technique analysis for 60 years monthly. We found the comparison of two periods show that the statistical distribution of all seasons does not show a significant shift in values, especially for the mean and mode. The monsoon rainfall pattern's maximum peak started from December to March, while the minimum peak

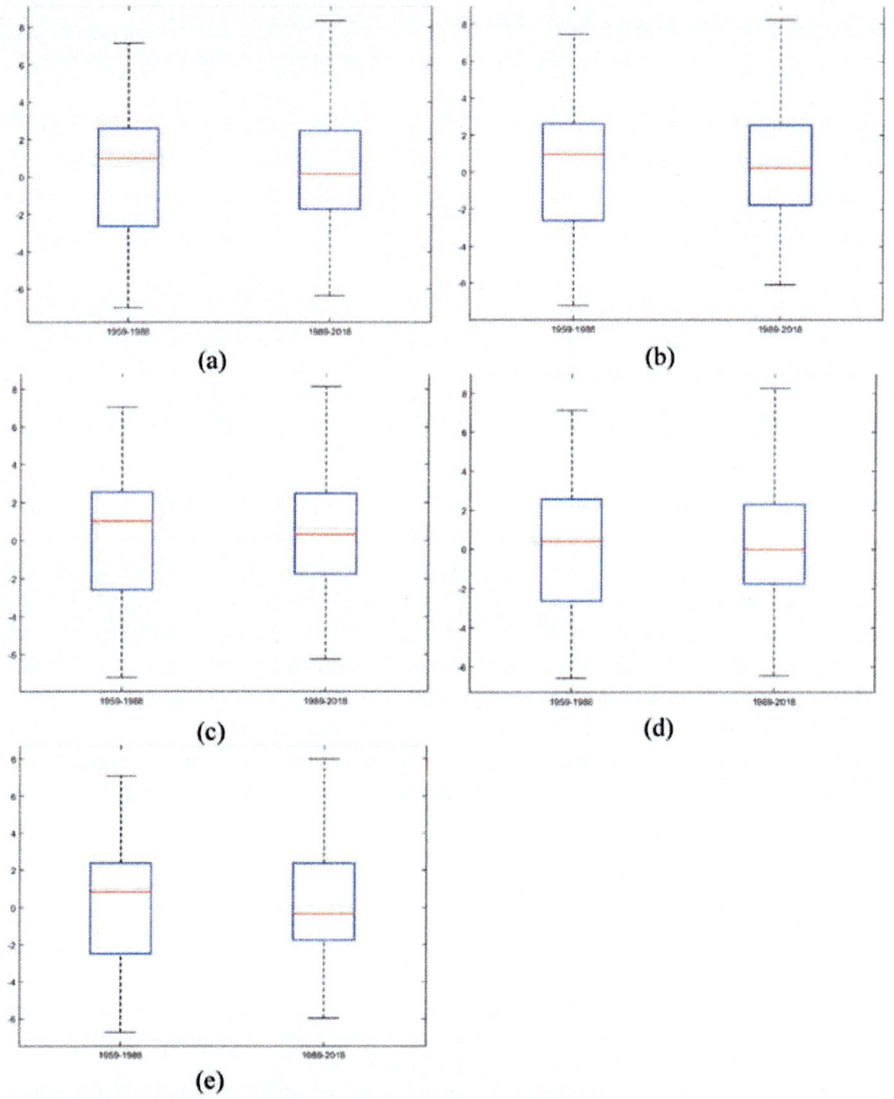

Fig. 2 Boxplot of PDSI in two decades in each district in DI Yogyakarta **a** Bantul, **b** Gunung Kidul, **c** Yogyakarta City, **d** Kulon Progo, **e** Sleman

Table 2 Rainfall accumulation during JJA and DJF season

Maximum rainfall (mm)	Bantul	Gunung Kidul	Yogyakarta City	Kulon Progo	Sleman
JJA 1958–1988	349,228	361,424	340,307	463,154	407,616
JJA 1989–2019	184,864	195,849	182,757	257,756	234,033
DJF 1958–1988	649,210	640,708	674,613	625,723	654,796
DJF 1989–2019	672,731	679,557	701,575	695,258	711,259

Table 3 PDSI values in the dry season JJA) and wet season (DJF)

	Bantul	Gunung Kidul	Kota Yogyakarta	Kulon Progo	Sleman
MIN PDSI					
JJA 1958–1988	−3.85	−3.87	−3.88	−3.89	−3.89
JJA 1989–2019	−4.09	−3.67	−3.88	−4.41	−3.72
DJF 1958–1988	−6.99	−7.21	−7.19	−6.56	−6.73
DJF 1989–2019	−6.36	−6.10	−6.24	−6.43	−5.96
MAX PDSI					
JJA 1958–1988	6.68	6.99	6.56	7.12	6.74
JJA 1989–2019	5.18	5.05	5.03	5.07	5.01
DJF 1958–1988	6.20	6.10	6.23	6.01	6.03
DJF 1989–2019	8.13	8.09	7.82	7.85	7.49

is generally between July and September; those patterns are similar to the other five area studies. By applying the PDSI it found that all study areas have the same pattern it indicating almost the Yogyakarta area potentially has a meteorological drought, especially for the Gunung Kidul area, a list of 50 mm. We found that the comparison of two periods in 30 years shows that the statistical distribution of all seasons does not show a significant shift in values, especially for the mean and mode. In the monsoon rainfall pattern, the maximum peak started from December to March, while the minimum peak is generally between July to September those patterns, are similar it is the other five area studies. By applying the PDSI, it found that all study areas have the same pattern indicating that almost the Yogyakarta area potentially has a meteorological drought, especially for the Gunung Kidul area, a list of 50 mm.

References

1. Abatzoglou, J.T., Dobrowski, S.Z., Parks, S.A., Hegewisch, K.C.: TerraClimate, a high-resolution global dataset of monthly climate and climatic water balance from 1958–2015. Sci Data **5**, 170191 (2018). https://doi.org/10.1038/sdata.2017.191
2. Burke, E.J., Brow, S.J.: Evaluating uncertainties in the projection of future drought. J. Hydrometeorol. **9**, 292–299 (2008). https://doi.org/10.1175/2007JHM929.1
3. Burke, E.J., Brown, S.J., Christidis, N.: Modeling the recent evolution of global drought and projections for the twenty-first century with the Hadley Centre climate model. J. Hydrometeor. **7**, 1113–1125 (2006). https://doi.org/10.1175/JHM544.1
4. Chang, T.J.: Investigation of precipitation droughts by use of Kriging method. J. Irrig. Drain. Eng. **117**(6), 935–943 (1992)
5. Christensen, J.H., Bruce, H., Busuioc, A.: Regional climate projections. In climate change 2007: the physical science basis. Contribution of working group 1 to the fourth assessment report of the intergovernmental panel on climate change (2007)
6. Cook, C.A., Woodhouse, C.A., Eakin, C.M., Meko, D.M., Stahle, D.W.: Long-term aridity changes in the western United States. Science **306**, 1015–1018 (2004). https://doi.org/10.1126/science.1102586.

7. Cook, R., Seager, R., Cane, M.A., Stahle, D.W.: North American drought: reconstructions, causes, and consequences. Earth Sci. Rev. **81**, 93–134 (2007). https://doi.org/10.1016/j.earscirev.2006.12.002
8. Dai, A.: Characteristics and trends in various forms of the Palmer Drought Severity Index (PDSI) during 1900–2008. J. Geophys. Res. **116**, D12115 (2011)
9. Dai, A., Trenberth, K.E., Karl, T.: Global variations in droughts and wet spells: 1900–1995. Geophys. Res. Lett. **25**, 3367–3370 (1998)
10. Dai, A., Trenberth, K.E., Qian, T.: A global data set of Palmer Drought Severity Index for 1870–2002: relationship with soil moisture and effects of surface warming. J. Hydrometeorol. **5**, 1117–1130 (2004)
11. DeGaetano, A.T.: A temporal comparison of drought impacts and responses in the New York City metropolitan area. Clim. Change **42**(3), 539–560 (1999)
12. Dore, M.H.I.: Climate change and changes in global precipitation patterns: what do we know? Environ. Int. **31**(8), 1167–1181 (2005)
13. Eltahir, E.A.B.: Drought frequency analysis of annual rainfall series in central and western Sudan. Hydrol. Sci. J. **37**(3), 185–1992 (1992)
14. Harris I., Jones P.D., Osborn T.J., Lister D.H.: Updated high-resolution grids of monthly climatic observations—The CRU TS3.10 dataset. Int. J. Climatol. **34**, 623–642 (2014)
15. Heim, R.R., Jr.: A review of twentieth-century drought indices used in the United States. Bull. Amer. Meteor. Soc. **83**, 1149–1165 (2002). https://doi.org/10.1175/1520-0477(2002)083%3c1149:AROTDI%3e2.3.CO;2
16. https://climate.nasa.gov/ diakses tanggal 11 April 2022
17. https://climatedataguide.ucar.edu/climate-data/palmer-drought-severity-index-pdsi diakses tanggal 11 April 2022
18. Huang, S.P., Li, Q., Huang, G., Leng, B.H., Ma, L.: The propagation from meteorological to hydrological drought and its potential influence factors. J. Hydrol. **547**, 184–195 (2017)
19. Hulme, M., Sheard, N.: Climate Change Scenarios for Indonesia. Leaflet CRU and WWF. Climatic Research Unit UEA, Norwich, UK (1999)
20. Keyantash, J., Dracup, J.A.: The quantification of drought: An evaluation of drought indices. Bull. Amer. Meteor. Soc. **83**(8), 1167–1180 (2002).
21. Kruger, A.C., Shongwe, S.: Temperature trends in South Africa: 1960–2003. Int. J. Climatol. **24**(15), 1929–1945 (2004)
22. Leng, G., Tang, Q., Rayburn, S.: Climate change impacts on meteorological, agricultural and hydrological droughts in China. Global Planet. Change **126**, 23–34 (2015)
23. Li, Z., Zhou, T., Zhao, X., Huang, K., Gao, S., Wu, H., Luo, H.: Assessments of drought impacts on vegetation in china with the optimal time scales of the climatic drought index. Int. J. Environ. Res. Public Health **12**(7), 7615–7634 (2015). https://doi.org/10.3390/ijerph120707615
24. Malisawa, M.S., Rautenbach, C.D.: Evaluating water scarcity in the Southern African Development Community (SADC) region by using a climate moisture index (CMI) indicator. Water Supply **12**(1), 45–55 (2012)
25. Measey, J., Hewitson, B., Busuioc, A., Chen, A., Gao, X., Held, I., Jones, R., Kolli, R.K., Kwon, W.T., Laprise, R., Magana, L., Menendez, C., Raisanen, J., Rinke, A., Sarr, A. Whetton, P.: Regional climate projections. Climate change 2007: The physical science basis. 847–940 (2007)
26. Measey, M.: Indonesia: a vulnerable country in the face of climate change, 2010: Glob. Majority E-J. **1**(1), 31–45
27. Mishra, A.K., Singh, V.P.: A review of drought concepts. J. Hydrol. **391**(1–2), 202–216 (2017)
28. Naylor, R.L., Battisti, D.S., Vimont, D.J., Falcon, W.P., Burke, M.S.: Assessing risks of climate variability and climate change for Indonesian rice agriculture. Proc. Natl. Acad. Sci. U.S.A. **104**(19), 7752–7757 (2007)
29. Nicholson, S.E., Tucker, C.J., Ba, M.B.: Desertification, drought, and surface vegetation: an example from the West African Sahel. Bull. Am. Meteor. Soc. **79**(5), 815–829 (1998)
30. Palmer, W.C.: Meteorological Drought. Research Paper No. 45, pp. 45–58. US Department of Commerce Weather Bureau, Washington, DC (1965)

31. Sheffield, J., Wood, E.F.: Projected changes in drought occurrence under future global warming from multi-model, multi-scenario, IPCC AR4 simulations. Clim. Dyn. **31**(1), 79–105 (2008)
32. Susandi, A.: Bencana perubahan iklim global dan proyeksi perubahan iklim Indonesia. Kelompok Keahlian Sains Atmosfer Fakultas Ilmu Kebumian dan Teknologi Mineral ITB: Bandung (2006)
33. Thornthwaite, C.W.: An approach toward a rational classification of climate. Geogr. Rev. **38**(1), 55–94 (1948)
34. Van der Schrier, G., Efthymiadis, D., Briffa, K.R., Jone, P.D.: European Alpine moisture variability for 1800–2003. Int. J. Climatol. **27**, 415–427 (2007). https://doi.org/10.1002/joc.1411
35. Van der Schrier, G., Briffa, K.R., Jones, P.D., Osborn, T.J.: Summer moisture variability across Europe. J. Clim. **19**, 2818–2834 (2006). https://doi.org/10.1175/JCLI3734.1
36. Van der Schrier, G., Briffa, K.R., Osborn, T.J., Cook, E.R.: Summer moisture availability across North America. J. Geophys. Res. **111**, D11102 (2006). https://doi.org/10.1029/2005JD006745
37. Vicente-Serrano, S.M., Beguena, S., Lopez-Moreno, J.I., Angulo, M., El Kenawy, A.: A new global 0.5 gridded dataset (1901–2006) of a multiscalar drought index: Comparison with current drought index datasets based on the Palmer drought severity index. J. Hydrometeor. **11**, 1033–1043 (2010)
38. Wang, L., Chen, W.: A CMIP5 multimodel projection of future temperature, precipitation, and climatological drought in China. Int. J. Climatol. **34**(6), 2059–2078 (2014)
39. Warburton Toucher, M., Schulze, R., Maharaj, M.: Is South Africa's temperature changing? An analysis of trends from daily records, 1950-2000, pp. 275-295 (2005)
40. Warburton, T.M., Schulze, R.E.: Potential impacts of climate change on the climatically suitable growth areas of Pinus and Eucalyptus: results from a sensitivity study in South Africa. Southern For.: J. For. Sci. **70**(1), 27–36 (2008)
41. Xu, H.-J.X.-P., Wang, C.-Y., Yang, X.-M.: Diverse responses of vegetation growth to meteorological drought across climate zones and land biomes in northern China from 1981 to 2014. Agric. For. Meteorol. **262**, 1–13 (2018)
42. Yusuf, A., Komarulzaman, A., Hermawan, W., Hartono, D., Sjahrir, K.: Scenarios for Climate Change Mitigation from the Energy Sector in Indonesia: The Role of Fiscal Instruments. Padjadjaran University, Department of Economics (2010)

Foraminiferal Paleoproductivity During Holocene in Sunda Strait

Eldian Yosua Budiarto, Marfasran Hendrizan, Rachmad Setijadi, Rainer Arief Troa, Rina Zuraida, and Eko Triarso

Abstract Sunda Strait is a part of southern Java Sea influenced by the South Java Current (SJC) and the Java Sea current system. Java Sea current system provides fresh water run-off from Sumatra and Java and decreases salinity which can influence foraminiferal productivity. This research aims to understand the foraminiferal productivity record influenced by the water mass condition during Holocene. This research is conducted based on spectrophotometry and magnetic susceptibility measurement to unravel the physical properties, and foraminiferal distribution to unravel the ecological properties. The results indicate that paleoproductivity in the study area can be divided into three facies including salinity, sea surface temperature, and thermocline depth, driven by variability in upwelling, run-off intensity, and monsoonal changes. Facies I shows a relatively high abundance of *Globigerina bulloides*, indicating that paleoproductivity is influenced by upwelling and southeast monsoon, followed by dry paleoecology. Facies II shows a relatively shallow thermocline reflected by the extinction of *Globorotalia crassaformis* and decreasing *Globorotalia hirsuta*. In this facies, paleoproductivity is influenced by run-off that increases terrigenous influx controlled by the northwest monsoon. Facies III shows decreasing *Globigerinoides* as an oligotrophic dweller, and this indicates that the water has become eutrophic caused by increasing run-off. The northwest monsoon influences this increasing run-off.

E. Y. Budiarto (✉)
Faculty of Earth Sciences and Technology-Institut Teknologi Bandung, Bandung 40132, Indonesia
e-mail: eldianyosuab@gmail.com

M. Hendrizan · R. Zuraida
Research Center for Climate and Atmosphere-National Research and Innovation Agency, Bandung 40135, Indonesia

R. Setijadi
Department of Geological Engineering, Universitas Jenderal Soedirman, Purbalingga 53371, Indonesia

R. A. Troa · E. Triarso
Ministry of Marine Affairs and Fisheries, Jakarta 10110, Indonesia

A. Basit et al. (eds.), *Proceedings of the International Conference on Radioscience, Equatorial Atmospheric Science and Environment and Humanosphere Science*, Springer Proceedings in Physics 290, https://doi.org/10.1007/978-981-19-9768-6_69

1 Introduction

Sunda Strait is a part of the Sunda Shelf. During late Pleistocene, the sea level was about 130 m lower than that of today, and Sunda Shelf was exposed [1]. During Holocene, Sunda Shelf was flooded due to sea level rise, opening Sunda Strait and connecting it with the Java Sea [1, 2]. Today, Sunda Strait's regional water is strongly influenced by South Java Current (SJC) [3]. The SJC-influenced water is characterized by changes in current and salinity. Near the surface (610 m), SJC's component changes its direction every 3 months, triggered by monsoonal changes [4]. The peak of easternward wind occurs in the end of northwest monsoon on May, and its transition to southeast monsoon occurs on November. This transition decreases salinity down to 32%. Northwest monsoon is characterized by the flow of warm and humid air from eastern Asia that increases rainfall in western part of Indonesia. In contrast, southeast monsoon is characterized by the flow of cold and dry air from Australia that causes upwelling along southern coast of Indonesia [5] which increases productivity in sea water [6].

Sunda Strait opening increased terrigenous input that might influence the Indian Ocean's productivity during Holocenece [7]. This increasing terrigenous input was influenced by freshwater run-off from Sumatra and Java [4]. As the result, it decreased salinity that might influence foraminiferal productivity. Previous researches in this study area concluded that sedimentation rate was around 19.1 cm kyr^{-1}, with sand content of 2.8%, silt content of 74.1%, and clay content of 23.1%. Clay minerals consist of 42.6% smectite, 25.2% kaolinite, and 18.9% illite. SiO_2 content showed declining trend upward and so is Fe_2O_3 content, in contrary, MgO and Na_2O showed inclining trend upward. Paleoproductivity was dominated by southeast monsoon and upwelling in the late Pleistocene, but then mainly controlled by the terrigenous materials input during Holocene [8].

This research is conducted on core BS-05 recovered from the southeastern part of Sunda Strait water. This research aims to understand the foraminiferal paleoproductivity record during Holocene. Foraminiferal records are used to identify paleoecological changes and their factors.

2 Study Area

The study area located at 7°1′38.83″N and 105°6′16.31″E (Fig. 1). The research is a part of BENTHIC cruise research conducted by the Ministry of Marine Affairs and Fisheries Republic of Indonesia and the First Institute of Oceanography Ministry of Natural Resources of the People's Republic of China, in collaboration with the former of National Research and Innovation Agency of Indonesia (Indonesian Institute of Sciences), and the Ministry of Energy and Mineral Resources Republic of Indonesia. This research took place on December 10–29th 2014. Study area is geographically located around 17 km to south from Java island, reached by research vessel Geomarin III.

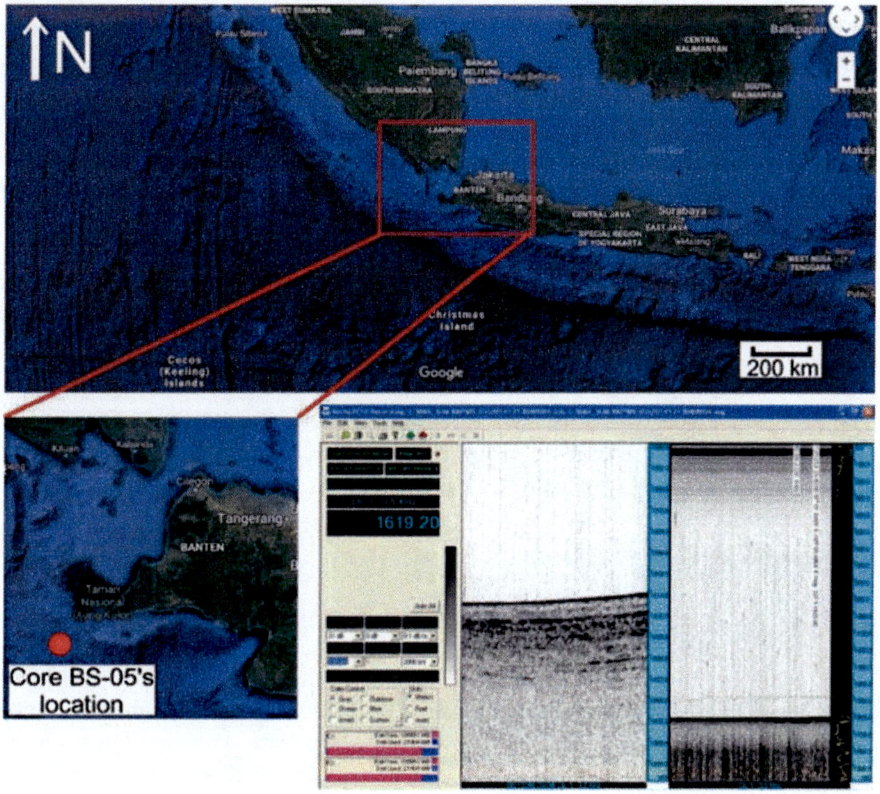

Fig. 1 Map of the study area [11] and subbottom profile of core BS-05 [10]

3 Analytical Methods

This research begins with the core recovery using the gravity core method. This research uses samples from the top 100 cm depths of the core. Several methods are used to detect changes in their physical and ecological attributes including spectrophotometry, magnetic susceptibility analysis, and foraminiferal analysis. Several pre-research studies about regional geology and previous researches are also conducted to gather more information about this study area.

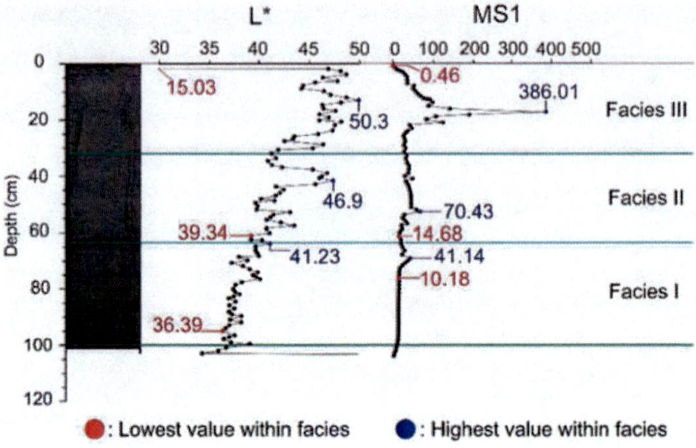

Fig. 2 Physical properties chart of core BS-05

4 Results and Discussion

4.1 Geomorphology

From its geomorphological aspects, this area is characterized by heights and steep slopes. This study area can be divided into three categories: volcanic remain hills, lava dome hills, and alluvial plains [9]. Sediment was deposited in Holocene-Recent, making it equivalent to alluvial unit that consists of finely-grained sediment (silt to clay). The subbottom profile acquisitioned by the BENTHIC research team also shows that this core is deposited in deep water with sloping morphology [10] (Fig. 1).

4.2 The Core's Physical Attributes

The core BS-05's physical attributes are observed using Multi-Sensor Core Logger (MSCL) device. The results consist of spectrofotometry's color lightness chart (L*) and magnetic susceptibility chart (MS1) which can be divided into 3 facies based on their values from older to younger strata (Fig. 2).

4.3 Spectrophotometry Analysis

This analysis is used to quantify the sediment's lightness (L*). The result can be divided into three facies, I, II, and III (Fig. 2).

Facies I

Facies I consists of grayish silt from 64 to 100 cm core depth. Facies I has a mean L* of 38.19. The lowest L* value of 36.39 is at 95 cm depth, while highest value of 41.23 is at 64 cm depth (Fig. 2). Among three facies, this facies has the lowest mean L* value, indicating the darkest color due to its relatively high clay mineral content.

Facies II

Facies II consists of grayish clay from 32 to 63 cm core depth. Facies II has mean L* of 42.17. The lowest L* value of 39.34 is at 61 cm depth, while the highest value of 46.9 is at 39 cm depth (Fig. 2). This facies has a higher mean L* than facies I, indicating its relatively brighter color. This brighter color suggests an increase of carbonate material in this facies.

Facies III

Facies III consists of grayish clay from 1 to 31 cm core depth. Facies III has a mean L* of 45.46. The lowest L* value of 16.03 is there at 1 cm depth, while the highest value of 50.3 is at 13 cm depth (Fig. 2). Among three facies, this facies has the highest mean L* value, indicating the lightest color due to its relatively high carbonate content.

4.4 Magnetic Susceptibility

Magnetic susceptibility is a measurement of metal-carrying components within sediments and can be used to identify sediment type and its metallic content. The higher the MS value the higher the metallic content within the sediment. Based on this analysis, core BS-05 can be divided into three facies which are facies I, II, and III (Fig. 2). Hereupon, magnetic susceptibility measurement results are pronounced as MS1.

Facies I

Facies I has a mean MS1 of 15.37. The lowest MS1 of 10.18 is at 76 cm depth, while the highest MS1 of 41.14 is at 69 cm depth (Fig. 2). Among three facies, this facies has the lowest mean MS, indicating that this facies has the lowest magnetic property. This lowest magnetic property reflects its relatively low metallic content.

Facies II

Facies II has a mean MS1 of 33.68. The lowest MS1 of 14.68 is at 61 cm depth, while the highest MS1 of 70.43 is at 53 cm depth (Fig. 2). This facies has higher a mean MS1 than facies I, indicating its relatively higher magnetic property. This higher magnetic property suggests an increase in metallic content in this facies.

Facies III

Facies III has a mean MS1 of 66.02. The lowest MS1 of 0.46 is at 1 cm depth, while the highest MS1 of 386.01 is at 17 cm depth (Fig. 2). Among three facies, this facies has the highest mean MS, indicating that this facies has the highest magnetic property. This highest magnetic property reflects its relatively high metallic content.

4.5 Foraminifera Analysis

Foraminifera analysis conducted on core BS-05 shows the planktonic and benthic foraminifera distribution. There are 17 species of planktonic foraminifera found in the samples such as *Neogloboquadrina dutertrei*, *Pulleniatina obliquiloculata*, *Pulleniatina finalis*, *Globorotalia crassaformis*, *Globorotalia hirsuta*, *Globorotalia menardii*, *Globorotalia tumida*, *Globigerinoides sacculifer*, *Globigerinella calida*, *Globorotalia ungulata*, *Globigerinoides trilobus*, *Globigerina bulloides*, *Hastigerina siphonifera*, *Globigerinoides ruber*, *Globigerinoides immaturus*, *Globigerinella praecalida*, *Globigerinoides elongatus*. There are 3 species of benthic foraminifera such as *Uvigerina peregrina*, *Bulimina marginata*, and *Cibicidoides wuellerstrofi*, which are not dominant in distribution.

Based on foraminifera analysis, paleocological changes in sediment depths can be divided into three facies, facies I (64–100 cm), facies II (32–64 cm), and facies III (0–32 cm) (Fig. 3). These facies are divided based on significant changes in specific species distribution which indicates variability in ecological factors (including thermocline depth, upwelling intensity, run-off, salinity, monsoonal changes, nutrient content, and terrigenous input) (Figs. 3 and 4).

Facies I

This facies is characterized by peak abundance of *Neogloboquadrina dutertrei* (48.29%), *Globorotalia menardii* (53.92%), *Globorotalia crassaformis* (100%), *Globigerina bulloides* (56.86%), *Globigerinoides ruber* (72.57%), *Globigerinella calida* (52.94%), *Globorotalia hirsuta* (70.15%), *Hastigerina siphonifera* (48.15%), and *Globigerinella praecalida* (85.29%) (Table 1). In this facies, thermocline depth is relatively deep, characterized by its high abundance of *Globorotalia crassaformis* [12]. Paleoproductivity is influenced by upwelling, characterized by *Globigerina bulloides* [13–15], *Globorotalia menardii* [16], *Neogloboquadrina dutertrei* [16, 17], and *Globorotalia hirsuta* [18]. The southeast monsoon controls the upwelling. Salinity is low, then increasing over time, it is characterized by the declining number of *Neogloboquadrina dutertrei*, because this species is typically abundant in warm water with lower salinity [19]. Because of the high intensity of upwelling, paleoecology becomes arid and colder and controlled by the southeast monsoon.

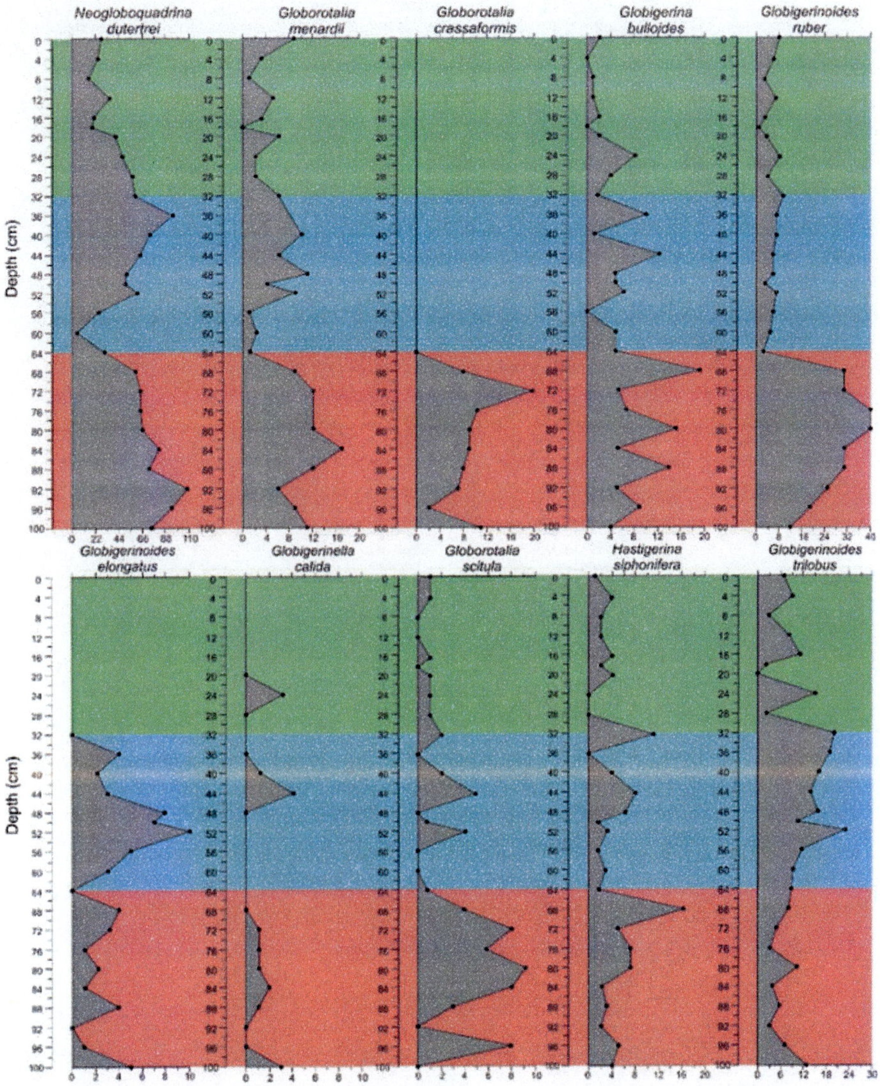

Fig. 3 Distribution charts of planktonic foraminifera of core BS-05 for ten species of *Neogloboquadrina dutertrei, Globorotalia menardii, Globorotalia crassaformis, Globigerina bulloides, Globigerinoides ruber, Globigerinoides elongatus, Globigerinella calida, Globorotalia hirsuta, Hastigerina siphonifera,* and *Globigerinoides trilobus. Red background indicates facies I, blue indicates facies II, and green indicates facies III*

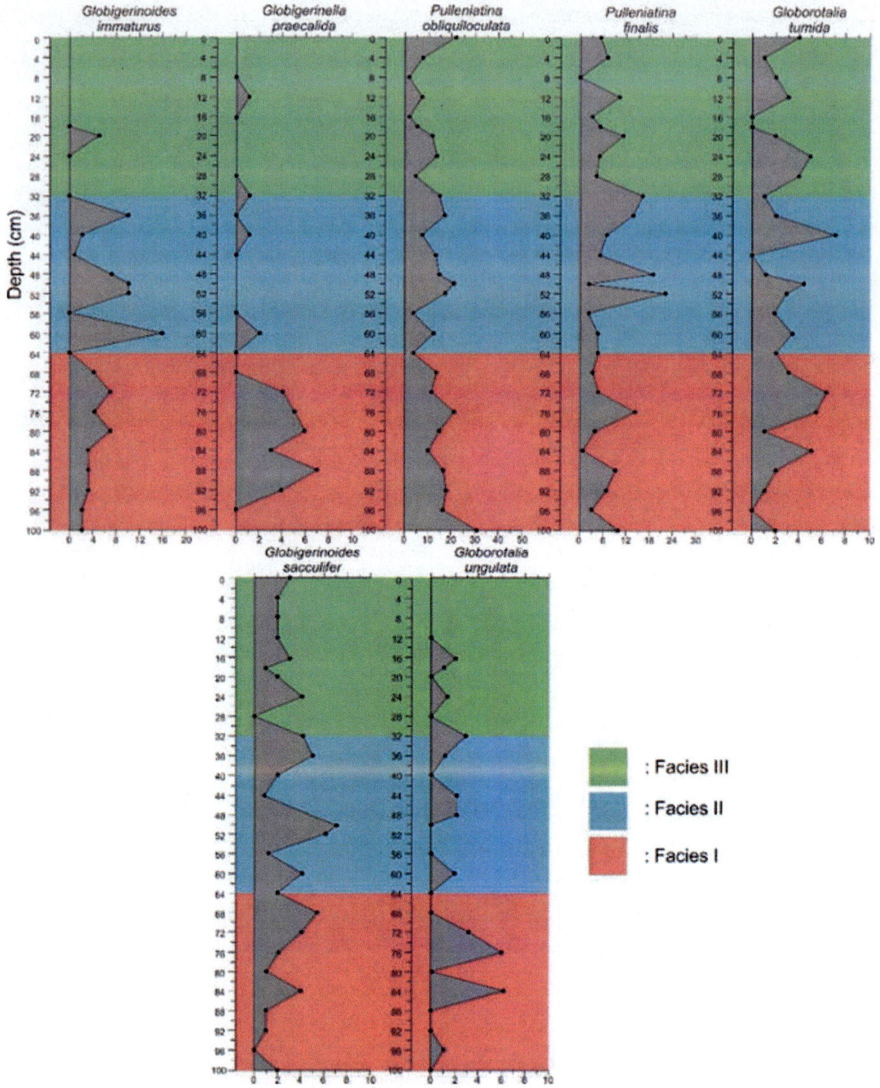

Fig. 4 Distribution charts of planktonic foraminifera of core BS-05 for seven species of *Globigerinoides immaturus*, *Globigerinella praecalida*, *Pulleniatina obliquiloculata*, *Pulleniatina finalis*, *Globorotalia tumida*, *Globigerinoides sacculifer*, and *Globorotalia ungulata*

Facies II

This facies is characterized by peak abundance of *Globigerinoides elongatus* (66.67%), *Globigerinoides trilobus* (47.49%), *Globigerinoides immaturus* (58.33%), *Pulleniatina finalis* (45.5%), and *Globigerinoides sacculifer* (45.45%) (Table 1). The result shows an extinction of one species *Globorotalia crassaformis*. *Globigerinella*

Table 1 The planktonic foraminiferal abundance of all facies, in percentage

Species	Percentage in Facies I (%)	Percentage in Facies II (%)	Percentage in Facies III (%)
Neogloboquadrina dutertreii	48.29	32.69	19.01
Globorotalia menardii	53.92	29.84	16.23
Globorotalia crassaformis	100	0	0
Globigerina bulloides	56.86	30.06	13.07
Globigerinoides ruber	72.57	15.51	11.91
Globigerinoides elongatus	33.33	66.67	0
Globigerinella calida	52.94	29.41	17.65
Globorotalia hirsuta	70.15	20.89	8.95
Hastigerina siphonifera	48.15	34.26	17.59
Globigerinoides trilobus	24.75	47.49	27.76
Globigerinoides immaturus	36.46	58.33	5.21
Globigerinella praecalida	85.29	11.76	2.94
Pulleniatina obliquiloculata	43.5	35.03	21.47
Pulleniatina finalis	29	45.5	25.5
Globorotalia tumida	38.03	32.39	29.58
Globigerinoides sacculifer	28.57	45.45	25.97
Globorotalia ungulata	54.84	32.26	12.9

calida once disappears at 68 cm depth (facies I), then reappears at 44 cm depth, then disappears again at 36 cm depth. In this facies, the thermocline is shallow, reflected by the extinction of *Globorotalia crassaformis* [12] and declining of *Globorotalia hirsuta* and *Globigerinella calida* [18]. Shallowing thermocline is caused by fresh water run-off. Sea surface temperature is declining reflected by increasing *Neogloboquadrina dutertrei* (within facies II) and increasing abundance of *Globigerinoides elongatus* compared to that in facies I. Upwelling intensity is decreasing, reflected by decreasing abundance of *Globigerina bulloides* [13–15], *Globorotalia menardii* [16], *Neogloboquadrina dutertrei* [16, 17], and *Globorotalia hirsuta* [18] compared to those of facies I. In this facies, paleoproductivity is influenced by northwest monsoon that increases terrigenous input, rather than upwelling. Previous research based on mean grain size analysis also indicates that increasing terrigenous input into Sunda Strait caused paleoproductivity to significantly increased during the late Holocene [8].

Facies III

This facies is characterized by decreasing abundance of various species, resulting in fewer individuals than the two other facies (Table 1). The result indicates an extinction of *Globigerinoides elongatus* followed by disappearance of *Globigerinella calida* at 28 cm depth, this species reappears at 24 cm depth, then disappears again at 20 cm depth upward. *Globigerinoides immaturus* disappears in 32 cm depth, reappears in 20 cm depth, and then disappears again at 8 cm depth upward. This facies is characterized by shallow thermocline, reflected by the absence of *Globorotalia crassaformis*. Other oligotrophic-dweller species like *Globigerinoides elongatus* also disappear. In this facies, *Globigerinoides* species which live in oligotrophic water are decreasing in abundance, indicating eutrophic water [20]. This implies that in this facies, water mass has excessive concentration of terrigenous material caused by run-off from increasing rainfall. The northwest monsoon causes increasing rainfall. Increasing run-off is also reflected by increasing MS1 within this facies (Fig. 2). This indicates increasing metallic content from terrestrial depositional processes in Holocene [7].

5 Conclusion

From the results and discussion, we can conclude that:

- From its geomorphological aspects, this area can be divided into 3 categories: volcanic remain hills, lava dome hills, and alluvial plains, with sediment deposited in sloping morphology.
- Core BS-05 is physically characterized by lighter color upward (from facies I to III) that reflects increasing carbonate material. Magnetic susceptibility value shows high number in facies III, indicating the increase of metallic content.
- There are 17 species of planktonic foraminifera and three species of benthic foraminifera found in the samples.
- Palaeoclimatology in this study area can be divided into three facies. In facies I, the thermocline is deeper, and paleoproductivity is influenced by upwelling, and increasing salinity, induced by arid climate. The southeast monsoon controls this arid climate. In facies II, the thermocline is shallower than facies I, followed by decreasing salinity, induced by run-off. Increasing run-off is controlled by northwest monsoon. In facies III, thermocline is shallow, followed by an excessive concentration of terrigenous material carried in by run-off. The northwest monsoon controls this increasing run-off.

Acknowledgements We would like to thank the Ministry of Marine Affairs and Fisheries Republic of Indonesia, the First Institute of Oceanography Ministry of Natural Resources of the People's Republic of China, the National Research and Innovation Agency of Indonesia, and the Ministry of Energy and Mineral Resources Republic of Indonesia for permitting us to use the core BS-05 for primary data of this study. We would also like to thank the R/V "Geomarin III" crew for their help with board sampling.

Author Contribution All authors are the main contributors to the article. EYB and MH conceived and wrote the initial draft of the paper; MH, RS, RAT, RZ, and EY discussed the results and modified the manuscript.

References

1. Hanebuth, T.J.J., Stattegger, K., Grootes, P.M.: Rapid flooding of the Sunda Shelf: a late-glacial sea-level record. Science **288**, 1033–1035 (2000)
2. Hanebuth, T.J.J., Voris, H., Yokoyama, K., Saito, Y.Y., Okuno, J.: Formation and fate of sedimentary depocentres on Southeast Asia's Sunda Shelf over the past sea-level cycle and biogeographic implications. Earth Sci. Rev. **104**, 92–110 (2011)
3. Gingele, F.X., Deckker, P.D., Hillenbrand, C.D.: Clay mineral distribution in surface sediments between Indonesia and NW Australia source and transport by ocean currents. Mar. Geol. **179**, 135–146 (2001)
4. Wijffels, S.E., Hautala, S., Meyers, G., Morawitz, W.: The WOCE Indonesian through flow repeat hydrography sections: I10 and IR6. Int. WOCE Newsl. **24**, 25–28 (1996)
5. Qu, T., Du, Y., Strachan, J., Myers, G., Slingo, J.: Sea surface temperature and its variability in Indonesian region. Journal Oceanography **18**, 50–61 (2005)
6. Webster, P.J., Moore, A.M., Loschnigg, J.P., Leben, R.R.: Coupled ocean-atmosphere dynamics in the Indian Ocean during 1997–98. Nature **401**, 356–360 (1999)
7. Setiawan, R.Y., Mohtadi, M., Southon, J., Groeneveld, J., Steinke, S., Hebbeln, D.: The consequences of opening the Sunda Strait on the hydrography of the eastern tropical Indian Ocean. Paleoceanography **30**, 1358–1372 (2015)
8. Xu, Y., Wang, L., Yin, X., Ye, X., Li, D., Liu, S., Shi, X., Troa, R.A., Zunaida, R., Triarso, E., Hendrizan, M.: The influence of the Sunda Strait opening on paleoenvironmental changes in the eastern Indian Ocean. J. Asian Earth Sci. **146**, 402–411 (2017)
9. Brahmantyo, B., Bandono: Klasifikasi bentuk muka bumi (landform) untuk pemetaan geomorfologi pada skala 1:25.000 dan aplikasinya untuk penataan ruang. Jurnal Geoaplika **1**(2), 71–78 (2006)
10. Troa, R.A., Zuraida, R., Hendrizan, M., Triarso, E., Liu., S., and all of BENTHIC Cruise participants: Cruise Report BENTHIC Phase I. RV. Geomarin III: Cirebon–Indian Ocean Off Western Sumatra–Cirebon, 1–29 Desember 2014. P3SDLP GEODEEP, Jakarta (2014)
11. Google Earth.: https://earth.google.com/web/. Last Accessed 12 Dec 2017
12. Huang, B.Q., Jian, Z.M., Cheng, X.R., Wang, P.X.: Foraminiferal responses to upwelling variations in the South China Sea over the last 220000 years. Mar. Micropaleontol. **47**, 1–15 (2002)
13. Bé, A.W.H.: An ecological, zoogeographic and taxonomic review of recent planktonic foraminifera. Oceanic Micropaleontol. **1**, 1–100 (1977)
14. Duplessy, J.C., Be´, A.W.H., Blanc, P.L.: Oxygen and carbon isotopic composition and biogeographic distribution of planktonic foraminifera in the Indian Ocean. Palaeogeography Palaeoclimatol. Palaeoecol. **33**, 9–46 (1981)
15. Brock, J.C., McClain, C.R., Anderson, D.M., Prell, W.L., Hay, W.W.: Southwest monsoon circulation and environments of recent planktonic foraminifera in the northwestern Arabian Sea. Paleoceanography **7**, 799–813 (1992)
16. Fairbanks, R.G., Sverdlove, M., Free, R., Wiebe, P.H., Be´, A.W.H.: Vertical distribution and isotopic fractionation of living planktonic foraminifera from the Panama Basin. Nature **298**, 841–844 (1982)
17. Curry, W.B., Thunell, R.C., Honjo, S.: Seasonal changes in the isotopic composition of planktonic foraminifera collected in Panama Basin sediment traps. Earth Planet. Sci. Lett. **64**, 33–43 (1983)
18. Shackleton, N.J., Vincent, E.: Oxygen and carbon isotope studies in recent foraminifera from the Southwest Indian Ocean. Mar. Micropaleontol. **3**, 1–13 (1978)
19. Thompson, P.R.: Planktonic foraminifera in the western north Pacific during the past 150,000 years: comparison of modern and fossil assemblages. Palaeogeogr. Palaeoclimatol. Palaeoecol. **35**, 241–279 (1981)
20. Seears, H.A., Darling, K.F., Wade, C.M.: Ecological partitioning and diversity in tropical planktonic foraminifera. BMC Evol. Biol. **12**, 1–15 (2012)

Microplastic Formation from Weathered Single-Use Plastic Straw in Panjang Island Beach, Banten Bay: Preliminary Result

Dwi Amanda Utami, Sri Yudawati Cahyarini, Ayu Utami Nurhidayati, Tubagus Solihuddin, and Marfasran Hendrizan

Abstract Plastic has been extensively produced and used in life due to its practicality. Plastic waste due to mismanagement is found in the environment and has become a global threat. Single-use plastic straws of various colors were observed stranded at the beach of Panjang Island, Banten Bay, Indonesia, showing signs of weathering. Plastic weathering due to photooxidation caused by UV irradiation, photothermal oxidation due to heat, and mechanical abrasion due to mechanic stress are the main process that will cause plastic to become brittle and break easily because it loses its elasticity, likely to produce microplastics. Wind is the most influential mechanism for horizontal transport of microplastic from the beach to the water surface. Weather events such as storms or high tides are likely to exchange plastic between the existing compartments. In area where monsoon occurs, such as at the tropical equator, high precipitation can cause a large flow of plastics from land to different marine environments. This work served as a preliminary result of a study to understand the process of plastic fragmentation into microplastic on a tropical beach environment.

1 Introduction

Plastics have been widely produced and used in life due to their durability, lightweight, low cost, strong plasticity and flexibility, and good insulation. In 2018, 360 million tons of plastic were produced globally [1]. Plastic waste has been found in freshwater [2], oceans [3], beaches [4, 5], and even Arctic polar water [6]. It is estimated that 8 million tons of plastic waste enter the world's oceans annually [7]. The input of plastic waste into the marine environment largely comes from the mismanagement of waste from industrial processes such as shipping and fish waste or runoff from rivers from municipal drainage systems [7].

D. A. Utami (✉) · S. Y. Cahyarini · A. U. Nurhidayati · T. Solihuddin · M. Hendrizan
Research Center of Climate and Atmosphere, Research Group of Paleoclimate and Paleoenvironment, National Research and Innovation Agency, Jakarta, Indonesia
e-mail: dwi.amanda.utami@brin.go.id

A. Basit et al. (eds.), *Proceedings of the International Conference on Radioscience, Equatorial Atmospheric Science and Environment and Humanosphere Science*, Springer Proceedings in Physics 290, https://doi.org/10.1007/978-981-19-9768-6_70

In the environment, plastics undergo extensive aging process such as light irradiation, wind erosion, and water abrasion. The physical, chemical, and biological process in the environment can slowly break down plastic waste, thus generating numerous smaller plastic debris [1]. Small fragments of plastic, known as microplastic (100 nm–5 mm), presently ubiquitous. Microplastic can enter the environment through two processes. The first process is the result of micro-sized plastic entering the environment, both from primary sources of microplastics such as microbeads or secondary sources like synthetic fibers from washing clothes. The second process of microplastic entering the environment is breakdown of larger plastic.

Microplastics have been found in the digestive organs of fish [8] as well as deposit and filter-feeding invertebrates such as crustaceans [9] and bivalves [10]. Laboratory experiments have shown that microplastic and related chemicals can cause adverse biological impacts on organisms [11]. Long-term exposure and adhesion of microplastics may have negatively effect on corals, such as bleaching and necrosis [12]. In addition, microplastics can increase pathogen transmission, which further increases the susceptibility of reef-building corals to disease [13, 14]. Microplastic can be transferred in the food web through ingestion. If ingested, microplastics can cause physical harm to organisms by blocking the gastrointestinal tract or by leaching of potentially toxic substances.

Studies have shown that the photooxidation process and mechanical abrasion can reduce the size of plastic fragments [15]. Photooxidation, photothermal oxidation, mechanical abrasion, hydrolysis, and biodegradation are the common process to weathered plastic polymer [16]. Nearly all synthetic polymers encounter photochemical reactions to absorb ultraviolet (UV) radiation that induces oxidation which causes plastic to brittle and broke easily because it has loses its elasticity [15]. Plastic surfaces can become weak due to exposure to UV radiation and oxidation, which is likely to produce microplastics through the friction of wind, waves, and sand. Other degradative mechanisms induced by the weathered process can cause small and large plastic that floats in the ocean or washes up on beaches to fragment into microplastics.

Photooxidation degradation begins when plastic polymers are affected by solar radiation (especially UV-B) [17]. Photooxidative and photothermal oxidative degradation may have a greater effect on coastal plastics than on the sea surface and seabed due to the higher availability of UV radiation and oxygen in combination with temperature [18]. In addition, the mechanical abrasion of plastic by sand may be greater at the beach due to the action of wind, waves, and tidal currents [15].

This work provides evidence of single-use plastic straw degradation into smaller plastic fragments in a tropical rubble coral beach. Plastic straws are the fifth largest category of items collected in marine waste during 2019 global coastal cleanup [19]. We aim to understand the process that governs the formation of microplastics from plastic straws by conducting field observation and microscopic examination.

2 Study Area

Field observation for this study was conducted in Panjang Island in the Puloampel district, Serang Regency, Banten, coordinated at 6° 25′ 18″ N–6° 28′ 12″ N and 106° 22′ 9″ E–106° 25′ 36″ E (Fig. 1). Panjang Island is approximately 8.2 km2 wide. This island is located north offshore of Banten Bay. The field observation was conducted in September 2022, during a field survey on the northeast beach of Panjang Island. The beach is a rubble beach that is mainly composed of branching coral rubble (Fig. 2). Field observation was following a transect ca. About 2–3 m above the natural high tide strandline. All samples were visually identified, photograph, carefully handled, and stored in Ziplock bags for further analysis. In the laboratory, the surface of the samples was observed and photograph using Olympus SZX16 microscope.

The shallow Banten Bay has an average depth of 7 m, characterized by muddy and sandy beaches composed of mud, silt, and sand [20]. The ocean current in Banten Bay is approximately 35 m/s, directed eastward during West Monsoon and westward during East Monsoon [21–23]. Due to its position, Panjang Island is protected from strong currents, high waves, and storms. Diverse benthic communities, including coral, seagrass, and mangrove, occur on Panjang Island [24]. A large portion of Panjang Island is covered with seagrass [24]. It is common to find mangroves along the coast where the substrate is muddy [25]. There are coral reefs on Panjang Island's north coast, but they are in relatively poor condition [25]. Similarly to other beaches in

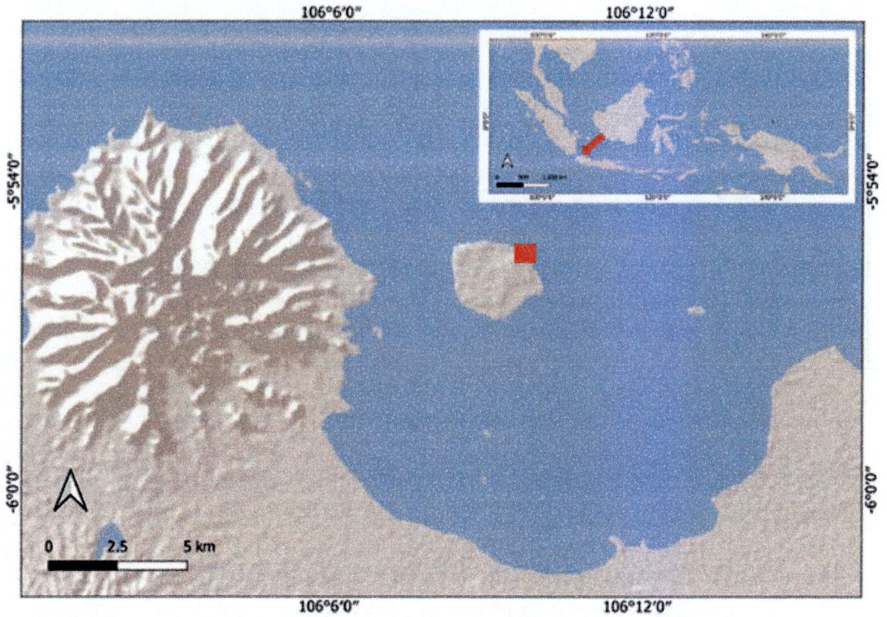

Fig. 1 Study area, located at the northeast beach of Panjang Island, Banten Bay, Indonesia as indicated in the red square

Fig. 2 Single-use plastic straws of various colors were stranded at the beach of Panjang Island, Banten Bay, showing signs of weathering and partly degraded into smaller plastic fragments

the Java Sea, Banten Bay's beaches contain coral and shell fragments of marine life, resembling coral reef ecosystems [26, 27]. A major concern for environmental pollution around Banten Bay is plastic litter, including larger plastic such as water bottles, food packaging, plastic packaging, cups/lids, plastic straw, and also microplastic [28, 29]. Hydrodynamic activity controls marine debris movement in the semi-enclosed Banten Bay [28].

3 Result and Discussion

The waters around Panjang Island, Banten Bay, have been shown to receive numerous waste originating from anthropogenic activities [28]. Disposed waste, waste from tourism, waste from shipping, and waste from fishery industries are included in the sources of anthropogenic waste that polluted Banten Bay. The most commonly found plastic waste in rivers is packaging and consumer products, while in the marine environment it is known to be dominated by fishery items [30]. Not only larger plastic such as polyethylene terephthalate bottles, polyethylene packaging, plastic packaging, and cups/lids, microplastics are also found in the sea of Banten Bay [28, 29].

Of the many types of plastic waste observed on Panjang Island, plastic straws of various colors are often found stranded on the beach (Fig. 2). Single-use plastic straws are usually made of polypropylene [31]. Among its main applications are the manufacture of plastic straw products that require hardness, flexibility, heat resistance, and low density [32, 33]. Plastic straws and stirrers were the fifth largest category of marine debris collected during the International Coastal Cleanup worldwide in 2019, reaching nearly 1 million [19]. In Indonesia, plastic straws are the third most common marine debris collected during the Coastal Cleanup, while number one is cigarette butts [19].

Plastic straws stranded at the beach of Panjang Island mostly show signs of weathering and are partly degraded into smaller plastic fragments (Fig. 2). Polymer weathering is commonly caused by photooxidation, photothermal oxidation, mechanical abrasion, hydrolysis, and biodegradation [16]. Although plastic materials are generally resistant to degradation, the very slow environmental weathering process will cause plastic breakdown and induces changes in their polymer properties [34]. Photodegradation is recognized as the most important process for initiating plastic degradation, being the most responsible are high-energy ultraviolet (UV) irradiation UV-B (290–315 μm) and medium energy UV-A (315–400 nm) [35]. The higher availability of UV radiation and oxygen in combination with higher temperature at the beach may exert a greater effect on photooxidative and photothermal oxidative plastic degradation process than on the sea surface and seabed [18, 36].

Polypropylene (PP) is a less stable polymer than polyethylene (PE) due to the presence of tertiary carbon, which is more susceptible to reaction with oxygen [34]. However, a lack of existing impurities or structural defects in polymers during manufacture or weathering can absorb UV radiation and generate free radicals [34]. Photodegraded plastics on the surface or in the photic zone of seawater can result in weight loss, changes in appearance and texture, deterioration in mechanical properties, signs of oxidation, and changes of physicochemical properties [37, 38]. Analysis of the surfaces of microplastic particles revealed signs of degradation and abrasion, indicating process of mechanical weathering (Fig. 3) [39]. Under accelerated weathering conditions, the formation of nano plastics increased over time during the degradation [40].

Photooxidized plastics on the beach can be subjected to various mechanical forces. Wind, tidal currents, and waves can cause friction with air, water, and sand on the beach. The influence of wind on plastics' horizontal transport mechanism is most influential on buoyant objects that extend above the water surface. During storms or oceanic high-tide events, plastics are more likely to be exchanged between the existing compartments, where high-wind-strength plastics are more affected than low-wind-strength plastic. Events with high precipitation, such as monsoons, can

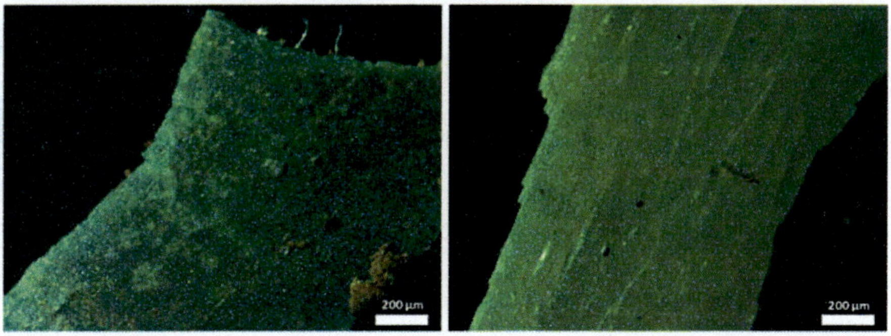

Fig. 3 The morphology of microplastic fragments from a single-use plastic straw in Panjang Island shows signs of weathering, such as changes in texture and color

cause a large flow of plastics from land to different aquatic environments [41]. Therefore, wind and rainfall can transport small plastic straw fragments stranded at the beach to the ocean on Panjang Island.

4 Conclusion

Dispose of plastic waste originating from anthropogenic activities was found at the beach of Panjang Island, including single-use plastic straws. Plastic straw stranded at the beach of Panjang Island mostly show signs of weathering and are partly degraded into smaller plastic fragments. Weathering of single-use plastic straws at the beach is primarily due to photooxidation, thermo-oxidation, and mechanical abrasion. This process caused plastic to become brittle and break easily because it loses its elasticity, likely to produce microplastics. This work served as a preliminary result of a study to understand the process of plastic fragmentation into microplastic on a tropical beach. Furthermore, more comprehensive analytical work is required to support this study, such as morphological screening using SEM–EDX and FTIR for polymer identification.

Acknowledgements This manuscript contributes to Pendanaan Ekspedisi dan Eksplorasi Deputi Bidang Fasilitasi Riset dan Inovasi BRIN No. 2860/II.01.00/8/2022 to DAU (batch 2 SK No. 53). Main contributor author DAU, contributor author member SYC, AUN, TS, MH.

References

1. Plastic Europe: Plastics—the Facts 2021. An analysis of European plastics production, demand and waste data. Brussel (2021)
2. McCormick, A., Hoellein, T.J., Mason, S.A., et al.: Microplastic is an abundant and distinct microbial habitat in an urban river. Environ. Sci. Technol. **48**, 11863–11871 (2014). https://doi.org/10.1021/es503610r
3. Utami, D.A, Reuning, L., Konechnaya, O., Schwarzbauer, J.: Microplastics as a sedimentary component in reef systems: a case study from the java sea. Sedimentology. https://doi.org/10.1111/sed.12879
4. Lusher, A.L., Welden, N.A., Sobral, P., Cole, M.: Sampling, isolating and identifying microplastics ingested by fish and invertebrates. Anal. Methods **9**, 1346–1360 (2017). https://doi.org/10.1039/C6AY02415G
5. Utami, D.A., Sujatmiko, K.A.: Unravelling the pathway of macro and micro debris in the beach of uninhabited Semak Daun reef platform, Kepulauan Seribu. IOP Conf Ser Earth Environ Sci **789**, 012047 (2021) https://doi.org/10.1088/1755-1315/789/1/012047
6. Lusher, A.L., Tirelli, V., O'Connor, I., Officer, R.: Microplastics in arctic polar waters: the first reported values of particles in surface and sub-surface samples. Sci. Rep. **5**, 14947 (2015). https://doi.org/10.1038/srep14947
7. Jambeck J.R., Geyer, R., Wilcox C, et al.: Plastic waste inputs from land into the ocean. Science **347**, 768–771 (2015) https://doi.org/10.1126/science.1260352

8. Rochman, C.M., Tahir, A., Williams, S.L., et al.: Anthropogenic debris in seafood: plastic debris and fibers from textiles in fish and bivalves sold for human consumption. Sci. Rep. **5**, 14340 (2015). https://doi.org/10.1038/srep14340
9. Murray, F., Cowie, P.R.: Plastic contamination in the decapod crustacean Nephrops norvegicus (Linnaeus, 1758). Mar. Pollut. Bull. **62**, 1207–1217 (2011). https://doi.org/10.1016/j.marpolbul.2011.03.032
10. Van Cauwenberghe, L., Janssen, C.R.: Microplastics in bivalves cultured for human consumption. Environ. Pollut. **193**, 65–70 (2014). https://doi.org/10.1016/j.envpol.2014.06.010
11. von Moos, N., Burkhardt-Holm, P., Köhler, A.: Uptake and effects of microplastics on cells and tissue of the blue mussel mytilus edulis L. after an experimental exposure. Environ. Sci. Technol. **46**, 11327–11335 (2012). https://doi.org/10.1021/es302332w
12. Reichert, J., Schellenberg, J., Schubert, P., Wilke, T.: Responses of reef building corals to microplastic exposure. Environ. Pollut. **237**, 955–960 (2018). https://doi.org/10.1016/j.envpol.2017.11.006
13. Zettler, E.R., Mincer, T.J., Amaral-Zettler, L.A.: Life in the "Plastisphere": microbial communities on plastic marine debris. Environ. Sci. Technol. **47**, 7137–7146 (2013). https://doi.org/10.1021/es401288x
14. Lamb, J.B, Willis, B.L., Fiorenza, E.A, et al.: Plastic waste associated with disease on coral reefs. Science **359**, 460–462. (2018) https://doi.org/10.1126/science.aar3320
15. Song, Y.K., Hong, S.H., Jang, M., et al.: Combined effects of UV exposure duration and mechanical abrasion on microplastic fragmentation by polymer type. Environ. Sci. Technol. **51**, 4368–4376 (2017). https://doi.org/10.1021/acs.est.6b06155
16. Carraher, C.E., Jr.: Introduction to Polymer Chemistry. CRC Press (2017)
17. Andrady, A., Amin, M.B., Xingzhou, H., Torikai, A.: Effects of increase solar ultraviolet radiation on materials. Ambio **24**, 191–196 (1995)
18. Andrady, A.L.: Microplastics in the marine environment. Mar. Pollut. Bull. **62**, 1596–1605 (2011). https://doi.org/10.1016/j.marpolbul.2011.05.030
19. Ocean Conservancy (2020) Together We Are Team Ocean. https://oceanconservancy.org/wp-content/uploads/202%0A0/10/FINAL_2020ICC_Report.pdf
20. Boer, W., van den Bergh, G.D., de Haas, H., et al.: Validation of accumulation rates in Teluk Banten (Indonesia) from commonly applied 210Pb models, using the 1883 Krakatau tephra as time marker. Mar. Geol. **227**, 263–277 (2006). https://doi.org/10.1016/j.margeo.2005.12.002
21. Mustikasari, E., Rustam, A., Adi, N., Purbani, D.: Pergerakan arus di Teluk Banten. In: Prosiding Seminar Nasional Teori dan Aplikasi Teknologi Kelautan (2012)
22. Satriadi, A.: Kajian Transpor Sedimen Tersuspensi Untuk Perencanaan Pembangunan Pelabuhan Bojonegoro Banten. Bul Oseanografi Mar **2**, 68–77 (2013)
23. Wisha, U.J., Husrin, S., Prihantono, J.: Hydrodynamics Banten Bay During Transitional Seasons (August-September) (Hidrodinamika Perairan Teluk Banten Pada Musim Peralihan (Agustus–September)). ILMU Kelaut Indones J Mar Sci **20**, 101. https://doi.org/10.14710/ik.ijms.20.2.101-112
24. Daud, M., Pin, T.G., Handayani, T.: The spatial pattern of seagrass distribution and the correlation with salinity, sea surface temperature, and suspended materials in Banten Bay. In: IOP Conf Ser Earth Environ Sci 243:012013. https://doi.org/10.1088/1755-1315/243/1/012013
25. Rustam, A., Adi, N.S., Mustikasari, E., et al.: Karakteristik Sebaran Sedimen dan Laju Sedimentasi Perairan Teluk Banten. J. Segara **14** (2018). https://doi.org/10.15578/segara.v14i3.7351
26. Utami, D.A., Reuning, L., Cahyarini, S.Y.: Satellite—and field-based facies mapping of isolated carbonate platforms from the Kepulauan Seribu complex, Indonesia. Depos. Rec. **4**, 255–273 (2018). https://doi.org/10.1002/dep2.47
27. Solihuddin, T., Utami, D.A., Salim, H.L., Prihantono, J.: Sedimentary environment of a modern carbonate platform of karimunjawa islands, central java. Indonesian J. Geosci. **6**, 57–72 (2019). https://doi.org/10.17014/ijog.6.1.57-72

28. Wisha, U.J., Gemilang, W.A., Wijaya, Y.J., Purwanto, A.D.: Model-based estimation of plastic debris accumulation in Banten Bay, Indonesia, using particle tracking—flow model hydrodynamics approach. Ocean Coast. Manag. **217**, 106009 (2022). https://doi.org/10.1016/j.ocecoaman.2021.106009
29. Falahudin, D., Cordova, M.R., Sun, X., et al.: The first occurrence, spatial distribution and characteristics of microplastic particles in sediments from Banten Bay, Indonesia. Sci. Total Environ. **705**, 135304. https://doi.org/10.1016/j.scitotenv.2019.135304
30. Schwarz, A.E., Ligthart, T.N., Boukris, E., van Harmelen, T.: Sources, transport, and accumulation of different types of plastic litter in aquatic environments: a review study. Mar. Pollut. Bull. **143**, 92–100 (2019). https://doi.org/10.1016/j.marpolbul.2019.04.029
31. Neto, A.M., Gomes, T.S., Pertel, M., et al.: An overview of plastic straw policies in the Americas. Mar. Pollut. Bull. **172**, 112813 (2021). https://doi.org/10.1016/j.marpolbul.2021.112813
32. Hwang, J., Choi, D., Han, S., et al.: An assessment of the toxicity of polypropylene microplastics in human derived cells. Sci. Total Environ. **684**, 657–669 (2019). https://doi.org/10.1016/j.scitotenv.2019.05.071
33. Zanghelini, G.M., Cherubini, E., Dias, R., et al.: Comparative life cycle assessment of drinking straws in Brazil. J. Clean Prod. **276**, 123070 (2020). https://doi.org/10.1016/j.jclepro.2020.123070
34. Zhang, K., Hamidian, A.H., Tubić, A., et al.: Understanding plastic degradation and microplastic formation in the environment: a review. Environ. Pollut. **274**, 116554 (2021). https://doi.org/10.1016/j.envpol.2021.116554
35. Liu, K., Wang, X., Wei, N., et al.: Accurate quantification and transport estimation of suspended atmospheric microplastics in megacities: implications for human health. Environ. Int. **132**, 105127 (2019). https://doi.org/10.1016/j.envint.2019.105127
36. Corcoran, P.L., Biesinger, M.C., Grifi, M.: Plastics and beaches: a degrading relationship. Mar. Pollut. Bull. **58**, 80–84 (2009). https://doi.org/10.1016/j.marpolbul.2008.08.022
37. O'Brine, T., Thompson, R.C.: Degradation of plastic carrier bags in the marine environment. Mar. Pollut. Bull. **60**, 2279–2283 (2010). https://doi.org/10.1016/j.marpolbul.2010.08.005
38. Iñiguez, M.E., Conesa, J.A., Fullana, A.: Recyclability of four types of plastics exposed to UV irradiation in a marine environment. Waste Manag. **79**, 339–345 (2018). https://doi.org/10.1016/j.wasman.2018.08.006
39. Furfaro, G., D'Elia, M., Mariano, S., et al.: SEM/EDX analysis of stomach contents of a sea slug snacking on a polluted seafloor reveal microplastics as a component of its diet. Sci. Rep. **12**, 10244 (2022). https://doi.org/10.1038/s41598-022-14299-3
40. Lambert, S., Wagner, M.: Characterisation of nanoplastics during the degradation of polystyrene. Chemosphere **145**, 265–268 (2016). https://doi.org/10.1016/j.chemosphere.2015.11.078
41. Lebreton, L., Andrady, A.: Future scenarios of global plastic waste generation and disposal. Palgrave Commun **5**, 6 (2019). https://doi.org/10.1057/s41599-018-0212-7

Comparison of CHAMP GPS Radio Occultation Dry Temperature Profiles Among Different Product Retrievals

Noersomadi and Nani Cholianawati

Abstract We investigated three different CHAllenging Minisatellite Payload (CHAMP) Global Positioning System Radio Occultation (GPS RO) dry temperature product retrievals in the upper troposphere and lower stratosphere layer, the layer that humidity is nearly zero in the atmosphere at about 1030 km altitude. The first product is released by University Corporation for Atmospheric Research (UCAR)—US, the second is provided by Research Institute for Sustainable Humanosphere (RISH), Kyoto University, Japan, and the third is available at Wegener Center (WEGC)-University of Graz, Austria. The RISH wave optics profiles have vertical resolutions of 0.1 km up to 30 km, which is able to indicate a small scale with an amplitude ± 0.3 K of a synoptic wave, a local or medium scale wave, in the lower stratosphere. The daily temperature perturbations of all three CHAMP products show downward propagation of planetary waves, a large-scale wave. Moreover, monthly temperature fluctuations, static stability, and potential energy variations describe the stratospheric Quasi-Biennial Oscillation associated with zonal wind anomaly in the tropics.

1 Introduction

Accurate information about the atmospheric conditions for weather prediction, mitigation of hydrometeorological disasters, and long-term climate projection is important for sustainability and human living. The atmospheric parameters could be obtained by conducting various observations such as the ground-based and satellite remote-sensing techniques [1–3]. Temperature is a key parameter that is related to pressure, humidity, and wind variation through the radiative transfer and thermodynamical process in the atmosphere [4]. It varies along latitude and longitude related to the energy transfer and dynamical process over the whole atmospheric fluid.

Noersomadi (✉) · N. Cholianawati
Research Center for Climate and Atmosphere, National Research and Innovation Agency (BRIN), Bandung, Indonesia
e-mail: noersomadi@brin.go.id

A. Basit et al. (eds.), *Proceedings of the International Conference on Radioscience, Equatorial Atmospheric Science and Environment and Humanosphere Science*, Springer Proceedings in Physics 290, https://doi.org/10.1007/978-981-19-9768-6_71

The radiosonde balloon is considered as the ground-based instrument for measuring vertical temperature, pressure, humidity, and wind velocity with high resolution up to about 10 m [3]. However, radiosonde measurement has limitations regarding the availability over the globe [5–7]. To solve this issue, the radio occultation technique, an observation of the atmosphere and ionosphere layer utilizing an excess phase of the signal from the navigation satellite, has been introduced [8]. When the electromagnetic signals from the navigation satellite pass through the ionosphere and atmosphere layers, the signals will bend at a certain point and altitude (tangent point) due to the density stratification [9, 10]. It is possible to obtain the refractive index from the measured bending angle of the Doppler shift of phase signal. Therefore, the atmospheric profiles could be determined from the inverse problem of the relationship between refractivity and atmospheric parameters [9, 11]. Hereafter, this observation technique is also known as global positioning system radio occultation (GPS RO) that retrieves high vertical resolution of global atmosphere pressure, temperature, and humidity [8].

The methods for retrieving atmospheric profiles from the observed refractive index are known as geometrical optics (GO), which determines the bending angle from the time derivative of the doppler phase shifting [9]. The second method is wave optics (WO), that analyzes the entire raw phase and complex amplitude signals by integral transforms [11, 12]. Both GO and WO assume a spherical symmetry [13] resulting the vertical resolution of about 1.4 km and 0.1 km, respectively [14, 15]. Therefore, the dry atmospheric profiles, where the refractive index depends on temperature and air pressure only, are obtained by utilizing hydrostatic equilibrium and ideal gas law. Meanwhile, the initial condition such as 1-D var of bending angle model is required to derive the wet profiles, when the refractive index also depends on both wet temperature and humidity components [10].

One of the early GPS RO missions was launched in 2001 by GeoForchungsZentrum (GFZ) German Geoscience named as CHAllenging Minisatellite Payload (CHAMP). CHAMP is the successful mission generation of the first long-term RO dataset [1]. The satellite is equipped with a single Low Earth Orbit payload with 87.2° polar orbit inclination at 454 km altitude [16]. For a single LEO satellite, it is expected to retrieve up to about 300 occultation profiles [14]. However, real observed atmospheric profiles obtained by CHAMP are ~150 profiles per day on average.

Since the temperature product depends on the retrieval processing method, it is important to elucidate recorded global temperature profiles, particularly related to accuracy and consistency issues [17]. In this work, we assess the consistency of temperature product retrievals from different datasets. The consistency of observed temperature product is important for model verification, including the role of synoptic and planetary-scale waves and long-term variations.

We explain the three different product retrievals and the method to reconstruct temperature fluctuations in the next section. We investigate the results of the comparison among the three retrievals in Sect. 3. Furthermore, the summary is in Sect. 4.

2 Data and Method

We analyze three product retrievals of CHAMP temperature profiles. The first product is released by the University Corporation for Atmospheric Research (UCAR). UCAR GPS RO data processing was described in [11, 18]. We used reprocessed CHAMP data version 2016 published by UCAR. CHAMP UCAR 2016 contains both dry and wet temperature profiles [19]. The second product is available in the Inter-university Upper atmosphere Global Observation NETwork (IUGONET) by Research Institute for Sustainable Humanosphere (RISH), Kyoto University, Japan [14, 19, 20]. RISH applied the wave optics method up to 30 km altitude to retrieve very high vertical dry temperature profiles in the lower stratosphere. Finally, the third product is CHAMP data version 5.6.2 processed using End-to-End Generic Occultation Performance Simulator (EGOPS) including both dry and wet profiles. The third CHAMP product is published by the Wegener Center (WEGC)-University of Graz, Austria [21, 22].

This work investigates CHAMP long-term observation from June 2001 to October 2008. All temperature products flagged as a bad profile by the three processing publishers were ignored. We compare an individual case of dry and wet temperature profiles at 060 km altitude range. Next, we focus on the typical dry profiles within five latitude bands, 60–90 S, 30–60 S, 20S–20 N, 30–60 N, 60–90 N, which are defined as polar south (PS), mid-latitude south (MS), tropics (TR), mid-latitude north (MN), and polar north (PN), respectively. We discuss different typical temperature profiles on four selected dates representing each season.

We assess the CHAMP products performance in detecting the temperature fluctuations by synoptic and planetary-scale waves. To analyze the fluctuations by synoptic or local atmospheric waves, we reconstruct temperature fluctuations from long and short vertical wavelength ($\lambda_z > 1$ km and $\lambda_z < 1$ km) ranges. We discuss the daily temperature perturbations, the deviation from the mean temperature profile, to evaluate the atmospheric planetary-scale waves.

The monthly mean temperature perturbations were performed to explore the radiative variations in the stratosphere, at 2030 km altitude, associated with the regular wind variations about 28 months high above the equatorial region known as the Quasi-Biennial Oscillation [23]. We analyze the temperature perturbations using CHAMP UCAR, the static stability using Brunt Vaisala frequency variations from CHAMP RISH, and potential energy variations with CHAMP WEGC. The static stability and potential energy are determined by the following formula:

$$N^2 = g/T[dT/dz + g/cp] \tag{1}$$

$$Ep = {}^1/_2(g/N)^2(T'/\overline{T})^2 \tag{2}$$

with N^2 is brunt vaisala frequency, g is gravitational acceleration, z is altitude, cp is specific heat constants for constant pressure, Ep is potential energy, T is temperature, and $\overline{T}$ is mean temperature.

To support our analysis, we utilize the zonal wind data from the NCEP reanalysis data provided by the NOAA Physical Science Laboratory, Boulder, Colorado, USA, from their website at https://psl.noaa.gov.

3 Results

3.1 Individual Comparison Among Three Products

Figure 1 shows the comparison of individual temperature product retrievals of CHAMP UCAR (dry and wet), CHAMP RISH (dry), and CHAMP WEGC (dry and wet). CHAMP UCAR provides the product reach to 60 km, while others were limited up to 40 km. The dry profiles indicate unrealistic values below ~8 km. We are able to see the RISH profile depicts very high vertical resolution in the stratosphere. Meanwhile, UCAR and WEGC products show smoother temperature profiles at 1530 km altitude intervals. All CHAMP temperature products describe an oscillation in stratosphere.

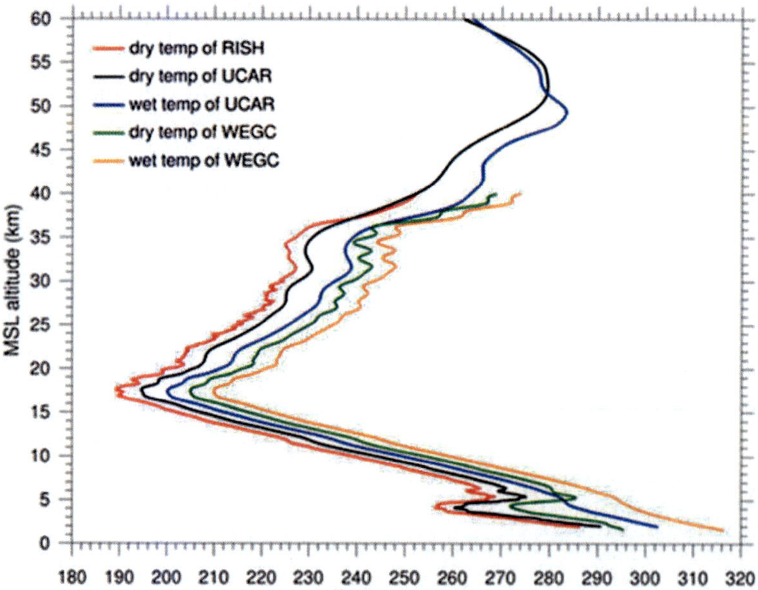

Fig. 1 Temperature profiles from the three product retrievals on January 15 2003, at 33.77 E and 11.39 N. Dry and wet profiles of UCAR and of WEGC are shifted by 5 K, 10 K, 15 K, and 20 K from the dry profile of RISH, respectively. The magnitudes of all dry profiles below about 8 km are unrealistic temperature data

Next, we re-plot profiles in Fig. 1 into Fig. 2 by focusing on the altitude of 15–22 km, which is called the tropopause layer and lower stratosphere regions. We observed detailed fluctuations in the CHAMP RISH profile, which indicates double minima temperature, 189.5 K at 16.8 km and 189 K at 17.6 km. On the other hand, UCAR and WEGC show a single minimum temperature at 17.6 km with slightly different values. This individual example resulting an issue regarding tropopause definition in the tropics. RISH wave optic profile suggested multiple minima, which may affect the vertical derivative of temperature and cold point temperature [24]. Both dry and wet profiles retrieved by WEGC almost coincided with each other in this altitude range. However, UCAR dry and wet describe as rather dissimilar though less than 1 K, particularly above 20 km. The wave optics applied by RISH could detect small amplitudes of short vertical scale fluctuations within 17–22 km.

For further discussion, we examined analyzing dry profiles only. We investigated typical temperature profiles in the region of (a) PS, (b) PN, (c) MS, (d) MN, and (e) TR, as shown in Fig. 3 to Fig. 6. We considered 15 January, 15 April, 15 July, and 15 October as the representative dates of mid-season of northern hemisphere winter, spring, summer, and autumn, respectively. Table 1 describes the details location and time of each occultation. In the northern winter season (summer in the southern), the minimum temperature of PS is observed at 215 K at ~9 km, while in PN the coldest temperature reaches 195 K at 22 km (Fig. 3a–b). The altitude of the minimum temperature in MS is between 15 and17 km, which seems close to the altitude of

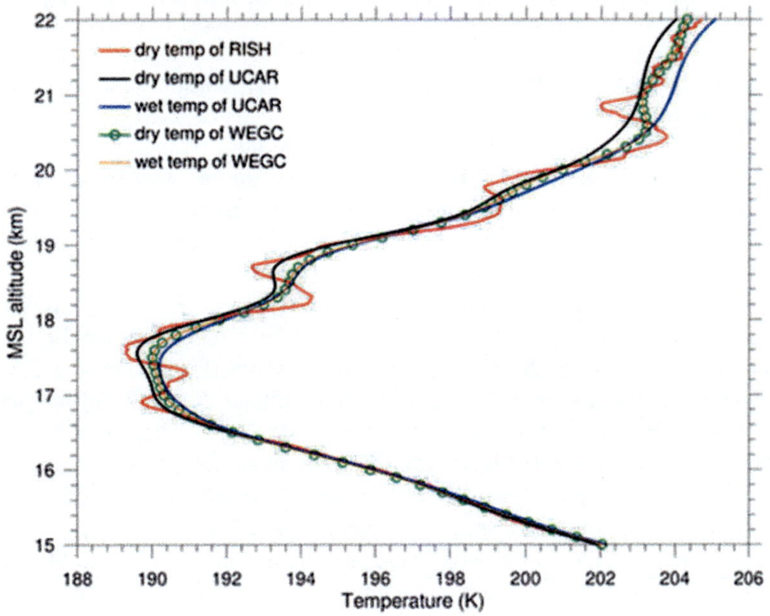

Fig. 2 Same profiles as in Fig. 1 concerning near the tropopause layer. No profile shifting is applied in this figure

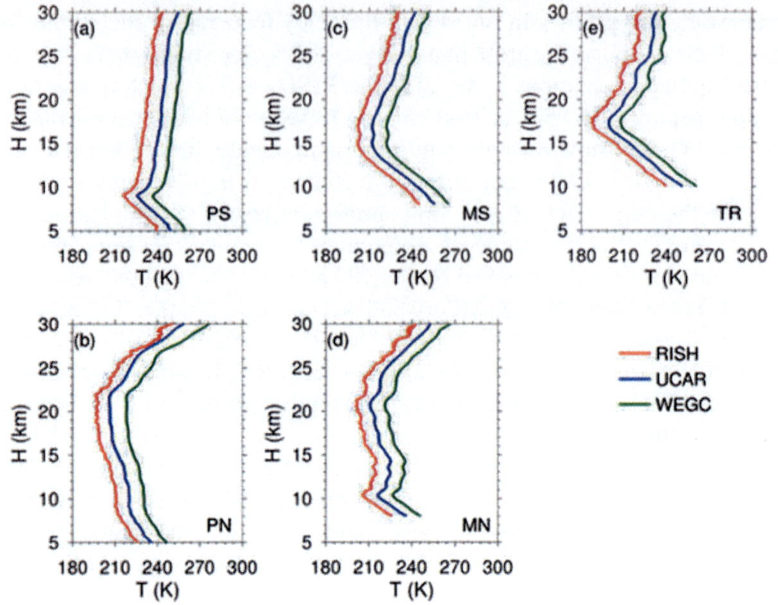

Fig. 3 Typical temperature profiles in the **a** polar south (PS); **b** polar north (PN); **c** mid-latitude south (MS); **d** mid-latitude north (MN); and **e** tropics (TR) of the three product retrievals on January 15, 2003 (northern hemisphere winter). Both UCAR and WEGC profiles are shifted by 10 K and 20 K, respectively. Details of occultation information of each profile is described in Table 1

minimum temperature in TR. Meanwhile, it is seen the minimum peak temperature 210 K at ~10 km in MN (Fig. 3c–e).

In the northern spring, we observed a local minimum temperature of 210 K at ~9 km in PS and 217 K at ~8 km in PN (Fig. 4a–b). On the other hand, the typical temperature profiles show similar patterns with slightly different minimum temperatures in the three regions of MS, MN, and TR (Fig. 4c–e). The minimum temperature in TR is the coldest temperature among other regions.

Figure 5 shows the opposite situation, as indicated in Fig. 3, particularly between the PS and PN, also among MS and MN. In the northern autumn, when the southern region experiences spring, the temperature pattern of PS (Fig. 6a) is still similar to the pattern in southern winter (Fig. 5a) with the altitude of minimum temperature reach ~20 km. The local minimum temperature profiles in PN are found in 9–11 km (Fig. 5b and 6b). Figures 5d and 6d describe an identical temperature pattern in the MN region during northern summer and autumn. The results show consistency with the global height and latitude temperature pattern by [25, 26].

Overall, we noted the RISH wave optics profile depicts small-scale fluctuations in the 17–30 km altitude interval in all samples. We also found the altitude of local minimum temperature profiles in the polar regions in the summer hemisphere is around 9–11 km. For the mid-latitude regions, the altitude of minimum temperature is observed at ~15 km during the spring and autumn seasons. Meanwhile, we noticed

Table 1 Time (HH:MM:SS UTC) and location of occultation event shown in Figs. 3, 4, 5 and 6

Date (2003)		15 Jan	15 Apr	15 Jul	15 Oct
Polar South (PS)	Time	00:07:51	02:20:48	18:12:32	01:59:38
	Lon	29.27E	121.57 W	146.76E	18.52E
	Lat	72.69S	89.76S	76.42S	66.07S
Mid-latitude South (MS)	Time	17:17:30	03:37:47	09:10:36	06:18:39
	Lon	136.3E	64.90E	7.42E	135.77E
	Lat	30.78S	32.11S	40.83S	43.77S
Tropics (TR)	Time	05:47:36	00:20:26	14:45:03	17:40:02
	Lon	129.60E	84.99E	108.32E	114.52E
	Lat	6.70 N	4.32 N	1.67S	6.62 N
Mid-latitude North (MN)	Time	02:11:07	12:08:56	11:29:29	16:17:45
	Lon	41.43E	114.77E	147.80E	114.10E
	Lat	49.51 N	39.06 N	32.68 N	37.56 N
Polar North (PN)	Time	11:44:16	10:47:18	22:03:50	01:01:48
	Lon	45.68E	142.57E	134.64E	31.87E
	Lat	70.08 N	80.57 N	83.67 N	67.53 N

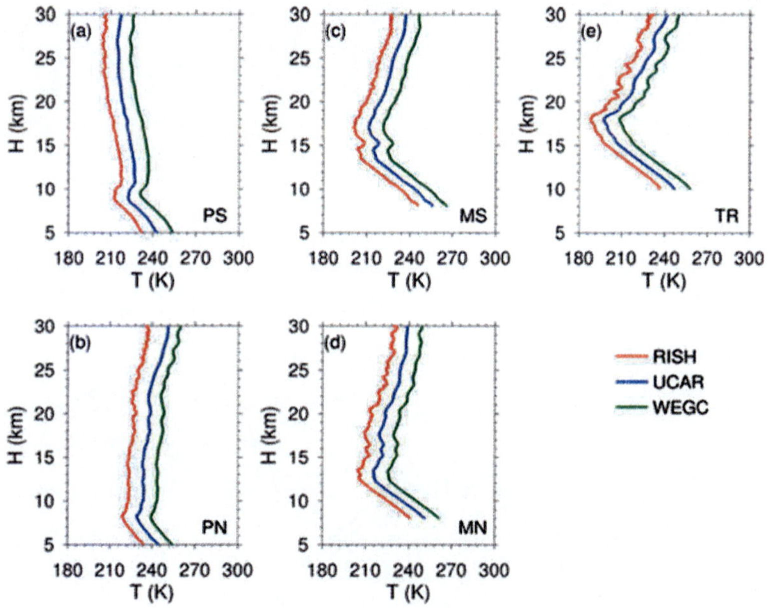

Fig. 4 Same as in Fig. 3 on April 15, 2003 (northern spring)

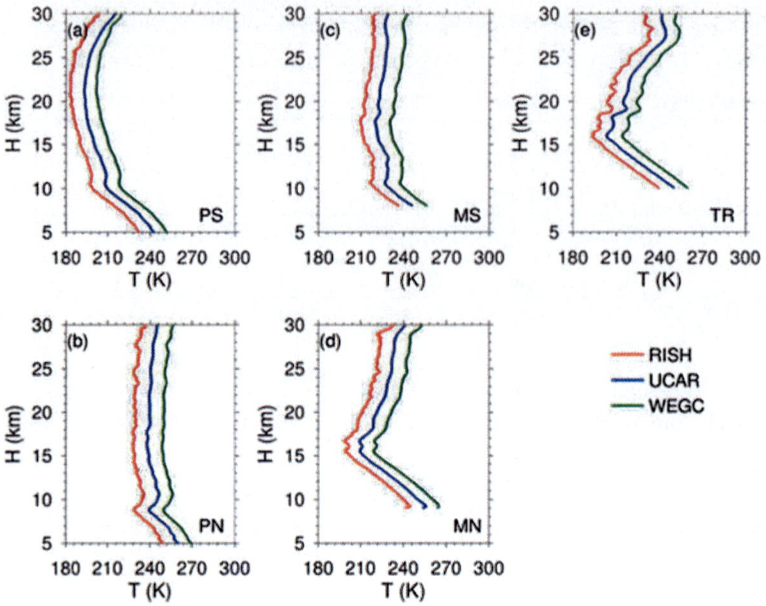

Fig. 5 Same Fig. 3 on July 15, 2003 (northern summer)

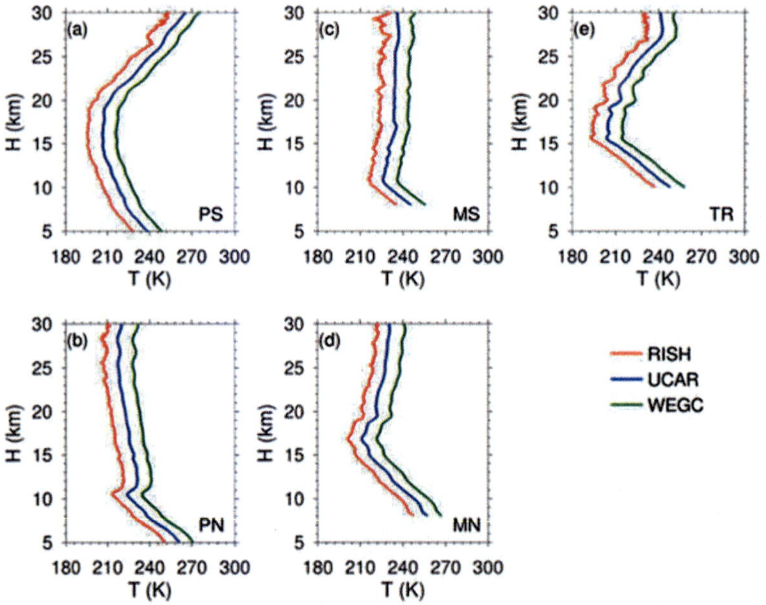

Fig. 6 Same as Fig. 3 on October 15, 2003 (northern autumn)

the altitude of minimum temperature in the tropics is relatively constant in all seasons at ~17 km.

3.2 Synoptic and Planetary Scale Waves

We collected all temperature profiles within 90–160 E and 15S–15 N in January 2004, then we calculated the daily mean temperature of all profiles at 20–32 km, which is known as the stable region of the stratosphere layer. The synoptic scale waves are often observed in this stable region [26, 27]. We demonstrated the temperature fluctuations in Fig. 7, derived by subtracting the monthly mean profile. The fluctuations consist of a mixture of multi-scale vertical wavelengths. We observed short vertical scales in the RISH profile up to 30 km related to the altitude limit of applying wave optics for the retrieval process [28].

To separate between long and short vertical scale waves, we used a fast fourier transform method for each profile. Therefore, we reconstructed temperature perturbations for $\lambda_z > 1$ km and $\lambda_z < 1$ km (Fig. 8). All three products indicate fluctuations from synoptic waves with $\lambda_z > 1$ km with amplitude ±2 K. CHAMP RISH product is able to demonstrate the amplitude of synoptic wave with λ_z ~2 km. For $\lambda_z <$ 1 km, the amplitude of UCAR and WEGC profiles are close to zero. However, the

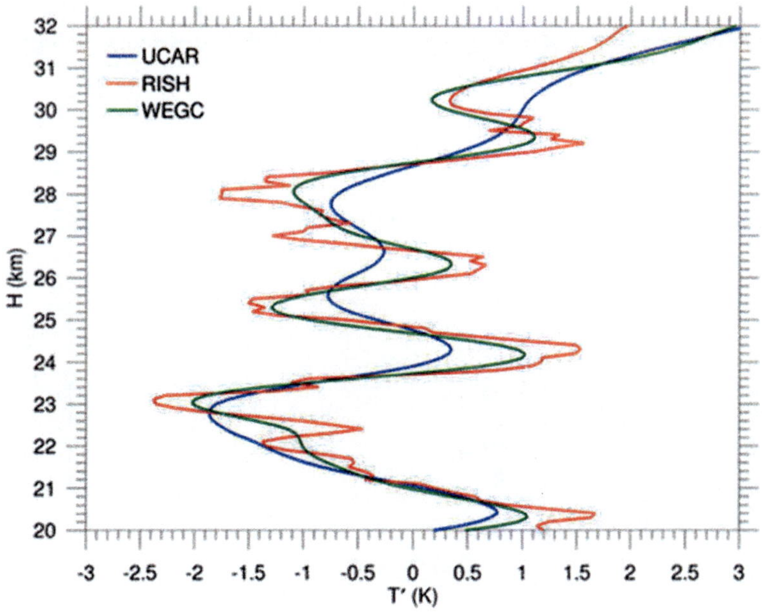

Fig. 7 Temperature fluctuations on January_1, 2004, over the 90–160 E, 15 S–15 N

RISH profile shows a significant amplitude of ±0.3 K. We conclude that wave optics applied by RISH could resolve long and short vertical scales of synoptic waves.

We analyzed the daily temperature profiles from all products within January 1, 2004, to April 25, 2004. The results of all three products are consistent. We show the height versus time variation of temperature perturbations from CHAMP UCAR in Fig. 9. The temperature perturbations were defined as the deviation from the mean temperature profile an entire period. To support the analysis, we also described the zonal wind anomaly over the same area. CHAMP temperature perturbations demonstrate clear downward cold and warm, indicating the propagation of planetary waves in March 2004. The downward temperature propagation is also associated with the westward and eastward wind. This propagation seems related to equatorial trapped wave activity [29, 30].

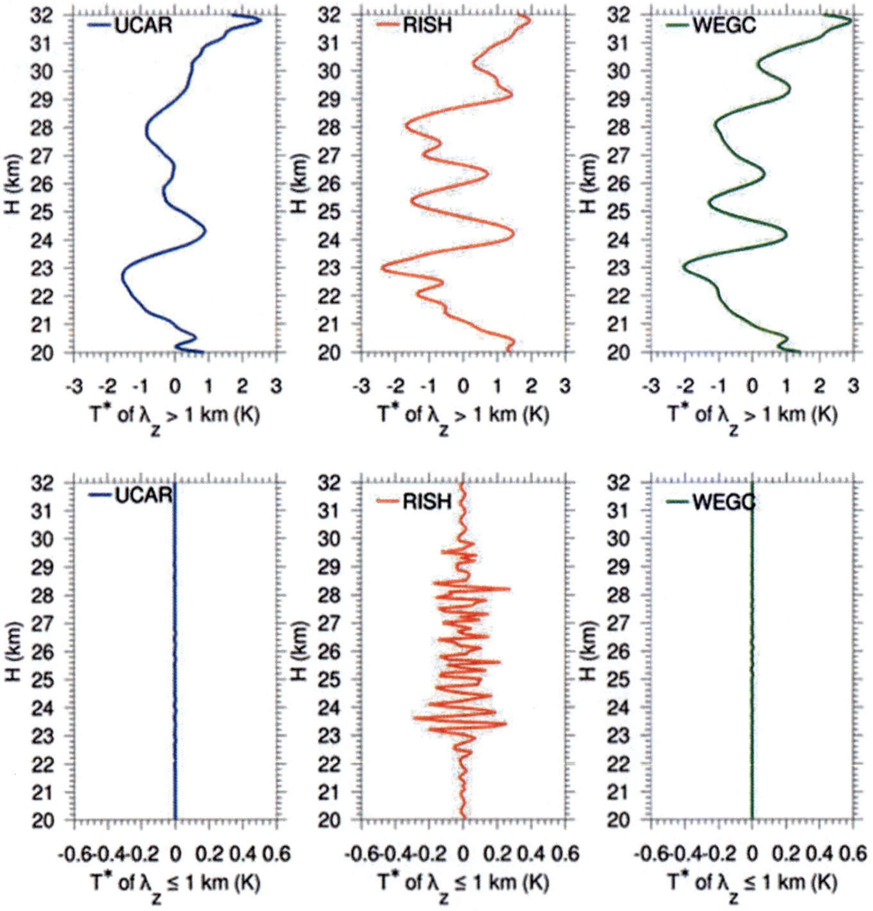

Fig. 8 Reconstructed temperature fluctuations T* of long ($\lambda_z > 1$ km) and short ($\lambda_z < 1$ km) vertical wavelength ranges from the temperature perturbations in Fig. 7

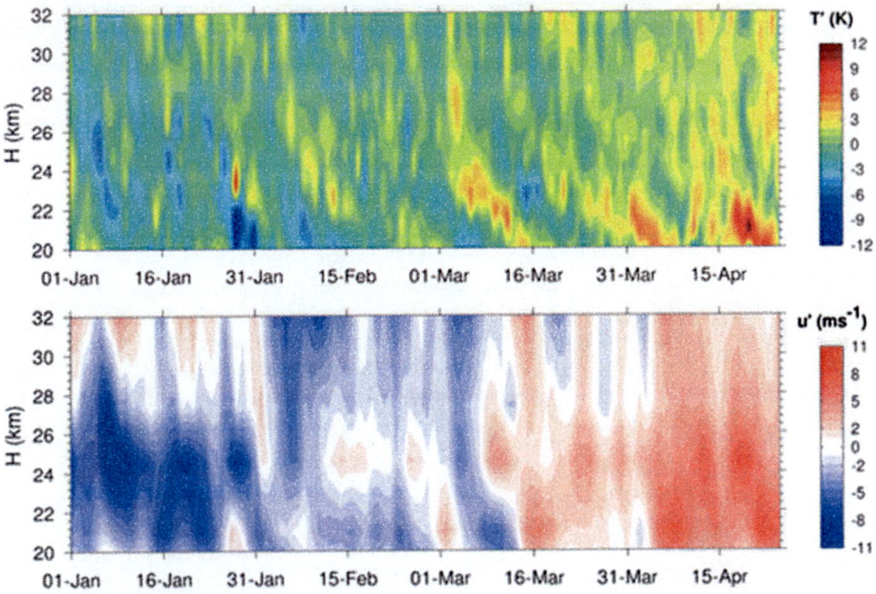

Fig. 9 Height versus time-cross-section of daily temperature perturbations (top) and zonal wind anomaly (bottom) over 90–160 E, 15S–15 N in the period of January 1 to April 25 2004

3.3 Zonal Mean Temperature Fluctuations in the Stratosphere

Figures 10, 11, and 12 show the height versus time cross-section of monthly temperature perturbations, static stability, and potential energy (Eq. 2) of synoptic-scale waves with three retrieval products. We overlaid the temperature perturbations and potential energy with the monthly zonal mean zonal wind at 70 hPa (~18.6 km) to 10 hPa (~32 km). For the static stability (Eq. 1), we superimposed it with the time derivative of zonal wind to analyze the relationship between vertical derivative of temperature and acceleration or deceleration of wind variation in the stratosphere. Top panel of each figure depicts the downward propagation of an oscillation between warm and cold temperature associated with an alternating westward and eastward wind. We noticed that a warm anomaly, particularly in the mid-2006, reached up to 6 K, and occurred when the wind direction changed from westward to eastward. On the other hand, a cold anomaly of about −6 to −4 K collocated with the change from eastward to westward.

The annual oscillation of N^2 is observed in the lower stratosphere as documented by [26, 28]. The downward propagation of the wind acceleration from 32 km altitude increased static stability starting from 28 km altitude. Meanwhile, when the wind decelerated, the strong static stability was seen trapped at ~22 km. We noted the high potential energy increased when the wind direction changed. All parameters shown

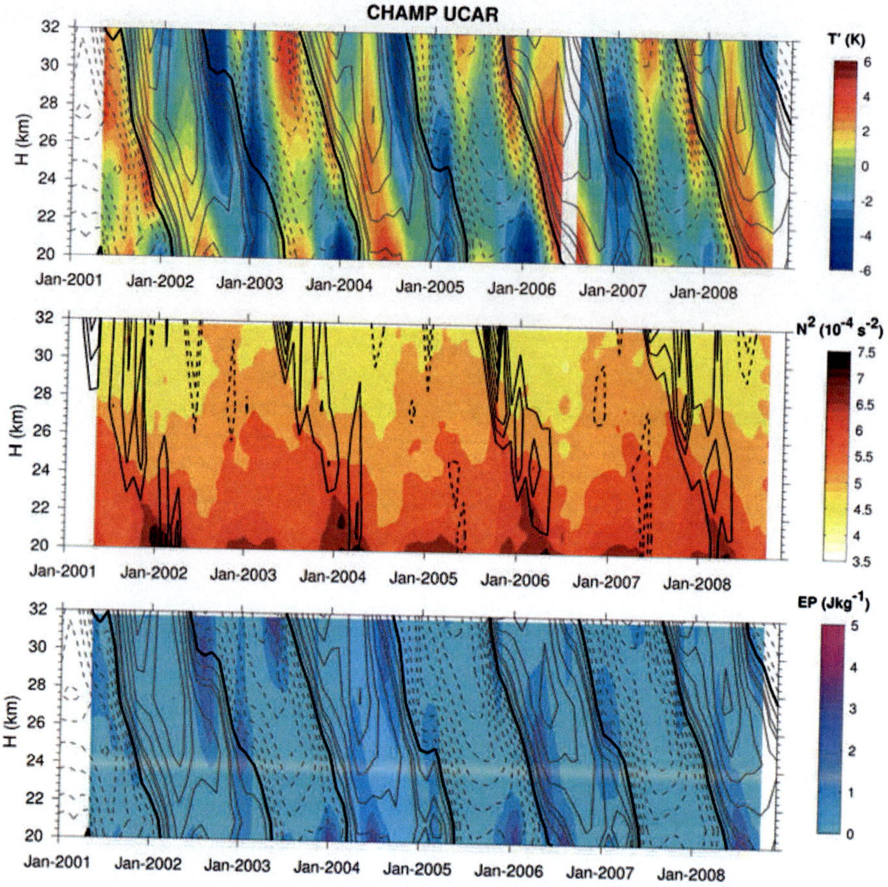

Fig. 10 Height versus time cross-section of CHAMP UCAR monthly temperature perturbations (top); Brunt Vaisala frequency square (middle); and potential energy of synoptic-scale waves (bottom). The top and bottom figures are overlaid by a contour line of zonal mean zonal wind anomaly with solid (eastward), dashed (westward), and thick solid (0 m.s^{-1}). The contour line ranges from –25 to 20 m.s^{-1} with an interval is 3 m.s^{-1}. The middle panel is overlaid by a contour line –0.2 to 0.4 m.s^{-2} of the time derivative of zonal mean zonal wind with solid (acceleration) and dashed (deceleration)

by three retrieval products describe consistency of clear downward propagation in the stratosphere, linked with the Quasi-Biennial Oscillation.

4 Summary

We investigated three CHAMP temperature product retrievals by UCAR, RISH, and WEGC. All CHAMP temperature products show fluctuations in the stratosphere

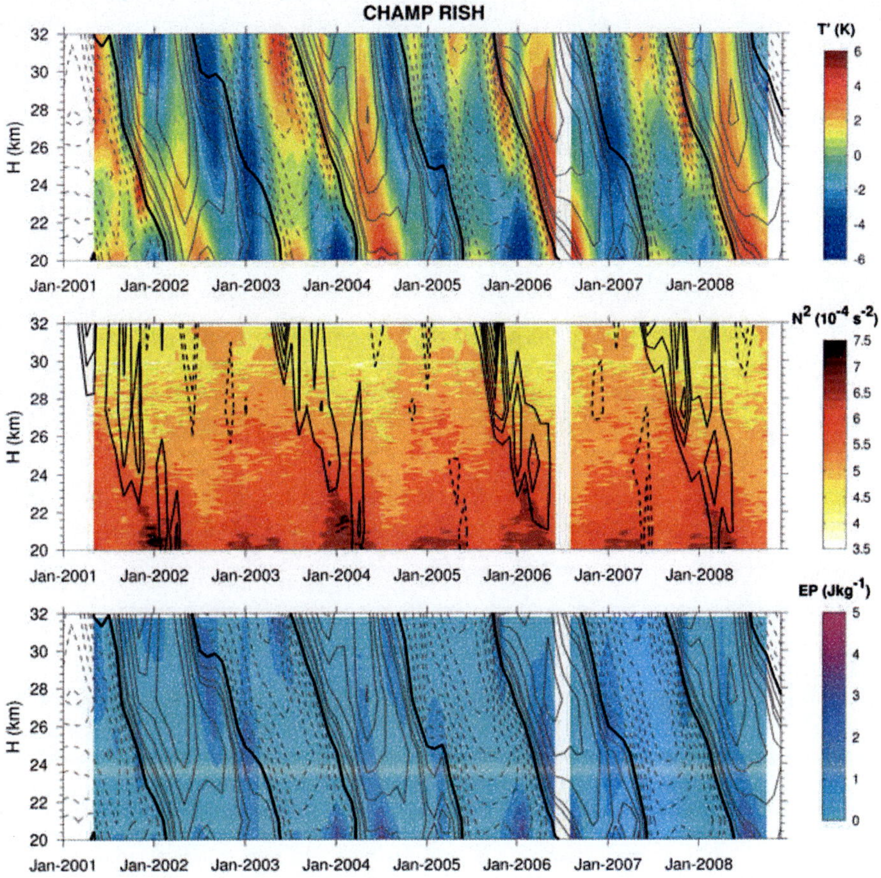

Fig. 11 Same Fig. 10 for CHAMP RISH

layer. The wave optics retrieval applied by RISH provides very high vertical resolution up to 30 km altitude. RISH wave optics profiles indicate small-scale fluctuations in the lower stratosphere. We found the altitude of local minimum temperature in the polar summer region is 9–11 km, in mid-latitude spring and autumn is ~15 km, and in the tropics is at ~17 km for all seasons.

We analyzed the vertical wavelengths of the temperature perturbations for $\lambda_z >$ 1 km (long) and $\lambda_z <$ 1 km (short) of synoptic scale waves. All three products show perturbation from the long vertical wavelengths. RISH wave optics profile was able to observe an amplitude $\pm$ 0.3 K of the short vertical scale. Hence, we summarize that wave optics profiles could resolve long and short vertical fluctuations. For the planetary-scale waves, CHAMP temperature describes downward cold and warm anomaly, associating with the westward and eastward wind.

The height versus time cross section of monthly temperature perturbations, static stability, and potential energy shows downward propagation in the stratosphere. A

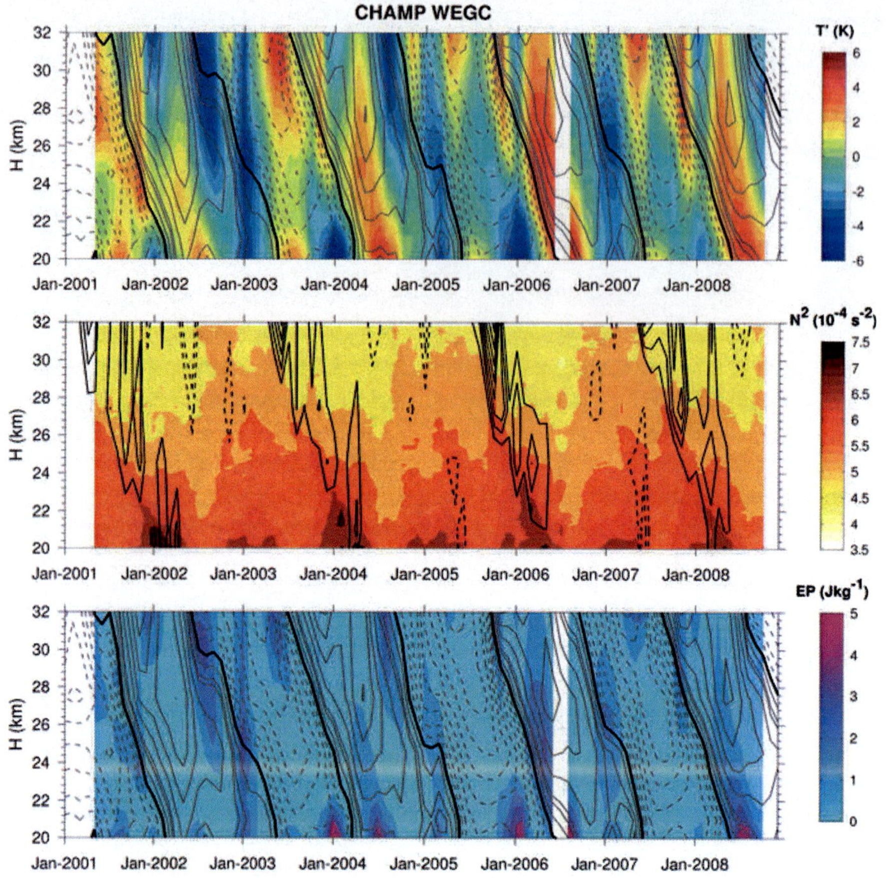

Fig. 12 Same Fig. 10 for CHAMP WEGC

warm (cold) anomaly of about ±6 K occurred when the wind direction changed from westward to eastward (eastward to the westward). Strong static stability is associated with downward wind acceleration. These are linked to the well-known phenomena in the tropical stratosphere called the Quasi-Biennial Oscillation.

Acknowledgements This research is supported by National Research and Innovation Agency of Indonesia (BRIN) Grant Number 15/III/HK/2022.

References

1. Wickert, J., et al.: GPS radio occultation with CHAMP: Initial results, pp. 4–7 (2001)
2. Hasebe, F., Aoki, S., Inai, Y., Nakazawa, T., Sugawara, S., Ikeda, C., Honda, H., Yamazaki, H., Halimurrahman., Komala, N., Putri, F.A., Budiyono, A., Soedjarwo, M., Ishidoya, S., Toyoda, S., Shibata, T., Hayashi, M., Eguchi, N., Nishi, N., Fujiwara, M., Coordinated upper-troposphere-to-stratosphere balloon experiment in Biak, (1213–1230) (2018)
3. Ho, S.-P., Zhou, X., Shao, X., Zhang, B., Adhikari, L., Kireev, S., He, Y., Yoe, J.G., Xia-Serafino, W., Lynch, E.: Initial assessment of the COSMIC-2/FORMOSAT-7 Neutral atmosphere data quality in NESDIS/STAR using in situ and satellite data. Remote Sens. **12**, 4099 (2020). https://doi.org/10.3390/rs12244099
4. Butchart, N.: The brewer-dobsoncirculation. Rev. Geophys. **52**, 157–184 (2014). https://doi.org/10.1002/2013RG000448
5. Ho, S., Peng, L., Vömel, H.: Characterization of the long-term radiosonde temperature biases in the upper troposphere and lower stratosphere using COSMIC and Metop-A/GRAS data from 2006 to 2014. (4493–4511) (2017)
6. Bell, T.M., Klein, P.M., Lundquist, J.K., Waugh, S.: Remote-sensing and radiosonde datasets collected in the San Luis Valley during the LAPSE-RATE campaign. Earth Syst. Sci. Data **13**, 1041–1051 (2021). https://doi.org/10.5194/essd-13-1041-2021
7. Yoneyama et al.: (2013). https://doi.org/10.1175/BAMS-D-12-00157.1
8. Anthes, R.A.: Exploring earth's atmosphere with radio occultation: contributions to weather, climate and space weather. Atmos. Meas. Tech. **4**(6), 1077–1103 (2011)
9. Gorbunov, M., Irisov, V., Rocken, C.: The influence of the signal-to-noise ratio upon radio occultation retrievals. Remote Sens. **14**(2742) (2022). https://doi.org/10.3390/rs14122742
10. Schreiner, W., Sokolovskiy, S., Hunt, D., Rocken, C.: Analysis of GPS radio occultation data from the FORMOSAT-3/COSMIC and Metop/GRAS missions at CDAAC. Atmos. Meas. Tech. **4**(10), 2255–2272 (2011)
11. Sokolovskiy, S., Rocken, C., Schreiner, W., Hunt, D.: On the uncertainty of radio occultation inversions in the lower troposphere. J. Geophys. Res. Atmos. **115**(22), 1–19 (2010)
12. Jensen, A.S., Lohmann, M.S., Benzon, H.H., Nielsen, A.S.: Full spectrum inversion of radio occultation signals. Radio Sci. **38**(3), 1040 (2003)
13. Melbourne, W.G.: Radio occultation using earth satellites. Wiley (2004)
14. Jensen, A.S., Lohmann, M.S., Nielsen, A.S., Benzon, H.H.: Geometrical optics phase matching of radio occultation signals. RadioSci. **39**(3), 1–8 (2004)
15. Tsuda, T., Lin, X., Hayashi, H., Noersomadi.: Analysis of vertical wave number spectrum of atmospheric gravity waves in the stratosphere using COSMIC GPS radio occultation data, Atmos. Meas. Tech. **4**(8) (2011)
16. Wickert et al.: GRAS SAF workshop on applications of GPSRO measurements, 16–18 (2008)
17. Noersomadi., Tsuda, T.: Comparison of three retrievals of COSMIC GPS radio occultation results in the tropical upper troposphere and lower stratosphere 2. Aeron. Earth Planet. Sp. **69**(1) (2017)
18. Schreiner, W., Rocken, C., Sokolovskiy, S., Hunt, D.: Quality assessment of COSMIC/FORMOSAT-3 GPS radio occultation data derived from single—and double-difference atmospheric excess phase processing. GPS Solut. **14**, 13–22 (2009). https://doi.org/10.1007/s10291-009-0132-5
19. https://data.cosmic.ucar.edu/gnss-ro/champ/repro2016/
20. http://database.rish.kyoto-u.ac.jp/arch/iugonet/data/GPS/champ/
21. Angerer, B., Ladstädter, F., Scherllin-Pirscher, B., Schwärz, M., Steiner, A.K., Foelsche, U., Kirchengast, G.: Quality aspects of the Wegener Center multi-satellite GPS radio occultation record OPSv5.6. Atmos. Meas. Tech. **10**, 4845–4863 (2017). https://doi.org/10.5194/amt-10-4845-2017
22. https://wegcwww.uni-graz.at/data-store/WEGC/OPS5.6:2019.1/

23. Baldwin, M.P., Gray, L.J., Dunkerton, T.J., Hamilton, K., Haynes, P.H., Randel, W.J., Holton, J.R., Alexander, M.J., Hirota, I., Horinouchi, T., Jones, D.B.A., Kinnersley, J.S., Marquardt, C., Sato, K., Takahashi, M.: The quasi-biennial oscillation. Rev. Geophys.
24. Zeng, Z., Sokolovskiy, S., Schreiner, W.S., Hunt, D.: 2019: Representation of vertical structures by radio occultation observations in the upper troposphere and lower stratosphere: comparison to high-resolution radiosonde profiles. J. Atmos. Oceanic Tech. (2019). https://doi.org/10.1175/JTECH-D-18-0105.1
25. Schmidt, T., Heise, S., Wickert, J., Beyerle, G., Reigber, C.: GPS radio occultation with CHAMP and SAC-C: global monitoring of thermal tropopause parameters. Atmos. Chem. Phys. **5**(6), 1473–1488 (2005)
26. Grise, K.M., Thompson, D.W.J., Birner, T.: A global survey of static stability in the stratosphere and upper troposphere. J. Clim. **23**(9), 2275–2292 (2010)
27. Duran, P., Molinari, J.: tropopause evolution in a rapidly intensifying tropical cyclone: a static stability budget analysis in an idealized axisymmetric framework. J. Atmos. Sci. **76**, 209–229 (2019)
28. Noersomadi and Tsuda, T.: Global distribution of vertical wavenumber spectra in the lower stratosphere observed using high-vertical-resolution temperature profiles from COSMIC GPS radio occultation. Ann. Geophys. **34**, 203–213. https://doi.org/10.5194/angeo-34-203-2016
29. Randel, W.J., Wu, F.: Kelvin wave variability near the equatorial tropopause observed in GPS radio occultation measurements. J. Geophys. Res. **110**, D03102 (2005). https://doi.org/10.1029/2004JD005006
30. Kim, J., Son, S.W.: Tropical cold-point tropopause: climatology, seasonal cycle, and intraseasonal variability derived from COSMIC GPS radio occultation measurements. J. Clim. **25**(15), 5343–5360 (2012)

Evaluation of ERA5 Precipitation Reanalysis Data in Indonesia

Sigit Kurniawan Jati Wicaksana and Iis Sofiati

Abstract Rainfall data is used widely in essential applications such as flood/drought monitoring, water management, and climate monitoring. This information is critical in Indonesia, where rainfall has a significant socioeconomic impact. The selection of a precipitation product can have a significant impact on the performance of such an application that mentioned above. Rainfall products are classified into three types: direct measurement, satellite based, and reanalysis. The purpose of this study is to evaluate the performance of ERA5 precipitation data against Global Satellite Mapping of Precipitation (GSMaP) and Climate Forecasting System (CFSv2) and analyze the annual trends and spatial patterns of reanalysis data in Indonesia to find out which areas have a better level of accuracy. The study area was in Indonesia, between 12°S and 10°N, and 90°E and 50°E. The dataset used in this study were ERA5, GSMAP, and CFSv2 from 2011to 2021. The research shows that ERA5 has a very good level of accuracy in the southern region of Indonesia. The results of this study can be used for further information for climate change and the hydrological cycle in Indonesia, especially in the use of precipitation data.

1 Introduction

Indonesia is a tropical country with a lot of rainfall each year. The availability of accurate precipitation data, both temporally and spatially, has become an essential requirement in various fields [1, 2]. In meteorology, precipitation is typically defined as rain, snow, sleet, or hail falling from a cloud toward the ground [3]. As a result, precipitation is one of the most critical variables in weather and climate studies. Accurate and reliable precipitation data are essential for determining climate trends

S. K. J. Wicaksana (✉)
Research Center for Oceanography, National Research and Innovation Agency, Jakarta, Indonesia
e-mail: Sigi020@brin.go.id

I. Sofiati
Research Center for Climate and Atmosphere, National Research and Innovation Agency, Bandung, Indonesia

A. Basit et al. (eds.), *Proceedings of the International Conference on Radioscience, Equatorial Atmospheric Science and Environment and Humanosphere Science*, Springer Proceedings in Physics 290, https://doi.org/10.1007/978-981-19-9768-6_72

and variability, including impact indices like floods and droughts, as well as water resource management in agriculture, forestry, and energy sectors, as well as weather, climate, and hydrologic forecasting [4–6].

A deep understanding of these processes and phenomena requires precipitation data with high precision and spatial–temporal resolution. At various spatial and temporal scales, various global rainfall products are available. To calculate rainfall estimates, these products take very different approaches. They have classified into three types: (1) reanalysis, which is based on a numerical weather prediction model and data assimilation; (2) gage-only products, which are derived solely from gage data; and (3) satellite-based products, which are based solely or partially on satellite data. All of these products differ in some way. They each have distinct advantages and disadvantages [7]. As a result, precipitation products based on satellite imagery and reanalysis have become the most promising data, outperforming calibrated meteorological data and weather radar data. Reanalysis datasets generally display greater variability than other types of datasets, such as satellite- or rain gage-based datasets, with the degree of variability varying by region [4, 6, 8, 9].

Satellite-based precipitation retrievals and precipitation estimates from models/reanalysis systems have become increasingly appealing and accessible in recent decades. Some of satellite based and reanalysis precipitation products, such as Tropical Rainfall Measuring Mission (TRMM) Multi-Satellite Precipitation Analysis (TMPA), the Global Satellite Mapping of Precipitation (GSMaP), the National Oceanic and Atmospheric Administration (NOAA), the Integrated Multi-satellite Retrievals for Global Precipitation Measurement, and a new global Multi-Source Weighted-Ensemble Precipitation (MSEWP), have been developed. Reanalysis precipitation data, which is simulated using new atmospheric models combined with advanced data assimilation systems, are an important alternative [4, 10, 11].

Because observed datasets are not always available or present, reanalysis data has emerged as a viable alternative. Reanalysis is not dependent on the density of surface observational networks and can provide surface variables in areas with little surface coverage. As a result, reanalysis systems provide nearly realistic atmospheric circulation fields, allowing many things to be done, including understanding precipitation changes through the lens of atmospheric dynamical mechanisms. There are several different reanalysis products available, but their quality is known to vary with frequent upgrades. The most popular reanalysis precipitation datasets include the National Centers for Environmental Prediction Remote Sens. 2020, 12, 2902 3 of 25 reanalysis (NCEP1 and NCEP2), NCEP Climate Forecast System Reanalysis (CFSR), ECMWF Reanalysis-5 (ERA-5) and ERA-Interim, the Japanese 55-year Reanalysis Project (JRA-55), and the National Aeronautics and Space Administration (NASA) Modern Era Reanalysis for Research and Applications (e.g., MERRA-2) [11, 12].

The purposes of this study are to (1) evaluate the performance of ERA5 precipitation data against GSMAP and CFSv2, (2) analyze the annual trends and spatial patterns of ERA5 precipitation reanalysis data in Indonesia to find out which areas have a better level of accuracy. The results are critical for the continued rational application of reanalysis data and information on climate change and the hydrological cycle in Indonesia.

2 Data and Methodology

2.1 *Study Area*

The study area was in Indonesia, between 12° S and 10° N, and 90° E and 50° E. Indonesia's tropical climate is influenced by monsoon winds that change every six months. Indonesia has a rainy season from November to April and a dry season from May to October [13].

2.2 *Dataset*

For this study, three satellite-based rainfall data products and reanalyzes were chosen: ERA5, CFSV2, and GSMAP. Because of the availability of data, the monthly precipitation from 2011 to 2021 was selected to be evaluated, and to prevent missing information and change the values of the other products, a nearest neighbor interpolation method with a spatial resolution of 0.25° was used. Precipitation data converted to one scale accumulations in monthly means of daily means has been scaled to include units such as "per day," respectively. The data used in this study are detailed below.

ERA5

ERA5 is the Copernicus Climate Change Service's latest generation ECMWF atmospheric reanalysis of global climate. Using advanced modeling and data assimilation systems, it transforms vast amounts of historical data into global estimates. Since 1979, and eventually, since 1950, estimates of a large number of atmospheric, land, and oceanic climate variables have been available on a 0.25° × 0.25° global grid. For the high-resolution realization of ERA5, which assimilates observations twice daily and provides hourly output forecast steps ranging from 0 to 18 h, the model time step is 12 min. Precipitation observations are not assimilated into the model; instead, forecast accumulations for hourly precipitation fields are generated. Data from ERA5 can be downloaded from (https://cds.climate.copernicus.eu/) [5, 14].

GSMaP

GSMap is one type of SPP developed by Japanese scientists. Initially, this project was developed by the Japan Science and Technology Agency (JST). However, the Japan Aerospace Exploration Agency is currently carrying it on (JAXA). GSMaP provides global rainfall data from microwave and infrared radiometers. These data cover the entire globe (from 60° N to 60° S) and have a temporal resolution of one hour. The data is available from 2000 to the present. GSMaP data can be downloaded from the JAXA website or by visiting (ftp://hokusai.eorc.jaxa.jp/standard/v7/) [4, 15, 16].

CFSv2

NCEP CFSv2 is the primary operational sub-seasonal forecast system used in the United States. CFSv2 is a fully coupled atmosphere–ocean modeling system. The spatial coverage spans the globe (90° N to 0° S) and generates hourly data with a ½° horizontal resolution (approximately 56 km). The data is available from 2011 to present. CFSv2 includes 16 runs per day, but only four runs were initialized daily during the reforecast period. These consist of four 9-month runs initialized (at 0000, 0600, 1200, and 1800 UTC) every five days, and a 0000 UTC one-season (3-month) run with three subsequent 45-day runs (0600, 1200, and 1800 UTC) initialized during the intervening four days. CFSv2 data can be downloaded from the NCAR website or by visiting (https://rda.ucar.edu/datasets/ds094.2/) [17].

2.3 Evaluation Method and Criteria

In this study, the evaluation of precipitation products was performed with calibration and validation for each month (January-December). The calibration phase uses regression analysis with a scatter plot. At this stage, it can be seen the relationship between reanalysis data and observation data. Meanwhile, the calibration model's accuracy is tested at the validation stage using the parameters root-mean-square error (RMSE 0, + ∞) and Pearson correlation coefficient (CC –1, 1). The correlation coefficient (CC) and root mean square error (RMSE) measure the quantitative accuracy of the precipitation products. The correlation coefficient (CC) represents the relationship between observed and estimated precipitation. The magnitude of the difference between observed and estimated rainfall is denoted by the RMSE, which assigns greater weight to significant errors. These performance statistics are expressed in the following ways:

$$\text{RMSE} = \sqrt{\frac{1}{N}\sum_{i=1}^{N}(S_i - G_i)^2} \tag{1}$$

$$CC = \frac{\sum_{i=1}^{N}(S_i - \overline{S})(G_i - \overline{G})}{\sqrt{\sum_{i=1}^{N}(S_i - \overline{S})^2}\sqrt{\sum_{i=1}^{N}(G_i - \overline{G})^2}} \tag{2}$$

where N is the total number of daily records, S_i and G_i are the estimated and gaged precipitation, respectively, $\overline{S}$ and $\overline{G}$ are their mean values.

3 Results

3.1 Comparison of Precipitation Products

Comparison of precipitation products with each other can help identify the consistency and differences in precipitation estimates among other precipitation products. Figure 1 shows the spatial–temporal distribution of the three precipitation products (ERA5 (A), GSMaP (B), and CFSv2 (C)). Overall, the spatial pattern of the mean annual rainfall of these products is similar. The results show that in general the performance of ERA5 with GSMaP, it looks underestimated in the waters of the Indian Ocean west of Sumatra, waters north of Kalimantan, and waters north of Papua. As for other regions, ERA5 can capture spatial patterns that are more similar to GSMaP than CFSv2. Furthermore, the ERA5 and CFSv2 data show underestimation (alongside the waters of Sumatra, critical Sulawesi, and maximum of the waters of Papua) and overestimation (withinside the Papuan plains).

Meanwhile, for CFSv2 compared to GSMaP, the dominant overestimate is in almost all mainland (large islands) such as Sumatra, Kalimantan, parts of Sulawesi, Java, and most of Papua. This phenomenon is closely related to the influence of the highlands on each island which will produce high-intensity orographic rainfall. However, for waters in the central region and some islands (such as West Nusa Tenggara, East Nusa Tenggara, and Maluku) the performance of CFSv2 looks closer to GSMaP [4].

3.2 Basic Statistics and Distribution of the Precipitation Products

The accuracy of ERA5 data against GSMAP and CFSv2 data was further analyzed statistically by calculating Root Mean Square Error (RMSE) and determination (R2) (Fig. 2). The linear correlation between ERA5 and GSMaP looks better (with a lower RMSE value of 3.949, a higher R2 of 0.740) compared to CFSv2 (RMSE 4.799, and R2 of 0.725). However, ERA5 has good (strong) linearity to (GSMaP) and (CFSv2) with the correlation coefficient (r) being (0.86) and (0.85), respectively. There are three categories of grouping on the R2 value, namely, the strong category, moderate category, and weak category. [18] stated that the R2 value of 0.75 belongs to the strong category, the R2 value of 0.50 belongs to the moderate category and the R2 value of 0.25 belongs to the weak category. R2 can not only be used in regression, but can use the R2 formula in all models to determine whether or not the model is in addition to strengthening the model that has been obtained [19].

Figure 3 shows the temporal pattern of annual rainfall from the three ERA5, GSMAP, and CFSv2 data were analyzed for 11 years, average for all of Indonesia, and the results are shown in Fig. 3. Generally, the annual pattern for ERA5, GSMAP, and CFSv2 look the same, with only slightly different intensity. The highest

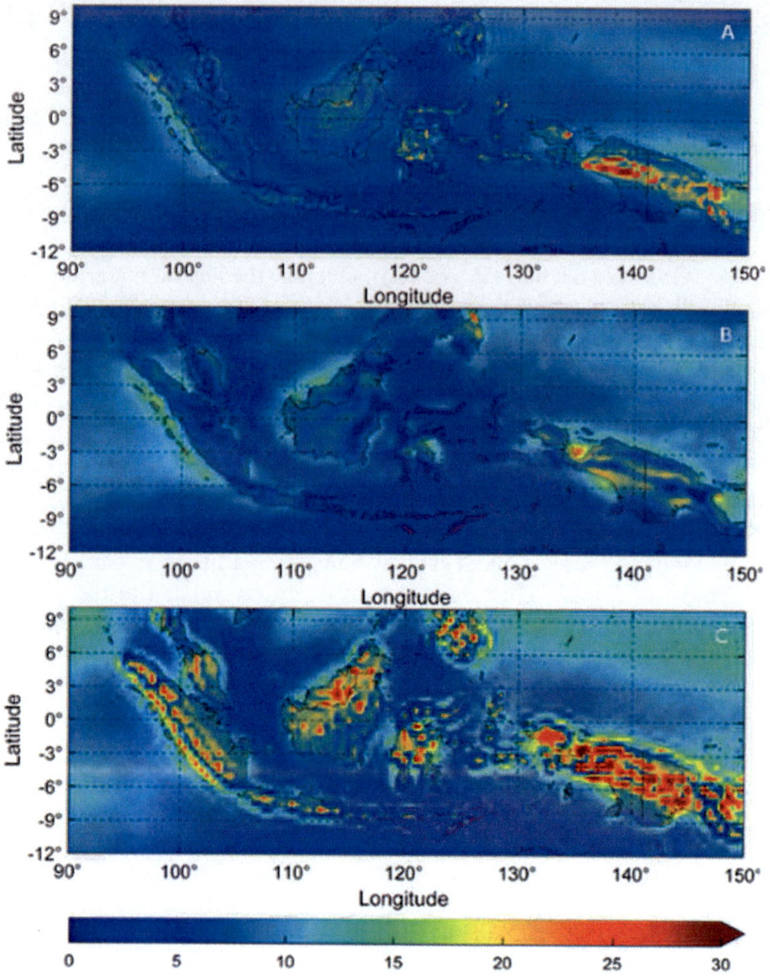

Fig. 1 Multiyear mean daily total precipitation from 2011 to 2021. (A) ERA5, (B) GSMAP, and (C) CFSv2

average rainfall occurs during the wet season (December-January–February, DJF) at 10.3 mm/day, and the lowest during the dry season (June-July–August) at about 6.2 mm/day. The annual cycle of ERA5 and GSMaP is nearly identical, except for January, June, July, September, October, and December, when ERA5 is slightly (1 mm/day) lower than GSMaP. Meanwhile, the annual cycle of CFSv2 is above ERA5 and GSMaP. Based on the results of the previous processing and analysis (Fig. 1), this also supports the finding that the CFSv2 data is overestimated for both ERA5 and GSMaP.

To make things apparent, annual cycle analysis was applied to several areas by following the type of rainfall from [20]. Three regions were studied marked with

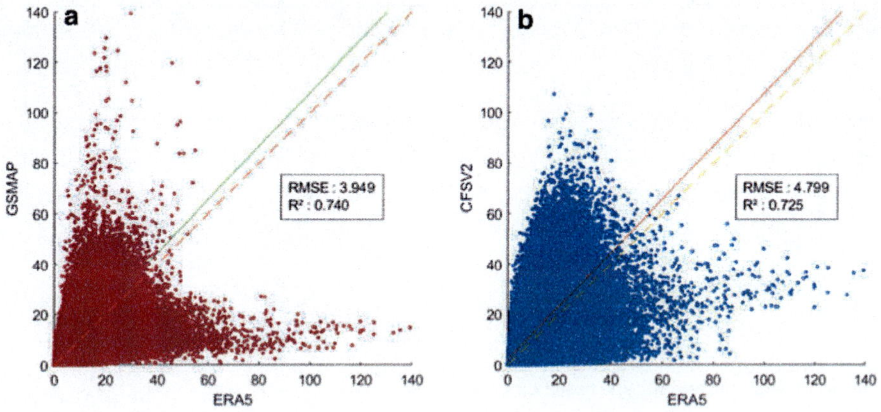

Fig. 2 Scatter plot of (**a**) ERA5 vs. GSMAP and (**b**) ERA5 versus. CFSv2 over the period 2011–2021

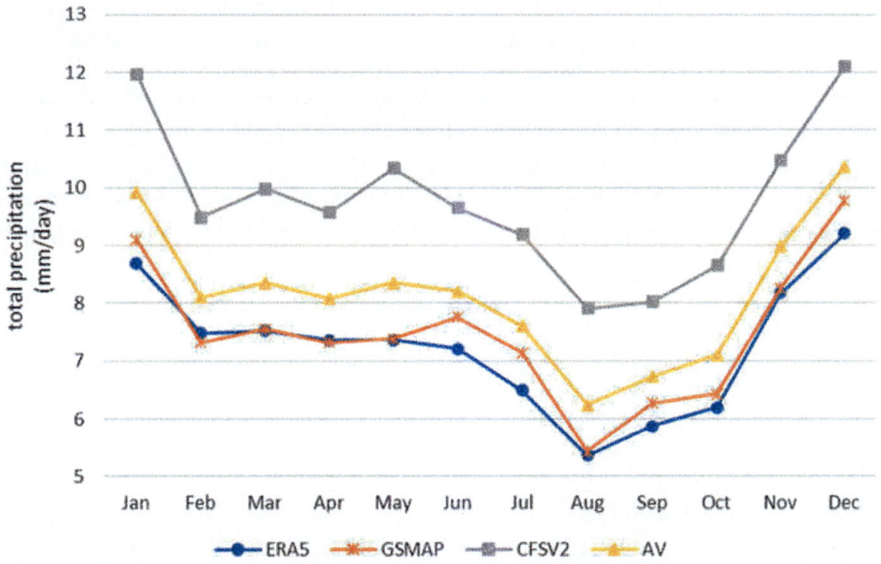

Fig. 3 Annual cycle of daily total precipitation rates in Indonesia for ERA5, GSMAP, and CFSv2 region averaged over the period 2011–2021

red boxes: Region.1 (6–10) °S, (105–110) °E, Region II (2–4) °N, (95–100) °E, and Region III (4–5) °S, (125–130) °E, as shown in Fig. 4.

In Region I, the highest average rainfall occurs during the wet season (December-January–February, DJF) at 14.1 mm/day, and the lowest during the dry season (June-July–August, JJA) at around 1.1 mm/day. The annual cycle of ERA5 and GSMaP is nearly identical, except for January, June, July, September, October, and December,

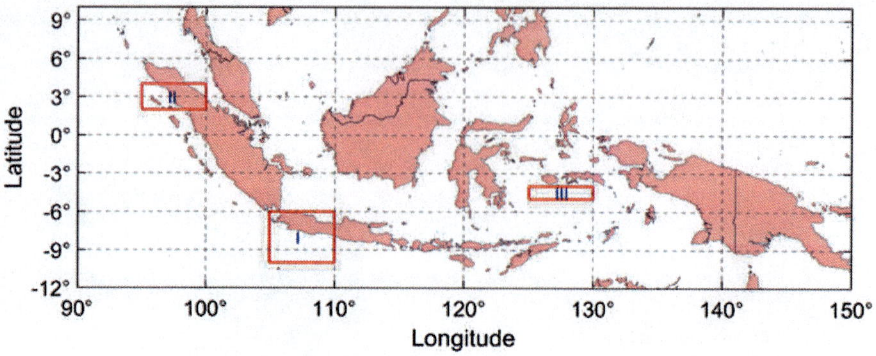

Fig. 4 location of rainfall type based on [20]

when ERA5 is slightly (1 mm/day) lower than GSMaP. Meanwhile, the annual cycle of CFSv2 is above ERA5 and GSMaP.

For Region II, there are two peaks of rainfall, namely in April and November as shown in Fig. 6. The highest average rainfall occurs twice, during the wet-to-dry transition season (March–April-May, MAM) of 13.1 mm/day, and the dry-to-wet transition (September–October-November, SON) was about 18.3 mm/day. Meanwhile, the lowest average was in June which only reached 6.7 mm/day.

Furthermore, for Region III as shown in Fig. 7, the highest average rainfall value occurs in the month (May, June, July, MJJ) at 13.8 mm/day, and the lowest average is in the month (August–September-October, ASO) about 2.5 mm/day. From the three

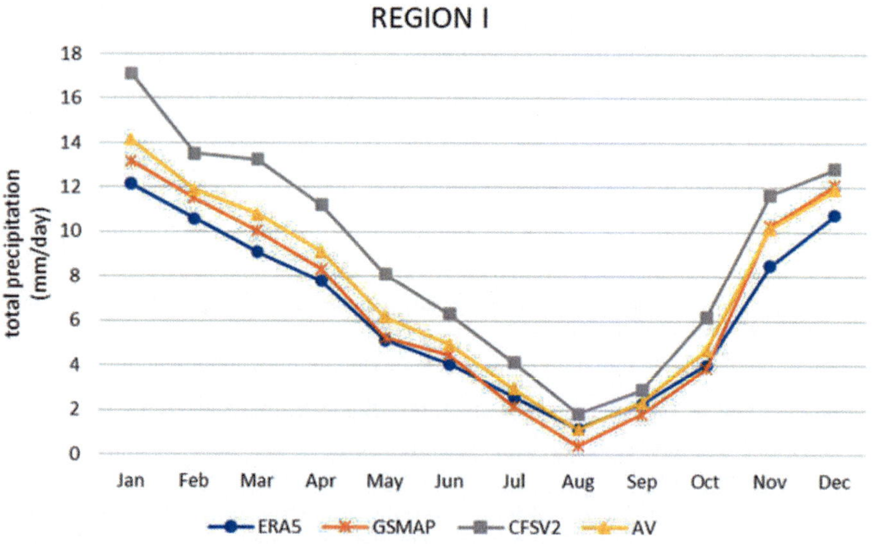

Fig. 5 Annual cycle of daily total precipitation rates in the region I (6–10) °S, (105–110) °E

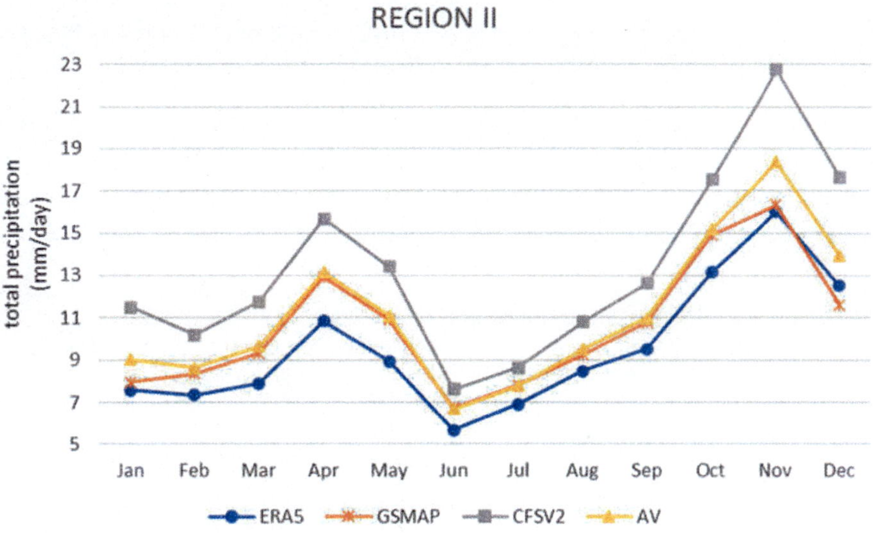

Fig. 6 Annual cycle of daily total precipitation rates in region II (2–4) °N, (95–100) °E

selected regions, the intensity of the annual cycle of ERA5 and GSMaP is almost the same, while the CFSv2 is above ERA5 and GSMaP.

In Fig. 8 can be seen that for both ERA5 versus. GSMAP and ERA5 versus. CFSv2, a solid and positive correlation (correlation coefficient value can reach one) mainly occurs in the southern and northern parts of Indonesia, for the equator area itself a strong correlation on average occurs in inter-island waters. Although there are some points in the Papua region that show a negative correlation, it is not as strong.

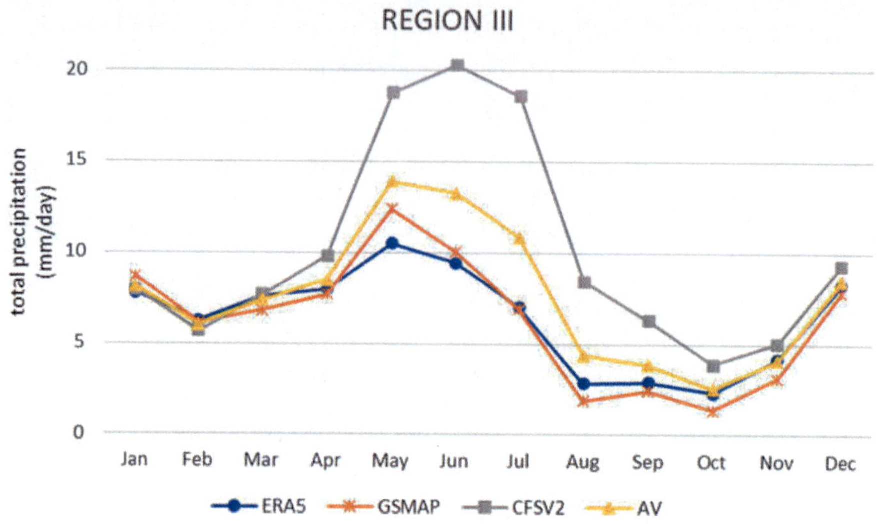

Fig. 7 Annual cycle of daily total precipitation rates in region III (4–5) °S, (125–130) °E

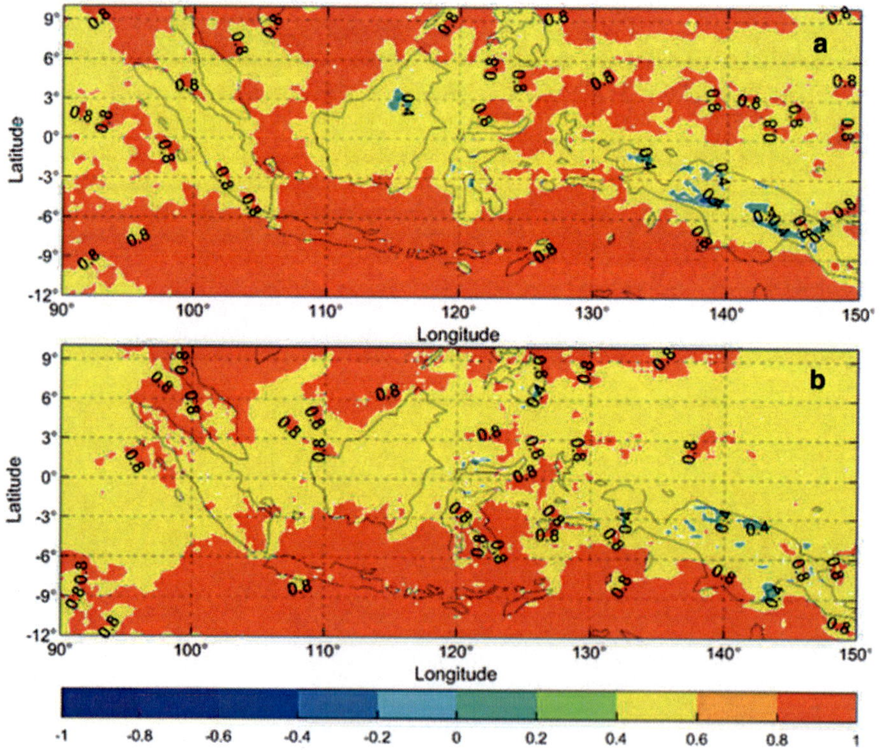

Fig. 8 Spatial Correlation total precipitation over the period 2011–2021 **a** ERA5 versus. GSMAP and **b** ERA5 versus. CFSv2

4 Conclusion

Three precipitation products were evaluated over the Indonesia region from 2011 to 2021. The main findings are as follows. In general, the spatial pattern for ERA5, GSMAP, and CFSv2 looks similar, only slightly different intensity, and the value of CFSv2 has a higher value (overestimate) than the other two products.

The evaluation results show that ERA5 is well-validated against two other products. The RMSE between ERA5 to GSMAP and CFSv2, respectively, is the value of 3.949 and 4.799 mm/day. The temporal pattern of annual rainfall from the three looks the same, only the intensity is slightly different. A comparison also made in three locations, and it shows that the three products had the same pattern as the research conducted by [20]. In general, the ERA5 precipitation product has a high level of accuracy and can be used as hydrological data in Indonesia, particularly in the southern region.

References

1. Arrokhman, N.A., Wahyuni, S., Suhartanto, E.: Evaluasi Kesesuaian Data Satelit untuk Curah Hujan dan Evaporasi Terhadap Data Pengukuran di Kawasan Waduk Sutami. J. Teknol. dan Rekayasa Sumber Daya Air. **1**(2), 904–916 (2021)
2. Lee, H.S.: General rainfall patterns in Indonesia and the potential impacts of local seas on rainfall intensity. Water (Switzerland). **7**(4), 1751–1768 (2015)
3. Siepielski, A.M., Morrissey, M.B., Buoro, M., Carlson, S.M., Caruso, C.M., Clegg, S.M., Coulson, T., DiBattista, J., Gotanda, K.M., Francis, C.D., Hereford, J., Kingsolver, J.G., Augustine, K.E., Kruuk, L.E.B., Martin, R.A., Sheldon, B.C., Sletvold, N., Svensson, E.I., Wade, M.J., MacColl, A.D.C.: Precipitation drives global variation in natural selection. Science (80-). **355** (6328), 959–962 (2017)
4. An, Y., Zhao, W., Li, C., Liu, Y.: Evaluation of six satellite and reanalysis precipitation products using gauge observations over the yellow river basin, China. Atmosphere (Basel), **11** (11) (2020)
5. Hassler, B., Lauer, A.: Comparison of reanalysis and observational precipitation datasets including era5 and wfde5. Atmosphere (Basel), **12** (11) (2021)
6. Sun, Q., Miao, C., Duan, Q., Ashouri, H., Sorooshian, S., Hsu, K.L.: A review of global precipitation data sets: data sources, estimation, and intercomparisons. Rev. Geophys. **56**(1), 79–107 (2018)
7. Le Coz, C., Van De Giesen, N.: Comparison of rainfall products over Sub-saharan Africa. J. Hydrometeorol. **21**(4), 553–596 (2020)
8. Bai, P., Liu, X.: Evaluation of Five Satellite-Based Precipitation Products in Two Gauge-Scarce Basins on the Tibetan Plateau. Remote Sens. 2018, Vol. 10, Page 1316. 10 (8), 1316 (2018)
9. Okamoto, K., Takahashi, N., Iwanami, K., Shige, S., Kubota, T.: High precision and high resolution global precipitation map from satellite data. 2008 Microw. Radiom. Remote Sens. Environ.—10th Spec. Meet. Proceedings, MICRORAD. (2008)
10. Dee, D.P., Uppala, S.M., Simmons, A.J., Berrisford, P., Poli, P., Kobayashi, S., Andrae, U., Balmaseda, M.A., Balsamo, G., Bauer, P., Bechtold, P., Beljaars, A.C.M., van de Berg, L., Bidlot, J., Bormann, N., Delsol, C., Dragani, R., Fuentes, M., Geer, A.J., Haimberger, L., Healy, S.B., Hersbach, H., Hólm, E.V., Isaksen, L., Kållberg, P., Köhler, M., Matricardi, M., Mcnally, A.P., Monge-Sanz, B.M., Morcrette, J.J., Park, B.K., Peubey, C., de Rosnay, P., Tavolato, C., Thépaut, J.N., Vitart, F.: The ERA-Interim reanalysis: configuration and performance of the data assimilation system. Q. J. R. Meteorol. Soc. **137**(656), 553–597 (2011)

11. Sun, S., Shi, W., Zhou, S., Chai, R., Chen, H., Wang, G., Zhou, Y., Shen, H.: Capacity of satellite-based and reanalysis precipitation products in detecting long-term trends across Mainland China. Remote Sens. **12**(18) (2020)
12. Wanzala, M.A., Ficchi, A., Cloke, H.L., Stephens, E.M., Badjana, H.M., Lavers, D.A.: Assessment of global reanalysis precipitation for hydrological modelling in data-scarce regions: a case study of Kenya. J. Hydrol. Reg. Stud. **41**, 101105 (2022)
13. Aldrian, E., Gates, L.D., Widodo, F.H.: Variability of Indonesian Rainfall and the Influence of ENSO and Resolution in ECHAM4 Simulations and in the reanalyses. MPI Rep. (346) (2003)
14. Qin, S., Wang, K., Wu, G., Ma, Z.: Variability of hourly precipitation during the warm season over eastern China using gauge observations and ERA5. Atmos. Res. **264**(September), 105872 (2021)
15. Kubota, T., Shige, S., Hashizume, H., Aonashi, K., Takahashi, N., Seto, S., Hirose, M., Takayabu, Y.N., Ushio, T., Nakagawa, K., Iwanami, K., Kachi, M., Okamoto, K.: Global precipitation map using satellite-borne microwave radiometers by the GSMaP project: production and validation. IEEE Trans. Geosci. Remote Sens. **45**(7), 2259–2275 (2007)
16. Pratama, A., Agiel, H.M., Oktaviana, A.A.: Evaluasi satellite precipitation product (GSMaP, CHIRPS, dan IMERG) di Kabupaten lampung selatan. J. Sci. Appl. Technol. **6**(1), 32 (2022)
17. Saha, S., Moorthi, S., Wu, X., Wang, J., Nadiga, S., Tripp, P., Behringer, D., Hou, Y.T., Chuang, H.Y., Iredell, M., Ek, M., Meng, J., Yang, R., Mendez, M.P., Van Den Dool, H., Zhang, Q., Wang, W., Chen, M., Becker, E.: The NCEP climate forecast system version 2. J. Clim. **27**(6), 2185–2208 (2014)
18. Joseph, F., Hair, J.R.E.A.R.L.T., W.C.B.: Multivariate Data Analysis Fifth Edition. Prentice Hall, Inc. (1998)
19. Ghozali, I.: Aplikasi Analisis Multivariate Dengan Program IBM SPSS 25. Universitas Diponegoro, Semarang (2018)
20. Aldrian, E., Dwi Susanto, R.: Identification of three dominant rainfall regions within Indonesia and their relationship to sea surface temperature. Int. J. Climatol. **23**(12), 1435–1452 (2003)
21. Liu, C.Y., Aryastana, P., Liu, G.R., Huang, W.R.: Assessment of satellite precipitation product estimates over Bali Island. Atmos. Res. **244**(January), 105032 (2020)
22. Wati, T., Hadi, T.W., Sopaheluwakan, A., Hutasoit, L.M.: Statistics of the Performance of Gridded Precipitation Datasets in Indonesia. Adv. Meteorol. **2022** (2022)

Utilization of ECMWF ERA-Interim Reanalysis Data for Analysis of Atmospheric Conditions During Tropical Cyclone Dahlia

Dendi Rona Purnama, I. Nyoman Agus Astina Putra, Dewangga Palguna, and Gandhi Mahendra

Abstract The decay of tropical cyclone Cempaka into a tropical depression was followed by the emergence of a new tropical low in the southwest of Bengkulu. This tropical low status was upgraded on 29th November 2017 at 19.00 LT into tropical cyclone Dahlia. In this study, an analysis of several weather parameters was carried out during the occurrence of tropical cyclone Dahlia, including wind conditions, divergence, sea surface temperature, geopotential height, air temperature, as well as the movement of water vapor transport on 30th November–1st December 2017. The influence caused by the presence of tropical cyclone Dahlia on atmospheric conditions is described in this study. This research utilizes ERA-Interim reanalysis data from the European Center for Medium-Range Weather Forecast (ECMWF). The method used is the descriptive method. The results showed that the sea surface temperature in the study area supports the formation of a tropical cyclone. The significant water vapor transport causes high water vapor intensity and low geopotential height during the research period. The cyclonic pattern is visible in the streamline, and the divergence conditions during the research period are generally negative in the lower tropospheric layers and positive in the upper tropospheric layers. The presence of MJO, Equatorial Rossby Waves, and Kelvin Waves in the same location and time supported the formation of TC Dahlia.

D. R. Purnama (✉)
Public Weather Services, Indonesian Agency for Meteorology Climatology and Geophysics, Angkasa I St. No. 2, Kemayoran, Central Jakarta 10720, Indonesia
e-mail: deronpurna@gmail.com

I. N. A. A. Putra
Kendari Marine Meteorological Station, Indonesian Agency for Meteorology Climatology and Geophysics, Jendral Sudirman St. No. 158, Kendari 93127, Indonesia

D. Palguna
Komodo Meteorological Station, Indonesian Agency for Meteorology Climatology and Geophysics, Yohanes Sehadun St., West Manggarai 86554, Indonesia

G. Mahendra
Sentani Meteorological Station, Indonesian Agency for Meteorology Climatology and Geophysics, Yabaso St., Sentani, Jayapura Region 99111, Indonesia

A. Basit et al. (eds.), *Proceedings of the International Conference on Radioscience, Equatorial Atmospheric Science and Environment and Humanosphere Science*, Springer Proceedings in Physics 290, https://doi.org/10.1007/978-981-19-9768-6_73

1 Introduction

One of the most destructive hydrometeorological disasters on Earth is tropical cyclones (TCs). In some cases, 90% of tropical cyclones have a terrible impact. Tropical cyclones are capable of causing enormous damage to the areas they pass through [1]. This damage can reach a radius of 250 miles from the center of a tropical cyclone [2]. Effects resulting from tropical cyclones can change geomorphology and ecology, especially in water areas [3]. Significant wave heights around tropical cyclones will increase as the intensity of the tropical cyclones increases [4].

Technically, a tropical cyclone is defined as a synoptic-scale non-frontal low-pressure system that grows over warm waters with areas of convective cloud and a maximum wind speed of at least 34 knots in more than half of the area surrounding its center, and lasts at least six hours [5, 6]. This strong cyclonic circulation is formed in the season when the water conditions warm, and this circulation is in the mature stage of the tropical disturbance [7, 8].

A cyclonic circulation can develop into a tropical cyclone due to the Coriolis force, which is supported by several parameters. Gray divides these parameters into two categories, namely thermal parameters (thermal energy of the ocean with sea temperatures > 26 °C to a depth of 60 m, the equivalent potential temperature difference between the surface layer with 500 mb layer > 10 °C, and wet air with humidity > 70% in the 700–500 mb layers) and dynamic parameters (strong low-level vorticity, weak vertical wind shear, and latitude Coriolis parameter > 3°) [9].

Most tropical cyclones form at latitudes 10°–20° from the equator (65%), and only ± 13% of tropical cyclones occur at latitude 22° [2]. With an average tropical cyclone occurrence of 10 times per year, the Indian Ocean region and the waters of Western Australia are the areas where tropical cyclone growth occurs most frequently in the world [10]. The early appearance of most tropical cyclones in the waters of Western Australia or the Southeast Indian Ocean can be found around the intertropical convergence zone (ITCZ). McBride and Keenan [11] estimated that approximately 85% of cyclones in the Southeast Indian Ocean have their origin location near the ITCZ. Tropical cyclone activity around the territory of Indonesia can also be influenced by the presence of El Nino and La Nina phenomena [1, 12] as well as influenced by Indonesian Throughflow conditions [13].

Tropical cyclones have an essential role in atmospheric dynamics, especially in the upper troposphere and lower stratosphere [14]. Several studies on the atmospheric conditions during tropical cyclones have been widely studied using various kinds of data. Fibriantika and Alhaqq [15] have analyzed the vertical profile of the atmosphere on the occurrence of tropical cyclones Cempaka and Dahlia using radiosonde data, with the results showing that tropical cyclone Cempaka has a more significant effect on the condition of the atmospheric vertical profile on Java Island compared to tropical cyclone Dahlia, especially at vertical profile parameters of air temperature, dew point temperature, and humidity. Analysis of tropical cyclone Nock-Ten by Ismail et al. [16] using Himawari satellite data shows that at the time of the occurrence, a high convective index and strong winds were recorded at category five on the

Saffir–Simpson scale. Using the Weather Research and Forecasting (WRF) model, the occurrence of tropical cyclone Haiyan was analyzed on the parameters of wind speed, air humidity, sea surface temperature, surface pressure, and rainfall [17]. In the study of Samrin et al. [18], ECMWF ERA-Interim reanalysis data, Himawari-8 satellite data infrared channel 1 (IR1), and Global Satellite Mapping of Precipitation (GSMaP) data were used to analyze atmospheric conditions and rainfall in tropical cyclone Cempaka.

Various types of modeling data have been widely used to simulate atmospheric conditions, both for analysis of past events, re-create past conditions (hindcast or back-cast), as well as for predicting future atmospheric conditions (forecast). The development of reanalysis data products started in the 1990s by the European Center for Medium-Range Weather Forecast (ECMWF) and the National Centers for Environmental Prediction (NCEP) has played a vital role in simulating future trends and scenarios related to climate change studies [19]. In de Lima and Alcântara's research [20], reanalysis data from the National Centers for Environmental Prediction/National Center for Atmospheric Research (NCEP/NCAR), reanalysis data from Climate Forecast System Reanalysis (CFSR), and ECMWF ERA-Interim reanalysis data were compared to find out how much error the three data generated against the observed data. The results show that the ERA-Interim ECMWF reanalysis data gives the smallest error values for the precipitation and temperature parameters.

The presence of tropical cyclone Dahlia seed also coincided with the collapse of tropical cyclone Cempaka on 29th November 2017 into a tropical depression. This tropical cyclone seed was upgraded to a tropical cyclone status at 19.00 LT and was at a position of 8.2 °S and 100.8 °E. Because the seeds of the cyclone appeared in the area of responsibility of the Tropical Cyclone Warning Center (TCWC) Jakarta, TCWC Jakarta has the right to give a name to the tropical cyclone, hereinafter referred to as tropical cyclone Dahlia.

The purpose of this study was to analyze the atmospheric conditions related to the tropical cyclone Dahlia phenomenon by utilizing the ERA-Interim ECMWF reanalysis data. The effect caused by the presence of tropical cyclone Dahlia on the conditions of the lower and upper layers of the atmosphere will be described in this study.

2 Data and Method

2.1 Time and Location

The research location is the waters of the Southeast Indian Ocean in the southern part of Sumatra and Java at coordinates 0°–20°S and 90°E–130°E (Fig. 1). Research time is limited to 30th November–1st December 2017.

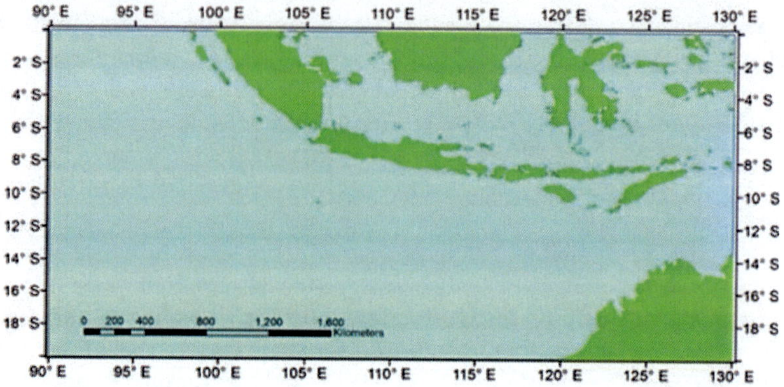

Fig. 1 Research location

2.2 Data

The data used in this study is the ECMWF ERA-Interim reanalysis data downloaded from https://www.ecmwf.int, which has a spatial resolution of 0.125° × 0.125° and a temporal resolution of 6 hours [21, 22]. Some of the parameters used to analyze the atmospheric conditions at the time of the tropical cyclone Dahlia were the zonal wind component (u), the meridional wind component (v), divergence, air temperature, and geopotential height at 850, 700, 500 and 200 mb layers. In addition, this study also used data on specific humidity of 1000, 700, 500 and 300 mb layers as well as sea surface temperature from ECMWF ERA-Interim. All data was downloaded to coincide with 30th November–1st December 2017.

2.3 Method

The descriptive method was used in this research. The ERA-Interim ECMWF reanalysis data was processed using the Grid Analysis and Display System (GrADS) application [23] to display weather parameters spatially during the tropical cyclone Dahlia. Furthermore, descriptive analysis was carried out on several weather parameters, including analyzing streamline, divergence values, sea surface temperature conditions, analyzing geopotential height and air temperature, and water vapor transport movements. The distribution of water vapor transport is calculated by dividing it into several layers, namely 1000–300, 1000–700, 700–500 and 500–300 mb.

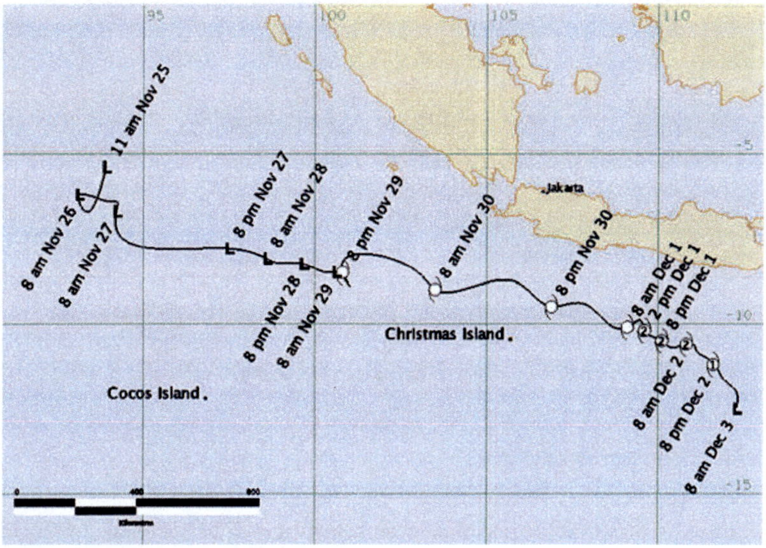

Fig. 2 Track of tropical cyclone Dahlia [24]

3 Result and Discussion

3.1 Tropical Cyclone Dahlia Movement

Tropical cyclone Dahlia began with the formation of a low-pressure pattern in the Indian Ocean west of Sumatera on 25th November 2017 and moved the southeast (Fig. 2). A tropical cyclone began to form on 29th November 2017, which was named tropical cyclone Dahlia by TCWC Jakarta. This tropical cyclone moved southeast to extinction in the waters south of Java on 3rd December 2017.

3.2 Sea Surface Temperatures Analysis

The sea surface temperature that supports the growth of a tropical cyclone is more than 26 °C [2, 9, 25]. In Fig. 3, the sea surface temperature on 30th November 2017 to 1st December 2017 in the southern waters of Java has temperatures ranging from 27.5–29 °C. This indicates that the region has warm sea surface temperatures and supports the occurrence of a tropical cyclone.

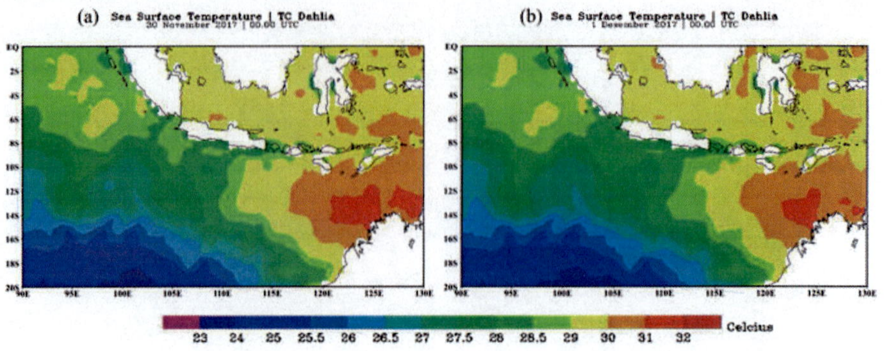

Fig. 3 Sea surface temperature on **a** 30th November 2017 and **b** 1st December 2017

3.3 Streamline Analysis

Streamlines are lines that are parallel to the horizontal wind speed vector at a certain height and time [26]. Figure 4 shows the streamline of tropical cyclone Dahlia that occurred in the southern hemisphere (SH) on 30th November 2017 and 1st December 2017 at 00.00 UTC or 07.00 LT at 850, 700, 500 and 200 mb layers.

On 30th November 2017 at the 850, 700 and 500 mb layers, a slightly elongated cyclonic circulation pattern was seen in the Southeast Indian Ocean (Fig. 4a). The wind speed around the center of the cyclone tends to increase with increasing altitude, while the wind speed at the center of the cyclone is relatively the same in all layers. At the 200 mb layer, the cyclonic wind circulation pattern has changed to anticyclonic winds. Because it is an anticyclonic region, the 200 mb layer is the layer where the air mass exits tropical cyclone Dahlia.

On 1st December 2017 which was the peak of activity for the tropical cyclone Dahlia, the cyclonic wind circulation pattern was not much different from the previous day, but became more visible (Fig. 4b). The intensity of the wind speed around the center of the cyclone also increased more than the previous date in all layers.

3.4 Geopotential Height and Air Temperature Analysis

Atmospheric pressure can be expressed as a function of altitude with the following equation [27]:

$$dp = \rho g dz \tag{1}$$

or

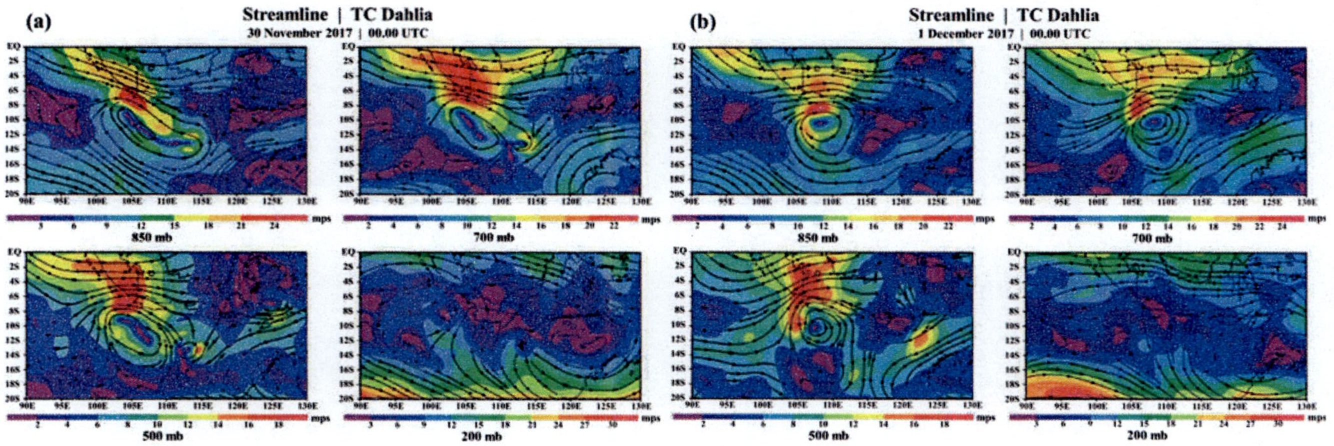

Fig. 4 Streamline in 850, 700, 500 and 200 mb layers at 00.00 UTC on **a** 30th November 2017 and **b** 1st December 2017. The unit of wind speed is meters per second (mps)

$$dp = -\rho d\emptyset \quad (2)$$

where $\emptyset = gz$ is called the geopotential and shows the magnitude of the potential energy at a height z. In the gas equation, $\rho = p/RT$, which shows that p is a function of pressure and T is a function of temperature. Thus, geopotential changes in pressure at a certain altitude layer are related to changes in temperature [27]. Figure 5 shows the geopotential height and air temperature on 30th November–1st December 2017.

On 30th November 2017, in the 850 mb and 700 mb layers, two nearly converging low geopotential areas were seen in the southwest of Java Island (Fig. 5a). In addition, there is another low geopotential area to the east. In the 500 mb layer, there is only one low geopotential area. Temperatures tend to be warmer around low geopotential areas. The low geopotential height indicates that the area is a low-pressure area, and the air movement tends to move up to the upper layers.

Two low geopotential areas in the 850 and 700 mb layers merged into one on 1st December 2017, with the geopotential height becoming progressively lower (Fig. 5b). Likewise, at the 500 mb layer, the geopotential height is getting lower. However, the temperature tends to be the same as the previous day. This condition can indicate the movement of air moving up, and the area is getting lower pressure. In contrast to the layer below, at the 200 mb layer, there is no closed isogeopotential lower than the surrounding area. This indicates that the air movement is no longer converging in this layer, but tends to show a strong divergence of air masses.

3.5 *Divergence Analysis*

The existence of divergence and convergence air mass flow patterns can cause vertical movement [28]. Divergence is an area of air mass distribution caused by air currents spreading over an area. The condition of the area with a negative divergence indicates the presence of air mass compression (convergence) in the area and supports the growth of convective clouds that cause heavy rains [29]. Figure 6 shows the condition of the divergence value on 30th November–1st December 2017.

Based on Fig. 6a, on 30th November 2017, in the 850 mb and 700 mb layers, the divergence value around the tropical cyclone Dahlia area, which is in the south of Java Island, is generally negative (−), which indicates the opportunity for the growth of rain-producing clouds in the region is quite large. At the 500 mb layer, there is a positive (+) divergence value which is squeezed by negative (−) divergence values, this indicates that there is a pattern of air movement that is starting to spread. The 200 mb layer is dominated by positive (+) divergence values, which are quite large and indicate that the air in this layer has been dominated by a diffused air pattern.

In the 850 mb and 700 mb layers on 1st December 2017, in the area where tropical cyclone Dahlia was formed, there was a low divergence value with a wider area coverage than the previous day (Fig. 6b). This indicates that there is a stronger compression of the air masses moving upwards than the previous day. Furthermore, in the 500 mb layer, the tropical cyclone Dahlia area begins to be dominated by

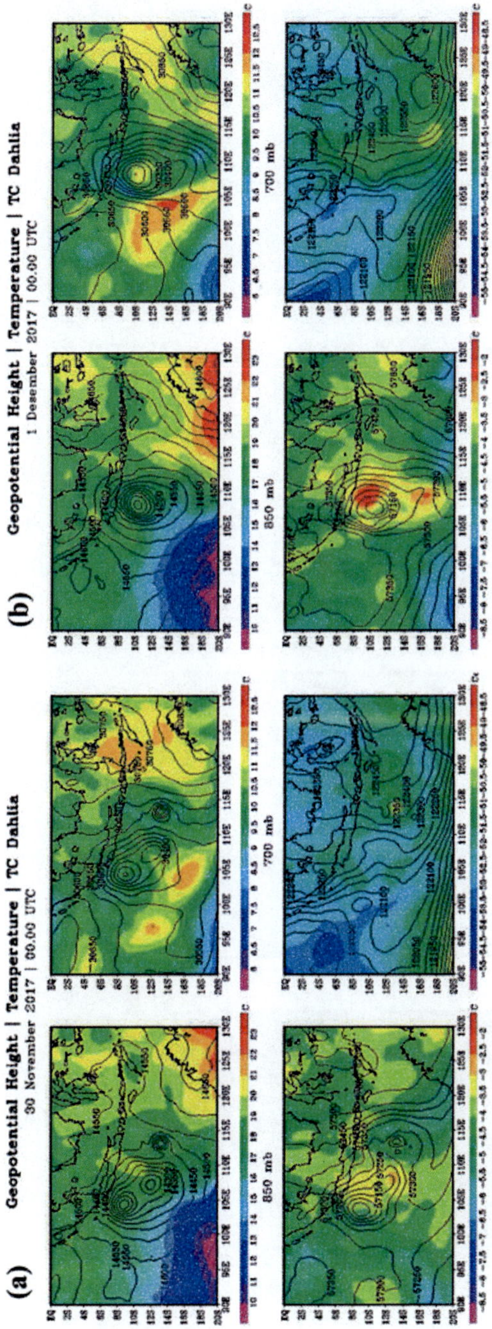

Fig. 5 Geopotential height (solid line) and air temperature (shaded) in the 850, 700, 500 and 200 mb layers at 00.00 UTC on **a** 30th November 2017 and **b** 1st December 2017. The unit of geopotential height is meter

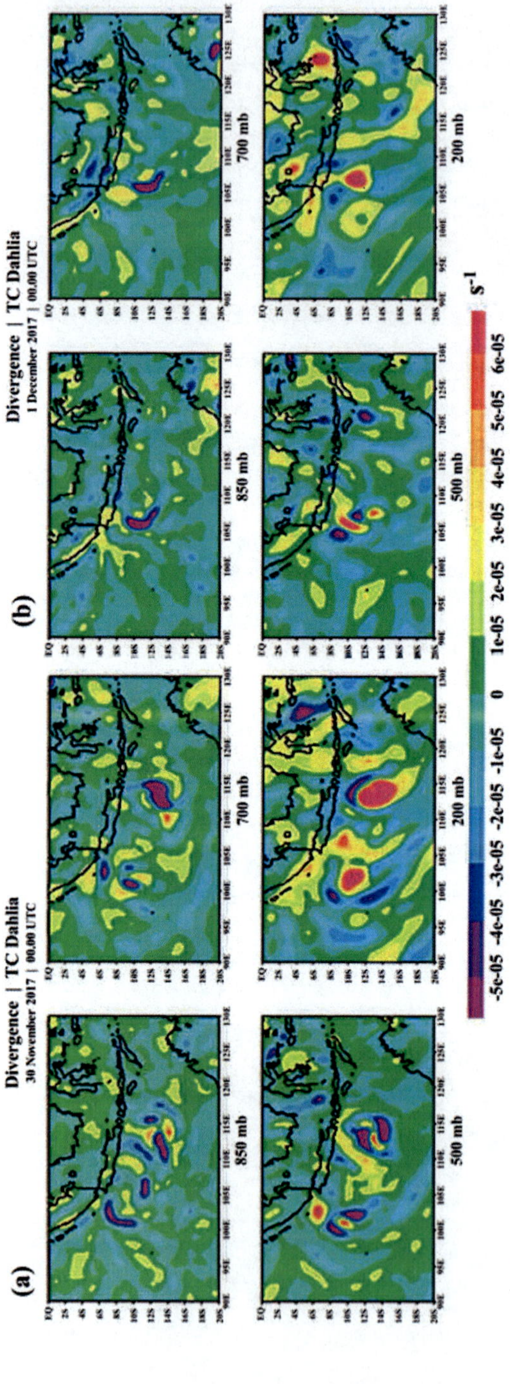

Fig. 6 Divergence of layers of 850 mb, 700 mb, 500 mb, and 200 mb at 00.00 UTC on **a** 30th November 2017 and **b** 1st December 2017

positive (+) divergence values. This indicates that the upward-moving air mass begins to spread in the 500 mb layer. In the 200 mb layer, the tropical cyclone Dahlia area shows a positive divergence value which indicates that in the 200 mb layer the air masses are spreading.

3.6 Water Vapor Transport Analysis

Water vapor transport can have different values and directions in each layer of the atmosphere [30]. Water vapor transport has an important role in analyzing the impact of some weather disturbances on rainfall. One widely adopted approach to determine the characteristics of water vapor transport is to calculate vertically integrated water vapor transport fluxes (IVT) between the pressure in the surface layer and the pressure limit at the highest altitude from a radiosonde measurement or other upper air condition measurement [31]. IVT can be calculated by

$$IVT = \sqrt{\left(\frac{1}{g}\int_{p_0}^{p} qudp\right)^2 + \left(\frac{1}{g}\int_{p_0}^{p} qvdp\right)^2} \quad (3)$$

where

q: specific humidity (kg/kg)
u: zonal wind component (m/s)
v: meridional wind component (m/s)
p_0: pressure in the lower boundary layer or surface layer pressure (mb)
p: pressure in the upper boundary layer or the pressure of the highest layer to be analyzed (mb).

The product components $q \times u \times dp/g$ and $q \times v \times dp/g$ at each grid point are summed vertically from the surface layer (bottom layer pressure limit) to a certain layer (top layer pressure limit) which are then combined into a horizontal transport vector, where the unit of IVT is kg m^{-1} s^{-1} [31, 32]. For example, if you want to calculate IVT in 1000–500 mb layers, then p_0 can be replaced with 1000, and p can be replaced with 500. In this study, the water vapor transport analyzed is between layers 1000–300 mb (overall troposphere), 1000–700 mb (lower troposphere), 700–500 mb (middle troposphere), and 500–300 mb (upper troposphere). Figure 7 shows water vapor transport on 30th November–1st December 2017.

In Fig. 7a, water vapor transport in the 1000–300 mb layer on 30th November 2017 shows a movement of water vapor from north to south which indicates negative meridional winds and positive zonal winds. This indicates the presence of Asian monsoon winds moving to SH carrying water vapor of 800 to more than 1600 kg m^{-1} s^{-1} to the southeastern part of Java Island. The high intensity of water vapor has an effect on increasing rainfall around the tropical cyclone Dahlia area. In the 1000–700

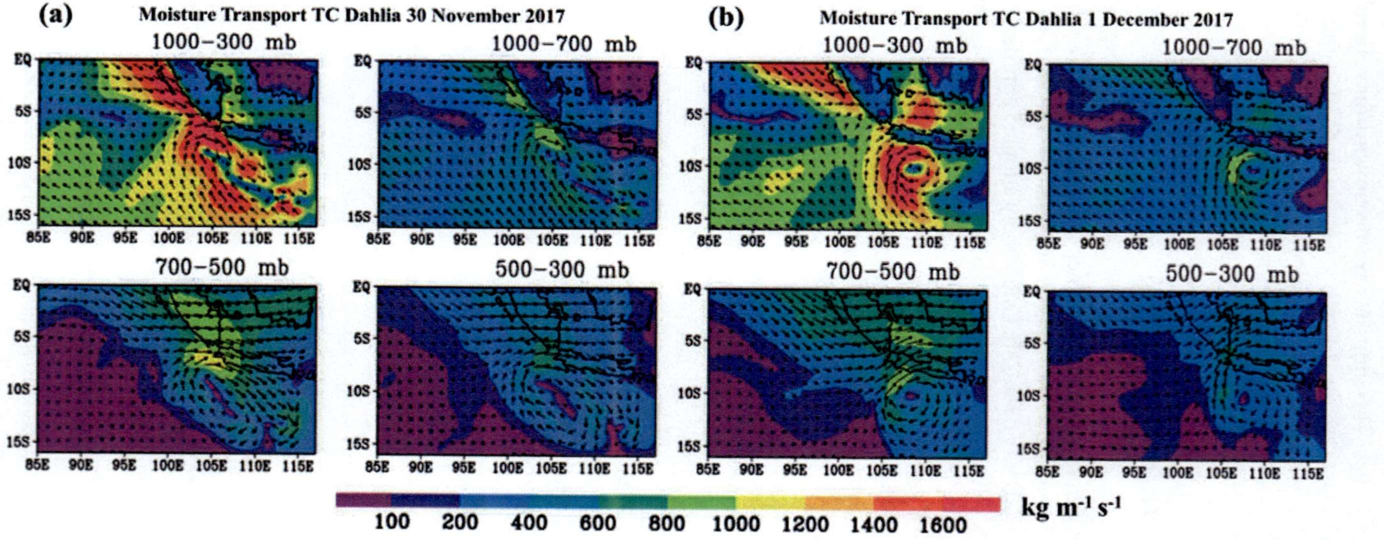

Fig. 7 Water vapor transport between 1000–300 mb, 1000–700 mb, 700–500 mb, and 500–300 mb at 00.00 UTC on **a** 30th November 2017 and **b** 1st December 2017

mb layer, the movement of water vapor tends to be the same as the 1000–300 mb layer, namely from the northwest of Sumatra Island to the southeast. Water vapor transport in this layer ranges from 400 to 1000 kg m^{-1} s^{-1}. The cyclonic pattern has been formed as in the previous layers. The distribution of water vapor in the 700–500 mb layers shows a fairly high water vapor, with the intensity of water vapor moving from along the northwest of Sumatra Island to the southeast in the range of 400–1200 kg m^{-1} s^{-1}. Likewise, in the 500–300 mb layer, the water vapor transport pattern is the same as the previous layer, but with less intensity (100–800 kg m^{-1} s^{-1}).

In Fig. 7b, in the 1000–300 mb layer, the water vapor is concentrated in several areas. Water vapor is concentrated mainly around the center of the cyclonic pattern (south of Java Island) and west of Sumatra Island with an intensity reaching more than 1600 kg m^{-1} s^{-1}. The water vapor mass movement pattern is the same as the previous day. The condition of the water vapor transport pattern in the 1000–700 mb layer is similar to the 1000–300 mb layer. The intensity in this layer tends to be lower than in the previous layer, which is 400–1000 kg m^{-1} s^{-1} in the tropical cyclone Dahlia area. In the 700–500 mb layer, the pattern water vapor transport moves from the northeast (in the South China Sea) and meets in the western area of Sumatra Island, flowing southeast to enter the territory of Indonesia. The cyclonic pattern is still visible in the south of Java Island. The concentration of water vapor occurs in the Sunda Strait and around the center of a tropical cyclone with an intensity of 400–1000 kg m^{-1} s^{-1}. In the 500–300 mb layer, the conditions are similar to the 700–500 mb layer with a lower intensity than the previous three layers, which is 200–1000 kg m^{-1} s^{-1} in the area of concentration of water vapor.

3.7 Influence of Tropical Waves

In addition to the thermal and dynamic parameters required for TCs growth [9], another potential factor causing TCs formation is the modulation by tropical waves. In conditions before, during, and after TC Dahlia, the Madden–Julian Oscillation (MJO) was active in quadrant 4 (maritime continent). Equatorial Rossby (ER) was also observed in the area where TC Dahlia grew, both before and during the occurrence of TC Dahlia. In addition, Kelvin Waves were also observed in the western part of Indonesia accompanied by the presence of low-frequency waves (Fig. 8).

The presence of MJO and ER is known to have a significant large modulation of TC genesis over the south Indian Ocean [35–37]. Meanwhile, Kelvin Waves have a significant small modulation [36]. With the presence of MJO and ER at the same time, the modulation was best attributable to its vorticity, convection, and shear fields. Thus, the presence of MJO, ER, and Kelvin Waves in the same location and time could be a supporting factor in the formation of TC Dahlia.

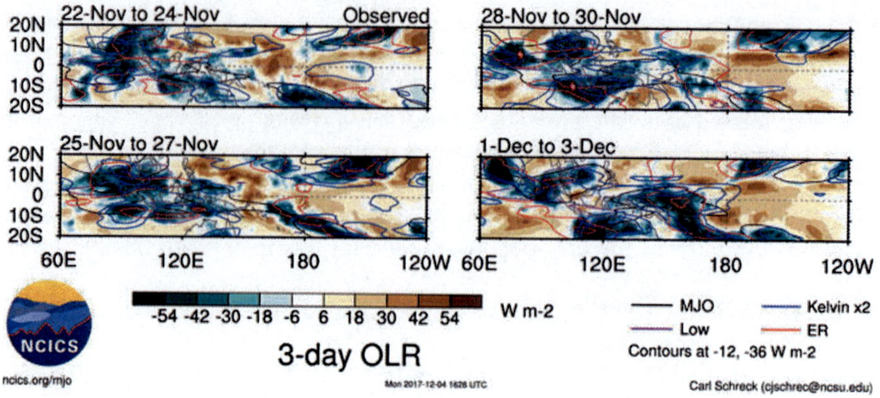

Fig. 8 Outgoing Longwave Radiation (OLR) in the 850 mb layer combines with line contours of MJO, Equatorial Rossby waves, Kelvin waves, and the low-frequency waves [33, 34]

4 Conclusion

The results of the analysis of the occurrence of tropical cyclone Dahlia using the ECMWF ERA-Interim reanalysis data showed that, in general, the data was able to represent atmospheric conditions that were in accordance with the conditions required for the formation of a tropical cyclone. The use of this data is able to describe atmospheric conditions, both vertically and horizontally, especially related to atmospheric conditions at the occurrence of the tropical cyclone Dahlia. The presence of MJO, ER, and Kelvin Waves in the same location and time supported the formation of TC Dahlia.

The analysis of the atmospheric conditions at tropical cyclone Dahlia showed that both cyclonic patterns and convergence conditions were found in the lower-to-middle troposphere layer (500 mb layer and below), both in streamline analysis, air temperature, isogeopotential, and divergence. Meanwhile, the upper troposphere shows an anticyclonic and divergence pattern. This means that in the upper troposphere, the movement of air masses is spread, especially in the 200 mb layer. In addition, the water vapor transport analysis showed a mass accumulation of water vapor around tropical cyclone Dahlia. The distribution of water vapor from the South China Sea indicates the activity of the Asian monsoon winds moving to the southern hemisphere.

References

1. Purnama, D.R., Zulistyawan, K.A., Christian, B., Veanti, D.P.O.: Dampak terjadinya el nino/la nina terhadap intensitas, masa hidup dan frekuensi siklon. Jurnal Meteorologi Klimatologi dan Geofisika **5**(2), 10–21 (2018)
2. Tjasyono, B.: Klimatologi Umum. Penerbit ITB, Bandung (2004)

3. Done, T.J.: Effects of tropical cyclone waves on ecological and geomorphological structures on the Great Barrier Reef. Cont. Shelf Res. **12**(7–8), 859–872 (1992)
4. Haryanto, Y.D., Riama, N.F., Purnama, D.R., Sigalingging, A.D.: The effect of the difference in intensity and track of tropical cyclone on significant wave height and wave direction in the Southeast Indian Ocean. The Scient. World J. **5492048**, 11 (2021)
5. Badan Meteorologi Klimatologi dan Geofisika.: http://web.meteo.bmkg.go.id/id/component/content/article/37-siklon-tropis/268-siklon-tropis. Last Accessed 01 Nov 2021
6. Zakir, A., Sulistya, W., Khotimah, M.K.: Perspektif operasional cuaca tropis. Pusat Penelitian dan Pengembangan BMKG, Jakarta (2010)
7. Saha, K.: Tropical Circulation Systems and Monsoons. Springer, Heidelberg (2010)
8. Kepert, J.D.: Tropical cyclone structure and dynamics. In: Global Perspectives on Tropical Cyclones, pp. 3–53. (2010)
9. Gray, W.M.: Hurricanes: their formation, structure and likely role in the tropical circulation. Meteorol. Over the Tropical Oceans **155**, 218 (1979)
10. Haryani, N.S., Zubaidah, A.: Dinamika siklon tropis di Asia Tenggara menggunakan data penginderaan jauh. Majalah Ilmiah Widya **29**(324), 54–58 (2012)
11. McBride, J.L., Keenan, T.D.: Climatology of tropical cyclone genesis in the Australian region. J. Climatol. **2**(1), 13–33 (1982)
12. Camargo, S.J., Sobel, A.H.: Western North Pacific tropical cyclone intensity and ENSO. J. Clim. **18**(15), 2996–3006 (2005)
13. Pradita, N., Ismah, S.F., Purnama, D.R.: Pengaruh arlindo terhadap durasi dan intensitas siklon tropis di Pasifik Barat Laut dan Samudra Hindia Selatan. In: Seminar Nasional Bumi dan Atmosfer, pp. 7–15. Sekolah Tinggi Meteorologi Klimatologi dan Geofisika, South Tangerang (2019)
14. Ratnam, M.V., Babu, S.R., Das, S.S., Basha, G., Krishnamurthy, B.V., Venkateswararao, B.: Effect of tropical cyclones on the stratosphere-troposphere exchange observed using satellite observations over the north Indian Ocean. Atmos. Chem. Phys. **16**(13), 8581–8591 (2016)
15. Fibriantika, E., Alhaqq, R.I.: Profil vertikal atmosfer selama aktifitas siklon tropis Cempaka dan Dahlia. Jurnal Meteorologi dan Geofisika **19**(2), 49–58 (2019)
16. Ismail, P., Hidayat, N.M., Siadari, E.L.: Analisis siklon tropis Nock-Ten berbasis data Satelit Himawari. Jurnal Meteorologi Klimatologi dan Geofisika **4**(3), 16–23 (2017)
17. Perawiska, E., Muliadi, M., Adriat, R.: Analisis unsur cuaca pada saat kejadian siklon tropis Haiyan menggunakan model WRF (weather research and forecasting). Prisma Fisika **6**(2), 129–136 (2018)
18. Samrin, F., Irwana, I., Hasanah, N.: Analysis of the meteorological condition of tropical cyclone Cempaka and its effect on heavy rainfall in Java Island. In: IOP Conference Series: Earth and Environmental Science vol. 303(1), pp. 012065. (2019)
19. Houle, D., Bouffard, A., Duchesne, L., Logan, T., Harvey, R.: Projections of future soil temperature and water content for three southern Quebec forested sites. J. Clim. **25**(21), 7690–7701 (2012)
20. de Lima, J.A.G., Alcântara, C.R.: Comparison between ERA Interim/ECMWF, CFSR, NCEP/NCAR reanalysis, and observational datasets over the eastern part of the Brazilian Northeast Region. Theoret. Appl. Climatol. **138**(3), 2021–2041 (2019)
21. Berrisford, P., Dee, D., Poli, P., Brugge, R., Fielding, K., Fuentes, M., Kållberg, P., Kobayashi, S., Uppala, S., Simmons, A.: The ERA-interim archive version 2.0, European Centre for Medium Range Weather Forecasts, Reading (2011)
22. Dee, D.P., Uppala, S.M., Simmons, A.J., Berrisford, P., Poli, P., Kobayashi, S., Andrae, U., Balmaseda, M.A., Balsamo, G., P. Bauer, Bechtold, P., Beljaars, A.C.M., van de Berg, L., Bidlot, J., Bormann, N., Delsol, C., Dragani, R., Fuentes, M., Geer, A.J., Haimberger, L., Healy, S.B., Hersbach, H., Hólm, E.V., Isaksen, L., Kållberg, P., Köhler, M., Matricardi M., McNally, A.P., Monge-Sanz, B.M., Morcrette, J.-J., Park, B.-K., Peubey, C., de Rosnay P., Tavolato, C., Thépaut, J.-N., Vitart, F.: The ERA-Interim reanalysis: Configuration and performance of the data assimilation system. Quart. J. The Royal Meteorol. Soc. **137**(656), 553–597 (2011)

23. Doty, B.E., Kinter III, J.L., Fiorino, M., Hooper, D., Budich, R., Winger, K., Schulzweida, U., Calori, L., Holt, T., Meier, K.: The grid analysis and display system (GrADS): An Update for 1997. In 13th International Conference on IIPS for Meteorology, Oceanography and Hydrology, American Meteorology Society, vol. 117 (1997)
24. Bureau of Meteorology.: http://www.bom.gov.au/cyclone/history/Dahlia.shtml. Last Accessed 01 Nov 2021
25. Wang, Z.: Tropical cyclogenesis. In: Encyclopedia of Atmospheric Sciences, vol. 6(2), pp. 57–64. (2015)
26. Wallace, J.M., Hobbs, P.V.: In: Atmospheric Science: an Introductory Survey. Elsevier (2006). 27. Wirjohamidjojo, S., Swarinoto, Y.S.: Meteorologi sinoptik. Pusat Penelitian dan Pengembangan BMKG, Jakarta (2013).
27. Wirjohamidjojo, S., Swarinoto, Y.S.: Meteorologi sinoptik. Pusat Penelitian dan Pengembangan BMKG, Jakarta (2013)
28. Trewartha, G.T., Horn, L.H.: An Introduction to Climate, 5th edn. McGraw-Hill Company, New York (1980)
29. Ariyanti, R.D., Sugihartati, F.M., Retharanita, W., Winarso, P.A.: Uji keakuratan data GFS menggunakan model WRF-ARW pada siklon tropis Flamboyan. J. Ilmu dan Inovasi Fisika **3**(2), 62–74 (2019)
30. Xiaoxia, Z., Yihui, D., Panxing, W.: Moisture transport in the Asian summer monsoon region and its relationship with summer precipitation in China. J. Meteorol. Res. **24**(1), 31–42 (2010)
31. Zhu, Y., Newell, R.E.: A proposed algorithm for moisture fluxes from atmospheric rivers. Mon. Weather Rev. **126**(3), 725–735 (1998)
32. Neiman, P.J., Ralph, F.M., Wick, G.A., Lundquist, J.D., Dettinger, M.D.: Meteorological characteristics and overland precipitation impacts of atmospheric rivers affecting the West Coast of North America based on eight years of SSM/I satellite observations. J. Hydrometeorol. **9**(1), 22–47 (2008)
33. North Carolina Institute for Climate Studies.: https://ncics.org/portfolio/monitor/mjo/. Last Accessed 01 Nov 2021
34. Wheeler, M., Weickmann, K.M.: Real-time monitoring and prediction of modes of coherent synoptic to intraseasonal tropical variability. Mon. Weather Rev. **129**(11), 2677–2694 (2001)
35. Camp, J., Wheeler, M.C., Hendon, H.H., Gregory, P.A., Marshall, A.G., Tory, K.J., Watkins, A.B., MacLachlan, C., Kuleshov, Y.: Skilful multiweek tropical cyclone prediction in ACCESS-S1 and the role of the MJO. Q. J. R. Meteorol. Soc. **144**(714), 1337–1351 (2018)
36. Bessafi, M., Wheeler, M.C.: Modulation of south Indian Ocean tropical cyclones by the Madden–Julian oscillation and convectively coupled equatorial waves. Mon. Weather Rev. **134**(2), 638–656 (2006)
37. Wang, B., Moon, J.Y.: An anomalous genesis potential index for MJO modulation of tropical cyclones. J. Clim. **30**(11), 4021–4035 (2017)

Evaluation of Detecting and Tracking Algorithms of Reflectivity Area Based on Rain Scanner Observation Data

Tiin Sinatra, Edy Maryadi, Syahrul, Syukri Darmawan, Ginaldi Ari Nugroho, and Asif Awaludin

Abstract A rain scanner for rainfall monitoring had been developed using marine radar. Rainfall monitoring as a part of atmospheric observation was conducted using this rain scanner in near real time. Rain scanner could produce spatial rain conditions in reflectivity every 2 min on a radius of 44 km. This study utilizes the output of the rain scanner and generates the rainfall movement trajectory. Preliminary processing using object detection is conducted to separate the rainfall object from its background. Six case studies of the single rainfall cell in Bandung in January 2021 are used. The results show that, in general, using edge detection, find contour, and frame difference methods can identify rain objects well. The tracking results can carry in the same direction. However, the detailed rain tracking results have differences. Verified with other data, this method could serve as a reliable and effective system for the trajectory of the movement of rain in urban areas.

1 Introduction

A weather radar is an essential instrument for providing rainfall distribution information over a large area within a certain time interval. Using the radar output, vectors showing the direction and speed of the precipitation trajectory can be determined [1]. The quantitative nowcast algorithm is mainly based on extrapolation of the rainfall distribution. The apparent movement of the rainfall distribution from the radar is extrapolated toward time at each grid or pixel. One of the extrapolation methods is Lagrangian persistence which allows extrapolation to obtain the next precipitation with the assumption that the precipitation features and movement are persistent [2]. The Lagrangian method consists of three steps: identifying, tracking, and forecasting [3]. The first step is identifying the rainfall object from the current radar data. The

T. Sinatra (✉) · G. A. Nugroho · A. Awaludin
BRIN—Research Organization for Earth Sciences and Maritime, Bandung, Indonesia
e-mail: tiin001@brin.go.id

E. Maryadi · Syahrul · S. Darmawan
BRIN—Research Organization for Electronics and Informatics, Bandung, Indonesia

A. Basit et al. (eds.), *Proceedings of the International Conference on Radioscience, Equatorial Atmospheric Science and Environment and Humanosphere Science*, Springer Proceedings in Physics 290, https://doi.org/10.1007/978-981-19-9768-6_74

second step is tracking the motion by identifying the same object in successive radar data. Later, forecasting using the extrapolation method [4].

The early steps of identification and tracking are essential to provide valuable information for the next forecasting calculation. Several methods have been proposed. A field-based method or cell-based method is often used in the identification method. The field-based advection considers all the pixels or grids of the radar data for calculating the motion vector. The cell-based method identifies a rainfall object from a connected domain within a certain threshold [5].

The tracking method based on cell-based identification (referred to as cell tracking) is often used since it is more efficient than field-based. Several studies of cell tracking have been done using various methods, such as cross-correlation [6], tracking barycenter of object cells [7], and motion vector calculation using optical flow [5]. The basic idea of this cross-correlation is to spatially shift the radar image taken at time t and overlay it with the image taken at time $t - \mathrm{d}t$. The next step is to look for the best possible relative position of the two images that correspond to the best cross-correlation coefficient [3]. On the barycenter method, it calculates the coordinate from the center line of the precipitation cell based on two consecutive radar data. Confidence factors were analyzed as a separate measurement with other neighborhood cells.

Another study tracked the results of radar monitoring using the optical flow method [8]. One method for tracking is moving object detection (MOV), a method for recognizing or determining the presence of an object and its scope and location on a moving image or video [9, 10] based on the pixel values over time [11] by separating an object from its background. The techniques that can be used to get the value of these pixels are temporal frame difference, background subtraction, optical flow, double difference, statistical methods, and others.

Identification and tracking are well-established and studied using common weather radar. Implementing those two methods into a rain scanner is different challenge. A rain scanner developed based on navigation radar is often used as the filling gap of the weather radar network [12–14]. The advantage of this rain scanner is small, compact size, and ease of installation. The rain scanner can detect rain with a daily pattern close to the daily pattern of rainfall observed by the rain gage [15, 16]. The output of the rain scanner is monitoring spatial rain conditions in a series of images every 2 min. This research utilizes the output of this rain scanner for further analysis in order to provide information in the form of a rain movement trajectory.

Tracking an object can be easier if the object does not change much in both feature and shape. This condition can be obtained if the temporal resolution between images from rain observations is very high and the observed rain comes from a single cloud. In fact, the observed rain can dissociate into several rain cells or, conversely, aggregate of several cells into a unique cell. Even for complex cases, the three conditions above can be observed in a one-time frame. A comprehensive method is needed to cover all three conditions. This study only tested the method for single-cell cases. The rain scanner data used in this study has 2 min temporal where the area and intensity change significantly within 2 min. This study aims to apply moving object detection to determine the movement of precipitation.

2 Methodology

Rain scanner is installed in Bandung (6.894941°S, 107.58648°E), West Java, with 44 km effective radius. It can cover rain observation over Bandung. Bandung is an urban area that consists of mountain area in the north and southern part (Fig. 1). The data used in this study are rain scanner observation data during January 2021, which was randomly selected. The characteristics of the rain object are summarized in Table 1. The rain scanner is able to produce rain observations spatially with a temporal resolution of 2 min and a spatial resolution of 120 m × 20 m at a radius of 44 km. Sequence data of rain scanner observations are then further processed for rain tracking.

Data processing begins with data collection for the six rain events. The rain object selected for each rain event is a single cell. In general, the steps in this study are shown in Fig. 2. In order to determine whether the trajectory result is good or not, we use three criteria. The first is the ability to identify the rain objects. The second is that the trajectory result could follow the rain object propagation based on Table 1. The third is that no backward trajectory path exist, only forward movement is captured.

Method-1

The first method uses the optical flow method. Optical flow Lukas-Kanade was used for estimating the direction of moving rain. Lucas-Kanade applies a local differential approach which calculate the flow a point by identifying the intersection of all flow path constraints to the image pixels within a frame. This method assumptions that

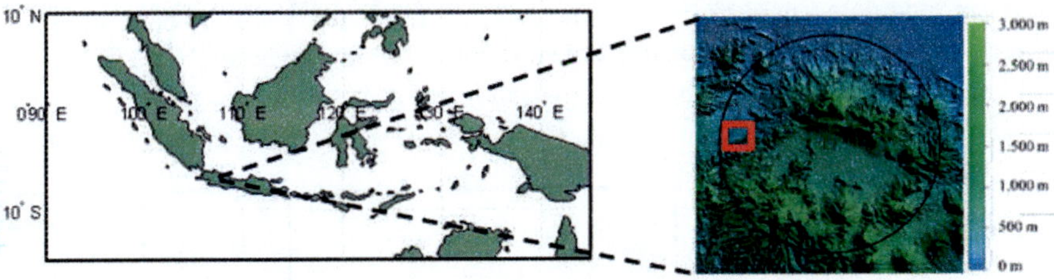

Fig. 1 Map of research location; black circle is coverage area of rain scanner

Table 1 Characteristics of the selected events

Event no	Start	End	Decaying type	Propagation
Event 1	1 Jan 2021 13:28	1 Jan 2021 14:20	Develop-decay	Southeast
Event 2	1 Jan 2021 14:38	1 Jan 2021 15:08	Develop	East
Event 3	18 Jan 2021 17:52	18 Jan 2021 18:34	Develop	Southeast
Event 4	20 Jan 2021 02:42	20 Jan 2021 03:18	Decay	Southeast
Event 5	30 Jan 2021 12:02	30 Jan 2021 12:28	Develop	Southeast
Event 6	30 Jan 2021 13:20	30 Jan 2021 13:38	Decay	Southeast

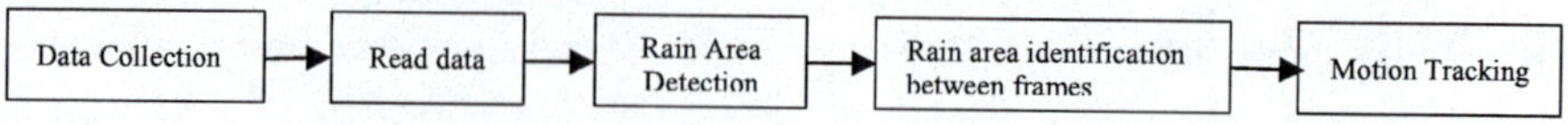

Fig. 2 Tracking rain algorithm

the brightness of the image is constant, where the movement at a point in the image is accompanied by no change in brightness level [17].

The first step in this method is to select objects in the frame using edge detection. Next, read the initial frame (t frame) to determine the interesting point formed on the detected rain object. The interesting point is determined using the Shi-Tomasi Corner Detection algorithm [18]. This algorithm determines the interesting point by comparing the eigenvalue with a certain value using (1).

$$R = \min(\lambda_1, \lambda_2) \tag{1}$$

Dot blue in Fig. 3a shows the interesting point determined. Then, estimate the movement of the interest point in the next frame ($t + 1$) until the frame-n ($t + n$) uses the Lukas-Kanade algorithm. The result of object tracking in frame-n is shown in Fig. 3b.

Method-2

Figure 4 shows algorithm of method 2. In this method, the first thing to do is to subtract the reflectivity value of the object in the frame-t with frame $t - 1$. The results obtained are positive value (new rain areas appeared-red color) and negative value (old rain areas disappeared—blue color). Position x and y for all red and blue areas are calculated using mean formula to get the center point of each area. The center point is then used to determine the direction of rain movement. The illustration of the application of this method is shown in Fig. 5.

Fig. 3 Simulation of rain detection and rain tracking uses Lukas-Kanade

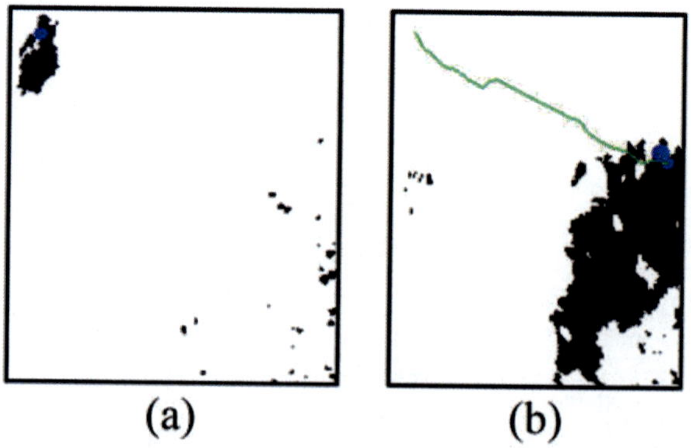

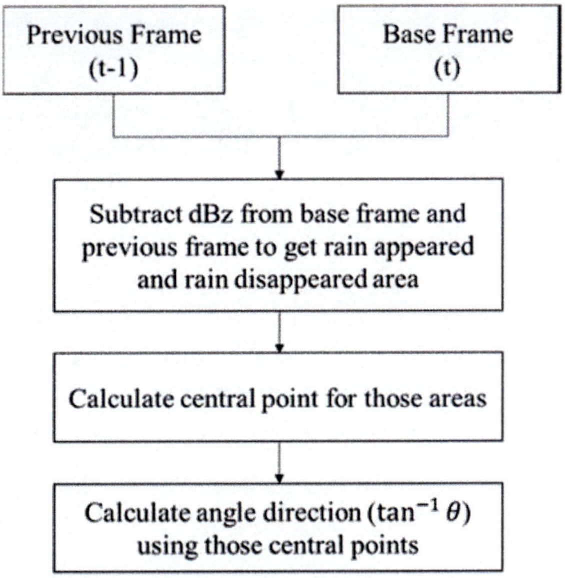

Fig. 4 Method 2 algorithm

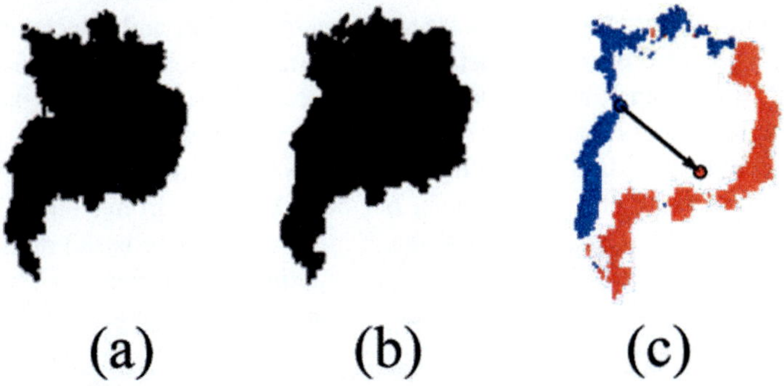

Fig. 5 **a** Previous frame ($t - 1$), **b** base frame (t), **c** the result of subtracted reflectivity from base frame and previous frame

Method-3 and 4

Methods 1 and 2 identify the selected rain object for tracking by cutting other rain objects that are not the focus of monitoring in one frame. On the other hand, methods 3 and 4 identify the focus rain objects without eliminating other rain objects in one frame. Figure 6b shows the results of tracking the position of the rain area on one rain scanner data using method 3. Rain area detection is a process to find the rain area in each spatial rain data automatically, with an algorithm depicted in Fig. 7. Rain area detection will mark areas in the spatial rain data and will treat them as rectangular

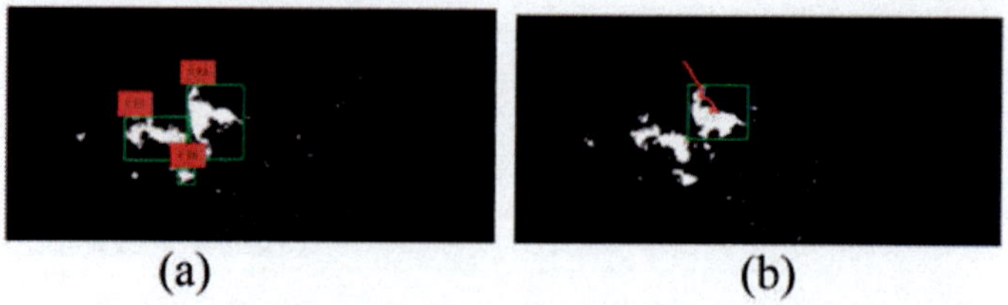

Fig. 6 **a** The result of rain detection. **b** The result of rain tracking. The rain area in the spatial rainfall data is marked with a green box

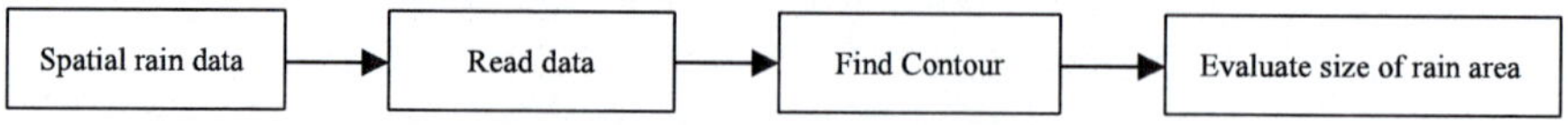

Fig. 7 Rain area detection algorithm of method 3

objects, with the center point as an attribute of the object. The rain area resulting from the rain area detection process is shown in Fig. 6a.

Identifying the rain area between the spatial rainfall data frame-t and frame $t + 1$ is calculated using the K-Nearest Neighbor (K-NN) method. Identification of the rain area is carried out to find the equivalent of each rain area on frame t with the rain area on spatial rain data to $t + 1$. K-NN calculates the distance between objects based on object attributes, in this case, is the center point, and selects as many as K objects that have the smallest distance or known as neighbors. Method 3 uses the value of $K = 1$ and the threshold large of rain area is 200 and 700, and the distance between objects is less than 3000. The improvement is conducted in this algorithm by applying a threshold on the Euclidean distance. Threshold of distance is used to prevent failed selected objects that can be occurred. Different from method 3, method 4 identifies the equivalent of the rain object in the next frame based on the closest distance without a specified range. In addition, method 4 identifies rain objects using edge detection, whereas method 3 uses one of shape identification techniques that is find contour.

3 Results and Discussion

Figure 8 is the result of observing rain with a rain scanner on January 18, 2021. It is seen that the object of rain changes in shape and reflectivity. Proper detection of rain objects needs to be considered for tracking rain.

Figure 9 shows the results of single-cell rain tracking with the four methods for each rain event. In general, it can be seen that the four methods show the movement of rain in the same direction, namely to the southeast for events 1, 3, 4, 5, and 6 and to the east for event 2.

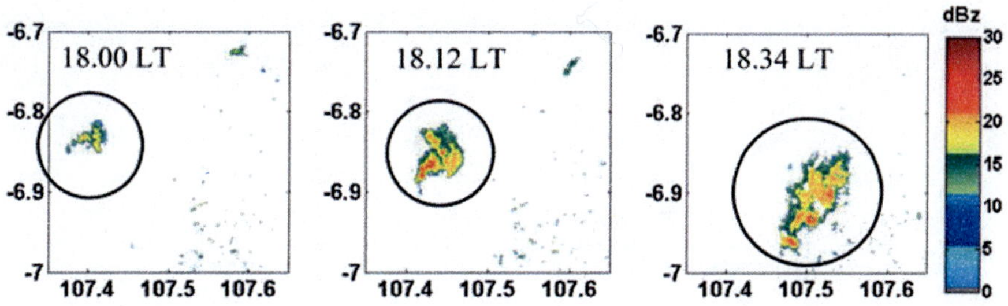

Fig. 8 Time variation of the rainfall event observed by rain scanner on January 18, 2021

Detailed analysis of the trajectory showed different characteristics. The trajectory from method 1 and method 2 in several events such as events 2, 3, 5, and 6 are quite different from methods 3 and 4. On the other hand, methods 3 and 4 show almost the same rain tracking (even coincide). This is because the tracking methods 3 and 4 are almost similar. Although, a small difference can be found between these two methods because of the difference in the identification method (edge detection and find contour), such as in event 2.

The deviation of the trajectory is very visible in the result of method 1 (event 2, 3, 5, and 6). Although in the identification process utilizes edge detection same as method 4, the tracking system depends on the determination of the interest point. The calculation of the interest point will affect the determination of the trajectory. All the trajectory results did not show any backward path, which shows that all of the methods are able to identify and track the rain object movement.

Another interesting characteristic is the start and endpoint of each method. The start points of the rain trajectory of method 2 is different from the other methods (events 1, 2, and 6), although the position of the endpoint is almost the same. This is because when determining the center point of rain, there is a part of another object (noise seen as speckle in Fig. 8) which is included in the calculation.

Only events 1 and 4 that give a better conformity of the start point (except method 2) and endpoint. This shows that the complexity of the rain object also affected the tracking result. Overall result confirms that the object identification still needs to be improved to increase tracking accuracy.

4 Conclusion

Detection and tracking applied on a single rain cell have been carried out for six different objects. The results show that in general, the four methods are able to identify rain objects well. Identification of rain objects using edge detection (method 1 and 4), frame difference (method 2), and find contour (method 3) gives almost the same results. The tracking results from the four methods as a whole are able to carry in the same direction. More detailed analysis shows that method 2 is less able to handle

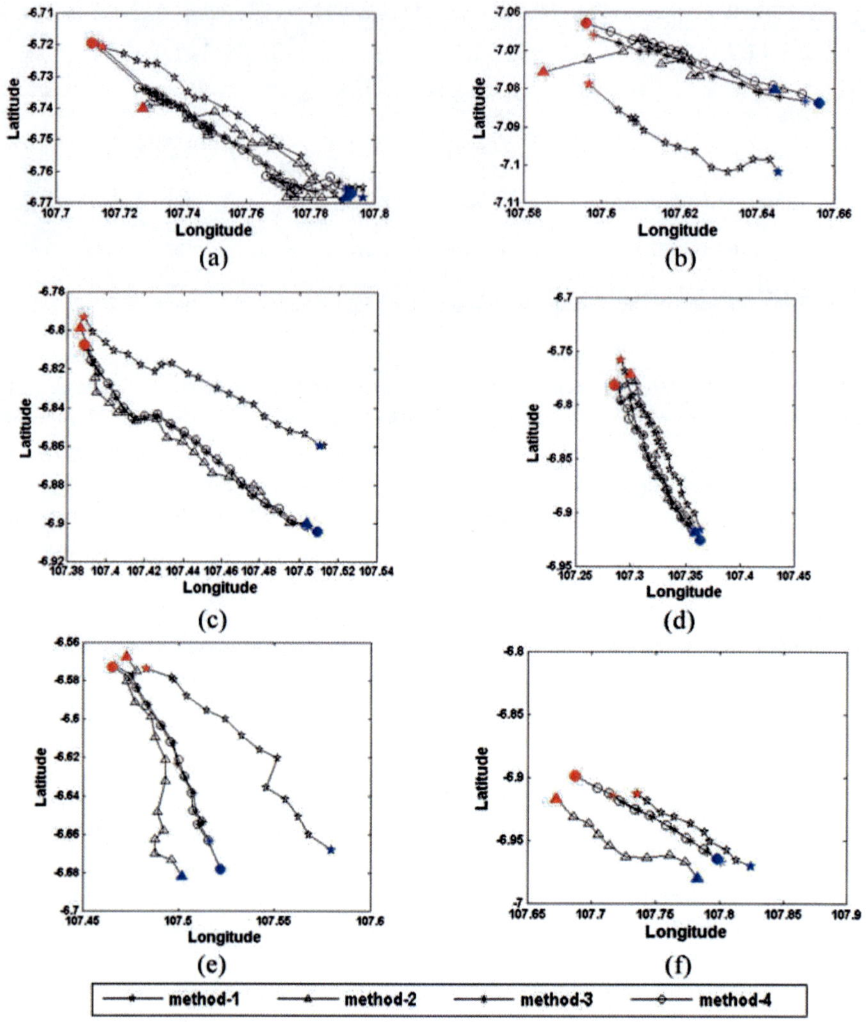

Fig. 9 Time variation of the rain tracking for **a** event 1, **b** event 2, **c** event 3, **d** event 4, **e** event 5, **f** event 6

speckle (noise) than other methods (1, 3, 4) because it is more sensitive to small/noise objects. In tracking rainfall, all methods are able to generate trajectory with almost similar direction to the rain object propagation. No backward trajectory paths were found within the result of those four methods. However, in the detailed analysis of the tracking result, Shi-Tomasi corner on method 1 showed a wider deviation of trajectory compared with other methods. Based on the conditions described, methods 3 and 4 give the optimum results for detecting and determining rain trajectory compared with other methods. Further evaluation is needed to compare the trajectory result with the reference data such as the automatic weather station network.

Acknowledgements The authors would like to thank the National Research and Innovation Agency for providing the funding number 15/III/HK/2022 for this research. This research also is partly funded by research grant of Riset Inovasi Untuk Indonesia Maju number 65/II 7/HK/2022. We would also like to express our gratitude to our senior researchers who gave us guidance and support.

References

1. Ayzel, G., Heistermann, M., Winterrath, T.: Optical flow models as an open benchmark for radar-based precipitation nowcasting (rainymotion v0.1). Geosci. Model Dev. **12**, 1387–1402 (2019)
2. Germann, U., Zawadzki, I.: Scale-dependence of the predictability of precipitation from continental radar images. Part I: description of the methodology. Mon. Weather Rev. **130**, 2859–2873 (2002). https://doi.org/10.1175/1520-0493(2002)1302.0.CO;2
3. Austin, G.L., Bellon, A.: The use of digital weather radar records for short-term precipitation forecasting. Q. J. Roy. Meteor. Soc. **100**, 658–664 (1974)
4. Austin, G. L., Bellon, A.: Guidelines for Nowcasting Techniques (WMO-No. 1198). Geneva (2017).
5. He, T., Einfalt, T., Zhang, J., Hua, J., Cai, Y.: New algorithm for rain cell identification and tracking in rainfall event analysis. Atmosphere **10**(9), 532 (2019)
6. Brémaud, P.J., Pointin, Y.B.: Forecasting heavy rainfall from rain cell motion using radar data. J. Hydrol. **142**(1–4), 373–389 (1993)
7. Raaf, O.: Efficient method for detecting and tracking rainfall clouds in non-Doppler radar images. J. Appl. Remote Sens. **8**(1), 083547 (2014)
8. Nugroho, G.A., Maryadi, E.: Identifikasi Pergerakan area Reflektivitas Hujan Menggunakan Metode optical flow Berdasarkan data Pengamatan RDH. In: Prosiding Seminar Nasional Sains Atmosfer 2015, pp. 93–99. LAPAN, Bandung (2015)
9. Dewi, S.R.: Deep Learning Object Detection pada Video Menggunakan Tensorflow dan Convolutional Neural Network. Universitas Islam Indonesia, Yogyakarta (2018)
10. Kumar, S., Agarwal, Y.K.: Computer vision based moving object detection and tracking. Int. J. Adv. Sci. Technol. **30**(1), 50–55 (2021)
11. Piccardi, M.: Background subtraction techniques: a review. In: 2004 IEEE International Conference on Systems, Man and Cybernetics (IEEE Cat. No.04CH37583), vol. 4, pp. 3099–3104 (2004)
12. Awaludin, A., Nugroho, G.A., Alam, S., Kurniawan, D.F. and Yuwono, R.: Development of marine radar signal acquisition and processing system. In: Proceeding of 2013 International Conference on Information Technology and Electrical Engineering (2013). https://doi.org/10.1109/ICITEED.2013.6676219
13. Sinatra, T., Nugroho, G.A., Awaludin, A.: Optimasi data radar Hujan dan Perluasan Jangkauannya Menggunakan Jaringan radar. Jurnal Sains Dirgantara **18**(1), 43–54 (2020)
14. Awaludin, A., Sinatra, T., Nugroho, G.A., Nauval F.: Clutter removal improvement of marine radar for weather observation. In: AIP Conference Proceedings, vol. 2366, p. 060023 (2021). https://doi.org/10.1063/5.0060062
15. Sinatra, T., Awaludin, A., Nauval, F., Purnomo, C.: Calibration of spatial rain scanner using rainfall depth of rain gauges. In: IOP Conference Series: Earth and Environmental Science, vol. 893, p. 012064 (2021). https://doi.org/10.1088/1755-1315/893/1/012064
16. Awaludin, A., Nugroho, G.A., Rahayu, S.A.: Analisis Kemampuan Radar Navigasi Laut Furuno 1932 Mark-2 untuk Pemantauan Intensitas Hujan. Jurnal Sains Dirgantara, vol. 10 (2013)

17. Lucas, B., Kanade, T.: An iterative image registration technique with an application to stereo vision. In: Proceedings of Seventh International Joint Conference on Artificial Intelligence, pp. 674–679. Vancouver, Canada (1981)
18. Abdulameer Kadhim, H., Abdullah Araheemah, W.: A method to improve corner detectors (Harris, Shi-Tomasi & FAST) using adaptive contrast enhancement filter. Period. Eng. Nat. Sci. **8**(1), 508–515 (2020)

Total Suspended Solids Concentration Estimation in Coastal Wasters Using Remote Sensing Data and Machine Learning Approach

D. N. Lintangsasi, A. Rahmadya, I. Ridwansyah, and F. Setiawan

Abstract Total suspended solids (TSS) concentration is an important water quality parameter generally used to investigate the watershed ecosystem health related to erosion and sedimentation process. It is crucial to routinely and rapidly monitor the TSS concentration, particularly in coastal regions. We propose a model for estimating TSS concentration from remote sensing data using a machine learning approach. We develop the model using in situ TSS data collected from three Indonesian coastal areas (TSS values 26.20–134.67 g/m^3, $n = 43$). First, we carried out Dark Object Subtraction (DOS) atmospheric correction on Landsat-8 OLI Level-1 images and distinguished water and non-water using the Normalized Different Water Index (NDWI). Second, we tested all possible band and band ratio combinations as TSS predictors and four machine learning methods, i.e., linear model (LM), k-nearest neighbor (k-NN), support vector machine (SVM), random forest (RF). Our results showed that the two-band ratios (Red/Ultra Blue and NIR/Ultra Blue) with the RF method gave the best performance (R^2 of 0.81 and the root mean square error of 10.02 g/m^3) and reasonable spatial TSS distribution. Our TSS estimation model well captured the TSS values and spatial distribution in three coastal regions, further valuable to support integrated watershed-river-coastal sustainable management.

1 Introduction

Sedimentation is a major problem often encountered in river estuary [1], which can trigger river siltation and pollution in water bodies. Sediment deposition in water bodies can be influenced by the concentration of suspended sediment, which is

D. N. Lintangsasi
Undergraduate School of Faculty of Geography, Gadjah Mada University, Yogyakarta, Indonesia

A. Rahmadya · I. Ridwansyah · F. Setiawan (✉)
Research Center for Limnology and Water Resources, National Research and Innovation Agency (BRIN), Jakarta Pusat, Indonesia
e-mail: faja006@brin.go.id

A. Basit et al. (eds.), *Proceedings of the International Conference on Radioscience, Equatorial Atmospheric Science and Environment and Humanosphere Science*, Springer Proceedings in Physics 290, https://doi.org/10.1007/978-981-19-9768-6_75

distributed by the river stream and finally deposited [2, 3]. Total suspended sediment or suspended solid (TSS) is a suspended material with a diameter of >1 μm which is retained on a 0.45 Millipore diameter filter [4]. TSS can consist of mud, fine sand, and microorganisms that most of which are formed from soil erosion entered into water bodies [5].

A conventional point-based TSS measurement on water bodies is limited to the number of sampling points; as the result, the spatial distribution of TSS remains unknown [6]. However, performing a routine monitoring of TSS and spatial distribution is urgently required to observe the possibility and direction of sedimentation in a water environment.

Satellite remote sensing provides high spatial and temporal resolution data of the earth's surface that has high potential to efficiently monitor the water quality such as TSS. Compared to manual monitoring, remote sensing technology has the advantages in the effectiveness of time, effort, and cost required. Several studies have proven that remote sensing imagery (Landsat-8 OLI imageries) can be used to observe TSS concentration in water. The image processing is carried out using a spectral transformation method using a TSS estimation algorithm [7–12]. However, some of these algorithms only effectively estimated TSS in water conditions within a specific TSS range as the model was built and further limited the application to certain water conditions.

Indonesia has various water conditions, as well as the TSS concentrations range. It is necessary to build a TSS estimation model that can cover a wide range of TSS concentrations. The development of the model can be done using a machine learning approach, which has the advantage of being able to handle more complex problems, especially when compared to physical models because the working principle does not require other information where limited assumptions are required such as physical model [13–15]. The objective of this study is to develop a model for estimating a TSS concentration from Landsat-8 OLI data. The developed model is expected to be more applicable to various water conditions in Indonesia.

2 Data and Method

2.1 In Situ TSS Data

We collected in situ TSS data from three conducted field surveys in three different coastal waters i.e., Batanghari Estuary (Jambi Province), Jakarta Bay (Jakarta Province), and Cimandiri Estuary (West Java Province) (Fig. 1a). In total, we have 43 points of data, with the minimum, maximum, and average values (mg/L) for Batanghari Estuary is 58.67, 134.67, and 85.80, respectively, for Jakarta Bay is 26.20, 100.80, and 58.59, respectively, and for Cimandiri Estuary is 58.00, 102.00, and 82.58, respectively, as shown in Fig. 1b.

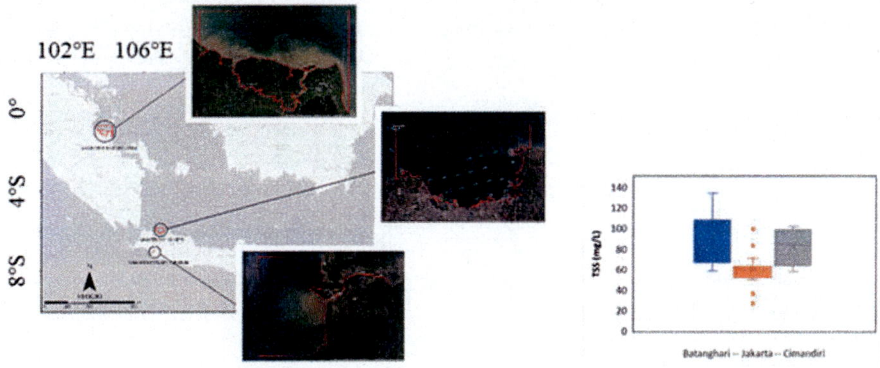

(a) (b)

Fig. 1 **a** Locations of the three Indonesian coastal areas, **b** distribution of TSS concentrations in the in situ dataset used in this study (yellow points are the outliers)

2.2 Landsat Data Collections and Pre-processing

We collected satellite images that were acquired by Landsat-8, from the United States Geological Survey (USGS) earth explorer website [https://earthexplorer.usgs.gov/]. Landsat-8 OLI Level-1 images were atmospherically corrected to get the surface reflectance (SR) value as input data for algorithm processing. We carried out atmospheric correction using Dark Object Subtraction (DOS) method [16]. The working principle of this atmospheric correction method is to eliminate the lower pixel value which is assumed as the darkest object in the image (Table 1).

We cropped the atmospherically corrected image using shapefile boundaries of each site that is digitized manually from the true color composite of the image. We removed the non-water pixels using the Normalized Different Water Index (NDWI) threshold [17], where a pixel with an NDWI value less than or equal to 0 is assumed as non-waters.

$$\text{NDWI} = \frac{(\text{blue} - \text{NIR})}{(\text{blue} + \text{NIR})} \tag{1}$$

Table 1 Three Landsat-8 OLI images corresponding with in situ TSS dataset

Location	Landsat Path and row	Acquisition date	Days gap (days)
Batanghari	125–61	September 21, 2021	17–22
Jakarta	122–65	November 3, 2021	15–17
Cimandiri	122–64	May 11, 2021	13–15

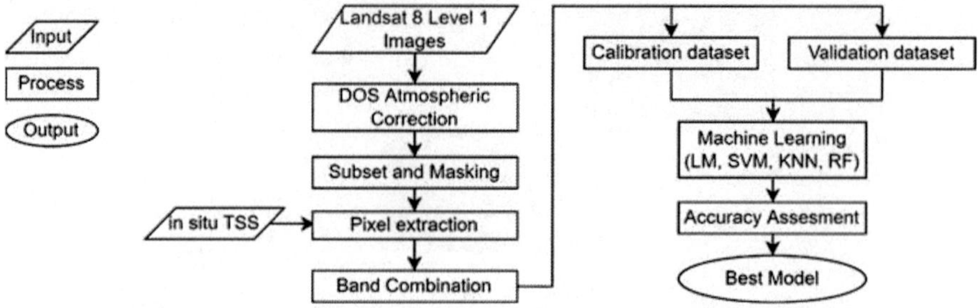

Fig. 2 Model development work flow

where blue and NIR are the atmospherically corrected values at the blue band and the near-infrared band, respectively.

2.3 Model Development

We used in situ TSS and the pre-processed Landsat Dataset to develop TSS estimation models. We obtained a total of 43 pairs. We then used the in situ TSS values as dependent variables and various combinations of the corresponding Landsat bands (e.g., using single bands, band ratios, band ratios and single bands, and two-band ratios) as independent variables. We used four machine learning approaches, i.e., linear model (LM), k-nearest neighbor (k-NN), support vector machine (SVM), and random forest (RF). We perform machine learning training and testing using the "caret" package in the *R* language [18]. We split the data into 75% for model training (calibration) and 25% for data testing (validation). We provided the workflow in Fig. 2.

2.4 Accuracy Assessment

We used the root means square error (RMSE) to assess the models' accuracy. The index is defined as follows:

$$\text{RMSE} = \sqrt{\frac{\sum_{i=1}^{N} (X_{\text{esti},i} - X_{\text{meas},i})^2}{n}} \quad (2)$$

where $X_{\text{esti},\,i}$ and $X_{\text{meas},\,i}$ are the estimated and measured TSS values, respectively; n is the number of samples. We also calculated the correlation between the measured and estimated values (R^2). We perform visual comparison between TSS estimation result image and the true color image to evaluate the TSS spatial distribution pattern.

3 Result and Discussion

3.1 The Machine Learning Models for Estimating the TSS from Landsat-8 OLI

We built and tested 68 models using our dataset, which consists of seven "single band-based models," 15 "two single band-based models," 15 "band ratio-based models," and 30 "two-band ratios-based models." None of the single band-based models has a high coefficient of determination, even though a complex machine learning approach has been applied. Either, the two single band-based model has a low coefficient of determination. This worse performance revealed that the use of a single band both as one band and two bands format are not robust enough as a TSS predictor.

Table 2 shows the selected developed TSS estimation models and their performances; we excluded the TSS estimation models with the worse performance by using thresholds of R^2 values <0.8. As the result, 12 TSS estimation models remained. One band ratio-based model (#14, a model with SVM and RF machine learning approach) and eleven two-band ratio-based models (#16, #18, #20, #22, #31, #32, #38, #41, #42, #43, #45 with RF machine learning approach).

Table 2 reveals a band ratio-based model with an RF machine learning approach outperformed the other machine learning approach, i.e., LM, k-NN, SVM (the R^2 ranged from 0.80 to 0.82), except for model #14, the SVM machine learning approach has the highest coefficient of determination (R^2) of 0.84. The root mean square error (RMSE) values show a similar pattern. The lowest RMSE value is 9.51 mg/L, obtained from model #45 with the TSS predictors B5/B1 and B6/B1, and the machine learning approach is random forest (RF).

Overall, the band ratio-based models and the two-band ratio-based models are more robust compared with the single band-based model and two single band-based models. This finding is in line with the result from Setiawan et al. [19]; band ratio provides a more robust water quality estimation. The band ratio has the benefit of mitigating the remained atmospheric effect, even though atmospheric correction (in this research: DOS) has been performed, as also mentioned by Doxaran et al. [20, 21].

3.2 The Reliability of the Estimated TSS Concentration

Based on the performance of the developed model, it was revealed the band ratio-based model combined with the RF machine learning approach produces a more accurate TSS estimation (shown by the higher R^2 value and relatively low RMSE value). However, it is necessary to visually check the spatial TSS distribution map resulting from machine learning. Generally, the TSS concentration will be higher on the shoreline and gradually lower onward to the open sea. In addition, as the TSS originated from the river channel to the estuary area, the result higher TSS concentration tends to be located near the estuary.

Table 2 Twelve selected developed TSS estimation models and their performances

Model No.	TSS predictor	R^2				RMSE (mg/L)			
		LM	k-NN	SVM	RF	LM	k-NN	SVM	RF
#14	B5/B1	0.17	0.29	**0.84**	**0.80**	18.66	17.33	**9.63**	**10.11**
#16	B1/B3 and B2/B3	0.25	0.27	0.37	**0.82**	17.75	17.56	17.02	**10.10**
#18	B1/B3 and B5/B3	0.25	0.25	0.28	**0.83**	17.73	17.71	18.31	**10.62**
#20	B2/B3 and B4/B3	0.26	0.27	0.33	**0.80**	17.61	17.53	17.83	**9.97**
#22	B2/B3 and B6/B3	0.25	0.28	0.29	**0.80**	17.72	17.39	18.22	**10.01**
#31	B3/B2 and B5/B2	0.23	0.30	0.43	**0.80**	17.95	17.24	17.24	**9.92**
#32	B3/B2 and B6/B2	0.24	0.29	0.37	**0.80**	17.89	17.29	17.51	**9.91**
#38	B2/B1 and B5/B1	0.22	0.23	0.29	**0.82**	18.13	18.00	18.16	**9.82**
#41	B3/B1 and B5/B1	0.24	0.31	0.46	**0.81**	17.89	17.14	17.38	**10.16**
#42	B3/B1 and B6/B1	0.24	0.28	0.34	**0.80**	17.86	17.57	17.67	**10.48**
#43	B4/B1 and B5/B1	0.24	0.31	0.41	**0.81**	17.82	17.18	16.95	**10.02**
#45	B5/B1 and B6/B1	0.17	0.39	0.54	**0.82**	18.65	16.22	15.49	**9.51**

[a] *LM* Linear Model, *k-NN* k-Nearest Neighbor, *SVM* Support Vector Machine, *RF* Random Forest

We used the above assumption to select the best TSS estimation model by visually checking the TSS concentration distribution maps. We observed that the most representative TSS concentration map was produced by model #43 with the random forest approach. The TSS concentrations gradually decrease from the estuary to the open sea. The TSS map resulting from other models looks less representative. For example, in models #38, #42, and #45, the low TSS is concentrated at the shoreline, while the high TSS is concentrated in the open sea.

3.3 TSS Concentration Map Using the Selected Model

We applied the #43 model and produced the TSS concentration map of Batanghari Estuary (Jambi Province), Jakarta Bay (Jakarta Province), and Cimandiri Estuary (West Java Province) as shown in Fig. 3. The TSS concentration in Batanghari Estuary is dominated by high TSS concentration on the shoreline and moderate

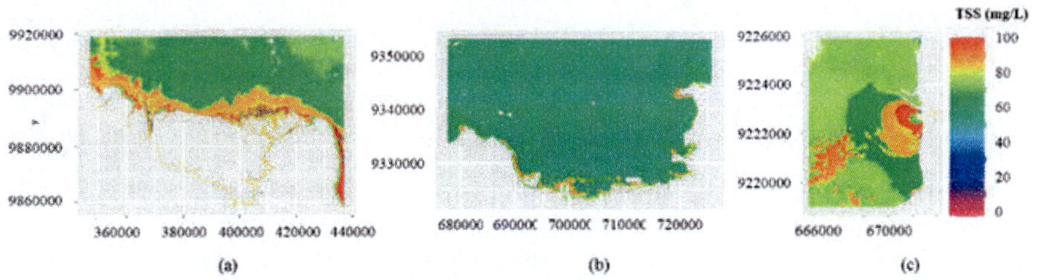

Fig. 3 Spatial distribution of TSS concentration of **a** Batanghari Estuary, **b** Jakarta Bay, and **c** Cimandiri Estuary

TSS concentration on the open sea. The high TSS concentration in Jakarta Bay tends to be low, where the high TSS concentration is not too wide toward the open sea.

The TSS concentration in Cimandiri Estuary forms a circular pattern. High TSS on the riverbank and changes gradually to low TSS concentration onward to the open sea area. However, based on image visualization, the circular pattern in the sea body has a lower TSS value that the area in the outermost sea body. We observed a moderate TSS in the open ocean of Cimandiri Estuary, which should have a low TSS concentration; it might be related to the image quality or due to atmospheric disturbances that have not been completely minimized during image correction.

4 Conclusion

We developed a model to estimate TSS concentration from Landsat-8 OLI data using a machine learning approach. The developed model suggests the use of the two-band ratios of band-4/band-1 and band-5/band-1 as the TSS predictor to reduce uncertainties and the random forest machine learning approach to improve the accuracies. Our TSS estimation model well captured the TSS values and spatial distribution in three coastal regions and seemly potential to be applicable to another water area in Indonesia. Thus, further valuable to support integrated watershed-river-coastal sustainable management.

Acknowledgements The authors thank the United States Geological Survey (USGS) for providing Landsat data. The Coremap-CTI (Coral Reef Rehabilitation and Management Program - Coral Triangle Initiative) dan Badan Riset Inovasi Nasional (BRIN) No. 17/A/DK/2021 for providing the research funding. The authors would also like to thank four anonymous reviewers for their valuable comments and suggestions for improving the quality of the manuscript. The R script for estimating TSS concentration are available upon request.

References

1. Van Rijn, L.C.: Estuarine and coastal sedimentation problems. In: Proceedings of the Ninth International Symposium on River Sedimentation, 18–21 Oct 2004, Yichang, China, pp. 105–120 (2004)
2. Yonar, et al.: Dinamika total suspended solid (TSS) Di Sekitar Terumbu Karang Pantai Damas, Trenggalek. J. Marine Coastal Sci. **10**, 48–57 (2021). (in bahasa)
3. Barus, B.S.: Keterkaitan sedimentasi terhadap kondisi ekosistem terumbu karang di perairan Teluk Lampung, Provinsi Lampung. Tesis. Institut Pertanian Bogor, Bogor (in bahasa) (2003)
4. Effendi, H.: Telaah Kualitas Air Bagi Pengelolaan Sumberdaya dan LingkunganPerairan. Fakultas Perikanan dan Ilmu Kelautan IPB. Bogor (2000) (in bahasa)
5. Budhiman, S.L.: Pemetaan Sebaran total suspended matter (TSM) Menggunakan data ASTER dengan Pendekatan bio-optical model. In: Pertemuan Ilmiah Tahunan MAPIN XIV "Pemanfaatan Efektif Penginderaan Jauh Untuk Peningkatan Kesejahteraan Bangsa", Surabaya, 14–15 Sept 2005, pp. 1–6 (2005) (in bahasa)
6. Saberioon, M., et al.: Retrieving Chl-a and total suspended solids in in-land waters using EnMAP simulated data. J. Hidrol. 1–33 (2022)
7. Budhiman, S.: Mapping TSM concentrations from multisensor satellite images in turbid tropical coastal waters of Mahakam Delta, Indonesia. Thesis. ITC, Netherlands (2004)
8. Doxoran, et al.: Use of reflectance band ratios to estimate suspended and dissolved matter concentrations in estuarine waters. Int. J. Remote Sens. **26**, 1763–1769 (2005)
9. Parwati, E., Trisakti, B., Carolila, I., Kartika, T., Harini, S., Dewanti, K.: Analisis Hubungan Penutup/Penggunaan Lahan dengan total suspended matter (TSM) Kawasan Perairan Segara Anakan Menggunakan data Inderaja. Jurnal Penginderaan Jauh dan Pengolahan Data Citra Digital **3**, 87–97 (2006). (In bahasa)
10. Laili, N., Arafah, F., Jaelani, L.M., Subehi, L., Pamungkas, A., Koenhardono, E.S., Sulisetyono, A.: Development of water quality parameter retrieval algorithms for estimating total suspended solids and chlorophyll-a concentration using Landsat-8 imagery at Poteran Island water. ISPRS Ann. Photogramm. Remote Sens. Spatial Inf. Sci. **2**(2W2), 55–62 (2015)
11. Jaelani, L.M., Limehuwey, R., Kurniadin, N., Pamungkas, A., Koenhardono, E.S., Sulisetyono, A.: Estimation of total suspended sediment and chlorophyll-a concentration from Landsat 8-OLI: the effect of atmosphere and retrieval algorithm. IPTEK J. Technol. Sci. **27**(1), 16–23 (2016)
12. Ouma, Y.O., et al.: Modelling reservoir chlorophyll-a, TSS, and turbidity using sentinel-2A MSI and Landsat-8 OLI satellite sensors with empirical multivariate regression. J. Sens. 1–21 (2020)
13. Chang, N.B., et al.: Exploring spatiotemporal patterns of phosphorus concentrations in a coastal bay with MODIS images and machine learning models. Remote Sens. Environ. **134**, 100–110 (2013)
14. Keller, S., et al.: Hyperspectral data and machine learning for estimating CDOM, chlorophyll a, diatoms, green algae, and turbidity. Int. J. Environ. Res. Public Health **15**, 1881 (2018)
15. Saberioon, M., et al.: Chlorophyll-a and total suspended solids retrieval mapping using sentinel-2A and machine learning for inland waters. Ecol. Indicators **113**, 1–11 (2020)
16. Chavez, P.S.: An improved dark-object subtraction technique for atmospheric scattering correction of multispectral data. Remote Sens. Environ. **24**, 459–479 (1988)
17. Gao, B.C.: NDWI—A normalized difference water index for remote sensing of vegetation liquid water from space. Remote Sens. Environ. **58**, 257–266 (1996)
18. Kuhn, M.: Building predictive models in R using the caret package. J. Stat. Softw. **28**(5), 1–26 (2008). https://doi.org/10.18637/jss.v028.i05
19. Setiawan, F., Matsushita, B., Hamzah, R., Jiang, D., Fukushima, T.: Long-term change of the Secchi disk depth in Lake Maninjau, Indonesia shown by Landsat TM and ETM+ Data. Remote Sens. **11**, 2875 (2019). https://doi.org/10.3390/rs11232875

20. Doxaran, D., Froidefond, J.M., Castaing, P.: A reflectance band ratio used to estimate suspended matter concentrations in sediment-dominated coastal waters. Int. J. Remote Sens. **23**, 5079–5085 (2002)
21. Doxaran, D., Froidefond, J.M., Castaing, P.: Remote-sensing reflectance of turbid sediment-dominated waters. Reduction of sediment type variations and changing illumination conditions effects by the use of reflectance ratios. Appl. Opt. **42**, 2623–2634 (2003)

Practical Approach to Designing Radar Linear and Nonlinear Frequency Modulation Chirp Waveforms

Harry Septanto and Djoko Suprijanto

Abstract A radar-based remote sensing system called synthetic aperture radar has many advantages over an optical-based one due to its independence from sunlight, clouds, and weather. Today's radar systems frequently use the pulse compression radar technique. Together with the transmitted pulse signal in the form of a chirp wave as the reference signal, signal processing with this technique will increase the range resolution and signal-to-noise ratio. This work deals with the design of linear and nonlinear frequency modulation (LFM and NLFM) chirp signals. The chirp signals are designed, tested, and evaluated using two free and open-source programs, GNU Octave and GNU Radio. This paper makes a contribution by outlining a step-by-step process for creating LFM and NLFM chirp waveforms that are expressed by a practical approach as opposed to numerous difficult mathematical equations and by contrasting the matching filtering results based on the chirp waveforms produced. In addition, this paper proposes a new rule for designing a type of piecewise linear nonlinear frequency modulation (NLFM) chirp called the Five-Stage Piecewise Linear NLFM (FS-PL-NLFM) chirp. Matched filtering results show that the designed FS-PL-NLFM chirp provides a lower peak sidelobe than the LFM's.

1 Introduction

The pulse compression technique with linear frequency modulation (LFM) waveforms has been widely employed for tracking and imaging radars since their development in the 1950s [1]. In imaging radars, synthetic aperture radar (SAR) is well

H. Septanto (✉)
Research Center for Smart Mechatronics, National Research and Innovation Agency,
KST Samaun Samadikun, Bandung 40135, Indonesia
e-mail: harry.septanto@brin.go.id

D. Suprijanto
Faculty of Mathematics and Natural Sciences, Combinatorial Mathematics Research Group,
Institut Teknologi Bandung, Jl. Ganeca 10, Bandung 40132, Indonesia
e-mail: djoko.suprijanto@itb.ac.id

A. Basit et al. (eds.), *Proceedings of the International Conference on Radioscience, Equatorial Atmospheric Science and Environment and Humanosphere Science*,
Springer Proceedings in Physics 290, https://doi.org/10.1007/978-981-19-9768-6_76

known as an imaging system that can be used in all types of weather and at any time of day or night [2]. These are its advantages over optical sensors, which are dependent on sunlight and cannot pass through clouds.

With a pulse compression technique, wide pulses with low peak power can be transmitted and still achieve the same detection range and resolution as narrow pulses with high peak power [2]. A matched filtering is often used to the received radar waveform for obtaining the maximum attainable signal-to-noise ratio (SNR) and for compressing the energy in the waveform into a relatively narrow pulse [3].

Although the LFM has good Doppler tolerance, the matched filtered result exhibits a relatively high peak sidelobe (approximately 13 dB), which may not be suitable for many radar applications [4]. A popular processing method for sidelobe suppression is the window-function technique [5, 6]. The SNR would be decreased, though, and a broader main lobe would result from this type of processing [7, 8]. Many studies have suggested various nonlinear frequency modulation (NLFM) chirp waveforms as an alternative approach for generating a low-peak sidelobe, e.g., [7, 8]. Unlike the LFM, the NLFM chirp waveform is sensitive to Doppler shift. Fortunately, it may not be a big problem for SAR-related applications [7], especially for airborne SAR [8], because the Doppler bandwidth in SAR is relatively small compared to the radar applications.

In addition to the field-programmable gate array (FPGA) [9, 10], another commonly utilized device for the implementation of LFM and NLFM chirps is the software-defined radio (SDR) [11]. GNU Radio is a free software development framework that is frequently used for SDR applications.

This research focuses on creating chirp signals using both linear and nonlinear frequency modulation (LFM and NLFM) for use with an SDR. This paper makes a contribution by outlining a step-by-step process for creating LFM and NLFM chirp waveforms that are expressed by a practical approach using two free and open-source software programs, GNU Octave (https://octave.org/) and GNU Radio (https://www.gnuradio.org/), as opposed to numerous difficult mathematical equations and by contrasting the matching filtering results based on the chirp waveforms produced.

In addition, this paper proposes a new rule to design a type of piecewise linear nonlinear frequency modulation (NLFM) class chirp called Five-Stage Piecewise Linear NLFM (FS-PL-NLFM) chirp. Matched filtering results show that the designed FS-PL-NLFM chirp provides a lower peak sidelobe than the LFM's.

2 Problem Definitions

The mathematical equation of LFM chirp waveform (baseband) can be represented by (1) and (2) (see [2, 3] for more information about LFM chirp). Meanwhile, the FS-PL-NLFM waveform can be represented by (1) and (3). In addition, Fig. 1 defines the equations' variables.

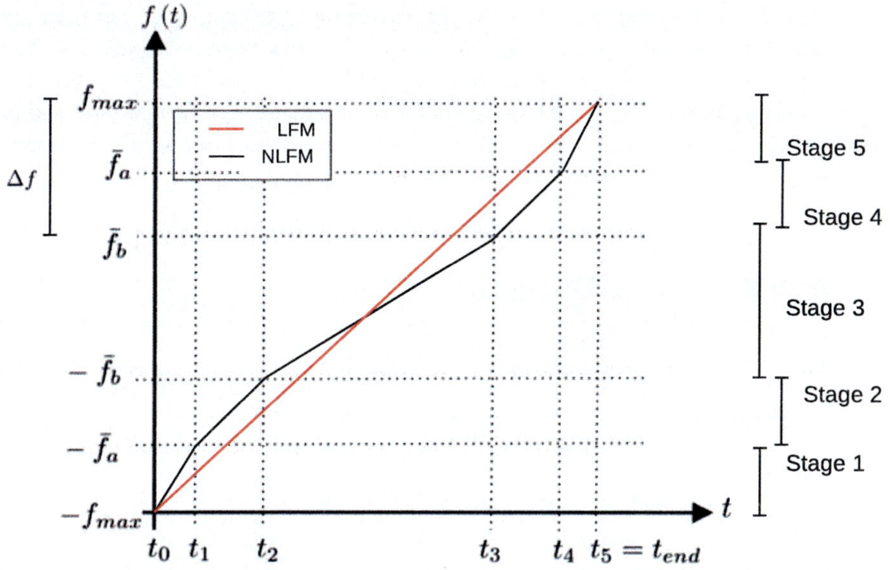

Fig. 1 Frequency function of LFM and FS-PL-NLFM

$$\begin{aligned} s(t) &= \mathrm{e}^{j2\pi f(t)t} \\ &= I(t) + jQ(t) \end{aligned} \tag{1}$$

where

$$\text{LFM: } f(t) = \frac{f_{\max} - (-f_{\max})}{t_{\text{end}} - t_0} t = \frac{2f_{\max}}{t_{\text{end}} - t_0} t, \; t_0 \le t < t_{\text{end}} \tag{2}$$

$$\text{FS-PL-NLFM: } f(t) = \begin{cases} \frac{-\bar{f}_a - (-f_{\max})}{t_1 - t_0} t, & t_0 \le t < t_1 \\ \frac{-\bar{f}_b - (-\bar{f}_a)}{t_2 - t_1} t, & t_1 \le t < t_2 \\ \frac{\bar{f}_b - (-\bar{f}_b)}{t_3 - t_2} t, & t_2 \le t < t_3 \\ \frac{\bar{f}_a - \bar{f}_b}{t_4 - t_3} t, & t_3 \le t < t_4 \\ \frac{f_{\max} - \bar{f}_a}{t_5 - t_4} t, & t_4 \le t < t_5 = t_{\text{end}} \end{cases} \tag{3}$$

Many references discuss the LFM and NLFM chirp waveform designs in the aforementioned mathematical equations. However, in a practical aspect, the realization of these mathematical equations is needed. Then, a natural research question is as follows: How can these mathematical formulas be realized in order to be prepared for implementation with an SDR? What types of software applications are required? These questions will be answered in the following sections.

The hardest part of using various NLFM waveform types is figuring out the right parameters so that the matched filtering result produces minimal sidelobes. The NLFM's peak sidelobe that is larger than the LFM's will be produced by choosing inappropriate parameters. Likewise, the choice of parameters is a difficulty in and of itself for the FS-PL-NLFM waveform. A new rule for determining the parameters proposed by this work will be provided in the later section.

3 Main Results and Discussion

The general steps for designing a chirp waveform for an SDR are the following:

1. GNU Octave: Produce a template signal in discrete-time form.
2. GNU Octave: Prepare a file of the template signal in a GNU Radio-compatible format.
3. GNU Radio: Conduct a simulation of detecting objects.
4. GNU Octave: Load saved data into GNU Octave, parse it, and conduct matched filtering of the simulation data.

The following descriptions elaborate on the steps mentioned above.

3.1 Template Signal

The template signal is the signal used as the reference in the matched filtering. This chirp waveform might be thought of as the ideal and desirable one. It is crucial to the process of matched filtering. The following snippet of code is to produce the template signal of the LFM and NLFM chirp. Instead of defining the sampling frequency first, we define the number of samples first.

```
%LFM
pkg load signal
fs = sig_samp/sig_dur; Ts = 1/fs;
t = 0:Ts:sig_dur;
I_all = chirp(t(1:end−1),−f_max,t(end),f_max,"linear", 0);
Q_all = chirp(t(1:end−1),−f_max,t(end),f_max,"linear", 90);
```

A new rule for determining the FS-PL-NLFM parameters that we propose is the following:

1. $\Delta f = \frac{2 f_{\max}}{5}$
2. $f_{\max} - \bar{f}_a = \bar{f}_a - \bar{f}_b = \frac{\Delta f}{2}$
3. Number of samples between t_0 and t_2 = number of samples between t_3 and t_{end} = rounded value to the nearest integer of $\frac{1}{2^3} \times$ the number of samples between t_0 and t_{end}

4. Number of samples between t_0 and t_1 = number of samples between t_4 and t_5 = rounded value to the nearest integer of $\frac{1}{1.9} \times$ the number of samples between t_0 and t_2

The following is a snippet of code for producing the FS-PL-NLFM chirp that satisfies the rule mentioned above.

```
%NLFM
pkg load signal
fs = sig_samp/sig_dur;
Ts = 1/fs;
t = 0:Ts:sig_dur;
delta_f = 2/5*f_max; %No. 1
delta_fmax_a = 1/2*delta_f; %No. 2
delta_fa_b = delta_f-delta_fmax_a;
fbar_a = f_max - delta_fmax_a;
fbar_b = fbar_a - delta_fa_b;
p02 = round(sig_samp/(2^3)); %No.3
p01 = round(p02/1.9); %No. 4
p12 = p02-p01; p23 = sig_samp-2*p02;
p34 = p12; p45 = p01; p35 = p02;
%t0-t1
Imax_a = chirp(t(1:p01),-f_max,t(p01+1),-fbar_a,"linear", 0);
Qmax_a = chirp(t(1:p01),-f_max,t(p01+1),-fbar_a,"linear", 90);
%t1-t2
Ia_b = chirp(t(1:p12),-fbar_a,t(p12+1),-fbar_b,"linear", 0);
Qa_b = chirp(t(1:p12),-fbar_a,t(p12+1),-fbar_b,"linear", 90);
%t2-t3
Ib_b = chirp(t(1:p23),-fbar_b,t(p23+1),fbar_b,"linear", 0);
Qb_b = chirp(t(1:p23),-fbar_b,t(p23+1),fbar_b,"linear", 90);
%t3-t4
Ib_a = chirp(t(1:p34),fbar_b,t(p34+1),fbar_a,"linear", 0);
Qb_a = chirp(t(1:p34),fbar_b,t(p34+1),fbar_a,"linear", 90);
%t4-t5
Ia_max = chirp(t(1:p45),fbar_a,t(p45+1),f_max,"linear", 0);
Qa_max = chirp(t(1:p45),fbar_a,t(p45+1),f_max,"linear", 90);
%real part
I_all = [Imax_a Ia_b Ib_b Ib_a Ia_max];
%imaginary part
Q_all = [Qmax_a Qa_b Qb_b Qb_a Qa_max];
```

3.2 GNU Radio-Compatible Format Data

It should be noted that this step requires a predefined value for the pulse repetitive interval (PRI) of the radar. One may refer to [2] for more detailed information about PRI. Besides, one has to prepare the LFM and FS-PL-NLFM chirp waveform data in order to satisfy the GNU Radio format by following the convention below:

- Generate a variable that consists of the chirp's real part $I(t)$ at the odd index and the imaginary part $Q(t)$ at the even index.

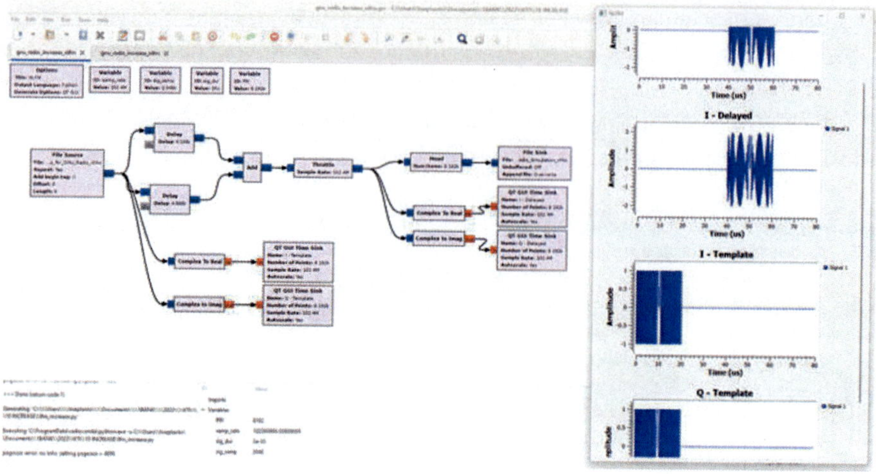

Fig. 2 Simulation on GNU radio

- Save/write the variable to a binary form. One may refer to the following link for more detailed: https://octave.sourceforge.io/octave/function/fwrite.html

3.3 *Simulation on GNU Radio*

Figure 2 illustrates a typical radar simulation realization. Radar back scatterings from two objects are simulated by "Delay" blocks.

3.4 *Matched Filtering*

In this step, matched filtering between the template chirp waveform and the simulation result from GNU Radio is calculated. Note that the simulation result is recorded in "File Sink" (see Fig. 2). However, first we have to load and parse the simulation result (recorded) data. The following link provides detailed information for loading/opening the binary data from the GNU Radio using GNU Octave: https://wiki.gnuradio.org/index.php/File_Sink. In addition, the following is a snippet of code for conducting the matched filtering between the NLFM's template chirp waveform and the simulation result from GNU Radio.

```
s_rec_nlfm_fft = fft(s_rec_nlfm); %Recorded chirp
s_temp_nlfm_fft = fft(s_temp_nlfm); %Template chirp
s_mf_nlfm = ifft(s_rec_nlfm_fft.*conj(s_temp_nlfm_fft));
s_mf_nlfm = 20*log10(abs(s_mf_nlfm));
s_mf_nlfm = s_mf_nlfm − max(s_mf_nlfm);
```

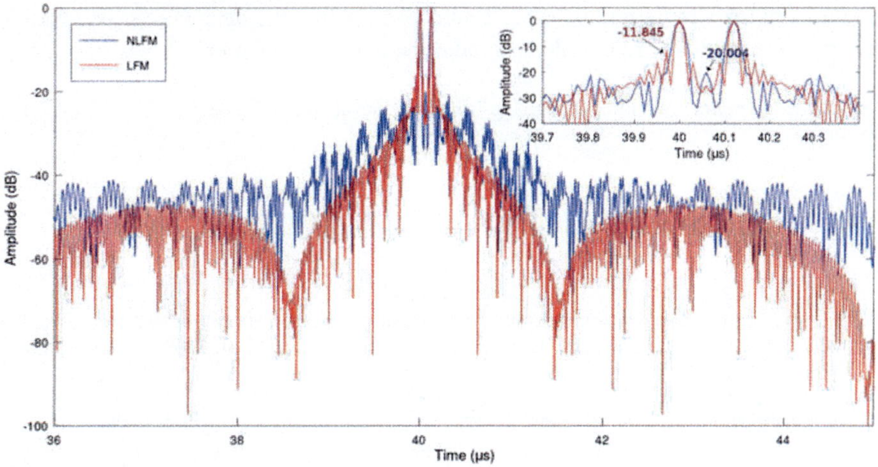

Fig. 3 Comparison of matched filtering results of the LFM and NLFM chirp waveforms

3.5 Discussion: Peak Sidelobe Comparison

Simulation that were carried out adopting the following parameters:

- Maximum frequency = 25 MHz.
- Pulse Duration = 20 μs.
- Chirp's number of samples = 2^{11}. Note that the selection of pulse duration and number of samples equivalent to the sampling frequency = 102.4 MHz
- PRI = 4 × Pulse Duration.
- Two objects: Delay from the first object = 2^{12} of samples and delay from the second object = $2^{12} + 12$ of samples.

Figure 3 shows the pulse compression results via matched filtering. It verifies that the proposed rule for the FS-PL-NLFM chirp yields a smaller peak sidelobe than the LFM's, i.e., $-20.004 < -11.845$. This peak sidelobe value is better than peak sidelobe value of the two-stages and tri-stages-PL-NLFM that have been reported in [12, 13]. In addition, the peak sidelobe resulted by the designed FS-PL-NLFM chirp is not better than the one shown in [7]. However, that work used 65 stages of piecewise linear, which is in contrast to the simplicity of the proposed FS-PL-NLFM.

4 Concluding Remarks

The step-by-step process for creating LFM and NLFM chirp waveforms using GNU Octave and GNU Radio is presented. The new rule for the five-stage piecewise linear NLFM chirp is presented. The rule comprises four points. The matched filtering result from the simulation on GNU Radio verifies that the designed FS-PL-NLFM chirp by the proposed rule yields a smaller peak sidelobe than the LFM's.

Acknowledgements We acknowledge the Riset dan Inovasi untuk Indonesia Maju (RIIM) grant for supporting this research. HS is the main contributor.

References

1. Cooke, C.: The early history of pulse compression radar-the history of pulse compression at Sperry Gyroscope Company. IEEE Trans. Aerosp. Electron. Syst. **24**(6), 825–833 (1988). https://doi.org/10.1109/7.18654
2. Özdemir, C.: Inverse Synthetic Aperture Radar Imaging With MATLAB Algorithms, 2nd edn. Wiley (2021)
3. Chan, Y.K., Koo, V.: An introduction to synthetic aperture radar (SAR). Progr. Electromagn. Res. B **2**, 27–60 (2008)
4. Alphonse, S., Williamson, G.A.: Evaluation of a class of NLFM radar signals. EURASIP J. Adv. Sig. Process. **2019**, 62 (2019). https://doi.org/10.1186/s13634-019-0658-9
5. Doerry, A.W.: Catalog of window taper functions for sidelobe control. Technical Report. Sandia National Laboratory (SNL-NM), Albuquerque, NM (United States) (2017)
6. Varshney, L., Thomas, D.: Sidelobe reduction for matched filter range processing. In: Proceedings of the 2003 IEEE Radar Conference (Cat. No. 03CH37474), pp. 446–451 (2003). https://doi.org/10.1109/NRC.2003.1203439
7. Saeedi, J., Faez, K.: Synthetic aperture radar imaging using nonlinear frequency modulation signal. IEEE Trans. Aerosp. Electron. Syst. **52**(1), 99–110 (2016). https://doi.org/10.1109/TAES.2015.140310
8. Wang, W., Wang, R., Zhang, Z., Deng, Y., Li, N., Hou, L., Xu, Z.: First demonstration of airborne SAR with nonlinear FM chirp waveforms. IEEE Geosci. Remote Sens. Lett. **13**(2), 247–251 (2016). https://doi.org/10.1109/LGRS.2015.2508102
9. Chua, M.Y., Koo, V.: FPGA-based chirp generator for high resolution UAV SAR. Progr. Electromagn. Res. **99**, 71–88 (2009). https://doi.org/10.2528/PIER09100301
10. Im, Y.T., Park, S.O.: An FMCW and chirp pulse-doppler radar system for surveillance in x-band. Microw. Opt. Technol. Lett. **55**(3), 474–479. https://doi.org/10.1002/mop.27343, https://onlinelibrary.wiley.com/doi/abs/10.1002/mop.27343 (2013)
11. Grabowski, A.: SDR-based LFM signal generator for radar/SAR systems. In: 2016 17th International Radar Symposium (IRS), pp. 1–3 (2016). https://doi.org/10.1109/IRS.2016.7497263
12. Chan, Y.K., Chua, M.Y., Koo, V.: Sidelobes reduction using simple two and tri-stages non linear frequency modulation (NLFM). Progr. Electromagn. Res. **98**, 33–52 (2009)
13. Widyantara, M., Sugihartono, Suratman, F., Widodo, S., Daud, P.: Analysis of non linear frequency modulation (NLFM) waveforms for pulse compression radar. Jurnal Elektronika dan Telekomunikasi **18**(1), 27–34 (2018). https://doi.org/10.14203/jet.v18.27-34

Ensembles Simulation on the Seasonal Rainfall Characteristics Over Indonesia Maritime Continent

Haries Satyawardhana, M. Arif Setyo Aji, Risyanto, Erma Yulihastin, Gammamerdianti, Candra N. Ihsan, Eka P. Wulandari, Lely Q. Avia, and Iis Sofiati

Abstract As a maritime continent, Indonesia is a complex land–sea composition and interaction, with seasons strongly influenced by the Australian–Asian monsoon system. Consequently, seasonal rainfall in this region has high variability. On the other hand, accurate seasonal prediction is significant for several sectors to support development in Indonesia, especially the agricultural sector. This seasonal prediction research uses the output of the Conformal Cubic Atmospheric Model (CCAM) for the Indonesia region, with the output of a spatial resolution of 32 km. The simulation period is carried out in two seasons: the wet season (December 2018) and the dry season (June 2019). Satellite rainfall data from GSMaP is used as a comparison. This study compares the probability of monthly rainfall >= 150 mm between model outputs and satellite. Rainfall characteristics/characteristics are obtained by comparing rainfall output from models and satellites, in the wet and dry seasons, with the average monthly rainfall for the 2014–2018 period. The results of seasonal predictions using CCAM include the values of rainfall reconstructed, rainfall probability >= 150 mm from time-lagged, and multi-sea surface temperature (multi-SST) ensemble. Seasonal predictions produce eight months of predictions, so the time-lagged ensemble uses predictions from eight members, while multi-SST ensembles are carried out using 11 simulations with 11 different SST inputs. The results show that the seasonal predictions output from rainfall reconstruction, probabilistic time-lagged, and multi-SST ensemble can capture monthly rainfall patterns >= 150 mm. However, in some parts of Indonesia, it is still far from being well predicted. Probabilistic rainfall >= 150 mm from a multi-SST ensemble has a better spatial pattern than the time-lagged ensemble compared to observation in almost all regions of Indonesia.

H. Satyawardhana (✉) · Risyanto · E. Yulihastin · Gammamerdianti · C. N. Ihsan · E. P. Wulandari · L. Q. Avia · I. Sofiati
Research Center for Atmospheric Science and Technology, Indonesia National Agency of Research and Innovation, Bandung, Indonesia
e-mail: hari031@brin.go.id

M. Arif Setyo Aji
Bandung Institute of Technology, Bandung, Indonesia

A. Basit et al. (eds.), *Proceedings of the International Conference on Radioscience, Equatorial Atmospheric Science and Environment and Humanosphere Science*, Springer Proceedings in Physics 290, https://doi.org/10.1007/978-981-19-9768-6_77

1 Introduction

Indonesia is a maritime country between two oceans (the Pacific and Indian) and two continents (Asia and Australia). This condition makes the Indonesian sea the only route connecting the Pacific and Indian Oceans in the Equatorial zone [1, 2]. This flow, known as the Indonesian Throughflow (ITF), carries warm water from the Pacific to the Indies and vice versa through the oceans in the inter-island region of Indonesia [3, 4]. This condition causes the movement of monsoon winds and affects the variability of rainfall and seasons in Indonesia both temporally and spatially [5].

Seasonal predictions are mid-range forecasts of the average weather on a time scale of several months to a year in the future in several regions [6]. The need for seasonal predictions, especially related to temperature and rainfall information, is increasing along with the public awareness of the effect of seasonal variability in important sectors [7]. This condition encourages innovation development to get more accurate seasonal predictions, one of which is by comparing and verifying the results of seasonal predictions to produce high predictions.

Among the many seasonal prediction models, one of the atmospheric models with a high resolution is the Conformal Cubic Atmospheric Model (CCAM) developed by CSIRO, Australia. CCAM is a conformal cubic grid-based global model that uses the Schmidt transform for regional and local forecasts. In addition, the multiple nesting technique is implemented for downscaling and integrates topographic and land use data [8, 9]. The conformal cubic coordinate system allows CCAM to be used as a global and regional prediction model. This feature is an advantage of CCAM compared to other global models in general.

In this study, Global Satellite Mapping of Rainfall (GSMaP) data from the Japan Aerospace Exploration Agency (JAXA) was used as a verifier for CCAM output rainfall prediction data due to its high spatial and temporal resolution [10, 11]. In addition, GSMaP products have been evaluated and verified through comparisons with observational data from several previous studies and showed promising results in capturing rainfall patterns and variability [12–16].

The purpose of this study is to compare the rainfall from the results of seasonal predictions using CCAM and satellites in the dry and wet seasons in the Indonesian Territory. In addition, the results were obtained to verify the rainfall prediction from several seasonal prediction simulations. Research on verifying seasonal prediction data is urgently needed, given the importance and challenge of seasonal predictions.

2 Data and Methods

2.1 Data

This study uses rainfall data from the results of seasonal predictions using CCAM and GSMaP. Prediction using CCAM is obtained with initial condition input from GFS and SST prediction from POAMA. CCAM seasonal prediction data contains eight months of prediction periods from the initial condition and has 32 km spatial and a daily temporal resolution, although processed monthly. The CCAM seasonal prediction data are time-lagged, multi-SST ensemble, and prediction rainfall reconstruction. These three data were obtained from the simulation of the seasonal prediction, which is explained further in subsection 2.2. Seasonal prediction simulations were carried out in two periods: June 2019, which represented the dry season, and December 2018 represented the wet season, with the research area being the Indonesian Maritime Continent (BMI) region.

CCAM is an atmospheric model developed by the Commonwealth Scientific and Industrial Research Organization (CSIRO), where CCAM was the development of DARLAM in 1994 [9]. This study uses CCAM to obtain medium-term predictions because CCAM can generate atmospheric prediction data using SST data guidance. The input data used are atmospheric data as initial conditions obtained from GFS data and SST prediction data obtained from the Predictive Ocean Atmosphere Model for Australia (POAMA) in the form of a surface temperature prediction ensemble. POAMA is a state-of-the-art seasonal to interannual forecast system based on a coupled ocean/atmosphere model developed in a joint project involving the Bureau of Meteorology Research Center (BMRC) and CSIRO Marine Research [18]. The comparison data for the rainfall parameter is GSMaP rainfall data. GSMaP rainfall data used is GSMaP_mvk which has 0.1° of spatial resolution. Before the comparative analysis, the two data (CCAM output and GSMaP) require spatial and temporal resolution adjustment.

2.2 Methods

This study uses CCAM to perform several simulations of seasonal predictions (medium-term forecasts). Initial condition data uses Global Forecasting System (GFS) data for the first date of each month and uses forcing SST predictions for the next eight months from POAMA. This simulation description had a similar scheme to our previous studies [17]. On post-processing, CCAM output data and simulations in this study were built by three ensemble prediction systems: time-lagged ensemble, multi-SST ensemble, and predicted rainfall reconstruction (Fig. 1).

The time-lagged ensemble system consists of an ensemble member, which results from CCAM seasonal predictions on the rainfall variable for the next eight months with differences in the time of the initial condition. The simulation results from this system are then cut (intersection) in June 2019 (representing the peak of the dry season) and December 2018 (the peak of the wet season), as described in Fig. 1a. Therefore, the time-lagged ensemble consists of eight members for each season simulation. While in Fig. 1b, the multi-SST ensemble system consists of an ensemble member that uses two initial conditions that have been given 11 different SST inputs. Finally, the predicted rainfall reconstruction is carried out using the output data from the multi-SST ensemble, whose work is depicted in Fig. 1c. Further explanation

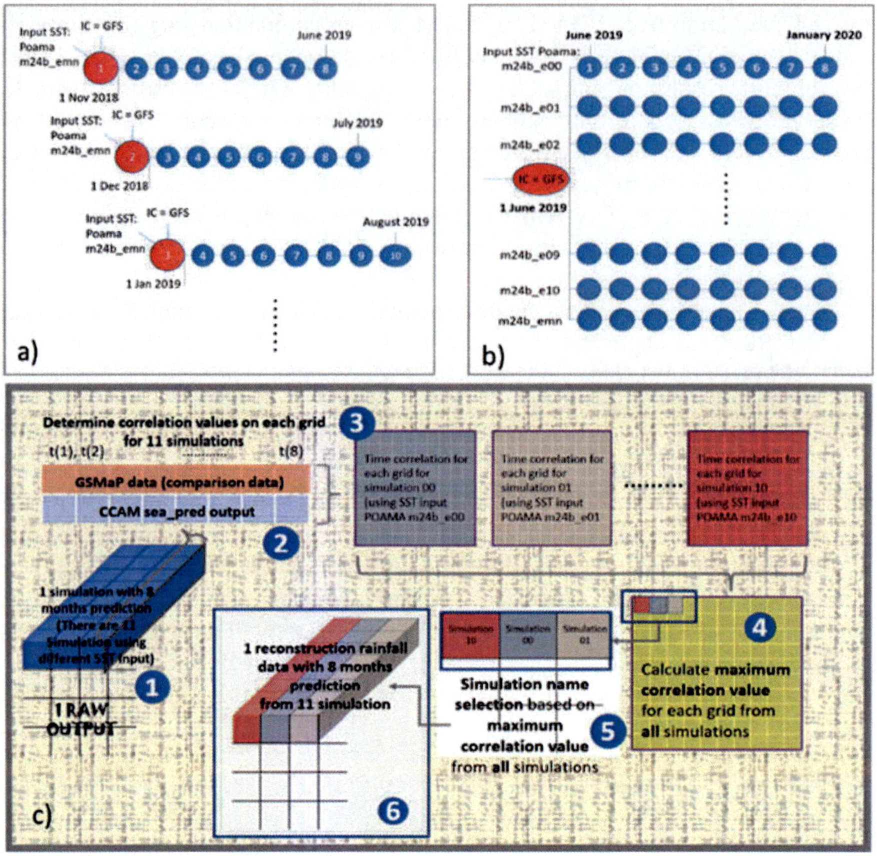

Fig. 1 **a** Seasonal prediction simulation scheme for time-lagged ensemble, **b** multi-SST ensemble, and **c** steps to reconstruct the rainfall prediction data. The date in the picture only illustrates the prediction scheme, not the date used for this study

regarding rainfall reconstruction can be seen in previous studies [20]. Thus, the time-lagged and multi-SST ensemble results were probabilistic values, and the rainfall reconstruction results were deterministic.

Rainfall data from these three simulations are classified based on the rainfall characteristics by the Indonesian Agency of Meteorology, Climatology, and Geophysics (BMKG). This rainfall characteristic is the result of comparing the rainfall intensity during the specified period and the average rainfall in the same period on a long-term scale (baseline) [21]. The long-term scale is defined as the arithmetic means of each climate (such as rainfall and temperature) for 30 years [22]. However, due to data availability, this study determined normal rainfall from the average rainfall only during 2014–2017. The characteristics of rain, according to BMKG, are divided into three categories, namely.

1. Above Normal (AN)—if rainfall values are more than 115% of baseline data.
2. Normal (N)—if rainfall values are between 85% - 115% of baseline data.
3. Below Normal (BN)—if rainfall values are less than 85% of baseline data [21].

3 Results and Discussion

This study compared the characteristics of monthly rainfall between the predicted results and satellite rainfall. The monthly rainfall characteristics used in this study describe rainfall above 150 mm/month. Figure 2a shows that rainfall above 150 mm/month in June 2019 mainly occurred north and around the equator, while rainfall in southern Indonesia, such as in Java, Bali, Nusa Tenggara, and Timor, was less than 150 mm/month. The rainfall of less than 150 mm/month is caused by the movement of the Australian monsoon, which brings dry air in from southern Indonesia through Java and its surroundings. The several prediction simulations show similar results, as shown in Figs. 2b and c, that the probability value of rainfall events of more than (less than) 150 mm/month occurring at the equator and northern BMI (southern BMI). It shows that the probabilistic value of rainfall of more than 150 mm/month generated by the simulation results can describe the occurrence of rainfall of more than 150 mm/month. Although the probability value of rainfall is more than 150 mm/month in some areas, it is still not well predicted, such as in the western coast of Sumatra, the Malacca Strait, the South China Sea, and the eastern part of the Philippine Sea. The rainfall probability value of more than 150 mm/month for prediction simulation using multi-SST ensembles is more similar to GSMaP data than time-lagged ensembles.

Figure 3 shows the spatial pattern of GSMaP rainfall of more than 150 mm/month and a simulated probability of rainfall of more than 150 mm/month, as predicted by CCAM in December 2018. In December 2018, most areas of Indonesia had entered the wet season, so rainfall of more than 150 mm/month occurs evenly in almost all parts of Indonesia (Fig. 3a). However, the probability of more than 150 mm/month rainfall shown by the CCAM seasonal prediction simulation (Fig. 3b and c) shows that only areas south of the equator get a high probability value for rainfall of more than 150 mm/month. It shows that the seasonal prediction simulation for the wet season in IMC was still not good enough for areas north of the equator following Schiemann et al. (2014) statement that the sensitivity model is not significantly associated with orographic conditions in the Indonesian Maritime Continent [23]. In addition, Satyawardhana et al. (2022) stated that the results of seasonal predictions in the areas on the west coast of Sumatra, Central Kalimantan, South Sulawesi, and the west coast of southern Papua have a negative correlation value with the observation data [20]. Therefore, research is still needed to get more accurate seasonal predictions by comparing and verifying the results of seasonal predictions to produce high-resolution seasonal predictions.

Figures 4 and 5 show a comparison of rainfall characteristics in Indonesia in June 2019 and December 2018 based on satellite data and model results. Figure 4a shows

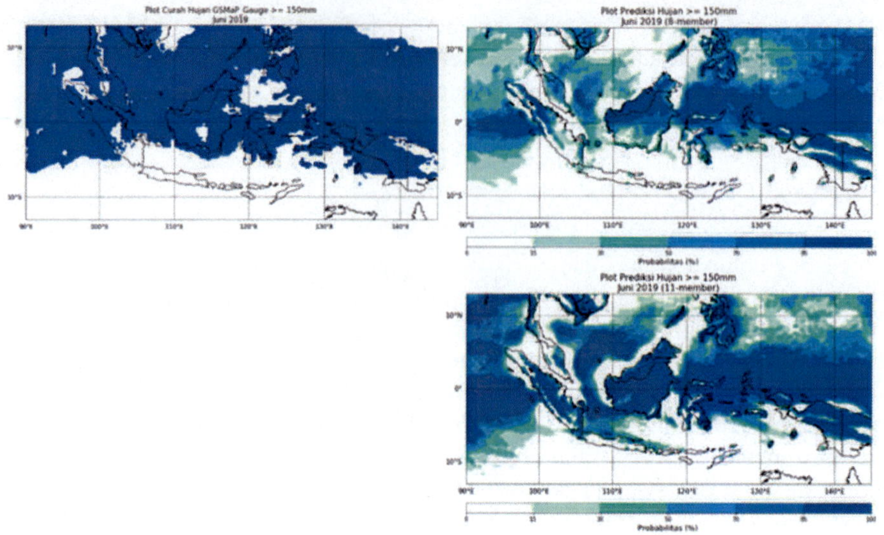

Fig. 2 Comparison of rainfall above 150 mm/month and probability of rainfall above 150 mm/month in June 2019

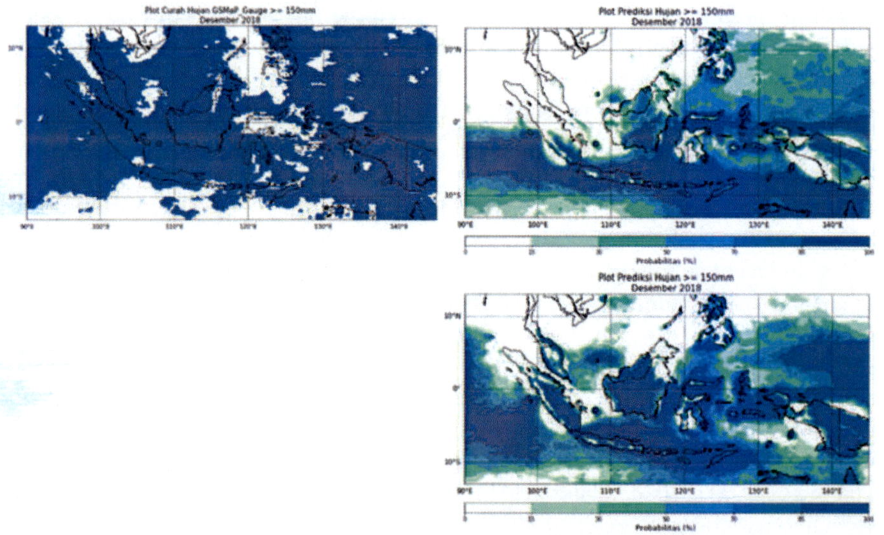

Fig. 3 Comparison of rainfall above 150 mm/month and probability of rainfall above 150 mm/month in December 2018

the rainfall characteristics using GSMaP in June 2019, which were below normal for southern Indonesia and the Sulawesi Sea. In addition, the western sea of Sumatra and the South China Sea were below-normal compared to the 5-year average rainfall. The results of the rainfall reconstruction (Fig. 4b) show that areas in the Java Sea have different results from the GSMaP data, which should be below normal. In addition, the below-normal rainfall characteristics described by satellite data in the Malacca Strait, West Pacific, and the South China Sea are also described reverse by the CCAM rainfall reconstruction results. Similar results are shown by the probabilistic value of time-lagged (Fig. 4c) and multi-SST ensemble (Fig. 4d). Even in the probability value of rainfall characteristics, the time-lagged ensemble has a low accuracy of rainfall characteristics in the Sumatra area. In contrast, the rainfall characteristics of the multi-SST ensemble are well predicted.

Figure 5a shows the spatial pattern of rainfall characteristics from GSMaP data which illustrates that the rainy season in Indonesia (December 2018) is classified as below-normal to normal. However, all the predictions in reconstructed rainfall, time-lagged, and multi-SST ensemble cannot capture normal rainfall (Fig. 5b, c, and d). Therefore, this seasonal prediction for December 2018 can only capture below and above-normal rainfall characteristics. This condition is due to the range of values for normal rainfall, which tends to be small compared to other categories.

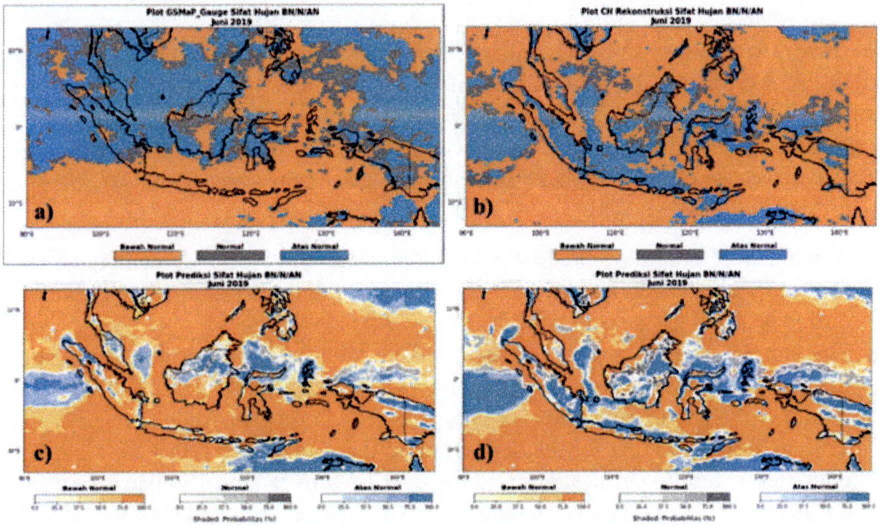

Fig. 4 Comparison of rainfall characteristics in June 2019 with the climatological average of rainfall in June, between **a** GSMaP, **b** CH reconstruction results, **c** time-lagged ensemble, and **d** multi-SST ensemble

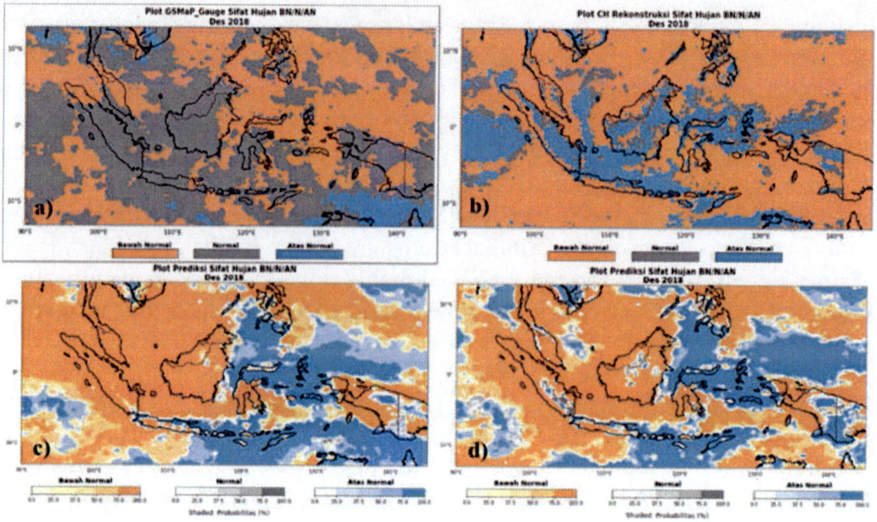

Fig. 5 Same as Fig. 2, but in December 2018

4 Summary and Conclusions

We investigated the ensemble simulations on the seasonal rainfall prediction during 2018–2019 over the Indonesia Maritime Continent. The wet (dry) seasons were represented by December 2018 (June 2019). To explore the methods, we compared a time-lagged ensemble, a multi-SST ensemble, and no-ensemble simulations, then validated the results using GSMaP satellite data. The main results could be described as follows:

- A multi-SST ensemble predicted shows a better qualitative agreement in rainfall characteristics compared with observed rainfall than a time-lagged ensemble during wet and dry season periods.
- The simulation results suggest that the methods have skillful predictions in the dry season, despite a low skill for the wet season. However, it might cause the model not to capture the normal rainfall observed characteristics.
- This study noticed that for the dry season in the southwestern IMC, a multi-SST consistently improved the rainfall characteristics over land (Sumatra-Java) and ocean (Java Sea) compared to a time-lagged ensemble. The southwestern IMC was mentioned in numerous studies [24–26] as a prominent region responsive to local and remote ocean forcing. In the dry season case study, the predicted improvement was depicted by a reduction of above-normal rainfall over the Java sea, whereas an enhancement of below-normal rainfall inland. It suggests that a multi-SST method that considers various SST from 11 ensemble members successfully elaborates physics and dynamics processes over boundary layers associated with increased rainfall over land during the dry season, which was proposed in a previous study [27].

- It should be noted that in the wet season, both ensembles method gave underestimated prediction results over southwestern IMC. Furthermore, it proposed that the simulations have a systematical bias due to meso-to-regional scales atmospheric processes that could not represent well in the CCAM model.

However, considering results that have almost similar predicted rainfall between a time-lagged and a multi-SST ensemble, this study suggests that a time-lagged ensemble should be chosen as a better strategy to reduce the time-running consumption of the model.

Acknowledgements The authors would like to thank the Japan Aerospace Exploration Agency (JAXA) for the GSMaP data and the Australian Bureau of Meteorology (BOM) for providing POAMA data. The authors also want to show our gratitude to CSIRO for developing CCAM that we can use to improve our capability in atmospheric modeling.

References

1. Koch-Larrouy, A., Madec, G., Bouruet-Aubertot, P., Gerkema, T., Bessières, L., Molcard, R.: On the transformation of Pacific water into Indonesian throughflow water by internal tidal mixing. Geophys. Res. Lett. **34**(4) (2007)
2. Sprintall, J., Révelard, A.: The Indonesian throughflow response to Indo-Pacific climate variability. J. Geophys. Res.: Oceans **119**(2), 1161–1175 (2014)
3. Wijffels, S., Schmidt, R.W., Bryden, H.L., Stigebrandt, A.: Transport of freshwater by the Oceans. J. Phys. Oceanogr **22**, 156–162 (1992)
4. Ganachaud, A., Wunsch, C., Marotzke, J., Toole, J.: Meridional overturning and large-scale circulation of the Indian Ocean. J. Geophys. Res. **105**, 26117–26134 (2000)
5. Dayem, K.E., Noone, D.C., Molnar, P.: Tropical western Pacific warm pool and maritime continent precipitation rates and their contrasting relationships with the walker circulation. J. Geophys. Res. Atmos. **112**(6), 1–12 (2007). https://doi.org/10.1029/2006JD007870
6. Jan van Oldenborgh, G., Balmaseda, M.A., Ferranti, L., Stockdale, T.N., Anderson, D.L.T.: Evaluation of atmospheric fields from the ECMWF seasonal forecasts over a 15-Year period. J. Clim. **18**(16), 3250–3269 (2005)
7. Schepen, A., Wang, Q.J., Robertson, D.E.: Seasonal forecasts of Australian rainfall through calibration and bridging of coupled GCM outputs. Mon. Weather Rev. **142**(5), 1758–1770 (2014). https://doi.org/10.1175/MWR-D-13-00248.1
8. Thatcher, M., McGregor, J.L.: A technique for dynamically downscaling daily-averaged GCM datasets using the conformal cubic atmospheric model. J. American Meteorol. Soc. **139**, 79–95 (2011)
9. Thatcher, M., McGregor, J., Dix, M., Katzfey, J.: A new approach for coupled regional climate modeling using more than 10,000 Cores. In: Denzer, R., Argent, R.M., Schimak, G., Hřebíček, J. (eds.) Environmental Software Systems. Infrastructures, Services, and Applications. ISESS 2015. IFIP Advances in Information and Communication Technology, vol. 448. Springer, Cham (2015). https://doi.org/10.1007/978-3-319-15994-2_61
10. Ushio, T., Sasashige, K., Kubota, T., Shige, S., Okamoto, K.I., Aonashi, K., Inoue, T., Takahashi, N., Iguchi, T., Kachi, M., Oki, R., Morimoto, T., Kawasaki, Z.I.: A Kalman filter approach to the global satellite mapping of precipitation (GSMaP) from combined passive microwave and infrared radiometric data. J. Meteorol. Soc. Japan Ser II **87A**, 137–151 (2009). https://doi.org/10.2151/jmsj.87A.137

11. Aonashi, K., Liu, G.: Passive microwave precipitation retrievals using TMI during the Baiu period of 1998. Part I: algorithm description and validation. J. Appl. Meteorol. **39**(12), 2024–2037 (2000)
12. Priyambodho, B.A., Kure, S., Yagi, R., Januriyadi, N.F.: Flood inundation simulations based on GSMaP satellite rainfall data in Jakarta Indonesia. Prog. Earth Planet Sci. **8**, 34 (2021). https://doi.org/10.1186/s40645-021-00425-8
13. Setiawati, M.D., Miura, F.: Evaluation of GSMaP daily rainfall satellite data for flood monitoring: case study–Kyushu Japan. J. Geosci. Environ. Prot. **4**(12), 101–117 (2016). https://doi.org/10.4236/gep.2016.412008
14. Admojo, D.D., Tebakari, T., Miyamoto, M.: Evaluation of a satellite-based rainfall product for a runoff simulation of flood event: a case study. J. Japan Soc. Civil Eng. Ser. B1 **74**(4), I_73–I_78 (2021)
15. Pakoksung, K., Takagi, M.: Effect of satellite-based rainfall products on river basin responses of runoff simulation on flood event. Model Earth Syst. Environ. **2**(3), 143 (2016). https://doi.org/10.1007/s40808-016-0200-0
16. Rusmanansari, A.H., Suwarman, R., Djamil, Y.S., Firdaus Permadhi, Y.: GSMaP seasonal rainfall verification over Western Java. In: Proceedings of the International Conference on Radioscience, Equatorial Atmospheric Science and Environment and Humanosphere Science, 2021. Springer Proceedings in Physics, vol. 275. Springer, Singapore (2022). https://doi.org/10.1007/978-981-19-0308-3_60
17. Satyawardhana, H., Gammamerdianti (2019) Comparison of seasonal prediction outputs based on dynamic atmosphere model and observations (case study: seasonal prediction in 2016−2017). In: IOP Conference Series: Earth Environment Science, vol. 303. pp. 012050. (2017)
18. Wang, G., et al.: POAMA: an Australian ocean-atmosphere model for climate prediction. Bull. Am. Meteorol. Soc. **1**, 4559–4563 (2004)
19. Kachi, M., Kubota, T., Aonashi, K., Ushio, T., Shige, S., Yamamoto, M.: Recent improvements in the global satellite mapping of precipitation (GSMaP). In: 2014 IEEE Geoscience and Remote Sensing Symposium, pp.3762−3765 (2014). https://doi.org/10.1109/IGARSS.2014.6947302. The GFDL Global Atmospheric Model Development Team.: The New GFDL Global Atmosphere and Land Model AM2-LM2: Evaluation with Prescribed SST Simulations. J. Climate **17**(24), 4641–4673 (2004)
20. Satyawardhana, H., Yulihastin, E., Gammamerdianti, Ihsan N.I., Wulandari, E.P.: Evaluation of CCAM seasonal prediction by GSMaP satellite rainfall data in Indonesia. In: Yulihastin, E., Abadi, P., Sitompul, P., Harjupa, W. (eds.) Proceedings of the International Conference on Radioscience, Equatorial Atmospheric Science and Environment and Humanosphere Science, 2021. Springer Proceedings in Physics, vol. 275, Springer, Singapore (2022). https://doi.org/10.1007/978-981-19-0308-3_14
21. BMKG: Prakiraan Musim Hujan 2022/2023 di Indonesia. BMKG 2022)
22. Muharsyah, R.: Penentuan Batas Atas Normal dan Bawah Normal Curah Hujan Bulanan Setara Tercile dengan Koefisien Regresi Linier Sederhana. Jurnal Meteorologi dan Geofisika **15**(1), (2014)
23. Reinhard, S., Marie-Estelle, D., Mizielinski, M., Malcolm, R., Len, S., Strachan, J., Vidale, P.L.: The sensitivity of the tropical circulation and Maritime Continent precipitation to climate model resolution. Climate Dynam. **42** (2013). https://doi.org/10.1007/s00382-013-1997-0
24. Hamada, J.-I., Mori, S., Kubota, H., Yamanaka, M.D.: Interannual rainfall variability over northwestern Jawa and its relation to the Indian Ocean Dipole and El Niño-Southern Oscillation events. SOLA **8**, 069–072 (2012)
25. Xu, Q., Guan, Z., Jin, D., Hu, D.: Regional characteristics of interannual variability of summer rainfall in the Maritime Continent and their related anomalous circulation patterns. J. Clim. **32**, 4179–4192 (2019)

26. Yulihastin, E., Putrantom M.F., Suaydhi, Sofiati, I.: Oceanic effect on precipitation development in the Maritime Continent during anomalously-wet dry seasons in Java. Indones. J. Geogr. **53**(3), 328–339 (2020)
27. Yulihastin, E., Nuryanto, D.E., Trismidianto, Muharsyah, R.: Improvement of heavy rainfall simulated with SST adjustment associated with mesoscale convective complexes related to severe flash flood in Luwu, Sulawesi, Indonesia. Atmosphere **12**,1445 (2021). https://doi.org/10.3390/atmos12111445.

Application and Analysis of Remote Sensing for the Initial Baseline of the Quantitative Comfort Index in Ibukota Nusantara (IKN) and Its Surroundings

Indah Susanti, Lilik S. Supriatin, Sinta B. Sipayung, Edy Maryadi, Adi Witono, Martono, Laras Toersilowati, and Amalia Nurlatifah

Abstract Convenience is an important aspect to optimize work productivity and performance, so it needs to be considered in development planning, especially in determining the location of the capital city of a country. This is important because the level of comfort will affect the level of energy use, especially electrical energy. The location of the capital city of a country should be chosen which can minimize the level of energy use in accordance with the principles of sustainable development (SDGs). Based on air temperature and humidity data from the Atmospheric Infrared Sounder (AIRS), the quantitative comfort level (Temperature–Humidity Index formulation) was processed and analyzed in Balikpapan City, Gunung Mas Regency, Katingan Regency, Kutai Kartanegara Regency, Palangkaraya City, and Tanah Bumbu Regency. The time span of the data processed and analyzed is January 2003 to December 2018. The results of statistical analysis show that the monthly average THI in the six regions is in the range of 24–27 °C, or in the partially comfortable category. Gunung Mas Regency is an area that has an average THI of 24.8, while other areas have a THI above 25. This means that the level of comfort in Gunung Mas Regency is the best compared to other areas analyzed. In addition, in the time span analyzed, the trend of THI in Gunung Mas Regency shows a decline, which means toward a better level of comfort. This has implications for the level of energy use which is also better.

L. S. Supriatin · S. B. Sipayung · A. Witono · Martono · L. Toersilowati · A. Nurlatifah
Research Center of Climate and Atmosphere, National Research and Innovation Agency, West Java, Indonesia

I. Susanti (✉)
Directorate of Repository, Multimedia and Scientific Publication, National Research and Innovation Agency, West Java, Indonesia
e-mail: indahpl@gmail.com

E. Maryadi
Research Center for Data and Information Sciences, National Research and Innovation Agency, West Java, Indonesia

A. Basit et al. (eds.), *Proceedings of the International Conference on Radioscience, Equatorial Atmospheric Science and Environment and Humanosphere Science*, Springer Proceedings in Physics 290, https://doi.org/10.1007/978-981-19-9768-6_78

1 Introduction

A few moments ago the issue of moving the capital city of Indonesia arose. The idea of relocating the nation's capital was based on traffic and flood that occurred in DKI Jakarta. Based on traffic conditions, Jakarta is ranked as the fourth worst city, and this has implications for real financial losses [2]. The loss in 2013 reached Rp 56 trillion per year, and in 2019 it was close to Rp 100 trillion [2]. From the hydrological aspect, the Jakarta area which is included in the flood-prone category reaches 50% and has a flood vulnerability level of under 10 years.

Moving the nation's capital is not an easy matter. In this case, careful consideration is needed in planning, both from the economic, social, and environmental aspects. The city is an ecosystem formed by various types of land cover, vegetation, and various types of land use from a very complex landscape [5]. Therefore, relocating the nation's capital means that an unavoidable urbanization process will occur. The trend is that urbanization puts a lot of pressure on the environment, including reducing the size of the habitat and changing the spatial pattern of the dynamics of living things. Several studies related to urbanization have shown changes in abiotic conditions in the remaining parts of the habitat, which are caused by increases in temperature, rainfall, humidity, and changes in the nitrogen cycle [6, 14].

Physical characteristics generally affect the ecological functions that support life. The level of primary productivity that determines the sustainability of an ecosystem is largely determined by temperature, rainfall, radiation, and carbon dioxide levels. Likewise, the level of human comfort is determined by temperature, water content in the air, wind, and several other factors. In areas with a low level of comfort, it requires higher facilities and energy to obtain an adequate level of comfort. The farther an area is from the optimal comfort level, the higher the additional energy required to obtain the required level of comfort. A clear example of this is the use of air conditioning (AC) in areas with low comfort levels. Planning the needs and use of facilities, infrastructure and energy in an area, needs to consider the carrying capacity and capacity of the environment in order to have sustainability. This consideration is in accordance with the demands of the development paradigm and international agreements that are currently developing, namely the sustainable development goals (SDGs).

Consideration of environmental comfort in urban development becomes important, because the level of health of humans and other living creatures is strongly influenced by environmental conditions, including the level of heat stress. Heat stress is influenced by air temperature, humidity, water movement, solar radiation, and precipitation [8]. Several studies have been conducted, such as by Basu [1], Hajat and Kosatky [7], Morabito, et al. [11], Tobias, et al. [15], Wu et al. [18] showed a relationship between high environmental temperatures and human mortality rates in some areas. Previous studies use air temperature variables such as daily average temperature, maximum, and minimum air temperature, as environmental predictors that affect mortality rates [11]. In addition, other authors use an alternative temperature metric that combines several meteorological parameters (humidity, temperature, wind speed, and solar radiation) into a single number, one of which is the Temperature–Humidity Index (THI). THI is a single value that describes the integrated effect of temperature and humidity on the degree of heat stress. Several studies use this index to show the heat load and the level of environmental comfort felt by humans that affect health and mortality rates, or indicate heat-free livestock that affects their productivity levels. The THI index has been used for various studies, one of which is to analyze the level of comfort in Jakarta as the current state capital [17]. The results of the analysis conducted by Wati and Fatkhuroyan [17], using data on air temperature and humidity from the Meteorology, Climatology, and Geophysics Agency (BMKG) meteorological stations in Kemayoran, Tanjung Priok, Halim Perdanakusuma, Cengkareng, and Pondok Betung, revealed that most areas of Jakarta have a THI of 25–27 C, which means that in the study area only some feel comfortable. This condition causes in general, in big cities such as Jakarta, the use of energy becomes relatively higher to obtain the optimum level of comfort.

Basically, there are difficulties in obtaining the appropriate accuracy and validity of heat measurements to analyze environmental influences on the level of ecosystem productivity or the level of human health. The significance of the application of an environmental indicator, whether using air temperature or thermal index, cannot be fully ascertained, especially if it is used in different geographical areas. However, an approach is needed to analyze the effect of environmental temperature pressure on certain aspects. In this study, the THI thermal index is used to analyze the level of comfort felt by humans to live in urban areas, although this is relative, because it is influenced by the psychological condition of the population. People who tend to live in cold areas will find it difficult to feel comfortable living in hot areas. In this case, it is necessary to adapt to a different time for each individual. However, in general, there are patterns of perception of comfort, as Nieuwolt has studied in assessing comfort level based on THI value. In the process of calculation and analysis, in this study, the application of THI is carried out for six cities that are candidates for the capital city of Indonesia. The selected city was carried out based on a literature study of several mass media which provided information related to the opinion of the Head of State and Bappenas regarding the candidate for the capital city. Analysis of the comfort level in this study is expected to be one of the considerations in the selection of the national capital in terms of the environment and can support sustainable development goals. This consideration is also expected to complement other considerations in terms of land availability, land status, disaster-prone status, geographical position, security and politics, as well as the presence of residents as supporting activities.

Remote sensing technology, that has developed today, has had a major impact on environmental studies. This is because remote sensing technology can provide data in a wide spatial range, including hard-to-reach areas. In addition, remote sensing data has a temporal resolution that can be adjusted to the needs of the analysis. The research uses Atmospheric Infrared Sounder (AIRS) sensor data which is an instrument on NASA's Aqua satellite. In general, studies of the air comfort index have used AWS data which provides limited spatial analysis. By using AIRS data, the comfort index can be analyzed spatially with a wider area coverage and more consistent time stages. In this case, the study of air comfort index using AIRS data is still very limited.

2 Data and Method

The data used in this study is remote sensing data from the Aqua satellite Atmospheric Infrared Sounder (AIRS) sensor with spatial resolution 1°. The working principle of this sensor is to capture infrared radiation emitted by the earth's surface and its atmosphere globally every day. The AIRS sensor can provide profile data for 20 levels of atmospheric altitude from the surface to the tropopause. The resulting data includes temperature, relative humidity, charge of carbon dioxide, methane, carbon monoxide, and several other parameters. At this time, with the development of existing algorithms, AIRS sensors can also detect volcanic smoke and forecast drought.

The AIRS data used for this study is monthly data for the period January 2003 to January 2018. The use of data with the length of the period is expected to show a more general trend of changes due to certain factors, including climate change. The coverage area analyzed is Kalimantan, which is then focused on several selected areas, namely Balikpapan City, Gunung Mas Regency, Katingan Regency, Kutai Kartanegara Regency, Palangkaraya City, and Tanah Bumbu Regency. The AIRS data, then, analyzed the monthly THI pattern spatially for the Kalimantan region and temporally for the selected area. In addition, it is also analyzed using a time series to see trends in changes that occur, so that future comfort conditions can be estimated.

The comfort index (THI) is calculated using the formula compiled by Nieuwolt [12], namely

$$\mathrm{THI} = 0.8\mathrm{T}\ \{(\mathrm{RH} \times \mathrm{T})/500\}$$

where T is the surface air temperature (°C), RH is the relative humidity, and THI is the comfort index (°C). As for the comfort limit, using a comfort interval based on Nieuwolt [12] and Emmanuel [4] was modified for a tropical climate [3]. Table 1 shows comfort index category based on Temperature–Humidity Index (THI)

Table 1 Comfort index category based on THI

THI (°C)	Categorized
21–24	Comfort
25–27	Comfort enough
>27	Not comfort

3 Results and Discussion

Figure 1 is a map of the island of Borneo with seven potential new capitals of Indonesia. Most of the areas on the island of Borneo are lowlands, which can be seen from the dominant green color on the topographic map of the island of Borneo (Fig. 1). The average temperature throughout the year in the lowlands is 23–28 °C. Several options circulating in the community regarding the selection of the national capital include Pontianak, Palangkaraya, Katingan, Gunung Mas, Tanah Bumbu, Balikpapan, and Kutai Kartanegara. Today, the Ministry of State Secretary announced and designated Kutai Kartanegara as a candidate for the state capital. The question is, when viewed from the climatic conditions, is Kutai Kartanegara Regency quite comfortable as the nation's capital? The analysis of this research will answer it.

The results of remote sensing data processing show that THI in Kalimantan has quite large temporal and spatial variations, with a range of values between 21 and 26.8 °C. This is an advantage of using AIRS data, especially for monthly data, because it shows good data coverage, both temporally and spatially. In situ measurement data at this time have not been able to provide data that is spatially and temporally consistent. In January which represents the rainy season, the average THI on the island

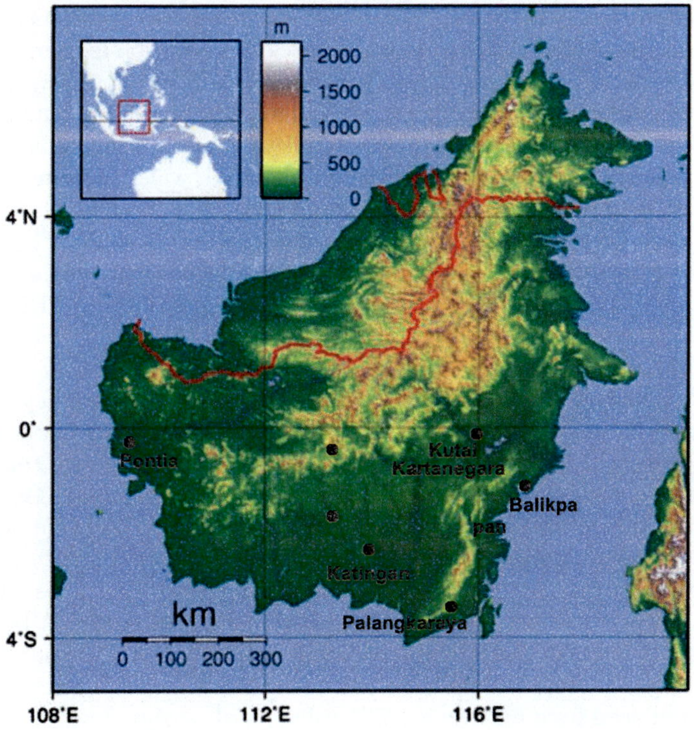

Fig. 1 Kalimantan map of topography and seven potential location, New Capital City, Indonesia

of Kalimantan is 24.5–26 °C where in this category the comfort index in Kalimantan is categorized as a partly comfortable area. In April which represents the transition season, the average THI in Kalimantan Island is 25–26.5 °C, including the category as a partly comfortable area with a comfort index of one level above January. The THI value in July and October which represents the dry season and relative transition is the same as the conditions in April, with the THI value being 25–26.5 °C which means that the conditions are partially comfortable. Kutai Kartanegara Regency itself, as the strongest candidate for the national capital, is in the range of THI values of 25.5–26.5 in four different months (January, April, July and October) which means it is in the partially comfortable category (Fig. 2).

The choice of moving the capital, of course, is expected to be an area with the best comfort. However, in most areas of Kalimantan in January, March, July, and October the conditions are partly comfortable. Efforts to optimize the THI value into the comfortable category must be done to be able to reduce air temperature and increase humidity.

Spatially, northern Kalimantan generally shows better comfort than southern Kalimantan (Fig. 2). This is influenced by the topography of the region, where the northern area tends to be higher than the southern area. Temporarily, variations in THI are influenced by monsoon patterns, where January which represents the rainy season period (December–January–February) is the period with the lowest THI. The difference in THI in January is about 1° lower compared to other periods. This difference has major implications for the various processes that occur on the earth's surface. Humans can tolerate temperature increases of up to 2–3 °C. However, some types of vegetation and animals have high sensitivity to changes in temperature, so that a difference of 1° in comfort level can affect the level of ecosystem productivity.

In general, seven (7) areas analyzed show high THI, which indicates a partially comfortable condition. Based on the cycle, the monthly pattern shows the highest value in May. The THI value in May is in the range of 25.4–26.2 °C, while the lowest THI value occurs in January with a range of values between 24.2 and 25.4 °C. However, Gunung Mas Regency is an area that has the lowest THI compared to six (6) other regions with differences about 1° for the average of all periods (January–December) (Fig. 3).

The difference of 1 °C for the THI value in Gunung Mas Regency compared to other areas has major implications in various aspects, especially implications for energy needs to achieve the level of comfort required for humans. The use of electrical energy for air conditioning and food and beverage cooling machines in Gunung Mas Regency will be lower than in the other six alternative areas. In fact, in the December–January period comfortable conditions in Gunung Mas Regency formed naturally due to the support of weather and climate factors, so that energy use can be minimized, although in this case it has not been analyzed from the topography of the area. In addition, areas with comfortable air conditions are areas that have potential in the tourism sector. The level of air comfort in Gunung Mas Regency is better, in general it can also be seen from the average THI value for all analyzed periods as can be seen in Table 1.

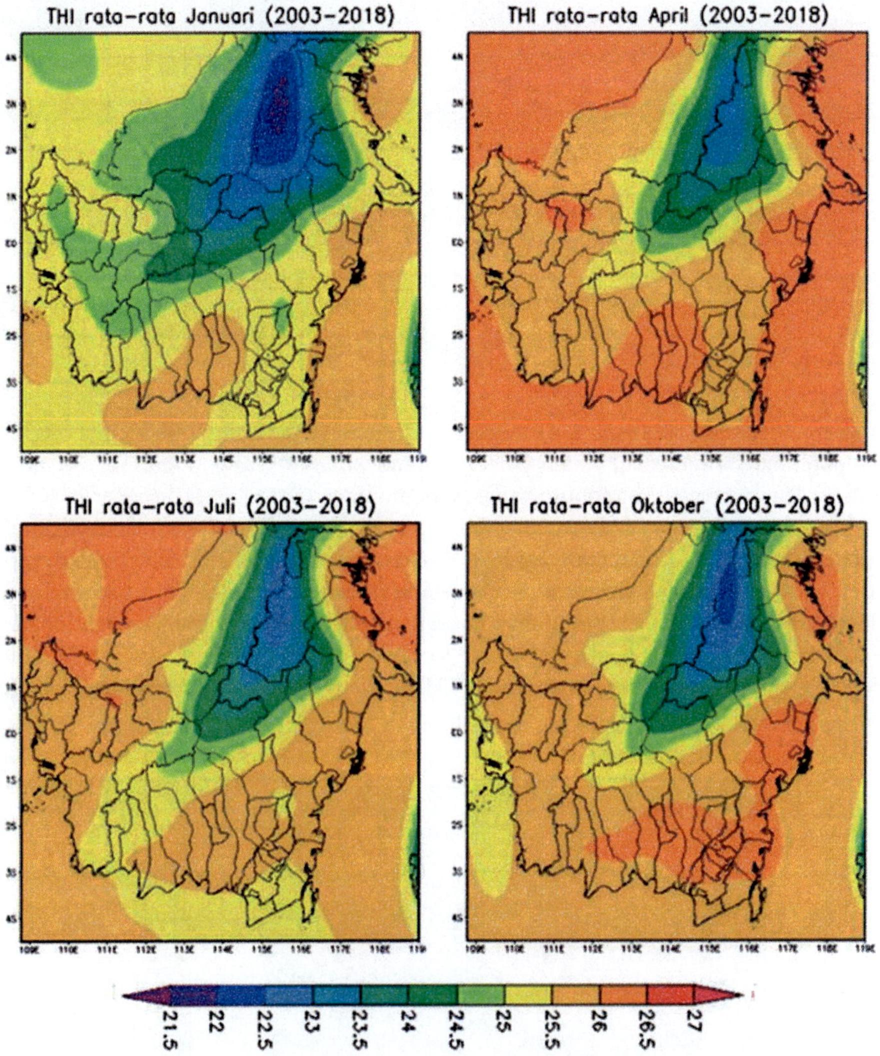

Fig. 2 THI in Kalimantan for monthly average in the range 2003–2018

Based on the trend of developing climatic conditions, in addition to showing a lower THI value compared to other regions, THI in Gunung Mas Regency also shows a decrease, which means that changes in temperature and humidity as elements that determine THI values lead to better comfort conditions. Figure 4 shows that from January 2003 to December 2018, the THI value in Gunung Mas Regency has decreased even though the decline is at a low rate, which is 2 × 10–5 per month or 2.4 × 10–4°/10 year. However, this is better than what happened in Pontianak City, which experienced a significant increase in the THI value, which was 9 × 10–5 per

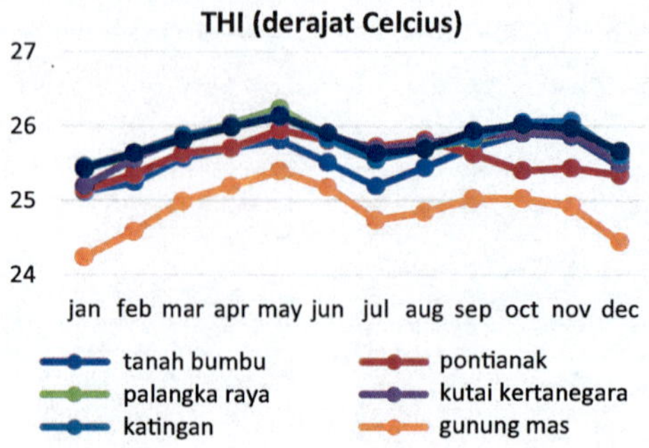

Fig. 3 Climatological average monthly THI (2003–2018) for the seven prospective capital cities

month or 1.08 °C/10 years. This tendency to increase the value of THI also shows the potential for increasing energy needs, in the value of seasonal rainfall, are more clearly seen when the maximum value is identified.

The THI value in Kutai Kartanegara as the strongest candidate for the national capital is predicted to decrease by 1 × 10–5 per month or 1.2 × 10–3 °C/10 years (Fig. 4). This trend of decreasing THI value is expected to be able to anticipate the energy needs of the community in adjusting to the comfortable conditions in Kutai Kartanegara. However, to be able to stay in perfect climatically comfortable conditions, it takes a long time for the THI value to be in the overall comfortable category (Table 2).

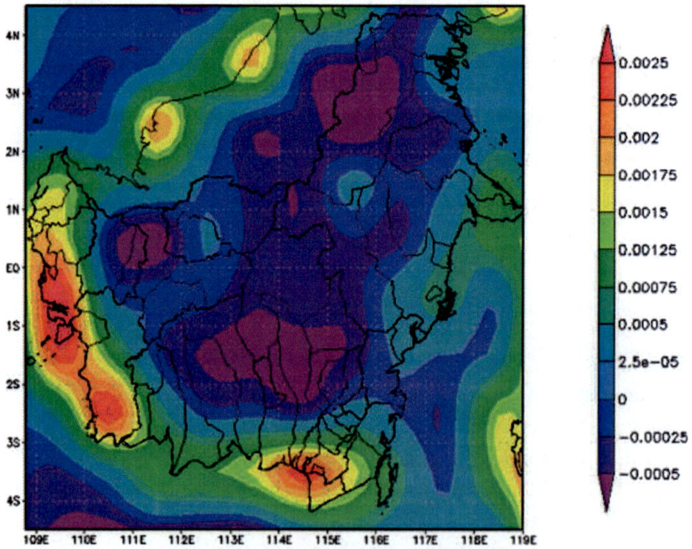

Fig. 4 Trend of change in THI value per month according to the slope value of the regression equation with respect to time from January 2003 to December 2018 in °C

Table 2 Average THI for the entire analysis period (2003–2018) for the seven prospective capital cities

Locations	Mean Temperature–Humidity Index (THI, °C)
Tanah Bumbu	25.6
Pontianak	25.6
Palangkaraya	25.8
Kutai Kartanegara	25.8
Katingan	25.8
Gunung Mas	24.9
Balikpapan	25.8

The downward trend in the value of THI in Gunung Mas is basically in line with the trend that occurred in the middle of the island of Borneo. In Fig. 4, it can be seen the trend of changes in the THI value per month according to the slope value of the regression equation with respect to time from January 2003 to December 2018 in °C. From the figure, it can be seen that in areas far from the coast, there is a trend of decreasing THI values, while in coastal areas, especially on the west coast of Kalimantan, there is a significant trend of increasing THI values, reaching more than 0.0025° per month, or 0.3°/10 years.

From the existing values and the changing trends that occur in Kalimantan, especially for areas targeted to become the prospective capital city of Indonesia, in terms of climate and comfort level, Gunung Mas Regency is the optimum area to be developed as the nation's capital. Apart from minimizing the use of energy, the level of comfort is related to the optimization of human resources. People will tend to be more productive if the surrounding environmental conditions are quite supportive.

The trend of environmental development that occurred in the analyzed period cannot be used as a benchmark that the change is a permanent trend. There is a possibility that the downward trend in the THI value will change if the character of the region changes which has implications for changes in the energy balance, including air temperature. The conversion of forest land into a built-up area will have a major impact on the character of the area, including changes in its ecosystem. These changes have been widely proven through various studies which state that rapid urbanization leads to an imbalance between needs and needs resource availability, infrastructure, and climate change [9]. This is the implication of the acceleration of population growth and its needs.

Consideration of environmental changes and availability of resources in determining the national capital is important. In the development of infrastructure and other functions as a supporter of the capital, it is unavoidable to avoid changes in the character of the region, which in general is a change from the natural environment to the built environment and built area. It also forms the dominance of artificial elements that have a positive role in promoting energy consumption [10,16]. Grimm et al. [6] and Pickett et al. [13] also demonstrated abiotic changes in the remaining parts of the habitat, caused by increases in temperature, rainfall, humidity, and changes in the nitrogen cycle [6, 13].

4 Conclusion

The AIRS data obtained shows consistent data coverage, both spatially and temporally, so that the full THI value can be obtained for the entire period and the entire scope of the analyzed area. Based on the calculation of the comfort index value represented in the Temperature–Humidity Index (THI), Gunung Mas Regency has the potential for better climatic comfort compared to the other six districts. Kutai Kartanegara Regency as a prospective capital city also has a fairly good comfort index in the category of partially comfortable (THI value 25.5–26.0), but it is also predicted that the THI value will decrease by 0.0012 °C per 10 years and will be more quickly if the flow of infrastructure development increases.

References

1. Basu, R.: High ambient temperature and mortality: a review of epidemiologic studies from 2001 to 2008. Environ. Health **8**(1), 40 (2009)
2. Bonasir, R.: Rencana pemindahan ibu kota dari Jakarta: Berapa lama waktu yang diperlukan dan apasajasyarat-syaratibu kota?. Internet:www.bbc.com/indonesia/indonesia-48093451. BBC News Indonesia (2019). Diakses Sept (2019)
3. Effendy, S.: Keterkaitan ruang terbuka hijau dengan urban heat island wilayah Jabotabek. IPB (Bogor Agricultural University), Disertasi (2007)
4. Emmanuel, R.: Thermal comfort implications of urbanization in a warm-humid city: the Colombo Metropolitan Region (CMR), Sri Lanka. Build. Environ. **40**(12), 1591–1601 (2005)
5. Foresman, T.W., Pickett, S.T., Zipperer, W.C.: Methods for spatial and temporal land use and land cover assessment for urban ecosystems and application in the greater Baltimore-Chesapeake region. Urban ecosystems **1**(4), 201–216 (1997)
6. Grimm, N.B., Faeth, S.H., Golubiewski, N.E., Redman, C.L., Wu, J., Bai, X., Briggs, J.M.: Global change and the ecology of cities. Science **319** (5864), 756–760 (2008)
7. Hajat, S., Kosatky, T.: Heat-related mortality: a review and exploration of heterogeneity. J. Epidemiol. Community Health **64**(9), 753–760 (2010)
8. Igono, M., Bjotvedt, G., Sanford-Crane, H.: Environmental profile and critical temperature effects on milk production of Holstein cows in desert climate. Int. J. Biometeorol. **36**(2), 77–87 (1992)
9. Liu, X., Peng, D.: Study on the threshold effect of urbanization on energy consumption. Theoret. Econom. Lett. **8**(11), 2220 (2018)
10. Ma, H., Xu, J., Wang, P.: Water resource utilization and China's urbanization. Resour. Sci. **36**(2), 334–341 (2014)
11. Morabito, M., Crisci, A., Moriondo, M., Profili, F., Francesconi, P., Trombi, G., Bindi, M., Gensini, G.F., Orlandini, S.: Air temperature- related human health outcomes: current impact and estimations of future risks in central Italy. Sci. Total Environ. **441**, 28–40 (2012)
12. Nieuwolt, S.: In: Tropical Climatology. An introduction to the Climates of the Low latitudes. Wiley (1977)
13. Pickett, S.T., Cadenasso, M.L., Grove, J.M., Boone, C.G., Groffman, P.M., Irwin, E., Kaushal, S.S., Marshall, V., McGrath, B.P., Nilon, C.H.: Urban ecological systems: scientific foundations and a decade of progress. J. Environ. Manage. **92**(3), 331–362 (2011)
14. Pickett, S.T.A., Burch, W.R., Dalton, S.E. et al.: A conceptual framework for the study of human ecosystems in urban areas. Urban Ecosys **1**, 185–199 (1997). https://doi.org/10.1023/A:1018531712889

15. Tobias, A., Armstrong, B., Zuza, I., Gasparrini, A., Linares, C., Diaz, J.: Mortality on extreme heat days using official thresholds in Spain: a multi-city time series analysis. BMC Public Health **12**(1), 133 (2012)
16. Wang, L., Wei, H.: An empirical study on the impact of urbanization on energy consumption in China. Resour. Sci. **36**(2), 1235–1243 (2014)
17. Wati, T., Fatkhuroyan, F.: Analisis Tingkat Kenyamanan Di DKI Jakarta Berdasarkan Indeks THI (Temperature Humidity Index). Jurnal Ilmu Lingkungan Undip **15**(1), 57–63 (2017)
18. Wu, W., Xiao, Y., Li, G., Zeng, W., Lin, H., Rutherford, S., Xu, Y., Luo, Y., Xu, X., Chu, C.: Mortality on extreme heat days using official thresholds in Spain: a multi-city time series analysis. Sci. Total Environ. **449**, 355–362 (2013)

A Model to Describe Elastic Light Scattering of a Single Sphere in a Non-homogeneous Illuminating Light Measured in an Optical Particle Counter

Moch S. Romadhon

Abstract Optical particle counters (OPCs) are widely used in the studies of atmospheric particles. The main principle of OPCs is to illuminate particles and retrieve particle size based on the light scattered by the particles. Ideally, illuminating light should be homogeneous so that scattered light is independent of particle position. However, this requirement is practically difficult. This paper demonstrates the effect of the non-homogeneity of illuminating light on the relation between scattered light and particle size. The effect was studied theoretically and experimentally. In this study, scattered light was measured by a Photomultiplier Tube (PMT). The scattered light can be characterised by two parameters: PMT pulse width and depth. The width indicates the residence time of particles in a sensing volume. The depth indicates the intensity of scattered light. The measurements result in multi-modal distributions of pulse depth as a function of particle size. Based on the distribution, the retrieval of particle size using PMT pulse depth can be misleading. This feature should be considered in designing an OPC or interpreting OPC outputs.

1 Introduction

Optical particle counters (OPCs) are widely used in the continuous measurements of atmospheric particle size distribution to study aerosols impact on climate and public health [1–3]. Their measurement principle is to illuminate particles in their sensing volume and develop relationship between the light scattered by the particles and the particle size [4]. Ideally, the illuminating light is homogeneous across the sensing volume so the scattering light is independent of particle position. However, one of the main problems undermining the accuracy of OPCs is the non-homogeneity of the illuminating light [5, 6]. In a particular case where the shape of illuminating light

M. S. Romadhon (✉)
Center for Climate and Atmosphere, National Research and Innovation Agency, Jakarta Pusat, Indonesia
e-mail: moch.syarif.romadhon@brin.go.id
URL: https://www.aerosols.web.id/

A. Basit et al. (eds.), *Proceedings of the International Conference on Radioscience, Equatorial Atmospheric Science and Environment and Humanosphere Science*, Springer Proceedings in Physics 290, https://doi.org/10.1007/978-981-19-9768-6_79

is Gaussian, the character of the light scattering has been studied theoretically [7]. However, there is no study for the case of arbitrary non-homogeneous plane wave of illuminating light either theoretically or experimentally. Pragmatically, OPCs manufacturers set their instruments sampling rate in a level that particles are aerodynamically focused on a specific position in the sensing volume. In a particular OPC design that requires relatively low sampling rate that particles can not be aerodynamically focused, measurement bias due the non-homogeneity can be significant. This paper presents the study for that case based on measurements using SPARCLE: an OPC to measure particle size and refractive index [8].

2 Design and Calibration

A part of SPARCLE design is shown in Fig. 1. One of the detectors used in the design is Photomultiplier Tube (PMT). The PMT can be used both to indicate the presence of particles in the sensing volume and to estimate particle size. This type of measurement is commonly used by commercial OPCs. The PMT is installed with its area vector direction along the x axis. The distance of the PMT from the centre of reference is around 6 mm. The diameter of the PMT window is around 25 mm. In SPARCLE, particles are transported from the sampling pipe to the sensing volume and removed through exhaust pipe. The path of particles in the sensing volume is modelled to be parallel to the y axis. In the sensing volume, particles are illuminated by a compressed laser beam. The beam intensity varies across the x–y plane. The residence time of particles in the sensing volume is expected around 1,000 ns by setting SPARCLE sampling flowrate around 0.2 $cm^3\,s^{-1}$.

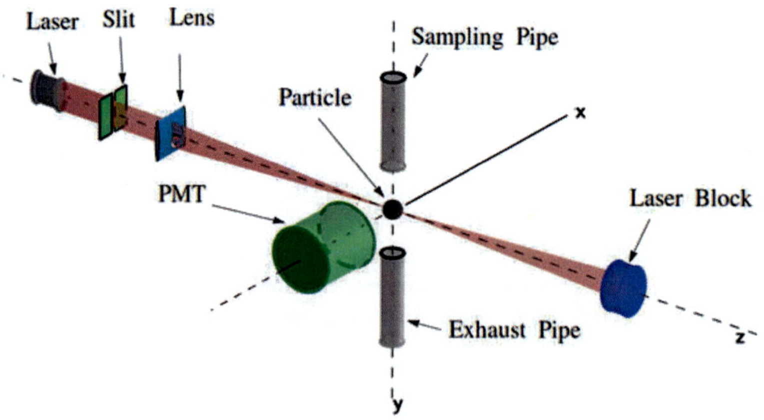

Fig. 1 Part of SPARCLE design. Particles are sampled through the sampling pipe and transported to the sensing volume. In the sensing volume, particles are illuminated by a compressed laser beam. The compression is done by a cylinder lens

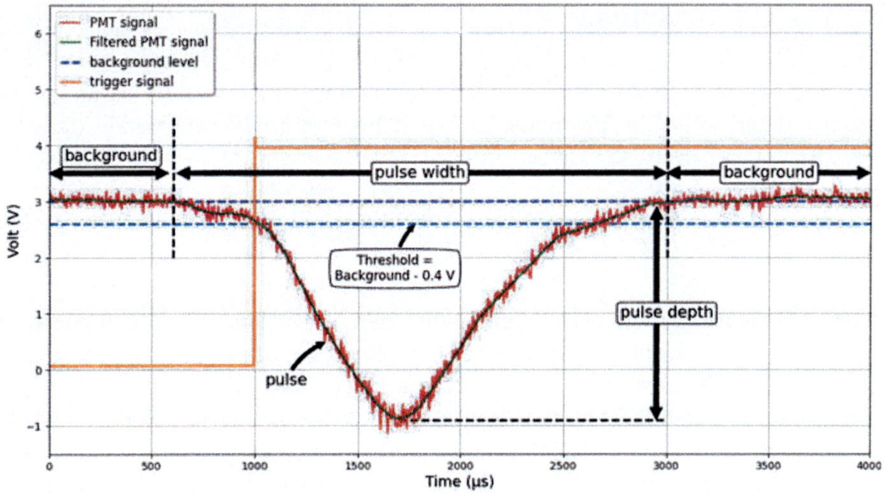

Fig. 2 One example of PMT signals. The pulse is plotted as a red line containing noises in the form of very high-frequency signals. The noise can be removed by applying a Savitzky–Golay filter. The filtered signals are plotted as a green line. Also shown in the figure, trigger signals which are generated when PMT signals reached a threshold level. The threshold level was set at 0.4 V lower than the mean background

The design was calibrated using four monodisperse solid aerosols. The aerosols were generated in two steps: first, by nebulising suspensions containing polystyrene latex (PSL) beads into droplets and, second, drying the droplets to form solid particles. The mean particle sizes of the four aerosols are 1100, 1800, 2000 and 3000 nm. The refractive index of the PSL beads is 1.59. One example of PMT pulses as a response to a PSL bead is shown in Fig. 2. As shown in the figure, PMT pulses can be characterised by two parameters: pulse depth and pulse width. The width indicates the residence time of particles in a sensing volume. The depth indicates the intensity of scattered light. Due to background in the sensing volume, a threshold level was applied to differentiate between scattering signals and the background. PMT pulses shallower than the threshold were omitted. In the suspension nebulisation, some droplets might not contain PSL beads and, when they were dried, residual particles were formed from solidification of surfactants contained in the suspensions. The surfactant was added by the suspension manufacturers to stabilise the suspension. Measurements by commercial GRIMM 1.108 OPC indicated that the size of residual particles were normally distributed peaked between 600 and 700 nm.

3 Model Description

The intensity of scattering light can be calculated using Mie theory assuming its shape is spherical [9]. In spherical coordinate, the intensity is expressed as

$$I = \frac{\lambda^2}{4\pi^2} \int_{\phi_1}^{\phi_2} \int_{\theta_1}^{\theta_2} \left[I_{\parallel}(\alpha, m, \theta, \phi) + I_{\perp}(\alpha, m, \theta, \phi) \right] G(\theta, \phi) \, \mathrm{d}\theta \, \mathrm{d}\phi, \tag{1}$$

where λ is the wavelength of the illuminating light; ϕ is azimuthal angles defined as the angle between the plane oscillation of the electric vector of the illuminating light and scattered light directions; θ is scattering angles defined as the angle between the propagation direction of the illuminating light and scattered light directions; α is a size parameter defined as $\alpha = \frac{\pi d_\mathrm{p}}{\lambda}$ with d_p as the particle diameter; $G(\theta, \phi)$ is geometrical factor specific to the instrument's design; $I_{\parallel}$ and $I_{\perp}$ are scattered light intensities propagated in parallel and perpendicular, respectively, relative to the plane oscillation of the electric vector of the illuminating light. The intensities are expressed as

$$\begin{aligned} I_{\parallel} &= |S_1(\alpha, m, \cos\theta_j)|^2 \sin^2\phi_j \\ I_{\perp} &= |S_2(\alpha, m, \cos\theta_j)|^2 \cos^2\phi_j \end{aligned} \tag{2}$$

where S_1 and S_2 are Mie amplitude functions [10]; scattered light I is collected over a solid angle represented by an arc of angle $\phi_1 - \phi_2$ swept around a sphere by $\theta_1 - \theta_2$. In a particular case where particles are illuminated in the centre of reference, scattered light collected by the PMT is the light over a solid angle represented by an arc of azimuth angles from $\phi_1 = 50°$ to $\phi_2 = -50°$ swept around a sphere from $\theta_1 = 20°$ to $\theta_2 = 160°$.

Since the beam intensity is distributed in the x and y axis, the intensity of scattering light measured by the PMT is a function of particle position along the x and y axis. The scattered intensity denoted as $I_1(x, y, \alpha, m)$ can be expressed as

$$I_1(x, y, \alpha, m) = f(x, y)\, I_0(\alpha, m), \tag{3}$$

where $f(x, y)$ is normalised distribution function of the beam intensity; $I_0(\alpha, m)$ is the intensity of scattering light when particles are illuminated by a beam with the intensity $\frac{P_\mathrm{b}}{A_\mathrm{b}}$ where P_b is the beam power and A_b is the beam cross section. The power of scattered light measured by the PMT, denoted as $P_0(\alpha, m)$, can be calculated numerically by dividing the area of the PMT window into N small areas over which I_1 is assumed to be homogeneous and adding up all the power of scattering light over the small areas. The power $P_0(\alpha, m)$ can be expressed mathematically as

$$P_0(\alpha, m) = \frac{P_b}{A_b k^2} \sum_{j=1}^{N} \mathrm{Re}\left[\frac{1}{r_j^2}\Big[|S_1(\alpha, m, \cos\theta_j)|^2 \sin^2\phi_j + |S_2(\alpha, m, \cos\theta_j)|^2 \cos^2\phi_j\Big]\hat{r}_j.\hat{A}_j\, A_j\right]. \tag{4}$$

where k is wave number calculated by $\frac{2\pi}{\lambda}$; A_j is the small area indexed by j; $\hat{A}_j$ is the unit vector of A_j; r_j is the distant of A_j from the centre of reference. The conversion of the power of scattered light into PMT outputs can be calculated as

$$V(x, y, \alpha, m) = P_1(x, y, \alpha, m)\left[\frac{\lambda\, \eta_{\mathrm{PMT}}}{1.24}\right] G_{\mathrm{PMT}}\, R_{\mathrm{lo}}\, G_{\mathrm{amp}}, \tag{5}$$

where η_{PMT} is the quantum efficiency of the photocathode on the PMT window; $\frac{\lambda\, \eta_{\mathrm{PMT}}}{1.24}$ is the conversion factor of light power into electrical currents with the unit of λ is nm; G_{PMT} is the gain of the PMT; R_{lo} is the load resistor installed on the PMT outlet; G_{amp} is the amplification factor of the amplifier.

The depth of PMT pulses can be regarded as the difference between PMT outputs when $y = 0$ and the background in SPARCLE scattering chamber. The depth is expressed mathematically as

$$V_{\mathrm{D}}(x, \alpha, m) = V(x, y = 0, \alpha, m) - V_{\mathrm{b}}, \tag{6}$$

where V_{D} is PMT pulse depth; $V(x, y = 0, \alpha, m)$ is PMT outputs when $y = 0$; V_{b} is the background. In the calibration, $V_{\mathrm{b}} = 3.1$ V. If PMT pulses can be assumed as normally distributed then the width of PMT pulses can be calculated as

$$T(x, z, \alpha, m) = \frac{2\sigma_y\sqrt{2\ln\left[\frac{V_{\mathrm{D}}(x,\alpha,m)}{V_{\mathrm{B}}}\right]}}{v(x, z)}. \tag{7}$$

where T is PMT pulse width; σ_y is the standard deviation of PMT pulses; $v(x, z)$ is the speed of air at particle position (x, z).

A modelling of the speed of air in the sensing volume using OpenFoam indicates that the speed is normally distributed around the symmetrical axis of the sampling pipe. The speed is expressed as

$$v(x, z) = \frac{Q}{\pi}\, \mathrm{e}^{\frac{-(x^2+z^2)}{2\sigma_v{}^2}}, \tag{8}$$

where Q is SPARCLE sampling flow rate; σ_v is the standard deviation of the normally distributed air speed and can be determined from OpenFoam simulation. The probability of a particle to be at a particular position is proportional to the speed of air in which the particle suspended. The probability can be expressed as

$$p(x, z) = B\, v(x, z), \tag{9}$$

where B is a normalisation coefficient. The probability of the PMT to generate pulses with the depth of V_{D}' as responses to particles of size parameter α_{o} and refractive index m can be calculated by integrating the probability of particle position where the light they scattering leads to PMT pulse depth equal to V_{D}'. The calculation can be expressed mathematically as

$$p(V_{\mathrm{D}}'(\alpha_{\mathrm{o}}, m)) = \sum_{i'} p(x_{i'}, z_{i'}), \tag{10}$$

where $(x_{i'}, z_{i'})$ is particle position where $V_{\mathrm{D}}(x_{i'}, \alpha, m) = V_{\mathrm{D}}'(\alpha_{\mathrm{o}}, m)$. Similarly, the probability of pulse width T'' as responses to the same particles can be calculated as

$$p(T''(\alpha_{\mathrm{o}}, m)) = \sum_{i''} p(x_{i''}, z_{i''}), \tag{11}$$

where $(x_{i''}, z_{i''})$ is particle position where $T(x_{i''}, z_{i''}, \alpha, m) = T''(\alpha_{\mathrm{o}}, m)$.

4 Results and Discussion

The distribution of PMT pulse depth and width as measured in SPARCLE calibration are shown as blue bars in Fig. 3. The average sampling rates of measurements are also indicated in the figure. As shown in the figure, the shape of pulse width distribution is skewed right with the range wider for slower sampling rate. The shape might be due to the variation of SPARCLE sampling flow rate resulted in various residence time of particles in the sensing volume. Meanwhile, the shape of pulse depth distributions is a multi-mode distribution with more than one peak identified. This results lead to a misleading interpretation when particle size is associated with PMT pulse depth.

Equations (10) and (11) are used to calculate the distributions of PMT pulse depth and width. The results of the calculation for PSL beads and residual particles are shown as orange and magenta dashed lines, respectively, in Fig. 3. Particle sizes used in the calculation are indicated in the figure. As shown in the figure, the shape of the calculated and measured distributions are similar with some discrepancies in fine structures of the distributions. The similarity indicates that the model is sufficient to describe the scattering light measured by the PMT.

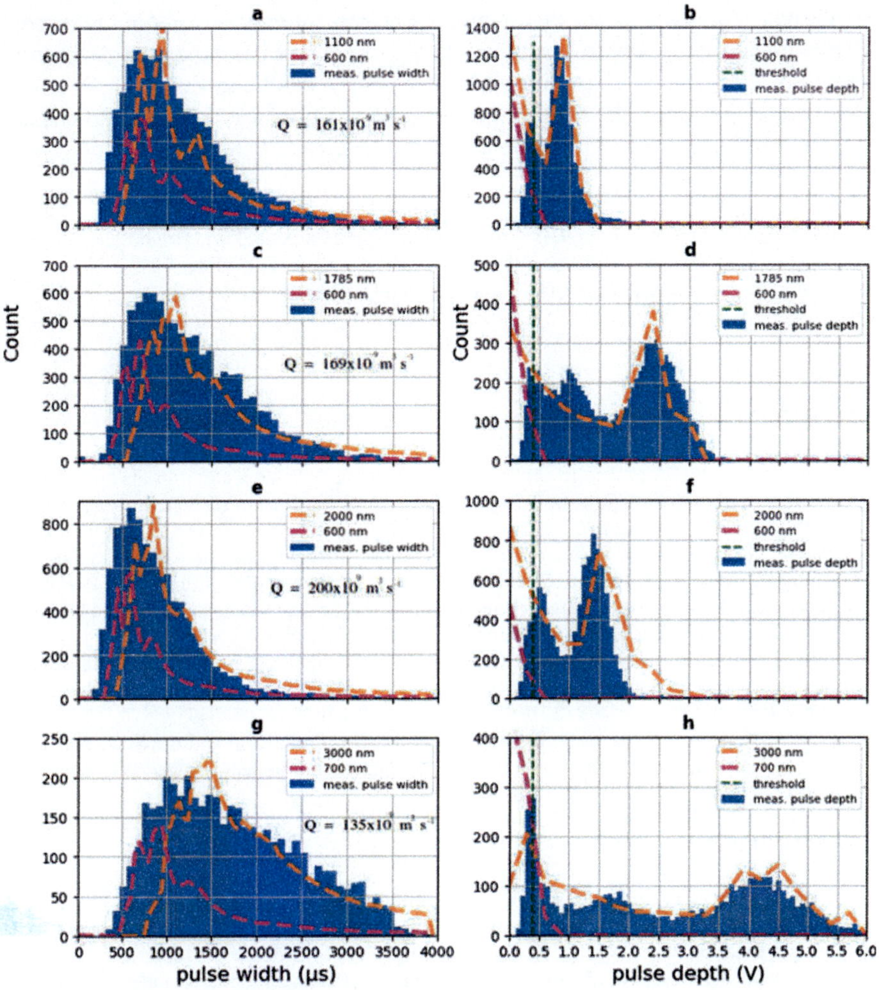

Fig. 3 Distribution of pulse widths and pulse depths. The blue areas are the measured distributions generated from the calibration data. The dashed lines are the distributions calculated from the simulations. The simulated distributions have been scaled up to fit with the measured distributions. The scaling factor for each simulation was searched manually and the fit was evaluated visually. The simulations assumed that the size distributions of aerosol particles are bimodal distributions which peaked at two sizes. The measured distributions can be approximated as the combination of the simulated distributions for the two sizes. The two sizes are listed in the figure

5 Conclusion

A model to describe the distribution of PMT pulse depth and width generated from the light scattered by particles illuminated in non-homogeneous beam intensity was developed. The comparison between the model and experimental results indicates the similarity of the shape of the distributions. The measurements result in multi-modal distributions of pulse depth as a function of particle size. Based on the distribution, the retrieval of particle size using PMT pulse depth can be misleading. Since most commercial OPCs use the relationship between the two, it is important to consider the bias that is introduced from the non-homogeneity of beam intensity used in the commercial OPCs.

Acknowledgements This research is funded by Lembaga Pengelola Dana Pendidikan, Kementerian Keuangan, Indonesia. This research was performed at Aerosol Lab. University of Oxford under supervision of Prof. Don Grainger, Dr. Dan Peters and Dr. Simon Proud.

References

1. Jianmin, C., Li, C., et al.: A review of biomass burning: Emission and impacts on air quality, health and climate in China. Sci. Total Environ. **579**, 1000–1034 (2017). https://doi.org/10.1016/j.scitotenv.2016.11.025
2. Myhre, G., et al.: Anthropogenic and natural radiative forcing. In: Stocker, T.F., et al. (eds.) Climate Change 2013: The Physical Science Basis. Contribution of Working Group I to the Fifth Assessment Report of the Intergovernmental Panel on Climate Change. Cambridge University Press, Cambridge, United Kingdom and New York, NY, USA (2013)
3. Wooster, M., et al.: New tropical peatland gas and particulate emissions factors indicate 2015 Indonesian fires released far more particulate matter (but less methane) than current inventories imply. Remote Sens. **10**, 495 (2018). https://doi.org/10.3390/rs10040495
4. Pramod, K., Baron, P.: Aerosol Measurement Principles, Techniques and Applications, chapter 5. https://doi.org/10.1016/0021-8502(80)90037-3, http://linkinghub.elsevier.com/retrieve/pii/0021850280900373 (2011). ISBN 9780470387412
5. Renliang, X.: Light scattering: a review of particle characterization applications. Particuology **18**, 11–21 (2015). https://doi.org/10.1016/j.partic.2014.05.002
6. Romanov, A.V., Yurkin, M.A.: Single-particle characterization by elastic light scattering. Laser Photonics Rev. **15** (2021). https://doi.org/10.1002/lpor.202000368
7. Zhai, C., Cao, Z.: Elastic light scattering of a single sphere in a Gaussian light sheet. Optik-Int. J. Light Electron. Opt. **194** (2019). https://doi.org/10.1016/j.ijleo.2019.163083
8. Romadhon, M.S.: SPARCLE: an instrument for aerosol size and refractive index measurement (2021). https://ora.ox.ac.uk/objects/uuid:3f586dca-7d46-42e9-a468-d8d26f894ce2
9. Mie, G.: Beiträge zur Optik trüber Medien, speziell kolloidaler Metallösungen. Ann. Phys. **330**(3), 377–445 (1908)
10. Hulst, H.C.: Light Scattering by Small Particles. Dover Publications Inc., New York (1957)

Analysis of Atmospheric Conditions on Hail Events in Bandung on March 8, 2022

Elfira Saufina, Tiin Sinatra, Anis Purwaningsih, Dita Fatria Andarini, Fahmi Rahmatia, Fadli Nauval, Ina Juaeni, Asif Awaludin, Aisya Nafiisyanti, Farid Lasmono, Adi Witono, Arief Suryantoro, Eddy Hermawan, and Acep Catur Nugraha

Abstract Hail is one of the atmospheric phenomena generally caused by weather anomalies due to convective storms with adequate updraft strength, sufficient availability of supercooled liquid water content, and conducive temperature and optimal time. This paper aims to analyze the atmospheric mechanism when hail occurred in Bandung on March 8, 2022. We used data analysis of the rain observation system by rain scanner to observe rain's distribution and temporal evolution. Also, we used analysis of satellite imagery data to observe the cloud type and cloud height during hail phenomenon in Bandung. Moreover, the atmospheric profile during the hail precipitation is analyzed using ERA5 data for several supporting parameters such as ice water content (ICW), liquid water content (LCW), moisture, vertical wind, and temperature profiles. The results indicate that heavy rain in the middle of Bandung area (study case: Antapani, Cicadas) was observed since 14:14 Local Time (LT) with rapid rain growth for approximately 1 h. The rain propagates and expands from south to north Bandung and converges with another cloud system developed in the northern area of Bandung. The rain is indicated to come from height clouds based on Himawari imagery. Persistent rainfall with a wide coverage lasts up to 15:30 LT and indicates cloud growth that reaches the height of the tropopause and stratosphere. This high cloud cluster is suspected to form ice precipitation falling to the study area at 14:40 LT. An indication of strong wind convergence since 14:00 LT is also one of the prominent features before the hail precipitation occurred over the study area.

E. Saufina (✉) · T. Sinatra · A. Purwaningsih · D. F. Andarini · F. Nauval · I. Juaeni · A. Awaludin · A. Nafiisyanti · A. Witono · A. Suryantoro · E. Hermawan · A. C. Nugraha
Research Center for Climate and Atmosphere, National Research and Innovation Agency (BRIN), Bandung, West Java 40173, Indonesia
e-mail: elfira.saufina@brin.go.id

F. Rahmatia
Agam Space and Atmospheric Observation Station, National Research and Innovation Agency, Bukit Kototabang, Koto Rantang Palupuh Agam, West Sumatera, Indonesia

F. Lasmono
Center for Data and Information, National Research and Innovation Agency (BRIN), Jakarta 10340, Indonesia

A. Basit et al. (eds.), *Proceedings of the International Conference on Radioscience, Equatorial Atmospheric Science and Environment and Humanosphere Science*, Springer Proceedings in Physics 290, https://doi.org/10.1007/978-981-19-9768-6_80

1 Introduction

Extreme weather is a natural disaster that has a detrimental impact on human life. Extreme weather can be characterized by heavy rain, lightning, and/or ice. An extreme weather event in the form of heavy rainfall with hail has been reported to occur on Tuesday, March 8, 2022, in the city of Bandung around the Cicadas and Antapani areas (central Bandung). This hail event occurred around 14:40–15:00 LT as reported by the media [1]. Based on World Meteorological Organization (WMO) definition, hail is the precipitation of either transparent or partly or completely opaque particles of ice (hailstones), usually spheroidal, conical, or irregular in form and of diameter generally between 5 and 50 mm which falls from a cloud either separately or agglomerated into irregular lumps. In general, hail occurs much in extra-tropical areas because it has a relatively lower freezing level. The freezing level is the height of the atmospheric layer with a temperature of 0 °C so that water drops can freeze [2]. Hail can occur in the tropics if the ice particles that fall from Cumulonimbus clouds are very large and fixed in the form of ice particles despite friction in the clouds. The temperature at the top of the cloud can reach −60 °C, and water vapor is generally found in the form of ice crystals [3]. Hail forms as a result of collision and coalescence processes from cloud drops that are supersaturated and produce large raindrops in the frozen phase [4]. Hail is usually local and uneven and occurs suddenly [5, 6]. In addition, hail generally occurs during the transition period and is accompanied by heavy rains accompanied by lightning and strong winds [7]. In Indonesia, the hail phenomenon rarely occurs because of the higher freezing level, so the ice crystals formed in this layer will melt into water when they fall to the ground [6]. Hail events can be triggered by overshooting phenomena or strong updraft activity in Cumulonimbus cloud cells. Cloud overshooting occurs when clouds that normally only grow in the troposphere then penetrate the stratosphere [2, 8].

There are several cases of hail in Indonesia that have been recorded and analyzed for atmospheric conditions, including the case of hail in Bogor on September 23, 2020 [9], heavy rain with ice in Bandung on March 17, 2017, which concluded that local convective activity and wind convergence in the study area during hail events [10], hail case in Jakarta on November 22, 2018, hail in Bandung on May 3, 2017 [2], and hail in Bandung on April 19 and 23, 2017 [3]. Of the various hail cases that occurred in the western part of Java Island, Bandung is the area that experiences the most hail events. Bandung is often one of the locations where extreme weather is possible due to its geographical location and topographical contours. Regional and local factors such as the effects of tropical cyclones, eddies, and wind bends (shear line) causing convective clouds to become one of the causes of extreme weather over Bandung [11].

Therefore, this paper aims to analyze more in the case of hail that occurred in Bandung, in particular, to determine the atmospheric conditions during the extreme weather phenomenon of hail in Bandung on March 8, 2022.

2 Data and Method

2.1 Precipitation Analysis

The temporal evolution of rain was analyzed descriptively based on the precipitation reflectivity data captured by the rain scanner developed from the x-band marine radar as part of atmospheric observation. The coverage of this radar can reach 44 km [12]. The temporal resolution of the rain scanner data is 2 min with a spatial resolution of $120 \times 120\ \text{m}^2$. With a sufficiently high temporal and spatial resolution, rain scanners can be used to see the evolution of rain during hail events. The reflectivity value indicates the presence of an object that produces an echo in the atmosphere. The higher the reflectivity value, the larger the size of the object [13]. In general, objects that can reflect high echo values are convective clouds containing large particles such as ice grains [14]. Rain scanner data installed in Sumedang (6.912°S, 107.838°E) was used to analyze hail occurrence on March 8, 2022.

2.2 Cloud Identification and Analysis

This study utilized the cloud properties/physical derived from Himawari level-2 products to identify the cloud characteristics. Himawari-8 is a satellite that was launched in 2015 with 10 min temporal resolution and has 16 channels as the latest generation of Multi Transpose Satellite-2 (MTSAT-2) observing the growth of convective clouds in more detail [15]. We obtained the cloud type, height, and temperature from Himawari product cloud properties with results based on the High-Resolution Cloud Analysis Information (HCAI) algorithm developed by the Japan Meteorological Agency.

Furthermore, the vertical structure of the cloud was analyzed by identifying the fraction of cloud cover (CC), cloud liquid water content (CLWC), and cloud ice water content (CIWC). We obtained those parameters from the European Center for Medium-Range Weather Forecasts (ECMWF) atmospheric reanalysis of the global climate (ERA5) on pressure-levels data, for twelve different altitudes (from 1000 to 100 hPa) over 6.6–7.2 S and 107.4–107.9 E [16]. The data can be downloaded from http://www.ecmwf.int/. In addition, the surface wind data at 925 hPa pressure level from ECMWF was used to analyze the wind convergence and wind direction before over the study area before and during ice precipitation.

3 Result and Discussion

3.1 *Temporal Evolution of Precipitation Based on Rain Scanner Observation*

Figure 1 shows the reflectivity distribution during the hail event on March 8, 2022, from a rain scanner located at Sumedang. The figure shows that it has rained around the city of Bandung including Antapani and Cicadas areas which were reported to experience heavy rain with ice on the day [1]. Rain scanner monitoring results show that light rain began to be detected in the east of Antapani and Cicadas at 13:52 LT (blue circle) and at 13:54 LT, it detected rain in the north of Antapani and Cicadas (red circle). Rainfall in both places is widespread as the intensity increases, which is indicated by an increase in the reflectivity value. At 14:14 LT the two rain objects began to unite (black circle). Meanwhile, rain objects in the southern area move closer to Antapani and Cicadas (green circle). At 14:24, the rain in the southern part split into two parts, one of these split parts began to merge with rain objects in the Antapani and Cicadas areas. At 14:30 LT, a fairly wide object of rain that extends vertically with a maximum reflectivity value (> 35 dBZ) is above the hail incident location. It was seen that the reflectivity value had started to decrease over Antapani and Cicadas at 14:46 LT. This continued until the rain stopped at around 15:28 LT.

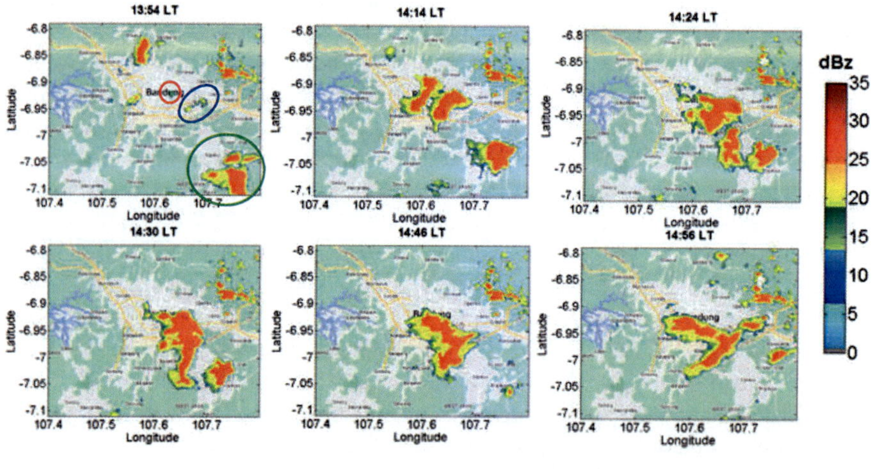

Fig. 1 Precipitation scanner observation during hail event in Bandung on March, 8, 2022

3.2 Wind Analysis

Wind convergence and wind direction over the study location were analyzed before and during the ice precipitation (Fig. 2). A convergence pattern shown by positive value (red contour) was detected from the morning at 10:00 LT (not shown). The convergence strengthened in time over the study area (black rectangle in Fig. 2) and reached its strongest magnitude (>9 × 10^{-5} s^{-1}) at 13:00 LT on March 8, 2022, approximately two hours before the hail event over Bandung (Fig. 2). The air masse's confluence forces air's upward motion over this area, thus supporting the formation of convective clouds. Relatively stronger convergence activities were still persistent over some locations, and this wind convergence continuously formed higher convective clouds that potentially resulted in ice precipitation. Moreover, during the ice precipitation (around 14:00–15:00 LT), the wind vectors supported the propagation of the rain object shown by the rain scanner in Fig. 1. Overall, convergence is one of the indications of the beginning of ice precipitation, as stated by BMKG that convergence is one of the causes of hail in Bandung on March 8, 2022 [17]. The identification of clouds formed by this process is discussed in the following subsection, by looking at cloud temperature, cloud top height, and the vertical structure of the clouds.

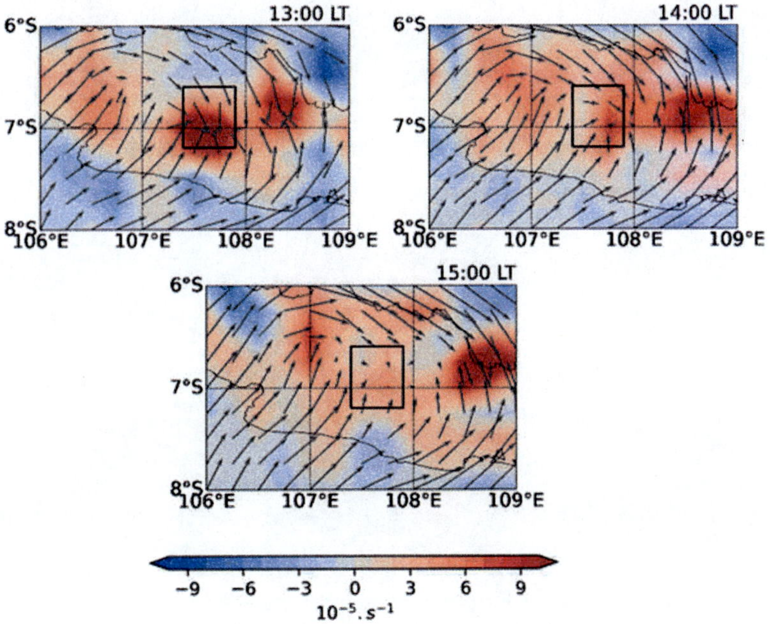

Fig. 2. 925 hPa wind convergence and direction at 12:00–15:00 LT March, 8, 2022

3.3 Cloud Identification and Vertical Analysis

Cloud Identification. The analysis of cloud identification was begun by investigating the cloud evolution from the spatial distribution of brightness temperature/temperature black body (TBB) of Himawari-8 Satellite over Bandung Area (black rectangle). Figure 3 shows the convective clouds observed in the Bandung area indicated by a cool TBB (<210 K) since 13:50 LT. This condition persisted until 14:20 LT and favored the environment to generate heavy precipitation. It is worth noting that at around 14:30 LT, the maximum value of TBB propagated to the west, while a higher value of TBB was found over the eastern region of Bandung City. It indicated that a hail event occurred, as observed from the maximum intensity of reflectivity in the rain scanner over Antapani and Cicadas (Fig. 1). Generally, a westward propagation of convective cloud was identified during this hail event, beginning from the convective cloud formation in the east of the Bandung area and continuing to grow persistently and eventually moving to the west of Bandung.

In addition, the small TBB value (<210 K) during hail events indicates that the presence of convective clouds in the study area is the cause of hail falling. It is

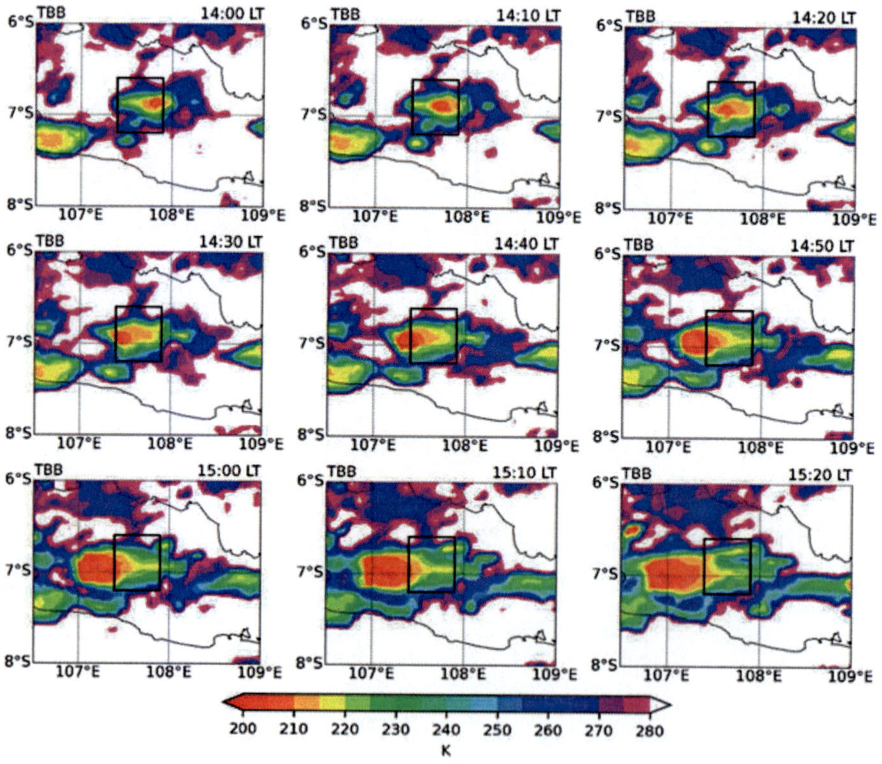

Fig. 3 TBB over study area from 14:00 LT to 15:20 LT on March 8, 2022

because of the TBB threshold for ice is around −75C. This result is consistent with previous research which showed that the cloud top temperature during hail events in Bandung on April 19, 2017, was −74 °C and on April 23, 2017, was −60 °C [3].

A persistent cloud formation (shown as lower TBB in Fig. 3) over the Bandung area before the hail event indicated intense convective activities. It is summarized from the cloud type depicted in Fig. 4 that the deep convection (DC) and Cirrostratus (Cs) dominated the cloud type in the study area from 14:10 to 14:50 LT. Those cloud types are categorized as high cloud according to the International Satellite Cloud Climatology Project (ISCPP) classification [18]. ISCCP defines nine cloud types according to the optical thickness and cloud top pressure (CTP) as measured by a global network of geostationary and polar orbit meteorological satellites. Deep convection has criteria in the form of high clouds, cloud top pressure in the range of 440–50 mb and cloud optical thickness in the range of 23–379 [19].

Convective clouds are divided into two types namely Cumulus and Cumulonimbus. Each of these types develops vertically depending on the available convective energy. The more convective energy, the higher the convective cloud grows is. If the convective cloud can reach a height above 6 km, it is referred to as deep convection, while if it is unable to reach that height, it is referred to as shallow convection. Only deep convection produces extreme rainfall including hail and thunderstorms [20]. Likewise, hail in Bandung on March 8, 2022, is produced by clouds with a deep convection type.

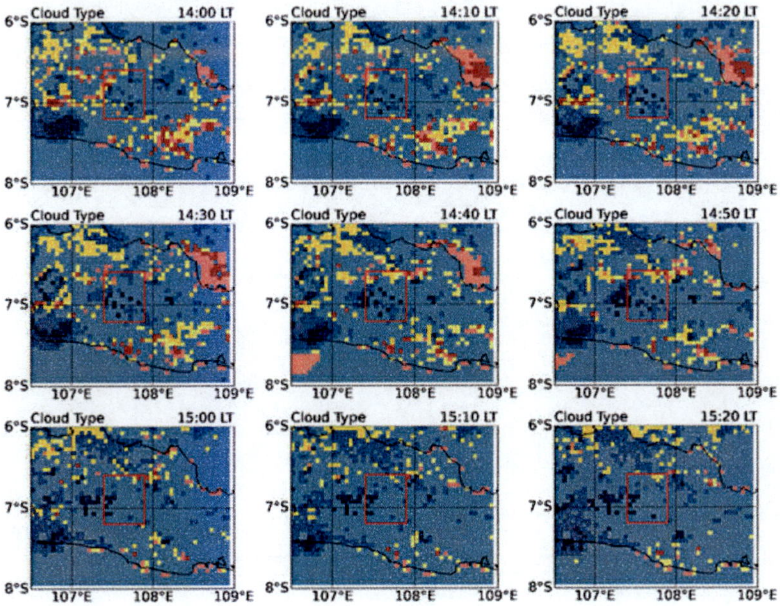

Fig. 4 Cloud type over study area from 14:00 LT to 15:20 LT on March 8, 2022. (Cirrus (Ci), Cirrostratus (Cs), deep convection (DC), Altocumulus (Ac), Altostratus (As), Nimbostratus (Ns), Cumulus (Cu), Stratocumulus (Sc), and Stratus (St))

The existence of convective clouds (deep convection) as a hail producer in the study area is also supported by cloud top height analysis as shown in Fig. 5. It was observed the presence of high clouds in the study area during the hail event as mentioned above. This convective cloud reaches heights above 12.5 km which is Cumulonimbus capillatus incus type (anvil top). An anvil top is a Cumulonimbus cloud that has developed the characteristic flat, anvil top form. An anvil top indicates a very strong updraft, the strongest among the Cumulonimbus species. Anvils usually top out at 10–12 km altitude, but they can reach as high as 16–18 km close to the equator [20]. This cloud causes extreme weather such as hail, strong winds, lightning, and very heavy rainfall.

Vertical Structure of Clouds. The vertical structure of clouds represents consistent results with the identified cloud type and altitude (Subsection 3.3). At 14:00 LT, clouds were identified at altitudes from 800 to 200 hPa, over the study location (along 107–109E and 6.75–7.75S) (see Fig. 6). This result supports the previous finding that deep convective cloud reached height above 200 hPa (or around 12 km). Moreover, the high ice water content (CIWC) can explain the location of the anvils

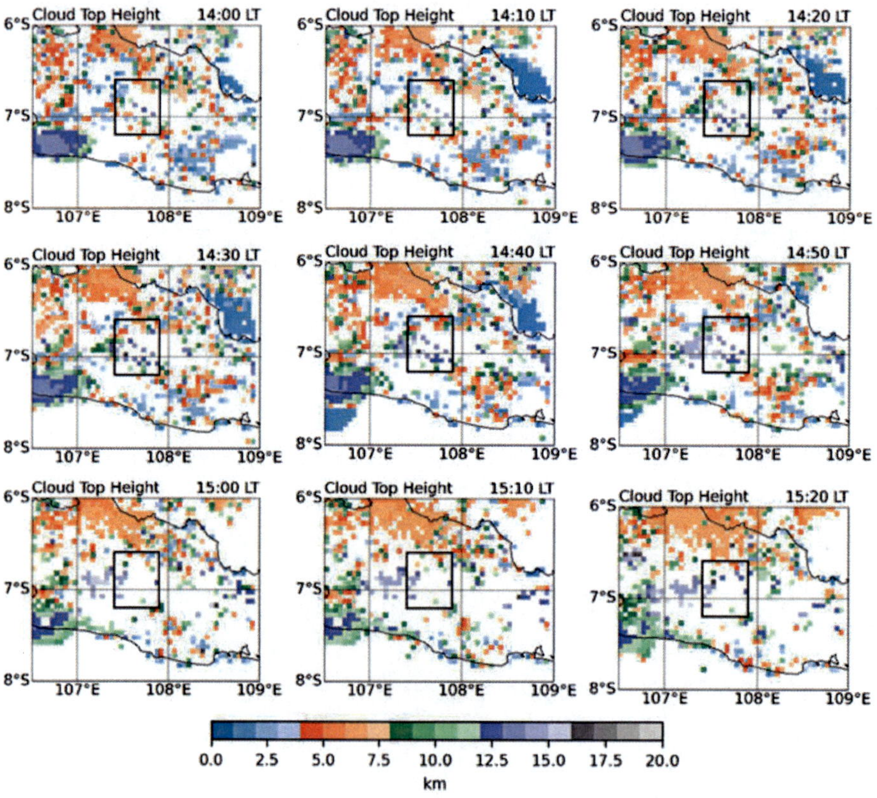

Fig. 5 Cloud top height over study area from 14:00 LT to 15:20 LT on March 8, 2022

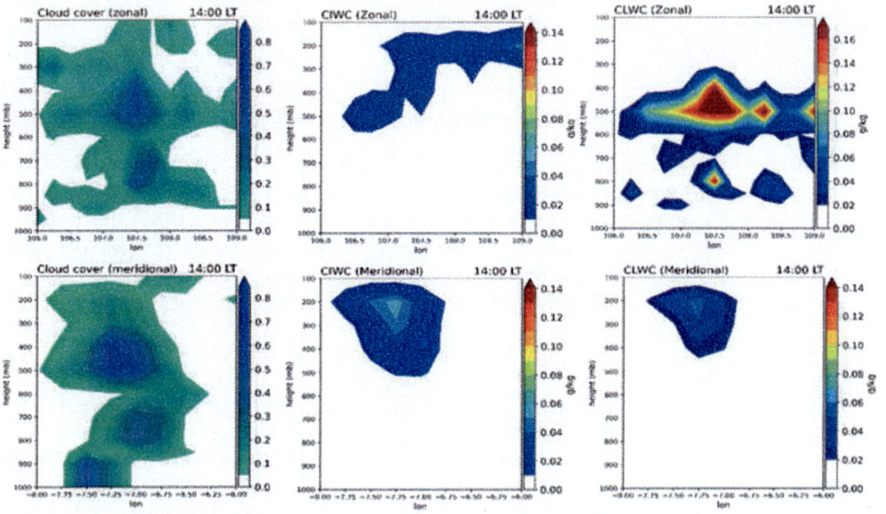

Fig. 6 Vertical cross section of cloud cover (%), cloud ice water content (CIWC), and cloud liquid water content (CLWC) (in g/kg) over area study

of this cloud. Higher CICW was at the same location, 107–109E and 6.75–7.75S, in the top of the cloud (around 200 hPa) or approximately 12 km altitude. Moreover, the liquid water content was reached around 500 hPa (around 6 km) which spread more in the zonal distribution than in meridional distribution. Overall, the deep convective cloud with anvil containing high ice content at 12 km altitude was identified during hail precipitation over the case study.

4 Conclusion

The case study of hail over Bandung on March 8, 2022, shows the presence of high-intensity rain as indicated by a large reflectivity value (>35dBZ) during hail events. This rain comes from convective clouds formed due to wind convergence around the study area. These clouds formed and grew around the eastern area of Bandung over Antapani and Cicadas which experienced hail as reported by the media. The convective clouds formed are dominated by deep convection clouds, which are tall clouds with a height above 12.5 km and capable of producing precipitation in the form of ice. The cloud top temperature from TBB also indicates that the cloud formation has passed over the ice threshold. Moreover, the vertical structure of clouds also represents that high ice water content, representing the anvil of clouds, was identified at an altitude of 11–12 km as high as DC clouds.

Acknowledgements This work is supported by TREAK, PRIMA BRIN. In addition, we thank for the availability of ECMWF ERA5 data downloaded from http://www.ecmwf.int/.

References

1. Ramadhan, A.S.: Beberapa Daerah di Bandung Diguyur Hujan Es. https://jabar.suara.com/read/2022/03/08/155853/beberapa-daerah-di-bandung-diguyur-hujan-es. Last Accessed 25 Aug 2022
2. Paski, et al.: Analisis Dinamika Atmosfer Kejadian Hujan Es Memanfaatkan Citra Radar dan Satelit Himawari-8 (Studi Kasus: Tanggal 3 Mei 2017 di Kota Bandung). Seminar Nasional Penginderaan Jauh ke-**4**, 371–381 (2017)
3. Hidayat, A.M., Efendi, U., Rahmadini, H.N.: Identifikasi Kejadian Hujan Es Berbasis Analisis Faktor Cuaca, Citra Satelit dan Model Numerik dengan Aplikasi GrADS (Studi Kasus: Kejadian Hujan Es Tanggal 19 dan 23 April 2017 di Bandung). Seminar Nasional Penginderaan Jauh ke-**4** 429–440 (2017)
4. Rogers, R.R., Yau, M.K.: In: A Short Course in Cloud Physics, 3rd edn, pp. 235. Elsevier Science, Burlington (2006)
5. Zakir, A.: Modul Praktis Analisa dan Prakiraan Cuaca. Pusat Pendidikan dan Latihan Meteorologi dan Geofisika, Jakarta (2008)
6. Fadholi, A.: Analisa Kondisi Atmosfer pada Kejadian Cuaca Ekstrem Hujan Es (hail). Simetri: Jurnal Ilmu Fisika Indonesia **1**(2) D (2012)
7. Madjid, F.M.: Studi Parameter Hujan Es ditinjau dari data radar, Pengamatan Udara Atas dan Pengamatan Udara Permukaan di Jawa Bagian Barat. [Thesis] Undergraduate Program in Applied Meteorology, Climatology and Geophysics, Tangerang Selatan (2018)
8. American Meteorology Society (AMS).: Glossary of Meteorology. htttp://glossary.ametsoc.org/. Last Accessed 15 Sept 2022
9. Prasetyo, S., Abdillah, S., Nugraheni, I.R., Sagita, N.: Studi Awan Konvektuf Penyebab Hujan Es Menggunakan Radar Cuaca Doppler Single Polarization di Bogor (23 September 2020). Jurnal Aplikasi Meteorologi **1**, 32–42 (2022)
10. Sinatra, T., Nugroho, G.A.: Analysis of heavy rainfall with hail using reflectivity and radial velocity derived from X-band radar in Bandung (Study Case: March 17, 2017). In: IOP Conference Series: Earth and Environmental Science, vol. 166, IOP Publishing, pp. 01203. (2018)
11. Paski, J.A.I.: Analisis Kejadian Cuaca Ekstrim Tanggal 20 April 2014 di Kota Bengkulu Memanfaatkan Citra Radar dan Satelit. Buletin Balwil II BMKG, vol. 4(9) (2014)
12. Nugroho, G.A., Awaludin, A., Rahayu, S.A.: Hasil pengamatan scanner hujan dan AWS pada kejadian hujan tanggal 2–3 Maret 2014 di daerah Bandung dan sekitarnya. Buku Variabilitas Cuaca dan Iklim di Indonesia (2014). ISBN 978-979-1458-818
13. Mutiasari, U.: Kajian Karakteristik dan Penentuan Ambang Batas Kejadian Hujan Es di Jawa Timur dengan Pemanfaatan Produk MAX dan VIL serta perhitungan VIL-Density. [Thesis] Undergraduate Program in Applied Meteorology, Climatology and Geophysics, Tangerang Selatan (2018)
14. Nugroho, et al.: Analisis Keadaan Atmosfer Kejadian Hujan Es Menggunakan Citra Radar Doppler CBand dan Citra Satelit Himawari 8 (Studi Kasus: Jakarta, 22 November 20118). Seminar Nasional Penginderaan Jauh ke **6** 183–19 (2019)
15. JMA.: Himawari User's Guide (2015). (http://www.jmanet.go.jp/msc/en/support/index.htm)
16. Hersbach, H., Bell, B., Berrisford, P., Biavati, G., Horányi, A., Muñoz Sabater, J., Nicolas, J., Peubey, C., Radu, R., Rozum, I., Schepers, D., Simmons, A., Soci, C., Dee, D., Thépaut, J-N.: ERA5 hourly data on pressure levels from 1959 to present. Copernicus Climate Change Service (C3S) Climate Data Store (CDS) (2018). Last Accessed 01 Oct 2022
17. Budianto, A.: Bandung Diguyur Hujan Es, Begini Penjelasan BMKG. https://daerah.sindonews.com/read/706797/701/bandung-diguyur-hujan-es-begini-penjelasan-bmkg-1646744610. Last Accessed 25 Aug 2022

18. Kurihana, T., Moyer, E., Willett, R., Gilton, D., Foster, I.: Data-driven cloud clustering via a rotationally invariant autoencoder. IEEE Trans. Geosci. Remote Sens. 1–25 (2021)
19. Rossow, W.B., Schiffer, R.A.: Advances in understanding clouds from ISCCP. Bull. Am. Meteor. Soc. **80**, 2261–2288 (1999)
20. World Meteorological Organization, Manual on the Observation of Clouds and other Meteors, International Cloud Atlas Volume 1 (1975). ISBN 92-63-10407-7

Microphysical Features During Rainfall Events in Bandung, West Java. Case Study: Weather Modification Technology in Citarum Basin, November 2021

Halda A. Belgaman, Sholehhudin A. Ayubi, Saraswati Dewi, Sopia Lestari, Findy Renggono, Edi Riawan, Neneng S. Juariah, and R. D. Goenawan

Abstract Supporting the Weather Modification Technology operating over Citarum Basin in October–November 2021, several rainfall observation instruments were installed around Bandung Area. Those instruments include X-band dual-polarization weather radar, Laser Precipitation Monitor (LPM) or laser disdrometer, and Automatic Weather Station (AWS). This study explored the rainfall characteristics around the Bandung area during November 2021. AWS was used to show accumulated rainfall amount; weather radar data was used to explore and analyze precipitation development mechanism both in temporal and spatial resolution; meanwhile, an LPM disdrometer was used to measure microphysics parameters of the rainfall such as the raindrop size and the number of raindrops, the velocity of the raindrops, reflectivity, and rainfall intensity. The temporal resolution of the weather radar was 5 min, the AWS was 5 min, and the LPM was 1 min. From the AWS data, the rainfall amount in November 2021 was recorded at around 491.5 mm, and the rainfall events mainly occurred in the afternoon until nighttime (11.00–22.00 LT). Data analysis from LPM (rainfall intensity, drop size, and number) showed that rainfall events from the convective or stratiform clouds over the Bandung area occurred in November 2021. Finally, cloud and precipitation development analyzed by weather radar data showed precipitation/cloud movement mainly propagated from North–North East to South, South–West, and West areas. Precipitation development from the south area also propagated to the Northern Bandung. Exploring and analyzing the microphysics characteristics of precipitation would help to determine the most suitable areas for cloud seeding activity.

H. A. Belgaman (✉) · S. A. Ayubi · S. Dewi · S. Lestari · F. Renggono · N. S. Juariah · R. D. Goenawan
National Research and Innovation Agency, Jakarta, Indonesia
e-mail: hald001@brin.go.id

E. Riawan
Bandung Institute of Technology, Bandung, Indonesia

A. Basit et al. (eds.), *Proceedings of the International Conference on Radioscience, Equatorial Atmospheric Science and Environment and Humanosphere Science*, Springer Proceedings in Physics 290, https://doi.org/10.1007/978-981-19-9768-6_81

1 Introduction

A brief definition of weather modification through cloud seeding activities is a deliberate intervention of the cloud development process for enhancing or suppressing the precipitation amount. In other parts of the world, cloud seeding activity was carried out at experimental and operational levels [1–3]. In Indonesia, Weather Modification Technology activities are carried out to manage water resources in the atmosphere by utilizing weather parameters to enhance or suppress the intensity of rainfall in certain areas in order to minimize the risk of disasters caused by climate and weather factors. The cloud seeding technique carried out in Indonesia is mainly in the form of warm cloud seeding using hygroscopic seeding material. A warm cloud is a cloud with water vapor as its condensation nuclei. The cloud seeding process is carried out by inserting large hygroscopic seeding material into the cloud that will act as giant cloud condensation nuclei (GCCN) and interact with natural cloud condensation nuclei inside; this interaction will accelerate the process of collision and coalescence within the cloud. It will ultimately accelerate and increase the efficiency of rain droplet growth [4]. There are two seeding methods currently applied in Indonesia. The first one is "dynamic" seeding using airplanes with powder and flares seeding materials. The second is "static" cloud seeding from the ground by utilizing orographic clouds using flare seeding materials [5].

In terms of purpose, cloud seeding activities in Indonesia are mainly divided into two main activities: mitigating hydrometeorological disasters, mitigating drought disasters, and haze disasters due to forest and land fires. The second was for water resources management to increase the lake water level for hydroelectric power plant operations and irrigation. One recently implemented was in the Upper Citarum Catchment Area (Citarum Basin). The Citarum Basin is mainly located over the West Java area, especially Bandung greater area shown in Fig. 1.

Studying weather characteristics, especially the mechanism of cloud growth and the rainfall development process occurring in the Citarum Basin area, is one of the essential aspects of water resource management activities through cloud seeding activities in this area. Previous studies had explored how the weather mechanism, especially local rainfall development, occurred in the Bandung Basin area and the West Java area using data from a network of an Automatic Weather Station (AWS) [6], weather radar [7, 8], and also using GPS measurement and numerical study [9].

During the implementation of weather modification operations through cloud seeding, [10] provided directions regarding the procedures that must be carried out, one of which is physical and statistical evaluation related to implementing cloud seeding activities. Unfortunately, the evaluation of cloud seeding activities in Indonesia has been carried out mainly using a statistical methodology, whether evaluating the rainfall amount or measuring the increase or decrease in lake water level in the target area during the operation period compared to historical value. The evaluation results are only limited to activity reports and are not entirely contained in scientific reports.

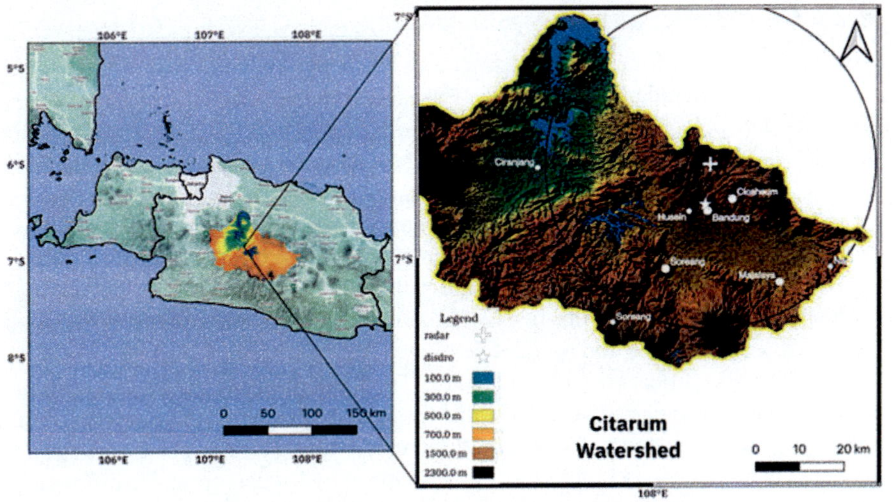

Fig. 1 Study area in the Citarum Basin in West Java, Indonesia. Plus (+) symbol represents weather radar data location, and the star (⋆) symbol represents LPM disdrometer location

Thus, the purpose of this study is as an initial study of the possibility of using X-band weather radar observation data and another cloud microphysics observation instrumentation to evaluate the cloud seeding activities in the Citarum watershed area.

2 Data and Methodology

The data used in this study came from rainfall observations from AWS and LPM disdrometers stationed at Bandung Institute of Technology Campus (Fig. 1 (⋆)) and also data from the Furuno WR-2100 X-band dual-polarization weather radar stationed in Lembang (Fig. 1, (+)). The data collection period was carried out throughout November 2021.

2.1 *Automatic Weather Station (AWS)*

The AWS data used in this study was rainfall amount data with a temporal resolution of 5 min, which was used for daily rainfall amount analysis and diurnal rainfall events. In addition, long-term hourly GSMaP rainfall amount data (from 2000 to 2020) was also used to analyze the climatology of monthly rainfall patterns in the target area.

2.2 Dual-Polarization Doppler Radar

Dual-polarization radar was often used for rainfall analysis with the benefits of better estimation of precipitation and the size distribution of hydrometeors, as well as the identification of the melting layer. This radar works by transmitting and receiving pulses in horizontal and vertical orientations. The returning frequencies provide measurements of horizontal and vertical dimensions of targets (hydrometeors), which provides a better estimation of the size and shape of hydrometeors.

In this study, the Furuno X-band dual-polarization Doppler radar data was used to analyze the rainfall events in November 2021. Three parameters of the radar binary data used include horizontal reflectivity (DBZ), differential reflectivity (ZDR), and the specific differential phase (KDP). The DBZ data can be used to estimate the precipitation intensity, and the value strongly depends on the number and the diameter of hydrometeors. Generally, the greater value corresponds to the stronger intensity of rain or hail [11]. To identify the shape of hydrometeors and to differentiate between water droplets and ice droplets, the ZDR parameter is used. The positive value indicates the water droplets, while the negatives represent the ice particles. However, the concentration of hydrometeors is also crucial in microphysical studies of rainfall. The raindrop concentration can be estimated using the KDP parameter from radar data, where the higher concentration gives a larger phase difference than the lower concentration.

2.3 Laser Precipitation Monitor

The amount, intensity, and type of precipitation data were also obtained from the Laser Precipitation Monitor (LPM), with 22 diameter and 20 falling speed classes. Using the laser-based measurement, ThiesClima LPM was operating in 1-min resolution.

The rain is classified using rain rate and the standard deviation for ten consecutive 1-min data for the convective and stratiform rain classification. Convective rain is characterized with rain rate >= 5 mm/h and the standard deviation > 1.5 mm/h, while stratiform rain is classified with standard deviation <= 1.5 mm/h [12]. Different research used the disdrometer to detect and observe the microphysical parameter, such as analyzing rainfall parameters in valley region [13], and other about the study of raindrops size distribution and its response to cloud microphysics [14].

3 Result and Discussion

3.1 AWS Data

Figure 2a shows the average monthly rainfall amount in the Citarum Basin area for the period 2000–2020 obtained from Global Satellite Mapping of Precipitation (GSMaP) data provided by Japan Aerospace Exploration Agency (JAXA), which shows the climatological pattern of rainfall was identified as an annual pattern which has one peak of rain as the wet season in November to March and one dry season with total minimum rain in the period June–October. In this type of annual pattern, rainfall is mainly determined by the movement patterns of the Asia and Australian monsoons [15]. The daily rainfall variability based on the data obtained from the automatic rain gauge shows the amount of daily rainfall between 0 and 72.3 mm/day (Fig. 2b). In November 2021, the monthly rainfall was 491.5 mm/month. The monthly rainfall amount in November 2021 was higher than the 2000–2020 average due to 2021 La-Nina episodes. La Nina episode tends to increase the rainfall amount over Indonesia in the dry season (SON) [16, 17] and affect Citarum River Basin [18].

Based on the rainfall intensity threshold issued by the BMKG, the rain that occurred throughout November consisted of 13 days of light rain (0.5–20 mm/day), nine days of moderate rain (20–50 mm/day), and two days of heavy rain (50–100 mm/day), also six days without rain. The daily rainfall pattern in November 2021 (Fig. 2c) shows that rain begins to occur in the afternoon, evening, and night. Figure 2c showed that there are four clusters of rain events that occurred on average in November, which were 13.00–14.30 LT, 15.00–16.00 LT, 16.30–17.30 LT, and 18.30–19.00. The average maximum rainfall during the day was 1.50 mm/5-min, while the average maximum rain at night was 0.49 mm/5-min.

3.2 Radar Data

To explore the rain process that occurred in the Citarum Basin, the conditions on November 4 (moderate rain), November 7 (light rain), November 10, and 13 (heavy rain) were chosen for radar data analysis.

Spatial analysis of weather radar data was used to analyze the cloud and rain growth process in the Citarum Basin. Figure 3 depicts the maximum reflectivity values (dbz) in a time—latitude/longitude graph (Hovmoller diagram) on 4 (Fig. 3a), 7 (Fig. 3b), 10 (Fig. 3c) and November 13, 2021 (Fig. 3d). The vertical dotted line indicates the location of the LPM disdrometer which is used to explore the rain microphysical analysis. On the 4th, 10th, and 13th, the rain grew in the East part of the Citarum Basin during the day and moved toward the west part of the basin until the afternoon or began to grow in the afternoon and continued to move westward until the evening.

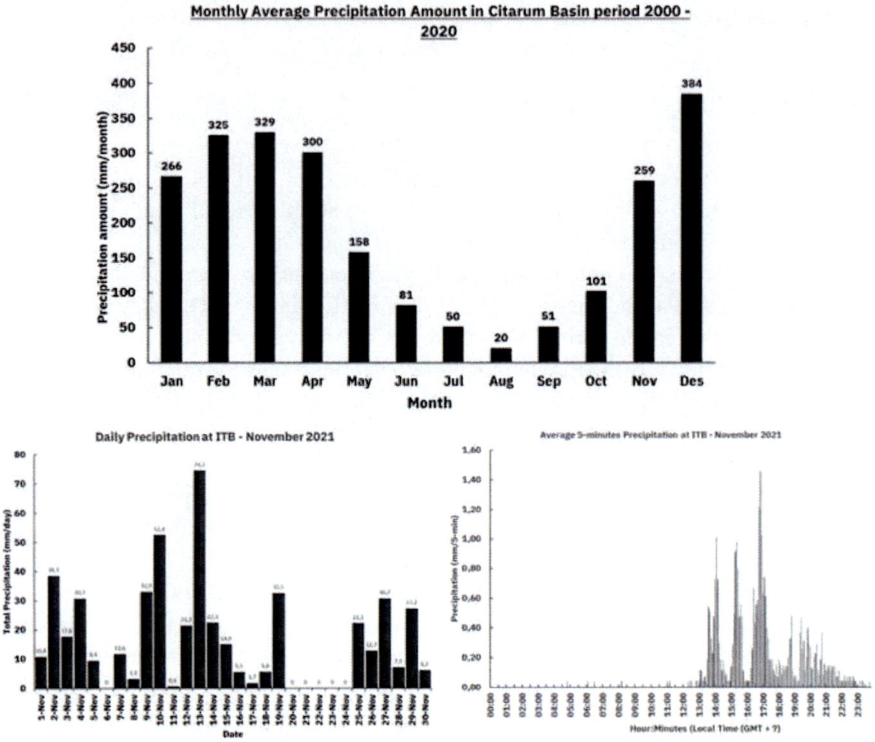

Fig. 2 Monthly rainfall amount for Citarum Basin area (top panel), daily rainfall amount (bottom left panel), and 5-min average rainfall amount (bottom right panel)

Meanwhile, on November 7, the condition was slightly different; clouds grew in the northern area of the Citarum Basin and then propagated to the southern area. [6] showed the same propagation in the Bandung area in February–March period.

The maximum reflectivity value ranges from 15 to 30 dBz representing light to moderate rain conditions. The regeneration of rainfall clouds in the Citarum Basin was also observed in the previous study [7, 8].

The subsequent analysis step was to distinguish between the types of rain produced by clouds. The method for identifying the convective or stratiform clouds/rain from the LPM disdrometer was mentioned in Sect. 2.3. Figure 4 shows the height-time weather radar reflectivity, drop size, and raindrop numbers collected from the LPM disdrometer on November 10, 2021. 10 November 2021 among all observation data was selected because it had relatively good quality data for all the instruments (AWS, disdrometer, and X-band weather radar).

The LPM disdrometer successfully identified the rain that occurred on November 10, 2021 at 13:30–15:30 LT as a convective rain, and at 15:40–16:20 LT and 19:10–20:50 LT was stratiform rain. The LPM disdrometer classified the types of convective and stratiform rain from the intensity of the rain, the size of the raindrops, and the

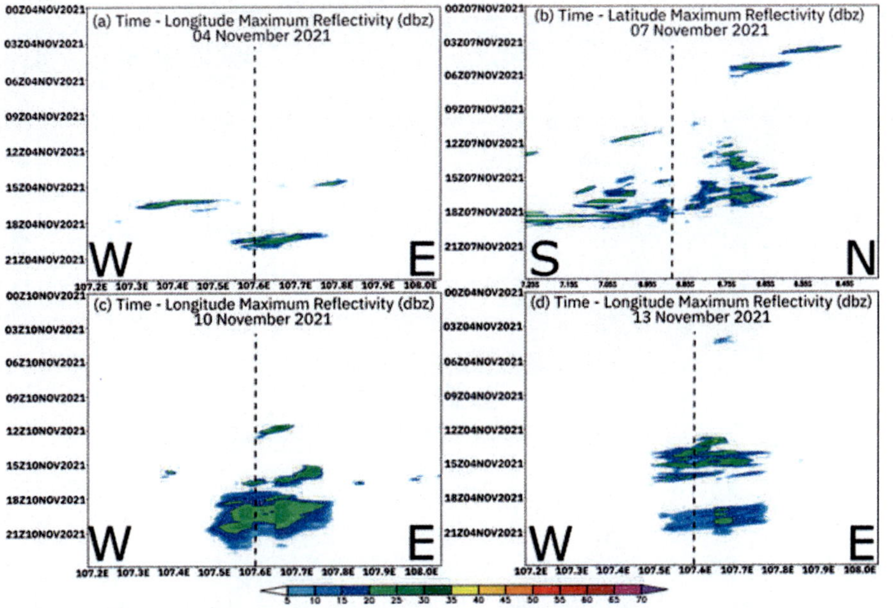

Fig. 3 Hovmoller diagram for maximum reflectivity on November 4, 2021, (top left), November 7, 2021, (top right), November 10, 2021, (bottom left), and November 13, 2021, (bottom right). The vertical dashed line indicated the LPM disdrometer location

number of raindrops. From Fig. 4, convective rain had a higher total number of grains than stratiform rain, both small and larger raindrops.

The X-band dual-polarization weather radar product can describe and identify the microphysical characteristics of rain events, as shown in Fig. 5. Using the LPM disdrometer, as previously discussed, we can identify convective and stratiform rainfall that occurs in the Citarum Basin. Focusing on rain on November 10, we confirmed how the response of microphysical variables such as size and number of raindrops generated by dual-polarized weather radar could reflect convective and stratiform rain.

Taking the November 10, 2021, 14:00 LT as a convective rain/cloud case compared to the November 10, 2021, 16:05 LT as a stratiform rain/cloud, the DBZ, ZDR, and KDP each showed significant difference signal to differentiate the convective and stratiform clouds. DBZ for the convective clouds having a value between 20–35 and 5–15 for stratiform clouds, DBZ value from 5 to 50 indicated light-moderate rain; in this case, both convective and stratiform, on November 10, showed an indication of light to moderate rain. The rainfall amount from the AWS data on November 10 convective rain was as much as 45.3 mm. Meanwhile, the rainfall mount for the stratiform cloud was 1.8 mm.

Furthermore, the ZDR value was between 1.5–2.5 DB for the convective and 0.5–1.5 DB for the stratiform cloud. Usually, the ZDR is used to differentiate the hydrometeor shape and form, as mentioned in Sect. 2.2. However, in this study, we

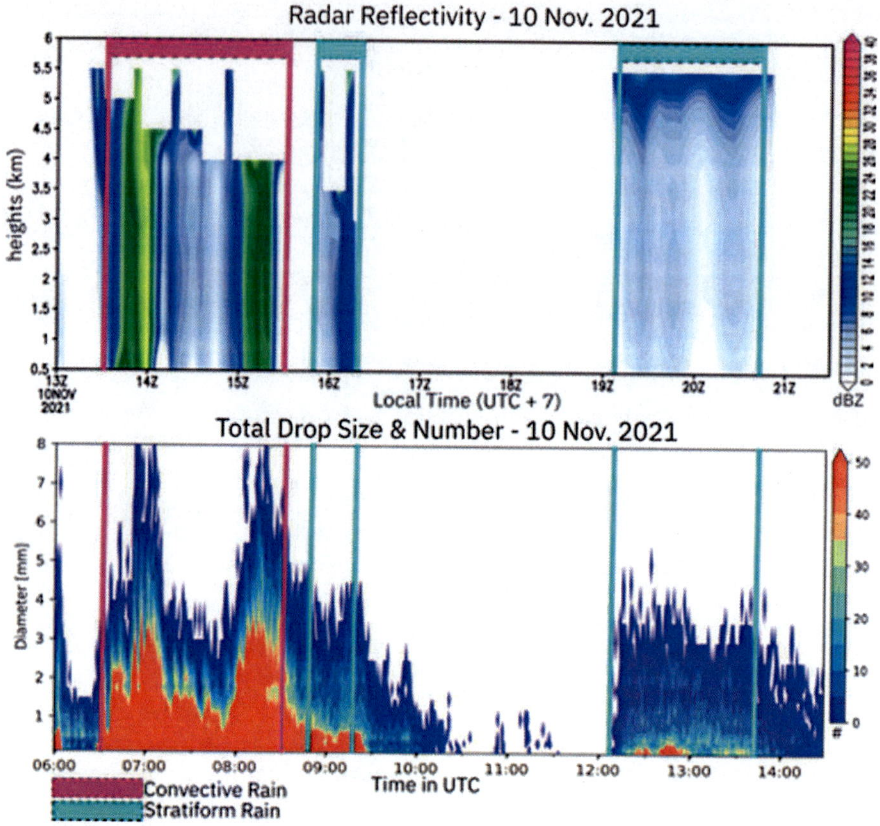

Fig. 4 November 10, 2021, height–time weather radar reflectivity (top panel) and drop size and raindrops numbers (bottom panel). The purple line indicates a convective rain event, and light green indicates stratiform rain

used ZDR to differentiate the size of the water droplets. A higher ZDR value indicates larger size droplets. In this study, the ZDR value confirms that a convective cloud will have larger droplets than a stratiform cloud.

Last, the KDP value for the convective cloud was around 2–4 deg/km; for the stratiform cloud, the KDP did not return any value. A larger KDP value indicated a higher concentration of droplets within the cloud.

4 Conclusion

In cloud seeding operations, the seeding targets are cumulus clouds with strong convective activity. This convective activity helped the Giant CCN put into the cloud to work efficiently. Cloud seeding carried out on stratiform rain clouds will not

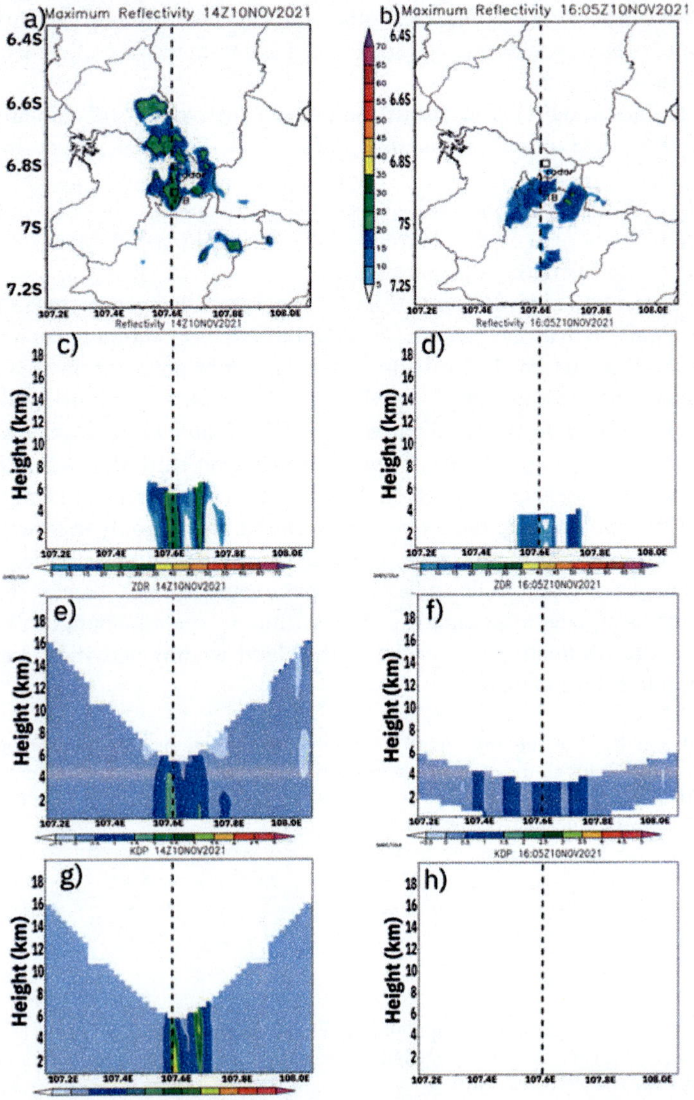

Fig. 5 X-band dual-polarization weather radar data of **a** maximum reflectivity for convective cloud, **b** same with **a** for stratiform cloud, **c** height–longitude reflectivity (dBz), **d** same with **c** for stratiform cloud, **e** height–longitude ZDR for convective cloud, **f** same with **e** for stratiform cloud, **g** height–longitude KDP for convective cloud, **h** same with **g** for stratiform cloud. Vertical dashed line indicated the location of LPM disdrometer longitude location

produce maximum results. Therefore, the study of determining cloud type using weather radar instrumentation and an LPM disdrometer will significantly assist the effective cloud seeding process.

Over the Indonesian region, stratiform and convective parts of rain are predominant and contribute to the total intensity of rainfall. The fundamental insight of a dominant type of rain in certain areas over Indonesia and the mechanism leading to the rain formation can be observed using dual-polarization radar. This information is beneficial for regional models currently facing a significant challenge in correctly simulating rainfall features over the region.

Dual-polarization radar is a powerful tool to observe changes in physical parameters in the cloud, such as the size and density of droplets. Combining the information from the weather station, LPM disdrometer, and dual polarization weather radar could advance the methodology for evaluating cloud seeding activities. DBZ, ZDR, and KDP could provide temporal variability of the cloud's rain droplet form, size, and concentration. Wang et al. [19] proposed Radar-Domain-Index (RDI) to evaluate cloud seeding activities. Unfortunately, the cloud seeding operation in Citarum Basin data was unavailable during radar observation, so this article only used weather data to present the possibility of physically evaluate the cloud seeding activities. Observation the changes of DBZ, ZDR and/or KDP parameters from X-band weather radar data before, during and after cloud seeding within a targeted cloud, can be used as methodology for evaluating how effective the cloud seeding activities in enhancing the rain droplets development.

Acknowledgements The authors would like to thank the Laboratory for Weather Modification Technology—BRIN, for all the resources used in this research. Author Contribution: Conceptualization, HAB, FR, and RDG; Radar data analysis, HAB, SAY, and SL; Disdrometer data analysis, FR and SD; rain gauge data analysis ER and NSJ.

References

1. Dong, X. et al.: Increase of precipitation by cloud seeding observed from a case study in November 2020 over Shijiazhuang, China. Atmos. Res. **262** (2021). https://doi.org/10.1016/j.atmosres.2021.105766
2. Kulkarni, J.R., Morwal, S.B., Deshpande, N.R.: Rainfall enhancement in Karnataka state cloud seeding program "Varshadhare" 2017. Atmos Res **219**, 65–76 (2019). https://doi.org/10.1016/j.atmosres.2018.12.020
3. Flossmann, A.I. et al.: Peer review report on global precipitation enhancement activities. [Online]. Available: https://hal.uca.fr/hal-01917801
4. Abshaev, A.M., Flossmann, A., Siems, S.T., Prabhakaran, T., Yao, Z., Tessendorf, S.: Rain enhancement through cloud seeding. In: Unconventional Water Resources, Cham, Springer International Publishing, pp. 21–49. (2022). https://doi.org/10.1007/978-3-030-90146-2_2
5. Renggono, F., et al.: Hygroscopic Ground-based generator cloud seeding design; a case study from the 2020 weather modification in Larona Basin Indonesia. Atmosphere (Basel) **13**(6), 968 (2022). https://doi.org/10.3390/atmos13060968

6. Dewi, E.S., Trilaksono, N.J.: Analysis of localized rainfall by using grid observation data of automatic weather station in Bandung basin case studies 28 February and 9 March 2017. pp. 020050. (2018). https://doi.org/10.1063/1.5047335
7. Nugroho, G.A., Sinatra, T., Risyanto.: Comparison and analysis of the first signal detection of two X-band radar with RGB Himawari satellite data before localized heavy rain march 17 2017 Bandung Basin Indonesia. IOP Conf. Ser. Earth Environ. Sci. **303**, 012056 (2019). https://doi.org/10.1088/1755-1315/303/1/012056
8. Syaraswati, R.A., Trilaksono, N.J., Nugroho, G.A.: Preliminary result of characteristic of convective cloud development observed by x-band meteorological radar in bandung basin. IOP Conf. Ser. Earth Environ. Sci. **303**(1), 012016 (2019). https://doi.org/10.1088/1755-1315/303/1/012016
9. Oigawa, M., Matsuda, T., Tsuda, T., Noersomadi.: Coordinated observation and numerical study on a diurnal cycle of tropical convection over a complex topography in West Java, Indonesia. J. Meteorol. Soc. Japan. Ser. II **95**(4), 261–281 (2017). https://doi.org/10.2151/jmsj.2017-015
10. Garstang, M., et al.: Weather modification: finding common ground. Bull. Am. Meteorol. Soc. **86**(5), 647–656 (2005). https://doi.org/10.1175/BAMS-86-5-647
11. Arbain, A.A., Sunarto, F., Mulyana, E.: Deteksi Es Dan Hail Di Atmosfer Dengan Radar Polarimetrik X-Band Furuno Wr-2100 (Studi Kasus: 24 Januari Dan 14 Februari 2016). Jurnal Sains and Teknologi Modifikasi Cuaca **19**(1), 21 (2018). https://doi.org/10.29122/jstmc.v19i1.2994.inbahasa
12. Wu, Z., Zhang, Y., Zhang, L., Zheng, H., Huang, X.: A Comparison of Convective and Stratiform Precipitation Microphysics of the Record-breaking Typhoon In-Fa (2021). Remote Sens (Basel) **14**(2), 344 (2022). https://doi.org/10.3390/rs14020344
13. González, S. et al.: Vertical structure and microphysical observations of winter precipitation in an inner valley during the Cerdanya-2017 field campaign. Atmos. Res. **264**, 105826 (2021). https://doi.org/10.1016/j.atmosres.2021.105826
14. Thomas, A., Kanawade, V.P., Chakravarty, K., Srivastava, A.K.: Characterization of raindrop size distributions and its response to cloud microphysical properties. Atmos. Res. **249** (2021). https://doi.org/10.1016/j.atmosres.2020.105292
15. Aldrian, E., Susanto, R.D.: Identification of three dominant rainfall regions within Indonesia and their relationship to Sea surface temperature. Int. J. Climatol. **23**(12), 1435–1452 (2003). https://doi.org/10.1002/joc.950
16. Kurniadi, A., Weller, E., Min, S., Seong, M.: Independent ENSO and IOD impacts on rainfall extremes over Indonesia. Int. J. Climatol. **41**(6), 3640–3656 (2021). https://doi.org/10.1002/joc.7040
17. Hamada, J.-I., et al.: Interannual rainfall variability over Northwestern Jawa and its relation to the Indian Ocean Dipole and El Niño-Southern oscillation events. SOLA **8**, 69–72 (2012). https://doi.org/10.2151/sola.2012-018
18. Sahu, N., Behera, S.K., Yamashiki, Y., Takara, K., Yamagata, T.: IOD and ENSO impacts on the extreme stream-flows of Citarum river in Indonesia. Clim. Dyn. **39**(7), 1673–1680 (2012). https://doi.org/10.1007/s00382-011-1158-2
19. Wang, F., et al.: Evaluation of hygroscopic cloud seeding in liquid-water clouds: a feasibility study. Atmos. Chem. Phys. **19**(23), 14967–14977 (2019). https://doi.org/10.5194/acp-19-14967-2019

The Role of Self-Organization Convective Clouds Resulting in Heavy Rainfall Over the Western Part of Java Island on July 15–16, 2022

Anis Purwaningsih, Trismidianto, Dita Fatria Andarini, Noersomadi, Teguh Harjana, Didi Satiadi, Fahmi Rahmatia, Elfira Saufina, Wendi Harjupa, Erma Yulihastin, Fadli Nauval, Ibnu Fathrio, Alfan Sukmana Praja, and Risyanto

Abstract Heavy rainfall during July 15–16, 2022 occurred over several western parts of Java Island. The recorded daily precipitation data reached about 155.2 mm/day on the West Java Climatology station, exceeding the extreme rainfall threshold (100 mm/day), while in this period, this area is generally experiencing a dry season. This paper aims to investigate the possible atmospheric mechanism resulting in extreme precipitation during this period using reanalysis and satellite data. We examined the atmospheric condition by analyzing cloud clusters, rainfall intensity, the vertical profile of temperature and humidity, and vertically integrated water vapor transport (IVT) during the heavy rainfall period. The results indicate that a convective cloud system developed along the western coast of Java Island from the night (July 15, 2022) until morning (July 16, 2022). The convective cloud initially developed over the western coast of Java Island and propagated eastward toward the center of the mainland, then dissipated on 00 Local Time (LT) July 16, 2022, over the mainland. The new clusters then developed again from the prior clusters from the sea area and propagated in an eastward development over the southern and northern coast of the land until the morning of July 16, 2022. Rainfall conditions exhibited a consistent result with convective clouds, shown by the double peaks of rainfall intensity during the disaster and the similar spatial pattern and propagation of precipitation. Moreover, we observed a warm temperature anomaly in the upper troposphere (9–12 km) in accordance with high moisture over the Sunda Strait, indicating the latent heat released due to convective cloud formation on July 15, 2022. Furthermore, the high moisture content transported from the Java Sea and the coastal area over the southern part of Java Island toward the flooding locations (IVT over 300 kg m^{-1} s^{-1}) supported the convective cloud formation during July 15–16, 2022.

A. Purwaningsih (✉) · Trismidianto · D. F. Andarini · Noersomadi · T. Harjana · D. Satiadi · F. Rahmatia · E. Saufina · W. Harjupa · E. Yulihastin · F. Nauval · I. Fathrio · A. S. Praja · Risyanto
Research Center for Climate and Atmosphere, National Research and Innovation Agency (BRIN), Bandung, Indonesia
e-mail: anispurwaningsih49@gmail.com

A. Basit et al. (eds.), *Proceedings of the International Conference on Radioscience, Equatorial Atmospheric Science and Environment and Humanosphere Science*, Springer Proceedings in Physics 290, https://doi.org/10.1007/978-981-19-9768-6_82

Overall, this study indicates that the convective clouds and their sustained development, due to supportive environmental conditions, triggered heavy rainfall events over study locations.

1 Introduction

Heavy rainfall occurred over several locations on the western part of Java Island from July 15–16, 2022, resulting in hydrometeorological disasters over three different cities: Bogor, Tangerang, and Garut. More specifically, flooding and landslides occurred in 16 regions of Bogor city, 13 regions of Tangerang, and two regions of Garut regency [1–6]. In addition, the disasters exerted economic and social loss on the victims of disasters in these locations [1–6]. During these hydrometeorological disasters, the recorded daily precipitation data reached about 155.2 mm/day on July 16, 2022, at the West Java Climatology station [7]. This value exceeds the extreme rainfall threshold determined by BMKG for extreme daily precipitation (100 mm/day). In addition, some other stations recorded heavy and extreme rainfall with intensity ranging from 50 mm/day to 110 mm/day on July 16, 2022 [7].

The recorded daily precipitation over the mentioned region in July 2022 is relatively higher compared to the normal condition since Java Island generally has a dry season over this period. However, BMKG stated that during this period in 2022, this region tends to have wet characteristics due to local and global atmospheric factors, such as the negative state of the Indian Ocean Dipole, weak La Nina, and the warming of sea surface temperature over the Indian Ocean [8]. Moreover, the outgoing longwave radiation (OLR) anomaly prediction indicates the high potency of cloud formation over the Maritime Continent (MC), specifically over the southern part of the MC. Furthermore, the indication of atmospheric waves, such as Madden Julian Oscillation and Kelvin and Rossby waves passing over the MC, trigger the intensification of convective activity during July 2022 [8].

From a large-scale point of view, around the time of the event, a synoptic disturbance was clearly visible in the form of low pressure and associated cyclonic circulation in the Indian Ocean (to the west of Sumatra Island), which induced a long convergence zone and rainband in the southeast of the depression that reached the West Java region (from https://earth.nullschool.net). The disturbance seemed to induce heavy rainfall and temporarily alter the usual dry season in the West Java area. Moreover, an initial investigation from the Global Satellite Mapping Precipitation (GSMaP) (https://sharaku.eorc.jaxa.jp/GSMaP/) indicates the formation of clouds clusters propagated from the Indian Ocean near the western part of Java Island to the western part of Java Island during the flooding episode (July 15–16, 2022). These cloud formations generally moved eastward with several phases of formation and dissipation. Those mentioned factors influence the increased precipitation in July over the western part of Java Island. However, as mentioned earlier, the hydrometeorological disasters occurred only in specific locations over this island, indicating that more specific factors are responsible for the disasters. Therefore, this paper investigates the specific atmospheric mechanism responsible for heavy rainfall over disaster locations.

One of the possible mechanisms is the interaction between Kelvin and Rossby waves which suppress the convection activity and develop more convection by interacting with other waves, background wind, and terrain profiles [9]. More specifically, the convergence zone is developed from an interaction between Kelvin waves modulating an eastward dry air propagation and Rossby waves modulating westward moisture propagation. This interaction results in the arising convection than the base-state convection [9], thus increasing precipitation in the convergence zone. Furthermore, the existing convection leads to the large-scale pattern that triggers more convection eastward due to the more convectively unstable characteristic of easterly warm and moist air [10]. Thus, the precipitation intensity over the convergence zone and its propagation area (east area of the convergence zone) is generally higher.

The interaction between Kelvin and Rossby appears as a Matsuno–Gill pattern cloud that can generally be identified over the equator and influence precipitation variability over the area [11, 12]. Moreover, a study reveals the association of the Matsuno–Gill pattern on deep convection and thunderstorm development [13]. The pattern is associated with Mesoscale Convective Complexes, which led to thunderstorm development [14]. Those developments are triggered by the unstable pattern that inhibits a self-sustain precipitation disturbance propagating eastward from the convergence zone [10]. In this paper, we investigated the irregular pattern of cloud formations (self-organization convective clouds) that led to heavy precipitation over the flooding areas.

This study investigates and analyzes the atmospheric condition concerning the cloud cluster formation resulting in heavy precipitation over the western part of Java Island on July 15–16, 2022. We analyzed the convection activity by identifying cloud formation, factors triggering the formation, and the precipitation condition over the study area during the period of hydrometeorological disaster events. In the next section, we explain about data and methodology used for this study. In the Result and Discussion section, we first identified cloud evolution and propagation over the study area, then examined rainfall's time series and spatial distribution. Finally, we analyzed the moisture and temperature before and during heavy rainfall. In the last part of this paper, we delivered and concluded the critical findings of this study.

2 Data and Methodology

2.1 Cloud Identification and Analysis

To investigate the cloud evolution and propagation associated with hydrometeorological disasters, we employed the cloud top brightness temperature/Temperature Black Body (TBB) of the Himawari-8 satellite with a 4 km spatial resolution and 1-h temporal interval [15]. The high temporal resolution data of the Advanced Himawari Imager (AHI) in Himawari-8 could be used to study the convective cloud evolution since it often develops in a relatively short period [15, 16]. We retrieved the infrared channel (Band 13) at 10.4 μm from http://weather.is.kochi-u.ac.jp/. We identified the convective cloud development from the spatial distribution of hourly TBB during 15 LT, July 15, 2022–12 LT, and July 16, 2022.

2.2 Precipitation Analysis

We observed rainfall characteristics during the flood event over the western part of Java Island. The observation utilized gridded hourly rainfall data from the Global Satellite Mapping of Precipitation gauge-corrected (GSMaP_gauge) version-7 for the period of July 15, 2022, to July 16, 2022 [17]. GSMaP_gauge is a standard product corrected by daily rain gauges from NOAA CPC Gauge-Based Analysis [18]. This data has a horizontal resolution of 0.1 × 0.1 degrees and can be accessed from https://sharaku.eorc.jaxa.jp/GSMaP.

We analyzed the temporal variation of rainfall before and during the flood event from a time series of hourly precipitation during two days in several flood areas. As a result, six locations over Bogor, Tangerang, and Garut were selected to represent the flood regions with higher impacts, as shown in Fig. 1.

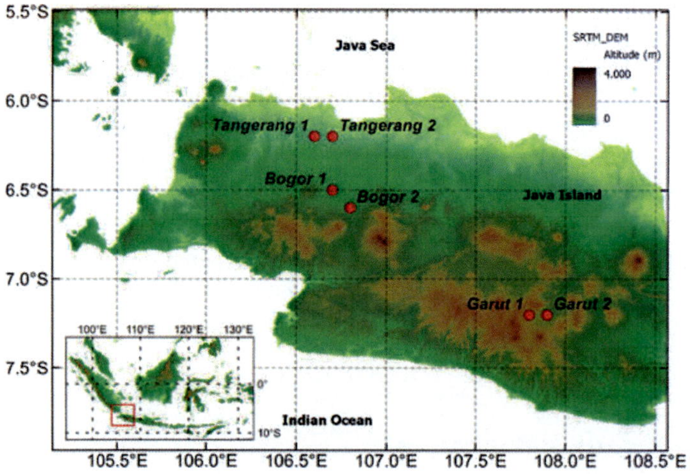

Fig. 1 Map of study locations. Red circles denote the disaster locations; Tangerang 1: 106.6E, 6.2S, Tangerang 2: 106.7E, 6.3S, Bogor 1: 106E, 6.5S, Bogor 2: 106.8E, 6.6S, Garut 1: 107.8E, 7.2S, Garut 2: 107.9E, 7.2S

2.3 Temperature and Moisture Analysis

We used the atmospheric data observed with the Constellation Observing Satellite for Meteorology, Ionosphere, and Climate Mission 2 (COSMIC-2) Global Navigation Satellite System (GNSS) Radio Occultation (RO) technique [19]. The RO measurement technique utilizes the GNSS signals which are bent when they pass through the atmospheric layers [20, 21]. In addition, we investigated temperature and humidity profiles to analyze the dynamic process induced by the convective cloud development resulting in extreme precipitation.

In addition, this study used the European Center for Medium-Range Weather Forecast reanalysis 5 (ERA5) data on a 0.25° × 0.25° latitude–longitude grid and hourly temporal resolution to analyze the moisture transport during heavy precipitation [22, 23]. We analyzed the specific humidity (q), zonal (u), and meridional (v) wind components from 12 LT, 15 July 2022, to 12 LT, 16 July 2022. We calculated the vertically integrated water vapor transport (IVT) at 1000–100 hPa pressure (p) level range, using the following equation —where g is the gravitational acceleration [24, 25]

$$IVT = \left[\left(\frac{1}{g}\int_{b}^{a} qudp\right)^2 + \left(\frac{1}{g}\int_{b}^{a} qvdp\right)^2\right]^{1/2}$$

3 Result and Discussion

3.1 Cloud Evolution and Propagation

Figure 2 denotes the spatial pattern of convective clouds indicated by the low magnitude of TBB (< 221 K [14]). The convective clouds initially developed (from 15 LT, July 15, 2022) over the western coast of Java Island and propagated eastward toward the center of the mainland. The maximum phase of the convective cloud system, denoted by the lowest TBB (<205 K), was detected at 19 LT over the mainland and supported by convergence wind toward the convective clouds. This convective cloud system then dissipated over the west part of the mainland on 00 LT, July 16, 2022. At the same time (00 LT, July 16, 2022), the convective cloud clusters were still persistent over the mainland (around 108E, 7S) and the ocean. These convective cells over the ocean around the southern and northern part of the coast (at 106E, 5.5S, and 8S) strengthened with TBB less than 205 K (00 LT, July 16, 2022). They initiated more convective development from midnight until morning.

From midnight through the morning of July 16, 2022, the new cloud clusters developed and aggregated from the prior clusters from the sea area. The more

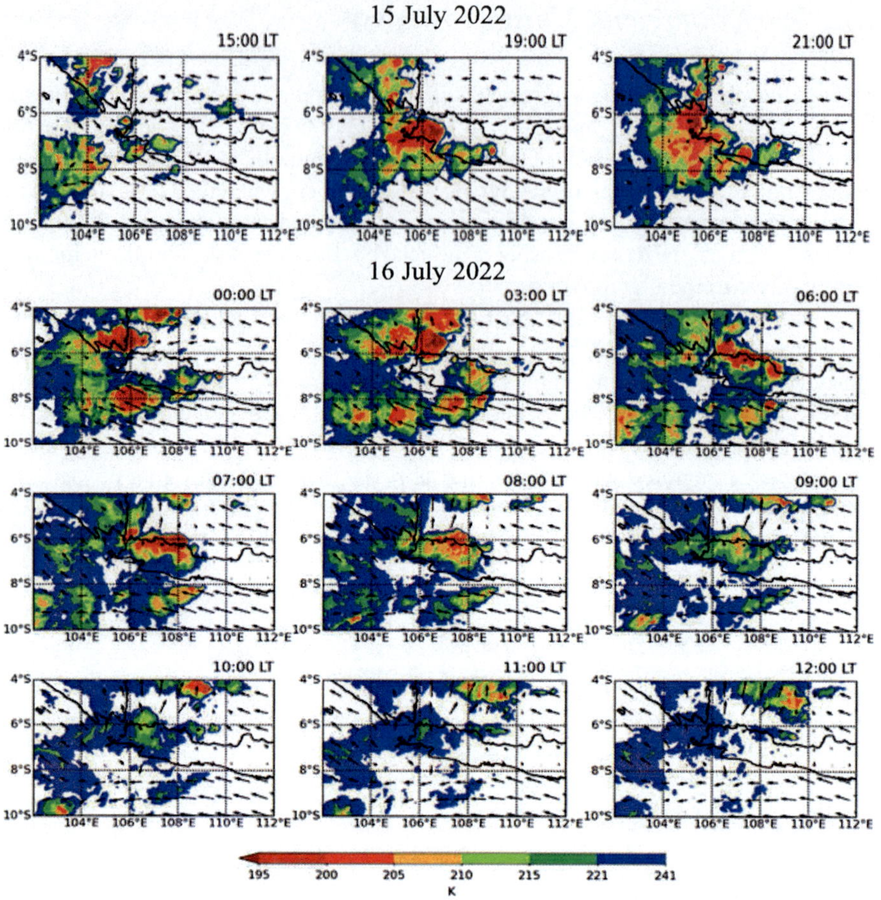

Fig. 2 Cloud top brightness temperature (TBB) (shaded) and 1000 hPa wind (vector) over the study area from 15 LT, July 15, 2022, to 12 LT, July 16, 2022

robust land breeze wind pattern over the northern coast supported the new convective cloud cluster development over the northern coast of Java. Furthermore, the clusters formed, then propagated eastward over the southern and northern coasts of the land. A different cluster from the north (around 4.5S and 107E) merged with the previous cloud cluster over the sea near the northern coast (00–03 LT, July 16, 2022), whereas over the southern coast, the strong cell from the ocean combined with a cell over the mainland (around 108E, 7S). Finally, the merging cells over the north and south coasts collided over the land. The expansion of these cells resulted in the cooler TBB (< 205 K) over the northern coast in the early morning of July 16, 2022. The decay of these convective cloud clusters started at 10 LT, July 16, 2022. The decay process was supported by wind divergence over the western part of Java Island and the surrounding sea (see Fig. 2).

Overall, the convective cloud formation over the mainland during flooding was divided into two phases. The first phase was during the evening until midnight of July 15, 2022, which was generally persistent over the center of the mainland. The second phase was initiated from the previous convective cloud and formed during the early morning of July 16, 2022. In this second phase, convective clouds propagated from the southern and northern coasts and moved inland. In the following subsection, an analysis of spatial distribution and temporal pattern of hourly precipitation is delivered concerning cloud development and propagation results.

3.2 Precipitation Analysis

To investigate the propagation of high rainfall intensity, we analyzed the spatial patterns of rainfall during the two periods of convective cloud formation over the mainland (Fig. 3). The spatial distribution of hourly rainfall indicates consistent results with the convective cloud formation of phase 1 and phase 2. Still, it shows a delay at the beginning of each phase. In the first stage, GSMaP data detected the eastward propagation of rainfall toward the coastal area at 17 LT, after convective clouds reached the coast at 15 LT. The high intensity of rainfall was initially detected over the coastal area of the western part of Java Island in the evening (17 LT, July 15, 2022). The inland and eastward propagation started from the southern coast of West Java from 19 LT and was followed by vast eastward propagation over the southern coast of West Java until midnight on July 16, 2022. Heavy rainfall was higher in West Java's northern and southern coast in the second period (from midnight until the morning of July 16, 2022). Over the northern area, heavy rainfall started to spread over the land area from 03LT and reached the highest intensity at 07LT in the location where the coolest TBB was indicated. During this second period, the eastward propagation of rainfall was observed over the coastal area, and then it spread westward over the center of the mainland.

Figure 4 presents the time series of hourly rainfall over six flooding locations that denote the timing of heavy rainfall's peak during July 15–16, 2022 (See Fig. 4). Generally, hourly precipitation data indicates two peaks of rainfall over the six flooding locations. The first peak was around 18–23 LT, July 15, 2022, and the second peak was during 02–12 LT, July 16, 2022. Bogor 1 and Bogor 2, in the center of the mainland, had the highest rainfall intensity on the first peak reaching 10 mm/hr (Bogor 1) and 25 mm/hr (Bogor 2) during the night (20LT, July 15, 2022). The higher rainfall intensity over the second peak was located more in the periphery of the land (Tangerang 1 and 2). The second rainfall peak over Tangerang ranged from 7 to 12 mm/hr, on the morning of July 16, 2022 (07 LT) and around 8 mm/hr on 10 LT, on July 16, 2022. The location and timing of the rainfall peaks are consistent with the propagation of clouds in which the lowest TBB intruded over the center mainland in the first peak. Clouds dissipated over the center of the mainland at 00 LT, July 16, 2022. The second convective aggregation started after the lowest TBB propagated from northern and southern coastal areas on July 16, 2022 in the morning.

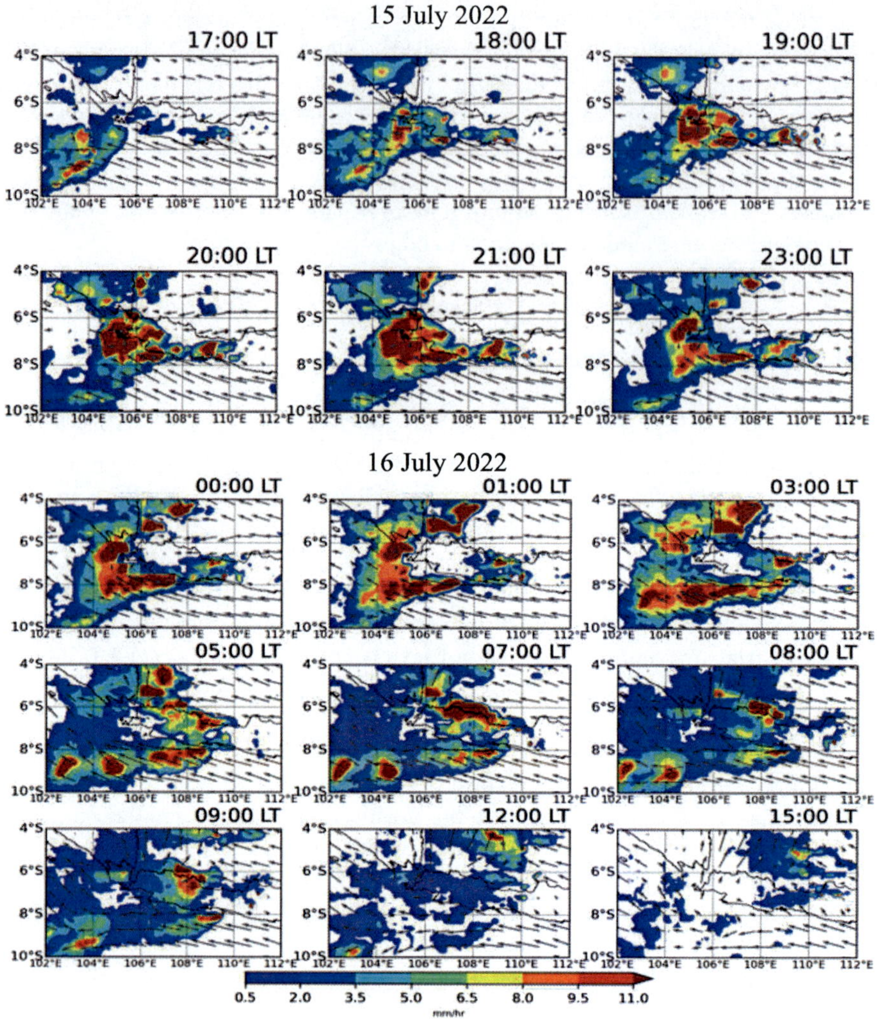

Fig. 3 Spatial distribution of hourly rainfall (mm/h) (shaded) and 1000 hPa wind (vector) from 12 LT, July 15, 2022 to 03 LT, July 16, 2022

Overall, two phases of cloud formation over the mainland (subsection 3.1) resulted in two peaks of high rainfall intensity over the flooding areas (subsection 3.2). The convective cloud formation and aggregation with eastward and inland propagation were from night until morning, when the land breeze is theoretically active. This pattern is similar to a case study revealed previously over Jakarta when the convective systems developed from night until morning and resulted in flash floods over Jakarta [26]. Moreover, another study indicated the same night–morning convective system development over the western coast of Sumatera, in which the combination

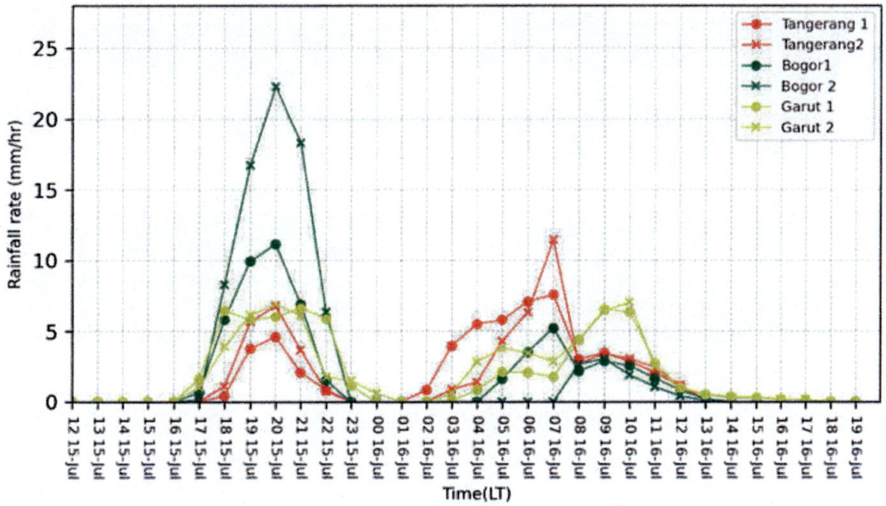

Fig. 4 Time series of hourly rainfall over six flooding locations from GSMaP data (see Fig. 1) during July 15–16, 2022

of land breeze and cold pool outflows of the convective system led to heavy precipitation over an area [27]. In this previous study, the divergent flow from the cold pool in combination with outland flow resulted in convergence flow that triggered new development of convective systems, thus resulting in continuous convective system development. Another study revealed the same pattern of self-aggregation of convective clouds and stated that diurnal forcing is critical in forming a persistent convective self-aggregation [28]. Those previous studies support the results of this paper, in which initial convection leads to more convective cloud development eastward during the evening until morning when the land breeze is active and potentially triggers extreme rainfall along the propagation of this convective cloud development.

3.3 Temperature and Moisture Analysis

We identified temperature and moisture conditions during the convective cloud formation, July 15–16, 2022. First, height versus longitude cross-sections of temperature observed with COSMIC-2 GNSS-RO show warming due to latent heat released during July 15–16, 2022, in the upper troposphere at 9–12 km (Fig. 5). A positive temperature anomaly (3 K) in the upper troposphere was detected in accordance with high moisture in the mid-troposphere, indicating the latent heat released due to convective cloud formation. We also notice the warming associated with an intensive cloud formation at lower altitudes (5–9 km) over the Sunda strait and Western Java

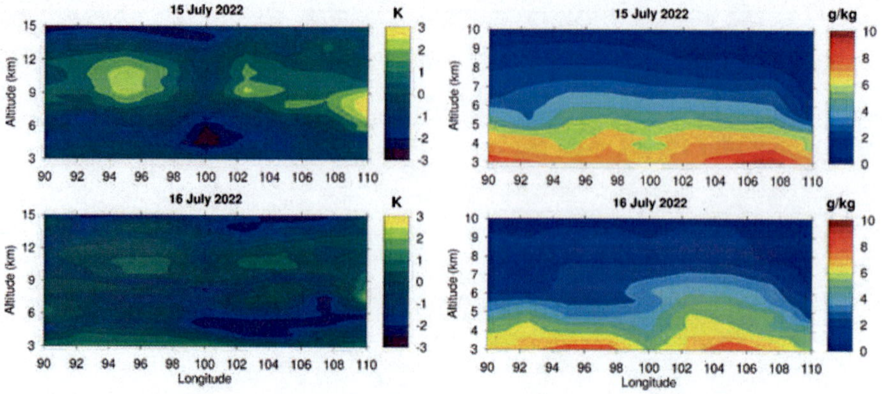

Fig. 5 Vertical cross-section of specific humidity (g/kg) and temperature (K) on July 15–16, 2022

Island (106–110E). Meanwhile, the high humidity and occurrence of precipitation in this period affect the atmospheric cooling by about −3 K at the lower troposphere (3–6 km) around 110E. Over the Sunda strait and Western Java Island, this cooling by about −1 K was identified.

Moreover, on July 16, 2022, a similar pattern at higher altitudes was shown with weaker warming than the previous day. In comparison, the temperature decreased and extended from 100 to 110 E, indicating cooling over the low altitude. Figure 5 shows that the latent heat released associated with high humidity related to cloud formation indicates warming in the upper troposphere and cooling in the lower altitude.

Furthermore, the spatial moisture analysis was conducted to identify the available moisture source supporting the convective cloud development. The zonal distribution of moist air represents a higher moisture content above the study area (102–110 E) both on July 15 and 16, 2022 (see Fig. 5 right). Moreover, the IVT plot shows the high moisture transport over the same location (the Java Sea and the southern coast of Western Java) with IVT values ranging from 300 to 500 kg/m/s (see Fig. 6). The high moisture flux (indicated by the arrow) moved toward the location where clouds formed. Furthermore, before the second peak of heavy rainfall, the air was moistening over the southern coast of Western Java at 03 LT, July 16, 2022 (see Fig. 6). The more abundant moisture over the southern coast intensified the development of convective clouds over this location. It contributed to the second episode of rainfall peaks starting from the southern area. The high moisture over those mentioned locations is supported by the warmer sea surface temperature (SST) over some areas (over the coastal area of Java Island, Sunda strait, and the Indian Ocean). The warmer SST intensifies evaporation, thus resulting in higher moisture in the atmosphere (the SST figure is not shown). The previous study emphasized that the formation of convective systems is more likely intensified and supported by high moisture over its surrounding location [29–32]. These findings support the result of this study that the high moisture content over the Java Sea and the southern coast of Western Java

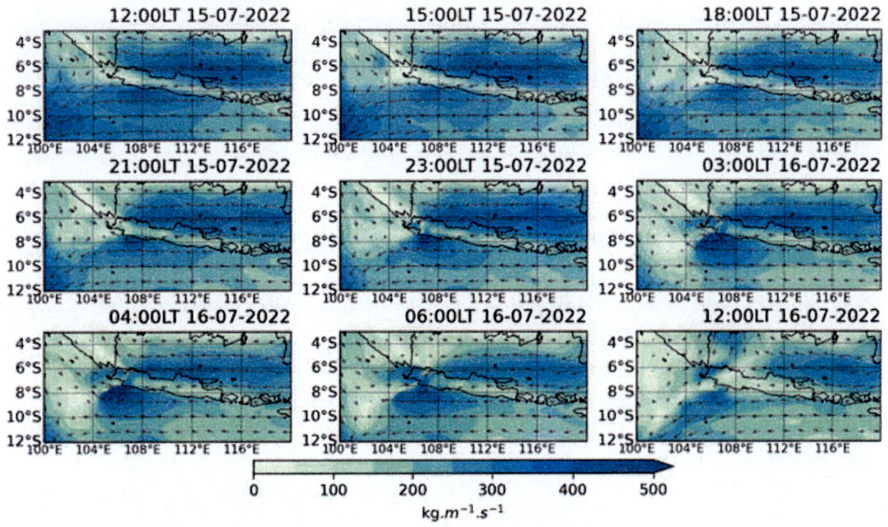

Fig. 6 Integrated water vapor transport (IVT) (shaded) and 1000 hPa wind (vector) from 12 LT, July 15, 2022, to 12 LT, July 16, 2022

played an essential role in cloud formation during flooding events over the study area.

4 Conclusions

A study has been carried out to understand the primary triggers of heavy rainfall events over three cities in the west part of Java Island (over six different locations: Tangerang 1, Tangerang 2, Bogor 1, Bogor 2, Garut 1, and Garut 2) from July 15 to 16, 2022. The heavy rainfall was triggered by the combination between favorable large-scale conditions and local convective development, organization, and propagation around the West Java region and its interaction with the local circulations (land breeze). This study emphasized the convective cloud self-aggregation as the primary trigger of heavy rainfall over the study areas. Convective organization over the mainland was first formed on the night of July 15, 2022. The propagation of convective clouds was from the ocean and moved eastward toward the center of the mainland (the west part of Java), resulting in the highest rainfall over Bogor (the first peak of rainfall intensity). This convective cloud dissipated over land at 00 LT, on July 16, 2022. During the dissipation over the mainland (00 LT), the existing convective cloud over the ocean initiated stronger convective clouds propagating eastward and inland and then merged with other convective cells during the early morning of July 16, 2022. Aggregation of the merging convective cloud triggered the second peak of rainfall during the morning of July 16, 2022, over Tangerang, Bogor, and Garut.

The latent heat released related to cloud formation indicates warming in the upper troposphere and cooling in the lower altitude. Therefore, the convective cloud development during July 15, 2022, can be identified by a positive temperature anomaly in the upper troposphere and the high moisture in the mid-troposphere, indicating the latent heat released due to convective cloud formation.

The more abundant moisture was identified over the Java Sea and the southern coast of Java Island, which can be explained by the warming of the sea surface temperature over the Sunda strait and its surroundings resulting in high evaporation activity. As a result, the higher moisture flux over the western part of Java Island and its vicinity moved toward the flooding areas from those locations. This high moisture is one factor that intensifies the convective cloud development over the study area. Overall, this study emphasized the roles of convective clouds and their sustainable development, potentially triggering heavy rainfall events over study locations.

Acknowledgements This research is supported by the Hydrometeorological Disaster Response and Analysis Team (TREAK), Research Center for Climate and Atmosphere, the National Research and Innovation Agency (BRIN). We gratefully acknowledge the Global Satellite Mapping of Precipitation gauge-corrected (GSMaP_gauge) version-7 for providing public access to rainfall data through https://sharaku.eorc.jaxa.jp/GSMaP, the Himawari-8 satellite infrared channel for the Cloud Top Brightness Temperature (TBB) data which was downloaded from http://weather.is.kochi-u.ac.jp/, and the European Center for Medium-Range Weather Forecast reanalysis 5 (ERA5) for the moisture transport data from Climate Data Storage (https://cds.climate.copernicus.eu/cdsapp#!/dataset/reanalysis-era5-pressure-levels?tab=overview). We thank you for the availability of the temperature and humidity profiles obtained by the Constellation Observing Satellite for Meteorology, Ionosphere, and Climate Mission 2 (COSMIC-2) Global Navigation Satellite System (GNSS) Radio Occultation (RO).

References

1. Ikhsan, A., Susanti, R.: Hujan Deras di Kota Bogor Picu Banjir dan Longsor, 1 Orang Tewas Tertimbun Halaman all—Kompas.com (2022). https://regional.kompas.com/read/2022/07/16/070928778/hujan-deras-di-kota-bogor-picu-banjir-dan-longsor-1-orang-tewas-tertimbun?page=all. Accessed 06 Oct 2022
2. Ramadhan, F.: Banjir Melanda Kabupaten Garut, Simak Prakiraan Cuaca untuk Hari Ini 16 Juli 2022—Pikiran-Rakyat.com (2022). https://www.pikiran-rakyat.com/jawa-barat/pr-015018855/banjir-melanda-kabupaten-garut-simak-prakiraan-cuaca-untuk-hari-ini-16-juli-2022. Accessed 06 Oct 2022
3. Naufal, M., Movanita, A.: 19 Titik di Kota Tangerang Terendam Banjir, Tertinggi sampai 1 Meter (2022). https://megapolitan.kompas.com/read/2022/07/16/11584381/19-titik-di-kota-tangerang-terendam-banjir-tertinggi-sampai-1-meter Accessed 06 Oct 2022
4. Nuryanti and Miftah.: Wilayah di Kota Tangerang yang Masih Tergenang Banjir, Ketinggian Air 50–100 Cm—Tribunnews.com (2022). https://www.tribunnews.com/metropolitan/2022/07/16/wilayah-di-kota-tangerang-yang-masih-tergenang-banjir-ketinggian-air-50-100-cm. Accessed 07 Oct 2022
5. Ghani, H.: Banjir Bandang Terjang Garut, Ratusan Warga Mengungsi (2022). https://www.detik.com/jabar/berita/d-6181827/banjir-bandang-terjang-garut-ratusan-warga-mengungsi. Accessed 07 Oct 2022

6. Mustika, I.: Update Titik Banjir di wilayah JABODETABEK, Mana Saja yang Terkena Banjir Hari ini 16 Juli 2022—Media Jabodetabek (2022). https://mediajabodetabek.pikiran-rakyat.com/jabodetabek/pr-1465020971/update-titik-banjir-di-wilayah-jabodetabek-mana-saja-yang-terkena-banjir-hari-ini-16-juli-2022. Accessed 07 Oct 2022
7. BMKG.: BMKG | Badan Meteorologi, Klimatologi, dan Geofisika (2022). https://www.bmkg.go.id/. Accessed 06 Oct 2022
8. BMKG,.: Analisis Dinamika Atmosfer Laut, Analisis and Prediksi Curah Hujan Update Dasarian Ii Juli 2022 (2022)
9. Roundy, P.E., Frank, W.M.: A climatology of waves in the equatorial region. J Atmos Sci **61**(17), 2105–2132 (2004). https://doi.org/10.1175/1520-0469(2004)061
10. Vallis, G.K., Penn, J.: Convective organization and eastward propagating equatorial disturbances in a simple excitable system. Q. J. R. Meteorol. Soc. **146**(730), 2297–2314 (2020). https://doi.org/10.1002/QJ.3792
11. Matsuno, T.: Quasi-geostrophic motions in the equatorial area. J. Meteorol. Soc. Japan. Ser. II **44**(1), 25–43 (1966). https://doi.org/10.2151/JMSJ1965.44.1_25
12. Gill, A.E.: Some simple solutions for heat-induced tropical circulation. Q. J. R. Meteorol. Soc. **106**(449), 447–462 (1980). https://doi.org/10.1002/QJ.49710644905
13. Sobel, A.H., Yuter, S.E., Bretherton, C.S., Kiladis, G.: Large-scale meteorology and deep convection during TRMM KWAJEX*—DocsLib (2004). https://docslib.org/doc/7863631/large-scale-meteorology-and-deep-convection-during-trmm-kwajex. Accessed 06 Oct 2022
14. Maddox, R.A.: Mesoscale Convective Complexes (1980)
15. Bessho, K. et al.: An Introduction to Himawari-8/9— Japan's new-generation geostationary meteorological satellites. J. Meteorol. Soc. Japan. Ser. II **94**(2), 151–183 (2016). https://doi.org/10.2151/JMSJ.2016-009
16. Putri, N.S., Iwabuchi, H., Hayasaka, T.: Evolution of mesoscale convective system properties as derived from Himawari-8 high resolution data analyses. J. Meteorol. Soc. Japan. Ser. II. **96B**, 2018–020 (2018). https://doi.org/10.2151/JMSJ.2018-020
17. Kubota, T., et al.: Global satellite mapping of precipitation (GSMaP) products in the GPM era. Adv. Global Change Res. **67**, 355–373 (2020). https://doi.org/10.1007/978-3-030-24568-9_20/COVER
18. Chen, M. et al.: Assessing objective techniques for gauge-based analyses of global daily precipitation. J. Geophys. Res.: Atmosp. **113**(D4) (2008). https://doi.org/10.1029/2007JD009132
19. Weiss, J.-P. et al.: COSMIC-2 mission summary at three years in orbit. Atmosphere **13**(9), 1409 (2022). https://doi.org/10.3390/ATMOS13091409
20. Kursinski, E.R., Hajj, G.A., Schofield, J.T., Linfield, R.P., Hardy, K.R.: Observing Earth's atmosphere with radio occultation measurements using the global positioning system. J. Geophys. Res.: Atmosp. **102**(D19), 23429–23465 (1997). https://doi.org/10.1029/97JD01569
21. Li, Y., Yuan, Y., Wang, X.: Assessments of the retrieval of atmospheric profiles from GNSS radio occultation data in moist tropospheric conditions using radiosonde data. Remote Sens. **12**(17), 2717 (2020). https://doi.org/10.3390/RS12172717
22. Hersbach, H., et al.: The ERA5 global reanalysis. Q. J. R. Meteorol. Soc. **146**(730), 1999–2049 (2020). https://doi.org/10.1002/QJ.3803
23. Hersbach, H. et al.: ERA5 hourly data on pressure levels from 1959 to present. Copernicus Climate Change Service (C3S) Climate Data Store (CDS) (2018). https://doi.org/10.24381/cds.bd0915c6
24. Lubis, S.W., Respati, M.R.: Impacts of convectively coupled equatorial waves on rainfall extremes in Java, Indonesia. Int. J. Climatol. **41**(4), 2418–2440 (2021). https://doi.org/10.1002/JOC.6967
25. Latos, B., et al.: Equatorial waves triggering extreme rainfall and floods in southwest sulawesi, indonesia. Mon. Weather Rev. **149**(5), 1381–1401 (2021). https://doi.org/10.1175/MWR-D-20-0262.1
26. Nuryanto, D.E., Pawitan, H., Hidayat, R., Aldrian, E.: Propagation of convective complex systems triggering potential flooding rainfall of Greater Jakarta using satellite data. IOP Conf. Ser. Earth Environ. Sci. **54**(1) (2017). https://doi.org/10.1088/1755-1315/54/1/012028

27. Trismidianto, T.W.H., Ishida, S., Moteki, Q., Manda, A., Iizuka, S.: Development processes of Oceanic convective systems inducing the heavy rainfall over the Western Coast of Sumatra on 28 October 2007. SOLA **12**(1), 6–11 (2016). https://doi.org/10.2151/SOLA.2016-002
28. Jensen, G.G., Fiévet, R., Haerter, J.O.: The diurnal path to persistent convective self-aggregation. J. Adv. Model Earth Syst. **14**(5), e2021MS002923 (2022). https://doi.org/10.1029/2021MS002923
29. Schumacher, R.S., Rasmussen, K.L.: The formation, character and changing nature of mesoscale convective systems. Nature Rev. Earth Environ. **1**(6), 300–314 (2020). https://doi.org/10.1038/s43017-020-0057-7
30. Nuryanto, D.E., Pawitan, H., Hidayat, R., Aldrian, E.: The occurrence of the typical mesoscale convective system with a flood-producing storm in the wet season over the Greater Jakarta area. Dyn. Atmosp. Oceans **96**, 101246 (2021). https://doi.org/10.1016/J.DYNATMOCE.2021.101246
31. Song, F. et al.: Crucial roles of Eastward propagating environments in the summer MCS initiation over the U.S. Great Plains. J. Geophys. Res.: Atmosp. **126**(16), e2021JD034991 (2021). https://doi.org/10.1029/2021JD034991
32. Hu, H., Leung, L.R., Feng, Z.: Early warm-season mesoscale convective systems dominate soil moisture–precipitation feedback for summer rainfall in central United States. Proc. Natl. Acad. Sci. USA **118**(43), e2105260118 (2021). https://doi.org/10.1073/PNAS.2105260118/SUPPL_FILE/PNAS.2105260118.SAPP.PDF

Characteristics of Marine Aerosol Deposition and Its Relation to Wind Condition in Three Locations of West Java

Dyah Aries Tanti, Arif Rachman, Opik Taopik, Amalia Nurlatifah, Asri Indrawati, Sumaryati, Wiwiek Setyawati, Nani Cholianawati, and Wilin Julian Sari

Abstract Sea salt is an aerosol emitted from the ocean. These particles are naturally distributed in nature. Studying the characteristics of sea salt is very important because sea salt has a lot of influence on the climate system, and its deposition is corrosive. This study aims to examine the characteristics of sea salt in West Java and determine how the wind influences the dynamics of its deposition concentration. Sulfates, nitrates, and ammonium in West Java are mostly distributed in urban areas (Cipedes). Meanwhile, chloride and alkaline compounds (Na^{+}, K^{+}, Ca^{2+}, and Mg^{2+}) are widely distributed in coastal areas (Pameungpeuk). Wind has a lot of influence on the concentration of sea salt deposition in West Java. Wind dynamics have the most influence on fluctuations in sea salt deposition concentrations in urban areas (Cipedes). The correlation is around 0.7. However, the wind has a strong and positive influence in this area which means that the concentration of sea salt deposition will increase if the wind speed also increases. This creates a tendency that sea salt in urban areas is mostly transported from other areas and not from the area itself.

1 Introduction

Sea salt is an aerosol emitted from the ocean. These particles are naturally distributed in nature. Sea spray aerosols or sea salt is important for a wide variety of processes.

D. A. Tanti (✉) · A. Nurlatifah · Sumaryati · W. Setyawati · N. Cholianawati
Research Center for Climate and Atmosphere, National Research and Innovation Agency, Jakarta, Indonesia
e-mail: dyahariest@gmail.com

A. Rachman · O. Taopik · A. Indrawati
Directorate of Laboratory Management, Research Facilities, and Science and Technology Park, National Research and Innovation Agency, Jakarta, Indonesia

W. J. Sari
Research Center for Quantum Physics, National Research and Innovation Agency, Jakarta, Indonesia

A. Basit et al. (eds.), *Proceedings of the International Conference on Radioscience, Equatorial Atmospheric Science and Environment and Humanosphere Science*, Springer Proceedings in Physics 290, https://doi.org/10.1007/978-981-19-9768-6_83

Part of the current interest is their role in climate [15]. The main effect of aerosols on the climate system is the scattering of solar radiation [4, 10]. Sulfate, nitrate, and sea salt aerosols can be considered purely scattering at visible wavelengths [18]. The scattering increases the fraction of solar radiation reflected back to space, cooling the climate system [10].

Aerosols in the form of Particulate Matter (PM) have a significant impact on human health. Particulate matter sizes below 2.5 microns (PM2.5) can settle in human alveoli and cause many respiratory problems [16]. In addition, aerosols in particulate matter form also play an important role in air pollution events and have significant influence on climatic conditions [19]. Therefore, knowledge of aerosols today is really important for many purposes such as the prevention of various lung diseases, reduction of air pollution, and control of climate change.

Water-soluble inorganic compounds such as SO_4^{2-} and NO_3^- are the main constituents of particulate matter in the air and account for 30–50% of PM2.5 [3, 8, 22]. This compound is usually the main cause of haze. Therefore, improving or reducing the concentration of these two compounds can significantly reduce air pollution and climate change caused by aerosols [7].

Knowing the characteristics of sea salt and researching the factors that influence it is important, so that effects of the transformation and transportation of it can be overcome. In addition, considering the important role and contribution of aerosols in the atmosphere and climate system, aerosol characteristics are important to be studied. As a compound that emits naturally in nature, meteorological factors are important and quite influential in determining the fate of sea salt.

Several previous studies have stated that wind direction and speed have significant effect on sea salt concentrations in an area [13]. Research on marine aerosols in South America provides strong evidence that marine aerosols from the South Atlantic Ocean reach the city of Buenos Aires [17]. An increase in sea salt concentration due to the influence of high wind speeds also occurred in Gdynia, a coastal city in Poland [9].

The southern coast of West Java has a coastline of 398 km. The long of the sea area around the southern region of West Java gives an indication that the incidence of sea salt deposition in the West Java area is quite high. This study aims to determine the characteristics of sea salt and its relation to wind conditions in West Java.

2 Data and Method

2.1 Research Locations

The sampling points used in this study include three points, namely Pameungpeuk (107.69E, 7.65S), Cipedes (107.59E, 6.88S), and Tanjungsari (107.72E, 6.87S). The location of the sampling point is presented in Fig. 1.

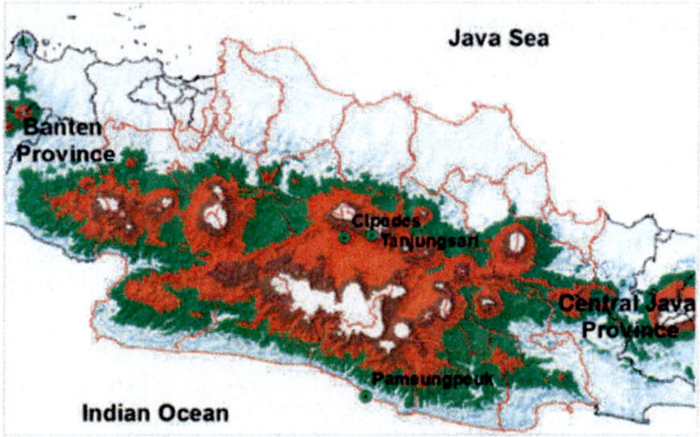

Fig. 1 Research locations

2.2 Data

Some of the data that is used in this study include.

- The concentration data of anions and cations dissolved in rainwater include SO_4^{2-},NO_3^-, Cl^-, NH_4^+, Na^+, K^+, Ca^{2+}, and Mg^{2+} ions from 2015 to 2018.
- Wind speed data taken from ERA5 Copernicus (https://cds.climate.copernicus.eu/) with spatial resolution of 0.25° × 0.25° and daily temporal resolution.

2.3 Method

- Rainwater sampling

Sampling was carried out by collecting rainwater every rainy day using an automatic rain sampler for the Cipedes location and a rain gauge for the Tanjungsari and Pameungpeuk locations. Furthermore, the collected rainwater samples were analyzed in the chemical laboratory using ion chromatography equipment to obtain concentrations of anions (SO_4^{2-}, NO_3^-, Cl^-) and cations (NH_4^+, Na^+, K^+, Ca^{2+}, Mg^{2+}).

- Pearson correlation

Pearson correlation is a measure of correlation or linear relationship between two variables (Pearson 1985). The values range from −1 to +1. The correlation value is calculated by the following formula:

$$r = \frac{n(\sum xy) - (\sum x)(\sum y)}{\sqrt{[n\sum x^2 - (\sum x)^2][n\sum y^2 - (\sum y)^2]}}$$

where x is observation data and y is model data. Table 1 provides an overview of the meaning of the Pearson correlation value.

Table 1 Meaning of Pearson correlation

r value	Meaning
$r > 0.5$	Both variables are positively and strongly correlated
$0 < r < 0.5$	Both variables are positively and weakly correlated
$r = 0$	Both variables are not correlated
$-0.5 < r < 0$	Both variables are negatively correlated (inversely) and weak
$r < -0.5$	Both variables are negatively (inversely) and strongly correlated

- Calculation of sea salts concentration

The calculation of sea salts (sulfate and calcium) concentration is estimated from sodium and non-sea salts concentration. The calculation of sea salts is determined in equations (1 and 2) [12].

$$\left[\mathrm{nss} - \mathrm{SO}_4^{2-}\right] = \left[\mathrm{SO}_4^{2-}\right] - 0.6028\left[\mathrm{Na}^+\right] \quad (1)$$

$$\left[\mathrm{nss} - \mathrm{Ca}^{2+}\right] = \left[\mathrm{Ca}^{2+}\right] - 0.2161\left[\mathrm{Na}^+\right] \quad (2)$$

$$\left[\mathrm{ss} - \mathrm{SO}_4^{2-}\right] = \left[\mathrm{SO}_4^{2-}\right] - \left[\mathrm{nss} - \mathrm{SO}_4^{2-}\right] \quad (3)$$

$$\left[\mathrm{ss} - \mathrm{Ca}^{2+}\right] = \left[\mathrm{Ca}^{2+}\right] - \left[\mathrm{nss} - \mathrm{Ca}^{2+}\right] \quad (4)$$

nss: concentration of non-sea salts ion
ss: concentration of sea salts ion

3 Results and Discussion

The conditions for the concentration of water-soluble inorganic compounds such as SO_4^{2-} and NO_3^- at three sampling stations in West Java are presented in Table 2. It can be seen that the highest value of SO_4^{2-} and NO_3^- were recorded at Cipedes station with an average concentration value of 36.95 mol/L for SO_4^{2-} and 27.38 mol/L for NO_3^-. This shows that in West Java, SO_4^{2-} and NO_3^- are more widely distributed in urban areas. Looking at the characteristics of its formation, many sulfate compounds in the atmosphere are formed due to anthropogenic activities such as burning of sulfur-containing fossil fuels, household wastes including detergents, and industrial effluents from tanneries, steel mills, sulfate-pulp mills, and textile plants [11]. Meanwhile, nitrate compounds can also be formed due to human activities such as runoff or seepage from fertilized agricultural lands, municipal and industrial waste water, refuse dumps, animal feedlots, septic tanks and private sewage disposal systems, urban drainage, and decaying plant debris [6]. Therefore, considering the dense population in urban areas of West Java, high concentrations of SO_4^{2-} and NO3- are more concentrated in West Java, is common.

The second largest SO_4^{2-} concentration was in Pameungpeuk. This shows that SO_4^{2-} in West Java is spread in coastal areas. Different things happened with NO_3^- where the second largest concentration was in rural areas (Tanjungsari) with an average concentration of 13.60 μmol/L. This difference happened caused by many people in rural area (Tanjungsari) cultivate crops and usually emitting more NO_3^- than the Pameungpeuk (beach) area. It is well known that anthropogenic sources of nitrate can come from runoff or seepage from fertilized agricultural lands, municipal and industrial waste water, refuse dumps, animal feedlots, septic tanks and private sewage disposal systems, urban drainage, and decaying plant debris.

Table 2 Mean of monthly concentration in West Java (in μmol/L)

		Pameungpeuk	Tanjungsari	Cipedes
SO_4^{2-}	Max	32.45	31.12	91.66
	Min	11.74	9.94	21.67
	Mean	17.78	18.29	36.95
NO_3^-	Max	12.10	26.13	74.78
	Min	4.22	4.90	15.43
	Mean	6.98	13.60	27.38
Cl^-	Max	395.37	42.43	42.87
	Min	86.89	7.47	9.43
	Mean	160.23	15.37	15.30
NH_4^+	Max	105.09	91.53	149.36
	Min	21.88	9.34	44,10
	Mean	43.01	31.46	73.29
Na^+	Max	297.26	36.05	52.58
	Min	76.27	9.15	13.25
	Mean	133.11	17.65	22.00
K^+	Max	26.48	12.13	14.33
	Min	8.53	2.08	3.22
	Mean	13.04	4.11	5.07
Ca^{2+}	Max	34.62	26.22	60.19
	Min	14.91	8.49	14.17
	Mean	20.27	13.66	22.12
Mg^{2+}	Max	38.58	11.74	15.06
	Min	12,61	2.39	3.36
	Mean	19.04	4.15	5.24

Chloride (Cl^-) ions are widely distributed in nature. However, the most dominant concentration of Cl^- is spread in the ocean with a value of up to 19.000 mg/L [21]. The distribution of Cl- in the three sample points representing coastal areas (Pameungpeuk), rural (Tanjungsari), and urban (Cipedes) shows that Cl- is widely distributed in coastal areas. The highest average concentration of Cl- occurred in Pameungpeuk exceeds a value of 160.23 μmol/L. The maximum concentration value in this area even reaches 395.37 μmol/L. This value is quite high when compared to the value of Cl- in the other areas which only reached 15.37 μmol/L in Tanjungsari and 15.30 μmol/L in Cipedes.

Ammonium is a form of dissolved inorganic nitrogen that is naturally occurring in low concentrations within groundwater [20]. The largest source of NH4 + emissions in the atmosphere is agriculture, including industrial processes, vehicular emissions and volatilization from soils and oceans [2]. It can be seen that the highest concentration of ammonium from the three sample points is occurred in Cipedes. This fact shows that a lot of ammonium in West Java is emitted in urban areas. When viewed from the source, ammonium can come from industrial processes or vehicle emissions and evaporation from the soil. Most likely these sources are the cause of the high concentration of ammonium in Cipedes. The second largest concentration is in Pameungpeuk which most likely comes from evaporation of sea water.

The concentration of alkaline cations such as sodium (Na^+), potassium (K^+), calcium (Ca^{2+}), and magnesium (Mg^{2+}) has the highest concentration in Pameungpeuk or coastal areas. This supports the fact that the cation compounds that are part of this sea salt are emitted from the sea. This compound is also quite reactive with chloride (Cl^-), and is often found in the form of salt compounds such as NaCl, KCl, $Ca(Cl)_2$, and $Mg(Cl)_2$. Therefore, although the concentration is large, the concentration of these cations is not greater than the concentration of Cl^-.

Monthly Characteristics of Sea Salt and Its Relation with wind

From the analysis of cations and anions concentration condition, we can determine the concentration of sea salts. In this research, the concentration of SO_4^{2-} and Ca^{2+} is figured out. Figure 2a shows the concentration value of sulfate (SO_4^{2-}), and Fig. 2b shows the concentration value of calcium (Ca^{2+}) in all sample points. From the two figures, it can be seen that in general the concentration of sea salts in the Pameungpeuk (coastal) area for both calcium and sulfate is the highest among the other two areas. This is quite understandable because the coastal area is the area closest to the sea salt emitter. Meanwhile, in general, in all regions it can be seen that the sulfate concentration is higher than the calcium concentration.

Figure 2 shows the characteristics of the monthly sea salt concentration in West Java and its relationship to the wind speed. While Table 3 shows the correlation value between wind speed and sea salt concentration value. It can be seen that in general, almost all sea salt concentrations in Cipedes experience high values in June–October or when the Australian monsoon occurs. In this season, the wind speed values tend to be high. This analysis is supported by the correlation value between sea salt concentration and wind speed in Cipedes which reaches above 0.45 (Table 3). In general, the wind speed factor is quite influential on the value of sea salt concentration in Cipedes.

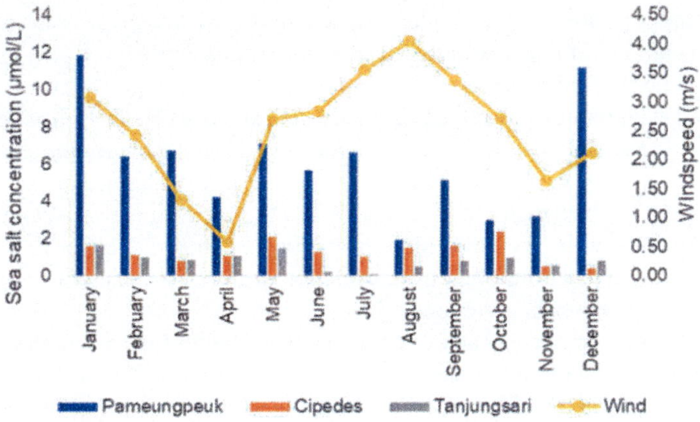

(a)

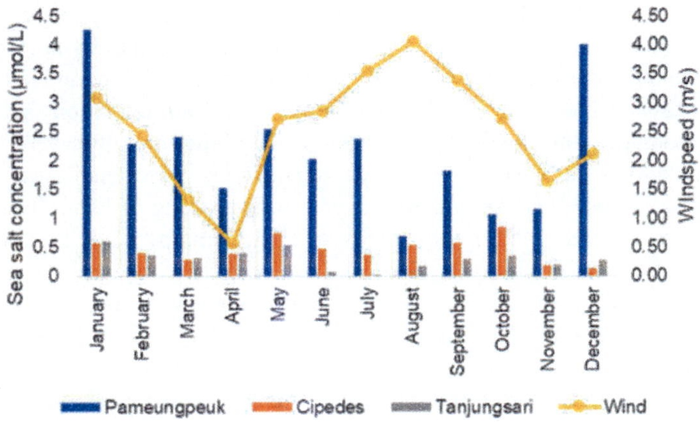

(b)

Fig. 2 Monthly concentration of sea salts and its overlay with monthly wind speed **a** SO_4^{2-} and **b** Ca^{2+}

Table 3 Pearson correlation coefficient between monthly sea salts concentration data and wind speed data

	Pameungpeuk	Cipedes	Tanjungsari
SO_4^{2-}	−0.02	0.45	−0.22
Ca^{2+}	−0.02	0.46	−0.24

For the Pameungpeuk area, in general the sea salt concentration has a very weak negative correlation with wind speed. This means that overall wind speed has little effect on the value of sea salt concentration in Pameungpeuk. This is probably because the Pameungpeuk area is already close to the emission source of sea salt (sea) so that the wind is not affect the sea salts concentration in this area.

Meanwhile in Tanjungsari, sea salt concentration has negative correlation with wind speed even though the value is in moderate category (Table 3). This means that when the wind speed is slow, there is tendency that the concentration of sea salt in this area is high and vice versa. The sea salt dispersion in Tanjungsari has an effect on the sea salt concentration in this area. When the incidence of sea salt dispersion due to wind is slower, the concentration of sea salt in this area is higher. This is possible in Tanjungsari although the correlation between sea salt concentration and wind speed in this area is quite low.

From all sample points, it can be concluded that the wind affects the concentration of sea salt in urban and rural areas. This is because the wind functions as a means of transporting pollutants that can spread pollutants. However, wind in urban areas has the most influence on the determination of sea salt concentrations compared to rural and coastal areas.

In addition to seasonal variability such as monsoons, inter-annual variability such as ENSO also greatly influences wind dynamics in Indonesia. During El Nino, the zonal wind that more dominant is the easterly wind, and the meridional wind that is more dominant is the southerly wind. This event can cause a reduction in the supply of water vapor in Indonesia [1]. This condition is also an indication that the wind speed in Indonesia during El Nino will tend to be higher than the average. On the other hand, during La Nina, the wind speed tends to be slower than usual.

2015 was the year that is recorded as a year with a strong El Nino. It can be seen that the average wind speed in this year is quite high compared to 2016 which was recorded as a weak La Nina year (Fig. 3). In the overall concentration of sea salts in Cipedes (urban), it can be seen that wind speed tends to be positively correlated with the concentration value. It can be seen that in 2016 where there was a weak La Nina and there was heavy rain throughout the year, the wind speed tended to be low and the overall concentration of sea salts was also low (Fig. 3). The same thing also happened in Tanjungsari where the low wind speed tends to also make the concentration value low. This indicates that the annual concentration values in Cipedes and Tanjungsari are the results of transportation from other areas due to high winds which actually give the effect of increasing this annual concentration value.

The difference occurred in Pameungpeuk. Figure 3 shows that wind speed has little effect on fluctuations in the average annual concentration of sea salt in Pameungpeuk. This indicates that the sea salts in Pameungpeuk are ions that originating come from the sea so that wind speed have little effect on the fluctuations of sea salts in this place.

Not only wind speed, wind direction also greatly affects the distribution of marine aerosols in Bandung, Sumedang, and Garut. Figure 4 shows the distribution of the dominant monthly winds in degrees. Overall, in the Pameungpeuk area, it can be seen that the dominant wind is heavily influenced by background winds where in the JJA season the dominant wind is from the east while at other times the wind is dominant from the west (Fig. 4c). This gives an indication that marine aerosols in the Pameungpeuk area are widely distributed to the west according to the movement of

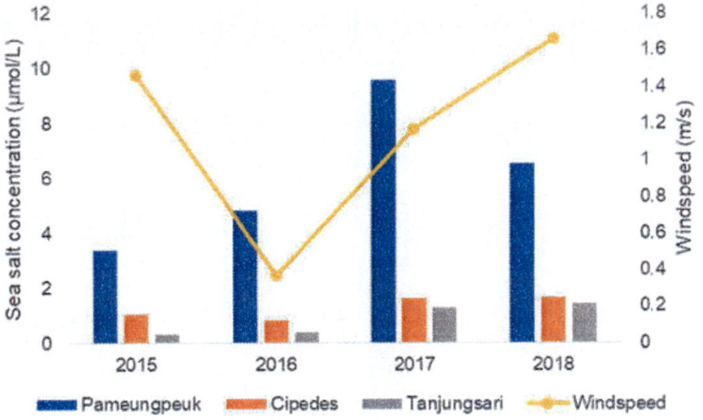

(a)

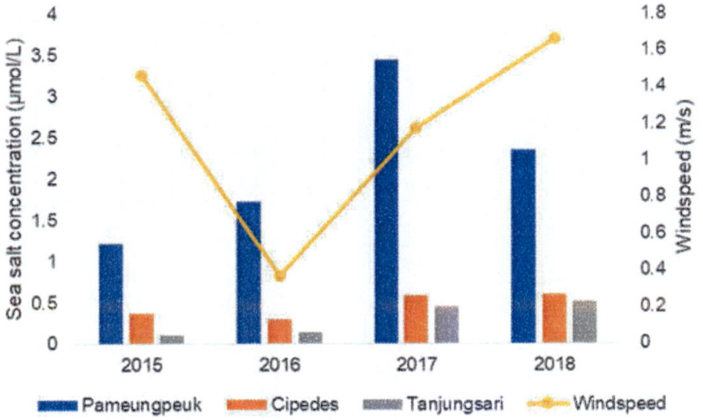

(b)

Fig. 3 Annual concentration of sea salt and its overlay with annual wind speed **a** SO_4^{2-} and **b** Ca^{2+}

the wind. In the Sumedang area, the dominant wind is still strong as evidenced by the distribution of the east and west winds according to the season, which are still clearly visible (Fig. 4b). However, the dominant wind is the easterly wind, and this is an indication that the sea salt is widely distributed to the western area of Sumedang. In the Bandung area, which is an urban area, inland and more local winds occur where the distribution and fluctuation of wind direction are more pronounced (Fig. 4a). This wind fluctuation gives an indication that indeed in the Bandung area the concentration of sea salt depends a lot on the wind and the various wind directions make the sea salt distributed in various directions.

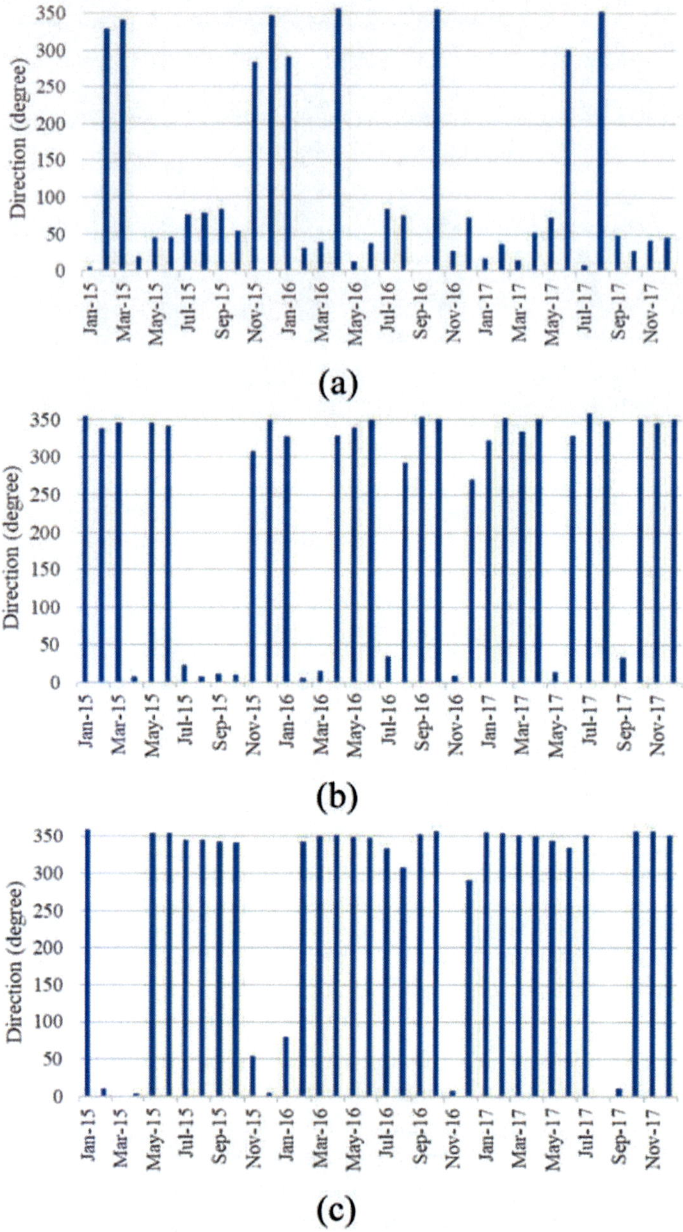

Fig. 4 Monthly wind direction from ERA5 data in **a** Cipedes, **b** Tanjungsari, and **c** Pameungpeuk

4 Conclusion

Sulfates, nitrates, and ammonium in West Java are mostly distributed in urban areas (Cipedes). Meanwhile, chloride and alkaline compounds (Na^+, K^+, Ca^{2+}, and Mg^{2+}) are widely distributed in coastal areas (Pameungpeuk). Wind has a lot of influence on the concentration of sea salt deposition in West Java. Wind dynamics have the most influence on fluctuations in sea salt deposition concentrations in urban area (Cipedes). However, the wind has a strong and positive influence in this area which means the concentration of sea salt deposition will increase if the wind speed also increases. This creates a tendency that sea salt in urban areas is mostly transported from other areas. Wind has low influence on fluctuations in sea salt deposition concentrations in coastal area (Pameungpeuk). It means that the sea salts in coastal area is come from that area itself.

Acknowledgements Authors would like to thank the Research Organization of Aeronautics and Space, Indonesia Research and Innovation Agency for moral and material support for this research.

References

1. Athoillah, I., Rini, S., Deassy, D.: Analisis spasial pengaruh kejadian el nino kuat tahun 2015 dan la nina lemah tahun 2016 terhadap kelembapan, angin dan curah hujan di Indonesia. Jurnal Sains and Teknologi Modifikasi Cuaca. **18**, 33 (2016). https://doi.org/10.29122/jstmc.v18i1.2140
2. Behera, S.N., Sharma, M., Aneja, V.P., Balasubramanian, R.: Ammonia in the atmosphere: a review on emission sources, atmospheric chemistry and deposition on terrestrial bodies. Environ Sci Pollut Res Int. **20**(11), 8092–8131 (2013). https://doi.org/10.1007/s11356-013-2051-9. Epub 2013 Aug 28 PMID: 23982822
3. Cao, J.J., Shen, Z.X., Chow, J.C., Watson, J.G., Lee, S.C., Tie, X.X., Ho, K.F., Wang, G.H., Han, Y.M.: Winter and summer PM2.5 chemical compositions in fourteen Chinese cities. J. Air Waste Manage. Assoc. **62**, 1214–1226 (2012)
4. Charlson, R.J., Schwartz, S.E., Hales, J.M., Cess, R.D., Coakley, J.A., Jr., Hansen, J.E., Hoffman, D.J.: Climate forcing by anthropogenic aerosols. Science **255**, 423–430 (1992). https://doi.org/10.1126/science.255.5043.423
5. https://kkp.go.id/djprl/artikel/21045-konservasi-perairan-sebagai-upaya-menjaga-potensi-kelautan-dan-perikanan-indonesia. Accessed in 24 Feb 2022
6. Illinois Department of Public Health.: https://dph.illinois.gov/topics-services/environmental-health-protection/private-water/fact-sheets/nitrates-drinking-water.html#:~:text=Illinois%20has%20adopted%20a%20drinking,guide%20for%20private%20water%20supplies. Accessed 23 Nov 2022
7. Jiang, F., Liu, F., Lin, Q., Fu, Y., Yang, Y., Peng, L., Lian, X., Zhang, G., Bi, X., Wang, X., Sheng, G.: Characteristics and formation mechanisms of sulfate and nitrate in size-segregated atmospheric particles from urban Guangzhou China. Aerosol Air Qual. Res. **19**, 1284–1293 (2019). https://doi.org/10.4209/aaqr.2018.07.0251
8. Lai, S.C., Zou, S.C., Cao, J.J., Lee, S.C., Ho, K.F.: Characterizing ionic species in PM2.5 and PM10 in four Pearl River Delta cities South China. J. Environ. Sci. **19**, 939–947 (2007)
9. Lewandowska, A.U., Falkowska, L.M.: Sea salt in aerosols over the southern Baltic. Part 2. The neutralizing properties of sea salt and ammonia**Parts of this paper were originally published

in Polish: Lewandowska, A., 2011, Chemizm aerozoli w rejonie Zatoki Gdańskiej, Wyd. UG, Gdańsk, pp.184. Oceanologia, **55**(2), 299–318 (2013). https://doi.org/10.5697/oc.55-2.299
10. Li, J., Carlson, B.E., Yung, Y.L., et al.: Scattering and absorbing aerosols in the climate system. Nat. Rev. Earth Environ. **3**, 363–379 (2022). https://doi.org/10.1038/s43017-022-00296-7
11. Malone Rubright, S.L., Pearce, L.L., Peterson, J.: Environmental toxicology of hydrogen sulfide. Nitric Oxide Biol. Chem. **71**, 1–13 (2017). https://doi.org/10.1016/j.niox.2017.09.011
12. Otosaka, S., Ueki, I., Sasano, D., Kumamoto, Y., Obata, H., Fukuda, H., Nishibe, Y., Maki, H., Goto, K., Ono, T., Aoyama, M. (eds).: Guideline of Ocean Observations, Volumes 1–10, 4th edn. Tokyo, Japan, The Oceanographic Society of Japan, pp. 666. (2020). ISBN 978-4-908553-67-7, https://doi.org/10.25607/OBP-772
13. Ovadnevaite, J., Ceburnis, D., Canagaratna, M., Berresheim H., Bialek, J., Martucci, G., Worsnop, D.R., O'Dowd, C.: On the effect of wind speed on submicron sea salt mass concentrations and source fluxes (2012). https://doi.org/10.1029/2011JD017379
14. Pearson, K.: Notes on regression and inheritance in the case of two parents. Proc. Royal Soc. London **58**, 240–242 (1895). Bibcode: 1895RSPS...58..240P
15. Penner, J.E., et al.: Aerosols, their direct and indirect effects. In: Houghton, J.T., et al. (eds.) Climate Change 2001: The Scientific Basis, Contribution of Working Group I to the Third Assessment Report of the Intergovernmental Panel on Climate Change, Cambridge University Press, Cambridge and New York, pp. 289–348. (2001)
16. Poschl, U.: Atmospheric aerosols: composition, transformation, climate and health effects. Angew. Chem. Int. Ed. **44**, 7520–7540 (2005)
17. Santos, M.D., Dawidowski, L., Smichowski, P., Ulke, A.G., Gómez, D.: Factors controlling sea salt abundances in the urban atmosphere of a coastal South American megacity. Atmos. Environ. **59**, 483–491 (2012). https://doi.org/10.1016/j.atmosenv.2012.05.019
18. Seinfeld, J.H., Bretherton, C., Carslaw, K.S., Coe, H., DeMott, P.J., Dunlea, E.J., Wood, R.: Improving our fundamental understanding of the role of aerosol−cloud interactions in the climate system. Proc. Natl. Acad. Sci. **113**(21), 5781–5790 (2016). https://doi.org/10.1073/pnas.1514043113
19. Seinfeld, J.H., Pandis, S.N.: Atmospheric Chemistry and Physics: From Air Pollution to Climate Change, 2nd edn. Wiley, New York (2006)
20. WHO.: Guidelines for Drinking-Water Quality: Fourth Edition Incorporating the First Addendum; World Health Organization: Geneva, Switzerland (2017)
21. Wu, D., Hu, Y., Liu, Y., Zhang, R.: Review of chloride ion detection technology in water. Appl. Sci. **11**, 11137 (2021). https://doi.org/10.3390/app112311137
22. Wu, X., Deng, J., Chen, J., Hong, Y., Xu, L., Yin, L., Du, W., Hong, Z., Dai, N., Yuan, C.S.: Characteristics of water-soluble inorganic components and acidity of PM2.5 in a coastal city of China. Aerosol. Air Qual. Res. **17**, 2152–2164 (2017)